Modern Techniques of Rice Crop Production

Naeem Sarwar • Atique-ur-Rehman •
Shakeel Ahmad • Mirza Hasanuzzaman
Editors

Modern Techniques of Rice Crop Production

Springer

Editors
Naeem Sarwar
Department of Agronomy
Bahauddin Zakariya University
Multan, Pakistan

Atique-ur-Rehman
Department of Agronomy
Bahauddin Zakariya University
Multan, Pakistan

Shakeel Ahmad
Department of Agronomy, Faculty of Agricultural Sciences and Technology
Bahauddin Zakariya University
Multan, Pakistan

Mirza Hasanuzzaman
Department of Agronomy
Sher-e-Bangla Agricultural University
Dhaka, Bangladesh

ISBN 978-981-16-4954-7 ISBN 978-981-16-4955-4 (eBook)
https://doi.org/10.1007/978-981-16-4955-4

This Springer imprint is published by the registered company Springer Nature Singapore Pte Ltd.
The registered company address is: 152 Beach Road, #21-01/04 Gateway East, Singapore 189721, Singapore

Preface

Being consumed by billions, rice becomes a more common food crop than other cereals. It ranks second in harvested or cultivated areas in the world. In terms of calorific values, rice occupies the top position over other cereals. Apart from its use as human food, straw is used as animal feed or in paper and other industries. Rice is grown on a wide range of soil under different climatic conditions. Different rice production systems include irrigated wet season, irrigated dry season, rainfed low land, rainfed upland, and deepwater rice. These ecosystems can be divided into subsystems based on hydrological or soil factors. Although rice is grown in a number of ways depending upon resource availability, its quality improves by ample supply of water and other resources. Improving rice yield and quality under changing climatic scenario is, therefore, vital to address food security. Numerous research works have been carried out in the past couple of decades to invent an eco-friendly integrated management approach. In recent decades, aerobic rice is well propagated to cope with water deficiency. Likewise, system of rice intensification is advocated as an advance in rice production. Organic cultivation methods are based more on knowledge of agronomic processes than input-based conventional production. In the era of climate change, rice is facing diverse abiotic stresses such as salt, drought, toxic metals, and environmental pollutants. Scientists are trying to develop stress tolerance cultivars using agronomic, genetic, and molecular approaches. A bulk of published work is available in this regard. However, invention and divisions occurring on every coming day under changing climate need to be disseminated to the reading community and researchers. This book is intended to improve the previously available literature about rice production. The chapters cover different modern aspects of rice production. The authors are experts in their fields of study and have a wide grasp of their assigned topics.

We, the editors, would like to give special thanks to the authors for their outstanding and timely work in producing such fine chapters. We are highly thankful to Dr. Aakanksha Tyagi, Associate Editor, Springer, India, for her prompt responses during the acquisition. We are also thankful to the project coordinator Veena

Perumal, Production Editor (Books), Springer Nature, and all other editorial staffs for their precious help in formatting and incorporating editorial changes in the manuscripts.

Multan, Pakistan Naeem Sarwar
Multan, Pakistan Atique-ur-Rehman
Multan, Pakistan Shakeel Ahmad
Dhaka, Bangladesh Mirza Hasanuzzaman

Contents

About the Editors

Naeem Sarwar is an Associate Professor in the Department of Agronomy, Faculty of Agricultural Sciences and Technology, Bahauddin Zakariya University, Multan, Pakistan. He completed his PhD with research work on irrigation and nutrient management in rice crop in the year 2011 from the University of Agriculture, Faisalabad, Pakistan. During his PhD, he was able to achieved scholarship under International Research Support Initiative Program (IRSIP) by Higher Education Commission (HEC) and joined as a research scholar in the University of California Davis, USA. He focused his research work on rice crop while staying there and published his research work in a well-reputed journal. Currently, he is working on different aspects of agronomy for sustainable crop production. He has published over 50 research articles in peer-reviewed international and national journals, written a couple of book chapters, and worked as a reviewer in many well-reputed journals. He has completed many research projects funded by Bahauddin Zakariya University Multan (BZU) and Higher education Pakistan (HEC). According to google scholar, his publications have received 736 citations with an h-index of 15. He has participated and presented his research work in many international/national conferences. Besides, he is member of different international/national professional societies like Australian Society of Agronomy, Pakistan Society of Agronomy etc. Moreover, many graduate students have successfully completed their research work under his supervision.

Atique-ur-Rehman is Associate Professor of Agronomy at Bahauddin Zakariya University, Multan, Pakistan. He received his Ph.D. from the University of Agriculture, Faisalabad, Pakistan, with main focus on the improvement of rice production systems by improving nutrition. He has completed several trainings on rice production and water savings from Hong Kong and China. He joined as an Assistant Professor in the Department of Agronomy, Pir Mehr Ali Shah Arid Agriculture University, Rawalpindi, in August 2012 and moved to Bahauddin Zakariya University, Multan, in October 2013. He has been devoting himself in teaching and researching the field of arable crops, especially focused on crop nutrition, water saving, and adaptation strategies. He has completed a number of projects in his field of expertise with main focus on crop nutrition under stress. He has published over 80 research articles, 10 book chapters, and many articles in daily newspapers. He is editor and reviewer of several peer-reviewed national and international journals. He is also member of number of professional societies like Pakistan Society of Agronomy and Soil Science Society of Pakistan. He has attended and presented his work in national and international conferences.

Shakeel Ahmad is Professor of Agronomy at the Department of Agronomy, Faculty of Agricultural Sciences (FAST), Bahauddin Zakariya University (BZU), Multan, Pakistan. He received his Ph.D. from the University of Agriculture (UAF), Faisalabad, Pakistan, during 2006. His Ph.D. thesis title was "Effect of Agro-environmental Factors Growth, Yield and Quality of Fine Basmati Rice." Then, he completed his Postdoctoral Fellowship with Prof. Gerrit Hoogenboom from the University of Georgia, USA, with financial support from Higher Education Commission (HEC), Islamabad. He joined Bahauddin Zakariya University in October 2002 as a Lecturer, and promoted to Assistant Professor, Associate Professor, and Professor in August 2007, August 2011, and July 2016, respectively. He has vast experience of working with a team of International Scientists as team member/Co-Principal Investigator of Agricultural Modeling Intercomparison and Improvement Project (AgMIP) during Phases-I and

II funded by UKaid through USDA and University of Columbia, USA. He won the Research Productivity Award (RPA) continuously for five years from the Pakistan Council for Science and Technology (PCST) through the Ministry of Science and Technology, Government of Pakistan, Islamabad. He also worked as Principal Investigator (PI) in many national projects. Currently, he serves as a Principal Investigator as well as Co-Principal Investigator in NRPU research grants from Higher Education Commission (HEC), Islamabad. Prof. Ahmad is a dedicated teacher and researcher in the field of agronomy, with major focus on crop modeling, climate change impact assessment, and adaptation strategies of agronomic crops. Prof. Ahmad has published over 200 articles in peer-reviewed journals and books. He has written over 100 book chapters on important aspects of plant physiology, plant stress responses, and climate change impact assessment and adaptation strategies. These book chapters were published by internationally renowned publishers. His publications received over 5000 citations with an h-index of 32 (according to Scopus). Prof. Shakeel Ahmad is also a research supervisor of undergraduate and postgraduate students and supervised 40 M.Phil. and 09 Ph.D. students so far. He is life member of professional societies like Pakistan Society of Agronomy (PSA), Pakistan Society of Botany, and Weed Science of Pakistan. He has attended and presented papers and posters in national and international conferences in different countries (USA, UK, Thailand, Sri Lanka, Nepal, Tanzania, China, Turkey, Kenya, etc.).

Mirza Hasanuzzaman is Professor of Agronomy at Sher-e-Bangla Agricultural University, Dhaka, Bangladesh. He completed his Bachelor of Science in Agriculture (Hons.) from Sher-e-Bangla Agricultural University, where he achieved First Class, received a gold medal for being first in his class, and earned a Sher-e-Bangla Agricultural University Award. He also completed a Master of Science in Agronomy from the same university, where he was once again at top of his class with a CGPA 4.0. In 2012, he received his Ph.D.

with a dissertation on "Plant Stress Physiology and Antioxidant Metabolism" from the United Graduate School of Agricultural Sciences, Ehime University, Japan, with a Japanese Government (MEXT) Scholarship. Later, he completed his postdoctoral research in the Center of Molecular Biosciences (COMB), University of the Ryukyus, Okinawa, Japan, with a "Japan Society for the Promotion of Science (JSPS)" postdoctoral fellowship. Subsequently, he became an Adjunct Senior Researcher at the University of Tasmania with an Australian Government's Endeavour Research Fellowship. Prof. Hasanuzzaman has over 180 publications in **Web of Science**. He has edited 10 books and written over 30 book chapters on important aspects of plant physiology, plant stress responses, and environmental problems in relation to plant species. His publications are cited over **10,000** times as per Scopus with an *h*-**Index of 52**. He is an editor and a reviewer of more than 100 ISI-indexed peer-reviewed international journals and the recipient of "Publons Peer Review Award 2017, 2018, and 2019." He is an active member of 40 professional societies and is the acting Research and Publication Secretary of the Bangladesh JSPS Alumni Association. He received the World Academy of Science (TWAS) Young Scientist Award 2014, UGC Gold Medal 2018 and several other awards and fellowships for his contribution to research. He is a fellow of the Bangladesh Academy of Sciences (BAS) and a foreign fellow of the Society for Science of Climate Change and Sustainable Environment.

Part I

Introduction

World Rice Production: An Overview

1

Atique-ur-Rehman, Naeem Sarwar, Shakeel Ahmad,
Muhammad Azam Khan, and Mirza Hasanuzzaman

Abstract

Rice is the important food crop for majority of the humans in the world with contribution more than many other cereal crops. However, with changes in world climatic conditions, rice production is adversely affected. Climate change has rapid effects, which can be seen in the form of extreme prevailing weather conditions. Among the adversities of climate change, the most important is the rise in global temperature, upon which agriculture is mostly dependent. Although, rice is vulnerable to climate imposed changes, however, more sensitive to high temperature. This necessitates shift of rice production systems and conventional technologies towards acclimatization of changing climate. Developing climate smart and integrated technologies and equipping rice crop with those is a possible strategy to cope with the changing climate.

Keywords

Rice · Climate change · Productivity · High temperature

Atique-ur-Rehman (✉) · N. Sarwar
Department of Agronomy, Bahauddin Zakariya University, Multan, Pakistan
e-mail: dr.atique@bzu.edu.pk

S. Ahmad
Department of Agronomy, Faculty of Agricultural Sciences and Technology, Bahauddin Zakariya University, Multan, Pakistan
e-mail: shakeelahmad@bzu.edu.pk

M. A. Khan
In-Service Agriculture Training Institute Sargodha, Sargodha, Pakistan

M. Hasanuzzaman
Department of Agronomy, Sher-e-Bangla Agricultural University, Dhaka, Bangladesh

N. Sarwar et al. (eds.), *Modern Techniques of Rice Crop Production*,
https://doi.org/10.1007/978-981-16-4955-4_1

1.1 Introduction

Rice is the sole food consumed by world population more than any other crop (Ahmad et al. 2008, 2009, 2012, 2013, 2015, 2019; Ahmad and Hasanuzzaman 2012; Ahmed and Ahmad 2019, 2020; Shahzad and Ahmad 2019). The main rice producers and consumers are in Asia, which are more than 90% of the world (Ahmed and Fayyaz-ul-Hassan 2017; Ahmed et al. 2020a, b; Akram et al. 2018, 2019; Hussain et al. 2018; Fahad et al. 2019; Fatima et al. 2020). It is a basic staple food for majority of the population living in this continent which distinguishes this from the rest of the world (FAO 2008) (Hafeez-ur-Rehman et al. 2019; Khan et al. 2019a, b; Naz et al. 2019; Peerzada et al. 2019; Razaq et al. 2019; Sarwar et al. 2013a, b; Tariq et al. 2020; Wasaya et al. 2019; Zahoor et al. 2019). In most of these rice-producing regions, rice is profoundly decorated in culture or traditions and is splendid for having exciting rice farming systems. Rice production boosted to more than three times, after the introduction of high-yielding varieties during green revolution of the late 1960s. This has been possible due to the introduction of improved rice cultivars with irrigation facilities along with subsidized inputs. With a total harvested area of approximately 158 M ha and production of more than 700 Mt. annually (470 Mt. milled rice), it is produced in more than hundred countries. Top ten rice-producing countries are China, India, Indonesia, Bangladesh, Viet Nam, Thailand, Myanmar, the Philippines, Japan, and Brazil (FAO; Fig. 1.1). From the total global production, almost 640 million tons of rice or 90% is produced in Asia, while 25 and 19 million tons are produced in Latin America and Sub-Saharan Africa, respectively. Figure 1.2 shows the area and production of rice from past years, while Fig. 1.3 represents current rice production and harvested area across regions. It can grow in a variety of environments and be productive under situations not suitable for other crops. Likewise, rice is grown on a wide range of soil with average yield less than 1 t ha^{-1} from very poor rainfed environments and more than 10 t ha^{-1} in rigorous temperate irrigated systems. Rice

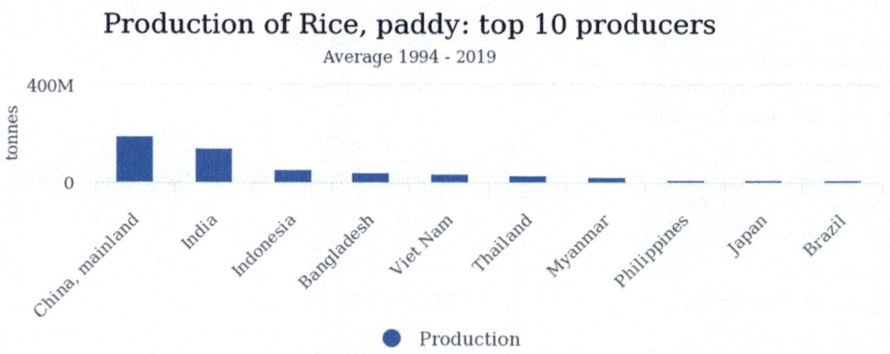

Fig. 1.1 Top ten rice-producing countries in the world (**Source:** FAOSTAT)

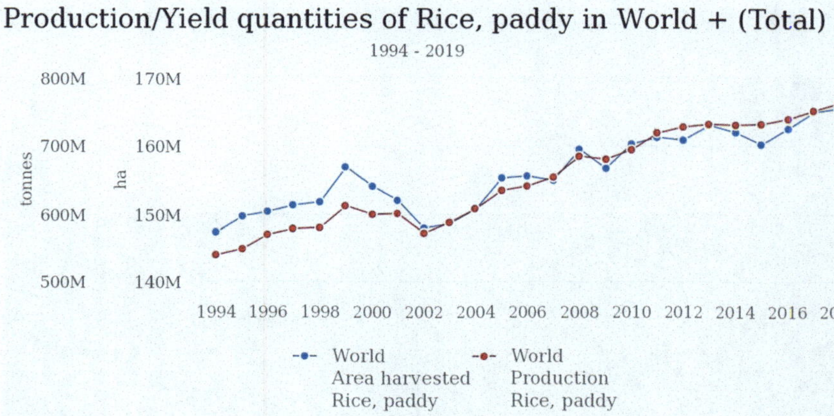

Fig. 1.2 Rice area and production in the world (**Source:** FAOSTAT)

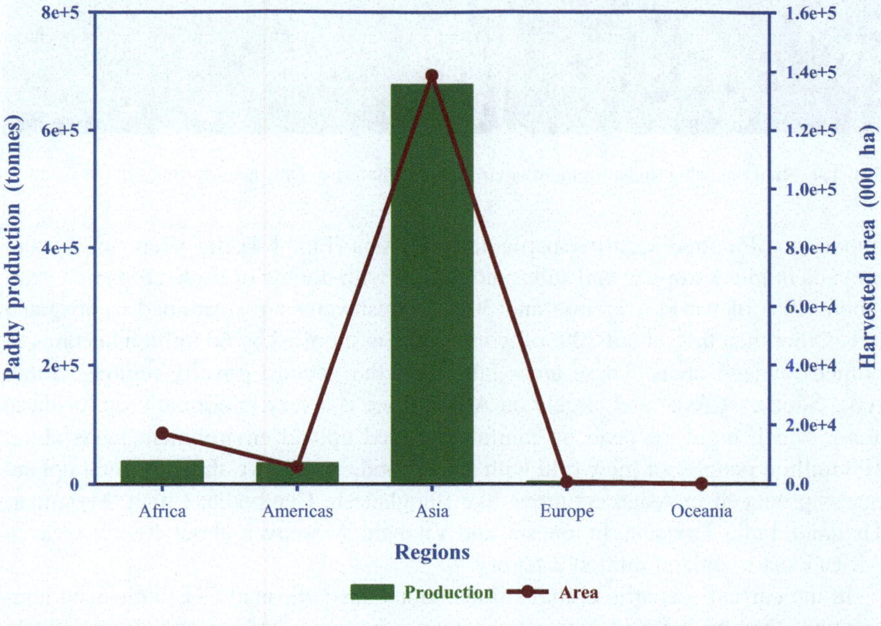

Fig. 1.3 Paddy productions and harvested area across regions in the world (**Source:** FAOSTAT 2021)

yields are relatively high in high-latitude areas with long day length under intensive farming techniques or in low-latitude desert areas with very high solar energy.

Rice can be cultivated in different ways all over the world, like lowland irrigated, rainfed lowland, and rainfed upland environment (Atique-ur-Rehman et al. 2014, 2018). Irrigated lowland rice contributes the largest share and provides about 75% of world's rice and is cultivated on 80 million hectares. This system seems to have more

Fig. 1.4 Rice maturing under ambient environment (Photo by: Dr. Shakeel Ahmad)

importance for food security, particularly in Asia (Fig. 1.4). Irrigated rice system prevails in many tropical and subtropical areas with double or triple crops in a year. About 40% of world irrigation and 30% of freshwater are consumed by irrigated rice. Other than this, about 20% of world's rice is supplied by 60 million hectares of rainfed lowland areas. These areas fall under the greatest poverty regions: South Asia, Southeast Asia, and largely on Africa. Rice is a very important crop for these areas, which highly depend on rainfall. Rainfed upland environment feeds about 100 million peoples of the world with staple food. About two-thirds of total upland rice is produced in Asian countries like Bangladesh, Cambodia, China, Myanmar, Thailand, India, Pakistan, Indonesia, and Vietnam. Moreover, about 40% of areas in Africa exist in upland rainfed category.

In the current scenario, climatic fluctuations have elevated CO_2; enhanced temperature, thereby creating extreme weather conditions; and enhanced pest attack, making things more complicated for the farmers. Depending upon the climatic variations among different regions, there is a huge difference in rice yield worldwide (Fig. 1.5). It is anticipated by rice researchers that global rice production, grain quality, and nutritional benefits are under severe threat by increased air temperature (Teixeira et al. 2013; Wang et al. 2011; Liu et al. 2010). Being C3 crop, rice may have benefited from increased atmospheric CO_2 concentration (Shimono et al. 2009); however, the benefit of CO_2 is countered by higher temperatures (Kadam et al. 2014). Rice grain yield and quality are adversely affected under higher

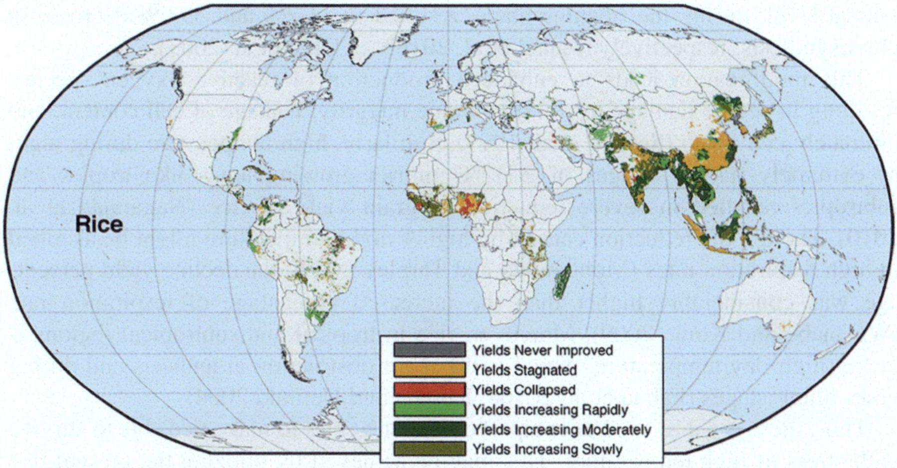

Fig. 1.5 Rice yield fluctuation in the world (**Source:** http://www.washingtonpost.com/blogs/worldviews/files/2012/12/PressRelease_Rice_HighRes.png)

temperatures, particularly at grain filling or reproductive stages (Bahuguna et al. 2015). Abiotic stresses are posing threat for rice but changing climate has triggered these with increased floods, drought, and salinity in many areas of the world. The annual loss of rice production is about 4 Mt. in just South Asia. Coastal areas are covered by about ten million hectares on which rice is single option to be grown, and poor are dependent on rice crop which is under threat. In addition to this, rising temperature especially at reproductive stage is impairing crop yield and is also a threat for future rice farmers. It is estimated that rice yield can be decreased from 7% to 1% on every rise of 1 °C of mean temperature. International Food Policy Research Institute said that about 12–14% of rice production would be reduced globally by 2050 due to changing climate and South Asia would be the worst-affected region. This would lead to increase in prices and less calorie intake, which ultimately increase child malnutrition. Food security is already a major issue in developing countries, so the climatic changes can further increase this problem. Proper management and planning are required to save the poor farmers and the rice-dependent population. Biotic stresses are other areas that need to be focused, as rice crops are already adversely affected by insect pests which can be triggered by changing climatic conditions of frequent extreme weather with erratic rainfall or water scarcity.

Diurnal temperature changes have significantly affected rice production by altering the physiological activities. For instance, from a certain level, day temperatures have direct effect on the activity of photosynthesis, which can be altered by varying the structure of thylakoid membrane and photosystem II (Zhang et al. 2005; Karim et al. 1997). Under stress conditions, the grain yield of rice depends on the cell membrane stability. It is noted that higher temperature of 3.6 °C and 7.0 °C than the critical level, during the ripening stage, resulted in 11.2% and 35.6% decrease in photosynthesis, respectively (Fahad et al. 2016b, c; Oh-e et al. 2007).

This modification leads to enhanced production of reactive oxygen species, resulting in severe damage to cell membrane integrity, leakage of cell content, and ultimately cell death (Schöffl et al. 1999). Similarly, high temperature during night or extremely narrow ranges of 2–3 °C in rice-growing areas like tropics and subtropics resulted in severe reduction in grain yield of rice (Nagarajan et al. 2010). Though the reduction caused by higher night temperature might be ascribed to high respiration rates (Mohammed and Tarpley 2009), the decline yield percentage was considerably higher than the increased percentage of respiration rate (Watanabe and Kume 2009). Moreover, rice in tropical and subtropical regions is facing high day temperature, which is the main obstruction at anthesis and during grain filling stages (Fahad et al. 2016a; Kobata and Uemuki 2004).

Thus, there is a dire need of comprehensive struggles to make rice able to survive with stress of high temperature. This may be achieved by utilizing the present rice genetic resources, targeting rice plants for improved tolerance to stress due to high temperature. Moreover, rice should be improved for tolerance to high temperature during grain filling stages to avoid losses of grain yield (Yamakawa et al. 2007). Likewise, adjustment of time of cultivation can regulate different critical stages like flowering and booting, to avoid rise of temperature stress, supporting rice plants to escape the unfavorable effects of heat stress (Shah et al. 2011).

Many different technologies or options are under study to manage climatic fluctuation, which needs to be documented for growers. Many varieties are introduced in different parts of the world for successful rice production by improving plants' tolerance to various biotic and abiotic stresses. Water-saving technologies like alternate wetting and drying, system of rice intensification, and direct seeding are being adopted by farmers to save the irrigation water.

With an up-surging world populace, the demands for rice are fostering, which seems to be precarious under changing climate with numerous emerging ambiguities. This includes a possible decline in area of production under rice with further strike on the growth and productivity. It is believed by many researchers that the rice production area is under stress and may drop owing to growing population, water deficiency, and other crops' competition. In addition to this, the enduring structural revolution in agriculture with conventional farming and lack of labor may affect rice production. Eventually, it is sole onus of rice scientists to either improve conventional production systems or devise such technologies, which are climate proof, enabling producers to grow more with the same or less water and other inputs.

Keeping in view the above worldwide importance of rice, the need to document the recent findings related to rice production becomes critical. Narration of each and every factor of rice production and its processing would be highly beneficial for the growers, researchers, and students. Rice production involves various important principles like seed, sowing, land preparation, insect pest management, irrigation management, fertilizer, harvesting, and processing. Lots of research have been done on these specific parameters, and much advancement is documented which needs to be filed in book format. Climate fluctuations have also forced the farmers to grow rice in altered ways; so many different technologies are under study, which also need to be summarized.

1.2 Major Issues

Integrated approaches and efforts have been employed to increase the production and productivity of rice to feed the burgeoning population of the world. Scientists have been successful to some extent in this regard. However, there are some major issues that are yet to be resolved, which need attention of all stakeholders and include (1) climate uncertainty and climate variability; (2) less nitrogen use efficiency for cereals; (3) availability of quality seeds; (4) nursery establishment; (5) water shortage; (6) water use and water use efficiency for transplanted or anaerobic rice; (7) weed problem for directed seeded or aerobic rice; (8) insect, pest, and disease management; (9) abiotic stresses at critical phenological stages; (10) local marketing problems; and (11) international trade barriers or issues.

1.3 Conclusion

No doubt the production and productivity of rice crop have been increased as compared to previous centuries. However, current issue of climate variability need solution. Due to climate uncertainty, the resource management (like water, land, nutrient, radiation) has totally changed. Likewise, insect, pest, and disease management needs new technologies to solve these issues, as insects' behavior and causal organisms' behavior have totally changed. Therefore, in order to overcome the above-mentioned old and current issues and to feed the burgeoning population of the world, integrated approaches are necessary on a sustained basis.

References

Ahmad S, Hasanuzzaman M (2012) Integrated effect of plant density, N rates and irrigation regimes on the biomass production, N content, PAR use efficiencies and water productivity of rice under irrigated semiarid environment. Not Bot Horti Agrobot Cluj-Napoca 40(1):201–211

Ahmad S, Zia-ul-Haq M, Ali H, Shad SA, Ammad A, Maqsood M, Khan MB, Mehmood S, Hussain A (2008) Water and radiation use efficiencies of transplanted rice (*Oryza sativa* L.) at different plant densities and irrigation regimes under semi-arid environment. Pak J Bot 40(1): 199–209

Ahmad S, Ahmad A, Zia-ul-Haq M, Ali H, Khaliq T, Anjum MA, Khan MA, Hussain A, Hoogenboom G (2009) Resources use efficiency of field grown transplanted rice (*Oryza sativa* L.) under irrigated semiarid environment. J Food Agric Environ 7(2):487–492

Ahmad S, Ahmad A, Soler CMT, Ali H, Zia-Ul-Haq M, Anothai J, Hussain A, Hoogenboom G, Hasanuzzaman M (2012) Application of the CSM-CERES-Rice model for evaluation of plant density and nitrogen management of fine transplanted rice for an irrigated semiarid environment. Precis Agric 13(2):200–218

Ahmad S, Ahmad A, Ali H, Hussain A, Garcia y Garcia A, Khan MA, Zia-Ul-Haq M, Hasanuzzaman M, Hoogenboom G (2013) Application of the CSM-CERES-Rice model for evaluation of plant density and irrigation management of transplanted rice for an irrigated semiarid environment. Irrig Sci 31(3):491–506

Ahmad A, Ashfaq M, Rasul G, Wajid SA, Khaliq T, Rasul F, Saeed U, Rahman MH, Hussain J, Baig IA, Naqvi AA, SAA B, Ahmad S, Naseem W, Hoogenboom G, Valdivia RO (2015) Impact of climate change on the rice–wheat cropping system of Pakistan. In: Hillel D, Rosenzweig C (eds) Handbook of climate change and agro-ecosystems: The agricultural modeling intercomparison and improvement project (AgMIP) integrated crop and economic assessments. Imperial College Press and The American Society of Agronomy, London, pp 219–258

Ahmad S, Abbas G, Ahmed M, Fatima Z, Anjum MA, Rasul G, Khan MA, Hoogenboom G (2019) Climate warming and management impact on the change of rice-wheat phenology in Punjab, Pakistan. Field Crop Res 230:46–61

Ahmed M, Ahmad S (2019) Carbon dioxide enrichment and crop productivity. In: Hasanuzzaman M (ed) Agronomic crops, vol 2. Springer, Singapore, pp 31–46

Ahmed M, Ahmad S (2020) Systems modeling. In: Ahmed M (ed) Systems modeling. Springer, Singapore, pp 1–44

Ahmed M, Fayyaz-ul-Hassan AS (2017) Climate variability impact on rice production: adaptation and mitigation strategies. In: Ahmed M, Stockle C (eds) Quantification of climate variability, adaptation and mitigation for agricultural sustainability. Springer, Switzerland, pp 91–111

Ahmed M, Ahmad S, Raza MA, Kumar U, Ansar M, Shah GA, Parsons D, Hoogenboom G, Palosuo T, Seidel S (2020a) Models calibration and evaluation. In: Ahmed M (ed) Systems modeling. Springer, Singapore, pp 151–178

Ahmed M, Ahmad S, Waldrip HM, Ramin M, Raza MA (2020b) Whole farm modeling: a systems approach to understanding and managing livestock for greenhouse gas mitigation, economic viability and environmental quality. In: Waldrip HM, Pagliari PH, He Z (eds) Animal manure. ASA Special Publication 67. American Society of Agronomy and Soil Science Society of America, Madison, WI, pp 345–371

Akram R, Turan V, Hammad HM, Ahmad S, Hussain S, Hasnain A, Maqbool MM, Rehmani MIA, Rasool A, Masood N, Mahmood F, Mubeen M, Sultana SR, Fahad S, Amanet K, Saleem M, Abbas Y, Akhtar HM, Hussain S, Waseem F, Murtaza R, Amin A, Zahoor SA, Sami ul Din M, Nasim W (2018) Fate of organic and inorganic pollutants in paddy soils. In: Hashmi MZ, Varma A (eds) Environmental pollution of paddy soils. Springer, Switzerland, pp 197–214

Akram R, Fahad S, Masood N, Rasool A, Ijaz M, Ihsan MZ, Maqbool MM, Ahmad S, Hussain S, Ahmed M, Kaleem S, Sultana SR, Mubeen M, Saud S, Kamran M, Nasim W (2019) Plant growth and morphological changes in rice under abiotic stress. In: Hasanuzzaman M, Fujita M, Nahar K, Biswas JK (eds) Advances in rice research for abiotic stress tolerance. Woodhead Publishing, Sawston, pp 69–85

Atique-ur-Rehman M, Farooq AN, Ahmad R (2014) Influence of boron nutrition on the rice productivity, grain quality and biofortification in different production systems. Field Crop Res 169:123–131

Atique-ur-Rehman M, Farooq A, Rashid F, Nadeem S, Stürz FA, Bel RW (2018) Boron nutrition of rice in different production systems. A review. Agron Sustain Dev 38:25

Bahuguna RN, Jha J, Pal M, Shah D, Lawas LM, Khetarpal S, Jagadish KSV (2015) Physiological and bio- chemical characterization of NERICA-L-44: a novel source of heat tolerance at the vegetative and reproductive stages in rice. Physiol Plant 154:543–559

Fahad S, Hussain S, Saud S, Hassan S, Chauhan BS, Khan F, Ihsan MZ, Ullah A, Wu C, Bajwa AA, Alharby H, Amanullah NW, Shahzad B, Tanveer M, Huang J (2016a) Responses of rapid viscoanalyzer profile and other rice grain qualities to exogenously applied plant growth regulators under high day and high night temperatures. PLoS One 11(7):e0159590

Fahad S, Hussain S, Saud S, Khan F, Hassan S, Amanullah NW, Arif M, Wang F, Huang J (2016b) Exogenously applied plant growth regulators affect heat-stressed rice pollens. J Agron Crop Sci 202:139–150

Fahad S, Hussain S, Saud S, Hassan S, Ihsan Z, Shah AN, Wu C, Yousaf M, Nasim W, Alharby H, Alghabari F, Huang J (2016c) Exogenously applied plant growth regulators enhance themorphophysiological growth and yield of rice under high temperature. Front Plant Sci 7: 1250

Fahad S, Noor M, Adnan M, Khan MA, Rahman I, Alam M, Khan IA, Ullah H, Mian IA, Hassan S, Saud S, Bakhat HF, Hammad HM, Ahmad S, Nasim W (2019) Abiotic stress and rice grain quality. In: Hasanuzzaman M, Fujita M, Nahar K, Biswas JK (eds) Advances in rice research for abiotic stress tolerance. Woodhead Publishing, Sawston, pp 571–583

FAO (2008) Food and Agriculture Statistics Global Outlook. Food and Agricultural Organization of the United Nations. http://faostat.fao.org/Portals/_Faostat/documents/pdf/world.pdf

Fatima Z, Ahmad M, Hussain M, Abbas G, Ul-Allah S, Ahmad S, Ahmed N, Ali MA, Sarwar G, Ehsan ul Haque, Iqbal P, Hussain S (2020) The fingerprints of climate warming on cereal crops phenology and adaptation options. Sci Rep 10:18013

Hafeez-ur-Rehman, Nawaz A, Awan MI, Ijaz M, Hussain M, Ahmad S, Farooq M (2019) Direct seeding in rice: problems and prospects. In: Hasanuzzaman M (ed) Agronomic crops. Production technologies, vol 1. Springer, Singapore, pp 199–222

Hussain M, Ahmad S, Hussain S, Lal R, Ul-Allah S, Nawaz A (2018) Rice in saline soils: physiology, biochemistry, genetics and management. Adv Agron 148:231–287

Kadam NN, Jagadish KS, Xiao G, Melgar RJ, Bahuguna RN, Quinones C, Tamilselvan A, Prasad PVV, Jagash KSV (2014) Agronomic and physiological responses to high temperature, drought, and elevated CO_2 interactions in cereals. Adv Agron 127:111–156

Karim MA, Fracheboud Y, Stamp P (1997) Heat tolerance of maize with reference of some physiological characteristics. Ann Bangladesh Agri 7:27–33

Khan MA, Ahmad S, Raza A (2019a) Integrated weed management for agronomic crops. In: Hasanuzzaman M (ed) Agronomic crops, vol 2. Springer, Singapore, pp 257–281

Khan SM, Ali S, Nawaz A, Bukhari SAH, Ejaz S, Ahmad S (2019b) Integrated pest and disease management for better agronomic crop production. In: Hasanuzzaman M (ed) Agronomic crops, vol 2. Springer, Singapore, pp 385–428

Kobata T, Uemuki N (2004) High temperatures during the grain-filling period do not reduce the potential grain dry matter increase of rice. Agron J 96:406–414

Liu ZJ, Guo YK, Bai JG (2010) Exogenous hydrogen peroxide changes antioxidant enzyme activity and protects ultrastructure in leaves of two cucumber ecotypes under osmotic stress. J Plant Growth Regul 29:171–183

Mohammed AR, Tarpley L (2009) Impact of high night time temperature on respiration, membrane stability, antioxidant capacity and yield of rice plants. Crop Sci 49:313–322

Nagarajan S, Jagadish SVK, Hari Prasad AS, Thomar AK, Anand A, Pal M, Agarwal PK (2010) Local climate affects growth, yield and grain quality of aromatic and non-aromatic rice in northwestern India. Agric Ecosyst Environ 138:274–281

Naz S, Fatima Z, Iqbal P, Khan A, Zakir I, Noreen S, Younis H, Abbas G, Ahmad S (2019) Agronomic crops: types and uses. In: Hasanuzzaman M (ed) Agronomic crops. Production technologies, vol 1. Springer, Singapore, pp 1–18

Oh-e I, Saitoh K, Kuroda T (2007) Effects of high temperature on growth, yield and dry-matter production of rice grown in the paddy field. Plant Prod Sci 10:412–422

Peerzada AM, Bukhari SAH, Dawood M, Nawaz A, Ahmad S, Adkins S (2019) Weed management for healthy crop production. In: Hasanuzzaman M (ed) Agronomic crops, vol 2. Springer, Singapore, pp 225–256

Razaq M, Shah FM, Ahmad S, Afzal M (2019) Pest management for agronomic crops. In: Hasanuzzaman M (ed) Agronomic crops, vol 2. Springer, Singapore, pp 365–384

Sarwar N, Ali H, Ahmad S, EhsanUllah AS, Mubeen K, Hill JE (2013a) Water wise rice cultivation on calcareous soil with the addition of essential micronutrients. J Anim Plant Sci 23(1):244–250

Sarwar N, Ali H, Maqsood M, EhsanUllah SAN, Shahzad M, Mubeen K, Shahid MA, Ahmad S (2013b) Phenological response of rice plants to different micronutrients application under water saving paddy fields on calcareous soil. Turk J Field Crops 18(1):52–57

Schöffl F, Prandl R, Reindl A (1999) Molecular responses to heat stress. In: Shinozaki K, Yamaguchi-Shinozaki K (eds) Molecular responses to cold, drought, heat and salt stress in higher plants. R.G. Landes Co, Austin, TX, pp 81–98

Shah F, Huang J, Cui KH, Nie LX, Shah T, Wu W, Wang K, Khan ZH, Zhu L, Chen C (2011) Physiological and biochemical changes in rice associated with high night temperature stress and their amelioration by exogenous application of ascorbic acid (vitamin C). Aust J Crop Sci 5: 1810–1816

Shahzad AN, Ahmad S (2019) Tools and techniques for nitrogen management in cereals. In: Hasanuzzaman M (ed) Agronomic crops, vol 2. Springer, Singapore, pp 111–126

Shimono H, Okada M, Yamakawa Y, Nakamura H, Kobayashi K, Hasegawa T (2009) Genotypic variation in rice yield enhancement by elevated CO_2 relates to growth before heading, and not to maturity group. J Exp Bot 60:523–532

Tariq M, Ahmed M, Iqbal P, Fatima Z, Ahmad S (2020) Crop phenotyping. In: Ahmed M (ed) Systems modeling. Springer, Singapore, pp 45–60

Teixeira EI, Fischer G, Van VH, Walter C, Ewert F (2013) Global hot-spots of heat stress on agricul- tural crops due to climate change. Agric For Meteorol 170:206–215

Wang Y, Frei M, Song Q, Yang L (2011) The impact of atmospheric CO_2 concentration enrichment on rice quality- a research review. Acta Ecol Sin 31:277–282

Wasaya A, Yasir TA, Ijaz M, Ahmad S (2019) Tillage effects on agronomic crop production. In: Hasanuzzaman M (ed) Agronomic crops, vol 2. Springer, Singapore, pp 73–99

Watanabe T, Kume T (2009) A general adaptation strategy for climate change impacts on paddy cultivation: Special reference to the Japanese context. Paddy Water Environ 7:313–320

Yamakawa H, Hirose T, Kuroda M, Yamaguchi T (2007) Comprehensive expression profiling of rice grain filling-related genes under high temperature using DNA microarray. Plant Physiol 144:258–277

Zahoor SA, Ahmad S, Ahmad A, Wajid A, Khaliq T, Mubeen M, Hussain S, Sami Ul Din M, Amin A, Awais M, Nasim W (2019) Improving water use efficiency in agronomic crop production. In: Hasanuzzaman M (ed) Agronomic crops. Springer, Singapore, pp 13–29

Zhang JH, Huang WD, Liu YP, Pan QH (2005) Effects of temperature acclimation pretreatment on the ultrastructure of mesophyll cells in young grape plants (*Vitis vinifera* L. cv. Jingxiu) under cross temperature stresses. J Integr Plant Biol 47:959–970

Climate Change and Global Rice Security 2

Allah Wasaya, Tauqeer Ahmad Yasir, Naeem Sarwar, Atique-ur-Rehman, Khuram Mubeen, Karthika Rajendran, Adel Hadifa, and Ayman E. L. Sabagh

Abstract

Rice is a first- or second-order staple food in many countries. It is mainly produced and consumed in Asia. There are so many causes of low yield of rice crops in the world. However, abiotic and biotic stresses are major factors for reducing rice yield and quality. The stresses' share in reducing rice yield across various regions varies greatly. So, in order to adapt to and mitigate climate change, both short-term and long-term approaches are needed for ensuring food security at regional and global levels.

A. Wasaya (✉) · T. A. Yasir
College of Agriculture, Bahauddin Zakariya University, Bahadur Sub-Campus, Layyah, Pakistan

N. Sarwar · Atique-ur-Rehman
Department of Agronomy, Bahauddin Zakariya University, Multan, Pakistan

K. Mubeen
Department of Agronomy, MNS University of Agriculture, Multan, Pakistan

K. Rajendran
School of Agricultural Innovations and Advanced Learning, Vellore Institute of 2 Technology (VIT), Vellore, Tamil Nadu, India

A. Hadifa
Rice Research and Training Center (RRTC), Field Crops Research Institute, Agricultural Research Center, Sakha, Kafr Elsheikh, Egypt

A. E. L. Sabagh
Faculty of Agriculture, Department of Field Crops, Siirt University, Siirt, Turkey

Faculty of Agriculture, Department of Agronomy, University of Kafrelsheikh, Kafrelsheikh, Egypt

N. Sarwar et al. (eds.), *Modern Techniques of Rice Crop Production*,
https://doi.org/10.1007/978-981-16-4955-4_2

Keywords

Rice · Abiotic · Biotic · Stresses · Climate change

2.1 Introduction

Crop production depends on soil characteristics, soil fertility, innovative technology, crop management practices, and climatic conditions (Ahmad et al. 2008, 2009, 2012, 2013, 2015, 2019; Ahmad and Hasanuzzaman 2012; Ahmed and Ahmad 2019, 2020; Shahzad and Ahmad 2019). Today, the agriculture sector is facing a serious challenge to provide food security for about 9 billion people worldwide by 2050 while protecting the environment and improving global ecosystems (Hafeez-ur-Rehman et al. 2019; Khan et al. 2019a, b; Naz et al. 2019; Peerzada et al. 2019; Razaq et al. 2019; Sarwar et al. 2013a, b; Tariq et al. 2020; Wasaya et al. 2019; Zahoor et al. 2019). These challenges will be further aggravated by climate change factors under changing climate scenarios in the near future (Cline 2007; IPCC 2007; Ahmed and Fayyaz-ul-Hassan 2017; Ahmed et al. 2020a, b; Akram et al. 2018, 2019; Hussain et al. 2018; Fahad et al. 2019; Fatima et al. 2020).

Climate change is a major threat to crop production and food security (Field 2012; Rezaei et al. 2015). Extreme weather conditions and uncertainty in rainfall patterns had an adverse effect on field crop production (Ullah et al. 2017; Ahmed et al. 2018). Variability in climatic factors plays a significant role in agriculture (Wang et al. 2009; Yang et al. 2011), mainly in grain production, which is the most important sector for feeding ever-increasing population throughout the world. Land and water resources are decreasing, while population is increasing daily, which demands more food to feed the increasing population. It is estimated that rice is the key staple food of 50% of the world's population (GRiSP 2013). With increasing population and to keep up with global population growth and demand, rice production must increase by 25% up to 2030 (Seck et al. 2012).

2.2 Global Rice Production

World's rice production has increased markedly since the beginning of the Green Revolution by almost 140%. World's rice cultivation and production are fluctuating every year, and it was grown in an area of 165 M ha with a total production of 755 Mt. (FAO 2019). From 1968 to 2010, the area under rice cultivation has been increased from 129 to 159.4 M ha. The world's largest rice producers are China and India. China's rice production is higher than other countries due to rice cultivation in irrigated areas. In fact, rice is well adapted to various agro-climatic conditions, and it is currently grown in both dry- and wet-land environmental conditions at high and low altitudes (Ahmad et al. 2019; Fatima et al. 2020).

Asia being the highest rice producer determines the future trend for rice production globally. In Asia, the rice area increased rapidly during the 1960s due to the

Green Revolution. About 75% of total rice is obtained from 85 to 90 M ha of paddy lowland areas in the world (IRRI, Africa Rice and CIAT 2010). This system produces about 56% of total rice area in Asia (Swain et al. 2005). Outside Asia, Brazil is the largest rice-producing country (Schwanck et al. 2015).

2.3 Global Rice Consumption

Rice plays a key role in the diet of many people on the earth, is consumed by approximately 3 billion people, and feeds people more than any other crops (Maclean et al. 2002). It is recorded that Asia is the largest rice producer and consumer (90%), where it is cultivated under irrigated and rainfed ecosystems. It ranks second in terms of harvested area after wheat globally (Garbach et al. 2014). It is used as primary source of calories by one-third world's population. South and Southeast Asia are the largest rice-growing regions, but most of the rice is grown in lowland and in primarily rainfed cultivation (Garbach et al. 2014) (Fig. 2.1).

Rice is used as staple food in many countries where its production exceeds one million t annually. Rice provides 50–80% calories of the total consumed in countries like Bangladesh, Cambodia, Indonesia, Myanmar, Thailand, and Vietnam (Rogers and Martyn 2009). Rice is becoming the fastest growing staple food in Africa.

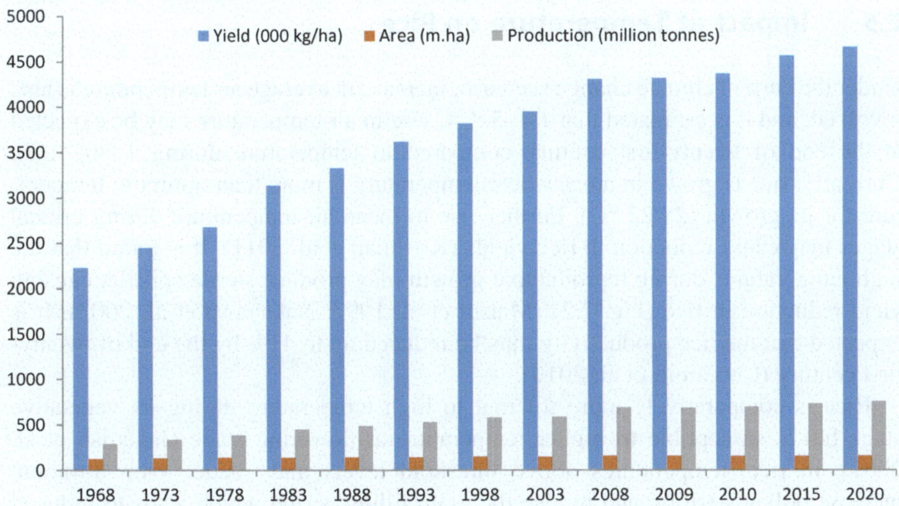

Fig. 2.1 Historical yield, area, and production of rice crop in the world (**Source:** FAOSTAT)

2.4 Effect of Climate Change on Rice Production

Rice production is expected to be influenced by current climate change due to increased temperature and CO_2 and change in rainfall pattern. The rapid change in climate can be noticed from extreme weather conditions and its effect on crop production and declining food security. Also, it will have an impact on freshwater systems, and reduction in the production of rice in some parts of Asia is recorded due to drought which was induced by increased temperature, decrease in number of rainy days, and increased dry spells (Cairns et al. 2012). High temperatures would cause an obvious reduction in world's rice production (Furuya and Koyama 2005). Furthermore, studies have demonstrated that global climate changes might affect rice food production and distribution in different parts of the world.

Understanding the impact of climate change is important on rice-based production systems to develop applicable strategies to adapt to and mitigate consequences of climate change and food security in the long term. Climate change effects on rice can be seen from germination period until the harvesting and post-harvesting stages. The occurrence of rainfall at harvesting and threshing time may cause reduction in grain quality (Brolley 2015). It is predicted that due to climate change, the temperature increase of 2 °C or more in the local condition without adaptation and mitigation will affect the overall agricultural production negatively by the mid-twenty-first century (IPCC 2014).

2.5 Impact of Temperature on Rice

Under the current climate change scenario, increase in average air temperature is also involved, and it is estimated that 1.4–5.8 °C rise in air temperature may be expected at the end of twenty-first century compared to temperature during 1980–2000. Currently, rice is grown in areas where temperature is more than optimum temperature for its growth (28/22 °C). Further rise in mean air temperature during critical stages may cause reduction in rice yields (Krishnan et al. 2011). It is found that the high temperatures during reproductive growth may produce sterile spikelet causing yield reduction in rice (Fig. 2.2) (Matsui et al. 1997; Nakagawa et al. 2003). It is expected that the rice productivity may be reduced up to 41% by the end of twenty-first century (Ceccarelli et al. 2010).

Rice is comparatively more tolerant to high temperature during its vegetative stage but is susceptible to higher temperature at flowering stage (Jagadish et al. 2007). In rice, temperatures above threshold levels may reduce crop duration, increase pollen sterility, and reduce the grain filling period which leads to reduced rice productivity and quality (Fitzgerald and Resurreccion 2009; Kim et al. 2011). It has been documented that an increase of 1 °C above the critical temperature (>24 °C) leads to 10% decrease in rice productivity (Peng et al. 2004; Welch et al. 2010). According to another field study, an increase of 3.6–7 °C could lead to photosynthetic reduction ranging from 12.2% to 35.6% (Oh-e et al. 2007; Wassmann et al. 2009a). It is observed that high temperature decreases the

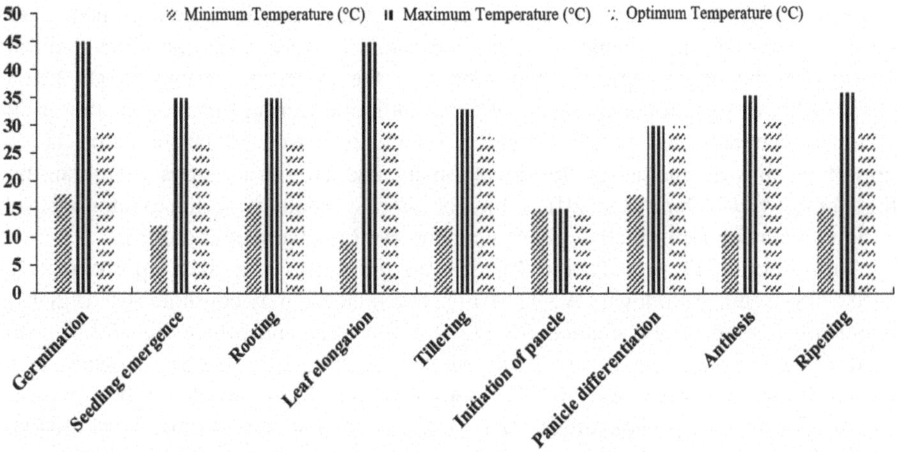

Fig. 2.2 Temperature ranges for the successful rice production at various growth stages (**Source:** Yoshida 1981)

photosynthetic efficiency and yield potential of rice (Wassmann et al. 2009a, b). This decrease in photosynthesis might be due to variations in organizational structure of thylakoid membrane (Karim et al. 1997) and loss of stacking of grana in chloroplast (Wahid and Close 2007).

On the whole, the reproductive phase is more sensitive to high temperature than vegetative phase in rice (Yoshida et al. 1981). Anthesis, which is identified with appearance of anthers and dehiscence of pollen grain, is more sensitive to higher temperature (Nakagawa et al. 2002). When compared to the pollen grains, pistil remains less affected under heat stress. Yoshida et al. (1981) demonstrated that pistil capability to be fertilized remained unaffected even more than a duration of 5 days at 41 °C in the reciprocal studies of manual pollen shedding from control plants onto stigma when exposed to higher temperatures and vice versa. Similarly, an increase of 30–80% in spikelet fertility was recorded by pollinating heat-stressed pistil with unstressed pollen in wheat (Saini and Aspinall 1982).

2.6 Effect of Solar Radiations, Day Length, and CO_2 Levels

The potential production is considered to be determined solely by the interaction of genotypic characteristics with the solar radiations, CO_2 level, and day length that it experiences during its growth cycle. Solar radiation provides the energy for the uptake of CO_2 utilized in the photosynthesis for the preparation of glucose, while temperature determines the crop growth duration for both vegetative and reproductive growth and rates of physiological and morphological processes. Day length can affect the rates of development at certain phases of crop life duration, besides to a lesser extent, the quantitative amount of solar radiation received by the crop.

Solar radiation is an important climatic factor which provides the light necessary for seed germination, expansion of leaf, and shoot growth. It also provides thermal energy for the physiological development of the plant. According to De Datta (1981), due to the higher intensity of solar radiation during dry season, the grain yield can be recorded more compared with wet season. Solar radiation is an important climatic factor for the development and growth of crops for obtaining higher crop yield (Deng et al. 2015). Increase in rice productivity due to intense solar radiations during reproductive phase was observed (Katsura et al. 2008).

Atmospheric CO_2 increased to 403.3 ± 0.1 μmol mol^{-1} in 2016, approximately 145% of pre-industrial level (WMO 2017). The increase may continue (IPCC 2014) which may increase the chances of climate warming and affect the global crop production (Rosenzweig et al. 2014; Melloy et al. 2014). Carbon dioxide is a growth-limiting factor mainly for C_3 crops like rice (Shimono et al. 2009), and it has direct effect on photosynthesis and stomatal conductance (Franks et al. 2013), which increase crop yield (Ainsworth 2008). An increased level of CO_2 may enhance growth and development in plants (Cai et al. 2015) and resulted in early flowering of rice (Li et al. 2013). It may increase the rate of chalky rice and impaired quality of rice (Liu et al. 2017).

2.7 Effect of Precipitation

Rainfall is one of the important factors affecting crop yield. Rainfall intensity and distribution has direct impact on yield of rice. Presently, around 40% of the total rice area is cultivated as lowland or upland, whereas around 3.5 M ha of rice is cultivated as flooded rice (Maclean et al. 2002). Both too much rainfall and too much dry spell are harmful for crop production. More rain leads to high water levels in rice fields which may cause drainage problems in roots and block the free movement of oxygen and form toxic compounds which damage plant roots. Second, high rain at reproductive phase may affect the fertilization and grain formation process and affect the rice quality. In addition, this excessive rainfall also affects the farming activities such as field preparation, crop harvesting, processing, and seed drying. It may also increase the chances of disease spread in plants (Basnayake et al. 2006). On the other hand, less rains adversely affect the rice yield in lowland areas. It may also decrease the average yield in such areas.

2.8 Effect of Drought Stress

Drought stress is the key climatic factor which limits agricultural production throughout the world (Passioura 1996). In Asia, drought is an important constraint to rice production affecting about 10 and 13 M ha of upland and rainfed lowland rice, respectively (Pandey et al. 2007). Around 50% of total rice production area in world is affected by drought (Bouman et al. 2005). According to an estimate, drought has affected about 55% of rice area in India, which resultantly affected 300 million people in 2002 (Pandey et al. 2007). Drought stress in rice-producing areas arises from the higher frequency of El Niño events and reduction in average rainfall coupled with high temperatures and increase in evapotranspiration (Tao et al. 2004).

Drought stress at early growth stages may affect vegetative growth rate besides productivity (Tao et al. 2006). Rice plants may die in case of very less rainfall of less than 200 in rainfed regions. High water stress at reproductive stages leads to reduction in flowering, grain filling, and final yield (Rana and Randhawa 2014). It was observed that rice yield has been reduced in past decades in Asia, owing to increasing drought stress (Tao et al. 2004; Fischer et al. 2002). Drought stress during flowering stage leads to reduction in rice compared to irrigation (Serraj and Atlin 2008). This situation may lead to reduced food security, especially in arid and semi-arid areas (Bates et al. 2008).

Drought stress is characterized by reduced soil moisture, less water potential, turgor pressure, stomatal activity in leaves, and a decrease in cell enlargement as well as growth in crop plants. Usually, severe drought reduced photosynthesis, disturbs plant metabolic pathways, and finally leads to plant death (Jaleel et al. 2008). It may also affect various physiological processes such as photosynthesis and respiration, ion uptake and translocation, carbohydrates, and metabolism of nutrients (Farooq et al. 2008).

Drought stress during flowering may cause sterility in spikelets and produce unfilled grains (Kamoshita et al. 2004) which might be due to early senescence and short grain filling period (Plaut et al. 2004). Drought stress at any critical stage affects crop yield (Bashir et al. 2016). Under drought stress, rice plants undergo leaf rolling and wilting which causes osmotic variations, reduced photosynthesis, as well as reduction in fertility and yield (Kamoshita et al. 2008; Pandey and Shukla 2015). Drought stress at vegetative growth causes reduced plant height, tillers, and biomass production in rice (Ji et al. 2012). Under drought stress, the plant releases hormones like ethylene that inhibit leaf and root growth on the primary phase (Basu et al. 2016), and stress occurring before flowering reduces yield. During grain filling, drought causes early senescence in the plant which shortens its filling period, lessens grain numbers, and reduces overall grain yield in rice (Plaut et al. 2004; Botwright et al. 2008).

2.9 Effect of Salinity

Salinity stress is a major limiting factor for agricultural crop production including rice cultivation as well (Velmurugan et al. 2016). Soil salinity has multiple effects on crops including stunted crop growth due to more osmotic potential. Inhibition of photosynthetic process, poor root growth, and reduced nutrient uptake ultimately resulted in reduction of crop yield (Velmurugan et al. 2016; Machado and Serralheiro 2017). Seed germination is the first stage which is severely affected due to salinity and resulted in poor crop germination (Ibrahim 2016). Mild salinity levels resulted in more production of abscisic acid which leads to seed dormancy, hence causes poor and delayed seed germination (Ibrahim 2016; He et al. 2019).

Higher salinity levels lead to cell damage, death of root cells, and reduced water and nutrient uptake (Ibrahim 2016; Thiam et al. 2013; Wang et al. 2011). In rice, it was noticed that salt stress leads to poor seed germination besides seedling (Pearson et al. 1966; Mondal and Borromeo 2016).

The salt stress affects vegetative and reproductive stages of crop plants (Senguttuvel et al. 2014). It adversely affects the various physiological processes like chlorophyll concentration and photosynthesis in plants (Senguttuvel et al. 2014). Furthermore, it affects the chlorophyll-a and chlorophyll-b ratio in plants (Senguttuvel et al. 2014). Salinity stress also affects the reproductive phase, especially flowering stage (Singh et al. 2004). At booting, it affects the pollen viability which leads to poor fertilization as well as grain filling, hence resulted in poor grain yield (Mohammadi-Nejad et al. 2010).

References

Ahmad S, Hasanuzzaman M (2012) Integrated effect of plant density, N rates and irrigation regimes on the biomass production, N content, PAR use efficiencies and water productivity of rice under irrigated semiarid environment. Not Bot Horti Agrobot Cluj-Napoca 40(1):201–211

Ahmad S, Zia-ul-Haq M, Ali H, Shad SA, Ammad A, Maqsood M, Khan MB, Mehmood S, Hussain A (2008) Water and radiation use efficiencies of transplanted rice (*Oryza sativa* L.) at different plant densities and irrigation regimes under semi-arid environment. Pak J Bot 40(1): 199–209

Ahmad S, Ahmad A, Zia-ul-Haq M, Ali H, Khaliq T, Anjum MA, Khan MA, Hussain A, Hoogenboom G (2009) Resources use efficiency of field grown transplanted rice (*Oryza sativa* L.) under irrigated semiarid environment. J Food Agric Environ 7(2):487–492

Ahmad S, Ahmad A, Soler CMT, Ali H, Zia-Ul-Haq M, Anothai J, Hussain A, Hoogenboom G, Hasanuzzaman M (2012) Application of the CSM-CERES-Rice model for evaluation of plant density and nitrogen management of fine transplanted rice for an irrigated semiarid environment. Precis Agric 13(2):200–218

Ahmad S, Ahmad A, Ali H, Hussain A, Garcia y Garcia A, Khan MA, Zia-Ul-Haq M, Hasanuzzaman M, Hoogenboom G (2013) Application of the CSM-CERES-Rice model for evaluation of plant density and irrigation management of transplanted rice for an irrigated semiarid environment. Irrig Sci 31(3):491–506

Ahmad A, Ashfaq M, Rasul G, Wajid SA, Khaliq T, Rasul F, Saeed U, Rahman MH, Hussain J, Baig IA, Naqvi AA, Bokhari SAA, Ahmad S, Naseem W, Hoogenboom G, Valdivia RO (2015) Impact of climate change on the rice–wheat cropping system of Pakistan. In: Hillel D, Rosenzweig C (eds) Handbook of climate change and agro-ecosystems: the agricultural modeling intercomparison and improvement project (AgMIP) integrated crop and economic assessments. Imperial College Press and The American Society of Agronomy, London, pp 219–258

Ahmad S, Abbas G, Ahmed M, Fatima Z, Anjum MA, Rasul G, Khan MA, Hoogenboom G (2019) Climate warming and management impact on the change of rice-wheat phenology in Punjab, Pakistan. Field Crop Res 230:46–61

Ahmed M, Ahmad S (2019) Carbon dioxide enrichment and crop productivity. In: Hasanuzzaman M (ed) Agronomic crops, vol 2. Springer, Singapore, pp 31–46

Ahmed M, Ahmad S (2020) Systems modeling. In: Ahmed M (ed) Systems modeling. Springer, Singapore, pp 1–44

Ahmed M, Fayyaz-ul-Hassan AS (2017) Climate variability impact on rice production: adaptation and mitigation strategies. In: Ahmed M, Stockle C (eds) Quantification of climate variability, adaptation and mitigation for agricultural sustainability. Springer, Switzerland, pp 91–111

Ahmed I, U-Rahman MH, Ahmed S, Hussain J, Ullah A, Judge J (2018) Assessing the impact of climate variability on maize using simulation modeling under semiarid environment of Punjab, Pakistan. Environ Sci Pollut Res 25(28):28413–28430

Ahmed M, Ahmad S, Raza MA, Kumar U, Ansar M, Shah GA, Parsons D, Hoogenboom G, Palosuo T, Seidel S (2020a) Models calibration and evaluation. In: Ahmed M (ed) Systems modeling. Springer, Singapore, pp 151–178

Ahmed M, Ahmad S, Waldrip HM, Ramin M, Raza MA (2020b) Whole farm modeling: a systems approach to understanding and managing livestock for greenhouse gas mitigation, economic viability and environmental quality. In: Waldrip HM, Pagliari PH, He Z (eds) Animal Manure. ASA Special Publication 67. American Society of Agronomy and Soil Science Society of America, Madison, WI, pp 345–371

Ainsworth EA (2008) Rice production in a changing climate: a meta-analysis of responses to elevated carbon dioxide and elevated ozone concentration. Glob Chang Biol 14(7):1642–1650

Akram R, Turan V, Hammad HM, Ahmad S, Hussain S, Hasnain A, Maqbool MM, Rehmani MIA, Rasool A, Masood N, Mahmood F, Mubeen M, Sultana SR, Fahad S, Amanet K, Saleem M, Abbas Y, Akhtar HM, Hussain S, Waseem F, Murtaza R, Amin A, Zahoor SA, Sami ul Din M, Nasim W (2018) Fate of organic and inorganic pollutants in paddy soils. In: Hashmi MZ, Varma A (eds) Environmental pollution of paddy soils. Springer, Switzerland, pp 197–214

Akram R, Fahad S, Masood N, Rasool A, Ijaz M, Ihsan MZ, Maqbool MM, Ahmad S, Hussain S, Ahmed M, Kaleem S, Sultana SR, Mubeen M, Saud S, Kamran M, Nasim W (2019) Plant growth and morphological changes in rice under abiotic stress. In: Hasanuzzaman M, Fujita M, Nahar K, Biswas JK (eds) Advances in rice research for abiotic stress tolerance. Woodhead Publishing, Sawston, pp 69–85

Bashir N, Mahmood S, Zafar Z, Rasul S, Manzoorand H, Athar H (2016) Is drought tolerance in maize (*Zea mays* L.) cultivars atthe juvenile stage maintained at the reproductive stage? Pak J Bot 48:1385–1139

Basnayake J, Inthavong T, Kam SP, Fukai S, Schiller JM, Chanphengxay M (2006) Climate diversity within the rice environments in Laos. In: Basnayake J et al (eds) Rice in Laos. International Rice Research Institute, Los Banos, Philippines, pp 47–64

Basu S, Ramegowda V, Kumar A, Pereira A (2016) Plant adaptation to drought stress. F1000. Resuscitation 5:1554

Bates BC, Kundzewicz ZW, Wu S, Palutikof JP (Eds) (2008) Climate Change and Water, 210 pp. Technical Paper of the Intergovernmental Panel on Climate Change, IPCC Secretariat, Geneva. http://www.ipcc.ch/ipccreports/tp-climate-change-water.htm

Botwright ATL, Lafitte HR, Wade LJ (2008) Genotype and environment interactions for grain yield of upland rice backcross lines in diverse hydrological environments. Field Crop Res 108:117–125

Bouman BAM, Peng S, Castaõeda A, Rand Visperas RM (2005) Yield and water use of irrigated tropical aerobic rice systems. Agric Water Manag 74(2):87–105

Brolley M (2015) Al Schmidley: rice security is food security for much of the world. Rice Today, 6 October 2015. http://ricetoday.irri.org/al- Singapore, July 2016 72 schmidley-rice-security-is-food-security-for-much-of-the-world/. Accessed 29 Mar 2016

Cai WW, Wan YF, Ai TC, You SC, Li YE, Wang B (2015) Impacts of elevated CO_2 concentration and temperature increasing on growth and yield of late rice. Chin J Agrometeorol 36(6): 717–723

Cairns JE, Sonder K, Zaidi PH, Verhulst N, Mahuku G, Babu R, Nair SK, Das B, Govaerts B, Vinayan MT, Rashid Z, Noor JJ, Devi P, San Vicente F, Prasanna BM (2012) Maize production in a changing climate: impacts, adaptation, and mitigation strategies. Adv Agron 114:1–58

Ceccarelli S, Grando S, Maatougui M, Michael M, Slash M, Haghparast R, Rahmanian M, Taheri A, Al-yassin A, Benbelkacem A, Labdi M, Mimoun H, Nachit M (2010) Plant breeding and climate changes. J Agric Sci (Camb) 148:627–637

Cline WR (2007) Global warming and agriculture: impact estimates by country. Center for Global Development, Peterson Institute for International Economics.

De Datta SK (1981) Principles and practices of rice production. Wiley, Canada

Deng NY, Ling XX, Sun Y, Zhang CD, Fahad S, Peng SB, Cui KH, Nie LX, Huang JL (2015) Influence of temperature and solar radiation on grain yield and quality in irrigated rice system. Eur J Agron 64:37–46

Fahad S, Noor M, Adnan M, Khan MA, Rahman I, Alam M, Khan IA, Ullah H, Mian IA, Hassan S, Saud S, Bakhat HF, Hammad HM, Ahmad S, Nasim W (2019) Abiotic stress and rice grain quality. In: Hasanuzzaman M, Fujita M, Nahar K, Biswas JK (eds) Advances in rice research for abiotic stress tolerance. Woodhead Publishing, Sawston, pp 571–583

FAO (2019) Statistic database. [2011-10-10]. http://faostat.fao.org/site/339/default.aspx

Farooq M, Aziz T, Basra SMA, Cheema MA, Rehamnh (2008) Chilling tolerance in hybrid maize induced by seed priming with salicyclic acid. J Agron Crop Sci 194:161–168

Fatima Z, Ahmad M, Hussain M, Abbas G, Ul-Allah S, Ahmad S, Ahmed N, Ali MA, Sarwar G, Ehsan Ul Haque, Iqbal P, Hussain S (2020) The fingerprints of climate warming on cereal crops phenology and adaptation options. Sci Rep 10:18013

Field CB (2012) Managing the risks of extreme events and disasters to advance climate change adaptation: special report of the intergovernmental panel on climate change. Cambridge University Press, London

Fischer G, Shah M, Velthuizen HV (2002) Climate change and agricultural vulnerability. International Institute for Applied Systems Analysis, Laxenburg, 152 pp.

Fitzgerald MA, Resurreccion AP (2009) Maintaining the yield of edible rice in a warming world. Funct Plant Biol 36:1037–1045

Franks P, Adams M, Amthor JS, Barbour M, Berry J, Ellsworth DS, Ghannoum O, Lloyd J, Lloyd J, McDowell NG, Norby RJ (2013) Sensitivity of plants to changing atmospheric CO_2 concentration: from the geological past to the next century. New Phytol 197:1077–1094

Furuya J, Koyama O (2005) Impacts of climatic change on world agricultural product markets: estimation of macro yield functions. JARQ 39(2):121–134

Garbach K, Vu Thanh TA, Buchori D, Ravanera R, Boualaphanh C, Ketelaar JW, Gemmill-Herren B (2014) The multiple goods and services of Asian rice production systems, 1st edn. Food and Agriculture Organization of the United Nations, Rome

Global Rice Science Partnership (GRiSP) (2013) Rice almanac. International Rice Research Institute. https://doi.org/10.1093/aob/mcg189

Hafeez-ur-Rehman, Nawaz A, Awan MI, Ijaz M, Hussain M, Ahmad S, Farooq M (2019) Direct seeding in rice: problems and prospects. In: Hasanuzzaman M (ed) Agronomic crops: production technologies, vol 1. Springer, Singapore, pp 199–222

He Y, Yang B, He Y, Zhan C, Cheng Y, Zhang J, Zhang H, Cheng J, Wang Z (2019) quantitative trait locus, qSE3, promotes seed germination and seedling establishment under salinity stress in rice. Plant J 97:1089–1104

Hussain M, Ahmad S, Hussain S, Lal R, Ul-Allah S, Nawaz A (2018) Rice in saline soils: physiology, biochemistry, genetics and management. Adv Agron 148:231–287

Ibrahim EA (2016) Seed priming to alleviate salinity stress in germinating seeds. J Plant Physiol 192:38–46

IPCC (2014) Climate change 2014: synthesis report. The Intergovernmental Panel on Climate Change, Geneva

IPCC (Inter-governmental Panel on Climate Change) (2007) Contribution of Working Group II to the fourth assessment report of the intergovernmental panel on climate change. In: Parry ML, Canziani OF, Palutikof JP, van der Linden PJ, Hanson CE (eds) Climate change 2007, fourth assessment report, intergovernmental panel on climate change. Cambridge University Press,

Cambridge, UK/New York, USA. http://www.ipcc.ch/publications_and_data/ar4/wg2/en/contents.html
IRRI (International Rice Research Institute) (2010) Scuba rice: breeding flood-tolerance to Asia's local mega rice varieties. International Rice Research Institute, Los Baños (Philippines). www.eiard.org/media/uploads/documents/case_studies/dfid_impact_case_study_sub1_rice_-_may_2010.pdf. Accessed 20 Nov 2011
Jagadish SVK, Craufurd PQ, Wheeler TR (2007) High temperature stress and spikelet fertility in rice (*Oryza sativa* L.). J Exp Bot 58(7):1627–1635
Jaleel CA, Gopi R, Sankar B, Gomathinayagam M, Panneerselvam R (2008) Alterations in morphological parameters and photosynthetic pigments responses of catharanthus roseus under soil water deficits. Colloids Surf B: Biointerfaces 62:312–318
Ji K, Wang Y, Sun W, Lou Q, Mei H, Shenand S, Chen H (2012) Drought-responsive mechanisms in rice genotypes with contrasting drought tolerance during reproductive stage. J Plant Physiol 169:336–344
Kamoshita A, Rodriguez R, Yamauchi A, Wade L (2004) Genotypic variation in response of rainfed lowland to prolonged drought and rewatering. Plant Production Science 7(4):406–420
Kamoshita A, Babu RC, Boopathi NM, Fukai S (2008) Phenotypic and genotypic analysis of drought-resistance traits for development of rice cultivars adapted to rainfed environments. Field Crop Res 109:1–23
Karim MA, Fracheboud Y, Stamp P (1997) Heat tolerance of maize with reference of some physiological characteristics. Ann Bangladesh Agri 7:27–33
Katsura K, Maeda S, Lubis I, Horie T, Cao WX, Shiraiwa T (2008) The high yield of irrigated rice in Yunnan, China-'A cross-location analysis. Field Crop Res 107:1–11
Khan MA, Ahmad S, Raza A (2019a) Integrated weed management for agronomic crops. In: Hasanuzzaman M (ed) Agronomic crops, vol 2. Springer, Singapore, pp 257–281
Khan SM, Ali S, Nawaz A, Bukhari SAH, Ejaz S, Ahmad S (2019b) Integrated pest and disease management for better agronomic crop production. In: Hasanuzzaman M (ed) Agronomic crops, vol 2. Springer, Singapore, pp 385–428
Kim J, Shon J, Lee CK, Yang W, Yoon Y, Yang WH, Kim YG, Lee BW (2011) Relationship between grain filling duration and leaf senescence of temperate rice under high temperature. Field Crop Res 122(3):207–213
Krishnan P, Ramakrishnan B, Reddy KR, Reddy VR (2011) High-temperature effects on rice growth, yield, and grain quality. Adv Agron 111:87–206
Li WC, Zhu QS, Wang YS, Wang SM, Chen XC, Zhang DW, Wang WL (2013) The relationships between physiological and biochemical indexes and the yield characteristics of rice under high temperature. Stress 29(9):5–10
Liu S, Waqas MA, Wang SH, Xiong XY, Wan YF (2017) Effects of increased levels of atmospheric CO_2 and high temperatures on rice growth and quality. PLoS One 12(11):e0187724
Machado RMA, Serralheiro RP (2017) Soil salinity: effect on vegetable crop growth. Management practices to prevent and mitigate soil salinization. Horticulture 3:30
Maclean JL, Dawe D, Hardy B, Hettel GP (eds) (2002) Rice almanac. International Rice Research Institute, Los Banos, Philippines, p 253
Matsui T, Omasa K, Horie T (1997) High temperature-induced spikelet sterility of japonica rice at flowering in relation to air temperature, humidity and wind velocity condition. Japanese J Crop Sci 66:449–455
Melloy P, Aitken E, Luck J, Chakraborty S, Obanor F (2014) The influence of increasing temperature and CO_2 on Fusarium crown rot susceptibility of wheat genotypes at key growth stages. Eur J Plant Pathol 140:19–37
Mohammadi-Nejad SRK, Arzani A, Rezaei A, Sabouri H, Gregorio GB (2010) Evaluation of salinity tolerance in rice genotypes. Int J Plant Prod 4:199–207
Mondal S, Borromeo TH (2016) Screening of salinity tolerance of rice at early seedling stage. J Biosci Agric Res 10:843–847

Nakagawa H, Horie T, Matsui T (2002) Effects of climate change on rice produc-tion and adaptive technologies. In: Mew TW, Brar DS, Peng S, Dawe D, Hardy B (eds) Rice science: innovations and impact for liveli-hood. International Rice Research Institute, China, pp 635–657

Nakagawa H, Horie T, Matsui T (2003) Effects of climate change on rice production and adaptive e technologies. In: Mew TW, Brar DS, Peng S, Dawe D, Hardy B (eds) Rice science: innovations and impact for livelihood. proceedings of the international rice research conference, Beijing, China, 16–19 September 2002. IRRI, Manila, The Philippines, pp 635–658

Naz S, Fatima Z, Iqbal P, Khan A, Zakir I, Noreen S, Younis H, Abbas G, Ahmad S (2019) Agronomic crops: types and uses. In: Hasanuzzaman M (ed) Agronomic crops: production technologies, vol 1. Springer, Singapore, pp 1–18

Oh-e I, Saitoh K, Kuroda T (2007) Effects of high temperature on growth, yield and dry-matter production of rice grown in the paddy field. Plant Production Science 10(4):412–422

Pandey V, Shukla A (2015) Acclimation and tolerance strategies of rice under drought stress. Rice Sci 22:147–161

Pandey S, Bhandari H, Hardy B (2007) Economic costs of drought and rice farmers' coping mechanisms: a cross-country comparative analysis. 203. International Rice Research Institute, Manila

Passioura JB (1996) Drought and drought tolerance. Plant Growth Regul 20(2):79–83

Pearson GA, Ayers AD, Eberhard DL (1966) Relative salt tolerance of rice during germination and early seedling development. Soil Sci 102:151

Peerzada AM, Bukhari SAH, Dawood M, Nawaz A, Ahmad S, Adkins S (2019) Weed management for healthy crop production. In: Hasanuzzaman M (ed) Agronomic crops, vol 2. Springer, Singapore, pp 225–256

Peng S, Huang J, Sheehy JE, Laza RC, Visperas RM, Zhong X, Centeno GS, Khush GS, Cassman KG (2004) Rice yield decline with higher night temperature from global warming. In: Redona ED, Castro AP, Llanto GP (eds) Rice integrated crop management: towards a RiceCheck syatem in the Philippines. Nueva Ecija, Philippines, PhilRice, pp 46–56

Plaut Z, Butow BJ, Blumenthal CS (2004) Transport of dry matter into developing wheat kernels and its contribution to grain yield under post-anthesis water deficit and elevated temperature. Field Crop Res 86(2–3):185–198

Rana DI, Randhawa DS (2014) Impact of rainfall on agriculture in Himachal Pradesh. State Centre on Climate Change, Himachal Pradesh, Himachal Pradesh

Razaq M, Shah FM, Ahmad S, Afzal M (2019) Pest management for agronomic crops. In: Hasanuzzaman M (ed) Agronomic crops, vol 2. Springer, Singapore, pp 365–384

Rezaei EE, Webber H, Gaiser T, Naab J, Ewert F (2015) Heat stress in cereals: mechanisms and modelling. Eur J Agron 64:98–113

Rogers TS, Martyn T (2009) Background paper: food. pacific economic survey 2009. FAO

Rosenzweig C, Elliott J, Deryng D, Ruane AC, Müller C, Arneth A, Boote KJ, Folberth C, Glotter M, Khabarov N, Neumann K (2014) Assessing agricultural risks of climate change in the 21st century in a global gridded crop model intercomparison. PNAS 111(9):3268–3273

Saini HS, Aspinall D (1982) Abnormal sporogenesis in wheat (*Triticum aestivum* L)induced by short periods of high-temperature. Ann Bot 49:835–846

Sarwar N, Ali H, Ahmad S, EhsanUllah AS, Mubeen K, Hill JE (2013a) Water wise rice cultivation on calcareous soil with the addition of essential micronutrients. J Anim Plant Sci 23(1):244–250

Sarwar N, Ali H, Maqsood M, EhsanUllah SAN, Shahzad M, Mubeen K, Shahid MA, Ahmad S (2013b) Phenological response of rice plants to different micronutrients application under water saving paddy fields on calcareous soil. Turk J Field Crops 18(1):52–57

Schwanck AA, Meneses PR, Farias CRJ, Funck GRD, Maia AHN, Del Ponte EM (2015) Bipolaris oryzae seed borne inoculum and brown spot epidemics in the subtropical lowland rice-growing region of Brazil. Eur J Plant Pathol 142:875–885

Seck PA, Diagne A, Mohanty S, Wopereis MCS (2012) Crops that feed the world 7: rice. Food Security 4(1):7–24

Senguttuvel P, Vijayalakshmi C, Thiyagarajan K, Kannanbapu JR, KOTA S, Padmavathi G, Geetha S, Sritharan N, Viraktamath BC (2014) Changes in photosynthesis, chlorophyll fluorescence, gas exchange parameters and osmotic potential to salt stress during early seedling stage in rice (*Oryza sativa* L.). SABRAO J Breed Genet 46:120–135

Serraj R, Atlin G (2008) Drought-resistant rice for increased rainfed production and poverty alleviation: a concept note. In: Serraj R, Bennett J, Hardy B (eds) Drought frontiers in rice: crop improvement for increased rainfed production. International Rice Research Institute, Los Baños, Philippines, pp 385–400

Shahzad AN, Ahmad S (2019) Tools and techniques for nitrogen management in cereals. In: Hasanuzzaman M (ed) Agronomic crops, vol 2. Springer, Singapore, pp 111–126

Shimono H, Okada M, Yamakawa Y, Nakamura H, Kobayashi K, Hasegawa T (2009) Genotypic variation in rice yield enhancement by elevated CO_2 relates to growth before heading, and not to maturity group. J Exp Biol 60:523–532

Singh RK, Mishra B, Singh KN (2004) Salt tolerant rice varieties and their role in reclamation programme in Uttar Pradesh. Indian Farming:6–10

Swain DK, Herath S, Pathirana A, Mittra BN (2005) Rain fed lowland and flood prone rice: a critical review on ecology and management technology for improving the productivity in Asia. Role of Water Sciences in Transboundry River Basin Management, Thailand

Tao F, Yokozawa M, Zhang Z, Hayashi Y, Grassl H, Fu C (2004) Variability in climatology ang agricultural production with the East Asia summer monsoon and El Niño South Oscillation. Clim Res 28:23–30

Tao H, Brueck H, Dittert K, Kreye C, Lin S, Sattelmacher B (2006) Growth and yield formation for rice (Oryza sativa L.) in the water-saving ground cover rice production system (GCRPS). Field Crop Res 95(1):1–12

Tariq M, Ahmed M, Iqbal P, Fatima Z, Ahmad S (2020) Crop phenotyping. In: Ahmed M (ed) Systems modeling. Springer, Singapore, pp 45–60

Thiam M, Champion A, Diouf D, Ourèye SYM (2013) NaCl effects on in vitro germination and growth of some Senegalese cowpea (*Vigna unguiculata* (L.) Walp.) cultivars. ISRN Biotechnol 1–11

Ullah A, Ahmad A, Khaliq T, Akhtar J (2017) Recognizing production options for pearl millet in Pakistan under changing climate scenarios. J Integr Agric 16(4):762–773

Velmurugan A, Swarnam TP, Ambast SK, Kumar N (2016) Managing waterlogging and soil salinity with a permanent raised bed and furrow system in coastal lowlands of humid tropics. Agric Water Manag 168:56–67

Wahid A, Close TJ (2007) Expression of dehydrins under heat stress and theirrelationship with water relations of sugarcane leaves. Biol Plant 51:104–109

Wang J, Mendelsohn R, Dinar A, Huang J, Rozelle S, Zhang L (2009) The impact of climate change on China's agriculture. Agric Econ 40:323–337

Wang Z, Wang J, Bao Y, Wu Y, Zhang H (2011) Quantitative trait loci controlling rice seed germination under salt stress. Euphytica 178:297–307

Wasaya A, Yasir TA, Ijaz M, Ahmad S (2019) Tillage effects on agronomic crop production. In: Hasanuzzaman M (ed) Agronomic crops, vol 2. Springer, Singapore, pp 73–99

Wassmann R, Jagadish SVK, Sumfleth K, Pathak H, Howell G, Ismail A, Serraj R, Redona E, Singh RK, Heuer S (2009a) Regional vulnerability of climate change impacts on Asian rice production and scope for adaptation. Adv Agron 102:91–133

Wassmann R, Jagadish SVK, Heuer S, Ismail A, Redona E, Serraj R, Singh RK, Howell G, Pathak H, Sumfleth K (2009b) Climate change affecting rice production: the physiological and agronomic basis for possible adaptation strategies. Adv Agron 101:59–122

Welch JR, Vincent JR, Auffhammer M, Moya PF, Dobermann A, Dawe D (2010) Rice yields in tropical/subtropical Asia exhibit large but opposing sensitivities to minimum and maximum temperatures. Proc Natl Acad Sci 107(33):14562–14567

WMO Greenhouse Gas Bulletin (2017) The State of Greenhouse in the atmosphere based on global observations through 2016: 13

Yang XG, Liu ZJ, Chen F (2011) The possible effects of global warming on cropping systems in China VI-Possible effects of future climate change on northern limits of cropping system in China. Sci Agric Sin 44:1562–1570; (in Chinese)

Yoshida S (1981) Fundamentals of rice crop science. International Rice Research Institute, Los Baños, Laguna

Yoshida S, Satake T, Mackill D (1981) High temperature Stress. IRRI Res Pap 67:1–15

Zahoor SA, Ahmad S, Ahmad A, Wajid A, Khaliq T, Mubeen M, Hussain S, Sami Ul Din M, Amin A, Awais M, Nasim W (2019) Improving water use efficiency in agronomic crop production. In: Hasanuzzaman M (ed) Agronomic crops, vol 2. Springer, Singapore, pp 13–29

Part II

Agronomic Management

Techniques of Rice Nursery Establishment and Transplanting

3

Ahmad Nawaz, Anees Ur Rehman, Zeeshan Haydar,
Hafeez Ur Rehman, Shakeel Ahmad, and Mubshar Hussain

Abstract

In rice, Nursery seedlings are grown in a specified area followed by transplanting under field conditions. Different techniques are being used by the rice farmers for raising the nursery seedlings followed by the transplanting across the world. However, each technique of raising and transplanting nursery has its merits and demerits. This chapter covers the comparison of manual and mechanical methods of rice nursery raising and transplanting techniques.

Keywords

Rice seed · Seedling · Transplanting · Manual · Mechanical

A. Nawaz
College of Agriculture, Bahauddin Zakariya University Multan, Layyah, Pakistan

A. U. Rehman · H. U. Rehman
Department of Agronomy, University of Agriculture, Faisalabad, Pakistan

Z. Haydar
Center for Agriculture and Biosciences International (CABI), Central and West Asia (CWA), Satellite Town Rawalpindi, Pakistan

S. Ahmad
Department of Agronomy, Faculty of Agricultural Sciences and Technology, Bahauddin Zakariya University, Multan, Pakistan
e-mail: shakeelahmad@bzu.edu.pk

M. Hussain (✉)
Department of Agronomy, Bahauddin Zakariya University, Multan, Pakistan

N. Sarwar et al. (eds.), *Modern Techniques of Rice Crop Production*,
https://doi.org/10.1007/978-981-16-4955-4_3

3.1 Introduction

Rice is an important cereal crop which is consumed as staple in many parts of globe (Ahmad et al. 2008, 2009, 2012, 2013, 2015, 2019; Ahmad and Hasanuzzaman 2012; Ahmed and Ahmad 2019, 2020; Shahzad and Ahmad 2019). In Pakistan, rice is mostly grown in Punjab, followed by Sindh, Balochistan, and KPK. Both Basmati and non-basmati rice varieties are grown in Pakistan (Hafeez-ur-Rehman et al. 2019; Khan et al. 2019a, b; Naz et al. 2019; Peerzada et al. 2019; Razaq et al. 2019; Sarwar et al. 2013a, b; Tariq et al. 2020; Wasaya et al. 2019; Zahoor et al. 2019). The Basmati rice plays an important role in foreign exchange earnings, thus contributing to country's economy (Ahmed and Fayyaz-ul-Hassan 2017; Ahmed et al. 2020a, b; Akram et al. 2018, 2019; Hussain et al. 2018; Fahad et al. 2019; Fatima et al. 2020). The rice residues also serve as an important feed source for animals. The rice husk is also used in paper industries as well as used for fuel purpose. The rice flour is mostly used for making bakery products. High-quality bran oil of unique properties is also extracted from rice bran. In contrast to other cereals' growth ecology, rice crop can grow well under flooded conditions. In most parts of the world including Pakistan, first the rice nursery is grown through various methods and then that nursery is transplanted in puddled flooded soil. In this chapter, we will discuss various rice nursery establishment methods common in Pakistan.

3.2 Methods of Nursery Establishment

The nursery methods common in Pakistan include dry method, wet method, Rab method, and tray method. Each method has been discussed in detail below.

3.2.1 Dry Method

Dry method is most suitable for the areas where soils are silty, and water cannot stay for long time on soil. In this method, pre-sowing irrigation is applied followed by 2–3 plowings when soil reaches workable soil moisture. Plowing is followed by planking. Then, the seeds are broadcasted at the seed rate of 1.5 kg per marla for non-basmati coarse varieties and 750 g per marla for Basmati varieties. It is followed by covering the broadcasted seed with a layer of one inch of dry farmyard manure, wheat straw, rich husk, or rice straw. After this, the field is irrigated. It is recommended to place some dry grass stubbles at the opening of water channel (Nakka) in order to prevent the overflow of rice seeds due to heavy water flow pressure. In this method, seed rate per unit area should be 1.5 times as compared to wet method. In this method, the nursery will be ready for transplanting after 35–40 days. Weeds in nursery can be controlled as per recommendation of agriculture department of each province depending on the type of weeds and weed infestation.

3.2.2 Wet Method

In this method, it is suggested to use recommended fungicide in 1.25 times more water as compared to seed quantity and keep seed dipped in fungicide contacting water for 24 h. Then, remove seed from water and put it under shade in heaps by covering heap with gunny bag. Water is sprinkled over seed after intervals, and heap is turned with hands about two to three times in a day for proper aeration and avoiding damage by heat due to suffocation. After 36–48 h, the seed sprouts and is ready for sowing. If registered seed is not available and we want to use our farm's seed, then add 500 g of salt in 20 L of water in order to separate the lightweight and bad quality seed which will float at the surface. After separating poor-quality seeds, wash the whole seed lot with water to remove excessive salts on seed.

After this, plow field one to two times when field is dry, and then irrigate the field. After irrigation, plow the field in standing water condition (i.e., puddling) one to two times followed by planking. After this, divide the whole field into small plots (10 marlas each). The spray of pre-emergence herbicides should be done to control weeds in nursery. The area where the nursery is going to be grown is watered for about 30 days before sowing. It facilitates weed germination which can be easily eradicated when plowing, and puddling is carried out during seedbed preparation. Then, the pre-germinated seeds, as discussed above, are sown in thoroughly puddled and leveled soil. The water level in field at the time of broadcasting of pre-germinated seeds should be 1.5–2 in. In this method, the recommended seed rate for non-basmati coarse varieties is 1 kg per marla and 500–750 g per marla for Basmati varieties. Broadcast the pre-germinated seed at evening time. Remove the applied irrigation water at next evening and again irrigate in next morning. Repeat the same pattern of irrigation for 1 week. The purpose of water removal is that the already standing water heats up rapidly at daytime, and it may damage the seeds due to suffocation. The water level in field should be increased according to the growth of seedlings, but it should not be more than 3 in. The seedlings become ready for transplanting after 25–30 days. However, if seedlings are not well established, broadcast 250-g urea or 400-g calcium ammonium nitrate per marla before 10 days of nursery transplanting. The nursery transplantation should be completed before the end of July.

3.2.3 Rab Method

This method is commonly used in districts of Dera Ghazi Khan and Muzaffargarh where generally soils are very hard and dry conditions. In this method, after initial land preparation, place farmyard manure, straw, or paddy straw having 2 in.' layer on surface of soil. Burn the straw in morning or afternoon when wind is not blowing. When ash cools down, mix it in soil. After this, broadcast the seed at a seed rate of 2 kg per marla for non-basmati coarse varieties and 1 kg per marla for Basmati varieties followed by a light irrigation. The nursery will be ready for transplantation after 35–40 days of sowing.

3.2.4 Tray Method

This method is used when paddy nursery needs to be transplanted through mechanical transplanter instead of manual transplanting. Mechanical paddy transplanting is the advanced form of paddy transplanting technology to shift the paddy seedling into the field by using paddy transplanter. This technology is more efficient than manual transplanting with the list of advantages, that is, less labor use, timely sowing, water saving, and ensure optimum plant population which enhance the yield and increase the more profit returns to the farmers which make them more prosper and happy.

In tray method, first the field is cultivated through rotavator (Fig. 3.1a) followed by the sieving of soil to get very fine soil particles (Fig. 3.1b). Then, the seeds are sown with seed cultivation machine in trays by using the sieved soil as shown in Fig. 3.2. These trays are made of special perforated plastic material (2 ft. × 1 ft. in dimension), 50 g in weight and are reusable. However, the nursery trays can be placed horizontally in Pakistan (Figs. 3.2c, 3.3c, and 3.4) and vertically in China (Fig. 3.5). The seedling machine has battery-operated power source to run the machine motor which gives drive to whole interconnected conveying belts through chain and sprocket mechanism. Machines have three sections: the first and third sections contain the soil, and middle section contains the seed during working as shown in Fig. 3.2b. Machine is fully automatic to fill the trays; just we need to enter empty tray from one side of the machine; it will automatically fill and lay out the trays on the land smoothly. During this mechanism, first a soil layer is spread on tray base for about ¾ inches, then, approximately 100 g per tray seed is distributed uniformly in whole tray, and finally, a topsoil layer of around ½ inch covers the seed layer. This machine

Fig. 3.1 Soil preparation through rotavator (**a**), soil sieving (**b**), and heaping (**c**) to use in plastic trays (Photo by: Zeeshan Haydar)

Fig. 3.2 Seedling tray (**a**), sections of seedling machine (**b**), and tray filling (**c**) (Photo by: Zeeshan Haydar)

Fig. 3.3 Horizontal method of raising rice seedlings in nursery in Pakistan (Photo by: Dr. Shakeel Ahmad)

continuously moves in one direction and makes a straight line of trays. Around 100 seedling trays are used for transplanting of one acre. After filling the required number of trays, it is wetted by the flooding method as shown in Fig. 3.3. To prepare the seedling, 2–4 kg of DAP and 4 kg of zinc are required for 100 trays, and 20-day seedlings can be shifted by paddy transplanter into the field.

3.3 Nursery Management Practices in Bangladesh

Rice is a widely sown crop in Bangladesh covering an area of about 75% of total cropped area. Climatic conditions of Bangladesh favor year-round rice sowing. Mainly, there are three rice sowing seasons in Bangladesh (i.e., Aus, Aman, and Boro seasons). The seed rate and fertilizer for rice nursery used in Bangladesh are presented in Tables 3.1 and 3.2.

3.4 Nursery Transplanting Methods

The following methods are being used for transplanting the rice nursery seedlings by the farmers into the fields. Mainly, the method of transplanting depends upon the conventional practices of an area or region, nature of soil, availability of labor force, economic conditions of the farmers, and the availability of advanced technological implements.

Fig. 3.4 Irrigation by flooding method, (**a**) 1 day after irrigation, (**b**) seedling growth at 8 days after sowing, (**c**) seedling growth at 15 days, and (**d**) rice crop after transplanting (Photo by: Zeeshan Haydar)

3.4.1 Manual Transplanting

This is a very old method of transplanting being used by the rice growers across the world. In this technique, the rice seedlings are manually dibbled in the prepared field after puddling (Fig. 3.6, Pakistan; Fig. 3.7, China; and Fig. 3.8, China). This technique is cheaper but needed much labor or manpower to transplant seedlings from the nursery area to the field. The major issue in this technique is to maintain the optimum planting density as the labor is not technical.

Fig. 3.5 Vertical method of raising rice seedlings in nursery in China (Photo by: Dr. Feng Ling Yang; Dr. Muhammad Ali Raza)

Table 3.1 Seed rate for rice sowing for different seasons in Bangladesh

Season	Seed rate (per acre)
1. Aus	
(a) Local Broadcast	28–37 kg in case of broadcasting
	14–19 kg in case of line sowing
(b) HYV Transplant	11–14 kg
(c) HYV Broadcast	8–9 kg
2. Aman	
(a) Local Transplant	9–11 kg
(b) Local Broadcast	33–37 kg
(c) HYV Transplant	8–9 kg
3. Boro	
(a) Local	8–11 kg
(b) HYV	8–11 kg
(c) Hybrid	8–11 kg

Source: Training module of Rice cultivation, Bangladesh Rice Research Institute, Gazipur, Bangladesh; Yearbook of Agricultural Statistics—2019, Bangladesh Bureau of Statistics (BBS), Dhaka, Bangladesh

3.4.2 Parachute Technology of Transplanting

Parachute technology follows the principle of parachute and was developed by scientists in China. In this method, nursery seedlings are being raised in trays. The

Table 3.2 Fertilizer rates for different seasons for raising rice nursery in Bangladesh

Season	Fertilizer rate (kg per hectare)				
	Nitrogen	Phosphorus	Potassium	Sulphur	Zinc
Boro	118.1	15.3	39.5	8.6	1.8
Transplanted Aus/Aman	79.1	10.1	39.5	7.2	1.8
Broadcast Aus	49.4	8.2	29.6	3.7	1.8

Source: Training module of Rice cultivation, Bangladesh Rice Research Institute, Gazipur, Bangladesh; Yearbook of Agricultural Statistics—2019, Bangladesh Bureau of Statistics (BBS), Dhaka, Bangladesh

Fig. 3.6 Manual transplanting of rice at Bahauddin Zakariya University Multan, Pakistan (Photo by: Dr. Shakeel Ahmad)

nursery seedling with an earth ball is being thrown above the ground in a parabolic curve and seedling following the principle of parachute fall on the ground and pushed into the soil. This method is applicable both manually (Fig. 3.9) and mechanically (Fig. 3.10).

3.4.3 Mechanical Transplanting

Mechanical transplanting is a technologically advanced technique to transplant the rice seedlings into the field (Figs. 3.11 and 3.12). Mechanical transplanters are being

Fig. 3.7 Manual transplanting of rice at farmer field in China (Photo by: Dr. Feng Ling Yang; Dr. Muhammad Ali Raza)

Fig. 3.8 Manual transplanting of rice at research farm in China (Photo by: Dr. Feng Ling Yang; Dr. Muhammad Ali Raza)

Fig. 3.9 Mechanical transplanting of rice by parachute technology by a farmer at farmer field at Sheikhupura, Punjab, Pakistan (Photo by: Dr. Shakeel Ahmad)

Fig. 3.10 Mechanical transplanting of rice by parachute technology by internship students of University of Agriculture students at a farmer field at Sheikhupura, Punjab, Pakistan (Photo by: Dr. Shakeel Ahmad)

Fig. 3.11 Mechanical transplanting of rice seedlings by transplanter in China (Photo by: Dr. Feng Ling Yang; Dr. Muhammad Ali Raza)

Fig. 3.12 Mechanical transplanting of rice seedlings by transplanter in China (Photo by: Dr. Feng Ling Yang; Dr. Muhammad Ali Raza)

used for transplanting. This technique is very effective to maintain optimum plant population in the field and required less labor as compared to manual method. This technique is very common in technologically advanced countries like China. However, in this technique, the nursery seedlings are raised in trays.

3.5 Conclusions

The nursery establishment and transplanting techniques are very important to harvest a good rice crop. A rice nursury establishment technique is more fruitful when it ensures healthy seedlings through better management options during establishment phase from sowing to transplanting. Transplanting technique also ensures the optimum plant population in the field.

References

Ahmad S, Hasanuzzaman M (2012) Integrated effect of plant density, N rates and irrigation regimes on the biomass production, N content, PAR use efficiencies and water productivity of rice under irrigated semiarid environment. Not Bot Horti Agrobot Cluj-Napoca 40(1):201–211

Ahmad S, Zia-ul-Haq M, Ali H, Shad SA, Ammad A, Maqsood M, Khan MB, Mehmood S, Hussain A (2008) Water and radiation use efficiencies of transplanted rice (*Oryza sativa* L.) at different plant densities and irrigation regimes under semi-arid environment. Pak J Bot 40(1): 199–209

Ahmad S, Ahmad A, Zia-ul-Haq M, Ali H, Khaliq T, Anjum MA, Khan MA, Hussain A, Hoogenboom G (2009) Resources use efficiency of field grown transplanted rice (*Oryza sativa* L.) under irrigated semiarid environment. J Food Agric Environ 7(2):487–492

Ahmad S, Ahmad A, Soler CMT, Ali H, Zia-Ul-Haq M, Anothai J, Hussain A, Hoogenboom G, Hasanuzzaman M (2012) Application of the CSM-CERES-Rice model for evaluation of plant density and nitrogen management of fine transplanted rice for an irrigated semiarid environment. Precis Agric 13(2):200–218

Ahmad S, Ahmad A, Ali H, Hussain A, Garcia y Garcia A, Khan MA, Zia-Ul-Haq M, Hasanuzzaman M, Hoogenboom G (2013) Application of the CSM-CERES-Rice model for evaluation of plant density and irrigation management of transplanted rice for an irrigated semiarid environment. Irrig Sci 31(3):491–506

Ahmad A, Ashfaq M, Rasul G, Wajid SA, Khaliq T, Rasul F, Saeed U, Rahman MH, Hussain J, Baig IA, Naqvi AA, SAA B, Ahmad S, Naseem W, Hoogenboom G, Valdivia RO (2015) Impact of climate change on the rice–wheat cropping system of Pakistan. In: Hillel D, Rosenzweig C (eds) Handbook of climate change and agro-ecosystems: the agricultural modeling intercomparison and improvement project (AgMIP) integrated crop and economic assessments. Imperial College Press and The American Society of Agronomy, London, pp 219–258

Ahmad S, Abbas G, Ahmed M, Fatima Z, Anjum MA, Rasul G, Khan MA, Hoogenboom G (2019) Climate warming and management impact on the change of rice-wheat phenology in Punjab, Pakistan. Field Crop Res 230:46–61

Ahmed M, Ahmad S (2019) Carbon dioxide enrichment and crop productivity. In: Hasanuzzaman M (ed) Agronomic crops, vol 2. Springer, Singapore, pp 31–46

Ahmed M, Ahmad S (2020) Systems modeling. In: Ahmed M (ed) Systems modeling. Springer, Singapore, pp 1–44

Ahmed M, Fayyaz-ul-Hassan AS (2017) Climate variability impact on rice production: adaptation and mitigation strategies. In: Ahmed M, Stockle C (eds) Quantification of climate variability, adaptation and mitigation for agricultural sustainability. Springer, Switzerland, pp 91–111

Ahmed M, Ahmad S, Raza MA, Kumar U, Ansar M, Shah GA, Parsons D, Hoogenboom G, Palosuo T, Seidel S (2020a) Models calibration and evaluation. In: Ahmed M (ed) Systems modeling. Springer, Singapore, pp 151–178

Ahmed M, Ahmad S, Waldrip HM, Ramin M, Raza MA (2020b) Whole farm modeling: a systems approach to understanding and managing livestock for greenhouse gas mitigation, economic viability and environmental quality. In: Waldrip HM, Pagliari PH, He Z (eds) Animal Manure. ASA Special Publication 67. American Society of Agronomy and Soil Science Society of America, Madison, WI, pp 345–371

Akram R, Turan V, Hammad HM, Ahmad S, Hussain S, Hasnain A, Maqbool MM, Rehmani MIA, Rasool A, Masood N, Mahmood F, Mubeen M, Sultana SR, Fahad S, Amanet K, Saleem M, Abbas Y, Akhtar HM, Hussain S, Waseem F, Murtaza R, Amin A, Zahoor SA, Sami ul Din M, Nasim W (2018) Fate of organic and inorganic pollutants in paddy soils. In: Hashmi MZ, Varma A (eds) Environmental pollution of paddy soils. Springer, Switzerland, pp 197–214

Akram R, Fahad S, Masood N, Rasool A, Ijaz M, Ihsan MZ, Maqbool MM, Ahmad S, Hussain S, Ahmed M, Kaleem S, Sultana SR, Mubeen M, Saud S, Kamran M, Nasim W (2019) Plant growth and morphological changes in rice under abiotic stress. In: Hasanuzzaman M, Fujita M, Nahar K, Biswas JK (eds) Advances in rice research for abiotic stress tolerance. Woodhead Publishing, Sawston, pp 69–85

Fahad S, Noor M, Adnan M, Khan MA, Rahman I, Alam M, Khan IA, Ullah H, Mian IA, Hassan S, Saud S, Bakhat HF, Hammad HM, Ahmad S, Nasim W (2019) Abiotic stress and rice grain quality. In: Hasanuzzaman M, Fujita M, Nahar K, Biswas JK (eds) Advances in rice research for abiotic stress tolerance. Woodhead Publishing, Sawston, pp 571–583

Fatima Z, Ahmad M, Hussain M, Abbas G, Ul-Allah S, Ahmad S, Ahmed N, Ali MA, Sarwar G, Ehsan ul Haque, Iqbal P, Hussain S (2020) The fingerprints of climate warming on cereal crops phenology and adaptation options. Sci Rep 10:18013

Hafeez-ur-Rehman, Nawaz A, Awan MI, Ijaz M, Hussain M, Ahmad S, Farooq M (2019) Direct seeding in rice: problems and prospects. In: Hasanuzzaman M (ed) Agronomic crops: production technologies, vol 1. Springer, Singapore, pp 199–222

http://dai.agripunjab.gov.pk/system/files/Rice%20Technology%202020-21.pdf (Urdu version translated into English)

http://www.knowledgebank.irri.org/step-by-step-production/growth/planting/how-to-prepare-the-seedlings-for-transplanting

https://agrinfobank.com.pk/improved-method-for-raising-rice-nursery/

Hussain M, Ahmad S, Hussain S, Lal R, Ul-Allah S, Nawaz A (2018) Rice in saline soils: physiology, biochemistry, genetics and management. Adv Agron 148:231–287

Khan MA, Ahmad S, Raza A (2019a) Integrated weed management for agronomic crops. In: Hasanuzzaman M (ed) Agronomic crops, vol 2. Springer, Singapore, pp 257–281

Khan SM, Ali S, Nawaz A, Bukhari SAH, Ejaz S, Ahmad S (2019b) Integrated pest and disease management for better agronomic crop production. In: Hasanuzzaman M (ed) Agronomic crops, vol 2. Springer, Singapore, pp 385–428

Naz S, Fatima Z, Iqbal P, Khan A, Zakir I, Noreen S, Younis H, Abbas G, Ahmad S (2019) Agronomic crops: types and uses. In: Hasanuzzaman M (ed) Agronomic crops. Production technologies, vol 1. Springer, Singapore, pp 1–18

Peerzada AM, Bukhari SAH, Dawood M, Nawaz A, Ahmad S, Adkins S (2019) Weed management for healthy crop production. In: Hasanuzzaman M (ed) Agronomic crops, vol 2. Springer, Singapore, pp 225–256

Razaq M, Shah FM, Ahmad S, Afzal M (2019) Pest management for agronomic crops. In: Hasanuzzaman M (ed) Agronomic crops, vol 2. Springer, Singapore, pp 365–384

Sarwar N, Ali H, Ahmad S, EhsanUllah AS, Mubeen K, Hill JE (2013a) Water wise rice cultivation on calcareous soil with the addition of essential micronutrients. J Anim Plant Sci 23(1):244–250

Sarwar N, Ali H, Maqsood M, EhsanUllah, Shahzad AN, Shahzad M, Mubeen K, Shahid MA, Ahmad S (2013b) Phenological response of rice plants to different micronutrients application under water saving paddy fields on calcareous soil. Turk J Field Crops 18(1):52–57

Shahzad AN, Ahmad S (2019) Tools and techniques for nitrogen management in cereals. In: Hasanuzzaman M (ed) Agronomic crops, vol 2. Springer, Singapore, pp 111–126

Tariq M, Ahmed M, Iqbal P, Fatima Z, Ahmad S (2020) Crop phenotyping. In: Ahmed M (ed) Systems modeling. Springer, Singapore, pp 45–60

Wasaya A, Yasir TA, Ijaz M, Ahmad S (2019) Tillage effects on agronomic crop production. In: Hasanuzzaman M (ed) Agronomic crops, vol 2. Springer, Singapore, pp 73–99

Zahoor SA, Ahmad S, Ahmad A, Wajid A, Khaliq T, Mubeen M, Hussain S, Sami Ul Din M, Amin A, Awais M, Nasim W (2019) Improving water use efficiency in agronomic crop production. In: Hasanuzzaman M (ed) Agronomic crops, vol 2. Springer, Singapore, pp 13–29

Rice Seed and Seedling Priming

4

Hafeez ur Rehman, Muhammad Farooq, Mubashir Hussain, and Shahzad M. A. Basra

Abstract

Earlier emergence and vigorous seedling stand are key indicators of crop performance. Seed priming as cost-effective hydration technique is central to enhance crop vigor to optimize input use in production and affect grain nutritional quality and food security in rice systems. Poor seedling growth and sub-optimal plant density associated with delayed transplanting of nursery seedling of low vigor is one of the major constraints in conventional flooded (CF) and water-saving aerobic (AR) and alternate wetting and drying (AWD) rice systems. Likely, poor and erratic stand restricts the success of direct seeded rice due to less weed competitiveness associated with low seed vigor. Seed hydropriming, osmopriming, and nutrient priming have been successfully employed in conventional transplanted system irrigated as AR or AWD and in direct seed rice systems to achieve healthy seedling stands, rapid crop development, high yields, and grain nutritional quality including input resource use efficiency. This chapter discusses the potential of priming for improving seed and seedling vigor, crop development, yields, grain nutritional quality, and their profitability in rice systems. This will help to reduce the yield gaps associated with crop vigor in actual and potential yields in rice production.

H. u. Rehman (✉) · S. M. A. Basra
Department of Agronomy, University of Agriculture, Faisalabad, Pakistan
e-mail: h.rehman@uaf.edu.pk

M. Farooq
Department of Agronomy, University of Agriculture, Faisalabad, Pakistan

Department of Plant Sciences, College of Agricultural and Marine Sciences, Sultan Qaboos University, Al-Khoud, Oman

M. Hussain
Department of Agronomy, Bahauddin Zakariya University, Multan, Pakistan

N. Sarwar et al. (eds.), *Modern Techniques of Rice Crop Production*,
https://doi.org/10.1007/978-981-16-4955-4_4

Keywords

Seed vigor · Wet nursery method · Crop stand · Water-saving rice cultivation · Grain biofortification · Profitability

4.1 Introduction

Rice (*Oryza sativa* L.) as staple feeds daily >3.5 billion people to fulfill their 20% daily calorie requirement (Ahmad et al. 2015; Rehman et al. 2019). More than 75% of world supplies are harvested from rice produced under flooded condition (CF) (Yang 2012; Van et al. 2001), and water as an important factor affects rice production (Ahmad and Hasanuzzaman 2012; Ahmad et al. 2008, 2009).

Irrigated rice in Asia is usually cultivated primarily by growing seedlings into nursery seedbeds and later their transplanting manually or mechanically into paddy fields maintained under CF or saturated condition with or without puddling. Depending on the freshwater availability, rice fields are maintained as flooded throughout crop growth cycle as in CF system, alternate wetting and drying (AWD) exposing soil to wet–dry cycles in AWD, at field capacity in aerobic rice (AR), and kept saturated under system of rice intensification (SRI) (Farooq et al. 2009a, b; Rehman et al. 2012). In each of these methods, wet bed method of nursery raising is mostly practiced by farmers in which rice seed is first soaked for 24 h in water. Then pre-germinated seeds incubated for 48 h are broadcasted uniformly on nursery raised bed resulting in poor and delayed emergence while producing nursery seedlings of uneven stand (Ahmad 1998; Farooq et al. 2008). These nursery seedlings of different ages ranging from 30 to 45 days are transplanted into the main rice fields (De Datta 1981; Singh and Singh 1999).

Nursery seedling with poor and delayed emergence raised by wet bed methods when transplanted results in sub-optimum planting density, and patchy and irregular crop stands subsequently have less growth rate. Nursery seedling of poor vigor is accompanied with delayed transplanting (>30 days) owing to scarce and high labor cost at critical time resulting in lower grain yields (Reddy 2004) and seed quality associated with poor seed setting owing to high temperature and high humidity at flowering (Rehman et al. 2019).

These effects are further increased on growth with transplanting shock when nursery of increased seedling age (Salam et al. 2001) is shifted. Nonetheless, seedling age is significant factor toward yield contribution in transplanting rice system by affecting tillers, dry matter production, and root traits, and several studies report transplanting of young nursery seedlings ($\leq$25 days) with positive effects on grain yield (Randriamiharisoa and Uphoff 2002; Horie et al. 2005; Pasuquin et al. 2008). Studies on SRI report better crop performance in terms of higher yields by transplanting 2–3 weeks or even more younger seedlings (Makarim et al. 2002).

Transplanting younger rice seedlings affects four phyllochron stages and produces more number and fertile tillers, with better capture resources including nitrogen, extended crop duration, higher 1000-grain weight, and grain yields (Ashraf et al. 1999; Mishra and Salokhe 2008) compared to transplanting aged nursery seedlings with more competition for resources.

Therefore, it is imperative to grow seedlings of high vigor, and transplanting at younger age is the primary factor to obtain uniform crop stand and increased rice production. Padalia (1980) reported that 50% success of rice cultivation depends upon the seedling, irrespective of method of nursery raising.

Seedling vigor in rice defines the plant characteristics such as survival, height, thickness and uniformity of stem, and establishment and development of new roots, and these traits vary with age, production system, and seedling hills before and after transplanting. Therefore, nursery seedling health with improved vigor plays a greater role in improving rice yields by affecting their establishment subsequent growth such as tillering in transplanted rice (TeKrony and Egli 1991; Himeda 1994; Ros et al. 2003; Sasaki 2004).

Nonetheless, sustainable rice production depends on efficient use of labor, water, and fertilizer to improve productivity, profitability, and resource use efficiency while reducing environmental footprints (Foley et al. 2011; Farooq et al. 2011a, b; Hoang et al. 2019). On the other hand, climate change-induced emissions of potent methane (CH_4), decreasing freshwater resources, high labor, and production costs are major challenges to conventional rice system (Linquist et al. 2012; Nawaz et al. 2019). Rice production under CF degrades soil physical and chemical properties by disintegrating soil aggregate, porosity, and permeability with increase in bulk density owing to the development of hardpan at shallow depth under puddle condition and decreases wheat productivity and delay its cultivation (Farooq et al. 2008; Nawaz et al. 2017, 2019; Nadeem et al. 2020). This suggests water-wise production by growing rice alternatively under direct seeding condition (DSR), AWD, and SRI (Farooq et al. 2009a, b, Farooq et al. 2011a, b, Rehman et al. 2012; Hoang et al. 2019).

Worldwide, these different methods of rice production have been adapted to sustain its productivity, and rice direct seeding is also being practiced as alternative to conventional transplanting in the United States, Western Europe including Italy and France, India, Russia, Japan, Cuba, Sri, Lanka, Malaysia, Vietnam, Thailand, Philippines, Pakistan, and in some parts of Iran (Farooq et al. 2011a, b; Kumar and Ladha 2011).

Direct seeded rice is practiced by broadcasting of pre-germinated seed on puddled soil in wet seeding, broadcasting, or drilling of seed in dry soil or at field capacity in dry seeding and broadcasting of seed in standing water in case of water seeding (Farooq et al. 2011a, b). Compared to wet and water seeding methods, dry DSR is more popular in areas with unpredictable water supply and rainfall such as for lowland rice cultivation and has advantages of less labor and water consuming, timely establishment, and earlier maturity, including reduced methane emissions (Ella et al. 2011; Gathala et al. 2011; Chauhan et al. 2012).

Among several factors, high weed pressure, nutrient management, and poor crop stand due to anoxic condition during germination, seed viability, and un-leveled fields affect rice production under dry DSR condition (Ladha et al. 2009; Tripathi et al. 2005; Manigbas et al. 2008). Previously, research has focused on reducing weed pressure and high emergence rate to improve its adaption and very less on developing cultivars of high early seedling vigor, a trait to determine successful crop establishment, improve weed competitiveness, and achieve high yield (Zhang et al. 2005a, b; Foolad et al. 2007; Mahender et al. 2015).

Seed germination, early seedling vigor, and uniform crop stands are key determinants of successful crop production and susceptible stages of plant growth cycle to adverse soil and environmental factors (Harris 1996; Hadas 2004). Availability of good quality seed and its cost influence both the quality and quantity of crop produce ultimately influencing food and nutritional security. Early seedling vigor is indicator of good quality seed which translates into quick, uniform germination, and development of crop stand with strong seedling growth detrimental to adverse soil and climatic condition. Earlier and uniform crop stands establish deeper and vigorous root systems to overcome seedbed constraints such as harden and drying upper soil layers, resist to sub-, supra-optimal temperature, and suppress weeds growth by reducing competition for water and nutrient sources (Farooq et al. 2018).

Nonetheless, early seedling vigor is an agronomical trait and indicator to improve speed and uniformity of emergence, seedling growth, and uniform crop stand in direct seeded (Foolad et al. 2007) and transplanted rice systems (Farooq et al. 2011a), in addition to breeding, developing cultivars of high seed vigor, seed priming as low cost along with effective technique for improving earlier, and better crop emergence for uniform stand establishment which flowers earlier and produce productivity (Harris et al. 2007; Ullah et al. 2019) in many crops including rice.

This chapter discusses the potential of priming for improvement of seed and seedling vigor, nursery seedling development and uniform stands, effects on crop growth, increase in yields, nutritional quality of harvested grains, and their economic benefits in conventional and water-saving rice systems. The major objective is potential application of priming to effect on seed and seedling vigor to optimize crop stands and shrink the yield gaps in different rice systems.

4.2 Rice Seed Priming

Priming of seed is a hydration treatment which involves soaking seed in simple water (hydropriming, on-farm priming), salts to lower water potential (osmopriming or osmohardening), crop growth regulators (hormonal priming), and crop nutrients (nutrient seed priming), along with organic biostimulants with or without aeration. Soaking is followed by drying to lower the moisture contents to the original dry weight for routine handling and safe storage of seed until use (Farooq et al. 2018). These are low cost, practicable, and effective techniques for improving seed and crop (Fig. 4.1) performance to address challenges of low seed quality, seedling vigor, late planting, lower and higher temperatures, nutrient deficiency, and salinity along with drought (Finch-Savage and Bassel 2016; Antonino et al. 2000; Farooq et al. 2009a, b, 2011a, b, 2014). Primed crops usually emerge earlier, produce uniform and healthy stands, vigorous root system, flower, and mature earlier relatively with higher crop yields (Table 4.1; Rehman et al. 2011a; Singh et al. 2015). Among priming treatments, on-farm priming, hydropriming, hardening osmopriming, osmohardening and hormonal priming with synthetic and natural biostimulants, and nutrient priming have been successfully employed for improving nursery seedling emergence along with development, produce uniform crop stands, improve growth, and yield performance in transplanted and dry seed rice systems (Table 4.1; Farooq et al. 2006a, b, c, d). Among different priming treatments, osmopriming and osmohardening with $CaCl_2$ and KCl have been extensively evaluated to improve germination besides uniform rice crop stand establishment (Farooq et al. 2006b, c, 2007a; Rehman et al. 2011a, b, 2014a, 2015a, b).

Priming treatments have been optimized for concentrations and duration for soaking of different osmotica, natural or synthetic growth regulators, plant-based biostimulants, micronutrients, and water. For example, rice seeds are hydroprimed in water for 24 h, osmopriming for 36 or 48 h, and on-farm priming for 12 h (Farooq et al. 2006a; Rehman et al. 2015b). Seed osmopriming with KCl has been found effectively to improve crop stand in coarse rice (Farooq et al. 2006a), seedling development in nursery and field, and yield attributes in transplanted rice (Farooq et al. 2007a, b).

4.3 Effects on Seedling Growth, Yield, and Resource Use Efficiency

Raising rice nursery seedlings by seed priming and their transplanting have several advantages including rapid crop development, early phenological growth, and productivity benefits including improved resources use efficiency in conventional and water-saving systems for rice (Tables 4.1 and 4.3). Early transplanted seedlings (<30 days) raised by primed seeds also reduced the time from transplanting to heading and maturity than seedlings raised by traditional method and their transplanting after 45 days (Farooq et al. 2007a, b). Likely, timely transplanted seedlings also result in earlier heading and maturity when raised by different priming methods and priming agents including osmohardening, hardening, and hormonal priming (Farooq et al. 2007a, b) than with delayed heading and maturity in seedlings raised after traditional method.

This improvement in crop stand and nursery seedling growth is attributed to earlier emergence, better seedling growth, increased root growth and its traits, and nutrient uptake contributing toward healthy and vigorous stands in direct seeded and transplanted rice systems (Farooq et al. 2018). Likely, higher yields in these rice systems are associated with increased total emergence, competitive advantage over weeds, productive tillers, number of panicles and growth attributes including leaf area and duration, crop growth rate, and increased dry matter production in aerobic as well as submerged condition (Mahajan et al. 2011). Reduced spikelet sterility and increased tillering in aerobic rice with AWD and SRI (Khalid et al. 2015; Das et al. 2021).

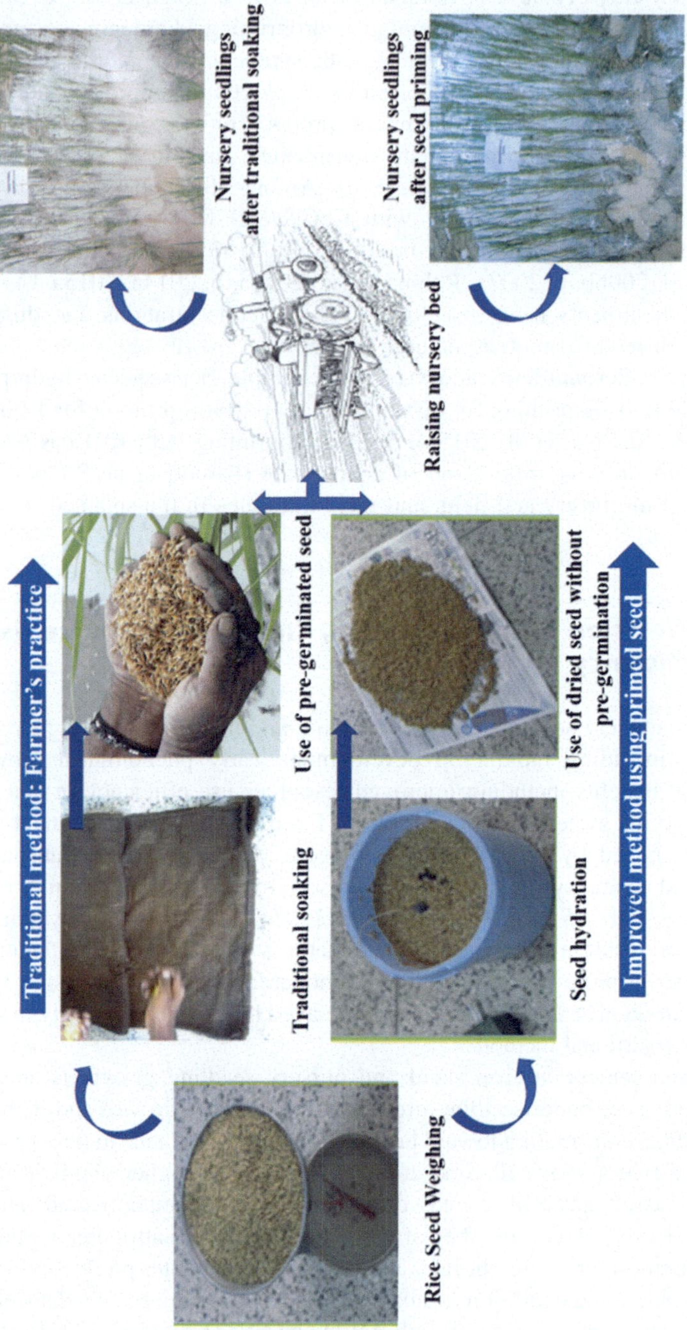

Fig. 4.1 Schematic presentation of improved nursery method "Rice Seed and Seedling Priming"

Table 4.1 Influence of seed priming on paddy yield in rice production systems

Seed priming type	Growing environment	Water-wise rice production system	% Increase in grain yield	References
Hydropriming	Field	CF	7.65	Rehman et al. (2016)
Hydropriming	Field	CF	−1.95	Rehman et al. (2014b)
Hydropriming	Field	CF	26.70	Farooq et al. (2007b)
Hydropriming	Field	AWD	−02.35	Rehman et al. (2016)
Hydropriming	Field	AWD	10.85	Rehman et al. (2014b)
Hydropriming	Field	DSR-SRI	36.00	
Hydropriming	Field	DSR	05.80	Rehman et al. (2016)
Hydropriming	Field	DSR	09.65	Rehman et al. (2014b)
Hydropriming	Field	DSR	−18.30	Rehman et al. (2011b)
Hydropriming	Field	DSR	03.70	Farooq et al. (2006d)
Osmopriming ($CaCl_2$)	Field	CF	42.80	Farooq et al. (2007b)
Osmopriming ($CaCl_2$)	Field	CF	42.80	Farooq et al. (2007b)
Osmopriming (KCl)	Field	DSR-AWD	09.55	Rehman et al. (2015b)
Osmopriming(CaCl2)	Field	DSR-AWD	14.10	Rehman et al. (2015b)
Osmopriming (CaCl2)	Field	DSR-SRI	31.50	
Osmopriming ($CaCl_2$)	Field	DSR	27.00	Rehman et al. (2011b)
Osmopriming (KCl)	Field	DSR	00.00	Rehman et al. (2011b)
Osmopriming (KCl)	Field	DSR	18.50	Farooq et al. (2006d)
Osmopriming ($CaCl_2$)	Field	DSR	14.80	Farooq et al. (2006d)
Osmopriming (KCl)	Field	DSR	10.50	Farooq et al. (2006c)
Nutripriming (B)	Field	CF	16.40	Rehman et al. (2014b)
Nutripriming (B)	Field	CF	27.00	Rehman et al. (2016)
Nutripriming (Zn)	Field	CF	41.10	Farooq et al. (2018)
Nutripriming (B)	Field	AWD	23.25	Rehman et al. (2014b)
Nutripriming (B)	Field	AWD	21.30	Rehman et al. (2016)
Nutripriming (B)	Field	DSR	17.80	Rehman et al. (2014b)
Nutripriming (B)	Field	DSR	17.55	Rehman et al. (2016)
Nutripriming (Zn)	Field	DSR and CF	02.90	Farooq et al. (2018)
Nutripriming (Zn)	Field	DSR	34.60	Farooq et al. (2018)

In addition to growth and yield advantages, seed priming has been reported to improve resource use efficiency regarding water productivity as by osmopriming with moringa leaf extracts (3%) (Rehman et al. 2015b) and osmopriming with Trichoderma and potassium nitrate under AWD (Das et al. 2021), reduce panicle sterility, and enhance gas exchange attributes by nutrient priming with micronutrients (Zn, B, Mn), thus affecting soil–plant water relationship in direct seeded and AWD rice systems (Rehman et al. 2014b, 2016). Likely, priming in rice genotypes efficient in purine permease 1 (*PUP1*) genes and low in seed phosphorus contents and also improvement in germination along with earlier seedling development in phosphorus-deficient soils (Pame et al. 2015), showing seed priming can be

Table 4.2 Influence of nutrient priming on increase in grain mineral concentration in rice production systems

Nutripriming (Nutrient and concentration)	Growing environment	Rice production system	% Increase in grain mineral concentration	References
H_3BO_3 (0.008 M)	Field	CF	700.00	Johnson et al. (2005)
B (0.001 and 0.01 B%)	Glasshouse	CF	33–47	Rehman et al. (2012)
B (0.1 mM)	Field	AWD	29.60	Rehman et al. (2014b)
B (0.1 mM)	Field	AR	27.50	Rehman et al. (2016)
Zn (0.5 M)	Field	DSR	26.67	Farooq et al. (2018)

combined with genetics to improve crop emergence in P-deficient soils. Weed competitive advantage by seed priming in direct seeded rice is owed to rapid emergence and increased seedling vigor at low seed rate reducing biomass of weeds which provide faster canopy development reducing 10% yield losses (Harris et al. 2002; Du and Tuong 2002; Anwar et al. 2012; Juraimi et al. 2012).

Seed nutrient priming by Zn can also reduce the soil application requirement especially under Zn-deficient soil by increased emergence, seedling growth, and crop stand producing better yields in rice (Table 4.2; Tehrani et al. 2003; Prom-u-thai et al. 2012).

4.4 Effects on Grain and Nutritional Quality Attribute

Seed priming induced improved seedling growth, and their transplanting at optimum age reduced mortality rate was associated with better capture resources of water and nutrients resulting in enhanced fertilization and less sterile spikelets. Moreover, increased pre- and post-anthesis net assimilation continued uniform supply of photosynthates throughout panicles producing maximum normal kernels, reducing kernel chalkiness, opaque, and abortive kernels in growing nursery seedlings (Table 4.2; Zheng et al. 2002; Farooq et al. 2007a, b, 2009a, b).

Similarly, Zn nutrient priming is promising strategy for agronomic biofortification in rice under transplanted and direct seeded water-saving rice systems (Farooq et al. 2018). Likely, Zn nutrient priming has been associated with decrease in antinutritional factors including grain phytic acid and Cd contents in grain and increase in protein contents (Seddigh et al. 2016; Rehman et al. 2018; Slamet-Loedin et al. 2015). Similarly, seed priming with boron (0.01 mM B) had been found to improve its grain concentration including panicle fertility under water-saving rice cultivation (Table 4.2; Johnson et al. 2005; Rehman et al. 2016).

Seed osmohardening with KCl and $CaCl_2$ has been observed to contain higher K and Ca contents in rice kernels under traditional and direct seeded rice systems. Likely, increases in seedling nitrogen are associated with increased number of secondary roots and reducing sugars with α-amylase activity in nursery transplanted rice (Farooq et al. 2007b; Rehman et al. 2011a).

4.5 Cost-Benefit Ratio (BCR) and Farmer's Practice

Success of seed priming depends on its cost-effectiveness, practicability, and adoption. The BCR varies among seed priming methods, and highest profit has been witnessed in rice under water-saving system, that is, hydropriming in AWD and DSR, and nutripriming with boron in AWD and with Zn in DSR (Table 4.3). These advantages of seed priming are associated with high yields and reduced inputs in terms of fertilizers and water. Nonetheless, seed priming has been practiced in various countries including Pakistan, Nepal, India, Bangladesh, China, and Australia in various crops including rice (Singh and Gill 1988; Harris et al. 2001; Farooq et al. 2006a, b, c; Hussain et al. 2013).

4.6 Rice Seedling Priming

In transplanted rice, rice seedlings are uprooted from the rice nursery area tagged into small-sized nursery bundles of 5–8 cm (or bunch) for transporting to the targeted field where transplanting is to be carried out. Before, uprooting the seedlings from the nursery, a short spell of stress is necessary to develop hardiness in the younger seedlings, so that these tender seedlings could bear the pulling or transplanting shock. In order to overcome this shock, the seedlings are being primed after uprooting and just before or prior to transplanting.

Seedling priming techniques involve the following, that is, (1) hydropriming (on-farm priming) as dipping the roots of uprooted seedlings in the standing water in a watercourse preferably under shade; (2) Zn priming as dipping the seedlings in Zn solution (35%) at the rate of 12.5 kg ha^{-1}, which is very effective for Zn application. The seedlings uptake the required quantity of Zn, which is required by the rice plant after transplanting; (3) nutripriming as application of biostimulants as booster dose to the younger seedlings; and (4) inoculation of rhizobacteria, which is carried out to enhance mineral nutrient uptake (N, P, K, etc.).

These above-mentioned seedling priming techniques are very cost-effective, practicable in nature, and very efficient to improve seedling performance in the field after transplanting to combat the issues of lower seed quality, seedling vigor, late planting, higher temperature, nutrient deficiency, salinity, and drought. Primed seedling re-start their re-growth after pulling or transplanting shock to perform better through producing uniform as well as healthy crops stands, vigorous root system, and mature earlier relatively with higher crop yields as compared to unprimed seedlings.

Table 4.3 Influence of seed priming on benefit-to-cost ratio in rice production systems

Seed priming type	Growing environment	Rice Production system	% Increase benefit: cost ratio over control	References
Hydropriming	Field	DSR-SRI	04.07	
Hydropriming	Field	CF	−02.30	Rehman et al. (2014b)
Hydropriming	Field	AWD	44.56	Rehman et al. (2014b)
Hydropriming	Field	DSR	07.48	Rehman et al. (2014b)
Hydropriming	Field	CF	05.82	Rehman et al. (2016)
Hydropriming	Field	AWD	09.78	Rehman et al. (2016)
Osmopriming ($CaCl_2$)	Field	DSR-SRI	05.88	
Nutripriming (B)	Field	DSR	12.63	Rehman et al. (2016)
Nutripriming (B)	Field	AWD	11.40	Rehman et al. (2016)
Nutripriming (B)	Field	CF	21.36	Rehman et al. (2016)
Nutripriming (B)	Field	DSR	16.82	Rehman et al. (2014b)
Nutripriming (B)	Field	AWD	57.60	Rehman et al. (2014b)
Nutripriming (B)	Field	CF	13.70	Rehman et al. (2014b)
Nutripriming (Zn)	Field	DSR	06.00	Farooq et al. (2018)
Nutripriming (Zn)	Field	DSR	18.18	Farooq et al. (2018)
Nutripriming (Zn)	Field	CF	15.38	Farooq et al. (2018)

4.7 Conclusion and Future Thrusts

Seed priming is viable and practicable solution to improve crop stand and seedling growth, productivity, nutritional quality, and profitability in traditional and water-saving rice systems. Seed priming can be integrated with genetics as evident from enhanced performance of rice varieties containing QTLs for *Sub1* such as Swarna and *Pup1* in IR74 under submerged and low soil P conditions, respectively (Ella et al. 2011; Sarkar 2012; Pame et al. 2015).

As seed vigor is less considered trait in traditional rice system, and priming had been found to induce stress memory in harvested progeny which needs to be investigated in case of traditional and water-saving rice systems. Such integration of stress-invoked memory in primed seed can be combined with molecular approaches to enhance seed vigor to translate this trait into next generations to address the challenges of seed and seedling vigor. With increasing nutritional deficiency of micronutrients, especially Zn, B, and Fe in human population worldwide, nutrient priming with these micronutrients can improve crop produce and grain micronutrient contents to help reduce malnutrition. In conclusion, as a cost-effective and practicable approach, seed priming can be effective technology to optimize yields using less resources, reduce the gaps between potential and actual yields, and improve socioeconomic condition of growers for sustainable food security.

References

Ahmad S (1998) Response of rice planted under varying management practices. Indian J Agric Sci 68:381–384

Ahmad S, Hasanuzzaman M (2012) Integrated effect of plant density, N rates and irrigation regimes on the biomass production, N content, PAR use efficiencies and water productivity of rice under irrigated semiarid environment. Not Bot Horti Agrobot Cluj-Napoca 40(1):201–211

Ahmad S, Zia-ul-Haq M, Ali H, Shad SA, Ammad A, Maqsood M, Khan MB, Mehmood S, Hussain A (2008) Water and radiation use efficiencies of transplanted rice (Oryza sativa L.) at different plant densities and irrigation regimes under semi-arid environment. Pak J Bot 40 (1):199–209

Ahmad S, Ahmad A, Zia-ul-Haq M, Ali H, Khaliq T, Anjum MA, Khan MA, Hussain A, Hoogenboom G (2009) Resources use efficiency of field grown transplanted rice (Oryza sativa L.) under irrigated semiarid environment. J Food Agric Environ 7(2):487–492

Ahmad A, Ashfaq M, Rasul G, Wajid SA, Khaliq T, Rasul F, Saeed U, Rahman MH, Hussain J, Baig IA, Naqvi AA, Bokhari SAA, Ahmad S, Naseem W, Hoogenboom G, Valdivia RO (2015) Impact of climate change on the rice–wheat cropping system of Pakistan. In: Hillel D, Rosenzweig C (eds) Handbook of climate change and agro-ecosystems: the agricultural modeling intercomparison and improvement project (AgMIP) integrated crop and economic assessments. Imperial College Press and The American Society of Agronomy, London, pp 219–258

Antonino ACD, Sampaio EVS, Dall'Olio A, Salcedo IH (2000) Balanço hídrico em solo com cultivos de subsistência no semi-árido do Nordeste do Brasil. Revista Brasileira de Engenharia Agrícola e Ambiental 4:29–34

Anwar MP, Juraimi AS, Puteh A, Selamat A, Rahman MM, Samedani B (2012) Seed priming influences weed competitiveness and productivity of aerobic rice. Acta Agricult Scand Sect B: Soil Plant Sci 62:499–509

Ashraf A, Khalid A, Ali K (1999) Effect of seedling age and density in growth and yield of rice in saline soil. Pak J Biol Sci 2:860–862

Chauhan BS, Mahajan G, Sardana V, Timsina J, Jat ML (2012) Productivity and sustainability of the rice-wheat cropping system in the Indo-Gangetic Plains of the Indian subcontinent: problems, opportunities, and strategies. Adv Agron 117:315–369

Das D, Noor Ul Basar, Ullah H, Attia A, Salin KR, Datta A (2021) Growth, yield and water productivity of rice as influenced by seed priming under alternate wetting and drying irrigation. Arch Agron Soil Sci. https://doi.org/10.1080/03650340.2021.1912320

De Datta SK (1981) Principle and practices of rice production. Wiley, New York
Du LV, Tuong TP (2002) Enhancing the performance of dry-seeded rice: effects of seed priming, seedling rate, and time of seeding. In: Pandey S, Mortimer M, Wade L, Tuong TP, Lopes K, Hardy B (eds) Direct seeding: research strategies and opportunities. International Rice Research Institute, Manila, Philippines, pp 241–256
Ella ES, Dionisio-Sese ML, Ismail AM (2011) Seed pre-treatment in rice reduces damage, enhances carbohydrate mobilization and improves emergence and seedling establishment under flooded conditions. AoB Plants, plr007
Farooq M, Basra SMA, Afzal I, Khaliq A (2006a) Optimization of hydropriming techniques for rice seed invigoration. Seed Sci Technol 34:507–512
Farooq M, Basra SMA, Hafeez K (2006b) Rice seed invigoration by osmohardening. Seed Sci Technol 34:181–187
Farooq M, Basra SMA, Wahid A (2006c) Priming of field-sown rice seed enhances germination, seedling establishment, allometry and yield. Plant Growth Regul 49:285–294
Farooq M, Basra SMA, Tabassum R, Afzal I (2006d) Enhancing the performance of direct-seeded fine rice by seed priming. Plant Prod Sci 9:446–456
Farooq M, Basra SMA, Khan MB (2007a) Seed priming improves growth of nursery seedlings and yield of transplanted rice. Arch Agron Soil Sci 53:315–326
Farooq M, Basra SMA, Ahmad N (2007b) Improving the performance of transplanted rice by seed priming. Plant Growth Regul 51:129–137
Farooq M, Basra SMA, Asad SA (2008) Comparison of conventional puddling and dry tillage in rice-wheat system. Paddy Water Environ 6:397–404
Farooq M, Kobayashi N, Wahid A, Ito O, Basra SMA (2009a) Strategies for producing more rice with less water. Adv Agron 101:351–388
Farooq M, Wahid A, Kobayashi N, Fujita D, Basra SMA (2009b) Plant drought stress: effects, mechanisms and management. Agron Sustain Dev 29:185–212
Farooq M, Rehman A, Aziz T, Habib M (2011a) Boron nutripriming improves the germination and early seedling growth of rice (Oryza sativa L.). J Plant Nutr 34:1507–1515
Farooq M, Siddique KHM, Rehman H, Aziz T, Lee DJ, Wahid A (2011b) Rice direct seeding: experiences, challenges and opportunities. Soil Tillage Res 111:87–98
Farooq M, Hussain M, Siddique KHM (2014) Drought stress in wheat during flowering and grain-filling periods. Crit Rev Plant Sci 33:331–349
Farooq M, Ullah A, Rehman A, Nawaz A, Nadeem A, Wakeel A, Nadeem F, Siddique KHM (2018) Application of zinc improves the productivity and biofortification of fine grain aromatic rice grown in dry seeded and puddled transplanted production systems. Field Crop Res 216:53–62
Finch-Savage WE, Bassel GW (2016) Seed vigour and crop establishment: extending performance beyond adaptation. J Exp Bot 67:567–591
Foley JA, Ramankutty N, Brauman K, Cassidy ES, Gerber JS, Johnston M, Mueller ND, O'Connell C, Ray DK, West PC et al (2011) Solutions for a cultivated planet. Nature 478: 337–342
Foolad MR, Subbiah P, Zhang L (2007) Common QTL affect the rate of tomato seed germination under different stress and nonstress conditions. Int J Plant Genomics 2007:97386
Gathala MK, Ladha JK, Kumar V, Saharawat YS, Kumar V, Sharma PK, Sharma S, Pathak H (2011) Tillage and crop establishment affects sustainability of South Asian rice-wheat system. Agron J 103(4):961–971
Hadas A (2004) Seedbed preparation—the soil physical environment of germinating seeds. In: Benech-Arnold RL, Sánchez RA (eds) Handbook of seed physiology: applications to agriculture. Haworth Press, New York, pp 3–49
Harris D (1996) The effect of manure, genotype, seed priming depth, and date of sowing on the emergence and early growth of sorghum (Sorghum bicolor L.) in semi-arid Botswana. Soil Tillage Res 40:73–88

Harris D, Raghuwanshi BS, Gangwar JS, Singh SC, Joshi KD, Rashid A, Hollington PA (2001) Participatory evaluation by farmers of on-farm seed priming in wheat in India, Nepal and Pakistan. Exp Agric 37:403–415

Harris D, Tripathi RS, Joshi A (2002) On-farm seed priming to improve crop establishment and yield in dry direct-seeded rice. Proceedings of the International workshop on direct seeding in asian rice systems: strategic research issues and opportunities. 25–28 January 2000. Pandey S, Mortimer M, Wade L, Tuong TP, Lopes K, Hardy B (Eds) (Bangkok, Thailand)

Harris D, Rashid A, Miraj G, Arif M, Shah H (2007) 'On-farm' seed priming with zinc sulphate solution—a cost-effective way to increase the maize yields of resource-poor farmers. Field Crop Res 102:119–127

Himeda M (1994) Cultivation technique of rice nursling seedlings: Review of research papers and its future implementation. Agric Hortic 69:679–683

Hoang TTH, Do DT, Tran TTG, Ho TD, Rehman H (2019) Incorporation of rice straw mitigates CH_4 and N_2O emissions in water saving paddy fields of Central Vietnam. Arch Agron Soil Sci 65(1):113–124

Horie T, Shiraiwa T, Homma K, Maeda Y, Yoshida H (2005) Can yields of lowland rice resume the increases that they showed in the 1980s? Plant Prod Sci 8:251–272

Hussain I, Ahmad R, Farooq M, Wahid A (2013) Seed priming improves the performance of poor quality wheat seed. Int J Agric Biol 15:1343–1348

Johnson SE, Lauren JG, Welch RM, Duxbury JM (2005) A comparison of the effects of micronutrient seed priming and soil fertilization on the mineral nutrition of chickpea (Cicer arietinum), lentil (Lens culinaris), rice (Oryza sativa) and wheat (Triticum aestivum) in Nepal. Exp Agric 41:427–448

Juraimi AS, Anwar MP, Selamat A, Puteh A, Man A (2012) The influence of seed priming on weed suppression in aerobic rice. Pak J Weed Sci Res 18:257–264

Khalid F, Ahmad AUH, Farooq M, Murtaza G (2015) Evaluating the role of seed priming in improving the performance of nursery seedlings for system of rice intensification. Pak J Agric Sci 52:27–36

Kumar V, Ladha JK (2011) Direct seeding of rice: recent developments and future research needs. Adv Agron 111:297–413

Ladha JK, Kumar V, Alam MM, Sharma S, Gathala M, Chandna P, Saharawat YS, Balasubramanian V (2009) Integrating crop and resource management technologies for enhanced productivity, profitability, and sustainability of the rice-wheat system in South Asia. In: Ladha JK, Singh Y, Erenstein O, Hardy B (eds) Integrated crop and resource management in the rice-wheat system of south Asia. International Rice Research Institute, Los Banos, pp 69–108

Linquist BA, Borbea MAA, Pittelkowa CM, Kessel CV, Groenigenb KJV (2012) Fertilizer management practices and greenhouse gas emissions from rice systems: a quantitative review and analysis. Field Crop Res 135:10–21

Mahajan G, Sarlach RS, Japinder S, Gill MS (2011) Seed priming effects on germination, growth and yield of dry direct-seeded rice. J Crop Improv 25:409–417

Mahender A, Anandan A, Pradhan SK (2015) Early seedling vigour, an imperative trait for direct-seeded rice: an overview on physio-morphological parameters and molecular markers. Planta 241:1027–1050

Makarim AK, Balasubramanian V, Zaini Z, Syamsiah I, Diratmadja IGPA, Handoko, Arafah, Wardana IP, Gani A (2002) Systems of rice intensification (SRI): evaluation of seedling age and selected components in Indonesia. In: Bouman BAM, Hengsdijk A, Hardy B, Bindraban PS, Tuong TP, Ladha JK (eds) Water-wise rice production. IIRI, Los Baños, Philippines, pp 123–139

Manigbas NL, Solis RO, Barroga WV, Noriel AJ, Arocena EC, Padolina TF, Cruz RT (2008) Development of screening methods for anaerobic germination and seedling vigor in direct wet-seeded rice culture. Philip J Crop Sci 33(3):34–44

Mishra A, Salokhe VM (2008) Seedling characteristics and the early growth of transplanted rice under different water regimes. Exp Agric 44:365–383

Nadeem F, Farooq M, Mustafa B, Rehman A, Nawaz A (2020) Residual zinc improves soil health, productivity and grain quality of rice in conventional and conservation tillage wheat-based systems. Crop Pasture Sci 71:322–333

Nawaz A, Farooq M, Lal R, Rehman A (2017) Comparison of conventional and conservation rice-wheat systems in Punjab, Pakistan. Soil Tillage Res 169:35–43

Nawaz A, Farooq M, Nadeem F, Siddique KHM, Lal R (2019) Rice– wheat cropping systems in South Asia: issues, options, and opportunities. Crop Past Sci 70:395–427

Padalia CR (1980) Effect of age of seedling on the growth and yield of transplanted rice. Oryza 81: 165–167

Pame AR, Kreye C, Johnson D, Heuer S, Becker M (2015) Effects of genotype, seed P concentration and seed priming on seedling vigor of rice. Exp Agric 51:370–381

Pasuquin E, Lafarge T, Tubana B (2008) Transplanting young seedlings in irrigated rice fields: Early and high tiller production enhanced grain yield. Field Crop Res 105:141–155

Prom-u-thai C, Rerkasem B, Yazici A, Cakmak I (2012) Zinc priming promotes seed germination and seedling vigor of rice. J Plant Nutr Soil Sci 175:482–488

Randriamiharisoa R, Uphoff N (2002) Factorial trials evaluating the separate and combined effects of SRI practices. In: Uphoff N et al (eds) Assessment of the system of rice intensification: proceedings of an international conference, Sanya, China, April 1–4, 2000. Cornell International Institute for Food, Agriculture and Development, Ithaca, pp 40–46

Reddy S (2004) Agronomy of field crops. Kalyani Publishers, New Delhi, India

Rehman H, Basra SMA, Farooq M, Ahmed N, Afzal I (2011a) Seed priming with $CaCl_2$ improves the stand establishment, yield and some quality attributes in direct-seeded rice (Oryza sativa L.). Int J Agric Biol 13:786–790

Rehman H, Basra SMA, Farooq M (2011b) Field appraisal of seed priming to improve the growth, yield and quality of directseeded rice. Turk J Agric For 35:357–365

Rehman HU, Aziz T, Farooq M, Wakeel A, Rengel Z (2012) Zinc nutrition in rice production systems: a review. Plant Soil 361:203–226

Rehman H, Nawaz Q, Basra SMA, Afzal I, Yasmeen A (2014a) Seed priming influence on early crop growth, phenological development and yield performance of linola (Linum usitatissimum L.). J Integr Agric 13:990–996

Rehman A, Farooq M, Nawaz A, Ahmad R (2014b) Influence of boron nutrition on the rice productivity, kernel quality and biofortification in different production systems. Field Crop Res 169:123–131

Rehman H, Iqbal H, Basra SMA, Afzal I, Farooq M, Wakeel A, Wang N (2015a) Seed priming improves early vigor, growth and productivity of spring maize. J Integr Agric 14:1745–1754

Rehman H, Kamran M, Basra SMA, Afzal I, Farooq M (2015b) Influence of seed priming on the performance and water productivity of direct-seeded rice in alternating wetting and drying. Rice Sci 22:189–196

Rehman A, Farooq M, Nawaz A, Ahmad R (2016) Improving the performance of short-duration basmati rice in water-saving production systems by boron nutrition. Ann Appl Biol 168:19–28

Rehman A, Farooq M, Naveed M, Nawaz A, Shahzad B (2018) Seed priming of Zn with endophytic bacteria improves the productivity and grain biofortification of bread wheat. Eur J Agron 94:98–107

Rehman H, Nawaz A, Awan MI, Ijaz M, Hussain M, Ahmad S, Farooq M (2019) Direct seeding in rice: problems and prospects. In: Hassanuzaman M (ed) Agronomic crops: production technologies, vol 1. Springer, Singapore, pp 199–222

Ros C, Bell RW, White PF (2003) Seedling vigour and the early growth of transplanted rice (Oryza sativa). Plant Soil 252:325–337

Salam MU, Jones JW, Kobayashi K (2001) Predicting nursery growth and transplanting shock in rice. Exp Agric 37:65–81

Sarkar RK (2012) Seed priming improves agronomic trait performance under flooding and non-flooding conditions in rice with QTL SUB1. Rice Sci 19:286–294
Sasaki R (2004) Characteristics and seedling establishment of rice nursing seedlings. Japan Agricult Res Quart 38:7–13
Seddigh M, Hossein A, Khoshgoftarmanesh Ghasemi S (2016) The effectiveness of seed priming with synthetic zinc-amino acid chelates in comparison with soil-applied $ZnSO_4$ in improving yield and zinc availability of wheat grain. J Plant Nutr 39:417–427
Singh H, Gill HS (1988) Effect seed treatment with salts on germination and yield of wheat. Agric Sci Dig 8:173–175
Singh RS, Singh SB (1999) Effect of age of seedlings, N-levels and time of application on growth and yield of rice under irrigated condition. Oryza 36(4):351–354
Singh H, Jassal RK, Kang JS, Sandhu SS, Kang H, Grewal K (2015) Seed priming techniques in field crops—a review. Agric Rev 36:251–264
Slamet-Loedin IH, Johnson-Beebout SE, Impa S, Tsakirpaloglou N (2015) Enriching rice with Zn and Fe while minimizing Cd risk. Front Plant Sci 6:121
Tehrani MM, Malakouti MJ, Savaghebi GR, Lotfollahi M, Zieyan A, Balali MR, Bybordi A, Majadi A (2003) Can the zinc fortified seed increase wheat grain yield in zinc deficient soils of Iran? In: '2nd International Congress of Plant Physiology'. New Delhi, India
TeKrony DM, Egli DB (1991) Relationship of seed vigour to crop yield: a review. Crop Sci 31: 816–822
Tripathi RP, Sharma P, Singh S (2005) Tillage index: an approach to optimize tillage in rice-wheat system. Soil Tillage Res 80:125–137
Ullah A, Farooq M, Hussain M, Ahmad R, Wakeel A (2019) Zinc seed priming improves stand establishment, tissue zinc concentration and early seedling growth of chickpea. J Anim Plant Sci 29:1046–1053
Van DW, Sakthivadivel R, Renshaw M, Silver JB, Birley MH, Konradsen F (2001) Alternate wet/dry irrigation in rice cultivation: a practical way to save water and control malaria and Japanese encephalitis? Research Report 47, IWMI: 1–30
Yang CM (2012) Technologies to improve water management for rice cultivation to cope with climate change. Crop Environ Bioinformat 9:193–207
Zhang ZH, Qu XS, Wan S, Chen LH, Zhu YG (2005a) Comparison of QTL controlling seedling vigor under different temperature conditions using recombinant inbred lines in rice (Oryza sativa). Ann Bot 95:423–429
Zhang ZH, Su L, Li W, Chen W, Zhu YG (2005b) A major QTL conferring cold tolerance at the early seedling stage using recombinant inbred lines of rice (Oryza sativa L.). Plant Sci 168:527–534
Zheng HC, Jin H, Zhi Z, Ruan SL, Song WJ (2002) Effect of seed priming with mixed- salt solution on germination and physiological characteristics of seedling in rice (Oryza sativa L.) under stress conditions. J Zhejiang Uni 2:175–178

Nursery Management of Transplanted Rice

5

Naeem Sarwar, Hakoomat Ali, Atique-ur-Rehman, Allah Wasaya, Omer Farooq, Khuram Mubeen, Muhammad Dawood, Muhammad Shehzad, and Shakeel Ahmad

Abstract

Rice is the food of more than half of the world's population which is mostly grown by transplanted method. Nursery management to get better or vigorous seedlings is very important to attain a good crop yield. Plants grown under well-managed fields with proper irrigation and fertilizer adjust easily in the main field. Nursey is uprooted manually from nursery field, and transplant in the main field is a common practice among farming community. This procedure may damage the seedlings, so the nursery plants with better re-rooting ability may have a great impact in the main field which may bear transplanting shock. Furthermore, the seedling age seems to be the main factor for uprooting and shifting in the field. In this chapter, we discuss the importance of inputs used in the nursery field and seedling age on transplanted rice.

N. Sarwar (✉) · H. Ali · Atique-ur-Rehman · O. Farooq · M. Dawood
Department of Agronomy, Bahauddin Zakariya University, Multan, Pakistan
e-mail: naeemsarwar@bzu.edu.pk

A. Wasaya
College of Agriculture, Bahauddin Zakariya University, Layyah, Pakistan

K. Mubeen
Department of Agronomy, Muhammad Nawaz Sharif University of Agriculture, Multan, Pakistan

M. Shehzad
Department of Agronomy, University of Poonch, Rawalakot, Pakistan

S. Ahmad
Department of Agronomy, Faculty of Agricultural Sciences and Technology, Bahauddin Zakariya University, Multan, Pakistan
e-mail: shakeelahmad@bzu.edu.pk

N. Sarwar et al. (eds.), *Modern Techniques of Rice Crop Production*,
https://doi.org/10.1007/978-981-16-4955-4_5

Keywords

Transplanted rice · Nursery · Fertilizer · Seedling age · Biochar

5.1 Introduction

Rice is grown as the second staple food of the world's population and fulfills their nutritional demand (Ahmad et al. 2008, 2009, 2012, 2013, 2015, 2019; Ahmad and Hasanuzzaman 2012; Ahmed and Ahmad 2019, 2020; Shahzad and Ahmad 2019). Population is increasing day by day, so the rice yield must increase 1% annually to meet the increasing demand (Khush 2005; Normile 2008; Ahmed and Fayyaz-ul-Hassan 2017; Ahmed et al. 2020a, b; Akram et al. 2018, 2019; Hussain et al. 2018; Fahad et al. 2019; Fatima et al. 2020; Hafeez-ur-Rehman et al. 2019; Khan et al. 2019a, b; Naz et al. 2019; Peerzada et al. 2019; Razaq et al. 2019; Sarwar et al. 2013a, b; Tariq et al. 2020; Wasaya et al. 2019; Zahoor et al. 2019). This goal can be achieved by reducing the yield losses. Transplanted rice is traditional method, but it still dominates among farmer community (Atique-ur-Rehman et al. 2014, 2018; Chen et al. 2007). Transplanting procedure damages the roots during uprooting and causes plant shock which can further affect the crop growth and yield (Li et al. 2016). It is therefore the need of the time to produce vigorous seedlings with better re-rooting ability to minimize the transplanting shock and yield loss (Ikeda et al. 2007; Sarwar et al. 2014). Generally, the high-quality seedlings produced in better environment have high re-rooting ability (Wu et al. 2016).

Poor nursery management is the principal cause of decreased yield, because after transplanting a poor nursery the yield attributes eventually get decreased (Padalia 1980). To acquire better outputs from paddy, nursery management is very important. Application of proper nutrients in optimum amount is a key to get desired nursery (Lal and Roy 1996). A proper amount of seed is necessary to get desired nursery because more seed rate in growing of nursery can have adverse effects. Among nutrients, nitrogen is most important because different studies revealed that application of nitrogen in nursery decreases chances of impermanency when transplanted in the field (Panda et al. 1991; Tekrony and Egli 1991). Although the nutrition is an important aspect of nursery management, the attributes of nursery vary with seedling age (Himeda 1994; Sasaki 2004). In lowland rice cultivation system, 25–50 days old seedling is transplanted mostly (Datta 1981; Wagh et al. 1988; Singh and Singh 1999).

There is a debate on the age of nursery, but it is also recommended by the outcomes of different studies that the 20 days old or less should not be transplanted. Above 25 days, seedling has enough capacity to withstand field conditions and give maximum yield (Wagh et al. 1988; Mandal et al. 1984; Singh and Singh 1998; Rao and Raju 1987; Ashraf et al. 1999; Nandini and Singh 2000; Thanunathan and Sivasubramanian 2002). The study of Khatun et al. (2002) disclosed that 45 days old contributes maximum in yield compared with 30, 60, and 75 days old nurseries. Kewat et al. (2002) described that 28 days old nursery has more contribution in yield than 14 days old nursery. Reddy and Reddy (1992) explicated that 30 days old nursery produces maximum yield as compared to 45 and 60 days old nurseries. Different other studies revealed that yield of paddy may increase with transplantation of 14 days old nursery. On the other hand, 21–23 days old nursery gave comparatively less yield (Makarim et al. 2002).

Nowadays, biochar has gained certain importance because of its role in improvement of soil (Thies and Rillig 2009) and can be applied in rice nursery for vigorous seedlings. To enhance soil characteristics, it is widely used as substitute for organic matter (Wolf 2008). There are several benefits of biochar including most commonly soil structure, water holding capacity, cation exchange capacity, soil pH, and soil strength, which also improves micro-biota of soil (Masulili et al. 2010). In biochar-amended soil, microbes secondarily enhance soil features like cation exchange capacity and lower the loss of nitrogen from soil (Chan and Xu 2009). Biochar increases soil efficiency as it compensates for the carbon belowground (Krishnakumar et al. 2014). Biochar can be a beneficial substance as its beneficial impacts including in physiological properties, acid substrate pH management, biological activity, and availability of nutrients in soil (Steiner and Harrtung 2014; Nemati et al. 2015; Kaudal et al. 2016).

5.2 Impact of Seedling Age

Age of seedling is a key factor in successful growth of transplanted rice. A lot of research has been done. According to studies of Makarim et al. (2002), 14 days old nursery performed well than 21–23 days old nursery. Another study explained that nursery was raised to assess different methods of sowing. 7–21 days old nursery was successful and participated more in yield by increasing the yield to 1 ton per ha (Pasuquin et al. 2008). Seedling can affect different parameters of paddy crop like dry matter accumulation, number of productive tillers, and plant height in shoot system, and it improves root length in root system and also provides better anchorage of root in the soil (Mishra and Salokhe 2008). It was obvious from the studies of Khakwani et al. (2005) that under consideration of physiological and morphological characteristics younger seedlings are more productive than older nursery. System of rice intensification emphasizes 14 days old nursery because it has more potential to grow better and yield enhancement (Balasubramanian 2004; Lehman 2007; Warnock et al. 2007; Woolf et al. 2010). Another study is outlined by McHugh (2002) to observe the age of seedling which explained that 8–15 days old nursery came out most productive in paddy yield. There is a competition of nutrients and space among seedlings when we allow more time for seedling to stay in nursery bed (Mandal et al. 1984). Reddy and Reddy (1992) worked on Warangal rice. 30, 45, and 60 days old nurseries was raised and transplanted to evaluate the results. It was indicated that 30 days old nursery had more potential than 45 and 60 days old nurseries because it gave more yield than others. The same results were observed by the study of Mohapatra and Kar (1991). Koshta et al. (1987) raised 20, 28, and 36 days old nurseries. The nursery which was transplanted after 20 days had maximum output than other nurseries. Interesting results were evaluated by Patel et al. (1987) that late sowing of younger nursery and early sowing of older nursery can provide higher yield trends. A study has been done on 3 and 5 leaf stages of nursery for panicle production. It was obvious that younger nursery can have higher ratio of spikelets. Moreover, the lower grain fertility was observed in younger nursery and higher grain fertility in older nursery (Tsai and Lai 1987). Das et al. (1988) experimented to assess age of nursery as they transplanted seedlings which were 2, 3, 4, and 5 weeks old. Higher outputs were observed in nursery at 4 weeks of age. Another study was conducted to transplant 25, 35, and 45 days older nurseries. It was observed that 25 days old nursery contributed more in kernel yield than others (Wagh et al. 1988). A 2-year study (1986–1987) was conducted in Sri Lanka by Joseph et al. (1989) to determine the seedling age. Better outcomes were observed in younger seedlings. The older seedling (35 days) gave low yield in both years as compared to younger seedlings (21 days). Singh and Singh (1998) conducted a field trial of transplanting 30, 40, 50, and 60 days old nurseries in Uttar Pradesh. High results were acquired from 50 days old seedlings. There is a certain decrease in 1000 grain weight of paddy by growing older seedlings (Kamadi et al. 1991).

On the other hand, there are different researches which emphasize on growing of older nurseries. Transplanting 45 days old nursery gives maximum output (Amina et al. 2002). Furuk et al. (2009) conducted trial on seedling age and concluded that transplanting of 4 weeks older nursery can give better yield attributes like paddy yield, straw yield, and panicle length than 2 weeks old nursery. Khatun et al. (2002) raised 30, 45, 60, and 75 days and then transplanted. Among all, 45 days old nursery outclassed other nurseries because it produces maximum yield. Kewat et al. (2002) concluded that there is a decrease in the yield by transplanting younger seedlings, as older seedling has a better capacity to withstand field conditions than younger seedlings. A 2-year study resulted that seedling age did not really matters but the yield can be obtained more by delayed sowing (20 July). But further delay can decrease the growth as well as yield of paddy. Gill and Sahi (1987) raised a 30, 45, and 60 days old nurseries and then transplanted, respectively. After evaluation of result, it was obvious that 60 days old nursery is best to obtain desired yield. Moreover, the yield parameters like 1000-grain weight, total dry matter, and kernel yield were observed to be enhanced in 60 days old nursery than other nurseries (30 and 45 days). Chandra and Manna (1988) raised 1 month and 20 months old nurseries. After transplanting both nurseries in 20, 25, 33, and 44 hills m^{-2}, it was concluded that age of nursery does not affect the yield but the density (33 hills m^{-2}) was more effective in yield enhancement.

5.3 Improving Seedling Vigor

To grow a successful transplanted rice crop, the nursery is the prime factor. To raise proper nursery, optimum amount of seed and nutrients is required. Paddy yield can be obtained by transplanting healthy nursery plants. Vigorous seedlings can be produced by providing better growing environment in nursery field. Proper fertilizer including organic, inorganic, and seed rate have a great impact to produce vigorous seedlings.

5.3.1 Management of Fertilizer and Seeding Density in Nursery Field

Studies revealed that younger transplantation of younger and vigorous seedlings produced with low seed rate recorded highest yield as well as net benefit (Sarwar et al. 2014). It is determined from previous studies that as the seedlings get older there is a competition observed of nutrients and space is observed (Mandal et al. 1984). The earlier scientists explained that germination rate and seedling vigor can be obtained by using an amended mat which is fertile so that it becomes a medium that provides proper nutrients with proper seed rate (2000–4000 seeds m^{-2}). After 2 weeks, it should be transplanted conventionally in the field. Panda et al. (1991) conducted experiment to raise nurseries with and without nitrogen. After evaluation of results, it was observed that plants treated with nitrogen application have better roots and better height. On the other hand, the plants which do not treat with nitrogen had poor roots and less height.

Healthy seedling can tolerate the environmental conditions as well as flooding in paddy field. Untreated nursery plants have low performance and less tiller production. A study was conducted in Kashmir during 2001–2002 on nutrients' application in nursery. NPK were used in optimum amount to grow nursery. It was observed that the application of nutrients had a significant impact on nursery growth. Moreover, after transplanting, different yield components were observed higher as compared to the plants with no treatments (Singh et al. 2005). Maskina et al. (1985) reported that yield aspects can be enhanced by optimum application of nutrients to the nursery. Lal and Roy (1996) studied seeding density and nutrients application to nursery. It was described that proper nutrient application grows perfect nursery to get desired yield.

Farooq et al. (2007) revealed that seed priming can also be an important practice in order to grow strong seedlings. Reddy (2004) elaborated that poor nursery results in less plant population, ultimately lowering paddy yield. Mishra and Salokhe (2008) raised nursery with and without nutrients. Adequate amount of nutrients was supplied to nursery. On the other side, no nutrients were applied. It was evaluated that nursery raised with nutrients has more vigorous growth and has more shoot and root growth as compared to nursery raise without nutrients. Pasuquin et al. (2008) considered different mediums for nursery including mat nursery, traditional wet method, seedling tray, and dapog. The 7–21 days old nursery was transplanted in field. Among all the mediums, congenitally wet bed nursery and dapog raised nursery had better results. Raju et al. (2001) experimented on application of weeds for green manuring in rice seedling development. It was observed that shoot characteristics (plant height, leaf, and biomass) and root system characteristics (root length and number of roots) were significantly improved by the application of green manure. The application of nutrients in nursery has significant role to boost plant growth which leads to increase the yield (Ros 1998; Ros et al. 1997). Shalaby et al. (1969) explicated that seedling grown with the nutrients results in higher yield. Tekrony and Egli (1991) evaluated that nutrients play a vital role in better crop establishment. Different features of plant like uniformity, viability, stem thickness, etc. are also affected by the application or the absence of nutrients (Matsuo and Hoshikawa 1993). Singh and Singh (1999) used different sources for nitrogen including Urea & DAP, Farmyard manure, and Gliricidia leaves. Two doses of nitrogen were used, that is, 60 and 120 kg N ha^{-1}. After result evaluation, it was noted that 120 kg N ha^{-1} had productive results. Yield parameters like productive tillers and panicle weight were boosted because the nutrients provide a better growth to seedlings. Singh et al. (1987) conducted 2 years of study on different aspects of nursery management to enhance the productivity of rice. Om et al. (1997) also studied different aspects of rice including plant height, panicle length, panicle weight, and productive tillers. After evaluation, it was concluded that the application of nitrogen proved helpful in increasing all the characteristics of seedling. An experiment was conducted with the application of weeds (Green manure) which are potassium rich to assess the growing capacity of rice seedlings. Different seedling characteristics were studied during this experiment like vigor, growth rate, fresh weight, dry weight, root length, shoot length, and biomass. After transplantation, results were evaluated that potassium application has significant impact to successfully grow the rice seedlings (Raju and Ganghwar 2004).

5.3.2 Biochar Application in Nursery Field

Seedling vigor is very important for healthy crop as it gets little transplanting shock when shift in the main field and easily adjusts to new environment. Success of transplanted rice heavily depends on nursey plant which can be achieved by proper nutrient and soil organic matter management (Sarwar et al. 2014). High organic matter not only provides nutrient, bind nutrient, but also creates porous soil structure which favors the uprooting of rice nursery. Biochar is a rich organic material which is produced through pyrolysis of organic plant waste which can be used in nursery field to get healthy and vigorous seedlings. Among its versatility, biochar can work with soil organisms like earthworms so that the nutrients may be available to plant (Noguera et al. 2010). As earlier, it is stated that biochar slows the rate of C emission; it also sequests 400 billion tons of carbon. Carbon dioxide concentration in atmosphere can be reduced up to 37 PPM (Tim Lenton 2009). So biochar has a long-lasting beneficial effect on soil.

Sohi et al. (2010) stated that biochar serves as conditioning material of soil because of its physical and chemical properties. It makes bonds with nutrients and water molecules to stay in the soil and is readily available to plant when required. Biochar is the biggest source of carbon and nitrogen. A large proportion of N can be available to plants from biochar. Firdaus and Syahirah (2017) also stated that major proportion of carbon and nitrogen can be obtained from biochar because it is rich in such compounds. Biochar also reduces the amount of synthetic fertilizers especially nitrogen which is beneficial in certain ways. Soil pollution can be minimized by the addition of biochar in the soil (Dong et al. 2014).

Helliwell (2015) emphasizes on the use of biochar, and organic matter can be added to soil as growing media. Chan and Xu (2009) illustrated that the application of biochar is very helpful for soil. It is an additive which supplies nutrients to the crop. Nitrogen is lost by volatilization in the process of pyrolysis, so biochar is used to retain other nutrients in the soil. The ash of biochar contains most of the mineral nutrients which are essential for plant growth. Cheng et al. (2008) also described that biochar retains other nutrients in the soil and makes easily available for plant. It is highly porous in nature and has a greater surface area due to which the minerals easily translocate from one place to other. It also neutralizes the effect of toxic compounds from the soil and improves the cation exchange capacity in the soil (Yu et al. 2006). Moreover, the application of biochar creates a nutrients zone which is very necessary for plant growth. Biochar prevents the flow of nutrients away from soil and has the ability to seize the nutrients so that they remain in soil profile and are easily transported from soil to other plant parts (Novak et al. 2009; Ding et al. 2010).

5.4 Conclusion

Studies revealed that nursery management is very important to get better rice crop yield. Nursery can be managed with proper fertilization, especially the nitrogen as well as by improving soil organic matter with the addition of biochar. Younger seedlings grown under proper nursery environment can easily adjust in the main field and fell less transplanting shock. Similarly, the seedlings grown under less seed rate showed better adjustment when transplanted in the main field.

References

Ahmad S, Hasanuzzaman M (2012) Integrated effect of plant density, N rates and irrigation regimes on the biomass production, N content, PAR use efficiencies and water productivity of rice under irrigated semiarid environment. Not Bot Horti Agrobot Cluj-Napoca 40(1):201–211

Ahmad S, Zia-ul-Haq M, Ali H, Shad SA, Ammad A, Maqsood M, Khan MB, Mehmood S, Hussain A (2008) Water and radiation use efficiencies of transplanted rice (*Oryza sativa* L.) at different plant densities and irrigation regimes under semi-arid environment. Pak J Bot 40(1): 199–209

Ahmad S, Ahmad A, Zia-ul-Haq M, Ali H, Khaliq T, Anjum MA, Khan MA, Hussain A, Hoogenboom G (2009) Resources use efficiency of field grown transplanted rice (*Oryza sativa* L.) under irrigated semiarid environment. J Food Agric Environ 7(2):487–492

Ahmad S, Ahmad A, Soler CMT, Ali H, Zia-Ul-Haq M, Anothai J, Hussain A, Hoogenboom G, Hasanuzzaman M (2012) Application of the CSM-CERES-Rice model for evaluation of plant density and nitrogen management of fine transplanted rice for an irrigated semiarid environment. Precis Agric 13(2):200–218

Ahmad S, Ahmad A, Ali H, Hussain A, Garcia y Garcia A, Khan MA, Zia-Ul-Haq M, Hasanuzzaman M, Hoogenboom G (2013) Application of the CSM-CERES-Rice model for evaluation of plant density and irrigation management of transplanted rice for an irrigated semiarid environment. Irrig Sci 31(3):491–506

Ahmad A, Ashfaq M, Rasul G, Wajid SA, Khaliq T, Rasul F, Saeed U, Rahman MH, Hussain J, Baig IA, Naqvi AA, Bokhari SAA, Ahmad S, Naseem W, Hoogenboom G, Valdivia RO (2015) Impact of climate change on the rice–wheat cropping system of Pakistan. In: Hillel D, Rosenzweig C (eds) Handbook of climate change and agro-ecosystems: the agricultural modeling intercomparison and improvement project (AgMIP) integrated crop and economic assessments. Imperial College Press and The American Society of Agronomy, London, pp 219–258

Ahmad S, Abbas G, Ahmed M, Fatima Z, Anjum MA, Rasul G, Khan MA, Hoogenboom G (2019) Climate warming and management impact on the change of rice-wheat phenology in Punjab, Pakistan. Field Crop Res 230:46–61

Ahmed M, Ahmad S (2019) Carbon dioxide enrichment and crop productivity. In: Hasanuzzaman M (ed) Agronomic crops, vol 2. Springer, Singapore, pp 31–46

Ahmed M, Ahmad S (2020) Systems modeling. In: Ahmed M (ed) Systems modeling. Springer, Singapore, pp 1–44

Ahmed M, Fayyaz-ul-Hassan AS (2017) Climate variability impact on rice production: adaptation and mitigation strategies. In: Ahmed M, Stockle C (eds) Quantification of climate variability, adaptation and mitigation for agricultural sustainability. Springer, Switzerland, pp 91–111

Ahmed M, Ahmad S, Raza MA, Kumar U, Ansar M, Shah GA, Parsons D, Hoogenboom G, Palosuo T, Seidel S (2020a) Models calibration and evaluation. In: Ahmed M (ed) Systems modeling. Springer, Singapore, pp 151–178

Ahmed M, Ahmad S, Waldrip HM, Ramin M, Raza MA (2020b) Whole farm modeling: a systems approach to understanding and managing livestock for greenhouse gas mitigation, economic viability and environmental quality. In: Waldrip HM, Pagliari PH, He Z (eds) Animal Manure. ASA Special Publication 67. American Society of Agronomy and Soil Science Society of America, Madison, WI, pp 345–371

Akram R, Turan V, Hammad HM, Ahmad S, Hussain S, Hasnain A, Maqbool MM, Rehmani MIA, Rasool A, Masood N, Mahmood F, Mubeen M, Sultana SR, Fahad S, Amanet K, Saleem M, Abbas Y, Akhtar HM, Hussain S, Waseem F, Murtaza R, Amin A, Zahoor SA, Sami ul Din M, Nasim W (2018) Fate of organic and inorganic pollutants in paddy soils. In: Hashmi MZ, Varma A (eds) Environmental pollution of paddy soils. Springer, Switzerland, pp 197–214

Akram R, Fahad S, Masood N, Rasool A, Ijaz M, Ihsan MZ, Maqbool MM, Ahmad S, Hussain S, Ahmed M, Kaleem S, Sultana SR, Mubeen M, Saud S, Kamran M, Nasim W (2019) Plant growth and morphological changes in rice under abiotic stress. In: Hasanuzzaman M, Fujita M, Nahar K, Biswas JK (eds) Advances in rice research for abiotic stress tolerance. Woodhead Publishing, Swaston, pp 69–85

Amina K, Mollah MIU, Rashid MH, Islam MS, Khan AH (2002) Seasonal effect of seedling age on the yield of rice. Pak J Biol Sci 5:40–42

Ashraf M, Khalid A, Ali K (1999) Effect of seedling age and density on growth and yield of rice in saline soil. Pak J Biol Sci 2:860–862

Atique-ur-Rehman M, Farooq AN, Ahmad R (2014) Influence of boron nutrition on the rice productivity, grain quality and biofortification in different production systems. Field Crop Research 169:123–131

Atique-ur-Rehman M, Farooq A, Rashid F, Nadeem S, Stürz FA, Bel RW (2018) Boron nutrition of rice in different production systems. A review. Agron Sustain Dev 38:25

Balasubramanian V (2004) Modified rice mat nursery for producing robust young seedlings in 15 days for enhanced productivity and profit in irrigated farming. Poster presented at the World Rice Research Conference Japan, November, 4–7

Chan KY, Xu ZH (2009) Chapter 5: biochar- nutrient properties and their enhancement. In: Lehmann J, Joseph S (eds) Biochar for environmental management: science and technology. Earthscan, London, UK, p 67

Chandra D, Manna GB (1988) Effect of planting date seedling age and plamting density on late planted wet season Rice. Int Rice Res Newslett 13:30–31

Chen S, Xia GM, Zhao WM, Wu FB, Zhang GP (2007) Characterization of leaf photosynthetic properties for no-tillage rice. Rice Sci 14:283–288
Cheng CH, Lehmann J, Engelhard M (2008) Natural oxidation of black carbon in soils: changes in molecular form and surface charge along a climosequence. Geochim Cosmochim Acta 72: 1598–1610
Das K, Biswal D, Pradhan T (1988) Effect of plant density and age of seedlings on the growth and yield of rice variety. Parjit Oryza 25:191–194
Datta SK (1981) Principles and practices of rice production. Wiley, Singapore
Ding Y, Liu Y-X, Wu W-X, Shi D-Z, Yang M, Zhong ZK (2010) Evaluation of biochar effects on nitrogen retention and leaching in multi-layered soil columns. Water Air Soil Pollut 213:47–55
Dong D, Feng Q, McGrouther K, Yang M, Wang V, Wu V (2014) Effects of biochar amendment on rice growth and nitrogen retention in a waterlogged paddy field. J Soils Sediments 15:153–162
Fahad S, Noor M, Adnan M, Khan MA, Rahman I, Alam M, Khan IA, Ullah H, Mian IA, Hassan S, Saud S, Bakhat HF, Hammad HM, Ahmad S, Nasim W (2019) Abiotic stress and rice grain quality. In: Hasanuzzaman M, Fujita M, Nahar K, Biswas JK (eds) Advances in rice research for abiotic stress tolerance. Woodhead Publishing, Swaston, pp 571–583
Farooq M, Basra SMA, Ahmad N (2007) Improving the performance of transplanted rice by seed priming. Plant Growth Regul 51:129–137
Fatima Z, Ahmad M, Hussain M, Abbas G, Ul-Allah S, Ahmad S, Ahmed N, Ali MA, Sarwar G, Ehsan ul Haque A, Iqbal P, Hussain S (2020) The fingerprints of climate warming on cereal crops phenology and adaptation options. Sci Rep 10:18013
Firdaus N, Syahirah N (2017) The effect of biochar application on nutrient availability of soil planted with MR219. J Microb Bioch Tech 9:583–587
Furuk MO, Rahman MA, Hassan MA (2009) Effect of seedling age and number of seedling per hill on the yield and yield contributing characters of BRRI Dhan-33. Int J Sustain Crop Prod 4:58–61
Gill PS, Sahi HN (1987) Effect of nitrogen levels in relation to age of seedlings and milling characteristics of rice. Indian J Agric Sci 57:630–634
Hafeez-ur-Rehman, Nawaz A, Awan MI, Ijaz M, Hussain M, Ahmad S, Farooq M (2019) Direct seeding in rice: problems and prospects. In: Hasanuzzaman M (ed) Agronomic crops: production technologies, vol 1. Springer, Singapore, pp 199–222
Helliwell R (2015) Effect of biochar on plant growth. Arboriculture J 37:238–242
Himeda M (1994) Cultivation technique of rice nurseling seedlings: review of research papers and its future implementation. Agric Hortic 69:679–683
Hussain M, Ahmad S, Hussain S, Lal R, Ul-Allah S, Nawaz A (2018) Rice in saline soils: physiology, biochemistry, genetics and management. Adv Agron 148:231–287
Ikeda H, Kamoshita A, Manabe T (2007) Genetic analysis of rooting ability of transplanted rice (*Oryza sativa* L.) under different water conditions. J Exp Bot 58:309–318
Joseph KDS, Hemachandra MPV, Jayawardena SDG (1989) Effect of seedling age on performance of the 3-month rice variety Bg-276-5. Trop Agric 145:93–101
Kamadi JT, Hatwar KG, Bobde GN, Patil SM (1991) Effect of age of seedlings at transplanting on the yield of rice varieties. J Soil Crop 1:154–156
Kaudal BB, Chen D, Madhavan DB, Downie A, Weatherly A (2016) An experiment of physiological and chemical properties of urban biochar for use as growing media substrate. Biomass Bioenergy 84:49–58
Kewat ML, Agrawal SB, Agrawal KK, Sharma RS (2002) Effect of divergent plant spacings and age of seedlings on yield and economics of hybrid rice (Oryza sativa). Indian J Agron 47:367–371
Khakwani AA, Masaki S, Muhammad Z, Safdar BM, Khalid N, Inayatullah A (2005) Effect of seedling age and water depth on morophological aspects of transplanted rice. J Zhejiang Univ (Sci) 5:389–395
Khan MA, Ahmad S, Raza A (2019a) Integrated weed management for agronomic crops. In: Hasanuzzaman M (ed) Agronomic crops, vol 2. Springer, Singapore, pp 257–281

Khan SM, Ali S, Nawaz A, Bukhari SAH, Ejaz S, Ahmad S (2019b) Integrated pest and disease management for better agronomic crop production. In: Hasanuzzaman M (ed) Agronomic crops, vol 2. Springer, Singapore, pp 385–428
Khatun A, Mollah MIU, Rashid IH, Islam MS, Khan AH (2002) Seasonal effect of seedling age on the yield of rice. Pak J Biol Sci 5:40–42
Khush GS (2005) What it will take to feed 5.0 billion rice consumers in 2030. Plant Mol Biol 59:1–6
Koshta LD, Sachidanad B, Raghu JS, Upadhyaya VB (1987) Effect of seedling age and soil submergemce on the performance of paddy. Oryza 24:226–230
Krishnakumar S, Rajalakshmi AG, Balaganesh B, Manikandan P, Vinoth C, Rajendran V (2014) Impact of biochar on soil health. Int J Adv Res 2:933–950
Lal M, Roy RK (1996) Effect of nursery seeding density and fertilizer on seedling growth and yield of rice (Oryza sativa). Indian J Agron 41:642–644
Lehmann J (2007) A handful of carbon. Nature 447:143–144
Lenton T (2009). http://climatechangepsychology.blogspot.com/2009/03/chris-goodall-johanneslehmanntim.htm
Li X, Zhong Q, Li Y, Li G, Ding Y, Wang S, Liu Z, Tang S, Ding C, Chen L (2016) Triacontanol reduces transplanting shock in machine-transplanted rice by improving the growth and antioxidant systems. Front Plant Sci 7:872
Makarim AK, Balasubramanian V, Zaini Z, Syamsiah I, Diratmadja IG, Handoko PA, Arafah A, Wardana IP, Gani A (2002) Systems of rice intensification (SRI): evaluation of seedling age and selected components in Indonesia. Int Rice Res Institute 129–139
Mandal BK, Sainik TR, Ray PK (1984) Effect of age of seedling and level of nitrogen on the productivity of rice. Oryza 21:225–232
Maskina MS, Sandhu PS, Meelu OP (1985) Effect of integrated use organic and inorganic nitrogen source on growth and nutrient composition of rice seedling. Oryza 22:11–16
Masulili A, Utomo WH, Syekhfani (2010) The characteristics of rice husk biochar and it's Influence on the properties of acid sulfate soils and rice growth. Geoderma 234:122–145
Matsuo T, Hoshikawa K (1993) Science of the rice plant, vol 1. Morphology. Food and Agricultural Policy Research Center, Tokyo, Japan, p 686
Mchugh O (2002) Farmer alternative wet/dry, non-flooded and continuously flooded irrigation practice in traditional and intensive system of rice cultivation in Madagascar. M. Sc. Thesis. Cornell University, Ithaca, NY, USA
Mishra A, Salokhe VM (2008) Seedling charactristics and the early growth of transplanted rice under different water regimes. Expt Agri 44:365–383
Mohapatra AK, Kar PC (1991) Effect of time of planting, age of seedling and level of nitrogen on yield and N-uptake of low land rice. Orissa J Agri Res India 4:23–26
Nandini DK, Singh AI (2000) Influence of seedling age and plant density on the performance of rice. Department of Agronic, College of Agriculture, Central Agriculture University, Imphal, Manipur, India
Naz S, Fatima Z, Iqbal P, Khan A, Zakir I, Noreen S, Younis H, Abbas G, Ahmad S (2019) Agronomic crops: types and uses. In: Hasanuzzaman M (ed) Agronomic crops: production technologies, vol 1. Springer, Singapore, pp 1–18
Nemati MR, Simard F, Fortin J, Beaudoin J (2015) Potential use of biochar in growing media. Vadose Zone J 14(6):1–8
Noguera D, Rondon M, Laossi KR, Hoyos V, Lavelle P, Cruz de Carvalho MH, Barot S (2010) Contrasted effect of biochar and earthworms on rice growth andresource allocation in different soils. Soil Biol Biochem 42:1017–1027
Normile D (2008) Reinventing rice to feed the world. Science 321:330–333
Novak JM, Busscher WJ, Laird DL, Ahmedna M, Watts DW, Niandou MAS (2009) Impact of biochar amendment on fertility of a Southeastern Coastal Plain soil. Soil Sci 174:105–112
Om H, Katyal SK, Dhiman SD (1997) Effect of nitrogen and seed rate in nursery on growth and yield of hybrid rice (Oryza sativa). Indian J Agron 42:275–277

Padalia CR (1980) Effect of age of seedling on the growth and yield of transplanted rice. Oryza 81: 165–167

Panda MM, Reddy MD, Sharma AR (1991) Yield performance of rainfed lowland rice as affected by nursery fertilization under conditions of intermediate deep water (15–50cm) and flash flood. Plant Soil 132:65–71

Pasuquin E, Lafarge T, Tubana B (2008) Transplanting young seedlings in irrigated rice fields: early and high tiller production enhanced grain yield. Field Crop Res 105:141–155

Patel DL, Patel ZG, Naik AG (1987) Effect of seeding date and seedling age on dry season yield international Rice Research Newsletter, 12: 46-47. (En) Dep. Agron. Gujrat Agric. Uni., Navsari, Gujrat India

Peerzada AM, Bukhari SAH, Dawood M, Nawaz A, Ahmad S, Adkins S (2019) Weed management for healthy crop production. In: Hasanuzzaman M (ed) Agronomic crops, vol 2. Springer, Singapore, pp 225–256

Raju AR, Ganghwar B (2004) Utilization of potassium-rich green-leaf manure for rice (Oryza sativa) nursery and their effect on crop productivity. Indian J Agron 49:244–247

Raju RA, Reddy MN, Gangwar B (2001) Nursery fertilization of rice (Oryza sativa) with native weed vegetation. Indian J Agron 46:94–100

Rao CP, Raju MS (1987) Effect of age of seedling, nitrogen and spacing on rice. Indian J Agron 32: 100–102

Razaq M, Shah FM, Ahmad S, Afzal M (2019) Pest management for agronomic crops. In: Hasanuzzaman M (ed) Agronomic crops, vol 2. Springer, Singapore, pp 365–384

Reddy S (2004) Agronomy of field crops. Kalyani Publishers, New Delhi, India

Reddy KS, Reddy BB (1992) Effect of transplanting time, plant density and seedling age on growth and yield of rice. Indian J Agron 37:18–21

Ros C (1998) Management of seedling nutrition in lowland rainfed rice. Ph. D. thesis, Murdoch University, Western Australia

Ros C, Bell RW, White PF (1997) Nursery applications of N and P increase Rice (Oryza sativa) yield. Kasetsart J (Nat Sci) 31:96–105

Sarwar N, Ali H, Ahmad S, EhsanUllah AS, Mubeen K, Hill JE (2013a) Water wise rice cultivation on calcareous soil with the addition of essential micronutrients. J Anim Plant Sci 23(1):244–250

Sarwar N, Ali H, Maqsood M, EhsanUllah SAN, Shahzad M, Mubeen K, Shahid MA, Ahmad S (2013b) Phenological response of rice plants to different micronutrients application under water saving paddy fields on calcareous soil. Turk J Field Crops 18(1):52–57

Sarwar N, Hakoomat A, Muhammad M, Ashfaq A, Ehsan U, Tasneem K, James EH (2014) Influence of nursery management and seedling age on growth and economic performance of fine rice. J Plant Nutr 37:1287–1303

Sasaki R (2004) Characteristics and seedlings establishment of rice nurseling seedlings. Japanese Agricultural Research Quarterly 38:7–13

Shahzad AN, Ahmad S (2019) Tools and techniques for nitrogen management in cereals. In: Hasanuzzaman M (ed) Agronomic crops, vol 2. Springer, Singapore, pp 111–126

Shalaby YY, Kobbia TE, Omar MA (1969) Rice yield as influenced by N- fertilization in nursery. J Soil Sci UAR 9:185–199

Singh RS, Singh SB (1998) Response of rice (Oryza sativa L.) to age of seedlings, and level and time of application of nitrogen under irrigated condition. Indian J Agron 43:632–635

Singh RS, Singh SB (1999) Effect of age of seedlings, N-levels and time of application on growth and yield of rice under irrigated condition. Oryza 36:351–354

Singh OP, Pal D, Om H (1987) Effect of seed rate in nursery and seedling/hill on yield of transplanted rice. Indian J Agron 32:96–97

Singh KN, Hassan B, Kanday BA, Bhat AK (2005) Effect of nursery fertilization on seedling growth and yield of rice (Oryza sativa). Indian J Agron 50:187–189

Sohi SP, Krull E, Lopez-Capel E, Bol R (2010) A review of biochar and its uses. Carbon 22:34–17

Steiner C, Harrtung T (2014) Biochar as growing media additives and peat substitute. Solid Earth 5: 995–999

Tariq M, Ahmed M, Iqbal P, Fatima Z, Ahmad S (2020) Crop phenotyping. In: Ahmed M (ed) Systems modeling. Springer, Singapore, pp 45–60

TeKrony DM, Egli DB (1991) Relationship of seed vigour to crop yield: a review. Crop Sci 31: 816–822

Thanunathan K, Sivasubramanian V (2002) Age of seedling and crop management practices for high density (HD) grain in rice. Crop Res 24:421–424

Thies JE, Rillig MC (2009) Characteristics of biochar: biological properties. In: Lehmann J, Joseph S (eds) Biochar for environmental management. Earthscan, Sterling, VA, pp 85–102

Tsai YZ, Lai LK (1987) Effect of seedling age on the growth and development of rice in different cropping seasons 2-studies on the effects on panicle formation and development. J Agri Association China 138:63–86

Wagh RG, Khanvilkar SA, Thorat ST (1988) Effect of age of seedlings at transplanting, plant densities and nitrogen fertilization on the yield of rice variety R711. Oryza 25:188–190

Warnock DD, Lehmann J, Kuyper TW, Rillig MC (2007) Mycorrhizal responses to biochar in soil - Concepts and mechanisms. Plant Soil 300:9–20

Wasaya A, Yasir TA, Ijaz M, Ahmad S (2019) Tillage effects on agronomic crop production. In: Hasanuzzaman M (ed) Agronomic crops, vol 2. Springer, Singapore, pp 73–99

Wolf D (2008) Biochar as a soil amendment: black carbon increases cation exchange in soils. Soil Sci Soc Am J 26:1719–1730

Woolf D, Amonette JE, Street-Perrott FA, Lehmann J, Joseph S (2010) Sustainable biochar to mitigate global climate change. Nat Commun 1:56

Wu W, Zhou Y, Zhang J, Chen G, Xu Y, Li S, Zuo Q, Ye W, Yu Y, Yang C, Sun X (2016) Analysis on population characteristics and yield formation advantages of pot-seedling mechanical transplanting middle-season indica hybrid rice. J Nucl Agric Sci

Yu XY, Ying GG, Kookana RS (2006) Sorption and desorption behaviors of diuron in soils amended with charcoal. J Agric Food Chem 54:8545–8550

Zahoor SA, Ahmad S, Ahmad A, Wajid A, Khaliq T, Mubeen M, Hussain S, Sami Ul Din M, Amin A, Awais M, Nasim W (2019) Improving water use efficiency in agronomic crop production. In: Hasanuzzaman M (ed) Agronomic crops, vol 2. Springer, Singapore, pp 13–29

Rice Cultivation Systems

6

Idrees Haider, Muhammad Arif Ali, Niaz Ahmed, Sajjad Hussain, Muhammad Arshad, Muhammad Bilal, Subhan Danish, Hassan Mehmood, Fariha Ilyas, and Shakeel Ahmad

Abstract

Globally, rice is grown in different types of cropping systems, and most common cropping systems for rice are upland rice, lowland rice, irrigated lowland rice, rainfed lowland rice, flood-prone rice and direct seeded rice. Every cropping system and planting technique has different soil management practices like puddling, furrow bed preparation, broadcasting in ploughed land or ploughing and drill seeding. Preparation and management of the soil for different planting techniques have key importance as soil management is further associated with nutrients management, water management and weed management that ultimately contributes to crop yield. In this chapter, we have reviewed the different rice planting techniques like flooded rice, cultivation by alternate wetting and drying and direct seeded rice along with all rice nurseries' planting techniques that can be used for transplanting in different cropping techniques. This chapter also covers

I. Haider · M. A. Ali (✉) · N. Ahmed · S. Danish · H. Mehmood · F. Ilyas
Department of Soil Science, Bahauddin Zakariya University, Multan, Punjab, Pakistan
e-mail: arif1056@bzu.edu.pk

S. Hussain
Department of Horticulture, Bahauddin Zakariya University, Multan, Punjab, Pakistan

M. Arshad
Institute of Environmental Science and Engineering, NUST, Islamabad, Pakistan

M. Bilal
Department of Environmental Science, COMSATS University Islamabad, Abbottabad, KPK, Pakistan

S. Ahmad
Department of Agronomy, Faculty of Agricultural Sciences and Technology, Bahauddin Zakariya University, Multan, Pakistan
e-mail: shakeelahmad@bzu.edu.pk

N. Sarwar et al. (eds.), *Modern Techniques of Rice Crop Production*,
https://doi.org/10.1007/978-981-16-4955-4_6

the other aspects of crop cycle management that are linked with soil preparation and management like crop nutrient management including site-specific nutrient management, crop water management and weed management in relation to the soil management. Our objective was to provide an overall summary of rice growth aspects associated with soil management for rice plantation. In the current water scarcity developing scenario of rice-producing regions like Africa and South Asia, there is a need to adopt the rice cultivation techniques with water-saving options and appropriately adopt the management practices for nutrients, water and weeds to meet the growing rice demand with increasing world's population.

Keywords

Rice · Soil management · Water management · Rice cultivation methods · Nutrients management · Puddling

6.1 Introduction

Rice (*Oryza sativa*), a member of Poaceae family, is a grassy plant with edible grains and used as a staple food in East Asia and Southeast Asia (Ahmad et al. 2008, 2009, 2012, 2013, 2015, 2019). Rice is cultivated globally in more than 100 countries, and total rice cultivated area is 158 million ha approximately and rice production is more than 700 million tonnes annually. Around 90% of the world's rice production is in Asia, that is, 640 million tonnes while 19 million tonnes are produced in Africa and 25 million tonnes in Latin America (GRiSP 2013). The more commonly worldwide cultivated species of rice is Asian rice, that is, *Oryza sativa,* while African rice (*Oryza glaberrima*) is also cultivated in some parts of West Africa. Rice is being grown in almost all parts of the world with different types of topographic locations and wide range of climatic conditions including areas with very frequent rainfalls exceeding 5100 mm with accessible water and wet conditions to very dry regions of the world receiving less than 100 mm rainfall (Ahmad and Hasanuzzaman 2012; Ahmed and Ahmad 2019, 2020; Shahzad and Ahmad 2019). Rice also tolerates a range of temperature and grows successfully in areas where average rice season temperature was witnessed as 33 °C in Sindh, Pakistan, while the average temperature of 17 °C was also evident in Otaru, Japan, rice-growing area (GRiSP 2013).

Globally, more than 50 million people consume rice in different forms as their regular food. In Bangladesh, Indonesia and Vietnam, 60% of calories are obtained from rice, while in China, Korea and Thailand 50–60% and 40–50% of calories are taken from rice in Japan and India (Wasim 2002). Multiple countries from Asia, Australia, America, Europe and Australia use the produce of rice crop as foreign exchange earnings (Prasad et al. 2017). Rice is considered as highly nutritious crop, and nutritional analysis of rice shows that it provides multiple essential nutrients like nitrogen (N), phosphorous (P), potassium (K), calcium (Ca), zinc (Zn) and sodium (Na). According to an estimate, 349–373 kcal energy is obtained from 100-g rice and additionally, rice grains also constitute 6.3–7.1-g protein, 0.3–0.5-g lipids, 77–78 g carbohydrates, 0.2–0.5-g glo fiber and 0.075–0.30-g vitamin E (Juliano 1993).

Rice is grown by different types of cropping systems around the globe, and frequently used cropping systems include upland rice cultivation, lowland rice cultivation, irrigated lowland rice, rainfed lowland rice, flood-prone rice and direct seeded rice (Ahmed et al. 2017, 2020a, b; Akram et al. 2018, 2019; Hussain et al. 2018; Fahad et al. 2019; Fatima et al. 2020). Seedbed preparation for rice cultivation is different from other crops which is also known as floating seedbed preparation as it is done in flooding conditions.

6.2 Methods of Rice Cultivation

Globally, rice is cultivated by flooding/transplanting method which is a conventional method. As a research on rice growth, development and requirements for rice growth progressed the new rice cultivation methods evolved. Nowadays, rice is being cultivated by different methods under the system of rice intensification. Following are the key methods which are in practice for rice cultivation.

6.2.1 Rice Cultivation in Flooding Conditions/Transplanting in Flooding Conditions

Rice cultivation in flooding conditions by transplanting the rice nursery is being practiced since rice crop evolution (Bouman 2003). About 70 million ha area in tropical Asia out of which around 20% (14 million ha) produce two to three rice crops per annum (which is thought to be most intensive soil management system of rice in tropical areas) is cultivated through flooding conditions/transplanting the rice seedlings which accounts for 40% of rice production around the globe (Buresh and Haefele 2010). For rice cultivation through transplanting, rice nursery is raised for 25–35 days and seedlings are transplanted in already puddled rice fields. In general, five different types of rice nurseries are raised to prepare rice seedlings that include the following as explained rice cultivation manual by Kega et al. (2015).

6.2.1.1 Wet-bed Nursery

This is the most common type of rice nursery for lowland rice. For such nursery, a fertile field is selected with easy access to irrigation and drainage. The field is irrigated, puddled and levelled followed by the addition of manure or fertilizer. Normally, 1- to 1.5-m-wide bed with manageable lengths is prepared. Pre-germinated seeds at the rate of 15–20 kg seed per 200 m^2 bed area are sown that are usually enough to produce seedling for one acre of field area (this may increase or decrease depending on soil conditions, rice variety or water availability for rice, etc.)

6.2.1.2 Dry-bed Nursery

Like wet nursery, 15–20 kg seed per 200 m^2 bed area is used for dry-bed nursery. Usually, light fertile soils with easy access to water are selected for dry-bed nursery. Appropriate quantity of organic manures is added on raised beds prepared for nursery sowing. Primed seeds are sown in dry nursery beds and covered lightly with light soil or rice husk. Nursery beds are then saturated after sowing. Nursery beds are watered periodically as per seedling growth, and seedlings get ready for transplantation in 25–30 days after sowing.

6.2.1.3 Mat Nursery

This type of nursery is similar to wet-bed nursery but in this type, plastic material or banana leaves are used for seed sowing that can be easily transported to fields for transplantation. Such nurseries are easy for mechanical transplantation. For such nursery, a levelled area is required near the water source that is divided into plots of 1-m width and 10–15-m length. Plastic material or banana leaves are spread in small plots, and 10–15 kg seed is sown in 25–35 m^2 area or 1 kg pre-germinated seeds are sown on 1.5 m^2 area. Mat nursery with pre-germinated seeds becomes ready for field transplantation in 10–15 days after sowing.

6.2.1.4 Hymeric Nursery

This is like wet-bed nursery, but biochar (carbonized rice husk) is used to grow seeds in nursery beds instead of soil. Biochar is used to pre-germinate the seeds and after 5–10 days, pre-germinated seeds are planted in the prepared fields. The rate of seed establishment in the field is usually reported high as a result of hymeric nursery.

6.2.1.5 Bubble Tray Nursery

For such type of nursery, plastic trays of 59 cm × 34 cm with more than 400 embedded holes are used. Around 300 nursery trays are required to plant one acre of field area, and these trays require around 100-m^2 area. Seeds are planted in bubble trays and watered regularly, and seedlings are transplanted after 25–30 days after sowing.

The rice cultivated through transplantation is one of the oldest techniques for growing rice, and this is also termed as "lowland rice", "wetland rice" or "paddy rice" and commonly cultivated in ploughed, levelled and puddled fields with surrounded earthen bunds for capturing and maximum utilizing the rainfall and irrigation water (Sanchez 2019). For growing lowland/paddy rice, rice nurseries are prepared by any of the suitable nursery methods (explained above) according to area suitability, transplanting method and rice varieties. The prepared seedlings are then transplanted in the levelled and puddled fields either through mechanical transplantation or manually by the labour force. Around 90% of lowland irrigated rice is transplanted by manual labour-driven method (Pandey et al. 2002). The main criticism for irrigated/transplanted rice is due to its high-water demand, especially in water-scarce regions. In many rice regions, either the main source of water is pumping underground water or irrigations are supplemented by pumping underground water in the times when canal or other surface water sources/canals are cannot meet rice water requirement.

6.2.2 Direct Seeded Rice

Direct seeded rice is one of the most mechanized rice cropping methods that was initially developed in the USA and Australia, where large and accurately levelled by laser technology are used for rice cultivation. For this technique, pre-germinated/partially germinated/primed seeds are either evenly broadcasted in saturated soil or standing water or seed drilling is done in dried fields, and then shallow flooding is done. Careful shallow flooding is required after 20 days of seeding to 20 days before maturity. If the fields are not properly levelled and have drainage problems, this may significantly reduce the rice yield in this rice cultivation technique (Sanchez 2019).

Rice is also being cultivated by direct seeding method which is thought to be a very effective substitute for transplanting in flooding conditions without paddling which reduces the labour costs (Mehmood et al. 2002). Rice cultivation through direct seeding method has potential for rice yield as conventional rice cultivation method by considering the effective management practices (Ikeda et al. 2008). By effective nutrient management through optimum nutrients, application at critical growth stages may produce significant results with uniform seedlings distribution in the field (Sarwar et al. 2013a, b, c), still nitrogen management (Yadav et al. 2007) and other cultural and agronomic management practices (Zhao et al. 2006) in direct seeded rice in the area of concern to be explored more. Proper management practices are required at the optimum time to attain higher yield and enhance nitrogen use efficiency in direct seeded rice as compared to transplanted rice (Jamil et al. 2017) with different fertilizer schemes for direct seeded rice (Yin et al. 2004).

6.2.3 Alternate Wet and Dry Method

The alternate wet and dry method is in practice nowadays in water-scarce or limited available water regions. This method allows some part of the topsoil to dry after being irrigated/saturated for many days. By evaporation of surface water, a part of topsoil gets dry, while soil holds water till 15 cm depth and capillary rise of water prevents the drought stress problem to rice crop (Sanchez 2019). Safe limits for wet and dry periods need careful monitoring for number of days without saturation condition attaining the soil moisture tension at −20 kPa (Bouman et al. 2007). According to the recommendation of the International Rice Research Institute, when water table lowers 15–20 cm in rice field, then rice needs irrigation. Dry spells have key importance to reduce methane emission and nitrous oxide emission. There is a need to carefully monitor the water table especially at critical crop growth stages like the tillering stage and panicle initiation stage to prevent the rice yield loss. At critical growth stages, the soil must be kept wet at 5 cm depth especially from flowering until 3–4 weeks before harvest to avoid grain yield reduction (Sanchez 2019).

The alternate wet and dry method saves about 30% of irrigation water if careful irrigation is done with a reliable source and can increase the yield. There is a potential risk of weeds growing in dry spells, and if appropriate weed control is practised and nutrient management is done accurately then rice may yield 30% higher production in clayey soils than the conventional method. In sandy or loamy soils with 40 cm deep water table, there could be 50% water saving that results in 20% decrease in rice production (Bouman et al. 2007). The alternate wet and dry method produces many encouraging results in puddled soils, and if soils are not puddled then soil cracking in dry periods raises water loss through evaporation (Bouman and Tuong 2001) and if irrigation is not precise and controlled then this method may increase rice water consumption.

Nutrient management especially nitrogen cycle experiences negative changes in this method. Nitrogen fertilizer losses are higher in the wet and dry method as compared with flooding method due to changes in nitrification as higher nitrification occurs upon drying and denitrification occurs upon wetting (Buresh et al. 2008). This usually happens to basal nitrogen fertilizer as plants are small to quickly take up ammonium ions and nitrogen application loss at transplanting stage may raise above 90% (Sanchez et al. 1973).

6.3 Field or Bed Preparation for Rice Cultivation

The field or bed preparation techniques for rice production vary with the method of rice cultivation. The most common technique for field or bed preparation is puddling in saturated or flooded fields. Puddling is a process of soil aggregates break down into uniform size mud by using any mechanical force, and this is done in the presence of higher moisture contents. In wetland rice cultivation system, puddling is one of the key soil management practices to manage the structure of topsoil (Sanchez 2019). Puddling is done through tillage operations which usually starts with moisture contents above saturation followed by tillage operations closer to field capacity as moisture level in soil determines the soil strength and aggregation. The puddling conditions may vary depending on soil texture, soil mineral type and contents and level of soil organic matter (Palm et al. 2007). Puddling results in many productive and yield-enhancing changes in the soil that includes the destruction of large soil aggregates, changes in the level of soil porosity, soil bulk density changes, increased moisture retention in soil, agronomic benefits like land levelling, incorporation of crop residues and weed control, changes in oxidation and reduction potential of soil, structural regeneration of soil and rice yield enhancement due to puddling (Sanchez 2019).

Rice sowing on either dry beds direct seed sowing or wet-bed nursery transplanting is also practised. For this purpose, beds in rice fields are prepared by mechanical implements and in dry seeding technique, rice seeds are spread on dry beds, then covered with clayey soil or mud and then irrigated in furrows, while for already prepared nurseries beds are prepared in the field, irrigated and then prepared rice nursery is transplanted in both edges of the wet beds (JICA 2014).

6.4 Nutrient Management in Different Rice Cultivation Techniques

Rice nutrient management depends on multiple factors including soil type, type of rice cultivation technique, type of rice variety and environmental conditions. According to area suitability, plant type, environmental conditions and rice cultivation technique, the most preferred nutrient management technique is site-specific nutrient management (SSNM) (Pampolino et al. 2007; Peng et al. 2010; Buresh et al. 2014). SSNM can overcome the large variability of rice fields/soils to supply mineral nutrients and may also maintain a balance of nutrient by avoiding nutrient depletion in soil (Sanchez 2019). SSNM especially focuses on the indigenous nutrient supply of soil, flexible nitrogen application and considering the management of soil P and K supplies. The best way to assess the indigenous nutrient supply of soil is to maintain the control fields without any nutrient application and assessing the nutrient uptake status of plants, while other aspects of SSNM can be assessed by the application of nutrients like P and K in different doses to crops (Buresh et al. 2014). The flexibility of nitrogen application timing is thought to be more effective when top dressing of nitrogen fertilizer is practised instead of different times of application of nitrogen (Fairhurst et al. 2007). The common method in practice to assess the rice nitrogen deficiency or need is by assessing the nitrogen contents in recent fully expanded leaf blade which can be done by leaf colour chart method (Alam et al. 2005; Singh et al. 2007).

In SSNM, the basal application of nitrogen fertilizer is minimized because young rice plants' nitrogen requirement is low and nitrogen losses are higher at the young plant stage. Therefore, top dressing of nitrogen is recommended at tillering and panicle initiation stage of rice and sometimes at the flowering stage (Cassman et al. 1998) depending on leaf colour (Yang et al. 2003). This ensures the nitrogen application at appropriate rate and due time when the plant requires nitrogen. In SSNM technique, the total nitrogen application varies from field to field or site to site, and sometimes lesser nitrogen is applied than the recommended dose. In terms of nitrogen management, flooded rice soils have three advantages compared to other rice planting techniques that include (1) no depletion of carbon and nitrogen by conserving organic carbon and organic nitrogen in the soil, (2) higher uptake level of nitrogen by rice plant by uptake of ammonium and nitrate ion by avoiding their fixation due to high mobility in water and (3) the higher sensitivity level of nitrogen to climatic conditions and management practices that allow the better planning for nitrogen application (Cassman et al. 1998).

Rice crop response to phosphorous is different from nitrogen. In flooding conditions, the soil solution phosphorous concentration increases, and most of the time no phosphorous application is required for lowland or wetland rice as compared to other crops growing in aerobic conditions that require a higher phosphorous application. Phosphorous availability decreases by an alternate wet and dry method which results in soil re-oxidation from reduction state (Sanchez and Briones 1973). According to Buresh and Haefele (2010), the approach for phosphorous application and response for crop under SSNM is different from nitrogen. SSNM focuses on preventing the phosphorus deficiency to plant and that can be done by adding the amount of phosphorous removed by grains yield removed from the soil, and if we keep on adding the same amount of phosphorous to the soil that is removed by plant, then phosphorus deficiency can be avoided and phosphorus depletion can also be avoided and phosphorus situation in the soil can be reassessed after 8–10 crop cycles (Fairhurst et al. 2007).

Potassium taken up by rice plant is almost in similar quantities as of nitrogen, but 80% of potassium accumulates in straw, while only 12% of nitrogen taken up by plants retains in straw (Sanchez et al. 1989). Irrigation water is also a good source of potassium to soil and adds about 22 kg/ha potassium in one crop cycle (Witt and Dobermann 2004). Potassium deficiency to rice may occur in calcareous soils due to calcium potassium imbalance. In intense cropping management system, the removal of potassium from soil exceeds indigenous K supplies, and poor recycling of rice straw also leads towards potassium losses and deficiency in the soil. Puddling is found to be a suitable operation for potassium recycling and addition to the soil after removal by crop (Dobermann et al. 2004). In SSNM, the calculation method for potassium is similar to phosphorus that focuses on the reduction of agronomic efficiency due to potassium deficiency by avoiding the potassium depletion from indigenous potassium supply (Buresh and Haefele 2010). The rate of potassium application is determined by the amount of potassium taken up by crop for grain yield and is also dependent on the way of straw management after harvest. If large quantities of straw are removed from the field, then potassium addition is unavoidable, and if most of the produced straw remains in the field then lesser potassium addition is required (Fairhurst et al. 2007). If higher doses of potassium application are necessary, then the basal application is not encouraged and splitting the potassium by half at planting and half at panicle initiation is a good option (Fairhurst et al. 2007). SSNM approach for other nutrients like zinc, sulphur, iron and Mn is also recommended for rice crops (Fig. 6.1).

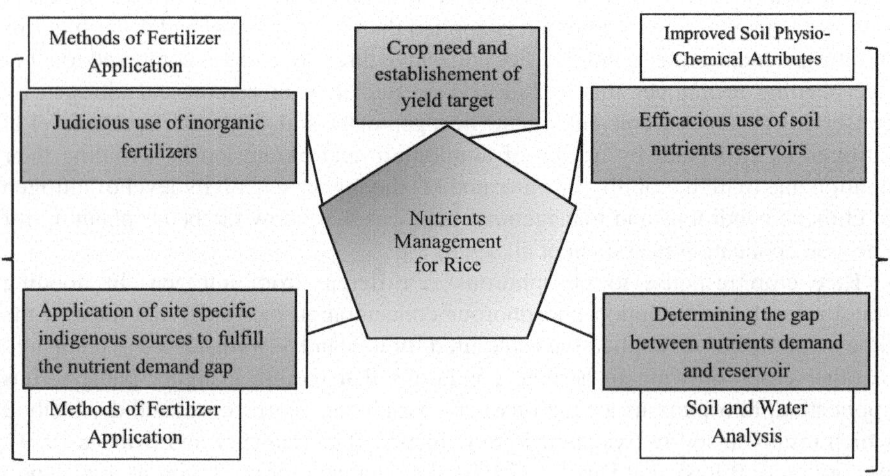

Fig. 6.1 Management of nutrients for the achievement of established yield target of rice (Adopted from Rizzo et al. 2012)

6.5 Water Management for Rice Cultivation

Rice is the only major crop that can grow and sustain in submerged/flooded conditions as it possesses the ability to oxidize its root zone. Rice contains aerenchyma cells/tissue that transports the oxygen from air to root and carbon dioxide and methane to roots to air (Kludze et al. 1993). A major part of oxygen transported through aerenchyma tissue is utilized by roots for respiration, and some of the oxygen is used to oxidize the rice rhizosphere.

In usual conditions, flooded/submerged rice does not face water/moisture stress as rice available moisture range is between flooding condition (0 kPa) and field capacity of the soil (−30 kPa), while other major crops face flooding stress between 0 kPa and − 10 kPa. If weather is hot, that is, sunny, rice crop may express water stress instead of flooding conditions because water transpiration rate from leaves exceeds water uptake by roots. Rice has fibrous root system that usually penetrates to the depth of 20 cm, and rice roots contain iron coating in flooding condition that may block some water uptake by rice crop. Severe water stress at any crop growth stage may decrease the rice yield which happens in rainfed systems (Sanchez 2019). Rice crop is found to have a higher yield in flooding conditions as compared to aerobic rice cultivation methods. In some cases, higher rice yields are obtained in the absence of flooding, but flooding is usually favourable for rice crop. Flooding condition of rice avoids any kind of water stress to rice crops, ensures easy weed management and enhanced availability of some nutrients like phosphorus and enhances soil pH. All three factors that improve flooding condition ultimately increase crop yield (Sanchez 2019).

Rice crop is one of the highest water-consuming crops of the globe and receives about 24–30 of global irrigation resource (Bouman et al. 2007). Keeping in view the highest water demand of the crop researchers has developed and adopted the new rice management systems with less/limited or comparatively reduced water requirement than conventional flooding methods. For water-saving options of rice crop, technologies were considered to lower water percolation, seepage and evaporation losses to increase the water productivity (Hafeez-ur-Rehman et al. 2019; Khan et al. 2019a, b; Naz et al. 2019; Sarwar et al. 2013a, b; Tariq et al. 2020; Wasaya et al. 2019; Zahoor et al. 2019). The water-saving methods include soil saturation without flooding, alternate wet and dry method, replacing puddle with possible dry tillage techniques, moving towards direct seeding from transplantation and introduction of aerobic rice cultivation method (Sanchez 2019) (Fig. 6.2).

6.6 Soil Management and Weeds Control

Weeds control is necessary as weeds create direct competition with rice crop for water, nutrients and sunlight and directly decrease the yield. Weeds also increase the cost of production and decrease grain quality. The most common and recommended practice for weed control is to start the land preparation 3–4 weeks before planting so that all weeds could grow before the crop and ploughing or use of harrow can

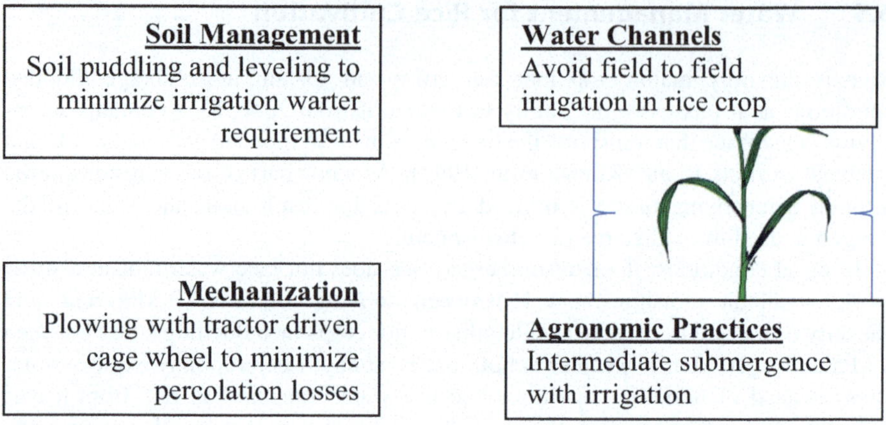

Fig. 6.2 Management of water in rice cultivation (**Source:** Conceptualized from Jansing et al. 2020)

eliminate the weeds from the field. Weeds management for rice in different rice cultivation techniques is different. In flooded rice, the weeds control is easy as most of the weeds cannot grow in flooding conditions, and soil preparation through puddling also eliminates the weeds (GRiSP 2013; Sanchez 2019). In saturated conditions, alternate wet and dry method and aerobic rice cultivation, the cultural control and use of weedicides are also the recommended methods for weeds control (Peerzada et al. 2019; Razaq et al. 2019).

6.7 Conclusion

Rice is one of the key staple crops that is grown by different types of cropping systems around the globe. The most practised and high-yielding cropping technique of rice in the world is flooded rice. As the water scarcity situation in many rice-producing regions like Africa and South Asia is developing, there is need to adopt the water-saving options for rice cultivations including alternate wetting and drying, from flooding to saturation, direct seeded rice and aerobic rice cultivation. Along with rice cultivation techniques, there is a need to focus on site-specific nutrient management along with water-saving options. As methods other than flooded rice have comparatively low yield potential and some associated risks like poor nutrient management, weed growth to reduce yield and re-oxidation of soil may occur that could negatively affect the rice yield. Besides the cropping techniques for water saving, effective nutrient management, efficient weed management, pest management and improved soil management practices are necessary to improve the yield potential of rice.

References

Ahmad S, Hasanuzzaman M (2012) Integrated effect of plant density, N rates and irrigation regimes on the biomass production, N content, PAR use efficiencies and water productivity of rice under irrigated semiarid environment. Not Bot Horti Agrobot Cluj-Napoca 40(1):201–211

Ahmad S, Zia-ul-Haq M, Ali H, Shad SA, Ammad A, Maqsood M, Khan MB, Mehmood S, Hussain A (2008) Water and radiation use efficiencies of transplanted rice (*Oryza sativa* L.) at different plant densities and irrigation regimes under semi-arid environment. Pak J Bot 40(1): 199–209

Ahmad S, Ahmad A, Zia-ul-Haq M, Ali H, Khaliq T, Anjum MA, Khan MA, Hussain A, Hoogenboom G (2009) Resources use efficiency of field grown transplanted rice (*Oryza sativa* L.) under irrigated semiarid environment. J Food Agric Environ 7(2):487–492

Ahmad S, Ahmad A, Soler CMT, Ali H, Zia-Ul-Haq M, Anothai J, Hussain A, Hoogenboom G, Hasanuzzaman M (2012) Application of the CSM-CERES-Rice model for evaluation of plant density and nitrogen management of fine transplanted rice for an irrigated semiarid environment. Precis Agric 13(2):200–218

Ahmad S, Ahmad A, Ali H, Hussain A, Garcia y Garcia A, Khan MA, Zia-Ul-Haq M, Hasanuzzaman M, Hoogenboom G (2013) Application of the CSM-CERES-Rice model for evaluation of plant density and irrigation management of transplanted rice for an irrigated semiarid environment. Irrig Sci 31(3):491–506

Ahmad A, Ashfaq M, Rasul G, Wajid SA, Khaliq T, Rasul F, Saeed U, Rahman MH, Hussain J, Baig IA, Naqvi AA, Bokhari SAA, Ahmad S, Naseem W, Hoogenboom G, Valdivia RO (2015) Impact of climate change on the rice–wheat cropping system of Pakistan. In: Hillel D, Rosenzweig C (eds) Handbook of climate change and agro-ecosystems: the agricultural modeling intercomparison and improvement project (AgMIP) integrated crop and economic assessments. Imperial College Press and The American Society of Agronomy, Swaston, pp 219–258

Ahmad S, Abbas G, Ahmed M, Fatima Z, Anjum MA, Rasul G, Khan MA, Hoogenboom G (2019) Climate warming and management impact on the change of rice-wheat phenology in Punjab, Pakistan. Field Crop Res 230:46–61

Ahmed M, Ahmad S (2019) Carbon dioxide enrichment and crop productivity. In: Hasanuzzaman M (ed) Agronomic crops, vol 2. Springer, Singapore, pp 31–46

Ahmed M, Ahmad S (2020) Systems modeling. In: Ahmed M (ed) Systems modeling. Springer, Singapore, pp 1–44

Ahmed M, Fayyaz-ul-Hassan, Ahmad S (2017) Climate variability impact on rice production: adaptation and mitigation strategies. In: Ahmed M, Stockle C (eds) Quantification of climate variability, adaptation and mitigation for agricultural sustainability. Springer, Switzerland, pp 91–111

Ahmed M, Ahmad S, Raza MA, Kumar U, Ansar M, Shah GA, Parsons D, Hoogenboom G, Palosuo T, Seidel S (2020a) Models calibration and evaluation. In: Ahmed M (ed) Systems modeling. Springer, Singapore, pp 151–178

Ahmed M, Ahmad S, Waldrip HM, Ramin M, Raza MA (2020b) Whole farm modeling: a systems approach to understanding and managing livestock for greenhouse gas mitigation, economic viability and environmental quality. In: Waldrip HM, Pagliari PH, He Z (eds) Animal Manure. ASA Special Publication 67. American Society of Agronomy and Soil Science Society of America, Madison, WI, pp 345–371

Akram R, Turan V, Hammad HM, Ahmad S, Hussain S, Hasnain A, Maqbool MM, Rehmani MIA, Rasool A, Masood N, Mahmood F, Mubeen M, Sultana SR, Fahad S, Amanet K, Saleem M, Abbas Y, Akhtar HM, Hussain S, Waseem F, Murtaza R, Amin A, Zahoor SA, Sami ul Din M, Nasim W (2018) Fate of organic and inorganic pollutants in paddy soils. In: Hashmi MZ, Varma A (eds) Environmental pollution of paddy soils. Springer, Switzerland, pp 197–214

Akram R, Fahad S, Masood N, Rasool A, Ijaz M, Ihsan MZ, Maqbool MM, Ahmad S, Hussain S, Ahmed M, Kaleem S, Sultana SR, Mubeen M, Saud S, Kamran M, Nasim W (2019) Plant growth and morphological changes in rice under abiotic stress. In: Hasanuzzaman M, Fujita M, Nahar K, Biswas JK (eds) Advances in rice research for abiotic stress tolerance. Woodhead Publishing, Swaston, pp 69–85

Alam MM, Ladha JK, Khan SR, Foyjunnessa HUR, Khan AH, Buresh RJ (2005) Leaf color chart for managing nitrogen fertilizer in lowland rice in Bangladesh. Agron J 97:949–959

Bouman BAM (2003) Rice and weeds: the coming crisis. Weed Sci 14(2):1–6

Bouman BAM, Tuong TP (2001) Field water management to save water and increase its productivity in irrigated lowland rice. Agric Water Manag 49:11–30

Bouman BAM, Humphreys E, Barker R (2007) Rice and water. Adv Agron 92:187–237

Buresh RJ, Reddy KR, van Kessel C (2008) Nitrogen transformations in submerged soils. In: Schepers JS, Raun WR (eds) Nitrogen in agricultural systems. Agronomy monograph, vol 49. ASA, CSSA, and SSSA, Madison, WI, pp 401–436

Buresh RJ, Haefele SM (2010) Changes in paddy soils under transition to water-saving and cropping systems. 19th World Congress of Soil Science, Brisbane, Australia. CSIRO Division of Soils, Canberra, published as CD ROM

Buresh RJ, Castillo R, van den Berg M, Gabinete G (2014) Nutrient management decision tool for small-scale rice and maize farmers. International Rice Research Institute, Los Baños

Cassman KG, Peng S, Olk DC, Ladha JK, Reichardt W, Dobermann A, Singh U (1998) Opportunities for increasing nitrogen-use efficiency from improved resource management in irrigated rice systems. Field Crop Res 56:7–39

Dobermann A, Wit C, Dawe D (2004) Increasing productivity of intensive rice systems through site-specific nutrient management. International Rice Research Institute, Los Baños

Fahad S, Noor M, Adnan M, Khan MA, Rahman I, Alam M, Khan IA, Ullah H, Mian IA, Hassan S, Saud S, Bakhat HF, Hammad HM, Ahmad S, Nasim W (2019) Abiotic stress and rice grain quality. In: Hasanuzzaman M, Fujita M, Nahar K, Biswas JK (eds) Advances in rice research for abiotic stress tolerance. Woodhead Publishing, Swaston, pp 571–583

Fairhurst TH, Witt C, Buresh RJ, Dobermann A (2007) Rice: a practical guide to nutrient management, 2nd edn. International Rice Research Institute (IRRI), Philippines, International Plant Nutrition Institute (IPNI) and International Potash Institute (IPI), Singapore

Fatima Z, Ahmad M, Hussain M, Abbas G, Ul-Allah S, Ahmad S, Ahmed N, Ali MA, Sarwar G, Ehsan ul Haque, Iqbal P, Hussain S (2020) The fingerprints of climate warming on cereal crops phenology and adaptation options. Sci Rep 10:18013

GRiSP (Global Rice Science Partnership) (2013) Rice almanac, 4th edn. International Rice Research Institute, Los Baños (Philippines), p 283

Hafeez-ur-Rehman, Nawaz A, Awan MI, Ijaz M, Hussain M, Ahmad S, Farooq M (2019) Direct seeding in rice: problems and prospects. In: Hasanuzzaman M (ed) Agronomic crops: production technologies, vol 1. Springer, Singapore, pp 199–222

Hussain M, Ahmad S, Hussain S, Lal R, Ul-Allah S, Nawaz A (2018) Rice in saline soils: physiology, biochemistry, genetics and management. Adv Agron 148:231–287

Ikeda H, Kamoshita A, Yamagishi J, Ouk M, Lor B (2008) Assessment of management of direct seeded rice production under different water conditions in Cambodia. Paddy Water Environ 6: 91–103

Jamil M, Hussain SS, Qureshi MA, Mehdi SM, Nawaz MQ, Javed Q (2017) Rice yield improvement through various direct seeding techniques on moderately salt affected soil. J Anim Plant Sci 27(3):848–854

Jansing MS, Mahichi F, Dasanayake R (2020) Sustainable irrigation management in paddy rice agriculture: a comparative case study of Karangasem Indonesia and Kunisaki Japan. Sustain For 12:1180

Japan International Cooperation Agency (JICA) (2014) Manual on rice cultivation by PROMPAC: The project for Rice Productivity Improvement in Chokwe Irrigation Scheme. National Directorate of Agricultural Extension, Ministry of Agriculture, Republic of Mozambique

Juliano BO (1993) Rice in human nutrition. Food and Agriculture Organization of the United Nations, Rome, Italy

Kega VM, Gikonyo EW, Muriithi CW, Macharia JMK, Muthoni L (2015) Rice cultivation manual. Kenya Agricultural and Livestock Research Organization (KALRO) Secretariat, Kenya

Khan MA, Ahmad S, Raza A (2019a) Integrated weed management for agronomic crops. In: Hasanuzzaman M (ed) Agronomic crops, vol 2. Springer, Singapore, pp 257–281

Khan SM, Ali S, Nawaz A, Bukhari SAH, Ejaz S, Ahmad S (2019b) Integrated pest and disease management for better agronomic crop production. In: Hasanuzzaman M (ed) Agronomic crops, vol 2. Springer, Singapore, pp 385–428

Kludze HK, DeLaune RD, Patrick WH (1993) Aerenchyma formation and methane and oxygen exchange in rice. Soil Sci Soc Am J 57:386–391

Mehmood N, Chatha ZA, Akhtar B, Saleem M (2002) Response of rice to different sowing methods. Asian J Plant Sci 1(2):144–145

Naz S, Fatima Z, Iqbal P, Khan A, Zakir I, Noreen S, Younis H, Abbas G, Ahmad S (2019) Agronomic crops: types and uses. In: Hasanuzzaman M (ed) Agronomic crops. Production technologies, vol 1. Springer, Singapore, pp 1–18

Palm CA, Sanchez PA, Ahamed S, Awiti A (2007) Soils: a contemporary perspective. Annu Rev Environ Resour 32:99–129

Pampolino MF, Manguiat IJ, Ramanathan S, Gines HC, Tan PS, Chi TTN, Rajendran R, Buresh RJ (2007) Environmental impact and economic benefits of site-specific nutrient management (SSNM) in irrigated rice systems. Agric Syst 93:1–24

Pandey S, Mortimer M, Wade L, Tuong TP, Lopez K, Hardy B (2002) Direct seeding: research strategies and opportunities. International Rice Research Institute, Los Baños

Peerzada AM, Bukhari SAH, Dawood M, Nawaz A, Ahmad S, Adkins S (2019) Weed management for healthy crop production. In: Hasanuzzaman M (ed) Agronomic crops, vol 2. Springer, Singapore, pp 225–256

Peng SB, Buresh RJ, Huang JL, Zhong XH, Zou YB, Yang JC, Wang GH, Liu YY, Hu RF, Tang QY, Cui KH, Zhang FS, Dobermann A (2010) Improving nitrogen fertilization in rice by site-specific N management. A review. Agron Sustain Dev 30:649–656

Prasad R, Shivay YS, Kumar D (2017) Current status, challenges, and opportunities in rice production. In: Chauhan BS, Jabran K, Mahajan G (eds) Rice production worldwide. Springer, Cham, pp 1–32

Razaq M, Shah FM, Ahmad S, Afzal M (2019) Pest management for agronomic crops. In: Hasanuzzaman M (ed) Agronomic crops, vol 2. Springer, Singapore, pp 365–384

Rizzo D, Marraccini E, Lardon S, Rapey H, Debolini M, Thenail C, Benoît M (2012) How can landscape management be enhanced by farming systems? A landscape agronomy perspective. Conference paper

Sanchez P (2019) Soil management in rice cultivation. In: Properties and management of soils in the tropics. Cambridge University Press, Cambridge, pp 492–543

Sanchez PA, Briones AM (1973) Phosphorus availability of some Philippine rice soils as affected by soil and water management practices. Agron J 65(2):226–228

Sanchez PA, Gavidia A, Ramírez GE, Vergara R, Minguillo F (1973) Performance of sulfur-coated urea under intermittently flooded rice culture in Peru. Soil Sci Soc Am Proc 37:789–792

Sanchez PA, Palm CA, Szott LT, Cuevas E, Lal R (1989) Organic input management in tropical agroecosystems. In: Coleman DC, Oades JM, Uehara G (eds) Dynamics of soil organic matter in tropical ecosystems. University of Hawaii, Honolulu HI, pp 125–152

Sarwar N, Ali H, Ahmad A, Ullah E, Ahmad S, Mubeen K, Hill JE (2013a) Water wise rice cultivation on calcareous soil with the addition of essential micronutrients. J Anim Plant Sci 23(1):244–250

Sarwar N, Ali H, Ahmad S, EhsanUllah AS, Mubeen K, Hill JE (2013b) Water wise rice cultivation on calcareous soil with the addition of essential micronutrients. J Anim Plant Sci 23(1):244–250

Sarwar N, Ali H, Maqsood M, EhsanUllah, Shahzad AN, Shahzad M, Mubeen K, Shahid MA, Ahmad S (2013c) Phenological response of rice plants to different micronutrients application under water saving paddy fields on calcareous soil. Turk J Field Crops 18(1):52–57

Shahzad AN, Ahmad S (2019) Tools and techniques for nitrogen management in cereals. In: Hasanuzzaman M (ed) Agronomic crops, vol 2. Springer, Singapore, pp 111–126

Singh SY, Singh B, Ladha JK, Bains JS, Gupta RK, Singh J, Balasubramanian V (2007) On-farm evaluation of leaf color chart for needbased nitrogen management in irrigated transplanted rice in northwestern India. Nutr Cycl Agroecosyst 78:167–176

Tariq M, Ahmed M, Iqbal P, Fatima Z, Ahmad S (2020) Crop phenotyping. In: Ahmed M (ed) Systems modeling. Springer, Singapore, pp 45–60

Wasaya A, Yasir TA, Ijaz M, Ahmad S (2019) Tillage effects on agronomic crop production. In: Hasanuzzaman M (ed) Agronomic crops, vol 2. Springer, Singapore, pp 73–99

Wasim MP (2002) A study of rice in the major growing countries of the world their growth instability and world share. Pak Econ Soc Rev 40(2):153–183

Witt C, Dobermann A (2004) Towards a decision support system for site-specific nutrient management. In: Dobermann A, Wit C, Dawe D (eds) Increasing productivity of intensive rice systems through site–specific nutrient management. International Rice Research Institute, Los Baños, pp 359–395

Yadav S, Gill MS, Kukal SS (2007) Performance of direct-seeded basmati rice in loamy sand in semi-arid sub-tropical India. Soil Tillage Res 97:229–238

Yang WH, Peng S, Huang J, Sanico AL, Buresh RJ, Wit C (2003) Using leaf color charts to estimate leaf nitrogen status of rice. Agron J 95:212–217

Yin XY, Xu YC, Shen QR, Zhou CL, Dittert K (2004) Absorption and apparent budget of nitrogen by direct-seeding rice cultivated in aerobic soil with or without mulching. Acta Ecol Sin 24: 1575–1581

Zahoor SA, Ahmad S, Ahmad A, Wajid A, Khaliq T, Mubeen M, Hussain S, Sami Ul Din M, Amin A, Awais M, Nasim W (2019) Improving water use efficiency in agronomic crop production. In: Hasanuzzaman M (ed) Agronomic crops, vol 2. Springer, Singapore, pp 13–29

Zhao DL, Atlin GN, Bastiaans L, Spiertz JHJ (2006) Developing selection protocols for weed competitiveness in aerobic rice. Field Crop Res 97:272–285

Smart Nutrient Management in Rice Crop

7

Naeem Sarwar, Atique-ur-Rehman, Hakoomat Ali, Allah Wasaya, Omer Farooq, Khuram Mubeen, Muhammad Dawood, Muhammad Shehzad, and Shakeel Ahmad

Abstract

Rice is a dominant food of almost half of the world population. Keeping in view its significance, it is important to develop management practices which are able to maintain higher yields, however, at the same time minimize adverse impacts to environment and optimizing positive benefits. Broader issues of water use, nutrient use efficiency, and emission of greenhouse gasses are all important drivers of long-term sustainability of rice systems. Another input for optimization of rice cultivation is the management of nitrogen fertilizer which is being used heavily. But in another way rice cultivation is also a cause of global warming as it produces greenhouse gasses. In rice cultivation, greenhouse gases (CH_4 and N_2O) are emitted due to anerobic conditions and heavy use of nitrogen fertilizers. Management of rice cultivation as with water-saving techniques and with judicious use of nitrogen fertilizer may become a mitigation strategy for greenhouse gas emissions as well as save costly inputs for sustainable rice agriculture. Lot of research work

N. Sarwar (✉) · Atique-ur-Rehman · H. Ali · O. Farooq · M. Dawood
Department of Agronomy, Bahauddin Zakariya University, Multan, Pakistan
e-mail: naeemsarwar@bzu.edu.pk

A. Wasaya
College of Agriculture, Bahauddin Zakariya University, Layyah, Pakistan

K. Mubeen
Department of Agronomy, Muhammad Nawaz Sharif University of Agriculture, Multan, Pakistan

M. Shehzad
Department of Agronomy, University of Poonch, Rawalakot, Pakistan

S. Ahmad
Department of Agronomy, Faculty of Agricultural Sciences and Technology, Bahauddin Zakariya University, Multan, Pakistan
e-mail: shakeelahmad@bzu.edu.pk

N. Sarwar et al. (eds.), *Modern Techniques of Rice Crop Production*,
https://doi.org/10.1007/978-981-16-4955-4_7

has been done on nitrogen sustainability, but still its efficiency is not more than 35% after application in the soil. Leaching and volatilizations are the major causes for the low nitrogen use efficiency. Many chemicals like nitrification inhibitors and urease inhibitors are being used to improve the nitrogen use efficiency which we discussed in this chapter. Moreover, some slow releasing fertilizers can also be developed and apply in the field to improve the nitrogen use efficiency. Biochar is porous structure which can also be loaded with nutrients and can act as slow releasing fertilizer. We reviewed the overall possibilities for improving nitrogen use efficiency. Integrated nutrient management is also an attractive and sustainable practice of applying nutrients to the soil. Last but not least, we discussed the nanofertilizers, which is one of the attractive advancements in the area of fertilizer.

Keywords

Rice · Greenhouse gasses · Integrated nutrient management · Slow releasing fertilizer · Nanofertilizer

Abbreviations

BGA	Blue-green algae
CEC	Cation exchange capacity
N	Nitrogen
OM	Organic matter
P	Phosphorus

7.1 Introduction

In Pakistan, rice is cultivated as second-order staple food and is also a good source of foreign exchange as the basmati rice is famous all over the world due to its specific fragrance (Ahmad et al. 2008, 2009, 2012, 2013, 2015, 2019; Rehman et al. 2014; Ahmad and Hasanuzzaman 2012; Ahmed and Ahmad 2019, 2020; Shahzad and Ahmad 2019). Most of the rice are cultivated with flooding technique in which flooding conditions are necessary to be maintained through the crop growth cycle starting from transplanting to physiological maturity (Ahmed et al. 2017, 2020a, b; Atique-ur-Rehman et al. 2018; Akram et al. 2018, 2019; Hussain et al. 2018; Fahad et al. 2019; Sarwar et al. 2011, 2014, 2019; Fatima et al. 2020). Farmers also mix farmyard manures like wheat straw and animal dung without following any recommendations. Flooded rice is a main source of CH_4 due to anerobic decomposition of organic matter. Nitrogen fertilizer is also the key component for increasing rice production to meet the domestic as well as export demand which in turn produce N_2O in the atmosphere. Use of unrecompensed dose of nitrogen in flooded rice is common practice at farmer level. Sustainable and economic crop productions are the basic needs of the present time which could only be possible with agronomic management practices (Hafeez-ur-Rehman et al. 2019; Khan et al. 2019a, b; Naz et al. 2019; Peerzada et al. 2019; Razaq et al. 2019; Sarwar et al. 2013a, b; Tariq et al. 2020; Wasaya et al. 2019; Zahoor et al. 2019).

Nitrogen fertilizer demand is increasing day by day and increased 1.4% in the year 2018. Among Asian countries, Pakistan is at fourth for increasing demand of nitrogen fertilizer (FAO 2015). Pakistan is already an energy-deficient country and has to invest lot of energy to manufacture fertilizer which needs to reduce for better economy. Most of Pakistan's cultivated area is deficient in nitrogen which is heavily fertilized with urea for successful crop production (Shah et al. 2012; Ehsanullah et al. 2012; Ali and Noorka 2013). The N is a key important nutrient applied for production of crops as it is limited in soils besides needs to be supplemented (Kawakami et al. 2013). But the nitrogen use efficiency of urea nitrogen is low as 20–50% in most of the soils. Some inorganic substrates like halloysite, montmorillonite, etc. have been also evaluated to enhance nitrogen use efficiency. These minerals are quite useful, but their extraction and purification are time consuming and costly affair (Pereira and Minussi 2012). This necessitated searching for an alternate substrate to regulate the release of nitrogen from urea. Biochar is microporous in nature, besides extensive surface area (120 m^2g^{-1}) that can be exploited for loading of nutrients (Spokas and Novak 2012).

Many products have been developed as nitrification inhibitors which inhibit or delay the nitrogen transformation process and improve the nitrogen availability to plants. These products include N-(n-butyl) thiophosphorictriamide (NBPT), phenyl phosphorodiamidate (PPD), phenylmercuric acetate (PMA), dicyandiamide (DCD), and hydroquinone (HQ) (Khan et al. 2013). Nanofertilizers are reacting similarly with this mechanism and so reduced their losses when applied in field (Yugandhar et al. 2015). Nutrient use efficiency under nanotechnologies can be attained by two means, by improving existing fertilizer having nano-properties or by developing new types of fertilizers (Pereira et al. 2015).

7.2 Fertilizer Application to Mitigate Greenhouse Gasses

Emission of nitrous oxide increases significantly with increasing nitrogen fertilization (Linquist et al. 2012). Production or accumulation of nitrous oxide/CH_4 also depends on the source of nitrogen applied (Cai et al. 2007; Bouwman et al. 2002; Burger and Venterea 2011). Nitrate-based fertilizer has been shown to reduce the CH_4 emission; it is because of lesser plant growth due to lower availability as much nitrogen is denitrified. Farmer prefers to use urea or ammonium sulfate other than nitrate-based fertilizer due to potential of higher availability. The cost factor and higher nitrogen contents in urea probably favor urea use over ammonium sulfate. Application of ammonium sulfate reduced the emission of CH_4 emission as compared to urea fertilization which is probably due to the addition of sulfate in the soil (Linquist et al. 2012). As far as nitrous oxide is concerned, studies showed that its emission increased with higher use of urea as compared with ammonium sulfate. The reason is difference in nitrification rates of both nitrogen sources. Burger and

Venterea (2011) explained that addition of urea and ammonium sulfate potentially affected the emission of nitrous oxide due to opposite reaction on soil pH.

Incorporation of nitrogen is often recommended to increase the nitrogen use efficiency as it limits the nitrification, denitrification, and volatilization. Studies revealed that placement of nitrogen in continuously flooded soil reduced the CH_4 emission as compared with broadcasting (Setyanto et al. 2000; Shang et al. 2011). Placement of nitrogen deep may encourage rice root growth which enhanced the CH_4 production. This CH_4 consumption is increased due to higher oxygen availability in the rhizosphere which overall decreases the CH_4 concentration (Kruger et al. 2001; Gilbert and Frenzel 1998), while in case of rainfed deep placement of nitrogen did not reduce the emission of CH_4 which might be due to changing water management year to year (Setyanto et al. 2000). Deep placement of nitrogen increased the emission of N_2O (Suranto et al. 1998; Fujinuma et al. 2011).

Higher cost and low fertilizer use efficiency are major reasons for low rhizosphere nutrients status to meet crop demand. Efficient nitrogen fertilizers like nitrification or urease inhibitor and slow releasing fertilizers are used to boost nitrogen use efficiency. These fertilizers minimized the losses linked with nitrification, ammonia volatilization besides leaching (Snyder et al. 2009). Such types of fertilizers are proved to be beneficial for higher nitrogen use efficiency in different rice production systems (Norman et al. 1989; Carreres et al. 2003). Nitrification inhibitors include compounds like dicyandiamide, thiosulfate, calcium carbide, etc., and urease inhibitor includes the compound hydroquinone that delays hydrolysis of urea. Some coated or encapsulated fertilizers having protective water-insoluble coating are also applied to reduce losses. As concerned with environmental impact, use of all types of efficiency enhancer products reduced the CH_4 emission which ranged from 15% to 18% (Linquist et al. 2012). In another study, encapsulated calcium carbide reduced the emission of CH_4 confirmed from both field and pot experiments (Bronson and Mosier 1991; Malla et al. 2005). Use of these compounds enhanced the oxidation of CH_4 in the root zone which in turn reduced the emission of CH_4 in the atmosphere. In case of N_2O emission, use of these fertilizers reduced the N-substrate for nitrification or denitrification which reduced the potential of its emission (Subbarao et al. 2006). Efficient nitrogen fertilizer reduced the emission of N_2O in the atmosphere ranged 28–36% in different rice cultivation systems (Akiyama et al. 2004; Arancon et al. 2004; Xuhui et al. 2010; Linquist et al. 2012).

7.3 Improving Nitrogen Use Efficiency

7.3.1 Use of Nitrogen Stabilizers

Population pressure has increased the food demand which in turn enhanced the flux of nitrogen application all over the world. Nitrogen fertilizer demand is increasing day by day and is expected to increase 1.4% in the year 2018. Among Asian countries, Pakistan is at fourth for increasing demand of nitrogen fertilizer (FAO 2015). Pakistan is already energy-deficient country and has to invest lot of energy to manufacture fertilizer which needs to reduce for better economy. Most of Pakistan's cultivated area is deficient in nitrogen which is heavily fertilized with urea for successful crop production (Shah et al. 2012; Ehsanullah et al. 2012; Ali and Noorka 2013). N is the key essential element applied for production of crops as it is limited in soils and so needs to be supplemented (Kawakami et al. 2013). Various nitrogenous sources are being applied like ammonium nitrate and nitrophosphate, but urea is extensively used because of its higher N contents (Soares et al. 2012). But applied nitrogen is not fully utilized by the crop plants as lot of portion is being wasted by leaching or volatilization. Nitrogen use efficiency is very low and reaches up to 33% in cereal production worldwide (Raun and Johnson 1999). Most dominant nitrogenous fertilizer is urea in most of the countries due to higher nitrogen percentage, low cost, no storage risk, and can be applied in variety of crops. But the nitrogen use efficiency of urea nitrogen is low as 20–50% in most of the soils (Bijay 2016).

At urea application in the soil, several chemical besides biological reactions take place that transforms the nitrogen into diverse forms and reduces availability to crops. Urea hydrolyzed into ammonium by urease enzyme and then by nitrifying bacteria into nitrate nitrogen. These urea products can be absorbed by plants or losses in the ammonia gas and nitrate as leaching. Loss of nitrogen not only decreases the nitrogen use efficiency but also enhances input cost of a climatic threat in global warming. Big portion of greenhouse gasses are emitted from agriculture which are 80% and 50%, respectively (IPCC 2007). Many products have been developed as nitrification inhibitors which inhibit or delay the nitrogen transformation process and improve the nitrogen availability to plants. These products include N-(n-butyl) thiophosphorictriamide (NBPT), phenyl phosphorodiamidate (PPD), phenylmercuric acetate (PMA), dicyandiamide (DCD), and hydroquinone (HQ) (Khan et al. 2013). These products are highly expensive which cannot be affordable for the farmers of developing countries like Pakistan. Some plant extracts having similar mode of action to enhance nitrogen use efficiency and also being used in developing countries like India. These products are cheaper and easily available which can attract the farming community to save the expensive fertilizer. Neem is widely growing tree in tropical and sub-tropical areas of Australia, America, Asia, and Africa, and has nitrification inhibitor properties (Schmutterer 1990). Studies showed that neem seed contains nitrification inhibitor compounds like Epinimbin, Deacetyl, Salannin, and Azadirachtin which enhances the nitrogen use efficiency when applied along with urea (Singh and Singh 1986). Neem oil and cake can be used as nitrification inhibitors in crop production to enhance nitrogen use efficiency which was reported many years back in 1970 (Bains et al. 1971). Neem oil or neem cake coated urea can enhance the crop productivity by minimizing nitrogen losses (Prasad et al. 2002). Complex behavior of N fertilizer in soil along with poor husbandry practices is mainly responsible for lower nutrient use efficiency in Pakistan. Such factors further increase N losses as nitrate leaching, volatilization of ammonium, and nitric oxide that are current concerning economic along with environmental issues and threats. On the basis of satisfied results of use of neem oil and neem cake along with urea fertilization, Indian government recently allowed fertilizer industries for urea coating (Bijay 2016).

7.3.2 Application of Enriched Biochar

Many studies have justified that soil application of biochar not only improved performance of major field crops like rice and wheat but also useful for the horticultural crops (Asai et al. 2009; Masulili et al. 2010; Utomo et al. 2017). It is also observed that increased crop performance was not due to nutrition added by biochar but indirectly with the improvement in soil quality. Many workers proved that its application improved the CEC, soil fertility, OM, and soil biology (Masulili et al. 2010; Liang et al. 2006; Rondon et al. 2007). The increase of soil cation exchange capacity would reduce nutrient loss caused by leaching (Laird et al. 2010) and hence increases fertilizer application efficiency (Utomo and Islami 2013).

Nitrogen fertilizer demand is increasing day by day and increased 1.4% in the year 2018. Among Asian countries, Pakistan is at fourth for increasing demand of nitrogen fertilizer (FAO 2015). Pakistan is already energy-deficient country and has to invest lot of energy to manufacture fertilizer which needs to reduce for better economy. Most of Pakistan's cultivated area is deficient in nitrogen which is heavily fertilized with urea for successful crop production (Shah et al. 2012; Ehsanullah et al. 2012; Ali and Noorka 2013). But the nitrogen use efficiency of urea nitrogen is low as 20–50% in most of the soils (Bijay 2016).

After a decade research work, scientists have just little improved the nitrogen use efficiency 30–35% with urea coating by using neem, tar, and starch (Smith and Harrison 1991). Other than organic substrates, some inorganic substrates like halloysite, montmorillonite, etc. have been also evaluated to enhance nitrogen use efficiency. These minerals are quite useful, but their extraction and purification are time consuming and a costly affair (Pereira and Minussi 2012). This necessitated searching for an alternate substrate to regulate the release of nitrogen from urea. Biochar is microporous in nature, besides extensive surface area (120 m^2g^{-1}) that can be exploited for loading of nutrients (Spokas and Novak 2012).

7.4 Integrated Nutrient Management

Integrated nutrient management is a strategy in which we can manage the soil fertility in a sustainable way with the combination of all possible sources. Currently, the farmers are focusing on the use of inorganic fertilizer for the crop production which is not sustainable methods as lot of fertilizer is wasted which is the cause of environmental pollution. In integrated method, some organic sources can be explored which are most efficient and can be used along with inorganic fertilizer. Some of those potential organic fertilizers are discussed below.

7.4.1 Vermicompost

Vermicompost is the way toward fertilizing the soil utilizing different worms. This manure is a supplement-rich natural compost and soil conditioner. Gomez and

Dominguez (2013) concluded that vermicomposting is a minimal effort strategy to change natural waste into biodegraded and balanced out side effects through the joined activity of worms and microorganisms. Contrasted and fertilizing the soil where microorganisms debase natural waste, associations among worms and microorganisms complete mineralization and humification of natural waste at a non-thermophilic stage (Pathma and Sakthivel 2012). Vermicompost is a uniform and unscented material having great actual design, plentiful labile assets, and high microbial movement (Pramanik et al. 2007; Ngo et al. 2011). It contains bigger amounts of mineral and minor components than customary fertilizer which are accessible for plant take-up (Tejada et al. 2009). It additionally contains plant development controlling substances like auxins, gibberellins, cytokinins, fulvic, and humic acids which are helpful for plant execution. Various examinations have demonstrated that vermicompost change could advance soil quality, by improving soil structure, expanding plant accessible supplements, and advancing microbial movement, in this way expanding plant creation comparative with show al substance treatment (Roy et al. 2010; Ngo et al. 2011).

Singh et al. (2008) and Wang et al. (2010) investigated that vermicompost joined with compound compost could advance plant development, productivity besides quality more than alone vermicompost. It suggested that vermicompost by itself could not fulfill every one of the requests of harvest creation for accessible supplements. In such a manner, microbial specialists could fill in for substance manures somewhat by upgrading soil supplement accessibility and plant development (Wu et al. 2012). Vermicompost, nonetheless, is better than manure regarding supplement and asset quality (Ngo et al. 2011; Song et al. 2014) along with its plentiful labile assets that could fill in as carbon along with fuel hotspots for microbial inoculants. Pathma and Sakthivel (2012) found that enormous particulate surface zones along with enhanced pH in vermicompost give good miniature natural surroundings to plant growth-promoting rhizobacteria (PGPR). Raja Sekar and Karmegam (2010) found that vermicompost can uphold endurance of helpful microbes *Azotobacter chroococcum, Rhizobium leguminosarum,* and *Bacillus megaterium* up till more than 10 months. However, a new study conducted by Wu et al. (2012) in pots demonstrated that vermicompost soil application can advance working of three PGPR species (A. chroococcum, B. super terium, and B. adhesive).

7.4.2 Biofertilizer

Nitrogen is considered to be a kingpin of fertilizers due to its demand for application. Nitrogen present in the atmosphere is in big percentage, and estimated that 80,000 tons nitrogen present over one hectare. This nitrogen is not available to plants as it is present in inert form which needs to be converted into available form. Fertilizer industries convert this inert nitrogen into urea, ammonium sulfate, etc. with the use of high temperature (400–500 C0) and pressure (200–1000 ATM). This process conserves large amount of energy obtained mainly from the fossil fuels. On average, energy requirement for 1 kg of nitrogen fertilizer is 11.2, for phosphorus 1.1, and for

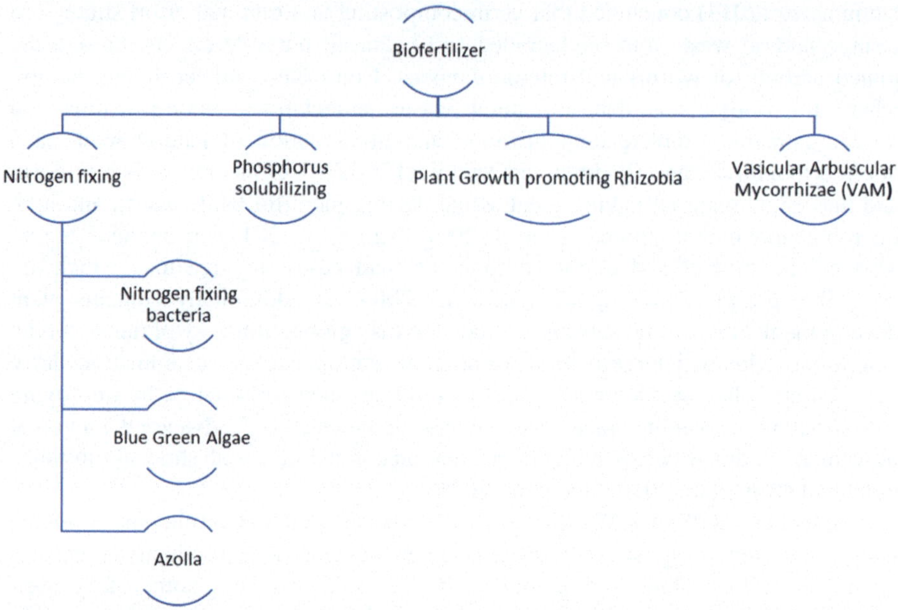

Fig. 7.1 Categories of biofertilizers

potash 1.0 KWH. Increasing cost of energy and diminishing of the fossils fuels are the major country's concerns. Biofertilizers are the big option of the time to meet the nutrition demand of the plants.

Biofertilizer is a substance which contains living microorganisms that colonizes rhizosphere or the interior of plant and promotes growth by increasing the supply or availability of nutrients or growth stimulate to target plants when applied to seed, plant surface, and soil.

Biofertilizer usage is the key component of integrated nutrient management, being cost-effective along with renewable plant nutrient source for supplementing inorganic fertilizers for sustained agricultural production and productivity. Many microorganisms along with their association for crops are being used during biofertilizer production, which can be categorized in diverse ways based on nature along with functionality. Biofertilizers can be categorized into different categories as shown in Fig. 7.1.

7.4.3 Nitrogen Fixing

7.4.3.1 Azotobacter

Azotobacter is a free-living microorganism which fixes the atmospheric nitrogen, present in neutral and alkaline soils. Azotobacter synthesizes growth substance as indole acetic acid, gibberellins, nicotinic acid, etc. Beneficial effects of Azotobacter biofertilizers on cereals, cotton, sugarcane, and vegetables for irrigated along with rainfed conditions are already documented. It can be applied through inoculation in which seed is first treated with sticky material and then treated with bacteria. In transplanted, crop seedlings are dipped in bacterial solution. Azotobacter application in different crops increases the root nitrogen contents as well as crop yield which might be because of the production of growth-promoting substances (Siddiqui et al. 2014).

7.4.3.2 Rhizobium

Generally utilized bio-compost is Rhizobium that colonizes the underlying foundations of explicit vegetables to shape tumor-like development, called root knobs. These knobs go about as plants of alkali creation. Rhizobium vegetable affiliation can fix up to 100–300 kg N/ha. It is noticed that rhizobia make a relationship with non-leguminous plants like rice, wheat, maize, and grain millets. Expanding the capacity of rhizobia in biofertilizer, crop upgrading movement in nonlegumes particularly oat grains would be a helpful innovation for expanded harvest yields among asset helpless ranchers. Ongoing discoveries showed both more harvest upgrading and biofertilizer credits in cereal yields due to rhizobia immunization (Mia and Shamsuddin 2010). Vaccination of the seed with a compelling strain of Rhizobium species alongside Azotobacter brought about huge expansion in nodulation, nitrogen content in the root, and grain yield over unimmunized controls. The helpful impact of test microbial inoculant and plant may be ascribed to the amalgamation of some development advancing substances (Siddiqui et al. 2014).

7.4.3.3 Azospirillum

These living beings structure acquainted beneficial interaction with numerous plants, especially with C4 components. The improvement of white, thick, and undulating fine pellide on semi-strong malate medium is the attribute of the Azospirillum. Notwithstanding nitrogen-fixing certain strains of Azospirillum denitrify under anerobic condition and likewise absorb NH_4, NO_3, or NO_2.

Inoculation of wheat, maize, and soybean is a broad farming practice that has end up being proficient in expanding creation and advancing sustenance of these yields. Plant growth-promoting bacteria (Pseudomonas + Azospirillum) inoculation expanded airborne biomass creation, collect record, and grain yield of rice by 4.7%, 16%, and 20.2%, separately (Ines et al. 2012).

7.4.3.4 Blue-Green Algae (BGA)

Utilization of BGA as a rice biofertilizer is very encouraging. In rice fields, BGA has contributed significantly to enriching and maintaining soil fertility. BGA is photosynthetically nitrogen-fixing organism which has nitrogen-fixing cells (heterocysts). These organisms mostly found in paddy field due to abundance of water. It uses sunlight and water for photosynthesis and nitrogen fixing. Most of BGA of rice field are filamentous comprising vegetative cell's chain including heterocysts as nitrogen-fixing machinery. Other than rice field, algal fungal association (lichens) occurs on the different soils, rocks, and trees. Blue-green algae synthesize growth-promoting substances such as auxin and amino acids which stimulate the rice growth when present in rice field.

7.4.3.5 Azolla

A little coasting water greenery, Azolla is ordinarily found in marsh fields and shallow new water bodies. Azolla makes relationship with nitrogen-fixing blue-green growth. Azolla anabaena affiliation is a live coasting N industrial facility utilizing energy from photosynthesis to fix barometrical N. Azolla normally structures the green tangle on water which frequently becomes reddish because of amassing of anthocyanin color and duplicate vegetative. It additionally delivers some natural acids which wretched the pH of soil.

7.4.3.6 Vesicular Arbuscular Mycorrhiza

Mycorrhiza is advantageous relationship of roots with foundations of vascular plants. Principle benefits of mycorrhiza to host plants lie in expansion of infiltration zone of root growth framework in dirt, working with an expanded P take-up. It not only solubilizes the phosphorus but also increases the root surface area for uptake. It assimilates nutrients for their own use and translocates them in plants in different forms. Furthermore, it also produces some substances like auxin and antibiotics in plants.

7.4.3.7 Phosphorus Solubilizing

Phosphorus is one of the major essential nutrients which require various functions in the plant. Phosphorus is present in the soil in organic and inorganic forms. Availability of the phosphorus to plant depends on its total concentration in the soil as well as on its solubility. Normally, most of the applied soil has been fixed in the soil and just a little fraction is available to plant. Many phosphorus-solubilizing bacteria have been identified which solubilizes soil phosphorus and improves its availability. These bacteria can be applied to crop by various methods like inoculation, seed treatment, soil application, or seedling dipping in solution.

7.4.3.8 Plant Growth-Promoting Rhizobia

They enhance the plant growth indirectly to reduce the deleterious effect of pathogen on plants. They also produce siderophore that chelate iron and increase its availability to plants. Furthermore, it also solubilizes the minerals and phosphorus which are fixed in the soil. Synthesize phytohormones and increase surface area of the plant roots. Plant growth-promoting rhizobia belong to many genera including Bacillus, Cellulomonas, Rhizobium, Azotobacter, etc.

7.5 Nanomaterials and Crop Nutrition

Study revealed that about 30–50% of crop productivity is attributed to fertilizer in temperate region, while it accounts for more than this under tropical climate (Stewarta and Robertsb 2012). In other words, we can say that it seems impossible to get optimum yield without fertilization. In green revolution, high yielding varieties were introduced which were highly responsive to input used, and so crop yield increased dramatically. Piccinno et al. (2012) noticed that fertilizer application was highly increased with largely extended arable land. Bouwman et al. (2013) concluded that quantity of N and P fertilizers increased around 20-fold and seven-fold, respectively, during 1950–2000. Unluckily, this crop production model was found unstable which did not guarantee safe crop production. It is even more disturbing if we consider projections on demographic increase expected till the end of ongoing century. Most of the applied fertilizer is going wasted under this system of crop production which not only increased investment in crops but also polluted the environment. Studies revealed that about 40–70% N, 80–90% P, along with 50–90% K supplied along fertilizer are not take up by plants due to many edaphic and fertilizer characteristic factors (Trenkel 2010; Solanki et al. 2015). Development along with utilization of nanotechnologies potential for crop fertilization is a new frontier in fertilizer research.

Key points of the nanofertilizers are to reduce the losses and synchronize their availability to crops in a smart way (Kim et al. 2018). Smart materials are the materials which can sense the surrounding environment and change its shape or behavior. Nanofertilizers are reacting similarly with this mechanism and so reduced their losses when applied in field (Yugandhar et al. 2015). Pereira et al. (2015) concluded that nutrient use efficiency under nanotechnologies can be attained by two methods: (1) by improvement of existing fertilizers with nano-properties and (2) by developing a new type of fertilizers. The first option concerns adoption of advanced technologies in order to produce formulations of nanosized fertilizer. The second option is more ambitious that is projected into future. It visualizes that appropriately designed nanocapsules comprising nutrients will be managed to release, when stimulated by environmental factors or man-induced pulses (Fig. 7.2).

7.6 Decision Support System Based for Rice Nutrition

The Decision Support System (DSS) has been successfully employed across the globe for improving the standards of agricultural research and to achieve higher production to feed the burgeoning population. Table 7.1 describes various studies mentioning the application of DSS technology for the management of nitrogen fertilizer for rice crop across diverse countries having different agro-climatic conditions. Ahmad et al. (2012) employed CERES-Rice model for the determination of the best nitrogen level for aromatic (Basmati rice) under agro-ecological conditions of Faisalabad, Pakistan (Fig. 7.3). The results revealed that this technique is the best technique for decision-making and can effectively be used by the stakeholders including scientists, researcher, policymakers, and farmers.

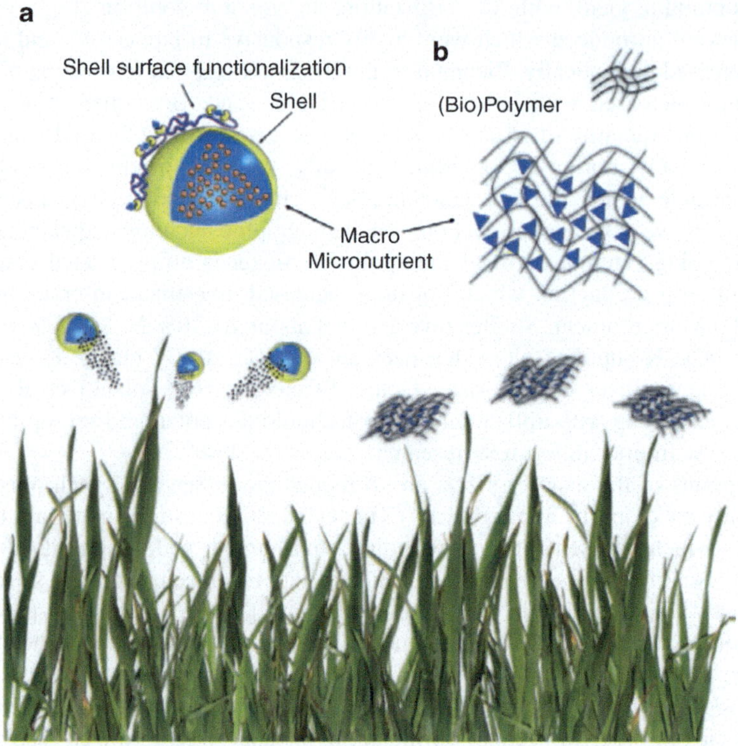

Fig. 7.2 Model of nanofertilizer [**Source:** Marchiol (2019)]

Table 7.1 Application of rice models for nitrogen management in rice

Country	Model application purpose	Model used	References
India	Planting date and nitrogen levels	CERES	Kumar et al. (2009)
Netherlands	Nitrogen management	ORYZA2000	Bouman and Van Laar (2006)
Pakistan	Seedlings hill^{-1} and nitrogen regimes	CERES	Ahmad et al. (2012)
Chile	Nitrogen fertilizer	ORYZA2000	Artacho et al. (2011)
India	Cultivars, nitrogen management	CERES	Shamim et al. (2012)
Pakistan	Nitrogen rates	ORYZA2000	Hameed et al. (2019)

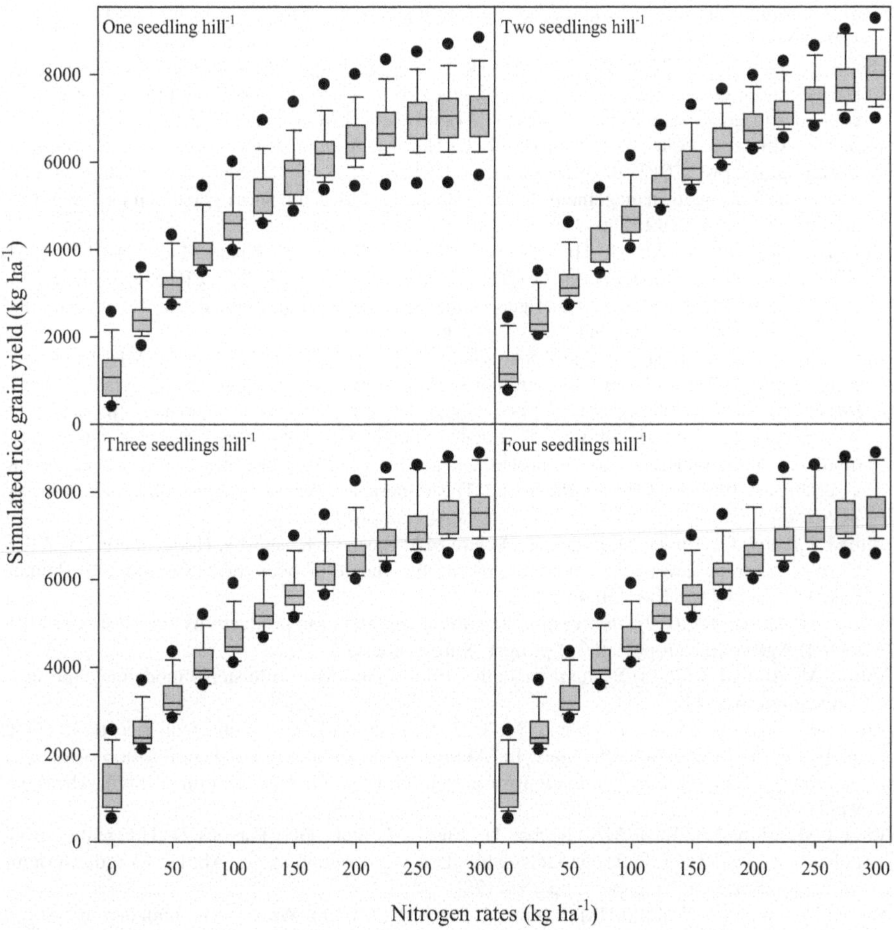

Fig. 7.3 Simulated grain yield of rice cultivar Basmati-385 at different planting densities and nitrogen application regimes. Box limits represent 25th and 75th percentiles, box central line represents median, and outliers represent minimum and maximum values. Simulated results were obtained from combination of historical weather data for 35 years, 4 plant densities, and 13 nitrogen application levels under irrigated conditions of Faisalabad, Punjab, and Pakistan (**Source:** Adopted from Ahmad et al. 2012)

References

Ahmad S, Hasanuzzaman M (2012) Integrated effect of plant density, N rates and irrigation regimes on the biomass production, N content, PAR use efficiencies and water productivity of rice under irrigated semiarid environment. Not Bot Horti Agrobot Cluj-Napoca 40(1):201–211

Ahmad S, Zia-ul-Haq M, Ali H, Shad SA, Ammad A, Maqsood M, Khan MB, Mehmood S, Hussain A (2008) Water and radiation use efficiencies of transplanted rice (*Oryza sativa* L.) at different plant densities and irrigation regimes under semi-arid environment. Pak J Bot 40(1): 199–209

Ahmad S, Ahmad A, Zia-ul-Haq M, Ali H, Khaliq T, Anjum MA, Khan MA, Hussain A, Hoogenboom G (2009) Resources use efficiency of field grown transplanted rice (*Oryza sativa* L.) under irrigated semiarid environment. J Food Agric Environ 7(2):487–492

Ahmad S, Ahmad A, Soler CMT, Ali H, Zia-Ul-Haq M, Anothai J, Hussain A, Hoogenboom G, Hasanuzzaman M (2012) Application of the CSM-CERES-Rice model for evaluation of plant density and nitrogen management of fine transplanted rice for an irrigated semiarid environment. Precis Agric 13(2):200–218

Ahmad S, Ahmad A, Ali H, Hussain A, Garcia y Garcia A, Khan MA, Zia-Ul-Haq M, Hasanuzzaman M, Hoogenboom G (2013) Application of the CSM-CERES-Rice model for evaluation of plant density and irrigation management of transplanted rice for an irrigated semiarid environment. Irrig Sci 31(3):491–506

Ahmad A, Ashfaq M, Rasul G, Wajid SA, Khaliq T, Rasul F, Saeed U, Rahman MH, Hussain J, Baig IA, Naqvi AA, Bokhari SAA, Ahmad S, Naseem W, Hoogenboom G, Valdivia RO (2015) Impact of climate change on the rice–wheat cropping system of Pakistan. In: Hillel D, Rosenzweig C (eds) Handbook of climate change and agro-ecosystems: the agricultural modeling intercomparison and improvement project (AgMIP) integrated crop and economic assessments. Imperial College Press and The American Society of Agronomy, Swaston, pp 219–258

Ahmad S, Abbas G, Ahmed M, Fatima Z, Anjum MA, Rasul G, Khan MA, Hoogenboom G (2019) Climate warming and management impact on the change of rice-wheat phenology in Punjab, Pakistan. Field Crop Res 230:46–61

Ahmed M, Ahmad S (2019) Carbon dioxide enrichment and crop productivity. In: Hasanuzzaman M (ed) Agronomic crops, vol 2. Springer, Singapore, pp 31–46

Ahmed M, Ahmad S (2020) Systems modeling. In: Ahmed M (ed) Systems modeling. Springer, Singapore, pp 1–44

Ahmed M, Fayyaz-ul-Hassan, Ahmad S (2017) Climate variability impact on rice production: adaptation and mitigation strategies. In: Ahmed M, Stockle C (eds) Quantification of climate variability, adaptation and mitigation for agricultural sustainability. Springer, Switzerland, pp 91–111

Ahmed M, Ahmad S, Raza MA, Kumar U, Ansar M, Shah GA, Parsons D, Hoogenboom G, Palosuo T, Seidel S (2020a) Models calibration and evaluation. In: Ahmed M (ed) Systems modeling. Springer, Singapore, pp 151–178

Ahmed M, Ahmad S, Waldrip HM, Ramin M, Raza MA (2020b) Whole farm modeling: a systems approach to understanding and managing livestock for greenhouse gas mitigation, economic viability and environmental quality. In: Waldrip HM, Pagliari PH, He Z (eds) Animal Manure. ASA Special Publication 67. American Society of Agronomy and Soil Science Society of America, Madison, WI, pp 345–371

Akiyama H, Mc TIP, Ball BC, Scott A (2004) N2O, NO and NH3 emissions from soil after the application of organic fertilizers, urea and water. Water Air Soil Pollut 156:113–129

Akram R, Turan V, Hammad HM, Ahmad S, Hussain S, Hasnain A, Maqbool MM, Rehmani MIA, Rasool A, Masood N, Mahmood F, Mubeen M, Sultana SR, Fahad S, Amanet K, Saleem M, Abbas Y, Akhtar HM, Hussain S, Waseem F, Murtaza R, Amin A, Zahoor SA, Sami ul Din M, Nasim W (2018) Fate of organic and inorganic pollutants in paddy soils. In: Hashmi MZ, Varma A (eds) Environmental pollution of paddy soils. Springer, Switzerland, pp 197–214

Akram R, Fahad S, Masood N, Rasool A, Ijaz M, Ihsan MZ, Maqbool MM, Ahmad S, Hussain S, Ahmed M, Kaleem S, Sultana SR, Mubeen M, Saud S, Kamran M, Nasim W (2019) Plant growth and morphological changes in rice under abiotic stress. In: Hasanuzzaman M, Fujita M, Nahar K, Biswas JK (eds) Advances in rice research for abiotic stress tolerance. Woodhead Publishing, Swaston, pp 69–85

Ali A, Noorka IR (2013) Nitrogen and phosphorus management strategy for better growth and yield of sunflower (Helianthus annuus L.). Soil Environ 32:44–48

Arancon NQ, Edwards CA, Bierman P, Welch C, Metzger J (2004) Influences of vermicomposts on field strawberries:Effects on growth and yields. Bioresour Technol 93:145–153

Artacho P, Meza F, Alcalde JA (2011) Evaluation of the Oryza2000 rice growth model under nitrogen-limited conditions in an irrigated Mediterranean environment. Chil J Agric Res 71(1): 23–33

Asai H, Samson BK, Stephan HM, Songyikhangsuthor K, Homma K, Kiyono Y, Inoue Y, Shiraiwa T, Horie T (2009) Biochar amendment techniques for upland rice production in Northern Laos 1. Soil physical properties, leaf SPAD and grain yield. Field Crop Res 111: 81–84

Atique-ur-Rehman M, Farooq A, Rashid F, Nadeem S, Stürz FA, Bel RW (2018) Boron nutrition of rice in different production systems. A review. Agron Sustain Dev 38:25

Bains SN, Prasad R, Bhatia PC (1971) Use of indigenous materials to enhance the efficiency of fertilizer nitrogen for rice. Fert News 16:30–32

Bijay S (2016) Agronomic benefits of neem coated urea -a review international fertilizer association. DOI: 10.13140/RG.2.2.10647.98722

Bouman BAM, Van Laar HH (2006) Description and evaluation of the rice growth model ORYZA2000 under nitrogen-limited conditions. Agric Syst 87(3):249–273

Bouwman AF, Boumans LJM, Batjes NH (2002) Emissions of N_2O and NO from fertilized fields: summary of available measurement data. Glob Biogeochem Cycles 16:6–13

Bouwman L, Kees KG, Klaas W, Van DH, Arthur H, Beusen W, Detlef P, Van V, Jaap W, Mariana CR, Elke S (2013) Exploring global changes in nitrogen and phosphorus cycles in agriculture induced by livestock production over the 1900–2050 period. PNAS 110:20882–20887

Bronson KF, Mosier AR (1991) Effect of encapsulated calcium carbide on dinitro-gen, nitrous oxide, methane, and carbon dioxide emissions from flooded rice. Biol Fertil Soils 11:116–120

Burger M, Venterea RT (2011) Effects of nitrogen fertilizer types on nitrous oxide emission. In understanding Greenhouse Gas Emissions from Agricultural Management. ACS Symposium Series. American Chemical Society, Washington, pp DC179–DC202

Cai Z, Shan Y, Xu H (2007) Effects of nitrogen fertilization on CH4 emissions from rice fields. Soil Sci Plant Nutr 53:353–361

Carreres R, Sendra J, Ballesteros R, Valiente EF, Quesada A, Carrasco D, Leganes F, Garcia CJ (2003) Assessment of slow release fertilizers and nitrification inhibitors in flooded rice. Biol Fertil Soils 39:80–87

Ehsanullah K, Jabran G, Asghar M, Hussain Rafiq M (2012) Effect of nitrogen fertilization and seedling density on fine rice yield in Faisalabad, Pakistan. Soil Environ 31:152–156

Fahad S, Noor M, Adnan M, Khan MA, Rahman I, Alam M, Khan IA, Ullah H, Mian IA, Hassan S, Saud S, Bakhat HF, Hammad HM, Ahmad S, Nasim W (2019) Abiotic stress and rice grain quality. In: Hasanuzzaman M, Fujita M, Nahar K, Biswas JK (eds) Advances in rice research for abiotic stress tolerance. Woodhead Publishing, Swaston, pp 571–583

FAO (2015) Fertilizer trends and outlook to 2018. P: 11

Fatima Z, Ahmad M, Hussain M, Abbas G, Ul-Allah S, Ahmad S, Ahmed N, Ali MA, Sarwar G, Ehsan ul Haque, Iqbal P, Hussain S (2020) The fingerprints of climate warming on cereal crops phenology and adaptation options. Sci Rep 10:18013

Fujinuma R, Venterea RT, Rosen C (2011) Broadcast urea reduces N(2)O but increases NO emissions compared with conventional and shallow-applied anhy-drous ammonia in a coarse-textured soil. J Environ Qual 40:1806–1815

Gilbert B, Frenzel P (1998) Rice roots and CH4 oxidation: the activity of bacteria, their dis tribution and the microenvironment. Soil Biol Biochem 30:1903–1916

Gomez BM, Dominguez J (2013) Recycling of solid organic wastes through vermicomposting: microbial community changes throughout the process and use of vermicompost as a soil amendment. Crit Rev Environ Sci Technol 44:1289–1312

Hafeez-ur-Rehman, Nawaz A, Awan MI, Ijaz M, Hussain M, Ahmad S, Farooq M (2019) Direct seeding in rice: problems and prospects. In: Hasanuzzaman M (ed) Agronomic crops: production technologies, vol 1. Springer, Singapore, pp 199–222

Hameed F, Xu J, Rahim SF, Wei Q, Liao Q, Ahmed S (2019) Rice growth and nitrogen uptake simulation by using ORYZA (v3) model considering variability in parameters. Pak J Agric Sci 56(1):245–259

Hussain M, Ahmad S, Hussain S, Lal R, Ul-Allah S, Nawaz A (2018) Rice in saline soils: physiology, biochemistry, genetics and management. Adv Agron 148:231–287

Ines E, Garcia de S, Juan MF, Luciana P, Di S, Jhovana S, Escobar O, Florencia D, Lucia F, Ana Fernandez S (2012) Inoculation of paddy rice with Azospirillum brasilense and Pseudomonas fluorescens: impact of plant genotypes on rhizosphere microbial communities and field crop production. Appl Soil Ecol 61:196–204

IPCC: Climate Change (2007) The physical science basis, contribution of working group I to the fourth assessment report of the intergovernmental panel on climate change. Cambridge University Press, Cambridge, New York, NY

Kawakami EM, Oosterhuis DM, Snider JL (2013) Nitrogen assimilation and growth of cotton seedlings under NaCl salinity and in response to urea application with NBPT and DCD. J Agron Crop Sci 199:106–117

Khan MA, Shah Z, Rab A, Arif M, Shah T (2013) Effect of urease and nitrification inhibitors on wheat yield. Sarhad J Agric 29:3

Khan MA, Ahmad S, Raza A (2019a) Integrated weed management for agronomic crops. In: Hasanuzzaman M (ed) Agronomic crops, vol 2. Springer, Singapore, pp 257–281

Khan SM, Ali S, Nawaz A, Bukhari SAH, Ejaz S, Ahmad S (2019b) Integrated pest and disease management for better agronomic crop production. In: Hasanuzzaman M (ed) Agronomic crops, vol 2. Springer, Singapore, pp 385–428

Kim DY, Kadam A, Shince S, Saratale RG, Patra J, Ghodake G (2018) Recent developments in nanotechnology transforming the agricultural sector: a transition replete with opportunities. J Sci Food Agric 98:849–864

Kruger M, Frenzel P, Conrad R (2001) Microbial processes influencing methane emission from rice fields. Global Change Bio 7:49–63

Kumar N, Tripathi P, Pal RK (2009) Validation of simulation modeling for growth parameters of rice genotypes using CERES 3.5v for Eastern Uttar Pradesh. Environ Ecol 27(4):1490–1494

Laird D, Fleming P, Wang B, Horton R, Karlen D (2010) Biochar impact on nutrient leaching from a midwestern agricultural soil. Geoderma 158:436–442

Liang B, Lehmann J, Kinyangi D, Grossman J, O'Neill B, Skjemstad JO, Thies J, Luizao FJ, Peterson J, Neves EG (2006) Black carbon increases cation exchange capacity in soils. Soil Sci Soc Am J 70:1719–1730

Linquist BA, Van Groenigen KJ, Adviento-Borbe MA, Pittelkow CM, Van Kessel C (2012) An agronomic assessment of greenhouse gas emissions from major cereal crops. Glob Chang Biol 18:194–209

Malla G, Bhatia A, Pathak H, Prasad S, Jain N, Singh J (2005) Mitigating nitrous oxide and methane emissions from soil in rice–wheat system of the Indo-Gangetic plain with nitrification and urease inhibitors. Chemosphere 58:141–147

Marchiol L (2019) Nanofertilizers: an outlook of crop nutrition in the fourth agricultural revolution. Ital J Agron 14:1367

Masulili A, Utomo WH, Syekhfani MS (2010) Rice husk biochar for rice based cropping system in acidic soils: the characteristics of rice husk biochar and its influence on the properties of acid sulfate soils and rice growth in West Kalimantan, Indonesia. J Agric Sci (Canada) 2:39–47

Mia MAB, Shamsuddin ZH (2010) Rhizobium as a crop enhancer and biofertilizer for increased cereal production. Afr J Biotechnol 9:6001–6009

Naz S, Fatima Z, Iqbal P, Khan A, Zakir I, Noreen S, Younis H, Abbas G, Ahmad S (2019) Agronomic crops: types and uses. In: Hasanuzzamean M (ed) Agronomic crops: production technologies, vol 1. Springer, Singapore, pp 1–18

Ngo PT, Rumpel C, Dignac MF, Billou D, Duc TT, Jouquet P (2011) Transformation of buffalo manure by composting or vermicomposting to rehabilitate degraded tropical soils. Ecol Eng 37: 269–276

Norman RJ, Wells BR, Moldenhauer KAK (1989) Effect of application method and dicyaniamide on urea-nitrogen-15 recovery in rice. Soil Sci Soc Am J 53:1269–1274

Pathma J, Sakthivel N (2012) Microbial diversity of vermicompost bacteria that exhibit useful agricultural traits and waste management potential. Springer Plus 1:1–19

Peerzada AM, Bukhari SAH, Dawood M, Nawaz A, Ahmad S, Adkins S (2019) Weed management for healthy crop production. In: Hasanuzzaman M (ed) Agronomic crops, vol 2. Springer, Singapore, pp 225–256

Pereira EL, Minussi FB (2012) Urea-montmorilloliteextruded nanocomposites: a novel slow release material. J Agric Food Chem 60:5273–5278

Pereira EI, Da Cruz CCT, Solomon A, Le A, Cavigelli MA, Ribeiro C (2015) Novel slow-release nanocomposite nitrogen fertilisers: the impact of polymers on nanocomposite properties and function. Ind Eng Chem Res 54:3717–3725

Piccinno F, Gottschalk F, Seeger S, Nowack B (2012) Industrial production quantities and uses of ten engineered nanomaterials in Europe and the world. J Nanopart Res 14:1109

Pramanik P, Ghosh G, Ghosal P, Banik P (2007) Changes in organic – C N, P and K and enzyme activities in vermicompost of biodegradable organic wastes under liming and microbial inoculants. Bioresour Technol 98:2485–2494

Prasad R, Sharma SN, Singh S, Devakumar C, Saxena VS, Shivay YS (2002) Neem coating of urea for environment and agriculture. Fert News 47:63–67

Raja SK, Karmegam N (2010) Earthworm casts as an alternate carrier material for biofertilizers: assessment of endurance and viability of Azotobacter chroococcum, Bacillus megaterium and Rhizobium leguminosarum. Sci Hortic 124:286–289

Raun WR, Johnson GV (1999) Improving nitrogen use efficiency for cereal production. Agron J 91: 357–363

Razaq M, Shah FM, Ahmad S, Afzal M (2019) Pest management for agronomic crops. In: Hasanuzzaman M (ed) Agronomic crops, vol 2. Springer, Singapore, pp 365–384

Rehman A, Farooq M, Nawaz A, Ahmad R (2014) Influence of boron nutrition on the rice productivity, grain quality and biofortification in different production systems. Field Crop Res 169:123–131

Rondon MA, Lehmann J, Ramirez J, Hurtado M (2007) Biological nitrogen fixation by common beans (*Phaseolus vulgaris* L.) increases with biochar additions. Biol Fertil Soils 43:699–708

Roy S, Arunachalam K, Dutta BK, Arunachalam A (2010) Effect of organic amendments of soil on growth and productivity of three common crops viz Zea mays Phaseolus vulgaris and Abelmoschus esculentus. Appl Soil Ecol 45:78–84

Sarwar N, Maqsood M, Wajid SA, Anwar-ul-Haq M (2011) Impact of nursery seeding density, nitrogen, and seedling age on yield and yield attributes of fine rice. Chilean J Agric Res 71 (3):343

Sarwar N, Ali H, Ahmad S, EhsanUllah, Ahmad S, Mubeen K, Hill JE (2013a) Water wise rice cultivation on calcareous soil with the addition of essential micronutrients. J Anim Plant Sci 23(1):244–250

Sarwar N, Ali H, Maqsood M, EhsanUllah SAN, Shahzad M, Mubeen K, Shahid MA, Ahmad S (2013b) Phenological response of rice plants to different micronutrients application under water saving paddy fields on calcareous soil. Turk J Field Crops 18(1):52–57

Sarwar N, Ali H, Maqsood M, Ahmad A, Ullah E, Khaliq T, Hill JE (2014) Influence of nursery management and seedling age on growth and economic performance of fine rice. J Plant Nutr 37 (8):1287–1303

Sarwar N, Rehman A, Farooq O, Wasaya A, Saliq S, Mubeen K (2019) Improved auxin level at panicle initiation stage enhance the heat stress tolerance in rice plants. In: Agronomy Australia conference, pp 2–4

Schmutterer H (1990) Properties and potential of natural pesticides from the neem tree, Azadirachta indica. Annu Rev Entomol 35:271–297

Setyanto P, Makarim AK, Fagi AM, Wassman R, Buendia LV (2000) Crop management affecting methane emissions from irrigated and rainfed rice in Central Java (Indonesia). Nutr Cycl Agroecosyst 58:85–93

Shah Z, Shah MZ, Tariq M, Bakht J, Rahman H (2012) Survey of citrus orchards for micronutrients deficiency in Swat Valley of Khyber Pakhtunkhwa Pakistan. Pak J Bot 44:705–710

Shahzad AN, Ahmad S (2019) Tools and techniques for nitrogen management in cereals. In: Hasanuzzaman M (ed) Agronomic crops, vol 2. Springer, Singapore, pp 111–126

Shamim M, Shekh AM, Pandey V, Patel HR, Lunagaria MM (2012) Simulating the phenology, growth and yield of aromatic rice cultivars using CERES-Rice model under different environments. J Agrometeorol 14(1):31–34

Shang Q, Yang X, Gao C, Wu P, Liu J, Xu Y, Shen Q, Zou J, Guo S (2011) Net annual global warming potential and greenhouse gas intensity in Chinese dou-ble rice-cropping systems: a 3-year field measurement in long-term fertilizer experiments. Glob Chang Biol 17:2196–2210

Siddiqui A, Ruchi S, Nidhi M, Kirti T (2014) Mixed effect of Rhizobium and Azotobacter as biofertilizer on nodulation and production of chick pea, Cicer arietinum. Biosci Biotech Res Comm 7:46–49

Singh M, Singh TA (1986) Leaching losses of nitrogen from urea as affected by application of neem cake. J Indian Soc Soil Sci 34:766–773

Singh R, Sharma R, Kumar S, Gupta R, Patil R (2008) Vermicompost substitution influences growth physiological disorders, fruit yield and quality of strawberry (Fragaria x ananassa Duch.). Bioresour Technol 99:8507–8511

Smith JD, Harrison HC (1991) Evaluation of polymers for controlled release properties when incorporated with nitrogen fertilizer solutions. Commun Soil Sci Plant Anal 22:559–573

Snyder CS, Bruulsema TW, Jensen TL, Fixen PE (2009) Review of greenhouse gas emissions from crop production systems and fertilizer management effects. Agric Ecosyst Environ 133:247–266

Soares JD, Cantarella H, De Campos MML (2012) Ammonia volatilization losses from surface-applied urea with urease and nitrification inhibitors. Soil Biol Biochem 52:82–89

Solanki P, Bhargava A, Chhipa H, Jain N, Panwar J (2015) Nanofertilisers and their smart delivery system. In: Rai M, Ribeiro C, Mattoso L, Duran N (eds) Nanotechnologies in food and agriculture. Springer, New York, pp 81–101

Song X, Liu M, Wu D, Qi L, Ye C, Jiao J, Hu F (2014) Heavy metal and nutrient changes during vermicomposting animal manure spiked with mushroom residues. Waste Manag 34:1977–1983

Spokas KA, Novak JM (2012) Biochar's role as an alternative N fertilizer: ammonia capture. Plant Soil 350:35–42

Stewarta WM, Robertsb TL (2012) Food Security and the Role of Fertilizer in supporting it. Procedia Engin 46:76–82

Subbarao GV, Ito O, Sahrawat KL, Berry WL, Nakahara K, Ishikawa T, Watanbe T, Suenaga K, Rondo M, Rao IM (2006) Scope and strategies for regulation of nitrification in agricultural systems – challenges and opportunities. Crit Rev Plant Sci 25:303–335

Suranto W, Murdiyarso D, Suratmo FG, Anas I, Saeni MS, Rambe A (1998) Nitrous oxide flux from irrigated rice fields in West Java. Environ Pollut 102:159–166

Tariq M, Ahmed M, Iqbal P, Fatima Z, Ahmad S (2020) Crop phenotyping. In: Ahmed M (ed) Systems modeling. Springer, Singapore, pp 45–60

Tejada M, Garcia MA, Parrado J (2009) Effects of a vermicompost composted with beet vinasse on soil properties, soil losses and soil restoration. Catena 77:238–247

Trenkel ME (2010) Slow-and controlled-release and stabilised fertilisers: an option for enhancing nutrient use efficiency in agriculture. International Fertiliser Industry Association, Paris, France, pp 1–162

Utomo WH, Islami T (2013) Nitrogen fertilizer requirement of maize (Zea mays L.) on biochar-treated soil. In: Hayashi K (ed) Biochar for future food security: learning from experiences and identifying research priorities. Proc. Int Workshop on "Biochar for Future Food Security: Learning From Experiences and Identifying Research Priorities". IRRI, Los Banos, ThePhilliphines

Utomo WH, Wisnubroto EI, Indrayatie ER (2017) Residual effect of biochar on growth and yield of red chili (Capsicum annum L.). J Adv Agric Technol 4:28–32

Wang D, Shi Q, Wang X, Wei M, Hu J, Liu J, Yang F (2010) Influence of cow manure vermicompost on the growth metabolite contents, and antioxidant activities of Chinese cabbage (Brassica campestris ssp. chinensis). Biol Fertil Soils 46:689–696

Wasaya A, Yasir TA, Ijaz M, Ahmad S (2019) Tillage effects on agronomic crop production. In: Hasanuzzaman M (ed) Agronomic crops, vol 2. Springer, Singapore, pp 73–99

Wu F, Wan JHC, Wu S, Wong M (2012) Effects of earthworms and plant growth–promoting rhizobacteria (PGPR) on availability of nitrogen, phosphorus, and potassium in soil. J Plant Nutr Soil Sci 175:423–433

Xuhui Z, Jinwei Z, David C (2010) Effect of biochar amendment on yield and methane and nitrous oxide emissions from a rice paddy from Tai Lake plain, China. Agric Ecol Environ 139:469–475

Yugandhar P, Vasu AT, Chavali M (2015) Smart materials in nanoscience and nanotechnology - An overview. Vfstr JSTEM 1:1–5

Zahoor SA, Ahmad S, Ahmad A, Wajid A, Khaliq T, Mubeen M, Hussain S, Sami Ul Din M, Amin A, Awais M, Nasim W (2019) Improving water use efficiency in agronomic crop production. In: Hasanuzzaman M (ed) Agronomic CROPS, vol 2. Springer, Singapore, pp 13–29

Irrigation Management in Rice

8

Khuram Mubeen, Naeem Sarwar, Muhammad Shehzad,
Abdul Ghaffar, and Mudassir Aziz

Abstract

Water issue in rice production systems can be resolved through either growing water-efficient rice varieties or using more efficient water management approaches under the resources available. The water-efficient rice cultivars should be more resistant to water shortage, short in duration having higher yield potential. The hybrid rice should be also grown in this context. The most common and traditional method of rice production has been the transplanted paddy rice (TPR). It, however, needs integration with agronomic management approaches to make the most of irrigation water applied. Aerobic rice systems in which crop stand is established by sowing rice directly in the non-puddled field without field ponding save water and labor. Once the direct seeded rice has emerged in the field, farmers can delay the irrigation for 7–15 days based on soil temperature and climatic condition. Delaying the irrigation can aid in making the rice seedlings drought resistant by improving the deeper roots. High-efficiency irrigation inclusive sprinkler or drip can significantly reduce the water use compared to TPR. Even flush irrigation can use less water than TPR. Water-saving technology whereby rice fields are periodically dried and watered known as alternate wetting and drying (AWD) can also be adopted. In AWD, water is ponded up to the depth of 5 cm once the water level in soil has fallen to below 15 cm for the whole growing season except at the flowering stage where

K. Mubeen (✉) · A. Ghaffar · M. Aziz
Department of Agronomy, Muhammad Nawaz Shareef University of Agriculture, Multan, Pakistan
e-mail: khurram.mubeen@mnsuam.edu.pk

N. Sarwar
Department of Agronomy, Bahauddin Zakaria University, Multan, Pakistan

M. Shehzad
Department of Agronomy, The University of Poonch, Rawalakot, AJK, Pakistan

N. Sarwar et al. (eds.), *Modern Techniques of Rice Crop Production*,
https://doi.org/10.1007/978-981-16-4955-4_8

5 cm water level is maintained. Global warming potential (GWP) is affected by systems of rice production through effects on methane, N_2O, and CO_2. Methane emissions were observed higher in conventional puddled transplanted rice as compared to dry direct seeded rice (DSR). The cost of water supply should also be given due consideration for making efficient water use in rice production on sustainable basis. Due to strong linkage of rice production with water resource, sustainable water policy requires cross-boundary application of decision support system, and cross-continental cooperation.

Keywords

Irrigation management · Transplanted rice · Aerobic rice · Alternate wetting · Drying

8.1 Introduction

More than half of the world population uses rice as a staple food (Ahmad et al. 2015, 2019, 2012, 2013, 2008, 2009; Ahmad and Hasanuzzaman 2012; Ahmed and Ahmad 2019, 2020; Shahzad and Ahmad 2019). However, rice sustainability is threatened by heavy underground water withdrawal and increased water usage (Hafeez-ur-Rehman et al. 2019; Khan et al. 2019a, 2019b; Naz et al. 2019; Peerzada et al. 2019; Razaq et al. 2019; Sarwar et al. 2013a, 2013b; Tariq et al. 2020; Wasaya et al. 2019; Zahoor et al. 2019). On contrary, increasing human population at the rapid pace demands a sharp rise in rice production (Fatima et al. 2020). More rice production with less water is therefore a huge challenge. Rice has the potential to grow in a varying range of hydrological conditions, climate, and soils. Based on the hydrology, rice can be grouped into irrigated lowland rice, rainfed lowland rice, flood-prone rice, and upland rice. Rice grown in lowlands is referred to as "paddy rice." Flood-prone rice is characterized by deepwater rice and floating rice. Uplands rice is grown in aerobic condition for more than 80% of the growing period. Water availability and use have become the main concerns of rice production worldwide even in irrigated areas owing to climate change in particular (Ahmed and Fayyaz-ul-Hassan 2017; Ahmed et al. 2020a, 2020b; Akram et al. 2018, 2019; Hussain et al. 2018; Fahad et al. 2019; Fatima et al. 2020).

8.2 Irrigation Management Methods in Rice

Rice is grown most commonly through transplanting; however, scarcity of water and labor along with higher labor wages demands looking for alternatives with higher water and crop productivity. It is noteworthy that only reducing the water use in puddled transplanted rice (PTR) results in gradual reduction in grain yield. Therefore, other pillars of production should also be accounted for improved water productivity. Depending on soil type, climatic situations, edaphic factors (pH, salinity, rainfall, and temperature), and water management methane emission may vary within PTR across regions and sites.

Farmers can also adopt alternate wetting and drying (AWD) system of water management involving the periodic drying and re-watering. After the water level goes down to 15 cm soil depth, the field is re-watered up to the depth of 5 cm. This practice is continued for most of the growing season except at the flowering stage, where 5 cm water level is maintained. Besides this, other management should also be accompanied by inclusive soil type, optimum plant population, timely nitrogen application, and maintenance of wetting and drying duration. Farmers can keep water ponded in the field for 2–3 weeks for suppressing weeds (Mubeen and Jabran 2019).

Various methods of direct seeded rice are compared with conventional transplanted paddy rice for performance based on economic yield, irrigation water use efficiency, effect on subsequent crops, labor use, GWP, and cost-effectiveness. At field level, farmer or farm manager should resolve water shortage and adapt to water availability to have economic gains (Table 8.1). Dry seeding and delayed flooding can be practiced where field is kept dry until tillering of rice, and then field is ponded for most of the growing season except drying duration needed for fertilizer application. Farmers can also practice dry seeding and intermittent irrigation. Hence, water saving in direct seeding varies based on soil type, precipitation, crop, and water management (Farooq et al. 2006a, 2006b). The other interventions that can be adopted are to reduce outflow in the form of leakage and percolation from rice fields. At the land preparation stage, reduced water usage can be associated with completion of land preparation in less time over PTR. Adoption of wet seeded rice system improved water productivity from 0.3 to 0.4 kg rice/m^3 water and increased yield from 6.3 to 6.9 t/ha. Total water use in wet seeded rice reduced from 2195 mm to 1700 mm.

However, it was observed that the irrigation schedule with drainage duration of 4–5 days is very damaging (Hussain et al. 2018). Higher rice yield can be obtained in fields with irrigation interval of 3 days in comparison to 2 days of irrigation interval (Gill and Singh 2008). However, other researchers have previously shown non-significant difference in grain yield of directly sown rice (Mahajan et al. 2006; Ramakrishna et al. 2007). DSR yield reduced more abruptly than PTR as water tension raised to 40 and 70 kPa (Yadav et al. 2011b). However, when soil was dried more to a water tension higher than 20 kPa, grain yield reduced rapidly in both DSR and PTR. DSR can have deeper roots compared to PTR (Biswas and Yamauchi 2007). Field applied with 5 cm irrigation depth on the day when ponded water vanished shows similar and highest dry matter and N, P, and K uptake with fields receiving irrigation 1 day after ponded water disappeared (Edwin and Anal 2008). Dry seeding of rice delayed growth duration by 13.5 days compared to continuous ponding, whereas intermittent irrigation in comparison to continuous ponded water delayed the tillering up to 7 days, reduced transpiration, and improved photosynthesis (Huang et al. 2008).

Sole water use reduction in PTR brought proportionate reduction in grain yield advocating the supplemental role of management practices for land, labor, and

Table 8.1 A comparative summary of benefits from different irrigation methods in rice over puddled transplanted rice

Sr. no.	Irrigation management approach	Benefits		References
01	Alternate wetting and drying	Saves water and maintains comparable grain yields in the rice farming		Mubeen and Jabran (2019)
		Reduces arsenic in the rice grains and methane emission from the rice fields		
		Improves growth of root and canopy structure.		
02	Direct seeded rice	Saves labor through avoiding nursery raising, uprooting seedlings, transplanting, and puddling		Wang et al. (2002)
		Well-spread demand for labor over time compared to paddy transplanted rice		
		Requires less irrigation water with more efficiency and low labor; can be highly mechanized		
		Irrigation water saving	35–57%	Sharma et al. (2002); Singh et al. (2002)
			28–33%	Kumar (2002)
			20%	Kaur (2004)
			13%	Mann and Ashraf (2004)
			30–50% with similar yields to flooded PTR	Yadav et al. (2011a, 2011b)
			30–50%, having yield loss of 20–30%	Bouman et al. (2005)
		Reduced water use by 9–24% with DSR (zero tillage or cultivated) in comparison to PTR		Jat et al. (2009)

nutrients in this context. Moreover, the rice crop should not face water stress at critical stage of growth. Water use was reduced by increasing the irrigation interval, however, water productivity improved in 8-day irrigation interval up to 60% without compromising yield when compared to continuous flooding (Rezaei et al. 2009). Moreover, AWD showed higher water productivity than continuous ponding (Yadav et al. 2011b). It is pertinent to note that rainfall occurrence time and amount also decide irrigation water use and savings (Saharawat et al. 2010). Reduction in water use up to 18% can be noticed through DSR while comparing with PTR.

Labor use was 37% higher in PTR than DSR mainly owing to transplanting practices. If weeds are managed through weedicides in an effective way, then considerable reduction in labor engagement can be obtained in DSR. Ponded rice fields which are puddled and nursery is transplanted are among the main sources of global methane emission (up to 20%) and CO_2 due to extended duration of water ponding creating anoxic soil situation. Combining DSR with intermittent irrigation reduced methane emission even more (up to 92%) compared to PTR. Combining reduced or zero tillage with dry DSR reveals highest cost reduction. The soil structure is damaged by ponding water for most of the growing season, whereas the direct sowing improves it which creates a favorable soil environment for succeeding crops like wheat whose yield was improved. Moreover, continuous water ponding hinders the root proliferation of winter crops in deeper soil layers because of developed hardpan. Aerobic rice cropping system productivity (DSR wheat, DSR chickpea, and DSR mustard) was higher than PTR-based cropping systems.

8.3 Decision Support System for Irrigation Management in Rice

The Decision Support System (DSS) can successfully be used across the world for decision-making along with agricultural research outputs to attain higher productivity under current and future climate uncertainty. Table 8.2 describes the various works of different research studies discussing the application of DSS technology for irrigation management for rice across various countries with notably diverse agro-environmental conditions. Ahmad et al. (2013) used CERES-Rice model for the determination of best irrigation regime for aromatic rice (cultivar; Basmati-385) at Faisalabad, Pakistan (Fig. 8.1). The results proved that DSS technique is the best option for decision-making and can effectively be used by the stakeholders including scientists, researcher, policy-makers, and farmers.

Table 8.2 Application of rice models for irrigation management in rice

Country	Rice model name	Application	References
India	ORYZA2000	Water use efficiency	Arora (2006)
Sri Lanka	ORYZA2000	Soil water contents	Soundharajan and Sudheer (2009)
South Korea	AquaCrop	Water use efficiency	Lee et al. (2017)
Thailand	CERES	Water use efficiency	Phakamas (2015)
Bangladesh	AquaCrop	Water use efficiency	Maniruzzaman and Nemes (2014)
Pakistan	CERES	Plant density and irrigation regimes	Ahmad et al. (2013)

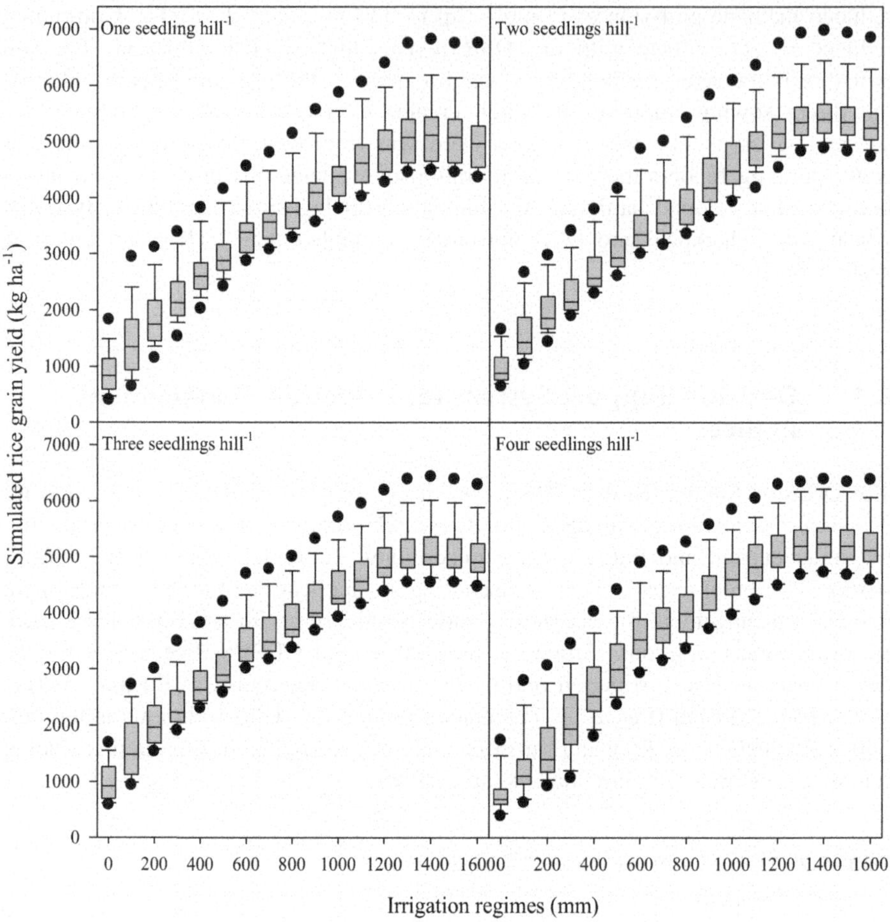

Fig. 8.1 Simulated grain yield for rice Basmati-385 at different planting densities and irrigation regimes. Box limits represent 25th and 75th percentiles, box central line represents median, and outliers represent minimum and maximum values. Simulated results were obtained from the combination of historical weather data for 35 years, 4 plant densities, and 17 irrigation levels for irrigated environment of Faisalabad, Pakistan (**Source:** Ahmad et al. 2013)

8.4 Future Research Need

More research is needed to finally quantify the trade-offs between economic and environmental concerns of rice production. An equal or higher yield is possible in direct seeding provided the harvest index of an aerobic rice variety for direct seeding is developed.

8.5 Summary

Irrigation water can be managed in either puddled transplanted rice or aerobic rice cultures (dry DSR, wet DSR, AWD) with a focus on reducing the water use, irrigation water need, improving water productivity, and making efficient use of available water. However, success of PTR or aerobic rice cultures is based on rainfall pattern and occurrence time, and soil type, crop, and water management in integration. Compared with PTR, nutrient uptake in straw and grain is also promoted to a reasonable extent in aerobic rice. Rice grain quality is also improved in medium water stress compared to sufficient water supply and more water-stressed conditions. It is also important to note that in PTR or aerobic rice, plants should not face water stress at critical growth stages. Keeping in mind the extent of resources available with farmers, the use of PTR or any of the feasible aerobic rice culture technology can be advocated. In areas with sufficient water availability over the growing season with reduced labor wages, it is better to recommend PTR system. However, if the labor wages are on the higher side with limited water availability during the course of the growing duration, then aerobic rice can be preferred.

8.6 Conclusion

While practicing puddled transplanted rice, farmer or farm manager should take care not to reduce the water use alone, as it results in marked decrease in grain yield. Moreover, farmers should account for the varying soil factors (pH, salinity, rainfall, temperature, etc.), climate, crop management, and water management for achieving the objective of improved water and crop productivity. Dry direct seeded rice in combination with reduced tillage or zero tillage reduces expenditures and improves profitability for the farmers. Wet direct seeded rice system uses less water than PTR either at land preparation stage or at irrigation stage. A pro-active implementation and coordination among stakeholders may assist to achieve the maximum benefits of AWD on sustainable basis to counter water scarcity without losing rice productivity. Moreover, under the backdrop of limited irrigation water availability, costing the water supply for irrigation to a reasonable level to sensitize the farmers for value of water can aid in making the efficient water use.

References

Ahmad S, Hasanuzzaman M (2012) Integrated effect of plant density, N rates and irrigation regimes on the biomass production, N content, PAR use efficiencies and water productivity of rice under irrigated semiarid environment. Not Bot Horti Agrobot Cluj-Napoca 40(1):201–211

Ahmad S, Zia-ul-Haq M, Ali H, Shad SA, Ammad A, Maqsood M, Khan MB, Mehmood S, Hussain A (2008) Water and radiation use efficiencies of transplanted rice (*Oryza sativa* L.) at different plant densities and irrigation regimes under semi-arid environment. Pak J Bot 40(1): 199–209

Ahmad S, Ahmad A, Zia-ul-Haq M, Ali H, Khaliq T, Anjum MA, Khan MA, Hussain A, Hoogenboom G (2009) Resources use efficiency of field grown transplanted rice (*Oryza sativa* L.) under irrigated semiarid environment. J Food Agric Environ 7(2):487–492

Ahmad S, Ahmad A, Soler CMT, Ali H, Zia-Ul-Haq M, Anothai J, Hussain A, Hoogenboom G, Hasanuzzaman M (2012) Application of the CSM-CERES-rice model for evaluation of plant density and nitrogen management of fine transplanted rice for an irrigated semiarid environment. Precis Agric 13(2):200–218

Ahmad S, Ahmad A, Ali H, Hussain A, Garcia Garcia A, Khan MA, Zia-Ul-Haq M, Hasanuzzaman M, Hoogenboom G (2013) Application of the CSM-CERES-Rice model for evaluation of plant density and irrigation management of transplanted rice for an irrigated semiarid environment. Irrig Sci 31(3):491–506

Ahmad A, Ashfaq M, Rasul G, Wajid SA, Khaliq T, Rasul F, Saeed U, Rahman MH, Hussain J, Baig IA, Naqvi AA, Bokhari SA, Ahmad S, Naseem W, Hoogenboom G, Valdivia RO (2015) Impact of climate change on the rice–wheat cropping system of Pakistan. In: Hillel D, Rosenzweig C (eds) Handbook of climate change and agro-ecosystems: the agricultural modeling intercomparison and improvement project (AgMIP) integrated crop and economic assessments. Imperial College Press and The American Society of Agronomy, London, pp 219–258

Ahmad S, Abbas G, Ahmed M, Fatima Z, Anjum MA, Rasul G, Khan MA, Hoogenboom G (2019) Climate warming and management impact on the change of rice-wheat phenology in Punjab, Pakistan. Field Crop Res 230:46–61

Ahmed M, Ahmad S (2019) Carbon dioxide enrichment and crop productivity. In: Hasanuzzaman M (ed) Agronomic crops, vol 2. Springer, Singapore, pp 31–46

Ahmed M, Ahmad S (2020) Systems modeling. In: Ahmed M (ed) Systems modeling. Springer, Singapore, pp 1–44

Ahmed M, Fayyaz-ul-Hassan AS (2017) Climate variability impact on rice production: adaptation and mitigation strategies. In: Ahmed M, Stockle C (eds) Quantification of climate variability, adaptation and mitigation for agricultural sustainability. Springer, Switzerland, pp 91–111

Ahmed M, Ahmad S, Raza MA, Kumar U, Ansar M, Shah GA, Parsons D, Hoogenboom G, Palosuo T, Seidel S (2020a) Models calibration and evaluation. In: Ahmed M (ed) Systems modeling. Springer, Singapore, pp 151–178

Ahmed M, Ahmad S, Waldrip HM, Ramin M, Raza MA (2020b) Whole farm modeling: a systems approach to understanding and managing livestock for greenhouse gas mitigation, economic viability and environmental quality. In: Waldrip HM, Pagliari PH, He Z (eds) Animal Manure. ASA Special Publication 67. American Society of Agronomy and Soil Science Society of America, Madison, WI, pp 345–371

Akram R, Turan V, Hammad HM, Ahmad S, Hussain S, Hasnain A, Maqbool MM, Rehmani MIA, Rasool A, Masood N, Mahmood F, Mubeen M, Sultana SR, Fahad S, Amanet K, Saleem M, Abbas Y, Akhtar HM, Hussain S, Waseem F, Murtaza R, Amin A, Zahoor SA, Sami ul Din M, Nasim W (2018) Fate of organic and inorganic pollutants in paddy soils. In: Hashmi MZ, Varma A (eds) Environmental pollution of paddy soils. Springer, Switzerland, pp 197–214

Akram R, Fahad S, Masood N, Rasool A, Ijaz M, Ihsan MZ, Maqbool MM, Ahmad S, Hussain S, Ahmed M, Kaleem S, Sultana SR, Mubeen M, Saud S, Kamran M, Nasim W (2019) Plant growth and morphological changes in rice under abiotic stress. In: Hasanuzzaman M, Fujita M, Nahar K, Biswas JK (eds) Advances in rice research for abiotic stress tolerance. Woodhead Publishing, Swaston, pp 69–85

Arora VK (2006) Application of a rice growth and water balance model in an irrigated semi-arid subtropical environment. Agric Water Manag 83(1–2):51–57

Biswas JK, Yamauchi M (2007) Mechanism of seedling of direct-seeded rice (Oryza sativa L.) under lowland condition. Rice Sci 60:191–226

Bouman BAM, Peng S, Castaneda AR, Visperas RM (2005) Yield and water use of tropical aerobic rice systems. Agric Water Manag 74:87–105

Edwin L, Anal PSM (2008) Effect of irrigation regimes and nitrogen management practices on uptake of nutrients and grain yield in hybrid rice (Oryza sativa L.). Environ Ecol 26:1146–1148

Fahad S, Noor M, Adnan M, Khan MA, Rahman I, Alam M, Khan IA, Ullah H, Mian IA, Hassan S, Saud S, Bakhat HF, Hammad HM, Ahmad S, Nasim W (2019) Abiotic stress and rice grain quality. In: Hasanuzzaman M, Fujita M, Nahar K, Biswas JK (eds) Advances in rice research for abiotic stress tolerance. Woodhead Publishing, Swaston, pp 571–583

Farooq M, Basra SMA, Wahid A (2006a) Priming of ûeld-sown rice seed enhances germination, seedling establishment, allometry and yield. Plant Growth Regul 49:285–294

Farooq M, Basra SMA, Tabassum R, Afzal I (2006b) Enhancing the performance of direct seeded rice by seed priming. Plant Prod Sci 9(4):446–456

Fatima Z, Ahmad M, Hussain M, Abbas G, Ul-Allah S, Ahmad S, Ahmed N, Ali MA, Sarwar G, Ehsan ul Haque, Iqbal P, Hussain S (2020) The fingerprints of climate warming on cereal crops phenology and adaptation options. Sci Rep 10:18013

Gill MS, Singh MP (2008) Grain yield and water productivity of direct seeded basmati rice (Oryza sativa) under various seed rates, weed control and irrigation schedules. Environ Ecol 26:594–597

Hafeez-ur-Rehman, Nawaz A, Awan MI, Ijaz M, Hussain M, Ahmad S, Farooq M (2019) Direct seeding in rice: problems and prospects. In: Hasanuzzaman M (ed) Agronomic crops: production technologies, vol 1. Springer, Singapore, pp 199–222

Huang DF, Xi LL, Wang ZQ, Liu LJ, Yang JC (2008) Effects of irrigation regimes during grain filling on grain quality and the concentration and distribution of cadmium in different organs of rice. Acta Agron Sin 34:456–464

Hussain M, Ahmad S, Hussain S, Lal R, Ul-Allah S, Nawaz A (2018) Rice in saline soils: physiology, biochemistry, genetics and management. Adv Agron 148:231–287

Jat ML, Gathala MK, Ladha JK, Saharawat YS, Jat AS, Kumar V, Sharma AS, Gupta R (2009) Evaluation of precision land leveling and double zero-till systems in the rice-wheat rotation: water use, productivity, profitability and soil physical properties. Soil Tillage Res 105:112–121

Kaur J (2004) Comparative performance of different planting techniques on growth, productivity and water saving in paddy (Oryza sativa L.). M.Sc. Thesis, Punjab Agricultural University, Ludhiana, India

Khan MA, Ahmad S, Raza A (2019a) Integrated weed management for agronomic crops. In: Hasanuzzaman M (ed) Agronomic crops, vol 2. Springer, Singapore, pp 257–281

Khan SM, Ali S, Nawaz A, Bukhari SAH, Ejaz S, Ahmad S (2019b) Integrated pest and disease management for better agronomic crop production. In: Hasanuzzaman M (ed) Agronomic crops, vol 2. Springer, Singapore, pp 385–428

Kumar A (2002) Growth and yield of different rice (Oryza sativa L) cultivars under various methods and time of sowing. M.Sc. Thesis, Punjab Agricultural University, Ludhiana, India

Lee KD, Park CW, So KH, Na SI (2017) Selection of optimal vegetation indices and regression model for estimation of rice growth using UAV aerial images. Korean J Soil Sci Fert 50(5): 409–421

Mahajan G, Sardana V, Brar AS, Gill MS (2006) Effect of seed rates, irrigation intervals and weed pressure on productivity of rice (Oryza sativa). Indian J Agric Sci 76:756–759

Maniruzzaman M, Nemes A (2014) Evaluation of the Aquacrop model to simulate rice growth under different water regimes in Bangladesh. Evaluation 159:331–340

Mann RA, Ashraf M (2004) Improvement of basmati and its production practices in Pakistan. In: Speciality rices of the world: breeding, production and marketing. Duffy et al. (Ed.), Food and agricultural organization of United Nations, Rome, Italy. pp. 134, 135, 137 and 140

Mubeen K, Jabran K (2019) Alternate wetting and drying system for water management in rice. In: Agronomic crops. Springer, Singapore, pp 101–110

Naz S, Fatima Z, Iqbal P, Khan A, Zakir I, Noreen S, Younis H, Abbas G, Ahmad S (2019) Agronomic crops: types and uses. In: Hasanuzzaman M (ed) Agronomic crops: production technologies, vol 1. Springer, Singapore, pp 1–18

Peerzada AM, Bukhari SAH, Dawood M, Nawaz A, Ahmad S, Adkins S (2019) Weed management for healthy crop production. In: Hasanuzzaman M (ed) Agronomic crops, vol 2. Springer, Singapore, pp 225–256

Phakamas N (2015) Performance of the CSM-CERES-rice model in evaluating growth and yield of rice in the farm level. J Agricul Techn 11(5):1285–1295

Ramakrishna Y, Singh S, Parihar SS (2007) Influence of irrigation regime and nitrogen management on productivity, nitrogen uptake and water use by rice (Oryza sativa). Indian J Agron 52: 102–106

Razaq M, Shah FM, Ahmad S, Afzal M (2019) Pest management for agronomic crops. In: Hasanuzzaman M (ed) Agronomic crops, vol 2. Springer, Singapore, pp 365–384

Rezaei M, Vahed HS, Amiri E, Motamed MK, Azarpore E (2009) The effects of irrigation and nitrogen management on yield and water productivity of rice. World Appl Sci J 7:203–210

Saharawat YS, Singh B, Malik RK, Ladha JK, Gathala M, Jat ML, Kumar V (2010) Evaluation of alternative tillage and crop establishment methods in a rice – wheat rotation in North Western IGP. Field Crop Res 116:260–267

Sarwar N, Ali H, Ahmad S, EhsanUllah AS, Mubeen K, Hill JE (2013a) Water wise rice cultivation on calcareous soil with the addition of essential micronutrients. J Anim Plant Sci 23(1):244–250

Sarwar N, Ali H, Maqsood M, EhsanUllah, Shahzad AN, Shahzad M, Mubeen K, Shahid MA, Ahmad S (2013b) Phenological response of rice plants to different micronutrients application under water saving paddy fields on calcareous soil. Turk J Field Crops 18(1):52–57

Shahzad AN, Ahmad S (2019) Tools and techniques for nitrogen management in cereals. In: Hasanuzzaman M (ed) Agronomic crops, vol 2. Springer, Singapore, pp 111–126

Sharma PK, Lav B, Ladha JK, Naresh RK, Gupta RK, Balasubramanian BV, Bouman BAM (2002) Crop-water relations in rice-wheat cropping under different tillage systems and water-management practicesin a marginally sodic, medium-textured soil. In: Bouman BAM, Hengsdijk H, Hardy B, Bindraban PS, Tuong TP, Ladha JK (eds) Water-wise rice production. International Rice Research Institute, Los Baños, Philippines, pp 223–235

Singh AK, Choudhury BU, Bouman BAM (2002) Effects of rice establishment methods on crop performance, water use, and mineral nitrogen. In: Bouman BAM, Hengsdijk H, Hardy B, Bindraban PS, Tuong TP, Ladha JK (eds) Water-wise rice production. Los Baños, Philippines, International Rice Research Institute, pp 237–246

Soundharajan B, Sudheer KP (2009) Deficit irrigation management for rice using crop growth simulation model in an optimization framework. Paddy Water Environ 7(2):135–149

Tariq M, Ahmed M, Iqbal P, Fatima Z, Ahmad S (2020) Crop phenotyping. In: Ahmed M (ed) Systems modeling. Springer, Singapore, pp 45–60

Wang H, Bouman BAM, Zhao D, Wang C, Moya PF (2002) Aerobic rice in northern China: opportunities and challenges. In: Bouman BAM, Hengsdijk H, Hardy B, Bindraban PS, Tuong TP, Ladha JK (eds) Water-wise rice production. Los Baños, Philippines, International Rice Research Institute, pp 143–154

Wasaya A, Yasir TA, Ijaz M, Ahmad S (2019) Tillage effects on agronomic crop production. In: Hasanuzzaman M (ed) Agronomic crops, vol 2. Springer, Singapore, pp 73–99

Yadav S, Gill G, Humphreys E, Kukal SS, Walia US (2011a) Effect of water management on dry direct seeded rice and puddled transplanted rice Part 1: crop performance. Field Crop Res 120: 112–122

Yadav S, Gill G, Humphreys E, Kukal SS, Walia US (2011b) Effect of water management on dry direct seeded rice and puddled transplanted rice Part 2: water balance and water productivity. Field Crop Res 120:123–132

Zahoor SA, Ahmad S, Ahmad A, Wajid A, Khaliq T, Mubeen M, Hussain S, Sami Ul Din M, Amin A, Awais M, Nasim W (2019) Improving water use efficiency in agronomic crop production. In: Hasanuzzaman M (ed) Agronomic crops, vol 2. Springer, Singapore, pp 13–29

Rice-Based Cropping Systems

9

Naeem Sarwar, Atique-ur-Rehman, Allah Wasaya, Omer Farooq,
Khuram Mubeen, Muhammad Dawood, Muhammad Shehzad,
and Shakeel Ahmad

Abstract

The annual crops are grown in succession with other crops due to different climatic requirements. Such differences make the basis of cropping system, and it appears at regional level. The rice cultivation is mainly followed by wheat and maize in Indo-Pak region and China. All these crops are the main players of food security in the world. On the other hand, the crops with similar climatic requirement may be intercropped with each other. The rice is also being intercropped with many other crops of summer season. The intercropping system not only improves the resource utilization but also increases farm income.

Keywords

Rice · Cropping system · Intercropping · Relay cropping · Resource use efficiency

N. Sarwar (✉) · Atique-ur-Rehman · O. Farooq · M. Dawood
Department of Agronomy, Bahauddin Zakariya University, Multan, Pakistan
e-mail: naeemsarwar@bzu.edu.pk

A. Wasaya
College of Agriculture, Bahauddin Zakariya University, Bahadur Sub-Campus, Layyah, Pakistan

K. Mubeen
Department of Agronomy, Muhammad Nawaz Sharif University of Agriculture, Multan, Pakistan

M. Shehzad
Department of Agronomy, University of Poonch, Rawalakot, Pakistan

S. Ahmad
Department of Agronomy, Faculty of Agricultural Sciences and Technology, Bahauddin Zakariya University, Multan, Pakistan
e-mail: shakeelahmad@bzu.edu.pk

N. Sarwar et al. (eds.), *Modern Techniques of Rice Crop Production*,
https://doi.org/10.1007/978-981-16-4955-4_9

9.1 Introduction

Rice (Oryza sativa L.) is a staple food consumed by the majority of the world's population (Emerick and Ronald 2019; Ahmad et al. 2008, 2009, 2012, 2013, 2015, 2019; Ahmad and Hasanuzzaman 2012; Ahmed and Ahmad 2019, 2020; Shahzad and Ahmad 2019; Ahmed et al. 2017, 2020a, b; Akram et al. 2018, 2019; Hussain et al. 2018; Fahad et al. 2019; Fatima et al. 2020) and one of the world's top three cereal crops (Yadav et al. 2019). Rice development is especially conspicuous in Asia, the Americas, and Africa (Hafeez-ur-Rehman et al. 2019; Khan et al. 2019a, b; Naz et al. 2019; Peerzada et al. 2019; Razaq et al. 2019; Sarwar et al. 2013a, b; Tariq et al. 2020; Wasaya et al. 2019; Zahoor et al. 2019). The Cerrado of Brazil is the world's largest producer of upland rice (Silva et al. 2020), a region with acidic soils and low soil fertilization (Allen et al. 2007). Small-level innovation use by ranchers and summer droughts incredibly limits plant improvement and yields of upland rice (Nascente et al. 2013). Investigations of intercropping frameworks have surveyed yields of soybean (Crusciol et al. 2013, 2014), corn (Crusciol et al. 2013; Borghi et al. 2013), and sorghum (Borghi et al. 2014; Crusciol et al. 2013), and in the mid-year season (Surve and Arvadia 2011) and biomass creation of tropical grass during the off-season (Pariz et al. 2017; Mateus et al. 2020). These examinations intend to grow better administration rehearses for expanding crop improvement as well as diminishing rivalry between intercropped species, in this manner expanding yields (Crusciol et al. 2014; Pariz et al. 2016; Moraes et al. 2019; Mateus et al. 2020). In either case, there is a scarcity of information on N, the executives of upland rice intercropped with forage grasses in tropical areas.

9.2 Rice-Based Cropping Systems

9.2.1 Cereal-Based Cropping Systems

The wheat–rice is one of the prominent rice-based cropping systems of South Asia. This system covers more than 90% area in Pakistan and is the main source of food in Asia. The rice is grown in Kharif, and wheat is grown during Rabi season. In South Asia, it covers about 13.5 million hectares with share of 10.0, 2.2, 0.8, and 0.5 million hectares from India, Pakistan, Bangladesh, and Nepal, respectively. In China, the rice–wheat cropping system occupies an area of 13 million hectares; however, it declined with the passage of time (Dawe et al. 2004). The decline occurred with the popularity of maize-rice cropping system. It was emerged in the late 1960 in India with the introduction of dwarf wheat varieties by CIMMYT (Mexico) (Mahajan and Gupta 2009). Both the crops are grown in rotation due to huge differences in climatic requirement. This system is exhausting the soil fertility, and yield of both the crops is stagnant, despite the significant application of synthetic inputs. The alternate wetting (rice) and drying (wheat) are severe threat for soil microbial population, adversely affecting the soil structure. The soil condition is converted to anaerobic from aerobic and then back to aerobic from anaerobic. The puddling conditions in rice badly affect the soil organic carbon dynamics, nutrient availability, and utilization efficiencies (Bhattacharyya et al.

2012). The changed soil physical properties due to puddling require additional energy for land preparation for sowing of succeeding crops. It breaks the soil capillaries and may also create hardpan which impedes the root growth of succeeding crop. Both the crops produce bulk residue which is sometimes burns in the field after combine harvesting, resulting in environmental pollution. The residue management is a major challenge for rice–wheat cropping system. The rice harvesting and wheat sowing are overlapped, and wheat sowing is mostly delayed in rice areas. The wheat grown following rice is less yielder due to unfavorable environmental conditions during reproductive development. The issue is being addressed by evolving early maturing rice varieties and zero tillage wheat planting. However, the fallow period (2–3 months) is available between wheat harvesting and rice nursery transplantation. The high-water requirement (5000 L per kg rice grains), deteriorating soil health, high labor, and energy requirement are certain constraints of rice–wheat cropping system. Despite these problems, the system has a pivotal role in global food security, effective weed control, and efficient utilization of excessive rains which are certain key benefits of the system. It can be exploited for cultivation of short-duration crops like mung bean, sesbania, and other fodder crops. Rice–rice is another prominent rice-based cropping system and practiced in West Coast Plains and Hills in India (Gill et al. 2008). It ranks second important rice-based cropping system after rice–wheat, and it occupies 2.12 million hectares in India (Yadav 1996). The rice–rice system occupies an area of 9.5 million ha in Nepal, India, and Bangladesh (Timsina et al. 2010). The rice–maize is another cropping system gaining popularity in Asia and occupies an area of 3.5 million ha (Timsina et al. 2010). The system also observes similar soil conditions to rice–wheat due to alternate wetting and drying conditions. The minor contribution is from rice–berseem, rice–lentil, rice–canola, rice–sunflower, rice–wheat–mung bean, rice–wheat–cowpeas, and rice–wheat–sesbania (Ali et al. 2012). The other cereals like barley, oat, fodder maize, millet, and sorghum may be other cereal-based cropping systems in rice-growing regions (Akanvou et al. 2002; Andrews and Kassam 1976; Dean et al. 2005; Ghanbari and Lee 2002; Li et al. 2009; Lou et al. 2013; Lu and Snow 2005; Patra and Chatterjee 1986).

9.2.2 Legume-Based Cropping System

The inclusion of legumes in rice-based cropping system is important for restoration of soil fertility in particular. Moreover, this practice will improve the biodiversity and could be the remedy for malnutrition through the availability of protein-based food grains. The legumes fixed the atmospheric nitrogen for plant use for plant use and improve the soil health through root penetration in deeper soil layers (Tariq et al. 2019). The short-duration legumes in the rice areas are beneficial with regard to cropping system diversity, biological nitrogen fixation, and nutritional security. The legumes have gained popularity as green legumes crops due to greater leaf proportion, and hence may be incorporated in soil to accelerate the process of restoration of soil fertility. The legumes shed their leaves on soil surface at maturity and contribute to soil fertility restoration. The legumes cultivation in rice areas may have potential for rainfed rice farming. The cropping systems are important for economic returns

Table 9.1 Proposed rice-based cropping systems in rice-growing countries

Rice-based cropping system	Reporting countries	Location	References
Legumes based			
Rice–berseem	Pakistan	Kala Shah Kaku (31.740655° N, 74.253146° E, 207 m asl)	Ali et al. (2012)
Rice–lentil	Pakistan	Kala Shah Kaku (31.740655° N, 74.253146° E, 207 m asl)	Ali et al. (2012)
Rice–pea	India	Wabagai (24° 32′ 45.276″ N, 93° 56′16.0188″ E, 778 m asl)	Meetei et al. (2020)
Rice–French Bean	India	Wabagai (24° 32′ 45.276″ N, 93° 56′16.0188 E and 778 m asl)	Meetei et al. (2020)
Rice–rice–black gram	India	Soil and Water Management Research Institute, Kattuthottam (10°45 N, 79° E and 50 m asl)	Porpavai et al. (2011)
Cereal based			
Rice–wheat	Pakistan	Kala Shah Kaku (31.740655° N, 74.253146° E, 207 m asl)	Ali et al. (2012)
Oil seed based			
Rice–canola/ mustard	Pakistan, India	Kala Shah Kaku (31.740655° N, 74.253146° E, 207 m asl) Wabagai (24° 32′ 45.276″ N, 93° 56′16.0188 E, 778 m asl)	Ali et al. (2012); Meetei et al. (2020)
Rice–sunflower	Pakistan	Kala Shah Kaku (31.740655° N, 74.253146° E, 207 m asl)	Ali et al. (2012)
Mixed cropping system			
Rice–wheat–mung bean	Pakistan	Kala Shah Kaku (31.740655° N, 74.253146° E, 207 m asl)	Ali et al. (2012)
Rice–wheat–cowpeas	Pakistan	Kala Shah Kaku (31.740655° N, 74.253146° E, 207 m asl)	Ali et al. (2012)
Rice–wheat–sesbania	Pakistan	Kala Shah Kaku (31.740655° N, 74.253146° E, 207 m asl)	Ali et al. (2012)
Rice–rice–onion	India	Soil and Water Management Research Institute, Kattuthottam (10°45 N, 79° E and 50 m asl)	Porpavai et al. (2011)
Groundnut–rice–blackgram	India	Soil and Water Management Research Institute, Kattuthottam (10°45 N, 79° E and 50 m asl)	Porpavai et al. (2011)
Rice–rice–green gram	India	Soil and Water Management Research Institute, Kattuthottam (10°45 N, 79° E and 50 m asl)	Porpavai et al. (2011)
Rice–onion–cowpea	India	Navsari Agricultural University	Jat et al. (2012)

and resource conservation and efficient utilization. Among the cropping system proposed by Ali et al. (2012) for Pakistan in Table 9.1, the maximum economic returns were gained from rice–lentil cropping system than rest of the treatments.

Growing peanut after rice is well-known cropping system in Thailand in which peanut completes its initial growth on residual moisture (Rehman et al. 2020; Ren et al. 2008; Singh et al. 2007; Umar et al. 2007; Zhang et al. 2011).

9.2.3 Mixed Cropping System

It is clear that rice is mainly grown in two crop rotations; however, it may be grown in three crop rotations like rice–wheat–mung bean, rice–wheat–cowpeas, and rice–wheat–sesbania in Pakistan. While in India, it may be grown in rice–mustard–green gram, rice–fenugreek–okra, rice–chickpea–sesamum, and rice–sorghum–groundnut rotation (Jat et al. 2012). The three crop rotations are possible with the inclusion of at least one short-duration crop. These are specific to certain areas and not widely practiced across the rice-growing regions. For example, rice–rice–pulse and rice–rice sesame are common cropping systems in Thanjavur district, India (Porpavai et al. 2011) and Maize-Jute-T. Aman is popular in Bogra district of Bangladesh (Ali et al. 2009). The potential legumes for rice-based cropping system are listed in Table 9.1.

9.3 Rice-Based Intercropping Systems

Intercropping is the development of at least two crops at the same time in a similar field. Numerous crops are intercropped with upland rice contingent upon the length of developing period and farmer choice. The basic frameworks include rice + maize, rice + maize + cassava, rice + cowpea, rice + sesame, rice + beniseed, rice + soybean, rice + mung bean, rice + pigeon pea, rice + Christian. Intercropping builds efficiency of integral, segment crops. Very much planned intercropping consolidates part crops that utilization assets more completely than would single y crops. Intercropping is the transcendent cropping framework in conventional and some degree present-day Agribusiness. Intercropping builds profitability through most extreme use of land, work, and capital assets and the board of pest irritations and inflectional diseases.

9.3.1 Rice and Cereal (Rice + Maize)

In Ghana, maize, a staple grain, is devoured in different customary dishes, for example, kenkey, banku, porridge, akpele, and TZ, to specify a couple. At the milk stage, maize can be eaten simmered or cooked. Mechanically, maize flour is utilized for making bread and liquor, and for domesticated animal feed. The straw might be utilized for taking care of animals or making covered rooftop houses, mats, and bushels, just as for mulch on crop ranches. Since the presentation of rice (Oryza sativa) into Ghana, its status has transformed from a "delicacy" to a staple food in northern Ghana, presently positioning among significant staple grain crops, like maize, pearl millet (Pennisetum glaucum), and (Sorghum bicolor). The adjustment of the situation with rice is a direct result of expanded utilization and maybe various homegrown and mechanical employments. By and large, in Ghana and especially in the northern piece of the country, laborer farmers for the most part produce crops in intercropping frameworks. The exhibition of these crops in intercropping frameworks is dictated by the proficiency of the part harvests to catch and use the accessible assets (Manu-Aduening 1999), and the similarity of the crops to contend successfully for daylight, supplements, and water. Willey (1979) saw that light is the main development asset for crop development and improvement and, in this manner, rivalry for light by plants regularly brings about plants getting slenderer and taller. Intercropping does not appear to be agreeable to the greater part of the cutting edge advances accessible to agriculturalists, yet most of the farmers in the jungle practice this framework (Onwueme and Sinha 1991). In Africa, cowpea cultivators, who do not approach insect sprays, intercrop cowpea, and maize. Furthermore, intercropping may likewise fence against all-out crop disappointment if there should arise an occurrence of antagonistic climate conditions. Maize/rice (cereal/grain) intercrop lately has become a typical practice among most farmers in northern Ghana, presumably due to the arrival of some upland rice assortments. Despite the fact that this training may guarantee productive and successful utilization of accessible resources, its impact on the yield of the staple harvests is generally secret.

9.3.2 Rice and Green Manure (Rice + Sasanian)

Other forage leguminous crops (cowpea, ricebean, pigeon pea, mung bean, and sesbania) are significant brief term summer crops that give more monetary re-visitation of cultivators (Ahmad et al. 2007; Iqbal et al. 2006). The region under these harvests cannot be expanded as they contend with rice, a significant Kharif grain crop of Pakistan. In addition, during sweltering late spring months, these fodder help to keep up creature well-being and milk creation other than improving fertilization of oil by means of natural nitrogen obsession (BNF) and organic matter accumulation (Wahla et al. 2010). In this way, one of the approaches to enhance late spring fodder creation is to develop these harvests as rummage in relationship with upland rice.

9.3.3 Rice and Pulses [Rice + Ricebean; Rice + Mung Bean; Rice + Pigeon Pea; Rice + Cowpea; Rice + Soybean]

Double cultivation of cowpea with cultivated rice could add around 12 t ha^{-1} of natural fertilizer and ensure 25% nitrogen saving (22.5 kg/ha^{-1}) with 11% yield upgrade and expansion in benefit. High land identical proportion, relative worth aggregate, high gross, and net return proportion were accounted for when maize was intercropped with red gram, green gram, dark gram, soybean, and groundnut. Higher net advantages in rice–cowpea intercrop were accounted for and the scientists additionally noticed that the intercrops of rice and cowpea were 14.03% higher than sole rice.

9.3.4 Rice and Grasses [Rice + Guinea Grass (Megathyrsus Maximus); Rice + Palisadegrass (Urochloa Brizantha)]

There is a scarcity of data on N the board of upland rice intercropped with forage grasses in tropical areas. Nitrogen (N) is the main supplement affecting turn of events and yield of rice, and its elements in the soil plant framework fluctuate as per soil conditions and manure the board technique (Nascente et al. 2013; Fageria et al. 2011). Expanding nitrogen use proficiency in agroecosystems is a continuous objective to improve farming maintainability, increment upland rice yield, and advance high income per region (Nascente et al. 2013). Cash crops and forage grasses are intercropped during mid-year, trailed by forage grass creation with creature munching in the off-season (Franzluebbers and Stuedemann 2014; Crusciol et al. 2014, 2016; Moraes et al. 2019). Progressive grass-no one but development can bargain the manageability of Integrated Crop-Livestock Systems (ICLS) because of soil nitrogen consumption through crop nitrogen evacuation (Garcia et al. 2016). In grass crops, low N manure recovery efficiency is common. The ideal N manure application rate varies depending on soil conditions, crop innovation stage, and type of harvest revolution (leguminous yields that repair atmospheric N vs. non-leguminous yields) (Borghi et al. 2014; Crusciol et al. 2016). Along these lines, improved N manure proposals in intercropped frameworks, especially those including upland rice, are required.

9.3.5 Rice and Water Spinach

The determination and plan of a proper intercropping framework enormously rely upon the idea of the segment harvests and area qualities, like the environment, soil conditions, nearby local area inclinations, and financial significance of the species. Numerous examinations have researched the chance of intercropping rice with dryland harvests like nut, mung bean, watermelon, and so on in upland fields (Ren et al. 2008; Li et al. 2009). Regardless, there is a scarcity of data on the effect of intercropping on pest concealment and yield profitability in irrigated rice farming systems. Water spinach is a herbaceous aquatic or semiaquatic enduring plant that is easy to grow and produces a high biomass yield rich in protein, the nutrients C and An, and minerals, especially Mn, K, Mg, and Fe. Therefore, water spinach is broadly developed and burned-through in southern China. By and large, spinach grown in water has a short development period and is impervious to regular pest irritations and diseases (Lin et al. 2014). Water spinach and rice are systematically various yields that do not have similar irritations and diseases, and both fill productively in a wetland climate. Those qualities are imperative to research the chance of intercropping these harvests fully intent on building up a privately adjusted differentiated rice editing framework in southern China.

9.3.6 Rice and Cotton Intercropping

Both rice and cotton are grown in summer, and their growth period is also common. The cotton may be grown with direct seeded or rainfed rice but the environment of flooded rice does not provide favorable conditions for cotton. In rainfed rice cultivation, the intercropped cotton may compensate for the negative impact of erratic or low rainfall. On the other hand, the bed furrow planting system has two different soil conditions, that is, the furrows are wet and beds are dry. There is sufficient furrow width for rice sowing and furrow environment is very suitable for rice, while beds have low moisture and thus support the cotton growth. Therefore, furrows will have rice and beds will have cotton. Almost no publication is found on this aspect; however, the technique may be very supportive for improving land use and resource use efficiency. The incidence of insect pests and other diseases in this system needs further investigation for proper recommendation. The pictorial view of rice–cotton intercropping is shown in Fig. 9.1.

9.4 Benefits of Rice-Based Intercropping Systems

With shrinking resources such as arable land, water systems, and electricity, there is an urgent need to plan and develop new harvesting strategies and methods to meet the growing demand for food, feed, and search through effective use of available rural information assets. Under the current arrangement of sole cropping, little farmers cannot address their expanded homegrown necessities to support typical livings from their restricted land, monetary assets, and water. This requires going for

Fig. 9.1 Cotton–rice intercropping in Bangladesh [adopted from Atique-ur-Rehman et al. (2020) and Ahmad and Hasanuzzaman (2020)]

fitting other option and more effective creation frameworks, for example, multicropping (inter/relay cropping) those can guarantee legitimate use of assets to get expanded creation per unit region and time on a manageable premise (Trenbath 1986). Intercropping, a unique feature of subtropical and tropical regions is gradually becoming more common among small farmers, as it provides a yield advantage over sole cropping through yield consistency and increased yield (Nazir et al. 2002; Bhatti et al. 2006). Via intercropping, the ability to raise different harvests, such as forage legumes and non-legumes, in relation to important staple food crops like rice, could be greatly improved (Saeed et al. 1999). It also aids in the preservation of soil fertility (Patra and Chatterjee 1986), utilizing supplements (Maingi et al. 2001; Nazir et al. 1997; Aggarwal et al. 1992; Ahmad and Saeed 1998) and guaranteeing monetary use of land, work, and capital (Singh et al. 1996; Jeyabal and Kuppuswamy 2001; Morris and Garrity 1993). All in all, non-legume yield is viewed as a stifling harvest in legume affiliations. Pakistan has dry to semi-parched subtropical environment with high-level light force and positive temperature ranges and a broad trench water system framework which converts into an enormous potential for raising at least two rural yields simultaneously.

9.4.1 Increased Productivity or Yield Advantage

Intercropping offers the chance of yield benefits comparative with sole-based cropping through yield steadiness and improved yield and subsequently giving differentiated necessities of small ranchers, soundness of yield over various seasons, weed control, creepy crawly irritations, and inflectional disorders just as control of soil disintegration (Willey 1979). In addition, it helps keeping up soil rich in fertility and productivity (Patra and Chatterjee 1986), utilizing supplements (Ahmad and Saeed 1998; Nazir et al. 1997; Aggarwal et al. 1992), and guaranteeing financial usage of labor, land, and capital (Singh et al. 1996; Morris and Garrity 1993).

9.4.2 Biological N Fixation (BNF)

Contrasted and other intercrop mixes, grain/legumes intercrop might be more effective on account of the capacity of the legume part to fix atmospheric nitrogen to serve succeeding yields developed on a similar real estate parcel.

9.4.3 Nitrogen Use Efficiency

Intercropping when utilized either independently or in mix with other viable techniques could be utilized in a practical way to reduce some of the issues in crop production and increase yield. Intercropping rice with soybean assists with expanding efficiency since soybean fixes nitrogen into the soil which is a significant supplement needed by rice. It is presumed that intercropping soybean and maize on level outcome in ideal yield and that the most extreme joined intercrop income from maize and soybean was from inter + intra-line planting course of action on level.

9.4.4 Land Use Efficiency

This conduct of farmers might be ascribed to the huge advantages they get from intercropping frameworks. Intercropping frameworks are utilized to augment creation and differentiate crops on a package of land on either schedule or space that would not be acquired in mono-cropping frameworks (Manu-Aduenning and Boa-Amponsem 2005). Intercropping guarantees the insurance of soil against vanishing and disintegration through the arrangement of sufficient ground cover. The spreading of collect advances a customary stockpile of nourishment for the family, accordingly guaranteeing food security. As per Rouanet (1995), the framework energizes proficient utilization of soil assets, which are misused to various profundities and various periods. Less use of technology by farmers and droughts in summer incredibly limits plant improvement and upland rice yields (Nascente et al. 2013). Upland rice has to deal with high soil acidity and interchangeable aluminum, both of which are permanent characteristics of degraded soils (Fageria 1998). Growing upland rice in degraded fields for two growing seasons before returning to pasture is a common rural practice in tropical areas (Kluthcouski et al. 2000). Notwithstanding, editing frameworks dependent on preservation the executives with crop revolution, intercropping, coordinated harvest animals framework, and no-culturing framework are prescribed to lessen soil corruption. These frameworks give incredible proficiency in saving normal assets and supporting high rural creation in tropical locales (Crusciol et al. 2015; Borghi et al. 2014; Moraes et al. 2019). In numerous locales of the world, including the jungles, ICLS might be a reasonable choice to improve food creation and abatement destitution (Food and Farming Association of the Assembled Countries (FAO 2017)). Since ICLS reduces the need to create new agrarian territories, these structures are seen as more sustainable (Surve and Arvadia 2011) and promote yield production diversity in the same region (Mateus et al. 2016; Crusciol et al. 2014). Upland rice intercropped with tropical search grasses could be a fantastic way to improve crop variety and soil quality, particularly in tropical soils with low fertility and dry winters (Wood and Mendelsohn 2014; Allen et al. 2007).

9.4.5 Water Use Efficiency

The units of yield produced per unit of available water are referred to as water use efficiency (WUE). Various methodologies are needed to enhance the production of rice and least utilization of water. Among various methodologies, water profitability or water use efficiency (WUE) is an effective methodology. Water use effectiveness can be controlled or improved by receiving rice-based intercropping designs.

9.4.6 Radiation Use Efficiency

The exhibition of these crops in intercropping frameworks is controlled by the proficiency of the segment crops to catch and use the accessible assets (Manu-Aduening 1999), and the similarity of the crops to contend adequately for daylight, supplements, and water. Willey (1979) saw that light is the main development asset for crop development and improvement and, thus, rivalry for light by plants frequently brings about plants getting more slender and taller.

9.4.7 Suppress Pests and Disease Incidence

China has the second biggest space of rice fields on the planet and has the most noteworthy rice creation (Zhang et al. 2011). In any case, especially stem borers, rice planthoppers, leaf folders, and leaf folders, just as the event of sicknesses, for example, rice blast and bacterial leaf blight, are not kidding challenge to rice creation. Farmers are also debating whether or not to use pesticides for better safety. Currently, the rice crop is responsible for the greatest amount of pesticide use in China; in reality, rice pesticide sales totaled $538 million in 2006. Southern China is among the significant rice-delivering areas in the country, with the majority of the rice developed under marsh conditions. The over the top utilization of agrochemical information sources like manure, herbicides, and pesticides expands the monetary weight on farmers as well as makes well-being dangers and ecological dangers. Techniques empowering a low agro-based chemical input are at present popular for practical vermin and infection, the executives in agroecosystems. Intercropping is the agricultural technique of all the while developing at least two crops in a similar space (Andrews and Kassam 1976). The training is for quite some time set up and still prevails in numerous tropical and subtropical locales, particularly in non-industrial nations. One of the significant benefits of intercropping is its capacity to diminish pest and disease spread. Francis (1989) announced that 18% of assessed tests showed that intercropping expanded the frequency of vermin contrasted with mono-cropping, while 9% tracked down no huge distinction. Comparative exploration likewise exhibited that broadening smothers pests yet at the same time lessens yield (Showler and Greenberg 2003; Schulthess et al. 2004). A lot of proof shows that intercropping supports biodiversity, which thus assists with smothering nuisances and disease. A meta-investigation of 45 articles containing 552 all-out correlations by Letourneau et al. (2011) showed that herbivore concealment, adversary upgrade, and harvest harm concealment were altogether more prominent in enhanced yield frameworks. All things considered, some different examinations contend that differentiated intercropping does not really diminish the occurrence of pests or enhance production. Such an expansion in nuisances can happen for a few reasons: (1) the subsequent harvest used may be a host for a particular irritation; (2) intercropping may increase the amount of shade provided by the covering, allowing pests and microbes to thrive; and (3) plant–plant competition can have negative effects on production that outweigh the benefits of pest concealment. Extra benefits of intercropping incorporate higher harvest yields and more prominent land use proficiency per unit land region. Willey (1990) concluded intercropping to be an efficient strategy for better return, with lower outer sources of info contrasted and mono-editing. Preventing the infection in crops is one of the important crop production practices to obtain high crop yields (Dordas 2008). There are numerous techniques for crop infectious prevention, including the use of synthetic, natural, physical, and cultural methodologies (Palti 1981). One of the compelling and natural well-disposed approaches to control infections is to apply biological methodologies in current agrarian frameworks (Risch et al. 1983), in which crop heterogeneity is made to give considerable infection concealment (Zhu et al. 2000; Garrett and Mundt 2000; Leung et al. 2003). An astounding model is the fruitful control of rice impact sickness exhibited by the enormous scope field to explore different avenues regarding blended planting of conventional and current rice assortments (Zhu et al. 2000). This model validated that "intraspecific harvest enhancement gives an environmental way to deal with infectious prevention that can be profoundly compelling over a huge territory and add to the supportability of yield creation" (Zhu et al. 2000). Rice (Oryza sativa) provides staple food to nearly one-third of the world's population and accounts for a large portion of the world's cereal yield (Lu and Snow 2005). Along these lines, the high yielding and supportable creation of rice is basic for the food security in this country. In any case, the practical rice creation is undermined by different parasite diseases, especially by the rice blast disease, which "spreads through various patterns of agamic conidiophores creation during the cropping season, causing necrotic spots on leaves and necrosis of panicles" (Zhu et al. 2000). Rice is also one of the most common food crops in China, and it is consumed throughout the country, despite its significant social qualities such as alcohol production (Luo et al. 2008) and strict social administrations (Zeng et al. 2012). Rice impact is the significant infection of the rice crop in many rice planting locales. Measurable information recommends that 10–20% of rice yield misfortunes are brought about by the serious assaults of rice blast disease in China (Sun et al. 1998; Liu et al. 2004). The generally utilized techniques to control rice impact illness are synthetic controls (Liu et al. 2004; Sun et al. 1998; Wen et al. 2013), which causes significant contamination in the rice biological systems and expands the expenses for rice production (Yang et al. 2012). Nonetheless, Zhu et al. (2000, 2003a) achieved a 94% reduction in rice blast disease and an 89% increase in grain yield for infection-resistant traditional rice varieties solely by combining these varieties with disease-resistant modern rice varieties. The methodology of Zhu et al. (2000, 2003a) utilizing the environmental strategy for intraspecific variety may give an effective choice to control these diseases in rice, notwithstanding its qualities for in situ protection of rice hereditary assets (Zhu et al. 2003b).

9.4.8 Economic Benefits

Small farmers establish over 70% of our cultivating local area in the Punjab territory, and their property possessions are persistently contracting, which clearly recommends that the arrangement of intercropping is their solitary wagered to guarantee effective usage of their assets for expanded production and family pay. Rice (Oryza sativa L.) being a staple food of the large numbers of individuals is the second significant wellspring of procuring unfamiliar trade after cotton in Pakistan. Consequently, its part in the financial improvement of Pakistan cannot be disregarded. As of now, rice is developed on an area of 2.62 million hectares in Pakistan with absolute yearly production of 5.54 million tones giving a normal yield of 2110 kg ha^{-1} (Govt. of Pakistan 2007). Maize and other forage legumes such as sesbania, cowpea, pigeon pea, ricebean, and mung bean are important short-term summer crops that provide farmers with more monetary return (Ahmad et al. 2007; Iqbal et al. 2006). The region under these crops cannot be expanded as they contend with rice, a significant Kharif grain of Pakistan. In addition, during sweltering summer months, these cereals help to keep up animal health and milk production other than improving soil productivity through biological nitrogen fixation (BNF) and adequate organic manure (Wahla et al. 2010). One strategy for increasing summer forage production is to cultivate these crops as a quest in conjunction with upland rice.

9.5 Conclusion

Rice crop provides food for millions of the people around the world and mainly grown in Asia in rice–wheat and rice–maize rotation. The share of legume-based rotation in total rice production is minor. The rice–wheat and rice–maize rotations are exhaustive and degrading the soil due to alternate wetting and drying practices. Even then, these are only rotations which are contributing major share in global food production to overcome the food crisis in the world. The rice is being intercropped with other legumes and fiber crops to bring diversity to farming system, improving resource use efficiency, and ultimately farmers' income. Intercropping rice is not extensively grown, but it can be potential tool to address the issue of low-income and poor ecosystem services.

References

Aggarwal PK, Garrity DP, Liboon SP, Morris RA (1992) Resource-use and plant interaction in a rice-mungbean intercrop. Agron J 84(01):71–78

Ahmad S, Hasanuzzaman M (2012) Integrated effect of plant density, N rates and irrigation regimes on the biomass production, N content, PAR use efficiencies and water productivity of rice under irrigated semiarid environment. Not Bot Horti Agrobot Cluj-Napoca 40(1):201–211

Ahmad S, Hasanuzzaman M (2020) Cotton production and uses. Springer Nature Singapore Pvt Ltd, p 641

Ahmad N, Saeed M (1998) Resource-use efficiency of some wheat-based intercropping systems at different patterns of wheat plantation. Pak J Agric Res 35(1–4):52–54

Ahmad AH, Ahmad R, Ahmad V, Tanveer A (2007) Performance of forage sorghum intercropped with forage legumes under different planting patterns. Pak J Bot 39(2):431–439

Ahmad S, Zia-ul-Haq M, Ali H, Shad SA, Ammad A, Maqsood M, Khan MB, Mehmood S, Hussain A (2008) Water and radiation use efficiencies of transplanted rice (*Oryza sativa* L.) at different plant densities and irrigation regimes under semi-arid environment. Pak J Bot 40(1): 199–209

Ahmad S, Ahmad A, Zia-ul-Haq M, Ali H, Khaliq T, Anjum MA, Khan MA, Hussain A, Hoogenboom G (2009) Resources use efficiency of field grown transplanted rice (*Oryza sativa* L.) under irrigated semiarid environment. J Food Agric Environ 7(2):487–492

Ahmad S, Ahmad A, Soler CMT, Ali H, Zia-Ul-Haq M, Anothai J, Hussain A, Hoogenboom G, Hasanuzzaman M (2012) Application of the CSM-CERES-Rice model for evaluation of plant density and nitrogen management of fine transplanted rice for an irrigated semiarid environment. Precis Agric 13(2):200–218

Ahmad S, Ahmad A, Ali H, Hussain A, Garcia y Garcia A, Khan MA, Zia-Ul-Haq M, Hasanuzzaman M, Hoogenboom G (2013) Application of the CSM-CERES-Rice model for evaluation of plant density and irrigation management of transplanted rice for an irrigated semiarid environment. Irrig Sci 31(3):491–506

Ahmad A, Ashfaq M, Rasul G, Wajid SA, Khaliq T, Rasul F, Saeed U, Rahman MH, Hussain J, Baig IA, Naqvi AA, Bokhari SAA, Ahmad S, Naseem W, Hoogenboom G, Valdivia RO (2015) Impact of climate change on the rice–wheat cropping system of Pakistan. In: Hillel D, Rosenzweig C (eds) Handbook of climate change and agro-ecosystems: the agricultural modeling intercomparison and improvement project (AgMIP) integrated crop and economic assessments. Imperial College Press and The American Society of Agronomy, London, pp 219–258

Ahmad S, Abbas G, Ahmed M, Fatima Z, Anjum MA, Rasul G, Khan MA, Hoogenboom G (2019) Climate warming and management impact on the change of rice-wheat phenology in Punjab, Pakistan. Field Crop Res 230:46–61

Ahmed M, Ahmad S (2019) Carbon dioxide enrichment and crop productivity. In: Hasanuzzaman M (ed) Agronomic crops, vol 2. Springer, Singapore, pp 31–46

Ahmed M, Ahmad S (2020) Systems modeling. In: Ahmed M (ed) Systems modeling. Springer, Singapore, pp 1–44

Ahmed M, Fayyaz-ul-Hassan, Ahmad S (2017) Climate variability impact on rice production: adaptation and mitigation strategies. In: Ahmed M, Stockle C (eds) Quantification of climate variability, adaptation and mitigation for agricultural sustainability. Springer, Switzerland, pp 91–111

Ahmed M, Ahmad S, Raza MA, Kumar U, Ansar M, Shah GA, Parsons D, Hoogenboom G, Palosuo T, Seidel S (2020a) Models calibration and evaluation. In: Ahmed M (ed) Systems modeling. Springer, Singapore, pp 151–178

Ahmed M, Ahmad S, Waldrip HM, Ramin M, Raza MA (2020b) Whole farm modeling: a systems approach to understanding and managing livestock for greenhouse gas mitigation, economic viability and environmental quality. In: Waldrip HM, Pagliari PH, He Z (eds) Animal Manure. ASA Special Publication 67. American Society of Agronomy and Soil Science Society of America, Madison, WI, pp 345–371

Akanvou R, Bastiaans L, Becker M (2002) Evaluating the use of two contrasting legume species as relay intercrop in upland rice cropping systems. Field Crop Res 74(1):23–36

Akram R, Turan V, Hammad HM, Ahmad S, Hussain S, Hasnain A, Maqbool MM, Rehmani MIA, Rasool A, Masood N, Mahmood F, Mubeen M, Sultana SR, Fahad S, Amanet K, Saleem M, Abbas Y, Akhtar HM, Hussain S, Waseem F, Murtaza R, Amin A, Zahoor SA, Sami ul Din M, Nasim W (2018) Fate of organic and inorganic pollutants in paddy soils. In: Hashmi MZ, Varma A (eds) Environmental pollution of paddy soils. Springer, Switzerland, pp 197–214

Akram R, Fahad S, Masood N, Rasool A, Ijaz M, Ihsan MZ, Maqbool MM, Ahmad S, Hussain S, Ahmed M, Kaleem S, Sultana SR, Mubeen M, Saud S, Kamran M, Nasim W (2019) Plant growth and morphological changes in rice under abiotic stress. In: Hasanuzzaman M, Fujita M, Nahar K, Biswas JK (eds) Advances in rice research for abiotic stress tolerance. Woodhead Publishing, Swaston, pp 69–85

Ali MY, Waddington SR, Timsina J, Hodson DP, Dixon J (2009) Maize-rice cropping systems in Bangladesh: status and research needs. J Agric Sci Technol 3(6):35–53

Ali RI, Awan TH, Ahmad M, Saleem MU, Akhtar M (2012) Diversification of rice-based cropping systems to improve soil fertility, sustainable productivity and economics. J Anim Plant Sci 22(1):108–112

Allen VG, Baker MT, Segarra E, Brown CP (2007) Integrated irrigated crop–livestock system in dry climates. Agron J 99:346–360

Andrews DJ, Kassam AH (1976) Importance of multiple cropping in increasing world food supplies. In: Papendick RT, Sachez PA, Triplet GB (Eds.), Multiple cropping, pp. 1–10

Atique-ur-Rehman AH, Sarwar N, Ahmad S, Farooq O, Nahar K (2020) Cotton-based intercropping systems. In: Ahmad S, Hasanuzzaman M (eds) Cotton production and uses. Springer, Singapore, pp 321–340

Bhattacharyya P, Roy KS, Neogi S, Chakravorti SP, Behera KS, Das KM, Rao KS (2012) Effect of long-term application of organic amendment on C storage in relation to global warming potential and biological activities in tropical flooded soil planted to rice. Nutr Cycl Agroecosyst 94(2–3):273–285

Bhatti IH, Ahmad R, Jabbar A, Nazir MS, Mahmood T (2006) Competitive behaviour of component crops in different sesame-legume intercropping systems. Int J Agric Biol 8(2):165–167

Borghi E, Crusciol CAC, Mateus GP, Nascente AS, Martins PO (2013) Intercropping time of corn and palisadegrass or guineagrass affecting grain yield and forage production. Crop Sci 53:629–636

Borghi E, Crusciol CAC, Trivelin PCO, Nascente AS, Costa C, Mateus GP (2014) Nitrogen fertilization (15NH4NO3) of palisadegrass and residual effect on subsequent no-tillage maize. R Bras Ci Solo 38:1457–1468

Crusciol CAC, Nascente AS, Mateus GP, Borghi E, Leles EP, Santos NCB (2013) Effect of intercropping on yields of corn with different relative maturities and palisadegrass. Agron J 105:599–606

Crusciol CAC, Nascente AS, Mateus GP, Pariz CM, Martins PO, Borghi E (2014) Intercropping soybean and palisade grass for enhanced land use efficiency and revenue in a no till system. Eur J Agron 58:53–62

Crusciol CAC, Nascente AS, Borghi E, Soratto RP, Martins PO (2015) Improving soil fertility and crop yield in a tropical region with palisadegrass cover crops. Agron J 107:2271–2280

Crusciol CAC, Marques RR, Carmeis FACA, Soratto RP, Costa CHM, Ferrari NJ (2016) Annual crop rotation of tropical pastures with no-till soil as affected by lime surface application. Eur J Agron 80:88–104

Dawe D, Frolking S, Li C (2004) Trends in rice–wheat area in China. Field Crop Res 87:89–95

Dean RA, Talbot NJ, Ebbole DJ, Farman ML, Mitchell TK, Orbach MJ, Thon M, Kulkarni R, Xu JR, Pan H, Read ND, Lee YH, Carbone L, Brown D, Oh YY, Donofrio N, Jeong JS, Soanes DM, Djonovic S, Kolomiets E (2005) The genome sequence of the rice blast fungus Magnaporthe grisea. Nature 434:980–986

Dordas C (2008) Role of nutrients in controlling plant diseases in sustainable agriculture: a review. Agron Sustain Dev 28:33–46

Emerick K, Ronald PC (2019) Sub1 rice: engineering rice for climate change. Cold Spring Harbor Perspect

Fageria NK (1998) "Manejo da Calagem e Adubação do Arroz," in Tecnologias Para Arroz de Terras Altas. Santo Antonio de Goiás: EMBRAPA-CNPAF, F Breseghello, LF Stone (eds) (Embrapa Arroz e Feijao). pp 67–78

Fageria NK, Moreira A, Coelho AM (2011) Yield and yield components of upland rice as influenced by nitrogen sources. J Plant Nutr 34:361–370

Fahad S, Noor M, Adnan M, Khan MA, Rahman I, Alam M, Khan IA, Ullah H, Mian IA, Hassan S, Saud S, Bakhat HF, Hammad HM, Ahmad S, Nasim W (2019) Abiotic stress and rice grain quality. In: Hasanuzzaman M, Fujita M, Nahar K, Biswas JK (eds) Advances in rice research for abiotic stress tolerance. Woodhead Publishing, Swaston, pp 571–583

Fatima Z, Ahmad M, Hussain M, Abbas G, Ul-Allah S, Ahmad S, Ahmed N, Ali MA, Sarwar G, Ehsan ul Haque, Iqbal P, Hussain S (2020) The fingerprints of climate warming on cereal crops phenology and adaptation options. Sci Rep 10:18013

Food and Agriculture Organization of the United Nations FAO (2017) Integrated Crop-Livestock Systems (ICLS)

Francis CA (1989) Biological efficiencies in multiple cropping systems. Adv Agron 42:1–42

Franzluebbers AJ, Stuedemann JA (2014) Crop and cattle production responses to tillage and cover crop management in an integrated crop– livestock systemin the southeastern USA. Eur J Agron 57:62–70

Garcia CMP, Costa C, Andreotti M, Meirelles PRL, Pariz CM, Freitas LA (2016) Wet and dry maize yield under intercrop culivation with marandu grass and/or dwarf pigeon pea and nutritional value of the marandu grass in succession. Aus J Crop Sci 10:1564–1571

Garrett KA, Mundt CC (2000) Host diversity can reduce potatolate blight severity for focal and general patterns of primary inoculum. Phytopathology 90:1307–1312

Ghanbari A, Lee HC (2002) Intercropped field beans (Vicia faba) and wheat (Triticum aestivum) for whole crop forage: effect of nitrogen on forage yield and quality. J Agric Sci 138:311–314

Gill MS, Shukla AK, Pandey PS (2008) Yield, nutrient response and economic analysis of important cropping systems in India. Indian J Fertilizers 4:11–48

Government of Pakistan (2007) Economic Survey of Pakistan 2006–2007. Ministry of Science and Technology, Government of Pakistan

Hafeez-ur-Rehman A, Nawaz A, Awan MI, Ijaz M, Hussain M, Ahmad S, Farooq M (2019) Direct seeding in rice: problems and prospects. In: Hasanuzzaman M (ed) Agronomic crops: production technologies, vol 1. Springer, Singapore, pp 199–222

Hussain M, Ahmad S, Hussain S, Lal R, Ul-Allah S, Nawaz A (2018) Rice in saline soils: physiology, biochemistry, genetics and management. Adv Agron 148:231–287

Iqbal I, Ayub M, Zaman H, Ahmad R (2006) Impact of nutrient management and legume association on forage qualitative traits of maize forage. Pak J Bot 38(4):1079–1084

Jat RA, Dungrani RA, Arvadia MK, Sahrawat KL (2012) Diversification of rice (*Oryza sativa* L.)-based cropping systems for higher productivity, resource-use efficiency and economic returns in south Gujarat, India. Arch Agron Soil Sci 58(6):561–572

Jeyabal A, Kuppuswamy G (2001) Recycling of organic wastes for the production of vermicompost and its response in rice-legume cropping systems and soil fertility. Heriana J Agron 15:153–170

Khan MA, Ahmad S, Raza A (2019a) Integrated weed management for agronomic crops. In: Hasanuzzaman M (ed) Agronomic crops, vol 2. Springer, Singapore, pp 257–281

Khan SM, Ali S, Nawaz A, Bukhari SAH, Ejaz S, Ahmad S (2019b) Integrated pest and disease management for better agronomic crop production. In: Hasanuzzaman M (ed) Agronomic crops, vol 2. Springer, Singapore, pp 385–428

Kluthcouski J, Fancelli AL, Dourado N, Ribeiro D, Ferraro LA (2000) Yield of soybean, corn, common bean and rice under no-tillage management. Sci Agric 57:97–104

Letourneau DK, Rivera BS, Lerma JM, Carmona EJ, Daza MC, Escobar S, Trujillo AR (2011) Does plant diversity benefit agroecosystems? A synthetic review. Ecol Appl 21(1):9–21

Leung H, Zhu Y, Revilla MI, Fan JX, Chen H, Pangga I, Cruz CV, Mew TW (2003) Using genetic diversity to achieve sustainable rice disease management. Plant Dis 87:1156–1169

Li YF, Zhang R, Sun S, Xu G (2009) Facilitated legume nodulation, phosphate uptake and nitrogen transfer by arbuscular inoculation in an upland rice and mung bean intercropping system. Plant Soil 315(1–2):285–296

Lin QC, Feng XY, Huang W, Cai LJ, Wang YX, Xu BQ, Dai YL (2014) Research progress on application of Ipomoea aquatica floating bed in eutrophication water. Meteorol Environ Res 5(10):52–55

Liu GQ, Meng ZH, Ren YJ, Li CG, Liu YW, Meng QX (2004) Study advances and countermeasures on blast resistance of rice. Chin Agric Sci Bull 20:211–214

Lou YG, Zhang WQ, Hu Y, Zhang J (2013) Biological control of rice insect pests in China. Biol Control 67(1):8–20

Lu BR, Snow AA (2005) Gene flow from genetically modified rice and its environmental consequences. Bioscience 55:669–678

Luo T, Fan WL, Xu Y (2008) Characterization of volatile and semi-volatile compounds in Chinese rice wines by headspace solid phase microextraction followed by gas chromatography-mass spectrometry. J Inst Brew 114:172–179

Mahajan A, Gupta RD (2009) The rice–wheat cropping system. In: Mahajan A, Gupta RD (eds) Integrated nutrient management (INM) in a sustainable rice—wheat cropping system. Springer, Dordrecht. https://doi.org/10.1007/978-1-4020-9875-8_7

Maingi MJ, Shisanya AC, Gitonga MN, Hornetz B (2001) Nitrogen fixation by common bean (Phaseolus vulgaris L.) in pure and mixed stands in semi-arid South East Kenya. Eur J Agron 14:1–12

Manu-Aduening JA (1999) Comparative assessment of component crop performance in a maize-FABA intercrop. In: Strategy for sustainable maize production in west and central Africa, Proceedings of a regional maize workshop

Manu-Aduenning JA, Boa-Amponsem K (2005) Influence of spatial arrangements on performance of a yam-maize-pepper intercrop. Ghana J Agric Sci 1:29–35

Mateus GP, Crusciol CAC, Pariz CM, Borghi E, Costa C, Martello JM (2016) Sidedress nitrogen application rates to sorghum intercropped with tropical perennial grasses. Agron J 108:433–447

Mateus GP, Crusciol CAC, Pariz CM, Costa NR, Borghi E, Costa C (2020) Corn intercropped with tropical perennial grasses as affected by sidedress nitrogen application rates. Nutr Cycl Agroecosyst 1:1–22

Meetei TT, Kundu MC, Devi YB (2020) Long-term effect of rice-based cropping systems on pools of soil organic carbon in farmer's field in hilly agroecosystem of Manipur, India. Environ Monit Assess 192(4):1–17

Moraes A, Carvalho PCF, Crusciol CAC, Lang CR, Pariz CM, Deiss L (2019) Integrated crop-livestock systems as a solution facing the destruction of Pampa and Cerrado biomes in South America by intensive monoculture systems. Agroecosyst Div 1:257–273

Morris RA, Garrity A (1993) Resource capture and utilization in intercropping: non-nitrogen nutrients. Field Crop Res 34:319–334

Nascente AS, Crusciol CAC, Cobucci T (2013) The no-tillage system and cover crops—alternatives to increase upland rice yields. Eur J Agron 45:134–131

Naz S, Fatima Z, Iqbal P, Khan A, Zakir I, Noreen S, Younis H, Abbas G, Ahmad S (2019) Agronomic crops: types and uses. In: Hasanuzzaman M (ed) Agronomic crops: production technologies, vol 1. Springer, Singapore, pp 1–18

Nazir MS, Elahi E, Jabbar A, Saeed M, Ahmad R (1997) Bio-economic assessment of different wheat-based intercropping systems. Pak J Agric Sci 34(1–4):62–64

Nazir MS, Jabbar A, Ahmad I, Nawaz S, Bhatti IH (2002) Production potential and economics of intercropping in autumn-planted sugarcane. Int J Agric Biol 4(1):140–142

Onwueme IC, Sinha TD (1991) Field crop production in tropical Africa. The Technical Centre for Agricultural and Rural Cooperation ACP-EU (CTA), Wageningen, The Netherlands

Palti J (1981) Cultural practices and infectious crop diseases. Springer-Verlag, Berlin, Heidelberg

Pariz CM, Costa C, Crusciol CAC, Meirelles PRL, Castilhos AM, Andreotti M (2016) Production and soil responses to intercropping of forage grasses with corn and soybean silage. Agron J 108: 1–13

Pariz CM, Costa C, Crusciol CAC, Meirelles PRL, Castilhos AM, Andreotti M (2017) Production, nutrient cycling and soil compaction to grazing of grass companion cropping with corn and soybean. Nutr Cycl Agroecosyst 108:35–54

Patra AP, Chatterjee (1986) Intercropping of soybean with rice, maize and pigeonpea in the plains of West Bengal. Indian J Agric Sci 56(6):413–417

Peerzada AM, Bukhari SAH, Dawood M, Nawaz A, Ahmad S, Adkins S (2019) Weed management for healthy crop production. In: Hasanuzzaman M (ed) Agronomic crops, vol 2. Springer, Singapore, pp 225–256

Porpavai S, Devasenapathy P, Siddeswaran K, Jayaraj T (2011) Impact of various rice based cropping systems on soil fertility. J Cereals Oilseeds 2(3):43–46

Razaq M, Shah FM, Ahmad S, Afzal M (2019) Pest management for agronomic crops. In: Hasanuzzaman M (ed) Agronomic crops, vol 2. Springer, Singapore, pp 365–384

Rehman AU, Ali H, Sarwar N, Ahmad S, Farooq O, Nahar K, Hasanuzzaman M (2020) Cotton-based intercropping systems. In: Ahmad S, Hasanuzzaman M (eds) Cotton production and uses. Springer, Singapore. https://doi.org/10.1007/978-981-15-1472-2_17

Ren LX, Yang X, Xu Y, Huang Q, Shen Q (2008) Intercropping with aerobic rice suppressed Fusarium wilt in watermelon. Soil Biol Biochem 40(3):834–844

Risch SJ, Andow D, Altieri MA (1983) Agro-ecosystem diversity and pest control: Data, tentative conclusions, and new research directions. Environ Entomol 12:625–629

Rouanet G (1995) The tropical agriculturalist series 1: maize. Macmillan, London, UK

Saeed MA, Ullah R, Ahmad, Jabbar A (1999) Bio-economic assessment of direct seeded rice-based intercropping systems under strip plantation. Pak J Biol Sci 2(3):980–983

Sarwar N, Ali H, Ahmad S, EhsanUllah AS, Mubeen K, Hill JE (2013a) Water wise rice cultivation on calcareous soil with the addition of essential micronutrients. J Anim Plant Sci 23(1):244–250

Sarwar N, Ali H, Maqsood M, EhsanUllah, Shahzad AN, Shahzad M, Mubeen K, Shahid MA, Ahmad S (2013b) Phenological response of rice plants to different micronutrients application under water saving paddy fields on calcareous soil. Turk J Field Crops 18(1):52–57

Schulthess F, Chabi-Olaye A, Gounou S (2004) Multi-trophic level interactions in a cassava-maize mixed cropping system in the humid tropics of West Africa. Bull Entomol Res 94(3):261–272

Shahzad AN, Ahmad S (2019) Tools and techniques for nitrogen management in cereals. In: Hasanuzzaman M (ed) Agronomic crops, vol 2. Springer, Singapore, pp 111–126

Showler AT, Greenberg SM (2003) Effects of weeds on selected arthropod herbivore and natural enemy populations, and on cotton growth and yield. Environ Entomol 32(1):39–50

Silva EC, Muraoka T, Bastos AVS, Franzin VI, Buzetti S, Soares FAL (2020) Biomass and nutrient accumulation by cover crops and upland rice grown in succession under no-tillage system as affected by nitrogen fertilizer rate. J Crop Sci Biotechnol 23:117–126

Singh G, Singh OP, Kumar M, Rajput AL, Murya S (1996) Effect of intercrops on yield and economics, weeds and pest infestation of deep water rice. Ann Agric Res 17(1):14–17

Singh S, Gupta RK, Bhushan L, Rao AN, Sivaprasad B (2007) Evaluation of mulching, intercropping with Sesbania and herbicide use for weed management in dry-seeded rice (Oryza sativa L.). Crop Prot 26(4):518–524

Sun GC, Du XF, Tao RX, Sun SY (1998) Control tactics and prospect of blast research in 21th century. Acta Phytopathol Sinica 28:289–292

Surve VH, Arvadia MK (2011) Performance of fodder sorghum (Sorghum bicolour L.), maize (Zea mays L.) and cowpea (Vigna unguiculata (L.) Walp.) under sole and intercropping systems. Int J Agric Res 2:28–31

Tariq M, Ali H, Hussain N, Nasim W, Mubeen M, Ahmad S, Hasanuzzaman M (2019) Fundamentals of crop rotation in agronomic management. In: Hasanuzzaman M (ed) Agronomic crops. Springer, Singapore. https://doi.org/10.1007/978-981-32-9151-5_24

Tariq M, Ahmed M, Iqbal P, Fatima Z, Ahmad S (2020) Crop phenotyping. In: Ahmed M (ed) Systems modeling. Springer, Singapore, pp 45–60

Timsina J, Buresh RJ, Dobermann A, Dixon J, Tabali J (2010) Strategic assessment of rice-maize systems in Asia. IRRI CIMMYT Alliance Project "Intensified Production Systems in Asia (IPSA)", IRRI-CIMMYT Joint Report, IRRI, Los Banos, Philippines

Trenbath BR (1986) Resource use efficiency by intercrops. In: Franic CA (ed) Multiple cropping system. MacMillan Pub. Co, New York, pp 57–81

Umar KJ, Hassan LG, Dangoggo SM, Ladan MJ (2007) Nutritional composition of water spinach (Ipomoea aquatica Forsk.) leaves. J Appl Sci 7(6):803–809

Wahla IHR, Ahmad, Ehsanullah, Ahmad A, Jabbar A (2010) Competitive functions of component crops in some barley based intercropping systems. Int J Agric Biol 11:69–72

Wasaya A, Yasir TA, Ijaz M, Ahmad S (2019) Tillage effects on agronomic crop production. In: Hasanuzzaman M (ed) Agronomic crops, vol 2. Springer, Singapore, pp 73–99

Wen XH, Xie MJ, Jiang J, Yang B, Shao YL, He W, Liu L, Zhao Y (2013) Advances in research on control method of rice blast. Chin Agric Sci Bull 29:190–195

Willey RW (1979) Intercropping: its importance and research needs. 1. Competition and yield advantages. Field Crop Abstract 32:2–10

Willey RW (1990) Resource use in intercropping systems. Agric Water Manag 17(1):215–231

Wood SA, Mendelsohn RO (2014) The impact of climate change on agricultural net revenue: a case study in the Fouta Djallon, West Africa. Environ Dev Econ 20:20–36

Yadav RL (1996) Cropping systems. In: 50 years of crop science research in India. ICAR Publications. pp 117–120

Yadav GSA, Lal R, Babu S, Datta M, Meena RS, Patil SB (2019) Impact of no-till and mulching on soil carbon sequestration under rice (Oryza sativa L.)-rapeseed (Brassica campestris L. var. rapeseed) cropping system in hilly agro-ecosystem of the Eastern Himalayas, India. Agric. Ecosyst Environ 275:81–92

Yang J, Shi ZF, Gao D, Liu L, Zhu YY, Li CY (2012) Mechanism on biodiversity managing crop diseases. Hereditas 34:1390–1398

Zahoor SA, Ahmad S, Ahmad A, Wajid A, Khaliq T, Mubeen M, Hussain S, Sami Ul Din M, Amin A, Awais M, Nasim W (2019) Improving water use efficiency in agronomic crop production. In: Hasanuzzaman M (ed) Agronomic crops, vol 2. Springer, Singapore, pp 13–29

Zeng YW, Du J, Pu XY, Luo X, Yang SM, Yang T (2012) Relationship between rice cultural diversity and ecological environment in Yunnan Province of China. Agric Sci Technol 13: 2247–2256

Zhang WJ, Jiang FB, Ou JF (2011) Global pesticide consumption and pollution: with China as a focus. Proc Int Acad Ecol Environ Sci 1(2):125–144

Zhu YY, Chen HR, Fan JH, Wang YY, Li Y, Chen JB, Fan JX, Yang SS, Hu LP, Leung H, Mew TW, Teng PS, Wang ZH, Mundt CC (2000) Genetic diversity and disease control in rice. Nature 406:718–722

Zhu YY, Chen HR, Fan JH, Wang YY, Li Y, Fan JX, Yang SS, Ma GL, Chen JB, Li ZS, Lu BR (2003a) The use of rice varietal diversity for rice blast control. Agric Sci China 2:400–408

Zhu YY, Wang YY, Chen HR, Lu BR (2003b) Conserving traditional rice through biodiversity management. Bioscience 53:158–162

Rice Ontogeny 10

Muhammad Tariq, Zeeshan Ahmed, Muhammad Habib Ur Rehman, Feng Ling Yang, Muhammad Hayder Bin Khalid, Muhammad Ali Raza, Muhammad Jawad Hassan, Tehseen Ahmad Meraj, Ahsin Khan, Atta Mohi Ud Din, Nasir Iqbal, and Shakeel Ahmad

Abstract

This chapter deals with morphological and metabolic changes in rice during its life span. Generally, crop development is completed in a sequence of events and characterized by the appearance of various organs in relation to crop stage. However, metabolic changes also occur in the background, and profiling these

M. Tariq
Central Cotton Research Institute, Multan, Pakistan

Department of Agronomy, Bahauddin Zakariya University, Multan, Pakistan

Z. Ahmed
Xinjiang Institute of Ecology and Geography, Chinese Academy of Sciences, Urumqi, Xinjiang, China

M. H. U. Rehman
Department of Agronomy, MNS-University of Agriculture, Multan, Pakistan

Institute of Crop Sciences and Resource Conservation (INRES), Crop Science Group, University Bonn, Bonn, Germany

F. L. Yang · M. H. B. Khalid
College of Agronomy, Sichuan Agricultural University, Chengdu, China

Sichuan Engineering Research Center for Crop Strip Intercropping System, Key Laboratory of Crop Ecophysiology and Farming System in Southwest China, Sichuan Agricultural University, Chengdu, China

M. A. Raza
College of Agronomy, Sichuan Agricultural University, Chengdu, China

Sichuan Engineering Research Center for Crop Strip Intercropping System, Key Laboratory of Crop Ecophysiology and Farming System in Southwest China, Sichuan Agricultural University, Chengdu, China

The Islamia University Bahawalpur, Bahawalpur, Pakistan

N. Sarwar et al. (eds.), *Modern Techniques of Rice Crop Production*,
https://doi.org/10.1007/978-981-16-4955-4_10

metabolites at various developmental stages is necessary for better understanding of morphology. The physiological and biochemical alterations are followed by morphological changes. In germination process, the macromolecules like starch, lipid, and protein are broken down into simple molecules. The nutrient uptake gradually increases with crop age, reaches peak, and then declines. The light duration stimulates the expression of genes responsible for biosynthesis of florigens to initiate flowering in short day length. The changes in metabolic concentrations in developing seeds are more prominent within 30 days of pollination. The hormone involvement in developing seed changes according to their associated role in seed development process. The hormones responsible for cell division are found during early seed development. Consequent upon these regulations and various metabolic pathways during germination and seed development, the seed the seed has been an important aspect of ontogenic studies. The compounds synthesize in seed development diminishes during seed germination.

Keyword

Biosynthesis · Development · Hormones · Regulation

10.1 Introduction

Multiple changes in the metabolic profile of the crop happen during growth and development (Ahmed et al. 2017, Ahmed et al. 2020a, b; Akram et al. 2018, 2019; Hussain et al. 2018; Fahad et al. 2019; Fatima et al. 2020). The morphological changes in various growth periods result in response to internal changes with regard to biochemistry, synthesis, and translocation of various metabolites and gene expressions. The changes in environmental conditions like temperature, light, and day length (photoperiod) are some of the key stimulants for activation of biosynthesis of the various molecules and hormones. The flower initiation in rice in response to short day length is one of the crop responses to environment changes (Hafeez-ur-Rehman et al. 2019; Khan et al. 2019a, b; Naz et al. 2019; Peerzada et al. 2019; Razaq et al. 2019; Sarwar et al. 2013a, b; Tariq et al. 2020a, b; Wasaya et al. 2019; Zahoor et al. 2019). The genes responsible for biosynthesis of metabolites are

M. J. Hassan
Shanghai Jiaotong University, Jiaotong, People's Republic of China

T. A. Meraj · A. Khan · A. M. U. Din
Sichuan Agricultural University, Yaan, People's Republic of China

N. Iqbal
School of Agriculture, Food and Wine, University of Adelaide, Waite, Australia

S. Ahmad (✉)
Department of Agronomy, Faculty of Agricultural Sciences and Technology, Bahauddin Zakariya University, Multan, Pakistan
e-mail: shakeelahmad@bzu.edu.pk

expressed in response to such changes. The morphological changes are classified as seedling stage, panicle initiation, and maturity (Ahmad et al. 2019). These are further subdivided into minor morphological changes, for example, inflorescence initiation is completed in nine further divisions (Ahmad et al. 2015; Ahmad et al. 2019; Ahmad et al. 2012, 2013; Ahmad et al. 2008, Ahmad et al. 2009a, b; Ahmad and Hasanuzzaman 2012; Ahmed and Ahmad 2019, 2020; Shahzad and Ahmad 2019).

The storage compounds in the germinating seeds are consumed to provide food at initial seedling growth. The emerging leaves synthesize food after germination process is over. The root initiation is completed for water, nutrients, and to provide support to plants. The main changes occur with the transformation of vegetative growth to reproductive in which process of seed development is completed. The dynamic changes with regard to seed water contents, hormones, and other metabolites occur in the process of seed development from fertilization to seed maturity (Wu et al. 2016; Dhatt et al. 2019; Sharma et al. 2009). Some compounds are translocated to seeds from other organs particularly from leaves, while others are consumed in seed development.

The ontogeny represents the history of structural changes, and the word "ontogeny" belongs to Greek, meaning mode of production (Tariq et al. 2020a, b). The morphogenesis and ontogenesis are other relevant terms which deal with key process of appearance of organs at particular growth stage. During developmental changes, many compounds are translocated, diminished, and added, and many new organs are added. The nutrients uptake and dry matter accumulation change at various growth stages. An overview of developmental changes as well as metabolic profile of rice will be discussed in this chapter.

10.2 Seed Germination

Seed germination is a fundamental phase in plant growth and can be considered as a basis for plant health and yield (Ali and Elozeiri 2017). Germination is marked as commencement of a new lifecycle that involves multifarious processes such as physiological and biochemical reactions. Mobilization of nutrients is from food reservoir of seed (endosperm) toward embryo of germinated seed (Xiong et al. 2021). Plant vigor is totally depending on the health and ability of the plant embryo, embedded inside the seed, to carry out its metabolic activity in a coordinated and sequential manner. The germination stages are illustrated in Fig. 10.1. The very first step of seed germination is "imbibition" in which several metabolic processes as synthesis of hydrolytic enzymes are stimulated. These enzymes hydrolyze the reserve food in endosperm, present in the form of proteins, carbohydrates, lipids, and phytic acid into simple accessible form for embryo uptake as food during germination process (Ali and Elozeiri 2017). Carbohydrates are hydrolyzed by amylase enzyme, and their activity increased gradually in the early days of germination process and soluble sugars are also increased while starch is decreased over time (Wang et al. 1988). In lipid hydrolysis, the ester carboxylate bonds are hydrolyzed by lipase enzyme, and released energy is used for the synthesis of different amino

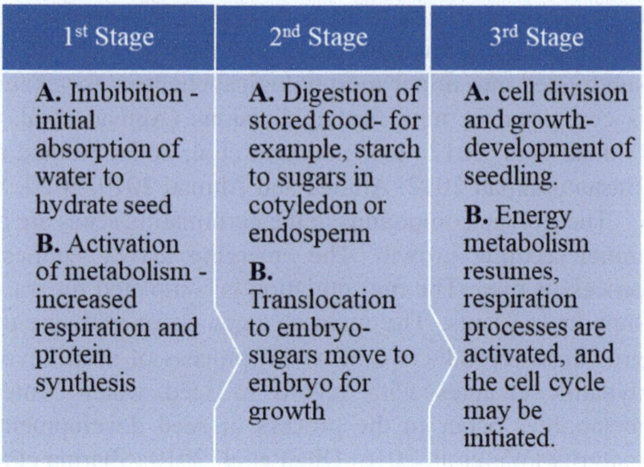

Fig. 10.1 The stages of seed germination process

acids and sugar components (Quettier and Eastmond 2009). Consequent to hydrolysis of lipids, the micromolecules such as fatty acids and organic alcohols are produced (Pereira et al. 2003; Leal et al. 2002). Two different species of lipase were detected, as their activity increased during germination. First is known as Lipase-I, and second is called Lipase-II, named on the basis of their relative mobility. Lipase-II was active during germination. Lipase-I actively participated in mobilization of lipid in germinating seed (Vijayakumar and Gowda 2012).

In germinating seeds, hydrolysis of stored phytin is done by acid phosphatase enzyme known as phytase. Hydrolysis reaction releases inositol, phosphate, and cations that are available for seedling growth (Hubel and Beck 1996). The plant growth-regulating hormones such as gibberellin (GA) and brassinosteroid (BR) and glutelin protein play important roles in regulation of seed germination and post-germination growth of seedling (Xiong et al. 2021). Transcription factor AP2/ERF encoded by SHB gene influences synthesis of gibberellic acid and meristem cells of root and also promotes elongation and proliferation of cells (Li et al. 2015).

10.3 "Omics" Studies of Germinated Seed

Modern approaches like "omics" support the new discoveries that successful germination depends on the quality of the mRNAs. In addition, physical appearance of seedling mainly depends on proteostasis and DNA integrity. It plays a crucial role in metabolic processes and takes part in the signaling pathways of hormones that regulate germination. Previous studies showed that seedling vigor depends on biochemical and molecular reactions that occurred during germination process (Rajjou et al. 2012). Metabolomic studies also play a pivotal role in understanding the metabolism reactions taking inside the germinating seed. Yang et al. (2019) studied the production of metabolites during seed germination, and they observed a total of 730 metabolites in their experiment. Among these metabolites, only 35 responded to low-temperature stress conditions. They also observed that only seven metabolites are involved in metabolism and biosynthesis of glutathione and inositol phosphate and amino acids and phenylpropanoids, respectively.

Furthermore, proteomic studies of germinating seeds revealed that approximately 148 proteins are produced differently during the process of seed germination. These include up-regulating and down-regulating proteins that affect the whole process. Out of 148 proteins, 69 were up-regulated (plus 20 induced proteins) and 63 were known to be down-regulated. Globulin and glutelin are characterized as down-regulated proteins that are mainly considered as storage protein of seed, in addition to a-amylase, the up-regulated protein that participates in glycolysis (Yang et al. 2007).

10.4 Crop Development

The rice growth and development are divided into vegetative (germination-panicle initiation, P1), reproductive (panicle initiation-heading), and grain filling and ripening (heading-maturity) stages (Ahmad et al. 2019). The days to panicle initiation, 50% heading, P1 to 50% heading, R7 and 50% heading to R7 range from 41–50 days, 66–108 days, 25–46 days, 84–125 days, and 12–23 days, respectively. The chlorophyll appearance in internodal distance between the fourth and fifth crown node triggers the transformation of vegetative to reproductive growth. The number of tillers increased during the first 25 days and continues to increase up to 120 days. The vegetative growth determines the overall growth season, and it is less in short maturing varieties and vice versa. However, there are also instances in which both vegetative and reproductive phases have been shortened for the development of early maturing varieties.

The general sequence of seedling development is designated as S0 (un-imbibed seed), S1 (coleoptile emergence), S2 (radicle emergence), and S3 (prophyll emergence from coleoptile). The seedling is supposed to be emerged along with the elongation of mesocotyl (first internode) and are followed by epiblast, a term describing the pushing of the tip of coleoptile from the soil surface. The duration of the first four leaf stages is presented by V1–V4 and is called pre-tillering. It is followed by tillering (V5) which starts from fifth leaf emergence and continues to six-leaf and secondary tiller emergence. The maximum tillering followed the sigmoid-shaped curve, and the duration between active tillering to reproductive growth initiation is known as the vegetative lag phase. The successive growth is included in reproductive growth and lasts for 1 month in most cases. It is characterized by R1 (panicle differentiation), R2 (booting), R3 (heading), R4 (anthesis), R5 (expansion of grains length and width), R6 (soft dough stage), R7 (hard dough stage), R8 (single grain maturity), and R9 (panicle maturity).

The root development is completed in seven stages (Cr1–Cr7), leaf in seven stages (P0–P6), spikelet in eight stages (Sp1–Sp8), and inflorescence in nine stages (In1–In9). The description of these stages is illustrated in Fig. 10.2. The leaves are originated as lateral organs of the shoot apical meristem. The interval between two successive leaves is known as plastochron and is denoted by P.

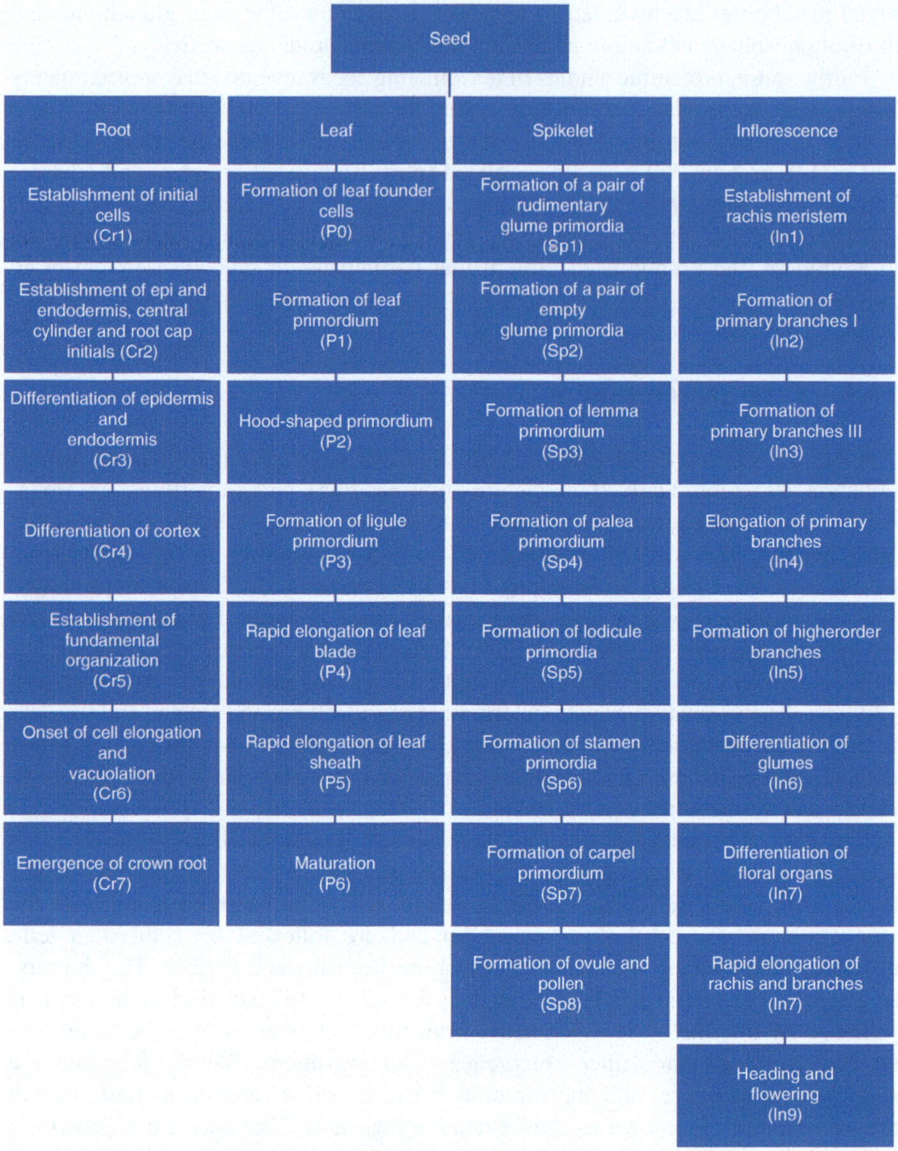

Fig. 10.2 Organs' development stages (Adapted from Itoh et al. 2005)

10.5 Dry Matter Accumulation and Nutrient Uptake

The dry weight is increased continuously; however, stem and leaves are the main parts of aboveground dry matter production at early stages, while spikelets also contribute to later stages. The tillers mainly contribute to shoot biomass, and tillering rate continuously increases up to 50–60 days after emergence followed by decline. All the tillers produced in the late season may not be effective for yield formation. The shoot length increases till full heading with no change at grains development to maturity. The dry matter increases till maturity; however, it may start to decrease from grains development to maturity in some varieties like Koshihikari (Hussain et al. 2014). The leaf area index increased up to 75 days and declined thereafter. The dry matter accumulation showed linear increase over crop age (Ahmad et al. 2009a, b; Ahmad and Hasanuzzaman 2012). The proportionate dry matter accumulation at various growth stages including before elongation (E), elongation to heading (E-H), heading (H)-25 DAH (days after heading), and 25 DAH-maturity (M) was analyzed, and it was concluded that maximum dry matter accumulation occurred during E-H followed by H-25 DAH, before E and 25 DAH-M, respectively (Fig. 10.3). The stem, leaves, and grains contribute 37.1%, 16.1%, and 39.4% in total biological yield at maturity (Puteh and Mondal 2013).

The daily nutrients uptake changes with crop age, and it followed a sigmoid pattern. The uptake of nitrogen, potassium, phosphorus, calcium, magnesium, and sulfur remains at the highest level during 70–90 days after emergence (García y García et al. 2003). The individual uptake pattern over growth period is illustrated in Figs. 10.4 and 10.5. Figure 10.4 shows maximum nitrogen, potassium, and sulfur uptakes that were recorded at 90, 60, and 80 days after emergence. These nutrients are required in large amount than other macronutrients like phosphorus, magnesium, and calcium. The maximum uptake of calcium and magnesium has been observed at 70 days, while phosphorus uptake is high at 60 days after emergence (Fig. 10.5).

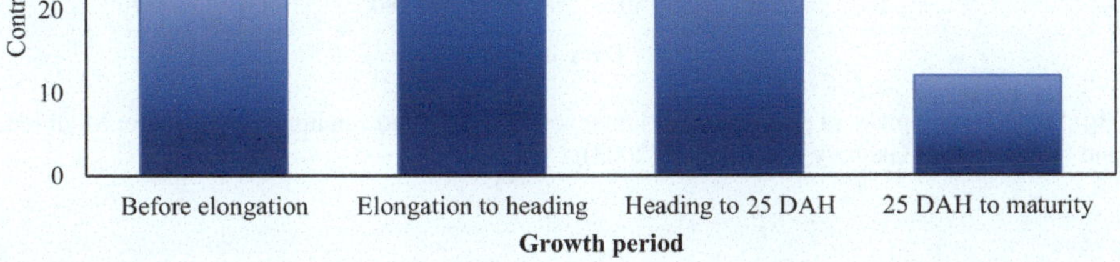

Fig. 10.3 Dry matter accumulation during various growth stages (Avg. of five rice hybrids) (**Source:** Wu et al. 2008)

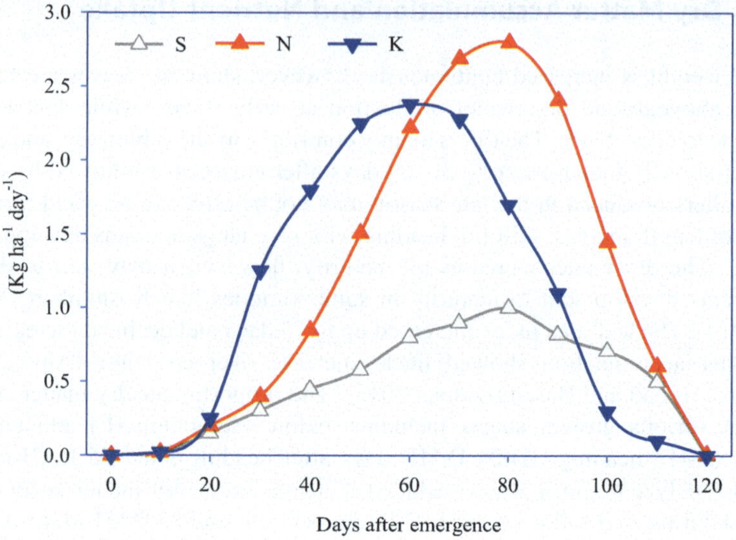

Fig. 10.4 Daily uptake of nitrogen (N), sulfur (S), and potassium (K) (**Source:** Modified and redrawn from García y García et al. (2003))

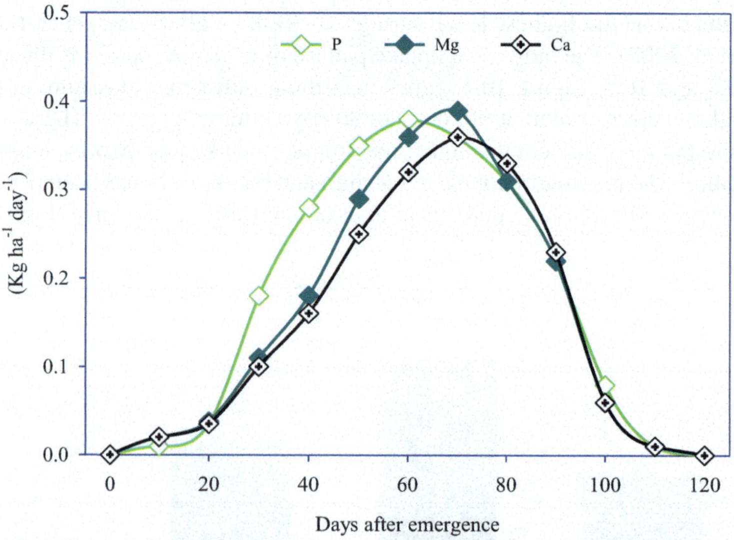

Fig. 10.5 Daily uptake of phosphorus (P), magnesium (Mg), and calcium (Ca) (**Source:** Modified and redrawn from García y García et al. (2003))

10.6 Flower Initiation

The floral initiation indicated the physiological transition from vegetative to reproductive growth which is a critical stage in the plant life cycle. The shoot apical meristem initially produces vegetative structures and then transits to the reproductive structure to initiate flowering. The flowering is triggered by both endogenous and environmental factors, especially light duration (photoperiod) and complex interplay of regulatory process with environment. Considering the crop flowering response is very important for adaptation to various environments. The flowering process is sequenced in inflorescence initiation, flower structure development, heading (panicle extrusion), and anthesis (flower opening) (Yoshida and Nagato 2011). The reproductive development of the rice starts with inflorescence meristem, branch meristem, spikelet meristem, and floral meristem. The regulation of flowering due to photoperiodic conditions involved the production of florigens.

Various studies investigated flowering regulation and its molecular mechanisms. The leaves perceive the day length (photoperiodic) condition and integrate it into the generation of florigens and transmitted to shoot apices (Itoh and Izawa 2013). The accumulation of florigens, that is, Hd3a/RFT1 proteins, induced the initiation of florescence meristem, and many genes are also involved in the maintenance of reproductive development. The Hd3a represents the HEADING DATE 3a and RFT1 denotes RICE FLOWERING LOCUS T 1. Both Hd3a and RFT1 are present on chromosome 6, just 11.5 kb away from each other (Komiya et al. 2008). The FT is synthesized in vascular leaf tissue and transported in phloem tissue. It enters the shoot apical meristem through cells located below the shoot apical meristem.

RFT1 and Hd3a have been regarded as important activators of flowering in long-day and short-day conditions, respectively (Komiya et al. 2009). During short day length, the Hd3a expression is higher than RFT1, indicating Hd3a is a more important florigen for flowering initiation in short day length. It is reduced during long-day conditions (Itoh and Izawa 2013). It was also observed that the accumulation of Hd3a increased gradually and reaches a peak about 1 month before flowering. The short day induced the expression of Hd3a by means of Ehd1 and Ghd7 proteins, and these both set the day length threshold. The blue light participates in the induction of Ehd1 for expression of Hd3a in the morning, and it has been confirmed by blue light pulses at various circadian phases in short and long day conditions (Itoh et al. 2010). It indicates that both morning and blue light are important for the expression of Ehd1, while Ghd7 expression in short days is induced in mid-night responding to light pulse (Itoh et al. 2010). The florigen is also known as flowering hormone due to its characters like mobility and universality in the flowering promotion (Tsuji et al. 2008). The CO (CONSTANS) protein activates the expression of FLOWERING LOCUS T (FT) which is involved in encoding florigenic protein for promoting flowering (An et al. 2004; Tiwari et al. 2010). The circadian clock regulates the expression of CO protein. The exposure to photoperiod shorter than 13.5 h induces the expression of Hd3a and Ehd1 protein (early heading) while reducing the expression of Ghd7 (Shrestha et al. 2014). Moreover, Ghd7, Hd1, Hd5, Hd6, and Hd16 promote inhibition of flowering in long day conditions. The expression level of these inhibitors starts decreasing at time of production of flowering signals.

10.7 Grain Development

The seed developments start with the double fertilization in embryo sac (Deng et al. 2013) and are followed by cell division, cell fate determination, tissue differentiation, and programmed cell death in developing zygote (Mahto et al. 2017). The remarkable changes in accumulation of various metabolites in the process of development of grains mainly occur within 30 days after pollination. The endosperm (triploid) is the major storage tissue during rice grain development (caryopsis). It nurtures the embryo during early seed development, acts food reserve for emerging seedlings in germination process, and is source of food for humans and animals. The endosperm constitutes most of the space in the seed coat which is composed of starchy endosperm, aleurone cells, transfer cells, and cells in vicinity of embryo (Olsen 2001). The husk is another important part of the seed which is formed as a result of drying lemma and the palea. The grain size increases with time after pollination along with regulation in water contents. The grain length initially increases from the first day and reaches at maximum of 6 days, while width and thickness (expansion) increased from 4 days up to 9 and 12 days after pollination. Small increase in fresh grains and caryopsis weight was observed during 3–6 days, and rapid increase was observed during 6–9 days after pollination followed by slight decline in subsequent days. The steady increase in dry caryopsis weight was observed during 6–21 days after pollination. The water contents were high during first 6 days followed by continuous decline (Wu et al. 2016). The key stages of endosperm development are given in Table 10.1.

Table 10.1 Description of endosperm development stage (Adopted from Wu et al. 2016)

Stages	Period (Days after pollination)	Characteristics
Coenocyte	1–2	Nuclear divisions, nuclei arrangement in periphery of embryo sac
Cellularization	3–5	Formation of anticlinal, periclinal, cell wall along periphery of central vacuole, filling of cavity of embryo sac
Storage product accumulation	6–21	Differentiation between aleurone and starchy endosperm, accumulation of storage protein in aleurone and subaleurone layers, the aleurone store proteins and lipids, while endosperm mainly accumulates proteins and starch
Maturation	22–30	No gains in caryopsis fresh and dry weight, minor morphological changes in cells of aleurone layers, loss of boundaries in cells of starchy endosperm, accumulation of protein bodies in subaleurone layers in large amount, crystallization of starch and storage proteins

10.8 Dynamics in Metabolites in Seed Development

The accumulation levels of various metabolites were investigated in developing rice grains at 3, 4, 7, and 10 days after fertilization (DAF) (Dhatt et al. 2019). The levels of glucose and fructose reached a peak after 10 days of fertilization and raffinose initially decreased up to 7 DAF followed by increase at 7 DAF, while proline, glycine, serine, threonine, glutamic acid, valine, tyrosine, and phenylalanine were lowest at 10 DAF. The β-alanine, aspartic acid, alanine, and shikimic acid improved with passage of time after fertilization, and the contents of β-alanine, aspartic acid, alanine, valine, tyrosine, and quinic acid were higher at 7 DAF than 10 DAF. Such patterns may further change in response to stressful environment, especially high night temperature.

The hormonal regulations also take place during the course of seed development. The genes participating in biosynthesis of auxin and gibberellins were found at early stages of seed development (Sharma et al. 2009). The auxin is involved in cell division in embryo and endosperm. The gibberellic acid, cytokinins, indole acetic acid, abscisic acid, putrescine, spermidine, and spermine are important regulators of grain development process. The cell divisions and filling rates were positively associated with hormones and polyamines. The concentration of gibberellic acids (GA1 and GA2) decreased with days after anthesis. Depending upon the varieties, the concentrations of cytokinins (Z + ZR), indole acetic acid, and abscisic acid in superior filled grains initially improved and attained maximum values during 10–15 days after anthesis and then declined with passage of time. The concentration of putrescine, spermidine, and spermine increased from 9 to 15 days after anthesis with minor role of varieties, while the concentrations of soluble- and insoluble-conjugated polyamines did not change with post-anthesis duration. The concentrations of these hormones in upper and lower spikelets of wild-type grains were maximum between 9–21 days and 24–27 days after anthesis, respectively. The superior grains exhibited great concentration of these hormones and vice versa in inferior spikelets. Moreover, the exogenous application of these hormones and polyamines further confirmed their role in grain filling process (Zhang et al. 2016). The grain filling rate initially increases up to 12–22 days followed by decline. The initial increase in cytokinin (Z + ZR), indole acetic acid, gibberellic acid, and abscisic acid up to certain duration after anthesis and subsequent decline has also been confirmed for developing grains in the results of study carried out by Yang et al. (2001).

The fertilized ovaries are transformed into caryopses, and the extent and rate of filling determine the final grain weight. The grains from lower spikelets are poorly filled as compared to upper spikelets. The delayed expression of genes responsible for transformation of sucrose to starch has been regarded as one of the major causes of poor grains filling in lower secondary panicle branches (Mahto et al. 2017). The photoassimilates are used for the synthesis of storage compounds (protein, starch, and lipids) during grain filling process.

References

Ahmad A, Ashfaq M, Rasul G, Wajid SA, Khaliq T, Rasul F, Saeed U, Rahman MH, Hussain J, Baig IA, Naqvi AA, SAA B, Ahmad S, Naseem W, Hoogenboom G, Valdivia RO (2015) Impact of climate change on the rice–wheat cropping system of Pakistan. In: Hillel D, Rosenzweig C (eds) Handbook of climate change and agro-ecosystems: the agricultural modeling intercomparison and improvement project (AgMIP) integrated crop and economic assessments. Imperial College Press and The American Society of Agronomy, London, pp 219–258

Ahmad A, Iqbal S, Ahmad S, Khaliq T, Nasim W, Husnain Z, Hussain A, Zia-ul-Haq M, Hoogenboom G (2009a) Seasonal growth, radiation interception, its conversion efficiency and biomass production of *Oryza sativa* L. under diverse agroenvironments in Pakistan. Pak J Bot 41(3):1241–1257

Ahmad S, Abbas G, Ahmed M, Fatima Z, Anjum MA, Rasul G, Khan MA, Hoogenboom G (2019) Climate warming and management impact on the change of rice-wheat phenology in Punjab, Pakistan. Field Crops Res 230:46–61

Ahmad S, Ahmad A, Ali H, Hussain A, Garcia y Garcia A, Khan MA, Zia-Ul-Haq M, Hasanuzzaman M, Hoogenboom G (2013) Application of the CSM-CERES-Rice model for evaluation of plant density and irrigation management of transplanted rice for an irrigated semiarid environment. Irrig Sci 31(3):491–506

Ahmad S, Ahmad A, Soler CMT, Ali H, Zia-Ul-Haq M, Anothai J, Hussain A, Hoogenboom G, Hasanuzzaman M (2012) Application of the CSM-CERES-Rice model for evaluation of plant density and nitrogen management of fine transplanted rice for an irrigated semiarid environment. Precis Agric 13(2):200–218

Ahmad S, Ahmad A, Zia-ul-Haq M, Ali H, Khaliq T, Anjum MA, Khan MA, Hussain A, Hoogenboom G (2009b) Resources use efficiency of field grown transplanted rice (*Oryza sativa* L.) under irrigated semiarid environment. J Food Agric Environ 7(2):487–492

Ahmad S, Hasanuzzaman M (2012) Integrated effect of plant density, N rates and irrigation regimes on the biomass production, N content, PAR use efficiencies and water productivity of rice under irrigated semiarid environment. Not Bot Horti Agrobot Cluj-Napoca 40(1):201–211

Ahmad S, Zia-ul-Haq M, Ali H, Shad SA, Ammad A, Maqsood M, Khan MB, Mehmood S, Hussain A (2008) Water and radiation use efficiencies of transplanted rice (*Oryza sativa* L.) at different plant densities and irrigation regimes under semi-arid environment. Pak J Bot 40(1): 199–209

Ahmed M, Ahmad S (2019) Carbon dioxide enrichment and crop productivity. In: Hasanuzzaman M (ed) Agronomic crops, vol 2. Springer, Singapore, pp 31–46

Ahmed M, Ahmad S (2020) Systems modeling. In: Ahmed M (ed) Systems modeling. Springer, Singapore, pp 1–44

Ahmed M, Ahmad S, Raza MA, Kumar U, Ansar M, Shah GA, Parsons D, Hoogenboom G, Palosuo T, Seidel S (2020a) Models calibration and evaluation. In: Ahmed M (ed) Systems modeling. Springer, Singapore, pp 151–178

Ahmed M, Ahmad S, Waldrip HM, Ramin M, Raza MA (2020b) Whole farm modeling: a systems approach to understanding and managing livestock for greenhouse gas mitigation, economic viability and environmental quality. In: Waldrip HM, Pagliari PH, He Z (eds) Animal Manure. ASA Special Publication 67. American Society of Agronomy and Soil Science Society of America, Madison, WI, pp 345–371

Ahmed M, Fayyaz-ul-Hassan, Ahmad S (2017) Climate variability impact on rice production: adaptation and mitigation strategies. In: Ahmed M, Stockle C (eds) Quantification of climate variability, adaptation and mitigation for agricultural sustainability. Springer, Switzerland, pp 91–111

Akram R, Fahad S, Masood N, Rasool A, Ijaz M, Ihsan MZ, Maqbool MM, Ahmad S, Hussain S, Ahmed M, Kaleem S, Sultana SR, Mubeen M, Saud S, Kamran M, Nasim W (2019) Plant growth and morphological changes in rice under abiotic stress. In: Hasanuzzaman M, Fujita M,

Nahar K, Biswas JK (eds) Advances in rice research for abiotic stress tolerance. Woodhead Publishing, Swaston, pp 69–85

Akram R, Turan V, Hammad HM, Ahmad S, Hussain S, Hasnain A, Maqbool MM, Rehmani MIA, Rasool A, Masood N, Mahmood F, Mubeen M, Sultana SR, Fahad S, Amanet K, Saleem M, Abbas Y, Akhtar HM, Hussain S, Waseem F, Murtaza R, Amin A, Zahoor SA, Sami ul Din M, Nasim W (2018) Fate of organic and inorganic pollutants in paddy soils. In: Hashmi MZ, Varma A (eds) Environmental pollution of paddy soils. Springer, Switzerland, pp 197–214

Ali AS, Elozeiri AA (2017) Metabolic processes during seed germination. Adv Seed Biol: 141–166

An H, Roussot C, Suarez-Lopez P, Corbesier L, Vincent C, Piñeiro M, Hepworth S, Mouradov A, Justin S, Turnbull C, Coupland G (2004) CONSTANS acts in the phloem to regulate a systemic signal that induces photoperiodic flowering of arabidopsis. Development 131:3615–3626

Deng ZY, Gong CY, Wang T (2013) Use of proteomics to understand seed development in rice. Proteomics 13:1784–1800

Dhatt BK, Abshire N, Paul P, Hasanthika K, Sandhu J, Zhang Q, Obata T, Walia H (2019) Metabolic dynamics of developing rice seeds under high night-time temperature stress. Front Plant Sci 10:1443. https://doi.org/10.3389/fpls.2019.01443

Fahad S, Noor M, Adnan M, Khan MA, Rahman I, Alam M, Khan IA, Ullah H, Mian IA, Hassan S, Saud S, Bakhat HF, Hammad HM, Ahmad S, Nasim W (2019) Abiotic stress and rice grain quality. In: Hasanuzzaman M, Fujita M, Nahar K, Biswas JK (eds) Advances in rice research for abiotic stress tolerance. Woodhead Publishing, Swaston, pp 571–583

Fatima Z, Ahmad M, Hussain M, Abbas G, Ul-Allah S, Ahmad S, Ahmed N, Ali MA, Sarwar G, Ehsan ul Haque, Iqbal P, Hussain S (2020) The fingerprints of climate warming on cereal crops phenology and adaptation options. Sci Rep 10:18013

García y García A, Dourado-Neto D, MDV B, López Ovejero RF, Favarin JL (2003) Logistic rice model for dry matter and nutrient uptake. Scientia Agricola 60(3):481–488

Hafeez-ur-Rehman, Nawaz A, Awan MI, Ijaz M, Hussain M, Ahmad S, Farooq M (2019) Direct seeding in rice: problems and prospects. In: Hasanuzzaman M (ed) Agronomic crops: production technologies, vol 1. Springer, Singapore, pp 199–222

Hubel F, Beck E (1996) Maize root phytase. Purification, characterization, and localization of enzyme activity and its putative substrate. Plant Physiol 112:1429–1436

Hussain M, Ahmad S, Hussain S, Lal R, Ul-Allah S, Nawaz A (2018) Rice in saline soils: physiology, biochemistry, genetics and management. Adv Agron 148:231–287

Hussain S, Fujii T, McGoey S, Yamada M, Ramzan M, Akmal M (2014) Evaluation of different rice varieties for growth and yield characteristics. J Anim Plant Sci 24(5):1504–1510

Itoh JI, Nonomura KI, Ikeda K, Yamaki S, Inukai Y, Yamagishi H, Kitano H, Nagato Y (2005) Rice plant development: from zygote to spikelet. Plant Cell Physiol 46(1):23-47

Itoh H, Izawa T (2013) The coincidence of critical day length recognition for florigen gene expression and floral transition under long-day conditions in rice. Molecular Plant 6(3):635–649

Itoh H, Nonoue Y, Yano M, Izawa T (2010) A pair of floral regulators sets critical day length for Hd3a florigen expression in rice. Nat Genet 42:635–638

Khan MA, Ahmad S, Raza A (2019a) Integrated weed management for agronomic crops. In: Hasanuzzaman M (ed) Agronomic crops, vol 2. Springer, Singapore, pp 257–281

Khan SM, Ali S, Nawaz A, Bukhari SAH, Ejaz S, Ahmad S (2019b) Integrated pest and disease management for better agronomic crop production. In: Hasanuzzaman M (ed) Agronomic crops, vol 2. Springer, Singapore, pp 385–428

Komiya R, Ikegami A, Tamaki S, Yokoi S, Shimamoto K (2008) Hd3a and RFT1 are essential for flowering in rice. Development 135(4):767–774

Komiya R, Yokoi S, Shimamoto K (2009) A gene network for long-day flowering activates RFT1 encoding a mobile flowering signal in rice. Develop 136(20):3443–3450

Leal MCM, Cammarota MC, Freire DMG, Sant'Anna JGL (2002) Hydrolytic enzymes as coadjuvants in the anaerobic treatment of dairy waste waters. Brazilian J Chemic Eng 19(2): 175–180

Li J, Zhao Y, Chu H, Wang L, Fu Y, Liu P, Upadhyaya N, Chen C, Mou T, Feng Y, Kumar P, Xu J (2015) SHOEBOX modulates root meristem size in rice through dose-dependent effects of gibberellins on cell elongation and proliferation. PLoS Genet 11:e1005464

Mahto A, Mathew IE, Agarwal P (2017) Decoding the transcriptome of rice seed during development. In: Jimenez-Lopez JC (ed) Advances in seed biology. IntechOpen Publishers, New York, p 25. https://doi.org/10.5772/intechopen.70574

Naz S, Fatima Z, Iqbal P, Khan A, Zakir I, Noreen S, Younis H, Abbas G, Ahmad S (2019) Agronomic crops: types and uses. In: Hasanuzzaman M (ed) Agronomic crops: production technologies, vol 1. Springer, Singapore, pp 1–18

Olsen OA (2001) Endosperm development: cellularization and cell fate specification. Annu Rev Plant Physiol Plant Mol Biol 52:233–267

Peerzada AM, Bukhari SAH, Dawood M, Nawaz A, Ahmad S, Adkins S (2019) Weed management for healthy crop production. In: Hasanuzzaman M (ed) Agronomic Crops, vol 2. Springer, Singapore, pp 225–256

Pereira EP, Zanin GM, Castro HF (2003) Immobilization and catalytic properties of lipase on chitosan for hydrolysis and etherification reactions. Brazilian J Chemic Eng 20(4):343–355

Puteh AB, Mondal MMA (2013) Salinity effect on dry mass partitioning in different plant parts and ion uptake in leaves of rice. J Environ Sci Nat Res 6(1):239–245

Quettier AL, Eastmond PJ (2009) Storage oil hydrolysis during early seedling growth. Plant Physiol Biochem 47(6):485–490

Rajjou L, Duval M, Gallardo K, Catusse J, Bally J, Job C, Job D (2012) Seed germination and vigor. Annual Review Plant Biol 63:507–533

Razaq M, Shah FM, Ahmad S, Afzal M (2019) Pest management for agronomic crops. In: Hasanuzzaman M (ed) Agronomic crops, vol 2. Springer, Singapore, pp 365–384

Sarwar N, Ali H, Ahmad S, EhsanUllah AS, Mubeen K, Hill JE (2013a) Water wise rice cultivation on calcareous soil with the addition of essential micronutrients. J Anim Plant Sci 23(1):244–250

Sarwar N, Ali H, Maqsood M, EhsanUllah, Shahzad AN, Shahzad M, Mubeen K, Shahid MA, Ahmad S (2013b) Phenological response of rice plants to different micronutrients application under water saving paddy fields on calcareous soil. Turk J Field Crops 18(1):52–57

Shahzad AN, Ahmad S (2019) Tools and techniques for nitrogen management in cereals. In: Hasanuzzaman M (ed) Agronomic crops, vol 2. Springer, Singapore, pp 111–126

Sharma R, Mohan SRK, Malik G, Deveshwar P, Tyagi AK, Kapoor S, Kapoor M (2009) Rice cytosine DNA methyltransferases–gene expression profiling during reproductive development and abiotic stress. FEBS J 276:6301–6311

Shrestha R, Gómez-Ariza J, Brambilla V, Fornara F (2014) Molecular control of seasonal flowering in rice, arabidopsis and temperate cereals. Annals Bot 114(7):1445–1458

Tariq M, Abbas G, Yasmeen A, Ahmad S (2020a) Cotton ontogeny. In: Ahmad S, Hasanuzzaman M (eds) Cotton production and uses. Springer, Singapore. https://doi.org/10.1007/978-981-15-1472-2_23

Tariq M, Ahmed M, Iqbal P, Fatima Z, Ahmad S (2020b) Crop phenotyping. In: Ahmed M (ed) Systems modeling. Springer, Singapore, pp 45–60

Tiwari SB, Shen Y, Chang CH, Hou Y, Harris A, Ma SF, McPartland M, Hymus GJ, Adam L, Marion C, Belachew A, Repetti PP, Reuber TL, Ratcliffe OJ (2010) The flowering time regulator CONSTANS is recruited to the FLOWERING LOCUS T promoter via a unique cis-element. New Phytol 187(1):57–66

Tsuji H, Tamaki S, Komiya R, Shimamoto K (2008) Florigen and the photoperiodic control of flowering in rice. Rice 1:25–35

Vijayakumar KR, Gowda LR (2012) Temporal expression profiling of lipase during germination and rice caryopsis development. Plant Physiol Biochem 57:245–253

Wang Y, Wang DM, Liang HG (1988) Effect on plumular axis on the amylase activity in cotyledons of germinating pea seeds. Acta Phytophysiol Sinica 14(3):244–249

Wasaya A, Yasir TA, Ijaz M, Ahmad S (2019) Tillage effects on agronomic crop production. In: Hasanuzzaman M (ed) Agronomic crops, vol 2. Springer, Singapore, pp 73–99

Wu W, Zhang H, Qian Y, Cheng Ye WG, Zhai C, Dai Q-g (2008) Analysis on dry matter production characteristics of super hybrid rice. Rice Science 15(2):110–118

Wu X, Liu J, Li D, Liu CM (2016) Rice caryopsis development II: dynamic changes in the endosperm. J Integr Plant Biol 58:786–798

Xiong M, Chu L, Li Q, Yu J, Yang Y, Zhou P, Zhou Y, Zhang C, Fan X, Zhao D, Yan C, Liu Q (2021) Brassinosteroid and gibberellin coordinate rice seed germination and embryo growth by regulating glutelin mobilization. Crop J (In Press). https://doi.org/10.1016/j.cj.2020.11.006

Yang J, Zhang J, Wang Z, Zhu Q, Wang W (2001) Hormonal changes in the grains of rice subjected to water stress during grain filling. Plant Physiol 127:315–323

Yang M, Yang J, Su L, Sun K, Li D, Liu Y, Guo T (2019) Metabolic profile analysis and identification of key metabolites during rice seed germination under low-temperature stress. Plant Sci 289:110282

Yang P, Li X, Wang X, Chen H, Chen F, Shen S (2007) Proteomic analysis of rice (*Oryza sativa* L.) seeds during germination. Proteomics 7(18):3358–3368

Yoshida H, Nagato Y (2011) Flower development in rice. J Exp Bot 62(14):4719–4730

Zahoor SA, Ahmad S, Ahmad A, Wajid A, Khaliq T, Mubeen M, Hussain S, Sami Ul Din M, Amin A, Awais M, Nasim W (2019) Improving water use efficiency in agronomic crop production. In: Hasanuzzaman M (ed) Agronomic crops, vol 2. Springer, Singapore, pp 13–29

Zhang W, Cao Z, Zhou Q, Chen J, Xu G, Gu J, Liu L, Wang Z, Yang J, Zhang H (2016) Grain filling characteristics and their relations with endogenous hormones in large- and small-grain mutants of rice. PLoS ONE 11(10):e0165321

Rice Phenotyping

11

Muhammad Tariq, Muhammad Habib Ur Rehman, Feng Ling Yang, Muhammad Hayder Bin Khalid, Muhammad Ali Raza, Muhammad Jawad Hassan, Tehseen Ahmad Meraj, Ahsin Khan, Atta Mohi Ud Din, Nasir Iqbal, Ahmed M. S. Kheir, and Shakeel Ahmad

M. Tariq
Central Cotton Research Institute, Multan, Pakistan

School of Agriculture, Food and Wine, University of Adelaide, Adelaide, SA, Australia

M. H. U. Rehman
Department of Agronomy, MNS-University of Agriculture, Multan, Pakistan

Institute of Crop Sciences and Resource Conservation (INRES), Crop Science Group, University Bonn, Bonn, Germany

F. L. Yang · M. H. B. Khalid
College of Agronomy, Sichuan Agricultural University, Chengdu, China

Sichuan Engineering Research Center for Crop Strip Intercropping System, Key Laboratory of Crop Ecophysiology and Farming System in Southwest China, Sichuan Agricultural University, Chengdu, China

M. A. Raza
College of Agronomy, Sichuan Agricultural University, Chengdu, China

Sichuan Engineering Research Center for Crop Strip Intercropping System, Key Laboratory of Crop Ecophysiology and Farming System in Southwest China, Sichuan Agricultural University, Chengdu, China

The Islamia University Bahawalpur, Bahawalpur, Pakistan

M. J. Hassan
College of Agronomy, Sichuan Agricultural University, Chengdu, China

T. A. Meraj
Shanghai Jiaotong University, Shanghai, People's Republic of China

N. Sarwar et al. (eds.), *Modern Techniques of Rice Crop Production*,
https://doi.org/10.1007/978-981-16-4955-4_11

Abstract

The conventional methods of screening desired traits from a population are slow and tedious, and their accuracy is highly influenced by working time and manpower professional skills. However, in current era of scientific developments, these tasks are being performed with modern phenotyping techniques. This chapter presents an overview of phenotyping various traits through imaging tools and other related technologies. The DJI Phantom 4, RootReader3D, and SmartGrain program are being used for recording plant height, root architecture, and automatic measurement of grains dimensions, respectively. The destructive sampling is sometimes compulsory for recording observations, and periodic measurement from the same plant often results in no seed availability for further selection process. The stress imposition induces morphological and physiological changes in plants, and monitoring these changes through phenotyping tools is very helpful for screening stress-tolerant germplasm. The phenotyping for stress-tolerant studies was performed with infrared camera for salt tolerance, LemnaTec 3D Scanalyzer for acquiring shoot images and infrared system, fluorescence, and DroughtSpotter for measuring drought-related parameters. Moreover, the Canon SX110 IS was used for the detection of rice hopper infestation and Micasense RedEdge for screening of tolerant genotypes against Hoja Blanca virus. The grain quality assessment is an important criterion of marketing rice at good prices. Such determination is also being carried out with modern phenotyping tools, and examples include RQS1.0 and CanoScan 9950FV for amylose content and chalkiness determination, respectively. We concluded that such technologies have great potential for accurate and fast measurement of various traits to accelerate the breeding process and monitoring growth, yield, and crop health.

Keywords

Imaging · Stress · Yield · Insects · Diseases

A. Khan · A. M. U. Din
Sichuan Agricultural University, Yaan, People's Republic of China

N. Iqbal
School of Agriculture, Food and Wine, University of Adelaide, Adelaide, SA, Australia

A. M. S. Kheir
Soil, Water and Environment Research Institute, Agriculture Research Centre, Giza, Egypt

S. Ahmad (✉)
Department of Agronomy, Faculty of Agricultural Sciences and Technology, Bahauddin Zakariya University, Multan, Pakistan
e-mail: shakeelahmad@bzu.edu.pk

11.1 Introduction

The phenotyping describes the quantitative measurement of physiological, biochemical, and developmental properties (Walter et al. 2015; Tariq et al. 2020). Conventional phenotyping requires lot of time, finance, and human resource. Hence, frequency of data recording remains very low with manual methods. The manual phenotyping process largely depends upon weather conditions to access the plot for observation recordings. The issue is more severe for flooded rice where standing water limits the human movements in the field. The modern phenotyping techniques thus enable the researchers to record observations on daily basis from large fields. The term phenotyping was known in 1960s, and image-based plant phenotyping was started in early twentieth century (Tariq et al. 2020). Currently, sensor-based technologies are being utilized to monitor the crop morphological development and to assess the crop responses to various stresses. The sensors are fixed with specific structure in the form of unmanned aerial vehicles and other platforms. The non-destructive approaches allow the researchers to record multiple observations, even from the same plant over time course. Furthermore, it also enables to study the dynamic plant response from individual plants against various environmental stresses. It also offers a solution to avoid destructive sampling from unique plants such as early generation and transgenic species for which seed collection is also crucial for future breeding. The modern phenotyping approaches are expensive; however, the cost is decreasing over time with the improvement in phenotyping knowledge.

Resulting upon internal movement, all the molecules emit infrared radiations (Kastberger and Stachl 2003). The devices for measurement of these radiations are known as near-infrared (NIR) and far-infrared (Far-IR). The infrared wavelength and far-red wavelength range from 0.9 to 1.7 μm and 7.5 to 13.5 μm, respectively. There are great variations in light reflection between healthy and unhealthy plants. The healthy plants reflect NIR light between 800 and 1400 nm, and red-light reflection from unhealthy plants is more than healthy plants (Yang et al. 2013). This phenomenon has great implication for measuring the overall health status of the crops.

The hyperspectral imaging techniques enable to divide the image into bands which can be used for assessing crop health and plant architecture. The scanning speed of hyperspectral imaging is slow, and its application is more effective for initial investigation of waveband. The plant visualization has been improved greatly with the advent of 3D structural tomography and functional imaging. The 3D tomography finds its application for phenotyping internal structure of multi-tiller crops like rice. The positron emission tomography (PET) and fluorescence imaging are two important technologies of functional imaging, and these are capable of investigating physiological changes in response to various stresses. These technologies have low spatial resolution and combine application of structural tomography, and functional imaging can accurately investigate the physiology activities (Yang et al. 2013) and detection of rice blast (Yang et al. 2012).

The visible image technology is being widely adopted in plant sciences since very past with the advent of digital camera in 1975. Initially, 2D imaging technology had

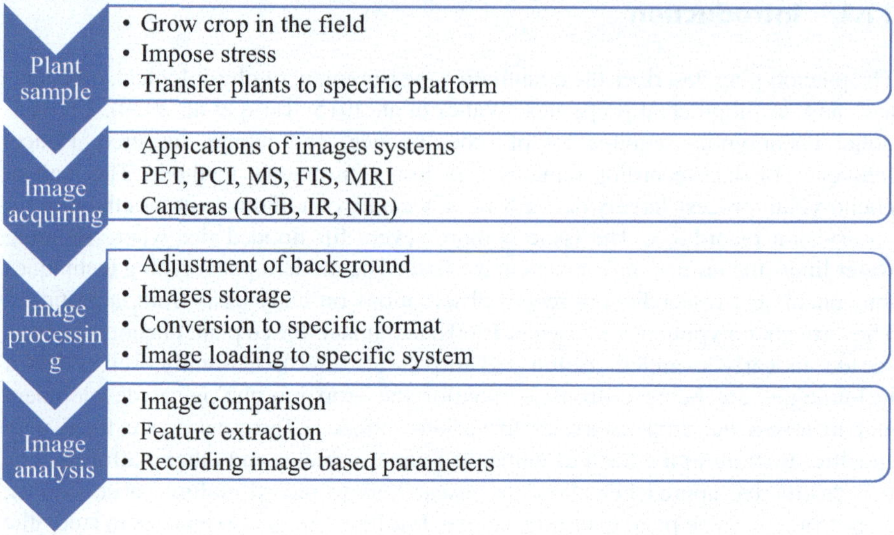

Fig. 11.1 An overview of processes of modern phenotyping. *PET* positron emission tomography, *PCI* phase-contrast X-ray imaging, *MS* mass spectrometry, *FIS* fluorescence imaging systems, *MRI* magnetic resonance imaging, *RGB* red green blue, *IR* infrared, *NIR* near-infrared

been used but replaced with 3D imaging. However, the use of 2D imaging technologies was suggested more reliable for high throughput than 3D imaging (Yang et al. 2013). An overview of modern phenotyping is illustrated in Fig. 11.1. The 3D imaging technologies such as RootReader3D for studying root architecture (Clark et al. 2011), 3D digitizer and L-system determination of growth and development system (Watanabe et al. 2005), visible light imaging rice phenology, near-infrared images for leaf area (Shibayama et al. 2011), hyperspectral technology for leaf growth and nitrogen status (Nguyen and Lee 2006), estimation of number of tillers through X-ray computed tomography (Yang et al. 2011), grain length and grains width determination through flatbed scanner (Igathinathane et al. 2009), use of visible light imaging, and X-ray digital radiography (DR) for estimation of filled and unfilled rice spikelets (Duan et al. 2011).

11.2 Phenotyping Measurement

11.2.1 Plant Height

The plant height measurement with UAVs (DJI Phantom 4) was performed for upland rice in Laos using canopy height model (CHM) and structure from motion (SfM) algorithm (Kawamura et al. 2020). The SfM is computer-based technique to generate 3D geometries from raw and unprocessed images. The digital surface model (DSMs) was generated by capturing field images prior to emergence and before flowering. The difference of DSMs at soil surface and DSMs of canopy of vegetation is used for measuring CHM. The pre-flowering stage is preferred time because further increase is rare, and genotypic plant height differences (except leaves to ear height) are very clear at this stage. The main steps in these techniques included acquiring RGB images, generation of dense point clouds and DSMs, calculation of differences of DSMs of pre-sowing and pre-flowering, and finally development of plant height estimation model. The image acquisition and generation of dense point clouds and DSMs were conducted with DJI Phantom 4 and Agisoft Metashape version 1.5.1, respectively. Measurement of plant height through phenotyping tool is more accurate at later growth stages than earlier. Furthermore, development of CHM should be from top (1–10%) among the largest pixel values. In this study, CHM3% was concluded the best metrics, and values showed better correlation with field observed values. By comparing the values, the root mean square error was 6.963% which was in acceptable range. Further comparison of manpower requirement indicated that manual data recording would be completed with three persons in 1 day; however, the UAVs completed this task with only one person in twelve minutes. In this way, the application of UAVs had great potential for recording plant height in minimum time.

11.2.2 Root Mass and Architecture

The knowledge of root diversity among genotypes is limited due to difficult methods of analyzing root system in soil media. The aboveground plant parts change in response to root growth in soil (Ayub et al. 2013). The screening of genotypes with respect to nutrient acquisition and drought tolerance is still to be worked out. The conventional and manual methods of studying root system are laborious, costly, and time-consuming. The phenotyping approaches including imaging techniques (2D and 3D), nuclear magnetic resonance imaging, X-ray computed tomography, visible light, and laser imaging are non-destructive measurement of the root architecture. The comparison was made between image base and manual root phenotyping in rice in seven different experiments. The imaging techniques were digital imaging of root traits (DIRT). The roots were placed on diffuse blackboard, and images were taken with digital camera. These images were further processed through DIRT software. The relative phenotypic variation between manual and image-based root phenotyping was assessed. The results demonstrated that image base root phenotyping was reliable for few traits but not useful for all traits (Bauw et al. 2020). Moreover, three-dimension root phenotyping of two rice types was performed with the help of RootReader3D. The performance of RootReader3D was then compared with parameters (27) recorded from 2D imaging. The images were captured with Nikon D300s Digital SLR Camera. It was concluded that RootReader3D is unique tool for characterizing root characteristics. The enhanced capacity of imaging 100 root systems per day makes its utility in breeding programs (Clark et al. 2011).

11.2.3 Yield Parameters

The determination of yield components is very important for the estimation of yield potential. The task is traditionally performed with the professional workers; however, it is tedious and requires huge human resource, especially for recording large datasets. Furthermore, the accuracy of the data is less and is mainly influenced by worker fatigue. The innovation in imaging technologies is being applied for the sake of recording yield-related traits, and few examples are quoted as follows. The automatic evaluation facility was used for estimating various yield attributes. The threshing unit, inspection unit, and packing weighing were the components of system. The threshing unit is used for threshing of spikelet through roller compaction process. The inspection unit comprised of different cameras and conveyers. The process of separation of spikelet and image analysis is completed here. The total spikelet vision and filled spikelet vision camera were used for acquiring images for total spikelet and filled spikelet. The process of image analysis was completed software which was programmed with NI Vision for LabVIEW 8.6. The software enables the image acquiring and analysis simultaneously in the computer. The filled spikelet is collected in tank which is poured into packing and weighing unit. The packet is coded and weighed on electric balance. The other parameters determined with this engineering prototype were grain length, width, and 1000-grain weight. The system was capable of analyzing 1440 plants in continuous 24 h working, and error values fall below 5%. The researchers further suggested its application for other cereals with certain modification (Duan et al. 2011).

The color images of grains spread on flatbed scanner were acquired for measuring length and width of grains. The black and white background was selected to develop better color contrast, while original color was converted into grayscale in Image J program for further conversion to binary image. The imageJ program has an add-on program, known as plugin. The plugin has facility of coding to fulfill the machine requirement. The system measures the orthogonal dimensions of an object by locating the boundary. The overall accuracy of plugin system was greater than 96.6% and capable of analyzing 254 ± 125 particles/s. Its accuracy can be improved by avoiding the presence of shade in the images. The dimensions (length and width) of irregular shapes can be easily measured in any orientation only through fitted ellipses' centroid coordinates and major axis inclination (Igathinathane et al. 2009).

The shape and size of the seed are important factors of yield formation and marketing process. The task of the measurement of seed size (length and width) is performed with calipers. However, the process is time-consuming, and quality of data is not well. The modern computer applications, particularly imaging technologies, are being made for such measurement. In this regard, the SmartGrain program is with the ability of automatic measurement of seed length, width, seed area, and perimeter length from image. The seeds were thoroughly spread on screen of Epson GT-X820 A4 scanner followed by scanning at 600 dpi. The seed area, perimeter length, and center of gravity are calculated with detection of outline of the seeds. The seed length was calculated through detection of longitudinal axis, while detection of transverse axis was used for calculation seed width, intersection of length and width, length-to-width ratio, circularity, and distance between intersection of length and width and center of gravity. Its application reduces the sampling error and to classify the lines with variable seed shape (Tanabata et al. 2012).

11.3 Phenotyping for Stresses

11.3.1 Salt Stress

The salinity is major threat for global rice production, especially in arid and semi-arid regions. It impairs normal growth and physiological mechanisms. The application of rapid and effective modern techniques for screening of tolerant plants is lacking. Therefore, phenotyping of rice genotypes for salt tolerance is crucial for breeding purposes. The screening on the basis of physiological traits is very common but such approaches are tedious and time-consuming. Furthermore, these involve destructive sampling, and sometimes it becomes very difficult to collect seeds from plants for which physiological measurements were made. The latest imaging techniques do not require destructive sampling and are very fast to record various observations. The osmotic and ionic stresses (Na and Cl accumulations) are very common consequences of the salinity stress. The phenotyping through various imaging techniques now enables us to differentiate these effects (Hairmansis et al. 2014). The infrared imaging techniques were used to evaluate the salt tolerance in rice using standard physiological parameters. The phenotyping with infrared camera is based on heat produced by the plants in stresses. It was further hypothesized that plant temperature depends upon physiological attributes which are also influenced by the presence of salts. The images of salt-treated and control plants were captured at 10.0 am with FLIR-SC-620 (FLIR Systems, USA) using a resolution of 640×480 pixels. The images were extracted to computer and analyzed using ThermaCAM Researcher Pro 2.10 software. The change in color temperature was used as indicator of relative salt stress. Four colors including blue < green < yellow < red color represent temperature in ascending order. The plants grown in saline environment displayed less blue color than their counterpart non-saline plants. Furthermore, leaf temperature of plants from saline environment was high than control. The average plant temperature in control plants was high than plants grown in 75 mM NaCl, whereas it was gradually increased from 27.9 °C to 29.5 °C with increasing NaCl concentration from 75 to 225 mM. The physiological parameters such as water content, stomatal conductance, performance index, and dark-adapted quantum yield were correlated with image temperature. The relative water contents and stomatal conductance showed negative correlation with image temperature with R^2 values of −0.852 and −0.612, respectively, while correlation between image temperature with performance index and dark-adapted quantum yield remained non-significant. It was concluded that variations in leaf temperature can be successfully applied to identify the stresses in plants (Siddiqui et al. 2014).

The measurement of plant biomass of rice genotypes (IR64 and Fatmawati) grown in various salinity levels was made through RGB images. LemnaTec 3D

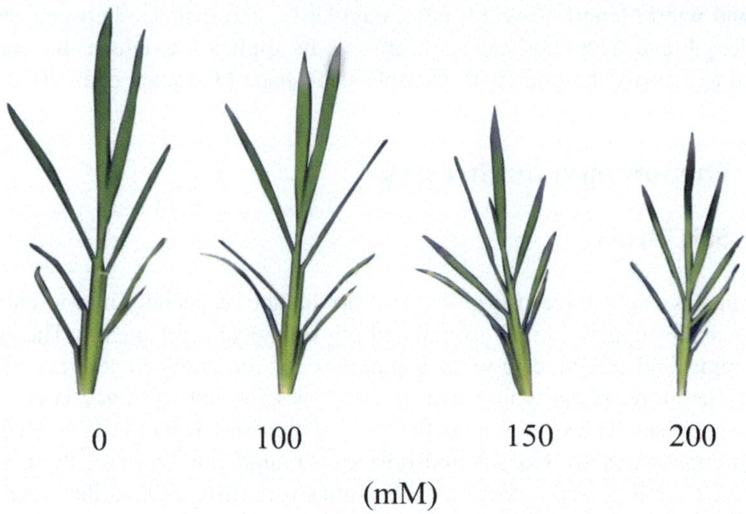

Fig. 11.2 The color changes in fluorescent images due to NaCl. The green indicates healthy and pink shows salt effects (Re-drawn following images of Hairmansis et al. (2014))

Scanalyzer system was used for capturing shoot images. In total, three images (two from side at 90° and one from top) were captured and analyzed through LemnaTec Grid software package. The areas of these three images were summed up for approximate measurement of shoot biomass. The shoot senescence was assessed with the help of fluorescent images from top. The green color of fluorescence image changed into pink with increasing salt concentration (Fig. 11.2). The authors suggested that image-based phenotyping may accelerate the process of development of salt-tolerant genotypes (Hairmansis et al. 2014).

11.3.2 Drought Stress

The severity and frequency of drought stress is becoming a major challenge of crop production in climate change scenarios. The upland rice on about 27 million hectares in the world is at the risk of drought (Zu et al. 2017). Phenotyping of large set of population is essential for the development of drought-tolerant genotypes. The application of imaging technologies in this regard may be helpful to accelerate the selection process for future breeding. The impact of drought is accompanied by numerous changes at cellular, physiological, and morphological levels. These techniques classify the population according to physiological and morphological responses to drought. For studying the genotypic response, the group of 20 wild-type and 20 *osphyb* (drought-tolerant mutant) seeds of rice was grown in normal and deficit moisture conditions. The plant area, color, and compactness were studied with RGB images. The water contents were determined with near-infrared image systems, while infrared system, fluorescence, and DroughtSpotter technology were

Table 11.1 Tools used for measuring drought-related parameters in studies conducted by Kim et al. (2020)

Phenotyping tools	Resolution	Manufacturer	Software	Phenotyping traits
RGB camera	6576 × 4384	LemnaTec, GmbH, Aachen, Germany	LemnaGrid software (LemnaTec GmbH, Aachen, Germany)	Projected plant area, plant color, convex hull area, compactness, eccentricity, object extent X and Y, and center of mass Y
NIR camera (Model: Goldeye G032	636 × 508 pixels	AVT Allied Vision, Exton, USA)	–	Water content
FLIR P620 camera	640 × 480 pixels	FLIR Systems Inc., North Billerica, MA, USA	FLIR Research IR 4.1 software	Plant temperature
PlantScreen™ Robotic XYZ System associated with RGB camera	1392 × 1040 pixels	Photons Systems Instruments, Brno, Czech Republic	ImageJ software	Photosynthetic efficiency
DroughtSpotter	–	(Phenospex, Herleen, the Netherlands)	–	Water use efficiency, plant water loss rate, and transpiration rate

used for measuring plant temperature, photosynthetic efficiency and water use efficiency, transpiration rate, and water loss, respectively. The detail of these tools is summarized in Table 11.1. The application of these tools had been very successful for analyzing drought-related traits in rice. These techniques showed clear differences between drought tolerance and drought-susceptible genotypes (Kim et al. 2020).

The different plant responses to drought with respect to xylem vessels of roots may be an important trait of selection for drought. Although such measurements were made manually in the past, now innovations in imaging technologies are getting in. However, the phenotyping of the response would require deep analysis of microscopic data. The novel automated framework was deployed for analysis of xylem vessel for characterization of drought-tolerant cultivars. In this process, the stele region is initially isolated from image background followed by detection of xylem vessels. Later on, feature extraction and unsupervised labeling were performed using different procedures. There were clear differences in phenotyping data of drought-susceptible and drought-tolerant genotypes, and the accuracy of the system was 98%. It suggested its potential application for breeding for drought tolerance (Bhugra et al. 2017).

11.3.3 Insect Pests and Diseases

The insect pests and diseases are major constraints of crop production, and both cause significant yield losses. The early-stage detection is thus very important to avoid yield losses through adopting various plant protection measures. The diseases and insect infestation are greatly influenced by environment and in-built plant resistance (Peerzada et al. 2019; Razaq et al. 2019).

The detection of stress caused by rice hopper infestation was estimated through analysis of visible images. The rice plants at tillering stage at 52 days were grouped into no infestation, mild, moderate, and severe infestation to monitor the rice hopper infestation using advanced phenotyping techniques. The digital camera (Canon SX110 IS) was used to capture the images of rice plants at 1600 × 1200 pixels with the objective to detect hopper infestation. The complete process was comprised of image processing, segmentation of infested areas, corner detection, and rice hopper quantity estimation. The images of rice stems were analyzed using MATLAB 2009a Image Processing Toolbox™. The results showed that the technique can be successfully used for the detection of rice hopper infestation (Zhou et al. 2011). Similarly, the hyperspectral imaging systems were applied to detect the incidence of rice blast at seedlings stage. For the purpose, the group of two hundred seedlings (100 for each healthy and infected) was scanned with this system in NIR wavelength range (900 to 1700 nm). The image spectral data were acquired through software kit, Isuzu Optics, Taiwan, China. There were great visual differences between images of healthy and infected seedlings. The results were satisfactory, and it was proposed that incidence of rice blast can be realized with the help of near-infrared hyperspectral image with approximate accuracy of up to 92% (Yang et al. 2012).

Real-time video detection system was developed to monitor the presence of rice sheath blight, stem borer, and brown spot. The video detection is highly different from still images, and extra care is required in its field application to avoid defocusing and motion blurring particularly in uneven fields and from leaves movement due to wind. Furthermore, the appearance of disease symptoms showed some variations which are another challenge for disease detection through this approach. The process was completed with transformation of videos into still frame which was sent to still image detector followed by processing of frames into videos. The image training models were used in this system for detection of untrained videos. The video evaluation metrics revealed satisfactory quality. The system detects the presence of lesions on the leaves in its first step application. The results showed that sheath blight was presented by red boxes, borer symptoms were indicated by purple boxes, and brown spot was denoted by blue boxes, and all these were detected simultaneously. The chances of confusion of detection of these three problems were controlled with image training models. The authors found that system can be applied for the detection of these problems (Li et al. 2020).

The unmanned aerial vehicles (UAVs) with Micasense RedEdge (multi-spectral camera) were used for screening of Hoja Blanca virus (RHBV)-resistant rice genotypes. The insect vectors were released for spreading diseases in the experimental area. The disease status was manually observed using FLAR methodology. For comparing visual observations with image acquisition, UAV at 20 m height was also performed during the same period. A semi-automated data analysis pipeline was developed using the following steps to analyze the images taken by UAV. The complete steps used by researchers include pre-processing images, reflection maps generation, radiometric calibration, crop masking, vegetation indices calculation, data extraction at plot level from vegetation index maps, data analysis, and image quality assessment. For conversion of digital number values into reflectance measurements, the radiometric calibration is very important. The application of these techniques will allow the rice breeders for earlier screening of RHBV-resistant genotypes (Delgado et al. 2019).

11.4 Phenotyping for Quality

The protein is an important constituent of animal diet in which protein acts as a major source of nitrogen. Mostly, the determination of protein is carried out indirectly through measuring the nitrogen with Kjeldahl method (Tariq et al. 2011; Ijaz et al. 2016; Bilal et al. 2017). The digital imaging technologies remain very important for inspection of grains quality. It basically detects the color changes in the treated solution. For instance, the fast measurement of amylose contents in rice grains can be made with the help of RQS1.0. The combination of chemical and digital imaging technology was applied for the determination of protein contents in rice grains. The process of color development was completed with the reaction of Biuret reagent with protein. The image chroma values were obtained through RQS1.0 system (image scanner, monitor, image analysis algorithms, a computer, and printer device). Upon reaction of Biuret reagent and protein, the resulting solution (5 ml) was poured into device in transparent bottles for capturing image. The chroma values were obtained through RQS1.0 system. The relationship curve between protein contents and chroma values was created. The concentration of bovine standard proteins was negatively associated with RGB chroma values. Among RGB, the G values showed greater sensitivity with protein contents than R and B values of chroma. The conclusion was drawn that digital chroma accurately measures the protein contents of grains treated with Biuret reagent (Sun et al. 2008).

The chalkiness in rice grains is the major issue for breeding for quality. The task is generally performed with naked eye due to lack of effective methods of classification. The opaque color of milled rice rather than translucent indicates chalkiness and rice with such appearance fetch lower price in the market. The chalkiness arose from loose packing of starch granules accompanied by development of air spaces. It alters the light reflection pattern; the scattered light is transmitted from translucent and prevented from opaque grains. The chalkiness in 12 rice genotypes was inspected with inexpensive image scanner CanoScan 9950FV. The scanner was operated in film positive scanning mode, and grains were placed screen and scanned at 300 dpi resolution. The color variation in images of normal and white back rice is presented

Fig. 11.3 The variations in image color of normal and white back rice grain (**Source:** Adapted and from Yoshioka et al. 2007)

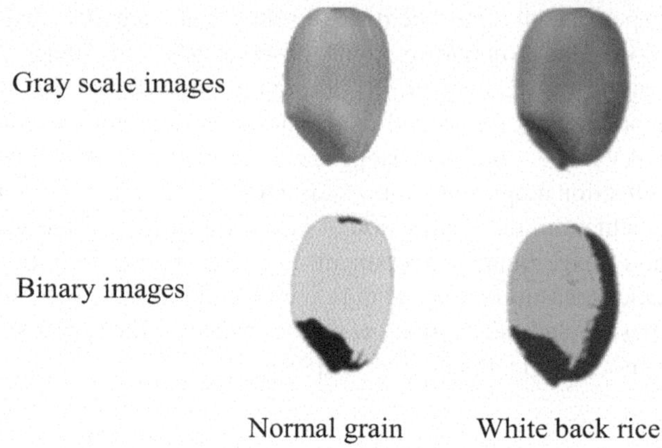

in Fig. 11.3. The results reported very high level of accuracy for the determination of chalkiness in rice grains using digital scanner (Yoshioka et al. 2007).

11.5 Conclusion

The deployment of modern tools for trait phenotyping is highly required particularly in flooded rice where manual observation recording remains a great challenge. Moreover, the techniques provide an alternative to destructive sampling which is a serious issue in trials where plants numbers are very small and seed harvesting is also crucial. Because of the ease of application, the researchers can now record observations on daily basis. We concluded that latest phenotyping tools can be successfully applied for grain quality assessment to monitoring growth characteristics, yield parameters, presence of pests, and diseases. There are variations between color and shape of healthy and affected plants from diseases, insect pests, and abiotic stresses. The quantification of these differences in images is further used monitoring stresses through phenotyping tools. Utilization of modern phenotyping tools in rice research may reduce the cost of data recording and save time.

References

Ayub M, Brook RM, Tariq M, Nadeem MA, Tahir M (2013) Genotypic diversity of perennial and annual wheat for root and shoot behavior. Romanian Agri Res 30:135–140

Bauw PD, Ramarolahy JA, Senthilkumar K, Rakotoson T, Merckx R, Smolders E, Van Houtvinck R, Vandamme E (2020) Phenotyping root architecture of soil-grown rice: a robust protocol 1 combining manual practices with image-based analyses. bioRxiv https://doi.org/10.1101/2020.05.13.088369

Bhugra S, Anupama A, Chaudhury S, Lall B, Chugh A (2017) Phenotyping of xylem vessels for drought stress analysis in rice. Fifteenth IAPR International Conference on Machine Vision Applications (MVA) Nagoya University, Nagoya, Japan, May 8–12, 2017

Bilal M, Ayub M, Tariq M, Tahir M, Nadeem MA (2017) Dry matter yield and forage quality traits of oat (*Avena sativa* L.) under integrative use of microbial and synthetic source of nitrogen. J Saudi Soc Agric Sci 16:236–241

Clark RT, MacCurdy RB, Jung JK, Shaff JE, McCouch SR, Aneshansley DJ, Kochian LV (2011) Three-dimensional root phenotyping with a novel imaging and software platform. Plant Physiol 156:455–465

Delgado C, Benitez H, Cruz M, Selvaraj M (2019) Digital disease phenotyping. IGARSS 2019 - 2019 IEEE International Geoscience and Remote Sensing Symposium, 28 July–2 Aug. 2019, Yokohama, Japan

Duan LF, Yang WN, Bi K, Chen SB, Luo QM, Liu Q (2011) Fast discrimination and counting of filled/unfilled rice spikelets based on bi-modal imaging. Comput Electron Agric 75:196–203

Hairmansis A, Berger B, Tester M, Roy SJ (2014) Image-based phenotyping for non-destructive screening of different salinity tolerance traits in rice. Rice 7:16

Igathinathane C, Pordesimo LO, Batchelor WD (2009) Major orthogonal dimensions measurement of food grains by machine vision using Image. J Food Res Int 42:76–84

Ijaz MK, Tariq M, Nadeem MA, Tahir M, Shah SA, Anwer A (2016) Forage production and pattern of forage quality indices during reproductive development of millet genotypes. J Environ Agric Sci 8:54–59

Kastberger G, Stachl R (2003) Infrared imaging technology and biological applications. Behav Res Methods Instrum Comput 35:429–439

Kawamura K, Asai H, Yasuda T, Khanthavong P, Soisouvanh P, Phongchanmixay S (2020) Field phenotyping of plant height in an upland rice field in Laos using low-cost small unmanned aerial vehicles (UAVs). Plant Prod Sci 23(4):452–465

Kim SL, Kim N, Lee H, Lee E, Cheon K, Kim M, Baek J, Choi I, Ji H, Yoon IS, Jung K, Kwon T, Kim K (2020) High-throughput phenotyping platform for analyzing drought tolerance in rice. Planta 252:38

Li D, Wang R, Xie C, Liu L, Zhang J, Li R, Wang F, Zhou M, Liu W (2020) A recognition method for rice plant diseases and pests video detection based on deep convolutional neural network. Sensors 20(578):1–21

Nguyen HT, Lee BW (2006) Assessment of rice leaf growth and nitrogen status by hyperspectral canopy reflectance and partial least square regression. Eur J Agron 24:349–356

Peerzada AM, Bukhari SAH, Dawood M, Nawaz A, Ahmad S, Adkins S (2019) Weed management for healthy crop production. In: Hasanuzzaman M (ed.), Agronomic crops; Volume 2 Springer Singapore, pp. 225–256

Razaq M, Shah FM, Ahmad S, Afzal M (2019) Pest management for agronomic crops. In: Hasanuzzaman M (ed) Agronomic crops, vol 2. Springer, Singapore, pp 365–384

Shibayama M, Sakamoto T, Takada E, Inoue A, Morita K, Takahashi W, Kimura A (2011) Estimating paddy rice leaf area index with fixed point continuous observation of near infrared reflectance using a calibrated digital camera. Plant Prod Sci 14:30–46

Siddiqui ZS, Cho JI, Park SH, Kwon TR, Ahn BO, Lee GS, Jeong MJ, Kim KW, Lee SK, Park SC (2014) Phenotyping of rice in salt stress environment using high-throughput infrared imaging. Acta Bot Croat 73(1):149–158

Sun J-P, Hou C-Y, Feng J, Wang X (2008) Determination of the protein content in rice by the digital chromatic method. J Food Qual 31:250–263

Tanabata T, Shibaya T, Hori K, Ebana K, Yano M (2012) SmartGrain: high-throughput phenotyping software for measuring seed shape through image analysis. Plant Physiol 160: 1871–1880

Tariq M, Ayub M, Elahi M, Ahmad AH, Chaudhary MN, Nadeem MA (2011) Forage yield and some quality attributes of millet (*Pennisetum americannum* L.) hybrid under various regimes of nitrogen fertilization and harvesting dates. Afr J Agric Res 6(16):3883–3890

Tariq M, Ahmed M, Iqbal P, Fatima Z, Ahmad S (2020) Crop Phenotyping. In: Ahmed M (ed) Systems modeling. Springer, Singapore. https://doi.org/10.1007/978-981-15-4728-7_2

Walter A, Liebisch F, Hund A (2015) Plant phenotyping: from bean weighing to image analysis. Plant Methods 11:14

Watanabe T, Hanan JS, Room PM, Hasegawa T, Nakagawa H, Takahashi W (2005) Rice morphogenesis and plant architecture: measurement, specification and the reconstruction of structural development by 3D architectural modeling. Ann Bot 95:1131–1143

Yang WN, Xu XC, Duan LF, Luo QM, Chen SB, Zeng SQ, Liu Q (2011) High-throughput measurement of rice tillers using a conveyor equipped with x-ray computed tomography. Rev Sci Instrum 82(2):025102

Yang Y, Chai RY, He Y (2012) Early detection of rice blast (Pyricularia) at seedling stage in Nipponbare rice variety using nearinfrared hyper-spectral image. Afr J Biotechnol 11:6809–6817

Yang W, Duan L, Chen G, Xiong L, Liu Q (2013) Plant phenomics and high-throughput phenotyping: accelerating rice functional genomics using multidisciplinary technologies. Curr Opin Plant Biol 16:180–187

Yoshioka Y, Iwata H, Tabata M, Ninomiya S, Ohsawa R (2007) Chalkiness in Rice: potential for evaluation with image analysis. Crop Sci 47:2113–2120

Zhou ZY, Zang Y, Luo XW, Wang P (2011) Color-based corner detection algorithm for rice planthopper infestation area on rice stem using the RGB color space. 2011 ASABE Annual International Meeting; Louisville, Kentucky, USA: 2011. Paper number 1111374

Zu X, Lu Y, Wang Q, Chu P, Miao W, Wang H, La H (2017) A new method for evaluating the drought tolerance of upland rice cultivars. Crop J 5:488–498

Rice Physiology Under Changing Climate 12

Rafi Qamar, Atique-ur-Rehman,
and Hafiz Muhammad Rashad Javeed

Abstract

There is an inseparable relation between agriculture and climate variables. The role of changing climate on food security has been at forefront of study and policy agendas in recent times. Agriculture and climate change are interrelated in a number of ways as climatic changes is one of the main cause of abiotic stresses, with unfriendly effects on agricultural production activities. Climate change impact is becoming fairly apparent, and improvements are even more visible in the world. As a result of climate change, intense abiotic factors such as frost damage, low and high temperatures, heavy rainfall, salinity, floods, and droughts are posing severe intimidation in production of rice and are also harmful to farmers earning from rice cultivation. Agricultural activities are influenced by climatic changes in numerous ways; for example, changes in plants growth, annual rainfall, heat waves, average temperatures, microbial activities, ozone depletion, momentous changes in atmospheric CO_2, and variations in sea levels. The threat of shifting global climate change has drawn a great deal of interest from scientists as these changes have adverse effects on global crop production and threaten food security issues worldwide. There is a serious need to build solutions against these pressures. In order to deal with those impacts, crop improvement may be a viable and successful counter. Advancement in molecular breeding can continue to harness the intrinsic ability of wild species by producing abiotic-tolerant lines. Wide tolerance screening in wild genotypes can be performed with the aid of

R. Qamar (✉)
Department of Agronomy, College of Agriculture, University of Sargodha, Sargodha, Pakistan

Atique-ur-Rehman
Department of Agronomy, Bahauddin Zakariya University, Multan, Pakistan

H. M. R. Javeed
Department of Environmental Sciences, COMSATS University Islamabad, Vehari, Pakistan

N. Sarwar et al. (eds.), *Modern Techniques of Rice Crop Production*,
https://doi.org/10.1007/978-981-16-4955-4_12

molecular markers to classify the underlying QTLs/genes. With the advancement of bioinformatics, DNA microarrays, mass spectrometry, RNA sequencing, or other new high-performance genomic techniques, it is now possible to decode the fundamental biochemical processes from a top-down approach. likewise, smart agricultural activities may be the best way to reduce the detrimental effects of climatic changes on crop adaptation until it could have a dramatic impact on global crop production. This chapter summarizes the causes of climatic changes, the physiological changes in crops, new breeding technology, the pressures generated by climate change, and biotechnological solutions to combat climatic changes in order to improve rice.

Keywords

Rice · Physiology · Climate change · Climate variability

12.1 Introduction

The atmosphere of the world has evolved across history. Natural structures, human well-being, and food production have been seriously impacted by catastrophic changes in the environment. Any of these temperature changes are due to very minor fluctuations in Earth's orbit that change the amount of solar energy our planet absorbs. With the exponential growth in the world's population, there is resulting increase in demand of food supply developing worries regarding the security of food (Ahmad et al. 2019; Ahmed and Ahmad 2020; Ahmed et al. 2020a, b; Fatima et al. 2020). Water supply, soil fertility, and air quality have a significant effect on agricultural production (Khan et al. 2019a, b). Climate change, also known as global warming, is an average rise in ambient temperature on Earth, which can be illustrated by some convincing facts (Atique-ur-Rehman 2018). According to evidence, climate change clearly influenced the world different regions, which will lead to droughts in the future due to decreased rainfall and increasing temperatures (Fig. 12.1). While the predicted changes in precipitation in the mid-to-late twenty-first century are unclear, increased incidence and intensity of extreme climatic events (severe storms, floods, droughts, etc.) are very much likely (Fig. 12.2). With sudden shifts in environmental conditions, attributable to the indirect and direct abiotic stress effects, the extreme impacts on plant production are increasing at considerable strength (Fig. 12.3). The ongoing deforestation and abundant use of fossil fuels cause the rise in the concentration of CO_2 in atmosphere, which has increased from 280 to 400 mol^{-1}. It is expected that the CO_2 concentration will increase twice, that is, to 800 mol^{-1}, by the end of this century. Emissions of toxic gases, in particular, CO_2, are the key reasons for the greenhouse effect and the colder average global temperatures. The impacts of change in climate and environmental variation are mainly calculated by sum of stress cycles, their effect on everyday life, and the damage to agricultural crops. In developed nations, crop yields are primarily impacted by unfavorable environmental conditions, with high temperatures and

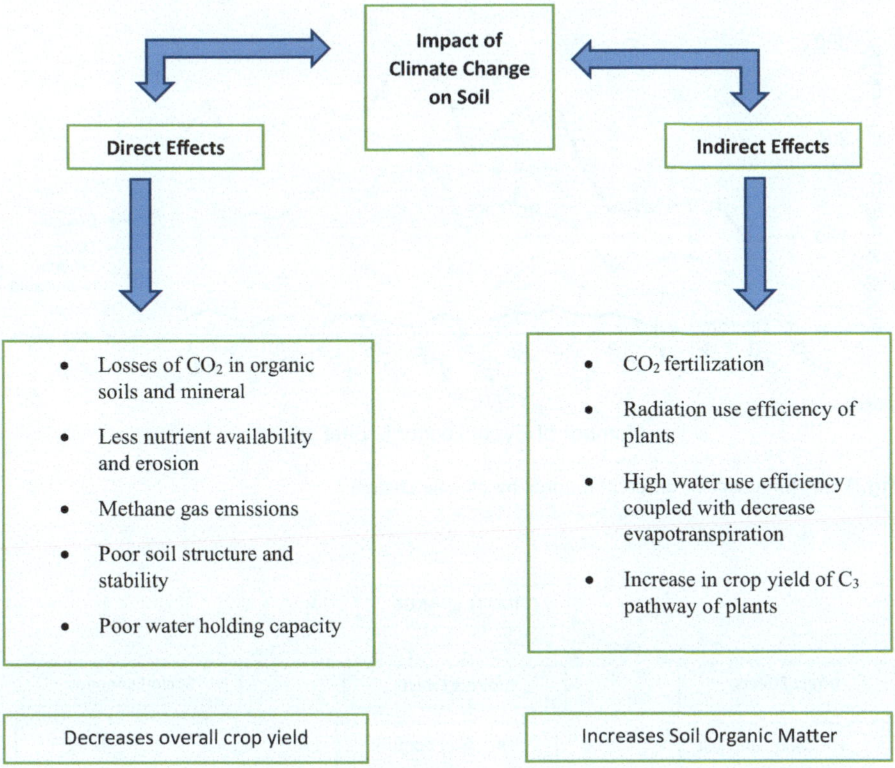

Fig. 12.1 Effect of climate change on soil fertility

excess CO_2 accumulation, pushing scientists to formulate new methods to overcome less predictable problems. In order to overcome these shortcomings and ensure food security, new climate-smart crop cultivars need to be developed. Abiotic stresses highly affect the growth of plant and its yield. Under the natural climatic conditions, plants also face several stresses such as cold, drought, waterlogging, salinity, and humidity (Fig. 12.4). Abiotic causes also include gas emissions, UV radiation, light intensity, unexpected flooding, and chemical and physical factors that create more stress.

Rice (*Oryza sativa* L.) is a staple food for hundred thousands of people across the globe (Atique-ur-Rehman 2018). Significant utilization of rice is occurring in subtropical and tropical Asia. Rice is cultivated worldwide as a soil and/or lowland crop. Though lowland irrigated rice is around 57% of total rice-growing land, it contributes to almost 76% of rice production globally. The period of rice development can be broken into three separate stages of growth:

1. Germination and tolerance.
2. Vegetative.
3. Reproduction, grain-filling, and grain ripening stages.

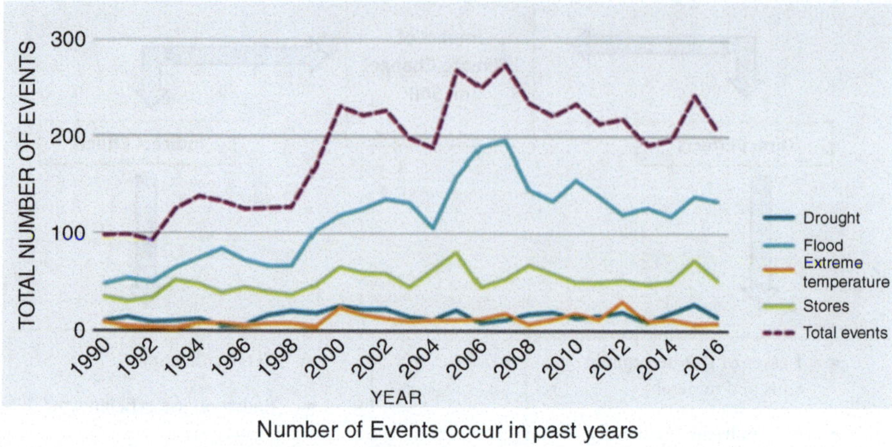

Fig. 12.2 Summary of different hazards by climate change

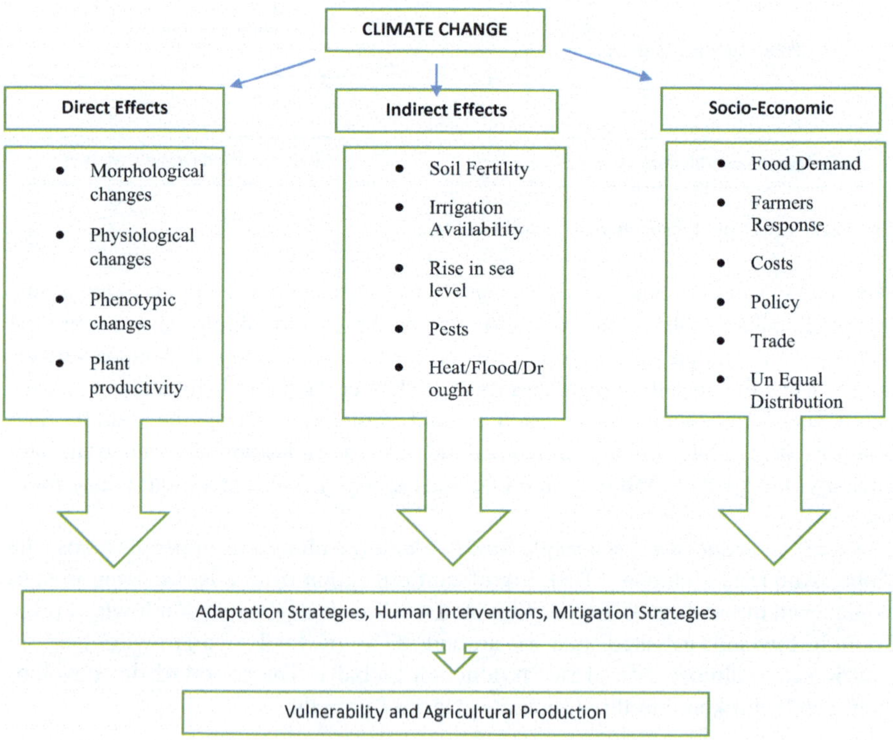

Fig. 12.3 Direct, indirect, and socioeconomic effects of climate change on agricultural production

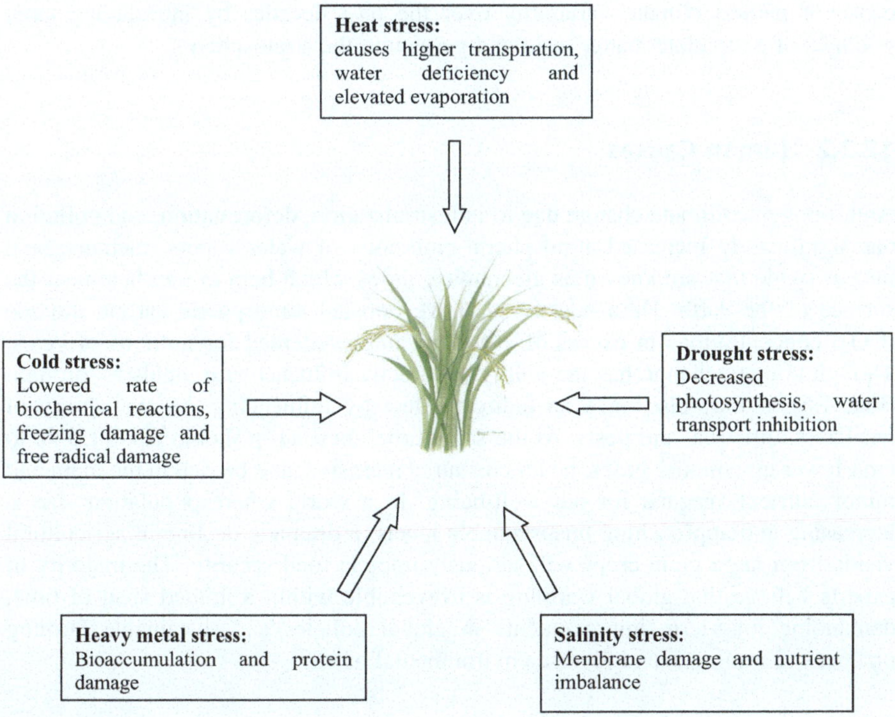

Fig. 12.4 Effect of different abiotic stresses on growth and morphological traits in rice

For a certain period of time, the rice plant remain in each growth stage then enter to the next growth stage. Tillering, root growing, stem growth, and growth of leaf are the main activities of the vegetative growth process. Booting, flowering, and panicle production (size of panicle and panicle spikelet's per panicle) are the primary events or stages of reproductive growth. During the grain–/spikelet-filling point, the weight of the grain/spikelet is determined. This chapter explores the physiology of the different stages of growth, formation of aerenchyma, and stomatal development in production of rice.

12.2 Causes of Climate Change

12.2.1 Natural Causes

The atmosphere of the world is complex and is always evolving by natural means, and there are a variety of natural causes responsible for climate change. Mainland drifts, volcanoes, ocean tides, world tilt, and meteorites are more dominant. Studies have demonstrated that massive explosive volcanic eruptions have been a significant cause of natural climate variability over the past decades by introducing large volumes of particulate matter (ash) and gases into the stratosphere.

12.2.2 Human Causes

Anthropogenic climate change due to industrialization, deforestation, and pollution has significantly increased atmospheric emissions of water vapors, methane, and nitrous oxide that are known as greenhouse gases which help to trap heat near the surface of the earth. Human activities have boosted atmospheric carbon dioxide (CO_2) concentrations in excess of 400 ppm, unprecedented for millions of years. Though climate change has the ability to directly influence rice yields by drought (Fig. 1.5), it may also have an indirect effect by influencing the availability of fertilizers, diseases, and pests. At the same time, these crop should also be able to emit fewer greenhouse gases, be less resource intensive, and be rich in the main and minor nutrients needed for our well-being. In a world where population size is increasing and approaching unsustainable levels, a dramatic decline in agricultural yields from large grain crops will seriously impede food security. The majority of experts believe that global warming is irreversible within a limited span of time, demanding long-term improvements in global policies and sustainable farming practices to minimize and reverse environmental harm.

12.3 Climate Change Impact

Global climate change has also had noticeable environmental consequences, and such lasting consequences include changes in growing seasons, precipitation levels, and extreme drought and heat waves. Climate change impact analyses can be performed on the basis of four main sectors of economy, including water, ecology, natural habitats, and agriculture. This is the reason why a minor climate change will adversely impact different farming practices. Surface air temperature and precipitation in the northeast area are projected to increase from 1.8 °C to 2.1 °C and 0.3–3%, respectively, in the 2030s to the 1970s. It will adversely impact paddy production, tea plantation, and other agricultural crop production, which will degrade the overall livelihoods of local residents. The major issue is paddy cultivation as rice is the most important food grain crop, followed by maize.

12.3.1 Influence of Climate Alteration on Rice Growth Stages

Climate change would decrease the yield of rice by 4.5–9 by 2039. Global warming forecasts suggested a rise in the duration of heat days and colder nights, with additional difficulties in producing higher crop yields. Monsoon rainfall is not the only environmental aspect affecting the yield of kharif rice. The yield of paddy and its response to climate change can be calculated using crop simulation models such as the GIS-based GEPIC model.

Relevant rice production parameters include climate variables such as rainfall, temperature, solar radiation, and atmospheric CO_2. The rise in temperature and rainfall variability was shown to be beneficial and detrimental to autumn and winter rice yields, but these variables were both favorable and negligible for summer rice. Rice spikelet affects yield reductions, while the rising pattern of atmospheric carbon dioxide concentration could increase rice yields and could decrease the daily maximum temperature trend. Threshold-bound temperature decrease the length of rice crops and also contribute to increased sterility of the spikelets, reduced duration of grain filling, and increased breathing rates, resulting in reduced production and reduced rice grains. Especially at the flowering stage, rice is very responsive to high temperature.

Rice yield is lowered by the increase in night temperature in cycle with global warming. In particular, throughout the latter part of the growing period, rice yield is typically decreased by higher temperature and lower solar radiation. Warming effects on rice phenology have been researched and it has been discovered that rice development in China has shortened over the past three decades. During the floral cycle of rice, temperature stress was correlated with sterility of the spikelets.

The increased CO_2 level of 340–680 ppm could increase major crop yields by 10–15%, in particular in C_3 plants like rice, but it is not appropriate that the incidence of photosynthetically active radiation (PAR) is likely to decrease by 1%.

12.3.2 Association of Flowering Time of Rice with Photoperiod and Temperature

Only part of rice life cycle, that is, between emergence and flourish, the rate of growth of rice generally is sensitive to photoperiod. In plants, three stage of the preflowering growth, namely young, inductive, and post-inductive stages, can be established. Just the length of the inductive process impacts photoperiod. The most significant exception in rice, which can last for a span of 50 days, which is very long, during the pre-inductive step also known as the simple vegetative process or BVP. Quantifying and reconstructing climate change reactions in reference to the use of the required temperature control unit and not just the atmospheric air temperature of utmost significance. It is, thus, shown that flowering is determined mainly by temperature and photoperiod responses in one season. Higher temperature in coming days rice cultivars with an increased tolerance of combined drought and heat stress during the flowering stage would be necessary to ensure the continuous implementation of water-saving technology to supplement success in the management of damage during other critical stages of growth, such as flowering.

12.3.3 Plant Yield and Climate Change

Plant physiology was affected by many forms owing to climatic fluctuations. Extremes in the atmosphere and variability in the ecosystem raised the risk of a variety of plant stresses. Climate change has a direct, secondary, and socioeconomic effect on crop production (Fig. 12.3). Furthermore, as the Food and Agriculture Organization has stated, climate change (drought, flooding, high temperatures, storms, etc.) has risen significantly (FAO). Increased greenhouse gas emissions and sudden changes in temperature, which would boost agriculture yields in the north, have been expected. The production of wheat is severely impacted by high temperatures in many countries due to climate change and can decrease crop yields by 6% with every 1 °C temperature increase. Drying and high temperatures are main stress factors, and Rubisco, the core photosynthesis enzyme, is disturbed if the temperature rises by 35 °C and slows the photosynthatic process. The negative effect of heat stress on antioxidant enzymes in *Zea mays* was stated by Gong et al. (1997). In sorghum, maize, and barley, the combined effect of drought and thermal stress on crop yield was measured. The cumulative influence of heat and drought stress has been found to have more detrimental effects than human stresses. Xu et al. (2005) were exposed to mixed drought and hot stresses and found that the photosystem II feature (PSII) decreased. The reproductive cycle of plant development is affected by climate change namely water deficits, and intense temperatures. The floral initiation and inflorescence were identified as badly influenced by water tension. Likewise, it can induce sterility in cereals if the temperature rises by about 30 °C during floret growth. In the meiotic process, the volume of wheat and rice was decreased by 35–75% due to the water deficit. In rice, the process of fertilization and anthesis is significantly influenced by dryness (Atique-ur-Rehman 2014, 2018). The harvest index is lowered to 60% and the crop set declines as a result of the water deficit. During the 1980s El Niño, the cocoa yield in West Africa was reduced drastically by the big drought events. It is proposed that agricultural produce will be the most impacted crop in Mexico by 2080 to be 25.7% due to climate change. Zhao et al. (2017) performed an experiment to examine the effect of climate change on major agricultural yields and showed substantial yield declines of 3.2%, 6%, 7.4%, and 3.1%, respectively, for corn, wheat, maize, and soya. New research in genomics allows for climate-smart agriculture through the production of climate-resilient plants to combat climate change. Drought stress affects wheat in every growth phase, but the most important of all is grain formation and reproductive phase. During moderate-drought stress in post-anthesis, the yield of wheat has been reduced by 1% to 30%, whereas in the event of a sustained mild-drought stress, it has risen to 92%. Dried stress decreased the yield of large grain legumes substantially. Drought stress decreased the yield of mash bean (*Vigna mungo* L.) from 31% to 57% in the floral stage, while drought stress during the reproductive stage was recorded a 26% decline. Maleki et al. (2013) also recorded a major drought stress impacting soybean yield and a 42% decrease in soybean grain filling was observed. Schlenker and Roberts (2009) identified the increase in maize yield at optimum 29 °C, but further temperature growth hindered maize yields. The maize yield was adversely affected by each 1 °C increase in temperature. Similarly, the maize yield decreased by 8.3% with an ideal growth temperature increase of 1 °C. Brown (2008) recorded a 10% decrease in wheat yield for each 1 °C temperature rise. Another study indicates that wheat yield falls by 3–4% with every 1 °C temperature rise. Easterling et al. (2007) defined that a 2 °C temperature increase induces a 7% yield decline, while an additional 4 °C temperature increases decreased yields of wheat by up to 34%. Similarly, with each temperature increase of 1PC, rice yield decreased by 2.6%. The threshold temperature for soybean has been found by Schlenker and Roberts (2009), but after this, additional changes in temperature have abrasively cut yields by 30 °C. This is to the maximum level. Soybean yields have risen at the same level. Eastburn et al. (2010) reported that increasing atmospheric CO_2 and ozone levels affected the kind of disease and influenced the kind of disease by an improvement in ozone and atmospheric CO_2 levels and increased the sensitivity of the disease to soybean by a steady increase in temperatures. Susceptibility to soybean was improved by persistent increase in temperature sickness. By applying different techniques; the number of scientific studies has significantly increased, in line with a biotic and biotic pressure in plants, by using various strategies (Fig 12.3).

12.3.4 Stem Physiology

The start of the tillering in rice begins at 3–4 leaf stage of seedlings. The secondary shoots that occur on the main shoot are called tillers. The completion of tillering takes 2–3 weeks. When rice plants having N contents >3.5 and no adequate light energy is present than rice stem phyiology continue without limitation (Evans et al. 1975). Concentration of P in the rice stem below 0.25% can limit tillage (Counce et al. 2003). The optimal water temperature for tiller emergence is 31 °C during day and 16 °C during night. The temperature of the water above 31 °C influences the emergence of the tiller. The tillering begins when adequate light hits the base of the plant during tillering period (Counce et al. 2003). During the rice growth cycle, some tillers can die and excess quantities or nutrients in tillers can be shifted to other good tillers (Evans et al. 1975). Sharma et al. (2017) isolated the MONOCULM 1 (MOC1) gene that regulates rice tillering and characterized it. The gene MOC1 codes regarded as GRAS belongs to nuclear protein families expressed in axillary buds thatinitiated the development and growth of the axillary buds.

Different hormones, such as auxin and cytokinin (CK and strigolactones (SL), play a significant role in the tillering method (Kebrom and Richards 2013). Auxins suppresses the axillary bud's outgrowth. These auxins are manufactured in young leaves and effectively but mainly transported through the phloem (Agusti and Greb 2013), which we call the Polar Auxin Transport (PAT). PAT is assisted by the familiar protein auxin and PIN-FORMED efflux carriers of ATP-binding cassette B, as stated by Zažímalová et al. (2010) (ABCB). Polar Auxin Transport helps auxin in passing through the epidermal layer of the outermost shooting apex and enters plant organ initiation positions. From here, auxins pass through the developing primordia toward the basipetal stream of the main shoot. PIN1, which is present in xylem parenchymal cells, supplies the addition of auxin to the basipetal main-shoot stream and plays a major role in PAT (Petrášek and Friml 2009). In terms of OPINlb (formerly called REH1), tillering in rice is improved (Xu et al. 2005; Chen et al. 2012). OsPIN2 overexpression improves tillering but reduces the size (Chen et al. 2012).

12.3.5 Leaf Physiology

Rice produces 10 leaves before the reproductive stage starts (Itoh et al. 2005). For each rice node, the leaf creation events take place as follows:

1. Leaf elongation.
2. Leaf initiation.
3. Collar formation.
4. Leaf sheath elongation.
5. Leaf blade maturation.
6. Node formation.
7. Internode elongation.

Just five internodes of the principal rice stem are extended (Moldenhauer et al. 2003). The sheets are continually shaped. The key activities that take place during leaf growth are tissue difference, cell division/expansion, tissue specification, and axis determination (Itoh et al. 2005). The mature rice leaf is strap like, and the proximal distal axis can easily be differentiated in three places. The leaf sharp edge is distal and is the primary site of photosynthesis. Leaf sheath is the next area which, therefore, protects the shoot apex while avoiding physical injury to the younger ones. There are three different sections, that is, a liger, atrium, and a lama joint (col), on the boundary of a leaf blade and leaf sheath (Itoh et al. 2005). The ligula is membranous and acuminated, usually broken between the mature leaves into two parts. The collar or lamina is a white region of the base of the leaf blade that makes the blade curve to the abaxial section. The atria (two in number) are thin, long-haired appendices on the margins of rice leaves (Itoh et al. 2005).

The adaxial/abaxial axis polarizes rice pads. There are many papillae and two separate forms of trichrome on the full surface except the adaxial surface of the sheath of the leaf. Cells bulliforms, grouped in vertical rows, hung in the epidermis of the leaf blade between vascular bundles. The leaves contain tiny and broad vascular bundles (Itoh et al. 2005). The phloem and xylem are found on the side of both abaxial vascular and adaxial vascular bundles. These packets are contained by the sheath cells in bundles. The adaxial/abaxial blade has vascular bundles that contain sclerotic fiber cells. The shape of the sheath and the blade is distinct with the margins pointing and membranous margin of the sheath (Itoh et al. 2005).

Leaf primordium appears like a swelling at the edge of the apical meristem of shoot. It develops opposite the spring apex and apical meristem and develops a primordium with the outline of a crown. In leaf primordium, the division of cells is many times greater than that of shoot apical meristem as restricted by the term histone H_4 (Itoh et al. 2000). The leaf cell process can be differentiated by using several markers from the leaf primordial stage. The rapid cells division/lengthening allows the primordial leaf to become the cap in apical and marginal regions and the start of the processor strand can be seen at this point in the middle of the leaf. In comparison, all primordial leaf margins cross and circle the apical meristems of the shoot (Itoh et al. 2005). At this point, the primordial form of the leaf tends to be cone like, and the border of the lame is evident. The protrusion of the primordium ligule originates at the border of the adaxial surface of a blade layer (Itoh et al. 2005).

The leaf blade spreads very quickly and exceeds the full length after ligula differentiation, but the elongation of the leaf blade remains the same. This leaf blade expansion is due to the increased activity of intercalary meristones in the basal area of the leaf blade (Kaufman 1959). At this stage, cell differentiation is decommissioned for the expression of genes (e.g., OsPNH1, OsSCR, and DL) (Nishimura et al. 2002). After leaf blade extension is ended, the sheath elongation becomes rapid. Unequal extension forces the blade to bend into the laminar joint in the adaxial and abaxial cells (Maeda 1965).

12.3.6 Root Physiology; Aerenchyma Formation

The supply of oxygen (O_2) is gradually reduced after flooding the rice field within 24 hours, O_2 is utlization by the soil bacteria (Pavanasasivam and Axley 1980). The roots of rice need to keep the oxygen alive and work properly. The roots of most flooded mineral soils are covered with iron. The siderophores are correlated with the iron and oxygen needs to be converted from ferric iron into ferrous. The leaves of rice can die in three to six phyllochron, so that they are not able to supply an oxygen. The knots are permanent and lead oxygen into the roots from the above flood water. Aerenchyma is a tissue which is able to lead oxygen, and produces broad intercellular spaces by orderly killing many plant tissues (programmed cell death) (Daneva et al. 2016). Three decades earlier, it has been reported that the acidification of cytoplasm and the lack of plasma membrane integrity showed the precedents of rice cell death and gas spaces radially scattered (Cong et al. 2017). Cell mortality is normally very fast in roots and coleoptiles. Tonoplast rupture is the first stage in the initiation of aerenchyma cells, preceded by the plasma membrane rupture, cytoplasm, wall loss, and cellular material harm (Kitambi et al. 2014). Aerenchyma is well adapted for sheathing the leaves, internodes, stems, and leaf midribs, which promote internal aeration between roots and shoots (Daneva et al. 2016; Ingram et al. 2015). In effect, the submerged leaves have developed gas films that promote the exchange of carbon dioxide (CO_2) between plant leaves and water, thus ubmarine net photosynthesis is improved by enhancing the absorption ofoxygen in night time and carbon in day time (Pincebourde and Casas 2016). This leaf gas filing, thus, facilitates photosynthesis in underwater sugar production in leaves and enables the springs and roots to be able to function properly (Ezquer et al. 2020).

For plants to live and work in submerged environments (e.g., rice), aerenchyma formation is essential. Aerenchymatic cells not only provide oxygen but also provide numerous gases, such as methane and carbon dioxide, from root to root and from soil to air (Colmer 2003; Lee et al. 2003). This ventilation is primarily due to gas delivery (Armstrong and Bruce 1996). Rice cells are called lysigenous aerenchymas, formed in root cortex, in pith cavity, and in stem cortex (Yang et al. 2011).

In rice, aerenchyma development starts in the apical root and then eventually extends into the basal components (El-Kereamy et al. 2015). Completely formed aerenchyma cells are isolated from the root's inner root by basal portions of the roots (Armstrong and Bruce 1996; El-Kereamy et al. 2015). In roots, the development of lysigenic aerenchymas can be improved under ventilated conditions by application of ethylene, and deoxygenated under stagnant conditions (0.1% agar) by silver ions (Wiengweera et al. 1997). During aerenchymatic development, cell wall degradation is carried out, and this mechanism is mediated by modulation of cell wall enzyme degradation through various enzyme activities such as xylanolytic, pectolytic and cellulolytic are engaged (Jackson and Armstrong 1999; Lee et al. 2003).

12.3.7 Leaf Physiology; Stomatal Formation

Stomatous structures are the microscopic opening structures in plant leaves epidermis. In rice, the stomachs are spread vertically on the surface of the leaf. The stomatous on the internerve region of the cells in adaxial face of the epidermis are primordia. The distribution of stomata cell rows is not spontaneous in rice leaf epidermis and these cells are situated at the borders of vascular groups. Both lines are further distinguished into two adjacent files (Itoh et al. 2005). Uniform and specific expression of the OsSCR gene in the stomach cell rows is expressed reliably and in particular in the stomatic cell rows. The rice stomata has two thin walled cell inwards and two neighborhood subsidiary cells. The development of stomatal cells occurs basipetally in leave ligule primordium on the blade epidermis. First, asymmetrically differentiated cells in the stomata comprise nonspecialized mother cells and epidermal cells. NEC is broad and mildly colored, while GMC is small and heavily colored (Itoh et al. 2005). The stomata on the adaxial surface of the sheath are primordial. The distribution of stomata cell rows is not spontaneous in rice leaf epidermis and these cells are situated at the borders of vascular groups. Both lines are further distinguished into two adjacent files (Itoh et al. 2005).

12.3.8 Grain Physiology

In rice, grain growth begins with double fertilization. Pollen germinates into a pollen tube after pollination that spreads to ovaries (Farooq et al. 2014). The development of pollen grain includes the energy supplied by the invertase activity of acid in the pollen tube growing. When fertilized, the endosperm and the emerging embryo mainly require nutrients supplied by sucrose via phloem transport. The extension of cells allows the caryopsis to stretch to the full lamb and palea space (the rice hull).

After 01 caryopsis is interrupted, the grain-filling process begins. At the end of the cell elongation, there is no accumulation of starch. Synthetic biochemistry is mainly done in two cell organelles, cytosol and plastid. The carbon path in the cytosol is from the sucrose imported. During grain production, the sucrose is separated by a saccharose synthase into UDP fructose and UDP glucose. The main mucolytic enzyme in the endosperm of rice is one or more of the isoforms of the sucrose synthase (Avigad and Dey 1997). Three isogens of sucrose synthase were associated with rice that coded different enzymes involved in various tissues and stages of development (Smith and Dilday 2002). The following step is to convert UDP glucose into glucose^{-1} phosphate.

The UDP glucose pyrophosphorylase is the enzyme concerned. Fructose can become glucose phosphates and eventually become starch by different enzymes. The next step is the transformation of G-1-P into G-6-P through phosphoglucose isomerase. At this point, the starch synthesis begins with the ADP glucose pyrophosphorylases in the cytosol and amyloplastic. Starch production in plastid (in cytosol) is regulated by AGP (Shannon et al. 1998; Farooq et al. 2014). The starch synthesis of ADP glucose and the further starch synthesis happen in the plastid. The single glucosyl unit adds straightened and branched chains made with starch synthase once the starch synthesis has begun. In the branching of starch chains, the additional starch synthesis is also occured through the starch branchingenzyme. For progressive formation, assembling, reassembling, and disassembling of developing endosperm, the branching, resizing, and debranching of the stub are essential. The enzymes involved are a starch branching enzyme, a depositary enzyme for starch synthesis (called starch synthase Myers et al. 2000). These events tear strongly formed granules, in which the starch is packed less and more branched amylopectin in alternating areas o1 (Myers et al. 2000). The stutter formation is very volatile in rice and other cereals and can fluctuate dramatically with temperature regimes. Starch rice granules are thinner than the majority of the crops. In contrast to other starch synthesis enzymes, the most sensitive enzyme is starch synthase (Keeling et al. 1994). High temperatures will decrease the activity of starch synthase during grain filling due to chalkiness. Potassium is essential for the correct operation for the starch synthase (Marschner et al. 1996). The creation of individual starch molecules, thus, represents the first step in grain growth. These molecules of starch later form the granules of starch (Myers et al. 2000). The aleurone cells are packed with protein and lipids after loading of endosperm cells with starch. Lipid, protein, and starch are found in subaleurone cells. The endosperm cells are not only starchy, but also contain a slight amount of protein (6–7%) (Juliano 1985). The genes involved in grain filling are common and well known in most cereals (Smith and Dilday 2002).

12.4 Existence of Abiotic Stress as a Result of Climate Change

Several abiotic stresses have arisen lately because of climate change. These impacts on rice's anatomy and physiology, which hinder the production and growth of rice plants overall. Below is some of the adverse conditions:

12.4.1 Drought

The adverse effect on rice crops caused by water shortages will remain largely below than higher atmospheric carbon emission levels, which demonstrates the relationship of CO_2 and drought (Fig. 12.5). The study showed that humidity stress affected rice morphologically (reduced height of plant, germination, number of tiller, biomass of plant species, sheet features, and various root) and physiologically (reduced transpiration, photosynthesis, stomatic behavior, water quality, chlorophyll content, relative water content, photosystem II behavior, membrane stability, and discrimination in carbon isotope and abscisic acid). Diminish grain production, the impaired growth

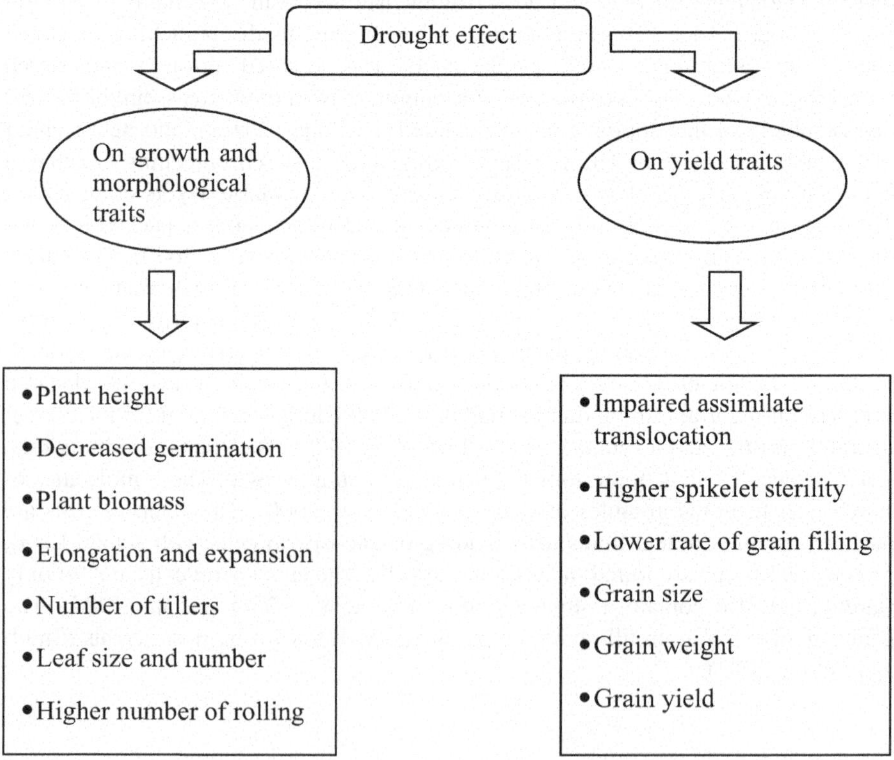

Fig. 12.5 Drought effect on various growth and developmental stages

of pollens during meiosis and panicle excretion, typically 70–75% of the spikelet's sterility under water stress, affect the reproductive level. It prevents processes including anther dehiscence, removal of pollen, germination of pollen, and fertilization.

Deficiency stress is extremely harmful to rice because it affects the rice's physiological, morphological, and efficiency parameters. The physiological basis of rice drought tolerance has been determined by many experiments. Ji et al. (2012) reported that peptidyl prolyl cis/trans isomerases (PPIases) catalyze, Cu/Zn-superoxide dismutase (CuZnSOD) and ascorbate peroxidase (APX) activity have been upregulated by drought stress.

Polyamine exogenous application increased water interaction, net photosynthesis, proline aggregation, anthocyanin, soluble phenolic accumulation, and also added to the oxidative harm to cell membranes (Farooq et al. 2009). In addition, improving the plant protection system for antioxidants can help to cope with drought stress in rice (Pandey and Shukla 2015). With increased drought stress in rice, ascorbate activity, glutathione (Selote and Khanna-Chopra 2004), dismutase of superoxide, reductase of monodehydroascorbate, reductase of the dehydroascorbate, reduction in glutathione (Sharma and Dubey 2005), ammonia lyase, and catalase phenylalanine (Shehab et al. 2010) are steadily improved and are very useful to the droughts' resistance. Abscisic acid can help trigger dryness tolerance in rice, strengthened status of the enzymes antioxidant, and enhanced transport of proteins and carbon metabolism, as well as expression of stress proteins (Latif 2014; Zhou et al. 2014). The overexpression of C4 photosynthesis enzymes including carboxylase phosphoenolpyruvate and orthophoto-dyskinesis pyruvate can be of benefit to rice plants in avoiding dryness (Zhou et al. 2011; Gu et al. 2013).

12.4.2 Crop Adaptation to Overall Extreme Climate Stresses

The atmosphere is severely altered and abiotically stressful by the rise in the Earth's temperature. Environmental changes are very negative and endanger crop species naturally. Overall, global warming and climate change both have some negative and positive effects on agricultural crops as well as on humans (Fig. 12.6). Drought and heat are the most common stress under field circumstances and have a major effect on plants. Plants need an ideal temperature for natural blooming and growth. It is stated that the changes in temperature greatly affect physiology of plants. Thermal stress influences the plant growth and grain yield, cold stress induces sterility, and drought stress has a negative impact on the morphology of sterility and on the plant morphophysiology (Table 12.1).

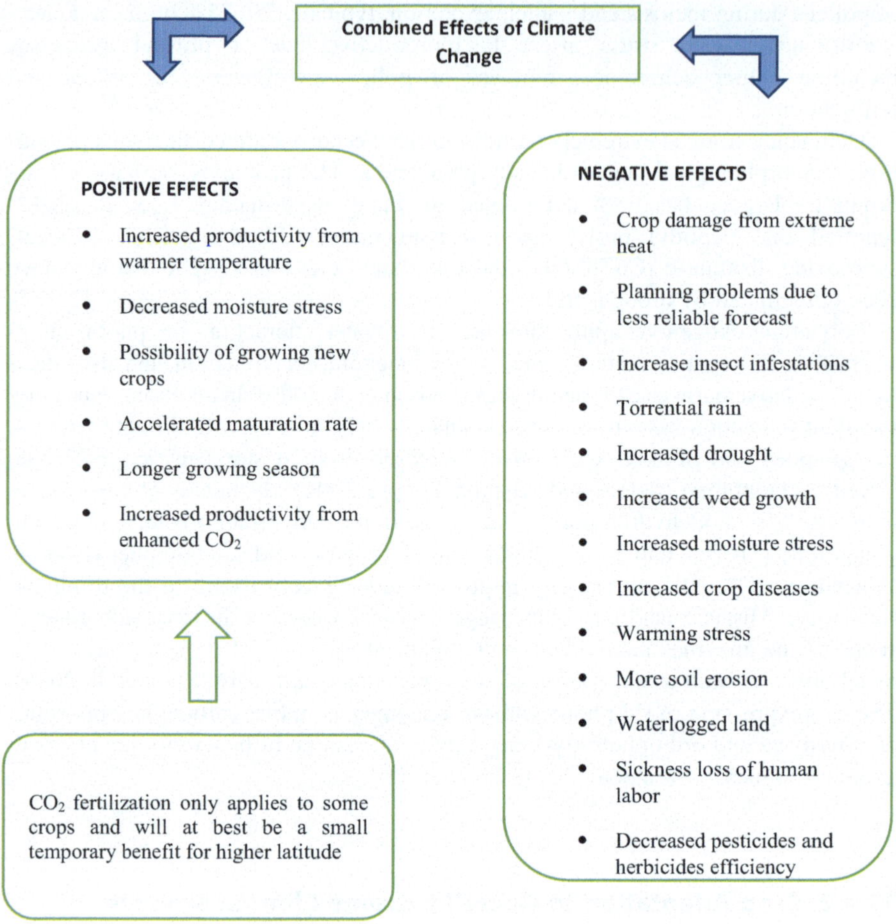

Fig. 12.6 Positive and negative effects of climate change on rice and humans

12.4.3 Plant Hormone Responses in Abiotic Stresses

Hormones are very essential for controlling several signaling reactions and pathways such as abscisic acid (ABA), salicylic acids (SA), and ethylene under various abiotic stresses (Fig. 12.7). The ABA's key function is to regulate stress reactions by associations with several other hormones, as shown in the crosstalk. In order to control climatic stress in the plant, ABA is the most important hormone. In various phases of plant growth, ABA has a major role, particularly in the closing and opening of stomata, drought stress, germination of seeds, and dormancy. PYR/PYL/RCAR-PP2C-SnRK2 is known as an ABA-generated signaling cascade and effectively tracks seed dormancy. Under conditions of drought, the development of a plant is severely slow down and the ABA levels in cells rise. Accumulation of

Table 12.1 Effect of abiotic stresses on rice crop's morphological, physiological, and yield parameters

Stress type	Stage of imposition	Trait	Cultivar	Decrease over control (%)
Drought	Grain filling and reproductive	Photosynthetic rate	Super Basinati	21.9
Drought	Grain filling and reproductive	Photosynthetic rate	Shaheen Basmati	15.0
Drought	Seedling stage	Transpiration rate	Mahsuri	173.3
Drought	Grain filling and reproductive	Transpiration rate	Super Basmati	36.0
Drought	Grain filling and reproductive	Transpiration rate	Shaheen Basmati	30.3
Drought	Grain filling and reproductive	Grains per panicle	Super Basmati	3.3
Drought	Grain filling and reproductive	Grains per panicle	Shaheen Basinati	3.9
Drought	Seedling stage	Chlorophyll contents	IR-29	113.4
Drought	Seedling stage	Chlorophyll contents	Pokkali	33.1
Drought	Reproductive stage	Chlorophyll contents	KDML105	9.3
Drought	Seedling stage	Chlorophyll contents	Pusa basmati	120.5
Drought	Seedling stage	Shoot weight	Mahsuri	33.9
Drought	Seedling stage	Leaf area	Mahsuri	211.6
Drought	Seedling stage	Sugar in leaf blade	Mahsuri	22.8
Drought	Seedling stage	Starch in leaf blade	Mahsuri	125.6
Drought	20 d before heading	Grain yield	Gangyou 527	22.1
Drought	20 d before heading	Grain yield	Yixiangyou 9	46.7
Drought	Whole cycle	Grain yield	Yangdao 6	3.14
Drought	20 d before heading	Grain yield	Gangyou I 8S	31.0
Heat		Grain weight	L-204	92.8
Heat		Grain yield	L-204	59.1
Heat	Booting	Grain yield	Hovaze	15.3
Heat	Booting	Grain yield	Fajr	37.5
Heat	Booting	Grain yield	Hashemi	30.4
Heat	Postheading	Grain yield	S-NMLYtz	1.5
Heat	Postheading	Grain yield	S-SWP	6.2
Heat	Postheading	Grain yield	DE-SMLYtz	9.7
Heat	Postheading	Grain yield	DE-SC	4.6
Salt		Root length	SS	13.2
Salt	–	Root length	RD6	25.5
Salt		Leaf area	ST	43.4
Salt		Leaf area	RD6	494.9

(continued)

Table 12.1 (continued)

Stress type	Stage of imposition	Trait	Cultivar	Decrease over control (%)
Salt	–	Chlorophyll contents	Ss	237.0
Salt		Chlorophyll contents	ST	259.0
Salt	–	Chlorophyll contents	RD6	268.0
Salt	–	Photosynthetic rate	RD6	40.7
Salt		Number of tillers	BR-II	5.5
Salt	–	Plant biomass	BR-11	4.32
Salt	–	Tiller per plant	M-202	161.7
Salt	–	Plant height	BR-11	11.6
Salt		Plant height	BBRI dhan44	16.7
Salt	–	Relative water contents	BBRI dhan44	25.5
Salt	–	Panicle number	BBRI dhan44	42.4

ABA in drought controls perspiration and prevents the disclosure of the stomach. In plants during water shortage, ABA also stimulates many physiological pathways, controls stomachs to close, and produces many genes that are stress intensive. Recently, ABA signaling machine has been studied and its operating mechanism has been explained. There were three units of signaling cascade, SnRK2/OST1 (protein kinase), PP2C (protein phosphatases), and PYR/PYL/RCAR proteins.

The ABA PYR/PYL/RCAR receptors were discovered by two separate groups of scientists. PP2C was first noted as the negatively regulated ABA in Arabidopsis knockout of abi1^{-1} and abi2^{-1}. Likewise, the protein kinase, which is the ABA activator, was extracted and isolated in SnRK2. Plants growing in stress weather conditions, salicylic acid has also controlled many physiological processes. Acetyl salicylic acid has been reported to promote the production of the protoplasmic clusters in corn, which regulate the cell cycle. SA's function was discovered by a group of scientists working on tobacco cell cultures, which was that SA controls bud growth and flora. Recent SA studies identified its effects on fruit production, nodulation of legumes, resistance to temperature, stomatic closure, breathing, genes related to senescence, and cell formation. The concentration of ABA in cells is increased. Accumulation of ABA in drought controls perspiration and prevents the disclosure of the stomata. In plants under water shortage condition, ABA also stimulates many physiological pathways, controls stomata opening and closing, and produces many genes which reduced stress intensity. A recent study was carried out on ABA signaling machines with an elucidation of their operating mechanism.

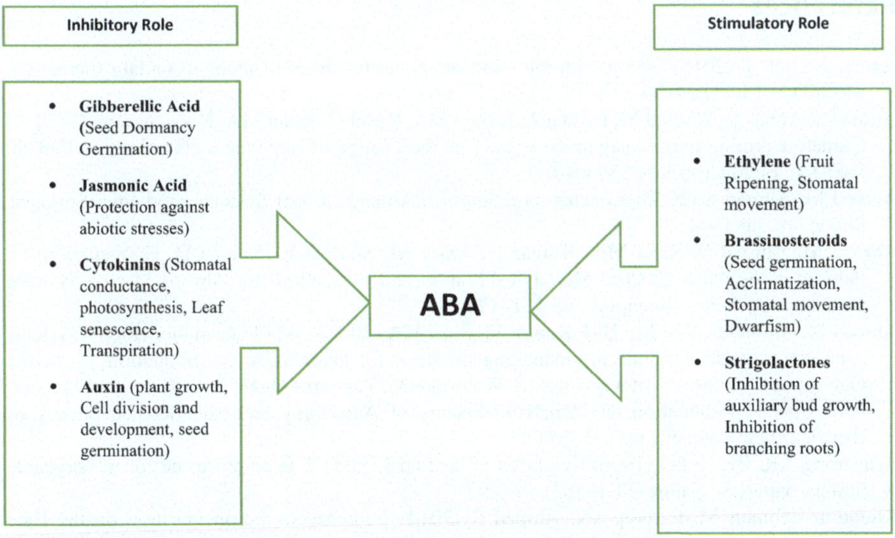

Fig. 12.7 Hormonal response under various stresses

12.5 Conclusion

The rice growth cycle can be separated into three different levels, namely germination/seedling, vegetative growth, and grain formation and development. The reproduction stage is the most critical step, in particular the production of grain, which determines its yield capacity. The expression of special genes and homeostasis of plant hormones control the growth of floral, gametogenesis, and grain development. Understanding the physiology of plants can help in increasing yields and productivity. The variations in gene expression between various rice species during the breeding phases must be studied. The expression of these genes in a number of growth conditions will also be the focus of future research discussions. But many abiotic stresses impair rice competitiveness. Inclusion, erosion, salinity, and heat stress are the most destructive factors among the abiotic stresses. Though physiological resistance mechanisms against abiotic stress in rice are reasonably well observed, further research will assess the physiological basis of assimilating division. Phenotypically stability in rice plant showed positive contribution against abiotic stress resistance further stress modulation genes are required to introduce in rice plants. The genomic tools combined with ecophysiological studies of rice plant will help to explain the behavior of introduced genotypes and plants physiological response to changing climatic conditions. This strategy will help inunderstanding the relationships between the environment and rice physiology and genes performance under less than ideal conditions.

References

Agusti J, Greb T (2013) Going with the wind–adaptive dynamics of plant secondary meristems. Mech Dev 130(1):34–44

Ahmad S, Abbas G, Ahmed M, Fatima Z, Anjum MA, Rasul G, Khan MA, Hoogenboom G (2019) Climate warming and management impact on the change of rice-wheat phenology in Punjab, Pakistan. Field Crop Res 230:46–61

Ahmed M, Ahmad S (2020) Systems modeling. In: Ahmed M (ed) Systems modeling. Springer, Singapore, pp 1–44

Ahmed M, Ahmad S, Raza MA, Kumar U, Ansar M, Shah GA, Parsons D, Hoogenboom G, Palosuo T, Seidel S (2020a) Models calibration and evaluation. In: Ahmed M (ed) Systems modeling. Springer, Singapore, pp 151–178

Ahmed M, Ahmad S, Waldrip HM, Ramin M, Raza MA (2020b) Whole farm modeling: a systems approach to understanding and managing livestock for greenhouse gas mitigation, economic viability and environmental quality. In: Waldrip HM, Pagliari PH, He Z (eds) Animal Manure. ASA Special Publication 67. American Society of Agronomy and Soil Science Society of America, Madison, WI, pp 345–371

Armstrong AR, Bruce PG (1996) Synthesis of layered LiMnO 2 as an electrode for rechargeable lithium batteries. Nature 381(6582):499–500

Atique-ur-Rehman M, Farooq AN, Ahmad R (2014) Influence of boron nutrition on the Rice productivity, grain quality and biofortification in different production systems. Field Crop Res 169:123–131

Atique-ur-Rehman M, Farooq A, Rashid F, Nadeem S, Stürz FA, Bel RW (2018) Boron nutrition of rice in different production systems. A review. Agron Sustain Dev 38:25

Avigad G, Dey PM (1997) Carbohydrate metabolism: Storage carbohydrates. In: Plant biochemistry, pp. 143–204

Brown LR (2008) Plan B 3.0: Mobilizing to save civilization (substantially revised). WW Norton & Company

Chen Y, Fan X, Song W, Zhang Y, Xu G (2012) Over-expression of OsPIN2 leads to increased tiller numbers, angle and shorter plant height through suppression of OsLAZY1. Plant Biotechnol J 10(2):139–149

Colmer TD (2003) Long-distance transport of gases in plants: a perspective on internal aeration and radial oxygen loss from roots. Plant Cell Environ 26(1):17–36

Cong X, Hubmayr RD, Li C, Zhao X (2017) Plasma membrane wounding and repair in pulmonary diseases. Am J Phys Lung Cell Mol Phys 312(3):L371–L391

Counce PA, Gealy DR, Sung SJS (2003) Rice physiology. In: Rice: origin, history, technology, and production. Wiley, New York, pp 129–152

Daneva A, Gao Z, Van Durme M, Nowack MK (2016) Functions and regulation of programmed cell death in plant development. Annu Rev Cell Dev Biol 32:441–468

Eastburn DM, Degennaro MM, Delucia EH, Dermody O, Mcelrone AJ (2010) Elevated atmospheric carbon dioxide and ozone alter soybean diseases at SoyFACE. Glob Chang Biol 16(1): 320–330

Easterling WE, Aggarwal PK, Batima P, Brander KM, Erda L, Howden SM, Tubiello FN (2007) Food, fibre and forest products. Climate Change 2007:273–313

El-Kereamy A, Bi YM, Mahmood K, Ranathunge K, Yaish MW, Nambara E, Rothstein SJ (2015) Overexpression of the CC-type glutaredoxin, OsGRX6 affects hormone and nitrogen status in rice plants. Front Plant Sci 6:934

Evans LT, Wardlaw IF, Fischer RA (1975) Wheat. p. 101–149. LT Evans (ed.) Crop physiology. Cambridge Univ. Press, New York. Wheat. In: Evans LT (ed) Crop physiology. Cambridge University Press, New York, pp 101–149

Ezquer I, Salameh I, Colombo L, Kalaitzis P (2020) Plant cell walls tackling climate change: Insights into plant cell wall remodeling, its regulation, and biotechnological strategies to improve crop adaptations and photosynthesis in response to global warming

Farooq M, Wahid A, Lee DJ (2009) Exogenously applied polyamines increase drought tolerance of rice by improving leaf water status, photosynthesis and membrane properties. Acta Physiol Plant 31(5):937–945

Farooq M, Wahid A, Siddique KH (2014) 14 Physiology of grain development in cereals

Fatima Z, Ahmad M, Hussain M, Abbas G, Ul-Allah S, Ahmad S, Ahmed N, Ali MA, Sarwar G, Ehsan ul Haque, Iqbal P, Hussain S (2020) The fingerprints of climate warming on cereal crops phenology and adaptation options. Sci Rep 10:18013

Gong M, Chen SN, Song YQ, Li ZG (1997) Effect of calcium and calmodulin on intrinsic heat tolerance in relation to antioxidant systems in maize seedlings. Funct Plant Biol 24(3):371–379

Gu JF, Qiu M, Yang JC (2013) Enhanced tolerance to drought in transgenic rice plants overexpressing C4 photosynthesis enzymes. Crop J 1(2):105–114

Ingram DS, Vince-Prue D, Gregory PJ (eds) (2015) Science and the garden: the scientific basis of horticultural practice. Wiley, New York

Itoh JI, Kitano H, Matsuoka M, Nagato Y (2000) Shoot organization genes regulate shoot apical meristem organization and the pattern of leaf primordium initiation in rice. Plant Cell 12(11): 2161–2174

Itoh JI, Nonomura KI, Ikeda K, Yamaki S, Inukai Y, Yamagishi H, Nagato Y (2005) Rice plant development: from zygote to spikelet. Plant Cell Physiol 46(1):23–47

Jackson MB, Armstrong W (1999) Formation of aerenchyma and the processes of plant ventilation in relation to soil flooding and submergence. Plant Biol 1(3):274–287

Ji K, Wang Y, Sun W, Lou Q, Mei H, Shen S, Chen H (2012) Drought-responsive mechanisms in rice genotypes with contrasting drought tolerance during reproductive stage. J Plant Physiol 169(4):336–344

Juliano BO (1985) The rice grain and its gross composition. Rice: Chem Technol

Kaufman PB (1959) Development of the shoot of Oryza sativa L. II Leaf histogenesis. Phytomorphology 9:277–311

Kebrom TH, Richards RA (2013) Physiological perspectives of reduced tillering and stunting in the tiller inhibition (tin) mutant of wheat. Funct Plant Biol 40(10):977–985

Keeling PL, Banisadr R, Barone L, Wasserman BP, Singletary GW (1994) Effect of temperature on enzymes in the pathway of starch biosynthesis in developing wheat and maize grain. Funct Plant Biol 21(6):807–827

Khan MA, Ahmad S, Raza A (2019a) Integrated weed management for agronomic crops. In: Hasanuzzaman M (ed) Agronomic crops, vol 2. Springer, Singapore, pp 257–281

Khan SM, Ali S, Nawaz A, Bukhari SAH, Ejaz S, Ahmad S (2019b) Integrated pest and disease management for better agronomic crop production. In: Hasanuzzaman M (ed) Agronomic crops, vol 2. Springer, Singapore, pp 385–428

Kitambi SS, Toledo EM, Usoskin D, Wee S, Harisankar A, Svensson R, Ernfors P (2014) RETRACTED: Vulnerability of glioblastoma cells to catastrophic vacuolization and death induced by a small molecule.

Latif HH (2014) Physiological responses of Pisum sativum plant to exogenous ABA application under drought conditions. Pak J Bot 46(3):973–982

Lee CH, Olson P, Evans RM (2003) Minireview: lipid metabolism, metabolic diseases, and peroxisome proliferator-activated receptors. Endocrinology 144(6):2201–2207

Maeda E (1965) Inhibition of lamina inclination by cytokinin in excised rice leaves. Plant Cell Physiol 6(4):653–660

Maleki A, Naderi A, Naseri R, Fathi A, Bahamin S, Maleki R (2013) Physiological performance of soybean cultivars under drought stress. Bull Env Pharmacol Life Sci 2(6):38–44

Marschner H, Kirkby EA, Cakmak I (1996) Effect of mineral nutritional status on shoot—root partitioning of photoassimilates and cycling of mineral nutrients. J Exp Bot 47:1255–1263

Moldenhauer KA, Gibbons JH, Smith CW, Dilday RH (2003) Rice morphology and development (pp. 103–127). John Wiley & Sons, Inc.: Hoboken, New Jersey

Myers AM, Morell MK, James MG, Ball SG (2000) Recent progress toward understanding biosynthesis of the amylopectin crystal. Plant Physiol 122(4):989–998

Nishimura EK, Jordan SA, Oshima H, Yoshida H, Osawa M, Moriyama M, Nishikawa SI (2002) Dominant role of the niche in melanocyte stem-cell fate determination. Nature 416(6883): 854–860

Pandey V, Shukla A (2015) Acclimation and tolerance strategies of rice under drought stress. Rice Sci 22(4):147–161

Pavanasasivam V, Axley JH (1980) Influence of flooding on the availability of soil zinc. Commun Soil Sci Plant Anal 11(2):163–174

Petrášek J, Friml J (2009) Auxin transport routes in plant development. Development 136(16): 2675–2688

Pincebourde S, Casas J (2016) Hypoxia and hypercarbia in endophagous insects: larval position in the plant gas exchange network is key. J Insect Physiol 84:137–153

Schlenker W, Roberts MJ (2009) Nonlinear temperature effects indicate severe damages to US crop yields under climate change. Proc Natl Acad Sci 106(37):15594–15598

Selote DS, Khanna-Chopra R (2004) Drought-induced spikelet sterility is associated with an inefficient antioxidant defence in rice panicles. Physiol Plant 121(3):462–471

Shannon JC, Pien FM, Cao H, Liu KC (1998) Brittle-1, an adenylate translocator, facilitates transfer of extraplastidial synthesized ADP-glucose into amyloplasts of maize endosperms. Plant Physiol 117(4):1235–1252

Sharma P, Dubey RS (2005) Drought induces oxidative stress and enhances the activities of antioxidant enzymes in growing rice seedlings. Plant Growth Regul 46(3):209–221

Sharma A, Saha TN, Arora A, Shah R, Nain L (2017) Efficient Microorganism compost benefits plant growth and improves soil health in calendula and marigold. Horticult Plant J 3(2):67–72

Shehab GG, Ahmed OK, El-Beltagi HS (2010) Effects of various chemical agents for alleviation of drought stress in rice plants (Oryza .sativa L.). Not Bot Horti Agr Cluj-Napoca 38:139–148

Smith CW, Dilday RH (eds) (2002) Rice: origin, history, technology, and production, vol 3. Wiley, New York

Wiengweera A, Greenway H, Thomson CJ (1997) The use of agar nutrient solution to simulate lack of convection in waterlogged soils. Ann Bot 80(2):115–123

Xu M, Zhu L, Shou H, Wu P (2005) A PIN1 family gene, OsPIN1, involved in auxin-dependent adventitious root emergence and tillering in rice. Plant Cell Physiol 46(10):1674–1681

Yang C, Zhang X, Zhou C, Seago JL Jr (2011) Root and stem anatomy and histochemistry of four grasses from the Jianghan Floodplain along the Yangtze River, China. Flora-Morpho Distrib Funct Ecol Plant 206(7):653–661

Zažímalová E, Murphy AS, Yang H, Hoyerová K, Hošek P (2010) Auxin transporters-why so many? Cold Spring Harb Perspect Biol 2(3):a001552

Zhao C, Liu B, Piao S, Wang X, Lobell DB, Huang Y, Asseng S (2017) Temperature increase reduces global yields of major crops in four independent estimates. Proc Natl Acad Sci 114(35): 9326–9331

Zhou B, Ding Z, Zhao M (2011) Alleviation of drought stress inhibition on photosynthesis by overexpression of PEPC gene in rice. Acta AgronomicaSinica 37(1):112–118

Zhou L, Xu H, Mischke S, Meinhardt LW, Zhang D, Zhu X, Fang W (2014) Exogenous abscisic acid significantly affects proteome in tea plant (Camellia sinensis) exposed to drought stress. Hortic Res 1(1):1–9

Water-Wise Cultivation of Basmati Rice in Pakistan 13

Amar Matloob, Khawar Jabran, Muhammad Farooq, Abdul Khaliq, Farhena Aslam, Tasawer Abbas, Ehsanullah, Umar Zaman, Sohail Irshad, and Bhagirath Singh Chauhan

Abstract

Basmati rice grown in Pakistan has a world over importance owing to its fragrant long grains and premium cooking quality. Further, the rice plays a significant role in fetching high foreign exchange and feeding nearly 200 million people in the country. The conventional rice system of Pakistan is characterized with puddling, enormous water supplies, and high energy input. Water-saving rice cultivation systems are highly desired in the wake of grievous water shortage and drought

A. Matloob (✉) · S. Irshad
Department of Agronomy, MNS University of Agriculture, Multan, Pakistan
e-mail: amar.matloob@mnsuam.edu.pk

K. Jabran
Plant Production and Technologies Department, Nigde Omer Halisdemir University, Nigde, Turkey

M. Farooq
Department of Agronomy, University of Agriculture, Faisalabad, Pakistan

Department of Plant Sciences, College of Agricultural and Marine Sciences, Sultan Qaboos University, Al-Khoud, Oman

A. Khaliq · Ehsanullah
Department of Agronomy, University of Agriculture, Faisalabad, Pakistan

F. Aslam
Department of Agronomy, Bahauddin Zakariya University, Multan, Pakistan

T. Abbas
Inservice Agriculture Training Institute, Sargodha, Pakistan

U. Zaman
Institute of Soil and Environmental Sciences, University of Agriculture, Faisalabad, Pakistan

B. S. Chauhan
Queensland Alliance for Agriculture and Food Innovation (QAAFI), The University of Queensland, Toowoomba, QLD, Australia

N. Sarwar et al. (eds.), *Modern Techniques of Rice Crop Production*,
https://doi.org/10.1007/978-981-16-4955-4_13

conditions with the subsequent objective of saving water for irrigating other crops. This would importantly help to ensure food security in the country. The water-saving rice cultivation methods including aerobic rice, alternate wetting and drying, system of rice intensification, and growing rice on raised beds can be the most suitable alternatives to the conventionally sown rice. Nonetheless, the high weed prevalence, quick moisture loss, and higher spikelet sterility are the major obstacles in widespread adoption of water-saving rice systems. Application of certain herbicides can be helpful in controlling the weeds; while mulching may not only conserve the soil moisture but also suppress the weeds in water-saving rice systems. Use of breeding and biotechnological approaches to develop cultivars which are well adapted to the water-saving rice cultivation conditions would help in improved yield and water productivity in these systems.

Keywords

Conventional rice system · Labor · Water-saving rice systems · Weed control · Mulch · Spikelet sterility

13.1 Introduction

All kinds of life require water for its maintenance and sustenance. Water has been ascribed as the Blue Gold of the twenty-first century. The available water resources in the developing world are either declining or not being properly utilized (Ahmad et al. 2004, 2008, 2009a, b, 2012, 2013, 2015, 2019). For example, the productivity of water is much lower in Pakistan compared with the rest of the world (Erenstein 2009; Ahmad and Hasanuzzaman 2012; Ahmed and Ahmad 2019, 2020; Shahzad and Ahmad 2019). Pakistan, once a water surplus country due to its plentiful surface water resources, is now a water-deficit country with acute water shortage and prolonged drought spells. The obvious reasons may be burgeoning population resulting in increased domestic and industrial needs of water, sedimentation of water reservoirs, erratic rainfalls, and uncertainty of climatic optima, especially in the last two decades (Hanjra and Qureshi 2010). Geographically, Pakistan is an arid and semiarid country where rainfall is neither regular nor sufficient to meet the current and future water demands. Pakistan is an agrarian economy wherein agriculture accounts for 19% of the gross domestic product and 60% of the total national export besides employing 45% of the country's labor force. Irrigated agriculture accounts for more than 90% of the country's agricultural production while only 10% is being produced from rainfed agriculture. Approximately, 25% of the country's total area is under cultivation, out of which 75% is irrigated by the world's largest contiguous gravity-run irrigation system. Irrigated agriculture is the principal consumer of freshwater, availing about 96% of it, quite higher than the world's average of 70% (Babel and Wahid 2008). It has been suggested that water availability to the agriculture sector across the globe has dropped from 72% in 1995 to 62% in 2020, while the corresponding drop in the developing countries has been estimated from

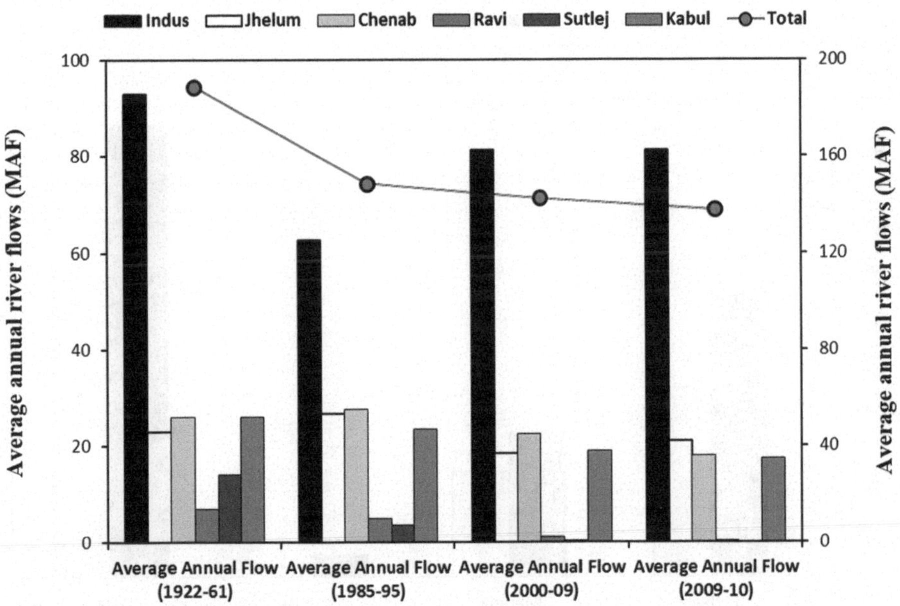

Fig. 13.1 Periodic changes in average annual flow of river Indus and its tributaries in the Indus Basin [**Source:** WRMD 2008, 2009]

87% to 73% (Khan et al. 2006). Estimates indicate that Pakistan irrigates three times more acreage than Russia. Nevertheless, the sustainability of the irrigated agriculture is threatened by food security issues, competition for water exerted by domestic and industrial sectors, and climate change. Increased water availability and its efficient utilization, thus, seem inevitable against the backdrop of augmenting the demands of agriculture and industrial sectors in a developing economy like Pakistan (Ahmed and Fayyaz-ul-Hassan 2017; Ahmed et al. 2020a, b; Akram et al. 2018, 2019; Hussain et al. 2018; Fahad et al. 2019; Fatima et al. 2020). The average annual flow of water in Indus Basin has decreased over years (Fig. 13.1). Moreover, sedimentation of reservoirs has also resulted in enormous loss in storage capacity (Figs. 13.2 and 13.3). The population, on the other hand, is increasing at a quick pace. The growth rate of the population in the country is 1.7%, which has increased the country's population by 25% within a span of 10 years. Per capita water availability in Pakistan has declined from 5600 m^3 to 1000 m^3 in 2004 (Kahlown et al. 2007) and the situation has become even worse in recent years (Kumar and Ladha 2011). The per capita water availability in Pakistan is already too low than other countries of the world (Fig. 13.4). This rise in population will affect water availability in two ways; firstly, by increasing the water demands for uses in domestic purposes and, secondly, by imposing stress on the availability of irrigation water by increasing its demand for food production (Fig. 13.5). Limited water availability for irrigation has been recognized as a major limiting factor that deters crop production (Tyagi et al. 2005). Thus, managing irrigation water in the wake of water productivity seems

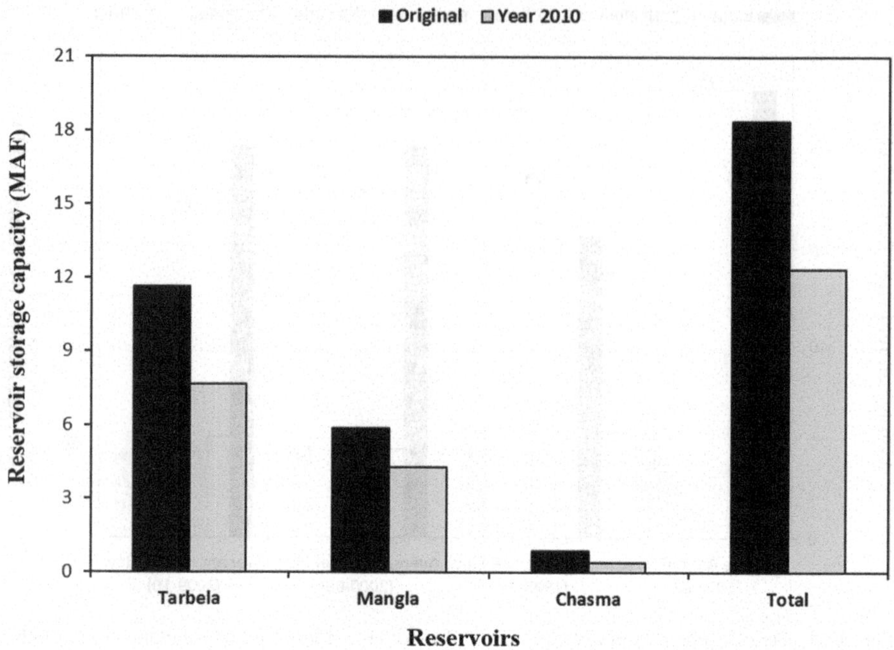

Fig. 13.2 Decline in storage capacity of major water reservoirs in Pakistan by the year 2010. [**Source:** National Water Policy (2003)] http://pc.gov.pk/mtdf/27-Water%20Sector/27-Water%20Sector.pdf. PILDAT Inter Provincial Water Issues in Pakistan Background Paper (draft) PILDAT-Pakistan Institute of Legislative Development and Transparency. http://pildat.org/Publications/publication/WaterR/Inter-ProvincialwaterissuesinPakistan-BackgroundPaper.pdf

imperative in a water-scarce country like Pakistan, where it remains a challenging task as the available groundwater is of marginal quality (Hafeez-ur-Rehman et al. 2019; Khan et al. 2019a, b; Naz et al. 2019; Peerzada et al. 2019; Razaq et al. 2019; Sarwar et al. 2013a, b; Tariq et al. 2020; Wasaya et al. 2019; Zahoor et al. 2019).

Surface water (138 million acre-foot; MAF), rainwater (6 MAF), and groundwater (50 MAF) are three water sources in Pakistan (Kahlown and Majeed 2003; Sufi et al. 2011). Accruing climatic changes are contingent on the intensity and occurrence of rainfall in the country (Chaudhary et al. 2009). The system for irrigational water provision in Pakistan is known as Indus Basin Irrigation System, having an annual flow of almost 103.5 MAF with 82 and 18% occurrence during Kharif (summer) and Rabi (winter) seasons, respectively (Kahlown and Majeed 2003; Sohail et al. 2014; Planning Commission 2020). During the recent decades, the water flow in the irrigational system has severely shrunk due to climatic changes and other reasons (Chaudhary et al. 2009). For example, during the years 2019–2020, the annual water flow in rivers was 94.4 MAF, which was 8.8% less than the average of the system (Government of Pakistan 2020). Earlier projections of Qutab and Nasiruddin (1994) suggested that the shortfall between water availability and demand during Rabi (winter season) will amplify from 3.5 MAF to 13 MAF by

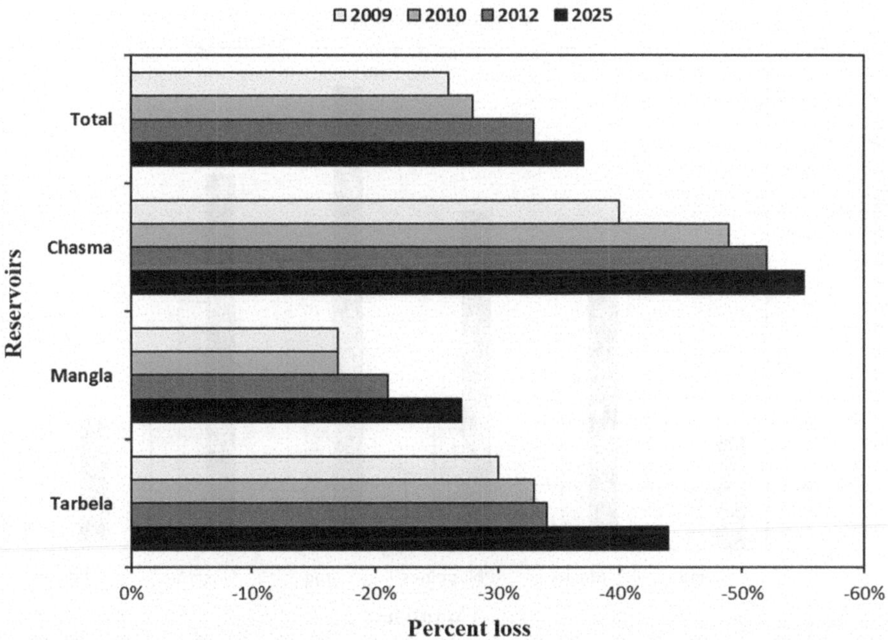

Fig. 13.3 Percent loss in storage capacity of major water reservoirs in Pakistan: current and future projections. [**Source:** WRMD (2008, 2009)]. National Water Policy (2003) http://pc.gov.pk/mtdf/27-Water%20Sector/27-Water%20Sector.pdf. PILDAT Inter Provincial Water Issues in Pakistan Background Paper (draft) PILDAT-Pakistan Institute of Legislative Development and Transparency. http://pildat.org/Publications/publication/WaterR/Inter-ProvincialwaterissuesinPakistan-BackgroundPaper.pdf

the year 2017. Pakistan has witnessed worse drought from 1999 to 2002 since its inception. The drought resulted in a high volume of groundwater use due to inevitable reliance and overexploitation of ground aquifers (Hussain et al. 2010). Pumping at the farm level has emerged as an essential resource that fulfills half of the crop water requirements. After the mid-1980s, groundwater has dominated the surface water as source of irrigation (Shah 2007) and is serving approximately 60% of the irrigated land (Shah et al. 2009). Nowadays, groundwater is used either as a sole source for irrigating crop fields or as a supplement to canal water and costs 16 times more than canal water. However, Erenstein (2009) argued that groundwater is no longer a supplemental to canal water, rather it has emerged as an integral and predominant source of water for irrigated agriculture in Punjab, Pakistan, ranging from 65% in the head- to over 90% in tail-end areas (Murray-Rust and Velde 1994). Currently, canal water fulfills only 35–40% of the total crop water requirement in the rice-wheat cropping system of the Indo-Gangetic Plains (IGP) while remaining requirements are being met from groundwater (Ladha et al. 2007). It seems that the first wave of irrigation development project aggravated the issues like salinity and waterlogging, while this second wave is heading toward overdraw of

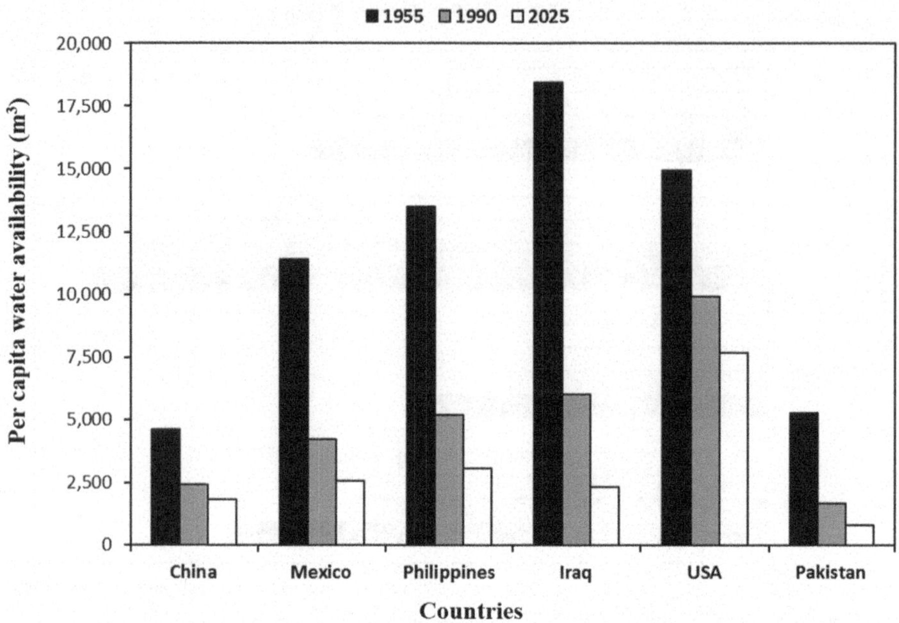

Fig. 13.4 Per capita water availability in Pakistan as compared to other countries. [**Source:** Engelman and LeRoy 1993]

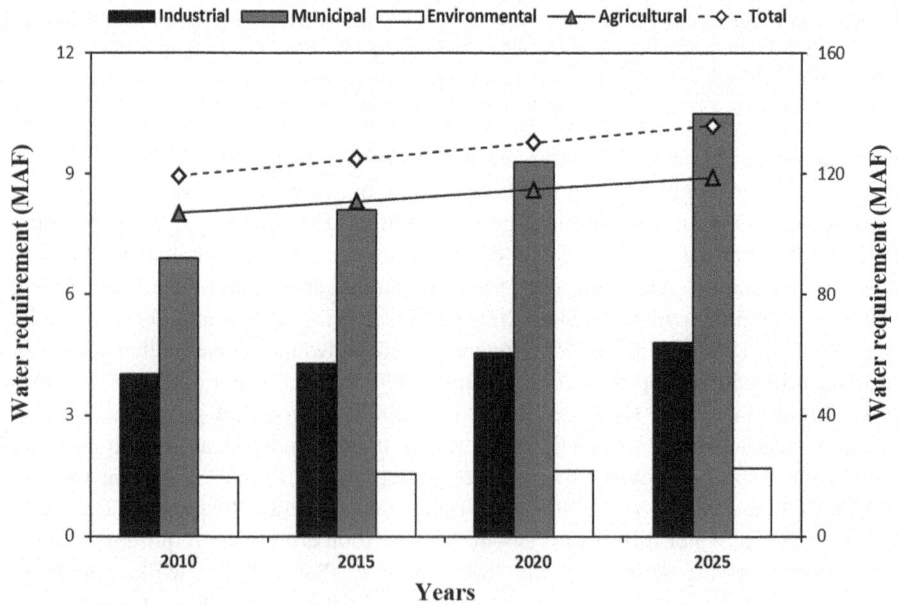

Fig. 13.5 Current and future water requirements of Pakistan (2010–2025). [**Source:** National Water Policy (2003)]

groundwater. Nevertheless, escalating energy crises and the resultant sky-rocketing prices of fossil fuels and electricity in Pakistan have deterred the luxurious use of this water resource (Erenstein 2009; Shah et al. 2006). Further, the groundwater table level is on downfall due to the huge water extractions (Jehangir et al. 2007). Groundwater is being depleted at an alarming rate of 1.0–2.0 m per year in the rice-wheat cropping system of IGP (Singh et al. 2002; Singh and Singh 2002; Tuong and Bouman 2003; Hira 2009; Humphreys et al. 2010). The water table is lowering steadily even in the areas getting a regular recharge from the canal system (Ahmad et al. 2007). Further, the annual groundwater recharge (53 mm) is much lower than the annual water extractions (82 mm) (Ahmad et al. 2005).

Rice (*Oryza sativa* L.) is among the most important food crop of the world as well as Pakistan. At least half the world's population is dependent on rice for acquiring their sustenance (Kumar and Ladha 2011). Rice ranks as the third largest crop after wheat and cotton in Pakistan. Rice contributes about 17% of the total cereal production of the country (Ahmad et al. 2005; Mann et al. 2016). Both fine and coarse rice are grown in Pakistan with a share of 40% and 60% in total rice production, respectively. Pakistan, being the house of aromatic fine rice, is world famous for its basmati rice (Akram 2009). Rice remains a major agricultural export commodity of the country (Government of Pakistan 2020) earning a significant amount of foreign exchange earnings (REAP 2020) as fine aromatic basmati rice is well known across the world for its exclusive aroma and excellent cooking quality. Pakistan is the fifth largest rice exporter and produces about 67% of the total basmati rice in the world. The basmati cultivars of indica rice type are generally grown in Pakistan, especially in the Punjab province (Jabran et al. 2017a, b). Pakistan also occupies fourth position in rice export after China, India, and Indonesia. It is grown on 11% of the total cropped area of Pakistan and, currently, 42% of its production potential is being harvested (Mann et al. 2016). The rice-growing area of Pakistan is shown in Fig. 13.6. As compared to the other countries like Laos (163 kg), Bangladesh (160 kg), Myanmar (157 kg), Cambodia (152 kg), Thailand (103 kg), China (77 kg), and South Korea (76 kg), per capita consumption of rice in Pakistan is only 14.5 kg (Anonymous 2012a, b) due to high cost of rice as compared to wheat flour (Sheikh and Kanasro 2003).

Despite the severe water scarcity, drought conditions, and unavailability of water at critical stages for many crops, a significant portion of available water resources is currently being used to irrigate the rice crop grown by conventional transplanted method (Jehangir et al. 2007). Almost all the rice in Pakistan is grown as irrigated rice (Soomro 2004; Swain et al. 2005) that consumes a substantial amount of water (Farooq et al. 2011; Jabran et al. 2015a, b). The development of water-saving rice cultivation systems seems imperative in the wake of looming water and energy crises, and ever increasing food demands of a burgeoning population. Lack of additional freshwater availability to sustain food production has warranted the need to avoid high volume of water use and promotion of water conservation measures in agriculture. The present chapter tends to summarize the pros and cons of conventionally transplanted rice (CTPR), bringing to the light the factors that necessitate a switch from CTPR to more appropriate practices, and highlights and

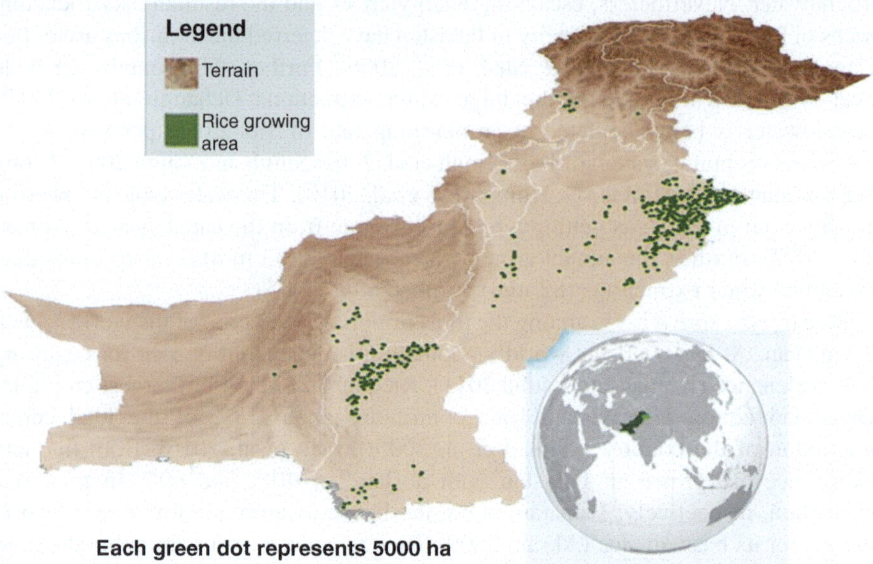

Fig. 13.6 Rice-growing areas of Pakistan (**Source:** http://ricepedia.org/pakistan)

critically evaluates the alternatives rice-growing strategies to cope with water scarcity in Pakistan.

13.2 Conventional Rice Cultivation System

Manual random transplanting of 30- to 35-day-old rice seedlings into a puddled field is the traditional rice cultivation method in Pakistan (Soomro 2004; Khaliq and Matloob 2011; Khaliq et al. 2014). This conventional system of rice transplanting is characterized by puddling of the land (repeated ploughing in standing water to a shallow depth of 10–20 cm to create a subsurface hardpan that restricts the downward water movement) and maintenance of a continuous layer of the standing water over the soil surface (Chauhan et al. 2017). This set of management practices is aimed at achieving certain objectives. Nevertheless, the cultivation of rice by puddling the soil and then transplanting rice seedlings carry some benefits besides several demerits. This method offers effective weed control due to standing water in the field and facilitates easy nursery uprooting and subsequent transplanting. The inherent size differential of transplanted rice seedlings in conjunction with flooded environments in CTPR provides a distinct competitive advantage, that is, a head start over a wide range of weed species that otherwise are quite problematic in aerobic rice culture. Puddling also benefits rice plants by reducing percolation of water (Kumar and Ladha 2011) and soil strength (Sharma and De Datta 1986). Ease in crop establishment, profuse tillering (Ampong-Nyarko and De Datta 1991), and increased water and nutrient availability are also extensively documented (Wade et al. 1998; Timsina and Connor 2001) in response to puddling. Wickham and Singh (1978) reported that puddling was effective in reducing percolation losses by 75% as compared to nonpuddled soil. Transplanting also provides farmers with flexibility to adjust to planting calendar and to accommodate rice crop to limited water supply since transplants are grown separately usually at a very high density. Another advantage of puddling is the possibility of green manuring that can significantly improve soil fertility. Flooding is considered responsible for soil nitrogen fertility when rice is grown without fertilizer addition. Submergence is believed to confer tolerance to lower temperatures in cooler climates while cooling the soil where high temperature is a problem.

13.2.1 Perils of Conventional Rice Cultivation System

The demerits of the conventional rice system surpass its benefits. Some of the salient demerits of growing rice by the conventional method are discussed in the subsequent paragraphs.

13.2.1.1 Water Issues

Looming water crises have been recognized as a major threat to agricultural productivity, in general (Sandhu et al. 2012), and irrigated rice, in particular (Soomro 2004; Farooq et al. 2011), with long-term consequences for regional and global food security (Seck et al. 2012). Irrigated rice is a gigantic user of fresh water and subject of debate as far as increasing water scarcity is concerned (Tuong and Bouman 2003). Water requirements of irrigated rice are approximately 2–3 times higher than for any other upland cereal (Bouman et al. 2007; Pathak et al. 2011a, b). In Pakistan, rice receives 5–6 times more water than the preceding wheat crop. Irrigated agriculture enjoys the largest share (90%) of fresh water in Asia and rice consumes more than 30 and 50% of the total diverted fresh water in the world and Asia, respectively (Barker et al. 1999). Rice consumes about 80% of the total irrigational water in rice-wheat cropping system (Bhushan et al. 2007). In Pakistan, a figure of 30% has been documented for rice by Mann et al. (2011). The CTPR needs water in field for wet tillage operations before it starts growing. Puddling alone accounts for 30% of the total water input in CTPR (Aslam et al. 2002; Gopal et al. 2010). In IGP, water is becoming increasingly scarce (Bhushan et al. 2007) and many Asian countries have witnessed a drop in the per capita water availability to the tune of 40–60% between 1955 and 1990 (Gleik 1993) and 34–76% from 1950 to 2005 (Kumar and Ladha 2011). This situation has become even worse in upcoming years (Matloob et al. 2015a, b) and is expected to aggravate further in near future just like a large snowball that becomes bigger and bigger as it rolls down. Pakistan is already shortlisted as a water-deficit country and per capita water availability has shown a tremendous drop in the last five decades (Fig. 13.7a). In Pakistan, 13–18 cm water is applied per irrigation that is far higher than consumptive use between two consecutive irrigations, that is, 8 cm (Kahlown et al. 2001). Malformed irrigation practices like

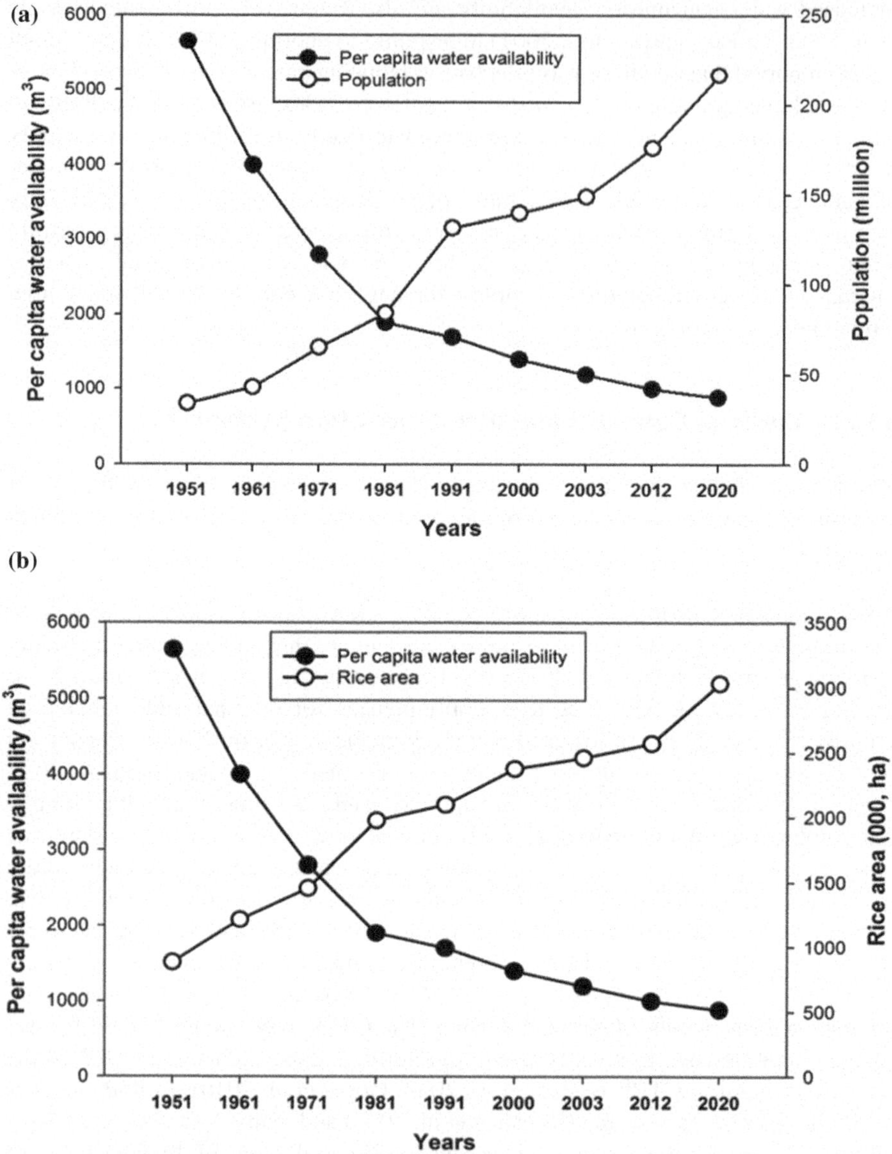

Fig. 13.7 Changes in per capita water availability (m^3) in Pakistan in relation to (a) population growth (million) and (b) rice area (000, ha) over time. Source: Pakistan Economic survey (various issues). State of the Environment Report (2005). Kahlown and Majeed (2003). http://pc.gov.pk/mtdf/27-Water%20Sector/27-Water%20Sector.pdf. WAPDA-Kalabagh Dam project (http://www.wapda.gov.pk/pdf/kbdam.pdf); Kumar and Ladha (2011)

overirrigation, lack of application uniformity, and prolonged irrigational events are also typical characteristics of rice growers in Pakistan (Kahlown and Kemper 2004). Seasonal water inputs for CTPR are estimated to range from 660 to 5280 mm (Bouman and Tuong 2001) that, however, can vary in response to agroecological, climatic, hydrological, and edaphic characteristics. Together with rice transplanting operations, puddling requires about 25% of the total seasonal water demand in rice (Chhokar et al. 2014). The CTPR, on an average, consumes as much as 2500 L of water to produce 1 kg of rice (Bouman 2009). This water is used for puddling and transplanting operations as well to account for evapotranspiration, seepage, and percolation losses in paddy fields during the crop growth period. The huge water requirement of CTPR is attributed merely to water inputs for puddling and nonproductive water loss by percolation and seepage (Hafeez et al. 2007). This nonproductive water use can range from 25 to 85% of the total seasonal water input but can vary in response to soil type and depth of water table (Kumar and Ladha 2011). For Pakistan, Soomro (2004) reported that nonbeneficial water loss by deep percolation may amount to 50% of the total water inflow that varies in the range of 33–64%. Globally, the area under rice is envisaged to decline as an outcome of water and labor shortage, and loss of arable land to urbanization and industrialization. Soomro (2004) documented an increase of 117, 368, and 116% in area, production, and yield of rice, respectively, in Pakistan considering 1950–1951 as a base year. Nevertheless, in Pakistan, a reverse trend is being observed and, despite of decreased per capita water availability, area under rice has expanded enormously (Fig. 13.7b). The CTPR on medium to heavy textured soil required about 1500 mm of water in a season. This is comprised of 150–250 mm for land preparation (puddling), 500–1200 to compensate for evapotranspiration to meet crop growth, and about 600 mm on account of seepage losses. Besides, 50 mm water is also required for raising rice nursery (Guerra et al. 1998). However, as a practice, farmers in Pakistani Punjab apply water more than this requirement (1850 mm), especially when rice is grown on light-textured soil (Bhatti and Kinje 1992) besides 500 mm of monsoon rain showers (Timsina and Connor 2001). However, Ghani et al. (1989) pointed out that even for heavy textured soils that are considered good for rice, water input for land preparation can be as higher as 1500 mm primarily due to soil cracking upon drying while the previous wheat crop is harvested. The water input for CTPR in Pakistan is more than 1400 mm while the actual water needed for the physiological maturity of rice crop is nearly 500 mm (Jehangir et al. 2007; Jabran et al. 2015a, b). Maintaining a water layer of 50–70 mm requires about 20 irrigations for 4–5 months rice crop (Ahmad et al. 2007; Jabran et al. 2015a, b). Freshwater resources of Pakistan have shown shrinkage in area and flow volume. Depletion of surface water resources has increased the burden on groundwater as an alternate source. Mushroom growth of tube wells (sixfold increase in the last two decades) has resulted in the mining of groundwater and a decrease in the water table. Pumping water from deep aquifers is not only costly but has also aggravated the risk and incidence of saline water intrusion into the rooting zone. In Pakistan, a number of groundwater structures (tube wells and dug wells) have been estimated around 0.80 million that are currently extracting 45–55 km^3 $year^{-1}$ of groundwater (Shah 2005).

The water productivity of CTPR in terms of total water input was estimated to be 0.17–0.38 kg m^{-3} that was quite lower than the water productivity (0.78–2.03 kg m^{-3}) observed for wheat (Ahmad et al. 2004; Jabran et al. 2015a, b). The difference in water productivities of both the crops was mainly attributed to the poor water management in the case of CTPR crop.

Jabran et al. (2012a) emphasized the need for water-efficient rice production in a country like Pakistan where rice is a conventional crop and water scarcity has endangered sustainable rice production. Matloob et al. (2015a, b) also mentioned the vulnerability of CTPR to severe water scarcity especially in Pakistan.

13.2.1.2 Labor Scarcity and Rising Wages

The CTPR also needs large labor inputs just like water. The labor requirements of CTPR are estimated to vary between 1400 and 1700 person-h ha^{-1} in a crop season. Out of this, the upper limit (600 h) is needed for land preparation operations (puddling), 100 h for nursery raising, 100–400 h for nursery uprooting and manual transplanting, 300 h for weeding and manual harvesting, and 100 h for drying (Kumar et al. 2008). The requirement of a large labor input within a short span of time is another critical issue as approximately 50% of labor is needed during first month for plowing, puddling, nursery pulling, and transplanting. A change in job preference has reduced the share of agriculture labor available for agricultural jobs (Dawe 2005; Kumar and Ladha 2011); hence, during the time of peak labor requirements, as is the case during manual transplanting of rice, skilled labor is either unavailable or wage rates are too high. A rapid growth of the economic sector has attracted agricultural laborers to nonfarm jobs in textile, garments, and food industries that involve less drudgery and wages are often higher than farm jobs. This situation has resulted in reduced farm profits due to the increased cost of production. During the last five decades, the share of the economically active population in agriculture has shown a marked reduction in Pakistan (Matloob et al. 2015a). The paucity of labor coupled with high wages has made CTPR a risky and costly enterprise as careless transplanting by hired labor results in suboptimal plant densities under field conditions (Khaliq et al. 2014).

13.2.1.3 Edaphic and Time Conflict

Deterioration of soil quality as a result of puddling is the other most important disadvantage of the CTPR. Puddling breaks soil capillaries, reduces voids, disperses clay particles, lowers soil strength, and destroys soil aggregation complex (Sharma and De Datta 1986). Puddling causes the clay particles to settle at the base of the tilled soil layer that impedes the percolation process. The loss of soil quality due to altered physical properties and formation of the subsurface hardpan due to drying of the compacted layer have been recognized as a serious threat to the productivity of ensuing upland crops like wheat (Kumar and Ladha 2011), especially in rice-wheat system (Timsina and Connor 2001). This reflects an edaphic conflict for the system (Farooq et al. 2006a, b; Sahrawat et al. 2010; Matloob et al. 2015a) as the subsequent wheat crop requires well-pulverized soil. This demands more time and energy for land preparation for ensuing wheat crop. Destruction of soil aggregates by puddling is conducive to the formation of cracks and crust upon drying that when broken by tillage leads to poor contact of seed with soil due to the formation of large clods (Timsina and Connor 2001). There is a substantive body of proof that supports reduction in wheat growth and; yield owing to poor soil structure, subsurface compaction, and suboptimal permeability (Kumar and Ladha 2011) and average yield reduction of 8% has been reported for wheat, when grown after CTPR (Kumar et al. 2008). Suboptimal soil environment due to puddling for CTPR is conducive to poor root growth of ensuing wheat crop and has been ascribed as the main reason for diminished wheat yields after CPTR rice due to limited water and nutrient uptake. Wheat yield was reduced by 0.4% for every 1 cm reduction in rooting depth (Sadras and Calvino 2001). Adding further, Ishaq et al. (2001) documented a considerable decrease in uptake of macronutrients by wheat crop owing to subsoil compaction that corresponded to 12–35% for N, 17–27% for P, and up to 24% for K. The magnitude of reduction in root length of wheat was far greater when it was grown after rice than after maize on sandy loam soil (Sur et al. 1981). Increased bulk density of the compacted soil layer reduced root growth of wheat and a strong negative relationship was observed (Hassan et al. 2007). The personal communication with many growers in the traditional rice belt revealed that the number of soil cultivation operations (and, hence, the tractor passes) typically ranges from 8 to 15, and may exceed this under special circumstances.

Besides edaphic conflict, there also exists a time clash that is especially apparent during wheat sowing after CTPR. The late-onset and erratic pattern of monsoon showers coupled with the drudgery of rice transplanting often delay rice planting, and fields remain reserved for a prolonged period of time (Farooq et al. 2006a, b). A delay in monsoon showers by few weeks would have consequences for the agricultural activity of the whole system (Joshi et al. 2013). Late vacation of paddy fields, especially of those planted to basmati rice varieties, is a major concern as it delays the wheat sowing beyond its optimum time, that is, mid-November (Hobbs 2001). A friable and well-pulverized seedbed is a unique requirement for a successful wheat crop that necessitates breaking of subsurface hardpan. To accomplish this goal, farmers have to plough their land at least six times with tractor-mounted cultivators followed each time by planking to get rid of the compacted soil layer and clods, respectively (Kumar et al. 2008). An average number of cultivations required to plant wheat after CTPR amounted to eight times in rice-wheat cropping system (Malik et al. 2004). Late maturing basmati varieties are usually harvested in the first week of November while the optimum time of sowing for wheat is the first fortnight of this month. Excessive residual moisture makes soil cultivation impossible and, hence, increases turnaround time (3–5 weeks) between rice harvesting and wheat planting. Waiting for the soil to reach working conditions results in delayed planting of wheat that in itself has negative implications for wheat grain yields (Pathak et al. 2003a, b). A typical response of wheat grain yield to sowing time in South-Asian countries has revealed a linear decline of 1–1.5% per day (35 kg ha^{-1} day^{-1}) after mid-November regardless of the maturity period of the wheat cultivar (Ortiz-Monasterio et al. 1994). In addition to the yield penalty associated with delayed wheat planting, reduction in input use efficiency of wheat crop is also a matter of concern (Hobbs 2001). The N response surface of wheat crop was much flatter in case of delayed wheat sowing than that of planted at the optimum time (Saunders 1990), suggesting that even additional N application cannot prevent yield loss owing to delayed sowing. Approximately, 80% of the total rice area in Pakistani Punjab is under the basmati group and delayed wheat sowing after basmati is an inevitable outcome of the current detrimental practice of puddling forcing 40–50% of the wheat to be planted beyond optimum time limit (Hobbs and Gupta 2004; Sahrawat et al. 2010).

13.2.1.4 Emission of Methane and Other Greenhouse Gases

Agriculture practices, in general, and CTPR, in particular, are known to exacerbate greenhouse gases (GHGs) emissions, thus, directly contributing toward global climate change (Hussain et al. 2014), and in turn, suffering from the negative implications of global warming (Pathak et al. 2011a, b). Hence, irrigated rice production can be considered as both perpetrator and victim of the climate change. Rice production is conducive to emission of major GHGs like methane (CH_4), nitrous oxide (N_2O), and carbon dioxide (CO_2). These GHGs contribute approximately 15, 5, and 60% to the total anthropogenic sources responsible for greenhouse effect (Rodhe 1990). Emission of these GHGs owing to agricultural activity has been estimated to be around 39, 60, and 1%, respectively (OECD 2000), with rice-based cropping systems being a potential contributor (Kumar and Ladha 2011; Pathak et al. 2011a, b). Recent estimates indicate that emission of CH_4, N_2O, and CO_2 from the agriculture sector is 45, 46, and 9%, respectively (Kasterine and Vanzetti 2010). Paddy fields are considered a potential source of CH_4 and N_2O and sink source of CO_2 as well. The share of rice fields toward global agricultural emission of CH_4 and N_2O is projected to be 30 and 11%, respectively. Rice cultivation accounts for about 10% of the total global methane emission (GMI 2010). The global warming potential (GWP) of these two GHGs is 21–25 and 298–310 times greater than CO_2 (IPCC 1997; 2007). Comparison of methane emission from different rice ecosystems reveals that rice culture with continuous flooding emits the highest fraction of methane (Fig. 13.8). Moreover, the average GWP of rice (3757 kg CO_2 eq ha^{-1} $season^{-1}$) was 5.7 and 2.7 times higher than that for wheat (662 kg CO_2 eq ha^{-1} $season^{-1}$) and maize (1399 kg CO_2 eq ha^{-1} $season^{-1}$), respectively (Linquist et al. 2012). Emission of GHGs from paddy fields is, however, sensitive to modification in cultural practices like tillage, irrigation, and fertilization regimes, and choice of cultivars (Hussain et al. 2014) and rice remains an important target to mitigate the emission of GHGs (Wassmann et al. 2004). Nevertheless, attempts to reduce the emission of one GHG may increase the emission of another; since a trade-off between CH_4 and N_2O is often observed while switching between rice establishment methods to mitigate GHGs and has been acknowledged as the major limitation in devising an effective mitigation strategy. Mid-season drainage in conjunction with direct seeding, or alternate wetting and drying regimes, has been proposed to cut short the CH_4 emissions by almost 50% (Lu et al. 2000; Wassmann et al. 2004). Avoiding flooding of rice fields may be a good solution to excessive methane emission from the rice fields (Li et al. 2004; Xu et al. 2000). Nevertheless, the net

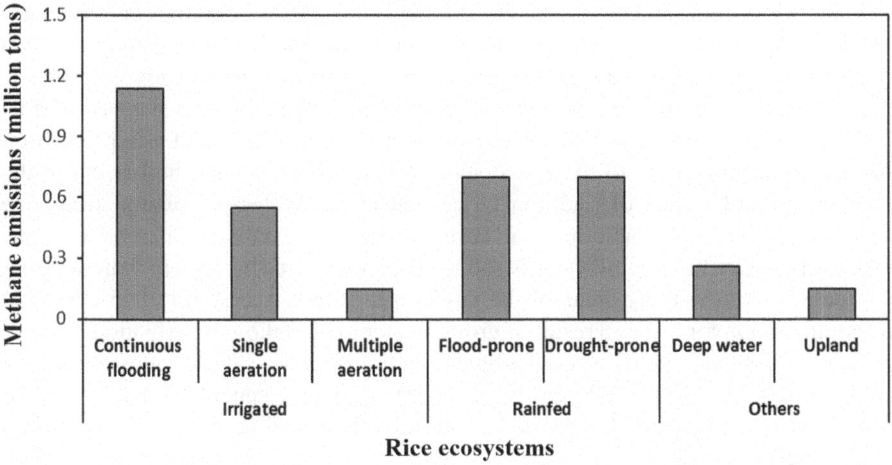

Fig. 13.8 Methane emission from different rice ecosystems under different moisture regimes [**Source:** Pathak et al. 2011a, b]

flux of GHGs emission under direct seeding is also dependent on N_2O emission that is usually higher under aerobic conditions (Farooq et al. 2011). Several authors reported that reduced emission of CH_4 is offset by increased emission of N_2O under aerobic conditions contrary to CTPR (Shang et al. 2011; Zhang et al. 2011). This warrants the need for diverse approaches that can furnish concurrent reduction in emission of both GHGs (Hussain et al. 2014), and no-till practices were found to benefit rice in this regard (Ahmad et al. 2009a, b). Another such approach is the irrigation management in such a way that soil redox potential is maintained between 100 and 200 mV because this range is high enough to avert CH_4 formation and low enough to favor the reduction of N_2O to N_2, since threshold redox potential identified for N_2O formation is 250 mV (Hou et al. 2000).

13.2.1.5 Other Environmental Hazards

Nitrogen losses into the environment as ammonia (NH_3) volatilization are greater under flooded conditions especially when the soil pH is higher as is the case with Pakistani soils. Ammonium ions (NH^{4+}) are loosely bound to water molecules and easily get converted into NH_3 that escapes as a gas. These volatilization losses may range from almost negligible to approximately 60% of the total nitrogen applied in a cropping season (Xing and Zhu 2000). Transportation or deposition of this NH_3 as gas or dissolved form may have negative implications for aquatic and terrestrial ecosystems causing eutrophication (Choudhury and Kennedy 2005). Denitrification losses on the other hand also represent a substantial loss of applied nitrogen. Studies at IRRI revealed that depending upon the method of crop establishment and urea application, denitrification losses may amount up to 46% (Buresh and De Datta 1990). Nitric oxide released as a result of the denitrification process leads to the formation of ozone that is a tropospheric pollutant and can adversely affect agricultural systems and human health (Matson et al. 1998). Moreover, both ammonium and nitrate forms of nitrogen are also vulnerable to leaching. Paddy rice is a major crop in Pakistan and it is suspected that nutrient loss from paddy fields to the groundwater by percolation through soil column under gravity, and subsurface flow to nearby watercourses, is a major environmental hazard. Irrigated rice production has led to accumulation of excessive sodium and other salts and higher pH in the shallow ground waters of traditional rice belt areas. Water percolation from rice fields raises the water table and can result in root zone salinity of nonrice crops, especially where the groundwater is saline. This also poses hazards of waterlogging and salinity to low-lying areas of the landscape (Bhutta and Alam 2005). Arsenic accumulation in topsoil as a result of groundwater exploitation for irrigation is also a critical issue as rice fields receive ample quantity of water fulfilled mostly by tube wells (Lu et al. 2009; Chakraborty et al. 2014). The potential bioavailability of arsenic is also higher under flooded conditions than aerobic conditions (Punshon et al. 2018).

Another very important disadvantage of conventionally flooded transplanted rice is the emission of CH_4 and other greenhouse gases posing serious threats to our environment in the form of elevated globe temperature (Cai et al. 1997). The decay of organic matter under anaerobic conditions of flooded rice fields causes the CH_4 emissions while the N_2O emission is accelerated under moistened conditions. Excessive nitrogen applications further accelerate the N_2O emissions from the rice fields. Microbial transformations of nitrogen are very important regarding methane emission. Yan et al. (2003) estimated the methane emission from the important rice-growing areas of the world including Pakistan. The total annual recorded methane emissions were 25.1 Tg per year out of which less than 1 Tg per year were emitted from the rice fields of Pakistan, which is significantly lower than China (7.67 Tg) and India (5.88 Tg). Nevertheless, a seasonal average of methane emission flux from wetland rice was higher for Pakistan compared to India.

13.3 Need for Alternative Rice Cultivation Systems

Our discussion regarding the conventional rice system characterized by puddling and inundation indicated that the demerits of this system surpass its benefits. Especially, the labor and water shortage accompanied by climatic changes, energy crises, increasing population, and escalating demand for food are important concerns. Problems of edaphic conflict and greenhouse gas emissions from the conventionally flooded rice fields not only challenge the sustainability of this system but these are also causing a considerable decrease in the productivity of this major cropping system. The situation demands water-saving rice cultivation systems with lower labor and energy inputs. Further, the need for a rice sowing system that does not deteriorate the soil structure after puddling and has an early crop maturity is highly desired. These objectives can be achieved by cultivating the rice by alternate methods such as aerobic rice, rice on raised beds, alternate wetting and drying, and system of rice intensification.

13.3.1 Aerobic Rice

Aerobic rice is considered most important among the water-saving rice cultivation systems (Jabran et al. 2015a, b). Aerobic rice does not desire soil puddling and flooding. Rather, the rice seeds are sown on a well-prepared seedbed (Fig. 13.9). The seed drilling machinery can be brought into use to make the sowing process easy. Subsequently, the processes such as sowing, uprooting, transporting, and transplanting of the nursery seedlings as in the case of conventional rice are avoided in aerobic rice. Owing to easiness in agronomic practices, aerobic rice is the preferred water-saving rice cultivation system. The salient benefits of aerobic rice include the saving of time, labor, energy, and water (Jabran et al. 2015a, b). Water saving may be considered as the major advantage of aerobic rice. The absence of a standing layer of water, the reduced percolation, and the seepage will result in considerable water saving under the aerobic rice system. The water required for aerobic rice is about half of the water needed for the conventional rice. For example, the rice sown under aerobic system would need 29% less water for land preparation. Total water input was 1240–1880 mm in flooded fields and 790–1430 mm in aerobic fields (Bouman et al. 2005). The water productivity of rice (with respect to rainfall and irrigation water input) under aerobic conditions was 32–88% higher than under flooded conditions (Bouman et al. 2005). The amount of water utilized to obtain comparable yields is much lesser in the nonflooded rice than in the conventionally flooded rice (Xu et al. 2007). Water used to raise the rice crop was almost 66% less in the nonflooded aerobic rice compared with the conventionally flooded rice. Aerobic rice reduces the amount of water that is lost through seepage in comparison with the CTPR. The water losses in the form of seepage were nearly two-third less in the aerobic rice than the water lost through seepage in the flooded rice. Also, higher water use efficiency was recorded in aerobic rice than the flooded rice. Under certain cases, the yield in aerobic rice may be less than conventional rice, however, water use efficiency is much higher than conventional system.

Conventionally transplanted flooded rice is more laborious compared with the rice grown under aerobic conditions. Balasubramanian et al. (2003) and Dawe (2005) commented on the labor requirements of the aerobic rice compared with transplanted rice and concluded that the labor requirements for the flooded rice were five times more than the aerobic rice (Balasubramanian et al. 2003; Dawe 2005). Singh et al. (2008), on the other hand, reported a 50% saving of the labor for aerobic rice than the transplanted rice. Unpuddled soils have a low bulk density than the puddled ones (Ishaq et al. 2001).

Some scientists have argued that aerobic rice is early maturing compared with the transplanted rice, thus, reducing the crop span and vacating the fields earlier for the sowing of the next crop (Farooq et al. 2006a, b). This is particularly important in rice-wheat system where wheat sowing is delayed due to late maturing and harvest of rice crop.

Aerobic rice sown on raised beds recorded 13–23% water savings over CTPR; however, rice yields under aerobic rice system were lower by 14–25%. Water use efficiency was also higher (0.45 g L^{-1}) for aerobic rice than CTPR (0.37–0.45 g L^{-1};

Fig. 13.9 Growth stages of basmati rice under aerobic conditions. (a) dry seeds, (b) emergence, (c) seedling stage, (d) tillering, (e) stem elongation, (f) panicle initiation, (g) soft dough, (h) hard dough, (g) physiological maturity, (i) physical maturity, and (j) harvesting

Bhushan et al. 2007). In India and Pakistan, reduction in grain yield under the aerobic rice systems ranged from 7.5 to 28.5% (Kumar and Ladha 2011). The trade-off between water savings and aerobic rice yields accompanied with high sterility has also been documented from field studies undertaken in Pakistan

Table 13.1 Yield comparison of directed seeded and conventional puddled transplanted rice in Pakistan

Direct seeded rice	Conventional puddled transplanted rice	References
2.56	3.34	Farooq et al. (2006a, 2007)
2.93	3.95	Farooq et al. (2006b, 2009)
3.56	4.24	Jabran et al. (2015a)
3.49	4.18	Khaliq et al. (2015)
2.82 (Basmati)	3.92 (Basmati)	Mann et al. (2016)
4.36 (Non-Basmati)	4.55 (Non-Basmati)	
5.35	4.65	Ishfaq et al. (2020)

(Table 13.1; Jabran et al. 2015a; Matloob 2014; Ishfaq et al. 2020). Farmer field trials undertaken at 25 locations in three major rice districts, for example, Gujranwala, Kasur, and Chiniot, revealed 25 water savings for aerobic rice over farmers' practice of CTPR. Moreover, productive tillers were more numerous (47% higher) and average rice yield of 5.4 t ha^{-1} was 27% higher than 4.2 t ha^{-1} recorded for CTPR (Mann et al. 2016). Furthermore, Awan et al. (2015) reported water productivity value of 0.38 g grain per kg total water input that was more than double of the country's average value of 0.16 for CTPR. Based on experimental evidence, there is a strong reason to believe that aerobic rice can be a potential alternative to CTPR, especially in the backdrop of looming water crises, provided the challenge of weed management is tacked in a sustainable manner and high yielding rice varieties adapted to this system are available (Ali et al. 2014; Jabran et al. 2015a, b). Improved water productivity matters for Pakistani rice farmers as the economic impact of adopting water-saving technologies and related agronomic practices was significant and manifested as higher net revenue per acre proving the economic viability of water-wise rice cultivation (Jabran et al. 2016; Ali et al. 2019). The water economy of rice cultivation has been recognized as the primary motive for adopting aerobic rice, and for resource-poor Pakistani farmers, aerobic rice has been proposed to enhance eco-efficiencies (Ahmad et al. 2014; Awan et al. 2015; Jabran et al. 2016). In conclusion, growing rice under aerobic conditions instead of CTPR would result in significant water saving. Further, the labor required to grow and mature the aerobic rice is far lower than the conventional rice. Reduced crop duration by almost 1 week is the other benefit of aerobic rice, which will allow vacating the rice field earlier for timely sowing of wheat crop.

13.3.2 Alternate Wetting and Drying

Under the alternate wetting and drying (AWD) system of growing rice, the crop is sown either by direct seeding or transplanting. The crop is neither flooded (as in the case of the conventional method) nor kept dry (as in the case of the aerobic rice). Rather the field is irrigated with heavy irrigation occasionally throughout the growing season (Jabran et al. 2015a, b). Once the crop is irrigated heavily, it is not reirrigated until the most of available moisture is consumed. Hence, avoiding wasteful irrigation application in AWD leads to considerable water savings. Further, the water losses in the form of deep percolation and evaporation are also minimized in AWD compared with the CTPR. Our studies indicated considerable water saving by growing rice using AWD (Jabran et al. 2015a, b). The yield outputs for AWD were comparable with the conventionally grown rice, in addition to water saving (Jabran et al. 2015a, b).

The AWD confronts water shortage in irrigated rice cultivated areas and it is an effective and sustainable way to save energy and water. Adoption of this technology can reduce water requirements in rice (Miah et al. 2009). In AWD, there is an effective use of optimum moisture by the crop. The use of this technology in rice production instead of traditional flooding practices alleviates water scarcity up to 30% (ACIAR 2013). The volume of irrigation water needed to yield a specific amount of rice decreases (water productivity is increased) in the case of AWD as compared to conventional flooding cultivation (Lampayan et al. 2009). The additional significant advantage of AWD is saving of fuel and energy usage. Conventional cultivation of rice is a major source of atmospheric CO_2 and CH_4; it is a well-known fact that reducing the standing water in rice fields will have a positive impact on environment quality (Bouman et al. 2007). Resultantly, the use of AWD in the rice production system will lead to efficient use of inputs like water besides saving fuel and energy.

The time interval during which soil remains nonflooded in AWD varies from 1 to more than 10 days depending upon the agroclimatic conditions. Drop in water level (15 cm) below the soil surface is considered as a criterion for applying irrigation. In Sheikhupura district of Punjab, Pakistan, it was found that canal water availability was 80, 50, and 20% of the seasonal water requirement at the head, middle, and tail, respectively, necessitating the remaining water requirements to be met through tube well (Nizami et al. 2020). Adoption of AWD and laser land leveling in this area reduced tube well operation hours by 19–24% (Nizami et al. 2020). Encouraging findings regarding the effectiveness of this technique for saving water in rice cultivation are also reported from Sindh province, wherein AWD improved rice yield by 13 and 17% over CTPR during 2013 and 2014 growing seasons, respectively. The corresponding improvements in water saving were 39 and 44% during the aforementioned period (Majeed et al. 2017). A recent study found that the water savings following AWD (−20 kPa soil moisture tension) were much greater in case of aerobic rice than CTPR system. In comparison to CTPR, the AWD under aerobic rice culture reduced water input by 27–29%, with a concurrent increase of 7–9% in yield and 44–50% in water productivity (Ishfaq et al. 2020).

13.3.3 System of Rice Intensification

System of rice intensification (SRI) can be practiced for higher yields and water saving for rice grown in Pakistan. The SRI is a sustainable way of optimizing different production elements like soil, water, light, etc. for permitting the plant to achieve its maximum potential, which is seldom achieved when inadequate practices are used (Zotoglo 2011). The SRI in contrast to CTPR comprises AWD technique in rice fields (Kepha et al. 2014). However, the SRI is a labor-intensive rice cultivation method and has been acknowledged as the main limiting factor in the wide-scale adoption of this innovative technology (Sharif 2011). The factors determining the success of SRI include the lower water input, low age rice nursery, plantation of single seedling, increased plant-plant distance, use of organic fertilizers, and non-chemical pest control. Low water input may be considered the major objective of growing rice by SRI, while transplanting low age nursery and increasing intrarow, intraplant distances help the profuse growth of rice plant. Root growth and dry biomass are increased abundantly resulting in healthier plants and luxuriant crop yields. However, a high labor requirement is the major drawback of growing rice by SRI. Further, the high temperature in rice-growing areas of Pakistan exceeding 40 °C during the months of rice transplanting (June–July) may result in seedling damage.

Rice grown using SRI can endure hostile impacts of torrential rains, drought conditions, and other economic failures as compared to conventional cultivation of rice. SRI is an eco-friendly and economic striking practice with minimum economic risk compared with other cultivation methods (Anthofer 2004). Water saving is one of the major benefits of SRI. Ceesay et al. (2006) reported an increase in water productivity from 2 to 6 times in case of SRI. Different trials on SRI reported water saving in Indonesia as 40%, 67% in the Philippines, and 25% in Sri Lanka as compared to CTPR (Sato 2006; Lazora 2004; Nemoto et al. 1995).

The SRI is an eco-friendly and valuable approach to farmers, that is also beneficial to the natural habitat and biodiversity. It does not entail chemical fertilizers and better use of organic manures with enhanced yield. This leads to improved soil and water quality which will ultimately enhance soil and human health.

Following six important elements of SRI were developed by Laulanie (Uphoff 2007):

1. Young rice seedlings should be transplanted.
2. Single seedling is planted per hill.
3. To increase soil fertility, organic fertilizers should be preferred.
4. AWD conditions are maintained in the field instead of continuous flooding.
5. Plant spacing should be more than conventional cultivation.
6. Instead of chemical weed control, rotary weeding should be preferred to enhance soil aeration.

The average rice yield is 8 t ha^{-1} in the case of SRI; while it is 3 t ha^{-1} with conventional cultivation (Uprety 2004). Similar findings were also reported by scientists, who found a higher average yield in SRI compared to the traditional method. Uphoff (2007) reported an increase in rice yield up to two- to three folds through the method of SRI. Husain et al. (2004) documented a 30% yield advantage for SRI in Bangladesh and Namara et al. (2003) showed an even larger benefit of 44% in Sri Lanka. While assessing the impacts of SRI in Punjab-Pakistan, the Director General of Agriculture (Water Management) reported a 30–50% increase in yield with a 15–20% improvement in grain weight. The SRI conserved 25–50% water and reduced seed input by 80–90%, whereas the cost of production was reduced by 20% (Gill 2007).

One of the major disadvantages of SRI is heavy weed infestation because fields are not flooded as is the case with CTPR. The use of herbicides is the most effective way to manage weeds but it does not affect the soil aeration. The method of cono weeders and rotary hoe is laborious but puts forth positive impacts on soil aeration (Satyanarayana et al. 2006). SRI is much labor intensive for many farmers and this is the major reason that about 40% of the farmers did not adopt this method in Madagascar (Husain et al. 2004). In Pakistan, mechanization of SRI resulted in 70% water savings with a significant reduction in labor requirement (25 man-h ha^{-1} compared to 85 in the case of CTPR; Sharif 2011). The innovations introduced were sowing nursery on mulched raised beds with a plastic sheet underneath to restrict root growth from penetrating deep into the soil. For precision planting, 137,500 seedlings were grown by using only 4 kg rice seed. In the field, raised beds were established using a machine that in a single-pass operation made raised beds with an open furrow on both sides, and placed fertilizer four inches away and deep from rice hill. This machine also placed compost exactly where rice seedlings will be transplanted. The transplanting was done into the dry soil that is a huge shift from conventional practice. For this purpose, a water-wheel rice transplanter was used to make pits (~ 6 cm) at a precise distance of 22.5 cm from each other. The machine automatically fills each pit with water when it passes over the bed and a single seedling is dropped per pit by hand. Immediately after completion of the transplanting task, irrigation was applied to cover top 2–3 cm of the raised bed in order to settle down the seedlings and cover their roots with fine soil particles. To overcome the major biological constraints, that is, weeds, a precision weeder was used that not only effectively controlled the weeds but also improved soil aeration (Sharif 2011).

13.3.4 Cultivation of Rice on Raised Beds

The use of raised beds and ridge sowing for growing vegetables has been a popular practice over the times. However, owing to several benefits of these techniques, these are getting popularity to grow many other field crops like soybean, potatoes, sugarcane, cotton, maize, wheat, sunflower, etc. Recently, some researchers from Pakistan have investigated the utility of raised beds and ridge sowing for growing rice crop with the aim of saving water (Jabran et al. 2012a). Water saving in raised beds and ridge sowing techniques is achieved by irrigating only the furrows among beds or ridges leaving the tops dry. The plant roots grasp water through seepage. Improved root growth is the important aspect of growing rice on either ridges or beds. Growing rice on raised beds saved irrigation water by 42% with a 16% increase in grain yield compared with the conventional rice sowing (Bhuyan et al. 2012). Further, not only the process of irrigation was accomplished much earlier but also the water use efficiency of rice was improved.

Recently, the construction of permanent raised beds in rice-wheat system of South Asia (including Pakistan) is getting the attention of researchers and growers. Such types of permanent raised beds are used to grow rice and wheat crop in sequence using a special zero-tillage machine. The fields are not required to prepare every time which results in significant decrease in cost of production for both rice and wheat crop. However, water saving is the major advantage of cropping rice and wheat on permanent raised beds. Only the furrows are irrigated and the crop staying on the bed acquires water reaching roots by seepage.

Raised beds are quite alike ridges or furrow beds that were first familiarized in mid-1990s (Ram et al. 2005). The implication of this technique was firstly reported in Haryana, India, in 2000 and Nepal in 2001 (Balasubramanian et al. 2003). It is supposed that this technique saves water as compared to conventional sowing techniques. It also helps to increase soil physical properties (Naresh et al. 2012). In contrast, conventional cultivation of rice through flooding deteriorates soil and reduces farm income owing to high use of water and labor (Hobbs and Gupta 2003). Multiple benefits of raised bed sowing are saving in irrigation water, better soil aeration, high WUE, and less lodging of crop (Majeed et al. 2015). Furthermore, the adoption of this technique is eco-friendly because of less CH_4 emission, and improved NUE resultantly reduce N_2O emission (Hobbs and Gupta 2003).

13.3.5 Mulching

Mulches can enhance the water retention in the soil for enhancing crop productivity (Jabran 2019). However, numerous additional benefits can be harvested through mulch application in crop systems. For example, the salient benefits of mulch application include improved weed control, reduced soil erosion, enhanced microbial activities, better insect pest and disease pathogen inhibition, and beneficially modified soil temperature (Jabran 2019). Mulch application has a particular significance under the water-scarce environments to benefit the crop plants with enhanced water availability through improved moisture retention (Jabran et al. 2015b, c). Further, recent studies prove the effectiveness of mulches for suppressing weeds in various field crops (Jabran 2019, 2020). Two important constraints of water-saving rice in Pakistan include the abundant weed prevalence and soil moisture losses at a quicker rate. Mulches can be importantly employed for controlling weeds and conserving soil water for enhanced rice productivity under water-saving systems. Here, the role of mulch application in soil moisture conservation and weed control is discussed. Mulch application has been found effective in increasing the soil moisture retention for rice crop under water-saving systems. Mulches of the synthetic materials, organic wastes, wheat straw, or soybean straw increase the interception of the rainfall water, enhance soil water contents, as well as improve the soil physical properties for better crop growth and development (Jabran and Farooq 2007; Jabran et al. 2015b, 2016).

Polyethylene sheets can be effectively utilized for water conservation in agroecosystems. The application of polyethylene sheets in the fields cultivated with the water-saving rice helped improving water use efficiency by more than 100%; nevertheless, the crop growth was also improved under this particular environment (Li et al. 2007). Plastic mulch fetched a respective yield and water-saving advantage of 28.6% and 30.6% over unmulched soil conditions in maize (Mahajan et al. 2007).

A 20.9–64.7% increase was noted in the soil moisture available for the rice plants when the field was covered with the plastic mulch instead without any mulch system. Elevated soil temperature, increased decomposition of organic matter, and more root growth were the results of using plastic mulch in the aerobic rice. Higher water use efficiency and comparable grain yields with the traditionally flooded rice were obtained from the rice grown with plastic mulch. Organic matter, nitrogen, and potassium contents in the soil were decreased when the rice was put in plastic mulching (Li et al. 2007). Rice grown with plastic mulch application attained more root dry mass, higher harvest index, and increased assimilate translocation to grains compared with the flooded rice, although the shoot biomass was lower compared with the flooded rice (Xu et al. 2007).

The activities of the antioxidant systems are enhanced in the aerobic rice grown with the plastic covering over the soil surface, thus, ameliorating the oxidative damage to water-deficient conditions compared with the one without any surface mulching. Activities of the enzymes including peroxidase and catalase were greater with plastic mulch than without mulch treatment at various growth stages of the crop. However, for the year when abundant soil moisture water was available to boost the leaf water status, the activities of antioxidant enzymes were almost similar for both the mulched and nonmulched treatments (Liang et al. 2002). The cost of plastic mulches is of important concern. These materials can be used only for one season and their cost may surpass the income gained from their usage. Fisher (1995) introduced the idea of "semi-permanent plastic mulching" which is usable in the field for more than 1 year and fetches a number of desired purposes.

The plastic films are nonbiodegradable, so the left over in the field may be environmentally hazardous and can cause certain problems for the cultural operations to be carried out for the next crop. Wang et al. (2004) introduced photobiodegradable polyethylene that can get decomposed after fulfilling the requirements of moisture conservation for one season before the onset of the next plantation.

Spreading straw mulch over the soil surface mainly reduces the evaporative water losses, thus, improving the soil water status compared with the bare soil (Jabran and Farooq 2007; Jabran 2019). Thus, the water storage efficiency of a mulched soil is higher compared with the unmulched (Shangning and Unger 2001). Statistically similar yields were obtained from the conventionally flooded and water-saving rice with straw mulch. Wheat straw has been reported to reduce the rice yields when scattered over the soil surface for water conservation. Rice yield was reduced by 14% in comparison with the conventionally transplanted rice. Also, nutrient absorption and dry matter accumulation were lower where wheat straw mulch was used compared with the transplanted rice. However, the long-term use of straw mulch posed positive effects on the nutrient balance of the system and also resulted in overall comparable yields of the whole system (Liu et al. 2003). Application of wheat straw mulch caused a 16% decrease in the rice yield at lower nitrogen addition; however, with higher nitrogen rates, the yield of rice with straw mulch was almost equal to the fields that were kept under conventional flooding or plastic mulch (Fan et al. 2005).

13.4 Challenges

13.4.1 Weeds

Aerobic systems are subject to far greater weed pressure than CTPR (Rao et al. 2007; Balasubramanian et al. 2003), in which weeds are controlled by the ponding of water and rice seedling transplantation, which have a "head start" over germinating weed seedlings (Matloob et al. 2015a). Aerobic rice fields are richer in terms of weed species diversity and density (Tomita et al. 2003a, b). Weed competition is the biggest among all challenges being faced by water-saving rice in Pakistan. Rice crop is occupied by the most troublesome weeds of the world such as barnyard grass (*Echinochloa crus-galli* L.), jungle rice (*Echinochloa colona* (L.) Link.), and sedges (*Cyperus* spp.) (Matloob et al. 2015a, b; Jabran and Chauhan 2015; Kraehmer et al. 2016). According to Matloob et al. (2015a, b), *Cyperus rotundus, Echinochloa colona, Cyperus difformis, Echinochloa crus-galli,* and *Cyperus iria* were major rice weeds in Pakistan. In addition to these weeds, some other weeds like *Paspalum distichum, Fimbristylis miliacea, Cynodon dactylon, Eleusine indica, Dactyloctenium aegyptium,* and *Leptochloa chinensis* were key weeds of rice in Pakistan. Years' long experiences lead to the development of control packages for the successful weed management in the conventionally sown rice (Jabran et al. 2012a, b; Jabran and Chauhan 2015). However, with the change in growing environment in the form of water-saving rice, the established weed control practices did not work properly (Jabran and Chauhan 2015). In addition, the weed flora of water-saving rice was more complex and difficult to control than the conventional rice system (Jabran et al. 2015a). Various weed species were recorded in water-saving rice, which are never seen in the CTPR. Horse purslane (*Trianthema portulacastrum* L.) and bermudagrass (*Cynodon dactylon* L.) are important examples in this regard (Khaliq et al. 2010; Matloob et al. 2015a, b).

In Pakistan, weeds are a major hindrance to attain optimum rice yield (Irshad and Cheema 2006; Jabran et al. 2012b; Matloob et al. 2019). In rice fields, weeds are major biological pests because environmental conditions are favorable for their growth, development, and dispersal (Rao et al. 2007; Matloob et al. 2015a, b). Dhakal et al. (2015) reported less attack of weeds and yield loss in transplanted rice as compared to aerobic rice. Singh et al. (2011) reported loss (12%) yield loss due to weed infestation in transplanted rice, and high yield losses were recorded for aerobic rice on a furrow-irrigated raised bed systems (Singh et al. 2008) and rice sown without tillage operations (Singh et al. 2011). The grain yield losses may be more than two-thirds in water-saving rice and uncontrolled weed infestation caused up to 80% yield loss in aerobic rice in Pakistan (Matloob et al. 2015b).

For weed control, farmers usually use preemergence herbicides such as acetochlor, butachlor, oxadiargyl, penoxsulam, ethoxysulfuron ethyl, and pyrazosulfuron in CTPR. A good weed management is possible in conventional cultivation system of PTR due to a wide choice of preemergence herbicides available in the country, which gives absolute control of all types of weed flora (Ashraf et al. 2006). However, fairly limited chemical weed control options are available for aerobic rice. Pendimethalin is used as a preemergence herbicide in aerobic rice but it is ineffective against sedges. The available postemergence herbicides, namely penoxsulam, and bispyribac sodium either alone or in combination with bensulfuron methyl, provide poor control of sedges and grasses such as crowfoot grass (Mann et al. 2007; Khaliq et al. 2012a, 2014). Therefore, weed control in aerobic rice is a real challenge for growers and is still a major obstacle in the large-scale adoption of this water-saving rice cultivation technique (Jabran and Chauhan 2015). However, some researchers reported some encouraging developments regarding the opportunities for weed management in water-saving rice (Table 13.2). Sequential application of pre- and postemergence herbicide was better for weed control and yield increase in aerobic rice (Table 13.2; Khaliq et al. 2011). Supplementing herbicides with other control measures such as increased seeding density (Khaliq et al. 2012b), narrow row spacing (Khaliq et al. 2014), competitive cultivars (Khaliq and Matloob 2011; Matloob et al. 2015a), the allelopathic crop extracts (Khaliq et al. 2012c), manual hoeing (Khaliq et al. 2013; Ihsan et al. 2014), and residue management (Haq et al. 2019) have been reported to be effective for weed control in aerobic rice in Pakistan.

Besides herbicides, mulches can also be used for weed control in aerobic rice. Weed suppression is one of the several benefits of applying mulches in the agricultural fields (Jabran 2019, 2020). The mulches originating from the residues of previous cops probably suppress the weeds in two ways. Firstly, these physically suppress the germination and growth of weeds. Secondly, these damage the weeds by exuding allelochemicals after decomposition. Plastic (black polyethylene) mulch, on the other hand, blocks the sunlight to prevent weed germination while the growth processes of already emerged weeds including photosynthesis, respiration, etc. are severely disturbed. Some studies indicate the successfulness of both the straw and plastic mulch for weed control in water-saving rice grown in Pakistan. For example, Jabran et al. (2010) evaluated mulches (each at 6 t ha^{-1}) from four different crops and plastic mulch for weed control in water-saving aerobic rice. Black plastic mulch was found to be the most effective in suppressing the weeds with a subsequent 36% increase in rice grain yield. Maize (*Zea maize* L.) was the most effective among straw mulches for reducing weed density and dry weight compared with control.

Table 13.2 Some studies on chemical weed control in direct seeded rice in Pakistan

Herbicides	Dose	Application window	Crop husbandry	DSR type and rice cultivar	Weed flora encountered	Reduction in weed density over control	Reduction in weed biomass over control	Yield increment over control	References
						(%)			
Ethoxysulfuron ethyl (Sunstar 15WDG)	200 g ha^{-1}	7 DAS	Pregerminated seeds, 100 kg ha^{-1} seed rate, R × R = 20 cm	Wet seeding, IR-6	Jungle rice	28–36	41–57	10–12	Baloch et al. (2005)
Butachlor (Machete 60EC)	2000 m ha^{-1}					33–43	30–51	7–33	
Pendimethalin (Stomp 330E) + 1 HW	750 g a.i ha^{-1}	Immediately after sowing flowed by flushing, and manual weeding at 40 DAS	Drill sowing, 30 kg ha^{-1} seed rate, R × R = 23 cm 120–65-60 kg NPK ha^{-1}	Dry seeded, Super Basmati	Purple nutsedge, rice flatsedge, small flower umbrella-sedge, false daisy, horse purslane, and common purslane	84	90	87	Mann et al. (2007)
Pendimethalin +2 HW	–	Immediately after sowing flowed by flushing, HW at 40 and 45 DAS				93	99	91	
Ethoxysulfuron+2,4-D (ester) +1 HW	18 g a.i ha^{-1}	35 DAS, HW at 40 DAS				75	83	85	
Pendimethalin >> > Ethoxysulfuron+2,4-D (ester)	–	Immediately after sowing flowed by flushing, 35 DAS				74	82	82	

(continued)

Table 13.2 (continued)

Herbicides	Dose	Application window	Crop husbandry	DSR type and rice cultivar	Weed flora encountered	Reduction in weed density over control	Reduction in weed biomass over control	Yield increment over control	References
Ethoxysulfuron (Sunstar Gold 60WG)	62.5 g ha^{-1}	15 and 25 DAS	Drill sowing, sown after 6 h soaking, 25 kg seed rate ha^{-1}, R × R = 22. 5 cm	Dry seeded, Super Basmati	Purple nutsedge, rice flatsedge, small flower umbrella-sedge, jungle rice, barnyard grass and goose weed	82	–	307	Hussain et al. (2018)
Bispyribac sodium (Nominee 100SC)	250 ml ha^{-1}					95	–	335	
Ethoxysulfuron+iodosulfuron (Stallion 13.75WG)	150 g ha^{-1}					84	–	217	
Ethoxysulfuron ethyl (Sunstar 15WDG)	200 g ha^{-1}					77	–	305	
Pendimethalin (Stomp 455CS)	1650 g a.i ha^{-1}		Osmo-hardened ($CaCl_2$) seeds, 75 kg ha^{-1} seed rate, drill sowing in 22.5 cm spaced rows, 125–55–40 kg NPK ha^{-1}	Dry seeded, Super Basmati	Horse purslane, jungle rice, crow foot grass, goose grass, barnyard grass, purple nutsedge and rice flatsedge, and corn spurrey	84	68	94	Khaliq et al. (2011, 2012a)
Bispyribac sodium (Nominee 100 SC)	30 g a.i ha^{-1}					62	29	37	
Penoxsulam (Ryzelan 240SC)	15 g a.i ha^{-1}					21	19	35	
Pendimethalin>>> > bispyribac sodium	–					87	80	98	
Pendimethalin>>> > penoxsulam	–					86	75	105	

Pendimethalin	1650 g a.i ha^{-1}	Immediately after sowing	Hydro-primed seed (24 h), 80 kg ha^{-1} seed rate, drill sowing in 20 cm apart rows, 100–67–63 kg NPK ha^{-1}	Dry seeded, Super Basmati	Purple nutsedge, jungle rice, crow foot grass, barn yard grass, and bitter weed	81	78	6	Akbar et al. (2011)
Butachlor	1800 g a.i ha^{-1}	4 DAS				81	74	18	
Pretilachlor	1250 g a.i ha^{-1}					87	87	19	
Pendimethalin (Stomp 330EC)	825 g a.i ha^{-1}	Immediately after sowing	Hydro-primed seed (12 h), 75 kg ha^{-1} seed rate, drill sowing in 30 cm apart rows, 143–88–68 kg NPK ha^{-1}	Dry seeded, Super Basmati	Horse purslane, crow foot grass, barn yard grass, and false daisy	51	61	52	Jabran et al. (2012a, b)
Bispyribac sodium (Nominee 100SC)	15 g a.i ha^{-1}	7 DAS				66	57	50	
Penoxsulam (Ryzelan 240SC)	25 g a.i ha^{-1}	15 DAS				91	90	62	
Bispyribac sodium (Nominee 100SC)	25 g a.i ha^{-1}	15 DAS	75 kg ha^{-1} seed rate, drill sowing in 22.5 cm apart rows, 150–85–67 kg NPK ha^{-1}	Dry seeded, Super Basmati	Purple nutsedge, jungle rice, barnyard grass, crow foot grass, Horse purslane, and false daisy	–	51	105	Saqib et al. (2012)

13.4.2 Blight and Blast

Bacterial blight of rice and rice blast caused by *Xanthomonas oryzae* pv. *oryzae* and *Pyricularia oryzae*, respectively, are the most important diseases of rice in Pakistan. One of the biggest challenges for rice production around the world is rice blast (Kuyek 2000). Observations indicated an increase in the severity of these diseases when rice was grown under water-saving conditions instead of conventional conditions (unpublished data). The exact reason for this increased weed severity is yet to be known. However, certain precautions can limit the diseases reasonably. Use of pathogen-free seed, adjustment of sowing time, and careful fertilizer management will reduce the severity of diseases. Generally, it is considered that an early sown crop provided with balanced nitrogen and luxuriant potassium will be better able to withstand the attack of these diseases. It accounts for yield loss of up to 10–30% every year (Wilson and Talbot 2009). In severe infestation or favorable conditions, this disease distress whole rice plants probably in 15–20 days; resultantly, 100% yield loss (Sakulkoo et al. 2018).

13.4.3 Lodging

Crop lodging, especially at vegetative stage, is the other critical issue of growing rice under the water-saving methods. Although this rice crop lodging is not specific only to water-saving rice, the lodging noted in water-saving rice is far more than that of conventional rice. Nevertheless, the strong winds in the months of August and September lead to this furious lodging while the crop is at the reproductive growth stage. Once the crop is lodged, the losses get irreversible. The processes of the reproductive phase such as pollination, fertilization, and grain formation are distorted. Further, the wedged plants are deprived of air circulation and light penetration. Hence, the process of food making is interrupted. Use of potassium fertilizer is supposed to reduce the lodging problem; however, experimenting is needed to confirm this assumption based on observations.

13.4.4 Panicle Sterility

An important drawback of the water-saving rice is the poor grain filling and spikelet sterility (Jabran et al. 2015b). Hormonal changes play an important role in grain filling and quality. Higher hormonal concentration has been found in well-filled grains (Xu et al. 2007). Increased concentrations of abscisic acid and decreased concentrations of indole-3-acetic acid and zeatin riboside were the reasons for shrinkage of grain-filling duration that resulted in substandard grains. Artificial application of indole-3-acetic acid proved helpful in enhancing the grain-filling period of poor grains (Xu et al. 2007). High weed infestation, lesser yield, and greater panicle sterility are biggest challenges for water-saving rice production technology as compared to conventional cultivation (Farooq et al. 2011; Jabran et al. 2015a, b).

13.4.5 Genotype Availability

Most of the breeding programs in Pakistan are purposed for developing rice varieties suitable for flooded environments. Provision of ample moisture is inevitable to harvest the optimum yields from the potential varieties. Nevertheless, very little or no work has been done to develop varieties suited to aerobic environments. Nonetheless, the existing rice germplasm can be passed through the process of "selection" to check their suitability for the aerobic environments (Jabran et al. 2015a). However, recently a lot of work is in progress for the development of rice cultivars suitable for growing by water-saving methods. The variety Basmati-515 is claimed to be suited to both the flooded and water-saving environments.

Importantly, the duration of a variety can serve as an important index to judge its water-saving capability. A long duration variety will stay longer in field causing more evaporative, percolation, and seepage losses, thus, having higher water demands and low water productivity. A short duration variety will, however, indirectly save water by its comparatively brief stay in the field, minimizing unavoidable losses, and increasing water productivity (Bouman et al. 2007).

13.5 Conclusions

Water-saving rice cultivation systems are needed in the wake of the edaphic conflict, and higher labor and water consumption of conventional rice systems in Pakistan. Aerobic rice, alternate wetting and drying, the system of rice intensification, rice on raised beds, and use of mulches are the attractive options for water-saving rice cultivation systems. Herbicides like pendimethalin, penoxsulam, and bispyribac sodium can be applied for weed management in water-saving rice systems. Mulch application in water-saving rice would help reduce moisture loss through evaporation, increasing soil moisture retention, water productivity, and suppressing the weeds. Further research is needed to develop the well-adapted cultivars for water-saving rice systems using conventional breeding and biotechnology approaches. Reduced water inputs (in the case of the water-saving rice) and improved water productivity can help to spare a significant amount of water to irrigate other crops to ensure food security in the country.

References

ACIAR (Australian Centre for International Agricultural Research) (2013) Progress report: Rice-Based Systems Research Program: Food Security in Lao PDR, Cambodia and Bangladesh, pp. 1–32

Ahmad S, Hasanuzzaman M (2012) Integrated effect of plant density, N rates and irrigation regimes on the biomass production, N content, PAR use efficiencies and water productivity of rice under irrigated semiarid environment. Not Bot Horti Agrobot Cluj-Napoca 40(1):201–211

Ahmad MD, Masih I, Turral H (2004) Diagnostic analysis of spatial and temporal variations in crop water productivity: a field scale analysis of rice-wheat cropping system of Punjab, Pakistan. J Appl Irrig Sci 39:43–63

Ahmad MD, Bastiaanssen WG, Feddes RA (2005) A new technique to estimate net groundwater use across large irrigated areas by combining remote sensing and water balance approaches, Rechna Doab, Pakistan. Hydrogeol J 13:653–664

Ahmad M-u-D, Turral H, Masih I, Giordano M, Masood Z (2007) Water-saving technologies: myths and realities revealed in Pakistan's rice-wheat systems. International Water Management Institute (IWMI), Colombo, Sri Lanka

Ahmad S, Zia-ul-Haq M, Ali H, Shad SA, Ammad A, Maqsood M, Khan MB, Mehmood S, Hussain A (2008) Water and radiation use efficiencies of transplanted rice (*Oryza sativa* L.) at different plant densities and irrigation regimes under semi-arid environment. Pak J Bot 40(1): 199–209

Ahmad S, Ahmad A, Zia-ul-Haq M, Ali H, Khaliq T, Anjum MA, Khan MA, Hussain A, Hoogenboom G (2009a) Resources use efficiency of field grown transplanted rice (*Oryza sativa* L.) under irrigated semiarid environment. J Food Agric Environ 7(2):487–492

Ahmad S, Li CF, Dai GZ, Zhan M, Wang JP, Pan SG, Cao CG (2009b) Greenhouse gas emission from direct seeding paddy field under different rice tillage systems in central China. Soil Tillage Res 106:54–61

Ahmad S, Ahmad A, Soler CMT, Ali H, Zia-Ul-Haq M, Anothai J, Hussain A, Hoogenboom G, Hasanuzzaman M (2012) Application of the CSM-CERES-Rice model for evaluation of plant density and nitrogen management of fine transplanted rice for an irrigated semiarid environment. Precis Agric 13(2):200–218

Ahmad S, Ahmad A, Ali H, Hussain A, Garcia y Garcia A, Khan MA, Zia-Ul-Haq M, Hasanuzzaman M, Hoogenboom G (2013) Application of the CSM-CERES-Rice model for evaluation of plant density and irrigation management of transplanted rice for an irrigated semiarid environment. Irrig Sci 31(3):491–506

Ahmad MD, Masih I, Giordano M (2014) Constraints and opportunities for water-savings and increasing productivity through Resource Conservation Technologies in Pakistan. Agric Ecosyst Environ 187:106–115

Ahmad A, Ashfaq M, Rasul G, Wajid SA, Khaliq T, Rasul F, Saeed U, Rahman MH, Hussain J, Baig IA, Naqvi AA, Bokhari SAA, Ahmad S, Naseem W, Hoogenboom G, Valdivia RO (2015) Impact of climate change on the rice–wheat cropping system of Pakistan. In: Hillel D, Rosenzweig C (eds) Handbook of climate change and agro-ecosystems: the agricultural modeling intercomparison and improvement project (AgMIP) integrated crop and economic assessments. Imperial College Press and The American Society of Agronomy, pp 219–258

Ahmad S, Abbas G, Ahmed M, Fatima Z, Anjum MA, Rasul G, Khan MA, Hoogenboom G (2019) Climate warming and management impact on the change of rice-wheat phenology in Punjab, Pakistan. Field Crop Res 230:46–61

Ahmed M, Ahmad S (2019) Carbon dioxide enrichment and crop productivity. In: Hasanuzzaman M (ed) Agronomic crops, vol 2. Springer, Singapore, pp 31–46

Ahmed M, Ahmad S (2020) Systems modeling. In: Ahmed M (ed) Systems modeling. Springer, Singapore, pp 1–44

Ahmed M, Fayyaz-ul-Hassan, Ahmad S (2017) Climate variability impact on rice production: adaptation and mitigation strategies. In: Ahmed M, Stockle C (eds.), Quantification of climate variability, adaptation and mitigation for agricultural sustainability. Springer Switzerland, pp. 91–111

Ahmed M, Ahmad S, Raza MA, Kumar U, Ansar M, Shah GA, Parsons D, Hoogenboom G, Palosuo T, Seidel S (2020a) Models calibration and evaluation. In: Ahmed M (ed) Systems modeling. Springer, Singapore, pp 151–178

Ahmed M, Ahmad S, Waldrip HM, Ramin M, Raza MA (2020b) Whole farm modeling: a systems approach to understanding and managing livestock for greenhouse gas mitigation, economic viability and environmental quality. In: Waldrip HM, Pagliari PH, He Z (eds) Animal Manure. ASA Special Publication 67. American Society of Agronomy and Soil Science Society of America, Madison, WI, pp 345–371

Akbar N, Ehsanullah, Jabran K, Ali MA (2011) Weed management improves yield and quality of direct seeded rice. Aust J Crop Sci 5:688–694

Akram M (2009) Aromatic rices of Pakistan. Pak J Agric Sci 22:154–160

Akram R, Turan V, Hammad HM, Ahmad S, Hussain S, Hasnain A, Maqbool MM, Rehmani MIA, Rasool A, Masood N, Mahmood F, Mubeen M, Sultana SR, Fahad S, Amanet K, Saleem M, Abbas Y, Akhtar HM, Hussain S, Waseem F, Murtaza R, Amin A, Zahoor SA, Sami ul Din M, Nasim W (2018) Fate of organic and inorganic pollutants in paddy soils. In: Hashmi MZ, Varma A (eds) Environmental pollution of paddy soils. Springer, Switzerland, pp 197–214

Akram R, Fahad S, Masood N, Rasool A, Ijaz M, Ihsan MZ, Maqbool MM, Ahmad S, Hussain S, Ahmed M, Kaleem S, Sultana SR, Mubeen M, Saud S, Kamran M, Nasim W (2019) Plant growth and morphological changes in rice under abiotic stress. In: Hasanuzzaman M, Fujita M, Nahar K, Biswas JK (eds) Advances in Rice Research for Abiotic Stress Tolerance. Woodhead Publishing, Swaston, pp 69–85

Ali A, Erenstein O, Rahut DB (2014) Impact of direct rice-sowing technology on rice producers' earnings: empirical evidence from Pakistan. Dev Studies Res 1:244–254

Ali J, Zulfiqar M, Nizami A (2019) Does improved water productivity matter for the farmers? A case of water-efficient rice production from Pakistan. Develop Countries Studies 9:44–51

Ampong-Nyarko K, De Datta SK (1991) A handbook for weed control in rice. International Rice Research Institute, Manila, Philippines, p 113

Anonymous (2012a) Bangladesh among top rice-eaters. http://www.defence.pk/forums/world-affairs/146344-bangladesh-among-top-rice-eaters.html. Accessed 28 Feb 2013

Anonymous (2012b) Rice in Pakistan. International Rice Research Institute, Los Baños, Philippines. http://www.irri.org/index.php?option=com_k2&view=item&id=12077:rice-in-pakistan&Itemid=100340&lang=en. Accessed 28 Feb 2013

Anthofer J (2004) Evaluation of the System of Rice Intensification (SRI) in Cambodia, February-April, 2004. Report to GTZ, Phnom Penh, summarized in paper presented to 2004. Deutscher Tropentag, Berlin. www.tropentag.de/2004/proceedings/node179.html

Ashraf MM, Awan TH, Manzoor Z, Ahmad M, Safdar ME (2006) Screening of herbicides for weed management in transplanted rice. J Anim Plant Sci 16:92–95

Aslam M, Qureshi AS, Horinkova VM (2002) Water-saving strategies for irrigated rice. J Drain Water Manage 6:25–36

Awan MI, Meinke H, Oort P, Bastiaans L (2015) Entry points for eco-efficient aerobic rice production system in Punjab, Pakistan. "*Building Productive, Diverse and Sustainable Landscapes*". Proceedings of the 17th ASA Conference, 20–24 September 2015, Hobart, Australia. www.agronomy2015.com.au.

Babel MS, Wahid SM (2008) Freshwater under threat: South Asia. Vulnerability assessment of freshwater resources to environmental change. UNEP and Asian Institute of Technology, Bangkok

Balasubramanian V, Ladha JK, Gupta RK, Naresh RK, Mehla RS, Bijay-Singh and Yadvinder-Singh (2003) Technology options for rice in the rice-wheat system in South Asia. In: Ladha JK, Hill JE, Duxbury JM, Gupta RK, Buresh RJ (eds.). Improving the productivity and sustainability of rice-wheat systems: issues and impacts, ASA Special Publication 65. ASA, CSSA, SSSA, Madison, Wisconsin. pp. 115–147.

Baloch MS, Hassan G, Morimoto T (2005) Weeding techniques in transplanted and wet-seeded rice in Pakistan. Weed Biolog Mang 5:190–196

Barker R, Dawe D, Tuong TP, Bhuiyan SI, Guerra LC (1999) The outlook for water resources in the year 2020: Challenges for research on water management in rice production. In: Assessment and

Orientation towards the 21st Century. Proceedings of the 19th session of the International Rice Commission. Cairo, Egypt. 7-9 September 1998. pp. 96–109

Bhatti MA, Kinje JW (1992) Irrigation management potential of paddy/rice production in Punjab in Pakistan. In: Soil and Water Engineering for Paddy Field Management. Proceedings of the International Workshop, January 1992, Bangkok. AIT, Bangkok, Thailand, pp. 355–366

Bhushan L, Ladha JK, Gupta RK, Singh S, Tirol-Padre A, Saharawat YS, Gathala M, Pathak H (2007) Saving of water and labor in a rice–wheat system with no-tillage and direct seeding technologies. Agron J 99:1288–1296

Bhutta MN, Alam MM (2005) Perspective and limits of groundwater use in Pakistan. In: Groundwater Research and Management: integrating science into management decisions. Roorkee, India, pp 105–113

Bhuyan MHM, Ferdousi MR, Iqbal MT (2012) Yield and growth response to transplanted aman rice under raised bed over conventional cultivation method. ISRN Agronomy Article ID 646859, 8 pages

Bouman BAM (2009) How much water does rice use? Rice Today 8:28–29

Bouman BAM, Tuong TP (2001) Field water management to save water and increase its productivity in irrigated rice. Agric Water Manag 49:11–30

Bouman BAM, Peng S, Castaneda AR, Visperas RM (2005) Yield and water use of irrigated tropical aerobic rice systems. Agric Water Manag 74:87–105

Bouman BAM, Lampayan RM, Tuong TP (2007) Water management in irrigated rice: coping with water scarcity. International Rice Research Institute, Los Baños, Philippines, 54p

Buresh RJ, De Datta SK (1990) Denitrification losses from puddled rice soils in the tropics. Biol Fertil Soils 9:1–13

Cai Z, Xing G, Yan X, Xu H, Tsuruta H, Yagi K, Minami K (1997) Methane and nitrous oxide emissions from rice paddy fields as affected by nitrogen fertilisers and water management. Plant Soil 196:7–14

Ceesay M, Reid WS, Fernandes E, Uphoff N (2006) The effect of repeated soil wetting and drying on lowland rice yield with System of Rice Intensification (SRI) methods. Int J Agric Sustain 4: 5–14

Chakraborty S, Alam MO, Bhattacharya T, Singh YN (2014) Arsenic accumulation in food crops: a potential threat in bengal delta plain. Water Qual Expo Health 6:233–246

Chaudhary Q, Mahmood A, Rasul G, Afzaal M (2009). Climate change indicators of Pakistan. Technical Report No. PMD-22:1–43

Chauhan BS, Jabran K, Mahajan G (eds) (2017) Rice production worldwide. Springer, Cham

Chhokar RS, Sharma RK, Gathala MK, Pundir AK (2014) Effects of crop establishment techniques on weeds and rice yield. Crop Prot 64:7–12

Choudhury ATMA, Kennedy IR (2005) Nitrogen fertilizer losses from rice soils and control of environmental pollution problems. Commun Soil Sci Plant Anal 36:1625–1639

Dawe D (2005) Increasing water productivity in rice–based systems in Asia–past trends, current problems, and future prospects. Plant Prod Sci 8:221–230

Dhakal M, Sah SK, McDonald A, Regmi AP (2015) Perception and economics of dry direct seeded rice in Terai of Nepal. J Agric Environ Sci 16:103–111

Engelman R, LeRoy P (1993) Sustainable Water; Population and the Future of Renewable Water Supplies. Population and Environment Programme. Population Action International, 1120 19th street NW Suite 550, Washington, DC, USA

Erenstein O (2009) Comparing water management in rice–wheat production systems in Haryana, India and Punjab, Pakistan. Agric Water Manag 96:1799–1806

Fahad S, Noor M, Adnan M, Khan MA, Rahman I, Alam M, Khan IA, Ullah H, Mian IA, Hassan S, Saud S, Bakhat HF, Hammad HM, Ahmad S, Nasim W (2019) Abiotic stress and rice grain quality. In: Hasanuzzaman M, Fujita M, Nahar K, Biswas JK (eds) Advances in rice research for abiotic stress tolerance. Woodhead Publishing, Swaston, pp 571–583

Fan M, Jiang R, Liu X, Zhnag F, Lu S, Zeng X, Christie P (2005) Interactions between non-flooded mulching cultivation and varying nitrogen inputs in rice–wheat rotations. Field Crop Res 91: 307–318

Farooq M, Basra SMA, Tabassum R, Afzal I (2006a) Enhancing the performance of direct seeded fine rice by seed priming. Plant Prod Sci 9:446–456

Farooq M, Basra SMA, Wahid A (2006b) Priming of field-sown rice seed enhances germination, seedling establishment, allometry and yield. Plant Growth Regul 49:285–294

Farooq M, Basra SMA, Ahmad N (2007) Improving the performance of transplanted rice by seed priming. Plant Growth Regul 51:129–137

Farooq M, Basra SMA, Ahmad N, Murtaza G (2009) Enhancing the performance of transplanted coarse rice by seed priming. Paddy Water Environ 7:55–63

Farooq M, Siddique KHM, Rehman H, Aziz T, Lee D, Wahid A (2011) Rice direct seeding: Experiences, challenges and opportunities. Soil Tillage Res 111:87–98

Fatima Z, Ahmad M, Hussain M, Abbas G, Ul-Allah S, Ahmad S, Ahmed N, Ali MA, Sarwar G, Ehsan ul Haque, Iqbal P, Hussain S (2020) The fingerprints of climate warming on cereal crops phenology and adaptation options. Sci Rep 10:18013

Fisher PD (1995) An alternative plastic mulching system for improved water management in dryland maize production. Agric Water Manag 27:155–166

Ghani MA, Bhuiyan SI, Hill RW (1989) Gravity irrigation management in Bangladesh. J Irrig Drain Eng 115:642–655

Gill M (2007) Increasing Water Productivity of Rice through Adoption of System of Rice Intensification (SRI), a presentation given to the SRI group at Cornell University, Ithaca, NY, by the Director General for Agriculture (Water Management) in Punjab, Pakistan

Gleik PH (ed) (1993) Water crisis: a guide to the world's fresh water resources. Pacific Institute for Studies in Development, Environment and Security. Stockholm Environment Institute. Oxford University Press, New York

GMI (Global Methane Initiative) (2010) Global methane emissions and mitigation opportunities. GMI. www.globalmethane.org. Accessed 17 August 2011

Gopal R, Jat RK, Malik RK, Kumar V, Alam MM, Jat ML, Mazid MA, Saharawat YS, McDonald A, Gupta R (2010) Direct dry seeded rice production technology and weed management in rice based systems. Technical Bulletin. International Maize and Wheat Improvement Center, New Delhi, India, 28p

Government of Pakistan (2020) Pakistan Economic Survey 2019–2020. Economic Adviser's Wing. Finance Division Government of Pakistan, Islamabad, p 21

Guerra LC, Bhuiyan SI, Tuong TP, Barker R (1998) Producing more rice with less water from irrigated systems. SWIM Paper 5. IWMI/IRRI, Colombo, Sri Lanka, 24p

Hafeez MM, Bouman BAM, Van de Giesen N, Vlek P (2007) Scale effects on water use and water productivity in a rice-based irrigation system (UPRIIS) in the Philippines. Agric Water Manag 92:81–89

Hafeez-ur-Rehman, Nawaz A, Awan MI, Ijaz M, Hussain M, Ahmad S, Farooq M (2019) Direct seeding in rice: problems and prospects. In: Hasanuzzaman M (ed) Agronomic crops: production technologies, vol 1. Springer, Singapore, pp 199–222

Hanjra MA, Qureshi ME (2010) Global water crisis and future food security in an era of climate change. Food Policy:365–377

Haq M, Khaliq A, Sheng Q, Matloob A, Hussain S, Fatima S, Aslam Z (2019) Weed growth, herbicide efficacy, and rice productivity in dry seeded paddy field under different wheat stubble management methods. J Integr Agric 18:907–926

Hassan FU, Ahmad M, Ahmad N, Abbasi MK (2007) Effects of subsoil compaction on yield and yield attributes of wheat in the sub-humid region of Pakistan. Soil Tillage Res 96:361–366

Hira GS (2009) Water management in northern states and the food security of India. J Crop Improv 23:136–157

Hobbs PR (2001) Tillage and crop establishment in South Asian rice-wheat systems. J Crop Prod 4: 1–22

Hobbs PR, Gupta RK (2003) Resource conservation technologies for wheat in the rice-wheat system. In: Ladha JK, Hill JE, Duxbury JM, Gupta RK, Buresh RJ (eds) Improving the Productivity and Sustainability of Rice-Wheat Systems: Issues and Impacts, ASA Special Publication 65. ASA, CSSA, SSSA, Madison, Wisconsin, pp 149–171

Hobbs PR, Gupta RK (2004) Problems and challenges of no-till farming for the rice–wheat systems of the Indo-Gangetic Plains in South Asia. In: Lal R, Hobbs P, Uphoff N, Hansen DO (eds) Sustainable agriculture and the rice–wheat system. Ohio State University, Marcel Dekker, Columbus, OH, New York, USA, pp 101–119

Hou AX, Chen GX, Wang ZP, Van Cleemput O, Jr Patrick WH (2000) Methane and nitrous oxide emissions form a rice field in relation to soil redox and microbiological processes. Soil Sci Soc Am J 64:2180–2186

Humphreys E, Kukal SS, Christen EW, Hira GS, Singh B, Yadav S, Sharma RK (2010) Halting the groundwater decline in north-west India: which crop technologies will be winners? Adv Agron 109:155–217

Husain AMM, Chowhan G, Barua P (2004) Completion Report on Verification and Refinement of the System of Rice Intensification (SRI) Project in Selected Areas of Bangladesh (SP 36 02), submitted to Poverty Elimination Through Rice Research Assistance (PETRRA), IRRI, Dhaka, June 2004

Hussain Z, Khan MA, Irfan M (2010) Water energy and economic analysis of wheat production under raised bed and conventional irrigation systems: a case study from a semi-arid area of Pakistan. Soil Tillage Res 109:61–67

Hussain S, Peng S, Fahad S, Khaliq A, Huang J, Cui K, Nie L (2014) Rice management interventions to mitigate greenhouse gas emissions: a review. Environ Sci Pollut Res 22: 3342–3360

Hussain M, Ahmad S, Hussain S, Lal R, Ul-Allah S, Nawaz A (2018) Rice in saline soils: physiology, biochemistry, genetics and management. Adv Agron 148:231–287

Ihsan MZ, Khaliq A, Matloob A, El-Nakhlawy FS, Abohassan RA, Daur I, Aslam Z (2014) Influence of herbicides applied alone or supplemented with manual weeding on weed growth, rice yield and grain quality in direct seeded rice (*Oryza sativa* L.). Philippine Agric Scientist 97: 377–384

IPCC (1997) Revised 1996 IPCC guidelines for national greenhouse gas inventories workbook, vol 2. Cambridge University Press, Cambridge

IPCC (2007) Summary for policymakers. In: Parry ML, Canziani OF, Palutikof JP, Van der Linden PJ, Hanson CE (eds) Climate Change 2007: Impacts, adaptation and vulnerability. Contribution of Working Group II to the Fourth Assessment Report of the Intergovernmental Panel on Climate Change. Cambridge University Press, Cambridge, pp 81–82

Irshad A, Cheema ZA (2006) Comparative efficiency of barnyard grass and rice to nitrogen under transplanted condition. J Tropical Agric Food Sci 34:385–392

Ishaq M, Hassan A, Saeed M, Ibrahim M, Lal R (2001) Subsoil compaction effects of crops in Punjab. Pakistan. I. Soil physical properties and crop yield. Soil Tillage Res 59:57–65

Ishfaq M, Akbar N, Anjum SA, Anwar-Ul-Haq M (2020) Growth, yield and water productivity of dry direct seeded rice and transplanted aromatic rice under different irrigation management regimes. J Integr Agric 19:2656–2673

Jabran K (2019) Role of mulching in pest management and agricultural sustainability. Springer, Cham. https://doi.org/10.1007/978-3-030-22-301-4

Jabran K (2020) Use of mulches for managing field bindweed and purple nutsedge, and weed control in spinach. Int J Agric Biol 23:1114–1120

Jabran K, Chauhan BS (2015) Weed management in aerobic rice systems. Crop Prot 78:151–163

Jabran K, Farooq M (2007) Mulching for resource conservation. Dawn, February 26, 2007. https://www.dawn.com/news/234678/mulching-for-resource-conservation. Assessed on 5 Dec 2020

Jabran K, Farooq M, Hussain M, Rehman H, Ali MA (2010) Wild oat (Avena fatua L.) and canary grass (Phalaris minor Retz.) management through allelopathy. J Plant Prot Res 50:41–44

Jabran K, Farooq M, Hussain M, Ehsanullah KMB, Shahid M, Lee DJ (2012a) Efficient weeds control with penoxsulam application ensures higher productivity and economic returns of direct seeded rice. Int J Agric Biol 14:901–907

Jabran K, Hussain M, Farooq M, Babar M, Doğan MN, Lee DJ (2012b) Application of bispyribac-sodium provides effective weed control in direct-planted rice on a sandy loam soil. Weed Biol Manag 12:136–145

Jabran K, Ullah E, Hussain M, Farooq M, Haider N, Chauhan BS (2015a) Water-saving, water productivity and yield outputs of fine-grain rice cultivars under conventional and water-saving rice production systems. Exp Agric 51:567–581

Jabran K, Ullah E, Hussain M, Farooq M, Zaman U, Yaseen M, Chauhan BS (2015b) Mulching improves water productivity, yield and quality of fine rice under water-saving rice production systems. J Agron Crop Sci 201:389–400

Jabran K, Ullah E, Akbar N (2015c) Mulching improves crop growth, grain length, head rice and milling recovery of basmati rice grown in water-saving production systems. Int J Agric Biol 17: 920–928

Jabran K, Hussain M, Fahad S, Farooq M, Bajwa AA, Alharrby H, Nasim W (2016) Economic assessment of different mulches in conventional and water-saving rice production systems. Environ Sci Pollut Res 23:9156–9163

Jabran K, Ullah E, Akbar N, Yasin M, Zaman U, Nasim W, Hussain M (2017a) Growth and physiology of basmati rice under conventional and water-saving production systems. Arch Agron Soil Sci 63:1465–1476

Jabran K, Riaz M, Hussain M, Nasim W, Zaman U, Fahad S, Chauhan BS (2017b) Water-saving technologies affect the grain characteristics and recovery of fine-grain rice cultivars in semi-arid environment. Environ Sci Pollut Res 24:12971–12981

Jabran K, Uludag A, Chauhan BS (2018) Sustainable weed control in rice. In: Corres NE, Burgos NR, Duke SO (eds) Weed Control: Sustainability, Hazards, and Risks in Cropping Systems Worldwide. CRC Press, Taylor & Francis Group, Boca Raton FL, pp 276–287

Jehangir WA, Masih I, Ahmed S, Gill MA, Ahmad M, Mann RA, Chaudhary MR, Qureshi AS, Turral H (2007) Sustaining crop water productivity in rice-wheat systems of South Asia: A case study from the Punjab, Pakistan. Colombo, Sri Lanka: International Water Management Institute. (IWMI Working Paper 115). 45p

Joshi E, Lal B, Kumar D, Gauyam P, Ak V (2013) Management of direct seeded rice for enhanced resource - use efficiency. Plant Know J 2:119–134

Kahlown MA, Kemper WD (2004) Seepage losses as affected by condition and composition of channel banks. Agric Water Manag 65:145–153

Kahlown MA, Majeed A (2003) Water-resources situation in Pakistan: challenges and future strategies. In: COMSATS' series of publications on science and technology 3. Water Resources in the South: Present Scenario and Future. Prospects Commission on Science and Technology for Sustainable Development in the South. pp. 21–39

Kahlown MA, Raoof A, Hanif M (2001) Plant population effect on paddy yield. J Drain Water Manage 5:1–5

Kahlown MA, Raoof A, Zubair M, Kemper WD (2007) Water use efficiency and economic feasibility of growing rice and wheat with sprinkler irrigation in the Indus Basin of Pakistan. Agric Water Manag 87:292–298

Kasterine A, Vanzetti D (2010) The effectiveness, efficiency and equity of market-based instruments to mitigate GHG emission from the agrifood sector, in UNCTAD Trade and Environment Review 2009/2010, Geneva. http://www.intracen.org/uploadedFiles/intracenorg/Content/Exporters/Sectors/Fair_trade_and_environmental_exports/Climate_change/TER_UNCTAD_KasterineVanzetti.pdf

Kepha GO, Bancy MM, Patrick GH (2014) Determination of the effect of the system of rice intensification (SRI) on rice Yields and water saving in Mivea irrigation scheme, Kenya. J Water Resour Prot 6:895–901

Khaliq A, Matloob A (2011) Weed crop competition period in three fine rice cultivars under direct seeded rice culture. Pak J Weed Sci Res 17:229–243

Khaliq A, Matloob A, Farooq M, Aslam F (2010) Weed occurrence and yield losses in dry direct seeded rice. 22nd Conf. of Asian Pacific Weed Science Society, March 8–12, 2010. Lahore, Pakistan. 12p

Khaliq A, Matloob A, Shafique HM, Cheema ZA, Wahid A (2011) Evaluating sequential application of pre and post emergence herbicides in dry seeded finer rice. Pak J Weed Sci Res 17:111–123

Khaliq A, Matloob A, Ahmad N, Rasul F, Awan IU (2012a) Post emergence chemical weed control in direct seeded fine rice. J Anim Plant Sci 22:1101–1106

Khaliq A, Matloob A, Mahmood S, Abbas RN, Khan MB (2012b) Seeding density and herbicide tank mixture furnishes better weed control and improve the growth, yield and quality of direct seeded fine rice. Int J Agric Biol 14:499–508

Khaliq A, Matloob A, Riaz Y (2012c) Bio-economic and qualitative impact of reduced herbicide usage in direct seeded fine rice through multipurpose tree water extracts. Chilean J Agric Res 72: 350–357

Khaliq A, Matloob A, Ihsan MZ, Abbas RN, Aslam Z, Rasul F (2013) Supplementing herbicides with manual weeding improves weed control efficiency, growth and yield of dry seeded rice. Int J Agric Biol 15:191–199

Khaliq A, Matloob A, Chauhan BS (2014) Weed management in dry seeded fine rice sown under varying row spacing in a rice-wheat system of Pakistan. Plant Prod Sci 17:321–332

Khaliq A, Matloob A, Hussain A, Hussain S, Aslam F, Zamir SI, Chattha MU (2015) Wheat residue management options affect productivity, weed growth and soil properties in direct-seeded fine aromatic rice. Clean Soil Air Water 43:1259–1665

Khan S, Tariq R, Yuanlai C, Blackwell J (2006) Can irrigation be sustainable? Agric Water Manag 80:87–99

Khan MA, Ahmad S, Raza A (2019a) Integrated weed management for agronomic crops. In: Hasanuzzaman M (ed) Agronomic crops, vol 2. Springer, Singapore, pp 257–281

Khan SM, Ali S, Nawaz A, Bukhari SAH, Ejaz S, Ahmad S (2019b) Integrated pest and disease management for better agronomic crop production. In: Hasanuzzaman M (ed) Agronomic Crops, vol 2. Springer, Singapore, pp 385–428

Kraehmer H, Jabran K, Mennan H, Chauhan BS (2016) Global distribution of rice weeds–a review. Crop Prot 80:73–86

Kumar V, Ladha JK (2011) Direct seeding of rice: recent developments and future needs. Adv Agron 111:297–413

Kumar V, Bellinder RR, Gupta RK, Malik RK, Brainard DC (2008) Role of herbicide-resistant rice in promoting resource conservation technologies in rice-wheat cropping systems of India: a review. Crop Prot 27:290–301

Kuyek D (2000) Blast, biotech and big business: implications of corporate strategies on rice research in asia. Barcelona, Spain, GRAIN

Ladha JK, Pathak H, Gupta RK (2007) Sustainability of the rice-wheat cropping system: issues, constraints, and remedial options. J Crop Improv 19:125–136

Lampayan RM, Palis FG, Flor RB, Bouman BAM, Quicho ED, De Dios JL, Espiritu A, Sibayan EB, Vicmudo VR, Lactaoen AT, Soriano JB (2009) Adoption and Dissemination of "Safe Alternate Wetting and Drying" in Pump Irrigated Rice Areas in the Philippines. In: 60th International Executive Council Meeting of the International Commission on Irrigation and Drainage (ICID), 5th Regional Conference, December 6-–11, 2009, New Delhi, India

Lazora RC (2004) Water-saving high-yielding TQPM and SRI trials runs and demonstrations in the Magballo-Balicotoc-Canlamay Intergrated Irrigation Sub Project. Report to National Irrigation Administration, Quezon City, form the Southern Philippines Irrigation Support Project

Li C, Mosier A, Wassmann R, Cai Z, Zheng X, Huang Y, Tsuruta H, Boonjawat J, Lantin R (2004) Modeling greenhouse gas emissions from rice-based production systems: Sensitivity and upscaling. Glob Biogeochem Cycles 18:GB1043. https://doi.org/10.1029/2003GB002045

Li YS, Wu LH, Zhao LM, Lu XH, Fan QL, Zhang FS (2007) Influence of continuous plastic film mulching on yield, water use efficiency and soil properties of rice fields under non-flooding condition. Soil Tillage Res 93:370–378

Linquist BA, Van Groenigen KJ, Adviento-Borbe MA, Pittelkow C, Van Kessel C (2012) An agronomic assessment of greenhouse gas emissions from major cereal crops. Glob Chang Biol 18:194–209

Liu XJ, Wanga JC, Lua SH, Zhanga FS, Zenga XZ, Aia YW, Penga SB, Christie P (2003) Effects of non-flooded mulching cultivation on crop yield, nutrient uptake and nutrient balance in rice-wheat cropping systems. Field Crop Res 83:297–311

Liang Y, Zhang CE, Guo DW (2002) Mulch types and their benefit in cropland ecosystems on the loess plateau in China. J Plant Nutr 25:945–955

Lu Y, Wassmann R, Neue HU, Huang C, Bueno CS (2000) Methanogenic responses to exogenous substrate in anaerobic rice soils. Soil Biol Biochem 32:1683–1690

Lu Y, Adomako EE, Solaiman ARM, Islam MR, Deacon C, Williams PN, Rahman GKMM, Meharg AA (2009) Baseline soil variation is a major factor in arsenic accumulation in Bengal delta paddy rice. Environ Sci Technol 43:1724–1729

Mahajan G, Sharda R, Kumar A, Singh KG (2007) Effect of plastic mulch on economizing irrigation water and weed control in baby corn sown by different methods. Afr J Agric Res 2: 19–26

Majeed A, Mahmood A, Niaz A, Javid S, Ahmad ZA, Shah SSH, Shah AH (2015) Bed planting of wheat improves nitrogen use efficiency and grain yield compared to flat planting. Crop J 3:118–124

Majeed A, Saleem M, Jalil S, Abbas SH, Hayat A (2017) Water saving rice production using alternate wetting and drying technique in rice based cropping system in Sindh, Pakistan. Sci Technol Dev 36:30–35

Malik RK, Yadav A, Gill GS, Sardana P, Gupta RK, Piggin C (2004) Evolution and acceleration of no-till farming in rice–wheat cropping system of the Indo-Gangetic plains. In: New Directions for a Diverse Planet. Proceedings of the Fourth International Crop Science Congress, 26 September–1 October 2004, Brisbane, Australia. Published on CDROM. http://www.cropscience.org.au.

Mann RA, Ahmad S, Hassan G, Baloch MS (2007) Weed management in direct seeded rice crop. Pak J Weed Sci Res 13:219–226

Mann RA, Hussain S, Saleem M (2011) Impact of dry seeding with alternate wetting and drying on rice productivity and profitability in Punjab, Pakistan. In: Resilient food system for a changing world. Proceedings of the 5th world congress on conservation agriculture, Brisbane, Australia. pp. 395–396

Mann RA, Ijaz M, Hussain S (2016) Aerobic and alternate wetting and drying rice production systems in Pakistan: progress and challenges. Regional: Development and Dissemination of Climate-Resilient Rice Varieties for Water-Short Areas of South Asia and Southeast Asia (Financed by the Climate Change Fund and the Government of Finland). International Rice Research Institute and partner organizations in Bangladesh, India, Nepal, Pakistan, and Philippines Los Baños, Laguna. pp. 211–243

Matloob A (2014) Weed Dynamics and Management Studies in Direct Seeded Rice (*Oryza sativa* L.). PhD dissertation. Department of Agronomy, University of Agriculture, Faisalabad, Pakistan

Matloob A, Khaliq A, Chauhan BS (2015a) Weeds of rice in Asia: problems and opportunities. Adv Agron 130:291–336

Matloob A, Khaliq A, Tanveer A, Husssain S, Aslam F, Chauhan BS (2015b) Weed dynamics in dry direct-seeded fine rice as influenced by tillage system, sowing time and weed competition duration. Crop Prot 71:25–38

Matloob A, Safdar ME, Abbas T, Aslam F, Khaliq A, Tanveer A, Rehman A, Chadhar AR (2019) Challenges and prospects for weed management in Pakistan: a review. Crop Prot 134:104724
Matson PA, Naylor R, Ortiz-Monasterio I (1998) Integration of environmental, agronomic, and economic aspects of fertilizer management. Science 280:112–115
Miah H, Sattar MA, Tuong TP (2009) Water saving irrigation practices in rice based canal system. International Rice Research Institute, Dhaka, Bangladesh
Murray-Rust DH, Velde EJV (1994) Conjunctive use of canal and groundwater in Punjab, Pakistan: management and policy options. Irrig Drain Syst 8:201–231
Namara RE, Weligamage P, Barker R (2003) Prospects for adopting system of rice intensification in Sri Lanka: a socio-economic assessment. Research Report: 75. IWMI, Colombo, Sri Lanka
Naresh RK, Singh B, Singh SP, Singh PK, Kumar A, Kumar A (2012) Furrow irrigated raised bed (FIRB) planting technique for diversification of ricewheat system for western IGP region. Int J Life Sci Biotechnol Pharm Res 9:134–141
National Water Policy (2003) Draft, prepared by the office of Chief Engineering Advisor. MW&P, GOP, Islamabad
Naz S, Fatima Z, Iqbal P, Khan A, Zakir I, Noreen S, Younis H, Abbas G, Ahmad S (2019) Agronomic crops: types and uses. In: Hasanuzzaman M (ed) Agronomic crops: production technologies, vol 1. Springer, Singapore, pp 1–18
Nemoto K, Morita S, Baba T (1995) Shoot and root development in rice related to the phyllochron. Crop Sci 35:24–29
Nizami A, Zulfiqar M, Ali J, Khan N, Sheikh I (2020) Improving water productivity in rice – a response to climate change and water stress in Pakistan. Sarhad J Agric 36:383–388
OECD (2000) Environmental Indicators for Agriculture Methods and Results. Executive summary. OECD, Paris
Ortiz-Monasterio J, Dhillon SS, Fischer RA (1994) Date of sowing effects on grain yield and yield components of irrigated spring wheat cultivars and relationships with radiation and temperature in Ludhiana, India. Field Crop Res 37:169–184
Pathak H, Ladha JK, Aggarwal PK, Peng S, Das S, Singh Y, Singh B, Kamra SK, Mishra B, Sastri ASRAS, Aggarwal HP, Das DK, Gupta RK (2003a) Trends of climatic potential and on-farm yields of rice and wheat in the Indo-Gangetic Plains. Field Crop Res 80:223–234
Pathak H, Prasad S, Bhatia A, Singh S, Kumar S, Singh J, Jain MC (2003b) Methane emission from rice-wheat cropping system of India in relation to irrigation, farmyard manure and dicyandiamide application. Agric Ecosyst Environ 97:309–316
Pathak H, Saharawat YS, Gathala M, Ladha JK (2011a) Impact of resource-conserving technologies on productivity and greenhouse gas emission in rice-wheat system. Greenhouse Gas Sci Technol 1:261–277
Pathak H, Tewari AN, Sankhyan S, Dubey DS, Mina U, Singh VK, Jain N, Bhatia A (2011b) Direct-seeded rice: Potential, performance and problems – A review. Curr Advan Agric Sci 3: 77–88
Peerzada AM, Bukhari SAH, Dawood M, Nawaz A, Ahmad S, Adkins S (2019) Weed management for healthy crop production. In: Hasanuzzaman M (ed) Agronomic crops, vol 2. Springer, Singapore, pp 225–256
Planning Commission (2020) Water–Background and Vision. 11th Five Year Plan. pp. 223–233. https://www.pc.gov.pk/uploads/plans/Ch20-Water1.pdf
Punshon T, Jackson BP, Meharg AA, Warczack T, Scheckel K, Guerinot ML (2018) Understanding arsenic dynamics in agronomic systems to predict and prevent uptake by crop plants. Sci Total Environ 581–582:209–220
Qutab SA, Nasiruddin (1994) Cost effectiveness of improved water management practices. In: Inayatullah C (ed) Water and community–an assessment of on farm water management programme. SDPI, Islamabad, Pakistan
Ram H, Yadvinder-Singh, Timsina T, Humphreys E, Dhillon SS, Kumar K, Kler DS (2005) Performance of upland crops on raised beds in northwestern India. In: Roth CH, Fischer RA, Meisner CA (eds) Evaluation and performance of permanent raised bed cropping systems in Asia, Australia and Mexico. Proceedings of a workshop held in Griffith, NSW, Australia, 1–3 March. CIMMYT, Mexico, pp 41–58

Rao AN, Johnson DE, Sivaprasad B, Ladha JK, Mortimer AM (2007) Weed management in direct-seeded rice. Adv Agron 93:153–255

Razaq M, Shah FM, Ahmad S, Afzal M (2019) Pest management for agronomic crops. In: Hasanuzzaman M (ed) Agronomic Crops, vol 2. Springer, Singapore, pp 365–384

REAP (Rice Export Association of Pakistan) (2020) Export of rice from Pakistan - 2010 to 2019. https://reap.com.pk/exporthistory

Rodhe H (1990) A comparison of the contribution of various gases to the greenhouse. Science 248: 1217–1219

Sadras VO, Calvino PA (2001) Quantification of grain yield response to soil depth in soybean, maize, sunflower, and wheat. Agron J 93:577–583

Sahrawat YS, Singh B, Malik RK, Ladha JK, Gathala M, Jat ML, Kumar V (2010) Evaluation of alternative tillage and crop establishment methods in a rice–wheat rotation in North Western IGP. Field Crop Res 116:260–267

Sakulkoo W, Osés-Ruiz M, Oliveira Garcia E, Soanes DM, Littlejohn GR, Hacker C, Correia A, Valent B, Talbot NJ (2018) A single fungal MAP kinase controls plant cell-to-cell invasion by the rice blast fungus. Science 359:1399–1403

Sandhu N, Jain S, Battan KR, Jain RK (2012) Aerobic rice genotypes displayed greater adaptation to water-limited cultivation and tolerance to polyethyleneglycol-6000 induced stress. Physiol Mol Biol Plants 18:33–43

Saqib M, Akbar N, Ehsanullah GA (2012) Development and appraisal of economical and sustainable approach for weed management in drill seeded aerobic rice (*Oryza sativa* L.). Pak J Agric Sci 49:281–287

Sarwar N, Ali H, Ahmad S, EhsanUllah AS, Mubeen K, Hill JE (2013a) Water wise rice cultivation on calcareous soil with the addition of essential micronutrients. J Anim Plant Sci 23(1):244–250

Sarwar N, Ali H, Maqsood M, EhsanUllah, Shahzad AN, Shahzad M, Mubeen K, Shahid MA, Ahmad S (2013b) Phenological response of rice plants to different micronutrients application under water saving paddy fields on calcareous soil. Turk J Field Crops 18(1):52–57

Sato S (2006) An evaluation of the System of Rice Intensification in Eastern Indonesia for its potential to save water while increasing productivity. Paper for International Dialogue on Rice and Water: Exploring Options for food security and Sustainable Environments. IRRI, Los Banos

Satyanarayana A, Thiyagarajan TM, Uphoff N (2006) Abstract on opportunities for saving with higher yield using the System of Rice Intensification. Acharya A.N. Ranga Agriculture University, Hyderabad, India; Tamil Nadu Agricultural University, Killikulam, India; and Cornell University, USA

Saunders DA (1990) Crop management research summary of results. Wheat Research Centre Monograph 5. Wheat-Rice Consortium, Nashipur. Bangladesh

Seck PA, Diagne A, Mohanty S, Wopereis MCS (2012) Crops that feed the world 7: rice. Food Security 4:7–24

Shah T (2005) Groundwater and human development: challenges and opportunities in livelihoods and environment. Water Sci Technol 51:27–37

Shah T (2007) The groundwater economy of South Asia: an assessment of size, significance and socio-ecological impacts. In: Giordano M, Vilholth KG (eds) The agricultural groundwater revolution: opportunities and threats to development. CABI Publication, Wallingford, UK, pp 7–30

Shah T, Singh O, Mukherji A (2006) Some aspects of South Asia's groundwater irrigation economy: analyses from a survey in India, Pakistan, Nepal Terai and Bangladesh. Hydrogeol J 14:286–309

Shah T, Ul H, Khattak MZ, Banerjee PS, Singh OP, Rehman SU (2009) Is irrigation water free? A reality check in the Indo-Gangetic Basin. World Dev 37:422–434

Shahzad AN, Ahmad S (2019) Tools and techniques for nitrogen management in cereals. In: Hasanuzzaman M (ed) Agronomic crops, vol 2. Springer, Singapore, pp 111–126

Shang QY, Yang XX, Gao CM, Gao CM, Wu PP, Liu JJ, Xu YC, Shen QR, Zou JW, Guo SW (2011) Net annual global warming potential and greenhouse gas intensity in Chinese double rice-cropping systems: a 3-year field measurement in long-term fertilizer experiments. Glob Chang Biol 17:2196–2210

Shangning J, Unger PW (2001) Soil water accumulation under different precipitation, potential evaporation, and straw mulch conditions. Soil Sci Soc Am J 65:442–448

Sharma PK, De Datta SK (1986) Physical properties and processes of puddled rice soils. Adv Soil Sci 5:139–178

Sharif A (2011) Technical adaptations for mechanized SRI production to achieve water saving and increased profitability in Punjab, Pakistan. Paddy Water Environ 9:111–119

Sheikh MA, Kanasro HA (2003) Production, consumption and export of rice. Press release the daily Dawn. http//Dawn.com, March 10

Singh DK, Singh AK (2002) Groundwater situation in India: Problems and perspective. Water Resour Dev 18:563–580

Singh S, Yadav A, Malik RK, Singh H (2002) Furrow irrigated raised bed planting system-A resource conservation technology for increasing wheat productivity in rice–wheat sequence. In: Herbicide Resistance Management and Zero Tillage in Rice–Wheat Cropping System, Proceedings International Workshop. CCS HAU, Haryana

Singh S, Ladha JK, Gupta RK, Bhushan L, Rao AN (2008) Weed management in aerobic rice systems under varying establishment methods. Crop Prot 27:660–671

Singh Y, Singh VP, Singh G, Yadav DS, Sinha RKP, Johnson DE, Mortimer AM (2011) The implications of land preparation, crop establishment method and weed management on rice yield variation in the rice-wheat system in the Indo-Gangetic plains. Field Crop Res 121:64–74

Sohail T, Delin H, Siddiq A (2014) Indus basin waters a main resource of water in Pakistan: an analytical approach. Curr World Environ 9:670–685

Soomro B (2004) Paddy and water environment related issues, problems and prospects in Pakistan. Paddy Water Environ 2:41–44

Sufi AB, Hussain Z, Sultan SJ, Tariq I (2011) Integrated water resource management in Pakistan. International Conference on Water Resources Engineering & Management (ICWREM-March 7–8, 2011). UET, Lahore, Pakistan

Sur HS, Prihar SS, Jalota SK (1981) Effect of rice–wheat and maize–wheat rotations on water transmission and wheat root development in a sandy loam of the Punjab, India. Soil Tillage Res 1:361–371

Swain D, Herath S, Pathirana A, Mittra BN (2005) Rainfed lowland and flood-prone rice: a critical review on ecology and management technology for improving the productivity in Asia. Role of Water Sciences in Transboundary River Basin Management, Proceedings of the International Symposium, 10–12 March 2005, Ubon Ratchathani, Thailand

Tariq M, Ahmed M, Iqbal P, Fatima Z, Ahmad S (2020) Crop phenotyping. In: Ahmed M (ed) Systems modeling. Springer, Singapore, pp 45–60

Timsina J, Connor DJ (2001) Productivity and management of rice-wheat cropping systems: issues and challenges. Field Crop Res 69:93–132

Tomita S, Nawata E, Kono Y, Nagata Y, Noichana C, Sributta A, Inamura T (2003a) Differences in weed vegetation in response to cultivating methods and water conditions in rainfed paddy fields in northeast Thailand. Weed Biol Manage 3:117–127

Tomita S, Nawata E, Kono Y, Inamura T, Nagata Y, Noichana C, Sributta A (2003b) Impact of direct dry seeding on rainfed paddy vegetation in north-east Thailand. Weed Biol Manage 3:68–76

Tuong TP, Bouman BAM (2003) Rice production in water-scarce environments. In: Kijne JW, Barker R, Molden D (eds) Water productivity in agriculture: limits and opportunities for improvement. CABI Publishing, UK, pp 53–67

Tyagi NK, Agrawal A, Sakhtivadivel R, Ambast SK (2005) Water management decisions on small farms under scarce canal water supply: a case study from NW India. Agric Water Manag 77: 180–195

Uphoff N (2007) The System of Rice Intensification: Using alternative cultural practices to increase rice production and profitability from existing yield potentials. International Rice Commission Newsletter, No. 55, Food and Agriculture Organization, Rome

Uprety R (2004) Performance of system of rice intensification in morang district. http://ciifad.cornell.edu/sri/countries/nepal/nepalrptuprety041.pdf

Wade LJ, George T, Ladha JK, Singh U, Bhuiyan SI, Pandy S (1998) Opportunities to manipulate nutrient-by-water interactions in rainfed lowland rice systems. Field Crop Res 56:93–112

Wang YZ, Yang KK, Wang XL, Zhou Q, Zheng CY, Chen ZF (2004) Agricultural application and environmental degradation of photobiodegradable polyethylene mulching films. J Polym Environ 12:7–10

Wasaya A, Yasir TA, Ijaz M, Ahmad S (2019) Tillage effects on agronomic crop production. In: Hasanuzzaman M (ed) Agronomic crops, vol 2. Springer, Singapore, pp 73–99

Wassmann R, Neue HU, Ladha JK, Aulakh MS (2004) Mitigating greenhouse gas emissions from rice–wheat cropping systems in Asia. Environ Dev Sustain 6:65–90

Wickham TH, Singh VP (1978) Water movement through wet soils. Soils and Rice. International Rice Research Institute, Los Baños, Philippines

Wilson RA, Talbot NJ (2009) Under pressure: Investigating the biology of plant infection by (*Magnaporthe oryzae*). Nat Rev Microbiol 7:185–195

WRMD (2008) Publication No. 382, 2008, Indus Basin Irrigation System, Historic Rivers and Canals Discharge Data 2007–2008, WRMD Directorate, WAPDA, WAPDA House Lahore, Pakistan

WRMD (2009) Publication No. 385,2009, Indus Basin Irrigation System, Historic Rivers and Canals Discharge Data 2008–2009, WRMD Directorate, WAPDA, WAPDA House Lahore, Pakistan

Xing GX, Zhu ZL (2000) An assessment of N loss from agricultural fields to the environment in China. Nutrient Cyc Agroecosys 57:67–73

Xu H, Cai ZC, Jia ZJ, Tsuruta H (2000) Effect of land management in winter crop season on CH_4 emission during the following flooded and rice-growing period. In: Wassmann R, Lantin R, Neue HU (eds) Methane emissions from major rice ecosystems in Asia. Springer, Netherlands, pp 327–332

Xu G, Chang ZZ, Hua J, Chang J (2007) Much improved water use efficiency of rice under non-flooded mulching cultivation. J Integr Plant Biol 49:1527–1534

Yan X, Ohara T, Akimoto H (2003) Development of region-specific emission factors and estimation of methane emission from rice fields in the East, Southeast and South Asian countries. Glob Chang Biol 9:237–254

Zahoor SA, Ahmad S, Ahmad A, Wajid A, Khaliq T, Mubeen M, Hussain S, Samiul Din M, Amin A, Awais M, Nasim W (2019) Improving water use efficiency in agronomic crop production. In: Hasanuzzaman M (ed) Agronomic crops, vol 2. Springer, Singapore, pp 13–29

Zhang W, Yu Y, Huang Y, Li T, Wang P (2011) Modeling methane emissions from irrigated rice cultivation in China from 1960 to 2050. Glob Chang Biol 17:3511–3523

Zotoglo K (2011) Training manual on system of rice intensification. Integrated Initiatives for Economic Growth (IICEM), Mali, pp 3–15

Rice Interactions with Plant Growth Promoting Rhizobacteria

14

Muhammad Baqir Hussain, Suleman Haider Shah, Amar Matloob, Rafia Mubaraka, Niaz Ahmed, Iftikhar Ahmad, Tanveer-ul-Haq, and Muhammad Usman Jamshaid

Abstract

Rice is an important cereal crop that serves as a staple food for half of the world population. The rhizosphere of the rice contributes to its nutrition where plant growth promoting rhizobacteria (PGPR) play a pivotal role. The PGPR are bacteria living in close vicinity of plant roots (3–5 mm) to positively influence plant growth via nitrogen fixation, nutrient solubilization, phytohormones, siderophores, exopolysaccharides, and enzymes production. The PGPR are inoculated to rice as single strain or in consortia with multiple benefits to the crop. Moreover, PGPR also contribute in alleviating the adverse impacts of salinity, heat, drought, heavy metals, and diseases on rice crop. Rice-PGPR interactions vary in time and space as per plant growth stage, growth conditions, management practices, and prevailing climatic and edaphic conditions. Rhizosphere of rice attracts or repels several soil microbes that modulate bacterial community composition. This chapter documents the interactions of rice with PGPR in the rhizosphere, PGPR-mediated growth improvements, underlying

M. B. Hussain (✉) · S. H. Shah · Tanveer-ul-Haq · M. U. Jamshaid
Department of Soil and Environmental Sciences, Muhammad Nawaz Shareef University of Agriculture, Multan, Pakistan
e-mail: baqir.hussain@mnsuam.edu.pk

A. Matloob
Department of Agronomy, Muhammad Nawaz Shareef University of Agriculture, Multan, Pakistan

R. Mubaraka
University of Koblenz-Landau, Koblenz, Germany

N. Ahmed
Department of Soil Science, Bahauddin Zakariya University, Multan, Pakistan

I. Ahmad
Department of Environmental Sciences, COMSATS University Islamabad, Vehari, Pakistan

N. Sarwar et al. (eds.), *Modern Techniques of Rice Crop Production*,
https://doi.org/10.1007/978-981-16-4955-4_14

mechanisms, and their possible utilization in alleviating suboptimal levels of growth factors. Current progress is documented based on the existing agronomic and experimental evidences; while missing links on various aspects of PGPR interaction with rice are pointed out.

Keywords

Rice · bacteria · growth · promoting

14.1 Introduction

It is anticipated that global human population will rise to 9.8 billion by 2050 (Harold and Reetz 2016), and food demand for this burgeoning population will also rise by 50% (IFPRI 2012). Crop production is the main source to supply food (Ahmad et al. 2008, 2009, 2012, 2013, 2015, 2019; Ahmad and Hasanuzzaman 2012; Ahmed and Ahmad 2019, 2020; Shahzad and Ahmad 2019). However, fertile land for crop production is gradually decreasing due to soil degradation, urbanization, and climate change. Soil degradation is the major factor affecting agricultural productivity of tropical and subtropical soils (Lamb et al. 2005). Intensive crop production system associated with unbalanced fertilizer use and greater nutrient removal by crops with very little residue recycling are deteriorating soil quality by organic matter depletion and nutrient loss. According to an estimate, about 24 billion tons of fertile soil worldwide is lost annually due to these anthropogenic, intensive agricultural activities (FAO 2011), and dwindling existing global food production capacity (Sarwar et al. 2013a, b, c). Continuous removal of essential nutrients is mining the agricultural fields, and depleted soil fertility has emerged as a serious threat to sustainable soil management necessitating the supplementation to overcome this nutrient loss to maintain productivity (Amundson et al. 2015; Ahmed et al. 2017, 2020a, b; Akram et al. 2018, 2019; Hussain et al. 2018; Fahad et al. 2019; Fatima et al. 2020). Hence, the use of synthetic fertilizers to replenish soil nutrient has increased manifold which is the major source for environmental pollution. Therefore, restoration of soil fertility with environment-friendly approaches is essential to ensure global food security without compromising soil, water, and environmental quality (Hafeez-ur-Rehman et al. 2019; Khan et al. 2019a, b; Naz et al. 2019; Peerzada et al. 2019; Razaq et al. 2019; Sarwar et al. 2013a, b; Tariq et al. 2020; Wasaya et al. 2019; Zahoor et al. 2019).

Rice (*Oryza sativa* L.) is the second major staple crop for human as half of the global population depends on it to meet their daily calories requirement (Ahmad et al. 2019; FAO 2020). Its contribution in global calorie intake is 23% (Joshi et al. 2020). However, the growth rate of rice yield has been reduced from 3.68% per year in 1980s to 0.74% per year in 1990s (Nguyen and Ferrero 2006). This yield increase rate is currently lower than increase in rice consumption rate. Many factors are responsible for decline in rice production like low or stagnant yields, high production costs, biotic and abiotic stresses, looming water shortage, and uncertainty of climatic optima, for example, temperature variation, drought, salinity, etc. (Nguyen and Ferrero 2006).

Utilization of plant growth promoting rhizobacteria (PGPR) for improving soil and plant health has been practiced for many decades. Almost all plants interact with microbes in their environment during all growth stages. The plant-microbe interaction could be beneficial, harmful, or neutral. A large number of microbial communities interact with plant roots in soil (Meena et al. 2012). These microbes include nitrogen fixing, nutrient solubilizing, phytohormone, and antibiotic- and enzyme-producing bacteria that have the ability to colonize plant roots for promoting plant growth and ameliorating different biotic and abiotic stresses (Glick 2012; Azcon et al. 2013). Various soil bacteria such as *Azospirillum, Rhizobium, Pseudomonas, Bacillus, Exiguobacterium, Chryseobacterium, Ralstonia, Kocuria, Serratia, Pantoea, Enterobacteria, Burkholderia,* and *Cyanobacteria* are most effective in rice-microbe interactions (Lucas et al. 2014; Pittol et al. 2016; Rêgo et al. 2018). The microbes in rice field perform various functions to increase nutrient availability and cycling of nutrient.

The community structure and population of microbes in rhizosphere vary with rice plant growth stage and genotype due to variation in quantity and composition of root exudates (Aulakh et al. 2001). The quantity of root exudates increases as rice plants grow from seedling to flowering stage and declines onward till maturity is reached. Organic acid contents in root exudates are increased but sugar contents show a decrease with the advancement in plant developmental stage (Aulakh et al. 2001). The bacterial communities that dominate the rice-soil-water system include proteobacteria, chloroflexi, actinobacteria, and acidobacteria (Hussain et al. 2011; Jiang et al. 2016), while dominating fungal communities include ascomycota, glomeromycota, and basidiomycota (Jiang et al. 2016; Yuan et al. 2018). However, some other microbial communities belonging to cyanobacteria, methanogens, methylotrophs, and methanotrophs have also been reported (Kumar 2017).

Plant growth and development are promoted by PGPR through different mechanisms. Some mechanisms involve direct stimulation, while others involve indirect stimulation. Direct stimulation involves phytohormones production, fixation of atmospheric nitrogen, reduction in ethylene level in plant body, enzymes production, siderophores production, and nutrient solubilization (Glick et al. 1999). Indirect stimulation involves antibiotic production, Fe chelation in rhizosphere, reduction in abiotic stress effects, production of fungal cell wall hydrolyzing extracellular enzymes, and change in rhizospheric microbial community structure (Kloepper and Schroth 1981; Glick 2012; Dey et al. 2014).

Status of rice in global food security demands understanding of the sustainable approaches for rice production. Use of PGPR in crop production system is emerging as a sustainable approach. So, the knowledge of microbial interaction with rice is of utmost importance. This knowledge includes understanding of the signals of rice roots resulting in molecular dialogue that attract microbes in rhizosphere, mechanisms of microbial-associated plant growth promotion, microbial community structure in rice rhizosphere, and variation associated with progressive growth stages. Such knowledge will enable us to explore and utilize this nature gifted microbial resource for rice production in a beneficial manner.

14.2 Plant Growth Promoting Mechanisms of Rhizobacteria

The PGPR employ different mechanisms to increase plant growth (Fig. 14.1) such as production of phytohormones, nutrient solubilization, biological nitrogen fixation, and protection of plants under biotic and abiotic stress conditions (Glick 2012; Dey et al. 2014).

14.2.1 Phytohormones Production

Root-associated PGPR affect plants through production of plant growth-regulating hormones like indole acetic acid (IAA), cytokinins, gibberellins, and abscisic acid in very low concentration that affect morphological and physiological activities within plant body (Arshad and Frankenberger 1998). Potential stimulatory effects of these growth regulators include: (i) IAA produced by the PGPR increases length and surface area of roots which favors plant's ability to access water and nutrients in soil more efficiently (Glick 2012); (ii) cytokinins are responsible for cell division, root

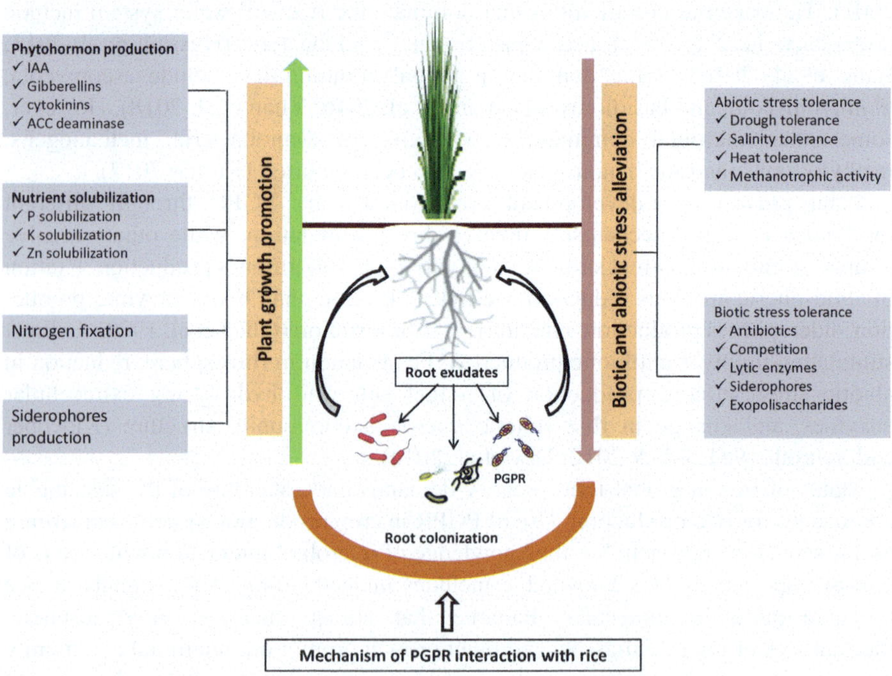

Fig. 14.1 Rice-PGPR interaction and plant growth promoting attributes

growth, and root hair development (Frankenberger and Arshad 1995); (iii) gibberellins stimulate leaf expansion and root elongation (Boiero et al. 2007); and (iv) abscisic acid controls stomatal closure, root growth and storage proteins, inhibits shoot growth, and produces proteinase inhibitors that are essential for pathogen defense (Mauseth 1991). However, various strains of *Pseudomonas aeruginosa* (GRC1, PSII, LES4, PRS4, PS15, and PSI), *Mesorhizobium loti* MP6, *Bacillus* sp. BPR7, *Bradyrhizobium* sp. BMP17, and *Sinorhizobium meliloti* PP3 and MSSP have been reported to produce IAA in the range of 24–100 μg/mL (Chandra et al. 2007; Pandey and Maheshwari 2007; Aeron et al. 2010; Kumar et al. 2012; Maheshwari et al. 2014). Many PGPR have been reported to synthesize phytohormones for promoting plant growth during interactions with rice (Malik et al. 1997; Mehnaz et al. 2001; Bhattacharjee et al. 2012; Vanegas and Uribe-Vélez 2014; Tiwari et al. 2017; Shen et al. 2019; Joshi et al. 2020).

14.2.2 Biological Nitrogen Fixation

Rice is generally cultivated in flooded conditions (85% of the total cultivated rice) where aerobic (*Azotobacter*) and anaerobic (*Clostridium*) bacteria can fix atmospheric nitrogen to supplement the nitrogen requirements of the crop. Flooded rice fields are favorable ecosystem for the Azolla-Anabaena association and blue green algae which can substantially contribute to atmospheric fixed nitrogen (Hashem 2001; Mian 2002). Certain nonsymbiotic bacteria such as *Azoarcus, Cyanobacteria, Azospirillum, Herbaspirillum, and Burkholderia* contribute biologically fixed nitrogen in rice field (Bohlool et al. 1992; Ladha and Reddy 1995; Baldani et al. 2000; Balandreau 2002). Nitrogen contribution from BNF is estimated about 30 kg N ha^{-1} per year in rice field (Herridge et al. 2008). Methanotrophic bacteria have also been reported to fix atmospheric nitrogen in addition to methane oxidation in flooded soils when supply of nitrogen is low in rice field (Minamisawa et al. 2016). *Azospirillum* N-4 inoculation to rice variety BAS-370 showed 70% nitrogen used by the plant derived from atmosphere through BNF (Malik et al. 1997). However, *Azoarcus* K-l, *Zoogloea* Ky-l, *Azospirillum brasilense* Wb-3, and *Pseudomonas* 96–51 are capable of BNF and IAA production showing about 29% nitrogen derived from atmosphere in rice cultivars NIAB-6 and BAS-370.

14.2.3 Nutrient Solubilization and Uptake

Soil contains large reservoirs of plant essential nutrients that are not readily available for plant use due to fixation in soil as an insoluble form. The PGPR promote bioavailability of these insoluble nutrients from soil and enhance uptake by plants leading to increased crop growth and yield. Many phosphorus solubilizing bacteria secrete organic acids that solubilize insoluble form of P and convert it into plant-available form by lowering the soil pH in microsites (Richardson et al. 2009; Joshi et al. 2020), chelation, or solubilization of Fe by siderophores production and exchange reactions with cations like Ca^{+2}, Mg^{+2}, Al^{+3}, and Fe^{+2} (Hameeda et al. 2008). *Bacillus paralicheniformis, B. haynesii,* and *B. licheniformis* have been identified as siderophores producers and solubilizer of phosphate and zinc in rice rhizosphere which improved the crop nutrition (Joshi et al. 2020). About twofold increase in the rice grain iron was observed by the inoculation of *P. putida* strains B17 and B19 (Sharma et al. 2013), whereas the phosphate solubilizing and nitrogen fixation ability in *B. arybhattai* MN1, isolated from rice seedlings, was a prominent character for enhanced nutrition in rice (Shen et al. 2019). Similarly, inoculation with *B. pumilus* was reported to enhance nutrition physiology of rice along with significant improvement in growth (Khan et al. 2016; Liu et al. 2018; Win et al. 2018). In another field experiment in China, Sun et al. (2020) observed that *B. mucilaginosus* and *Aspergillus niger* can solubilize potassium, silicon, and phosphorus in soil and increase rice yields significantly.

14.2.4 Methanotrophic Activity

Methane as a greenhouse gas is released in substantial amount from flooded rice fields. According to an estimate, the rice paddies contribute about 10% of total methane produced and 1.5% of the total greenhouse gases produced anthropogenically at a global scale (Nazaries et al. 2013; FAO 2014; Jhala et al. 2014). The paddies methane emissions have been increased from 17.4 Tg in 1961 to 24.4 Tg in 2016 with an average increased rate of 1.2 Tg per decade (FAO 2016). The methane gas is produced in rice fields as a result of organic debris and root exudates decomposition by the action of methanogenic microbes, and contributes 9–19% toward total global production (Conrad 2009; Han et al. 2016). However, the oxidation of the methane in rice rhizosphere can reduce the levels, but little is known about methanotrophs and rice interaction (Takeda et al. 2008). Methane gas is converted into biomass by methanotrophs and its level is reduced in flooded soil. Methane monooxygenase enzyme catalyzes this methanotrophic activity in the presence of oxygen that is released by the algal community grown at water surface in flooded rice field (Jhala et al. 2014; Kumar 2017). *Bacillus aerius, Rhizobium sp., B. subtilis, P. illinoisensis, B. megaterium,* and *Methylobacterium extrorquens* have showed soluble methane monooxygenase (22.5–37 nmol min^{-1} mg of $protein^{-1}$) and methane dehydrogenase (35–75 nmol min^{-1} mg of $protein^{-1}$) activity with a significant reduction of methane from 5% to 1% methane solution in 10 days (Jhala et al. 2014).

14.2.5 Enzymes Production

Rhizobacteria produce an enzyme 1-aminocyclopropane-1-carboxylic acid (ACC) deaminase that degrades the precursor of ethylene to lower its increased concentration in plants to improve plant growth under biotic and abiotic stress conditions (Ghosh et al. 2003; Govindasamy et al. 2008; Duan et al. 2009; Bhattacharjee et al. 2012; Nascimento et al. 2014). Similarly, reactive oxygen species scavenger antioxidant enzymes like catalase, peroxidase, and superoxide dismutase are produced by PGPR to dilute the oxidative stress on rice plant (Laxa et al. 2019). The cell wall of the plant pathogens is composed of chitin, lipids, glucan, cellulose, and proteins. Microbes inhibit the growth of phytopathogens through the production of cell wall degrading hydrolytic enzymes like peroxidase, chitinase, cellulase, proteases, and glucanase (Chet et al. 1990; Gupta et al. 1995; Kobayashi et al. 2002; Chater et al. 2010). It has been recorded that the inoculation of rice genotypes ADT43 and IR50 with *Pseudomonas* strains PF1 and TDK1 showed a higher activity of antioxidant enzymes catalase, peroxidase, and nitrate reductase to alleviate the impact of 100 mM NaCl stress (Sen and Chandrasekhar 2015).

14.2.6 Antibiotics Production

Antibiotics like polyenes, macrolides, aminoglycosides, nucleosides, benzoquinones, etc. and many other metabolites are produced by bacteria and actinobacteria that show antibiosis by suppression of phytopathogenic activities (Berdy 2012; Wani and Gopalakrishnan 2019). Besides antibiotic production, the soil microbes inhibit the growth of potential plant pathogens through competition for space and nutrients and production of bacteriocins and lytic enzymes (Leclere et al. 2005; Dutta and Podile 2010; Palaniyandi et al. 2013). Some microbes like *Pseudomonas aeruginosa* and *P. fluorescens* have been confirmed for the production of antibiotics and inhibition of fungal growth (Haas and Défago 2005; Wong et al. 2015). *Bacillus amyloliquefaciens* BAS23 reduced the incidence of dirty panicle fungal pathogens of rice by the production of antifungal lipopeptide iturin A (Saechow et al. 2018).

14.2.7 Siderophores Production

Iron in soil exists as hydroxides and oxyhydroxides that are insoluble and inaccessible forms for plant uptake. Rhizobacteria secrete low molecular weight, iron-chelating compounds called siderophores that trap iron and make it accessible to plants under low iron supply conditions (Rajkumar et al. 2010). The siderophores produced by rhizobacteria are of high affinity for iron while those produced by phytopathogens are of low affinity; therefore, rhizobacteria-produced siderophores efficiently colonize the plant roots and suppress pathogenic activity via induced iron deficiency. This is the way microbes serve as biocontrol agents against phytopathogens and fulfil the iron nutritional requirement of the plant (Glick 2012; Johnson et al. 2013). The most efficient microbes for siderophores production include *Pseudomonas aeruginosa* and *P. fluorescens* (Haas and Défago 2005; Wong et al. 2015). Similarly, endophytic bacteria including Burkholderia, Pseudomonas, Enterobacter, Pantoea, Bacillus, and Sphingomonas have been reported to produce siderophores predominantly within the plant body to support rice plant growth (Loaces et al. 2011; Kumar et al. 2021).

14.2.8 Exopolysaccharides Production

All soil physical, chemical, and biological activities depend on soil structure. The microbially produced exopolysaccharides (EPS) like polysaccharide lipids and lipopolysaccharide proteins improve soil structure, play a vital role for survival of microbes under adverse environmental conditions (Orlandelli et al. 2016), and support microbes for root colonization (Oades 1993). These EPS maintain water potential in soil that maintain nutrient uptake by plants under stressed conditions (Alami et al. 2000; Munns 2002). It has been observed that *Bacillus pumilus* strain JPVS11 produced EPS to alleviate the ill effects of Na cations on rice plant growth and improved soil health by developing biofilm around the plant roots (Kumar et al. 2021).

14.3 Rice Inoculation with PGPR

The PGPR inoculation (Table 14.1) to seed surface or dipping of seedling roots is an efficient method for introducing microbes in rice rhizosphere to colonize plant roots (Hussain et al. 2009; Gomez-Ramirez and Uribe-Velez 2021). The inoculation can be divided into two type: (i) single strain inoculation and (ii) consortia inoculation, on the basis of number of strains used for preparing and delivering inoculants. In order to develop inoculants, the selection of microbial strains is a very crucial step (Bhattacharjee et al. 2008). Colonization of PGPR with roots is not uniform and it varies not only in different parts of root but also associated with the quantity and nature of root exudates (Compant et al. 2010). Root colonization with bacteria is also regulated by a mechanism of "quorum sensing" that is a bacterial population density-induced gene expression which occurs in inter/intrabacterial species (Miller and Bassler 2001). Quorum sensing can change pattern of root colonization by affecting bacterial competitiveness and host-bacteria specificity (Compant et al. 2010).

14.3.1 Single Strain Inoculation

Seed inoculation with microbes was started in 1896 on commercial scale and initially single efficient microbial strain was selected to be used for inoculation (Hartmann et al. 2008). It is more effective in most of the cases and many biofertilizers commercially available contain single strain for inoculation (Nuti and Giovannetti 2015). Single-strain inoculation with three isolates of *Bacillus* spp. (IBUN-02704, IBUN-02724, and IBUN-02755) to rice showed a significant increase in shoot length, dry biomass, and uptake of phosphorus as compared to uninoculated control (Gomez-Ramirez and Uribe-Velez 2021). In another study, the single-strain inoculation in rice showed significant improvement in various parameters as compared to uninoculated control. *Rhizobium leguminosarum* (LSI-29) showed 45%, 43%, and 25% increase in straw dry weight, paddy yield, and thousand grain weight, respectively, and *Rhizobium phaseoli* (A2) showed 28% and 29% increase in shoot

Table 14.1 List of some important plant growth promoting rhizobacteria (PGPR) showing positive interaction with rice

PGPR	Study type	Plant growth promotion mechanism	Plant response	References
Bacillus paralicheniformis strain EN107	Pot study	– Phosphorus solubilization – Zinc solubilization	Increased root length, shoot length, plant biomass, and drought stress tolerance	Joshi et al. (2020)
Bacillus paralicheniformis strain EN121		– IAA production – Siderophore production		
Bacillus licheniformis strain EN108		– Phosphorus solubilization		
Bacillus licheniformis strain A21		– Siderophore production – Phosphorus solubilization		
Bacillus haynesii strain EN43		– Siderophore production – Zinc Solubilization		
Bacillus haynesii strain EN124		– Siderophore production		
Pseudacidovorax intermedius strain E1, *Mycobacterium vanbaalenii* strain E10, *Bacillus sporothermodurans* strain M1, *Bacillus fortis strain* T9, and *Bacillus marisflavi* strain T3	Pot study	– N fixation – Fungicide stress tolerance	Not studied	Shen et al. (2019)
Rhizobium larrymoorei strain E2, *Bacillus aryabhattai* strain E7, and *Bacillus aryabhattai* strain MN1		– N fixation – IAA production – Fungicide stress tolerance	Promoted root growth. However, only *R. larrymoorei* promoted shoot growth	
Sphingomonas aquatilis strain E4, *Mycobacterium sphagni* strain E9, *Mycobacterium aromaticivorans* strain EN1, *Pseudomonas*		– N fixation – P solubilization – Fungicide stress tolerance	Only *Pseudomonas granadensis* promoted root and shoot growth	

(continued)

Table 14.1 (continued)

PGPR	Study type	Plant growth promotion mechanism	Plant response	References
granadensis strain T6, *Pseudomonas baetica* strain TN1, and *Pseudomonas plecoglossicida* strain TN3				
Flavobacterium ginsengiterrae strain E5, *Brevundimonas vesicularis* strain E6, and *Bacillus fortis* strain M7		– Fungicide stress tolerance	Not studied	
Azospirillum lipoferum N-4, *Azospirillum brasilense* Wb-3, *Azoarcus* K-1, *Pseudomonas* 96–51, and *Zoogloea* Ky-1	Pot and field	– IAA production – N fixation	Increased plant biomass	Malik et al. (1997)
Herbaspirillum seropedicae and *Burkholderia* spp.	Pot	– N fixation	Increased plant fresh and dry weight and N uptake	Baldani et al. (2000)
Pseudomonas monteilii, P. plecoglossicida, Brevibacterium altitudinis, B. antiquum, Enterobacter ludwigii, and *Acinetobacter tandoii*	Field study	–	Enhanced root weight, root length and volume, tiller numbers, panicle numbers, stover yield, and grain yield	Gopalakrishnan et al. (2012)
Enterobacter sp., *Chryseobacterium* sp., Lactococcus sp., *Enterobacter* sp., and *Acinetobacter* sp.	Pot study	– N fixation – IAA production	Increase root and shoot biomass	Vanegas and Uribe-Vélez (2014)
Pseudomonas sp. and Chryseobacterium sp.	Field study	–	Increased rice production and protection against rice blast fungus	Lucas et al. (2009)

(continued)

Table 14.1 (continued)

PGPR	Study type	Plant growth promotion mechanism	Plant response	References
Bacillus amyloliquefaciens	Growth chamber study	– Increased proline, total soluble sugar, and lipid peroxidation levels under abiotic stress conditions – IAA production – ACC deaminase production	–	Tiwari et al. (2017)
Bacillus Cereus	Pot study	– Cadmium toxicity tolerance	Increase in seedling growth	Jan et al. (2019)
Rhizobium Leguminosarum bv. trifolii SN10	Pot study	– IAA production – ACC deaminase production	– Enhances rice growth	
Aeromonas veronii and *Enterobacter cloacae*	Pot study	– IAA production – N fixation	– Increase in seedling growth	Mehnaz et al. (2001)
Bacillus pumilus	Pot study and field study	– IAA production – Gibberellins production – P solubilization – Siderophore production – NaCl and high boron stress tolerance	– Increased rice growth and production – Increased nitrogen uptake	Win et al. (2018) Liu et al. (2020) Khan et al. (2016) Liu et al. (2018) Wang et al. (2019)
Bacillus mucilaginosus and Aspergillus niger	Field study	Not studied	– Increase in nutrients (K, N, P, and Si) availability and uptake – Increased plant growth and yield	Sun et al. (2020)
Azospirillum brasilense	Growth room study	– Root colonization	Increased plant growth	Thomas et al. (2019)
Paenibacillus polymyxa Sx3	Growth room study	– Nitrogen fixation – P solubilization	– Increased plant growth	Abdaullah et al. (2019)

(continued)

Table 14.1 (continued)

PGPR	Study type	Plant growth promotion mechanism	Plant response	References
		– IAA production – Bacterial leaf blight suppression		
Burkholderia pyrrocinia BRM 32113	Green house study	– Leaf blast suppression	– Plant growth promotion.	Arriel-Elias et al. (2019)

length and grain number per panicle, respectively. However, both strains improved the NPK nutrition in rice significantly (Hussain et al. 2009).

14.3.2 Consortia Inoculation

Recent trends have shifted from utilization of single strain to consortia inoculation with rationale that combination of diverse strains with different plant growth promoting traits would be more effective (Vorholt et al. 2017; Woo and Pepe 2018). Compatible microbial strains with different modes of action can be combined to compose consortia for inoculation with the objective of harvesting complementary or synergistic benefits (Lopez-Cervantes and Thorpe 2013). For developing consortia, microbial strains are selected on the basis of genetic diversity and their adaptation to tolerate variations in soil environmental conditions, like temperature, pH, salinity, or moisture, and plant growth promoting attributes, like phytohormone production, nutrient solubilization, exopolysaccharides production antibiotics, or stress-related enzymes production (Sekar et al. 2016; Fouda et al. 2021). Basic assumption is that under variable soil conditions, different strains of microbial consortium produce effects on plant growth in response to their activation by rhizospheric signals and ecophysiological responses by host plants (Lopez-Cervantes and Thorpe 2013; Bradacova et al. 2019). Many studies have reported that consortia inoculation is more efficient as compare to single-strain inoculation (Bashan 1998; Nuti and Giovannetti 2015; Sekar et al. 2016). In a study, it was observed that the bacterial strains *Enterobacter hormaechei* (AM122) and *Lysinibacillus xylanilyticus* (DB25), isolated from rice varieties, Ambemohar-157 and Dehradun Basmati, respectively were capable of synthesizing 2-acetyl-1-pyrroline (2AP) for improving aroma and yield of rice. The strains performed significantly higher in consortia inoculation as compared to single-strain inoculation for improving growth, yield, and aroma of the rice (Dhondge et al. 2021). In another study, consortia (consisting of bacterial strains *Mycobacterium senegalense* LM1, *Rhizobium rhizoryzae* BMU, *Providencia stuartii* LM18, and *Bacillus methylotrophicus* N2P4) inoculation in rice under reduced inorganic fertilizer recommended dose (50%, 75%, and 100% of recommended NPK levels) about 33.55% increase in rice production, and 25% fertilizer use efficiency was increased over uninoculated control was observed by Pratiwi et al. (2021).

14.4 Rice Inoculation with PGPR Under Stress Conditions

Rice crop faces several stress conditions from sowing to maturity. These stresses may be abiotic like salinity, drought, metal, and heat or biotic like pathogens. The estimated loss to rice crop yield due to abiotic stresses is about 70% (Kaur et al. 2008; Mantri et al. 2012). As a consequence of global warming and soil degradation, soil salinity and drought are becoming main environmental challenges for agriculture productivity in arid and semiarid regions (Tuteja 2007; Kompas et al. 2018).

Microbes are more abundant with low diversity in rhizosphere as compared to bulk soil. The role of PGPR is indispensable for inducing stress tolerance and plant growth promotion under stress conditions. Seed inoculation with efficient PGPR is done to reinforce the microbial population in the rice rhizosphere. As a result of PGPR interaction with plants, growth performance and resistance to biotic and abiotic stresses are enhanced. Various mechanisms are involved in plant growth promotion by PGPR under stress conditions.

14.4.1 Salinity Stress

Accumulation of salts in plant rhizosphere results in osmotic stress in plants. Salt stress leads to physiological drought condition in the rhizosphere. The PGPR produce auxins, 1-aminocyclopropane-1-carboxylic acid (ACC) deaminase, and accumulate proline that confer salt stress and help to maintain normal growth cycle (Nautiyal et al. 2013; Srivastava et al. 2016). Other mechanism involved in PGPR-induced salt tolerance includes production of Na^+-binding exopolysaccharides (Ashraf et al. 2004) and improving ionic homeostasis (Hamdia et al. 2004). Khan et al. (2016) indicated that *Bacillus pumilus* increased plant tolerance to high level of NaCl and boron, and enhanced rice growth.

14.4.2 Drought Stress

A major abiotic constraint to global crop production is drought. It has the potential to become a serious problem on 50% of arable global lands (Vinocur and Altman 2005; Kasim et al. 2013). Severe growth reduction has been reported in rice due to drought stress (Lafitte et al. 2007). The mechanism adopted by PGPR for drought tolerance in plants includes production of growth hormones, ACC deaminase production to reduce ethylene level in roots, and production of bacterial exopolysaccharides (Timmusk and Nevo 2011; Kim et al. 2013; Timmusk et al. 2014) to hold water and nutrients in the rhizosphere. Rice inoculation with PGPR induces drought tolerance and increase crop yield and biomass production (Redman et al. 2011; Singh et al. 2020). A PGPR strain *Bacillus amyloliquefaciens* NBRI-SN13 (SN13) showed a significant increase in the seedling growth of rice cv. Saryu-52 under drought in hydroponic conditions. The results from the study suggested that the strain NBRI-SN13 ameliorated the impact of drought in rice through phytohormone production, nutrient solubilization, and stimulation of stress tolerance gene expressions (Tiwari et al. 2017). Two bacterial strains belonging to *Mycobacterium* sp. and *Bacillus* sp. were isolated from the rhizosphere of rice and tested for improving drought tolerance in rice (cultivar IR64). The results revealed significant improvement in the germination and seedling growth of rice under drought as compared to uninoculated control which was attributed to plant growth promoting character of strains as phosphate solubilization, auxins production, and ACC deaminase production (Karmakar et al. 2021).

14.4.3 Metal Stress

Excessive level of heavy metals in soil or irrigation water is considered not only as pollutant but also hinders plant growth (Nagajyoti et al. 2010). Numerous studies have reported positive effects of PGPR on plant growth through metal uptake reduction (Luo et al. 2011; Shah et al. 2017; Khan et al. 2017). The PGPR ameliorate metal stress by production of 1-aminocyclopropane-l-carboxylic acid (ACC) deaminase and exopolysaccharides. Microbially produced ACC deaminase decreases the plant ethylene levels produced in response to metal stress while exopolysaccharides bound to form complexes with the metal cations and reduced plant uptake (Han and Lee 2005; Saharan and Nehra 2011). In rice, PGPR inoculation alleviated metal stress by reducing heavy metal plant uptake and accumulation in plant body, especially Cd^{+2} in rice grains (Rizwan et al. 2016; Awasthi et al. 2018; Jan et al. 2019). A strain of *Pseudomonas* sp. K32 (isolated from rice rhizosphere contaminated with metals) showed its tolerance for multimetals including As^{3+} (3700 μg/mL), Pb^{2+} (3800 μg/mL), and Cd^{2+} (4000 μg/mL) and capability to produce indole acetic acid, solubilize phosphorus, and nitrogen fixation. The strain K32 showed a significant improvement in rice seedling growth and reduction in uptake of Cd^{2+} in the seedling under high Cd^{2+} stress (Pramanik et al. 2021).

14.4.4 Heat Stress

Heat stress is one of the major abiotic stresses that limit rice yield (Wahid et al. 2007). Current trends of climate change-mediated rise in atmospheric temperature indicate an increase of 0.8 °C in average global temperature since mid-of-last century and predict 1.8–4.0 °C rise in average temperature by the end of this century (Solomon et al. 2007). Heat stress induces changes in pattern of hormone production and increases ethylene level that impairs plant normal growth (Saleem et al. 2007). Prolonged heat stress at reproduction stage changes balance between abscisic acid and cytokinin that results in grain abortion (Cheikh and Jones 1994; Sulaiman et al. 2018). Some researchers reported ACC deaminase production ability of PGPR when inoculated to rice (Tiwari et al. 2017) that helped rice plant to tolerate heat stress. Similarly, Tiwari et al. (2017) reported induced tolerance in rice against heat stress (45 °C for 24 h) when inoculated with *Bacillus amyloliquefaciens* NBRI-SN13 where about 5% increase in proline content of rice seedling was recorded as compared to control. Moreover, the heat tolerance gene expression was induced in rice by the *B. amyloliquefaciens* NBRI-SN13 inoculation.

14.4.5 Diseases

A rice crop confronts some plant diseases during its physiological growth periods from germination to maturity. Rhizospheric PGPR *Bacillus amyloliquefaciens* has been identified for biocontrol of infection caused by *Rhizoctonia solani* in rice (Srivastava et al. 2016). Similarly, Abdaullah et al. (2019) showed that *Paenibacillus polymyxa* Sx3 suppressed bacterial leaf blight and enhanced plant growth. In addition, Arriel-Elias et al. (2019) reported significant suppression of leaf blast in rice when inoculated with *Burkholderia pyrrocinia* BRM 32113. A PGPR strain K32 of *Pseudomonas* sp. inhibited the growth of six rice fungal pathogens including *Cladosporium herbarum, Aspergillus flavus, Alternaria alternata, Paecilomyces sp., Aspergillus parasiticus,* and *Rhizopus stolonifer* by the production of biocontrol enzymes such as β-1,3-glucanase (4.38 ± 0.35 U/mg protein), protease (7.72 ± 0.28 U/mg protein), and chitinase (8.17 ± 0.44 U/mg protein) (Pramanik et al. 2021).

14.5 Rhizosphere Microbial Community of Rice

Root exudates play an important role in shaping microbial community in rhizosphere (Chaparro et al. 2014) and it vary further with plant growth stages (Li et al. 2014) and health condition of the plant. Major difference in rhizospheric environment of rice from all other major crops originates from the fact the rice is grown under flooded conditions. Resultantly, rice rhizosphere is divided into oxic and anoxic zones that favor specific microbial groups with aerobic, anaerobic, or facultative metabolism (Brune et al. 2000; Breidenbach et al. 2016). Bacterial species belonging to *Herbaspirillum* and *Geobacter* are abundant in rhizosphere as compared to bulk soil, especially during early growth stages of rice (Hassan and Mathesius 2012). Rice rhizosphere is enriched with iron reducers (*Geobacter* and *Anaeromyxobacter*) and fermenters (*Clostridiaceae* and *Opitutaceae*) communities (Breidenbach et al. 2016). Microbial communities of fermenters (*Clostridiaceae* and *Opitutaceae*) and acetogens are responsible for breakdown of complex carbon compounds including root material that serves as a substrate for methanogensis (Watanabe et al. 1999; McInerney et al. 2008; Breidenbach et al. 2016). Population of *Archea* is more in rice rhizosphere than bulk soil while its community composition remains the same (Breidenbach et al. 2016).

14.6 Conclusions and Future Outlook

The PGPR are playing substantial role in crop production since their discovery. They are being used as a tool in biofertilization, biocontrol, and biostimulation to enhance crop productivity in sustainable agriculture. Application of methanotrophs can reduce methane emission from paddy field which is produced as a result of methanogensis. Their plant growth promoting traits are due to specific mechanisms that they adopt when interacting with plant-soil system. Understanding and exploitation of this natural resource can enhance rice growth and yield under biotic and abiotic stresses. The effects and functions of PGPR in plant and soil are not completely defined and well understood. This may play a pivotal role in rice production system by devising strategies for reducing fertilizer requirement and lowering greenhouse gaseous emissions. Nanotechnology is an emerging technology in the field of agriculture, being used as a carrier for genetic material delivery and plant transformation. Its application in exploiting PGPR resource may be explored. The major challenge is management of microbial communities in rice rhizosphere in such a way to get maximum colonization of rice roots with PGPR.

References

Abdaullah Y, Yang M, Zhang M, Masum MM, Ogunyemi SO, Hossain A, An Q, Yan C, Li B (2019) Plant growth promotion and suppression of bacterial leaf blight in rice by *Paenibacillus polymyxa* Sx3. Lett Appl Microbial 68:423–429

Aeron A, Pandey P, Maheshwari DK (2010) Differential response of sesame under influence of indigenous and non-indigenous rhizosphere competent rhizosphere fluorescent pseudomonads. Curr Sci 99(2):166–168

Ahmad S, Hasanuzzaman M (2012) Integrated effect of plant density, N rates and irrigation regimes on the biomass production, N content, PAR use efficiencies and water productivity of rice under irrigated semiarid environment. Not Bot Horti Agrobot Cluj-Napoca 40(1):201–211

Ahmad S, Zia-ul-Haq M, Ali H, Shad SA, Ammad A, Maqsood M, Khan MB, Mehmood S, Hussain A (2008) Water and radiation use efficiencies of transplanted rice (*Oryza sativa* L.) at different plant densities and irrigation regimes under semi-arid environment. Pak J Bot 40(1): 199–209

Ahmad S, Ahmad A, Zia-ul-Haq M, Ali H, Khaliq T, Anjum MA, Khan MA, Hussain A, Hoogenboom G (2009) Resources use efficiency of field grown transplanted rice (*Oryza sativa* L.) under irrigated semiarid environment. J Food Agric Environ 7(2):487–492

Ahmad S, Ahmad A, Soler CMT, Ali H, Zia-Ul-Haq M, Anothai J, Hussain A, Hoogenboom G, Hasanuzzaman M (2012) Application of the CSM-CERES-Rice model for evaluation of plant density and nitrogen management of fine transplanted rice for an irrigated semiarid environment. Precis Agric 13(2):200–218

Ahmad S, Ahmad A, Ali H, Hussain A, Garcia y Garcia A, Khan MA, Zia-Ul-Haq M, Hasanuzzaman M, Hoogenboom G (2013) Application of the CSM-CERES-Rice model for evaluation of plant density and irrigation management of transplanted rice for an irrigated semiarid environment. Irrig Sci 31(3):491–506

Ahmad A, Ashfaq M, Rasul G, Wajid SA, Khaliq T, Rasul F, Saeed U, Rahman MH, Hussain J, Baig IA, Naqvi AA, Bokhari SAA, Ahmad S, Naseem W, Hoogenboom G, Valdivia RO (2015) Impact of climate change on the rice–wheat cropping system of Pakistan. In: Hillel D, Rosenzweig C (eds) Handbook of climate change and agro-ecosystems: the agricultural modeling intercomparison and improvement project (AgMIP) integrated crop and economic assessments. Imperial College Press and The American Society of Agronomy, London, pp 219–258

Ahmad S, Abbas G, Ahmed M, Fatima Z, Anjum MA, Rasul G, Khan MA, Hoogenboom G (2019) Climate warming and management impact on the change of rice-wheat phenology in Punjab, Pakistan. Field Crop Res 230:46–61

Ahmed M, Ahmad S (2019) Carbon dioxide enrichment and crop productivity. In: Hasanuzzaman M (ed) Agronomic crops, vol 2. Springer, Singapore, pp 31–46

Ahmed M, Ahmad S (2020) Systems modeling. In: Ahmed M (ed) Systems modeling. Springer, Singapore, pp 1–44

Ahmed M, Fayyaz-ul-Hassan, Ahmad S (2017) Climate variability impact on rice production: adaptation and mitigation strategies. In: Ahmed M, Stockle C (eds) Quantification of climate variability, adaptation and mitigation for agricultural sustainability. Springer, Switzerland, pp 91–111

Ahmed M, Ahmad S, Raza MA, Kumar U, Ansar M, Shah GA, Parsons D, Hoogenboom G, Palosuo T, Seidel S (2020a) Models calibration and evaluation. In: Ahmed M (ed) Systems modeling. Springer, Singapore, pp 151–178

Ahmed M, Ahmad S, Waldrip HM, Ramin M, Raza MA (2020b) Whole farm modeling: a systems approach to understanding and managing livestock for greenhouse gas mitigation, economic viability and environmental quality. In: Waldrip HM, Pagliari PH, He Z (eds) Animal Manure. ASA Special Publication 67. American Society of Agronomy and Soil Science Society of America, Madison, WI, pp 345–371

Akram R, Turan V, Hammad HM, Ahmad S, Hussain S, Hasnain A, Maqbool MM, Rehmani MIA, Rasool A, Masood N, Mahmood F, Mubeen M, Sultana SR, Fahad S, Amanet K, Saleem M, Abbas Y, Akhtar HM, Hussain S, Waseem F, Murtaza R, Amin A, Zahoor SA, Sami ul Din M, Nasim W (2018) Fate of organic and inorganic pollutants in paddy soils. In: Hashmi MZ, Varma A (eds) Environmental pollution of paddy soils. Springer, Switzerland, pp 197–214

Akram R, Fahad S, Masood N, Rasool A, Ijaz M, Ihsan MZ, Maqbool MM, Ahmad S, Hussain S, Ahmed M, Kaleem S, Sultana SR, Mubeen M, Saud S, Kamran M, Nasim W (2019) Plant growth and morphological changes in rice under abiotic stress. In: Hasanuzzaman M, Fujita M, Nahar K, Biswas JK (eds) Advances in rice research for abiotic stress tolerance. Woodhead Publishing, Swaston, pp 69–85

Alami Y, Achouak W, Marol C, Heulin T (2000) Rhizosphere soil aggregation and plant growth promotion of sunflowers by exopolysaccharide producing *Rhizobium sp.* strain isolated from sunflower roots. Appl Environ Microbiol 66:3393–3398

Amundson R, Berhe AA, Hopmans JW, Olson C, Sztein AE, Sparks DL (2015) Soil and human security in the 21st century. Science 348:1261–1271

Arriel-Elias MT, de Carvalho Barros Côrtes MV, de Sousa TP, Chaibub AA, de Filippi MCC (2019) Induction of resistance in rice plants using bioproducts produced from *Burkholderia pyrrocinia* BRM 32113. Environ Sci Pollut Res Int 26:19705–19718

Arshad M, Frankenberger WT (1998) Plant growth regulating substances in the rhizosphere: microbial production and function. Adv Agron 62:45–151

Ashraf M, Hasnain S, Berge O, Mahmood T (2004) Inoculating wheat seedlings with exopolysaccharide producing bacteria restricts sodium uptake and stimulates plant growth under salt stress. Biol Fertil Soils 40:157–162

Aulakh MS, Wassmann R, Bueno C, Kreuzwieser J, Rennenberg H (2001) Characterization of root exudates at different growth stages of ten rice (*Oryza sativa* L.) cultivars. Plant Biol (Stuttg) 3: 139–148

Awasthi S, Chauhan R, Dwivedi S, Srivastava S, Srivastava S, Tripathi RD (2018) A consortium of alga (*Chlorella vulgaris*) and bacterium (*Pseudomonas putida*) for amelioration of arsenic toxicity in rice: a promising and feasible approach. Environ Exp Bot 150:115–126

Azcon R, Medina A, Aroca R, Ruiz-Lozano JM (2013) Abiotic stress remediation by the arbuscular mycorrhizal symbiosis and rhizosphere bacteria/yeast interactions. In: de Brujin FJ (ed) Molecular microbial ecology of the rhizosphere. Wiley, Hoboken, pp 991–1002

Balandreau J (2002) The spermosphere model to select for plant growth promoting rhizobacteria. In: Kennedy IR, Choudhury ATMA (eds) Biofertilisers in action. Rural Industries Research and Development Corporation, Canberra, Australia, pp 55–63

Baldani VLD, Baldani JI, Dobereiner J (2000) Inoculation of rice plants with the endophytic diazatrophs *Herbaspirillum seropedicae* and *Burkholderia spp*. Biol Fertil Soils 30:485–491

Bashan Y (1998) Inoculants of plant growth-promoting bacteria for use in agriculture. Biotechnol Adv 16:729–770

Berdy J (2012) Thoughts and facts about antibiotics: where we are now and where we are heading. J Antibiot 65:385–395

Bhattacharjee RB, Sin A, Mukhopadhyay SN (2008) Use of nitrogen-fixing bacteria as biofertiliser for non-legumes: prospects and challenges. Appl Microbiol Biotechnol 80:199–209

Bhattacharjee RB, Jourand P, Chaintreuil C, Dreyfus B, Singh A, Mukhopadhyay SN (2012) Indole acetic acid and ACC deaminase-producing *Rhizobium leguminosarum*bv. *trifolii* SN10 promote rice growth, and in the process undergo colonization and chemotaxis. Biol Fertil Soils 48:173–182

Bohlool BB, Ladha JK, Garrity DP, George T (1992) Biological nitrogen fixation for sustainable agriculture: a perspective. Plant Soil 141:1–11

Boiero L, Perrig D, Masciarelli O (2007) Phytohormone production by three strains of *Bradyrhizobium japonicum* and possible physiological and technological implications. Appl Microbiol Biotechnol 74:874–880

Bradacova K, Floria AS, Bar-Tal A, Minz D, Yerminyahu U, Shawahna R, Kraut-Cohen J, Zolti A, Erel R, Dietel K, Weinmann M, Zinnerman B, Berger N, Ludewig U, Neumann G, Poster G (2019) Microbial consortia versus single-strain inoculants: an advantage in PGPM-assisted tomato production? Agronomy 9:1–25

Breidenbach B, Pump J, Dumont MG (2016) Microbial community structure in the rhizosphere of rice plants. Front Microbiol 6:1537

Brune A, Frenzel P, Cypionka H (2000) Life at the oxic-anoxic interface: microbial activities and adaptations. FEMS Microbiol Rev 24:691–710

Chandra S, Choure K, Dubey RC, Maheshwari DK (2007) Rhizosphere competent *Mesorhizobium loti* MP6 induces root hair curling, inhibits *Sclerotiorum* and enhances growth of Indian mustard (*Brassica campestris*). Braz J Microbiol 38(1):124–130

Chaparro JM, Badri DV, Vivanco JM (2014) Rhizosphere microbiome assemblage is affected by plant development. ISME J 8:790

Chater KF, Biro S, Lee KJ, Palmer T, Schrempf H (2010) The complex extracellular biology of Streptomyces. FEMS Microbiol Rev 34:171–198

Cheikh N, Jones RJ (1994) Disruption of maize kernel growth and development by heat stress (role of cytokinin/abscisic acid balance). Plant Physiol 106:45–51

Chet I, Ordentlich A, Shapira R, Oppenheim A (1990) Mechanisms of biocontrol of soil-borne plant pathogens by rhizobacteria. Plant Soil 129:85–92

Compant S, Clément C, Sessitsch A (2010) Plant growth-promoting bacteria in the rhizo- and endosphere of plants: their role, colonization, mechanisms involved and prospects for utilization. Soil Biol Biochem 42:669–678

Conrad R (2009) The global methane cycle: recent advances in understanding the microbial processes involved. Environ Microbiol Rep 1:285–292

Dey R, Pal KK, Tilak KVBR (2014) Plant growth-promoting rhizobacteria in crop protection and challenges. In: Goyal A, Manoharachary C (eds) Future challenges in crop protection against fungal pathogens. Springer, New York, pp 31–58

Dhondge HV, Pable AA, Barvkar VT, Dastager SG, Nadaf AB (2021) Rhizobacterial consortium mediated aroma and yield enhancement in basmati and non-basmati rice (*Oryza sativa* L.). J Biotechnol 328:47–58

Duan J, Muller KM, Charles TC, Vesely S, Glick BR (2009) 1-Aminocyclopropane-1-carboxylate (ACC) deaminase genes in rhizobia from Southern Saskatchewan. Microb Ecol 57:423–436

Dutta S, Podile AR (2010) Plant growth promoting rhizobacteria (PGPR): the bugs to debug the root zone. Crit Rev Microbiol 36:232–244

Fahad S, Noor M, Adnan M, Khan MA, Rahman I, Alam M, Khan IA, Ullah H, Mian IA, Hassan S, Saud S, Bakhat HF, Hammad HM, Ahmad S, Nasim W (2019) Abiotic stress and rice grain quality. In: Hasanuzzaman M, Fujita M, Nahar K, Biswas JK (eds) Advances in rice research for abiotic stress tolerance. Woodhead Publishing, Swaston, pp 571–583

FAO (2011) The state of the world's land and water resources for food and agriculture (SOLAW) managing systems at risk. Food and Agriculture Organization of the United Nations/Earthscan, Rome, London

FAO (2014) Emissions - Agriculture. Emissions of Methane and Nitrous Oxide Produced From Agricultural Activities. Average 1990–2013

FAO (2016) FAO FAOSTAT. http://www.fao.org/faostat/en/#home. Accessed 21 Feb 2018

FAO (2020) Food Outlook - Biannual Report on Global Food Markets: June 2020. Food Outlook, 1. Rome

Fatima Z, Ahmad M, Hussain M, Abbas G, Ul-Allah S, Ahmad S, Ahmed N, Ali MA, Sarwar G, Ehsan ul Haque, Iqbal P, Hussain S (2020) The fingerprints of climate warming on cereal crops phenology and adaptation options. Sci Rep 10:18013

Fouda A, Eid AM, Elsaied A, El-Belely EF, Barghoth MG, Azab E, Gobouri AA, Hassan SED (2021) Plant growth-promoting endophytic bacterial community inhabiting the leaves of *Pulicaria incisa* (Lam.) DC inherent to arid regions. Plants 10:76

Frankenberger WT, Arshad M (1995) Phytohormones in soil: microbial production and function. Dekker, New York, 503p

Ghosh S, Penterman JN, Little RD, Chavez R, Glick BR (2003) Three newly isolated plant growth-promoting *Bacilli* facilitate the seedling growth of canola, *Brassica campestris*. Plant Physiol Biochem 41:277–281

Glick BR (2012) Plant growth-promoting bacteria: mechanisms and applications. Australas Sci 2012:963401

Glick BR, Patten CL, Holguin G, Penrose DM (1999) Biochemical and genetic mechanisms used by plant growth-promoting bacteria. Imperial College Press, London

Gomez-Ramirez LF, Uribe-Velez D (2021) Phosphorus solubilizing and mineralizing Bacillus spp. contribute to rice growth promotion using soil amended with rice straw. Curr Microbiol 78:932–948

Gopalakrishnan S, Upadhyaya HD, Vadlamudi S, Humayun P, Vidya MS, Alekhya G, Singh A, Vijayabharathi R, Bhimineni RK, Seema M, Rathore A, Rupela O (2012) Plant growth promoting traits of biocontrol potential bacteria isolated from rice rhizosphere. Springerplus 1:71p

Govindasamy V, Senthilkumar M, Gaikwad K, Annapurna K (2008) Isolation and characterization of ACC deaminase gene from two plant growth-promoting rhizobacteria. Curr Microbiol 57: 312–317

Gupta R, Saxena RK, Chaturvedi P, Virdi VS (1995) Chitinase production by *Streptomyces viridificans*: its potential in fungal cell wall lysis. J Appl Bacteriol 78:378–383

Haas D, Défago G (2005) Biological control of soil-borne pathogens by fluorescent pseudomonads. Nat Rev Microbiol 3:307–319

Hafeez-ur-Rehman A, Nawaz A, Awan MI, Ijaz M, Hussain M, Ahmad S, Farooq M (2019) Direct seeding in rice: problems and prospects. In: Hasanuzzaman M (ed) Agronomic crops: production technologies, vol 1. Springer, Singapore, pp 199–222

Hamdia MAES, Shaddad M, Doaa MM (2004) Mechanisms of salt tolerance and interactive effects of *Azospirillum brasilense* inoculation on maize cultivars grown under salt stress conditions. Plant Growth Regul 44:165–174

Hameeda B, Harini G, Rupela OP, Wani SP, Reddy G (2008) Growth promotion of maize by phosphate-solubilizing bacteria isolated from composts and macrofauna. Microbiol Res 163: 234–242

Han H, Lee S (2005) Physiological responses of soybean-inoculation of *Bradyrhizobium japonicum* with PGPR in saline soil conditions. Res J Agric Biol Sci 1:216–221

Han X, Sun X, Wang C, Wu M, Dong D, Zhong T, Thies JE, Wu W (2016) Mitigating methane emission from paddy soil with rice-straw biochar amendment under projected climate change. Sci Rep 6:24731

Harold F, Reetz J (2016) Fertilizers and Their Efficient Use. International Ferlizers Industry Association. www.fertilizers.org

Hartmann A, Rothballer M, Schmid M, Hiltner L (2008) A pioneer in rhizosphere microbial ecology and soil bacteriology research. Plant Soil 312:7–14

Hashem MA (2001) Problems and prospects of cyanobacterial biofertilizer for rice cultivation. Aust J Plant Physiol 28:881–888

Hassan S, Mathesius U (2012) The role of flavonoids in root-rhizosphere signalling: opportunities and challenges for improving plant-microbe interactions. J Exp Bot 63:3429–3444

Herridge DF, Peoples MB, Boddey RM (2008) Global inputs of biological nitrogen fixation in agricultural systems. Plant Soil 311:1–18

Hussain MB, Mehboob I, Zahir ZA, Naveed M, Asghar HN (2009) Potential of *Rhizobium* spp. for improving growth and yield of rice. Soil Environ 28(1):49–55

Hussain Q, Liu Y, Zhang A, Pan G, Li L, Zhang X, Song X, Cui L, Jin Z (2011) Variation of bacterial and fungal community structures in the rhizosphere of hybrid and standard rice cultivars and linkage to CO_2 flux. FEMS Microbiol Ecol 78:116–128

Hussain M, Ahmad S, Hussain S, Lal R, Ul-Allah S, Nawaz A (2018) Rice in saline soils: physiology, biochemistry, genetics and management. Adv Agron 148:231–287

IFPRI (2012) Global food policy report. International Food Policy Research Institute, Washington, DC

Jan M, Shah G, Masood S, Shinwari KI, Hameed R, Rha ES, Jamil M (2019) *Bacillus cereus* enhanced phytoremediation ability of rice seedlings under cadmium toxicity. Biomed Res Int 1: 1–12

Jhala YK, Vyas RV, Shelat HN, Patel HK, Patel HK, Patel KT (2014) Isolation and characterization of methane utilizing bacteria from wetland paddy ecosystem. World J Microbiol Biotechnol 30: 1845–1860

Jiang YJ, Liang YT, Li CM, Wang F, Sui YY, Suvannang N, Zhou JZ, Sun B (2016) Crop rotations alter bacterial and fungal diversity in paddy soils across East Asia. Soil Biol Biochem 95:250–261

Johnson LJ, Koulman A, Christensen M, Lane GA, Fraser K, Forester N, Johnson RD, Bryan GT, Rasmussen S (2013) An extracellular siderophore is required to maintain the mutualistic interaction of *Epichloë festucae* with *Lolium perenne*. PLoS Pathog 9:e1003332

Joshi B, Chaudhary A, Singh H, Kumar PA (2020) Prospective evaluation of individual and consortia plant growth promoting rhizobacteria for drought stress amelioration in rice (*Oryza sativa* L.). Plant Soil 457:225–240

Karmakar J, Goswami S, Pramanik K, Maiti TK, Kar RK, Dey N (2021) Growth promoting properties of *Mycobacterium* and *Bacillus* on rice plants under induced drought. Plant Sci Today 8(1):49–57. https://doi.org/10.14719/pst.2021.8.1.965

Kasim WA, Osman ME, Omar MN, Abd El-Daim IA, Bejai S, Meijer J (2013) Control of drought stress in wheat using plant growth promoting bacteria. J Plant Growth Regul 32:122–130

Kaur G, Kumar S, Nayyar H, Upadhyaya HD (2008) Cold stress injury during the pod-filling phase in chickpea (*Cicer arietinum* L.): effects on quantitative and qualitative components of seeds. J Agron Crop Sci 194:457–464

Khan A, Zhao XQ, Javed MT, Khan KS, Bano A, Shen RF, Masood S (2016) *Bacillus pumilus* enhances tolerance in rice (*Oryza sativa* L.) to combined stresses of NaCl and high boron due to limited uptake of Na^+. Environ Exp Bot 124:120–129

Khan MHU, Khattak JZK, Jamil M, Malook I, Khan SU, Jan M, Din I, Saud S, Kamran M, Alharby H, Fahad S (2017) *Bacillus safensis* with plant-derived smoke stimulates rice growth under saline conditions. Environ Sci Pollut Res 24:23850–23863

Khan MA, Ahmad S, Raza A (2019a) Integrated weed management for agronomic crops. In: Hasanuzzaman M (ed) Agronomic crops, vol 2. Springer, Singapore, pp 257–281

Khan SM, Ali S, Nawaz A, Bukhari SAH, Ejaz S, Ahmad S (2019b) Integrated pest and disease management for better agronomic crop production. In: Hasanuzzaman M (ed) Agronomic crops, vol 2. Springer, Singapore, pp 385–428

Kim YC, Glick B, Bashan Y, Ryu CM (2013) Enhancement of plant drought tolerance by microbes. In: Aroca R (ed) Plant responses to drought stress. Springer, Berlin, pp 383–413

Kloepper JW, Schroth MN (1981) Relationship of in vitro antibiosis of plant growth promoting rhizobacteria to plant growth and the displacement of root microflora. Phytopathology 71:1020–1024

Kobayashi DY, Reedy RM, Bick J, Oudemans PV (2002) Characterization of a chitinase gene from *Stenotrophomonas maltophilia* strain 34S1 and its involvement in biological control. Appl Environ Microbiol 68:1047–1054

Kompas T, Pham VH, Che TN (2018) The effects of climate change on GDP by country and the global economic gains from complying with the Paris climate accord. Earth's Future 6:1153–1173

Kumar M (2017) Bacterial community interaction with plants: with special reference to paddy field. J Agric Sci Bot 1:34–35

Kumar P, Dubey RC, Maheshwari DK (2012) *Bacillus* strains isolated from rhizosphere showed plant growth promoting and antagonistic activity against pathogens. Microbiol Res 167(8): 493–499

Kumar A, Singh S, Mukherjee A, Rastogi RP, Verma JP (2021) Salt-tolerant plant growth-promoting *Bacillus pumilus* strain JPVS11 to enhance plant growth attributes of rice and improve soil health under salinity stress. Microbiol Res 242:126616

Ladha JK, Reddy PM (1995) Extension of nitrogen fixation to rice - necessity and possibilities. GeoJournal 35:363–372

Lafitte HR, Yongsheng G, Yan S, Lil ZK (2007) Whole plant responses, key processes, and adaptation to drought stress: the case of rice. J Exp Bot 58:169–175

Lamb D, Erskine PD, Parrotta JA (2005) Restoration of degraded tropical forest landscapes. Science 310:1628–1632

Laxa M, Liebthal M, Telman W, Chibani K, Dietz KJ (2019) The role of the plant antioxidant system in drought tolerance. Antioxidants 8(4):94. https://doi.org/10.3390/antiox8040094

Leclere V, Bechet M, Adam A, Guez JS, Wathelet B, Ongena M, Thonart P (2005) Mycosubtilin overproduction by *Bacillus subtilis* BBG100 enhances the organism's antagonistic and biocontrol activities. Appl Environ Microbiol 71:4577–4584

Li X, Rui J, Mao Y, Yannarell A, Mackie R (2014) Dynamics of the bacterial community structure in the rhizosphere of a maize cultivar. Soil Biol Biochem 68:392–401

Liu ZP, Wang ZG, Xu WH, Chen WJ, Lv ZH, Long WC, Shi YR (2018) Screen, identification and analysis on the growth-promoting ability for the rice growth-promoting rhizobacteria. J Agric Resour Environ 35:119–125

Liu H, Wang Z, Xu W, Zeng J, Li L, Li S, Gao Z (2020) *Bacillus pumilus* LZP02 promotes rice root growth by improving carbohydrate metabolism and phenylpropanoid biosynthesis. Mol Plant Microbe Interact 33:1222–1231

Loaces I, Ferrando L, Scavino AF (2011) Dynamics, diversity and function of endophytic siderophore-producing bacteria in rice. Microb Ecol 61:606–618

Lopez-Cervantes J, Thorpe DT (2013) Microbial composition comprising liquid fertilizer and processes for agricultural use. WO Patent, WO2013148278A1

Lucas JA, Ramos-Solano B, Montes F, Ojeda J, Megias M, Gutierrez Mañero FJ (2009) Use of two PGPR strains in the integrated management of blast disease in rice (*Oryza sativa*) in Southern Spain. Field Crop Res 114:404–410

Lucas JA, García-Cristobal J, Bonilla A, Ramos B, Gutierrez-Ma~nero J (2014) Beneficial rhizobacteria from rice rhizosphere confers high protection against biotic and abiotic stress inducing systemic resistance in rice seedlings. Plant Physiol Biochem 82:44–53

Luo S, Wan Y, Xiao X, Guo H, Chen L, Xi Q, Zeng G, Liu C, Chen J (2011) Isolation and characterization of endophytic bacterium LRE07 from cadmium hyperaccumulator *Solanum nigrum* L. and its potential for remediation. Appl Microbiol Biotechnol 89:1637–1644

Maheshwari DK, Aeron A, Dubey RC, Agrwal M, Dheeman S, Shukla S (2014) Multifaceted beneficial associations with Pseudomonas and rhizobacteria on growth promotion of *Mucuna pruriens* L. J Pure Appl Microbiol 8(6):4657–4667

Malik KA, Bilal R, Mehnaz S, Rasul G, Mirza MS, Ali S (1997) Association of nitrogen-fixing, plant promoting rhizobacteria (PGPR) with kallar grass and rice. Plant Soil 194:37–44

Mantri N, Patade V, Penna S, Ford R (2012) Abiotic stress responses in plants: present and future. In: Ahmad P, Prasad MNV (eds) Abiotic stress responses in plants. Springer, New York, USA, pp 1–19

Mauseth JD (1991) Botany: an introduction to plant biology. Saunders, Philadelphia, pp 348–415

McInerney MJ, Struchtemeyer CG, Sieber J, Mouttaki H, Stams AJ, Schink B, Rohlin L, Gunsalus RP (2008) Physiology, ecology, phylogeny, and genomics of microorganisms capable of syntrophic metabolism. Ann N Y Acad Sci 1125:58–72

Meena KK, Kumar M, Kalyuzhnaya MG, Yandigeri MS, Singh DP, Saxena AK, Arora DK (2012) Epiphytic pink-pigmented methylotrophic bacteria enhance germination and seedling growth of wheat (*Triticum aestivum*) by producing phytohormone. Antonie Leeuwenhoek 101:777–786

Mehnaz S, Mirza MS, Haurat J, Bally R, Normand P, Bano A, Malik KA (2001) Isolation and 16S rRNA sequence analysis of the beneficial bacteria from the rhizosphere of rice. Can J Microbiol 472:110–117

Mian MH (2002) Azobiofer: a technology of production and use of Azolla as biofertiliser for irrigated rice and fish cultivation. In: Kennedy IR, Choudhury ATMA (eds) Biofertilisers in action. Rural Industries Research and Development Corporation, Canberra, Australia, pp 45–54

Miller MB, Bassler BL (2001) Quorum sensing in bacteria. Annu Rev Microbiol 55:165–199

Minamisawa K, Imaizumi-Anraku H, Bao Z, Shinoda R, Okubo T, Ikeda S (2016) Are symbiotic methanotrophs key microbes for N acquisition in paddy rice root? Microbes Environ 31:4–10

Munns R (2002) Comparative physiology of salt and water stress. Plant Cell Environ 25:239–250

Nagajyoti PC, Lee KD, Sreekanth TVM (2010) Heavy metals, occurrence and toxicity for plants: a review. Environ Chem Lett 8:199–216

Nascimento FX, Rossi MJ, Soares CR, McConkey BJ, Glick BR (2014) New insights into 1-aminocyclopropane-1-carboxylate (ACC) deaminase phylogeny, evolution and ecological significance. PLoS One 6:e99168

Nautiyal CS, Srivastava S, Chauhan PS, Seem K, Mishra A, Sopory SK (2013) Plant growth-promoting bacteria *Bacillus amyloliquefaciens* NBRISN13 modulates gene expression profile of leaf and rhizosphere community in rice during salt stress. Plant Physiol Biochem 66:1–9

Naz S, Fatima Z, Iqbal P, Khan A, Zakir I, Noreen S, Younis H, Abbas G, Ahmad S (2019) Agronomic crops: types and uses. In: Hasanuzzaman M (ed) Agronomic crops: production technologies, vol 1. Springer, Singapore, pp 1–18

Nazaries L, Murrell JC, Millard P, Baggs L, Singh BK (2013) Methane, microbes and models: fundamental understanding of the soil methane cycle for future predictions. Environ Microbiol 15:2395–2417

Nguyen N, Ferrero A (2006) Meeting the challenges of global rice production. Paddy Water Environ 4:1–9

Nuti M, Giovannetti G (2015) Borderline products between bio-fertilizers/bio-effectors and Plant protectants: the role of microbial consortia. J Agric Sci Technol 5:305–315

Oades JM (1993) The role of biology in the formation, stabilization and degradation of soil structure. Geoderma 56:182–186

Orlandelli RC, Vasconcelos AFD, Azevedo JL, Silva MLC, Pamphile JA (2016) Screening of endophytic sources of exopolysaccharides: preliminary characterization of crude exopolysaccharide produced by submerged culture of *Diaporthe* sp. JF766998 under different cultivation time. Biochim Open 2:33–40

Palaniyandi SA, Yang SH, Zhang L, Suh JW (2013) Effects of actinobacteria on plant disease suppression and growth-promotion. Appl Microbiol Biotechnol 97:9621–9636

Pandey P, Maheshwari DK (2007) Two-species microbial consortium for growth promotion of *Cajanus cajan*. Curr Sci 92(8):1137–1142

Peerzada AM, Bukhari SAH, Dawood M, Nawaz A, Ahmad S, Adkins S (2019) Weed management for healthy crop production. In: Hasanuzzaman M (ed) Agronomic crops, vol 2. Springer, Singapore, pp 225–256

Pittol M, Durso L, Valiati VH, Fiuza LM (2016) Agronomic and environmental aspects of diazotrophic bacteria in rice fields. Ann Microbiol 66:511–527

Pramanik K, Mandal S, Banerjee S, Ghosh A, Maiti TK, Mandal NC (2021) Unraveling the heavy metal resistance and biocontrol potential of Pseudomonas sp. K32 strain facilitating rice seedling growth under Cd stress. Chemosphere 274:129819

Pratiwi E, Akhdiya A, Purwani J, Husnain, Syakir M (2021) Impact of methane-utilizing bacteria on rice yield, inorganic fertilizers efficiency and methane emissions. 1st International Conference on Sustainable Tropical Land Management. IOP Conf Ser: Earth Environ Sci 648:012137. https://doi.org/10.1088/1755-1315/648/1/012137

Rajkumar M, Ae N, Prasad MNV, Freitas H (2010) Potential of siderophore-producing bacteria for improving heavy metal phytoextraction. Trends Biotechnol 28:142–149

Razaq M, Shah FM, Ahmad S, Afzal M (2019) Pest management for agronomic crops. In: Hasanuzzaman M (ed) Agronomic Crops, vol 2. Springer, Singapore, pp 365–384

Redman RS, Kim YO, Woodward CJDA, Greer C, Espino L, Doty SL, Rodriguez RJ (2011) Increased fitness of rice plants to abiotic stress via habitat adapted symbiosis: a strategy for mitigating impacts of climate change. PLoS One 6:e14823p

Rêgo MCF, Cardoso AF, Ferreira TC, de Filippi MCC, Batista TFV, Viana RG, da Silva GB (2018) The role of rhizobacteria in rice plants: growth and mitigation of toxicity. J Integr Agric 17: 2636–2647

Richardson AE, Barea JM, McNeil AM, Prigent-Combaret C (2009) Acquisition of phosphorus and nitrogen in the rhizosphere and plant growth promotion by microorganisms. Plant Soil 321:305–339

Rizwan M, Ali S, Adrees M, Rizvi H, Rehman MZ, Hannan F, Qayyum MF, Hafeez F, Ok YS (2016) Cadmium stress in rice: toxic effects, tolerance mechanisms and management: a critical review. Environ Sci Pollut Res 23:17859–17879

Saechow S, Thammasittirong A, Kittakoop P, Prachya S, Thammasittirong SNR (2018) Antagonistic activity against dirty panicle rice fungal pathogens and plant growth-promoting activity of *Bacillus amyloliquefaciens* BAS23. J Microbiol Biotechnol 28(9):1527–1535

Saharan B, Nehra V (2011) Plant growth promoting rhizobacteria: a critical review. Life Sci Med Res 21:1–30

Saleem M, Arshad M, Hussain S, Bhatti AS (2007) Perspective of plant growth promoting rhizobacteria (PGPR) containing ACC deaminase in stress agriculture. J IndustMicrobiol Biotechnol 34:635–648

Sarwar MH, Sarwar MF, Sarwar M, Qadri NA, Moghal S (2013a) The importance of cereals (*Poaceae: Gramineae*) nutrition in human health: a review. J Cereals Oilseeds 4:32–35

Sarwar N, Ali H, Ahmad S, EhsanUllah, Ahmad S, Mubeen K, Hill JE (2013b) Water wise rice cultivation on calcareous soil with the addition of essential micronutrients. J Anim Plant Sci 23(1):244–250

Sarwar N, Ali H, Maqsood M, EhsanUllah, Shahzad AN, Shahzad M, Mubeen K, Shahid MA, Ahmad S (2013c) Phenological response of rice plants to different micronutrients application under water saving paddy fields on calcareous soil. Turk J Field Crops 18(1):52–57

Sekar J, Raj R, Prabavathy VR (2016) Microbial consortial products for sustainable agriculture: commercialization and regulatory issues in India. In: Singh HB, Sarma BK, Keswani (eds) Agriculturally important microorganisms: commercialization and regulatory requirements in Asia. Springer, Singapore, pp 107–133

Sen S, Chandrasekhar CN (2015) Effect of PGPR on enzymatic activities of rice (*Oryza sativa* L.) under salt stress. Int J Adv Agric Res 3:37–41

Shah G, Jan M, Afreen M, Anees M, Rehman S, Daud MK, Malook I, Jamil M (2017) Halophilic bacteria mediated phytoremediation of salt-affected soils cultivated with rice. J Geochem Explor 174:59–65

Shahzad AN, Ahmad S (2019) Tools and techniques for nitrogen management in cereals. In: Hasanuzzaman M (ed) Agronomic crops, vol 2. Springer, Singapore, pp 111–126

Sharma A, Shankhdhar D, Shankhdhar SC (2013) Enhancing grain iron content of rice by the application of plant growth promoting rhizobacteria. Plant Soil Environ 59(2):89–94

Shen FT, Yen JH, Liao CS, Chen WC, Chao YT (2019) Screening of rice endophytic biofertilizers with fungicide tolerance and plant growth promoting characteristics. Sustain For 11:1133p

Singh DP, Singh V, Gupta VK, Shukla R, Prabha R, Sarma BK, Patel JS (2020) Microbial inoculation in rice regulates antioxidative reactions and defense related genes to mitigate drought stress. Sci Rep 10:4818p

Solomon S, Qin D, Manning M, Averyt K, Marquis M (2007) Climate change: the physical science basis: working group I contribution to the fourth assessment report of the IPCC. Cambridge University Press, London

Srivastava S, Bist V, Srivastava S, Singh PC, Trivedi PK, Asif MH, Chauhan PS, Nautiyal CS (2016) Unraveling aspects of *Bacillus amyloliquefaciens* mediated enhanced production of rice under biotic stress of *Rhizoctonia solani*. Front Plant Sci 7:587

Sulaiman C, Moung-Ngam P, Arikit S, Vanavichit A, Malumpong C (2018) Effects of heat stress at vegetative and reproductive stages on spikelet fertility. Rice Sci 25:218–226

Sun T, Liu YYN, Wu S, Zhang JZ, Qu B, Xu JG (2020) Effects of background fertilization followed by co-application of two kinds of bacteria on soil nutrient content and rice yield in Northeast China. Int J Agric Biol Eng 13:154–162

Takeda K, Tonouchi A, Takada M, Suko T, Suzuki S, Kimura Y, Matsuyama N, Fujita T (2008) Characterization of cultivable methanotrophs from paddy soils and rice roots. Soil Sci Plant Nutr 54(6):876–885

Tariq M, Ahmed M, Iqbal P, Fatima Z, Ahmad S (2020) Crop phenotyping. In: Ahmed M (ed) Systems modeling. Springer, Singapore, pp 45–60

Thomas J, Kim HR, Rahmatallah Y, Wiggins G, Yang Q, Singh R, Glazko G, Mukherjee A (2019) RNA-seq reveals differentially expressed genes in rice (*Oryza sativa*) roots during interactions with plant growth promoting bacteria, *Azospirillum brasilense*. PLoS One 14:e0217309

Timmusk S, Nevo E (2011) Plant root associated biofilms. In: Maheshwari DK (ed) Bacteria in agrobiology. Plant nutrient management. Springer, Berlin, pp 285–300

Timmusk S, Abd El-Daim IA, Copolovici L, Tanilas T, Kännaste A, Behers L, Nevo E, Seisenbaeva G, Stenström E, Niinemets U (2014) Drought-tolerance of wheat improved by rhizosphere bacteria from harsh environments: enhanced biomass production and reduced emissions of stress volatiles. PLoS One 9:1–13

Tiwari S, Prasad V, Chauhan PS, Lata C (2017) *Bacillus amyloliquefaciens* confers tolerance to various abiotic stresses and modulates plant response to phytohormones through osmoprotection and gene expression regulation in rice. Front Plant Sci 8:1510

Tuteja N (2007) Abscisic acid and abiotic stress signaling. Plant Signal Behav 2:135–138

Vanegas J, Uribe-Vélez D (2014) Selection of mixed inoculants exhibiting growth-promoting activity in rice plants from undefined consortia obtained by continuous enrichment. Plant Soil 375:215–227

Vinocur BA, Altman (2005) Recent advances in engineering plant tolerance to abiotic stress: achievements and limitations. Curr Opin Biotechnol 16:123–132

Vorholt JA, Vogel C, Carlström CI, Mueller DB (2017) Establishing causality: opportunities of synthetic communities for plant microbiome research. Cell Host Microbe 22:142–155

Wahid A, Gelani S, Ashraf M, Foolad MR (2007) Heat tolerance in plants: an overview. Environ Exp Bot 61(3):199–223

Wang H, Liu Z, Xu W, Wang Z (2019) Determination of growth-promoting ability of several strains to rice. Jiangsu Agr Sci 47:94–99

Wani SP, Gopalakrishnan S (2019) Plant growth-promoting microbes for sustainable agriculture. In: Sayyed R, Reddy M, Antonius S (eds) Plant growth promoting rhizobacteria (PGPR): prospects for sustainable agriculture. Springer, Singapore, pp 19–45

Wasaya A, Yasir TA, Ijaz M, Ahmad S (2019) Tillage effects on agronomic crop production. In: Hasanuzzaman M (ed) Agronomic crops, vol 2. Springer, Singapore, pp 73–99

Watanabe A, Takeda T, Kimura M (1999) Evaluation of origins of CH_4 carbon emitted from rice paddies. J Geophys Res 104:23623–23629

Win KT, Oo AZ, Ohkama-Ohtsu N, Yokoyama T (2018) *Bacillus Pumilus* strain TUAT-1 and nitrogen application in nursery phase promote growth of rice plants under field conditions. Agronomy 8:216p

Wong WS, Tan SN, Ge L, Chen X, Yong JWH (2015) The importance of phytohormones and microbes in biofertilizers. In: Maheshwari DK (ed) Bacterial metabolites in sustainable agroecosystem. Springer, New York, pp 105–158

Woo SL, Pepe O (2018) Microbial consortia: promising probiotics as plant biostimulants for sustainable agriculture. Front Plant Sci 9:1801

Yuan C, Zhang L, Hu H, Wang J, Shen J, He J (2018) The biogeography of fungal communities in paddy soils is mainly driven by geographic distance. J Soils Sediments 18:1795–1805

Zahoor SA, Ahmad S, Ahmad A, Wajid A, Khaliq T, Mubeen M, Hussain S, Sami Ul Din M, Amin A, Awais M, Nasim W (2019) Improving water use efficiency in agronomic crop production. In: Hasanuzzaman M (ed) Agronomic Crops, vol 2. Springer, Singapore, pp 13–29

Plant Growth Regulators for Rice Production in Changing Environment

15

Tauqeer Ahmad Yasir, Allah Wasaya, Wasif Azhar, Saima Kanwal, Naeem Sarwar, Muhammad Ishaq Asif Rehmani, and Abdul Wahid

Abstract

The productivity of rice is greatly affected by changing climate. Due to climate change and climate variability, the insect, pests, and diseases pressure have been shifted. Plant growth regulators could play an important role to boost rice growth and yield. In order to ensure food security, climate change adaptation and mitigation strategies are very much needed at regional and global scales.

Keywords

RICE · Climate · Change · Variability

15.1 Introduction

Agriculture is facing an exigency situation globally due to changing climate (Ahmed et al. 2017, 2020a, b; Ahmed and Ahmad 2019; Akram et al. 2018, 2019; Hussain et al. 2018; Fahad et al. 2019; Fatima et al. 2020). Crops yield increased due to successful green revolution, which has now reached to the ultimate potential, and,

T. A. Yasir (✉) · A. Wasaya · W. Azhar · S. Kanwal
College of Agriculture, Bahauddin Zakariya University, Bahadur Sub-Campus, Layyah, Pakistan
e-mail: tauqeer.yasir@bzu.edu.pk

N. Sarwar
Department of Agronomy, Bahauddin Zakariya University, Multan, Pakistan

M. I. A. Rehmani
Department of Agronomy, Ghazi University, Dera Ghazi Khan, Pakistan

A. Wahid
Department of Plant Breeding and Genetics, Sindh Agriculture University, Tandojam, Pakistan

N. Sarwar et al. (eds.), *Modern Techniques of Rice Crop Production*,
https://doi.org/10.1007/978-981-16-4955-4_15

finally, resulting in deceleration of agricultural productivity and stagnant growth (Mitchell and Sheehy 2006). Nearly about 50% of calories are provided by cereal crops, mainly wheat, rice, and maize, in underdeveloped countries (Reynolds et al. 2012). Cereal production must be improved to a great extent to feed the fast growing population and to sustain global food demand under climate change scenario (Hafeez-ur-Rehman et al. 2019; Khan et al. 2019a, b; Naz et al. 2019; Peerzada et al. 2019; Razaq et al. 2019; Sarwar et al. 2013a, b; Tariq et al. 2020; Wasaya et al. 2019; Zahoor et al. 2019). It is alarming that changing climatic precursors like high temperature, water stress, salinity, and unpredictable rains are resulting in decreased yield of cereals as well other crops, whereas an increase in yield proliferation is either not observed or denied by increasing temperature (Prasad et al. 2006).

Rice (*Oryza sativa* and *Oryza japonica*) is a monocotyledonous cereal crop that belongs to family Poaceae, and is considered as an extensive food crop in many regions of the world (Ahmad et al. 2015, 2019, 2012, 2013, 2008, 2009; Ahmad and Hasanuzzaman 2012; Ahmed and Ahmad 2019, 2020; Shahzad and Ahmad 2019). One-third of the human population use rice as the primary source of food. It is estimated that its production and consumption is exceeding 90% in the Asian countries than rest of the world. By adopting advanced field production practices along with using improved seed of high yielding rice cultivar, the production of rice has been enhanced all over the world. Although rice production fulfills the present needs of food security, there is an immediate requirement to multiply the yield as the world population of 6.4 billion has already reached 7.7 billion in 2020 and predicted to be 9.5 billion by 2050.

Higher plants produce phytohormones known as plant growth regulators (PGRs) that are organic-based substances. PGRs can control plant's life functions at a point apart from their production place and perform predominately even in small amounts. Plant growth regulators significantly affect plant growth characters, yield, and its related parameters, photosynthetic indicators, biochemical properties, as well as seed quality traits. Plant growth regulators contain cytokinins, gibberellins (gibberellic acid), auxin, ethylene, growth retardants, and growth inhibitors (Gollagi et al. 2019). Several types of research display that exogenous implementation of plant growth regulators affects the plant response. In plants, the internal hormonal configuration can be affected through the use of plant growth regulators by using both the supplementary suboptimized levels or via the interaction by means of their synthesis, translocation, or silencing of previous hormonal intensities (Arshad and Frankenberger Jr 1993).

15.2 Plant Growth Regulators and Rice Growth

15.2.1 Gibberellins: GA3

Early emergence and rooting can be promoted by the application of suitable plant growth regulator and it also help in providing a good harbor of the plant and good seedling establishment. According to Bevilaqua et al. (1993), germination of various seeds and shoot growth can be increased by using gibberellic acid (GA3), a plant growth regulator.

15.2.2 Effect of Gibberellins on Seed Germination

Germination percentage, seedling emergence, and seedling height considerably improved in rice by seed treatment with use of GA3 (Asborno et al. 1999). Under salinity stress condition, priming with GA3 is very efficient, which can change ions uptake and their accumulation arrangement among plant roots and shoots by producing grains in different quantities. Improved vegetative growth has been observed by treating the seeds with GA3 growth regulator in drilled seeded rice (Helms et al. 1990).

With the application of GA3, speedy and vigorous seed emergence has been achieved (Bevilaqua et al. 1996). Rice seeds treated with GA3 followed by IBA caused highest emergence of rice seeds at field Madurai (India). Additionally, GA3-treated rice seeds also showed higher germination percentage and seedlings with increased coleoptile length (Devasagayam et al. 1996). GA3 is more effective than IBA in terms of increasing shoot growth, as IBA is unable to enhance the activity of enzyme α-amylase. Conversely, water potential increased in cell by increasing softness of cell wall to cause water movement inside and resulting to produce elongated cells, when seeds were treated with GA3 followed by hydrolysis (Arteca 1995). It has been recorded that during cold stress, particularly under cold-sensitive cultivars, action of α-amylase or GA3 was considered as a controlling factor to the emergence process (Chen et al. 2005). The ability and productiveness of GA3 can be enhanced by enhancing concentration and soaking period of GA3 (Viera et al. 2002).

15.2.3 Effect of Gibberellins on Seed Yield

Panicle exertion of male sterile lines can be improved by foliage applied GA3, when emergence of panicles gets started in hybrid rice seed production, which is generally used as essential technology. Furthermore, it was found with the help of various investigations that farmers carried on spraying GA3 of very low strength for several days. For the reason that postanthesis use of GA3 can influence properties of stigma (Tian 1991; Zhou et al. 2017) which, affect the rate of fertilization and final crop yield. Applying GA3 increases crossing rate, quantity of spikelets, seed weight, and production of hybrid rice (Cheng et al. 2011). It was founded by Zheng et al. (2018) that the involvement of the panicle can enhance the yield by 32.3% with the application of GA3 after anthesis; and in hybrid rice seed production, carbon is synthesized in the panicle used for seed filling. These results showed that duration of seed filling could be elongated and increased by application of GA3 after anthesis. Spraying GA3 could efficiently postpone the process of senescence in plants for all types of rice. In another study, it was revealed that the subordinate grain fleshiness in rice was increased by 5.5% while spraying with GA3 after flowering. Likewise, with the application of GA3, the duration of seed filling period can be extended which results in enhancing sink capacity of the seeds to attain more grain weight the grain weight with increasing the sink capacity by application of GA3 (Dong et al. 2016). Various postharvest fruit properties can be enhanced by preharvest GA3 applications and have clear effects during storage on the shelf life of the fruits (Erogul and Sen 2016). Application of GA3 after flowering in hybrid rice seed production significantly increases the strength of stigma and similarly enhance crossing rate and grain weight.

Application of GA3 alone or in combination with ethephon significantly increased the plant height (Watanabe and Siagusa 2004). Dong et al. (2016) revealed that different rice cultivars respond differently to exogenous application of growth regulator in terms of quality parameters. Application of GA3 (57.7 μmol per liter) during early grain filling stage in rice also effects the grain positions. Chalkiness and amylase content were enhanced by using GA3 exogenously, however, yield parameters and protein content were reduced. Some findings also proved that the application of GA3 has a potential to improve yield-related traits in early-flower producing spikelets than in late-flower producing spikelets. It was recorded by Sheng et al. (1993) that water-soaked rice seeds in 50 ppm MET for 24 h before sowing produce 246,000 more panicles per hectare than nonsoaked treated seeds. Rio de Grande and Sao Paulo (Brazil) revealed that number of panicles increased dominantly by seed treatment with 1.0 g GA3 per 50 kg seed (Bevilaqua et al. 1996). Wahyuni et al. (2003) noted that there was no considerable impact on yield-contributing characters observed by using treated rice seeds with 25 ppm, 50 ppm, and 100 ppm GA3 and 10 ppm, 20 ppm and, 40 ppm IBA. Although in China, in hybrid rice, the use of different concentrations of uniconazole highly improved grain filling on lower panicle site, which multiplied grains on each and, hence, enhancing grain weight per panicle (Xiang et al. 2003). Spraying of GA3 at 60 ml/ha in rice resulted in 13% higher seed yield (Elankavi et al. 2009). Moreover, with the application of phytohormones, growth and yield of rice crop can be increased by 50.5% (Elankavi et al. 2009). Shi-Hua et al. (2006) researched that application of plant growth regulators increases biological yield with larger panicle that is connected with a greater number of grain panicles, which culminate to high productivity.

15.3 Cytokinins: Kinetin, Zeatin, and 6-Benzylaminopurine (6-BA)

Hutchison and Kieber (2002) reported that cytokinins (CKs) have been used to regulate various developmental processes and environmental plant responses, involving leaf senescence, apical dominance, chloroplast development, and to control cell division. CKs exhibit a key role in multiple developmental functioning related to growth happening in the plants. Hare et al. (1997) found these results due to the interactions of CKs with other plant hormones and environmental signals.

6-Benzylaminopurine (6-BA) is a synthetic compound of cytokinins, which can profoundly decrease the ethylene sensitivity of cut flowers (Yuan et al. 2012). Exogenous application of 6-BA checks synthesis of ethylene, invokes the production of 1-aminocyclopropane-1-carboxylate oxidase (ACO) and 1-aminocyclopropane-1-carboxylate synthase (ACS), for prompt controlling of signaling transduction linked with cytokinins receptors (Hall et al. 2001). Zhang et al. 2007 observed that, when leaves of rice treated with 6-BA during later stages of growth, delay in leaf senescence takes place and due to this the rate of seed setting becomes higher which ultimately increased seed yield and grain yield.

15.3.1 Effect of Cytokinins on Seed Germination

Adenine (Ade) is regarded as the common precursor for the manufacturing of cytokinins in the tissues of plants as well as to culture microbes. Nieto and Frankenberger Jr (1989) investigated that some microbial isolates of Pseudomonas and Azotobacter were introduced to produce cytokinins actively when isopentyl alcohol (IA) and Ade were supplied to the growing medium. It was reported that under salt stress condition, the use of CKs to prime the seeds is noticed to be advantageous over unprimed seeds to germination, growth, and yield of some crop species grown.

15.3.2 Effect of Cytokinins on Seed Yield

Treating roots with different strengths of Ade or kinetin have been found effective to achieve higher production of rice (Zahir et al. 1999). Leaf senescence can be decreased by applying kinetin to the source structure (flag leaf) in rice crop, which increases rice yield (Biswas and Mondal 1986). In transplanted rice (TPR), scientists reported that leaf area index and plant height, concentration of leaf nitrogen during flowering and milky growth, as well as seed and straw production in rice can be significantly enhanced by cytokinins or triacontanol application on foliage at early-to-late growth stages in rice. Effect of four treatments was investigated by Sakeena and Salam (1989), including water spray, 10 ppm of kinetin spray at flowering stage, 10 ppm of kinetin applied at 10 days (after flowering), and the same concentration sprayed twice at flowering as well as at 10 days after producing flowers, in rice. In a result, protein content was noted to be lesser in entire leaf parts, while higher in the grains due to application of kinetin. Furthermore, highest amount of protein in the grains was analyzed in the treatment which received kinetin twice at different growth stages. Grain protein and functions of protein were also managed by applying uniconazole (Han and yang 2009).

Macronutrients like N, P, and K in the grains together with rice straw were improved by application of kinetin and Ade + Isopentyl alcohol. Furthermore, gradual conversion of precursors into cytokinins was observed to be more efficient, perhaps bring about in a constant stream for a longer period of time (Zahir et al.

2001). It was discovered that the traits responsible to produce higher outcome in rice might be interconnected and results in positive response of rice by the application of cytokinins. In a field study, it was observed that the rice yield was enhanced to 45.8% by spraying cytokinin as compared to controlled condition (Métraux and Kende 1983). Maksimov et al. (1979) found that the uptake of plant essential nutrients was influenced by the use of synthetic cytokinins, by greatly influencing the growth of roots.

Nitrogen content of rice grain can also be enhanced by cytokinins application (Yoshida 1987). He further revealed that a synthetic cytokinins-active phenyl urea with a physiologic activity surpasses that of zeatin. Pandey et al. (2001) recorded that panicle weight can be enhanced with the application of cytokinin alone and in combination with triacontanol and DAT on foliage or on soil. Similarly, paclobutrazol when applied at 50 mg per liter prominently enhances the quantity of filled grains on rice panicle, when compared with the control treatment (Thuc et al. 2016). It was recorded by Gurmani et al. (2006) that seed yield and other vegetative and panicle-related traits of paddy also increased by exogenous treatment of cytokinins as compared to treatment with other plant growth products like BA, ABA, and CCC.

15.4 Auxins

15.4.1 Effect of Auxins on Seed Germination

Advance root initiation, adventitious root formation, and early root development are commonly practiced by auxins like IAA, IBA, and NAA (Pan et al. 1999). Immersing the effects of two plant growth enhancers, gibberellic acid (GA3) and indole-3-acetic acid (IAA), on unhusked rice seeds vulnerable to low levels of salinity stress evidenced that total soluble and decreasing sugars in endosperms of rice seeds soaked with GA3 (10 μM) accumulated slowly under low NaCl stress as germination progressed (Kim et al. 2006). Rice seeds soaked with IAA have more clear effects than GA3. Furthermore, germinating rice seeds soaked with IAA were much more efficient in growth due to the stimulating effect of by α-amylase than that germinating rice seeds soaked with GA3. During salt stress condition, rice seeds soaked with IAA and GA3 exhibited comparatively higher amounts of endogenous production of IAA. Particularly, in GA3-soaked rice seeds, IAA content is more enhanced when compared with IAA soaked rice seeds. Moreover, the maximum concentrations among all the treatments were shown by gibberellin content in IAA-soaked rice seeds.

15.4.2 Effect of Auxins on Callus Formation

The callus induction efficiency can be considerably enhanced by use of higher dose of auxin, though such calli are poor in green plant regeneration and less in embryogenic. The anthers from some rice cultivars generated white color in Z2 media; fragile and consolidate callus texture consisting of a combined mixture of kinetin, NAA, and 2, 4-D (Shahnewaz et al. 2003). MS media having several concentrations of tryptophan, kinetin and auxin, when applied on paddy cultivar Swat-II, various colors (white, yellow, and brown) were noted in several embryogenic and nonembryogenic callus (Bano et al. 2005). In another study, compact callus texture was produced, in some rice genotypes, having white color in the media, when treated with NAA and 2, 4-D of several strengths (Roly et al. 2014). Double haploid lines are produced by the anthers of BC2F3 of *Oryza sativa* L. × *Oryza rufipogon* via fragile texture of callus having light green color in the media like N6 (Ambarawati et al. 2009). In rice cultivar Swarna, various types of callus texture with many callus colors were recorded on the N6 media having different combinations of kinetin, 2, 4-D, BAP, and NAA (Shukla et al. 2014). Callus induced by 2, 4-D in combination with kinetin growth enhancer in rice cultivars, one treatment carried maximum callus growth and caused minimal callus growth by the remaining treatments. However, minimal growth of callus growth was invoked in treatments T2, T3, and T4.

Elevated callus growth was induced by the anthers of Kyoto Asahi rice cultivar by media-type N6 media having kinetin and 2, 4-D (Hoffmann-Benning and Kende 1992). Anyhow, combination of NAA and kinetin growth enhancer induces minimal production of callus in both of the genotypes. In rice variety Azucena, different combinations of kinetin, NAA, and 2–4-D growth enhancers were used to test callus growth, color and type and found that the treatment consisting of 2, 4-D 2 mg L^{-1} + NAA 2 mg L^{-1} + Kinetin 1 mg L^{-1} caused early callus production (Sharma 2018). Japonica rice variety Azucena induced the callus growth with media N6 having kinetin, NAA, and 2–4-D with similar level of callus production. Consequently, Azucena and Moroberekan varieties have predominant embryogenic calli with compact type. Various kinds of callus, color of callus, and callus growth can be produced by growth regulators together with their concentrations. So, to obtain high embryogenic calli and callus production, various natures of growth regulators and their strengths must be studied.

15.4.3 Effect of Auxins on Seed Yield

Prominent results of several growth regulators on yield of rice and its accrediting characters have been reported. NAA at the rate of 500 ppm ha −1 influence the grains formation and number of spikelet positively (Misra and Sahu 1957). NAA indicated positive results by increasing crop yield with their maximum concentrations (Alam et al. 2002). Development of female panicles out of leaf sheath in rice can be raised by GA3 to ameliorate the capability to receive pollens coming out of male part (Gavino et al. 2008). Low amount of IAA and ABA content identified the low grain filling in rice (Yang et al. 1999; Wang et al. 1998). Furthermore, during rice grain filling, the ABA production in large-sized grains was found greater than in the small-sized grains (Kato et al. 1993). During early stages of grain filling, there is a positive correlation of ABA content in wheat grain with grain filling rate (Bai et al. 1989). Percentage of problematic grains can be decreased by using NAA along with 10 μmolL-1 and 20 μmolL-1 concentrations. According to Sakata et al. (2010), NAA commonly use exogenous application of auxin in plants to enhance auxin level in plants. Chen and Zhao (2008) proved that exogenous application of auxin in plants behaves as molecular signal for pollen tube growth.

15.5 Ethylene

To support gas exchange with atmosphere, flooded rice enhances development of young tillers headed to air and water areas (Jackson and Ram 2003). It was reported that during water logging condition, stem elongation promotion was mimicked under normal aerobic condition by ethylene and ACC treatment on seedlings and intermodal parts of rice (Van der Straeten et al. 2001; Raskin and Kende 1984; Métraux and Kende 1983; Yang and Choi 2006). Opposite to that, ethylene production was decreased and internodes elongation was also blocked in deepwater rice during puddling which inhibit ethylene production with the effect of aminooxyacetic acid and aminoethoxyvinylglycine (Raskin and Kende 1984; Métraux and Kende 1983). Further, ethylene activates the division of cells and their elongation in their corresponding regions of intermodal sections of rice under puddled condition (Métraux and Kende 1984). Ethylene is important to promote appearance and growth of the roots other than seminal. It has been observed that, at the nodal sections of stem, the establishing of adventitious root can be enhanced by the use of ethylene precursors, 2- ethephon and chloroethyl phosphonic acid whereas, under submergence, ethylene intercedes the loss of epidermal cells around the adventitious roots, which sustenance the permeation of the roots (Mergemann and Sauter 2000; Steffens and Sauter 2005). Under internodes of deepwater rice, the levels of ABA were reduced by ethylene application (Yang and Choi 2006; Hoffmann-Benning and Kende 1992). Furthermore, during submergence, reduction in ABA levels was checked by 1-methylcyclopropene (1-MCP), which inhibits ethylene production. These evidences show that during stress situation, ethylene induced under submergence can control the decline in ABA. In the latest studies, it was recorded that ABA 80-hydroxylase-1 pertained to the reduction in ABA under puddling in deepwater and lowland rice (Yang and Choi 2006; Saika et al. 2007). ABA 80-hydroxylase transforms ABA to 80-hydroxy ABA, which is quickly isomerized to form phaseic acid (PA). It was observed that mRNA levels of this enzyme in internodes increased by submergence and ethylene treatments in deepwater as well as lowland rice (Yang and Choi 2006; Saika et al. 2007).

Under salinity conditions, the germination of rice seed was regulated by ethylene, and the internal ethylene production differs with growth stages. The treatment of seedlings of rice with the precursors that can enhance ET resulted in enhancing plant tolerance against anaerobic water stress; anyhow, suppressing the signaling of ET, which is responsible to stimulate growth. According to Sajid et al. (2020), initial germination and further growth of rice were boosted by priming rice seedlings with ethephon. Priming rice seedlings with ethephon resulted in upgrading levels of ET. Generally, higher ET rate in rice may be favorable for the germination process of seeds, but may not be suitable for the growing seedlings of rice to produce healthy plants and higher seed yield.

15.6 Abscisic Acid (ABA)

Abscisic acid (ABA) is an important plant growth regulator or phytohormone that plays the significant role in adaptive reaction of plants to many environmental stresses, such as cold (Chen et al. 1983), high temperature (Robertson et al. 1994), dried condition (Lu et al. 2009), and salinity stress (Zhang et al. 2006). Tolerance against salinity stress can be increased by foliage treatment of ABA in rice (Bohra et al. 1995; Gurmani et al. 2013; Sripinyowanich et al. 2013).

In rice, tolerance of rice calli against osmotic, saline, and freezing stresses was increased by pretreatment with ABA (Martinez-Beltran and Manzur 2005). Recent studies suggested that exogenous application of ABA induces antioxidant system to upgrade rice chilling tolerance (Wang et al. 2013). It was also noted by Gurmani et al. (2013), during salinity condition, that rice survival and growth can be promoted by ABA pretreatment. Increased K/Na ratio following exogenous ABA application is interconnected with the promotion of salinity tolerance in rice (Bohra et al. 1995). Gurmani et al. (2013) reported that tolerance against salinity stress can be increased with exogenous ABA application by reducing Na transport and, thus, the Na/K ratio in rice. It was observed by Li-Xing et al. (2015) that increase in survival rate and seedling growth of rice under alkalinity stress was considerably improved with ABA seed priming.

Gurmani et al. (2006) reported that presoaking of rice seeds in ABA-containing solution increases tolerance to salinity and grain yield by 21% in saline fields. Further, viability rate and growth under salinity stress conditions can be enhanced by pretreatment of rice seedlings with ABA (Li et al. 2010; Gurmani et al. 2013; Sripinyowanich et al. 2013). In the previous research, it was showed that ABA priming has significant effect to address alkaline stress in rice; further, it shows significant improve in the survival rate, root growth, and biomass production under alkalinity stress situation due to pretreatment of rice seedlings with ABA. Under alkaline fields, priming effect of ABA was attributed by physiological analysis to minimize Na^+/K^+ ratio, water loss, and cell membrane injury. Exogenously applied AM1/quinabactin was helpful to induce plant tolerance against cold stresses and dried conditions.

Viable alternatives are provided by ABA agonists to ABA for agricultural use, while their effects on the plants to respond to SA stress remained unclear. Additionally, to strengthen the ABA signaling pathway, genetic and molecular alteration of regulatory components could also have valuable approach in improving the tolerance of plants against stresses (Wani et al. 2016). It is reported that occurrence of leaf withering reduced by 22%–50% and seedling death decreased by 11%–17% with exogenous application of ABA compared to the control treatment. A more prominent effect of ABA application was found in the harder SA field, which revealed that there is significant improve in plant growth and the final grain yield by 8%–55% with ABA application. ABA treatment induces the expressions of the ABA-responsive genes SalT and OsWsi18, and strengthens the induced levels until 8d after transplanting. The ABA enhanced tolerance against acidity in *Oryza sativa* by increasing the photosynthetic rate.

It was found by Luo et al. (2011) that the ABA content can be enhanced by applying ABA (10 μM) on the roots of rice. It also contributes to rectify rice ability to tolerate in the environments prone to acidic rain, as there is a substantial positive interconnection between plant tolerance mechanism and ABA deposition. According to López et al. (2008), exogenous application of 10 μM ABA could enhance cell division via cell proliferation, cell extension, and cell differentiation through enhancing the levels of plant growth enhancers like GA3, IAA, and ZT. Exogenously used ABA in an amount of 10 μM alleviates SAR-induced delay on the growth of roots. However, the levels of GA3, IAA, and ZT in the roots of rice declined by exogenous 100 μM ABA under SAR (pH 4.5 or 3.5). These results showed that elevated quantity of ABA may speed up the reduction in these hormones by optimizing the efficiency of some antioxidant enzymes (Seo et al. 2006; Zentella et al. 2007).

Low concentration of exogenous ABA (10 μM) increases endogenous hormonal contents (ABA, IAA, GA3, and ZT) by mitigating the deleterious effects of acidic rains on morphological and yield-related characteristics of rice, which elevates the uptake of ions of macronutrients and reduces the ammonium and Ca contents. Exogenous application of 10 μM ABA helped to maintain the uptake of nutrients to the roots of rice through optimizing the functioning of plasma membrane H + -ATPase.

It was identified that endogenous ABA content is higher in salinity-tolerant lines than in salinity-sensitive lines (Perales et al. 2005). ABA-responsive genes improve tolerance in plants significantly to salinity stress by overexpressing some transgenic rice plants (Xiang et al. 2008). Proteomic evidences showed that increase in ABA results in improving tolerance in rice seedlings against salinity (Li et al. 2010). Pons et al. (2013) demonstrated that rice cell lines selected for their various tolerance to salt to measure the response to ABA of H+ pumps and Na+/H+ antiporters correlated with the plasma membrane and the tonoplast. In salt stress condition, ABA acts synergistically with salinity on H+ pumping and antagonistically with stimulation of Na+/H+ antiport. It seems that cytoplasmic ion homeostasis in rice cell lines is managed by ABA application, which involves in tolerance responses to salinity.

15.7 Paclobutrazol (PBZ)

Paclobutrazol (PBZ) comes under the category of plant growth retarder and belongs to the group triazole. Different triazoles, like other plant growth regulators, exhibit a number of growth regulatory effects. Tolerance of various plant species can be improved with triazoles in all kinds of stresses (biotic and abiotic), involving infections due to fungi, water moisture stress, pollution in the air, and high or low temperature stress, through decreasing the damages to plants from oxidative stress and by regulating the functioning of antioxidant (Bahram 2009).

Generally, foliar application is adopted while treating the plants with PBZ. Pan (2013) reported that there was an increase in spikelets number, rate of seed setting, and yield in the studied rice varieties when treated with PBZ (50 mg per liter) during crop heading stage. Moreover, head rice rate and amylase content were also improved by PBZ treatment of the studied rice genotypes Peizataifeng and Huayou 86. Peng et al. (2011) observed that PBZ-treated rice plants performed better in terms of producing more grains, more seeds set, and higher 1000-seed weight when matched with the control treatment. Moreover, PBZ showed resistance against lodging; its application obviously increases root biomass and root activity to enhance phosphorus and potassium accumulation in rice stem, leaves, and grains. It was demonstrated by Pan (2013) that SOD and POD activities can be promoted by spraying PBZ and can be reduced by the accumulation of MDA in flag leaves at late growth stages.

Application of PBZ moderately alleviated the damaging effects of rice senescence by improving the activity of enzymatic antioxidants and modifying antioxidant system, which assisted in sustaining plant growth. It was also recorded that vegetative growth of rice plants can effectively decrease by application of paclobutrazol, which can also increase chlorophyll content in rice plants. Less photosynthates were supplied for vegetative growth of rice plants and more photosynthates allocated for seed development of rice when rice seedlings were treated with paclobutrazol compared to control plants or those plants treated with gibberellin (Dewi et al. 2016). Syahputra et al. (2013) reported that gibberellin contents in the leaves were decreased by applying paclobutrazol with concentrations of 200 mg/L to 600 mg/L to rice plant during preanthesis stage compared to that of control. According to Dewi et al. (2016), black rice plants treated with either 25 or 50 ppm paclobutrazol reduce leaf senescence, and more green leaves were found in treated plants compared to control. As a result, increase in the activity of oxidative enzymes prevented the cell maturation. Previous observation showed higher root activity in rice and wheat treated with plant growth regulators (Zhao et al. 2006).

The plant performance might be improved by the application of paclobutrazol under stressful condition through intriguing root activity of the plants. Spraying 200 ppm paclobutrazol on the rice seedlings resulted in 50% increment in survival percentage as compared to control during submerged conditions. The elevated seedling survival was apparently due to less energy utilized for seedling elongation. The similar kind of regulatory effect of paclobutrazol was also observed in moisture stress conditions for production of anaerobic proteins and for the repair of membrane integrity (Chon et al. 2000).

References

Ahmad A, Ashfaq M, Rasul G, Wajid SA, Khaliq T, Rasul F, Saeed U, Rahman MH, Hussain J, Baig IA, Naqvi AA, Bokhari SAA, Ahmad S, Naseem W, Hoogenboom G, Valdivia RO (2015) Impact of climate change on the rice–wheat cropping system of Pakistan. In: Hillel D, Rosenzweig C (eds.), Handbook of Climate Change and Agro-ecosystems: The Agricultural Modeling Intercomparison and Improvement Project (AgMIP) Integrated Crop and Economic Assessments, Published by Imperial College Press and the American Society of Agronomy, pp. 219–258

Ahmad S, Abbas G, Ahmed M, Fatima Z, Anjum MA, Rasul G, Khan MA, Hoogenboom G (2019) Climate warming and management impact on the change of rice-wheat phenology in Punjab, Pakistan. Field Crops Res 230:46–61

Ahmad S, Ahmad A, Ali H, Hussain A, Garcia A, Khan MA, Zia-Ul-Haq M, Hasanuzzaman M, Hoogenboom G (2013) Application of the CSM-CERES-Rice model for evaluation of plant density and irrigation management of transplanted rice for an irrigated semiarid environment. Irrig Sci 31(3):491–506

Ahmad S, Ahmad A, Soler CMT, Ali H, Zia-Ul-Haq M, Anothai J, Hussain A, Hoogenboom G, Hasanuzzaman M (2012) Application of the CSM-CERES-Rice model for evaluation of plant density and nitrogen management of fine transplanted rice for an irrigated semiarid environment. Precis Agric 13(2):200–218

Ahmad S, Ahmad A, Zia-ul-Haq M, Ali H, Khaliq T, Anjum MA, Khan MA, Hussain A, Hoogenboom G (2009) Resources use efficiency of field grown transplanted rice (*Oryza sativa* L.) under irrigated semiarid environment. J Food Agric Environ 7(2):487–492

Ahmad S, Hasanuzzaman M (2012) Integrated effect of plant density, N rates and irrigation regimes on the biomass production, N content, PAR use efficiencies and water productivity of rice under irrigated semiarid environment. Not Bot Horti Agrobot Cluj-Napoca 40(1):201–211

Ahmad S, Zia-ul-Haq M, Ali H, Shad SA, Ammad A, Maqsood M, Khan MB, Mehmood S, Hussain A (2008) Water and radiation use efficiencies of transplanted rice (*Oryza sativa* L.) at different plant densities and irrigation regimes under semi-arid environment. Pak J Bot 40(1): 199–209

Ahmed M, Ahmad S (2019) Carbon dioxide enrichment and crop productivity. In: Hasanuzzaman M (ed) Agronomic crops, 2nd edn. Springer, Singapore, pp 31–46

Ahmed M, Ahmad S (2020) Systems modeling. In: Ahmed M (ed) Systems modeling. Springer, Singapore, pp 1–44

Ahmed M, Ahmad S, Raza MA, Kumar U, Ansar M, Shah GA, Parsons D, Hoogenboom G, Palosuo T, Seidel S (2020a) Models calibration and evaluation. In: Ahmed M (ed) Systems modeling. Springer, Singapore, pp 151–178

Ahmed M, Ahmad S, Waldrip HM, Ramin M, Raza MA (2020b) Whole farm modeling: a systems approach to understanding and managing livestock for greenhouse gas mitigation, economic viability and environmental quality. In: in: Waldrip HM, Pagliari PH, He Z (eds.), Animal manure. ASA special publication 67. Madison, WI: American Society of Agronomy and Soil Science Society of America, pp. 345–371

Ahmed M, Fayyaz-ul-Hassan M, Ahmad S (2017) Climate variability impact on rice production: adaptation and mitigation strategies. In: Stockle C (ed) Ahmed M. Adaptation and Mitigation for Agricultural Sustainability. Springer International Publishing Switzerland, Quantification of Climate Variability, pp 91–111

Akram R, Fahad S, Masood N, Rasool A, Ijaz M, Ihsan MZ, Maqbool MM, Ahmad S, Hussain S, Ahmed M, Kaleem S, Sultana SR, Mubeen M, Saud S, Kamran M, Nasim W (2019) Plant growth and morphological changes in rice under abiotic stress. In: Fujita M, Nahar K, Biswas JK (eds) Hasanuzzaman M. Woodhead Publishing, Advances in Rice Research for Abiotic Stress Tolerance, pp 69–85

Akram R, Turan V, Hammad HM, Ahmad S, Hussain S, Hasnain A, Maqbool MM, Rehmani MIA, Rasool A, Masood N, Mahmood F, Mubeen M, Sultana SR, Fahad S, Amanet K, Saleem M,

Abbas Y, Akhtar HM, Hussain S, Waseem F, Murtaza R, Amin A, Zahoor SA, Sami Ul Din M, Nasim W (2018) Fate of organic and inorganic pollutants in paddy soils. In: Hashmi MZ, Varma A (eds.), Environmental pollution of paddy soils. Springer Nature Switzerland AG, pp. 197–214

Alam SM, Shereen A, Khan M (2002) Growth response of wheat cultivars to naphthalene acetic acid (NAA) and ethrel. Pak J Bot 34(2):135–137

Ambarawati AD, Somantri IH, Utami DW, Apriana A, Moeljopawiro S (2009) Rice anther culture to develop double haploid population and blast resistant lines. J Agron Biogen 5(2):71–77

Arshad M, Frankenberger Jr WT (1993) Microbial production of plant growth regulators. Soil Microbial Echol. F.B. Metting Jr. (Ed), Marcel Dekker Inc. New York, pp. 307–347

Arteca RN (1995) Plant growth substances: principles and applications. Chapman and Hall, New York, p 332

Asborno MD, Vidal AA, Bezus R, Beltrano J (1999) Rice: temperature and gibberelic acid effect on initial growth stages. Agro Ciencia 15(1):47–53

Bahram B (2009) Amelioration of chilling stress by paclobutrazol in watermelon seedlings. Sci Hortic-Amsterdam 121:144–148

Bai XF, Cai YP, Nie F (1989) Relationship between abscisic acid and grain filling of rice and wheat. Physiol Commun (China) 3:40–41

Bano S, Jabeen M, Rahim F, Ilahi I (2005) Callus induction and regeneration in seed explants of rice (*Oryza sativa* cv. Swat-II). Pak J Bot 37(3):829–836

Bevilaqua GAP, Peske ST, Santos-Filho BG, Baudet LML (1993) Performance of irrigated rice seed treated with growth regulator. I. Effect on emergence in the field. Revista Brasileira de Sementes 15(1):67–74

Bevilaqua GAP, Scwengber JE, Peske ST (1996) Cultivar differences in treatment of rice seeds with gibberellic acid. Lavoura Arrozeira 49(428):8–12

Biswas AK, Mondal SK (1986) Regulation by kinetin and abscisic acid of correlative senescence in relation to grain maturation, source-sink relationship and yield of rice (Oryza sativa L.). Plant Growth Regul 4(3):239–245

Bohra JS, Dörffling H, Dörffling K (1995) Salinity tolerance of rice (Oryza sativa L.) with reference to endogenous and exogenous abscisic acid. J Agron Crop Sci 174:79–86

Chen D, Gunawardena TA, Naidu BP, Fukai S, Basnayake J (2005) Seed treatment with gibberellic acid and glycinebetaine improves seedling emergence and seedling vigour of rice under low temperature. Seed Sci Technol 33(2):471–479

Chen D, Zhao J (2008) Free IAA in stigmas and styles during pollen germination and pollen tube growth of Nicotiana tabacum. Physiol Plantarum 134:202–215

Chen HH, Li PH, Brenner ML (1983) Involvement of abscisic acid in potato cold acclimation. Plant Physiol 71:362–365

Cheng JP, Zhao F, You AB, You AQ, Wu JP (2011) Effects of spraying exogenous gibberellin (GA_3) at heading stage on heading uniformity and yield of rice (Oryza sativa L.). J Huazhong Agric Univ 30:657–662

Chon NM, Koseki NN, Hirata Y, Saka H, Abe H (2000) Effects of brassinolide on coleoptile and leaf growth in rice seedlings. Plant Prod Sci 3(4):360–365

Devasagayam MM, Jayapaul P, Balasubramanian R (1996) Effect of seed treatment with growth regulators on seedling characters of wet seeded rice. J Ecotoxicol Environ Monit 6(3/4): 199–202

Dewi K, Agustina RZ, Nurmalika F (2016) Effects of blue light and paclobutrazol on seed germination, vegetative growth and yield of black rice (*Oryza sativa* L. 'Cempolreng'). Biotropia 23(2):85–96

Dong CF, Xu NX, Ding CL, Gu HR, Zhang WJ, Li XL (2016) Suitable gibberellic acid treatment for double-purpose rice (Oryza sativa L.) varieties at different harvest times. Field Crops Res 193:178–185

Elankavi S, Kuppuswamy G, Vaiyapuri V, Raman R (2009) Effect of phytohormones on growth and yield of rice. Oryza 46:310–313

Erogul D, Sen F (2016) The effect of preharvest gibberellic acid applications on fruit quality of 'Angelino' plums during storage. Sci Hortic 202:111–1116

Fahad S, Noor M, Adnan M, Khan MA, Rahman I, Alam M, Khan IA, Ullah H, Mian IA, Hassan S, Saud S, Bakhat HF, Hammad HM, Ahmad S, Nasim W (2019) Abiotic stress and rice grain quality. In: Fujita M, Nahar K, Biswas JK (eds) Hasanuzzaman M. Woodhead Publishing, Advances in Rice Research for Abiotic Stress Tolerance, pp 571–583

Fatima Z, Ahmad M, Hussain M, Abbas G, Ul-Allah S, Ahmad S, Ahmed N, Ali MA, Sarwar G, Ehsan Ul Haque ME, Iqbal P, Hussain S (2020) The fingerprints of climate warming on cereal crops phenology and adaptation options. Sci Rep 10:18013

Gavino RB, Pi Y, Abon CC (2008) Application of gibberellic acid (GA3) in dosage for three hybrid rice seed production in the Philippines. J Agric Technol 4:183–192

Gollagi SG, Lokesha R, Dharmpal S, Sathish BR (2019) Effects of growth regulators on growth, yield and quality of fruit crops: a review. Journal of Pharmacognosy and Phytochemistry 8(4): 979–981

Gurmani AR, Banno A, Saleem M (2006) Effect of growth regulator on growth, yield and ion accumulation of rice. Pak J Bot 38(5):1415–1424

Gurmani AR, Bano A, Ullah N, Khan H, Jahangir M, Flowers TJ (2013) Exogenous abscisic acid (ABA) and silicon (Si) promote salinity tolerance by reducing sodium (Naþ) transport and bypass flow in rice ('Oryza sativa' indica). Aust J Crop Sci 7:1219–1226

Hafeez-ur-Rehman WT, Nawaz A, Awan MI, Ijaz M, Hussain M, Ahmad S, Farooq M (2019) Direct seeding in rice: problems and prospects. In: Hasanuzzaman M (ed.), Agronomic crops, Volume 1: Production technologies. Springer Nature Singapore Pte Ltd, pp. 199–222

Hall MA, Moshkov IE, Novikova GV, Mur LAJ, Smith AR (2001) Ethylene signal perception and transduction: multiple paradigms? Biol Rev 76:103–128

Han H, Yang W (2009) Influence of uniconazole and plant density on nitrogen content and grain quality in winter wheat in South China. Plant Soil Environ 55(4):159–166

Hare PD, Cress WA, van Staden J (1997) The involvement of cytokinins in plant responses to environmental stress. Growth Regul 23:79–103

Helms RS, Dilday RH, Carlson RD (1990) Using GA3 seed treatment in direct seeded rice in southern U.S.A. In: Direct seeded flooded rice in the tropics: selected papers from international rice research conference. International Rice Research Institute, Manila, Philippines, pp 113–114

Hoffmann-Benning S, Kende H (1992) On the role of abscisic acid and gibberellin in the regulation of growth in rice. Plant Physiol 99:1156–1161

Hussain M, Ahmad S, Hussain S, Lal R, Ul-Allah S, Nawaz A (2018) Rice in saline soils: physiology, biochemistry, genetics and management. Adv Agron 148:231–287

Hutchison CE, Kieber JJ (2002) Cytokinin signaling in Arabidopsis. Plant Cell (Suppl) 14:S47–S59

Jackson MB, Ram PC (2003) Physiological and molecular basis of susceptibility and tolerance of rice plants to complete submergence. Ann Bot 91:227–241

Kato T, Sakurai N, Kuraishi S (1993) The changes of endogenous abscisic acid in developing grains of two rice cultivars with different grain size. Jpn J Crop Sci 62:456–461

Khan MA, Ahmad S, Raza A (2019a) Integrated weed management for agronomic crops. In: Hasanuzzaman M (ed.), Agronomic crops; volume 2 Springer nature Singapore Pte ltd., pp. 257–281

Khan SM, Ali S, Nawaz A, Bukhari SAH, Ejaz S, Ahmad S (2019b) Integrated pest and disease management for better agronomic crop production. In: Hasanuzzaman M (ed.), Agronomic crops; volume 2 Springer Nature Singapore Pte ltd. pp. 385–428

Kim SK, Son TK, Park SY, Lee IJ, Lee BH, Kim HY, Lee SC (2006) Influences of gibberellin and auxin on endogenous plant hormone and starch mobilization during rice seed germination under salt stress. J Environ Biol 27(2):181–186

Li F, Guo S, Zhao Y, Chen D, Chong K, Xu Y (2010) Overexpression of a homopeptide repeat-containing bHLH protein gene (*OrbHLH001*) from Dongxiang wild rice confers freezing and salt tolerance in transgenic *Arabidopsis*. Plant Cell Rep 29:977–986

Li-Xing W, Bing-Sheng L, Ming-Ming W, Hong-Yuan M, Hao-Yu Y, Xiao-Long L, Chang-Jie J, Zheng-Wei L (2015) Priming effect of abscisic acid on alkaline stress tolerance in rice (Oryza sativa L.) seedlings. J Plant Physiol Biochem 90:50–57

López MA, Bannenberg G, Castresana C (2008) Controlling hormone signaling is a plant and pathogen challenge for growth and survival. Curr Opin Plant Biol 11(4):420–427

Lu SY, Su W, Li HH, Guo ZF (2009) Abscisic acid improves drought tolerance of triploid bermudagrass and involves H_2O_2 and NO induced antioxidant enzyme activities. Plant Physiol Biochem 47:132–138

Luo L, Xu F, Weng H, Hong S, Duan L, Li Z (2011) Inducing effects and its biological mechanisms of ABA on the chilling resistance of sweet pepper seedlings. Acta Bot Boreali-Occident Sin 31(1):94–100

Maksimov GB, Medvedev SS, Alamgir ANM (1979) Effect of kinetin on corn seedling roots. Mysore J Agric Sci 29:208–212

Martinez-Beltran J, Manzur CL (2005) Overview of salinity problems in the world and FAO strategies to address the problem. In: proceedings of the international salinity forum. Riverside, California, p 311e313

Métraux J-P, Kende H (1983) The role of ethylene in the growth response of submerged deep water rice. Plant Physiol 72:441–446

Métraux J-P, Kende H (1984) The cellular basis of the elongation response in submerged deep-water rice. Planta 160:73–77

Mergemann H, Sauter M (2000) Ethylene induces epidermal cell death at the site of adventitious root emergence in rice. Plant Physiol 124:609–614

Misra G, Sahu G (1957) Effect of plant growth substances on an early variety. Buliton of the Torrey Botanical Club 84(6):142–149

Mitchell P, Sheehy J (2006) Supercharging rice photosynthesis to increase yield. New Phytol 171: 688–693

Naz S, Fatima Z, Iqbal P, Khan A, Zakir I, Noreen S, Younis H, Abbas G, Ahmad S (2019) Agronomic crops: types and uses. In: Hasanuzzaman M (ed.), Agronomic crops. Volume 1: Production technologies. Springer nature Singapore Pte. Ltd., pp. 1–18

Nieto KF, Frankenberger WT Jr (1989) Biosynthesis of cytokinins in soil. Soil Sci Soc Am J 53: 735–740

Pan S (2013) Roles of plant growth regulators on yield, grain qualities and antioxidant enzyme activities in super hybrid rice (Oryza sativa L.). Rice 6:9

Pan R, Zhao RJ, Tian X (1999) Comparative effect of IBA, BSAA, and 5,6- cl-IAA-me on the rooting of hypocotyls in mung bean. Plant Growth Regul 27:91–98

Peerzada AM, Bukhari SAH, Dawood M, Nawaz A, Ahmad S, Adkins S (2019) Weed management for healthy crop production. In: Hasanuzzaman M (ed.), Agronomic crops; volume 2 Springer Nature Singapore Pte ltd., pp. 225–256

Peng ZP, Huang JC, Yu JH, Yang SH, Li WY (2011) Effects of PP333 and nutrient elements applied on yields and root growth of rice. Chin Agric Sci Bull 27(05):234–237

Perales L, Arbona V, Gomez-Cadenas A, Cornejo MJ, Sanz A (2005) A relationship between tolerance to dehydraton of rice cell lines and ability for ABA synthesis under stress. Plant Physiol Biochem 43:786–792

Pons R, Cornejo MJ, Sanz A (2013) Is ABA involved in tolerance responses to salinity by affecting cytoplasm ion homeostasis in rice cell lines? Plant Physiol Biochem 62:88–94

Prasad PVV, Boote KJ, Allen LH, Sheehy JE, Thomas JMG (2006) Species, ecotype and cultivar differences in spikelet fertility and harvest index of rice in response to high temperature stress. Field Crops Res 95:398–411

Raskin H, Kende (1984) Regulation of growth in stem sections of deep-water rice. Planta 160:66–72

Razaq M, Shah FM, Ahmad S, Afzal M (2019) Pest management for agronomic crops. In: Hasanuzzaman M (ed.), agronomic crops; volume 2 springer nature Singapore Pte ltd. pp. 365–384

Reynolds M, Foulkes J, Furbank R, Griffiths S, King J, Murchie E, Parry M, Slafer G (2012) Achieving yield gains in wheat. Plant Cell Environ 35:1799–1823

Robertson AJ, Ishikawa M, Gusta LV, MacKenzie SL (1994) Abscisic acid induced heat tolerance in Bromus Inermis Leyss cell-suspension cultures (heat-stable, abscisic acid-responsive polypeptides in combination with sucrose confer enhanced thermostability). Am Soc Plant Biol 105:181–190

Roly ZY, Mahmudul I, Pallob ES, Saiful IA, Shah S, Dipesh D, Mujjammil EH, Khalekuzzaman B (2014) In vitro callus induction and regeneration potentiality of aromatic rice (Oryza sativa L.) cultivars in differential growth regulators. Int J Appl Sci Biotechnol 2(2):160–167

Saika H, Okamoto M, Miyoshi K, Kushiro T, Shinoda S, Jikumaru Y, Fujimoto M, Arikawa T, Takahashi H, Ando M, Arimura S, Miyao A, Hirochika H, Kamiya Y, Tsutsumi N, Nambara E, Nakazono M (2007) Ethylene promotes submergence-induced expression of OsABA8ox1, a gene that encodes ABA 80 -hydroxylase in rice. Plant Cell Physiol 48:287–298

Sajid H, Chunquan Z, Jie H, Jing H, Lianfeng Z, Xiaochuang C, Satyabrata N, Maqsood AK, Qingduo L, Yali K, Qianyu J, Junhua Z (2020) Ethylene response of salt stressed rice seedlings following Ethephon and 1-methylcyclopropene seed priming plant growth regulation. https://doi.org/10.1007/s10725-020-00632-1

Sakata T, Oshino T, Miura S, Tomabechi M, Tsunaga Y, Higashitani N, Miyazawa Y, Takahashi H, Watanabe M, Higashitani A (2010) Auxins reverse plant male sterility caused by high temperatures. Proc Natl Acad Sci 107:8569–8574

Sakeena I, Salam MA (1989) Influence of potassium and kinetin on protein partitioning in rice. Int Rice Res Newsl 14(3):29–30

Sarwar N, Ali H, Ahmad S, Ehsan U, Ahmad S, Mubeen K, Hill JE (2013a) Water wise rice cultivation on calcareous soil with the addition of essential micronutrients. J Anim Plant Sci 23(1):244–250

Sarwar N, Ali H, Maqsood M, Ehsan U, Shahzad AN, Shahzad M, Mubeen K, Shahid MA, Ahmad S (2013b) Phenological response of rice plants to different micronutrients application under water saving paddy fields on calcareous soil. Turk J Field Crops 18(1):52–57

Seo M, Hanada A, Kuwahara A, Endo A, Okamoto M, Yamauchi Y, North H, Marion-Poll A, Sun T, Koshiba T, Kamiya Y, Yamaguchi S, Nambara E (2006) Regulation of hormone metabolism in Arabidopsis seeds: phytochrome regulation of abscisic acid metabolism and abscisic acid regulation of gibberellin metabolism. Plant J 48(3):354–366

Shahnewaz S, Bari MA, Siddiqu NAE, Khatun N, Rahman MH, Haque ME (2003) Induction of haploid rice plants through in vitro anther culture. Pak J Biol Sci 6(14):1250–1252

Shahzad AN, Ahmad S (2019) Tools and techniques for nitrogen management in cereals. In: Hasanuzzaman M (ed.), agronomic crops; volume 2 springer nature Singapore Pte ltd., pp. 111–126

Sharma A (2018) Morphology response of growth regulators on callus of Japonica rice through anther culture. Clin Biotechnol Microbiol:424–442

Sheng YH, Shen YQ, Sheng MZ, Cao HF, Wu YJ, Fan HW (1993) The biological effect of treating seeds with multi-effect triazole (MET) in raising rice seedlings with plastic sheet mulch. Acta Agric Shanghai 9:59–63

Shi-Hua V, Ben-Yi C, Jian-Li W, Wei-Feng S, Shi-Hua C (2006) Review and prospects in rice breeding and extension in China. Rice Sci 13:1–8

Shukla R, Dube A, Koshy EP (2014) Production of high quality embryogenic callus of rice. The Bioscan 9(3):1077–1080

Sripinyowanich S, Klomsakul P, Boonburapong B, Bangyeekhun T, Asami T, Gu H et al (2013) Exogenous ABA induces salt tolerance in indica rice (Oryza sativa L.): the role of OsP5CS1 and OsP5CR gene expression during salt stress. Environ Exp Bot 86:94–105

Steffens B, Sauter M (2005) Epidermal cell death in rice is regulated by ethylene, gibberellin, and abscisic acid. Plant Physiol 139:713–721

Syahputra BAS, Sinniah UR, Syed Rastan SO, Ismail MR (2013) Changes in gibberellic acid (GA) content in 3 Oryza sativa due to paclobutrazol treatment. J Food Pharmac Sci 1:14–17

Tariq M, Ahmed M, Iqbal P, Fatima Z, Ahmad S (2020) Crop phenotyping. In: Ahmed M (ed.), Systems Modeling, Springer Nature Singapore Pte Ltd. pp. 45–60

Thuc LV, Duy MV, Hoi N, TandNhung MT (2016) The effects of a potential retardant on the stem's hardness, growth and yield of IR50404 rice. Vietnam J Agri Sci 14(10):1557–1563

Tian DC (1991) Out crossing cultivation of Rice. Sichuan Science and Technology Press, Chengdu, China

Van der Straeten D, Zhou Z, Prinsen E, van Onckelen HA, van Montagu MC (2001) A comparative molecular-physiological study of submergence response in lowland and Deepwater rice. Plant Physiol 125:955–968

Viera AR, Viera MDG, Viera GC, Fraga AC, Oliveira JA, Santos CD (2002) Action of gibberellic acid on dormancy and activity of a-amylase in rice seeds. Rev Bras Sementes 24(2):43–48

Wahyuni S, Sinniah UR, Yusop MK, Amarthalingam R (2003) Improvement of seedling establishment of wet seeded rice using GA3, IBA as seed treatment. Indones J Agric Sci 4(2):56–62

Wang GJ, Miao W, Wang JY, Ma DR, Li JQ, Chen WF (2013) Effects of exogenous abscisic acid on antioxidant system in weedy and cultivated rice with different chilling sensitivity under chilling stress. J Agron Crop Sci 199:200–208

Wang Z, Yang J, Zhu QZ, Zhang Y, Lang X, Wang (1998) Reason for poor grain plumpness in inter subspecific hybrid rice. Acta Agron Sin 24(6):782–787

Wani SH, Kumar V, Shriram V, Sah SK (2016) Phytohormones and their metabolic engineering for abiotic stress tolerance in crop plants. Crop J 4:162–176

Wasaya A, Yasir TA, Ijaz M, Ahmad S (2019) Tillage effects on agronomic crop production. In: Hasanuzzaman M (ed.), agronomic crops; volume 2 springer nature Singapore Pte ltd., pp. 73–99

Watanabe H, Siagusa M (2004) Effects of combine application of ethephon and gibberellin on growth and nutrient uptake of rice seedling growing under direct seedling conditions. Proc. 4 th Int'l crop Sci, Brisbane, Australia. Available at: http://www.cropscience.com

Xiang Y, Tang N, Du H, Ye H, Xiong LZ (2008) Characterization of OhsZIP23 as a key player of the basic leucine zipper transcription factor family for conferring abscisic lacid sensitivity and salinity and drought tolerance in rice. Plant Physiol 148:1938–1952

Xiang ZF, Yang WY, Ren WJ, Wang XC (2003) Effects of uniconazole (5% WP) on grain filling and grain yield of rice. Hybrid Rice 19(5):50–55

Yang J, Wang Z, Zhu Q, Lang Y (1999) Regulation of ABA and GA to rice grain filling. Acta Agron Sin 25:341–348

Yang S-H, Choi D (2006) Characterization of genes encoding ABA 80 -hydroxylase in ethylene-induced stem growth of Deepwater rice (Oryza sativa L.). Biochem Biophys Res Commun 350: 685–690

Yoshida R (1987) Effect of Zeatin and abscisic acid on grain ripening in rice. Proceedings of the 14th Annual Meeting Plant Growth Regulator Society of America, 2±6 August 1987, Honolulu, HI, USA, pp 91–96

Yuan Y, Qian HM, Wang Y, Shi YM, Tang DQ (2012) Hormonal regulation of freesia cutflowers and FhACS1. Sci Hortic-Amsterdam 143:75–81

Zahir ZA, Asghar HN, Arshad M (2001) Cytokinin and its precursors for improving growth and yield of rice. Soil Biol Biochem 33:405–408

Zahir ZA, Asghar HN, Malik MAR, Arshad M (1999) Effect of cytokinin precursors on growth and yield of rice (Oryza sativa L.). J Anim Plant Sci 9:53–56

Zahoor SA, Ahmad S, Ahmad A, Wajid A, Khaliq T, Mubeen M, Hussain S, Sami Ul Din M, Amin A, Awais M, Nasim W (2019) Improving water use efficiency in agronomic crop production. In: Hasanuzzaman M (ed.), Agronomic crops; Volume 2 Springer Nature Singapore Pte Ltd., pp. 13–29

Zentella R, Zhang ZL, Park M, Thomas SG, Endo A, Murase K, Fleet CM, Jikumaru Y, Nambara E, Kamiya Y, Sun TP (2007) Global analysis of DELLA direct targets in early gibberellin signaling in Arabidopsis. Plant Cell 19(10):3037–3057

Zhang JH, Jia WS, Yang JC, Ismail AM (2006) Role of ABA in integrating plant responses to drought and salt stresses. Field Crop Res 97:111–119

Zhang WX, Peng CR, Sun G, Zhang FQ, Hu SX (2007) Effect of different external phytohormones on leaves senescence in late growth period of late-season rice. Acta Agric Jiangxi 19(2):11–13

Zhao XF, Fang ZG, Gao ZM (2006) Effects of paclobutrazol (PP333) on root vigor and IAA oxidase and peroxidase activities in leaf of rice and wheat seedlings. Guangxi Agric Sci 37:379–381

Zheng HB, Wang XM, Li YX, Huang GF, Tang QY, Tang JW (2018) Contributions of photosynthetic organs to the seed yield of hybrid rice: the effects of gibberellin application examined by carbon isotope technology. Seed Sci Technol 46:533–546

Zhou H, Li PB, Xie WB, Hussain S, Li YB, Xia D, Zhao H, Sun SY, Chen JX, Ye H (2017) Genome-wide association analyses reveal the genetic basis of stigma exsertion in rice. Mol Plant 10:634–644

Salinity Tolerance in Rice 16

Usman Khalid Chaudhry, Niaz Ahmed, Muhammad Daniyal Junaid, Muhammad Arif Ali, Abdul Saboor, Subhan Danish, Sajjad Hussain, and Shakeel Ahmad

Abstract

Rice is the main staple food crop across the globe. Among abiotic stresses, salinity stress is increasing at an alarming rate. It inhibits rice growth and yield as rice is a sensitive crop to salinity. It influences various physiological functioning of the rice, which results in retarded growth and ultimately gives poor yield. In this chapter, we highlighted influence of physiological changes and effect on rice grain in response to salinity stress and their adaptation strategies. Moreover, currently numerous studies have explored the molecular response/changes in rice to cope with salinity stress. In this regard, we explained the abscisic acid and signaling under salinity stress along with the functions of transcription factors. Final part of this chapter covers the importance of modern breeding techniques to screen and develop salt tolerant cultivars within a short period of time as compared to conventional breeding approaches.

U. K. Chaudhry · M. D. Junaid
Department of Agricultural Genetic Engineering, Faculty of Agricultural Sciences and Technologies, Niğde Ömer Halisdemir University, Niğde, Turkey

N. Ahmed · M. A. Ali (✉) · A. Saboor · S. Danish
Department of Soil Science, Faculty of Agricultural Sciences and Technology, Bahauddin Zakariya University, Multan, Punjab, Pakistan
e-mail: arif1056@bzu.edu.pk

S. Hussain
Department of Horticulture, Faculty of Agricultural Sciences and Technology, Bahauddin Zakariya University, Multan, Punjab, Pakistan

S. Ahmad
Department of Agronomy, Faculty of Agricultural Sciences and Technology, Bahauddin Zakariya University, Multan, Pakistan
e-mail: shakeelahmad@bzu.edu.pk

N. Sarwar et al. (eds.), *Modern Techniques of Rice Crop Production*,
https://doi.org/10.1007/978-981-16-4955-4_16

Keywords

Rice · Salinity · Abscisic acid · Transcription factors · Physiology · Growth · QTL mapping · Molecular markers

16.1 Introduction

Plants as being sessile grow under natural environmental conditions where so many factors are involved for their nurturing. So, any deviation from their favorable growth conditions at different growth stages exerts pressure on plant growth and yield (Ahmad et al. 2015, 2019, 2012, 2013, 2008, 2009; Ahmad and Hasanuzzaman 2012; Ahmed and Ahmad 2019, 2020; Shahzad and Ahmad 2019). Plants as being immutable suffer from multifarious environmental adversities in field (Chaudhry et al. 2021a). Various abiotic stress factors are involved for curtailing growth of plants. The prime climatic pressure that is under thorough investigation in context of its effect on plants is salinity stress (Junaid et al. 2021). It include excessive accumulation of salt concentrations in soil mainly due to anthropogenic activities and poor management practices (Shahzad et al. 2017). Severity of salinity stress is lethal for plant. Currently it is the main challenge for maintaining plant growth and crop productivity under such stress scenario for sustainable agriculture (Chaudhry et al. 2020). Soil salinity is the major threat to the agricultural crops especially rice, as it is native to freshwater marshes. Salinization is the main hurdle for rice growth due to inherent sensitivity of rice to salinity stress (Hasanuzzaman et al. 2018). According to Bohn et al. (1985), normal soil has pH = 4.5–7.5, electrical conductivity (EC) <4 dSm-1, sodium absorption ratio (SAR) <15, and exchangeable sodium percentage (ESP) <15 for optimum plant growth. However, during salt stress, soil consists of high levels of soluble salts, which inhibits optimum plant growth. More than 6% of world's land is saline or sodic and 20% of cultivable land has become salt stressed (Munns 2005). Numerous studies have been reported the behavior of rice to varying salinity stress at morpho-physiological and molecular levels. Beside this there are still research gaps that need to be explored about rice to exploit the potential of the local rice cultivars on saline soils. Among all the major abiotic stresses, salinity stress is one of the most brutal stress factors that results in complicated physiological changes, starting from ionic disequilibrium hyperosmotic stress, hyperionic, and ultimately disrupting the metabolic functioning of plants (Tomaz et al. 2020). Moreover, all these disruptions in physiological functioning of rice in response to salinity stress cause morphological changes in rice. Salinity stress leads severe reduction of plant growth and development, damage to membrane, ionic imbalance due to higher accumulation of Na^+ and Cl^- ions in plant, elevated lipid peroxidation, and finally higher production of reactive oxygen species, that is, superoxide radicals, hydrogen peroxide, and hydroxyl radicals (Chaudhry et al. 2021b). Area under salinity stress is increasing on yearly basis due to several factors, that is, climatic changes, rise in sea water levels, unnecessarily excessive irrigation without adopting drainage practices, and presence of underlying rocks possessing harmful salts. It is predicted that if the current situation of salinity stress would continue, there may be 50% loss to current cultivated agricultural land by the year 2050 (Jamil et al. 2011). Rice production in all over the world is greatly affected by salt stress. After drought, it is widely spread and a major factor in decreased rice productivity (Gregorio et al. 1997). Salinity hinders almost all the important processes occurring in the plant hence decreasing vegetative growth, fruit and flowering, vigor, and as a result economic yield (Sairam and Tyagi 2004). In various plant species, there are several mechanisms to counter salt stress including accumulation of compatible solutes (proline, etc.) and mineral ion homeostasis. Additionally, intensity of salt stress also defines the extent of plant response depending upon its genotype, concentration, and type. In order to differentiate salt tolerant and susceptible genotypes, screening for salt tolerant genotypes can play important role for future rice production. Though rice is considered as saline tolerant crop, various genotypes cannot withstand high levels of salinity. Rice crop has a wide genetic variability regarding saline tolerance, and as it is a complex trait, there are a number of salt responsive genes which control this trait (Zhu et al. 2001).

16.2 Salinity Stress Effect on Physiological Changes in Rice

Salt stress causes number of physiological disruptions in rice including photosynthesis rate, stomatal conductance, photosynthetically active radiation, relative water content, transpiration rate, and degradation of pigment (Cattivelli et al. 2008). According to Ramezani et al. (2012), water use efficiency (WUE) of rice is also significantly affected by salinity. Recent literature shows that WUE in rice is negatively correlated with increased salinity (Gholipoor et al. 2002). This may be due to the negative pleiotropic effects on the development and physiology of rice at biochemical and molecular level (Tester and Davenport 2003), which may result into abnormal functioning or plant death (Nishimura et al. 2011).

16.2.1 Stomatal Closure

In plants, immediate closure of stomata after sensing salt stress is the first response (Asim et al. 2021). Concentration of salts if increased (0–20 dSm-1) may decrease CO_2 assimilation and reduce plant growth which resultantly result causes low intracellular CO_2 and hence reducing photosynthesis rate (Amirjani 2010). When NaCl concentration increases in rhizosphere, the transpiration rate reduces, which causes lowered water potential in roots. Furthermore, this leads to lower ABA transportation from root to shoot and it induces stomatal closure to prevent dehydration of leaf tissues (Zheng et al. 2001). Zeinolabedin (2012) also concluded that under salinity stress, ABA biosynthesis occurs that closes the stomata. Stomatal action is controlled by redox action of ascorbate as it regulates the concentration of H_2O_2 (Chen et al. 2014). In most of plant species, salinity stress hinders photosynthetic activity (Dionisio-Sese and Tobita 2000). On the other hand, in monocot species like rice, it is believed that stomatal closure, rubisco inefficiency, degeneration of important cations, and reduction of sink activity are the main reasons behind reduced photosynthetic activity (Flowers and Yeo 1981). Salinity may affect stomatal conductance, with reduction in guard cell turgidity and CO_2 partial pressure in cell (Dionisio-Sese and Tobita 2000). Stomatal closure plays a key role in plants against salt stress.

16.2.2 Photosynthesis and Transport

Previous studies suggest that rice plants exhibit lower photosynthetic efficiency with disruption to photosystem II. Higher Na^+ and Cl^- damages the chlorophyll contents that may also hinder electron transport in PS II (Munns et al. 2006). Salt stress significantly decreases carotenoids and chlorophyll contents in rice leaves (Ahmed et al. 2020). According to Garcia et al. (2012), high salinity also reduces PS II efficiency and lowers K^+/Na^+ ratio. A study showed that under high salinity, there is higher nonphotochemical quenching in salt susceptible plants; however, reduced quantum yield of PS II and low photochemical quenching was observed (Dionisio-Sese and Tobita 2000). High photosynthetic rate has direct influence in sink strength of spikelets (Chen et al. 2013; Li et al. 2012).

16.3 Effect on Ionic Balance

Salinity may lead to ionic imbalance in plants, and it is may be due to competition of various ions like Na^+ and Cl^- with K^+, Ca^{2+}. NaCl causes specific ion toxicity in plants (Na^+ and Cl^-), and it leads to decrease of N, P, K, Ca, Mg and increase in Na^+/K^+, Na^+/Ca^{2+}, and Ca^{2+}/Mg^{2+} and Cl^-/NO_{3-} in plants and causes nutrient imbalance (Grattan and Grieve 1999; Abd El-Wahab 2006; Zeinolabedin 2012). In rice, it is also observed that salinity negatively influences nutrient uptake (Chakrabarti and Mukherji 2003; Abdelgadir et al. 2005). High accumulation of ions like Na^+ and Cl^- causes ionic imbalance and ultimately it reduces nutrient uptake in plant cell and tissues. As a result, ionic competition between K^+, Ca^{2+}, and Mn^{2+} causes decrease in their concentrations (Sudhir and Murthy 2004). Previous studies also suggest that concentration of Na^+, Ca^{2+}, K^+, and Mg^{2+} in shoot and root is significantly affected by high salinity (Abdur et al. 2011). According to Wimmer et al. (2001), Boron and Silicon (B and Si) availability reduces under high salt concentration in plants. It is suggested that high NaCl concentrations in rice decrease Zinc (Zn) availability and increase the influence of cadmium (Cd) toxicity (Amanullah 2016).

16.4 Hormonal Production and its Positive Role for Conferring Salinity Tolerance

Plant hormones are chemical messengers within plant parts, and they play important role in development, growth, and response to biotic and abiotic stress so that's why they are also known as plant growth regulators (PGRs) (Javid et al. 2011). PGRs also help plant in adaptation to changing environmental conditions by nutrient balance and source sink transformation (Fahad et al. 2014). Auxin is the main hormone that plays significant role in the development of plant and confers shape to the plant body, due to its control on cell division. However, its action rigorously associated with its concentration along the direction of growth of developing organs. Its optimum concentration is also dependent upon the perception, signaling, biosynthetic regulation, and finally transportation to the different organs. It is well studied that variation of auxin gradient stimulates the growth patterns and ultimately assists in salinity tolerance in rice (Du et al. 2013). Cytokinins is another essential plant hormone that influences meristem tissues of the plant, so it is extremely vital for plant under salinity stress conditions to maintain its production for normal growth functioning. It has been also reported that it interplays with abscisic acid signal transduction that interferes the stomatal closure related to the production of abscisic acid (Daszkowska-Golec and Szarejko 2013). Gibberellins plays pivotal role during the recovery phase of rice by interacting with reactive oxygen species to alleviate the oxidative stress. Different hormonal exposure showed positive behavior of rice against salinity stress and helped in improved salinity tolerance (Formentin et al. 2018).

16.5 Effects of Salinity Stress on Rice Growth

At tillering stage, rice is less susceptible to salt stress as compared to early seedling stage (Shereen et al. 2005). Kazemi and Eskandari (2011) demonstrated that fresh weight and seedling growth decreases with an increase of 5 to 7.5 dS m^{-1} salt stress. Greenhouse studies also suggest that rice stand density and biomass production decrease under salt stress (Yong et al. 2020). Abiotic stress factors directly affect plant root in the environment (Smet et al. 2012). Transportation of water and solutes in root is governed by different pathways such as apoplastic, symplastic, and transcellular pathways. In rice, apoplastic pathway plays key role in transportation of water and solutes (Ochiai and Matoh 2002; Kronzucker and britto 2011). Apoplastic pathway is also responsible for Na^+ mobility which passages through solvent drags and Casparian bands (Gong et al. 2006). During salt stress, rice plant shows significant reduction in mean root length, number, and shoot length (Jamil et al. 2006; Jiang et al. 2010). Thus, one can assume the response of rice plant to salt stress by observing its root and shoot lengths. According to Munns (2002), cell elongation and cell division in rice seriously deteriorates under salinity stress, as a result it negatively effects economic yield. Genome duplication and ploidy levels of plant also affect resistance to salt stress in rice. It was observed in tetraploid rice

(HN2026-4x and Nipponbare-4x) with better root growth, organelles, membrane, epidermis cell frequency, H+ efflux on root tip, and nuclei stability. It is due to the exposure of protective gap between pericycle cells and cortex which escalates H^+ transport to root surface and plays role in salt resistance (Tu et al. 2014). At early seedling stage, salt stress significantly increases leaf mortality rate in rice (Shereen et al. 2005). Furthermore, leaf mortality also leads to reduced photosynthesis rate (Amirjani 2010). Leaf senescence is much prone to salt stress, and cells inside the leaves which are exhibiting transpiration are also damaged under salinity stress (Munns et al. 2006).

16.6 Effects on Rice Grain

Various rice cultivars showed panicle sterility at fertilization and pollination stages; it is due to the changes in genetic mechanisms or nutrient deficiencies caused by salinity stress (Khatun and Flowers 1995). Multiple studies suggest that salinity causes panicle sterility during fertilization which declines grain setting, reduced stigmatic surface, and pollen bearing capacity (Abdullah et al. 2001). Moreover, salt stress hinders transfer of carbohydrates for vegetative growth and development of spikelet. In another study, it was observed that major yield traits such as tillers per plant, number of spike per panicle, and percentage of sterile florets decreased with increased salinity (Zheng et al. 2001). According to Abdullah et al. (2001), starch synthetase activity declines under salt stress; moreover, decline in transportation of soluble sugars to spikelet also occurs under salinity. Economic yield components including number of florets per plant, tillers per plant, panicle length, 1000 grain weight were drastically reduced under heavy salinity stress (Farshid and Hassan 2012). Noticeable reduction in yield-related traits was observed with an increase of 1.9 to 6.1 dS m^{-1} salt concentration (Zheng et al. 2001). Farshid and Hassan (2012) demonstrated that with increase in salt stress (2–8 dS m^{-1}), the grain number per panicle decreases. All yield traits are associated with one another and there is a need to unfold the mechanisms so that yield losses to salt stress could be minimized.

16.7 Adaptation Responses to Salt Stress

Ability to tolerate salt stress is an important trait for the development of sustainable rice cultivars (Momayezi et al. 2009). Tolerance of rice plant depends on cellular, molecular, and physiological levels. Osmotic adjustment is an important physiological phenomenon as it enables a plants to cope with toxic ion concentrations. Compartmentalization of Nacl in vacuoles or its accumulation in cytoplasm is osmotic adjustment (Asch and Wopereis 2001). Accumulation of compatible solutes, glycine betaine, proline, and free sugars is an important characteristic for physiologists and plant scientists (Jampeetong and Brix 2009). According to Nounjan and Theerakulpisut (2012), trehalose or carbohydrates are better than proline under osmotic stress. Salt stress causes ionic and osmotic stress in plants, which leads to disruption in normal cell growth and division. As a response to biotic and abiotic stresses, plants exhibit osmotic and ionic signaling and maintain homeostasis in osmosis/ion (Kumar et al. 2013). Plant leaf area in rice also plays an important role in lowering the effect of Na^+ concentration by diluting it in plant leaves and transpiration force (Akita and Cabuslay 1990).

16.8 Salinity Stress and ABA Signaling

Salinity stress induces ionic and osmotic stress that hinders rice normal cellular growth and cell division. High salt accumulation in the vicinity of rice roots exerts pressure to activate various signaling molecules for adaptation to harsh conditions. In order to cope the salinity stress, rice plant needs to maintain ionic and osmotic homeostasis with quick activation of osmotic signaling. Moreover, osmotic stress also rapidly increases the production of abscisic acid (ABA) synthesis with the regulation of ABA-dependent and independent pathways. ABA is an imperative plant hormone due to its critical role in response to salinity stress. Exogenous application of ABA mimics stress effect on plant which results in desiccation of cell and creates osmotic imbalance. ABA also overlaps for gene expression patterns of stress-related genes after exposure to any abiotic stress; therefore, it suggests ABA shares common factor with numerous stress signals during signaling pathways. Furthermore, these common factors crosstalk with each other for maintaining homeostasis in cell during salinity stress conditions.

16.8.1 ABA-Dependent and ABA-Independent Pathways

Higher salinity stress induces osmotic stress with the synthesis of ABA. Previous studies have reported that the imposition of osmotic stress due to high salt environment for rice is transmitted by two pathways, which includes ABA-dependent and ABA-independent pathways (Fig. 16.1). Genetic analysis indicated that there is no evident demarcation that creates difference between ABA-dependent and independent pathways, thereby the components participating might often crosstalk and even converge during signaling pathways. Calcium is considered the secondary messenger for multifarious stresses, which is a strong candidate for mediating such crosstalks. Moreover, numerous studies have reported the positive correlation of ABA for higher increase in calcium in plant cells (Chinnusamy et al. 2004; Xiong et al. 2002). In some of the gene transcript levels such as *RD29A* gene, its accumulation was reported to be regulated in both ABA-dependent and ABA-independent (Yamaguchi-Shinozaki and Shinozaki 1993). Proline accumulation in response to salinity stress was also regulated by both ABA signaling pathways. Calcium role in ABA-dependent pathway was reported for the induction of *P5CS* genes under salinity stress conditions (Knight et al. 1997). The expression level of *P5CS*, *RD22*, *RD29A*, *COR47*, and *COR15A* genes was decreased in los5 mutant; however, the mechanism for signal regulation is not well known but transcriptional regulation

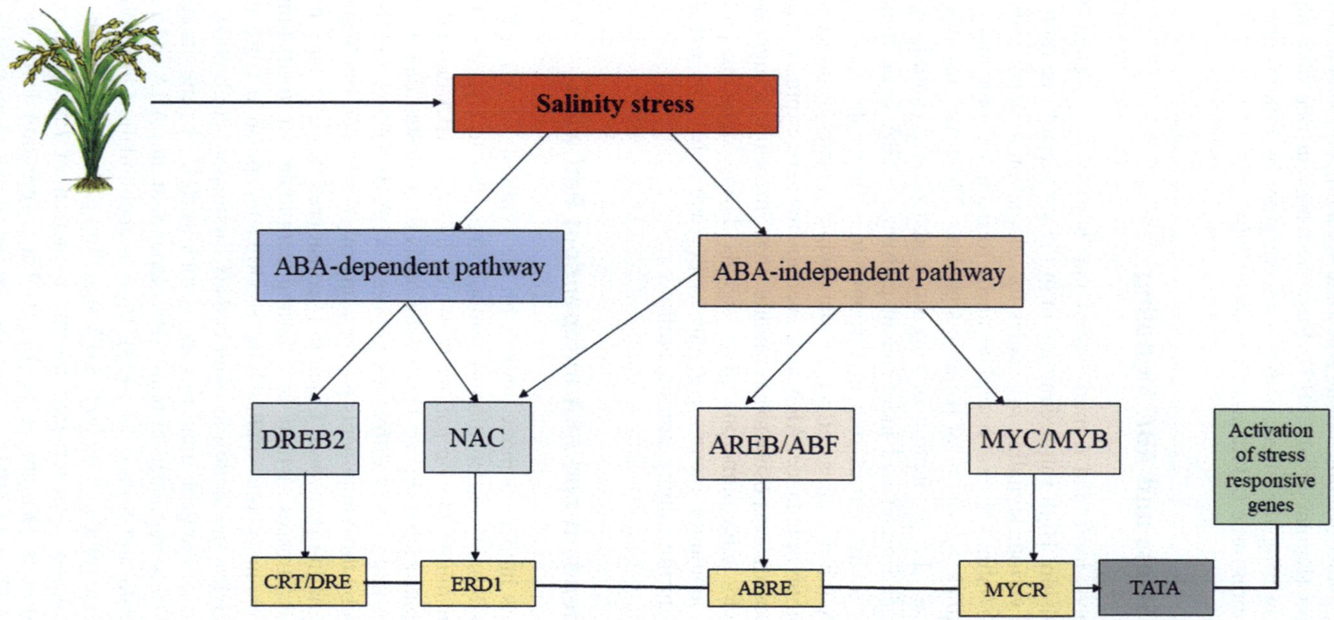

Fig. 16.1 Rice responses to salinity stress via ABA-dependent and ABA-independent pathways. ABA-independent pathways signaling is regulated by dehydration responsive element binding protein (DREB2), and NAC activates the expression of C-repeat elements/dehydration responsive element and ERD1 (early responsive to dehydration stress). The transcription factors involved for the expression of ABA-responsive element (ABRE) and MYC/MYB activate the responsive gene

of *RD29A* is known to some extent (Xiong et al. 2002). Numerous genes are involved in the ABA pathway in response to salinity stress in rice. The first genes that regulate within couple of hours exposure to salinity stress have been reported as 9-cis-epoxycarotenoid dioxygenases genes (OsNCED3) phytoene synthase in rice (OsPSY3), and the expression of this gene depends upon the level of ABA accumulation in the roots of rice (Welsch et al. 2008). It also regulates the expression of various other genes involved in inducing signaling in response to salinity stress.

16.8.2 Regulation of Transcription Factors Via ABA

Transcription factors are the core regulators that play role in turning on/off specific genes expression. It also enhances the transcription of gene. It also regulates, assimilates, and balances hormonal concentration and environmental signals in plant (Shah et al. 2021). Only one transcription factors can regulate the gene expression of several genes via specific binding to cis-acting elements of target genes. Approximately 7% of the coding sequence of the plant genome is allocated to transcription factors that offer complicated transcriptional regulation (Udvardi et al. 2007). Transcription factors are important for regulating the network of ABA-related genes, that is, ABA-responsive binding element proteins (AREB)/ABRE binding factor (ABF), myelocytomatosis (MYC), and no apical meristem (NAM: NAC).

AREB/ABF belongs to the subfamily of basic leucine zipper (bZIP) transcription factor. ABF regulates changes in plants behavior when plant is under the influence of stress. Salinity stress disrupts the plant functioning such as speed-up flowering stage and shortens life cycle of plant. ABF function in stress tolerance is conserved as it results in upregulation of bZIP which assists in alleviating the oxidative stress and activates antioxidant system of plant (Li et al. 2020). This suggests bZIP-positive functions in stress response, which is good strategy for developing rice with improved stress tolerance. In rice, *OsbZIP23* showed the enhanced sensitivity to ABA with improved salinity tolerance (Xiang et al. 2008). Myelocytomatosis transcription factor which is categorized by helix-loop-helix domain is the member of subfamily of bHLH transcription factor. Its conserved function in plant increased tolerance to oxidative stress. In rice, *OsbHLH035* interacts with JAZ protein which is responsible for the activation of jasmonic acid signaling. Its higher expression reported to confer higher tolerance to stress in rice (Chen et al. 2018). NAC transcription factors are specific transcription factors, and it constitutes more than 100 NAC making it the largest family of transcription regulators in plants. It indicated that NAC transcription regulation is involved in numerous biological processes of plant (Olsen et al. 2005). It is involved in transcriptional network of ABA. Its higher expression reported for positive behavior of rice growth towards stress tolerance. Several studies have been reported elucidating the overexpression of this transcription for salinity tolerance in rice (Hu et al. 2006; Zheng et al. 2009; Hong et al. 2016).

16.9 Role of Transporters in Rice

In rice, several channels and pumps have been reported for its role in ionic homeostasis in plant cells. Approximately, 1200 different protein transporters have been annotated by using the genome sequence of rice and 84% of them are actively involved in response to salinity stress (Nagata et al. 2008). Glycophytes restrain the uptake of Na^+ and compartmentalize it to older plant tissues, for example, leaves which serve as a sink to Na^+ that ultimately abscised. The removal of Na^+ ion from cytoplasm to vacuole was observed by salt-inducible enzyme (Apse et al. 1999).

16.9.1 Na^+/H^+ Antiporters

The Na^+ transporters are crucial for Na^+ tolerance in rice which was studied for the very first time in sugar beet (Blumwald and Poole 1987). Ion transporters function is important for the maintenance of physiologic concentration of plant. The Na^+/H^+ antiporters is crucial for ionic homeostasis in cell to ensure proper cellular functioning of rice plant. It shift it to its survival capability to cope with the stress conditions (Wang et al. 2003). The Na^+/H^+ antiporter plays positive role in the exchange of Na^+ with H^+ ion in the membrane which directly effects the other functions such as cytoplasmic regulation of pH and decreases Na^+ level in plants (Serrano et al. 1999). Moreover, four different vacuolar Na^+/H^+ antiporters were reported to confer salinity tolerance in rice (Bassil et al. 2012).

16.9.2 Na^+/K^+ Symporter

Several ions are present in soil solution available for plant uptake. Both sodium (Na^+) and potassium (K^+) ions existed long ago in the ocean during early evolution of life, whereas only K^+ ions are the beneficial for maintaining osmotic and electrolyte balance in the cells of plants (Garciadeblás et al. 2003). Salinity has diverse range of effects, but toxicity of Na^+ is the primary element that causes damage to cell of salt-sensitive plants such as rice. Contrarily K^+ is an essential ion as Na^+ and K^+ compete for their uptake. Moreover, K^+ uptake assists in several physiological changes to alleviate the negative effects of higher Na^+ accumulation in cells. The normal concentration of K^+ was found to be 100–200 mM, while the normal range of Na^+ is 1–10 mM for optimal metabolic functions in cell. Therefore, it suggests that plants should have higher K^+/Na^+ ratio for its normal growth functions. Contrarily salinity stress disturbs the ionic concentration in the cytosol with the higher uptake of Na^+ ions resulting lower K^+ ions (Kader et al. 2006). Plant needs to inhibit Na^+ uptake and favors the uptake of K^+ ions at intermediate levels. For this purpose, plant modulates ionic balance for normal metabolism. Plants regulate ionic balance to maintain normal metabolism by restricting the uptake of Na^+ and Cl^- ions and enhancing the uptake of K^+ ions. The uptake of K^+ ions is observed to be increased with the higher gene expression of K^+ transporters and H^+ pumps that generate the force for ionic transport. It was reported in rice that higher degree of salt tolerance was attributed to effective system for the selection of K^+ ion over Na^+ ion (Yang et al. 2014). Differential response regarding salinity tolerance among rice genotypes was due to obstruction of Na^+ in leaves. It was due to higher expression levels of Na^+/K^+ transport proteins that assisted in higher K^+ influx in rice genotypes reported differential response (Ali et al. 2021; Hussain et al. 2018).

16.10 QTL Mapping

Briefly it is a genome-wide speculation of genotypic and phenotypic relationship at several genomic locations for quantitative trait about genomic positions, numbers, effects, and QTL interaction. It is efficient approach for seeing genetic complexity of plant at genomic level (Yousaf et al. 2021). It also helps in digging the specific location of loci to find resistance genes to cope abiotic stress conditions (Gökçe et al. 2021). With the invention of QTL mapping, it provided ease and swiftness in abiotic stress breeding program with the selection of phenotypic trait from a plant population. It utilizes statistical models for the prediction of association between desired phenotypic trait and genetic markers (Naeem et al. 2021). These developed models serve as a basis to deduce progenies of plants within breeding population for abiotic stress breeding. After confirmation of association, it can be used for screening of plants for stress tolerance by focusing on specific plant trait from a several plant population and within a short period of time (Wassan et al. 2021).

16.10.1 QTLs for Salinity Tolerance in Rice

It is well reported through several studies that salinity stress regulates the function of group (Islam et al. 2019). Therefore, it is important to dissect regions by exploiting QTL mapping approach that helps rice in salinity tolerance. It is a modern genetic tool to localize and identify molecular marker linked to a specific trait (Gökçe et al. 2021). QTL mapping also helped in marking a region that selectively favors the higher uptake of K^+ over Na^+, which is important for the maintenance of ionic homeostasis as mentioned above during salinity stress (Gökçe and Chaudhry 2020). The major QTL reported is *saltol* which is exclusively responsible for salinity tolerance by maintaining Na^+/K^+ homeostasis (Thomson et al. 2010). Moreover, it was also explained through studies that *SKC1* gene exists in the QTL region of *saltol*, whose primary function is to restrict Na^+ entry and transportation to different cells in plants. This gene (*SKC1*) also encodes HKT transporter for unloading of excessive Na^+ from the shoots to roots. It also helps in unloading of excessive Na^+ from xylem for protecting plants aerial parts (Zang et al. 2008). The identification of *saltol* locus in rice helped in developing the salt tolerant cultivars within a short period of time as compared to conventional breeding techniques (Thomson

Table 16.1 QTLs identified for salinity tolerance in rice

QTLs for salinity tolerance	Traits	References
*SalTol*10–1	Salinity tolerance	Islam et al. (2011)
qSDM-5	Dry matter contents	Prasad et al. (2000)
Q_{NaK}	Na^+:K^+ discrimination	Flowers et al. (2000)
qSKC-1	Shoot potassium concentration	Lin et al. (2004)
qSNC-7	Shoot sodium concentration	
qSDS-1	Survival of rice seedlings	
qST3	Increased K^+ accumulation over Na+	Lee et al. (2006)
qRKC1	Higher K+ uptake	Thomson et al. (2010)
qSGEM-6	Increased germination percentage	Prasad et al. (2000)
qGY2.1 s	Grain yield	Mohammadi et al. (2013)
qTGW10.1 s	Grain weight	
qSPFR2.1 s	Fertility of spikelet	
qPL1.1 s	Length of panicle	
qPH1.1 s	Height of plant	
QTn10	Total number of tillers	Zang et al. (2008)
QFw11b	Shoot fresh weight	
QPh9	Plant height	
qNAK-6	Na^+/K^+ ratio	Sabouri et al. (2009)
qRL6.1	Increased root length	Bizimana et al. (2017)
qSTR-1, *qSTR*-5	Salt tolerance rating	Ming-zhe et al. (2005)
qDWS-8	Dry weight of shoot	

et al. 2010). Several other QTLs associated for salinity stress tolerance in rice have been reported, given in Table 16.1.

Higher Na^+ accumulation under salinity stress causes osmotic stress which induces oxidative stress. As described above in physiological changes due to salinity stress that salinity causes stomatal closure and other physiological with decreased influx of CO_2 in rice leads to the formation of ROS species. Lipid peroxidation in rice occurs to higher level of ROS species. Malondialdehyde (MDA) indicates the rate of lipid peroxidation that rice counters during salinity stress; it is accumulated in higher amount in sensitive cultivars resulting in membrane damage (Frukh et al. 2020). In rice QTLs (*qMDA-1a* and *qMDA-1b*) for MDA have been mapped to elucidate the behavior of rice/selection of rice cultivars for salinity tolerance (Jiang et al. 2009). Salinity stress effects the growth of rice with a stunted plant height, reduced leaf growth, and shorter root length and these morphological changes result in decreased rice panicles leading to lower yield. In order to ensure higher yield by coping with salinity stress, genes responsible for improved growth and yield have been marked by utilizing QTL mapping (Gökçe and Chaudhry 2020). QTLs for various morphological and yield traits have been reported by Mohammadi et al. (2013) which helped in marker-assisted backcrossing and to find useful genes for salinity tolerance in rice. In another study, different QTLs for plant height, tiller number, and fresh weight of shoot in rice have been reported showing the salinity tolerance (Zang et al. 2008).

16.11 Molecular Markers for Salinity Tolerance in Rice

At specific genes loci, a major technique under use for the visualization of DNA sequence variations is the use of molecular markers. Before DNA sequencing, molecular markers were the only means to produce abundant data about DNA variations for the past 30 years. The first markers used to identify differences in DNA were used for restriction enzymes on recognition sites. RFLP marker explains polymorphic pattern and changes in length between restriction patterns. In the field of molecular genetics, invention of PCR machine is a huge breakthrough. A series of methods for identifying the variations by analyzing PCR product sizes among genetic sequences were developed. On the other hand, there are other techniques in which variations in the sequence lengths in PCR product can be identified by using primer annealing site or restriction site polymorphism. With the passage of time, single locus marker techniques are replaced by multiple locus marker techniques. This is the reason that Amplified Fragment Length Polymorphism (AFLP) marker is widely accepted in molecular genetics. The easiest way to find salt tolerant rice cultivars is by doing screening of large number of breeding populations by using molecular markers. Screening of the rice genotypes at field level is relatively difficult because of heterozygous nature of the soil and several environmental factors that influence physiological functioning of rice. In contrast under laboratory conditions, it is relatively easy and advantageous as compared to screening in field (Quijano-Guerta and Kirk 2002).

Molecular markers include (Karp et al. 1996)

1. Restriction fragment length polymorphism.
2. Random amplified polymorphic DNA.
3. Inter simple sequence repeats.
4. Simple sequence repeats.
5. Amplified fragment length polymorphism.
6. Microsatellite markers.

All the above-mentioned molecular markers generate more accurate and reliable information compared to morpho-physiological markers. Microsatellite markers are found to be the more accurate/effective way to identify genetic variations in rice genotypes (Islam 2004; Bhuiyan 2005). Rice cultivars were screened for salinity tolerance and four rice genotypes were found to be tolerant to salinity stress by using SSR markers (Ali et al. 2014). Another study also endorsed authenticity of the SSR markers for the screening rice genotypes (Tahjib-Ul-Arif et al. 2018). Recently another study successfully screened the rice genotypes with the use of RAPD marker technology (Mazumder et al. 2020).

References

Abd El-Wahab MA (2006) The efficiency of using saline and freshwater irrigation as alternating methods of irrigation on the productivity of Foeniculum vulgare mill subsp. Vulgare var. vulgare under North Sinai conditions. Res J Agric Biol Sci 2:571–577

Abdelgadir EM, Oka M, Fujiyama H (2005) Nitrogen nutrition of rice plants under salinity. Biol Plant 49:99–104

Abdullah Z, Khan MA, Flowers TJ (2001) Causes of sterility in seed set of rice under salinity stress. J Agron Crop Sci 187:25–32

Abdur RM, Talukderb NM, Tofazzal IM, Duttad RK (2011) Salinity effect on mineral nutrient distribution along roots and shoots of rice (*Oryza sativa* L.) genotypes differing in salt tolerance. Arch Agron Soil Sci 57:33–45

Ahmad A, Ashfaq M, Rasul G, Wajid SA, Khaliq T, Rasul F, Saeed U, Rahman MH, Hussain J, Baig IA, Naqvi AA, Bokhari SAA, Ahmad S, Naseem W, Hoogenboom G, Valdivia RO (2015) Impact of climate change on the rice–wheat cropping system of Pakistan. In: Hillel D, Rosenzweig C (eds.), Handbook of Climate Change and Agro-ecosystems: The Agricultural Modeling Intercomparison and Improvement Project (AgMIP) Integrated Crop and Economic Assessments, Published by Imperial College Press and the American Society of Agronomy, pp. 219–258

Ahmad S, Abbas G, Ahmed M, Fatima Z, Anjum MA, Rasul G, Khan MA, Hoogenboom G (2019) Climate warming and management impact on the change of rice-wheat phenology in Punjab, Pakistan. Field Crops Res 230:46–61

Ahmad S, Ahmad A, Ali H, Hussain A, Garcia A, Khan MA, Zia-Ul-Haq M, Hasanuzzaman M, Hoogenboom G (2013) Application of the CSM-CERES-Rice model for evaluation of plant density and irrigation management of transplanted rice for an irrigated semiarid environment. Irrig Sci 31(3):491–506

Ahmad S, Ahmad A, Soler CMT, Ali H, Zia-Ul-Haq M, Anothai J, Hussain A, Hoogenboom G, Hasanuzzaman M (2012) Application of the CSM-CERES-Rice model for evaluation of plant density and nitrogen management of fine transplanted rice for an irrigated semiarid environment. Precis Agric 13(2):200–218

Ahmad S, Ahmad A, Zia-ul-Haq M, Ali H, Khaliq T, Anjum MA, Khan MA, Hussain A, Hoogenboom G (2009) Resources use efficiency of field grown transplanted rice (*Oryza sativa* L.) under irrigated semiarid environment. J Food Agric Environ 7(2):487–492

Ahmad S, Hasanuzzaman M (2012) Integrated effect of plant density, N rates and irrigation regimes on the biomass production, N content, PAR use efficiencies and water productivity of rice under irrigated semiarid environment. Not Bot Horti Agrobot Cluj-Napoca 40(1):201–211

Ahmad S, Zia-ul-Haq M, Ali H, Shad SA, Ammad A, Maqsood M, Khan MB, Mehmood S, Hussain A (2008) Water and radiation use efficiencies of transplanted rice (*Oryza sativa* L.) at different plant densities and irrigation regimes under semi-arid environment. Pak J Bot 40(1): 199–209

Ahmed M, Ahmad S (2019) Carbon dioxide enrichment and crop productivity. In: Hasanuzzaman M (ed.), Agronomic crops; volume 2 Springer Nature Singapore Pte ltd., pp. 31–46

Ahmed M, Ahmad S (2020) Systems modeling. In: Ahmed M (ed.), Systems modeling, Springer Nature Singapore Pte Ltd. pp. 1–44

Ahmed N, Khalil A, Gulshan AB, Bashir S, Saleem M, Hussain R, Ali MA, Iqbal J, Bashir S (2020) The efficiency of magnesium (Mg) on rice growth, biomass partitioning and chlorophyll contents in alkaline soil condition. Pure Appl Biol 10:325–333

Akita S, Cabuslay GS (1990) Physiological basis of differential response to salinity in rice cultivars. Plant Soil 123:277–294

Ali MN, Yeasmin L, Gantait S, Goswami R, Chakraborty S (2014) Screening of rice landraces for salinity tolerance at seedling stage through morphological and molecular markers. Physiol Mol Biol Plants 20(4):411–423

Ali A, Raddatz N, Pardo JM, Yun DJ (2021) HKT sodium and potassium transporters in Arabidopsis thaliana and related halophyte species. Physiol Plant 171:546–558

Amanullah I (2016) Dry matter partitioning and harvest index differ in rice genotypes with variable rates of phosphorus and zinc nutrition. Rice Sci 23:78–87

Amirjani MR (2010) Effect of NaCl on some physiological parameters of rice. Eurpoean Journal of Biological Science 3:6–16

Apse MP, Aharon GS, Snedden WS, Blumwald E (1999) Salt tolerance conferred by overexpres Sion of a vacuolar Na+/H+ antiport in Arabidopsis. Science 285:1256–1258

Asch F, Wopereis MCS (2001) Responses of field grown irrigated rice cultivars to varying levels of flood water salinity in a semi-arid environment. Field Crops Res 70:127–137

Asim A, Gökçe ZNÖ, Bakhsh A, Çayli IT, Aksoy E, Çalişkan S, Çalişkan ME, Demirel U (2021) Individual and combined effect of drought and heat stresses in contrasting potato cultivars overexpressing miR172b-3p. Turk J Agric For 45:651–668

Bassil E, Coku A, Blumwald E (2012) Cellular ion homeostasis: emerging roles of intracellular NHX Na+/H+ antiporters in plant growth and development. J Exp Bot 63(16):5727–5740

Bhuiyan MA (2005) Efficiency in evaluating salt tolerance in rice using phenotypic and marker assisted selection. Mymensingh, Bangladesh

Bizimana JB, Luzi-Kihupi A, Murori RW, Singh RK (2017) Identification of quantitative trait loci for salinity tolerance in rice (Oryza sativa L.) using IR29/Hasawi mapping population. J Genet 96:571–582

Blumwald E, Poole RJ (1987) Salt tolerance in suspension cultures of sugar beet: induction of Na+/ H+ antiport activity at the tonoplast by growth in salt. Plant Physiol 83:884–887

Bohn HL, Brain LM, O'Connor GA (1985) Soil chemistry, 2nd edn. John Wiley and Sons, New York, pp 234–248. Graham W (1994) Water Resources Manager, Jefferson County PU#1, MS Geology, Easton Washington University, 133–147

Cattivelli L, Rizza F, Badeck FW, Mazzucotelli E, Mastrangelo AN, Francia E, Mare C, Tondelli A, Stanca AM (2008) Drought tolerance improvement in crop plants. An integrated view from breeding to genomics. Field Crops Res 105:1–14

Chakrabarti N, Mukherji S (2003) Effect of phytohormone pretreatment on nitrogen metabolism in Vigna radiata under salt stress. Biol Plant 46:63–66

Chaudhry UK, Gökçe ZN, Gökçe AF (2020) Effects of salinity and drought stresses on the physio-morphological attributes of onion cultivars at bulbification stage. Int J Agri Biol 24:1681–1689

Chaudhry UK, Junaid MD, Gökçe AF (2021a) Influence of environmental adversities on physiological changes in plants. In: Developing climate-resilient crops. CRC Press, pp 853–110

Chaudhry UK, Gökçe ZN, Gökçe AF (2021b) Drought and salt stress effects on biochemical changes and gene expression of photosystem II and catalase genes in selected onion cultivars. Biologia 76:3107–3121

Chen HC, Cheng WH, Hong CY, Chang YS, Chang MC (2018) The transcription factor OsbHLH035 mediates seed germination and enables seedling recovery from salt stress through ABA-dependent and ABA-independent pathways, respectively. Rice 11(1):50

Chen T, Xu Y, Wang J, Wang Z, Yang J, Zhang J (2013) Polyamines and ethylene interact in rice grains in response to soul drying during grain filling. J Exp Bot 64:2523–2538

Chen Y, Cothren JT, Chen D, Amir MH, Ibrahim LL (2014) Effect of 1-MCP on cotton plants under abiotic stress caused by ethephon. Am J Plant Sci 5:3005–3016

Chinnusamy V, Schumaker K, Zhu JK (2004) Molecular genetic perspectives on cross-talk and specificity in abiotic stress signalling in plants. J Exp Bot 55(395):225–236

Daszkowska-Golec A, Szarejko I (2013) The molecular basis of ABA-mediated plant response to drought. In: Abiotic stress-plant responses and applications in agriculture. IntechOpen, pp 103–134

Dionisio-Sese ML, Tobita S (2000) Antioxidant responses of rice seedlings to salinity stress. Plant Sci 135:1–9

Du H, Liu H, Xiong L (2013) Endogenous auxin and jasmonic acid levels are differentially modulated by abiotic stresses in rice. Front Plant Sci 4:397

Fahad S, Hussain S, Matloob A, Khan FA, Khaliq A, Saud A, Hassan S, Shan D, Khan F, Ullah N, Faiq M, Khan MR, Tareen AK, Khan A, Ullah A, Ullah N, Huang JL (2014) Phytohormones and plant responses to salinity stress: a review. Plant Growth Regul 75:391–404

Farshid A, Hassan ER (2012) Physiological characterization of rice under salinity stress during vegetative and reproductive stages. Indian J Sci Technol 5:2578–2586

Flowers TJ, Koyama ML, Flowers SA, Sudhakar C, Singh KP, Yeo AR (2000) QTL: their place in engineering tolerance of rice to salinity. J Exp Bot 51:99–106

Flowers TJ, Yeo AR (1981) Variability in the resistance of sodium chloride salinity within rice (*Oryza sativa* L.) varieties. New Phytol 88:363–373

Formentin E, Sudiro C, Perin G, Riccadonna S, Barizza E, Baldoni E, Lavezzo E, Stevanato P, Sacchi GA, Fontana P, Toppo S (2018) Transcriptome and cell physiological analyses in different rice cultivars provide new insights into adaptive and salinity stress responses. Front Plant Sci 9:204

Frukh A, Siddiqi TO, Khan MIR, Ahmad A (2020) Modulation in growth, biochemical attributes and proteome profile of rice cultivars under salt stress. Plant Physiol Biochem 146:55–70

Garcia MS, Tellez LI, Merino FCG, Caldana C, Victoria ED, Cabrera HEB (2012) Growth, photosynthetic activity, and potassium and sodium concentration in rice plants under salt stress. Acta Sci Agron 34:317–324

Garciadeblás B, Senn ME, Bañuelos MA, Rodríguez-Navarro A (2003) Sodium transport and HKT transporters: the rice model. Plant J 34(6):788–801

Gholipoor M, Soltani A, Shekari F, Shekari FB (2002) Effects of salinity on water use efficiency and its components in chickpea (*Cicer arietinum* L.). Acta Agron Hung 50:127–134

Gökçe AF, Chaudhry UK (2020) Use of QTL in developing stress tolerance in agronomic crops. In: Hasanuzzaman M (ed) Agronomic crops. Springer Publishing, Singapore pp, pp 527–556

Gökçe AF, Junaid MD, Chaudhry UK (2021) Mapping QTLs for abiotic stress. In: Developing climate-resilient crops. CRC Press, pp 175–201

Gong HJ, Randall DP, Flowers TJ (2006) Silicon deposition in the root reduces sodium uptake in rice (*Oryza sativa* L.) seedlings by reducing bypass flow. Plant Cell Environ 29:1970–1979

Grattan SR, Grieve CM (1999) Salinity mineral nutrient relations in horticultural crops. Sci Hortic 78:127–157

Gregorio GB, Dharmawansa S, Mendoza RD (1997) Screening rice for salinity tolerance. IRRI discussion paper series NO.22, international rice research institute, Manila, Philippines, pp. 2–23

Hasanuzzaman M, Fujita M, Nahar K, Biswas JK (2018) Advances in rice research for abiotic stress tolerance. Woodhead Publishing

Hong Y, Zhang H, Huang L, Li D, Song F (2016) Overexpression of a stress-responsive NAC transcription factor gene ONAC022 improves drought and salt tolerance in rice. Front Plant Sci 7:4

Hu H, Dai M, Yao J, Xiao B, Li X, Zhang Q, Xiong L (2006) Overexpressing a NAM, ATAF, and CUC (NAC) transcription factor enhances drought resistance and salt tolerance in rice. Proc Natl Acad Sci 103(35):12987–12992

Hussain M, Ahmad S, Hussain S, Lal R, Ul-Allah S, Nawaz A (2018) Rice in saline soils: physiology, biochemistry, genetics, and management. Adv Agron 148:231–287

Islam M, Ontoy J, Subudhi PK (2019) Meta-analysis of quantitative trait loci associated with seedling-stage salt tolerance in rice (*Oryza sativa* L.). Plants 8:33

Islam MR, Hassan L, Salam MA, Collard BCY, Singh RK, Gregorio GB (2011) QTL mapping for salinity tolerance at seedling stage in rice. Emir J Food Agr:137–146

Islam, M.M., 2004. Mapping salinity tolerance genes in rice (Oryza sativa L.) at reproduction stage

Jamil A, Riaz S, Ashraf M, Foolad MR (2011) Gene expression profiling of plants under salt stress. Crit Rev Plant Sci 30:435–458

Jamil M, Bae LD, Yong JK, Ashraf M, Chun LS, Shik RE (2006) Effect of salt (NaCl) stress on germination and early seedling growth of four vegetables species. J Cent Eur Agric 7:273–282

Jampeetong A, Brix H (2009) Effects of NaCl salinity on growth, morphology, photosynthesis and proline accumulation of Salvinia natans. Aquat Bot 91:181–186

Javid MG, Ali S, Foad M, Seyed AMMS, Iraj A (2011) The role of phytohormones in alleviating salt stress in crop plants. Aust J Crop Sci 5:726–734

Jiang J, Zhuang JY, Fan YY, Bo SHEN (2009) Mapping of QTLs for leaf malondialdehyde content associated with stress tolerance in rice. Rice Sci 16(1):72–74

Jiang XJ, Zhang S, Miao L, Tong T, Liu Z, Sui Y (2010) Effect of salt stress on rice seedling characteristics, effect of salt stress on root system at seedling stage of rice. North Rice 40:21–24

Junaid MD, Chaudhry UK, Gökçe AF (2021) Climate change and plant growth–South Asian perspective. In: Climate Change and plants: biodiversity, growth and interactions. CRC Press, pp 37–53

Kader MA, Seidel T, Golldack D, Lindberg S (2006) Expressions of OsHKT1, OsHKT2, and OsVHA are differentially regulated under NaCl stress in salt-sensitive and salt-tolerant rice (*Oryza sativa* L.) cultivars. J Exp Bot 57(15):4257–4268

Karp A, Seberg OLE, Buiatti M (1996) Molecular techniques in the assessment of botanical diversity. Ann Bot 78(2):143–149

Kazemi K, Eskandari H (2011) Effects of salt stress on germination and early seedling growth of rice (*Oryza sativa*) cultivars in Iran. Afr J Biotechnol 10(77):17789–17792

Khatun S, Flowers TJ (1995) Effects of salinity on seed set in rice. Plant Cell Environ 18:61–67

Knight H, Trewavas AJ, Knight MR (1997) Calcium signalling in Arabidopsis thaliana responding to drought and salinity. Plant J 12(5):1067–1078

Kronzucker JK, Britto TD (2011) Sodium transport in plants: acritical review. New Phytol 189: 54–81

Kumar K, Kumar M, Seong-Ryong K, Ryu H, Cho YG (2013) Insights into genomics of salt stress response in rice. Rice 6:6–27

Lee SY, Ahn JH, Cha YS, Yun DW, Lee MC, Ko JC, Lee KS, Eun MY (2006) Mapping of quantitative trait loci for salt tolerance at the seedling stage in rice. Mol Cells 21(2):192–196

Li G, Meng X, Wang R, Mao G, Han L, Liu Y (2012) Duallevel regulation of ACC synthase activity by MPK3/MPK6 cascade and its downstream WRKY transcription factor during ethylene induction in Arabidopsis. PLoS Genet 8:1002767

Li Q, Zhao H, Wang X, Kang J, Lv B, Dong Q, Li C, Chen H, Wu Q (2020) Tartary buckwheat transcription factor FtbZIP5, regulated by FtSnRK2. 6, can improve salt/drought resistance in transgenic Arabidopsis. Int J Mol Sci 21(3):1123

Lin HX, Zhu MZ, Yano M, Gao JP, Liang ZW, Su WA, Hu XH, Ren ZH, Chao DY (2004) QTLs for Na+ and K+ uptake of the shoots and roots controlling rice salt tolerance. Theor Appl Genet 108:253–260

Mazumder SR, Hoque H, Sinha B, Chowdhury WR, Hasan MN, Prodhan SH (2020) Genetic variability analysis of partially salt tolerant local and inbred rice (*Oryza sativa* L.) through molecular markers. Heliyon 6(8):e04333

Ming-zhe Y, Jian-fei W, Hong-you C, Hu-qu Z, Hong-sheng Z (2005) Inheritance and QTL mapping of salt tolerance in rice. Rice Sci 12:25–32

Mohammadi R, Mendioro MS, Diaz GQ, Gregorio GB, Singh RK (2013) Mapping quantitative trait loci associated with yield and yield components under reproductive stage salinity stress in rice (*Oryza sativa* L.). J Genet 92(3):433–443

Momayezi MR, Zaharah AR, Hanafi MM, Mohd Razi I (2009) Agronomic characteristics and proline accumulation of Iranian rice genotypes at early seedling stage under sodium salts stress. Malaysian J Soil Sci 13:59–75

Munns R (2002) Comparative physiology of salt and water stress. Plant Cell Environ 25:239–250

Munns R (2005) Genes and salt tolerance: bringing them together. New Phytol 167:645–663

Munns R, James RA, Lauchli A (2006) Approaches to increasing the salt tolerance of wheat and other cereals. J Exp Bot 57:1025–1043

Naeem M, Demirel U, Yousaf MF, Caliskan S, Caliskan ME (2021) Overview on domestication, breeding, genetic gain and improvement of tuber quality traits of potato using fast forwarding technique (GWAS): a review. Plant Breed 140:519–542

Nagata T, Iizumi S, Satoh K, Kikuchi S (2008) Comparative molecular biological analysis of membrane transport genes in organisms. Plant Mol Biol 66:565–585

Nishimura T, Cha-um S, Takagaki M, Ohyama K (2011) Survival percentage, photosynthetic abilities and growth characters of two indica rice (*Oryza sativa* L. spp. indica) cultivars in response to isosmotic stress. Span J Agric Res 9:262–270

Nounjan N, Theerakulpisut P (2012) Effect of exogenous proline and trehalose on physiological responses in rice seedlings during salt-stress and after recovery. Plant Soil Environ 58:308–315

Ochiai K, Matoh T (2002) Characterization of the Na+ delivery from roots to shoots in rice under saline stress: excessive salt enhances apoplastic transport in rice plants. Soil Sci Plant Nutr 48: 371–378

Olsen AN, Ernst HA, Leggio LL, Skriver K (2005) NAC transcription factors: structurally distinct, functionally diverse. Trends Plant Sci 10:79–87

Prasad SR, Bagali PG, Hittalmani S, Shashidhar HE (2000) Molecular mapping of quantitative trait loci associated with seedling tolerance to salt stress in rice (*Oryza sativa* L.). Curr Sci 78(2): 162–164

Quijano-Guerta C, Kirk GJD (2002) Tolerance of rice germplasm to salinity and other soil chemical stresses in tidal wetlands. Field Crops Res 76(2–3):111–121

Ramezani M, Seghatoleslami M, Mousavi G, Sayyari-Zahan MH (2012) Effect of salinity and foliar application of iron and zinc on yield and water use efficiency of Ajowan (*Carum copticum*). Int J Agric Crop Sci 4:421–426

Sabouri H, Rezai AM, Moumeni A, Kavousi A, Katouzi M, Sabouri A (2009) QTLs mapping of physiological traits related to salt tolerance in young rice seedlings. Biol Plant 53:657–662

Sairam RK, Tyagi A (2004) Physiology and molecular biology of salinity stress tolerance in plants. Curr Sci 86(3):407–421

Serrano R, Mulet JM, Rios G, Marquez JA, de Coo IF, Leube MP, Mendizabal I, Pascual A, Zamora-Ros R, Montesinos C, Proft M (1999) A glimpse of the mechanisms of ion homeostasis during salt stress. J Exp Bot 50:1023–1036

Shah WH, Rasool A, Saleem S, Mushtaq NU, Tahir I, Hakeem KR, Rehman RU (2021) Understanding the integrated pathways and mechanisms of transporters, protein kinases, and transcription factors in plants under salt stress. Int J Gen 2021:5578727

Shahzad AN, Ahmad S (2019) Tools and techniques for nitrogen management in cereals. In: Hasanuzzaman M (ed.), agronomic crops; volume 2 springer nature Singapore Pte ltd., pp. 111–126

Shahzad S, Khan MY, Zahir ZA, Asghar HN, Chaudhry UK (2017) Comparative effectiveness of different carriers to improve the efficacy of bacterial consortium for enhancing wheat production under salt affected field conditions. Pak J Bot 49:1523–1530

Shereen A, Mumtaz S, Raza S, Khan MA, Solangi S (2005) Salinity effects on seedling growth and yield components of different inbred rice lines. Pak J Bot 37:131–139

Smet ID, White PJ, Bengough AG, Dupuy L, Parizot B, Casimiro I, Heidstra R, Laskowski M, Lepetit M, Hochholdinger F, Draye X, Zhang H, Broadley MR, Peret B, Hammond JP, Fukaki H, Mooney S, Lynch JP, Nacry P, Schurr U, Laplaze L, Benfey P, Beeckman T, Bennett M (2012) Analyzing lateral root development: how to move forward. Plant Cell 24:15–20

Sudhir P, Murthy SDS (2004) Effects of salt stress on basic processes of photosynthesis. Photosynthetica 42:481–486

Tahjib-Ul-Arif M, Sayed MA, Islam MM, Siddiqui MN, Begum SN, Hossain MA (2018) Screening of rice landraces (*Oryza sativa* L.) for seedling stage salinity tolerance using morpho-physiological and molecular markers. Acta Physiol plant 40(4):70

Tester M, Davenport R (2003) Na+ tolerance and Na+ transport in higher plants. Ann Bot 9: 503–527

Thomson MJ, de Ocampo M, Egdane J, Rahman MA, Sajise AG, Adorada DL, Tumimbang-Raiz E, Blumwald E, Seraj ZI, Singh RK, Gregorio GB, Ismail AM (2010) Characterizing the Saltol quantitative trait locus for salinity tolerance in rice. Rice 3:148–160

Tomaz A, Palma P, Alvarenga P, Gonçalves MC (2020) Soil salinity risk in a climate change scenario and its effect on crop yield. In: Climate Change and Soil Interactions. Elsevier, pp. 351–396

Tu Y, Jiang A, Gan L, Hossain M, Zhang J, Bo P, Xiong Y, Song Z, Cai D, Xu W, Zhang J, He Y (2014) Genome duplication improves rice root resistance to salt stress. Rice 7:2–15

Udvardi MK, Kakar K, Wandrey M, Montanari O, Murray J, Andriankaja A, Zhang JY, Benedito V, Hofer JM, Chueng F, Town CD (2007) Legume transcription factors: global regulators of plant development and response to the environment. Plant Physiol 144:538–549

Wang W, Vinocur B, Altman A (2003) Plant response to drought, salinity and extreme temperatures: towards genetic engineering for stress tolerance. Planta 218:1–14

Wassan GM, Khanzada H, Zhou Q, Mason AS, Keerio AA, Khanzada S, Solangi AM, Faheem M, Fu D, He H (2021) Identification of genetic variation for salt tolerance in Brassica napus using genome-wide association mapping. Mol Genet Genomics 296:391–408

Welsch R, Wust F, Bar C, Babili SA, Beyer P (2008) A third phytoene synthase is devoted to abiotic stress-induced abscisic acid formation in rice and defines functional diversification of phytoene synthase genes. Plant Physiol 147:367–380

Wimmer MA, Muehling KH, Lauchli A, Brown PH, Goldbach HE (2001) Interaction of salinity and boron toxicity in wheat (*Triticum aestivum* L.). In: Horst WJ, Schenk MK, Burkert A, Claassen N, Flessa H, Frommer WB, Goldbach H, Olfs HW, Romheld V (eds) Plant nutrition: food security and sustainability of agro-ecosystems through basic and applied research. Springer, Netherlands, pp 426–427

Xiang Y, Tang N, Du H, Ye H, Xiong L (2008) Characterization of OsbZIP23 as a key player of the basic leucine zipper transcription factor family for conferring abscisic acid sensitivity and salinity and drought tolerance in rice. Plant Physiol 148(4):1938–1952

Xiong L, Schumaker KS, Zhu JK (2002) Cell signaling during cold, drought, and salt stress. Plant Cell 14(suppl 1):S165–S183

Yamaguchi-Shinozaki K, Shinozaki K (1993) Characterization of the expression of a desiccation-responsive rd29 gene of Arabidopsis thaliana and analysis of its promoter in transgenic plants. Mol Gen Genet 236(2–3):331–340

Yang T, Zhang S, Hu Y, Wu F, Hu Q, Chen G, Cai J, Wu T, Moran N, Yu L, Xu G (2014) The role of a potassium transporter OsHAK5 in potassium acquisition and transport from roots to shoots in rice at low potassium supply levels. Plant Physiol 166(2):945–959

Yong MT, Solis CA, Rabbi B, Huda S, Liu R, Zhou M, Shabala L, Venkataraman G, Shabala S, Chen ZH (2020) Leaf mesophyll K^+ and Cl^- fluxes and reactive oxygen species production predict rice salt tolerance at reproductive stage in greenhouse and field conditions. Plant Growth Regul 92:53–64

Yousaf MF, Demirel U, Naeem M, Çalışkan ME (2021) Association mapping reveals novel genomic regions controlling some root and stolon traits in tetraploid potato (Solanum tuberosum L.). 3 Biotech 11:1–16

Zang J, Sun Y, Wang Y, Yang J, Li F, Zhou Y, Zhu L, Jessica R, Mohammadhosein F, Xu J, Li Z (2008) Dissection of genetic overlap of salt tolerance QTLs at the seedling and tillering stages using backcross introgression lines in rice. Sci China Life Sci 51(7):583–591

Zeinolabedin J (2012) The effects of salt stress on plant growth. Tech J Eng Appl Sci 2:7–10

Zheng L, Shannon MC, Lesch SM (2001) Timing of salinity stress affecting rice growth and yield components. Agric Water Manag 48:191–206

Zheng X, Chen B, Lu G, Han B (2009) Overexpression of a NAC transcription factor enhances rice drought and salt tolerance. Biochem Biophys Res Commun 379(4):985–989

Zhu GY, Kinet JM, Lutts S (2001) Characterization of rice (*Oryza sativa* L.) F3 populations selected for salt resistance. I. Physiological behaviour during vegetative growth. Euphytica 121:251–263

Rice Pollination **17**

Wali Muhammad, Munir Ahmad, and Shahid Hussain Shahid

Abstract

Rice (*Oryza sativa* L.) is an important and most nutritional cereal crop of the world and known as the second largest crop in Pakistan earning foreign exchange. Pollination of rice crop is important for the variability in genetic setup and its most crucial factor for the plant breeders as genetic stability and variability are important for the development of new rice varieties and increase in seed quantity and quality. Mostly, rice is known as the self-pollinating crop but crossing of rice pollens is also possible through wing, natural gravity, and natural pollinator agents (insects). Among the insects providing pollination services in rice crop, bees are most important as bees visit the rice flower for nectar and pollen collection and serve for pollen transfer among different flowers in the rice fields. Impact of insect pollinators in genetic flow of rice has been scientifically proved in many studies. About 510 insect species are found engaged in rice pollination of nectar and pollen feeding activities in rice crop. The role of honeybee was also identified in transgene flow in rice crop from a distance of 500 meters. Moreover, it is also emphasized that the natural yield and gene flow of rice crop can be successfully maintained through engaging the honeybee colonies in or near the rice fields along with associated benefits of apiculture industry. The risk of transgene flow in rice crop should also be considered as honeybees are reported to transfer viable pollen from GM rice varieties to their non-GM parents.

W. Muhammad (✉) · S. H. Shahid
Agriculture Pest Warning & Quality Control of Pesticides, Government of Punjab, Multan, Pakistan

M. Ahmad
Department of Entomology, PMAS-Arid Agriculture University Rawalpindi, Rawalpindi, Pakistan

N. Sarwar et al. (eds.), *Modern Techniques of Rice Crop Production*,
https://doi.org/10.1007/978-981-16-4955-4_17

Keywords

Rice · Pollination · Anemophily · Insect pollination

Abbreviations

FOT Flower opening time
GM Genetically modified
MP Mature pollen

17.1 Introduction of Pollination in Rice Crop

Rice (*Oryza sativa* L.) is an important cash and food crop in the world providing about 21% of global human per capita energy and 15% of per capita proteins to human (Ahmad et al. 2015, 2019, 2012, 2013, 2008, 2009). In Pakistan, rice is the second largest crop to generate foreign exchange (Ahmed et al. 2017, 2020a, b; Akram et al. 2018, 2019; Hussain et al. 2018; Fahad et al. 2019; Fatima et al. 2020; Hafeez-ur-Rehman et al. 2019; Khan et al. 2019a, b; Naz et al. 2019; Peerzada et al. 2019; Razaq et al. 2019; Sarwar et al. 2013a, b; Tariq et al. 2020; Wasaya et al. 2019; Zahoor et al. 2019). Rice has also many nutritional values including proteins, minerals, fiber, and vitamins (Memon 2013; Ahmad and Hasanuzzaman 2012; Ahmed and Ahmad 2019, 2020; Shahzad and Ahmad 2019).

The development of new varieties is very important and continuous process for the rice development and research centers to main the genetic viability and its nutritional values. There is a lengthy process for the development of new varieties, for which genetic pools are developed to maintain genetic variations. For this purpose, the natural mechanism of pollination is adopted, and crossing is also made for new variations in genetic makeup (Pu et al. 2014). Two types of crossing mechanisms in rice crop are reported for crossing of male and female flowers.

17.1.1 Self-Pollination

Rice is known as self-pollinating crop as it has a perfect flower containing pistil and stamens for its own pollination, hence there is no need to cross the flower for rice production, but for seed production crossing is essential before anthesis, and selective male parents are pollinated (Moon and Jung 2020; Win et al. 2020).

17.1.2 Cross Pollination

Rice is known as self-pollinating crop, but supplementary pollination is considered necessary for enhancing the out-crossing rate for increase in seed setting. Usually, supplementary pollination is carried out by shaking the parent pollen through ropes and sticks. Seed yield can be increased through continuous supplementary pollination after each 3–4 h for 10 to 12 days during the flowering at flower opening time (FOT) (Kobayasi et al. 2009). This management is considered important for improving the seed quality and quantity (Win et al. 2020; Hayes et al. 1955). Rice is also called as autogamous crop as the crops have natural crossing less than 5% (Destro and Montalván 1999).

17.2 Mode of Rice Pollination

17.2.1 Wind Pollination

Wind pollination is also reported in rice crop which is specifically termed as anemophily. Rice plants produce a high quantity of dry and light mature pollens (MP) which can be carried through wind. Stigma of rice plants is longer in size creating easiness to capture the pollens. This phenomenon is helpful in rice pollination that acquires no extra resources like insects such as bees as pollinators.

17.2.2 Animal Pollination

Animals are bigger group that help pollination of the crop plant for increasing in number of seeds and for healthier crop yields. When there are flowering plants or crops in our surroundings, there are also some bees, birds, butterflies, and bats working hard for vital crop yields as an unnoticed work force. These abovementioned groups work alone or in groups for maintaining the plant's diversity. The animals that are helpful for rice pollination are discussed below:

1. Insect Pollinators
   However, many animals help in crop pollination, but in case of rice crop, only insects are reported to help in pollination, and other animals like mammals and bats are not reported for the pollinating services in rice crop. As insects are considered responsible for every third bite of food we eat, native bees and some other insects are also reported to pollinate the rice crop as shown in Fig. 17.1.

Fig. 17.1 A bee visiting the rice crop on blooming

17.3 Biodiversity of Rice Pollinators

Rice flower has the greater quantiles of dry and viable pollen for pollen feeding insects and insects are reported to visit the rice field at the time of anthesis affecting the gene flow in rice variety development process. Many insect species are observed visiting the rice fields, about 510 insect species are noted as insect pollinators of rice crop, and these include the hover flies, honeybees, and some other insect species carrying rice pollens. *Apis mellifera* is reported as highly visiting insect in rice crop and collecting the pollen on a daily basis. The maximum visiting frequency of *Apis mellifera* was observed between 12.00 and 13.00 h.

17.4 Importance and Scope of Insect Pollination in Rice Cultivation

Studies proved that the higher diversity of rice visiting insects are carrying the rice pollens from a longer distance, which cannot be covered through wind pollination only. The presence of insect pollinators is also a risk factor for the breeders working on GM rice varieties as vector of viable pollens and increases the frequency of transgene flow. It was earlier assumed that there are no factors affecting natural gene flow in rice are changing the consideration of ecological risk assessment of gene flow in rice crop especially GM rice varieties.

Rice flowers are helpful for maintaining the honeybees' colonies in starvation days as risk of bees' mortalities increases with changing climate.

17.5 Future Scope of Rice Pollination Management

Currently the concern of genetic flow in rice gene due to cross pollination is not considered a serious issue. The natural crossing rate of the rice may be altered due to change in climate or the environment including insects and other pollinating agents of rice crop (Silva et al. 2005). The occurrence of genetic flow in rice due to crossing may contaminate the genetic and certified seed production. Economic losses are also expected if there is no proper management of crossing agents of rice crop. Exact sizing of barriers and hindering the insect pollinators are required to stop these contaminations in new developing seed lines of rice.

Global climate change is also detrimental for rice pollination (Piao et al. 2019). Increased heat event due to changing climate is characterized with pollen destruction and reduction of pollination capacity. This can greatly affect the rice production in developing countries where climate is changing with increased heat (Zhao et al. 2017). About 30% of rice spikelet are reported to be sterile due to heat wave in 2013 leading to severe yield losses of rice crop (Tan and Shen 2016). Intensified pollination services may address these changing and crop pollinations may be carried out through crossing through insect pollinators and use of wind sources (Wu et al. 2020).

Global environmental problems are being addressed with novelty of techniques and use of artificial intelligence in agriculture. Similar approach may be used for the improvement of rice pollination through development of robotic insect swarms (Srinivasan 2011; Wyss Institute 2016). Chemical sprays and hormones may also be used for increasing the fruit setting in unpollinated flowers in rice crop (Wright 2013). Golden rice development is best example of the rice variety which can produce grains without pollination or intensified insect pollinators are used. Breeding of genetic bees, resistant to pesticides, pathogens, and parasites are also suitable in future for sustainable crop yields (Schulte et al. 2014).

17.6 Summary

Rice (*Oryza sativa* L.) is most important cereal crop of the world and several transgenic varieties are developed to combat the climate change. GM rice varieties are considered as low risk for the environment as rice is a self-pollinating crop. Impact of insect pollinators in genetic flow of rice has been scientifically proved in many studies. About 510 insect species are found engaged in rice pollination of nectar and pollen feeding activities in rice crop. The role of honeybee was also identified in transgene flow in rice crop from a distance of 500 meters. Moreover, it is also emphasized that the natural yield and gene flow of rice crop can be successfully maintained through engaging the honeybee colonies in or near the rice fields along with associated benefits of apiculture industry. The risk of transgene flow in rice crop should also be considered as honeybees are reported to transfer viable pollen from GM rice varieties to their non-GM parents. Robotic and genetically mediated insect pollinators may be used for the improvement of rice pollination where pollen is destroyed due to climate change.

References

Ahmad A, Ashfaq M, Rasul G, Wajid SA, Khaliq T, Rasul F, Saeed U, Rahman MH, Hussain J, Baig IA, Naqvi AA, Bokhari SAA, Ahmad S, Naseem W, Hoogenboom G, Valdivia RO (2015) Impact of climate change on the rice–wheat cropping system of Pakistan. In: Hillel D, Rosenzweig C (eds.), Handbook of Climate Change and Agro-ecosystems: The Agricultural Modeling Intercomparison and Improvement Project (AgMIP) Integrated Crop and Economic Assessments, Published by Imperial College Press and the American Society of Agronomy, pp. 219–258

Ahmad S, Abbas G, Ahmed M, Fatima Z, Anjum MA, Rasul G, Khan MA, Hoogenboom G (2019) Climate warming and management impact on the change of rice-wheat phenology in Punjab, Pakistan. Field Crops Res 230:46–61

Ahmad S, Ahmad A, Ali H, Hussain A, Garcia A, Khan MA, Zia-Ul-Haq M, Hasanuzzaman M, Hoogenboom G (2013) Application of the CSM-CERES-Rice model for evaluation of plant density and irrigation management of transplanted rice for an irrigated semiarid environment. Irrig Sci 31(3):491–506

Ahmad S, Ahmad A, Soler CMT, Ali H, Zia-Ul-Haq M, Anothai J, Hussain A, Hoogenboom G, Hasanuzzaman M (2012) Application of the CSM-CERES-Rice model for evaluation of plant density and nitrogen management of fine transplanted rice for an irrigated semiarid environment. Precis Agric 13(2):200–218

Ahmad S, Ahmad A, Zia-ul-Haq M, Ali H, Khaliq T, Anjum MA, Khan MA, Hussain A, Hoogenboom G (2009) Resources use efficiency of field grown transplanted rice (*Oryza sativa* L.) under irrigated semiarid environment. J Food Agric Environ 7(2):487–492

Ahmad S, Hasanuzzaman M (2012) Integrated effect of plant density, N rates and irrigation regimes on the biomass production, N content, PAR use efficiencies and water productivity of rice under irrigated semiarid environment. Not Bot Horti Agrobot Cluj-Napoca 40(1):201–211

Ahmad S, Zia-ul-Haq M, Ali H, Shad SA, Ammad A, Maqsood M, Khan MB, Mehmood S, Hussain A (2008) Water and radiation use efficiencies of transplanted rice (*Oryza sativa* L.) at different plant densities and irrigation regimes under semi-arid environment. Pak J Bot 40(1): 199–209

Ahmed M, Ahmad S (2019) Carbon dioxide enrichment and crop productivity. In: Hasanuzzaman M (ed.), Agronomic crops; Volume 2 Springer Nature Singapore Pte ltd., pp. 31–46

Ahmed M, Ahmad S (2020) Systems modeling. In: Ahmed M (ed.), Systems modeling, Springer Nature Singapore Pte Ltd. pp. 1–44

Ahmed M, Ahmad S, Raza MA, Kumar U, Ansar M, Shah GA, Parsons D, Hoogenboom G, Palosuo T, Seidel S (2020a) Models calibration and evaluation. In: Ahmed M (ed), Systems modeling. Springer Nature Singapore Pte Ltd. pp. 151–178

Ahmed M, Ahmad S, Waldrip HM, Ramin M, Raza MA (2020b) Whole farm modeling: a systems approach to understanding and managing livestock for greenhouse gas mitigation, economic viability and environmental quality. In: Waldrip HM, Pagliari PH, He Z (eds.), Animal manure. ASA special publication 67. Madison, WI: American Society of Agronomy and Soil Science Society of America, pp. 345–371

Ahmed M, Fayyaz-ul-Hassan W, Ahmad S (2017) Climate variability impact on rice production: adaptation and mitigation strategies. In: Ahmed M, Stockle C (eds) Quantification of climate variability, adaptation and mitigation for agricultural sustainability. Springer, Cham, pp 91–111

Akram R, Fahad S, Masood N, Rasool A, Ijaz M, Ihsan MZ, Maqbool MM, Ahmad S, Hussain S, Ahmed M, Kaleem S, Sultana SR, Mubeen M, Saud S, Kamran M, Nasim W (2019) Plant growth and morphological changes in rice under abiotic stress. In: Hasanuzzaman M, Fujita M, Nahar K, Biswas JK (eds) Advances in rice research for abiotic stress tolerance. Woodhead Publishing, pp 69–85

Akram R, Turan V, Hammad HM, Ahmad S, Hussain S, Hasnain A, Maqbool MM, Rehmani MIA, Rasool A, Masood N, Mahmood F, Mubeen M, Sultana SR, Fahad S, Amanet K, Saleem M, Abbas Y, Akhtar HM, Hussain S, Waseem F, Murtaza R, Amin A, Zahoor SA, Sami Ul Din M, Nasim W (2018) Fate of organic and inorganic pollutants in paddy soils. In: Hashmi MZ, Varma A (eds), Environmental pollution of paddy soils. Springer Nature Switzerland AG, pp. 197–214

Destro D, Montalván R (1999) Modo de reprodução das plantas superiores In: Destro D, Montalván R. Melhoramento genético de plantas. Londrina: UEL, pp. 9–25

Fahad S, Noor M, Adnan M, Khan MA, Rahman I, Alam M, Khan IA, Ullah H, Mian IA, Hassan S, Saud S, Bakhat HF, Hammad HM, Ahmad S, Nasim W (2019) Abiotic stress and rice grain quality. In: Hasanuzzaman M, Fujita M, Nahar K, Biswas JK (eds) . Woodhead Publishing, Advances in rice research for abiotic stress tolerance, pp 571–583

Fatima Z, Ahmad M, Hussain M, Abbas G, Ul-Allah S, Ahmad S, Ahmed N, Ali MA, Sarwar G, Haque EU, Iqbal P, Hussain S (2020) The fingerprints of climate warming on cereal crops phenology and adaptation options. Sci Rep 10:18013

Hafeez-ur-Rehman WT, Nawaz A, Awan MI, Ijaz M, Hussain M, Ahmad S, Farooq M (2019) Direct seeding in rice: problems and prospects. In: Hasanuzzaman M (ed.), Agronomic crops, Volume 1: Production technologies. Springer Nature Singapore Pte Ltd, pp. 199–222

Hayes HK, Immer FR, Smith DC (1955) Methods of plant breeding. McGraw Hill Book Co., New York, p 551

Hussain M, Ahmad S, Hussain S, Lal R, Ul-Allah S, Nawaz A (2018) Rice in saline soils: physiology, biochemistry, genetics and management. Adv Agron 148:231–287

Khan MA, Ahmad S, Raza A (2019a) Integrated weed management for agronomic crops. In: Hasanuzzaman M (ed.), Agronomic crops; volume 2 Springer Nature Singapore Pte ltd., pp. 257–281

Khan SM, Ali S, Nawaz A, Bukhari SAH, Ejaz S, Ahmad S (2019b) Integrated pest and disease management for better agronomic crop production. In: Hasanuzzaman M (ed.), Agronomic crops; volume 2 Springer Nature Singapore Pte ltd. pp. 385–428

Kobayasi K, Masui H, Atsuta Y, Matsui T, Yoshimoto M, Hasegawa T (2009) Flower opening time in rice–cultivar difference and effect of weather factors. In: Proceedings of MARCO symposium

Memon NA (2013) Rice: important cash crop of Pakistan. In: Exclusive on Rice. Pak Food J pp. 21–23

Moon S, Jung KH (2020) First steps in the successful fertilization of rice and Arabidopsis: pollen longevity, adhesion and hydration. Plan Theory 9(8):956

Naz S, Fatima Z, Iqbal P, Khan A, Zakir I, Noreen S, Younis H, Abbas G, Ahmad S (2019) Agronomic crops: types and uses. In: Hasanuzzaman M (ed.), Agronomic crops. Volume 1: Production technologies. Springer Nature Singapore Pte. Ltd., pp. 1–18

Peerzada AM, Bukhari SAH, Dawood M, Nawaz A, Ahmad S, Adkins S (2019) Weed management for healthy crop production. In: Hasanuzzaman M (ed.), Agronomic crops; Volume 2 Springer Nature Singapore Pte ltd., pp. 225–256

Piao S, Liu Q, Chen A, Janssens IA, Fu Y, Dai J, Liu L, Lian X, Shen M, Zhu X (2019) Plant phenology and global climate change: current progresses and challenges. Glob Change Biol 25: 1922–1940

Pu DQ, Shi M, Wu Q, Gao MQ, Liu JF, Ren SP, Yang F, Tang P, Ye GY, Shen ZC, He JH (2014) Flower-visiting insects and their potential impact on transgene flow in rice. J Appl Ecol 51(5): 1357–1365

Razaq M, Shah FM, Ahmad S, Afzal M (2019) Pest management for agronomic crops. In: Hasanuzzaman M (ed.), Agronomic crops; Volume 2 Springer Nature Singapore Pte ltd. pp. 365–384

Sarwar N, Ali H, Ahmad S, Ehsan U, Ahmad S, Mubeen K, Hill JE (2013a) Water wise rice cultivation on calcareous soil with the addition of essential micronutrients. J Anim Plant Sci 23(1):244–250

Sarwar N, Ali H, Maqsood M, Ehsan U, Shahzad AN, Shahzad M, Mubeen K, Shahid MA, Ahmad S (2013b) Phenological response of rice plants to different micronutrients application under water saving paddy fields on calcareous soil. Turk J Field Crops 18(1):52–57

Schulte C, Theilenberg E, Müller-Borg M, Gempe T, Beye M (2014) Highly efficient integration and expression of piggyBac-derived cassettes in the honeybee (Apis mellifera). Proc Natl Acad Sci 111(24):9003–9008

Shahzad AN, Ahmad S (2019) Tools and techniques for nitrogen management in cereals. In: Hasanuzzaman M (ed.), Agronomic crops; Volume 2 Springer Nature Singapore Pte ltd., pp. 111–126

Silva EFD, Silva LMD, Montalván R (2005) Crossing rate and distance in upland rice. Bragantia 64(2):197–201

Srinivasan MV (2011) Honeybees as a model for the study of visually guided flight, navigation, and biologically inspired robotics. Physiol Rev 91(2):413–460

Tan SQ, Shen SH (2016) Distribution of high thermal damage to rice in middle and lower reaches of the Yangtze River in recent 32 years. Jiangsu Agric Sci 44:97–101

Tariq M, Ahmed M, Iqbal P, Fatima Z, Ahmad S (2020) Crop phenotyping. In: Ahmed M (ed), Systems modeling. Springer Nature Singapore Pte Ltd. pp. 45–60

Wasaya A, Yasir TA, Ijaz M, Ahmad S (2019) Tillage effects on agronomic crop production. In: Hasanuzzaman M (ed.), Agronomic crops; Volume 2 Springer Nature Singapore Pte ltd., pp. 73–99

Win A, Tanaka TST, Matsui T (2020) Panicle inclination influences pollination stability of rice (Oryzasativa L.). Plant Prod Sci 23(1):60–68. https://doi.org/10.1080/1343943X.2019.1698971

Wright CJ (2013) Manipulation of fruiting. Butterworths, London

Wu C, Cui K, Tang S, Li G, Wang S, Fahad S, Nie L, Huang J, Peng S, Ding Y (2020) Intensified pollination and fertilization ameliorate heat injury in rice (Oryza sativa L.) during the flowering stage. Field Crops Res 252:107795

Wyss Institute (2016) The RoboBees project and colony collapse disorder. http://wyss.harvard.edu/viewpage/536

Zahoor SA, Ahmad S, Ahmad A, Wajid A, Khaliq T, Mubeen M, Hussain S, Sami Ul Din M, Amin A, Awais M, Nasim W (2019) Improving water use efficiency in agronomic crop production. In: Hasanuzzaman M (ed), Agronomic crops; Volume 2 Springer Nature Singapore Pte Ltd., pp. 13–29

Zhao C, Liu B, Piao S, Wang X, Lobell DB, Huang Y, Huang M, Yao Y, Bassu S, Ciais P, Durand JL, Elliott J, Ewert F, Janssens IA, Li T, Lin E, Liu Q, Martre P, Müller C, Peng S, Peñuelas J, Ruane AC, Wallach D, Wang T, Wu D, Liu Z, Zhu Y, Zhu Z, Asseng S (2017) Temperature increase reduces global yields of major crops in four independent estimates. Proc Natl Acad Sci U S A 114:9326–9331

Part III

Plant Protection

Plant Protection Strategies in Rice Crop 18

Omer Farooq, Naeem Sarwar, Hafiz Muhammad Aatif, Muqarrab Ali, Atique-ur-Rehman, Azhar Abbas Khan, Muhammad Mazhar Iqbal, Muhammad Zeeshan Manshaa, and Shakeel Ahmad

Abstract

Rice is an important cereal crop in Asia. It is also used as staple food in various countries. It is mainly produced and consumed in Asia. Its yield is affected by insects, pests, and diseases in the rice growing regions in the world. Several approaches for controlling insects, pests, and diseases have been reported in scientific literature. However, modern scientific results revealed that integrated insect pest management, integrated disease management, and integrated weed management are ideal options for controlling insects, pests, and diseases.

Keywords

Oryza sativa L. · Insects · Pests · Diseases · Integrated management

O. Farooq (✉) · N. Sarwar · S. Ahmad
Department of Agronomy, Faculty of Agricultural Sciences and Technology, Bahauddin Zakariya University, Multan, Pakistan
e-mail: shakeelahmad@bzu.edu.pk

H. M. Aatif · A. A. Khan · M. Z. Manshaa
College of Agriculture, Bahauddin Zakariya University, Layyah, Pakistan

M. Ali
Department of Agronomy, MNS University of Agriculture, Multan, Pakistan

Atique-ur-Rehman
Departmentof Agronomy, Bahauddin Zakariya University, Multan, Pakistan

M. M. Iqbal
Soil and Water Testing Laboratory Chiniot, Department of Agriculture, Government of Punjab, Multan, Punjab, Pakistan

N. Sarwar et al. (eds.), *Modern Techniques of Rice Crop Production*,
https://doi.org/10.1007/978-981-16-4955-4_18

18.1 Introduction

Plant protection is an imperative aspect in agriculture that concerns with preventing crop yield losses caused by insects, diseases, and weeds. Among the large numbers of available options to deal with these yield reducing pests, selection of an accurate measure or method is sometimes very difficult and also requires professional skills, experience, and specialized knowledge. No doubt, almost all products for plant protection are categorized as poisons that are harmful for the human beings and environment also but at the same time are supposed to be effective means of controlling pests (Ozkara et al. 2016; Hong-xing et al. 2017; January et al. 2018). International Rice Research Institute has suggested an average of 37% loss in rice crops because of pests, whereas by using modern skills farmers can be supported in categorizing the type and class of pests present in their paddy fields. Modernized plant protection measures deal with existing problems of insect/pests by integrated use of all possible ways of controlling these pests, and these options could be administrative, preventive, chemical, mechanical, and biological means in order to minimize the loss of yields and also causing minimal damage to the environment. Practically all plants are affected by these pests that cause significant economic loss. Hence, plant protection has equal significance in all areas, in landscapes, water flows, channels, urban spaces, sport objects, and industrial spaces. Plant protection measures have also a very crucial role in storage of plant harvest and their products or by products.

At present chemical measures of controlling rice insects and diseases are more common and no doubt, contributing a lot in achieving the best targets. Worldwide, about two million tons of pesticide is utilized yearly for the plant protection (Alavanja 2009). However, there are several concerns with these chemicals like killing of beneficial soil microbes, human health issues, environmental pollution, and increasing resistant development in pests (Venter et al. 2006; Wasim et al. 2009; Köhler and Triebskorn 2013). The health disorders like nausea, blurred vision, and respiratory disorders are common in farmers who use pesticides than those who do not use (Ross et al. 2001; Sankoh et al. 2016). Ideally, plant protection strategies involve good practices to avoid the occurrence of pests in the first place, timely detection and close monitoring of potential losses caused, the use of suitable treatments, and assessment of this treatments in relation to environment disturbance.

18.2 Insect Management

18.2.1 Stem Borers

Stem borers include white stem borer (*Scirpophaga innotata* Wlk), pink stem borer (*Sesamia inferens* Wlk), and yellow stem borer (*Scirpophaga incertulas* Wlk). Nearly all insects have tremendous inherent ability to survive under unfavorable conditions (Misof et al. 2014). Rice insects are usually exogenic, therefore; the original size of population directly affects the development of the following generation (Hong-xing et al. 2017).

These borers are responsible for dead heart and white head. Stem borers make tiny hole in leaf sheaths and live there for some days. Further they bore into the stem and feed on it, and after that cut the central stem, this condition is called dead heart. Dead heart condition continued before reproductive stage. But when borers attack at flowering stage and allow the panicle to cut it from base then this condition is known as white head. The white head has white color with empty spike. In their management, following measures are taken.

The larvae of these insect hibernate in the crop leftovers and stubbles that remain after harvesting. So, destruction of crop stubbles is necessary for borer management. Rice nursery should not be sown before 20th May. If nursery is sown earlier than 20th May, then hibernating larvae start feeding on it and multiply their number. Transplanting of nursery should be completed before end of July. Plant only healthy seedlings, while discard unhealthy and affected seedlings from rice fields. Critical period of borer attack occurs between ends of August to start of September. Application of nitrogen fertilizers also affects these borers; higher amount application of nitrogenous fertilizers enhanced the rates of dead heart and white head of rice (Tan 1986).

18.2.2 Leaf Folder (*Cnaphalocrocis Medinalis* Gn)

On the leaf surface, the female leaf folder lays eggs usually in clusters of 10–12. Eggs hatch after 6 days. The newly born larvae attach to the leaf edges from tip downward and stay attach to the edges and make leaf like cylinder. After that larvae start feeding inside the plant leaf, damaging it and inserting pathogens inside plant body. Theses insects attack the crop from second week of August to October. In order to control leaf folder, sowing of less susceptible varieties should be ensured. Application of fertilizers (especially N, P, and K) and other essential nutrient elements in balanced amount can not only improve their utilization efficiency but also rice plant vigor, further, enhance the resistant ability to leaf folder and other pests of rice plant (de Kraker et al. 2000).

18.2.3 White-Backed Plant Hopper (*Sogatella Furcifera*)

The female of white back plant hopper laid eggs on leaf in groups. Each group contains 4–19 eggs. Nymph and adult damage the plant by sucking plant sap and causing injury to plants.This insect attacks the crop which has more vegetative and flashy growth. Their excreta are wax like which help the mold to develop and establish. The chlorosis at lower leaves and presence of mold at leaves are signs of attack. This starts from a patch and spread and covers large area having crop burn; therefore, this is called as hopper burn.

18.2.4 Brown-Plant Hopper (*Nilaparvata Lugens*)

It is a secondary pest for rice crop that causes potential damage (Fahad et al. 2015). Attack of brown plant hopper is occurred after mild winter followed by very hot summer having a little rain while monsoon season is highly favorable for insect to promote its population. Female of this insect lays egg at leaf sheath. Nymph and adults suck the sap of plant from stem and leafs. In this condition, plants become more vulnerable to disease and turn into yellow. Sowing of resistant varieties is found helpful against the brown plant hopper. Cultivation/transplanting materials should be clean. Injured and disease-infected plants should not be selected for cultivation. Avoid continuous standing of much water. Proper plant to plant distance should be maintained. Remove affected plants from rice field. Remove the crop stubbles when ploughing for next crop. Mechanized harvesting in stubble removing can play an important role (Wu et al. 2014). Don't apply nitrogen in excess amount. Nitrogen fertilizer should be applied in splits.

18.2.5 Green Leaf Hopper (*Nephotettix Nigropictus* or *N. Virescens*)

Green leaf hopper mainly feeds on foliage. The nymph and adults feed on sap of plant leaves. Resistance of host-plant is a desirable character for pest management (Cheng et al. 2008), so ensure the sowing of resistant varieties of rice. Clean cultivation/transplanting should be a helpful strategy to reduce the insect attack. Injured and disease-infected plants should not be cultivated. Avoid continuous standing of water in water. Remove crop stubbles. Transplanting should be completed in July.

18.2.6 Gall Midge *(Orseolia Oryzae)*

The insect remains active both in dry and wet seasons. Adult gall midge is small like mosquito. Female lays eggs on leaf. Plant defense mechanism against this insect is not generally reported; however, some wild species were reported to show some limited resistance (Khan et al. 1991). Insect starts attacking crop from its nursery stage and continues attacking till tillering stage. The maggot infects stem, reaches to its apical point, and starts feeding on its central shoot. Attacked tiller looks shining and round/tubular in shape; this condition is named silver shoot. Attacked plant cannot produce panicle. Insect attacks at early stages of crop result in stunted growth.

Acquiring natural resistance is possibly the most effective way of developing gall midge resistant varieties of rice (Bentur et al. 2016). Planting should be completed in July. Properly maintain plant to plant distance. Remove the affected plants and destroy them. Don't apply nitrogen in excess. Apply nitrogen fertilizer in more splits.

18.2.7 Rice Hispa (*Dicladispa Armigera*)

The pest is active from May to October. The female scrapes the soft/delicate portions of leaf and lays the eggs. The adults and grubs affect plant by feeding on it, particularly beetle suck chlorophyll of leaf and produce parallel white streaks and grub feed by mining into leaf and sucking chlorophyll.

At the time of nursery transplantation, clip the leaf. By this way, eggs of insect are removed. Collect and destroy the beetles of rice hispa. Completely dry the field for 3–4 days and then irrigate. Crop rotation is best in managing the disease. After harvesting remove the stubbles of crop because after crop season insect feed on stubbles.

18.2.8 Grass Hopper (*Hieroglyphus* Spp.)

Grasshopper female lays egg inside the soil and along the borders of field. Nymph and adult of this insect feed on different grasses. Their adults mostly eat on plant leaves and stem. Controlling all weed plants from the field and border around the field regularly. At the end of crop, plough field to expose eggs of insect. Their population extent and epidemic frequency can be effectively reduced by agro-ecosystem regularization through ecological engineering (Zhu et al. 2015; Gurr et al. 2016). Rice farming with animal husbandry pattern makes a good farming system. In rice-duck system, ducks feed on these hoppers and ineffective paddy tillers (Dai et al. 2004; Long et al. 2013).

18.2.9 Integrated Insect Management

The integrated insect management (IIM) is a recent concept that is gaining acceptance over the past couple of decades as an eco-friendly approach (Wu and Guo 2005).

IIM involves following measures:

- Always prefer resistant varieties for sowing.
- Sowing within time.
- Maintain proper planting geometry with optimum plant population.
- Crop rotation must be done to control insect. By crop rotation crop specific insects cannot increase their number. Also insect resistance cannot develop.
- Timely removal of affected plants and their destruction.
- Ploughing of field exposes the insects and their eggs.
- Biological control also effective against insect management. This method uses natural enemies of insects.

18.3 Economic Importance of Diseases

Many fungal, bacterial, viral, and nematode diseases attack on rice plant, and it has estimated that 36.5% average losses are caused by these diseases (Agrios 2005). Losses reported due to fungal diseases include rice blast, brown leaf spot, sheath blight of rice, sheath rot, grain rot, false smut, tungro disease, and bakanae disease of rice (Gupta et al. 2021*)*. The important bacterial diseases on rice have losses as bacterial leaf streak (8–17%) and bacterial blight (70%) that cause much damages to rice on different stages under the influence of diversity of environmental conditions. Viral diseases are rice yellow mottle virus (20–100%), rice stripe virus (70%) disease while root knot disease caused by nematodes has also severe losses (Chen et al. 2016). Rice is staple food and these rice diseases have drastic effect not only on the economy of farmers, but also cause the death, that is, momentous Bengal famine was occurred in 1842–1843 only due to Brown spot disease, in which 50–90% rice yield losses lead to the death of about two million people due to famishment (Agrios 2005).

Rice blast (*Magnaporthe oryzae*), major threat to food security worldwide, malady decreased the world rice production around 30% on which 60 million population could feed. Globally price of rice becomes higher, due to paramount losses, resulting in lower welfare of consumer and food security (Nalley et al. 2016). Rice production decreased due to rice blast resulting in lowering the yield and higher cost of production, directly or indirectly (Skamnioti and Gurr 2009). Among rice diseases, one of the most devastating maladies is rice blast which is most prevailing and pricy in temperate areas of world (Wang and Valent 2009). In 1910, sheath blight was initially documented from Japan. After that it was reported from all rice cultivating regions worldwide. Economically, this malady has become the one among the foremost vital rice maladies around the globe.

18.4 Rice Blast

18.4.1 Importance

Main barrier in battling the world's insecurity of food is rice blast caused by *Magnaporthe oryzae* resulting in 30% losses in production of rice worldwide which is enough for 60 million people. Prices go up due to these production losses in rice which reduces the food security and consumer welfare (Nalley et al. 2016). Rice production decreased due to rice blast resulting in lower yield and higher costs of production directly or indirectly (Skamnioti and Gurr 2009).

18.4.2 Pathogen Biology

Pyriform macroconidia, in *Magnaporthe oryzae*, are formed upon conidiophores projected on plants from lesions. The tag *Pyricularia* denotes the pyriform structure of the conidia. Where the conidiogenous cells of *Pyricularia* are polyblastic and integrated on the conidiophores, thus are sympodial, cylindrical, denticulate, and geniculate. Appressorium develops at tip point of germ tube after germination of conidia, which stick to the tissues of plant surface; an infection peg originates from appressorium and enters into tissues of plant. Melanin is the key pigment of appressorium and conidiophores wall.

18.4.3 Disease Symptom

Leaves along with nodes, panicles, and collar showed lesions or spots as a result of fungal attack.

Spindle-shaped lesions appeared on leaves with wider center and pointed towards both ends. Lesions become diamond shaped as the larger exhibiting grayish center with brown margins.

Susceptible lines leaves, in favorable environment, showed lesions which rapidly expanded, become amalgamate, and lead to be necrotic completely. Leaves showed lesions of various shapes viz. diamond to elongate along with pointed and tapered ends.

18.4.4 Disease Cycle

Initially, spore of blast infects rice plant and indices lesions, while completed its cycle in 20 days where fungus sporulates many folds and scatters its newly induced airborne spores. The infection cycle continued, under favorable moisture and temperature conditions.

18.4.5 Management Strategies

Since the pathogen of rice blast has a wide range of host so the crop rotation and eradication has little or no value. Lots of work is done to develop effective management strategies during this century.

18.4.5.1 Cultural Practices

Cultural practices play a vital role in the management of disease, including balance supply of nutrients, water, planting time and spacing, etc.

18.4.5.2 Chemical Control

In the utmost current field, timely use chemicals are more effective against blast problem. There are number of fungicides available to control this problem, but the imperative thing is selection of most suitable and safe fungicides, that is, difenoconazole and tebuconazole + trifloxystobin, which are found effective against the blast. However, it has some hazards, if not used wisely.

18.4.5.3 Breeding for Resistance and Marker Assisted Selection

Marker assisted selection is an efficient strategy through which varieties with high yield can be inserted with R genes of blast. But the barrier in keeping the advantages of yield is linkage drag (Wang et al. 2015). The utmost concrete result of any breeding program is higher production; however, maintenance breeding (biotic and abiotic stresses) commonly consequences in pathogen resistance, which can be observed as vindicating potential losses of crop.

18.4.5.4 Biological Control

Approaches related to biological control are getting more attention now a day due to its usefulness and less health problems. Scientist and researchers have given more attention in developing novel and advanced natural products-based fungicides and trying to introduce biocontrol agents to manage plant diseases. In the recent past, bacteria were proved as significant biocontrol agents against phytopathogenic fungi especially rice blast viz. 1Re14, 2R37, 1Pe2, *B. subtillis* strain B-332, and *S. sindenius* isolate 263.

18.5 Brown Leaf Spot of Rice

18.5.1 Importance

Bengal famine was occurred in 1842–1843 only due to Brown spot disease, in which 50–90% rice yield losses lead to the death of about two million people due to famishment (Agrios 2005).

18.5.2 Pathogen

Different scientists explained the pathogen differently. In 1900, it was first described by B. Haan as *Helminthosposium oryzae*. Drechsler (1934) considered that the fungus belongs to the genus Cochliobolus but Dastur (1942) formally transferred it to that genus. The name *C. miyabeanus* (Ito and Kuribayashi 1927). In Iran, *B. victoriae* (*C. victoriae*) has been found to be the most predominant species associated with brown spot followed by *B. oryzae* (*C. miyabeanus*), *B. bicolor*, and *B. indica* (Motlagh and Behzad 2008). Considerable variation has been observed among the pathogen isolates with respect to size, color, number of cells per conidium (Harish et al. 2007), and virulence (Misra 1985; Harish et al. 2007). Viability of the pathogen is directly influenced by temperature and R.H. of soil (Dallagnol et al. 2011).

18.5.3 Symptoms

The initial symptoms appear on areal parts. On leaves, typical brown spots with whitish or grey center, oval or cylindrical in shape appear. Latter on these spots merge and due to this leaf dries up. On susceptible cultivar, the larger spots are produced and may reach up to 1 cm. The similar symptoms are appeared on leaf sheaths and coleoptiles. Ramakrishnan and Subramanian (1977) observed that brown, minute spots are caused by *H. rostratum* (*D. rostrata*), rectangular brown spot by *H. oryzae* (*C. miyabeanus*), and leaf tip blight by *H. halodes*.

The pathogen produces lesions and these lesions extend downward beneath the sheath resulting in severe wet rotting, partially filled to chaffy dull grains and occasional hanging down of panicles. In severe cases, greyish mycelial growth is seen between sheath and stalk. Dry rot symptoms are observed when infection extends upward (Sunder et al. 2005).

18.5.4 Disease Cycle

The pathogen can survive 2–3 years in soil and infected plant parts. The host epidermis is penetrated through infection pegs arising from the appressoria (Ou 1985). Factors such as plant age, soil and fertilization, seeding methods, sowing time, temperature, light, and moisture have been observed to affect the disease development. In the early stage of rice plant, only minute spots are formed but after ear formation, large spots develop. The rice kernels are observed to be more susceptible at the flowering and milk stages than at the soft dough or mature stages (Ou 1985). Brown spot is known to occur in nutrient deficient soils (Ou 1985). Low organic matter in soil with high pH favors this disease (Holanda et al. 2002).

18.5.5 Disease Management

18.5.5.1 Chemical Control

Foliar spraying with edifenphos, aureofungin, and dithane M-45 has been observed effective in avoiding airborne infection and secondary spread of the disease (Ou 1985), whereas treating seed with 0.4% tricyclazole solution and further spraying with 0.25% mancozeb plus 0.08% tricyclazole (Viswanathan and Narayanasamy 1990) and treating seed with 2 g/kg thiram with 3 sprays of Companion or Ridomyl MZ (Ved et al. 2005) proved highly effective against brown spot.

18.5.5.2 Host Resistance

Host plant resistance to the disease is most economical way to manage brown spot of rice. In the past, breeding efforts have been more emphasized on diseases such as blast and bacterial blight (Savary et al. 2011); however, much efforts are now needed for brown spot disease of rice. For identifying resistance sources, inoculation methods have to be full proof (Vu-Van and Sangchote 2006).

18.5.5.3 Recent Advances

Indispensable mean in management approaches of disease includes understanding of diversity, variability, and mechanisms that impact the genotypical variation in population of pathogen. Information regarding variation in pathogen is helpful in recognizing and characterizing of germplasm resistance, but this consideration is restricted to brown spot pathogen. Resistance in host plant is an efficient mean in management of brown spot. Currently, URP markers have been used for analysis of *B. oryzae* genetic diversity. Comparative analysis of genome of Cochlibolus species tends to understand the genes role in secondary metabolism, miner protein secretion in pathogenesis, and diversity present at various levels viz. inter- and intraspecific.

18.5.5.4 Biological Control

Commercially existing antagonistic species (Pseudomonas and Trichoderma sp.) can suppress diseases by direct effect on the pathogen through antibiosis and mycoparasitism or by enhancing immunity of plant through induced resistance. In field trials, two sprays of extract of neem cake and leaf extract of *N. oleander* and *T. viride* minimize occurrence of brown spot disease by 70, 53, and 48%, respectively, whereas increase the grain yield significantly (Harish et al. 2008).

18.6 Sheath Blight of Rice

18.6.1 Importance

In 1910, sheath blight was initially documented from Japan. After that it was reported from all rice-cultivating regions worldwide. Economically, this malady has become one among the foremost vital rice maladies around the globe. Sheath blight can cause complete failure of crop under favorable climatic conditions (Singh 2016).

18.6.2 Biology

Among foremost maladies, sheath blight (*Rhizoctonia solani*) of rice is more damaging one (Molla et al. 2020). The pathogen *R. solani* belongs to genus *Rhizoctonia* and is soilborne, abundant pathogen that causes damage on a wide variety of economically important crops (Ajayi-Oyetunde and Bradley 2018).

18.6.3 Disease Symptoms

Initially, symptoms appear on sheath of leaf at or just overhead the water stripe as circular, oval or ellipsoid, greenish-gray colored spots which become water soaked. Lesions become large coalesce with centers of grayish white color encircled by irregular outline which shown tanned to dark brown color (Dolar et al. 2016). Leaf blades become infected and exhibit uneven lesions showing brown, green, or yellow orange outlines (Manjunat et al. 2021). Blades of leaf show extensive and coalesce growth of lesion completely or partially, which may reveal rattlesnake pattern of skin. Upward flow of water and nutrients interrupted due to tissue damage, canopy turn out to be dense, as plant moves to heading stage forming a humid microenvironment which supports the disease development (Singh et al. 2016). Under adverse situation, panicles and flag leaves of plant become infected as the malady moves up. Early infections, induced by the fungus, of sheath can expand and weaken the culms followed by lodging as well as tiller collapse.

18.6.4 Disease Cycle

Asexual stage of *R. solani* belongs to Fungi-imperfecti, *hyphomycetes, Agonomycetales* was called *R. solani* whereas, sexual stage belongs to *Basidiomycota, hymenomycetes, homobasidiomycetida,* and *Tulasnellale* was called as *Thanatephorus cucumeris*. This stage of Sclerotium belongs to *Rhizoctonia, Tulasnellaceae*, and *ThanatepHorus*. It is a type of filamentous fungi (Shahid et al. 2014). Sheath blight starts when a piece of infected plant stem or sclerotium from a foregoing season floats to the flooded water in paddy field and has contact with rice stem (Lanoiselet et al. 2005).

18.6.5 Management

Optimizing nitrogen fertilizer use in combination with spacing of plant can decrease the infection advancement. Replacement of fungicides with the natural ones or adoption of bio-agents can offer efficient control of sheath blight malady as well as lower the environmental influence. Genetic methods that exhibit promising sheath blight control comprise of exogenous treatment with dsRNA to silence pathogen gene expression and genome editing to develop rice lines that have lesser susceptibility to this disease and developing transgenic rice lines overexpressing or silencing pathogenesis-related genes (Ajayi-Oyetunde and Bradley 2018).

18.6.5.1 Host Plant Resistance

Microbial pathogens and parasitic fungi are restricted by the plants through initiation of systemic and local attributes. Signals induced at the area of infection trigger different defense-related responses against various pathogenic entities viz. *R. solani*. Association of various proteins with ROS regulation exhibits the key function in induction of systemic acquired resistance (SAR) and induced systemic resistance (ISR). Superoxide dismutase (SOD), catalase (CAT), ascorbate peroxidase (APx), and glutathione peroxidase (GPx) significantly minimize the ROS. In rearrangement of the plants defense actions, rhizosphere microbiome perform a vital part (Maignien et al. 2014). Soil microbial community plays an important role in primary ecosystem viz. nutrient cycling and decomposition, but their role is repeatedly ignored despite their services (Singh et al. 2005). The key barriers that were recognized in operative plant protection against this sheath blight disease are the pathogen adaptive flexibility, deficiency of resistant varieties of rice, single resistance genes absence for use in breeding, and lower contact of growers to informative events for best control strategies (Chethana 2018).

18.6.5.2 Chemical Control

In the recent past, a number of chemical fungicides are applied to control this disease. Different strategies were applied to manage this problem. Presently carbendazim (BCM), a systemic fungicide, is used against *R solani*. Thiamine priming was also used against *R solani.*, on rice (Bahuguna et al. 2012). Internationally, chemical control is considered the most affective practice to decrease yield loss caused by sheath blight (Aktaruzzaman et al. 2015).

18.6.5.3 Biological Control

Biological control is an efficient disease management strategy gaining momentum in recent times (Jayaprakashvel et al. 2014). Biocontrol application of *P. fluorescens* PF-08 and *T. harzianum* UBSTH-501 alone or in combination not only helps in control of the disease but also increases plant growth along with reduction in application of toxic chemical pesticides (Dai et al. 2004). The most effective and long-term strategy for the management of sheath blight is to reduce the concentration of initial inoculum by killing sclerotia or to inhibit their germination (Prasanna et al. 2021). Several studies have examined the potential use of biocontrol agents for control of sheath blight through the application of antifungal bacteria isolated from the soil, sclerotia, or other habitats.

18.7 Ufra Rice Diseases

18.7.1 Disease History

Ufra is a disease of deep water and lowland rice which is caused by the nematode *Ditylenchus angustus* in parts of Asia (Neog and Chauhan 2015) and is regarded as the fourth worst disease of rice in Bangladesh according to Islam and Catling (Islam and Rice 2012). Yield losses induced by this nematode are severe and range from 20 to 90% (Bridge et al. 1990). Butler was first reported this disease from Noakhali in 1913. Generally, this disease is observed in deepwater rice growing areas. Susceptible varieties of rice are more vulnerable to Ufra disease.

18.7.2 Pathogen and its Biology

Ditylenchus angustus is an obligate parasite and is liable for considerable yield losses especially in Asian countries, that is, 15% in India (Rao et al. 1986), almost 100% in Thailand (Hashioka 1963), and Bangladesh (Miah and Bakr 1977). It was observed that all life stages of the nematode are infectious and have capability to invade the plant, whereas Juvenile stage 4 (J4 stage) cause maximum damage to the plant. *Ditylenchus angustus* nematodes mostly feed ectoparasitically at the collar region of the plant and complete its life cycle within 10–20 days at 27–30 °C (Bridge and Starr 2007).

18.7.3 Host Range

Ditylenchus angustus nematode is mostly restricted to deeply flooded cultivated rice, particularly deepwater rice and the Oryza wild rice. It has also been reported on grassy weeds including *Echinochloa colona* and *Leersia hexandra.*

18.7.4 Symptoms

Ditylenchus angustus develops and reproduces in the leaf sheath and stalk, where it feeds ectoparasitically on meristematic tissues (CAB International. *Ditylenchus angustus*). This disease is characterized by first whitish and then brownish leaf tips of plant, stem distortion above the last node, and stunted development of the ear and in severely attacked plants, by decay. Mostly nematodes are spread by rain splashes and through irrigation water.

18.7.5 Management

18.7.5.1 Cultural Control

Cultural practices are the most effective and inexpensive approaches to control this problem, such as deep plowing, fallowing during summer, destruction of weeds and burning the stubbles of earlier crops, use of certified planting materials, and cultivation of resistant, tolerant or nonhost crops, which may substantially prevent the crop losses in cereals which is caused by nematodes.

The resistance mechanism of different rice cultivar against nematodes has been investigated in many plant nematode interactions experiments by a number of scientists. It was observed that some cultivars of *Oryza sativa* have different levels of resistance against *Ditylenchus angustus* (Das et al. 2000; Latif et al. 2011). Miah and Bakr (1977) reported that wild rice species *O. subulata* also showed resistance against these nematodes. In the recent past, "Manikpukha" cultivar had shown highly resistant response against *D. angustus* (Khanam et al. 2018).

18.7.5.2 Biological Control

Use of seed treatment with biopesticides of *Trichoderma* spp. or *P. fluorescens* as a general treatment may prove quite effective in growing healthy crop with considerably higher yields.

18.7.5.3 Chemical Control

In the present scenario, different companies have different nematicides which are used to suppress or control the population of nematodes. Application of nematicides before sowing or transplanting of rice nursery has better result.

18.8 Khaira Disease

This problem is caused by Zn deficiency. Sandy soil and soil with low organic matter are deficient in Zn. Susceptible rice varieties, high soil pH ($\geq$7 under anaerobic conditions), and higher availability of Fe, Mn, P, Cu, Ca and Mg after flooding and excessive liming are some of the major reasons for Zn deficiency in rice plant.

18.8.1 History

Khaira was first spotted in rice (*Oryza sativa*) on calcareous soils of Tarai region of Uttar Pradesh of India (Nene 1966). Widespread phenomenon of this disease in low land rice and in growing area of Asia, next to nitrogen (N) and phosphorous (P) deficiency yield loss may be up to 25%.

18.8.2 Distribution

Mostly this disease is common in densely or intensively planted rice field. The saline-sodic type of soil is another important factor of Khaira disease. Calcareous soils and soils having high content of organic matter are also facing this problem. And this problem is also occurring in those soils where ratio of total available phosphorus is higher than normal soils. Higher PH level in soils is also contributing to cause this disease.

18.8.3 Symptoms

Deficiency symptom in rice appears when tissue levels fall below 20 mg/kg (Takkar 1991). Deficiency of Zn causes numerous symptoms which appeared 2 to 3 weeks after transplanting. Preliminary deficiency of Zn may cause interveinal chlorosis, size of young leaves may reduce, and characteristic brown rusty spots may develop. Acute deficiency of Zn may cause stunted plant growth with brown rusty appearance.

18.8.4 Favorable Conditions

Deficiency of available Zn in the soil favors this disease. Plant varieties susceptible to Zn deficiency (Zn inefficient cultivars) are favorable for this disease. When pH of the soil increases, Zn is precipitated as sparingly soluble $Zn(OH)^2$.

18.8.5 Management

Diverse management strategies are used to manage this disease. As "Khaira disease" appears on leaf at the early stages of growth and is directly related to Zn deficiency, its early diagnoses play important role for its management. After sowing or transplantation of a crop, the disease is preferentially cured by three foliar sprays of Zn at regular intervals (one week). Foliar spray of micronutrients is more appropriate as compared to the soil application, because it decreases the soil toxicity by decreasing elements accumulation in soil, easy to apply. Foliar application of zinc sulfate ($ZnSO_4.7H_2O$) at late growth stages can also be helpful in reducing this problem (Abd El-Baky et al. 2010; Yosefi et al. 2011). In order to get maximum benefits of Zn treatment, it is strongly recommended its application before sowing or transplanting.

18.8.6 Integrated Disease Management

Integrated disease management (IPM) has been an important measure for crop protection since the 1960s (Horgan 2017). Integrated disease management deals with the management of multiple pests simultaneously, regular monitoring of pests as well as their natural enemies, application of treatment and economic threshold when using pesticides, and the use of multiple and suppressive tactics in an integrated manner (Duke and Powles 2008). The following strategies are preferred during integrated disease management.

- Select healthy seeds for cultivation.
- Always grow disease resistant varieties.
- Timely sowing of nursery.
- Maintain optimum plant population.
- Don't irrigate field to field. Irrigate using water channel.
- Remove burn and destroy the affected plants and their stubbles.
- Weed control should be ensured as certain weeds act as host for some pathogens.
- Need-based application of chemical control.

18.9 Weeds

Since the last century, more than 1800 plant species have been recorded in paddy fields (Kamoshita et al. 2014). Weeds are identified as one of major biological constraints that hinders the attainment of optimal rice productivity in major rice producing countries of South Asia like India (Rao et al. 2015); Bangladesh (Rashid et al. 2012), Sri Lanka (Weerakoon et al. 2011), and Pakistan (Farooq and Cheema 2013; Irshad and Cheema 2006). In rice critical weed crop competition is for about 30 days after direct sowing or transplanting nursery. Generally sedges and grasses mostly dominate in rice crop. Among these, *Cyperus rotundus* and *Echinochloa crusgalli* are the most competitive weeds. Most of weeds that infest rice crop are following:

Echinochloa colonum	*Echinochloa crusgalli*
Panicum sp	*Setaria glauca*
Cyperus rotundus	*Cyperus deformis*
Fimbristylis miliacea	*Cynodon doctylon*
Digitaria sp	*Commelina benghalensis*
Ageratum conyoides	*Ammonia braccifera*
Monochoria vaginalis	*Seirpus sp*

For insect, pests, and disease management, weed control is important as certain weeds act as host for certain insect and pests. For example, Carolina foxtail is a new host for blast pathogen and earlier it was unrecognized (Jia et al. 2008).

18.9.1 Weed Management

Manual removal of weed is mostly used in weed management but in rice manual weeding is difficult. Manual weeding depends on nature and intensity of weed infestation. Moreover, identification between grassy weeds and rice plant is difficult at early stage. Intercultivation can be done with rotary weeder if rice is sowed by maintaining proper plant distance. Preparatory tillage, flooding, and puddling are most effective ways in controlling the rice weeds. Soil compaction and continuous submergence reduce weeds growth. Plant population should be optimum; there should be no space enough for weeds to grow. Crop rotation plays an important role in weed management as Mollah et al. (2015) found lower rice yields in a Wheat-Fallow-Rice cropping program due to high weed infestation when the field was left empty after wheat harvest. Weed control was quite impoverished in the Wheat-Mungbean-Rice cropping system. In comparison to the Wheat-Fallow-Rice cropping trend, infestation in DSR was lower, implying that including mungbean in the rice-wheat cropping method is an important weed control measure in DSR.

Competitive crop cultivar usage is a low-input cultural weed control approach that should be used. Crop competition can be improved by planting weed-resistant cultivars (e.g., allelopathic nature, early growth vigor, rapid ground coverage, and more light interception). Preventative measures should be implemented at all levels of crop development, from planting to harvesting. Using clean crop seeds and machinery in the field is a much cheaper and easier choice. Furthermore, nonchemical weed control measures are gaining great attention among the sustainable growers around the world (Farooq et al. 2019).

18.9.2 Integrated Weed Management

18.9.2.1 Monitoring of Weeds

Crop field should be monitored and emerging weeds, their growth stage, appearance of new weeds, and resistant weeds should be noted.

18.9.2.2 Ploughing

Ploughing is effective in controlling weeds particularly perennial weeds by stopping their emergence.

18.9.2.3 Water Management

Continuous standing water can reduce weeds. Most of broad leaf weeds suppress their growth at submergence or field capacity.

18.9.2.4 Crop Rotation

The moisture-loving weeds can be controlled by proper crop rotation.

18.9.2.5 Stale Seedbed Technique

This method includes the successively removal of weeds before sowing. First land is prepared and weeds are allowed to germinate. After weed germination, weeds are removed mechanically, manually, or even chemically.

18.9.2.6 Crop Stand

Crop plant should be sown at optimum plant population. There should be no enough empty space allowing weeds to grow and compete with crop.

18.9.2.7 Fertilizer Management

Nitrogen fertilizer should be applied in splits. Excessive nitrogen applications at early stages allow weeds to grow (Table 18.1).

Table 18.1 Cultural management practices for some rice pest (Adapted from Pathak et al. 1998)

Cultural practices	Hispa	Yellow stem borer	Gall midge	Brown plant hopper	Green leaf hopper	Leaf folder	Rodents
Early planting	*	*	*	*	*	–	–
Synchronized planting	–	*	*	*	*	*	*
Crop rotation	*	–	–	–	–	–	–
Direct seeding	–	*	*	–	–	–	–
Intercropping	–	*	–	*	–	–	–
Removal of affected plants	–	–	–	*	*	–	–
Burning/ ploughing	–	*	–	*	*	–	–
Stubble removal	*	–	–	*	*	*	–
Proper water management	–	*	–	*	*	*	–
Proper fertilizer management	–	*	*	–	–	*	*
Plant density	*	–	–	*	*	–	–

*Shows that crop is affected by this management by respective pest.
–Shows that crop is not affected by this management.

References

Abd El-Baky MMH, Ahmed AA, El-Nemr MA, Zaki MF (2010) Effect of potassium fertilizer and foliar zinc application on yield and quality of sweet potato. Res J Agric Biol 6:386–394

Agrios G (2005) Plant pathology, vol 26–27, 5th edn. Elsevier Academic Press, Amsterdam, pp 398–401

Ajayi-Oyetunde OO, Bradley CA (2018) Rhizoctonia solani: taxonomy, population biology, and management of rhizoctonia seedling disease of soybean. Plant Pathol 67:3–17

Aktaruzzaman MD, Islam B, Rashid D (2015) Fungicidal management of sheath blight of rice. doi: https://doi.org/10.13140/RG.2.1.3055.3443

Alavanja MC (2009) Pesticides use and exposure extensive worldwide. Rev Environ Health 24(4): 303–309

Bahuguna RN, Joshi R, Shukla A, Pandey M, Kumar J (2012) Thiamine primed defense provides reliable alternative to systemic fungicide carbendazim against sheath blight disease in rice (*Oryza Sativa* L.). Plant Physiol Biochem 57:159–167

Bentur JS, Rawat N, Divya D, Sinha DK, Agarrwal R, Atray I, Nair S (2016) Rice-gall midge interactions: Battle for survival. J Insect Physiol 84:40–49

Bridge J, Starr JL (2007) Cereals: rice (Oryza sativa). In: Bridge J, Starr JL (eds) Plant nematodes of agricultural importance: a colour handbook. Manson Publishing, London, pp 52–60

Bridge J, Luc M, Plowright RA (1990) Nematode parasites of rice. In: Luc M, Sikora RA, Bridge J (eds) Plant parasitic nematodes in subtropical and tropical agriculture. CAB International, Wallingford, pp 69–108

Chen GH, Zhu PY, Zheng XS, Yao XM, Zhang FC, Sheng XQ, Xu HX, Lu ZX (2016) The practice of ecological engineering on the control of rice insect pests in Jinhua. China Plant Prot 36(1): 31–36

Cheng JA, Zhu JL, Zhu ZR, Zhang LG (2008) Rice planthopper outbreak and environment regulation. J Environ Entomol 30(2):176–182

Chethana B (2018) A new combination fungicide for the management of sheath blight and neck blast diseases of paddy. Int J Pure Appl Biosci 6:651–655

Dai ZM, Yang HS, Zhang X, Zhou XZ, Han ZH, Yu Y, Qian GP, Zhang SQ, Masaharu M, Yoshitaka N, Wei HJ (2004) The research on the benefit of Yunnan rice-duck intergrowth model and its comprehensive evaluation (part 3). Chin Agric Sci Bull 4:265–267

Dallagnol LJ, Rodrigues FA, Da Matta FM (2011) Brown spot of rice as affected by phyton irradiance and temperature. J Phytopathol 159:630–634

Das D, Sarma N, Borgohain R, Das P (2000) Rayada B3—a high-yielding, ufra-resistant Deepwater rice for Assam, India. Int Rice Res Notes 25:17

Dastur JF (1942) Notes on some fungi isolated from'black point' effected kernels in the central provice. Indian J Agr Res 12:731–742

Dolar P, You MP, Vincent L, Barbetti MJ (2016) Reservoir of cultivated rice pathogens in wild rice in Australia. Eur J Plant Pathol 147:295–311

Drechsler C (1934) Phytopathological and taxonomical aspects of *Ophilobolus, Pyrenophora, Helminthosporium* and a new genus *Cochliobolus*. Phytopathology 24:973

Duke S, Powles S (2008) Perspective integrated pest management (IPM): definition, historical development and implementation, and the other IPM Lester. Pest Manag Sci 63(11):1100–1106

Fahad S, Nie L, Hussain S, Khan F, Khan FA, Saud S, Muhammad H, Li L, Liu X, Tabassum A, Wu C, Xiong D, Cui K, Huang J (2015) Rice pest management and biological control. October 2020:85–106. https://doi.org/10.1007/978-3-319-16988-0_4

Farooq O, Cheema ZA (2013) Impact of planting time and seedbed conditions on little seed canary grass and lambsquarters dynamics in wheat (*Triticum aestivum*). Int J Agric Biol 15:1003–1007

Farooq O, Mubeen K, Ali HH, Ahmad S (2019) Non-chemical Weed Management for field crops. In: Hasanuzzaman M (ed) Agronomic crops, volume 2: management practices. Springer Nature Singapore Pte Ltd. https://doi.org/10.1007/978-981-32-9783-8_16

Gupta L, Vermani M, Ahluwalia SK, Vijayaraghavan P (2021) Molecular virulence determinants of Magnaporthe oryzae: disease pathogenesis and recent interventions for disease management in rice plant. Mycology. https://doi.org/10.1080/21501203.2020.1868594

Gurr GM, Lu ZX, Zheng XS, Xu HX, Zhu PY, Chen GH, Yao XM, Cheng JA, Zhu ZR, Catindig JL, Villareal S, Chien HV, Cuong LQ, Channoo C, Chengwattana N, Lan LP, Hai LH, Chaiwong J, Nicol HI, Perovic DJ, Wratten SD, Heong KL (2016) Multi-country evidence that crop diversification promotes ecological intensification of agriculture. Nat Plants 2:16014

Harish S, Saravanakumar D, Kamalakannan A, Vivekananthan R, Ebenezar EG, Seetharaman K (2007) Phylloplane microorganisms as a potential biocontrol agent against *Helminthosporium oryzae* Breda de Haan, the incitant of rice brown spot. Arch Phytopathol Plant Prot 40:148–157

Harish S, Duraiswamy S, Ramalingam R, Ebenezar EG, Seetharaman K (2008) Use of plant extracts and biocontrol agents for the management of brown spot disease in rice. BioControl 53:555–567

Hashioka Y (1963) The rice stem nematode angustus in Thailand. FAO Ditylenchus Plant Prot Bull 11:97–102

Holanda FSR, da Mendonca MC, Melo AS, Cassetari ND, de Carvalho JG, Bertoni JC (2002) Influence of the availability of nutrients and irrigation water on the severity of brown spot and leaf scald in rice crops. Revista Ensaios-e-Ciencia;-Serie Ciencias Biologicas. Agrariase-da-Saude 6:85–96

Hong-xing X, Ya-jun Y, Yan-hui L, Xu-song Z, Jun-ce T, Feng-xiang L, Qiang F, Zhong-xian L (2017) Sustainable management of rice insect pests by non-chemical-insecticide technologies in China. Rice Sci 24(2):61–72

Horgan FG (2017) Integrated pest management for sustainable rice cultivation: a holistic approach, pp. 309–342

Irshad A, Cheema ZA (2006) Comparative efficiency of barnyard grass and rice to nitrogen under transplanted condition. J Trop Agric Food Sci 34:385–392

Islam Z, Rice CD (2012) Pests of Bangladesh: their ecology and management. The University Press, Ltd., Dhaka, Bangladesh, p 422

Ito S, Kuribayashi K (1927) Production of ascigerous stage in culture of *Helminthosporium oryzae*. Ann Phytopathol Soc Jpn 2:1–8

January B, Rwegasira GM, Tefera T (2018) Farmers' perceptions of rice production constraints and stem borers management practices in Tanzania. J Agric Sci 10:57–71

Jayaprakashvel M, Sharmika N, Vinothini S, Venkatramani M, Muthezhilan R, Hussain JA (2014) Biological control of sheath blight of rice using marine associated fluorescent pseudomonads. Biosci Biotechnol Res Asia 11(spl. Edn. 1): 115–121

Jia Y, Gealy D, Lin MJ, Wu L, Black H (2008) Carolina foxtail (*Alopecurus carolinianus*): susceptibility and suitability as an alternative host to rice blast disease (Magnaporthe oryzae [formerly M. grisea]). Plant Dis 92:504–507

Kamoshita A, Araki Y, Nguyen Y (2014) Weed biodiversity and rice production during the irrigation rehabilitation process in Cambodia. Agric Ecosyst Environ 194:1–6

Khan ZR, Litsinger JA, Barrion AT, Villanueva FFD, Fernandez NJ, Taylo LD (1991) World bibliography of rice stem borers 1794–1990. International Rice Research Institute, Los Baños, Philippines

Khanam S, Bauters L, Singh RR, Verbeek R, Haeck A, Sultan SMD, Dameestere K, Kyndt T, Gheysen G (2018) Mechanisms of resistance in the rice cultivar Manikpukha to the rice stem nematode *Ditylenchus angustus*. Mol Plant Pathol 19:1391–1402

Köhler HR, Triebskorn R (2013) Wildlife ecotoxicology of pesticides: can we track effects to the population level and beyond? Science 341(6147):759–765

de Kraker J, Rabbinge R, van Huis A, van Lenteren JC, Heong KL (2000) Impact of nitrogenous-fertilization on the population dynamics and natural control of rice leaffolders (Lep.: Pyralidae). Int J Pest Manag 46(3):225–235

Lanoiselet VM, Cother EJ, Ash GJ, Harper JDI (2005) Yield loss in rice caused by Rhizoctonia oryzae-sativae in Australia. Aust Plant Pathol 34:175–179

Latif M, Rafii M, Jahan AHQ, Hossain M (2011) Cost-effective management of ufra disease of rice and identification of resistant landraces. Sci Res Essays 6:2668–2675

Long P, Huang H, Liao XL, Fu ZQ, Zheng HB, Chen AW, Chen C (2013) Mechanism and capacities of reducing ecological cost through rice-duck cultivation. J Sci Food Agric 93(12): 2881–2891

Maignien L, DeForce EA, Chafee ME, Murat Eren A, Simmons SL (2014) Ecological succession and stochastic variation in the assembly of Arabidopsis thaliana phyllosphere communities. mBio 5:e00682–13

Manjunat O, Sagar BV, Prakasham V, Reddy CN (2021) Variability studies on sheath blight of Rice in Karnataka. Int J Curr Microbiol Appl Sci 7(10):724–736

Miah S, Bakr M (1977) Chemical control of ufra disease of rice. PANS 23:412–413

Misof B, Liu S, Meusemann K, Peters RS, Donath A, Mayer C, Frandsen PB, Ware J, Flouri T, Beutel RG et al (2014) Phylogenomics resolves the timing and pattern of insect evolution. Science 7:763–767

Misra AK (1985) Variability in *Drechslera oryzae*- the causal organism of brown spot disease of rice. Indian Phytopath 38:168–169

Molla KA, Karmakar S, Molla J, Bajaj P, Varshney RK, Datta SK, Datta K (2020) Understanding sheath blight resistance in rice: the road behind and the road ahead. Plant Biotechnol J 18:895–915

Mollah MIU, Bhuiya MSU, Khatun A, Hossain SMA (2015) Increasing crop diversity and productivity of rice (Oryza Sativa L.)-wheat (Triticum aestivum L.) cropping system through bed planting. Bangladesh Rice J 19:37–46

Motlagh MRS, Behzad KL (2008) Characterization of new Bipolaris Spp.: the causal agent of rice brown spot disease in the north of Iran. Int J Agric. Biol 6:10

Nalley L, Tsiboe F, Durand-Morat A, Shew A, Thoma G (2016) Economic and environmental impact of rice blast pathogen (Magnaporthe oryzae) alleviation in the United States. PLoS One 11(12):e0167295

Nene YL (1966) Symptoms, cause and control of Khaira disease of paddy. Bull Indian Phytopathol Soc 3:97–191

Neog PP, Chauhan MK (2015) Incidence and management of ufra disease caused by Ditylenchus angustus in Deepwater rice in Assam. Annals Plant Protection Sciences 23:189–190

Ou SH (1985) Rice diseases, 2nd edn. CMI, Kew, England, 370 pp

Ozkara A, Akyil D, Konuk M (2016) Pesticides, environmental pollution and health. In: Larramendy ML, Soloneski S (eds) Environmental health risk- hazardous factors to living species. Intech. https://doi.org/10.5772/63094

Pathak MD, Rao YRVJ, Kameswar Row KV, Mukhopadhyay SK (1998) Integrated pest management in rice. In: CRRI (ed) Rainfed rice for sustainable food security. Central Rice research institute, Cuttack, India

Prasanna R, Reddy KRN, Reddy M, Vijay K, Kumar K, Sudini H (2021) Biological control of rice sheath blight disease by pseudomonas fluorescens isolates. PhD thesis

Ramakrishnan G, Subramanian CL (1977) Studies on helminthosporiose of rice: involvement of different species in disease development. Madras Agric J 64:60–61

Rao Y, Prasad J, Panwar M (1986) Stem nematode (Ditylenchus angustus) a potential pest of rice in Assam and West Bengal, India. Inter Nematol Network Newslett 3:24–26

Rao AN, Wani SP, Ladha JK (2015) Weed management research in India- An analysis of past and outlook for future. pp. 1–26. In: DWR - Souvenir, Celebrating Silver Jubilee (1989-2014). Directorate of Weed Research, Jabalpur, India

Rashid MH, Alam M, Rao AN, Ladha JK (2012) Comparative efficacy of pretilachlor and hand weeding in managing weeds and improving the productivity wet-seeded rice in Bangladesh. Field Crops Res 128:17–26

Ross JH, Driver JH, Cochran RC, Thongsinthusak T, Krieger RI (2001) Could pesticide toxicology studies be more relevant to occupational risk assessment? Ann Occup Hyg 45(1):5–17

Sankoh AI, Whittle R, Semple KT, Jones KC, Sweetman AJ (2016) An assessment of the impacts of pesticide use on the environment and health of rice farmers in Sierra Leone. Environ Int 94: 458–466

Savary S, Nelson A, Sparks AH, Willocquet L, Duveiller E, Mahuku G, Forbes G, Garrett KA, Hodson D, Padgham J, Pande S, Sharma M, Yuen J, Djurle A (2011) International agricultural research tackling the effects of global and climate changes on plant diseases in the developing world. Plant Dis 95:1204–1216

Shahid AA, Shahbaz M, Ali M (2014) A comparative study of the commercially available fungicides to control sheath blight of rice in Lahore. J Plant Pathol Microb 5:240

Singh R (2016) Sheath blight of Rice: a review. Int J Agric Invent 1:161–169

Singh RK, Singh CV, Shukla VD (2005) Phosphorus nutrition reduces brown spot incidence in rainfed upland rice. Intern Rice Res Notes 30:31–32

Singh DB, Malviya D, Wasiullah SS, Pradhan JK, Singh BP, Roy M, Imram M, Pathak N, Baisyal BM, Rai JP, Sarma BK, Singh RK, Sharma PK, Kaur SD, Manna MC, Sharma SK, Sharma AK (2016) Bio-protective microbial agents from rhizosphere eco-systems trigger plant defense responses provide protection against sheath blight disease in rice (Oryza sativa L.). Microbiol Res 192:300–312

Skamnioti P, Gurr SJ (2009) Against the grain: safeguarding rice from rice blast disease. Trends Biotechnol 27(3):141–150

Sunder S, Singh R, Dodan DS, Mehla DS (2005) Effect of different nitrogen levels on brown spot (*Drechslera oryzae*) of rice and its management through host resistance and fungicides. Pl Dis Res 20:111–114

Takkar PN (1991) Zinc deficiency in Indian soils and crops. (in) zinc in crop nutrition. International Lead zinc research organization Inc. and Indian Lead zinc information center, New Delhi pp. 55–64

Tan YC (1986) Preliminary study on the relationship between fertilizer application and occurrence of striped stem borer, Chilo suppressalis. Entomol Knowl 23(3):101–102

Ved R, Biswas SK, Kumar K, Yadav MD (2005) Evaluation of fungicides for the management of brown leaf spot (*Drechslera oryzae*) and blast (*Pyricularia oryzae*) of paddy. Farm Sci J 14:38–39

Venter O, Brodeur NN, Nemiroff L, Belland B, Dolinsek IJ, Grant JWA (2006) Threats to endangered species in Canada. Bioscience 56(11):903–910

Viswanathan R, Narayanasamy P (1990) Chemical control of brown leaf spot of rice. Indian J Mycol Pl Pathol 20:139–143

Vu-Van BA, Sangchote S (2006) Seed borne and transmission of *Bipolaris oryzae*, the causal pathogen of brown spot of rice. Kasetsart J Natural Sci 40:353–360

Wang G-L, Valent B (2009) Advances in genetics, genomics and control of rice blast disease. Springer Science+Business Media B.V

Wang X, Jia M, Ghai P, Lee F, Jia Y (2015) Genome-wide association of rice blast disease resistance and yield-related components of rice. Mol Plant-Microbe Interact 28:1383–1392

Wasim MD, Sengupta D, Chowdhry A (2009) Impact of pesticides use in agriculture: their benefits and hazards. Interdiscip Toxicol 2(1):1–12

Weerakoon WMW, Mutunayake MMP, Bandara C, Rao AN, Bhandari DC, Ladha JK (2011) Direct-seeded rice culture in Sri Lanka: lessons from farmers. Field Crops Res 121:53–63

Wu KM, Guo YY (2005) The evolution of cotton pest management practices in China. Annu. Rev Entomol 50:31–52

Wu SW, Peng ZP, Jiang GF, Qin GQ, He CY (2014) Effects of harvest method and treatment of rice stubble on over-wintering of Chilo suppressalis Walker. Hunan Agric Sci 1:48–50

Yosefi K, Galavi M, Ramrodi M, Mousavi SR (2011) Effect of bio-phosphate and chemical phosphorus fertilizer accompanied with micronutrient foliar application on growth, yield and yield components of maize (single cross 704). Aus J Crop Sci 5(2):175–180

Zhu PY, Wang GW, Zheng XS, Tian JC, Lu ZX, Heong KL, Xu HX, Chen GH, Yang YJ, Gurr GM (2015) Selective enhancement of parasitoids of rice Lepidoptera pests by sesame (Sesamum indicum) flowers. BioControl 60(2):157–167

Weed Management Strategies in Direct Seeded Rice

19

Muhammad Mansoor Javaid, Athar Mahmood,
Muhamamad Ather Nadeem, Naeem Sarwar,
Muhamamd Ehsan Safdar, Masood Ahmad, Mirza Hasanuzzaman,
and Shakeel Ahmad

Abstract

Rice is an important cereal in the world. It is a first-order or second-order staple food across the globe. Its productivity and production has a major role for ensuring food security. However, its productivity and production is hampered due to the presence of abundant weeds. Huge losses in yield and quality have been reported in literature. Different approaches have been adopted to control these weeds. However, integrated weed management is an appropriate strategy to harvest a bumper crop and ultimately ensuring food security at national, regional, and global levels.

Keywords

Cereal · Rice · Staple · Food · Weeds management

M. M. Javaid · M. A. Nadeem · M. E. Safdar
Department of Agronomy, College of Agriculture, University of Sargodha, Sargodha, Pakistan

A. Mahmood
Department of Agronomy, University of Agriculture, Faisalabad, Pakistan

N. Sarwar
Department of Agronomy, Faculty of Agriculture, Bahauddin Zakariya University, Multan, Pakistan

M. Ahmad
Department of Botany, University of Agriculture, Faisalabad, Pakistan

M. Hasanuzzaman
Department of Agronomy, Bangladesh Agriculture University, Dhaka, Bangladesh

S. Ahmad (✉)
Department of Agronomy, Faculty of Agricultural Sciences and Technology, Bahauddin Zakariya University, Multan, Pakistan
e-mail: shakeelahmad@bzu.edu.pk

N. Sarwar et al. (eds.), *Modern Techniques of Rice Crop Production*,
https://doi.org/10.1007/978-981-16-4955-4_19

19.1 Introduction

Among the cereals, rice possesses an inevitable role in food supply in the world (Gadal et al. 2019). About 90% of the Asian population and half of the world population rely on rice to meet the daily calories (Bandumula 2018). The most common method of rice establishment in irrigated low land areas is transplanting (Sathish et al. 2017), which consumes more water and degrades the land (Kumar and Rajitha 2019). Water resource limitations, labor shortage, and increasing cost of labor make rice transplanting uneconomical and these caused yield reduction (Sangeetha and Baskar 2015). In puddled transplanting rice, out of the total required water (150 cm), about 20–25 cm water is being used in the process of puddling. Puddling disturbs the soil aggregated and makes the soil hard which leads to the development of cracks. These cracks facilitate the deep water penetration and water requirement increases manyfold (Nagargade et al. 2018). Physical condition of the soil for succeeding crops is not favorable after puddling. Main limitations of transplanted flooded rice system like water resources and shortage of labor make it unsustainable and uneconomical (Nagargade et al. 2018). Further, increasing input cost like diesel and electricity and climate change further aggravated this issue. Due to aforementioned issues, there is a dire need to shift from conventional transplanting to alternate method of rice production (Jabran et al. 2015; Thakur et al. 2014).

For sustainable agriculture production, alternate methods of rice production are considered sustainable, eco-friendly, and more productive as compared to the conventional techniques of rice cultivation (Jabran et al. 2015; Thakur et al. 2013). Among the alternate methods of rice production, system of rice intensification (SRI) and direct seeded rice (DSR) are the most common methods. Direct seeded rice is the most promising approach for water saving which can reduce water application by 44% by reducing percolation, seepage, and evaporation losses relative to the conventional transplanted system (de Borja Reis et al. 2018). Globally, out of the 161 M ha rice cultivation, 33 M ha rice is being cultivated as DSR (Shekhawat et al. 2020). However, DSR is facing much more pressure of weed infestation than conventional puddled transplanted system (Matloob et al. 2015). Weeds are suppressed in conventional puddled transplanted rice seedling by flooded condition and quick establishment of transplanted seedling over germinating weed seeds, whereas in DSR early germination and establishment of weeds reduced the yield up to 90% depending on weed flora and climatic conditions (Nie et al. 2012; Prasad et al. 2017).

19.2 Yield Losses in Direct Seeded Rice

In DSR, aerobic conditions facilitate the weed seed germination relative to transplanted flooded rice which caused yield losses up to 90% (Singh et al. 2008; Jabran et al. 2015). Table 19.1 represents the losses due to weed with varied weed composition. Low production under DSR is reported due to the weed management

Table 19.1 Yield losses with the varied composition of weed

Weeds composition	Yield losses %	References
Dactyloctenium aegyptium, Echinochloa crus-galli, Echinochloa colona, Leptochloa chinensis, Commelina benghalensis L., Caesulia axillaris, Eclipta prostrata, Euphorbia hirta, P. oleracea L., T. portulacastrum, Lindernia spp.	73	Singh et al. (2007)
Cyperus rotundus, Cyperus iria, Digitaria sanguinalis, Echinochloa colona, E. crus-galli, Sphenoclea spp., *Euphorbia hirta, T. portulacastrum, Ammannia* spp. *Ludwigia* spp., *Eclipta alba, Eclipta prostrata*	77–82	Mahajan et al. (2009)
Cyperus rotundus, Cyperus difformis, Cyperus iria, S. zeylanica, Echinochloa colona, Echinochloa crus-galli	80	Hussain et al. (2015)
Echinochloa crus-galli, Echinochloa colona, Dactylaeium aegyptium, L. chinensis, Eleusine indica, Cyperus rotundus, Cyperus iria, T. portulacastrum, Ipomoea aquatica, P. oleracea	82–86	Khaliq et al. (2012)
Echinochloa crus-galli, Echinochloa colona, Cyperus iria, Cyperus difformis, I. rugosum, Commelina spp., *F. miliacea, Caesulia axillaris*	85–98	Singh et al. (2011)
Echinochloa crus-galli, D. aegyptium, Echinochloa colona, C. rotundus, Eclipta alba, Euphorbia granulata, T. Portulacastrum	47	Mohtisham et al. (2013)
Eclipta prostrate, T. portulacastrum, E. crus-galli, D. aegyptium, C. rotundus	70–75	Jabran et al. (2012)

issues in this system. Aerobic conditions in DSR conserve water, but the absence of flooded conditions favors the germination of weed seeds (Rao et al. 2007; Joshi et al. 2013). In DSR, most critical time of weed-crop competition is 41 DAS; however, weed-free conditions for 70 DAS are required for maximum production (Shekhawat et al. 2020). It is estimated that DSR yield may reduce from 15% to 99% (Bajwa et al. 2020; Xu et al. 2019). Xu et al. (2019) reported that yield losses were reported up to 48% and 74% in transplanted and DSR, respectively. Table 19.2 represents the common weed flora in DSR. Weed composition varies to crop establishment method, water and soil management, cultural method, crop rotation, weed seed bank, and type of weeds in that area. *Ischaemum rugosum, Echinochloa colona* and *E. crus-galli, Cyperus iria, C. difformis, Commelina* spp., and *Fimbristylis milaiacea* are the notorious weeds of DSR, many of which cause 85–98% yield losses (Jabran et al. 2015; Nagargade et al. 2018). Moreover, *E. crus-galli* and *E. colona* grow best in saturated soil, so these are more abundant in DSR (Nagargade et al. 2018). Extent of yield losses due to weeds is dependent on the emergence time of weeds, and losses are higher with early emergence or at the time of crop emergence (Aldrich 1987). Weeds emerged early in DSR occupy space quickly and compete with crop for resources like nutrients, water, and carbon dioxide. Under this situation, weed biomass production is higher in DSR relative to conventional transplanted rice system, which leads to lower yield (Nagargade et al. 2018;

Table 19.2 Most common weed species associated with DSR

Family	Country	Weeds	Reference
Compositae	India	*Ageratum conyzoides* L.	Khan et al. (2003)
	India	*Caesulia axillaris* Roxb.	Singh et al. (2007)
	India	*Eclipta alba* (L.) Hassk.	Mahajan et al. (2009)
	India; Pakistan	*Eclipta prostrate* (L.) L.	Singh et al. (2009); Jabran et al. (2012)
	Malaysia	*Emilia sonchifolia* (L.) DC. Ex DC. e	Anwar et al. (2013)
	India	*Parthenium hysterophorus* L.	Singh et al. (2009)
Poaceae	Malaysia	*Axonopus compressus* (Sw.) Beauv.	Anwar et al. (2013)
	USA	*Brachiaria platyphylla* (Munro ex C.Wright) Nash	Riar and Norsworthy (2011)
	Sri Lanka	*Cynodon dactylon* (L.) Pers.	Weerakoon et al. (2011)
	Thailand; Pakistan; India Philippines	*Dactyloctenium aegyptium* (L.) Willd.	Sanusan et al. (2010); Khaliq et al. (2012); Zhao et al. (2007); Singh et al. (2006)
	Philippines; Malaysia	*Digitaria ciliaris* Retz. Koeler	Zhao et al. (2007); Anwar et al. (2013)
	USA	*Digitaria sanguinalis* (L.) Scop.	Tindall et al. (2005)
	Pakistan; Philippines; India	*Echinochloa colona* (L.) link	Khaliq et al. (2012); Zhao et al. (2007); Chauhan and Opeña (2013)
	India; Pakistan; USA	*Echinochloa crus-galli* (L.) P. Beauv	Singh et al. (2006); Khaliq et al. (2012); Tindall et al. (2005)
	Philippines	*Echinochloa glabrescens* Munro ex hook.	Chauhan (2013)
	Malaysia	*Echinochloa oryzicola* Vasinger	Azmi et al. (2005)
	Malaysia	*Echinochloa apicta* (J.Koenig) Kunth	Azmi et al. (2005)
	Malaysia	*Echinochloa stagnina* (Retz.) P.Beauv	Azmi et al. (2005)
	Philippines; Malaysia; Pakistan	*Eleusine indica* (L.) Gaertn	Zhao et al. (2007); Anwar et al. (2010); Khaliq et al. (2012)

(continued)

Table 19.2 (continued)

Family	Country	Weeds	Reference
	India	*Eragrostis japonica* (Thunb.) Trin.	Singh et al. (2009)
	Malaysia; Thailand; Sri Lanka	*Ischaemum rugosum* Salisb.	Weerakoon et al. (2011); Sanusan et al. (2010)
	India; Pakistan; Philippines	*Leptochloa chinensis* (L.) Nees	Singh et al. (2006); Khaliq et al. (2012); Zhao et al. (2007)
	USA	*Leptochloa fascicularis* (lam.)	Pittelkow et al. (2012)
	India	*Leptochloa panacea* (Retz.)	Singh et al. (2006)
	USA	*Leptochloa panicoides* Hitchc.	Tindall et al. (2005)
	Sri Lanka; Thailand	*Panicum repens* L.	Weerakoon et al. (2011); Sanusan et al. (2010)
	Malaysia	*Paspalum conjugatum* P.J. Bergius	Anwar et al. (2010)
	India	*Paspalum distichum* L.	Singh et al. (2009)
	India	*Saccharum spontaneum*	Singh et al. (2009)
	India	*Setaria glauca* (L.) P. Beauv.	Midya et al. (2005)
	India	*Sorghum halepense* (L.) Pers	Singh et al. (2009)
	USA	*Urochloa platyphylla* Nash.	Riar and Norsworthy, (2011)
Cyperaceae	Malaysia	*Cyperus aromaticus* (Ridl.) Mattf	Anwar et al. (2013)
	USA	*Cyperus difformis* L.	Pittelkow et al. (2012)
	USA	*Cyperus esculentus* L.	Riar and Norsworthy (2011)
	Thailand	*Cyperus imbricatus* Retz.	Sanusan et al. (2010)
	Thailand; Philippines; Pakistan	*Cyperus iria* L	Sanusan et al. (2010); Chauhan and Opeña (2013); Khaliq et al. (2012)
	Philippines; Pakistan	*Cyperus rotundus* L.	Khaliq et al. (2012); Chauhan (2013)
	Malaysia	*Cyperus sphacelatus* Rottb.	Anwar et al. (2013)
	Thailand	*Fimbristylis dichotoma* L.	Sanusan et al. (2010)
	Thailand; Malaysia	*Fimbristylis miliacea* (L.) Vahl.	Sanusan et al. (2010); Anwar et al. (2010)

(continued)

Table 19.2 (continued)

Family	Country	Weeds	Reference
	India	*Fimbristylis quinquangularis* Kunth	Singh et al. (2009)
	USA	*Schoenoplectus mucronatus* (L.) Palla	Pittelkow et al. (2012)
	Philippines	*Scirpus mucronatus* L.	Caton et al. (2002)
	India	*Scirpus supinus* L.	Singh et al. (2009)
Amaranthaceae	India	*Digera alternifolia* (L.) Asch.	Midya et al. (2005)
	India	*Digera arvensis* (L.) Mart.	Mahajan and Chauhan (2013a)
	India	*Digera muricata* (L.) Mart.	Singh et al. (2009)
	Malaysia	*Alternanthera sessilis* (L.) DC.	Anwar et al. (2013)
Leguminosae	Thailand	*Alysicarpus vaginalis* (L.) DC.	Sanusan et al. (2010)
	USA	*Sesbania herbacia* (mill.) McVagh	Godara et al. (2012)
Euphorbiaceae	USA	*Caperonia palustris* (L.)	Godara et al. (2012)
	Pakistan	*Euphorbia granulate* Forsk.	Mohtisham et al. (2013)
	India	*Euphorbia hirta* L.	Singh et al. (2006)
	India	*Euphorbia microphylla* lam.	Singh et al. (2009)
Portulacaceae	India; Pakistan; Philippines	*Portulaca oleracea* L.	Singh et al. (2006); Khaliq et al. (2012); Zhao et al. (2007)
Lythraceae	India	*Ammannia auriculata* Willd	Singh et al. (2009)
	Philippines	*Ammannia baccifera* L.	Chauhan (2013)
	USA; Philippines	*Ammannia* spp.	Pittelkow et al. (2012); Caton et al. (2002)
Sphenocleaceae	Pakistan	*Sphenoclea zeylanica* Gaertn.	Hussain et al. (2015)
Portulacaceae	India; Pakistan; Philippines	*Portulaca oleracea* L.	Singh et al. (2006); Khaliq et al. (2012); Zhao et al. (2007)
Convolvulaceae	Pakistan	*Ipomoea aquatic* Forssk.	Khaliq et al. (2012)

Shekhawat et al. 2020). It is concluded that rice yield losses are highly dependent on competition duration. The study depicted that the total biomass of weeds in DSR was 84–194% higher compared to conventional flooded rice and yield losses were 33% more in DSR relative to transplanted flooded rice (Mahajan et al. 2009).

19.3 Method of Direct Seeded Rice

Direct seeded rice is the establishment of rice crop with sowing of seed directly in field rather than seedling transplanting (Kaur and Singh 2017). After the completion of germination, the crop can be subsequently flooded or the crop can remain rainfed by providing fluctuating aerobic and anaerobic conditions. Direct seeded rice is classified into (1) water-seeded rice, (2) dry-seeded rice, and (3) wet-seeded rice based on the soil conditions (Rao et al. 2007).

19.4 Weed-Crop Competition and Weed Flora Shift in DSR

Weed emergence time with respect to emergence of crop has significant role in weed-crop competition. Early weed emergence had adverse effects on crop yield especially when weeds grow at the time of crop emergence. Losses due to weeds are dependent on the duration of weed infestation. Although DSR is grown at a seed rate to provide optimum plant population, weed infestation can cause total loss of yield. A study conducted by Gibson et al. (2001) showed 99% losses in DSR with weed infestation. At the early stage of development, all annual crops are the most susceptible to weeds; however, critical period of weed-crop competition is reliant on overlap period of two separate components: (1) period in which crops remain weed free so that weeds emerged at lateral stage do not reduce yield and (2) presence of weeds before weed interference with crop growth (Rao et al. 2007). So removing weeds at early stage and keeping the field free of weeds till the end of the critical period are mandatory to avoid yield reduction.

The occurrence, growth, and development of weeds are greatly influenced with the method of rice cultivation. More weed species are reported in DSR relative to conventional flooded rice, and control measures are more difficult in complex flora (Pathak et al. 2011). Rice is cultivated in tropical and subtropical areas and water conditions in these areas favor the weed growth. Furthermore, soil in DSR system remains moist rather than flooded, which provides favorable condition for the germination of weed seeds (Jabran and Chauhan 2015). Hence, more seed germination with favorable climatic conditions aids in weed-crop competition in DSR. For instance, *T. portulacastrum* L. are unable to grow in flooded rice but rapidly grow and develop in DSR (Jabran and Chauhan 2015). There are pieces of evidence that rice-weed competition in DSR occurs at the early stage of crop development and there is a dire need for early weed control (Rao et al. 2007; Chauhan 2012; Khaliq et al. 2012). Generally, DSR has more diversity of weed species relative to transplanted flooded rice. About 21 species, 18 genera, and 13 families are reported in transplanted puddled rice, whereas 50 species, 18 genera, and 22 families were found in DSR (Kaur and Singh 2017). Agronomic management, crop establishment, and moisture conditions in DSR alter the weed flora and weed diversity. A shift to the DSR method doubled the diversity index of weed flora (Singh et al. 2016). According to Singh et al. (2016), there is 2.5–4.7 times increase in grassy and broadleaf weeds in India. Grasses and sedges were shifted more aggressively in DSR in Southeast Asia (Yaduraju and Mishra 2008). Initially, broadleaf weeds were dominant in DSR, whereas after a decade some grass weeds like *Ischaemum rugosum*, *E. crus-galli*, and *L. chinensis* became prominent in most DSR growing areas (Shekhawat et al. 2020).

19.5 Weed Management in DSR

19.5.1 Preventive Weed Management Strategies

Prevention of weed production is a long-term successful tool of weed management. Weed seed bank can be reduced by eliminating weeds for few years or increase if weeds are allowed to grow and produce their seeds (Schwartz-Lazaro and Copes 2019). The following preventive strategies are helpful in effective weed management program.

- Preventing reproduction of weed during crop growing season provides a chance to reduce weed seed bank densities
- Weed infestation can be greatly reduced by eliminating reproduction on bunds and the border of DSR production areas. A strong campaign among the farmer community cultivating DSR is direly needed to demolish weed reproduction on the bund and border in fields.
- Reduction of weed reproduction in fellow fields in DSR growing region is a very low cost-effective technique to decrease crop production losses. It is very easy and convenient to adopt any nonchemical or chemical weed control to stop weed reproduction in fellow field
- Dispersal of weed seed is a major problem caused by various abiotic and biotic factors such as winds, water, insect, birds, and human activities. Limiting dispersal of weed seed by understanding the mechanism of dispersal and ecological management can be effective for weed management in DSR areas
- Weed infestation via seed dispersal due to contaminated rice seed with various species of weed seed can be effectively controlled and achieved by educating and motivating the farmer community to use noncontaminated and cleaned rice seed for DSR cultivation
- Farm machinery and equipment items such as seed drills, fertilizer spreaders, combined harvesters are other sources of weed seed dispersal in DSR area. These causes of weed seed dispersal can be easily avoided by using cleaned and noncontaminated machinery and equipment during different field practices

- Among various sources, irrigation water is of prime importance which could readily disperse various species of weeds. The management strategies to prevent dispersal via irrigation water may include filtering water, proper cleaning of water channels, and removing floating weeds
- Application of manures and immature composting has viable weed seed which readily spread in fields of DSR systems and considered significant sources for weed dispersal. Pretreatment and preventive measures are needed to be followed, that is, direct manure must be avoided and composting process exposed at high temperature (55–70 °C) lethal for weed seed could help to prevent dispersal and weed infestation. Grazing in the fellow field also should be encouraged to eliminate weed before seed production deposited by manures
- Other weed seed dispersal agents include migratory birds, ducks, winds, and earthworms which spread weed in DSR field. As dispersal of weed seed is a multidimensional process, no specific method can be recommended. To control weed in this way, integrative management strategies must be adopted to prevent crop losses
- In the DSR system, weed seed predation can play a considerable role in weed mortality. Although little information is available on weed seed predation, previous researchers suggested that predation may have 60–78% mortality in a short duration of 14 days. Management strategies in seed predation include careful identification of specific predators and the creation of circumstances favorable for predators and predations.
- The persistence of weed seed directly is linked with weed population in fields. Decay of weed seed in DSR field could be a tool to assess the persistence of weed seed. Many factors determine the rate of decay of weed seeds including microorganism, physiological death, decline in seed viability soil moisture, gas exchange, soil nitrogen contents, and temperature. Creating an environment (moisture and nitrogen addition, etc.) that facilitates microbes to boost up seed decay and declining viability could be effective in reducing weed population in DSR system
- In addition, decay of seed can be triggered by manipulating tillage of field as it could promote rapid predations, and nontillage practice may cause more seed decay consequently reduce the weed densities
- Soil organic amendments like composting and cover crops may change the microbial community of soil. This organic application could increase the microbial biomass of the soil and in return decrease the weed emergence and increase weed seed decay.
- Irrigation management can also be an effective tool to promote condition for preventing weed population such as high moisture content increases rate of decay and negatively influences seed viability of weeds
- Promoting fatal germination strategies weed population could be reduced such as by promoting conditions to create dormancy and inhibiting the emergence of weed seeds during crop growing season.
- Another effective approach to mitigate the weed losses in DSR could be stale seedbed preparation, that is, in the first step weed might be allowed to grow before the cultivation of crop and killed by cultural or chemical method before the establishment of the crop

- Weed densities could be reduced by inhibiting their germination and emergence by spreading mulches. Some types of mulches have allelopathic impacts and thus prevent the seedling establishment of weeds. In the DSR system, this could be an effective strategy to reduce weed population in the field
- Emergence, germination, and growth of weeds can also be controlled by the flooding process using anaerobic tolerant rice varieties. Many weed species in DSR are highly affected by depth, duration, and timing of flooding application
- Crop rotation with keeping residue in the field could give proper solution to keep weed densities under threshold level and to reduce yield losses in DSR system.
- Integrative preventive strategies could be highly cost-effective to control weed in the DSR system

19.5.2 Cultural Weed Management in Direct Seeded Rice (DSR)

It is very difficult to eradicate weeds under direct seed rice (DSR) system as compared to transplanted rice. Due to the abundance of weeds in DSR, the average yield loss in rice is much higher. Herbicides control or kill the weeds completely in DSR system but serious health and environmental issues increased. Therefore, it is essential to control the weeds in an effective and manageable manner through integrated weed management methods. In DSR system focuses on reducing the use of herbicides and emphasis on cultural methods to control weeds is direly needed (Shekhawat et al. 2020). However, limited information is present about the cultural prevention of weed management in DSR (Table 19.3). Therefore, it is direly needed to develop integrated weed management methods to control weeds that will

Table 19.3 Cultural weed control methods and their properties

Cultural weed control methods	Properties
Water management	This practice ensures better root and seedling development as well as encourages the emergence of weed seeds
Land preparation	Uprooting and burial of weeds at planting stage provide weed-free field in which tillage practices give a competitive advantage to crops and inhibit germination and growth of weeds
Weed competitive cultivars	Combination of rice cultivars with other management practices may help to minimize the herbicide resistance among various species of weeds
Soil solarization	Solarization technique has a promising potential for weed management when combined with other technique to suppress weed, that is, herbicide, hoeing, hand weeding, etc.
Intercropping	By employing intercropping, land utility could be increased by sowing of more than one crop at same piece of land at the same time
Mulching	From previous investigation, it is suggested that mulches have commendable role to inhibit weeds in DSR systems. With this respect, residue of previous crop is very crucial. A combined effort of eliminating weed by mulching and herbicide spray can be very effective

give a high yield of rice. For obtaining effective, manageable, and long-term weed control in the DSR system, it is direly needed to replace the use of herbicides with other cultural strategies, like use of suitable fertilizers, high seedling rate, stale seedbed practices, cropping weed-competitive cultivars, optimum sowing time, and water inputs and mechanical weeding (Nagargade et al. 2018).

19.5.2.1 Water Management

Weed management by water suppression is one of the key cultural methods in direct seeded rice system (DSR). Floodwater management considerably impacts on vigor, density and uniform establishment of rice stand, intensity of weed competition, and herbicides efficacy. The reason that forced the farmers of California to shift from dry seeding to water seeding in twentieth century was mainly to suppress *E. crus-galli* (Nagargade et al. 2018). Depth of water could exert a clear impact on the structure of weed populations and existence of specific weed species into the growing rice crop. Observation showed that in comparison with 12 DAS, flooding up to 5 cm at 4–5 days after sowing (DAS) significantly reduced the growth and density of weed species *E. crus-galli* in DSR system. Precise and accurate water management, especially at the crop establishment phase (7–15 days DAS), is of prime importance in dry drill-seeded rice (Sharda et al. 2017). Another important parameter to be considered is that the field must be provided moisture but not saturated to prevent rotting of seed. It is compulsory to apply irrigation after sowing seed of rice in dry soil if it is unlikely to rain followed by saturating the field. This will not only encourage seedling development but also gear up the emergence of weed seed. Early weed control using preemergence (PE) herbicide is crucial to prevent weed emergence which may limit yield.

A researcher concluded that flood watering after application of herbicide or manual weeding can markedly reduce subsequent establishment of weeds in DSR system (Saha et al. 2021). Water management has a prime importance in suppression of weed in DSR (Raj and Syriac 2017; Rathika et al. 2020). Enough moisture should be ensured in the field during the preemergence herbicides application in direct seeded rice system. It is observed that herbicides provide excellent control when applied into water compared to the absence of standing water. Rational water management combined with chemical weed management gives viable and cost-effective opportunity for conserving moisture and high productivity of rice.

19.5.2.2 Land Preparation and Tillage

Burial and uprooting of weeds at the time of planting ensured a weed-free field in which tillage practices provide a huge competitive advantage to crops and inhibit weed emergence. Tillage application causes mortality, dormancy of weed seeds due to unfavorable conditions like water gaseous regime, as well as suppression of the germination of weed seed present at the upper soil layers (Swanton et al. 2008). Contrary to that, tillage is the cause of 85% of the total weed seed dispersal, and on the other hand, no tillage encourages weed predation by birds, rodents, and insects (Shekhawat et al. 2020). Previous research elaborated that 78–90% of *E. colona* seeds were removed (in 2 weeks) by the process of predation only (Chauhan et al.

2010). Various factors such as depth, frequency, and timing of soil manipulations determine the emerging weed density in the field.

In DSR, good seedbed preparation coupled with laser leveling provides a weed-free field, suppresses weed by 50%, enhances the efficacy of herbicide application, and reduces demand of labor for manual weed management (Shekhawat et al. 2020). It was observed that soil inversion using deep tillage machinery inhibits germination of *L. chinensis* which hides its seeds below 5 cm depth (Benech-Arnold et al. 2000). Early crop stand is also a very effective factor affecting weed densities in DSR.

19.5.2.3 Enhancing Crop Competitiveness

Implementation of nonchemical weed management practices, such as cultivars with high competitive potential against weeds, seeding rates and row spacing, adjusting plant geometry, can give potential advantage for competing against weeds (Dass et al. 2017).

19.5.2.4 Weed Competitive Cultivars

Among various cultural weed management strategies, use of highly competitive rice cultivars with other management practices may help to minimize the herbicide resistance among weeds in DSR system. The result of previous study elaborated that traits like tolerance against early submergence for uniform crop stand and seed emergence under anaerobic conditions are considered to be very effective regarding weed management in DSR system (Mahajan and Chauhan 2013a, b; Rao et al. 2007). Cultivars in which faster canopy developed with allelopathic potential may be favored smother weeds in DSR system.

Those rice cultivars that expand canopy earlier compared to others can control weed without environmental damage or any other damage to plants (Zhao et al. 2007). However, using more seed rate could recompense slow canopy developer cultivars to occupy enough space to overcome weeds. In smothering rice cultivars, DM accumulation and LAI can be used as most important standards for weed selection. For example, Apo rice cultivar was reported to be a high-yielding cultivar competitive against weeds in DSR systems (Zhao et al. 2006). The cultivars of African rice included ACC 102257 and IG 10 which had the ability of weed suppression due to its speedy growth than other tested cultivars. The IG 10 cultivars showed 70% less biomass of weed as compared to weedy plots (Fofana and Rauber 2000).

19.5.2.5 Soil Solarization

In soil solarization technique, energy from sunlight is used in order to suppress crop pests and weeds. Sunlight is captured by spreading coverings of transparent plastic sheet on the soil. In this method, the entrapped solar energy enhances the temperature of soil and kills the seeds near soil surface and germinated and germinating weed in DSR field. Soil solarization technique is also being applied to kill pathogens and pests and improve soil properties. Previous observation showed that rice yield was increased in solarized field in comparison with the non-solarized field. Although the use of soli solarization to control weeds has not been explored limitedly in DSR system, its positive impacts on soil properties have been well documented (Jabran and Chauhan 2015). Observation of previous study suggested that soil solarization application enhanced temperature by 20–25% and decreased weed density up to 25–40% and also helped in reducing the soil pathogen causing various diseases in plants (Culman et al. 2006).

As a result useful impacts of solarization considerably increased rice growth and yield. Regarding this aspect, previous researcher carried out a study to explore the effect of solarization on weed control in aerobic rice system. In this experiment, various weeds were included such as *E. colona*, *Euphorbia hirta*, *Eclipta alba*, *Cyperus iria*, *Phyllanthus niruri*, Hassk., *Cynodon dactylon*, and *Ageratum conyzoides* (Khan et al. 2003). Before sowing of crop, covering of field with black sheet for 1 week enhanced soil temperature by 50 °C and decreased weed population of grasses by 64% and broad leaved weed by 50% in comparison with unweeded field. Interestingly, grain yield of rice was approximately double in plots employed with solarization as compared to nontreated plots. It can be concluded that solarization technique could be highly effective for reducing weed pressure in DSR system areas where weeds are the main yield-limiting factor. This technique can be applied with other integrative weed management methods like tillage, hand weeding, and herbicide application. However, further research is direly needed to find out more beneficial aspects of solarization in DSR system.

19.5.2.6 Intercropping

Weed suppression could be achieved by intercropping, that is, mungbean (*Vigna radiata* L.), *Sesbania* spp., corn (*Z. mays* L.), watermelon (*Citrullus lanatus*), and peanuts (*Arachis hypogaea* L.), in aerobic rice system used as intercrops (Ahmad et al. 2007; Ren et al. 2008; Singh et al. 2007). Intercrops effectively suppress weeds by decreasing distance among them and using resources which were destroyed by weeds previously. Intercropping practices increased land utility in which more than one crops are cultivated at the same piece of land at same time. However, intercropping must be practiced by keeping in view well-defined principles. Experimental data of many studies proved that overall productivity of DSR system could be enhanced when intercropped with other crops. Observations of various studies exhibited the effectiveness of intercropping in DSR system for weed management (Ren et al. 2008; Singh et al. 2007).

19.5.2.7 Mulching

Covering crop with mulch is a very effective and promising multipurpose method to get many benefits in agro-ecosystem. Among those effects include soil moisture retention, weed inhibition, soil improvement, and fertility enhancement. Therefore, use of mulch is highly recommended in DSR system to achieve high moisture retention and weed suppression. Application of mulch in DSR area could be used to reduce weed pressure by sunlight blockage, releasing of allelochemicals, and physical suppression. Weeds that are considered highly problematic in DSR system such as *E. colona*, *E. crus-galli*, and *Eclipta alba* could be effectively suppressed by wheat residues mulch application (Kumar et al. 2013). Additionally plants have been identified with allelochemical properties to suppress weeds, for example, mulch of *Tagetes minuta* L. proved to be effective in decreasing rice weeds, including *C. rotundus* and *E. crus-galli* in field, glass house, and laboratory trials (Batish et al. 2007). Other plants that have allelopathic ability were also identified including *Datura stramonium* L., *Clerodendrum trichotomum*, *Melia azedarach* L., and *Desmodium triflorum* L. having capability to reduce weed by 70–90% when applied to rice soils at 1–2 Mg ha^{-1} (Hong et al. 2004).

19.6 Manual Weeding

In spite of extensive use of herbicide to control weeds in DSR, hand weeding or pulling of weeds is either partially or extensively used worldwide. However, this method relies on labor availability. Hand weeding normally initiates at 15–30 days after sowing and carries on for many days (Rao et al. 2007). Sometimes dense weed population and few labor men delay the weeding, and in the last days of weeding, weeds compete with crops and reduce the yield. Studies suggested that hand weeding is most effective when weed density is lower and availability of labor is easy along with normal expenditure (Singh et al. 2005). It is easy to pull weeds when they have sufficient size (Khaliq et al. 2011). Labor scarcity, high labor cost, presence of perennial weeds, repeated hand weeding, and weather conditions reduced the efficacy of manual weeding in DSR. Furthermore, manual weeding becomes more important when suitable herbicides is not available to control some specific weeds or when herbicides are beyond the reach of farmers due to high cost, especially in case of small farmers. It is reported that about 100 person-day ha^{-1} are required to control weeds in DSR (Trung et al. 1995). In terms of preventive measures, hand pulling of rice weeds at seed producing stage is most effective. Manual weeding is only easy when weed plants are larger than rice plant. Further, patches of weeds in field can be pulled or cut manually to prevent seeds dispersal.

19.7 Mechanical Weeding

Use of simple implements to control weeds is one of the ancient methods and this method remains practical for small farmers. This method is practical line sowing rice where row cultivation either with hand pulled seed drill or with animal-drawn implement reduced the time in weeding. Reports showed the reduction in weed density up to 72% with three times mechanical weeding done after 4, 6, and 8 weeks after sowing (Akbar et al. 2011). With labor shortage, mechanical weeding proved to be efficient and suitable in DSR. Recently, some mechanical seeders like straight-spike weeder, ring hoe, and two-row spike weeder are being used effectively for weed management in DSR. Similarly, motored mechanical weeders are introduced for weed management in conventional and DSR systems. Mechanical weeding is the most effective early stage of weed emergence with minimum biomass.

19.8 Chemical Weed Control in DSR

Nonchemical weed management strategies include manual or mechanical hoeing or hand weeding which have limitations like labor unavailability and increasing labor cost. As profit is major drivers for rice farmers; hence, they seek alternatives of manual weeding. Despite the un-availability of broad-spectrum herbicides for DSR, chemical weed control is necessary for managing weeds (Chauhan 2012). Khaliq et al. (2012) reported that a chemical method to control weeds in DSR is most effective when weed density is high. Several preemergence (PRE), early postemergence (POST), and POST have been reported effective in DSR, but their efficacy is conditional to correct application of herbicides. Common herbicides used in alternate rice systems are presented in Table 19.4. For example, many weed seeds remain in upper soil surface (2–3 cm) in dry-direct seeded rice, and PRE application provides satisfactory weed control (Chauhan and Johnson 2009). Pedinmethaline, oxdiazon, oxadiargyl, and pretilachlor are commonly used PRE herbicides, while penoxsulam, 2–4-D, metsulfuron methyl, bispyribac-sodium, and fenoxaprop are commonly used POST herbicides in DSR (Singh et al. 2019). During chemical weed management in DSR, herbicide resistance, management, and mitigation should also be focused. Another aspect of spraying herbicide in DSR is toxic effects on crop. For

Table 19.4 Common herbicides used in alternate rice systems

Herbicide	Target weed	Dose (g. a.i. ha^{-1})
Preemergence herbicides		
Pendimethalin	Complex	825
Pretilachlor	Grasses and broadleaved	500
Imazosulfuron	*C. palustris*, *S. herbacia*	168
Oxadiazon	Grasses and broadleaved	1000
Butachlor	Grasses and broadleaved	500
Postemergence herbicide		
Triclopyr	Broadleaved	500
Ethoxysulfuron	Broadleaved	18
Bispyribac-sodium	Grasses, sedges, broadleaved	30
Bispyribac-sodium	*C. palustris*, *S. herbacia*	17.6
Bensulfuron	Broadleaved	60
2,4-D (ester)	Broadleaved	500
Ethoxysulfuron	*Echinochloa* spp., *Cyperus* spp., and *S. zeylanica*	37.5
Metsulfuron	Grasses	15
Triclopyr	Grasses and broadleaved	500
Cyhalofop butyl	Grasses and broadleaved	120
Fenoxyprop-p-ethyl	Grasses and broadleaved	50
Propanil	Broadleaved and grasses	1750
Ethoxysulfuron	Broadleaved and grasses	18
Cyhalofop butyl	Broadleaved and grasses	120
Anilofos	Grasses	400

example, oxadiazon and metolachlor are used for weedy rice but should be applied well before the sowing of crop as these herbicides are toxic to emerging crop if applied at the time of swing (Eleftherohorinos and Dhima 2002).

In DSR, herbicide persistence with long residual activity resulted in herbicide resistance and weed shift. For instance, over-reliance on herbicides and untimely application of herbicides with higher doses reduced the efficacy of some herbicides like butachlor, anilofos, and pretilachlor (Mahajan and Chauhan 2008). About 30 resistant weed species are reported against propanil, 2–4-D, and sulfonylurea. Use of proper spray technique and weed species-specific herbicide enhances the herbicide efficacy (Gressel 2005). For example, bispyribac-sodium is effective against most common grasses but cannot control *Cyperus* spp., *L. chinensis*, and *D. aeguptium* (Busi et al. 2017). Other cultural practices like tillage and irrigation also affect the herbicide efficacy. Use of safeners with POST herbicides also enhanced the herbicide selectivity. For example, pendimethalin does not control *C. rotundus* and *I. triloba*, but co-application with safeners increased the efficacy and selectivity (Webster et al. 1999).

Nano-herbicides characterized with high penetration and efficacy prevent the overuse of herbicide and reduced environmental contamination (Pereira et al. 2014). Further, a precise dose reduced the chance of herbicide resistance. Nanoparticles of atrazine enhanced the entry, translocation, and bio-availability of herbicide molecule (Mudhoo and Garg 2011). Another study showed that nanoparticles of carboxymethyl facilitate atrazine degradation and reduce the residues in the environment (Susha et al. 2008).

19.9 Herbicide-Resistant (HR) Rice

Herbicide-resistant rice has a great potential to increase the efficacy of weed management. Rice crops remain resistant to specific herbicide or group of herbicides to which rice is otherwise sensitive. Glyphosate- and glufosinate-resistant rice varieties have been developed with transgenic technology. Imidazolinone-resistant rice were developed with mutagenesis and conventional breeding method (Olofsdotter et al. 2000). Despite the potential benefit of HR rice, there are several risks associated with this technology. For instance, this technology offers the control of weedy rice, but gene flow can occur from HR rice to nonresistant rice which is a major threat in future (Kumar et al. 2008).

19.10 Integrated Weed Management

Since no single control measure can produce appropriate levels of weed control, substantial progress in weed regulation can be achieved if different ingredient is combined in a logical series (Rajashekar et al. 2017). Enhanced agronomic efficiency methods, optimal fertilization and water regulation, integration of crop residues in soil, and the addition of remaining crop residues help to enhance the herbicides (Rao et al. 2017). According to Brar and Walia (2008), plant density combined with high nitrogen rate resulted in superior weed control. In rice and green gram intercropping, already present sprinkle of pendimethalin supplied successful weed control and resulted in a substantial increase in the production of two crops (Kumar et al. 2020). Jabran and Chauhan (2015) concluded that relying on single weed control methods led to not only case weed control failure but also herbicide resistance. Similarly application of higher dose of herbicide or repeated application of herbicides leads to environment damage. Beltran et al. (2012) suggested that using multiple weed control techniques is more effective and economical. Singh et al. (2008) depicted that use of PRE and POST herbicides and hand weeding resulted in very effective weed control in rice. Under DSR, stale seed followed by mulching followed by spraying of early and late POST herbicide reduced weed density significantly. Integration of pretilachlor as PRE herbicide along with one hand weeding at 30 DAS provided good weed control in dry DSR. Crop residues of previous wheat crop at 4 t ha^{-1} in DSR remained effective for grasses and broad leaf weeds. So integration of various available options to control weeds in alternate rice systems will not only be effective and economical but also helpful to tackle the problem of herbicide resistance.

19.11 Future Research Needs

Weed invasion is a chief risk to the development and productivity of DSR across the world. In many developed countries, DS is the only technique for rice establishment and relies on mechanization along with close attention to weed control and management. In Asia, due to shortage of agricultural labor, DS is encouraged. There is a need to economize the water use and increase water-use efficiency (WUE) that may decrease water need up to 50%. In DSR system, water management plays a key role for management of weed; however, greater understanding is still required to be able to increase role of water in management of weed in itself and in combination with herbicides. Significant research work for understanding relationship between water depth/drainage with weed removal and identifying rice germplasm with practices to improve survival rice-seedling and emergence under flooding are still needed.

References

Ahmad R, Jabbar A, Ahmad A, Ullah E, Bhatti I (2007) Evaluation of direct seeded upland rice-based intercropping systems under strip planting geometry. Pak J Agric Sci 44:189–193

Akbar N, Jabran K, Ali MA (2011) Weed management improves yield and quality of direct seeded rice. Aust J Crop Sci 5:688–694

Aldrich RJ (1987) Predicting crop yield reductions from weeds. Weed Technol 1:199–206

Anwar M, Juraimi A, Mohamed M, Uddin M, Samedani B, Puteh A et al (2013) Integration of agronomic practices with herbicides for sustainable weed management in aerobic rice. Sci World J 2013:916408

Anwar MP, Juraimi AS, Man A, Puteh A, Selamat A, Begum M (2010) Weed suppressive ability of rice (*Oryza sativa*'L.) germplasm under aerobic soil conditions. Aust J Crop Sci 4:706–717

Azmi M, Chin D, Vongsaroj P, Johnson D (2005) Emerging issues in weed management of direct-seeded rice in Malaysia, Vietnam, and Thailand. In: Rice is life: Scientific Perspectives for the 21st Century, pp 196–198

Bajwa AA, Ullah A, Farooq M, Chauhan BS, Adkins S (2020) Competition dynamics of *Parthenium hysterophorus* in direct-seeded aerobic rice fields. Exp Agric 56:196–203

Bandumula N (2018) Rice production in Asia: key to global food security. Proc Natl Acad Sci India Sect B: Biol Sci 88:1323–1328

Batish DR, Arora K, Singh HP, Kohli RK (2007) Potential utilization of dried powder of Tagetes minuta as a natural herbicide for managing rice weeds. Crop Prot 26:566–571

Beltran JC, Pannell DJ, Doole GJ, White B (2012) A bioeconomic model for analysis of integrated weed management strategies for annual barnyardgrass (*Echinochloa crus-galli* complex) in Philippine rice farming systems. Agric Syst 112:1–10

Benech-Arnold RL, Sánchez RA, Forcella F, Kruk BC, Ghersa CM (2000) Environmental control of dormancy in weed seed banks in soil. Field Crops Res 67:105–122

Brar AS, Walia US (2008) Effect of rice residue management techniques and herbicides on nutrient uptake by Phalaris minor Retz. and wheat (Triticum aestivum L.). Indian J Weed Sci 40 (3–4):121–127

Busi R, Nguyen NK, Chauhan BS, Vidotto F, Tabacchi M, Powles SB (2017) Can herbicide safeners allow selective control of weedy rice infesting rice crops? Pest Manag Sci 73:71–77

Caton BP, Hill JE, Mortimer AM, Foin TC, Lubigan RT (2002) Canopy development of direct-seeded rice and some important grass and sedge weeds in response to water management. Agric Forest Meteorol 111:39–53

Chauhan B, Johnson D (2009) Ecological studies on *Cyperus difformis*, *Cyperus iria* and *Fimbristylis miliacea*: three troublesome annual sedge weeds of rice. Ann Appl Biol 155: 103–112

Chauhan B, Migo T, Westerman P, Johnson D (2010) Post-dispersal predation of weed seeds in rice fields. Weed Res 50:553–560

Chauhan B, Opeña J (2013) Weed management and grain yield of rice sown at low seeding rates in mechanized dry-seeded systems. Field Crops Res 141:9–15

Chauhan BS (2012) Weed ecology and weed management strategies for dry-seeded rice in Asia. Weed Technol 26:1–13

Chauhan BS (2013) Shade reduces growth and seed production of *Echinochloa colona*, *Echinochloa crus-galli*, and *Echinochloa glabrescens*. Crop Prot 43:241–245

Culman S, Duxbury J, Lauren J, Thies J (2006) Microbial community response to soil solarization in Nepal's rice–wheat cropping system. Soil Biol Biochem 38:3359–3371

Dass A, Shekhawat K, Choudhary AK, Sepat S, Rathore SS, Mahajan G et al (2017) Weed management in rice using crop competition-a review. Crop Prot 95:45–52

de Borja Reis AF, de Almeida REM, Lago BC, Trivelin PC, Linquist B, Favarin JL (2018) Aerobic rice system improves water productivity, nitrogen recovery and crop performance in Brazilian weathered lowland soil. Field Crops Res 218:59–68

Eleftherohorinos IG, Dhima KV (2002) Red rice (*Oryza sativa*) control in rice (*O. sativa*) with preemergence and postemergence herbicides. Weed Technol 16:537–540

Fofana B, Rauber R (2000) Weed suppression ability of upland rice under low-input conditions in West Africa. Weed Res Oxford 40:271–280

Gadal N, Shrestha J, Poudel MN, Pokharel B (2019) A review on production status and growing environments of rice in Nepal and in the world. Arch Agric Environ Sci 4:83–87

Gibson KD, Hill JE, Foin TC, Caton BP, Fischer AJ (2001) Water-seeded rice cultivars differ in ability to interfere with watergrass. Agron J 93:326–332

Godara RK, Williams BJ, Webster EP, Griffin JL, Miller DK (2012) Evaluation of imazosulfuron for broadleaf weed control in drill-seeded rice. Weed Technol 26:19–23

Gressel J (2005) Crop ferality and volunteerism. CRC Press, Boca Raton

Hong N, Xuan T, Tsuzuki E, Terao H, Matsuo M, Khanh T (2004) Weed control of four higher plant species in paddy rice fields in Southeast Asia. J Agron Crop Sci 190:59–64
Hussain S, Ramzan M, Akhter M, Aslam M (2015) Weed management in direct seeded rice
Jabran K, Chauhan BS (2015) Weed management in aerobic rice systems. Crop Prot 78:151–163
Jabran K, Hussain M, Farooq M, Babar M, Doğan MN, Lee DJ (2012) Application of bispyribac-sodium provides effective weed control in direct-planted rice on a sandy loam soil. Weed Biol Manag 12:136–145
Jabran K, Ullah E, Hussain M, Farooq M, Haider N, Chauhan B (2015) Water saving, water productivity and yield outputs of fine-grain rice cultivars under conventional and water-saving rice production systems. Exp Agric 51:567
Joshi E, Kumar D, Lal B, Nepalia V, Gautam P, Vyas A (2013) Management of direct seeded rice for enhanced resource-use efficiency. Plant Knowl J 2:119
Kaur J, Singh A (2017) Direct seeded rice: prospects, problems/constraints and researchable issues in India. Curr Agric Res J 5:13
Khaliq A, Matloob A, Ahmad N, Rasul F, Awan I (2012) Post emergence chemical weed control in direct seeded fine rice. J Anim Plant Sci 22:1101–1106
Khaliq A, Riaz MY, Matloob A (2011) Bio-economic assessment of chemical and non-chemical weed management strategies in dry seeded fine rice (*Oryza sativa* L.). J Plant Breed Crop Sci 3: 302–310
Khan A, Srivastava R, Ghorai A, Singh S (2003) Efficient soil solarization for weed control in the rain-fed upland rice ecosystem. International agrophysics 17
Kumar A, Kadam S, Arif M, Meena R, Verma T (2020) Legumes an alternative land use options for sustaining soil health. Agriculture & Food e-newsletter 6
Kumar KA, Rajitha G (2019) Alternate wetting and drying (AWD) irrigation-a smart water saving technology for rice: a review. Int J Curr Microbiol Appl Sci 8:2561–2571
Kumar V, Bellinder R, Brainard D, Malik R, Gupta R (2008) Risks of herbicide-resistant rice in India: a review. Crop Prot 27:320–329
Kumar V, Singh S, Chhokar RS, Malik RK, Brainard DC, Ladha JK (2013) Weed management strategies to reduce herbicide use in zero-till rice–wheat cropping systems of the indo-Gangetic Plains. Weed Technol 27:241–254
Mahajan G, Chauhan B, Johnson D (2009) Weed management in aerobic rice in northwestern indo-Gangetic Plains. J Crop Improv 23:366–382
Mahajan G, Chauhan BS (2008) Performance of penoxsulam for weed control in transplanted rice. Pest Technol 2:114–116
Mahajan G, Chauhan BS (2013a) Herbicide options for weed control in dry-seeded aromatic rice in India. Weed Technol 27:682–689
Mahajan G, Chauhan BS (2013b) The role of cultivars in managing weeds in dry-seeded rice production systems. Crop Prot 49:52–57
Matloob A, Khaliq A, Chauhan BS (2015) Weeds of direct-seeded rice in Asia: problems and opportunities. Adv Agron 130:291–336
Midya A, Bhattacharjee K, Ghose S, Banik P (2005) Deferred seeding of blackgram (*Phaseolus mungo* L.) in rice (*Oryza sativa* L.) field on yield advantages and smothering of weeds. J Agron Crop Sci 191:195–201
Mohtisham A, Ahmad R, Ahmad Z, Aslam MR (2013) Effect of different mulches techniques on weed infestation in aerobic rice (*Oryza sativa* L.). Am-Eur J Agric Environ Sci 13:153–157
Mudhoo A, Garg V (2011) Sorption, transport and transformation of atrazine in soils, minerals and composts: a review. Pedosphere 21:11–25
Nagargade M, Singh M, Tyagi V (2018) Ecologically sustainable integrated weed management in dry and irrigated direct-seeded rice. Adv Plants Agric Res 8:319–331
Nie L, Peng S, Chen M, Shah F, Huang J, Cui K et al (2012) Aerobic rice for water-saving agriculture. A review. Agron Sustain Develop 32:411–418
Olofsdotter M, Valverde B, Madsen K (2000) Herbicide resistant rice (*Oryza sativa* L.): global implications for weedy rice and weed management. Ann Appl Biol 137:279–295

Pathak H, Tewari A, Sankhyan S, Dubey D, Mina U, Singh VK et al (2011) Direct-seeded rice: potential, performance and problems–a review. Curr Adv Agric Sci 3:77–88
Pereira AE, Grillo R, Mello NF, Rosa AH, Fraceto LF (2014) Application of poly (*epsilon-caprolactone*) nanoparticles containing atrazine herbicide as an alternative technique to control weeds and reduce damage to the environment. J Hazard Mater 268:207–215
Pittelkow C, Fischer A, Moechnig M, Hill J, Koffler K, Mutters R et al (2012) Agronomic productivity and nitrogen requirements of alternative tillage and crop establishment systems for improved weed control in direct-seeded rice. Field Crops Res 130:128–137
Prasad R, Shivay YS, Kumar D (2017) Current status, challenges, and opportunities in rice production. In: Rice production worldwide. Springer, Cham, pp 1–32
Raj SK, Syriac EK (2017) Weed management in direct seeded rice: a review. Agric Rev 38:41–50
Rajashekar B, Sudharani V, Parveen SN, Shivacharan G (2017) Knowledge of farmers about integrated weed management (IWM) practices in major crops. Int J Farm Sci 7:33–36
Rao A, Johnson D, Sivaprasad B, Ladha J, Mortimer A (2007) Weed management in direct-seeded rice. Adv Agron 93:153–255
Rao AN, Brainard DC, Kumar V, Ladha JK, Johnson DE (2017) Preventive weed management in direct-seeded rice: targeting the weed seedbank. Adv Agron 144:45–142
Rathika S, Ramesh T, Shanmugapriya P (2020) Weed management in direct seeded rice: a review. IJCS 8:925–933
Ren L, Su S, Yang X, Xu Y, Huang Q, Shen Q (2008) Intercropping with aerobic rice suppressed fusarium wilt in watermelon. Soil Biol Biochem 40:834–844
Riar DS, Norsworthy JK (2011) Use of imazosulfuron in herbicide programs for drill-seeded rice (Oryza sativa) in the mid-South United States. Weed Technol 25:548–555
Saha S, Munda S, Singh S, Kumar V, Jangde HK, Mahapatra A et al (2021) Crop establishment and weed control options for sustaining dry direct seeded rice production in eastern India. Agronomy 11:389
Sangeetha C, Baskar P (2015) Influence of different crop establishment methods on productivity of rice–a review. Agric Rev 36:113–124
Sanusan S, Polthanee A, Audebert A, Seripong S, Mouret JC (2010) Suppressing weeds in direct-seeded lowland rainfed rice: effect of cutting dates and timing of fertilizer application. Crop Prot 29:927–935
Sathish A, Kumar KA, Reddy PRR, Devi MU (2017) Effect of different crop establishment methods and irrigation regimes on rice (*Oryza sativa* L.) yield and water use efficiency. Int J Curr Microbiol Appl Sci 6:90–95
Schwartz-Lazaro LM, Copes JT (2019) A review of the soil seedbank from a weed scientists perspective. Agronomy 9:369
Sharda R, Mahajan G, Siag M, Singh A, Chauhan BS (2017) Performance of drip-irrigated dry-seeded rice (*Oryza sativa* L.) in South Asia. Paddy Water Environ 15:93–100
Shekhawat K, Rathore SS, Chauhan BS (2020) Weed management in dry direct-seeded rice: a review on challenges and opportunities for sustainable rice production. Agronomy 10:1264
Singh K, Singh S, Pannu R (2019) Efficacy of pendimethalin and cyhalofop-butyl+ penoxsulam against major grass weeds of direct-seeded rice. Indian J Weed Sci 51:227–231
Singh S, Bhushan L, Ladha J, Gupta R, Rao A, Sivaprasad B (2006) Weed management in dry-seeded rice (*Oryza sativa*) cultivated in the furrow-irrigated raised-bed planting system. Crop Prot 25:487–495
Singh S, Chhokar R, Gopal R, Ladha J, Gupta R, Kumar V et al (2009) Integrated weed management: a key to success for direct-seeded rice in the indo-Gangetic plains. In: Integrated crop and resource Management in the Rice–Wheat System of South Asia. Los Banos, Philippines: International Rice Research Institute, Los Banos, pp 261–278
Singh S, Ladha J, Gupta R, Bhushan L, Rao A (2008) Weed management in aerobic rice systems under varying establishment methods. Crop Prot 27:660–671

Singh S, Ladha J, Gupta R, Bhushan L, Rao A, Sivaprasad B et al (2007) Evaluation of mulching, intercropping with *Sesbania* and herbicide use for weed management in dry-seeded rice (Oryza sativa L.). Crop Prot 26:518–524

Singh S, Singh G, Singh V, Singh A (2005) Effect of establishment methods and weed management practices on weeds and rice in rice-wheat cropping system. Indian J Weed Sci 37:51–57

Singh VP, Dhyani VC, Singh SP, Kumar A, Manalil S, Chauhan BS (2016) Effect of herbicides on weed management in dry-seeded rice sown under different tillage systems. Crop Prot 80:118–126

Singh Y, Singh V, Singh G, Yadav D, Sinha R, Johnson D et al (2011) The implications of land preparation, crop establishment method and weed management on rice yield variation in the rice–wheat system in the indo-Gangetic plains. Field Crops Res 121:64–74

Susha VS, Chinnamuthu CR, Pandian K (2008) Remediation of herbicide atrazine through metal nano particle. In: International Conference on Magnetic Materials and their Applications in the 21st Century

Swanton CJ, Mahoney KJ, Chandler K, Gulden RH (2008) Integrated weed management: knowledge-based weed management systems. Weed Sci 56:168–172

Thakur AK, Mohanty RK, Patil DU, Kumar A (2014) Impact of water management on yield and water productivity with system of rice intensification (SRI) and conventional transplanting system in rice. Paddy Water Environ 12:413–424

Thakur AK, Rath S, Mandal KG (2013) Differential responses of system of rice intensification (SRI) and conventional flooded-rice management methods to applications of nitrogen fertilizer. Plant Soil 370:59–71

Tindall K, Williams B, Stout M, Geaghan J, Leonard B, Webster E (2005) Yield components and quality of rice in response to graminaceous weed density and rice stink bug populations. Crop Prot 24:991–998

Trung HM, Tan NT, Cung HA (1995) Present status and prospect of weed control in rice in Vietnam. In: Proceeding of the 5th Asian-pacific Weed Science Society Conference, Tsukuba, Japan

Webster EP, Baldwin FL, Dillon TL (1999) The potential for clomazone use in rice (*Oryza sativa*). Weed Technol 13:390–393

Weerakoon W, Mutunayake M, Bandara C, Rao A, Bhandari D, Ladha J (2011) Direct-seeded rice culture in Sri Lanka: lessons from farmers. Field Crops Res 121:53–63

Xu L, Li X, Wang X, Xiong D, Wang F (2019) Comparing the grain yields of direct-seeded and transplanted rice: a meta-analysis. Agronomy 9:767

Yaduraju N, Mishra J (2008) Sedges in rice culture and their management. In: Direct seeding of rice and weed management in the irrigated rice-wheat cropping system of the Indo-Gangetic Plains, pp 191–203

Zhao D, Atlin G, Bastiaans L, Spiertz J (2006) Developing selection protocols for weed competitiveness in aerobic rice. Field Crops Res 97:272–285

Zhao D, Bastiaans L, Atlin G, Spiertz J (2007) Interaction of genotype× management on vegetative growth and weed suppression of aerobic rice. Field Crops Res 100:327–340

Contemporary Management of Insect Pests in Rice 20

Farhan Mahmood Shah, Muhammad Razaq, and Yasir Islam

Abstract

Rice, *Oryza sativa* L., is an essential food for the survival of human society. Rice production is challenged by several biotic and abiotic constraints, including insect pests. From sowing until harvesting, about 217 insect species reportedly feed on rice plant, including planthoppers, stem borers, and leaf folders, altogether causing losses of 37%. Chemical pesticides are perceived to be the most promising solutions to cope with rice pest challenges, but concerns of pest resistance and environment pollution necessitate their replacement with safer, efficient and more resiliant approaches. Integrated pest management (IPM) is a globally endorsed paradigm that benefits from ecological principles and creates new avenues to manage pest challenges, for example, it promotes the use of sowing date adjustment, intercropping, plant spacing, flooding, managing crop residues, and host plant resistance (HPR) to discourage pest development and severity to subsequently reduce pressure on agrochemicals. Manipulating habitat management practices, for example, banker plant systems, bund crops, trap plants to support biological control is another promising way of enhancing natural pest suppression and reducing insecticide load. Sensible use of inundative releases of *Trichogramma* wasps to manage lepidopteran pests has proved to be equally effective as chemical control with respect to grain yield. Recent advancements to reduce pesticide use include internet of things (IoT) systems, which are

F. M. Shah · M. Razaq (✉)
Department of Entomology, Faculty of Agricultural Sciences and Technology, Bahauddin Zakariya University, Multan, Pakistan
e-mail: muhammadrazaq@bzu.edu.pk

Y. Islam
Hubei Insect Resources Utilization and Sustainable Pest Management Key Laboratory, College of Plant Science and Technology, Huazhong Agricultural University, Wuhan, China

N. Sarwar et al. (eds.), *Modern Techniques of Rice Crop Production*,
https://doi.org/10.1007/978-981-16-4955-4_20

automated programmed systems based on artificial intelligence (AI). Automated sensor based systems mounted on ground vehicles or aerial vehicles visualizing pest (through visual imaging), or devices able to sense chemical/physical cues from stressed plants (under herbivore attack) are tested effective in timely implementation of precise control. Machine learning based automated traps can mitigate conventional pest monitoring challenges for timely pest detection. This chapter discusses recent advances made with IPM in typical context of rice pest management.

Keywords

Artificial intelligence · Losses · Insect pests · Banker plant · Trap plant · Bund crops

20.1 Introduction

The current world's population is 7 billion, which is projected to reach 9.1 billion by 2050. As a result of which, a consumption growth of 70% is also expected. However, the constant depletion of agriculture resources from increased urbanization and industrilization are among the key constraints limiting agricultural production, making it difficult for the global food system to balance between supply and demand. It is expected, the poor world implementing ineffective production/protection technologies is more likely to suffer from food shortage (Prosekov and Ivanova 2018). Cereals, legumes, and tubers/roots, including, rice *Oryza sativa* L.; maize *Zea mays* L.; wheat *Triticum aestivum* L.; millet *Digitaria exilis* K.; barley *Hordeum vulgare* L.; rye *Secale cereale* L.; sorghum *Sorghum bicolor* L.; and oats *Avena sativa* L. (all Poales: Poaceae) are major staple diets that upon consumption provide essential nutrients, vitamins, fibers, oils, etc. (Naerstad et al. 2012) and altogether provide 90% of world's calories (Bender and Smith 1997). Rice supplies about 20% of the globe's dietary energy supply, whereas 19% and 5% are supplied by wheat and maize, respectively (Alexandratos and Bruinsma 2012).

Rice is a staple diet for nearly half of world's inhabitants (Ahmad et al. 2015, 2019, 2012, 2013, 2008; 2009; Ahmad and Hasanuzzaman 2012; Ahmed and Ahmad 2019, 2020; Shahzad and Ahmad 2019). Rice occupies one-fifth of total world's cereal cropland (Fairhurst and Dobermann 2002) and its cultivation is reported from 110 countries worldwide. Rice grains are rich sources of consumable, allergen-free starch and proteins (Juliano 1992; Shih and Daigle 2000), fulfilling 35–60% calorie requirements (Yorobe Jr et al. 2016), thus are important for achieving food security (Long-ping 2014). Global leaders in rice production are mainly Asian countries with 90% contribution, and the rest comes from Africa, Australia, Europe, Latin America, and the USA (Rejesus et al. 2013). The annual production of rice in Asia stood at 675 million tons, while it was only 29 million tons in Africa and 36 million tons in the Americas (Elert 2014). China is the top rice-producing nation and is followed by the India as the second nation. The last decade has witnessed a marked increase in rice production. It has increased from 439 million tons in 2010 to 496 million tons in 2020, which, by 2035, expectedly reach 555 million tons (FAO 2017).

Insect pests are prominent biotic constraints limiting rice production. In Asian rice agroecosystems, rice pests are reported to cause huge crop losses that, on an annual basis, are estimated to reach 37%. Rice pests reportedly destroyed 120–200 million tons of rice in tropical Asia (Willocquet et al. 2004). As grain feeders, leaf defoliators, stem borers, and root feeders, rice pests assault rice plant from sowing until harvesting. According to Grist and Lever (1969), about 800 insect species feed on this crop. Although, this research has been referenced in almost all researches describing rice pest species, studies often conclude that most of these pests do very little damage, are no-longer pests or no longer present. We made a research to compile information on rice insect pests using some of the highly cited published references since 1990s onward (Pathak and Khan 1994; Bambaradeniya et al. 2004; Heinrichs and Muniappan 2017; Heinrichs et al. 2017), based on which, we could conclude overall 217 species of insects to be pests of rice crop, globally.

The frequently reported global rice pests are brown planthopper (hereafter referred to as BPH) *Nilaparvata lugens* Stål; white-backed planthopper (hereafter referred to as WBPH), *Sogatella furcifera* Horváth); small brown planthopper, *Laodelphax striatellus* Fallen (Hemiptera: Delphacidae) (Savary et al. 2012); rice leaf folder, *Cnaphalocrocis medinalis* (Guenée) (Lepidoptera: Pyralidae), rice striped stem borer, *Chilo suppressalis* (Guenée) (Lepidoptera: Crambidae), and rice yellow stem borer, *Scirpophaga incertulas* (Walker) (Lepidoptera: Pyralidae) (Heinrichs et al. 2017). Images of some of the important rice pests from Pakistan are shown in Fig. 20.1. *Chilo suppressalis* that started to spread across northern Taiwan since 1960 (Cheng and Chiu 1999) has now become widely dispersed as a pest of wheat, rice, and maize across Asia, Europe, and Oceania (Luo et al. 2019). In larval stages, it attacks rice plant's leaves, stems, and panicles, and causes white heads and dead hearts (Cheng et al. 2010). In Japan, it caused dead hearts (10–20%) and white heads (>5%) (Liu 1990). *Cnaphalocrocis medinalis* is another damaging, obnoxious, cosmopolitan insect pest of rice in Asia, Northeastern Australia, Madagascar, and Oceania (Khan et al. 1988). Its larva stitches the blades of the leaf longitudinally with silk strands and feeds on the mesophyll tissue within a longitudinally folded leaf blade, which also assist this pest avoiding direct contact with pesticides (Wang et al. 2009). The outbreaks of *C. medinalis* have been witnessed by many Asian countries (Heong 1993), including China, where in 2003, this pest attacked 20 million hectares and destroyed 760 million (kg) grains (Li et al. 2012). Leaf hoppers (including BPH) are phytophagous arthropods that have resurged since 1960 as a result of pesticide-induced resistance (Wu et al. 2020). The nymph/adult hoppers remove phloem sap from vascular tissues, leading to a condition called "hopperburn," which disturbs photosynthesis (Padmavathi et al. 2013). Some species of rice hoppers are also known to transmit plant viruses (e.g., ragged stunt virus/grassy stunt virus) (Heong and Hardy 2009). Indeed, BPH is now regarded a global challenge limiting rice production (Jia-An 2011; Hereward et al. 2020).

Fig. 20.1 Images of some frequent rice pests from fields of rice, located in Punjab, Pakistan (Photographs by Noor Abid Saeed; Nuclear Institute for Agriculture & Biology (NIAB); Faisalabad – Pakistan)

Like many other economically important crops, rice pest management has been dominated by synthetic chemical pesticides. Chemical pesticides became part and parcel of pest management worldwide since the Paul Muller discovered the insecticidal properties of an organochlorine insecticide, dichlorodiphenyltrichloroethane (DDT) around World War II (Osteen and Szmedra 1989). The chemical pesticides reduced yield gaps and helped to ensure food security and sustainability, but problems began with their irrational use (Abdollahzadeh et al. 2015) that resulted in serious challenges of 3R problems (Residues, Resistance, Resurgence) (Razaq et al. 2019a, 2019b; Naeem et al. 2021), environmental pollution, and off-target effects (Chiu et al. 2018). Pesticide-induced challenges are now global concerns, generally existent across territories where pesticides are being used. The chemical control of rice pests evolved parallel to the issues of pest resistance (Zhao 2000). The history started with conventional pesticides (e.g., chlorinated hydrocarbons, carbamates, pyrethroids), the latter substituted by neonicotinoids and miscellaneous molecules (Ogah and Nwilene 2017). Some common chemicals used in rice pest management worldwide include lambda-cyhalothrin, thiamethoxam, dinotefuran, acephate, pymetrozine, triflumezopyrim against rice hoppers, and chlorpyrifos, carbofuran, abamectin, and fipronil against rice stem borers (Chen and Klein 2012; Seni and Naik 2017; Jun et al. 2020; Zhao et al. 2021). Irrational use led to resistance selection in rice pests, making its management through these means challenging (Nakao 2017) because *N. lugens* has developed resistance to 29 compounds from major pesticide classes, including organochlorines (Davies et al. 2007), organophosphates (Lu et al. 2017; Zhang et al. 2017), carbamates (Chung et al. 1982; Min et al. 2014), pyrethroids (Sun et al. 2017), neonicotinoids (Matsumura and Sanada-Morimura 2010; Jin et al. 2019), to fipronil (Harris 2006; Khoa et al. 2018) and pymetrozine (Zhang et al. 2014). Likewise, insecticide resistance reports are common in lepidopterans infesting rice (Yao et al. 2017; Mao et al. 2019; Zhao et al. 2021). Resistance is a long-standing and ever-expanding challenge for crop protection, udermining crop protection prolonged sustainability, as its reduces pesticide efficacy and necessitates need for new molecules, which is not only costly but also a very time taking process due to want of prescreening tests to understand cross-resistance and safety risks to pest and pollinators, timely availability of relevant individuals, and high production costs (McLeman et al. 2020). Integrated pest management (IPM) strategies provide chemical alternative strategies (e.g., biopesticides) and decision-guiding approaches (e.g., action threshold that aims to prevent an increasing population below damaging levels) as viable solutions to phasing out large-scale use of pesticides (Shah et al. 2017; Shah et al. 2019; Islam et al. 2020; Shah et al. 2020). Action threshold levels have been established in rice pest control (Morgan et al. 1989) to prevent economic loss (Litsinger et al. 2011). Threshold levels of 3.5% egg masses (1 m × 1 m) (Appert 1970), 5% white head, and 10% dead heart (Nwilene et al. 2008) have been tested effective in suppressing stem borers with reduced reliance on synthetic chemicals without sacrificing marketable yields.

IPM considers multiple nonchemical methods (e.g., cultural, biological, host plant resistance) to lower pesticide loads. As a last resort, chemical pesticides can be applied keeping in view their selectivity, safety, and resistance risks (Abdulla 2007; Savary et al. 2012; Sparks and Nauen 2015; Sparks et al. 2020). The history of IPM in rice production parallels with the introduction of chemical pesticides, fertilizers, and semi-dwarf rice varieties in tropical agriculture (Cohen et al. 1997) which attracted rice cultivation on large areas than before. This also increased insecticide use that subsequently killed biocontrol agents and irrational use selected for pest resistance (e.g., BPH in Asia) (Kenmore 1991; Wu et al. 2020). Rice IPM program was initiated to educate farmers to adapt IPM by developing necessary skills (Matteson et al. 1994). As this strategy benefited from ecological principle to counter pest challenges, so is more sustainable due to reduced health and environmental risks (Pandey et al. 2010). This IPM included intellectual integration of nonchemical but effective control methods (cultural, biological, host plant resistance). Here, we discuss rice insect pests and their management in a typical context of contemporary pest management. The recent advances made with biological control through ecological engineering or habitat management practices those that have proved to be at par in producing rice grain yields, and the scope of precision pest management are also discussed.

20.2 Cultural Practices

It is a nonchemical method, implemented to improve cropping environment for a better plant growth to discourage pest buildup, and help plant escape susceptible crop stage by breaking herbivore/host phenological synchronization (Ahmed et al. 2017, 2020a, 2020b; Akram et al. 2018; Akram et al. 2019; Hussain et al. 2018; Fahad et al. 2019; Fatima et al. 2020). In general, cultural control benefits crop over pest by creating an unfavorable environment for herbivore. Soil fertilization with zinc/potassium helped plants enforce vigor and defense against stem borer injury (Sarwar 2011; Sarwar 2012). Managing crop residues (e.g., ploughing, disking) proved an efficacious strategy to kill diapausing larvae of stem borers (e.g., *Chilo* spp.) (Kfir et al. 2002; Sarwar 2012; Togola et al. 2020). Burning crop residues also kills hidden larvae but it is not safe due to causing environmental pollution (Goswami et al. 2020; Singh et al. 2020). Flooding the fields led to the killing of overwintering larvae of *C. suppressalis* and *S. incertulas*, with a greater success against the latter species due to its weak crawling ability (Zhu et al. 2007; Luo et al. 2020). The optimal plant spacing of rice (i.e., 33 × 33 cm) proved to be more promising in suppressing rice planthoppers and to protect yield losses better than by 18 × 18 cm, 23 × 23 cm, and 28 × 28 cm spacings (Asghar et al. 2020). Using trap crops (i.e., plants attracting oviposition but not allowing hatching larvae survive) is another promising method. The use of a dead-end trap plant, vetiver grass (*Vetiveria zizanioides*) caused significant suppression of *C. suppressalis* abundance in rice fields (Lu et al. 2019). The use of intercropping or strip cropping could also be valuable for rice pest management (Ning et al. 2017; Lan et al. 2018). Adjusting sowing time is another efficient strategy to break phenological synchronization between the host and herbivore. This sowing time adjustment creates an asynchrony between pest and susceptible crop stage, hence discourages pest population buildup during susceptible crop stage, allowing chemical control reduction without sacrificing harvestable yield (Shah et al. 2017; Shah et al. 2019; Shah et al. 2020). The plantation of rice during May suppressed infestation and damage of leaf miner (*Hyderllia griseola* Fallen) and stem borer (*Chilo agamamnon* Bles) better than June-plantation (Shalaby 2018). Late-plantation of rice (i.e., August 30) suffered the greatest damage from rice leaf folder (15.42%) than early-plantation (August 1) with 7.83% damage (Rautaray et al. 2019). Against BPH, planting of the rice early (July 5–8) and middle (July 18–23) allowed pest suppression and yield production to be better than late-plantation (August 5–7) (Chander and Mohan 2020). Cultural practices are more convenient for farmers with small land holdings particularly in developing countries; however, implementation of these strategies can be complex as these require continuous and collective efforts at farm level and require through knowledge of pest ecology. Even a well-culturally managed field can become infested, once the pest move from poorly managed fields in the surroundings to the well-managed field (Hafeez-ur-Rehman et al. 2019; Khan et al. 2019a, b; Naz et al. 2019; Peerzada et al. 2019; Razaq et al. 2019; Sarwar et al. 2013a, b; Tariq et al. 2020; Wasaya et al. 2019; Zahoor et al. 2019).

20.3 Host Plant Resistance

Host plant resistance (HPR) crops are the central theme of IPM programs, implemented in order to prevent pest damage in a viable manner. The HPR crops are modified plants incorporated with heritable traits to express defense, tolerance, or avoidance against phytophagous pests. Conventional breeding for resistance in rice plants started against rice pests in 1976 at the International Rice Research Institute (IRRI) (Wu and Khush 1985). At that time, a number of screening programs and germplasm collection identified several resistant genes/sources against BPH, WBPH, green leafhopper, stem borers, and gall midges (Khush 1980). However, the durability of HPR crops is compromised by many reasons such as pest adaptiveness as a result of dynamic nature of the host plant/pest complex interactions allowing transitioning of pests from minors to majors, or with the selection of virulent biotypes and with the introduction of high-yielding rice cultivars, which means that novel resistance sources/genes need to be continuously identified and incorporated to deal with adapting pests (Horgan 2018; Loc and Van Hoa 2019; Makkar et al. 2019). Though, the conventional breeding method has made considerable progress with the development of resistant rice cultivars, breeding cultivated *O. sativa* gene pool with the conventional method has become difficult due to several reasons, such as with the scarcity of resistance sources (typically against the yellow stem borer), complex varietal screening protocols, and complex genetic nature of resistance (Makkar and Bentur 2017; Loc and Van Hoa 2019). The novel approaches include biotechnology, genomics, and RNA interference (RNAi) have become important in identifying or understanding the underlying resistance mechanism in rice pests (Kathuria et al. 2007; Makkar et al. 2019; Du et al. 2020). The recent advances made with molecular biology and biotechnology (e.g., mutagenesis, single gene/gene pyramiding, transgene) have been integrated with conventional breeding methods to quicken identification and assist with transfer of novel resistance sources into plant materials for enhancing tolerance/resistance (Gupta et al. 2019; Rasool et al. 2020; Togola et al. 2020). Against rice hoppers, about 70 resistant genes have been identified using molecular marker tagging and about six genes expressing BPH resistance have been cloned in different lines using map-based cloning techniques. Gene pyramiding and marker-assisted selection and using transgenic approaches (e.g., RNAi) targeting plant volatiles and plant lectins contribute to rice plant resistance against rice hoppers (Rao et al. 1998; Foissac et al. 2000; Sarao et al. 2017; Thulasinathan et al. 2020). Against BPH, some of the developed resistant rice varieties are Bw453, Ld371, Bw367, 62_355, Bg310, Bg454, Bg250, Bg403, At306, At307, At362, At402, and Ld253 (Ananda et al. 2020). Against lepidopterans, the *Bacillus thuringiensis* derived Cry genes have been shown to have biocidal potential against many pest larvae including those of *C. suppressalis*, *S. inferens*, and *C. medinalis* (Makkar et al. 2019; Estiati 2020; Majumder et al. 2020; Talakayala et al. 2020).

20.4 Light Traps/Sex Pheromones

Against nocturnal pests, particularly lepidopterans and coleopterans, the use of light traps is an economical way of monitoring pest populations (Cheng 2009). Light traps are mechanically designed lamp-based traps that are networked in the field to attract insects. They are designed in such a way as to discourage trapped insects' evasion. The trapped individuals can be used to provide assistance with species identification and density estimates that can then guide pest management decisions (Baehaki 2011). In rice pest management, various types of light traps have been developed, including frequency-vibration-based-lamps (Lu et al. 2015), or rustic light traps (Bentley 2009), frequency trembler grid lamps, and fan-inhaling lamps (He et al. 2013). Against adult BPH, WBPH, and stem borers, light traps are used to support both the appropriate time of sowing (Bentley 2009) and implement control timely under field and nursery environments (Baehaki 2011; Baskaran et al. 2019). Though light traps are economical but can be disadvantageous for nontarget insects, especially during pest non-immigrant period (Lu et al. 2015). Attempts are being made to improve functioning of light traps through artificial intelligence (see Discussion below), or through the application of species-specific pheromones to prevent damage to nontarget organisms. Sex pheromones are species-specific chemicals that can trigger mating behavior in the conspecifics. The female insects are emitters of sex pheromones and conspecific male partners are receivers: preparing synthetic blends of actual compounds can be used to deceive male partners to pass optimum mating period or capturing them. Since the 1990s, the synthetic blends of sex pheromones (mimic natural pheromones structurally but function differently) have been used in IPM for flight monitoring, pest forecasting, mating disruption, and mass trapping of several male pests, mainly lepidopterans (Chen and Klein 2012; Chen et al. 2014; Furlan et al. 2020; Liang et al. 2020; Noeth et al. 2020; Sammani et al. 2020). This method as being highly species-specific requires small amounts and is considered safe for plants/nontarget organisms (McNeil 1991; Lucchi and Benelli 2018; Lucchi et al. 2018; Lucchi et al. 2019). In rice pest management, pheromone-baited traps have been mainly used against moths (Abdulla 2007), including *C. suppressalis*, and *C. medinalis* to monitor pest population and timely implement control (Wilson et al. 2015), but this method fails to help decision-making against immatures.

20.5 Biological Control

Biological control includes the action of predators, parasitoids, and pathogens to reduce the populations of the pests. With the concept of integrated control, the biological control method became an integral part of pest management strategies; it complements other control tactics, particularly habitat management or ecological engineering, because these practices reduce populations of insects by enhancing activities of the natural enemies. Successful implementation of IPM with biological control agents needs research regarding potential strain identification, rearing, maintenance, inundative release programs, and the assessment of impact evaluation regarding insect suppression, pesticide load reduction, and crop production enhancement. Research projects developed rarely take into account these research areas mainly due to financial constraints (Babendreier et al. 2019). Here, we review some recent advances in developing biological control in rice production systems and also discuss challenges.

Rice ecosystems are rich in beneficial arthropod communities, and more than 1200 species of predators and parasitoids of arthropods have been reported from China on this crop (Lu et al. 2015). *Trichogramma* parasitizing eggs of lepidopteran pests constitute a major group of biocontrol agents that have been targeted greatly for their host specificity and performance. Six *Trichogramma* strains collected from rice fields in the Greater Mekong Subregion were evaluated alongside three other strains for egg parasitism of the striped stem borer, *C. suppressalis*, for possible incorporation in the pest management program. More parasitism rates and emergence rate were noted on 24 h old eggs than on older age groups (24–72 h) in the laboratory vial tests. *Trichogramma chilonis* Ishii (Hymenoptera: Trichogrammatidae) CJ (strain abbreviation; originating host: Sentinel *Corcyra cephalonica* (Stainton); originating location: Cuijia, XingÕan, China (34 5546 N, 109 3751 E)) strain provided a significantly higher parasitism rate among the tested strains. This strain was released at the rate of 100,000 wasps per hectare to open fields (Ko et al. 2014). In China, *Trichogramma japonicum* Ashmead (Hymenoptera: Trichogrammatidae) is another important egg parasitoid of lepidopterans, *Tryporyza incertulas* Walker (Lepidoptera: Pyralidae) (100% egg parasitism), *C. suppressalis*, and *C. auricilius* (Chen et al. 2010). The other important parasitoid species of lepidopterans include *Cotesia chilonis* (Matsumura) (Hymenoptera: Braconidae) and *Apanteles cypris* Nixon (Hymenoptera: Braconidae). *Anagrus nilaparvatae* (Pang et Wang) (Hymenoptera: Mymaridae) is an important egg parasitoid of *N. lugens*, *S. furcifera*, and *L. striatellus* (Ren and Xian 2017). Implementing biological control through promising strains, rearing methods, temperatures, and selective pesticides increased rice yields by 2–10% and reduced pesticide use by 1.5 applications per season (Ko et al. 2014; Babendreier et al. 2020). This implies that a strategic use of biocontrol agents can enhance food production and reduce chemical load, and may help with restoration of agroecosystems supporting natural enemies to make the pest control more resilient and environmentally sustainable.

20.6 Ecological Engineering or Landscape Management in Pest Management

Governing insect seasonality or abundance over time were the fundamental aspects of an integrated control concept (Stern et al. 1959). Insect pest populations are regulated by many interacting density-independent (mainly abiotic) and density-dependent (biotic) factors. The latter factors mostly operate naturally, but anthropogenic changes (caused by humans) disturbed natural ecosystem functions. Monoculture and chemicals coupled with the high-yielding varieties were the main factors responsible for population escalations of pests due to depletion of predators, parasitoids, and pathogens. Improper pesticide application, landscape design, and irrigation patterns in rice crops are shown to disrupt the biological control. In the 1990s, these facts were experimentally proven in rice production systems in Indonesia. In early-planted rice, generalist predators develop on ample populations of detritivores and plankton-feeding insects, adding organic matter can thus favors soil biota which then increases the resource for predators to develop and suppress pests later in the season. Early-season applications with pesticides killed pre-existing predators and hampered biological control, which then led to pest resurgence in the later season. Conserving pre-season predators is therefore recommended to improve pest control with major reductions in pesticide load (Settle et al. 1996). Restoring agroecosystems to support natural enemies thus represents a feasible means of reducing pesticide load.

Many concepts, such as sustainable pest management, ecosystem resilience, and regenerative agriculture, were subsequent developments since the concept called integrated control was introduced. The ecological or environmental pest management concepts were developed as alternative terms to IPM. Taking advantage from soil health and weeds against the pest underlies the philosophy of these IPM concepts (Coll and Wajnberg 2017; Michaud 2018). For conservation biological control, the more relevant terms are habitat management and ecological engineering, the latter being the fine-tuned form of plant diversity that is feasible agronomically at the farm level and suppresses the pests by increasing natural enemies of the insect pests (Zhu et al. 2014). Species of the plants other than the main crop provide food (nectars and pollens), shelter, and alternate prey host to natural enemies of insect pests, increasing biocontrol diversity, population, functions, making farming systems more resilient and sustainable (Lu et al. 2015; Gurr et al. 2016).

In developing habitat management programs, determining complex interactions of plant species, pests, and their natural enemies is crucial to manipulate cropping habitat (Gurr et al. 2017). The information resulting from interaction strength can be intellectually integrated to support biological control and suppress pest problems. Against the pests, such approaches are reported to be more efficacious than chemical control alone (Bottrell and Schoenly 2018), which has made habitat management a hot topic in the context of pest management. The last decade has witnessed an increasing interest in research developing habitat management technologies for the rice pest control almost across all rice-growing regions worldwide. In this part, we detail some important implications of habitat management in rice pest management.

In Asian rice production systems, 23 plant species were evaluated using Y-tube olfactometer assays to determine suitable candidates for incorporation as ecological engineering plants through the response of parasitoids, *Anagrus optabilis* (Perkins) and *Anagrus nilaparvatae* (Pang et Wang) (Hymenoptera: Mymaridae) to rice planthoppers. Both parasitoids were attractive to *Sesamum indicum* L. (Lamiales: Pedaliaceae), *Emilia sonchifolia* L. (Asterales: Asteraceae), and *Impatiens balsamena* L. (Ericales: Balsaminaceae). Female *A. nilaparvatae* and *A. optabilis* parasitized significantly more eggs of *N. lugens* with about 40% reduced handling time and had increased longevity when fed with sesame flowers in adult stage (Zhu et al. 2013). In another study, the influence of flowers of plant species viz., *Tagetes erecta*, *Trida procumbens*, *Emilia sonchifolia*, and *S. indicum* on biological traits of an important predator of rice planthoppers, mirid bug, *Cyrtorhinus lividipennis* Reuter (Hemiptera: Miridae), revealed increased survival of adult males and females. Nectar sourced as food from flowers of all tested species increased the consumption of *N. lugens*; however, *S. indicum* was the most favorable for the females of *C. lividipennis*. In separate experiments, feeding *C. lividipennis* with *S. indicum* flowers strongly promoted predation, increased search rate, and decreased handling time, suggesting the suitability of *S. indicum* for use as an ecological engineering plant in the rice pest management (Zhu et al. 2014). In another study, the laboratory evaluation of *S. indicum* flowers as the nectar source proved to favor the biology of egg and larval parasitoids of rice borers but became biologically deficit for adult borers. Flowers of sesame increased the survival of adults *Apanteles ruficrus*, *C. chilonis*, and *T. chilonis* but no such effects were evident on two stem borers, *S. inferens* and *C. suppressalis*. Females of *T. chilonis* laid more eggs when their adults were fed with sesame flowers but the feeding of adult borers did not enhance egg production. These results support the suitability of *S. indicum* to be manipulated against rice stem borers in strengthening the biological control of rice pests (Zhu et al. 2015). Against rice pests, other than arthropod predators, the effect of insectivorus birds was evaluated with respect to vegetation density in rice agroecosystem as habitat management/ecological engineering in Philippines. Foraging response of insectivorous birds increased in these ecologically engineered fields to arthropods and snails on the rice plants, as compared to fields where such vegetation was not provided (Horgan et al. 2017b).

In the context of ecological engineering, another approach receiving particular attention is the growing of rice bunds (lvees) with flowering plants to attract nectar-feeding parasitoids. The selected plant species should have positive effects on enemy foraging and life-table responses. While selecting bund crops, it is also recommended to consider crops discouraging pest feeding and development (Gurr et al. 2012). In China, a banker plant system to control the BPH has been developed to enhance biological control. In the landscape design of this system, a grass species, *Leersia sayanuka* Ohwi, is planted near the rice fields. *Nilaparvata muiri* is a sibling species of BPH and distributed widely across many Asian rice-growing countries. *Leersia sayanuka* is the host of *N. muiri*; however in the laboratory, it was proved that BPH cannot complete the life cycle on this plant species. Likewise, *N. muiri* cannot complete its life cycle on rice; resultantly, *L. sayanuka* cannot support population developments of BPH. An egg parasitoid, *A. nilaparvatae* parasitizes eggs of both the species of hoppers. Experiments in rice fields depicted that BPH populations were significantly lower in fields where *L. sayanuka* was grown as the banker plant system, as compared to those rice fields where the banker plant system was not established (Zheng et al. 2017). A recent study in Philippines compared the communities of insect pests with their natural enemies in rice fields and also on levees with and without mung bean, okra, sesame, and weeds. All the vegetation types around the rice fields provided habitat to parasitoids, predatory bugs and spiders, subsequently nurturing lower densities of the pests than fields without vegetation. The higher densities of the predator *C. lividipennis* and *N. lugens* egg parasitism by *Anagrus* sp. were recorded in field plots with okra or sesame bunds. The fields surrounded with mung bean had more populations of grain-sucking pests and lepidopterans; however, their damage did not increase (Horgan et al. 2019).

In China, Thailand, and Vietnam, planting cotyledon plants around rice fields on levees was determined to be successful in enhancing densities and predation of natural enemies in the rice production system. Employing this approach helped suppress the *N. lugens* and *S. furcifera* with 70% reduction in insecticide applications, hence economical. This intervention increased rice yields by 5% and growers' profit by 7.52% (Gurr et al. 2016). In India, the planting of *Vigna unguiculata* (L.) Walpers as a border crop around the rice fields attracted the highest populations of the generalist predator *Coccinella septempunctata* L. to control *N. lugens*, as compared to other crops viz., *Helianthus annuus* L., *Solanum lycopersicum* L., *S. indicum*, *Abelmoschus esculentus* L., and *Solanum melongena* L. bordered around the rice field (Chandrasekar et al. 2016). Similarly, in Indian rice production systems the cultivation of aromatic basmati varieties on bunds with rice crop enhanced the densities of *C. lividipennis*, an important predator of *N. lugens*, in field studies as well as in the laboratory in choice experiments with olfactometer (Chandrasekar et al. 2017). Rice growers in Philippines usually plant rice bunds (levees) with vegetables, namely string beans (*V. unguiculata*), to enhance their incomes, but they do so without knowing the role of this landscape in pest management. Farmers apply insecticides as standard practice to manage rice pests. To determine the effect of planting beans on the side of the rice fields on pest management, rice fields were planted with string beans on bunds and applied with insecticides (grower's standard practice) and in other conditions, rice fields were planted without string beans but managed with insecticides, and fields planted with string beans but not managed with insecticide treatment. The rice yield was similar across all treatments, but fields planted with beans produced 3.6 kg of fresh beans for each meter of the bund. Insecticide-treated fields had maximum leaf damage, and lowest densities of odonatans and minimum parasitism of planthopper eggs (Horgan et al. 2017a). In another study, rice bunds called diversified fields that were established by planting with 40 patches of 2 m^2 per hectare on rice bund fields at three sites in Philippines were established for comparison with conventionally grown fields regarding insect pests' and parasitoids' abundance and yield attributes. Diversification fields did not affect the densities of *N. lugens*, *C. medinalis*, and damage of stem borers at all the sites but had abundant *S. incertulas* moths. Both types of landscapes resulted in similar rice yields but diversified fields provided range of vegetables. Egg parasitism of planthoppers was similar in both the treatments however; it was more near vegetable patches. Egg parasitism on stem borer by *T. japonicum* and other parasitoids was typically greater in diversified fields (Vu et al. 2018).

20.7 Challenges in the Adoption Rate of Biological Control

The adoption rate of biological control is still low particularly in developing countries, despite its unequivocal advantages. In the multicountry study in the Greater Mekong Subregion to incorporate *Trichogramma* in rice pest management systems, growers from China were ready to pay for parasitoid cards. Moreover, Chinese farmers with high acreage started establishing their own facilities for producing natural enemies. But rice growers in Myanmar were not ready to pay for the cards even less than the production costs (Babendreier et al. 2020). One of the important motivational factors in the developing nations is the education level of the farmers which hinders farmer interest in the innovative practices like biological control for their highly technical nature. In a recent study to determine the motivations to accept biological control in Iran among rice growers, it was shown that the health maintenance was the motivation for the educated rice growers (who had received college education) to adapt to biological control (Abdollahzadeh et al. 2016). Another factor, which also determines the adoption of biological control/IPM in developing countries, is the lack of collective action by the farmers. *Trichogramma* release proved more effective in the villages where all the farmers participated. Moreover, farmers who did not attend all the training sessions were not aware of natural enemies (Babendreier et al. 2020). It is a well-known fact that biological control is a knowledge-intensive method desiring time, money, skills, and effort to be implemented efficiently. Farmers are mostly reluctant to implement it, as small negligence can cause biocontrol failure, for instance, prolonged storage of eggs at 4 °C reduced susceptibility to parasitism by *Trichogramma*. Literature review does not provide a desired level of research on biological control in many developing countries to provide alternatives to chemical control which are not only cheaper but also knowledge-intensive. The stakeholders can play their role in developing region-specific biocontrol technologies to reducing pesticide load and concerns.

20.8 Internet of Things (IoT) and Rice Pest Management

Agriculture 4.0 refers to the digitalization of agriculture through smart technologies like Internet of Things (IoT) and artificial intelligence (AI). IoT considers smart electronic devices to receive and transfer data through wireless networks without human assistance. AI is the simulation of human intelligence that can be used to program smart devices to think and act like humans. The intellectual integration of such digital techniques with farming practices allowed optimization of agriculture inputs (e.g., fuel, water, agrochemicals). Smart devices are increasingly being developed and proposed in rice pest management to overcome challenges of conventional pest management including monitoring, detection, and decision implementation. Here, we provide a brief overview of some recent advances made with AI-programmed IoT systems in a typical context of rice pest management along with conventional IPM challenges.

Timely pest detection is important for an effective pest management to control pest before they are well established and accrue losses. Accurate pest detection requires efficient sampling and identification techniques. As distributions of pests are often patchy and unpredictable, the use of complex sampling methods is recommended. The conventional monitoring methods are difficult and often do not comply with nonuniform and unpredictable distribution of pest across fields. Rice planthoppers' monitoring through stem tapping for identification or counting of pests on enamel plates coated with glue to avoid hopper escape has been a complex methodology to apply. When pests are abundant (>50), the density estimate based on personal experience is preferred over counts, which can affect the detection accuracy. The capturing of rice hoppers using sweep nets to visually count the number of pests, and using digital imaging of trapped individuals for subsequent inspection by human experts, are complex methods due to the small size, large densities, and a complex background of BPH. Machine learning is a feasible alternative method of recognizing and counting pests by an automation process (Skawsang et al. 2019). Machine learning based systems have been proposed for pest identification, classification, and data acquisition, the latter uses image as a source, which is captured from a wireless camera continuously observing the sticky trap and subsequently processed and analyzed by the server. Most devices use intact specimens and detection of imperfect specimens, with missing body parts and distorted shapes being difficult to recognize than intact specimens. PENYEK, an automated, deep-learning based detection pipeline, is used to identify BPH specimens using images taken from readily available sticky pads made of plastic plates clipped on steel plates and sprayed with glue. This device is shown to detect imperfect specimens with 95% accuracy level (Nazri et al. 2018). Several other devices like single-layer detection algorithm and two-layer detection algorithm, based on deep-learning technology, have been proposed to rapidly and accurately detect rice BPH. These devices are reported to be efficacious in detecting pest populations as well as in counting their numbers, but results have been comparatively better for a two-layer detection algorithm than a single-layer detection algorithm (He et al. 2020). The monitoring devices, including IoT-assisted unmanned aerial vehicle (UAV) or machine vision devices, have been developed to detect rice pests. An IoT-assisted UAV-based rice detection model using Imagga cloud was applied to identify rice pests. This UAV model focuses on the AI mechanism and Python programming paradigm to send rice pest images to Imagga cloud, which then processes the information and upon detection of the pest based on tags, it sends the information to the owner for further action. This type of device can detect any pest affecting rice during production (Bhoi et al. 2021).

Automated monitoring systems like rice light-trap based on machine vision have been developed (Mohammed et al. 2018; Bjerge et al. 2020). An intelligent light trap, a cloud server, and a mobile phone or computer client platform are the main components. The piling up of insects inside the trap is avoided by their dispersal through vibrating action of a vibration plate and a moving rotary conveyor belt. This system works by trapping, killing, and dispersing insects within the trap, followed by the imaging of trapped insects and communication with the server for the identification and counting of trapped individuals, which have been tested under field conditions with satisfactory results (Qing et al. 2020). Manual detection of stem borer is complex, time-consuming, and ineffective. The use of hyperspectral imaging (visible/near-infrared hyperspectral imaging, spectral range of 380–1030 nm) combined with chemometrics (based on the successive projection algorithm and backpropagation neural network) was proposed to detect striped stem-borer early pest infestation and degree of infestation with a classification accuracy of >95% (Fan et al. 2017). For an accurate detection of *C. medinalis* and *S. incertulas*, a light trap based on a four-layer deep neural network, with the search/rescue optimization (DNN-SAR) method has been proposed. These traps are designed to attract and trap rice pests, and the images are analyzed with 98.29% pest detection accuracy (Muppala and Guruviah 2020). A video detection system (on the basis of deep learning) with custom backbone to build real-time pest detection video has been proposed for detecting pests of rice plants. This system that involved still-image detector having faster-RCNN (region-based convolutional neural network) framework, video-based evaluation metrics to reflect video detection quality (based on a machine-learning classifier) and to detect relatively blurry videos that used image-training models, was found to be more suitable than ResNet-50, VGG16, YOLOv3, and ResNet-101 backbone system in the detection of untrained rice videos (Li et al. 2020).

Herbivore-induced plant volatiles are emitted by attacked plant hosts to inform natural enemies (Dicke 1994; Arimura et al. 2009) and induct defense in neighboring plants (Kost and Heil 2006). Also, leaf reflectance traits can change as a result of herbivory because the phloem sap drainage induces desiccation, nutrient and chlorophyll loss, the latter in particular results in the change of color or leaf reflectance traits eventually before host death. The reflectance trait change with respect to responsible species is difficult to determine when using naked eye but possible with digital technologies. These volatile compounds and leaf reflectance change are important cues for pest management. The color change from plant hoppers (*N. lugens* and *S. furcifera*) feeding on rice phloem juice was examined using digital images captured with a digital camera across 39 rice cultivars during seedling box-test, with reflectance values: for red (580 nm), green (540 nm), and blue (550 nm). Reflectance (red and blue) gradually increased as the herbivore damage increased, with greater red reflectance for plants infested with *N. lugens* and blue/green reflectance for plants infested with *S. furcifera*. This distinct impact on color change (phenotyping) by the two planthopper species was obvious (Horgan et al. 2020). The hyperspectral reflectance data of rice panicles under *N. lugens* attack and infestation with fungal pathogen, *Ustilaginoidea virens*, through near-infrared spectroscopy allowed discrimination between healthy/damaged panicles (Liu et al. 2010). Electric-nose technologies based on sensors, developed using hardware and software, are important advancements in detecting cues from insect/plant hosts. Some of these are smart sprayers that could be mounted on ground vehicles (e.g., tractor) or others are unmanned rotary wing aircrafts (e.g., aerial drones), the former type uses ground base-sensing whereas the latter type uses satellite base-sensing. The ground vehicles are drone- mounted sprayers that consider imaging system to detect density/unit area and apply metered amount of pesticides (Chang et al. 2014). In wild berries, the use of smart sprayers reduced 78.5% herbicides and increased yield by 137.8% (Esau et al. 2018). The drone-mounted sprayer mainly consists of LiPo (lithium polymer) batteries, brushless direct current (BLDC) motors, pump, pesticide tank, and supporting frame. Using these technologies is very important for spraying metered amount of pesticides in orchards and other crops including rice where pesticide application is very difficult with human intervention (Yallappa et al. 2017; Li et al. 2019). Some drones rely on chemical cues (insect/plant volatiles) for detecting pest and implementing precise control. A monitoring drone surveying crop collects information and passes to another drone, which processes and implements control according to the information received (i.e., release biological control or apply chemicals) in the hotspot area (Li et al. 2019; Wang et al. 2019; Filho et al. 2020). In addition to detecting plant chemical cues, electric-nose technology has been implemented in detecting pest, including BPH. Bionic e-nose technology was used to sample BPH volatiles with respect to BPH age and amount. The main ingredients of BPH volatiles are aromatics, sulfur-containing organics, nitrogen oxides, and chlorine-containing organics, which may be exploited in developing specialized sensors for BPH detection (Xu et al. 2014). Though volatiles themselves from BPH are important for its detection, how phloem feeding affects BPH volatile composition remains an important question to address for an accurate pest detection. Electric-nose technology applied to study the similarity and differences between volatiles from BPH and rice stem concluded a certain similarity between the volatiles from BPH and infested BPH rice stem, meaning the use of electric-nose technology can help strengthen BPH predictions (Xu et al. 2018).

Challenges and limitations are still due even with such novel methods, for example, tractor-mounted smart sprayers need thorough coverage for an accurate detection, but moving through fields leads to soil compaction and main-crop destruction. Aircraft sensing could be limited under cloudy/windy weather, and detecting through various heights of plant canopy can also be difficult, for instance, imaging through small-heighted plants could visualize pests residing in the lower plant area, but such pests become difficult to detect when plants grew large and the canopy further expands. Further, pest classification can be difficult in the case of small-sized arthropods, as a result, most technologies are laboratory limited (Bieganowski et al. 2020). Developing fine-tuned sensors, with the ability to accurately detect and classify pests, are much needed to implement control precisely and efficiently.

20.9 Conclusion

Rice is an important cereal grain and staple diet for nearly half of the world's population. Being an economically important crop, sustainable rice production is most desired but sustainability is challenged by rice pests that reduce crop yields and attract heavy use of pesticides in their management, subsequently creating issues of resistance and environmental pollution (Wu et al. 2004). IPM considers intellectual integration of several ecological approaches to benefit from natural ecosystem services to lowering pest pressure and thus the pesticide loads. The major challenge for successful IPM implementation is its collective execution over large-area within-farming community (Parsa et al. 2014), but poor farmers worldwide hesitate to adopt IPM due to many reasons, such as technical difficulties, labor-intensive nature, high cost, and less-perceived benefits. In this chapter, we discussed several economical and sustainable methods, their advancement, and implications in rice pest management. The pest control potential of sowing time adjustment, managing crop residue, plant spacing, intercropping, and flooding has been obvious and suitable in phasing out large-scale use of pesticides. The latest research in the context of biological control creates new avenues to strengthen biological control by creating natural enemy-supportive and pest discouraging environments, through manipulation of cropping habitat/environment. Ecological engineering/landscape management through banker plant system, bund crop, or intercropping with trap crops demonstrated promising potential toward lowering pest pressure by favoring the development of natural enemies. Eliminating weed supportive environment for rice pest development could be another way of suppressing rice pest (Shimada and Sugiura 2021). In Asian countries, research is relatively scarce on habitat management and ecological engineering approaches providing sustained management of rice pests. To deal with challenges related to conventional IPM involving complex sampling procedures for monitoring and decision-making the contemporary IPM introduces artificial intelligence approaches that are more efficient in implementing precise control are being used by the developed world. The machine-learning and automated devices (e.g., smart sprayers, drones, traps) have shown promising potential toward precise-control implementation, and efforts are being concentrated to improving these techniques. Technical/financial aid at national/global levels to support research should be given supreme attention (Rahman 2012), typically to introduce these expensive technologies in the less-developed world, where pesticide are most used and their concerns are rapidly growing (Wiyono 2020). The abundant, locally available, and accessible plant materials with a multitude of new chemistries' offering novel action sites represent novel alternatives to synthetic pesticides typically in pesticide-dominated-pest-control-system of poor world (Shah et al. 2017; Shah et al. 2019; Shah et al. 2020).

Acknowledgments We are grateful to Rizwan Ahmad Pasha (RAM) for collection of data about rice pest and to Muhammad Usman (MU) for cross-checking the information gathered. Both RAM and MU are members of the IPM laboratory at the Department of Entomology, Faculty of Agricultural Sciences and Technology, Bahauddin Zakariya University, Pakistan. We acknowledge Noor Abid Saeed (Senior Scientist at the Plant Protection Division; Nuclear Institute for Agriculture and Biology (NIAB); Faisalabad – Pakistan) for photographing rice pests.

References

Abdollahzadeh G, Sharifzadeh MS, Damalas CA (2015) Perceptions of the beneficial and harmful effects of pesticides among Iranian rice farmers influence the adoption of biological control. Crop Prot 75:124–131

Abdollahzadeh G, Sharifzadeh MS, Damalas CA (2016) Motivations for adopting biological control among Iranian rice farmers. Crop Prot 80:42–50

Abdulla NR (2007) Diversity of stem borer egg parasitoids and patterns of parasitism in Morogoro and coast regions in Tanzania. Sokoine University of Agriculture, Morogoro

Ahmad A, Ashfaq M, Rasul G, Wajid SA, Khaliq T, Rasul F, Saeed U, Rahman MH, Hussain J, Baig IA, Naqvi AA, Bokhari SAA, Ahmad S, Naseem W, Hoogenboom G, Valdivia RO (2015) Impact of climate change on the rice–wheat cropping system of Pakistan. In: Hillel D, Rosenzweig C (eds.), Handbook of climate change and agro-ecosystems: the agricultural modeling intercomparison and improvement project (AgMIP) integrated crop and economic assessments, Published by Imperial College Press and the American Society of Agronomy: London, pp. 219–258

Ahmad S, Abbas G, Ahmed M, Fatima Z, Anjum MA, Rasul G, Khan MA, Hoogenboom G (2019) Climate warming and management impact on the change of rice-wheat phenology in Punjab, Pakistan. Field Crops Res 230:46–61

Ahmad S, Ahmad A, Ali H, Hussain A, Garcia y Garcia A, Khan MA, Zia-Ul-Haq M, Hasanuzzaman M, Hoogenboom G (2013) Application of the CSM-CERES-Rice model for evaluation of plant density and irrigation management of transplanted rice for an irrigated semiarid environment. Irrig Sci 31(3):491–506

Ahmad S, Ahmad A, Soler CMT, Ali H, Zia-Ul-Haq M, Anothai J, Hussain A, Hoogenboom G, Hasanuzzaman M (2012) Application of the CSM-CERES-Rice model for evaluation of plant density and nitrogen management of fine transplanted rice for an irrigated semiarid environment. Precis Agric 13(2):200–218

Ahmad S, Ahmad A, Zia-ul-Haq M, Ali H, Khaliq T, Anjum MA, Khan MA, Hussain A, Hoogenboom G (2009) Resources use efficiency of field grown transplanted rice (*Oryza sativa* L.) under irrigated semiarid environment. J Food Agric Environ 7(2):487–492

Ahmad S, Hasanuzzaman M (2012) Integrated effect of plant density, N rates and irrigation regimes on the biomass production, N content, PAR use efficiencies and water productivity of rice under irrigated semiarid environment. Not Bot Horti Agrobot Cluj-Napoca 40(1):201–211

Ahmad S, Zia-ul-Haq M, Ali H, Shad SA, Ammad A, Maqsood M, Khan MB, Mehmood S, Hussain A (2008) Water and radiation use efficiencies of transplanted rice (*Oryza sativa* L.) at different plant densities and irrigation regimes under semi-arid environment. Pak J Bot 40 (1):199–209

Ahmed M, Ahmad S (2019) Carbon dioxide enrichment and crop productivity. In: Hasanuzzaman M (ed) Agronomic crops, vol 2. Springer Nature Singapore Pte Ltd., Singapore, pp 31–46

Ahmed M, Ahmad S (2020) Systems modeling. In: Ahmed M (ed) Systems modeling. Springer Nature Singapore Pte Ltd, Singapore, pp 1–44

Ahmed M, Ahmad S, Raza MA, Kumar U, Ansar M, Shah GA, Parsons D, Hoogenboom G, Palosuo T, Seidel S (2020b) Models calibration and evaluation. In: Ahmed M (ed) Systems modeling. Springer Nature Singapore Pte Ltd., pp 151–178

Ahmed M, Ahmad S, Waldrip HM, Ramin M, Raza MA (2020a) Whole farm modeling: a systems approach to understanding and managing livestock for greenhouse gas mitigation, economic viability and environmental quality. In: Waldrip HM, Pagliari PH, He Z (eds) Animal manure. ASA special publication 67. American Society of Agronomy and Soil Science Society of America, Madison, pp 345–371

Ahmed M, Fayyaz-ul-Hassan, Ahmad S (2017) Climate variability impact on rice production: adaptation and mitigation strategies. In: Ahmed M, Stockle C (eds) Quantification of climate variability, adaptation and mitigation for agricultural sustainability. Springer International Publishing Switzerland, Cham, pp 91–111

Akram R, Fahad S, Masood N, Rasool A, Ijaz M, Ihsan MZ, Maqbool MM, Ahmad S, Hussain S, Ahmed M, Kaleem S, Sultana SR, Mubeen M, Saud S, Kamran M, Nasim W (2019) Plant growth and morphological changes in rice under abiotic stress. In: Hasanuzzaman M, Fujita M, Nahar K, Biswas JK (eds) Advances in rice research for abiotic stress tolerance. Woodhead Publishing, Duxford, pp 69–85

Akram R, Turan V, Hammad HM, Ahmad S, Hussain S, Hasnain A, Maqbool MM, Rehmani MIA, Rasool A, Masood N, Mahmood F, Mubeen M, Sultana SR, Fahad S, Amanet K, Saleem M, Abbas Y, Akhtar HM, Hussain S, Waseem F, Murtaza R, Amin A, Zahoor SA, Sami Ul Din M, Nasim W (2018) Fate of organic and inorganic pollutants in paddy soils. In: Hashmi MZ, Varma A (eds) Environmental Pollution of Paddy Soils. Springer Nature Switzerland AG, Cham, pp 197–214

Alexandratos N, Bruinsma J (2012) World agriculture towards 2030/2050: the 2012 revision. FAO Agricultural Development Economics Division. Food and Agriculture Organization of the United Nations

Ananda GKS, Wijesundara WWMUK, Herath HMVPK, Karunrathne SI, Kannangara SK, Ranaweera LT, Rathnayake PGRG, Abeysundara SP, Udawela UAKS, Weerakoon WMW, Weebadde CK, Sooriyapathirana SDSS (2020) Establishment of a scheme of DNA fingerprinting and sequencing to distinguish the caryopsis samples of the exportable rice cultivars from a panel of mega-varieties. Sri Lankan J Agricul Ecosyst 2(2):31–55

Appert J (1970) M. Separatella (the white rice borer). New observations and a review of the entomological problems of rice in Madagascar. Agron Trop 25(4):329–367

Arimura G-I, Matsui K, Takabayashi J (2009) Chemical and molecular ecology of herbivore-induced plant volatiles: proximate factors and their ultimate functions. Plant Cell Physiol 50: 911–923

Asghar M, Hassan T-U, Arshad M, Aziz A, Latif MT, Sabir AM (2020) Effect of plant spacing on incidence of rice planthoppers in transplanted rice crop. Int J Trop Insect Sci 41:1–11

Babendreier D, Hou M, Tang R, Zhang F, Vongsabouth T, Win KK, Kang M, Peng H, Song K, Annamalai S, Horgan FG (2020) Biological control of lepidopteran pests in rice: a multi-nation case study from Asia. J Integr Pest Manag 11(1):5

Babendreier D, Wan M, Tang R, Gu R, Tambo J, Liu Z, Grossrieder M, Kansiime M, Wood A, Zhang F (2019) Impact assessment of biological control-based integrated pest management in rice and maize in the greater Mekong subregion. Insects 10:226

Baehaki S (2011) Strategi fundamental pengendalian hama wereng batang coklat dalam pengamanan produksi padi nasional. Pengembangan Inovasi Pertanian 4:63–75

Bambaradeniya C, Edirisinghe J, De Silva D, Gunatilleke C, Ranawana K, Wijekoon S (2004) Biodiversity associated with an irrigated rice agro-ecosystem in Sri Lanka. Biodivers Conserv 13:1715–1753

Baskaran RM, Sridhar J, Sharma K, Senthil-Nathan S (2019) Influence of summer weather on prevalence of rice yellow stem-borer in Central India: monitoring and biocontrol strategy. Biocatal Agric Biotechnol 21:101340

Bender W, Smith M (1997) Population, food, and nutrition. Popul Bull 51:5–25

Bentley JW (2009) Impact of IPM extension for smallholder farmers in the tropics. In: Integrated pest management: dissemination and impact. Springer, pp 333–346

Bhoi SK, Jena KK, Panda SK, Long HV, Kumar R, Subbulakshmi P, Jebreen HB (2021) An internet of things assisted unmanned aerial vehicle based artificial intelligence model for rice pest detection. Microprocess Microsyst 80:103607

Bieganowski A, Dammer K-H, Siedliska A, Bzowska-Bakalarz M, Bereś P, Dąbrowska-Zielińska K, Pflanz M, Schirrmann M, Garz A (2020) Sensor-based outdoor monitoring of insects in arable crops for their precise control. Pest Manag Sci 77(3):1109–1114

Bjerge K, Sepstrup MV, Nielsen JB, Helsing F, Hoye TT (2020) A light trap and computer vision system to detect and classify live moths (Lepidoptera) using tracking and deep learning. bioRxiv

Bottrell D, Schoenly K (2018) Integrated pest management for resource-limited farmers: challenges for achieving ecological, social and economic sustainability. J Agric Sci 156:408–426

Chander S, Mohan B (2020) Effect of planting date of rice on brown planthopper. Indian J Entomol 82:277–279

Chandrasekar K, Muthukrishnan N, Soundararajan R (2017) Ecological engineering cropping methods for enhancing predator, *Cyrtorhinus lividipennis* (Reuter) and suppression of planthopper, *Nilaparvata lugens* (Stal) in rice-effect of intercropping system. J Entomol Zool Stud 5:1778–1782

Chandrasekar K, Muthukrishnan N, Soundararajan R, Robin S, Prabhakaran N (2016) Ecological engineering cropping method for enhancing predator *Coccinella septempunctata* and suppression of Planthopper, *Nilaparvata lugens* (Stal) in Rice. Adv Life Sci 5:1288–1294

Chang YK, Zaman QU, Esau TJ, Schumann AW (2014) Sensing system using digital photography technique for spot-application of herbicide in pruned wild blueberry fields. Appl Eng Agric 30: 143–152

Chen HF, Huang SS, Zhang YZ, Zeng X, Huang ZH (2010) Control efficacy of *Trichogramma japonicum* against *Chilo suppressalis* and *Chilaraea auricilia*. J Appl Ecol 21:743–748

Chen RZ, Klein MG (2012) Efficacy of insecticides against the Rice stem-borer, Chilo suppressalis (Walker) (Lepidoptera: Crambidae), and use of sex pheromones to time accurately the yearly application. Int J Pest Manag 58:354–360

Chen RZ, Klein MG, Sheng CF, Li QY, Li Y, Li LB, Hung X (2014) Mating disruption or mass trapping, compared with chemical insecticides, for suppression of *Chilo suppressalis* (Lepidoptera: Crambidae) in northeastern China. J Econ Entomol 107:1828–1838

Cheng C, Chiu Y (1999) Review of changes involving rice pests and their control measures in Taiwan since 1945. Plant Prot Bull (Taipei) 41:9–34

Cheng J (2009) Rice planthopper problems and relevant causes in China. In: Heong KL, Hardy B (eds) Planthoppers: new threats to the sustainability of intensive rice production systems in Asia. Los Banos (Philippines) International Rice Research Institute, Los Banos, pp 157–178

Cheng X, Chang C, Dai SM (2010) Responses of striped stem borer, *Chilo suppressalis* (Lepidoptera: Pyralidae), from Taiwan to a range of insecticides. Pest Manag Sci 66:762–766

Chiu YH, Williams PL, Gillman MW, Gaskins AJ, Mínguez-Alarcón L, Souter I, Toth TL, Ford JB, Hauser R, Chavarro JE (2018) Association between pesticide residue intake from consumption of fruits and vegetables and pregnancy outcomes among women undergoing infertility treatment with assisted reproductive technology. JAMA Intern Med 178:17–26

Chung TC, Sun CN, Hung CY (1982) Resistance of *Nilaparvata lugens* to six insecticides in Taiwan. J Econ Entomol 75:199–200

Cohen MB, Alam SN, Medina EB, Bernal CC (1997) Brown planthopper, *Nilaparvata lugens*, resistance in rice cultivar IR64: mechanism and role in successful *N. lugens* management in Central Luzon, Philippines. Entomol Exp Appl 85:221–229

Coll M, Wajnberg E (2017) Environmental pest management: challenges for agronomists, ecologists, economists and policymakers. John Wiley & Sons, Hoboken

Davies T, Field L, Usherwood P, Williamson M (2007) DDT, pyrethrins, pyrethroids and insect sodium channels. IUBMB Life 59:151–162

Dicke M (1994) Local and systemic production of volatile herbivore-induced terpenoids: their role in plant-carnivore mutualism. J Plant Physiol 143:465–472

Du B, Chen R, Guo J, He G (2020) Current understanding of the genomic, genetic, and molecular control of insect resistance in rice. Mol Breed 40:1–25

Elert E (2014) Rice by the numbers: a good grain. Nature 514:S50–S51

Esau T, Zaman Q, Groulx D, Farooque A, Schumann A, Chang Y (2018) Machine vision smart sprayer for spot-application of agrochemical in wild blueberry fields. Precis Agric 19:770–788

Estiati A (2020) Development of Bt rice potential for yellow stem borer control. J Crop Sci Biotechnol 23:395–403

Fahad S, Noor M, Adnan M, Khan MA, Rahman I, Alam M, Khan IA, Ullah H, Mian IA, Hassan S, Saud S, Bakhat HF, Hammad HM, Ahmad S, Nasim W (2019) Abiotic stress and rice grain quality. In: Hasanuzzaman M, Fujita M, Nahar K, Biswas JK (eds) Advances in rice research for abiotic stress tolerance. Woodhead Publishing, pp 571–583

Fairhurst T, Dobermann A (2002) Rice in the global food supply. Better Crops Intl 16:3–6

Fan Y, Wang T, Qiu Z, Peng J, Zhang C, He Y (2017) Fast detection of striped stem-borer (*Chilo suppressalis Walker*) infested rice seedling based on visible/near-infrared hyperspectral imaging system. Sensors 17:2470. https://doi.org/10.3390/s17112470

FAO (2017) Agriculture Organization of the United Nations (2017). The future of food and agriculture. Trends and challenges, Rome

Fatima Z, Ahmad M, Hussain M, Abbas G, Ul-Allah S, Ahmad S, Ahmed N, Ali MA, Sarwar G, ul Haque E, Iqbal P, Hussain S (2020) The fingerprints of climate warming on cereal crops phenology and adaptation options. Sci Rep 10:18013

Filho FHI, Heldens WB, Kong Z, de Lange ES (2020) Drones: innovative technology for use in precision pest management. J Econ Entomol 113:1–25

Foissac X, Loc NT, Christou P, Gatehouse AM, Gatehouse JA (2000) Resistance to green leafhopper (*Nephotettix virescens*) and brown planthopper (*Nilaparvata lugens*) in transgenic rice expressing snowdrop lectin (*Galanthus nivalis* agglutinin; GNA). J Insect Physiol 46:573–583

Furlan L, Contiero B, Chiarini F, Benvegnù I, Tóth M (2020) The use of click beetle pheromone traps to optimize the risk assessment of wireworm (Coleoptera: Elateridae) maize damage. Sci Rep 10:1–12

Goswami SB, Mondal R, Mandi SK (2020) Crop residue management options in rice–rice system: a review. Arch Agron Soil Sci 66:1218–1234

Grist DH, Lever RJAW (1969) Pests of rice. Lever London, Longmans, p 520

Gupta MK, Vadde R, Donde R, Gouda G, Kumar J, Nayak S, Jena M, Behera L (2019) Insights into the structure–function relationship of brown plant hopper resistance protein, Bph14 of rice plant: a computational structural biology approach. J Biomol Struct Dyn 37:1649–1665

Gurr GM, Lu Z, Zheng X, Xu H, Zhu P, Chen G, Yao X, Cheng J, Zhu Z, Catindig JL (2016) Multi-country evidence that crop diversification promotes ecological intensification of agriculture. Nat Plants 2:1–4

Gurr GM, Read DM, Catindig JLA, Cheng J, Liu J, Lan LP, Heong KL (2012) Parasitoids of the rice leaffolder *Cnaphalocrocis medinalis* and prospects for enhancing biological control with nectar plants. Agric For Entomol 14:1–12

Gurr GM, Wratten SD, Landis DA, You M (2017) Habitat management to suppress pest populations: progress and prospects. Annu Rev Entomol 62:91–109

Hafeez-ur-Rehman NA, Awan MI, Ijaz M, Hussain M, Ahmad S, Farooq M (2019) Direct seeding in rice: problems and prospects. In: Hasanuzzaman M (ed) Agronomic crops, Production technologies, vol 1. Springer Nature Singapore Pte Ltd, Singapore, pp 199–222

Harris R (2006) Monitoring of neonicotinoid resistance in *Nilaparvata lugens* and subsequent management strategies in Asia Pacific. Proceedings of the international workshop on ecology and management of rice planthoppers, pp.16–19

He C, Fang B, Zhang Y, Qing X (2013) Comparison of pest-controlling effects in rice field between fan-inhaling lamp and frequency trembler grid lamp. Hybrid Rice 28:58–63

He Y, Zhou Z, Tian L, Liu Y, Luo X (2020) Brown rice planthopper (*Nilaparvata lugens* Stal) detection based on deep learning. Precis Agric 21:1385–1402

Heinrichs E, Muniappan R (2017) IPM for tropical crops: rice. CAB Rev 12:1–30

Heinrichs E, Nwilene FE, Stout M, Hadi BA, Freitas T (2017) Rice insect pests and their management. Burleigh Dodds Science Publishing Limited

Heong K (1993) Rice leaffolders: are they serious pests. Research on rice leaffolder management in China. Proc. Int. workshop on economic threshold for Rice Leaffolder in China, pp. 8–11

Heong KL, Hardy B (2009) Planthoppers: new threats to the sustainability of intensive rice production systems in Asia. International Rice Research Institute, Los Baños

Hereward JP, Cai X, Matias AMA, Walter GH, Xu C, Wang Y (2020) Migration dynamics of an important rice pest: the brown planthopper (*Nilaparvata lugens*) across Asia—insights from population genomics. Evol Appl 13:2449–2459

Horgan F, Ramal A, Villegas J, Jamoralin A, Bernal C, Perez M, Pasang J, Naredo A, Almazan M (2017a) Effects of bund crops and insecticide treatments on arthropod diversity and herbivore regulation in tropical rice fields. J Appl Entomol 141:587–599

Horgan FG (2018) Integrating gene deployment and crop management for improved rice resistance to Asian planthoppers. Crop Prot 110:21–33

Horgan FG, Jauregui A, Cruz AP, Martínez EC, Bernal CC (2020) Changes in reflectance of rice seedlings during planthopper feeding as detected by digital camera: potential applications for high-throughput phenotyping. PLoS One 15:e0238173

Horgan FG, Martínez EC, Stuart AM, Bernal CC, de Cima ME, Almazan MLP, Ramal AF (2019) Effects of vegetation strips, fertilizer levels and varietal resistance on the integrated management of arthropod biodiversity in a tropical rice ecosystem. Insects 10:328

Horgan FG, Ramal AF, Villegas JM, Almazan MLP, Bernal CC, Jamoralin A, Pasang JM, Orboc G, Agreda V, Arroyo C (2017b) Ecological engineering with high diversity vegetation patches enhances bird activity and ecosystem services in Philippine rice fields. Reg Environ Chang 17: 1355–1367

Hussain M, Ahmad S, Hussain S, Lal R, Ul-Allah S, Nawaz A (2018) Rice in saline soils: physiology, biochemistry, genetics and management. Adv Agron 148:231–287

Islam Y, Shah FM, Shah MA, Khan MM, Rasheed MA, Rehman S, Ali S, Zhou X (2020) Temperature-dependent functional response of *Harmonia axyridis* (Coleoptera: Coccinellidae) on the eggs of *Spodoptera litura* (Lepidoptera: Noctuidae) in laboratory. Insects 11:583

Jia-An LY-GC (2011) Basic research on the outbreak mechanism and sustainable management of rice planthoppers. Chin J Appl Entomol 2

Jin R, Mao K, Liao X, Xu P, Li Z, Ali E, Wan H, Li J (2019) Overexpression of CYP6ER1 associated with clothianidin resistance in *Nilaparvata lugens* (Stål). Pestic Biochem Physiol 154:39–45

Juliano BO (1992) Structure, chemistry, and function of the rice grain and its fractions. Cereal Foods World 37:772–772

Jun Z, Sun W-Q, Yao L, L-Q GE, Yang G-Q, J-X XU, Fang L (2020) Effects of a novel mesoionic insecticide, triflumezopyrim, on the feeding behavior of rice planthoppers, *Nilaparvata lugens* and *Sogatella furcifera* (Hemiptera: Delphacidae). J Integr Agric 19:2488–2499

Kathuria H, Giri J, Tyagi H, Tyagi AK (2007) Advances in transgenic rice biotechnology. Crit Rev Plant Sci 26:65–103

Kenmore PE (1991) How rice farmers clean up the environment, conserve biodiversity, raise more food, make higher profits. In: Indonesia's IPM-a model for Asia

Kfir R, Overholt W, Khan Z, Polaszek A (2002) Biology and management of economically important lepidopteran cereal stem borers in Africa. Annu Rev Entomol 47:701–731

Khan MA, Ahmad S, Raza A (2019a) Integrated weed management for agronomic crops. In: Hasanuzzaman M (ed) Agronomic crops, vol 2. Springer Nature Singapore Pte Ltd., Singapore, pp 257–281

Khan SM, Ali S, Nawaz A, Bukhari SAH, Ejaz S, Ahmad S (2019b) Integrated pest and disease management for better agronomic crop production. In: Hasanuzzaman M (ed) Agronomic crops, vol 2. Springer Nature Singapore Pte Ltd., Singapore, pp 385–428

Khan Z, Barrion A, Litsinger J, Castilla N, Joshi R (1988) A bibliography of rice leaffolders (Lepidoptera: Pyralidae). Int J Trop Insect Sci 9:129–174

Khoa DB, Thang BX, Liem NV, Holst N, Kristensen M (2018) Variation in susceptibility of eight insecticides in the brown planthopper *Nilaparvata lugens* in three regions of Vietnam 2015-2017. PLoS One 13:e0204962

Khush GS (1980) Breeding for multiple disease and insect resistance in rice. MP-Texas Agricultural Experiment Station (USA), Mount Pleasant

Ko K, Liu Y, Hou M, Babendreier D, Zhang F, Song K (2014) Evaluation for potential *Trichogramma* (hymenoptera: Trichogrammatidae) strains for control of the striped stem borer (Lepidoptera: Crambidae) in the greater Mekong subregion. J Econ Entomol 107:955–963

Kost C, Heil M (2006) Herbivore-induced plant volatiles induce an indirect defence in neighbouring plants. J Ecol 94:619–628

Lan N, Xiang H, Zhang JE, Wang F, Luo H (2018) Effects of rice and *Canna indica* L. intercropping on rice growth, disease/pest incidence and yield. Chin J Eco-Agric 26:1170–1179

Li D, Wang R, Xie C, Liu L, Zhang J, Li R, Wang F, Zhou M, Liu W (2020) A recognition method for rice plant diseases and pests video detection based on deep convolutional neural network. Sensors 20:578. https://doi.org/10.3390/s20030578

Li SW, Yang H, Liu YF, Liao QR, Du J, Jin DC (2012) Transcriptome and gene expression analysis of the rice leaf folder, *Cnaphalocrosis medinalis*. PLoS One 7:e47401

Li X, Andaloro JT, Lang EB, Pan Y (2019) Best management practices for unmanned aerial vehicles (UAVs) application of insecticide products on rice. 2019 ASABE annual international meeting. Am Soc Agric Biol Eng, p 1

Liang YY, Luo M, Fu XG, Zheng LX, Wei HY (2020) Mating disruption of *Chilo suppressalis* from sex pheromone of another pyralid rice pest *Cnaphalocrocis medinalis* (Lepidoptera: Pyralidae). J Insect Sci 20(3):1–8

Litsinger J, Bandong J, Canapi B (2011) Effect of multiple infestations from insect pests and other stresses to irrigated rice in the Philippines: II. Damage and yield loss. Int J Pest Manag 57:117–131

Liu T (1990) A survey on the occurrence of rice stem borers and their damage in Taichung areas. Bull Taichung District Agric Improv Station 29:39–47

Liu ZY, Shi JJ, Zhang LW, Huang JF (2010) Discrimination of rice panicles by hyperspectral reflectance data based on principal component analysis and support vector classification. J Zhejiang Univ Sci B 11:71–78

Loc HT, Van Hoa N (2019) Techniques for determining the brown plant hopper *Nilaparvata lugens* (Stål) resistance in rice in Vietnam. In: Experimental techniques in host-plant resistance. Springer, Singapore, pp 73–81

Long-ping Y (2014) Development of hybrid rice to ensure food security. Rice Sci 21:1–2

Lu K, Wang Y, Chen X, Zhang Z, Li Y, Li W, Zhou Q (2017) Characterization and functional analysis of a carboxylesterase gene associated with chlorpyrifos resistance in *Nilaparvata lugens* (Stål). Comp Biochem Phys C 203:12–20

Lu Y-H, Zheng X-S, Lu Z-X (2019) Application of vetiver grass *Vetiveria zizanioides*: Poaceae (L.) as a trap plant for rice stem borer *Chilo suppressalis*: Crambidae (Walker) in the paddy fields. J Integr Agric 18:797–804

Lu Z, Zhu P, Gurr GM, Zheng X, Chen G, Heong KL (2015) Rice pest management by ecological engineering: a pioneering attempt in China. In: Rice planthoppers. Springer, Dordrecht, pp 161–178

Lucchi A, Benelli G (2018) Towards pesticide-free farming? Sharing needs and knowledge promotes integrated Pest management. Environ Sci Pollut Res 25:13439–13445

Lucchi A, Ladurner E, Iodice A, Savino F, Ricciardi R, Cosci F, Conte G, Benelli G (2018) Eco-friendly pheromone dispensers—a green route to manage the European grapevine moth? Environ Sci Pollut Res 25:9426–9442

Lucchi A, Suma P, Ladurner E, Iodice A, Savino F, Ricciardi R, Cosci F, Marchesini E, Conte G, Benelli G (2019) Managing the vine mealybug, *Planococcus ficus*, through pheromone-mediated mating disruption. Environ Sci Pollut Res 26:10708–10718

Luo GH, Luo ZX, Zhang ZL, Sun Y, Lu MH, Shu ZL, Tian ZH, Hoffmann AA, Fang JC (2020) The response to flooding of two overwintering rice stem borers likely accounts for their changing impacts. J Pest Sci 94:451–461

Luo M, Wang Z, Yang B, Zheng L, Yao Z, Seyrek UA, Chung H, Wei H (2019) Effects of winter cover crops on rice pests, natural enemies, and grain yield in a rice rotation system. J Insect Sci 19:25

Majumder S, Datta K, Satpathy S, Datta SK (2020) Development and evaluation of lepidopteran insect resistant jute expressing the fused Bt-Cry1Ab/ac toxin driven by CaMV35S promoter. Ind Crop Prod 156:112873

Makkar GS, Bentur J (2017) Breeding for stem borer and gall midge resistance in rice. In: Breeding insect resistant crops for sustainable agriculture. Springer, Singapore, pp 323–352

Makkar GS, Bhatia D, Suri K, Kaur S (2019) Insect resistance in rice (*Oryza sativa* L.): overview on current breeding interventions. Int J Trop Insect Sci 39:259–272

Mao K, Li W, Liao X, Liu C, Qin Y, Ren Z, Qin X, Wan H, Sheng F, Li J (2019) Dynamics of insecticide resistance in different geographical populations of *Chilo suppressalis* (Lepidoptera: Crambidae) in China 2016–2018. J Econ Entomol 112:1866–1874

Matsumura M, Sanada-Morimura S (2010) Recent status of insecticide resistance in Asian rice planthoppers. Jpn Agric Res Q JARQ 44:225–230

Matteson PC, Gallagher KD, Kenmore PE (1994) Extension of integrated pest management for planthoppers in Asian irrigated rice: empowering the user. In: Planthoppers. Springer, Dordrecht, pp 656–685

McLeman A, Troczka BJ, Homem RA, Duarte A, Zimmer C, Garrood WT, Pym A, Beadle K, Reid RJ, Douris V (2020) Fly-Tox: a panel of transgenic flies expressing pest and pollinator cytochrome P450s. Pestic Biochem Physiol 169:104674

McNeil JN (1991) Behavioral ecology of pheromone-mediated communication in moths and its importance in the use of pheromone traps. Annu Rev Entomol 36:407–430

Michaud J (2018) Problems inherent to augmentation of natural enemies in open agriculture. Neotrop Entomol 47:161–170

Min S, Lee SW, Choi B-R, Lee SH, Kwon DH (2014) Insecticide resistance monitoring and correlation analysis to select appropriate insecticides against *Nilaparvata lugens* (Stål), a migratory pest in Korea. J Asia Pac Entomol 17:711–716

Mohammed ME, El-Shafie HA, Al-Hajhoj MBR (2018) Design of an automated solar-powered light trap for monitoring and mass trapping of major date palm pests. Ecol Environ Conserv 24: 177–185

Morgan D, Tugwell N, Bernhardt J (1989) Early rice field drainage for control of rice water weevil (Coleoptera: Curculionidae) and evaluation of an action threshold based upon leaf-feeding scars of adults. J Econ Entomol 82:1757–1759

Muppala C, Guruviah V (2020) Detection of leaf folder and yellow stemborer moths in the paddy field using deep neural network with search and rescue optimization. Inf Process Agric. https://doi.org/10.1016/j.inpa.2020.09.002. (In Press; Corrected Proof)

Naeem A, Hafeez F, Iftikhar A, Waaiz M, Güncan A, Ullah F, Shah FM (2021) Laboratory induced selection of pyriproxyfen resistance in *Oxycarenus hyalinipennis* costa (Hemiptera: Lygaeidae): cross-resistance potential, realized heritability, and fitness costs determination using age-stage, two-sex life table. Chemosphere 269:129367. DOI: 129310.121016/j.chemosphere.122020.129367

Naerstad R, Dees MW, Le VH, Holgado R, Hermansen A (2012) Occurrence of skin blemish diseases (scab and scurf) in Norwegian potato production. Potato Res 55:225–239

Nakao T (2017) Insecticide resistance in rice planthoppers. In: Advances in agrochemicals: ion channels and G protein-coupled receptors (GPCRs) as targets for Pest control, GPCRs and ion channels, vol 2. ACS Publications, Washington, pp 23–39

Naz S, Fatima Z, Iqbal P, Khan A, Zakir I, Noreen S, Younis H, Abbas G, Ahmad S (2019) Agronomic crops: types and uses. In: Hasanuzzaman M (ed) Agronomic crops, Production technologies, vol 1. Springer Nature Singapore Pte. Ltd., Singapore, pp 1–18

Nazri A, Mazlan N, Muharam F (2018) PENYEK: automated brown planthopper detection from imperfect sticky pad images using deep convolutional neural network. PLoS One 13:e0208501

Ning C, Qu J, He L, Yang R, Chen Q, Luo S, Cai K (2017) Improvement of yield, pest control and Si nutrition of rice by rice-water spinach intercropping. Field Crops Res 208:34–43

Noeth K, Verleur P, Bouwer M, Crous J, Roux J, Hurley B, Slippers B (2020) Mass trapping of *Coryphodema tristis* (Lepidoptera: Cossidae) using a sex pheromone in *Eucalyptus nitens* compartments in Mpumalanga, South Africa. South For J For Sci 82:271–279

Nwilene F, Agunbiade T, Togola M, Youm O, Ajayi O, Oikeh S, Ofodile S, Falola O (2008) Efficacy of traditional practices and botanicals for the control of termites on rice at Ikenne, Southwest Nigeria. Int J Trop Insect Sci 28:37–44

Ogah E, Nwilene F (2017) Review article incidence of insect pests on rice in Nigeria: a review. J Entomol 14:58–72

Osteen CD, Szmedra PI (1989) Agricultural pesticide use trends and policy issues. US Department of Agriculture, Economic Research Service, Washington

Padmavathi C, Katti G, Padmakumari AP, Voleti SR, Rao LVS (2013) The effect of leaffolder *Cnaphalocrocis medinalis* (Guenee) [Lepidoptera: Pyralidae] injury on the plant physiology and yield loss in rice. J Appl Entomol 137:249–256

Pandey S, Byerlee D, Dawe D, Dobermann A, Mohanty S, Rozelle S, Hardy B (2010) Rice in the global economy. In: Strategic research and policy issues for food security. IRRI, Los Baños

Parsa S, Morse S, Bonifacio A, Chancellor TC, Condori B, Crespo-Pérez V, Hobbs SL, Kroschel J, Ba MN, Rebaudo F (2014) Obstacles to integrated pest management adoption in developing countries. Proc Natl Acad Sci 111:3889–3894

Pathak MD, Khan ZR (1994) Insect pests of rice. International Rice Research Institute, Los Banos, p 19

Peerzada AM, Bukhari SAH, Dawood M, Nawaz A, Ahmad S, Adkins S (2019) Weed management for healthy crop production. In: Hasanuzzaman M (ed) Agronomic crops, vol 2. Springer Nature Singapore Pte Ltd., Singapore, pp 225–256

Prosekov AY, Ivanova SA (2018) Food security: the challenge of the present. Geoforum 91:73–77

Qing Y, Jin F, Jian T, Xu W-G, Zhu X-H, Yang B-J, Jun L, Xie Y-Z, Bo Y, Wu S-Z (2020) Development of an automatic monitoring system for rice light-trap pests based on machine vision. J Integr Agric 19:2500–2513

Rahman M (2012) Problems and suggestions for farmers adoption of IPM practices in rice (*Oryza sativa* L) cultivation. Bangladesh J Agr Res 37:121–128

Rao K, Rathore KS, Hodges TK, Fu X, Stoger E, Sudhakar D, Williams S, Christou P, Bharathi M, Bown DP (1998) Expression of snowdrop lectin (GNA) in transgenic rice plants confers resistance to rice brown planthopper. Plant J 15:469–477

Rasool A, Akbar F, Rehman A, Jabeen H (2020) Genetic engineering of rice for resistance to insect pests. In: Rice research for quality improvement: genomics and genetic engineering. Springer, Singapore, pp 129–148

Rautaray B, Bhattacharya S, Panigrahi D, Panda T, Dash S (2019) Effect of planting dates on incidence of rice leaf folder and its impact on grain yield in north eastern coastal plains of Odisha

Razaq M, Mensah R, Athar H-U-R (2019) Insect pest management in cotton. In: Jabran K, Chauhan BS (eds) Cotton production. John Wiley & Sons, Inc., Washington

Razaq M, Shah FM, Ahmad S, Afzal M (2019a) Pest management for agronomic crops. In: Hasanuzzaman M (ed) Agronomic crops, vol 2. Springer Nature Singapore Pte Ltd, Singapore, pp 365–384

Razaq M, Shah FM, Ahmad S, Afzal M (2019b) Pest management for agronomic crops. In: Hasanuzzaman M (ed) Agronomic crops. Springer, Singapore, pp 365–384

Rejesus R, Martin AM, Gypmantasiri P (2013) Meta-impact assessment of the irrigated rice research consortium. IRRI, Los Banos

Ren L, Xian Z (2017) Survey of Cnaphalocrocis medinalis (Guenée) (Lepidoptera: Pyralidae) parasitoids in Guangxi, China1. J Agric Urban Entomol 33:71–76

Sammani AM, Dissanayaka DM, Wijayaratne LK, Bamunuarachchige TC, Morrison WR (2020) Effect of pheromones, plant volatiles and spinosad on mating, male attraction and burrowing of *Cadra cautella* (walk.) (Lepidoptera: Pyralidae). Insects 11:845

Sarao P, Bhatia D, Brar D (2017) Advances in breeding for resistance to hoppers in rice. In: Breeding insect resistant crops for sustainable agriculture. Springer, Singapore, pp 101–130

Sarwar M (2011) Effects of zinc fertilizer application on the incidence of rice stem borers (*Scirpophaga* species) (Lepidoptera: Pyralidae) in rice (*Oryza sativa* L.) crop. J Cereals Oilseeds 2:61–65

Sarwar M (2012) Effects of potassium fertilization on population buildup of rice stem borers (lepidopteron pests) and rice (*Oryza sativa* L.) yield. J Cereals Oilseeds 3:6–9

Sarwar N, Ali H, Ahmad S, EhsanUllah AS, Mubeen K, Hill JE (2013b) Water wise rice cultivation on calcareous soil with the addition of essential micronutrients. J Anim Plant Sci 23(1):244–250

Sarwar N, Ali H, Maqsood M, EhsanUllah SAN, Shahzad M, Mubeen K, Shahid MA, Ahmad S (2013a) Phenological response of rice plants to different micronutrients application under water saving paddy fields on calcareous soil. Turk J Field Crops 18(1):52–57

Savary S, Horgan F, Willocquet L, Heong KL (2012) A review of principles for sustainable pest management in rice. Crop Prot 32:54–63

Seni A, Naik BS (2017) Evaluation of some insecticides against Brown plant hopper, *Nilaparvata lugens* (Stal) in rice, *Oryza sativa*. L. Int J Bio-resour Stress Manag 8:268–271

Settle WH, Ariawan H, Astuti ET, Cahyana W, Hakim AL, Hindayana D, Lestari AS (1996) Managing tropical rice pests through conservation of generalist natural enemies and alternative prey. Ecology 77:1975–1988

Shah FM, Razaq M, Ali A, Han P, Chen J (2017) Comparative role of neem seed extract, moringa leaf extract and imidacloprid in the management of wheat aphids in relation to yield losses in Pakistan. PLoS One 12:e0184639

Shah FM, Razaq M, Ali Q, Ali A, Shad SA, Aslam M, Hardy ICW (2020) Action threshold development in cabbage pest management using synthetic and botanical insecticides. Entomol Gen 40:157–172

Shah FM, Razaq M, Ali Q, Shad SA, Aslam M, Hardy IC (2019) Field evaluation of synthetic and neem-derived alternative insecticides in developing action thresholds against cauliflower pests. Sci Rep 9:1–13

Shahzad AN, Ahmad S (2019) Tools and techniques for nitrogen management in cereals. In: Hasanuzzaman M (ed) Agronomic crops, vol 2. Springer Nature Singapore Pte Ltd., Singapore, pp 111–126

Shalaby SE (2018) Influence of sowing dates and insecticides on the rice stem borer (*Chilo agamamnon* Bles) and rice leafminer (*Hydrellia griseola* Fallen). J Entomol Res 42:195–200

Shih FF, Daigle KW (2000) Preparation and characterization of rice protein isolates. J Am Oil Chem Soc 77:885–889

Shimada N, Sugiura S (2021) Indirect effects of weeds on rice plants via shared heteropteran herbivores. J Appl Entomol 145:117–124

Singh R, Yadav DB, Ravisankar N, Yadav A, Singh H (2020) Crop residue management in rice–wheat cropping system for resource conservation and environmental protection in North-Western India. Environ Dev Sustain 22:3871–3896

Skawsang S, Nagai M, Tripathi NK, Soni P (2019) Predicting rice pest population occurrence with satellite-derived crop phenology, ground meteorological observation, and machine learning: a case study for the central plain of Thailand. Appl Sci 9:4846

Sparks TC, Crossthwaite AJ, Nauen R, Banba S, Cordova D, Earley F, Ebbinghaus-Kintscher U, Fujioka S, Hirao A, Karmon D (2020) Insecticides, biologics and nematicides: updates to IRAC's mode of action classification-a tool for resistance management. Pestic Biochem Physiol 167:104587

Sparks TC, Nauen R (2015) IRAC: mode of action classification and insecticide resistance management. Pestic Biochem Physiol 121:122–128

Stern V, Smith R, van den Bosch R, Hagen K (1959) The integration of chemical and biological control of the spotted alfalfa aphid: the integrated control concept. Hilgardia 29:81–101

Sun H, Yang B, Zhang Y, Liu Z (2017) Metabolic resistance in *Nilaparvata lugens* to etofenprox, a non-ester pyrethroid insecticide. Pestic Biochem Physiol 136:23–28

Talakayala A, Katta S, Garladinne M (2020) Genetic engineering of crops for insect resistance: An overview. J Biosci 45:1–12

Tariq M, Ahmed M, Iqbal P, Fatima Z, Ahmad S (2020) In: Ahmed M (ed) Crop phenotyping, Systems modeling, Springer Nature Singapore Pte Ltd, Singapore, pp 45–60

Thulasinathan T, Nallathambi J, Rahman H, Kambale R, Ayyenar B, Venkatasamy B, Muthurajan R (2020) Marker assisted introgression and validation of resistance genes Bph20 and Bph21 for brown plant hopper (*Nilaparvata lugens* stal) into a popular rice variety of CO51. J Pharmacogn Phytochem 9:939–944

Togola A, Boukar O, Tamo M, Chamarthi S (2020) Stem borers of cereal crops in Africa and their management. In: Pests control and acarology. IntechOpen, London. https://doi.org/10.5772/intechopen.83719

Vu Q, Ramal AF, Villegas JM, Jamoralin A, Bernal CC, Pasang JM, Almazan MLP, Ramp D, Settele J, Horgan FG (2018) Enhancing the parasitism of insect herbivores through diversification of habitat in Philippine rice fields. Paddy Water Environ 16:379–390

Wang F, Hu G, Chen X, Shen H, Luo S, Xin D, Xu S, Zhang X, Zhai B (2009) Analysis on the causes of recent outbreaks of *Cnaphalocrocis medinalis* in Nanning, China. Chin J Rice Sci 23: 537–545

Wang G, Lan Y, Qi H, Chen P, Hewitt A, Han Y (2019) Field evaluation of an unmanned aerial vehicle (UAV) sprayer: effect of spray volume on deposition and the control of pests and disease in wheat. Pest Manag Sci 75:1546–1555

Wasaya A, Yasir TA, Ijaz M, Ahmad S (2019) Tillage effects on agronomic crop production. In: Hasanuzzaman M (ed) Agronomic crops, vol 2. Springer Nature Singapore Pte Ltd., Singapore, pp 73–99

Willocquet L, Elazegui FA, Castilla N, Fernandez L, Fischer KS, Peng S, Teng PS, Srivastava R, Singh H, Zhu D (2004) Research priorities for rice pest management in tropical Asia: a simulation analysis of yield losses and management efficiencies. Phytopathology 94:672–682

Wilson B, Hardy T, Beuzelin J, VanWeelden M, Reagan T, Miller R, Meaux J, Stout M, Carlton C (2015) Expansion of the Mexican rice borer (Lepidoptera: Crambidae) into rice and sugarcane in Louisiana. Environ Entomol 44:757–766

Wiyono S (2020) Building new phase of IPM in Indonesia. E&ES 468:011003

Wu CF, Khush G (1985) A new dominant gene for resistance to whitebacked planthopper in rice. Crop Sci 25:505–509

Wu J, Ge L, Liu F, Song Q, Stanley D (2020) Pesticide-induced planthopper population resurgence in rice cropping systems. Annu Rev Entomol 65:409–429

Wu JC, Qiu HM, Yang GQ, Liu JL, Liu GJ, Wilkins R (2004) Effective duration of pesticide-induced susceptibility of rice to brown planthopper (*Nilaparvata lugens* Stål, Homoptera: Delphacidae), and physiological and biochemical changes in rice plants following pesticide application. Int J Pest Manag 50:55–62

Xu S, Zhou Z, Lu H, Luo X, Lan Y, Zhang Y, Li Y (2014) Estimation of the age and amount of brown rice plant hoppers based on bionic electronic nose use. Sensors 14:18114–18130

Xu S, Zhou Z, Tian L, Lu H, Luo X, Lan Y (2018) Study of the similarity and recognition between volatiles of brown rice plant hoppers and rice stem based on the electronic nose. Comput Electron Agr 152:19–25

Yallappa D, Veerangouda M, Maski D, Palled V, Bheemanna M (2017) Development and evaluation of drone mounted sprayer for pesticide applications to crops. In: 2017 IEEE global humanitarian technology conference (GHTC). IEEE, Piscataway, pp 1–7

Yao R, Zhao DD, Zhang S, Zhou LQ, Wang X, Gao CF, Wu SF (2017) Monitoring and mechanisms of insecticide resistance in *Chilo suppressalis* (Lepidoptera: Crambidae), with special reference to diamides. Pest Manag Sci 73:1169–1178

Yorobe JM Jr, Ali J, Pede VO, Rejesus RM, Velarde OP, Wang H (2016) Yield and income effects of rice varieties with tolerance of multiple abiotic stresses: the case of green super rice (GSR) and flooding in the Philippines. Agric Econ 47:261–271

Zahoor SA, Ahmad S, Ahmad A, Wajid A, Khaliq T, Mubeen M, Hussain S, Sami Ul Din M, Amin A, Awais M, Nasim W (2019) Improving water use efficiency in agronomic crop production. In: Hasanuzzaman M (ed) Agronomic crops, vol 2. Springer Nature Singapore Pte Ltd., Singapore, pp 13–29

Zhang X, Liu X, Zhu F, Li J, You H, Lu P (2014) Field evolution of insecticide resistance in the brown planthopper (*Nilaparvata lugens* Stål) in China. Crop Prot 58:61–66

Zhang Y, Yang B, Li J, Liu M, Liu Z (2017) Point mutations in acetylcholinesterase 1 associated with chlorpyrifos resistance in the brown planthopper, *Nilaparvata lugens* Stål. Insect Mol Biol 26:453–460

Zhao S (2000) Plant chemical protection. China Agricultural Press, Beijing. [in Chinese]

Zhao X, Xu X, Wang X-g, Yin Y, M-y L, Wu Y-q, Y-h L, Cheng Q-h, C-w G, Shen L-t (2021) Mechanisms for multiple resistances in field populations of rice stem borer, *Chilo suppressalis* (Lepidoptera: Crambidae) from Sichuan Province, China. Pestic Biochem Physiol 171:104720

Zheng X, Lu Y, Zhu P, Zhang F, Tian J, Xu H, Chen G, Nansen C, Lu Z (2017) Use of banker plant system for sustainable management of the most important insect pest in rice fields in China. Sci Rep 7:1–8

Zhu P, Gurr GM, Lu Z, Heong K, Chen G, Zheng X, Xu H, Yang Y (2013) Laboratory screening supports the selection of sesame (*Sesamum indicum*) to enhance *Anagrus* spp. parasitoids (hymenoptera: Mymaridae) of rice planthoppers. Biol Control 64:83–89

Zhu P, Lu Z, Heong K, Chen G, Zheng X, Xu H, Yang Y, Nicol HI, Gurr GM (2014) Selection of nectar plants for use in ecological engineering to promote biological control of rice pests by the predatory bug, *Cyrtorhinus lividipennis*, (Heteroptera: Miridae). PLoS One 9:e108669

Zhu P, Wang G, Zheng X, Tian J, Lu Z, Heong KL, Xu H, Chen G, Yang Y, Gurr GM (2015) Selective enhancement of parasitoids of rice Lepidoptera pests by sesame (*Sesamum indicum*) flowers. BioControl 60:157–167

Zhu Z, Cheng J, Zuo W, Lin X, Guo Y, Jiang Y, Wu X, Teng K, Zhai B, Luo J (2007) Integrated management of rice stem borers in the Yangtze Delta, China. Area-wide control of insect pests. Springer, Dordrecht, pp 373–382

21 Insect Chitin Biosynthesis and Regulation in *Cnaphalocrocis Medinalis* Using RNAi Technology

Muhammad Shakeel, Naeem Sarwar, Omer Farooq, Juan Du, Shang-Wei Li, Yuan-Jin Zhou, Xiaolan Guo, and Shakeel Ahmad

Abstract

Rice leaf folder (*Cnaphalocrocis medinalis*) is one of the most destructive insect pests of rice crop. It is becoming difficult to control due to insecticidal resistance; therefore, chitin can be considered a soft target to control this notorious insect pest. Chitin is an abundant biopolymer and widespread amino polysaccharide in nature. It is also believed to be the second order of magnitude after cellulose that is mainly produced by fungus, nematodes, and in arthropods. In insects, chitin formation is an important mechanism that plays a vital role for their survival. It helps in insect growth (tracheal and gut epithelium, epidermal cuticle formation) and deals with metamorphosis. For this reason, insects continuously produce chitin synthesis enzymes in different body tissues. Chitin biosynthesis pathway starts from trehalase (TRE) and ends up with the utilization of chitin synthesis (CHS). Eight enzymes are involved in the formation of cuticle in insects including *C. medinalis*. However, insect chitin biosynthesis mechanism can be inhibited

M. Shakeel · J. Du · S.-W. Li (✉)
Provincial Key Laboratory for Agricultural Pest Management of Mountainous Regions, Institute of Entomology, Guizhou University, Guiyang, Guizhou, China

N. Sarwar · O. Farooq
Department of Agronomy, Bahauddin Zakariya University, Multan, Pakistan

Y.-J. Zhou
Baihuahu Town Agricultural Service Center, Guiyang, Guizhou Province, China

X. Guo
College of Forestry, Guizhou University, Guiyang Shi, Guizhou, China

S. Ahmad
Department of Agronomy, Faculty of Agricultural Sciences and Technology, Bahauddin Zakariya University, Multan, Pakistan
e-mail: shakeelahmad@bzu.edu.pk

N. Sarwar et al. (eds.), *Modern Techniques of Rice Crop Production*,
https://doi.org/10.1007/978-981-16-4955-4_21

by silencing these eight enzymes using RNA interference. In this chapter, we will discuss the silencing of chitin biosynthesis pathway in *C. medinalis* using RNAi technology.

Keywords

Chitin · Rice leaf folder · Chitin biosynthesis · Gene silencing · RNA interference

21.1 Introduction

Rice (*Oryza sativa* L.) serves as a staple food for a large number of the world's population (Faostat FAO 2020; Kumar et al. 2008) especially in Asia, where about 90% of people use rice to meet their dietary requirements (Khush et al. 2002). In commercial production, the productivity and quality of rice are affected by many biotic stresses, especially pests. Pests damage rice crops at different growth and developmental stages. Global rice yield losses due to pests (diseases, animal/insects, and weeds) range from 20% to at least 30% of the achievable yield (Savary et al. 2000). Many insects like stem pitting, rice gundhi worm, brown plant hopper, white backed plant hoppers, rice leaf folders, etc. damage rice fields. Among them, leaf insects, because of the ability to remove the chlorophyll content of the leaves or remove the leaves, significantly reduce crop yield. The rice leaf folder is one of the most injurious insect pests of rice (Gunathilagaraj and Gopalan 1986).

Rice leaf folder, *Cnaphalocrocis medinalis* Guenée (Lepidoptera), is commonly known as rice leaf folding machine. It is a major pest in rice fields, which is widespread in China, Pakistan, Sri Lanka, Bangladesh, Japan, India, Korea, Philippines, and Indonesia. *C. medinalis* is a migratory pest with 1–11 generations per year. *C. medinalis* is polyphagous insect pest of *Avena sativa L.*, *Eleusine coracana* L., *Hordeum vulgare* L., *O. sativa* L., *Panicum miliaceum* L., *Pennisetum glaucum* L., *Susu bicolor* L., *Zea mays* L., and *Triticum aestivum* L. (www.plantwise.org/KnowledgeBank).

Existence of alternative hosts around rice fields provide shelter for insects and create a favorable environment (de Kraker 1996) and complete multiple generations in one growth period (Baskaran et al. 2017). By producing different varieties, application of fertilizers, and continuous use of pesticides produced resistance in it (Kaushik 2010). Excessive use of pesticides and fertilizers ratio appears to be favorable for the development of *C. medinalis* populations (Li et al. 2020). Rice cultivation and resurrection by pesticides are important factors in the abundance of leaves (Yang et al. 2020). Favorable environment such as high humidity with high temperature, and shady areas of the field are favorable for growth and development of leaf folder (Gangwar 2015). The *C. medinalis* larvae are nocturnal and hide them in rice leaves during the day to avoid predation. In recent decades, rice plantations, especially in areas with modern and fragrant varieties, grew up extensively (Gangwar 2015) but leaf folder seems a potential threat for quality production. New irrigation system, multiple paddy crops, variability of modern semi-dwarf varieties, application of high-level nitrogenous fertilizers, and increased pesticide resistance and resurrection lead to increased threat to *C. medinalis*.

In order to manage *C. medinalis* population, it is very important to know its life cycle. It plays an important role in managing the *C. medinalis* population in the field.

The adult female moth is golden yellow with brown edges on both wings. While resting, the shape of the body is the same as that of an equilateral triangle. There are three bands that run across the entire forewing. It is short, comma-shaped, and curved outward. The length of an adult moth is 10–12 mm. The wing extension is 13–15 mm. Male adults are smaller than female and have noticeable dark brown scale fragments along the mid-costa. They usually mate between twilight and midnight. Adults appear in the evening and mating occurs at night (Gangwar 2015). The egg on the leaf is transparent, yellowish-white, 0.90 mm long, 0.39 mm wide, and almost flat with slightly convex surface. Egg hatching starts 3–4 days after egg laying. Laid eggs are alone or in cluster of 3–8. The female lays 135–175 eggs along the central ribs of the young leaves (Gangwar 2015).

The *C. medinalis* larvae are yellow and later on turn into yellow-green with brown hair that is about 12–25 mm long. Newly hatched larvae migrate at the center of the fresh leaves and start feeding. Second instar larvae fold the leaves by stitching with mouth and start scraping the green material inside the leaves. The larva feeds green pigment inside the folded leaves which causes linear and pale white stripes. Second instar larvae feed on growing paddy leaves resulting in paper-dry leaves (Khan et al. 1996). Typically, one larva feeds on each folded leaf, but later on, it migrates to another leaf. Therefore, a single larva that eats multiple leaves can damage many rice plants. A single larva damages several rice leaves interfering with photosynthesis and reduced rice yield (Alvi et al. 2003). Scratched leaves become membranous, whitish, and wither that seize photosynthesis and reduce the crop productivity (Mishra et al. 1998). At the plant stage, crops usually recover from larval damage, but damage to flag leaves during the reproductive stage significantly reduces crop yield. Severe invasion can completely annihilate plants (Rao and Ramasubbaiah 1988). The losses with leaf folder attack during crop growth are insurmountable (Singh et al. 2003). Previous research suggested that the 17.5% of damaged leaves caused in a yield loss of 16.5%. *C. medinalis* are active throughout the year, and adults moths fly short distances when disturbed (Han et al. 2015). Depending on the reported agroecological situation, the extent of the loss can extend from 63% to 80% (Rajendran et al. 1986).

C. medinalis larvae take 14–18 days to become pupae. The fifth instar larvae show dark brown head, 20–25 mm long, and 1.5–2 mm wide (Gangwar 2015). The fifth instar larvae stich all sides and produce white silk webbing of the folded leaves to become pupae. Newly emerged pupae are light brown in color. The light brown pupae turn into reddish brown before developing adult. Adult emergence normally starts on the sixth to tenth day and are active from May to October (Gangwar 2015).

At present, *C. medinalis* attack is mainly controlled by using various insecticides. Behavioral and physiological modification study showed the detoxification of the pesticides against *C. medinalis*. Therefore, scientists are looking for other strategies to control this pest. Chitin synthetic pathway can be a soft target site to control *C. medinalis* which can adopt.

21.2 Insect Chitin Biosynthesis

Chitin is a second most important polysaccharide after cellulose which is mainly composed of N-acetyl-β-D–glucosamine residues linked by β 1, 4 glyosidic linkages (Merzendorfer 2006).

Chitin is widely present in fungi, pharynx, eggshell of nematodes, cyst wall of protozoan (*E. histolytica*), mollusk shells, and in cuticle and peritrophic matrix of arthropods (Van Dellen et al. 2006).

It is a polysaccharide nitrogen-containing compound. Chitin is a linear polysaccharide with crystal structures of α-chitin, β-chitin, and γ-chitin. Different polymers such as α, β, and γ form are converted into microfibrils during chitin synthesis (Peters 1992). The α-chitin is composed of polymer chains in an antiparallel manner; β-chitin is composed of polymer chains in a parallel manner; γ-chitin is composed of polymer chains. The α-chitin is widely present in insect exoskeletons and shells of crustaceans. Due to the strong flexibility of β-chitin and γ-chitin, they are widely distributed in insect periphagus cocoon (Kenchington 1976).

Chitin is an important component in the formation of embryonic cuticle, tracheal and epidermal cuticle, and extracellular linings that protect the inside and outside of insect's body (Muthukrishnan et al. 2012). Insects utilize chitin as a remarkable polymer for the production of various anatomical structures. In addition, chitin deposition also found in foregut, hindgut, tracheae, salivary glands, and in mouthparts of the insects.

Chitin synthesis plays a crucial role in growth and development and also responsible for ecdysis (Merzendorfer and Zimoch 2003). The body wall of insects contains a large amount of chitin, which can reduce the persecution of insects by the environment and enable insects to complete normal molting and metamorphosis. Chitin also acts as a barrier against invading the different harmful microorganisms (Langer and Vinetz 2001).

The synthesis of chitin is a complex process in insects and requires a series of enzymes to complete it (Merzendorfer 2006). Synthesis of chitin in insects starts with trehalase and ends with chitin. Eight enzymes are involved followed by trehalase and hexokinase phosphoglucose isomerase, phosphofructose glutaminase, phosphoglucosamine transferase, phosphoglucosamine acetylmutase, UDP-N-acetylglucosamine pyrophosphorylase, and chitin synthase (Cohen 1987). Cohen (2001) studied that chitin biosynthesis has a tremendous biological importance, but still less research was done about the chitin biosynthetic pathway in insect and other vertebrates. Jaworski et al. (1963) finally established the pathway from UDP-GlcNAc to chitin by using southern armyworm (*Spodoptera eridania*). Trehalase is the first key enzyme in insect chitin synthesis pathway in which chitin synthase is the last key enzyme and is currently the two most widely studied enzymes such as *Drosophila melanogaster* (Gagou et al. 2002a, b), *Aedes aegypti* (Kato et al. 2005), and *Manduca sexta* (Zhu et al. 2002).

The chitin synthesis sugar is being synthesized from body fat glycogen, which is catalyzed by an enzyme glycogen phosphorylase that converts glucose-1-phosphate into trehalose, which is finally released into blood stream. Trehalose is a source of energy in many insects, which is widely distributed in the gut and in the epidermis to yield intracellular glucose (Becker et al. 1996). Two glycolytic enzymes that are acted upon in the cytosol are involved to convert glucose into fructose-6-phosphate. Hexokinase and glucose-6-phosphate isomerase are two enzymes that are acted upon glucose and converted into fructose-6-phosphate. Glutamine-6-P aminotransferase and Glucosamine-6-P-N-acetyltransferase are enzymes which act upon fructose-6-phosphate that converts glucosamine-6-P into N-acetyl glucosamine 6-P in the presence of glutamine and acetyl CoA as coenzymes. The conversion of sugar derivatives follows the synthesis pathway leads them to the formation of UDP-GlcNAc serve as the action site for CHS and this product finally transferred into respective speculative chitin (N-acetylglucosamine) $_{n}$. Glycogen biosynthesis pathway based on the model which require glycognin as a primer and CHS may fulfill the needs of priming function (Fig. 21.1). At early studies demonstrated that by using as whole insect or isolated different bodied parts like larvae and pupae (epidermal and hypodermal cells, integuments, gut, leg regenerates, oocytes, abdomen, and discs cells) were have capabilities of chitin synthesis. For this reason (trehalose, glucose, glycogen, glucosamine, fructose, GlcNAc) and other compounds could be serving as precursors of isolated bodied tissues.

21.3 Regulation of Insect Biosynthetic Genes Using RNAi

Chitin synthesis acts as a foundation for growth and development in insects. In the line of investigation, chitin synthase disruption leads to severe developmental disorders and deformities ultimately causing death. RNA interference (RNAi) technology is used to control gene expression as "posttranscriptional gene silencing" at mRNA level, remarkably well conserved and regulated in several eukaryotic organisms (Zhuang and Hunter 2012). RNAi is a highly specific defense system found in plants (Baulcombe 2004) as well as in animals (Aouadi et al. 2009), and triggered by various small RNAs including short-interfering RNAs (siRNAs), microRNAs (miRNAs), piwi-interacting RNAs (piRNAs), exogenous RNAs, and endogenous siRNAs (Brodersen and Voinnet 2009). Basically, RNAi mechanism can be divided into three steps, that is, at first, dsRNA molecule is being introduced into the cell and processed into the small RNA duplexes by Dicer and RNase III enzymes. Dicer numbers can be varied from one to many based on individual to individual that are responsible for different types of short dsRNA product (Meister and Tuschl 2004), and in the second step, ssRNA is loaded in the form of single strand after unwinding, called as guide strand loaded into a protein complex called RISC, and third is guide strand directs to target the mRNA and RISC bound to specific protein slicer is an "Argonaute protein" that leads to target the mRNA (Whangbo and Hunter 2008) (Fig. 21.2).

RNAi is divided into cell-autonomous RNAi and non-cell-autonomous RNAi. Cell autonomous is in which RNAi is triggered by siRNAs, is a short double-stranded RNA (dsRNA) fragment of about 20–25 bp, and is incorporated into RNA-induced silencing complex (RISC) processed by Dicer, an RNase III (Price

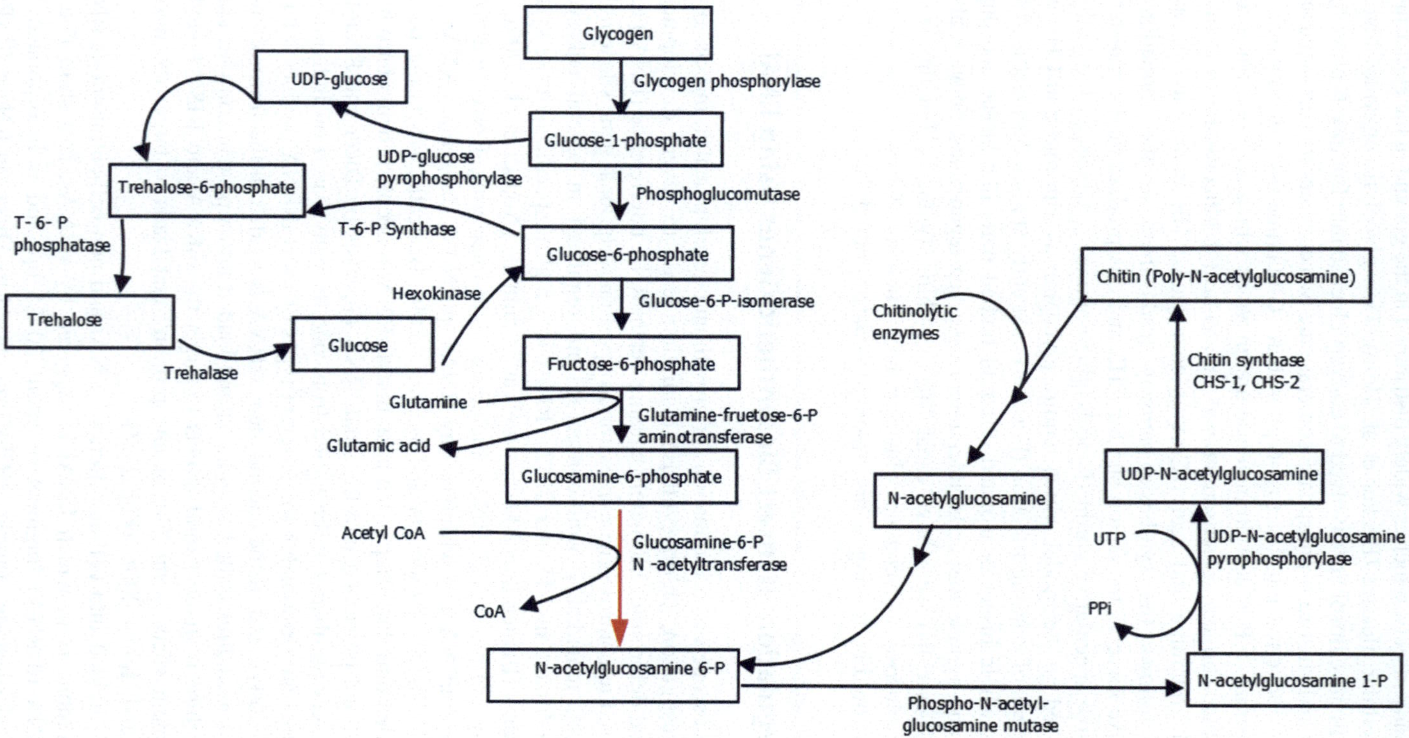

Fig. 21.1 The diagrammatic representation of biosynthesis of chitin in insects that starts from glycogen that stored energy in body tissues occurring in most insects which ends with the chitin polymers

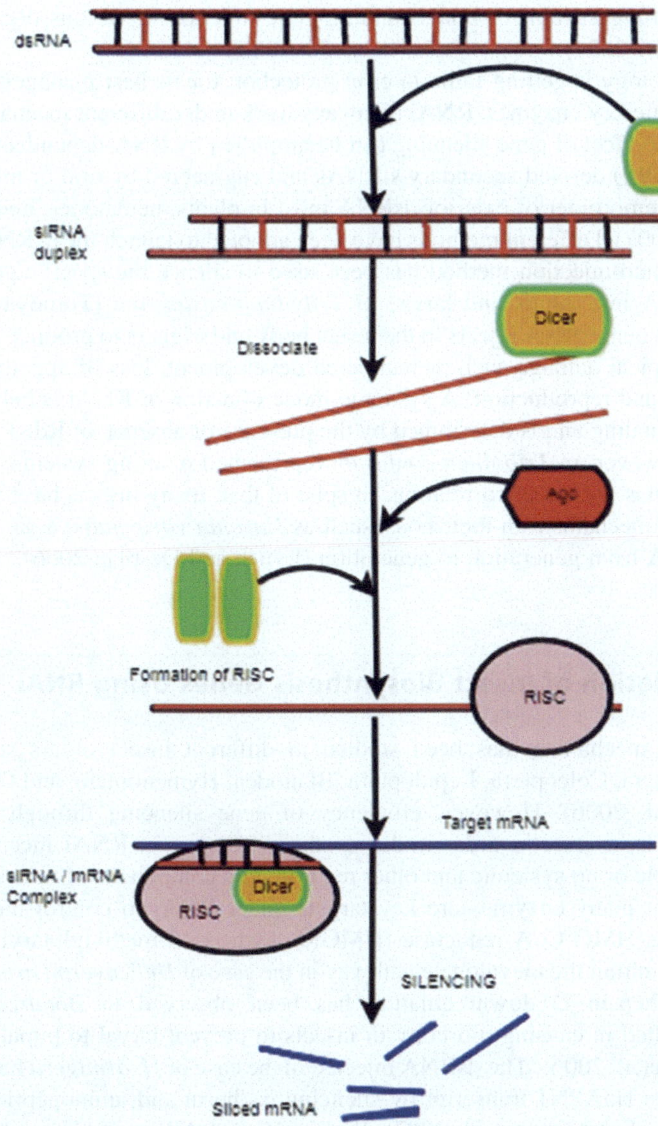

Fig. 21.2 RNAi pathway for sequence-specific gene silencing mechanism

and Gatehouse 2008). Two different kinds of non-cell-autonomous RNAi are evolved, that is, systemic RNAi and environmental RNAi (Grishok 2005). The specificity depends on sequence based on dsDNA on the part or as a whole gene transcription (Borgio 2010). RNAi was first discovered in nematodes and later on it was known in many organisms including plants, fungi, and insect possessing RNAi pathway for sequence-specific gene silencing mechanism and it is also rapidly evolved as a strongest tool for understanding gene function in various organisms (Mello and Conte 2004).

RNAi technology is getting fame in crop protection for its best management in inhibiting specific key enzymes. RNAi pathways work under different mechanisms; for example, the effect of gene silencing can be amplified by RNA-dependent RNA polymerase (RdRP) derived secondary siRNAs and engineered by oral or injection administration, embroider of exterior dsRNA most in plants, nematodes, and fungi (Pak and Fire 2007). Different methods have been adopted to launch the dsRNA into the host. The microinjection method has been used to silence the specific gene by injecting dsRNA into adults and larvae of *Tribolium castaneum* (Tomoyasu and Denell 2004). When dsRNA injects in the insect body and triggers to produce RNAi, causes phenotypical damage such as hampered development, loss of appetite, and stunted growth and reproduction. A systemic mode of action of RNAi is helpful to silence gene signaling and is determined by the presence or absence of RdRP. RNA polymerase, however, in *Tribolium confusum* represented a strong systemic RNAi response but it has not direct correlation. In spite of that, many insects have shown systemic RNAi mechanism in their body such as *Nasonia vitripennis*; a wasp also transfers dsRNA from generation to generation (Lynch and Desplan 2006).

21.4 Regulation of Insect Biosynthesis Genes Using RNAi

Gene silencing mechanism has been studied in different insect orders such as Hemiptera, Diptera, Coleoptera, Lepidoptera, Blattodea, Hymenoptera, and Orthoptera (Cruz et al. 2006). However, efficiency of gene silencing through RNAi technology differ in insect's order is due to the difference in RNAi mechanism included systemic or no systemic and other regarding its components. Some studies have shown that many enzymes are key targets under RNAi to control the gene function such as HMG-CoA reductase (HMGR), hydroxy-3-methylglutaryl coenzyme is used to inhibit the mevalonate pathway in the case of *Helicoverpa armigera*, as well as cathepsin D downregulation has been observed in *Bombyx mori* (BmCatD) resulted in causing mortality in insects to prevent larval to pupal transformation (Gui et al. 2006). The dsRNA injected in the case of *H. armigera* has been known to reduce HaAPN1 transcript by silencing cadherin and aminopeptidase-N (APN) enzymes (Sivakumar et al. 2007). Walker III and Allen (2011) worked for silenced the PG gene after treatment of *L. lineolaris* bugs with PG1dsRNA resulting in reduced function and expression of PG1 enzyme. Similarly Kumar et al. (2008) studied and observed stunted larval growth by feeding siRNA to silenced acetylcholinesterase gene of *H. armigera* and ultimately caused death. Gene malfunction can also be seen in all parts of the body of bed bug after injection of dsRNA that helps to prevent the ClCPR gene expression (Zhu et al. 2012). Quan et al. (2002) have found the effect of injected dsRNA in *B. mori* embryos for recognition of different patterns of protein, and later on similar studies have been performed in embryos of *Hyalophora cecropia* by its own RNAi effects (Bettencourt et al. 2002). The dsRNA response is also detected in *Spodoptera litura* and *Epiphyas postvittana*, but no effect is known against homopteran pests (Price and Gatehouse 2008).

To understand the efficiency of RNAi is generally assumed to depend upon different "delivery methods" occurrence and different stages of the host, if dsRNA has been assumed to be injected inside of the cell of pre-differentiated embryos and mainly depends upon RNAi machinery inside of the cell. Similarly, cationic lipids stimulated by introducing the nucleic acid into the tissue cell act as a delivery agent. Introduction of dsRNA has mostly been done through microinjection and found successful method employed in *B. mori* varied (Pan et al. 2009), and two studies about transgenic *B. mori* have also been described about high level of silencing gene expressing hairpin RNAs (Dai et al. 2008). Transportation of dsRNA into the host body is notably a different method in which microinjection delivery method in one of the most important for injecting dsRNA into the embryos of several lepidopteron insects has been reported mainly in *Mamestra brassicae, S. exigua,* and *Plodia interpunctella,* and silencing of gene has known to be found and observed in these specimen except in *S. exigua* (Tsuzuki et al. 2005). Basically, two different approaches applied in RNAi technology are systemic RNAi as well as environmental RNAi functioned when dsRNA can be induced by injecting and through feeding of dsRNA into insect. Wonderful differences potentially exist in different lepidopteran insects involving high or low silencing gene with respect to the effectivity and level of systemic and environmental RNAi (Terenius et al. 2011). Bettencourt et al. (2002) reported the heritable effect of dsRNA in next generation, treating the pupae of *H. cecropia,* suggesting that dsRNA could entering in the gonads of the immature pupae, leads the success of silencing of target immunity regarding genes. Gandhe et al. (2007) investigated the higher level of silencing effect on injecting of same routine concentration in small and large species. He further found that calculated effects of RNAi appear in higher species rather than lower species.

Surprising results were achieved by injecting dsRNA into the host in *Manduca sexta* and *M. mori*. However, lower and no effect was found in *Spodoptera littoralis, Chrysodeixis,* and *Bicyclus anynana* when high concentration of dsRNA was applied (Iga and Smagghe 2010). This approach was generally applied to *M. configurata* where silencing the target gene was found successful in the midgut tissues cultured in the presence of dsRNA when applied in very high concentration. Similarly, high level of gene silencing was observed in *S. littoralis*, where dsRNA injection was applied to vas deferens tissues cultured in vitro (Gvakharia et al. 2003). Application of feeding dsRNA is also a powerful tool and most attractive approach that helps to open more possibility to access new strategies of insect pest control in different transgenic crops engineering by specific hairpin RNAs, and this approach was also applied and showed feasible results in several insect pests species including lepidopteron *H. armigera*. Surprisingly, high level of gene silencing was achieved by applying the siRNA as compared to dsRNA, suggesting the big difference between siRNA and dsRNA in same specie (Kumar et al. 2008). Feeding of dsRNA gave comfortable results in many lepidopteron insect pests including *M. sexta, S. exigua, Plutella xylostella,* and *Ostrinia nubilalis* while low silencing gene effects were seen in other insect pest of same insect order including *Diatraea saccharalis, Trichoplusia ni,* and *Epiphyas postvittana* (Khajuria et al. 2015).

21.4.1 Trehalose and its Regulation Using RNAi

Trehalose serves as a disaccharide and non-reducing sugar which is difficult to hydrolyze by acid in which two glucose molecules are joined together in the form of α/α-1, 1-glycosidic linkage. Trehalose is found almost in 80 different species including yeast, fungi, plants and bacteria, insects, and other invertebrates (Elbein 1974). Elbein et al. (2003) reported trehalose as an essential part of fungal spore that provides carbon during growth. But in the case of mycobacteria, trehalose incorporated by itself into glycolipids and serves as building material as well as structural component of the body (Elbein and Mitchell 1973). Trehalose plays a multidimensional role in different organisms and also provides energy while performing various physical mechanisms, such as flight (Asano 2003). Trehalose was firstly studied in insects that were found in larval hemolymph, pupae, and adults. A major role of trehalose is to supply power in certain energy-requiring activities such as flight (Elbein et al. 2003). Trehalose is not examined in mammals but trehalases (TREs) have been found in various body parts including small intestine in different species (Richards and Renandya 2002). Cabib and Leloir (1958) reported that trehalose biosynthesis pathway in yeast was catalyzed by two enzymes; the first enzyme namely TPS1 (trehalose-6-phosphate synthase) that catalyzes the transfer of glucose into UDP-glucose forming the G6P yielded into UDP+T6P (trehalose-6-phosphate) and the second that is TPS2 (trehalose-6-phosphate phosphatase) hydrolyses T6P into trehalose and inorganic phosphate. Later on, it has been demonstrated in various organisms, including in plants (Eastmond et al. 2002) and in insects (Wyatt 1967). In insects, trehalase enzymes are also involved in the synthesis of cuticle during molting, flight, stress recovery, growth, and development (Tatun et al. 2008). Trehalose releases the reserved energy catalyzed by trehalases to meet the energy demands for flight and structural developments in insects (Reyes-DelaTorre et al. 2012). Chitin biosynthesis pathway starts with the production of trehalose that is reduced into glucose catalyzes by the specific enzyme trehalase (Merzendorfer and Zimoch 2003). Regulation of *ApTPS* and *ApTRE* using RNAi caused chitin degradation in *Acyrthosiphon pisum*. Chitin metabolism genes were severely affected after silencing trehalose-6-phosphate synthases gene in *Nilaparvata lugens*. In case of *C. medinalis*, silencing of *CmTre* caused significant deformities and mortalities (Zhang et al. 2020: Unpublished data) (Fig. 21.3).

21.4.2 Hexokinase and its Regulation Using RNAi

It is involved in transcriptional regulations and production of multifunctional proteins (Bryson et al. 2002). HK converts glucose to glucose-6-phosphate in the glycolysis pathway. Furthermore, HK acts as a substrate for conversion of glucose

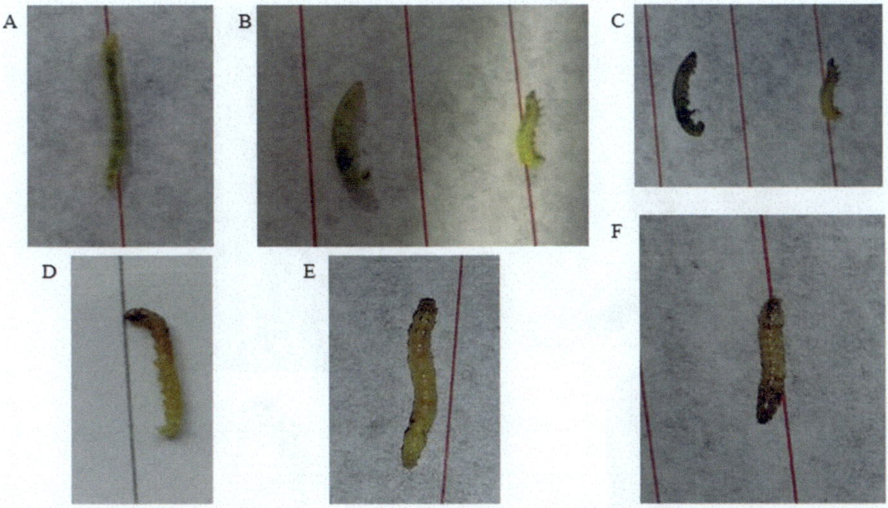

Fig. 21.3 Phenotype changes of *C. medinalis* after injection of dsRNA

6-phosphate into pyruvate. In chitin biosynthetic pathway, HK acts as a second most important enzyme for chitin metabolism (Zhang et al. 2011). Firstly, it was investigated in *Nematocida parisii* (Cuomo et al. 2012). Later on, it was further studied in *Paranosema locustae* (Senderskiy et al. 2014; Timofeev et al. 2017; Reinke et al. 2017). In insect species, HK was found in *Bombyx mori, Aedes togoi, Anopheles stephensi,* and *Nilaparvata lugens* (Yanagawa 1978). *HK* was also studied in *Cnaphalocrocis medinalis* (Shakeel et al. 2020). Post-injection of HK inhibitor deoxy-2-glucose (DOG) showed HK dysfunction that leads to stunted growth, abnormality, and delayed pupation in *H. armigera*. Knockdown of *CmHK* caused larval weight loss, less oviposition, disrupted transformation of larva-pupa-adult, and less mRNA expression pattern (Shakeel et al. 2020) (Fig. 21.4).

21.4.3 Fructose-6-Phosphate Aminotransferase and its Regulation Using RNAi

Fructose-6-phosphate amidotransferase (GFAT) is hexosamines that play a crucial role in the biosynthesis pathway. According to prokaryotic and in lower or higher eukaryotic cells, GFAT was treated GLMS, GFA or GFAT activity by uridine 50-diphosphate N-acetyl glucosamine (UDP-GlcNAc) (Tourian and Sidbury 1983). In the form of N-acetyl glucosamine, it makes this amino sugar into several membrane molecules. The activity of some eukaryotic GFAT is modulated by Cam P-dependent protein kinase A (PK A). GFAT forms new cuticles at different sites during molting development (Kato et al. 2006). GFAT has been isolated and characterized from a number of sources such as *A. aegypti* (Kato et al. 2002), yeast (Watzele and Tanner 1989), mice (Hu et al. 2004), and *Homo sapiens*

Fig. 21.4 Regulation of chitin biosynthesis using RNAi (Shakeel et al. 2020)

(Li et al. 2007). Knockdown of GFAT caused severe deformities and mortality in *N. lugens*. Knockdown of *CmGFAT* caused phenotypic changes, abnormalities, or even death in *C. medinalis* (Zhou et al. 2021; unpublished data) (Fig. 21.5).

21.4.4 UDP-GlcNAc Pyrophosphorylase and Its Regulation Using RNAi

N-acetyl glucosamine (GlcNAc) is a substrate required for GPI anchor formation. It is an important component of the glycosyl that helps in the modification of extracellular proteins (Moussian 2008). Sugar N-acetyl glucosamine (GlcNAc) linked by b-1, 4-linkage. In insects, chitin is the cuticle, organ, and peripheral nutritional matrix (PM) (Merzendorfer 2006). Chitin synthase converts UDP-N-acetyl glucosamine (UDP-GlcNAc) against growing chitin polymer (Cohen 2001). In cells, it is catalyzed by UDP-GlcNAc pyrophosphorylase (UAP). UAP is also important for glycosylation. Of proteins, sphingolipids and secondary metabolites with N-acetyl glucosamine (GlcNAc) or GPI anchor Protein, or 7-b-hydroxylation bond to cell membrane bile acids (Eisenhaber et al. 2003). In case of insects, three types of UAP act as regulatory enzymes in chitin synthesis pathway (Muthukrishnan et al. 2012). However, most of insects have only one *UAP* gene, except for *Locusta migratoria*, *T. castaneum*, and *L. decemlineata* that have two *UAP* genes. *C. medinalis* also has

Fig. 21.5 Phenotype of *C. medinalis* after silencing of *CmGFAT* gene

one UAP regulatory gene in chitin synthesis pathway (Zhou et al. 2021). RNAi experiments revealed that knockdown of *LdUAP2* in *L. decemlineata* and *TcUAP1* in *T. castaneum* causes reduction of chitin synthesis in the peritrophic matrix (PM) (Arakane et al. 2011). Knockdown of *LmUAP1* gene causes higher mortality in *L. migratoria* (Arakane et al. 2011). Zhou et al. (2021) showed that silencing of *CmUAP* causes slow growth, reduces feeding, stops excretion, and leads to weight loss and even death, in *C. medinalis* (Zhou et al. 2021) (Fig. 21.6a,b,c).

21.4.5 Chitin Synthase and its Regulation Using RNAi

Insect chitin biosynthesis catalyzed by highly conserved enzyme known as chitin synthases (CHSs) was found in chitin synthesizing organisms (Merzendorfer and Zimoch 2003). CHSs enzyme encodes by genes that varied among species to species, one gene in parasitic nematodes (Harris and Fuhrman 2002) and in some fungi (Roncero 2002). Furthermore, a nematode named as *Caenorhabditis elegans* has these genes (Gagou et al. 2002b) that encodes CHSs which resembled a protein sequence (Arakane et al. 2004a, b). TcCHS1 is expressed predominantly in *Tribolium*, and TcCHS2 is mainly present in later stages of larvae and pupae (Arakane et al. 2004a, b). *MsCHS1* in *Manduca*, concentration level relatively remains constant in the epidermis during feeding period, but simultaneously drops during molting stage and gradually increases at maximum level during pupal molting stage. Two types of cell, tracheal and columnar cells, expressing a chitin

Fig. 21.6 (**a**) Abnormal development of *C. medinalis* larvae after dsRNA injection (Zhou et al. 2021). (**b**) Abnormal development of *C. medinalis* pupae after dsRNA injection (Zhou et al. 2021). (**c**) Abnormal development of *C. medinalis* adults after dsRNA injection (Zhou et al. 2021). A, wing

synthase in the larval midgut represent two different genes (Zimoch and Merzendorfer 2002), and both of the genes are expressed during growth and development (Zimoch et al. 2005). During molting, MsCHS-1 are notably detected at the stage of tracheal growth but MsCHS-2 are present during molting stage because of the highest concentration level of MsCHS-1 found during molting. When the ecdysteroid level is high, the MsCHS-2 become at low stage represents the gene expression in columnar cells by the ecdysteroids. Same study has been done, level of mRNA and enzyme during chitin synthase activity at larval stage is achieved at maximal level indicate chitin synthesis is absolutely dependent on MsCHS-2 regulation (Zimoch et al. 2005).

Ostrowski et al. (2002) studied the mutation in the CHS1 gene which guides head cuticle deformation in the case of Drosophila (precise data obtained from mutant Drosophila flies represents the integrity, stability of procuticle and epicuticle, and enzymatic activity) (Moussian and Roth 2005). Miscarried abnormal growth in the fifth instar larvae of *Manduca* was measured after the application of CHS1 dsRNA that leads to a miscarried metamorphosis. Both mutation and RNAi are precisely tuned the effects of chitin synthase genes as well as ecdysterone hormone secreted by the prothoracic gland into the hemolymph of Drosophila (Harshman et al. 1999). In *Tribolium castaneum*, buffer injected dsRNA for *TcCHS1* and *TcCHS2* produced a lethal phenotypic effect on larvae, pupae, and adult developmental stages of *Tribolium castaneum* along with untreated insects with normal phenotypic expression. Highest mortality was observed in all lethal phenotypic conditions at all developmental stages instead of adult stages (Arakane et al. 2005). It was reported that RNAi-based *TcCHS1* disrupted all larval, pupal, and adult stages and greatly damaged the whole of chitin synthetic contents, while in *TcCHS2,* no effects were observed (Arakane et al. 2005). In case of *C. medinalis*, silencing of *CmCHSB* caused larval growth and caused larval lethality (Zhang et al. 2020).

21.5 RNAi Research Progress and Application

Napoli first discovered gene silencing in plants (Napoli et al. 1990), but did not find out the real cause. Later, in 1995, *C. elegans* was successfully blocked by using antisense RNA (*Caenorhabditis elegans*) Cepar-1 gene expression; this phenomenon is called RNAi phenomenon (Guo and Kemphues 1995) whereas, in 1998, double-chain RNA-mediated RNA interference can lead to gene silencing. At this point, people gradually uncovered RNAi mystery. Due to the widespread existence of RNAi in eukaryotes, researchers are now deepening their research on RNAi. By using this RNAi mechanism, Andrew and Cranig have won the Nobel Prize. The discovery of RNAi promotes the development of the entire biomolecular sciences.

Fig. 21.6 (continued) folding; b, abnormal wing spreading; c, wings failing to cover abdomen; d, wing curling deformed development

RNA interference technology can enable eukaryotes to resist virus infection, control transposons translocation, and regulate gene expression and function. At present, RNA interference technology is widely used in the research of gene function and control of pests by improving the resistance in crops. It has a very important position in the field of crop breeding and has also been applied to medical fields such as controlling infectious diseases. RNA interference (RNAi) is a phenomenon in which homologous mRNA is induced by double-stranded RNA (dsRNA) to produce specific silent degradation (Dong and Friedrich 2005). As the transcribed mRNA is specifically degraded, the gene cannot be successfully expressed as a protein (Hannon 2002). RNAi technology has certain characteristics such as high efficiency, specificity, length restriction, PTGS mechanism, and its delivery. RNAi can generally be divided into cell-autonomous RNAi, environmental RNAi, and systemic RNAi. Cell-autonomous RNAi mainly occurs in specific cells or introduces dsRNA into cells and tissues; environmental RNAi refers to cells or tissues that absorb dsRNA from the environment. For example, researchers inject dsRNA into nematodes to cause gene silencing in nematodes and their offspring (Tabara et al. 1998). Systemic RNAi refers to the transmission of RNAi effects from one cell to another cell or one tissue to another tissue, such as the RNAi effect caused by the introduction of dsRNA into a desert locust (*Schistocerca gregaria*) (Dong and Friedrich 2005) and RNAi effects induced by the introduction of dsRNA in German small cockroach (*Blattella germanica*) (Martín et al. 2006). All such species produce systemic RNAi effect.

The technical principle of RNAi can be briefly described as the endogenous double-stranded RNA that is cleaved by a specific endonuclease into short-interfering RNA (siRNA) with a size of 21–25 bp (Hutvágner and Zamore 2002). Then, the siRNA reacts with enzymes in living cells and finally degrades this mRNA with the same sequence of siRNA, hence causing interference effects. The silencing effect caused by RNA interference technology mainly goes through three stages (Ushida 2001). In the initial stage, dsRNA is recognized and cleaved by a specific endonuclease (Dicer). The process of small interfering RNA with a size of 21–25 bp. Dicer enzyme is a nucleotides family that is very conserved and needed in the evolutionary process (Kim et al. 2005). This enzyme was first discovered in Drosophila, and its domain mainly includes RNase catalytic region, helicase domain, PAZ structural domain, dsRNA binding region, etc. After the siRNA is formed, some proteins and Dicer enzyme are combined with siRNA. The final composition has RNA-induced silencing complex (Filipowicz 2005). The RNA helicase in the cell relies on the energy of ATP to open the helix structure of the siRNA on the RISC to form two single strands. The antisense strand of the siRNA catalyzes the binding of the Argonaute protein in the RISC and recognizes by the principle of base complementary pairing target mRNA, and finally siRNA 3′-end 12 bases at the place. The target mRNA is cleaved by endonuclease to degrade the RNA and silence the corresponding gene (Hood 2004). When the last step is completed, RNA polymerases (RdRPs) in the antisense strand and target of siRNA work after combination. To Antisense strand as Primer, target RNAs a primer to synthesize new double-stranded RNA. Then enter the initial stage again, it is cleaved by Dicer enzyme to produce new siRNA, which produces new of siRNA is called the second siRNA. New siRNA and again with RICS effect to identify the target mRNA. This endogenous polymerase only exists in some species such as nematodes and has not been found in insects, but as long as there is RdRP, when the foreign dsRNA is introduced into the insect body, RNAi will be present throughout the growth and development of the insect.

21.6 Application of RNAi in Insect Control

RNA interference technology has high efficiency and specificity so that is why researchers are continuously exploring it in modern researches. This technology has wide adaptation in the fields of gene functioning, infectious diseases control, biological pest control, and many other fields. Ordinary pest control methods are very time-consuming, costly, and not very efficient; similarly spraying of pesticides can also cause environmental pollution that is not only harmful to human health but also leads to resistance development against such pesticides. This necessitates the development of sustainable and high-efficiency prevention and control measure approach, whereas RNAi technology is gradually appearing as a reasonable and reliable alternative. By using the mechanism of RNAi, researchers can introduce exogenous dsRNA into insects through feeding, injection, soaking, and other methods to silence endogenous genes and allow the pests to die, thereby achieving the goal of green and efficient pest control. The feeding method is to put exogenous dsRNA into the feed of the insects to eat. It is very simple and convenient to use and will not cause any mechanical damage to the insects. For the purpose, Epiphyas postvittana was introduced with the target gene dsRNA into their body, thereby triggering the effect of RNAi; similarly, *Rhodnius prolixus* was introduced with RPNP2 dsRNA in its saliva, which successfully caused gene silencing (Turner et al. 2006; Araujo et al. 2006). Under normal circumstances, when dsRNA is introduced into insects by this method, the intestine first absorbs dsRNA, and then it is absorbed by other tissues. This method has successfully been applied in controlling of pests such as beet armyworm, cotton bollworm, and yellow beetle. Injecting dsRNA into the insects is one of the more widely used methods, and the RNAi effect caused by the injection method is excellent. Usually, dsRNA is injected into the front chest or abdomen of the test insect to silence the expression of the gene. Relish-dsRNA was injected into the body of Italian bee *Anpis melliferan*, causing RNAi effects (Schlüns and Crozier 2007). This method is used to study the function of genes and the effect of RNAi is very obvious. For larger insects, the injection is very simple and easy to observe. The injection can also be quantified, and the mortality of insects is low. For example, the study of *Bombyx mori* can use injection. Whereas, for small insects, the mechanical damage caused by the injection method is greater, the mortality rate is higher. For example, *Tetranychus urticae* has a small body and a higher mortality rate by injection. Therefore, it is very important to choose a suitable injection instrument according to the size of the insect. Genetic engineering mediation refers to the production of dsRNAi in plants through technical means such as genetic modification, and the RNAi effect occurs after pests use this plant as food. Generally, the plasmid of the hairpin structure of the target gene is first constructed and introduced into the plant through genetic engineering technology, so that the plant itself can produce the dsRNA of the target gene of the pest. Likewise, genetically modified corn was used to produce *Dianbrotican virgiferan* ANTPanse dsRNA, and then the corn root firefly beetle was allowed to feed on it. It turns out that the insect has an RNAi effect. In another case, a similar method was used to introduce the dsRNA of the cotton bollworm CYP gene into tobacco plants; later cotton bollworms were allowed to feed on such plants and it was found that the expression of the cotton bollworm CYP gene was inhibited (Mao et al. 2007). In addition, specific cells of insects can be immersed in a culture medium containing dsRNA, which can also effectively inhibit the expression of target genes. The immersion method has also been applied to insects such as fruit flies and *Aedes aegypti*. The use of RNAi technology to control pests is mainly to allow insects to feed on dsRNA. The dsRNA reaches the midgut of insects through the oral cavity and is absorbed by epithelial cells. Through genetic engineering, the dsRNA of the target gene of the insect body is introduced into the plant body and transformed into transgenic plants to control pests. The dsRNA of the target gene of the insect body is added to the feed, which can inhibit the expression of the gene after the insect eats and achieve the purpose of controlling pests. RNAi technology can also be used in insect viruses, allowing the dsRNA of insect target genes to be infected by recombinant viruses, inhibiting the expression of target genes, and achieving the Rani effect. Uhlirova et al. (2003) used the BR-C-dsRNA recombinant virus to infect the silkworm *B. mori*, so that the gene BR-C in the silkworm could not be expressed normally, which caused difficulties in the emergence of the silkworm, which could not successfully emerge as adults, specifically, the eyes and wings of the adults. The use of viruses is specific to the host, and this technical method can be applied to pest control.

21.7 Conclusion

Insecticidal resistance has become a major challenge due to the use of commercially applied insecticides for the control of insect pest in different crops. A long-term pest control strategy may be achieved with the introduction of latest techniques to overcome the issues evolving from insufficient insect's pest control. As we discuss thoroughly in this chapter, chitin synthesis and its regulation are crucial for growth and development. With understanding the chitin synthesis pathway, many enzymes are playing important roles in chitin formation. As we discussed, trehalose regulates homeostasis and is involved in different physiological mechanisms and chitin synthesis during molting stage; and chitin synthases (CHSs) are involved in chitin synthesis in epidermal cells of the chitin and peritrophic matrix in the midgut. As a part of this study, many questions that have been raised to control insect pests need to be answered. These are enzymes such as TREs, HKs, UAPs, and GFAT that are regulated by using RNAi technique which provides a safe and better direction to control insect pests by silencing its gene expression mechanism. RNAi technique is molecular based and has also been adopted as an emerging insect pest control technology, an effective tool for gene silencing, and dsRNA is considered a wonderful strategy for gene function. The RNAi-based insect pest control strategy provides a vast platform to better understand the gene function, and different approaches adopted to introduce dsRNA into the host vary from host to host and for different purposes. For its application in pest control, many scientists have to find answers to various questions, such as finding suitable delivery method of incorporation of dsRNA into the host, long-lasting effects of dsRNA after introduction and stabilization during delivery, and a cheaper method of its installation for its large-scale use. At last, we believe that with better understanding of these enzyme target genes and with an increase in our knowledge on RNAi mechanism and dsRNA delivery, RNAi technology will become a wonderful and efficient approach in insect-pest research in coming years.

References

Alvi SM, Ali MA, Chaudhary SU, Iqbal S (2003) Population trends and chemical control of rice leaf folder, *Cnaphalocrocis medinalis* on rice crop. Int J Agri Biol 5:615–617

Aouadi M, Tesz GJ, Nicoloro SM, Wang M, Chouinard M, Soto E, Czech MP (2009) Orally delivered siRNA targeting macrophage map 4k4 suppresses systemic inflammation. Nature 458: 1180–1184

Arakane Y, Hogenkamp DG, Zhu YC, Kramer KJ, Specht CA, Beeman RW, Muthukrishnan S (2004a) Characterization of two chitin synthase genes of the red flour beetle, *Tribolium castaneum*, and alternate exon usage in one of the genes during development. Insect Biochem Mol Biol 34:291–304

Arakane Y, Hogenkamp DG, Zhu YC, Kramer KJ, Specht CA, Beeman RW, Muthukrishnan S (2004b) Characterization of two chitin synthase genes of the red flour beetle, *Tribolium castaneum*, and alternate exon usage in one of the genes during development. Insect Biochem Mol Biol 34:291–304

Arakane Y, Muthukrishnan S, Kramer KJ, Specht CA, Tomoyasu Y, Lorenzen MD, Beeman RW (2005) The Tribolium chitin synthase genes TcCHS1 and TcCHS2 are specialized for synthesis of epidermal cuticle and midgut peritrophic matrix. Insect Mol Biol 14:453–463

Arakane Y, Baguinon MC, Jasrapuria S, Chaudhari S, Doyungan A, Kramer KJ, Muthukrishnan S, Beeman RW (2011) Both UDP N-acetylglucosamine pyrophosphorylases of Tribolium castaneum are critical for molting, survival and fecundity. Insect Biochem Mol Biol 41:42–50

Araujo RN, Santos A, Pinto FS, Gontijo NF, Lehane MJ, Pereira MH (2006) RNA interference of the salivary gland nitrophorin 2 in the triatomine bug Rhodnius prolixus (Hemiptera: Reduviidae) by dsRNA ingestion or injection. Insect Biochem Mol Biol 36:683–693

Asano N (2003) Glycosidase inhibitors: update and perspectives on practical use. Glycobiology 13: 93–104

Baskaran RM, Sharma KC, Kumar J (2017) Seasonal and relative abundance of stem-borer and leaf-folder in wet land rice eco-system. J Entomol Zool Stud 5:879–884

Baulcombe D (2004) RNA silencing in plants. Nature 431:356–363

Becker A, Schlöder P, Steele JE, Wegener G (1996) The regulation of trehalose metabolism in insects. Experientia 52:433–439

Bettencourt R, Terenius O, Faye I (2002) Hemolin gene silencing by ds-RNA injected into Cecropia pupae is lethal to next generation embryos. Insect Mol Biol 11:267–271

Borgio JF (2010) RNAi mediated gene knockdown in sucking and chewing insect pests. J Biopest 3:386

Brodersen P, Voinnet O (2009) Revisiting the principles of microRNA target recognition and mode of action. Nat Rev Mol Cell Biol 11:141–148

Bryson JM, Coy PE, Gottlob K, Hay N, Robey RB (2002) Increased hexokinase activity, of either ectopic or endogenous origin, protects renal epithelial cells against acute oxidant-induced cell death. J Biol Chem 277:11392–11400

Cabib E, Leloir LF (1958) The biosynthesis of trehalose phosphate. J Biol Chem 231:259–275

Cohen E (1987) Chitin biochemistry: synthesis and inhibition. Annu Rev Entomol 32:71–93

Cohen E (2001) Chitin synthesis and inhibition: a revisit. Pest Manag Sci 57:946–950

Cruz J, Mané-Padrós D, Bellés X, Martín D (2006) Functions of the ecdysone receptor isoform-a in the hemimetabolous insect Blattella germanica revealed by systemic RNAi in vivo. Dev Biol 1: 158–171

Cuomo CA, Desjardins CA, Bakowski MA, Goldberg J, Ma AT, Becnel JJ, Didier ES, Fan L, Heiman DI, Levin JZ, Young S (2012) Microsporidian genome analysis reveals evolutionary strategies for obligate intracellular growth. Genome Res 22:2478–2488

Dai H, Ma L, Wang J, Jiang R, Wang Z, Fei J (2008) Knockdown of ecdysis-triggering hormone gene with a binary UAS/GAL4 RNA interference system leads to lethal ecdysis deficiency in silkworm. Acta Biochim Biophys Sin 40:790–795

Dong Y, Friedrich M (2005) Nymphal RNAi: systemic RNAi mediated gene knockdown in juvenile grasshopper. BMC Biotechnol 5:1–7

Eastmond PJ, Van Dijken AJ, Spielman M, Kerr A, Tissier AF, Dickinson HG, Jones JD, Smeekens SC, Graham IA (2002) Trehalose-6-phosphate synthase 1, which catalyses the first step in trehalose synthesis, is essential for Arabidopsis embryo maturation. Plant J 29:225–235

Eisenhaber B, Maurer-Stroh S, Novatchkova M, Schneider G, Eisenhaber F (2003) Enzymes and auxiliary factors for GPI lipid anchor biosynthesis and post-translational transfer to proteins. BioEssays 25:367–385

Elbein AD (1974) Interactions of polynucleotides and other polyelectrolytes with enzymes and other proteins. Adv Enzymol Relat Areas Mol Biol 40:29–64

Elbein AD, Mitchell M (1973) Levels of glycogen and trehalose in mycobacterium smegmatis and the purification and properties of the glycogen synthetase. J Bacteriol 113:863–873

Elbein AD, Pan YT, Pastuszak I, Carroll D (2003) New insights on trehalose: a multifunctional molecule. Glycobiology 13:17–27

Faostat FAO (2020) Food and agriculture Organization of the United Nations

Filipowicz W (2005) RNAi: the nuts and bolts of the RISC machine. Cell 122:17–20

Gagou ME, Kapsetaki M, Turberg A, Kafetzopoulos D (2002a) Stage-specific expression of the chitin synthase DmeChSA and DmeChSB genes during the onset of drosophila metamorphosis. Insect Biochem Mol Biol 32:141–146

Gagou ME, Kapsetaki M, Turberg A, Kafetzopoulos D (2002b) Stage-specific expression of the chitin synthase DmeChSA and DmeChSB genes during the onset of drosophila metamorphosis. Insect Biochem Mol Biol 32:141–146

Gandhe AS, John SH, Nagaraju J (2007) Noduler, a novel immune up-regulated protein mediates nodulation response in insects. J Immunol 179:6943–6951

Gangwar RK (2015) Life cycle and abundance of rice leaf folder, *Cnaphalocrocis medinalis* (Guenee)-a review. Life 5(15)

Grishok A (2005) RNAi mechanisms in Caenorhabditis elegans. FEBS Lett 579:5932–5939

Gui ZZ, Lee KS, Kim BY, Choi YS, Wei YD, Choo YM, Jin BR (2006) Functional role of aspartic proteinase cathepsin D in insect metamorphosis. BMC Dev Biol 6:1–11

Gunathilagaraj K, Gopalan M (1986) Rice leaffolder (LF) complex in Madurai, Tamil Nadu, India. Int Rice Res Notes

Guo S, Kemphues KJ (1995) A gene required for establishing polarity in C. elegans embryos, encodes a putative Ser/Thr kinase that is asymmetrically distributed. Cell 81:611–620

Gvakharia BO, Bebas P, Cymborowski B, Giebultowicz JM (2003) Disruption of sperm release from insect testes by cytochalasin and β-actin mRNA mediated interference. Cell Mol Life Sci 60:1744–1751

Han Y, Lei W, Wen L, Hou M (2015) Silicon-mediated resistance in a susceptible rice variety to the rice leaf folder, Cnaphalocrocis medinalis Guenée (Lepidoptera: Pyralidae). PLoS One 10: 20557

Hannon GJ (2002) RNA interference. Nature 418:244–251

Harris MT, Fuhrman JA (2002) Structure and expression of chitin synthase in the parasitic nematode Dirofilaria immitis. Mol Biochem Parasitol 122:231–234

Harshman LG, Loeb AM, Johnson BA (1999) Ecdysteroid titers in mated and unmated Drosophila melanogaster females. J Insect Physiol 45:571–577

Hood E (2004) RNAi: What's all the noise about gene silencing? Environ Health Perspect 112: A224–A229

Hu Y, Riesland L, Paterson AJ, Kudlow JE (2004) Phosphorylation of mouse glutamine-fructose-6-phosphate amidotransferase 2 (GFAT2) by cAMP-dependent protein kinase increases the enzyme activity. J Biol Chem 279:29988–29993

Hutvágner G, Zamore PD (2002) RNAi: nature abhors a double-strand. Curr Opin Genet Dev 12: 225–232

Iga M, Smagghe G (2010) Identification and expression profile of Halloween genes involved in ecdysteroid biosynthesis in Spodoptera littoralis. Peptides 31:456–467

Jaworski E, Wang L, Marco G (1963) Synthesis of chitin in cell-free extracts of Prodenia eridania. Nature 198:790

Kato N, Dasgupta R, Smartt CT, Christensen BM (2002) Glucosamine: fructose-6-phosphate aminotransferase: gene characterization, chitin biosynthesis and peritrophic matrix formation in Aedes aegypti. Insect Mol Biol 11:207–216

Kato N, Mueller CR, Wessely V, Lan Q, Christensen BM (2005) Mosquito glucosamine-6-phosphate N-acetyltransferase: cDNA, gene structure and enzyme kinetics. Insect Biochem Mol Biol 35:637–646

Kato N, Mueller CR, Fuchs JF, Wessely V, Lan Q, Christensen BM (2006) Regulatory mechanisms of chitin biosynthesis and roles of chitin in peritrophic matrix formation in the midgut of adult Aedes aegypti. Insect Biochem Mol Biol 36:1–9

Kaushik C (2010) Extent of damage by leaf folder, *Cnaphalocrocis medinalis* (Guenee) in paddy cultivars at Raiganj, Uttar Dinajpur, West Bengal. Curr Biotica 4:365–367

Kenchington W (1976) Adaptation of insect peritrophic membranes to form cocoon fabrics. The Insect Integument 3:497–513

Khajuria C, Vélez AM, Rangasamy M, Wang H, Fishilevich E, Frey ML, Carneiro NP, Gandra P, Narva KE, Siegfried BD (2015) Parental RNA interference of genes involved in embryonic development of the western corn rootworm, *Diabrotica virgifera virgifera LeConte*. Insect Biochem Mol Biol 63:54–62

Khan ZR, Abenes ML, Fernandez NJ (1996) Suitability of graminaceous weed species as host plants for rice leaffolders, Cnaphalocrocis medinalis and Marasmia patnalis. J Crop Prot 15: 121–127

Khush RS, Cornwell WD, Uram JN, Lemaitre B (2002) A ubiquitin-proteasome pathway represses the drosophila immune deficiency signaling cascade. Curr Biol 12:1728–1737

Kim DH, Behlke MA, Rose SD, Chang MS, Choi S, Rossi JJ (2005) Synthetic dsRNA dicer substrates enhance RNAi potency and efficacy. Nat Biotechnol 23:222–226

de Kraker J (1996) The potential of natural enemies to suppress rice leaf folder populations

Kumar NS, Tang B, Chen X, Tian H, Zhang W (2008) Molecular cloning, expression pattern and comparative analysis of chitin synthase gene B in Spodoptera exigua. Comp Biochem Physiol B: Biochem Mol Biol 149:447–453

Langer RC, Vinetz JM (2001) Plasmodium ookinete-secreted chitinase and parasite penetration of the mosquito peritrophic matrix. Trends Parasitol 17:269–272

Li Y, Roux C, Lazereg S, LeCaer JP, Laprévote O, Badet B, Badet-Denisot MA (2007) Identification of a novel serine phosphorylation site in human glutamine: fructose-6-phosphate amidotransferase isoform 1. Biochemistry 46:13163–13169

Li Y, Hallerman EM, Wu K, Peng Y (2020) Insect-resistant genetically engineered crops in China: development, application, and prospects for use. Annu Rev Entomol 65:273–292

Lynch JA, Desplan C (2006) A method for parental RNA interference in the wasp Nasonia vitripennis. Nat Protoc 1:486

Mao YB, Cai WJ, Wang JW, Hong GJ, Tao XY, Wang LJ, Chen XY (2007) Silencing a cotton bollworm P450 monooxygenase gene by plant-mediated RNAi impairs larval tolerance of gossypol. Nat Biotechnol 25:1307–1313

Martín D, Maestro O, Cruz J, Mane-Padros D, Belles X (2006) RNAi studies reveal a conserved role for RXR in molting in the cockroach *Blattella germanica*. J Insect Physiol 52:410–416

Meister G, Tuschl T (2004) Mechanisms of gene silencing by double-stranded RNA. Nature 431: 343–349

Mello CC, Conte D (2004) Revealing the world of RNA interference. Nature 431:338

Merzendorfer H (2006) Insect chitin synthases: a review. J Comp Physiol 176:1–15

Merzendorfer H, Zimoch L (2003) Chitin metabolism in insects: structure, function and regulation of chitin synthases and chitinases. J Exp Biol 206:4393–4412

Mishra BK, Senapati B, Mishra PR (1998) Chemical control of rice leaf folder, Cnaphalocrocis medinalis (Guenee) in Orissa. J Insect Sci 11:137–140

Moussian B (2008) The role of GlcNAc in formation and function of extracellular matrices. Comp Biochem Physiol B: Biochem Mol Biol 149:215–226

Moussian B, Roth S (2005) Dorsoventral axis formation in the drosophila embryo—shaping and transducing a morphogen gradient. Curr Biol 15:887–899

Muthukrishnan S, Merzendorfer H, Arakane Y, Kramer KJ (2012) Chitin metabolism in insects. Insect Biochem Mol Biol:193–235

Napoli C, Lemieux C, Jorgensen R (1990) Introduction of a chimeric chalcone synthase gene into petunia results in reversible co-suppression of homologous genes in trans. Plant Cell 2:279–289

Ostrowski S, Dierick HA, Bejsovec A (2002) Genetic control of cuticle formation during embryonic development of Drosophila melanogaster. Genetics 16:171–182

Pak J, Fire A (2007) Distinct populations of primary and secondary effectors during RNAi in C. elegans. Science 315:241–244

Pan MH, Wang XY, Chai CL, Zhang CD, Lu C, Xiang ZH (2009) Identification and function of abdominal-a in the silkworm, *Bombyx mori*. Insect Mol Biol 18:155–160

Peters W (1992) Peritrophic membranes. Zoophysiology 30:1–232

Price DR, Gatehouse JA (2008) RNAi-mediated crop protection against insects. Trends Biotechnol 26:393–400

Quan GX, Kanda T, Tamura T (2002) Induction of the white egg 3 mutant phenotype by injection of the double-stranded RNA of the silkworm white gene. Insect Mol Biol 11:217–222

Rajendran R, Rajendran B, Sundara Babu PC (1986) Varietal resistance of rice leaf folder to *Cnaphalocrocis medinalis*. Int Rice Res 11:17–18

Rao BN, Ramasubbaiah K (1988) Dissipation of monocrotophos and carbaryl in tomato Lycopersicon lycopersicum L. Indian J Entomol 50:209–214

Reinke AW, Balla KM, Bennett EJ, Troemel ER (2017) Identification of microsporidia host-exposed proteins reveals a repertoire of rapidly evolving proteins. Nat Commun 8:1–11

Reyes-DelaTorre A, Peña-Rangel MT, Riesgo-Escovar JR (2012) Carbohydrate metabolism in drosophila: reliance on the disaccharide trehalose. Carbohydrates-comprehensive studies on glycobiology and glycotechnology. InTech, Winchester, pp 317–338

Richards JC, Renandya WA (2002) Methodology in language teaching: an anthology of current practice. Cambridge University Press, Cambridge

Roncero C (2002) The genetic complexity of chitin synthesis in fungi. Curr Genet 41:367–378

Savary S, Willocquet L, Elazegui FA, Castilla NP, Teng PS (2000) Rice pest constraints in tropical Asia: quantification of yield losses due to rice pests in a range of production situations. Plant Dis 84:357–369

Schlüns H, Crozier RH (2007) Relish regulates expression of antimicrobial peptide genes in the honeybee, Apis mellifera, shown by RNA interference. Insect Mol Boil 16:753–759

Senderskiy IV, Timofeev SA, Seliverstova EV, Pavlova OA, Dolgikh VV (2014) Secretion of Antonospora (Paranosema) locustae proteins into infected cells suggests an active role of microsporidia in the control of host programs and metabolic processes. PLoS One 9:93585

Shakeel M, Du J, Li SW, Zhou YJ, Sarwar N, Bukhari SAH (2020) Characterization, knockdown and parental effect of hexokinase gene of Cnaphalocrocis medinalis (Lepidoptera: Pyralidae) revealed by RNA interference. Genes 11:1258

Singh HM, Srivastava RK, Rizvi SMA, Elazegui FA, Castilla NP, Savary S (2003) Yield reduction due to brown spot and leaf folder injuries at varying levels of fertilizers and water supply to rice crop. Annal Plant Prot Sci 11:16–19

Sivakumar S, Rajagopal R, Venkatesh GR, Srivastava A, Bhatnagar RK (2007) Knockdown of aminopeptidase-N from Helicoverpa armigera larvae and in transfected Sf21 cells by RNA interference reveals its functional interaction with *bacillus thuringiensis* insecticidal protein Cry1Ac. J Biol Chem 282:7312–7319

Tabara H, Grishok A, Mello CC (1998) RNAi in C. elegans: soaking in the genome sequence. Science 282:430–431

Tatun N, Singtripop T, Tungjitwitayakul J, Sakurai S (2008) Regulation of soluble and membrane-bound trehalase activity and expression of the enzyme in the larval midgut of the bamboo borer Omphisa fuscidentalis. Insect Biochem Mol Biol 38:788–795

Terenius O, Papanicolaou A, Garbutt JS, Eleftherianos I, Huvenne H, Kanginakudru S, Smagghe G (2011) RNA interference in Lepidoptera: an overview of successful and unsuccessful studies and implications for experimental design. J Insect Physiol 57:231–245

Timofeev SA, Senderskiy IV, Tsarev AA, Tokarev YS, Dolgikh VV (2017) Heterologous expression of Paranosema (Antonospora) locustae hexokinase in lepidopteran, Sf9, cells is followed by accumulation of the microsporidian protein in insect cell nuclei. J Invertebr Pathol 143:104–107

Tomoyasu Y, Denell RE (2004) Larval RNAi in Tribolium (Coleoptera) for analyzing adult development. Dev Genes Evol 214:575–578

Tourian AY, Sidbury JB (1983) Phenylketonuria and hyperphenylalaninemia. J Inherit Metab Dis. [edited by] John B. Stanbury..[et al.]

Tsuzuki S, Sekiguchi S, Kamimura M, Kiuchi M, Hayakawa Y (2005) A cytokine secreted from the suboesophageal body is essential for morphogenesis of the insect head. Mech Dev 122:189–197

Turner CT, Davy MW, MacDiarmid RM, Plummer KM, Birch NP, Newcomb RD (2006) RNA interference in the light brown apple moth, Epiphyas postvittana (Walker) induced by double-stranded RNA feeding. Insect Mol Boil 15:383–391

Uhlirova M, Foy BD, Beaty BJ, Olson KE, Riddiford LM, Jindra M (2003) Use of Sindbis virus-mediated RNA interference to demonstrate a conserved role of broad-complex in insect metamorphosis. PNAS 100:15607–15612

Ushida C (2001) RNA interference (RNAi): gene suppression by dsRNA. *Tanpakushitsu kakusan koso*. Protein, Nucleic Acid, Enzyme 46:1381–1386

Van Dellen KL, Bulik DA, Specht CA, Robbins PW, Samuelson JC (2006) Heterologous expression of an Entamoeba histolytica chitin synthase in Saccharomyces cerevisiae. Eukaryot Cell 5: 203–206

Walker WB III, Allen ML (2011) RNA interference-mediated knockdown of IAP in Lygus lineolaris induces mortality in adult and pre-adult life stages. Entomologia Experimentalis et Applicate 138:83–92

Watzele G, Tanner W (1989) Cloning of the glutamine: fructose-6-phosphate amidotransferase gene from yeast: pheromonal regulation of its transcription. J Biol Chem 264:8753–8758

Whangbo JS, Hunter CP (2008) Environmental RNA interference. Trends Genet 24:297–305

Wyatt GR (1967) The biochemistry of sugars and polysaccharides in insects. Adv Insect Phys 4: 287–360

Yanagawa HA (1978) Tissue distribution, purifications, and properties of multiple forms of hexokinase in the silkworm, Bombyx mori. Insect Biochem 8:293–305

Yang Y, Liu X, Xu H, Liu Y, Ali P, Bodlah MA, Lu Z (2020) The abundance and diversity of gut bacteria of rice leaffolder Cnaphalocrocis medinalis (Guenée) across life stages. J Asia Pac Entomol 23:430–438

Zhang WQ, Chen XF, Tang B, Tian HG, Chen J, Yao Q (2011) Insect chitin biosynthesis and its regulation. Chin J Appl Entomol 48:475–479

Zhang ZJ, Xia L, Du J, Li SW, Zhao F (2020) Cloning, characterization, and RNAi effect of the chitin synthase B gene in Cnaphalocrocis medinalis. J Asia Pac Entomol

Zhou YJ, Du J, Li S-W, Shakeel M, Li J-J, Meng X-G (2021) Cloning, characterization, and RNA interference effect of the UDP-N-Acetylglucosamine Pyrophosphorylase gene in Cnaphalocrocis medinalis. Genes (Basel) 12(4):464

Zhu YC, Specht CA, Dittmer NT, Muthukrishnan S, Kanost MR, Kramer KJ (2002) Sequence of a cDNA and expression of the gene encoding a putative epidermal chitin synthase of Manduca sexta. Insect Biochem Mol Biol 32:1497–1506

Zhu F, Sams S, Moural T, Haynes KF, Potter MF, Palli SR (2012) RNA interference of NADPH-cytochrome P450 reductase results in reduced insecticide resistance in the bed bug, Cimex lectularius. PloS One 7:31037

Zhuang JJ, Hunter CP (2012) RNA interference in Caenorhabditis elegans: uptake, mechanism, and regulation. Parasitology:139–560

Zimoch L, Merzendorfer H (2002) Immunolocalization of chitin synthase in the tobacco hornworm. Cell Tissue Res 308:287–297

Zimoch L, Hogenkamp DG, Kramer KJ, Muthukrishnan S, Merzendorfer H (2005) Regulation of chitin synthesis in the larval midgut of Manduca sexta. Insect Biochem Mol Biol 35:515–527

Integrated Management of Rice Diseases 22

Muhammad Imran Hamid and Muhammad Usman Ghazanfar

Abstract

The advances in rice production and research have reached a high level, and diseases and abiotic disorders remain the major cause of yield loss and less profit. The major diseases of rice and suitable integrated management practices are reviewed in this chapter. The diseases reduce the quality and yield and also increase the cost of production. The main issues regarding the practical identification of diseases based on symptoms in the field condition and presence of pathogen are discussed. The proper and accurate identification of diseases is utmost important to adopt suitable management practices. The principles of integrated disease management and utilization of recent technologies in combination with different control strategies, including resistance cultivars, boosting the host defense mechanisms, intercropping, and crop rotation, are considered in this chapter. The economically important rice diseases are blast, sheath blight, brown spot, bacterial blight, sheath rot, stem rot, false smut, and abiotic disorders. The impact of diseases on rice production is increasing day by day, and the only possible solution is to adopt integrated disease management strategies to overcome these issues and improve production.

Keywords

Rice · Diseases · Management · Integrated

M. I. Hamid (✉) · M. U. Ghazanfar
Department of Plant Pathology, College of Agriculture, University of Sargodha, Sargodha, Pakistan
e-mail: imran.hamid@uos.edu.pk

N. Sarwar et al. (eds.), *Modern Techniques of Rice Crop Production*,
https://doi.org/10.1007/978-981-16-4955-4_22

22.1 Introduction

Agriculture is the backbone of most of the nations to feed large populations and also contributes in the development of countries (Awan and Alam 2015; Liu et al. 2014). Despite of the advances in the production technology of rice, diseases are the major cause of yield reduction and increase the management cost at rice farms. Diseases are responsible to reduce the production and quality and to lower the profit of the produce. The impact of diseases has increased over time on the rice production in Pakistan (Ashfaq et al. 2017; Shaheen et al. 2019). The use of high-yield producing cultivars and application of more nitrogen (N) fertilizer have improved rice yields (Ahmad et al. 2008, 2009, 2012, 2013, 2015, 2019; Ahmad and Hasanuzzaman 2012; Ahmed and Ahmad 2019, 2020; Shahzad and Ahmad 2019) but also increased the disease incidence significantly. Rice is grown on large scale with short rotations, or no rotations on low profile soils with less fertility and with low water-holding capacity. All these factors contribute toward the development of disease conducive conditions.

Rice (*Oryzae sativa*) is the main crop of the region (Ahmed and Fayyaz-ul-Hassan 2017; Ahmed et al. 2020a, b; Akram et al. 2018, 2019; Hussain et al. 2018; Fahad et al. 2019; Fatima et al. 2020; Hafeez-ur-Rehman et al. 2019; Khan et al. 2019a, b; Naz et al. 2019; Peerzada et al. 2019; Razaq et al. 2019; Sarwar et al. 2013a, b; Tariq et al. 2020; Wasaya et al. 2019; Zahoor et al. 2019), and multiple efforts are being considered by the government agencies and private sector to improve the varietal response toward high yielding potential and resistance to insect and diseases (Naureen et al. 2009). With the lack of formal education and trainings at farm level, farmers are struggling to manage and cure crops from diseases with diverse biology and infection process (Sarma et al. 2010). These factors are substantially responsible for the lower crop productivity and development of disease at any stage of plant growth over large areas (Naureen et al. 2009; Government of Sindh 2018). In this situation, disease covers large areas in the form of epidemics and remains unmanaged throughout the cropping season, and pathogen survives to infect next season crop.

The disease development in plant is strictly ruled by three factors, exemplified by "Plant Disease Triangle" (Moore et al. 2020). The factors are susceptible variety, virulent pathogen, and favorable atmosphere that resulted in rigorous epidemics when all causes are exist for a considerable time period (Chappelka and Grulke 2016). When single factor did not exist or when not all three factors are managed enough longer, disease will be less severe or not exist (Moore et al. 2020). The understanding of these disease-causing factors needs to be demonstrated at all levels of crop cultivation.

Rice crop goes through different developmental stages from nursery plants to maturity (Mohan et al. 2016). During this time period, plants face diverse biotic and abiotic stresses that ultimately lead toward the yield loss if not managed timely and properly (Zhang and Xie 2014). Moreover, cultivation of susceptible cultivars toward different diseases is another threatening issue that assemble the pathogen inoculums to spread far distant especially most dangerous disease, known as rice blast (Li et al. 2011). The cultivation of blast susceptible cultivars in the fields with blast history will favor the development of fungus during the transfer of nursery plants. The virulent blast pathogen attacks on plant leaves, causing leaf blast disease that is favored by repeated light rainfalls in June to early-July. During the booting stage to early heading stage, farmers are unable to maintain the deep irrigation (4 or more inches) due to low pumping capability in summers besides reduced water-holding capability of soils that increase the stress on plants and favor the development of blast pathogen (Miah et al. 2017). The environmental conditions favor the disease development with light rains and warm days with extended dew at nights and ultimately result in neck blast through heading besides early seed filling stage. This leads toward complete crop failure and up to 80% yield loss. The hot and dry weather at booting, if nitrogen addition was not extreme besides deep flood irrigation, was continued after the mid-season, and crop losses may be less due to the neck blast. The unfavorable environment resists the pathogen to cause disease, and field shows more resistance to neck blast at the time of panicle emergence even though susceptible cultivar was present (Gohel and Chauhan 2015). Several pathogens attack on rice during the cropping season, which include fungi, bacteria, viruses, and nematode (Moore et al. 2020). The proper management practices especially integrated disease management strategies can lead toward high yield (Wang et al. 2014). The concerns over losing the effectiveness of conventional use of pesticides and development of resistance in plant pathogens by various reasons, such as improper and indiscriminate use of chemical pesticides, substandard pesticides, no observations on prescribed waiting periods, incorrect advice besides promotion of some specified pesticides through dealers, and incorrect discarding of pesticide bottles besides cleaning of equipment, have promoted development and application of integrated disease management.

The focus of this chapter is to elaborate integrated disease management strategies to overcome the overall impact of disease on crop to gain high produce.

22.2 Disease Identification and Diagnosis

The accurate identification and diagnosis of a disease is of utmost importance based on symptoms and presence of pathogen for successful management (Sharma and Bambawale 2008). The proper management practices can be adopted only when précised disease diagnosis is carried out under field conditions at farm level and may send to laboratory for an expert opinion (Shurtleff and Averre 1997). The steps involved in the diagnosis of a disease are to view at affected crop plant in field, distribution then also consider the existing environmental aspects. The presence and spreading of pathogen on plant parts will assist to know nature and pathogen dissemination, viz. seed, water, soil, or air borne (Horst 2001). Air-borne pathogens in the rice ecosystem are blast, brown spot, sheath blight, and sheath rot. However, sheath blight, stem rot, and damping off disease of rice are soil borne (Carlile et al. 2001). Moreover, if diseased plants are distributed randomly, occurring repeatedly across large fields, this suggests contribution of air-borne pathogens. If the disease-infected crop plants present in smaller round spots and extending slowly toward healthy plants, this suggests that causal organism may be soil borne. In rice, tungro and grassy stunt virus may be suspected as air-borne pathogens or vector-based diseases that may affect large areas of field and even across broad geographical areas.

Secondly, the location and symptoms on plant parts provide more clues regarding kind of disease (Sharma and Bambawale 2008). In most of the cases, lower leaves of plants are affected that show symptoms and slowly dry up that indicates the distribution of nutrients to other organisms that may be possibly soil-borne pathogens. The origin and biology of pathogens need to be considered to understand the life cycle and infection process. The water-borne pathogens disseminate through irrigation and spread far distant to infect large number of plant population.

During the 1970s era, there was dire necessity to increase the food productivity to meet demands of burgeoning population from inadequate soil resources that required usage of intensive cropping schemes. There is no doubt that production of major cereal crops was increased to triple and met national food requirements based on higher inputs of manures, insecticides, and irrigation water besides multiple cropping patterns with narrow genetic bases. Narrowing of genetic bases, introduction of nonlocal crops, and changes in cropping patterns favored the rise of diseases, especially mycological novelties. Several less important diseases developed a potential threat, besides in certain cases implicit epidemic status.

The main practice to manage plant diseases is with chemical pesticides with various active compounds that are losing effectiveness with passage of time by the phenomenon of resistance development in pathogens, that's why need to convert towards alternative technologies such as Integrated Disease Management (IDM) (Sharma and Bambawale 2008; Fry 2012). The integrated management of plant diseases mainly works on six principles that include avoidance, exclusion, eradication, protection, resistance, and therapy.

22.3 Concept and Principles of Integrated Disease Management

Basic concept of integrated disease management is a decision-based process involving coordinated use of multiple tactics to optimize management of pathogens in eco-friendly and economical ways. Implications of this concept are regular assessment of pathogen effects, their natural enemies and antagonists, simultaneous management of multiple pathogens, use of treatment and economical thresholds while applying chemicals, and integrated use of multiple and pathogen suppressive tactics.

22.4 Principles of Plant Disease Management

1. Avoidance: The disease prevention by selecting time in a year or site where there is no pathogen or environment is not suitable for infection (Lovic and Hopkins 2003).
2. Exclusion: Control introduction of inoculums (Sharma et al. 2015).
3. Eradication: Destroy the inoculums (Fry 2012).
4. Protection: Stops the infection by using toxic compounds or some other barrier (O'Brien 2017).
5. Resistance: Plant the cultivars with resistance to infection (Arriel-Elias et al. 2019).
6. Therapy: Infected plants should be cured (Ghorbanpour et al. 2018).

The main factors affecting plants are microorganisms, including fungi, bacteria, nematodes, viruses, and mycoplasmas, or may be provoked by physiological orders that include high or low temperatures, lack or excess of soil moisture, soil aeration, deficiency of macro and micro nutrients, and other soil-related issues, such as acidic or alkaline soils. The factors limiting the disease progress are the relatively low inoculum levels and less availability of susceptible plants during the infection phase of pathogens. The causative agents of disease in green plants are tens of thousands in numbers and include almost every form of life, but primary agents of disease may also be unresponsive. The abiotic agents of disease include mineral deficiencies and excesses, biologically produced toxicants, air pollutants, improperly used pesticide chemicals, and other environmental factors such as water, wind, temperature, and sunlight. The abiotic factors may play pivotal role in disease development by stressing the plants and distracting the physiological mechanisms, that may induce the pathological responses and express in the form of characterizitic symptoms. The biotic agents or living organisms directly cause infection in plants and are responsible for decline and yield losses of crops (Fry 2012). The combination of disease management practices such as integrated strategies allows us to control the biotic and abiotic factors to cause disease. The integrated disease management practices are vital to control pathogens at any stage of life and crop development.

22.4.1 Blast

Blast a ubiquitous disease that caused enormous productivity losses in favorable environmental circumstances. The pathogen causing the disease is air-borne fungus named *Pyricularia oryzae,* which is a universal issue in rice-growing areas. Fungus overwinters on diseased rice straw or seeds and is responsible to cause early infection. The fungus initially attacks on leaves and later on infects collars, nodes, and panicles of rice plants (Fig. 22.1). The blast fungus is present as different races in rice-producing areas that are genetically different biological variants. These pathogenic strains infect only some rice varieties but not all cultivars. The sowing of particular and regular resistant varieties may overcome the resistance, and new races

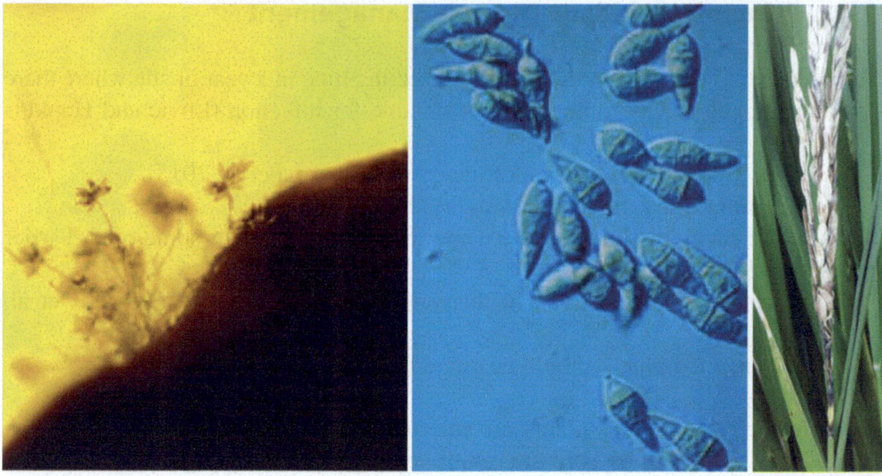

Fig. 22.1 Rice blast fungus growth on infected seed surface. Magnified spores of rice blast fungus and neck blast symptoms

of pathogen may emerge from time to time. For instance, Pi-ta resistance initially occurs in Katy rice besides usage broadly for the development of new cultivars since 1989. Later on, this resistance was overwhelmed by a race of blast fungus recognized as IE-1k that was first observed during 1994 and which has broken the resistance of several rice cultivars considered as resistance to blast pathogen.

Blast pathogen forms lesions on leaves that are typically spindle to diamond-shaped spots. Lesions vary in size from small to large and are mostly based on plant susceptibility, with most commonly noted lesions in field with reddish brown boundary besides off-white to tan center. The pathogen completes the life cycle and produces abundant air-borne spores from lesions during favorable circumstances. On heading stage, spores transfer to node region and infect the node below the panicle and cause "neck blast" which is the most damaging type of blast. Free moisture favors the spore dispersal and development of disease (Fig. 22.2).

Epidemiology

Blast pathogen is more virulent when temperatures are slightly cooler and moisture contents are high. This is favored by long dew periods and cool nights. Late transplanting of rice plants in field increases the probability of blast infection and serves yield loses.

Management

- The rice cultivars should be spot checked for resistance against blast pathogen. Select durable resistant varieties in the fields with blast history or problematic to irrigate.

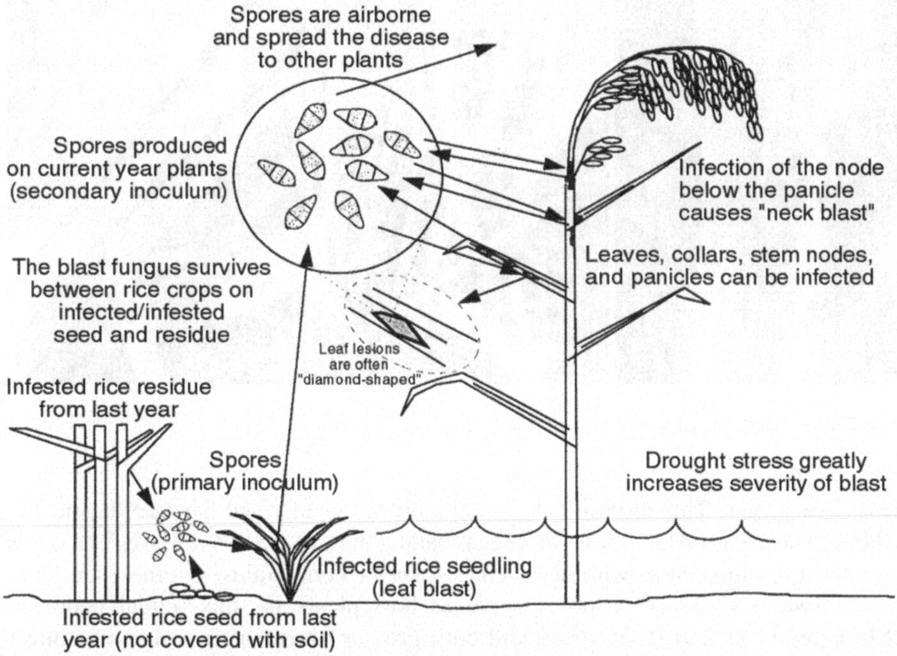

Fig. 22.2 Disease cycle of rice blast disease

- Treat the seeds with proper fungicides before sowing.
- N fertilizer should be used at recommended rate, and avoid high N rates or high organic matter soils, especially in fields with susceptible cultivars.
- Timely development of nursery plants and transplantation is necessary to avoid heavy pressure of blast disease late during season.
- Need to maintain a consistent and deep flood ($\geq$4 in.) to prevent the plants from blast and especially for susceptible cultivars.
- Scouting of fields of leaf-blast indications on every variety needs to be considered even though cultivars are resistant to blast. The best time to scout the fields is before the crop enters booting stages. If the blast symptoms are observed in the field, then maintain deep flood irrigation and prepare to apply preventive fungicides at early heading stage. Fungicides may limit the development of fungus and spores for secondary infection.

22.4.2 Sheath Blight

Sheath blight is among the common diseases of rice and also one of the most significant diseases of rice-growing areas that is widespread and infecting all rice cultivars (Fig. 22.3). Sheath blight exists in each rice field besides causing constant

Fig. 22.3 Sheath blight symptoms on rice plants

loses every year. This disease behaves like hidden enemy and damage drastically. *Rhizoctonia solani* AG1-1A is the fungal pathogen to cause this disease. The crops adopted as rotation crops with rice such as soybean, corn, grains, legumes, sorghum, and grasses for fodder purposes serve as host plants for this fungal pathogen (El-Shafey et al. 2019). Soybean and corn provide abundant inoculums to infect rice crop.

The fungus survives in the soil for at least two to three years in the form of "sclerotia." Sclerotia are hard and brownish structures having 2–3 mm diameter that stay dormant in the soil. The fungal masses can survive between infected straw and other crop residues that have not so longer survival due to exposure to outer environmental conditions. The sclerotia float out from the soil with flood irrigation water and move to all around the field. Moreover, rainfall and soil cultural practices could also spread sclerotia in all field. The sclerotia start germinating and infect rice crop as permanent flood is established and rice plants establish somehow in the field. The fungus infects the sheath of rice just above the waterline and develops symptoms. More severe attack is seen during late tillering to earlier reproductive phases. Sheath blight produces oval to long, purple-bordered lesions on deceased sheaths besides dying tissues bands on leaf blades. The fungus infects the rice tissue throughout the plant body by developing microscopic hyphae that grow fast under conducive conditions. The disease cycle of sheath blight pathogen is shown in Fig. 22.4.

The fungus develops sclerotia on dead or dying plant tissue in latter stages of rice season. Sclerotia fall into soil prior or while rice harvesting and assure the continuous fungus survival in field. Some further sheath diseases are also infecting rice and usually confused with sheath blight. The watchful identification and diagnosis are essential to develop appropriate management practices.

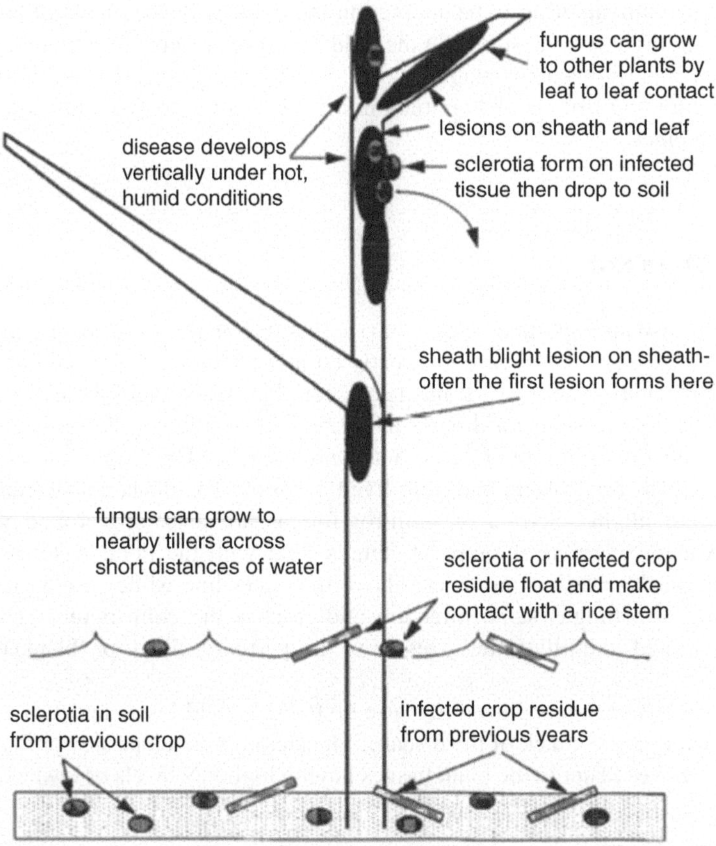

Fig. 22.4 Disease cycle of sheath blight pathogen

Epidemiology

The conditions that favor the disease development of sheath blight are the use of short, leafy rice cultivars, hot and humid weather with temperature ranges between 80° and 92 °F during the day and 74 °F and above at night, and high rates of N fertilizer. The dry and hot weather and cooler temperature reduce disease incidence besides sheath blight severity disease. Severe productivity loses may occur from heavy death of leaf sheaths before the emerging of panicles. Moreover, susceptible semidwarf cultivars are also more vulnerable to sheath disease, and loses may reach up to 50% (El-Shafey et al. 2019).

Management

- Recommended rate of N should be used for the fields. High N rates at preflood stage increase sheath blight activity.
- Thick stand of the crop favors the disease development.

- Resistant cultivars need to be adopted in the fields with sheath blight history.
- If sheath blight is wide spread in the field, use proper fungicides that may reduce the losses during heading stage.
- Fields scouting from panicle initiation to 50% heading to avoid unnecessary use of fungicides.

22.4.3 Stem Rot

Stem rot is also an important disease of rice that infects rice at all stages and causes severe yield losses. The disease is directly correlated with the availability of potassium (K) in sandy loam soils. It has been seen that fields with low levels of K are more susceptible to stem rot disease as a result of insufficient K fertilization. The causal organism of stem rot of rice is *Sclerotium oryzae*. The fungus survives in soil among crops as tiny, black, and long-lived sclerotia (Chethana and Kumar 2019). Same sheath blight sclerotia of stem-rot fungus are dispersed around with soil besides water. Sclerotia of stem rot fungus float with the flood water and cause small and blackish lesions at or just above the water line while coming in contact with plant. The fungus starts infection and reaches the culm before grain fill is completed, and the tiller dies, resulting in partially filled or blanked grains (Fig. 22.5).

Stem rot has been recognized to cause up to 70% yield losses and 100% lodging in fields with severe attack of the disease. The stem rot develops slower than sheath blight but moves faster in the plant tissues having higher N levels or lower Kl levels.

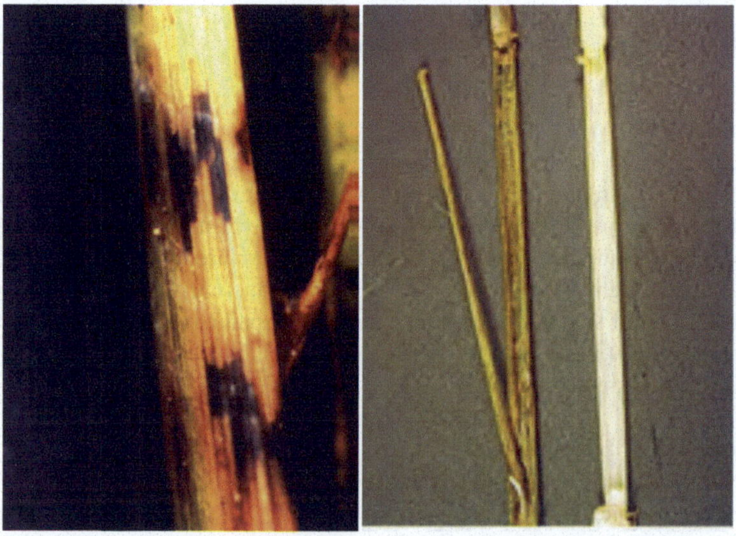

Fig. 22.5 Stem rot symptoms on rice plants

The N:K ratio of 3:1 or above afterward the mid-season of crop favors the stem rot. There are no resistant rice varieties known as stem rot in the soils lacking to K.

Management

- Recommended K fertilizer application after the soil testing.
- Soil sample should be collected on regular basis before the sowing of crop to check the fungal infestation.
- Only recommended rate of N fertilizer should be used and avoid excess N levels.

22.4.4 Crown Sheath Rot

Crown sheath rot or black sheath rot is a widespread disease in major rice-growing fields. The disease is more prevalent in the fields where rice was not planted on regular basis or in new rice fields where other crops were sown. Crown or black sheath rot is caused by the fungus *Gaeumannomyces graminis* var. *graminis* that infects rice sheath above the waterline and overwinters on rice straw and other members of grass family (Peixoto et al. 2013). It is observed that black sheath disease is less noticed if sheath blight or stem rot is already present on plant.

The symptoms appear on sheath by forming a green ring to inter node elongation stages (Fig. 22.6). The disease cycle of this disease is not well understood. The symptoms usually occur in the form of dark brown to grayish black up to 1 in. long having irregular shape. Lesions are developed just above water line. The main distinguishing symptom of black sheath rot fungus is the development of mycelial mat under the sheath that shape like a fan. Later on, sheath of the rice plant start rotted and blackish spore development on the lesions. The yield losses due to black

Fig. 22.6 Crown sheath rot disease symptoms on rice plants

sheath rot disease were recorded as 20% in research plots, but losses in commercial fields are not accurately recorded and may be minor. High-yield losses may occur if susceptible cultivars are planted in field, and overdoses of N fertilizers may be used.

Management

- Resistant cultivars should be planted in the field with consistent history.
- The plant-to-plant distance should be maintained to avoid thick stands in the field.
- N fertilizers should be used properly and over fertilization should be avoided.
- Fungicides may provide some control at early stages of disease development but mostly no need to apply.

22.4.5 Kernel Smut

The disease became more important in 1990s and considered an emerging disease since 2010. The losses in yield due to this disease are 10–30% and can be more severe if environmental conditions favor the disease development. The rice infected with smut disease is not desired for boiling purposes as it turns color into gray. The pathogen that causes kernel smut is a fungus, presently identified as *Neovossia horrida* or *Tilletia barclayana* (El-Kazzaz et al. 2015). Fungus survives as black teliospores, which are microscopic and black in color and turn the grains into black masses. The teliospores are very common in all type of soils and water and can be found in every rice field.

The smut fungus floats on water and develops mycelial mat on the surface of water and produces sporidia that blow into air to infect flowers of rice (Fig. 22.7). The spores enter into flowers and infect the endosperm of rice kernel and convert it into black masses of teliospores. The teliospores release from kernels with drying and rupturing.

Fig. 22.7 Typical symptoms and spores on the spike of rice plant

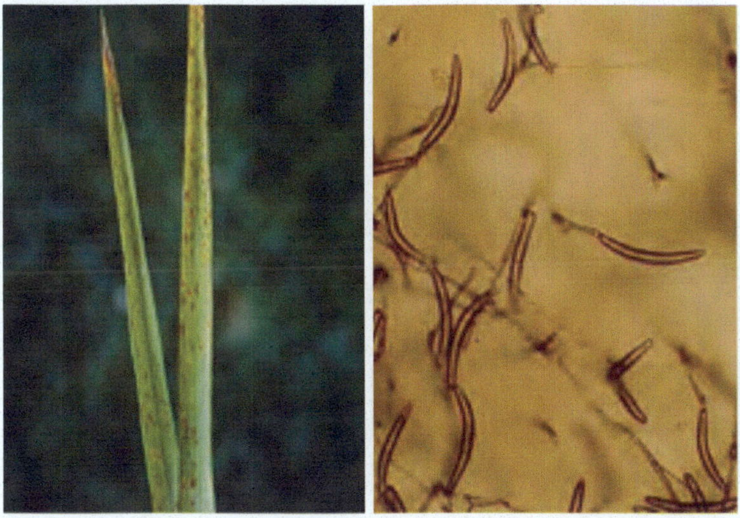

Fig. 22.8 Symptoms of brown spot disease on rice plant leaves and magnified spores of fungus

Management

- Resistant varieties in fields having history of kernel smut disease.
- Apply a fungicide with active ingredient of propiconazole as appropriate dose and at the time of booting stage.
- Recommended rate of N fertilizer should be used and avoid high rate of N.

22.4.6 Brown Spot

The brown spot disease is an old rice disease, which is almost prevalent in all rice-growing areas of the world. The fungus known as *Bipolaris oryzae* (*Helminthosporium oryzae*) is the causal agent of rice brown spot. Fungus survives on infected grain besides most likely on diseased residues. It commonly attacks at seedling stage but can also infect leaves and panicles of weak rice plants (Fig. 22.8). The field with deficiencies of N, K, and P is more susceptible against brown spot disease.

Fungus disseminates as air-borne spores that can infect the plant in the presence of free moisture. The fungus forms oval shape lesions on leaves and panicles. The resistant cultivars behave to form small and dark brown spots. On susceptible cultivars, spots are larger with dark border and gray center and often confused with leaf blast lesions.

Management

- Use clean and fungicide-treated seed.
- Plant resistant cultivars to brown spot.
- Recommended rates and timing for the application of N fertilizers.

Fig. 22.9 Symptoms of bacterial blight on spike of rice plants and complete damage of rice ears

- Soil sample collection and testing from laboratory to make sure the availability of nutrient and especially K.
- Fungicides are not commonly recommended for brown spot disease control.

22.4.7 Bacterial Panicle Blight

Bacterial panicle blight or panicle blight is caused because of bacteria *Burkholderia glumae* and *B. gladioli*. Symptoms on panicles appear at heading stage without showing any initial symptoms (Fig. 22.9). The pathogen is responsible to cause seedling rot and seed rot (Mulaw et al. 2018). At heading stage, if disease is severe, panicle clusters do not fill out and turn over due to the blank heads. Later on, color of panicles turns into tan or grayish and other saprophytes may attack on it. The flag leaf of plant often shows reddish brown lesions that may result in the death of flag leaf.

The bacterial panicle blight was greatly noticed in very hot years of 1990s and later on widespread during the years of 2010 and 2011. The disease was responsible to cause up to 50% losses. The disease was an unidentified issue in many rice-growing regions of the world. Later on, pathogen was isolated and characterized to study the life cycle and losses. The environmental circumstances like high night temperatures, high N fertility, water stress, higher seeding rates besides late sowing may be responsible to upsurge the disease incidence besides severity. Most varieties of rice are susceptible against disease, and no labeled chemicals are developed to control this disease.

Fig. 22.10 Symptoms of straight head disease on rice plant

Management

- Plant rice early, as late-planted rice may be more vulnerable to bacterial blight disease.
- No water stress and avoid excessive N rates.
- Plant moderately resistant cultivars to manage the disease of bacterial blight.

22.4.8 Straight Head

Straight head is considered a physiological disorder, and cause of this disease is unknown. In research plots or controlled conditions, alike indications can be caused with arsenic, but in field circumstances, straight head symptoms can be observed (Li et al. 2017). Straight head is an ancient problem in rice and usually in low textured soils of rice. "Drain and dry" irrigation approaches are presently used to tackle straight heads that were used by growers during early 1900s.

Symptoms are observed as blanked or moderately blanked panicles and distorted panicles or less seeds (Fig. 22.10). If the disease is severe and straight heads are formed, the panicles may not emerge from the boot and new tillers can emerge from nodes below the panicle. The affected plants may suffer badly and are dark green in color and may appear any time in the fields with consistent rice plantations.

Management

- Plant resistant to moderately resistant cultivars with history of straight head problem.
- Draining and drying strategy of the soil may provide best control of the problem. The cultivars with dual resistant to blast and straight head should be considered.
- The fields with history of cotton production and silty or sandy-loam soils containing higher organic matter favor straight head.

Fig. 22.11 False smut disease development and spore balls on the panicle of rice plant

22.4.9 False Smut

False smut or orange or green smut is reported from various regions of the world from rice fields. False smut disease is caused by fungus *Ustilaginoidea virens* that survives in soil or in contaminated rice seed as spore balls. The favorable environmental conditions play crucial role in the development of disease. False smut replaces the rice kernel with unseen gall and blank kernels, a certain environmental circumstances (Cartwright et al. 2017).

Disease galls progress inside diseased kernels and emerge from glumes as silvery color, and later on, spore balls appear as bright orange to olive green or brown in color. The microscopic spore balls are released on disturbing the kernels which appear like orange dust (Fig. 22.11). Spores contaminate other panicles, and life cycle of fungus carries on till end of rice season. It is observed that late maturity of crop favors the disease of false smut. It is a common problem in rice, but heavy yield losses are not reported to date.

Management

- Resistant varieties in fields with false smut history.
- Early planting of rice results in early maturity that can reduce the false smut incidence and severity.
- Proper N fertilizer application is recommended as high N rates favor the disease greatly.
- Fungicides with active ingredient of propiconazole control false smut if applied at booting to boot split stage.

Fig. 22.12 Comparison of healthy and diseased roots of rice plants

22.4.10 Autumn Decline

The autumn decline is considered another rice physiological disorder and considered to be induced by hydrogen sulfide (H_2S) toxicity in soils which are highly anaerobic. Rice roots turn black, rot, and ultimately die that result the yellowing, wilting, and stunting besides occasionally plant death (Wamishe et al. 2017). Fields with this disorder were tested, and high levels of sulfate in the soil were recorded. The cold water areas of rice production show symptoms of this disorder. It has been observed that undecomposed crop residues at flooding reduce the sulfur process and aggregate the autumn decline (Fig. 22.12).

Management

- Drain and dry strategy can work well for this disorder as suggested for straight head problem. Proper draining of field should be done if black root rotting is observed during the scouting process.
- The anaerobic conditions favor disease development, so aerobic conditions need to be established. The fields may not need to dry too much, as soon white roots start to appear, light flood can be maintained.
- Alternative water source should be used if routine water has high sulfur contents.
- The soil and water testing is recommended to manage the problem.

References

Ahmad S, Hasanuzzaman M (2012) Integrated effect of plant density, N rates and irrigation regimes on the biomass production, N content, PAR use efficiencies and water productivity of rice under irrigated semiarid environment. Not Bot Horti Agrobot Cluj-Napoca 40(1):201–211

Ahmad S, Zia-ul-Haq M, Ali H, Shad SA, Ammad A, Maqsood M, Khan MB, Mehmood S, Hussain A (2008) Water and radiation use efficiencies of transplanted rice (*Oryza sativa* L.) at different plant densities and irrigation regimes under semi-arid environment. Pak J Bot 40(1): 199–209

Ahmad S, Ahmad A, Zia-ul-Haq M, Ali H, Khaliq T, Anjum MA, Khan MA, Hussain A, Hoogenboom G (2009) Resources use efficiency of field grown transplanted rice (*Oryza sativa* L.) under irrigated semiarid environment. J Food Agric Environ 7(2):487–492

Ahmad S, Ahmad A, Soler CMT, Ali H, Zia-Ul-Haq M, Anothai J, Hussain A, Hoogenboom G, Hasanuzzaman M (2012) Application of the CSM-CERES-Rice model for evaluation of plant density and nitrogen management of fine transplanted rice for an irrigated semiarid environment. Precis Agric 13(2):200–218

Ahmad S, Ahmad A, Ali H, Hussain A, Garcia y Garcia A, Khan MA, Zia-Ul-Haq M, Hasanuzzaman M, Hoogenboom G (2013) Application of the CSM-CERES-Rice model for evaluation of plant density and irrigation management of transplanted rice for an irrigated semiarid environment. Irrig Sci 31(3):491–506

Ahmad A, Ashfaq M, Rasul G, Wajid SA, Khaliq T, Rasul F, Saeed U, Rahman MH, Hussain J, Baig IA, Naqvi AA, Bokhari SAA, Ahmad S, Naseem W, Hoogenboom G, Valdivia RO (2015) Impact of climate change on the rice–wheat cropping system of Pakistan. In: Hillel D, Rosenzweig C (eds) Handbook of climate change and agro-ecosystems: the agricultural modeling intercomparison and improvement project (AgMIP) integrated crop and economic assessments. Imperial College Press and the American Society of Agronomy, pp 219–258

Ahmad S, Abbas G, Ahmed M, Fatima Z, Anjum MA, Rasul G, Khan MA, Hoogenboom G (2019) Climate warming and management impact on the change of rice-wheat phenology in Punjab, Pakistan. Field Crops Res 230:46–61

Ahmed M, Ahmad S (2019) Carbon dioxide enrichment and crop productivity. In: Hasanuzzaman M (ed) Agronomic crops, vol 2. Springer Nature Singapore Pte Ltd, pp 31–46

Ahmed M, Ahmad S (2020) Systems modeling. In: Ahmed M (ed) Systems modeling. Springer Nature Singapore Pte Ltd, pp 1–44

Ahmed M, Fayyaz-ul-Hassan, Ahmad S (2017) Climate variability impact on rice production: adaptation and mitigation strategies. In: Ahmed M, Stockle C (eds) Quantification of climate variability, adaptation and mitigation for agricultural sustainability. Springer International Publishing Switzerland, pp 91–111

Ahmed M, Ahmad S, Raza MA, Kumar U, Ansar M, Shah GA, Parsons D, Hoogenboom G, Palosuo T, Seidel S (2020a) Models calibration and evaluation. In: Ahmed M (ed) Systems modeling. Springer Nature Singapore Pte Ltd, pp 151–178

Ahmed M, Ahmad S, Waldrip HM, Ramin M, Raza MA (2020b) Whole farm modeling: a systems approach to understanding and managing livestock for greenhouse gas mitigation, economic viability and environmental quality. In: Waldrip HM, Pagliari PH, He Z (eds) Animal manure. ASA Special Publication 67. American Society of Agronomy and Soil Science Society of America, Madison, WI, pp 345–371

Akram R, Turan V, Hammad HM, Ahmad S, Hussain S, Hasnain A, Maqbool MM, Rehmani MIA, Rasool A, Masood N, Mahmood F, Mubeen M, Sultana SR, Fahad S, Amanet K, Saleem M, Abbas Y, Akhtar HM, Hussain S, Waseem F, Murtaza R, Amin A, Zahoor SA, Sami ul Din M, Nasim W (2018) Fate of organic and inorganic pollutants in paddy soils. In: Hashmi MZ, Varma A (eds.), Environmental pollution of paddy soils. Springer, pp. 197–214

Akram R, Fahad S, Masood N, Rasool A, Ijaz M, Ihsan MZ, Maqbool MM, Ahmad S, Hussain S, Ahmed M, Kaleem S, Sultana SR, Mubeen M, Saud S, Kamran M, Nasim W (2019) Plant growth and morphological changes in rice under abiotic stress. In: Hasanuzzaman M, Fujita M, Nahar K, Biswas JK (eds) Advances in rice research for abiotic stress tolerance. Woodhead Publishing, pp 69–85

Arriel-Elias MT, Côrtes MVDCB, de Sousa TP, Chaibub AA, de Filippi MCC (2019) Induction of resistance in rice plants using bioproducts produced from *Burkholderia pyrrocinia* BRM 32113. Environ Sci Pollut Res 26(19):19705–19718

Ashfaq M, Mubashar U, Haider MS, Ali M, Ali A, Sajjad M (2017) Grain discoloration: an emerging threat to rice crop in Pakistan. J Anim Plant Sci 27:696–707

Awan AG, Alam A (2015) Impact of agriculture productivity on economic growth: a case study of Pakistan. Indus Eng Lett 5(7):27–33

Carlile MJ, Watkinson SC, Gooday GW (2001) The fungi, 2nd edn. Academic Press, New York, NY

Cartwright RD, Groth DE, Wamishe YA, Greer CA, Calvert LA, Cruz CM, Verdier V, Way MO (2017) Diseases of rice. American Phytopathological Society, St. Paul

Chappelka AH, Grulke NE (2016) Disruption of the 'disease triangle' by chemical and physical environmental change. Plant Biol 18:5–12

Chethana BS, Kumar MP (2019) Rice stem rot *Sclerotium hydrophilum* Sace isolated and characterized in Southern Karnataka. Oryza 56(2):242–245

El-Kazzaz MK, Salem EA, Ghoneim KE, Elsharkawy MM, El-Kot GAN, Kalboush ZA (2015) Biocontrol of *Tilletia barclayana*, the causal agent of kernel smut disease in rice. Egyptian J Biol Pest Control 25(3):535–544

El-Shafey RAS, Elamawi RM, Saleh MM, Tahoon AM, Emeran AA (2019) Morphological, pathological and molecular characterization of rice sheath blight disease causal organism *Rhizoctonia solani* AG-1 IA in Egypt. Arch Phytopathol Plant Protec 52(5–6):507–529

Fahad S, Noor M, Adnan M, Khan MA, Rahman I, Alam M, Khan IA, Ullah H, Mian IA, Hassan S, Saud S, Bakhat HF, Hammad HM, Ahmad S, Nasim W (2019) Abiotic stress and rice grain quality. In: Hasanuzzaman M, Fujita M, Nahar K, Biswas JK (eds) Advances in rice research for abiotic stress tolerance. Woodhead Publishing, pp 571–583

Fatima Z, Ahmad M, Hussain M, Abbas G, Ul-Allah S, Ahmad S, Ahmed N, Ali MA, Sarwar G, Ehsan ul Haque, Iqbal P, Hussain S (2020) The fingerprints of climate warming on cereal crops phenology and adaptation options. Sci Rep 10:18013

Fry WE (2012) Principles of plant disease management. Academic Press

Ghorbanpour M, Omidvari M, Abbaszadeh-Dahaji P, Omidvar R, Kariman K (2018) Mechanisms underlying the protective effects of beneficial fungi against plant diseases. Biol Control 117: 147–157

Gohel NM, Chauhan HL (2015) Integrated management of leaf and neck blast disease of rice caused by *Pyricularia oryzae*. Afr J Agric Res 10(19):2038–2040

Government of Sindh (2018) Diseases of rice crop and their control. Agriculture, Supply and Prices Department, Government of Sindh, Karachi, Pakistan. http://agri.sindh.gov.pk/estimate-of-rice-crop-for-theyear-of-2013

Hafeez-ur-Rehman, Nawaz A, Awan MI, Ijaz M, Hussain M, Ahmad S, Farooq M (2019) Direct seeding in rice: problems and prospects. In: Hasanuzzaman M (ed) Agronomic crops, Production technologies, vol 1. Springer Nature Singapore Pte Ltd, pp 199–222

Horst RK (2001) Westcott's plant disease handbook, 6th edn. Kluwer Academic Publishers, Boston, MA

Hussain M, Ahmad S, Hussain S, Lal R, Ul-Allah S, Nawaz A (2018) Rice in saline soils: physiology, biochemistry, genetics and management. Adv Agron 148:231–287

Khan MA, Ahmad S, Raza A (2019a) Integrated weed management for agronomic crops. In: Hasanuzzaman M (ed) Agronomic crops, vol 2. Springer Nature Singapore Pte Ltd, pp 257–281

Khan SM, Ali S, Nawaz A, Bukhari SAH, Ejaz S, Ahmad S (2019b) Integrated pest and disease management for better agronomic crop production. In: Hasanuzzaman M (ed) Agronomic crops, vol 2. Springer Nature Singapore Pte Ltd, pp 385–428

Li J, Yang J, Liu L, He H, He X, Zhu Y, Li C (2011) Geographic distribution of avirulence genes in rice blast fungus in Yunnan province, China. J Resources Ecol 2(2):181–185

Li X, Hu B, Pan X, Zhang N, Wu D (2017) Association analysis of arsenic-induced straight head in rice (*Oryza sativa* L.) based on the selected population with a modified model. Biomed Res Int 2017:e7641362

Liu W, Liu J, Triplett L, Leach JE, Wang GL (2014) Novel insights into rice innate immunity against bacterial and fungal pathogens. Annu Rev Phytopathol 52:213–241

Lovic BR, Hopkins DL (2003) Production steps to reduce seed contamination by pathogens of cucurbits. HortTechnology 13:50–54

Miah G, Rafii MY, Ismail MR, Sahebi M, Hashemi FSG, Yusuff O, Usman MG (2017) Blast disease intimidation towards rice cultivation: a review of pathogen and strategies to control. J Anim Plant Sci 27:1058–1066

Mohan KJ, Balasubramanian M, Palanivel S (2016) Detection and recognition of diseases from paddy plant leaf images. Int J Comput Appl 144(12):34–41

Moore D, Robson GD, Trinci AP (2020) 21st century guidebook to fungi. Cambridge University Press, London

Mulaw T, Wamishe Y, Jia Y (2018) Characterization and in plant detection of bacteria that cause bacterial panicle blight of rice. Am J Plant Sci 9(04):667

Naureen Z, Price AH, Hafeez FY, Roberts MR (2009) Identification of rice blast disease-suppressing bacterial strains from the rhizosphere of rice grown in Pakistan. Crop Prot 28(12):1052–1060

Naz S, Fatima Z, Iqbal P, Khan A, Zakir I, Noreen S, Younis H, Abbas G, Ahmad S (2019) Agronomic crops: types and uses. In: Hasanuzzaman M (ed) Agronomic crops, Production technologies, vol 1. Springer Nature Singapore Pte Ltd, pp 1–18

O'Brien PA (2017) Biological control of plant diseases. Australas Plant Pathol 46(4):293–304

Peerzada AM, Bukhari SAH, Dawood M, Nawaz A, Ahmad S, Adkins S (2019) Weed management for healthy crop production. In: Hasanuzzaman M (ed) Agronomic crops, vol 2. Springer Nature Singapore Pte Ltd, pp 225–256

Peixoto CN, Ottoni G, Filippi MC, Silva-Lobo VL, Prabhu AS (2013) Biology of *Gaeumannomyces graminis* var. *graminis* isolates from rice and grasses and epidemiological aspects of crown sheath rot of rice. Trop Plant Pathol 38(6):495–504

Razaq M, Shah FM, Ahmad S, Afzal M (2019) Pest management for agronomic crops. In: Hasanuzzaman M (ed) Agronomic crops, vol 2. Springer Nature Singapore Pte Ltd, pp 365–384

Sarma SK, Singh KR, Singh A (2010) An expert system for diagnosis of diseases in rice plant. Int J Artif Intell 1(1):26–31

Sarwar N, Ali H, Ahmad S, Ullah E, Ahmad S, Mubeen K, Hill JE (2013a) Water wise rice cultivation on calcareous soil with the addition of essential micronutrients. J Anim Plant Sci 23(1):244–250

Sarwar N, Ali H, Maqsood M, Ullah E, Shahzad AN, Shahzad M, Mubeen K, Shahid MA, Ahmad S (2013b) Phenological response of rice plants to different micronutrients application under water saving paddy fields on calcareous soil. Turk J Field Crops 18(1):52–57

Shaheen R, Sharif MZ, Amrao L, Zheng A, Manzoor M, Majeed D, Kiran H, Jafir M, Ali A (2019) Investigation of bacterial leaf blight of rice through various detection tools and its impact on crop yield in Punjab, Pakistan. Pak J Bot 51(1):307–312

Shahzad AN, Ahmad S (2019) Tools and techniques for nitrogen management in cereals. In: Hasanuzzaman M (ed) Agronomic crops, vol 2. Springer Nature Singapore Pte Ltd, pp 111–126

Sharma OP, Bambawale OM (2008) Integrated management of key diseases of cotton and rice. In: Integrated management of diseases caused by fungi, phytoplasma and bacteria. Springer, Dordrecht, pp 271–302

Sharma KK, Singh US, Sharma P, Kumar A, Sharma L (2015) Seed treatments for sustainable agriculture-a review. J Appl Nat Sci 7(1):521–539

Shurtleff MC, Averre CW (1997) The plant disease clinic and field diagnosis of abiotic diseases. American Phytopathological Society, St. Paul, MN

Tariq M, Ahmed M, Iqbal P, Fatima Z, Ahmad S (2020) Crop phenotyping. In: Ahmed M (ed) Systems modeling. Springer Nature Singapore Pte Ltd, pp 45–60

Wamishe YA, Hardke JT, Roberts TL, Fryer JM, Gebremariam T, Mulaw T, Belmar SB (2017) Identifying management strategies to reduce autumn decline in rice. Res Ser Arkan Agricult Exper Stat 643:128–136

Wang X, Lee S, Wang J, Ma J, Bianco T, Jia Y (2014) Current advances on genetic resistance to rice blast disease. In: Yan W, Bao J (eds) Rice-germplasm, genetics and improvement. InTech publishers, pp 195–217

Wasaya A, Yasir TA, Ijaz M, Ahmad S (2019) Tillage effects on agronomic crop production. In: Hasanuzzaman M (ed) Agronomic crops, vol 2. Springer Nature Singapore Pte Ltd, pp 73–99

Zahoor SA, Ahmad S, Ahmad A, Wajid A, Khaliq T, Mubeen M, Hussain S, Sami Ul Din M, Amin A, Awais M, Nasim W (2019) Improving water use efficiency in agronomic crop production. In: Hasanuzzaman M (ed) Agronomic crops, vol 2. Springer Nature Singapore Pte Ltd, pp 13–29

Zhang F, Xie J (2014) Genes and QTLs resistant to biotic and abiotic stresses from wild rice and their applications in cultivar improvements. In: Yan W, Bao J (eds) Rice-germplasm, genetics and improvement. InTech publishers, pp 59–78

Diversity and Management of Plant Viruses Infecting Rice

23

Zafar Iqbal, Muhammad Naeem Sattar,
and Muhammad Nadir Naqqash

Abstract

Rice (*Oryza sativa* L., family Poaceae) is the leading cereal crop that is widely cultivated across the globe. Asia is the largest producer of rice with over 7.05 billion tons production in 2018, followed by the United States (38 million tons [MT]), Africa (33 MT), Europe (4 MT), and Oceania (0.65 MT). While the leading rice-producing countries are China (148.5 million MT) followed by India (116.42 million MT), Indonesia (36.7 million MT), Bangladesh (34.91 MT), and Vietnam (27.77 million MT). Rice supplies 21% of energy and 15% of protein to humans and plays a crucial role in the global food chain. However, rice cultivation is under continuous stress due to several biotic and abiotic constraints. Among the biotic constraints, rice-infecting viruses (RIVs) and their insect vectors cause enormous yield losses to worldwide rice production. RIVs encompass huge genomic diversity and include single-stranded, double-stranded, negative-sense single-stranded, negative-sense double-stranded, positive-sense single-stranded, and ambisense viruses. More than 15 RIVs are known and 10 of these RIVs pose a significant threat to Asian rice production. To sustain the global food security, it is of dire need to curb the RIVs and their insect vectors simultaneously. Several conventional to modern approaches have been employed to sustain the rice production against RIVs. Nonetheless, the contemporary CRISPR-Cas-based approaches and its expanding toolkit can offer unlimited utilities to improve rice yield and control RIVs and their insect vectors via

Z. Iqbal · M. N. Sattar (✉)
Central Laboratories, King Faisal University, Al-Ahsa, Saudi Arabia
e-mail: zafar@kfu.edu.sa; mnsattar@kfu.edu.sa

M. N. Naqqash
Department of Plant Production and Technologies, Faculty of Agricultural Science and Technology, Nigde Omer Halisdemir University, Niğde, Turkey

N. Sarwar et al. (eds.), *Modern Techniques of Rice Crop Production*,
https://doi.org/10.1007/978-981-16-4955-4_23

transgene-free genome-editing capabilities. The importance of rice, RIVs, and their control strategies are discussed.

Keywords

Rice (*Oryza sativa*) · Rice-infecting viruses · Sustainable production · Insect vectors · CRISPR-Cas · RNA interference

23.1 Introduction

Rice (*Oryza sativa* L., family Poaceae) is the world's largest cereal crop that is widely cultivated around the globe. The maximum rice production is in Asia with over 7.05 billion tons in 2018, followed by the United States (38 million tons [MT]), Africa (33 MT), Europe (4 MT), and Oceania (0.65 MT). The leading rice-producing countries are China (148.5 million MT) followed by India (116.42 million MT), Indonesia (36.7 million MT), Bangladesh (34.91 MT), and Vietnam (27.77 million MT) (https://www.statista.com/statistics/255945/top-countries-of-destination-for-us-rice-exports-2011/). China, Pakistan, India, Indonesia Vietnam, and rest of other Asian countries produced ~92% of the rice and it is widely cultivated and consumed in Asia (ca 650 MT) and therefore directly supplies ca 36% of the daily required calories (Normile 2008). In 57th session of the United Nations General Assembly, the importance of rice as a staple food has been accepted. Rice has influenced millions of people's cultures, lifestyles, and economies of many countries. In 2004, after realizing its importance, the UN named 2004 as Rice's International Year.

The precise origin of rice is difficult to map and has been lost in antiquity, but vividly the rice domestication is an important development in the history of the mankind. Rice cultivation is believed to begin simultaneously across different regions about 6500 years ago, but the first-ever rice crop was witnessed around 5000 BC in China and around 4500 BC in Thailand. Subsequently it spread to Cambodia, South India, and Vietnam and then japonica- and indica-derived species expanded to other Asian countries such as Japan, Korea, Pakistan, Philippines, Sri Lanka, and Indonesia. The cultivation of Asian rice started around 800 BC in the Middle East and Mediterranean Europe. In 1964, most likely, rice reached to North America (in South Carolina) via Madagascar. While Spanish took it to South America in the beginning of eighteenth century (source: UNCTAD.org). Rice can play a role in ensuring food security, fighting hunger, meeting nutritional demands, and eliminating poverty. Two types of rice crops are domesticated, *Oryza sativa* and *Oryza glaberrima*, and are native to Southern Asia and Southeast Africa, respectively (Crawford and Shen 1998). Rice can be grown in diverse environments, even in those areas where cultivation of other crops is not suitable (Breviario and Genga 2013). Initially rice was cultivated without submersion, but it is thought that mutations made rice a semiaquatic plant. As mentioned earlier, rice can be cultivated in diverse environments, but most suitable environmental conditions for rice are

Table 23.1 Nutritional values of raw, unenriched, long grain, white and brown rice (per 100 g) according to the estimates of the USDA nutrient database

	Amounts	
Nutrient	White rice	Brown rice
Carbohydrates	79.9 g	76 g
Calcium	28 mg	9 mg
Dietary fiber	1.3 g	3.6 g
Fat	0.6 g (saturated fat 0.2 g)	3.2 g (saturated 0.6 g)
Iron	0.65 mg	1.29 mg
Magnesium	25 mg	116 mg
Manganese	25 mg	2.853 mg
Niacin	1.6 mg	6.494 mg
Protein	7 g	7.5 g
Pantothenic acid	1.01 mg	1.065 mg
Phosphorous	115 mg	311 mg
Potassium	115 mg	250 mg
Sodium	5 mg	5 mg
Sugar	0.12 g	0.66 g
Thiamin	0.07 mg	0.541 mg
Vitamin B6	0.16 mg	0.477 mg
Zinc	1.09 mg	2.13 mg

warm and humid climate, where it grows faster. The structure of rice plant is very intriguing, it develops multiple tillers on a main stem, depending on the variety rice height ranges from 0.6 to 6 m (floating rice). The tiller has a ramified panicle measuring 20–30 cm long, while 50–300 grain flowers are produced in each panicle and the fruit is a caryopsis (https://unctad.org/en/Pages/Home.aspx). Rice supply 21% of energy and 15% of protein for humans, but the nutritional value differs in different types of rice (Gnanamanickam 2009); the nutritional values of long grain, unenriched, raw, white and brown rice are shown in Table 23.1.

Besides the worldwide production and economical importance of rice, the key constraints to rice production include insects, pests, and microbial diseases that cause significant annual losses to rice production. In addition, climate change and adverse environmental conditions such as salinity, drought, and unpredictable thermal acclimations have impacted rice production. The population of the world is expected to exceed 9 billion by 2050 (rising by 34%) and according to the estimates of FAO, this will increase the agricultural products demand by ca 70%. To feed this estimated population, the demand for cereal production will rise from 2.1 billion tons to around 3.0 billion tons by 2050. The mounting scenario demands extraordinary measures to ensure the food security, nonetheless, the rice-infecting viruses (RIV) are becoming major constraints to rice production. The management of RIVs and their insect vectors should be a high-end priority to ensure the global food security.

23.2 Major Viral Diseases of Rice and Their Impact

RIVs encompass huge genomic diversity and include single-stranded (ss), double-stranded (ds), negative-sense ss and ds, and positive-sense ss viruses. These RIVs include rice black-streaked dwarf virus (RBSDV), southern rice black-streaked dwarf virus (SRBSDV), rice dwarf virus (RDV), rice grassy stunt virus (RGSV), rice ragged stunt virus (RRSV), rice stripe virus (RSV), rice necrosis mosaic virus (RNMV), rice yellow mottle virus (RYMV), rice transitory yellowing virus (RTYV), rice hoja blanca virus (RHBV), rice tungro bacilliform virus (RTBV), rice tungro spherical virus (RTSV), rice stripe necrosis virus (RSNV), rice stripe mosaic virus (RSMV), rice gall dwarf virus (RGDV), rice bunchy stunt virus (RBSV), rice giallume virus (RGV), and rice necrosis virus (RNV). Although RIVs are generally considered to be of less importance as they cause <1.5% crop losses across the globe annually, sporadic epidemics could cause devastating damage (ca 80–100%) to a particular area or region (Ramasamy and Jatileksono 1996). In Southern Vietnam, RGSV alone (or may be with RRSV coinfection) has seriously affected more than 485,000 ha of rice fields, leading to a loss of US$ 120 million worth for 828,000 tons of rice (Cabauatan et al. 2009). The biology and epidemiology of leading RIVs are discussed in the subsequent section.

23.2.1 Rice Black-Streaked Dwarf Virus

Rice black-streaked dwarf virus (RBSDV) is a fijivirus (genus *Fijivirus*, family Reoviridae) and comprises 10 dsRNA segmented genome. RBSDV is a nonenveloped, double-shelled, icosahedral shape virus having spikes on 80 nm diameter virion. The RBSDV genome consists of 10 linear dsRNAs of ca 30 kb and designated from S1 to S10 based on their molecular weight in descending order; the smallest is of 1.8 kb, while the largest is of 4.5 kb (Zhang et al. 2001). RBSDV encodes 13 different size proteins ranging from 203 to 1464 amino acids. Two open reading frames (ORFs) are encoded by S5, S7, and S9 and these are further categorized as ORF1 and ORF2, for example, S5-encoded ORFs are referred to as ORF5-1 and ORF5-2. While the remaining seven segments encode a single ORF and designated as ORF1 to ORF10, respectively (Firth and Atkins 2009; Azuhata et al. 1993). Each segment has genus-specific conserved terminal sequences at 5′ end (AAGUUUUU) and 3′ end (CAGCUNNNGUC) and a conserved inverted repeat of 7–11 nucleotides (nts) adjacent to these conserved sequences. The segment S1 encodes RNA-dependent RNA polymerase (RdRP), S2 is a core protein, S3 is a capping enzyme, and S4 is an outer spike protein, respectively. P5-1 and P9-1 are viroplasm components, P5-2 localized subcellularly in the chloroplast, P7-1 localized at plasmodesmata, and have a role in the development of tubular structure in both insects and plant cells (Sun et al. 2013). P9-1 binds to ssRNA, P6 is an RNA silencing suppressor, and interacts and recruits P9-1 to viroplasm-like structures (Xu et al. 2015). P8 is a minor capsid and acts as a transcriptional repressor, while P10 is an outer capsid protein (Liu et al. 2007).

RBSDV infects a number of different cereal crops such as rice, barley, maize, sobal, wheat, and some grasses (Shikata 1974) causing severe yield losses, particularly in Asia. Upon infection, it induces leaves and stem darkening, growth stunting, leaf margin splitting, and twisting of the leaf tips symptoms. It also produces different colored (ranging from white to black) waxy galls on abaxial sides of the leaves and along the veins. The galls are the outcome of hypertrophy and hyperplasia of the phloem cells due to the localization of RBSDV (Uyeda et al. 1995).

RBSDV is transmitted by small brown planthopper (SBPH), *Laodelphax striatellus*, in a circulative, persistent, and propagative manner. The active transmittance ratio of *L. striatellus* was ca 30% and the shortest acquisition time was 30 min, while incubation period was recorded (ranging 4–35 days). During feeding, RBSDV is acquired via stylet; after ingestion, the virus particles move to insect hemolymph via the gut, enter and replicate into the epithelial cells of insect midgut, exit and then enter into salivary glands, and eventually transmitted to rice plants during feeding (Hogenhout et al. 2008). As RBSDV infects multiple host plants, so in the endemic areas, after the rice harvesting, the planthopper begin to feed on grass and weeds then switch to barley and wheat, follow ovipository, and transmit the virus (Lee et al. 1977). Nymphs emerge out in September and October, diapause overwinter, feed on RBSDV-infected wheat and barley plants in March and November, acquire the virus, and switched to newly sowing rice plants in May and June. Many adults of the first generation are macropterous and can fly long distances. Second- and third-generation adults switch from early-sown to late-sown rice and spread RBSDV. Generally in a rice monocropping system, the incidence of RBSDV is high, while in a rice polycropping system, RBSDV incidence is high on late-planted rice. Typically a higher RBSDV incidence is recorded along irrigation canal fields, and the use of high-nitrogen fertilizer increases disease incidence and severity (Park et al. 1982).

RBSDV cause significant losses to rice productivity in the Asian countries, especially in China, Japan, Korea, and a few other Asian countries. The spread of RBSDV has been sporadic and generally remain limited to small areas. Nonetheless, the first major outbreak of RBSDV was recorded in Japan during 1941, while the second outbreak was recorded on maize crops between 1957 and 1961, which then spread to maize and rice crops between 1965 and 1967. In China, the outbreaks of RBSDV occurred on rice, maize, wheat, and barley in 1963 and 1965–1967. The highest incidence of RBSDV in Korea was recorded in 1975–1976 (Lee et al. 1977). But during the high-incidence era, RBSDV was proliferating indigenously and remained endemic to small areas in these three countries. In rice and maize cocultivation, RBSDV caused more damage to maize than rice, while quite opposite is recorded if both were planted in different localities. A severe disease incidence occurred in rice early plantation in Japan, while a decrease in incidence was observed after 1968 when the area for barley and wheat cultivation was reduced. The exact opposite was observed in China during the cropping years of 1963–1967 when wheat cultivation was increased (Hui 1986). The occurrence of major RBSDV outbreaks simultaneously in Japan and China pinpointed toward a possible correlation, and the principal cause seemed to be the migration of the planthopper from southern China to Japan or vice versa (Kisimoto 1979) (Table 23.2).

Table 23.2 Virus names, genus and family, host, genome, genome size, insect vector, and occurrence

Virus name	Genus and family of the virus	Host(s)	Genome	Genome size	Insect vector	Occurrence
Rice black-streaked dwarf virus (RBSDV)	Genus *Fijivirus*, family Reoviridae	Rice, wheat, barley, maize, sobal, grass	dsRNA (10 segments)	ca 29.1 kb	*Laodelphax striatellus*	China, Japan, Korea, and Asia
Southern rice black-streaked dwarf virus (SRBSDV)	Genus *Fijivirus*, family Reoviridae	Rice	dsRNA (10 segments)	29.1 kb	*Sogatella furcifera*	China, Vietnam, and Japan
Rice dwarf virus (RDV)	Genus *Phytoreovirus*, family Reoviridae	Rice, barely, wheat, grass weeds	dsRNA (12 segments)	25.13 kb	*Nephotettix* spp.	China, Japan, Korea, Nepal, and Philippines
Rice grassy stunt virus (RGSV)	Genus *Tenuivirus*	Rice	Ambisense RNA (six segments)	25.1 kb	*Nilaparvata lugens* and *Nilaparvata* spp.	China, Japan, India, Philippines, Taiwan, and southeastern Asia
Rice ragged stunt virus (RRSV)	Family Reoviridae	Rice	dsRNA (10 segments)	26 kb	*Nilaparvata lugens* and *Nilaparvata* spp.	Indonesia, Malaysia, Sri Lanka, China, Japan, Thailand, India, Vietnam, and southeastern Asia
Rice stripe virus (RSV)	Genus *Tenuivirus*	Rice, wheat, oats, millet, barley, maize	ssRNA (4 strands)	16 kb	*Laodelphax striatellus*	China, Japan, Korea, Siberia, and Taiwan
Rice necrosis mosaic virus (RNMV)	Genus *Bymovirus*, family Potyviridae	Rice	Bipartite (+)sense ssRNA	10.7 kb	*Polymyxa graminis*	Japan
Rice yellow mottle virus (RYMV)	Genus *Sobemovirus*, family Solemoviridae	Rice	(+)sense/messenger-sense RNA, single, polypeptide and 5 ORFs	4.45 kb	By many beetles and mechanically	Africa

Rice Transitory Yellowing Virus (RTYV)	Genus *Nucleorhabdovirus*, family Rhabdoviridae	Rice	(−)sense, nonsegmented ssRNA, 6 ORFs	14.02 kb	*Nephotettix cincticeps*, *N. apicalis*, and *N. impicticeps*	China, Japan, Vietnam, and Thailand
Rice hoja blanca virus (RHBV)	Genus *Tenuivirus*, family Phenuiviridae	Rice, barley, oats	(−)sense, tetrapartite, 7 ORFs	16.92 kb	*Tagosodes orizicolus* Müir	Latin America, United States, Caribbean, Venezuela
Rice tungro bacilliform virus (RTBV)	Genus *Tungrovirus*, family Caulimoviridae	Rice	Circular double-stranded DNA	8 kb	*Nephotettix virescens*	South, Southeast Asia, and Southern China
Rice tungro spherical virus (RTBV)	Genus *Waikavirus*, family Secoviridae	Rice	(+)sense ssRNA genome, single polypeptide	12.18 kb	*Nephotettix virescens*	South, Southeast Asia, and Southern China
Rice stripe necrosis virus (RSNV)	Genus *Benyvirus*, family Benyviridae	Rice	RNA, RNA1 and RNA2, multipartite	11.25 kb	*Polymyxa graminis*	Colombia, Ivory coast, Ecuador, Liberia, Nigeria, Mali, and Brazil
Rice stripe mosaic virus (RSMV)	Genus *Cytorhabdovirus*, family Rhabdoviridae	Rice	(−)sense ssRNA	12.4 kb	*Recilia dorsalis*	China

23.2.2 Southern Rice Black-Streaked Dwarf Virus

Southern rice black-streaked dwarf virus (SRBSDV) associated disease was first observed in 2001 in Guangdong Province, China. SRBSDV is sometimes also referred to as RBSDV-2. Although RBSDV and SRBSDV shared strikingly same resemblances in their respective disease phenotype, both have different genomes and insect vectors. The principal cause of SRBSDV disease is white-backed planthopper (WBPH; *Sogatella furcifera*). SRBSDV particle size is about 75–80 nm in diameter, and it is multilayered, nonenveloped, and isometric shaped equipped with B-spikes (Xie et al. 2014). The genome of SRBSDV consists of 10 linear dsRNA segments labeled as S1 to S10 based on their molecular weight. Full-length SRBSDV genome sequences have been known from China and Vietnam and genome sizes vary from 29,106 to 29,115 nts, while the segment size ranges from 1.8 to 4.5 kb (Wang et al. 2010). Like RBSDV and other fijiviruses, SRBSDV also shared conserved features and have conserved sequences at 5′ and 3′ ends of each segment. A total of 13 different proteins are encoded by 10 genome segments, while S5, S7, and S9 encode two ORFs. The two ORFs encoded by S5 and S9 do not overlap, but the S5 ORFs overlap partially (Xue et al. 2014). The S1-encoded P1 is an RdRP protein that localizes in viroplasm and functions as a structural protein. P5-1 and P9-1 are also structural proteins and detected from the filamentous and granular matrix (Mao et al. 2013; Li et al. 2013). P7-1 is a tubular component and serves as a virus movement mediator across basal lamina in insect cells (Jia et al. 2014). P6 interacts with the host translation factor eEF-1A and recruits P5-1 and P9-1 during viroplasm formation (Li et al. 2013; Li et al. 2015a, b). The functions of the remaining proteins are yet to be elucidated.

Until 2008, SRBSDV remained sporadic and limited to southern China (Zhou et al. 2008; Wang et al. 2010). Nonetheless, in 2009, SRBSDD spread to nine provinces in southern China and 19 provinces in northern Vietnam, and cumulatively damaged ca 340,000 ha of paddy fields, of which 300,000 ha were damaged in China. In the following year, it became more prevalent and spread to 13 Chinese provinces and affected 1,300,000 ha, while in 29 provinces of Vietnam, it affected 60,000 ha of rice fields, leading to complete crop failure in many parts (Wang et al. 2010). Although concerted efforts were taken to curb the SRBSDD, yet ca 700,000 ha in China and more than 500,000 ha in Vietnam got infected in 2011. The disease also spread to some areas of Japan in 2010 and subsequently infected some other parts of Japan by 2013 (Matsumura and Sakai 2011).

Although rice is susceptible at all the stages of its life span, nonetheless, the symptoms, yield losses, and severity of SRBSDD depend on the initial infection stage. Generally early-stage infection leads to severe yield losses (Zhou et al. 2010). Symptoms typically induced by SRBSDV vary at different stages of rice plants: at early stage, it causes stiffness in the leaves and dwarfism; at tillering stage, it causes severe stunting in growth, excessive tillering, and dwarfism; at elongation stage, it do not cause dwarfism, but barren grains, small spikes, and reduced grain weight; and at the booting stage, infected rice plants usually exhibit no apparent symptoms. After elongation, the infected plants show dwarfism, white to black waxy galls along the stem and veins, and dark green foliar leaves (Zhou et al. 2013). SRBSDV also infects sorghum, maize, and some grassy weeds (Zhang et al. 2008a).

WBPH transmit SRBSDV in a propagative and persistent manner, interestingly another plant hopper, SBPH, is capable of acquiring but has no role in the transmission of SRBSDV. During sap feeding from the phloem, virions are acquired within 5 min of feeding; after passing from stylet and foregut, the virions crossed the cell membrane of the midgut epithelial cells and start propagation (Jia et al. 2012). The virions then cross the basal lamina into the muscle cells and eventually move to the salivary glands where they are egested with saliva during feeding. WBPH can transmit the virus after 6–14 days of the acquisition period, and WBPH may transmit the virus to maize after acquiring it from rice, but rarely acquires SRBSDV from infected maize plants, as maize plants are not an appropriate host (Hoang et al. 2011).

23.2.3 Rice Dwarf Virus

Rice dwarf virus (RDV), a principal cause of rice dwarf disease (RDD), is a phytoreovirus (genus *Phytoreovirus*, family Reoviridae) (Zheng et al. 2000). RDV was the first studied plant virus and RDD was identified from Japan during 1890–1930, which became an epidemic after 1967 in Japan (Kiritani 1983). The virion of RDV is a double-shelled icosahedral shape of 70 nm diameter, consisting 12 dsRNA segments, classified from S1 to S12 on the basis of decreasing molecular weight, and the segment size range from 4.4 to 0.83 kb (Omura and Yan 1999). Of these 12 segments, seven segments (S1–S3, S5, and S7–S9) encode structural proteins, while remaining encode nonstructural proteins (Zhong et al. 2003). S1-encoded protein (P1) is an RdRP. P3 protein forms a thin layer of capsid protein and encapsidate it (Nakagawa et al. 2003). P5 is a guanylyltransferase, while P8 mainly along with smaller number of P2 form the outer layer of the virion. The proteins P3 (114 kDa) and P8 (46 kDa) represent approximately 29% and 52%, respectively, of the total RDV protein mass. Pns4 is a phosphoprotein that forms tubules in the insect vector and localized around the viroplasm matrix. The S6-encoded protein, Pns6, plays a part in the cell-to-cell dissemination to the nonhost plants. Pns10 is a silencing suppressor of RNA and forms tubules that package virions to increase viral spread, and Pns11 binds to the nucleic acids (Xu et al. 1998).

RDV is mainly prevalent in China, Korea, Japan, Nepal, Mindanao, and Philippines (Cabauatan et al. 1993b). In central and southern Japan, the occurrence of RDV was limited to just some parts and was occasionally observed during the period 1889–1930. Nonetheless, it began to spread to the surrounding vicinities, and the level of RDD increased from 1967 to 1978 then declined. Meanwhile in this period, it also spread to Chejiang, China, and a higher disease incidence was observed (Li et al. 1979). The Japanese epidemic that began around 1955 was due to early cropping of paddy rice, which supported first generation of green leafhoppers to replicate, raised their population density, and dispersed the RDV.

While the reduction of RDD in 1979 was due to excessive use of insecticides to rice seedlings via machine transplant nursery boxes and plowing adjoining rice fields in early spring or winters (Nakasuji 1974). Generally the incidence of RDV is high at the field edge. Excessive use of nitrogen fertilizer not only increases the insect vector population but also the susceptibility of rice plants, leading to an increase in the RDV incidence.

RDV-infected rice plants exhibit stunting, delayed and incomplete panicle exertion, and chlorotic leaf specks symptoms. Three different strains of RDV designated as O, D84, and S have been identified. Among these, RDV-S induces the worst, while RDV-O induces the mildest effects. At molecular level, a substitution of 20 amino acids was found in 10 of 12 virus proteins among three strains. RDV has an unusual ability to multiply in both plants and leafhopper vector and its transovarial mode of transmission (Honda et al. 2007). During phloem feeding, RDV particles are ingested via stylet by *Nephotettix* spp., and initially infects epithelium filter chamber, replicates, assemble, and then spreads to anterior midgut, crosses the basal lamina of the midgut epithelium, then passes into the hemolymph, and eventually reached to the salivary glands, then transmitted. Besides rice, RDV infects grassy weeds and these weeds did not act as an important virus reservoir but as nursery for insect vector (Ishii and Yoshimura 1973). After harvesting late rice, infective *Nephotettix cincticeps* start feeding on weeds or grass and overwinter as nymphal diapauses. Overwintered adults come in March–April and feed on weeds. Some of these overwintered insects switch to rice and transmit RDV in warm areas in early April or May. In June, oviposit first-generation adults emerge out and transmit RDV to newly planted rice fields.

23.2.4 Rice Grassy Stunt Virus

Rice grassy stunt virus (RGSV) caused grassy stunt disease of rice (RGSD) and categorized as a major menace to rice crops in Southeast Asian countries including China, Japan, India, Philippines, and Taiwan (Shikata et al. 1980). RGSV is a tenuivirus (genus *Tenuivirus*) and has a filament-shaped virion of 60 nm size. However, RGSV particles are pleomorphic and can form a thin filamentous, circular, or spiral configuration. RGSV genome comprises six ambisense ssRNA segments with a cumulative length of 25,142 nts and all segments have identical terminal sequences over 17 nts. RGSV genome segments are classified as RNA1 to 6, these six segments encode 12 ORFs (Ramirez 2008; Toriyama et al. 1998). RGSV RNA1 encodes a 339-kDa protein on the complementary sense RNA (cRNA1) and function as RdRP. The RNA2 segment of tenuivirus encodes 23 kDa protein in its virus sense (v) sequence and 94 kDa protein in its complementary (c) sequence and the amino acid sequences across tenuiviruses. Although the function of these proteins has not yet been experimentally determined, the 94 kDa protein vaguely resembles to membrane glycoproteins. Two proteins are encoded on RNA5, a c-sense 36-kDa protein encoded by cRNA5 is a capsid protein (Chomchan et al. 2003), while a virion-sense (vRNA 5) encoded 22-kDa P5 protein accumulates in both rice leaves and BPH and plays a key role in RGSV infection (Chomchan et al. 2002). A v-sense 23-kDa protein (P2) encoded on vRNA2 and another 21-kDa (P6) protein encoded by vRNA6 are expressed only in the leaves of infected rice leaves.

RGSV typically affects rice crops where year-round rice cultivation is practiced and it induces yellowing (chlorosis), growth stunting, excess tillering (branching), erect and narrow leaves, mottling symptoms in the infected rice plants (Satoh et al. 2013). Although some other tenuivirus also induced chlorosis and stunting phenotype in infected plants, but excessive tillering is a hallmark of RGSV infection. At the cellular level, rice cells infected with RGSV showed the formation of fibril in the cytoplasm and nucleus. Based on the symptomatology, cultivars reaction, and insect transmission, different strains of RGSV have been reported. These strains have been reported from different countries; three strains—wilted stunt virus isolate (GSW), grassy stunt B (GSB), and grassy stunt Y (GSY)—were reported from Taiwan strain (Ching-Chung 1982), and fourfrom Philippines and referred to as severe (S2), moderately severe (SC), moderately mild (M2), and very mild (M3) (Jonson et al. 1990). Some of these strains causing yellow-orange leaf discoloration and premature plant death have been observed in Taiwan (Ching-Chung 1982), Philippines and Thailand (Hibino et al. 1985), and India (Mariappan et al. 1984).

High-RGSD incidence was observed in Indonesia from 1970 to 1977, in Philippines from 1982 to 1983, in India from 1973 to 1974 and 1981 to 1984, and in Japan from 1977 to 1978. Since 1984, RGSV occurrence has been low in Asia (reviewed by Hibino 1996); while not well known, the low incidence is attributed to a change in efficiency of insect transmission and low population density. High incidences of grassy stunt and ragged stunt disease were also reported in Vietnam's Mekong Delta from 2000 to 2007 (Du et al. 2005). RGSV and BPH affected ca 8000 ha of rice in central Java, Indonesia in 1971, causing an estimated yield loss of 77.8% (Tantera and Satomi 1973); in the subsequent years, from 1974 to 1977, RGSD progressed further to Java, Bali, Sumatra, and Sulawesi, and affected a total of 1.2 million ha led to yield losses of ca three million tons of paddy rice over $510 million (Palmer and Rao 1981). A major outbreak of RGSV occurred in 1973 in Philippines causing a massive yield loss of ca $26 million. In India, RGSV infected 15,000 ha and reported yield losses were of ca $20 million (Dyck and Thomas 1979). In Vietnam, during 2005–2006, ca 485,000 ha of paddy fields were damaged by BPH and caused losses of worth US$ 120 million and yield losses of 828,000 tons (Van Du et al. 2007).

RGSV is transmitted by BPH (*Nilaparvata lugens*) and by two other *Nilaparvata* spp. in a propagative and persistent manner, but not via eggs (Hibino 1996). *Nilaparvata lugens* is monophagous to rice, so after cultivation of one paddy field it travels to remote areas to new paddy fields. RGSV and *N. lugens* are generally endemic in tropical areas, especially in those areas where year-round rice is grown. Nymph and adult *N. lugens* are common vectors for RGSV and typically they acquire RGSV after feeding 20–30 min on an infected plant.

23.2.5 Rice Ragged Stunt Virus

Rice ragged stunt virus (RRSV) associated disease was first recognized in Indonesia in 1976 (Hibino 1979). RRSV is an oryzavirus (genus *Oryzavirus*, family Reoviridae) having an icosahedral shape and double-shelled particle of about 75–80 nm in diameter (Boccardo and Milne 1984). The virus genome comprises 10 dsRNA segments (referred to as S1–S10), each segment has genus-specific nt sequences at 5′ GAUAAA and at 3′ GUGC (Yan et al. 1992). A total of 15 proteins are encoded on RRSV genome, of these five are structural while remaining are nonstructural proteins. RRSV genome has been fully sequenced, and its largest segment, RNA1, encodes a putative RdRP. Segment 2 encodes a major outer capsid protein (VP2) and segment S3 encodes a major inner capsid protein (VP3), while other minor structural proteins are encoded by S1, S5, and S8, respectively. The molecular weight of structural protein ranges from 33 to 120 kDa, while the molecular weight of three nonstructural proteins ranges from 31 to 88 kDa. The genome segments S5 encodes ca 91 kDa protein and have a minor structural protein and did not share any significant sequence homology to other reovirus proteins (Li et al. 1996). The segment S9 encodes a ca 38 kDa protein that is a highly immunogenic structural protein (Upadhyaya et al. 1995).

Rice plants infected with RRSV exhibit growth stunting, twisted or serrated edges of the leaves, galls on the abaxial side of the leaves, and plants look ragged. The gall forms due to hyperplasia and phloem tissue hypertrophy (Hibino 1996). The newly emerging leaves reveal symptoms at the heading stage. During infection, RRSV localized inside the phloem tissues and induce gall formation. Infected rice cells contain large inclusion of viroplasm matrix and many virions. RRSV reduces rice yield by causing panicles, unfilled seeds, and loss of plant weight. RRSV infection is typically high in tropical regions where rice is planted all around the year, making it easier and providing suitable growth conditions for the BPH. Early-instar nymphs are more efficient transmitters of RRSV than mature BPHs. The BPH can acquire RRSV within 24 h of feeding on infected plants, and after 6 h of acquisition, can transmit the virus to other plants and remain infective for a lifetime.

RRSV is transmitted by BPH, *N. lugens*, and *Nilaparvata* spp. in a persistent and propagative fashion but not via eggs. RRSV along with its insect vectors can cause up to 80% yield losses. In Indonesia, field trials showed that RRSV incidence ranged from 34% to 76% and can reduce grain yield by 53–66% (Palmer and Rao 1981). RRSD was initially observed in Indonesia in 1977 and then subsequently recognized in Philippines, Malaysia, and Thailand. By the following year, it spread to China, Sri Lanka, and India (Hibino 1979); to Taiwan in 1978 (Chen et al. 1979a); and to Japan in 1979 (Shinkai et al. 1980). The origin of RRSD is vague; immediately after the first onset of disease incidence, it became epidemic in Indonesia and Philippines in 1977–1981, in Thailand in 1980–1982, and the second wave appeared during 1989–1990. However, the RRSD incidence has remained low in many countries since 1982. The occurrence of RRSV in Cambodia, Lao PDR (Laos), and Burma is suspected but not proven. Major RRSV and RGSV outbreaks occurred in China and Vietnam in 2006–2007 (Cabunagan and Choi 2009).

23.2.6 Rice Stripe Virus

Rice stripe virus (RSV) is a type of the genus *Tenuivirus* and categorized as a leading rice pathogen. RSV is filamentous in shape having 8 nm diameter and 500–2000 nm length. The genome of RSV is ssRNA and comprises four RNA segments designated as RNA1 to 4, these segments have genus-specific and conserved 5′ and 3′ terminal sequences. A total of seven proteins are encoded in both senses, that is, negative and ambisense (Ramírez and Haenni 1994). The RdRP is encoded by RNA1 (~9 kb) in the viral complementary strand and found associated with filamentous ribonucleoprotein complex of RSV. All other RNA segments (RNA2, 3, and 4) encode two proteins on both sense strands (ambisense), for example, one ORF at the 5′ viral RNA terminus (vRNA) and the second ORF at the 5′ viral complementary RNA terminus (vcRNA) and both these ORFs are separated by a nonoverlapping and noncoding intergenic region (Hamamatsu et al. 1993). RNA2 (~3.5 kb) encodes two proteins: vRNA-NS2 (p2) of 22.8 kDa with an unknown function, and vcRNA-NSvc2 (pc2) of 94 kDa which is a potential membrane glycoprotein. RNA3 (~2.5 kb) encodes vRNA silencing suppressor (NS3 of ~23 kDa) and vcRNA nucleocapsid protein (NCP of 35 kDa) (Takahashi et al. 1993). RNA4 encodes nonstructural disease-specific protein (SP of 21.5 kDa) and vcRNA-cell to cell protein (NSvc4 of 32.5 kDa) (Zhu et al. 1992). RSV-encoded SP protein localized in the cytoplasm, chloroplast, and nucleus of the infected rice cells and this localization is directly correlated with RSV disease symptom development.

Although rice is the major host plant for RSV, it also infects maize, wheat, oats, foxtail millet, several weeds, and *Arabidopsis* (Falk and Tsai 1998; Sun et al. 2011). Rice stripe disease (RSD) is characterized by yellowish white stripes, mottling, and necrotic streaks on the leaves. Nonetheless, the severity of the disease varies at different stages of the plant, but earlier sown seedlings or young seedlings are more susceptible to RSV. At the seedling stage, plants showed twisted, wilted, folded, and droopy leaves. Later plants get stunted, have just a few tillers, made a few panicles, and die prematurely. Infected plant panicles have whitish to brown deformed and unfilled spikelets, and may not be fully exerted. While at the tillering stage, leaves showed less mottling or chlorosis and delayed panicle exertion and ripening.

RSD is more prevalent in temperate regions of East Asia, including China, Korea, Japan, Philippines, Taiwan, and has also been reported from far-eastern Russia (Wang et al. 2008). In 1963, RSV was first recorded in just small area of China and then subsequently it spread to 16 provinces of China (Wei et al. 2009). The prevalence of RSD was high in Shanghai, Jiangso, and Zhejiang in 1964, during 1975–1976 in Beijing, in 1984 in Shandong, and during 1974–1990 in Yunnan (Lin et al. 1990). In Japan, RSV incidence was very high from 1960 to 1972 and increased again from 1977 to 1986 (Kiritani 1983). In Korea, it was high during 1964–1965 and 1973–1974 (Chon et al. 1975). RSV remained endemic in western and central Japan, but it recently reemerged in Zhejiang and eastern China and reduced the yield up to 40% (Wu et al. 2009). In 2008, RSV occurred in western Japan due to viruliferous planthoppers migrated from China. In recent decades, the frequent occurrence of RSV has been from China, Japan, and Korea. Previously RSV was limited to southern part of Korea, but has recently been observed across different altitudes, suggestive of a rapid spread of RSV (Lee et al. 2008).

SBPH (*L. striatellus*) and some other planthoppers transmit RSV in a persistent and propagative fashion. RSV is also transmitted via eggs of adult females to their progeny and has extremely low mechanical transmission ability. The ability of small BPH to acquire RSV is highly variable, experiments showed that small BPH can acquire and transmit RSV from frozen plant samples.

23.2.7 Rice Necrosis Mosaic Virus

Rice necrosis mosaic virus (RNMV) belongs to the family Potyviridae (genus *Bymovirus*) and has a nonenveloped, filamentous, and flexuous-shaped particle of 205–550 nm in length and 13–14 nm in width. RNMV has a bipartite, linear, and (+) sense ssRNA genome which has two segments, RNA1 (7.7 kb) and RNA2 (~3.7 kb). The 3′ terminus has poly(A) tail and 5′ terminus has a covalently bound genome-linked protein (VPg) (Adams et al. 2005). Genomic RNA is infectious and both segments encode polyproteins which are cleaved by virus-encoded protease (Nla-Pro) to several functional proteins. Bymoviral RNA1-encoded large ORF subdivided into eight proteins. The order of genes from N-terminus is P3, 7K1, protein (CI), 7K2, VPg, NIa-Pro, protein b (NIb), and CP (Adams et al. 2005). Bymoviral RNA2-encoded ORF is processed into two proteins, P1 and P2 (Adams et al. 2005). The 6K2 protein of *Potyvirus* is considered as a functional homolog of RNMV-encoded 7K2 protein which has a role in viral replication by anchoring the viral replication complex on the endoplasmic reticulum, systemic movement, and symptoms induction (Spetz and Valkonen 2004).

RNMV was first identified in Okayama, Japan in 1959 and then in southwestern and central Japan in 1962–1970 (Fujii 1978) and later reported from India in 1978–1979 (Ghosh 1980). Although RNMV naturally infects rice, the Indian isolate also infects some weeds, the role of the weeds has not been elucidated yet in the disease cycle. Some other crops such as barley, wheat, quinoa, *Avena sativa* (oats), maize, *Chenopodium amaranticolor*, and *Nicotiana tobacco* did not develop symptoms or infection when grown in infectious soil or after infectious sap.

RNMV is transmitted by a fungus, *Polymyxa graminis*, to rice plants (Inouye and Fujii 1977) and it induced mosaic symptoms characterized by streaks on the lower leaves, moderate stunting, spindle-shaped yellow flecks, reduction in tillers, necrotic fleck lesions on basal portions of sheath and stems, and plants are somewhat prostrate (Wagh et al. 2016). The zoospores of *P. graminis* penetrate into the plant roots and then pass through soil or seeds to the next generation, though seed transmission is almost negligible (Ghosh 1980). RNMV-affected soil retains long-term infectivity and a very high transmission rate through the soil at 25–30°C and low moisture. Wet soil does not favor the RNMV infection, so rice seeded directly in flooded fields have rare chances to get infected.

23.2.8 Rice Yellow Mottle Virus

Rice yellow mottle virus (RYMV) is a sobemovirus (genus *Sobemovirus*, family Solemoviridae). RYMV virions are polyhedral in shape of ca 30 nm in diameter and contain one small (+)sense/messenger-sense RNA genome of ca 4450 nts which is not polyadenylated (Yassi et al. 1994). RYMV genome has five overlapping ORFs. Protein (P1) encoded by ORF1 suppresses the host defense by downregulating the DCL4, controls RYMV accumulation within host cells, and maintains a subtle balance between host defense and RYMV counterdefense strategies (Lacombe et al. 2010; Opalka et al. 1998). The overlapping ORF2a and ORF2b encode a polyprotein that is subsequently cleaved into serine protease, VPg, and RdRP. ORF2b is thought to be translated as a ribosomal frameshifting mechanism and have a role in the resistance breakdown. ORF4 encodes (26 kDa) coat protein which have nuclear localization signal (NLS) at N-terminal and have a role in cell-to-cell and systemic movement of the RYMV (Yassi et al. 1994).

RYMV causes substantial losses to rice production from eastern to western Africa. Initially RYMV was observed in Kenya in 1966, and then in 1970, it was reported from province Nyanza, Kenya (Bakker 1974). Soon after first identification, RYMV spread to many Western, Central, and East African countries and was identified from Liberia, Nigeria, Sierra Leone, and Tanzania in 1976 (Raymundo and Buddenhagen 1976). Subsequently RYMV was identified from Côte d'Ivoire (Fauquet and Thouvenel 1977); later in 1980 it was reported from Ghana, Koba, Niger, Burkina Faso, Mali, Malawi, and Rwanda (John et al. 1984). RYMV was reported from Madagascar in 1989 and the severity of infection was quite high that the rice cultivation had to be abandoned in Marovoay and Lake Alaotra within just a few years (Reckhaus and Randrianangaly 1990). In 2001, RYMV was reported from Cameroon and Chad (Central Africa) (Traoré et al. 2001). In 1983 and 1986, RYMV infected a wide area of rice cultivation in different regions of Africa, it infected ca 75% area of Sahel, 40% of total rice area in Sudan, 18% of Guinea savanna, and 7.5% area in the tropical rainforest (Awoderu 1991). The yield losses ranged from 1% to 49% in a susceptible cultivar Ita212 and 10% to 78% in Ngoyumaboi (depending on isolates) (Kouassi et al. 2005). In Kenya, the RYMV incidence was 54% and 96% of the 2015 and 2016 collected samples, respectively (Adego et al. 2017). Generally yield losses vary from 10% to 100% depending on preinfection plant age, rice susceptibility, and environmental factors. In addition, RYMV occurrence is more prevalent in irrigated paddies, although it occurs in inland swamps, deep rice, and upland rice (Raymundo and Konteh 1980).

RYMV is transmitted semipersistently by several insect vectors belonging to the coleopteran order such as *Chaetocnema pulla* (Halticinae), *Sesselia pusilla* (Galerucinae), *Dicladispa* (Chrysispa) *viridicyanea* (Hispinae), and *Trichispa sericea* (Hispinae) (Bakker 1971, 1974). Some other insects are also considered as potential vectors of RYMV such as *Conocephalus merumontanus* (Sjost.) and long-horned grasshopper (Abo et al. 2000). RYMV is also mechanically transmissible by farm machinery such as scythes, contaminated hands, close contact among plants during planting (Abo et al. 2000, 1997), grazing or trampling animals (Sarra et al. 2004), and by irrigation water (Abo et al. 2000).

RYMV is generally limited to rice plants and initially causes yellow-green linear or oblong spot at the base of the youngest leaves. Later, these spots spread to the leaf veins, appearing as yellow or orange streaks. Leaves emerged after the onset of symptoms became twisted and mottled. In addition, RYMV causes dark-brown discoloration, poor panicle exertion, growth stunting, reduction of tillers, and spikelets sterility (Bakker 1970, 1974; Awoderu 1991). Infected plants after 20–50 days of transplantation may show yellow stripes and spots, produce flowers and seeds, but have stunted growth.

RYMV is a major biotic constraint to rice cultivation in Africa and serological studies have shown that six strains of RYMV are present. RYMV diversity has shown that the virus originated 200 years ago in East Africa and spread across the continent.

23.2.9 Rice Transitory Yellowing Virus

Rice transitory yellowing virus (RTYV) is a nucleorhabdovirus (genus *Nucleorhabdovirus*, family Rhabdoviridae) (Hiraguri et al. 2010). RTYV virions have 94–98 × 129–194 nm bullet-shaped particles containing nonsegmented and (−)sense ssRNA genome of 14.02 kb size (Huang et al. 2003). RTYV genome has 3′ leader sequences followed by five genes (N, P, M, G, and L) and then 5′ trailer sequences. The virions of RTYV comprise a host-derived lipid envelope and a ribonucleocapsid core (made of N protein), while phosphoprotein (P) and polymerase (L) are bounded to it. The glycoprotein (G) bulges out from the lipid envelope and matrix protein (M) binds the envelope to the ribonucleocapsid core. RTYV's genome organization is unique, having two genes besides basic gene order (Huang et al. 2003; Hiraguri et al. 2010); a small protein P6 is encoded between G and L ORFs. The gene 3-encoded protein (P3) presumably have movement function; its secondary structure analysis suggested that it resembles to the "30K" superfamily and found to be associated with the ribonucleocapsid core (Hiraguri et al. 2012).

RTYV is transmitted persistently and preoperatively by three species of rice green leafhoppers (RGH), *Nephotettix cincticeps*, *N. apicalis*, and *N. impicticeps* (Chiu et al. 1968; Shikata 1972). However, RTYV is not transmissible via eggs. The transmission efficiency of RTYV by *N. cincticeps* and *N. apicalis* is high, but low by *N. impicticeps*. In infected RGH cells, RTYV virions localize in the cytoplasmic vacuolate structures and remain inside the insect vectors for a lifetime after acquisition period of 21–30 days (*N. cincticeps*) and 3–29 days (*N. apicalis*). Insects may acquire RTYV while feeding on infected plants for 15–240 min (*N. cincticeps*) and 5–600 min (*N. apicalis*) (Hibino 1996).

RTYV-infected plants exhibit growth stunting, discoloration, reduced tillering, leaf yellowing (two lower leaves). As the infection progressed (usually up to 3 weeks), these leaves turn bright yellow or orange (Chiu et al. 1968). Plants exhibit severe stunting in extreme situations, and yellow (or orange) leaves open wider, rusty flecks grow on the leaves, and ultimately leaves get wilted. The initial symptoms get recovered gradually and infected plants tend to produce new and symptom-free leave. RYSV yield losses are 70–90% in the pandemic areas (Chen 1984). Round inclusion bodies are found in RYSV-infected rice plant parenchyma cells (Su and Huang 1965) due to accumulation of RYSV particles (Shikata 1972).

RTYV occurs in China (central and southern regions), Japan, Taiwan, Thailand, Vietnam, and Laos (Hibino 1990). Three major epidemics of RTYV have occurred in Taiwan during the years 1960–1962, then 1973–1975, and the last between 1977 and 1980 (Chiu et al. 1968; Su 1969). In China, province Guandong has witnessed two outbreaks, one in 1964–1966 and the other in 1979 (Faan and Pui 1980), while three outbreaks in 1966, 1969, and 1973 have been noticed in province Fujian (Xie and Lin 1980), and one in Chejiang province during 1970–1972 (Chen et al. 1979b). In 1979, RTYV was found in Thailand (Inoue et al. 1986). Generally the RTYV incidence has been low to endemic levels since 1980s.

Naturally infections of RTYV are limited to rice plants only. After harvesting late-planted rice, RTYV overwinters in insect or rice stubble. RTYV overwinter in rice stubble in Fujian, southern China, and Taiwan (Faan and Pui 1980; Uyeda et al. 1995), while it overwinters in *N. cincticeps* in the lower and middle basins of Yangtze River basins (Li et al. 1979). Upon harvesting early rice, second- or third-generation *N. cincticeps* adults switch to late-planted rice. The incidence of RTYV is typically low in the first rice crop and high in the second. Nonetheless, during epidemic conditions, RTYV incidence is directly related to *N. cincticeps* population and winter temperatures during second crop (Lin et al. 1989; Chen et al. 1980).

23.2.10 Rice Hoja Blanca Virus

Rice hoja blanca virus (RHBV) belongs to the family Phenuiviridae (genus *Tenuivirus*). Virions of RHBV adopt different structures such as circular, spiral, or branched and are nonenvelope. The genome of RHBV is tetrapartite, (−)sense ssRNA, and consists of four RNA components, RNA1 is in (−) sense, whereas the remaining RNA2 to RNA4 have complementary (ambisense) translation strategy. All RNA components have a conserved terminal nucleotide sequence (5′ACACAAGTC3′) (Jimenez et al. 2018). RNA1 is of ca 9 kb and encodes RdRP, RNA2 (3.6 kb) encodes two proteins: a nonstructural (NS2) protein at 5′ end and a glycoprotein at 3′ end ORF. Similarly RNA3 (2.3 kb) encodes two ORFs: NS3 protein at 5′ end and nucleocapsid protein (NC3) at 3′ proximal end ORF. RNA4 (ca 2 kb) encodes the major nonstructural protein (NCP) at 5′ end and NS4 protein at 3′ proximal end (Ramirez et al. 1992; Jimenez et al. 2018).

RHBV infects rice plants and its vector, *Tagosodes orizicolus* (*Müir*), also feed on wild rice species, barley, oats, and rye (Cordero and Newsom 1962; Zeigler and Morales 1990). Agricultural records revealed that RHBV was present in Colombia's Cauca Valley area in 1935 50 years before its first biological characterization in 1980 (Nance and United States Agricultural Research Service, Crops Research 1958). RHBV has been mainly prevalent in Central and Southern America, the Caribbean, and Latin America from 1957 to 1959 (Atkins 1960). The first RHBV outbreak occurred in 1956–1967 in Latin America, however, during this outbreak, *T. orizicolus* caused more severe damage than RHBV (Jennings and Pineda 1971). RHBV caused significant losses to rice crop in Cuba, Costa Rica, Colombia, Panama, and Venezuela during 1956–1958 (Zeigler et al. 1994). Later in 1980–1985, the disease reached to an epidemic level in some Latin American countries (Zeigler and Morales 1990). *Tagosodes orizicolus* was observed in southern United States rice-growing areas during 1957, 1959, and 1962. After the introduction of early-maturing rice cultivars, there was an outburst of *T. orizicolus* and its population density increased much. Recently the presence and full-length genome of RHBV have been reported from Peru and Colombia (Jimenez et al. 2018; Bolaños et al. 2017).

Generally RHBV-induced different symptoms depend on the infectivity, the rice strain, and the plant age. Typically mature plant tissues after infection remain asymptomatic, while young leaves exhibit different symptoms, for example, after 4 days of infection, creamy-colored spots (up to 5 mm) appear on young leaves then lead to chlorosis, the spots turn white and then entire leaves eventually succumb to infection, becoming chlorotic and dwarf. After the successful onset of symptoms, newly emerging leaves exhibit heavy stripes or complete chlorosis. The tillers of RHBV-infected plants are stunted and often have discolored or malformed grains. The size of the infected plant roots get reduced and the roots eventually turn brown (Zeigler et al. 1994). In addition, RHBV also induce symptoms in *T. orizicolus* by infecting its organs and glands, the severity of symptoms ranges from mild to severe, and in severe cases it leads to digestive and respiratory track failure, thus reducing the fertility or longevity of infected female (Jennings and Pineda 1971).

RHBV is transmitted via *T. orizicolus* in a circulative, propagative, and transovarially manner. A potential *T. orizicolus* vector requires up to 12 h to acquire the virus from a systemically infected rice plant, although some can acquire within an hour, with an optimal acquisition feeding time of 8 h (Mcmillian et al. 1962). However, for transmission, a minimum of 1 week to 1 month is required to complete the incubation period by a vector. During feeding, RHBV virions are acquired via stylet, transferred to midgut, then penetrated into the hemolymph to finally reach the salivary glands to be transmitted back after 3–7 h feeding by nymph, female and male adults with ca 90% efficiency (Mcmillian et al. 1962; Galvez 1968). RHBV is mainly dispersed by congenitally infectious virus *T. orizicolus*.

23.2.11 Rice Tungro Bacilliform Virus

Rice tungro bacilliform virus (RTBV) is classified into the genus *Tungrovirus*, family Caulimoviridae (Azzam and Chancellor 2002). RTBV transcribes their genome via reverse transcription and virions are elongated icosahedron of a diameter of 30 × 130 nm. The RTBV genome comprises circular dsDNA of ca 8 kb size and contains two conserved discontinuities in genome which resulted from reverse transcription during replication (Bao and Hull 1992). Four ORFs are encoded into corresponding proteins, P1 to P4, from the pregenomic RNA using specialized translation mechanisms (Fütterer et al. 1997). The function of P1 (24 kDa) protein is still unknown. The ORF2 and 3 get translated by leaky scanning and encode P2 and P3 proteins, respectively. The P2 interacts with the core domain of P3 polyprotein and potentially have a role in RTBV capsid assembly. ORF3 encodes a 196-kDa P3 polyprotein that processed into different functional proteins, the order of proteins from N to C terminus include movement (MP), coat protein (CP), protease (PR), reverse transcriptase (RT), and RNase H (RH) (Herzog et al. 2000; Qu et al. 1991; Marmey et al. 2005). The P4 protein is a suppressor of RNA silencing.

Two major RTBV strains are present in Indian subcontinent and Southeast Asia. Both strains shared up to 70% sequence homology; the Indian strain is easily distinguishable from the Southeast Asian strain by the absence of a noncoding ca 65 nts region (Hull 2008). The infected rice plants with RTBV alone or in association with rice tungro spherical virus (RTSV) cause characteristic tungro disease symptoms that include stunting, yellowing, leaf discoloration (yellow-orange), and reduction in tillers (Hibino et al. 1988). The newly emerging young leaves may exhibit striping or mottling, but plants infected only with RTBV exhibit relatively milder symptoms. In infected plants, RTBV is found in the vascular tissues and it usually failed to form panicles in the infected plants, however, if panicles formed, they remained unfilled or with discolored grains. Although rice tungro disease was first described in 1964–1965 (Rivera and Ou 1965), tungro, which means "degenerated growth" in a Filipino dialect, has a long-standing history and traced back to 1859 to mentek disease in Indonesia (Ou 1985). RTBV was first described in 1975 (Saito and Roechan 1975) and its role as a member of the virus complex causing tungro disease was reported in 1978 (Hibino et al. 1978). Tungro disease associated with RTBV alone causes severe economic losses of worth US$ 1.5 billion in South and Southeast Asia (Herdt 1991).

RTBV is conditionally transmitted by *Nephotettix virescens*, a leafhopper, in a semipersistent fashion, only after *N. virescens* fed on the plants infected with both RTSV and RTBV, or when RTSV is acquired first and then RTBV (Cabauatan and Hibino 1988). RBTV can be acquired easily by *N. virescens* after feeding for 30 min on a susceptible infected plant and RBTV can be transmitted just after a few minutes (Ling 1974). Once RBTV was acquired, then *N. virescens* can transmit it for approximately 4 and 7 days, respectively. The nymphs of *N. virescens* can also transmit RTBV, but lose after molting (Ling 1974).

23.2.12 Rice Tungro Spherical Viruses

RTSV is a waikavirus (genus *Waikavirus*, family Secoviridae) (Murphy et al. 1995). The RTSV genome comprises a (+)sense ssRNA of 12.18 kb and genome has protection at both ends, with a VPg cap at the 5′ end and poly(A) tail at the 3′ end. RTSV virion has a isometric, nonenveloped, and encapsidated icosahedron of 30 nm diameter (Jones et al. 1991). Only a single, large ORF of more than 390 kDa is encoded on the genome and the putative large ORF is expected to start from the first in-frame start (AUG) codon at nt position 515 (Shen et al. 1993). Besides a large ORF, two smaller ORFs, ORF2 and ORF3, have been identified at 3′ end, however, these ORFs did not get translated. RTSV thus has an unusually long 3′ noncoding region of ca 1240 nts (Thole and Hull 1996). The long ORF is translated into a huge polypeptide that is subsequently processed by its own protease into eight functional proteins, and these proteins include (from 5′ to 3′) a leader protein (P1), three different capsid proteins (CP1–CP3), a helicase, VPg protein, a protease, and an RdRP. The replication of RTSV took place in the cytoplasm within ER-derived membrane vesicles.

Tungro disease is collectively caused by both RTBV and RTSV in South Asia, Southeast Asia, and Southern China and caused significant damage to rice productivity (Hibino et al. 1988; Cabauatan and Hibino 1988). Typically tungro disease incidence is low in rainfed or upland rice fields, however, when it reaches to an epidemic level then it is disastrous to the rice fields on vast acreage, even in rainforests or upland areas. RTSV remains restricted to phloem tissue during infection and induces mild symptoms, however, rice disease in Kyushu, Japan is an exception (Cabauatan et al. 1993a). RTSV alone causes independent infections to rice plants in Japan (Hibino 1996), possibly in central China, and in South and Southeast Asia (Bajet et al. 1986). In Japan, RTSV was also referred to as rice waikavirus (Shinkai 1977). RTSV alone induces mild symptoms including growth stunting, discoloration of grain, and leaf yellowing leading to a 20–30% reduction in the yield in Japan (Teng et al. 1988). Rice yield losses due to RTSV alone are not estimated, however, experimental conditions showed 2–40% yield losses (Hasanuddin and Hibino 1989). In Kyushu, Japan, RTSV reached to an epidemic level in 1971–1973 (Shinkai 1977), and 24,825 ha (ca 7.4% of total cultivated) area was affected. RTSV, as the RTBV, is transmitted in a semipersistent way by *N. virescens*.

23.2.13 Rice Stripe Necrosis Virus

Rice stripe necrosis virus (RSNV) is a nonenveloped, rod-shaped, multipartite benyvirus (genus *Benyvirus*, family Benyviridae). RSNV virions have a bimodal particle having a length of 110–160 nm and 270–380 nm diameter (Morales et al. 1999; Lozano and Morales 2009). The genome organization of RSNV is almost similar to beet necrotic yellow vein virus (BNYVV) genome. As demonstrated for BNYVV, RNA1 (6614 nts) of RSNV encodes a putative replicase-associated motifs, while RNA2 (4631 nts) encodes six different ORFs and include CP, read-through protein, triple gene block, and a 17-kDa cysteine-rich protein (Lozano and Morales 2009).

Naturally RSNV infects rice plants, and infected plants exhibit stunting, chlorosis, yellow streaks on leaves, reduction of tillers, and necrosis. In RSNV-infected cells, virions form aggregates in the cytoplasm and form paracrystalline inclusion bodies. The severity and effects of RSNV on rice production generally depend on the different factors such as plant age at time of infection, number of infected tillers, and the rice variety. During 1981 and 1983, 73 rice varieties were tested against RSNV in Côte d'Ivoire and 0–100% infection rates were observed depending on the variety (Fauquet et al. 1988). In Colombia and Ecuador, 20–40% yield losses were recorded in the last decade (Paz et al. 2009).

RSNV is a soil-borne virus and transmitted mechanically through a fungus *Polymyxa graminis* (order Plasmodiophorales). This fungus is an obligate pathogen and resides in the roots of infected rice plants (Fauquet et al. 1988; Dick 2001). RSNV was initially identified in Ivory Coast in 1977 (Louvel and Bidaux 1977). Since first identification, RSNV has also been found to infect rice in Nigeria, Liberia, and Sierra Leone (Fauquet et al. 1988). In disease-affected areas, season-to-season stripe necrosis occurs erratically. The occurrence of RSNV disease is higher in upland rice. In the forest, the disease occurs in clusters that grow gradually.

23.2.14 Rice Stripe Mosaic Virus

Rice stripe mosaic virus (RSMV) belongs to the genus *Cytorhabdovirus* (family Rhabdoviridae) (Yang et al. 2017). Currently RYSV is known only to infect rice species (Huang et al. 2003). RSMV virions form bacilliform particles 45- to 55-nm wide and 300- to 375-nm long, which form translucent granular-fibrillar viroplasm inside the cytoplasm of the infected cells. RSMV genome consists of (−)sense ssRNA and encodes seven ORFs. ORF1-, ORF5-, and ORF7-encoded proteins showed the highest sequence homology to known rhabdoviruses-encoded nucleocapsid (N), glycoprotein (G), and large polymerase (L) proteins, respectively. The ORF2 is a potent phosphoprotein (P) because it contains multiple phosphorylation sites. The ORF4 product is thought to be a viral matrix (M) protein because of its sequence homology (10.3–14.3%) with M protein of other rhabdoviruses.

RSMV occurs frequently in province Guangdong, China and transmitted by a leafhopper *Recilia dorsalis*, additionally, a leafhopper *N. virescens* also has the ability to acquire and transmit RSMV. Initially RSMV particles accumulate in epithelial cells of *R. dorsalis*, then proceed to the visceral muscles, subsequently RSMV quickly disperse to the salivary glands throughout the suspensory ligament. In the meantime, RSMV disperse from the filter chamber to the central nervous system, midgut, hindgut, esophagus, and hemolymph (Zhao et al. 2019).

RSMV-infected plants exhibit striped mosaic, dwarf, increased number of tillers with unfilled grains (Yang et al. 2017). RSMV induced the same phenotype in different rice cultivars, indica, hybrid and japonica. The induced symptoms were striped mosaic, dwarfing, crinkled or twisted leaves, increased number of tillers, shortening of panicles, and mostly unfilled grains. Nonetheless, induced symptoms vary slightly under different environmental conditions. The indica cultivar showed the highest reduction in yield, followed by hybrid indica and then japonica (Chen et al. 2019).

23.3 Diagnostics of Rice-Infecting Viruses

Accurate diagnosis of viruses in the plants and/or insect vectors plays a key role to monitor and estimate imminent disease outbreaks. Development of quick, precise, and highly sensitive techniques for the accurate detection of RIVs is very important to establish the control strategies, accordingly. Serological tests have been used extensively around the globe for the detection of RIVs at insect-pest control centers, agricultural research, and plant protection centers. Both serological and molecular techniques that have been used widely are discussed here.

23.3.1 Latex Agglutination Reaction

The latex agglutination reaction (LAR) is a classical, simple, rapid, and robust method for the detection of an antigen or an antibody via immunochemical reactions. The LAR consists of a process of attachment of carrier particles (latex) with the antibody or the antigen through adsorption. Even in developed countries such as Japan, LAR has been very common due to its frequent use in the detection of RSV in SBH owing to the fact that it is very sensitive as compared to other serology-based assays and required far less time, facilities, and expertise. However, the double antibody sandwich (DAS)-ELISA has started replacing the LAR due to availability of its commercially available kits, sensitivity, accuracy, and cost-effective application (Takahashi et al. 1991).

23.3.2 ELISA

Enzyme-linked immunosorbent assay (ELISA) can be broadly divided into two types: direct ELISA/dot ELISA and ID-ELISA. Dot/direct-ELISA is a quick, simple, and accessible technique to screen more samples in less time (Manoharan et al. 2004). However, standardization, optimization of ingredients, and procedure for the detection of RIVs are essential for validation of assays, quality control, and reproducibility.

23.3.2.1 Direct ELISA

Direct ELISA (ID-ELISA) has been used for the isolation and detection of SRBSDV polyclonal S10 antibodies. However, dot-ELISA technique can be successfully utilized to identify new serologically reactive SRBSDV antigens isolated from the infected rice plants (Esen et al. 1983). The ingredients commonly used for dot-ELISA are carbonate buffer, sodium carbonate, sodium bicarbonate, TBS, TBST, blocking buffer, antibody buffer, and alkaline phosphatase buffer. Color development solution can be developed via 5-bromo-4-chloro-3-indolyl phosphate p-toluidine salt and alkaline phosphatase containing nitro blue tetrazolium. BCIP solution and NBT solution should be prepared in *N*, *N*-dimethylformamide.

Virus antigen is usually obtained from the extract of infected rice, which is finely powdered in carbonate buffer, while the debris are separated by centrifugation. The supernatant can be removed and stored at −20°C (Wang et al. 2012).

Based on the S10 sequence, the S10 polyclonal antibody can be easily designed using BLAST in NCBI GenBank. Keyhole limpet hemocyanin (KLH) acts as the surface for the conjugation of the synthetic peptides viz. S10. The anti-SRBSDV rabbit antiserum is usually used for the purification of polyclonal antibodies (Chen and Song 2011).

23.3.2.2 Indirect ELISA (ID-ELISA)

The main step of this technique is binding of primary and secondary enzyme-labeled antibodies. Hence, the target antigen is indirectly detected by the secondary antibody that is labeled with the enzyme, therefore the name indirect ELISA is assigned. The surface of microtiter plate is blocked with the blocking proteins by the immobilized antigen attached to the surface of the microtiter plate. The enzyme-labeled secondary antibodies reacts with the immobilized antigen bound to the primary antibodies (in antisera), followed by the development of color. Higher the immobilized target antigen, higher would be the detection signals and the vice versa. Macromolecules can be easily detected with the ID-ELISA and a disease-associated antibody can be easily diagnosed in the sample antisera or sap with the help of primary antibody. So, indirect ELISA can be efficiently used for the detection of various diseases (Kruseman 1983; Matyjaszek-Matuszek et al. 2013).

Preparation of antiserum includes the eluted P10 and NaCl solution. An efficient ID-ELISA recipe consists of extraction buffers viz. carbonic acid buffer, PBS buffer, and citrate buffer for the extraction of RBSDV from respective samples. For antiserum incubation, protein alkaline phosphatase conjugate can be added to the wells (Sakamoto et al. 2018). This technique has been successfully deployed for the detection of the rice black-streaked dwarf virus (RBSDV).

23.3.3 Loop-Mediated Isothermal Amplification

Loop-mediated isothermal amplification (LAMP) is an economical and robust technique for the detection of certain diseases or the pathogens associated with them. It is a single-tube DNA amplification technique developed by Notomi et al. (2000). LAMP has alleviated many of the problems associated with the serological techniques such as need of a virus-specific antiserum. Additionally the sensitivity of ELISA is not enough to differentiate between different viral species. On the other hand, advanced molecular techniques like RT-PCR are more expensive as they require lab cycler and expensive ingredients. So, the LAMP include an unmatched sensitivity and can be performed rapidly without requiring any special equipment. A total of four primers (two primers sets), having an ability to recognize six different regions of the target sequence, are used which amplify the target sequence at 60–65°C in an hour. Thus, LAMP assay alleviated the need of a thermal cycler and can be performed either in a water bath or a dry thermal block. This technique has been deployed in various crops for the accurate detection of different species of viruses including the RIVs (Fukuta et al. 2003; Varga and James 2006; Boubourakas et al. 2009). Nine species of viruses have been diagnosed via RT-LAMP viz. RDV, RGDV, RBSDV, RTYV, RRSV, RSV, RTSV, RGSV, and RTBV (Le et al. 2010). As strains of viruses contain highly conserved regions, the RT-LAMP primers should be prepared from the conserved region of the nucleotide sequences. A specific set of primers targeting the S10 sequence of the SRBSDV genome could distinguish between closely related SRBSDV and RBSDV genomes in infected rice plants (Ichiki et al. 2013). Meanwhile, another specific set of primers targeting the S9 of SRBSDV genome have the ability to identify and distinguish between SRBSDV and RBSDV genome both in plants and insect vectors (Zhou et al. 2012). Ingredients required for RT-LAMP technique are: NaOH, Tris–Cl buffer, two outer primers and two inner primers, RNA amplification kit, KCl, Tris–HCl, $MgSO_4$, $(NH_4)_2SO_4$, betaine, Tween 20, Bst DNA polymerase, dNTPs, AMV reverse transcriptase, the template RNA, and a fluorescent detection reagent (Le et al. 2010).

23.3.4 Reverse Transcription PCR

Serological techniques need a virus-specific antiserum, which are difficult to obtain. Using a reverse transcription polymerase chain reaction (RT-PCR) for the detection is more effective as compared to ELISA (Takahashi et al. 1993; Ichiki et al. 2013) and has been used to diagnose viral diseases in rice (Periasamy et al. 2006; Zhang et al. 2008a, b). RT-PCR is a modified version of a common PCR. In RT-PCR, ssRNA (mRNA) is employed as a template to synthesize complementary DNA (cDNA) via reverse transcriptase enzyme. RT-PCR is being used widely across the globe to detect, quantify RNA viruses, and to measure the gene expression. RT-PCR is suitable for the detection of viruses at larger scale and has been successfully deployed for the detection and forecasting of epidemics in rice (Wang et al. 2004; Zhang et al. 2008a, b). RT-PCR is not a cost-effective technique, however, to make it cost-effective, a multiplex RT-PCR can be opted. By opting multiplex RT-PCR, the simultaneous detection of RSV and RBSDV has been achieved. The simultaneous detection of both RSV and RBSDV is almost impossible through ELISA due to the limited quantity and quality of specific antibodies. In addition, RT-PCR has been employed successfully to detect three RIVs—RBSDV, RSV, and RDV—together (Cho et al. 2013). Such multiplexing require three sets of primers that required many efforts to optimize the PCR conditions, nonetheless, the use of three sets of primers has been executed successfully by achieving a reliable result (Hamamatsu et al. 1993). Such executions are not only meaningful in distinguishing symptoms and associated principal pathogen in plants, but also in insect vector SBPH. RT-PCR has also been used to decipher the disease etiology between RSV and RBSDV. This technique requires three sets of primers, total RNA, NanoDrop (spectrophotometer), cDNA template, TBE buffer, dNTP mix, Taq DNA polymerase, and ddH_2O (Li et al. 2015a, b).

23.3.5 Real-Time PCR

Real-time PCR (also referred as qPCR) includes the amplification of template DNA/cDNA, which is detected and quantified by the fluorescence signal. The signal strength increases in a direct proportion with the increase in PCR amplicon in the reaction. qPCR offers the precise and accurate quantification of the target nucleic acid, even the starting amount of the nucleic acid and this can be monitored in a real-time manner. This method can also be used for the quantification of target RNA.

Generally three steps are involved in qRT-PCR: in the first step, cDNA is made from RNA template then PCR is performed using specific primers. During second step, fluorescent reporter signal related to the amount of PCR product is observed for each cycle. In the third and final step, amplicons are incubated sequentially to get a melting curve to verify the presence of a specific amplicon followed by a cooling step. The expression is always measured in relation to the housekeeping gene. This technique has been successfully deployed for the diagnosis of SRBSDV (Matsukura et al. 2013).

RT-PCR-based techniques have been employed for the detection and quantification of diverse RIVs. To amplify and quantify the RSV in individual *L. striatellus* was demonstrated by using RdRP-specific primers. As an internal positive control, a set of primers targeting the actin gene of small BPH was designed across the boundary of an intron and an exon. So, a duplex RT-PCR was successfully executed for the simultaneous detection of RSV and actin in *L. striatellus* (Lijun et al. 2003). Another study used RSV-polymerase and CP gene for RT-PCR-mediated detection of RSV in rice and *L. striatellus* (Lee et al. 2004).

Zhang et al. (2008a, b) developed a one-step RT-PCR method using the TaqMan chemistry for RSV detection in rice and *L. striatellus.* A conserved region of CP was exploited in the RT-PCR for the simultaneous detection of RSV in rice and insect. The developed assay was sensitive *enough* to detect the 20 viral copies with a high reproducibility. The findings of this study revealed that RSV accumulated to different levels in different plant and insect tissues, highest level of RSV was observed in rice leaf and SBPH thoraco-abdominal tissue (Zhang et al. 2008b).

The absolute quantification of RTBV and RTSV was achieved via RT-PCR using SYBR green chemistry. The titer of RTBV DNA and RTSV RNA in infected rice plants was quantified simultaneously and results were evaluated on the basis of melt curve analysis. The sensitivity of the developed assay was 10^3 and 10^5 folds higher than standard PCR and dot-blot hybridization, respectively (Sharma and Dasgupta 2012).

A unique, but simple RT-PCR assay requiring no RNA isolation was demonstrated for the detection of RSV and RBSDV in a single SBPH. In this assay, a single SBPH in sterile water was grinded and then crude extract was directly used as a template. The results of the assay were further verified by determining the sequences of the amplicon and dot blot. The sensitivity of assay was quite high and RSV can be detected after the 10^3-fold dilution of the extract (Xu et al. 2017).

23.4 Management Strategies to Curb RIVs

The productivity of rice across the globe is often severely compromised by a number of biotic factors, and viruses impart a key role in reducing the productivity and quality of rice grain. To sustain worldwide rice productivity, several conventional and modern approaches have extensively been practiced to limit the rice-infecting viruses. Many resistance genes/QTLs have been successfully introgressed through breeding and various molecular approaches have been opted to mitigate the virus stress in rice. In the subsequent section, conventional and molecular approaches of resistance are described.

23.4.1 Conventional Approaches

Plant viruses are internal obligatory parasitic agents, therefore, curative measurements to cure an infected plant is not possible. Under such circumstances, adoption of prophylactic approaches is crucial to combat viral epidemics in rice. These approaches are mainly based on agronomic or cultural practices, biosecurity, and managing viral vectors through chemical or biological control (Fig. 23.1).

23.4.1.1 Regular Monitoring of Viral Pathogens

Combination of modern molecular biology techniques and conventional assessment of viral symptoms can be very helpful to continuously characterize the spectrum of virus etiology in the field during a specific period of time. Such systems require sensitivity, specificity, and rapidity to accurately diagnose and monitor viral pathogens (Boonham et al. 2014). However, the reliability of a pathogen diagnosis test is the key to viral disease management in rice. Such infected plants should be eradicated from the rice field once detected positive to minimize the virus spread.

23.4.1.2 Integrated Pest Management (IPM) Approach to Counter Vector Populations

The RIVs exclusively need an insect vector for their successful transmission from one plant to another plant or across the fields. Thus, RIVs can be efficiently controlled by curbing the insect vector population in a field. This can be practiced through judicial use of pesticides and other appropriate pest control strategies. The transmission of RIV-associated disease can also be limit with the use of viral nonhost "trap plants" around the rice fields (Bragard et al. 2013).

23.4.1.3 Vigorous Measures for Weed and Alternate Host Plants Management in the Field

The presence of reservoir plants (such as weed and alternate host plants) in the vicinity of rice plantation can be crucial for viral epidemics because most of the epidemics are reported to be associated with the emergence of new viruses or their variants, which somehow escaped from the weed or alternate host reservoirs into the crop plants. Although evolution of new viral species is a complex evolutionary

Fig. 23.1 Conventional approaches to limit the rice-infecting viruses and their insect vectors

phenomenon, which is affected by multiple controlling factors, the RIVs can be effectively controlled through spatial field management that may have durable impact on resistance (Fabre et al. 2012).

23.4.1.4 International Phytosanitary Measures

The worldwide trade of crop plants and related commodities are under international legislations to minimize the spread of viral epidemics and trading virus-free agriculture commodities. Such legislations should be considered during the developmental stage of crop plants and decontamination of agronomic tools.

23.4.1.5 Transplanting Viral-Resistant Cultivars

The most efficient, sustainable, and frequent strategy against RIVs can be the use of resistant crop cultivars to curtail virus infection in the field. The agronomic selection of resistant plants is in practice from centuries. However, integrated crop genetic improvement programs have been in use from the middle of twentieth century, which rely upon the knowledge of complex interactions between plant-virus-vector to develop resistant crop cultivars (Fig. 23.1).

23.4.2 Molecular/Modern Approaches

Recent advances in biotechnology and genetic engineering helped to develop rice plants that have improved resistance against viral diseases and insect pathogens. In pathogen-derived resistance (PDR), viral sequences (or genes) are expressed in plants to prevent or reduce infections of the plant virus in the last couple of decades. RNA interference (RNAi) is another resistance approach that has offered wonderful utilities to confer resistance against rice-infecting viruses.

23.4.2.1 RNA Interference

RNA interference (RNAi) is a conserved mechanism of gene regulation that is triggered by dsRNA which is subsequently processed into small (21–26 nts) RNAs by the dicers. In the downstream process, RNA-induced silencing complexes (RISC) recruit these small RNAs to mediate gene silencing based on sequence homology at either the transcriptional (TGS) or posttranscriptional (PTGS) levels. RNAi is described as a first antiviral mechanism (Waterhouse et al. 2001) especially against RNA viruses, as these are both activators and targets of RNA silencing. A majority of the plant viruses have ds secondary structures in their genomes or during replication RdRP produces ds RNA intermediates. These dsRNA structures led to the production of small RNAs that trigger the RNA silencing machinery to degrade the viral genome and initiate systemic signals via plasmodesmata to the neighboring and long-distance cells (Pumplin and Voinnet 2013). RNAi performs exceptionally well against RNA viruses and rarely against DNA viruses, however, it poses some limitations such as sequence homologies, viral titer, selection of target and environmental temperature, therefore its efficacy may vary (Szittya et al. 2003). On the other hand, the pros of RNAi include an endogenous mechanism triggered by dsRNA, which induces extremely specific RNA degradation and increased gene silencing efficiency at various cellular levels (Schwarz et al. 2002).

To confer RNAi-mediated resistance against plant viruses, two approaches have been opted. First, RNAi has been applied to various insects that serve as vectors for plant viruses, and second, viruses have been targeted directly to limit their replication and spread. RNAi has been applied against ca 30 insect species belonging to different orders to limit the spread of pathogens which they transmit (Fereres and Raccah 2015). RNAi-mediated resistance was achieved against RYMV by transforming the RYMV-susceptible rice cultivars with the RYMV-encoded RdRP (Pinto et al. 1999). The ORF4 of RTBV, both in sense and antisense, was used in transgenic rice plants to confer RNAi-mediated resistance against RTBV. The results revealed the degradation of the target transgene transcript and varying resistance responses were demonstrated by two different transgenic rice plants. Initially RTBV accumulated to a higher level after inoculation in the line, RTBV-O-Ds1, followed by a quick reduction up to 50-folds till 40 days postinoculation (dpi). This line exhibited comparable tungro symptoms to nontransgenic inoculated plants. However, in the other line, RTBV-ODs2, the virus titer gradually increased by 40 dpi but remained lesser than nontransgenic plants, and plants showed extremely mild symptoms (Tyagi et al. 2008). These transgenic rice cultivars exhibited resistance to RYMV strains at different African locations, nonetheless, later the transgenic resistance was found to be only partial and temporary and may explain why PTGS-mediated rice transgenic resistance is not high or durable.

To confer RNAi-mediated resistance against RDV, two of its ORFs, Pns12 and Pns4, were expressed in rice plants. Transgenic rice plants expressing Pns12 showed a higher degree of resistance to RDV infection. In contrast, transgenic rice plants with Pns4 exhibited less resistance to RDV and some plants just showed delayed onset of symptoms, and symptoms were comparable to the nontransgenic rice plants. These findings suggested that the use of viral replication proteins to limit the viruses could be more practical and effective (Shimizu et al. 2009). Sasaya et al. (2014) designed several constructs to confer resistance to rice viruses and showed that not all the designed construct worked effectively against the rice viruses and concluded that identification of the "Achilles' heel" gene is important to limit the target virus via RNAi. The transgenic rice plants harboring Pns6, P8, and Pns12 genes showed resistance, were comparable to the control healthy plants, and did not contain detectable amounts of the virus. Nonetheless, transgenic plants bearing constructs for the P2, P5, P7, P9, or Pns10 failed to show resistance against RDV. The transgenic plants developed symptoms that were indistinguishable from the control RDV-infected plants (Sasaya et al. 2014). Transgenic rice lines carrying a hairpin of 500 bp fragment of different RTSV regions produced small interfering RNAs, hence can show improved resistance against RTSV (Le et al. 2015).

To target the RSV via RNAi, an inverted repeat construct targeting the gene NCP was engineered in the rice plants and three transgenic rice lines were developed. The expression of small interfering RNAs (siRNAs) was quantified in three lines that showed varied expression levels of siRNAs corresponding to different segments of the NCP gene. Approximately 47% of NCP-derived siRNAs belonged to the fragment from 594 to 832 nt in line 4, suggestive of a hotspot region for RNAi silencing. This study concluded that the massive abundance of siRNA derived from the inverted repeat of NCP is the major reason for RSV resistance (Li et al. 2016). Ahmed et al. (2017) developed transgenic rice plants harboring RNAi constructs targeting the S7-2 or S8 genes of RBSDV. The homozygous T5 transgenic lines harboring either S7-2-RNAi or S8-RNAi construct exhibited a higher level of resistance when infected through viruliferous small BPH under field infections (Ahmed et al. 2017). In another study, an RNAi gene construct of highly conserved RTSV partial coat protein 3 (CP3) gene sequences was used to develop the transgenic rice plants (cultivar Taipei-309). Two homozygous transgenic lines of the T4 generation were inoculated via viruliferous insect vectors and inoculated plants not only showed highly resistant phenotypes against tungro disease, but also showed an inability to transmit the virus complex. This resistance may likely be due to suppression of RTSV proteins having role in virus transmission (Malathi et al. 2019).

RNAi has an unparalleled advantage in controlling insect pests and has been successfully exploited to control insects (Gordon and Waterhouse 2007). BPH (*N. lugens*) is a typical rice-specific phloem sap feeder and transmit RGSV and RRSV (Table 23.2). The potential of RNAi to target *N. lugens* was exploited by targeting the ubiquitous hexose transporter gene *NlHT1*, the carboxypeptidase gene *Nlcar*, and the trypsin-like serine protease gene *Nltry*. These three genes are present in *N. lugens* and expressed in midgut of the insect. These genes were isolated for the development of dsRNA constructs for the transformation of rice plants. When nymphs were fed on the rice plants expressing dsRNA, transcript levels of targeted midgut genes were reduced; however, lethal phenotypic effects were not observed after feeding the dsRNA. When mature *N. lugens* were fed on the transgenic plants, RNAi gets triggered and transcript levels of target genes were suppressed. These results opened up new ways to control *N. lugens* in the fields (Zha et al. 2011).

The *N. lugens* ecdysone receptor (NlEcR) was targeted to control it via RNAi. In rice plants, a 360-bp fragment of EcR-c sharing a common region between NlEcR-A and NlEcR-B was introduced to generate transgenic rice plants. Upon feeding on these transgenic lines, a significant reduction of NlEcR mRNA associated with reduced number of offspring per pair of *N. lugens* was observed. The survival rate of the nymphs was almost 90% in all lines, and the average number of offspring per pair reduced up to 66% in the treated groups as compared to the control (Yu et al. 2014). Matsumoto and Hattori (2016) demonstrated an RNAi-based insect control strategy to limit the green rice leafhopper *N. cincticeps*, which cause tungro disease in rice plants. A dsRNA targeting the *Nclac2* gene caused a strong knockdown of the *laccase-2* gene in the injected female leafhoppers and first instar nymphs. These effects were observed for up to 2 weeks and these RNAi constructs also induced silencing in the salivary glands, leading to high mortality rates and body sideline depigmentation (Matsumoto and Hattori 2016). Coronatine insensitive1 (COI1) is an F-box protein vital to all jasmonate reactions whose function is unknown in rice. In rice plants, RNAi-mediated silencing of OsCOI1 was achieved to determine the role of OsCOI1 in rice defense against *N. lugens* and *Cnaphalocrocis medinalis*. Upregulation of OsCOI1 and methyl jasmonate (MeJA) was witnessed by

C. medinalis infestation, but not by *N. lugens*. The transgenic rice plants showed reduced *C. medinalis* resistance, but no change against *N. lugens* was observed, which is suggestive of no role OsCOI1 in controlling *N. lugens* (Ye et al. 2012).

The knockdown of RDV-encoded Pns10 (a nonstructural protein) inhibited the formation of tubules in leafhoppers, which ultimately prevented the RDV intercellular spread and transmission via *N. cincticeps* (Chen et al. 2012). Similarly RNAi-mediated knockdown of Pns4 in *N. cincticeps* culture cells prevented the viroplasm accumulation, inhibited the formation of minitubules, and inhibited the viral infection (Chen et al. 2015). Another study has shown that silencing the planthopper (*L. striatellus*) cuticular protein (CPR1) reduced the vector's ability to transmit RSV (Liu et al. 2015). The upregulation of three *L. striatellus* genes, peroxiredoxin, cathepsin B, and cytochrome P450, was reported during RSV infection. A rice leaf-mediated method was used to suppress the replication of RSV in *L. striatellus*. RNAi-based silencing of the three genes severely downregulated the RSV-encoded NS3 gene over 73.6% and compromised RSV replication. These results suggest effective RNAi-based approaches to RSV control and provide insight into RSV-*L. striatellus* interactions (Fang et al. 2020).

23.4.2.2 CRISPR-Cas and Rice Viruses

CRISPR-Cas is an adaptive immune system that degrades exogenous DNA discovered in *Escherichia coli* in 1987. Other bacteria including *Shigella dysenteriae*, *Salmonella enterica*, and *Mycobacterium tuberculosis* were later identified (Nakata et al. 1989). But the potential to exploit this system for genome editing was not popular until 2012 (Jinek et al. 2012). This system's immune memory is preserved as spacer sequences from foreign genomes incorporated into CRISPR arrays (Koonin et al. 2017). These spacers and Cas proteins act as a monitoring system to recognize and degrade foreign nucleic acids. This process has three steps. The first step, called adaptation, immunization, or spacer acquisition, is to identify and incorporate foreign DNA (spacers) into the CRISPR locus. The spacer sequence of viral or plasmid DNA is called protospacer. Typically there is a short-conserved sequence in the protospacer's immediate vicinity, called the adjacent protospacer motif (PAM). The second step consists of machine speech. A primary transcript (precrRNA) is transcribed from CRISPR locus to small CRISPR RNAs (crRNA). In the final step, called interference or immunity, crRNAs along with transactivating crRNA (tracrRNA) form a Cas-protein ribonucleoprotein complex. It recognizes and degrades foreign DNA by base pairing (Bhaya et al. 2011). CRISPR/Cas was developed and used for genome editing in a wide range of species such as mammalian cells, bacteria, fungi, and plants (Krappmann 2017). The genome-editing method includes two components: a DNA endonuclease (the most widely used is *Streptococcus pyogenes* Cas9 protein) and a customizable single guide RNA (sgRNA) (Fig. 23.1). Cas9 has a bilobed architecture with a large REC lobe and a small nuclease lobe (NUC) with two nuclease domains, RuvC and HNH, each cutting a different DNA strand. SgRNA is a small noncoding RNA, fusing crRNA and tracrRNA.

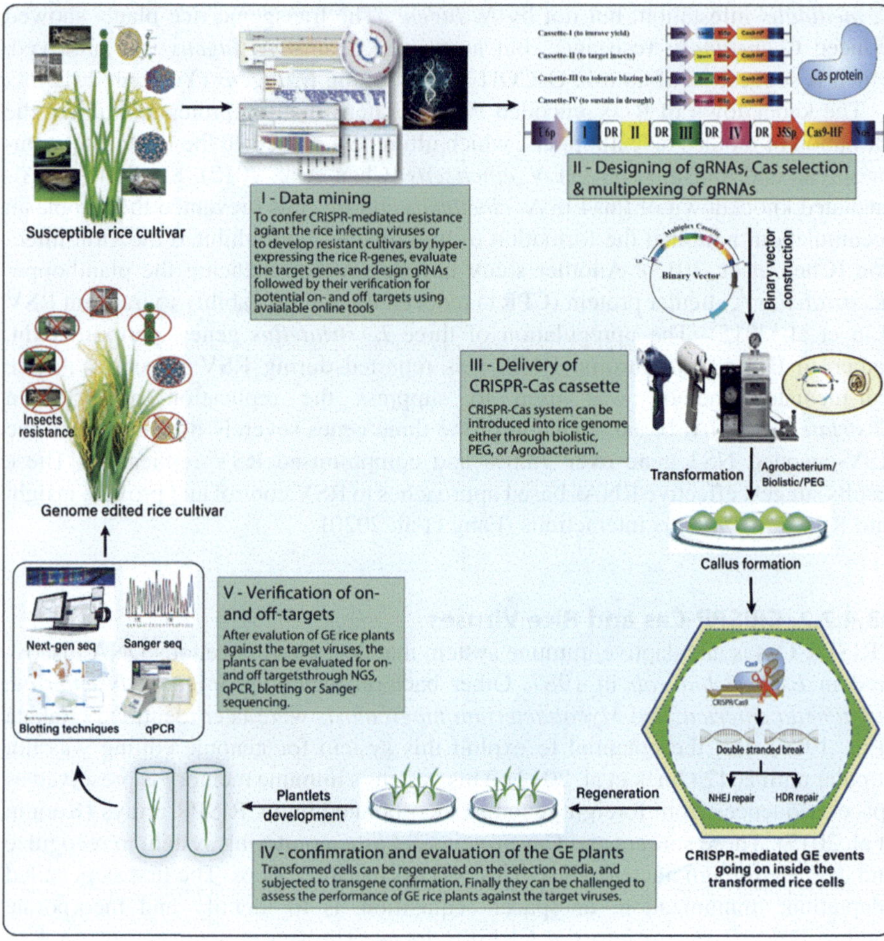

Fig. 23.2 A generalized layout for the successful execution of CRISPR-Cas system in rice genome engineering to engineer elite characteristics

The contemporary CRISPR-Cas-based approaches and its expanding toolkit can offer unlimited utilities to improve rice yield and control RIVs and their insect vectors via its transgene-free genome-editing abilities. Currently many CRISPR-Cas systems are available for editing/targeting different types of the genome, nonetheless, just a few of these have been executed in rice plants including CRISPR-Cas9, CRISPR-Cas12 (cpf1), and CRISPR-Cas13. All these CRISPR-Cas systems require DNA breakage as double-stranded breaks (DSBs). These breaks are then repaired by homology-dependent repair (HDR; for insertions or replacement) or nonhomologous end joining (NHEJ; to knockout a gene) leading to changes at targeted sites. A detailed overview for the successful execution of CRISPR-Cas system has been depicted in Fig. 23.2.

In CRISPR-Cas9 systems, the specificity of the Cas9-directed DNA cleavage is determined by the single guide RNA (sgRNA) base pairing to the target sites. To cleave the target genomes, Cas9 requires the protospacer adjacent motif (PAM) which in most cases is 5′NGG3′. The CRISPR-Cas9 system initially executed in rice plants to verify the proof of concept by targeting three genes (ROC5, SPP, and YSA), the execution of CRISPR-Cas9 led to the production of albino phenotype, indicating that this system can be used successfully for the editing of rice genome (Feng et al. 2013). After this, CRISPR-Cas9 system was opted for multiplexing to engineer the rice genome at multiple positions of a gene or different loci. To achieve multiplex genome editing in rice, tandemly arrayed tRNA–gRNA architecture was built, and this system was referred to as polycistronic tRNA–gRNA (PTG) comprising multiple gRNAs separated by tRNAs. The mature gRNAs were cleaved by the endogenous RNase P and RNase Z and then recruited by Cas9 to edit multiple genome locations with 100% efficiency (Xie et al. 2015). This tRNA processing system has also been exploited in rice plants to achieve herbicide resistance. To achieve this, the homology-dependent repair (HDR) by donor repair template (DRT) approach was opted by designing a dual purpose chimeric sgRNA. This system can further be extended to improve other rice traits via targeted genome editing (Butt et al. 2017).

The Cas9 and Cas12 proteins have been engineered to substitute a single nucleotide at the specific target locus without inducing strand breaks, and such systems are referred to as base editors (BE). Such systems were engineered by fusing Cas9 nickase and a deaminase to facilitate the substitution of cytidine (C) to uridine (U), adenine (A) to guanine (G), and C to thymine (T) (Komor et al. 2016). Such BEs allow precise and more efficient base substitution than other techniques to achieve desirable mutations (Hess et al. 2017). BE system has been executed in rice plants to target the OsCDC48 gene, which regulates senescence and cell death, to explore the feasibility of BE systems in rice and the results revealed a successful nucleotide substitution with 43.48% mutation frequency and no off-targets or indels (Zong et al. 2017). Shimatani et al. (2017) engineered the rice gene, OsALS, to achieve herbicide resistance by substituting the C to T to change alanine to valine at amino acid 96 position (Shimatani et al. 2017).

Recently a new CRISPR-Cas-mediated RTSV resistance source was developed through the mutagenesis of eIF4G alleles in rice plants. The genome-edited transgenic rice plants after inoculation with RTSV led to a significant reduction in RTSV titer and plants showed improved height and grain yield compared to their wild-type counterparts under glasshouse conditions. In addition, the mutation generated via CRISPR-Cas in eIF4G gene are heritable with no off-targets (Macovei et al. 2018).

Three crRNAs (SA, SB, and SC) targeting the dsRNA genome of SRBSDV were inserted into three independent pCR12 vectors and stably transformed into rice plants. Viruliferous insect vectors of SRBSDV were fed on T1 transgenic lines, along with control wild-type rice plants. Typical symptoms were observed in control plants 40 days later, including major dwarfing and yellow-streaked leaves, however, transgenic plants showed mild symptoms and pCR12-SB lines had no obvious symptoms. All of these infections were associated with reduced SRBSDV titer and

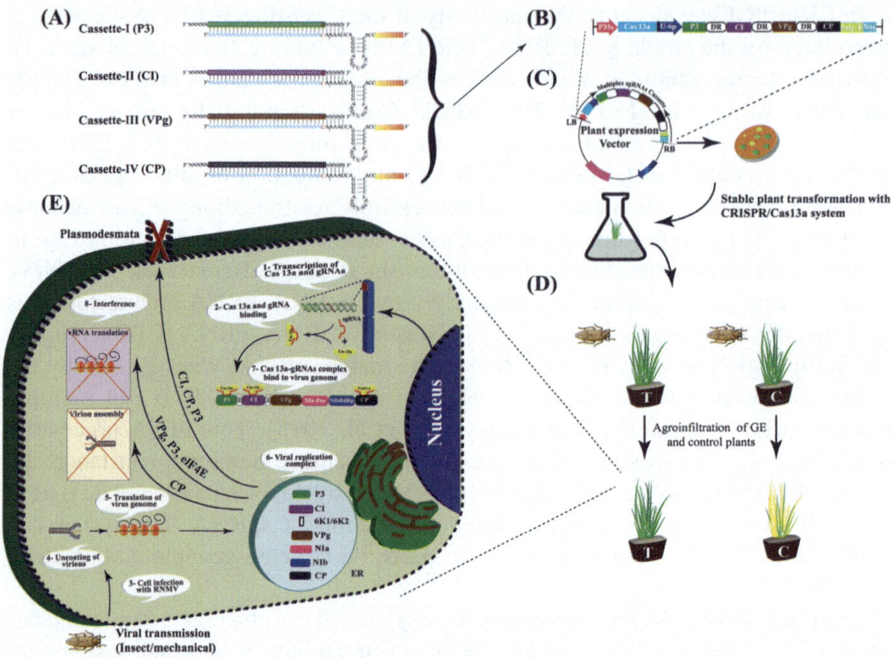

Fig. 23.3 An exemplary application of CRISPR-Cas9-based GE technique in Hassawi rice against rice necrosis mosaic virus (RNMV). (**a**) Individual sgRNA cassettes to target four genes (P3, CI, VPg, and CP) simultaneously. Each gene is represented in a specific color by a 26- to 28-nt long, direct repeat sequence. The poly-T tail sequences (represented in yellow background color) assist the homologous recombination. (**b**) Multiplexed sgRNA cassette of the individual sgRNAs expressed under a common promotor in a single binary vector. (**c**) Hassawi rice plants transformed via *Agrobacterium*-mediated plant transformation protocol with the recombinant bacterial plasmids carrying the multiplexed sgRNA cassette(s). (**d**) Aphids-mediated RNMV inoculation of the successfully genome-edited Hassawi rice plants. The figure was produced using CS5 version of Adobe Illustrator software

virus accumulated at low levels in the transgenic rice plants. In the same study, the CRISPR-Cas13 system was also executed to confer resistance against the ssRNA genome virus, RSMV. Transgenic rice plants were developed and challenged with RSMV, and these plants exhibited very mild symptoms associated with less viral RNA accumulation. To test the resistance inheritability, homozygous T3 lines of both rice plants were infected with SRBSDV or RSMV, respectively. All the tested lines showed stable resistance against these viruses (Zhang et al. 2019). A multiplexed CRIPSR/Cas 9-based technique to confer resistance in Hassawi rice (an indigenous rice cultivar to Saudi Arabia) against RNMV is depicted (Fig. 23.3).

23.5 Rice Plant Host-Virus Interactions

Rice-infecting viruses are obligatory parasites and necessarily require host cell for replications, gene expression, and development of a successful infection, respectively. To establish their infection successfully, rice-infecting viruses profoundly persuade the plant physiology and distinctly disturb several internal processes leading to disease development. The canonical roles of virus-encoded proteins beyond virus replication and virion assembly are crucial for infection progression. The consequences of an extensive array of interactions during their biological cycle lead phenotypic and developmental anomalies followed by virus-induced symptoms in plants. Symptom development is suggestive of a successful proliferation and pathogenesis. The most commonly occurring symptoms induced by different rice-infecting viruses include dwarfed plant growth, leaf chlorosis, and induction of plant sterility. The movement protein (MP) of rice strip virus (RSV), also called pc4, has a crucial role in intra- and intercellular virus movement besides foliar symptom development (Zhang et al. 2012). Moreover, RSV-encoded disease-specific protein (SP) also actively takes part to disrupt the structure and function of rice chloroplast. Nevertheless, the capsid protein (P2) induces the disruption of gibberellins (Gas) and thus promotes dwarfed phenotype of the rice plants infected with rice dwarf virus (RDV). It seems that the viral proteins prominently target the rice plant hormonal pathways for modulating plant development. A series of defense and counterdefense are depicted (Fig. 23.4).

The host-virus arm race is a multilayered interaction process which is recurrently evolving. The virus-encoded proteins not only interact with the rice plant hormonal pathways, but also disrupt the host microRNA (miRNA) profile. For example, the miR319 expression in the rice plants infected with rice ragged stunt virus (RRSV) disturb the jasmonic acid (JA) levels, which increase the rice plant susceptibility (Zhang et al. 2016). Several physiological, developmental, and stress-related genes are regulated by miRNAs, and thus, it is not surprising that rice-infecting viruses regulate and interact with endogenous miRNAs.

The virus-host plant interactions are based on several other cellular mechanisms in rice plants hijacked by the virus. These interactions involve either a single viral-encoded protein or complex interaction between several virus proteins and host factors (Zheng et al. 2017). In turn, the infected rice plants also deploy multiple defense strategies to disrupt viral replication and in planta movement. Several classical pathways can be listed such as RNA interference for RNA silencing, N-gene mediated resistance, phytohormonal response, dominant/recessive gene-mediated resistance, and cellular autophagy. During rice plant response to the invading viruses, different defense signaling molecules interfere with rice host plant-virus interactions (Zhang et al. 2019). The cellular RNA silencing mechanism (RNAi) is a crucial baseline for host plant resistance against invading plant viruses. Rice-infecting viruses also have several counterdefense strategies against cellular RNAi mechanism. The RNA-dependent RNA polymerase 6 (RDR6) produces viral dsRNA, which is processed into viral siRNAs. Probably that is the reason that RDV downregulates the expression of RDR6 in the susceptible rice plants. In rice,

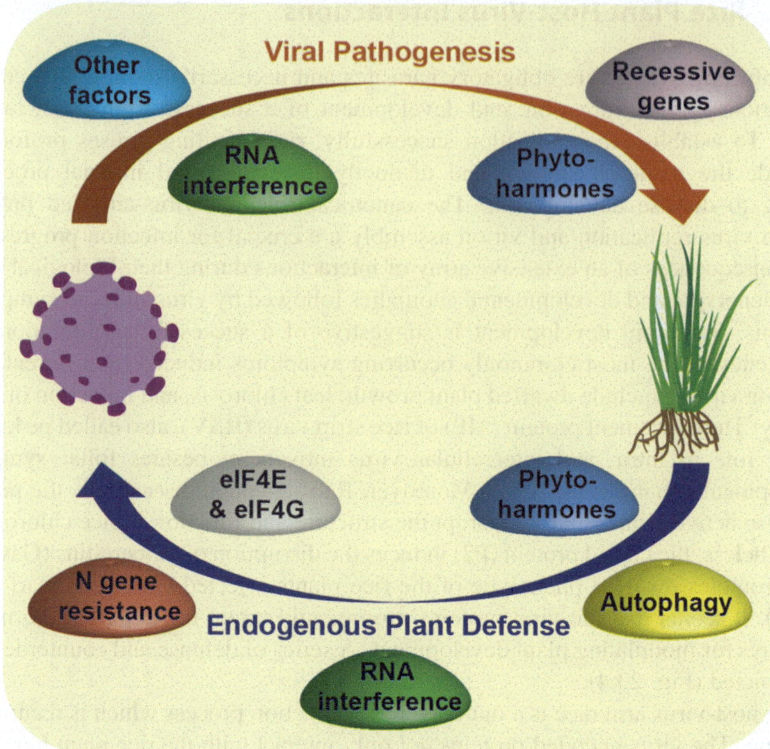

Fig. 23.4 Defense and counterdefense strategies between rice-infecting viruses and rice plants

S-acylation of group 1 remorin (REM1) inhibits the long-distance movement of viruses. It has been found that the NSvc4 protein of RSV blocks the S-acylation of REM1 and successfully translocates to the neighboring cells. However, the nonacylated REM1 can be accumulated in the endoplasmic reticulum periphery and induces autophagy to restrict the virus movement.

The rice antiviral defense pathway is a multilayered process, which comprised different types of RNAi, hormones, and resistance-related genes. These different pathways may be interconnected through a complex cross-talk during immune response that requires further investigations. Advanced insights into the host plant-virus interactions can expedite innovative crop protection strategies through high-throughput biotechnological, genetic, and breeding approaches.

23.6 Conclusions

Rice plays a core role in global food security and it is consumed by more than half the people around the globe. In Asia, where word "food" is substituted for rice and "rice" for agriculture, rice is not only a staple food, but a main source of livelihood.

The world population is increasing exponentially, so agricultural production is needed to be increased exponentially, nonetheless, certain biotic and abiotic constraints are limiting the agricultural production. Among biotic factors, RIVs and their insect vectors are posing a huge challenge for global food security. RIVs are causing significant losses to rice production in Asia. A laconic control against insect vectors is use of insecticides, but it is not cost-effective and can lead to resistance development and also pose an environment challenge.

Resistance breeding is another promising area to introgress the elite agronomic traits into the rice plants. Although wonderful achievements have been met, this technique is time-consuming; however, molecular breeding approaches have substituted the conventional breeding, but still failing to meet the pace. Modern molecular-based approaches like RNAi offers an alternative approach to develop RIV-resistant rice plants. The execution of RNAi is an effective, more promising approach for conferring resistance against RIVs. Several RNAi-based rice varieties have performed exceptionally well against RIVs and their insect vectors. In such rice plants, viral genes were targeted that have key roles in viral infection and proliferation. Beside wonderful utilities, RNAi-based crops have not been commercialized because of the transgenic nature. The contemporary CRISPR-Cas-based genome-editing technologies hold great potential for rice crop improvement. The unmatched advantage of such genome-editing technologies is to develop transgene free plants. In rice plants, different versions of CRISPR-Cas system, including Cas9, Cas12a, and Cas13, have been employed successfully, so can be opted to develop resistance against RIVs.

The diagnosis of RIV is a basic and prime important step to decipher the precise disease etiology. Choosing a most suitable diagnostic method is of prime importance to determine incidence, for screening, and to track disease transmission in the field. Conventional detection approaches only provide information on the existence of target pathogens. ELISA-based detection methods have been practiced since a long time and these are reliable, rapid, and accurate, but processing multiple samples is difficult in short period of time. RT-LAMP and RT-PCR techniques are simple, accessible, robust, and more accurate techniques to analyze the RIVs both qualitatively and quantitatively in less time. For the identification of closely related viruses, such as SRBSDV and RBSDV, and early detection of disease before symptoms, both techniques are matchless. In addition, for the screening of virus resistance and the study of viral population dynamics, viral replication and virus-host interactions are feasible, especially real-time RT-PCR. Nonetheless, the execution of such technique is not suitable as these are not cost-effective techniques and not suitable for large surveys.

During a course of successful infection, RIVs profoundly persuade the plant physiology and dramatically disrupt a variety of internal processes leading to disease growth. In addition, RIVs hijack the internal machinery of rice plant to execute their replication, translation, assembly, and dissemination. Canonical functions of virus-encoded proteins outside virus replication and virion assembly are essential to the progression of infection. Such diverse arrays of interaction lead to physiological abnormalities and contribute to phenotypic and developmental anomalies accompanied by virus-induced phenotype.

To sustain the rice productivity and meet the global food security, accurate detection of RIVs, better genome engineering techniques, and elucidating the molecular mechanisms underlying virus-host interactions are very important. The use of CRISPR-based technologies can precisely target the rice plant genome to confer resistance against RIV. Until now, the majority of genome-edited crops reported to date have not yet been commercialized due to biosafety and GMO regulations. However, the GE plants should be classified as non-GMO, because transgene-free GE plants can be achieved through the modern approaches, which would be comparable to the plants generated through traditional breeding.

References

Abo ME, Sy AA, Alegbejo MD (1997) Rice yellow mottle virus (RYMV) in Africa: evolution, distribution, economic significance on sustainable rice production and management strategies. J Sustain Agric 11(2–3):85–111

Abo ME, Alegbejo MD, Sy AA, Misari SM (2000) An overview of the mode of transmission, host plants and methods of detection of rice yellow mottle virus. J Sustain Agric 17(1):19–36

Adams MJ, Antoniw JF, Beaudoin F (2005) Overview and analysis of the polyprotein cleavage sites in the family Potyviridae. Mol Plant Path 6(4):471–487

Adego AK, Were HK, Muoma J, Mukoye B, Hébrard E, Pinel-Galzi A, Poulicard N, Fargette D (2017) Current status of rice yellow mottle disease in Western Kenya. Res Journali's J Agricult 4(9):1–12

Ahmed MMS, Bian S, Wang M, Zhao J, Zhang B, Liu Q, Zhang C, Tang S, Gu M, Yu H (2017) RNAi-mediated resistance to rice black-streaked dwarf virus in transgenic rice. Transgenic Res 26(2):197–207

Atkins J (1960) Occurrence of hoja blanca and its insect vector, *Sogata orizicola Muir*, on rice in Louisiana. Plant Dis Report 44(6):390–393

Awoderu VA (1991) The rice yellow mottle virus situation in West Africa. J Basic Microbiol 31(2): 91–99. https://doi.org/10.1002/jobm.3620310204

Azuhata F, Uyeda I, Kimura I, Shikata E (1993) Close similarity between genome structures of rice black-streaked dwarf and maize rough dwarf viruses. J Gen Virol 74(7):1227–1232

Azzam O, Chancellor TCB (2002) The biology, epidemiology, and management of rice tungro disease in Asia. Plant Dis 86(2):88–100

Bajet N, Aguiero V, Daquioag R, Jonson G, Cabunagan R, Mesina E, Hibino H (1986) Occurrence and spread of rice tungro spherical virus in the Philippines. Plant Dis 70(10):971–973

Bakker W (1970) Rice yellow mottle, a mechanically transmissible virus disease of rice in Kenya. Neth J Plant Pathol 76(2):53–63

Bakker W (1971) Three new beetle vectors of rice yellow mottle virus in Kenya. Neth J Plant Pathol 77(6):201–206

Bakker W (1974) Characterization and ecological aspects of rice yellow mottle virus in Kenya. Pudoc

Bao Y, Hull R (1992) Characterization of the discontinuities in rice tungro bacilliform virus DNA. J Gen Virol 73(5):1297–1301

Bhaya D, Davison M, Barrangou R (2011) CRISPR-Cas systems in bacteria and archaea: versatile small RNAs for adaptive defense and regulation. Annu Rev Genet 45:273–297

Boccardo G, Milne R (1984) Plant reovirus group. CMI/AAB descriptions of plant viruses no. 294. Commonwealth Microbiology Institute and Association of Applied Biology Kew

Bolaños C, Leiva AM, Saavedra J, Bruzzone C, Cruz M, Cuellar WJ (2017) Occurrence and molecular detection of rice hoja blanca virus (genus *Tenuivirus*) in Peru. Plant Dis 101(6): 1070–1070

Boonham N, Kreuze J, Winter S, van der Vlugt R, Bergervoet J, Tomlinson J, Mumford R (2014) Methods in virus diagnostics: from ELISA to next generation sequencing. Virus Res 186:20–31

Boubourakas IN, Fukuta S, Kyriakopoulou PE (2009) Sensitive and rapid detection of peach latent mosaic viroid by the reverse transcription loop-mediated isothermal amplification. J Virol Methods 160(1–2):63–68

Bragard C, Caciagli P, Lemaire O, Lopez-Moya JJ, MacFarlane S, Peters D et al (2013) Status and prospects of plant virus control through interference with vector transmission. Annu Rev Phytopathol 51:177–201

Breviario D, Genga A (2013) Stress response in rice. Rice Research: Open Access

Butt H, Eid A, Ali Z, Atia MAM, Mokhtar MM, Hassan N, Lee CM, Bao G, Mahfouz MM (2017) Efficient CRISPR/Cas9-mediated genome editing using a chimeric single-guide RNA molecule. Front Plant Sci 8:1441

Cabauatan P, Hibino H (1988) Isolation, purification, and serology of rice tungro bacilliform and rice tungro spherical viruses. Plant Dis 72(6):526–528

Cabauatan P, Kobayashi N, Ikeda R, Koganezawa H (1993a) *Oryza glaberrima*: an indicator plant for rice tungro spherical virus. Int J Pest Manag 39(3):273–276

Cabauatan P, Koganezawa H, Cabunagan R, Cruz F (1993b) Rice dwarf (RD), a new virus disease in the Philippines. Int Rice Res Notes 18(1)

Cabauatan PQ, Cabunagan RC, Choi I-R (2009) Rice viruses transmitted by the brown planthopper *Nilaparvata lugens* Stål. In: Heong KL, Hardy B (eds) Planthoppers: new threats to the sustainability of intensive rice production systems in Asia. International Rice Research Institute, Los Baños, pp 357–368

Cabunagan R, Choi I (2009) Why outbreaks of virus diseases occurred in south Vietnam? Ricehoppers

Chen C (1984) Effects of rice transitory yellowing virus on yield and yield components. Plant Prot Bull (Taiwan) 26(1):11–22

Chen Z, Song BA (2011) The prevention and control technology of southern rice 0black-streaked dwarf virus disease, 1st ed.; China Chemical Industry Press: Guiyang, pp. 3–5

Chen C, Chiu R, Wang E (1979a) Rice ragged stunt: a virus disease new to Taiwan. Plant Prot Bull (Taiwan) 21(4):447

Chen S, Yuan Y, King D, Chen K, Lin R, Kao T (1979b) Studies on the development and epiphytotics of the rice yellow stunt disease. Acta Phytopathol Sinica 9(1):41–71

Chen C, Ko W-H, Wang E-S, Yu S-M, Hu D (1980) Epidemiological studies on rice transitory yellowing with special reference to its transmission by the rice green leafhoppers. TDAIS Bull (Taiwan, ROC) 4:1–61

Chen Q, Chen H, Mao Q, Liu Q, Shimizu T, Uehara-Ichiki T, Wu Z, Xie L, Omura T, Wei T (2012) Tubular structure induced by a plant virus facilitates viral spread in its vector insect. PLoS Pathog 8(11):e1003032

Chen Q, Zhang L, Chen H, Xie L, Wei T (2015) Nonstructural protein Pns4 of rice dwarf virus is essential for viral infection in its insect vector. Virol J 12(1):211

Chen S, Li W, Huang X, Chen B, Zhang T, Zhou G (2019) Symptoms and yield loss caused by rice stripe mosaic virus. Virol J 16(1):145

Ching-Chung C (1982) Three symptomatologic types of rice virus diseases related to grassy stunt in Taiwan. Plant Dis 66(1):15

Chiu R-j, Jean J-h, Chen M-h, Lo T-c (1968) Transmission of transitory yellowing virus of rice by two leafhoppers. Phytopathology 58(6)

Cho SY, Jeong RD, Yoon YN, Lee SH, Shin DB, Kang HW, Lee BC (2013) One-step multiplex reverse transcription-polymerase chain reaction for the simultaneous detection of three rice viruses. J Virol Methods 193(2):674–678

Chomchan P, Miranda GJ, Shirako Y (2002) Detection of rice grassy stunt tenuivirus nonstructural proteins p2, p5 and p6 from infected rice plants and from viruliferous brown planthoppers. Arch Virol 147(12):2291–2300

Chomchan P, Li S-F, Shirako Y (2003) *Rice grassy stunt Tenuivirus* nonstructural protein p5 interacts with itself to form oligomeric complexes in vitro and in vivo. J Virol 77(1):769–775

Chon T, Hyun J, Park C (1975) A study on the population dynamics of overwintered small brown planthopper, *Laodelphax striatellus* (Fallen). Korean J Entomol 5:21–32

Cordero A, Newsom L (1962) Suitability of *Oryza* and other grasses as hosts of *Sogata orizicola Muir*. J Econ Entomol 55(6):868–871

Crawford GW, Shen C (1998) The origins of rice agriculture: recent progress in East Asia. Antiquity 72(278):858–866

Dick MW (2001) Straminipilous fungi: systematics of the peronosporomycetes including accounts of the marine straminipilous protists, the plasmodiophorids and similar organisms. Kluwer Academic

Du PV, Cabunagan RC, Choi IR (2005) Rice "yellowing syndrome" in Mekong river delta. Omonrice 13:135–138

Dyck V, Thomas B (1979) Brown planthopper: threat to rice production in Asia. International Rice Research Institute, Manila, pp 3–17

Esen A, Conroy JM, Wang SZ (1983) A simple and rapid dot-immunobinding assay for zein and other prolamins. Anal Biochem 132(2):462–467

Faan H, Pui W (1980) A preliminary investigation on the affecting factors and disease management of rice yellow stunt disease in Kwangtung. J South Chin Agricult Coll 1(3):1–15

Fabre F, Rousseau E, Mailleret L, Moury B (2012) Durable strategies to deploy plant resistance in agricultural landscapes. New Phytol 193:1064–1075

Falk BW, Tsai JH (1998) Biology and molecular biology of viruses in the genus *Tenuivirus*. Annu Rev Phytopathol 36(1):139–163

Fang Y, Choi JY, Park DH, Park MG, Kim JY, Wang M, Kim HJ, Kim WJ, Je YH (2020) Suppression of rice stripe virus replication in *Laodelphax striatellus* using vector insect-derived double-stranded RNAs. Plant Pathol J 36(3):280–288

Fauquet C, Thouvenel J-C (1977) Isolation of the rice yellow mottle virus in Ivory Coast. Plant Dis Report 61(6):443–446

Fauquet C, Thouvenel J-C, Fargette D, Fishpool L (1988) Rice stripe necrosis virus: a soil-borne rod-shaped virus. In: Developments in applied biology II viruses with fungal vectors, pp 71–82

Feng Z, Zhang B, Ding W, Liu X, Yang DL, Wei P, Cao F, Zhu S, Zhang F, Mao Y (2013) Efficient genome editing in plants using a CRISPR/Cas system. Cell Res 23(10):1229–1232

Fereres A, Raccah B (2015) Plant virus transmission by insects. eLS:1–12

Firth AE, Atkins JF (2009) Analysis of the coding potential of the partially overlapping 3′ ORF in segment 5 of the plant fijiviruses. Virol J 6(1):32

Fujii S (1978) Studies on rice necrosis mosaic. Special Bulletin, Okayama Prefectural Agric Exp Stn, 81

Fukuta S, Iida T, Mizukami Y, Ishida A, Ueda J, Kanbe M, Ishimoto Y (2003) Detection of Japanese yam mosaic virus by RT-LAMP. Arch Virol 148(9):1713–1720

Fütterer J, Rothnie HM, Hohn T, Potrykus I (1997) Rice tungro bacilliform virus open reading frames II and III are translated from polycistronic pregenomic RNA by leaky scanning. J Virol 71(10):7984–7989

Galvez GE (1968) Transmission studies of the hoja blanca virus with highly active, virus-free colonies of *Sogatodes oryzicola*. Phytopathology 58(6):818–821

Ghosh SK (1980) Rice necrosis mosaic. Proceed Plant Sci 89(4):291–299

Gnanamanickam SS (2009) Rice and its importance to human life. In: Gnanamanickam SS (ed) Biological control of rice diseases. Springer Netherlands, Dordrecht, pp 1–11

Gordon KHJ, Waterhouse PM (2007) RNAi for insect-proof plants. Nat Biotechnol 25(11): 1231–1232

Hamamatsu C, Toriyama S, Toyoda T, Ishihama A (1993) Ambisense coding strategy of the rice stripe virus genome: in vitro translation studies. J Gen Virol 74(6):1125–1131
Hasanuddin A, Hibino H (1989) Grain yield reduction, growth retardation and virus concentration in rice plants infected with tungro-associated viruses. Trop Agric Res 22:56–73
Herdt RW (1991) Research priorities for rice biotechnology. Rice Biotechnol 6:19–54
Herzog E, Guerra-Peraza O, Hohn T (2000) The rice tungro bacilliform virus gene II product interacts with the coat protein domain of the viral gene III polyprotein. J Virol 74(5):2073–2083
Hess GT, Tycko J, Yao D, Bassik MC (2017) Methods and applications of CRISPR-mediated base editing in eukaryotic genomes. Mol Cell 68(1):26–43
Hibino H (1979) Rice ragged stunt, a new virus disease occurring in tropical Asia. Rev Plant Prot Res 12:98–110
Hibino H (1990) Insect-borne viruses of rice. In: Advances in disease vector research. Springer, pp 209–241
Hibino H (1996) Biology and epidemiology of rice viruses. Annu Rev Phytopathol 34(1):249–274
Hibino H, Roechan M, Sudarisman S (1978) Association of two types of virus particles with penyakit habang (tungro disease) of rice in Indonesia. Phytopathology 68(10):1412–1416
Hibino H, Cabauatan P, Omura T, Tsuchizaki T (1985) Rice grassy stunt virus strain causing tungro-like symptoms in the Philippines. Plant Dis 69(6):538–541
Hibino H, Daquioag RD, Cabauatan PQ, Dahal G (1988) Resistance to rice tungro spherical virus in rice. Plant Dis 72:843–847
Hiraguri A, Hibino H, Hayashi T, Shimizu T, Uehara-Ichiki T, Omura T, Sasaya T (2010) Complete sequence analysis of rice transitory yellowing virus and its comparison to rice yellow stunt virus. Arch Virol 155(2):243–245
Hiraguri A, Hibino H, Hayashi T, Netsu O, Shimizu T, Uehara-Ichiki T, Omura T, Sasaki N, Nyunoya H, Sasaya T (2012) The movement protein encoded by gene 3 of rice transitory yellowing virus is associated with virus particles. J Gen Virol 93(10):2290–2298
Hoang AT, Zhang H-m, Yang J, Chen J-p, Hébrard E, Zhou G-h, Vinh VN, Cheng J-a (2011) Identification, characterization, and distribution of southern rice black-streaked dwarf virus in Vietnam. Plant Dis 95(9):1063–1069
Hogenhout SA, Ammar E-D, Whitfield AE, Redinbaugh MG (2008) Insect vector interactions with persistently transmitted viruses. Annu Rev Phytopathol 46(1):327–359
Honda K, Wei T, Hagiwara K, Higashi T, Kimura I, Akutsu K, Omura T (2007) Retention of rice dwarf virus by descendants of pairs of viruliferous vector insects after rearing for 6 years. Phytopathology 97(6):712–716
Huang Y, Zhao H, Luo Z, Chen X, Fang R-X (2003) Novel structure of the genome of rice yellow stunt virus: identification of the gene 6-encoded virion protein. J Gen Virol 84(8):2259–2264
Hui XL (1986) Research on rice virus diseases in China. In: International symposium on virus diseases of rice and leguminous crops in the tropics: proceedings of a symposium on tropical agriculture research, Tsukuba, October 1–5, 1985. Tropical Agriculture Research Center, Ministry of Agriculture, Forestry, p 45
Hull R (2008) Rice tungro disease. In: Mahy BWJ, Van Regenmortel MHV (eds) Encyclopedia of virology, 3rd edn. Academic Press, Oxford, pp 481–485
Ichiki TU, Shiba T, Matsukura K, Ueno T, Hirae M, Sasaya T (2013) Detection and diagnosis of rice-infecting viruses. Front Microbiol 4:289
Inoue H, Omura T, Hayashi T, Morinaka T, Putta M, Chettanachit D, Parejarearn A, Disthaaporn S, Kadokao S (1986) Rice transitory yellowing virus. In: Virus diseases of rice and legumes in the tropics, Technical Bulletin of the Tropical Agriculture Research Center of Japan, vol 21, pp 65–77
Inouye T, Fujii S (1977) Rice necrosis mosaic virus. CMI/AAB Descriptions of plant viruses, No. 172. Commonwealth Mycological Institute, Kew; Association of Applied Biologists, Warwick
Ishii M, Yoshimura S (1973) Epidemiological studies on the rice black-streaked dwarf virus in Kanto-Tosan district, Japan. J Cent Agricult Exp Stat 17:61–121

Jennings P, Pineda TA (1971) The effect of the hoja blanca virus on its insect vector. Phytopathology 61(2):142–143

Jia D, Chen H, Zheng A, Chen Q, Liu Q, Xie L, Wu Z, Wei T (2012) Development of an insect vector cell culture and RNA interference system to investigate the functional role of fijivirus replication protein. J Virol 86(10):5800–5807

Jia D, Mao Q, Chen H, Wang A, Liu Y, Wang H, Xie L, Wei T (2014) Virus-induced tubule: a vehicle for rapid spread of virions through basal lamina from midgut epithelium in the insect vector. J Virol 88(18):10488–10500

Jimenez J, Carvajal-Yepes M, Leiva AM, Cruz M, Romero LE, Bolaños CA, Lozano I, Cuellar WJ (2018) Complete genome sequence of rice hoja blanca tenuivirus isolated from a susceptible rice cultivar in Colombia. Genome Announc 6(7):e01490–e01417

Jinek M, Chylinski K, Fonfara I, Hauer M, Doudna JA, Charpentier E (2012) A programmable dual-RNA–guided DNA endonuclease in adaptive bacterial immunity. Science 337:816–821

John V, Thottapilly G, Awoderu V (1984) Occurrence of rice yellow mottle virus in some Sahelian countries in West Africa. FAO Plant Protect Bull 32(3):86–87

Jones M, Gough IK, Dasgupta I, Rao BS, Cliffe J, Qu R, Shen P, Kaniewska M, Blakebrough M, Davies J (1991) Rice tungro disease is caused by an RNA and a DNA virus. J Gen Virol 72(4): 757–761

Jonson G, Koganezawa H, Mew T, Bajet N, Calilung V (1990) Variants of rice grassy stunt disease in the Philippines. In: 21. Pest control council of the philippines annual convention. Bacolod City (Philippines). 7–10 May 1990

Kiritani K (1983) Changes in cropping practices and the incidence of hopper-borne diseases of rice in Japan. In: Changes in cropping practices and the incidence of hopper-borne diseases of rice in Japan, pp 239–247

Kisimoto R (1979) Brown planthopper migration. In: Brown planthopper: threat to rice production in Asia, pp 113–124

Komor AC, Kim YB, Packer MS, Zuris JA, Liu DR (2016) Programmable editing of a target base in genomic DNA without double-stranded DNA cleavage. Nature 533(7603):420–424

Koonin EV, Makarova KS, Zhang F (2017) Diversity, classification and evolution of CRISPR-Cas systems. Curr Opin Microbiol 37:67–78

Kouassi N, N'guessan P, Albar L, Fauquet C, Brugidou C (2005) Distribution and characterization of rice yellow mottle virus: a threat to African farmers. Plant Dis 89(2):124–133

Krappmann S (2017) CRISPR-Cas9, the new kid on the block of fungal molecular biology. Med Mycol 55(1):16–23

Kruseman AN (1983) Application of ELISA for assessment of antiserum immunoreactivity in endocrine immunocytochemical studies. J Clin Pathol 36(4):406–410

Lacombe S, Bangratz M, Vignols F, Brugidou C (2010) The rice yellow mottle virus P1 protein exhibits dual functions to suppress and activate gene silencing. Plant J 61(3):371–382

Le DT, Netsu O, Uehara-Ichiki T, Shimizu T, Choi IR, Omura T, Sasaya T (2010) Molecular detection of nine rice viruses by a reverse-transcription loop-mediated isothermal amplification assay. J Virol Methods 170(1–2):90–93

Le DT, Chu HD, Sasaya T (2015) Creation of transgenic rice plants producing small interfering RNA of rice tungro spherical virus. GM Crops Food 6(1):47–53

Lee JY, Lee SH, Chung BJ (1977) Studies on the occurrence of rice black-streaked dwarf virus in Korea. Korean J Appl Entomol 16(2):121–125

Lee B-C, Hong Y-K, Kwak D-Y, Oh B-G, Park S-T, Kim S-C (2004) Detection of rice stripe virus using RT-PCR. Res Plant Dis 10(1):30–33

Lee B-C, Young-Nam Y, Sung-Jun H, Yeun-Gyu H, Yeon KD, Jong-Hee L, Won KH, Hwang H-G (2008) Analysis on the occurrence of rice stripe virus (Analysis on the occurrence of rice stripe virus). Res Plant Dis 14(3):210–213

Li D, Wang G, Cheng F (1979) Epidemiological study on rice virus diseases and their control in Zhejiang Province. Acta Phytopathol Sinica 9(2):73–82

Li Z, Upadhyaya NM, Kositratana W, Gibbs AJ, Waterhouse PM (1996) Genome segment 5 of rice ragged stunt virus encodes a virion protein. J Gen Virol 77(12):3155–3160
Li J, Xue J, Zhang H-M, Yang J, Lv M-F, Xie L, Meng Y, Li P-P, Chen J-P (2013) Interactions between the P6 and P5-1 proteins of southern rice black-streaked dwarf fijivirus in yeast and plant cells. Arch Virol 158(8):1649–1659
Li J, Xue J, Zhang H-M, Yang J, Xie L, Chen J-P (2015a) Characterization of homologous and heterologous interactions between viroplasm proteins P6 and P9-1 of the fijivirus southern rice black-streaked dwarf virus. Arch Virol 160(2):453–457
Li S, Wang X, Xu J, Ji Y, Zhou Y (2015b) A simplified method for simultaneous detection of rice stripe virus and rice black-streaked dwarf virus in insect vector. J Virol Methods 211:32–35
Li L, Guo C, Wang B, Zhou T, Lei Y, Dai Y-h, He W, Liang C, Wang X-f (2016) RNAi-mediated transgenic rice resistance to rice stripe virus. J Integr Agric 15(11):2539–2549
Lijun C, Xizhi M, Lin K, Kejing D, Shouyuan Z, Changben L (2003) Detecting rice stripe virus (RSV) in the small brown planthopper (*Laodelphax striatellus*) with high specificity by RT-PCR. J Virol Methods 112(1):115–120
Lin Q, Lechen T, Lianhui X (1989) The epidemic forecasting and path analysis on rice transitory yellowing. J Fujian Agricult Coll 17:38–41
Lin Q, Xie L, Zhou Z, Xie L, Wu Z (1990) Studies on rice stripe: I. distribution of and losses caused by the disease. J Fujian Agricult Coll 19(4):421–425
Ling K (1974) Capacity of *Nephotettix virescens* to infect rice seedlings with tungro. Philipp Phytopathol 10:42–49
Liu H, Wei C, Zhong Y, Li Y (2007) Rice black-streaked dwarf virus outer capsid protein P10 has self-interactions and forms oligomeric complexes in solution. Virus Res 127(1):34–42
Liu W, Gray S, Huo Y, Li L, Wei T, Wang X (2015) Proteomic analysis of interaction between a plant virus and its vector insect reveals new functions of hemipteran cuticular protein. Mol Cell Proteomics 14(8):2229–2242
Louvel D, Bidaux J (1977) Observations of new pathological symptoms on early varieties of rice in Côte d'Ivoire. L'Agronomie Tropicale 32(3):257–261
Lozano I, Morales F (2009) Molecular characterisation of rice stripe necrosis virus as a new species of the genus *Benyvirus*. Eur J Plant Pathol 124(4):673–680
Macovei A, Sevilla NR, Cantos C, Jonson GB, Slamet-Loedin I, Čermák T, Voytas DF, Choi I-R, Chadha-Mohanty P (2018) Novel alleles of rice eIF4G generated by CRISPR/Cas9-targeted mutagenesis confer resistance to rice tungro spherical virus. Plant Biotechnol J 16(11): 1918–1927
Malathi P, Muzammil SA, Krishnaveni D, Balachandran SM, Mangrauthia SK (2019) Coat protein 3 of rice tungro spherical virus is the key target gene for development of RNAi mediated tungro disease resistance in rice. Agri Gene 12:100084
Manoharan S, Parthiban M, Prabhakar TG, Ravikuma G, Koteeswaran A, Chandran NDJ, Rajavelu G (2004) Rapid serological profiling by an immunocomb-based dot-enzyme-linked immunosorbent test for three major poultry diseases. Vet Res Commun 28(4):339–346
Mao Q, Zheng S, Han Q, Chen H, Ma Y, Jia D, Chen Q, Wei T (2013) New model for the genesis and maturation of viroplasms induced by fijiviruses in insect vector cells. J Virol 87(12): 6819–6828
Mariappan V, Hibino H, Shanmugam N (1984) A new rice virus disease in India. Int Rice Res Newsl 9(6):9–10
Marmey P, Rojas-Mendoza A, de Kochko A, Beachy RN, Fauquet CM (2005) Characterization of the protease domain of rice tungro bacilliform virus responsible for the processing of the capsid protein from the polyprotein. Virol J 2(1):33
Matsukura K, Towata T, Sakai J, Onuki M, Okuda M, Matsumura M (2013) Dynamics of Southern rice black-streaked dwarf virus in rice and implication for virus acquisition. Phytopathology 103(5):509–512

Matsumoto Y, Hattori M (2016) Gene silencing by parental RNA interference in the green rice leafhopper, *Nephotettix cincticeps* (hemiptera: *cicadellidae*). Arch Insect Biochem Physiol 91(3):152–164

Matsumura M, Sakai J (2011) Occurrence of new disease caused by southern rice black-streaked dwarf virus transmitted by the whitebacked planthopper. Plant Prot 65:46–48

Matyjaszek-Matuszek B, Pyzik A, Nowakowski A, Jarosz MJ (2013) Diagnostic methods of TSH in thyroid screening tests. Ann Agric Environ Med 20(4):731–735

Mcmillian WW, Mcguire JU, Lamey HA (1962) Hoja blanca transmission studies on rice. J Econ Entomol 55(5):796–797

Morales FJ, Ward E, Castaño M, Arroyave JA, Lozano I, Adams MJ (1999) Emergence and partial characterization of rice stripe necrosis virus and its fungus vector in South America. Eur J Plant Pathol 105(7):643–650

Murphy FA, Fauquet CM, Bishop DH, Ghabrial SA, Jarvis AW, Martelli GP, Mayo MA, Summers MD (1995) The viruses. In: Virus taxonomy. Springer, pp 15–507

Nakagawa A, Miyazaki N, Taka J, Naitow H, Ogawa A, Fujimoto Z, Mizuno H, Higashi T, Watanabe Y, Omura T, Cheng RH, Tsukihara T (2003) The atomic structure of *rice dwarf virus* reveals the self-assembly mechanism of component proteins. Structure 11(10):1227–1238

Nakata A, Amemura M, Makino K (1989) Unusual nucleotide arrangement with repeated sequences in the Escherichia coli K-12 chromosome. J Bacteriol 171:3553–3556

Nakasuji F (1974) Epidemiological study on rice dwarf virus transmitted by the green rice leafhopper *Nephotettix cincticeps*. JARQ 8:84–91

Nance NW, United States Agricultural Research Service, Crops Research (1958) Index to Plant disease reporter supplements 243–248, vol 249. Crops Research Division, Agricultural Research Service, United States Department of Agriculture, Beltsville, MD

Normile D (2008) Reinventing rice to feed the world. Science 321(5887):330–333

Notomi T, Okayama H, Masubuchi H, Yonekawa T, Watanabe K, Amino N, Hase T (2000) Loop-mediated isothermal amplification of DNA. Nucleic Acids Res 28(12):e63–e63

Omura T, Yan J (1999) Role of outer capsid proteins in transmission of Phytoreovirus by insect vectors. In: Maramorosch K, Murphy FA, Shatkin AJ (eds) Advances in virus research, vol 54. Academic Press, pp 15–43

Opalka N, Brugidou C, Bonneau C, Nicole M, Beachy RN, Yeager M, Fauquet C (1998) Movement of rice yellow mottle virus between xylem cells through pit membranes. Proc Natl Acad Sci 95(6):3323–3328

Ou SH (1985) Rice diseases. 2nd edition, Commonwealth Agricultural Bureau, Wallingford., p 380

Palmer L, Rao P (1981) Grassy stunt, ragged stunt and tungro diseases of rice in Indonesia. Int J Pest Manage 27(2):212–217

Park R, Jung Y, Kim D, Lee D (1982) Studies on the occurrence of rice virus diseases in Yongnam area. Res Rep ORD 24:98–106

Paz CL, Espinoza MA, Amano KY (2009) El virus del "entorchamiento" del arroz en Ecuador

Periasamy M, Niazi FR, Malathi VG (2006) Multiplex RT-PCR, a novel technique for the simultaneous detection of the DNA and RNA viruses causing rice tungro disease. J Virol Methods 134(1–2):230–236

Pinto YM, Kok RA, Baulcombe DC (1999) Resistance to rice yellow mottle virus (RYMV) in cultivated African rice varieties containing RYMV transgenes. Nat Biotechnol 17(7):702–707

Pumplin N, Voinnet O (2013) RNA silencing suppression by plant pathogens: defence, counter-defence and counter-counter-defence. Nat Rev Micro 11(11):745–760

Qu R, Bhattacharyya M, Laco GS, De Kochko A, Subba Rao BL, Kaniewska MB, Scott Elmer J, Rochester DE, Smith CE, Beachy RN (1991) Characterization of the genome of rice tungro bacilliform virus: comparison with Commelina yellow mottle virus and caulimoviruses. Virology 185(1):354–364

Ramasamy C, Jatileksono T (1996) Intercountry comparison of insect and disease losses. In: Evenson RE, Herdt RW, Hossain M (eds) Rice research in Asia: progress and priorities. CAB International and IRRI, Wallingford

Ramirez BC (2008) Tenuivirus. In: Encyclopedia of virology, 3rd edn. CNRS, Paris
Ramírez B-C, Haenni A-L (1994) Molecular biology of tenuiviruses, a remarkable group of plant viruses. J Gen Virol 75(3):467–475
Ramirez B-C, Macaya G, Calvert LA, Haenni A-L (1992) Rice hoja blanca virus genome characterization and expression in vitro. J Gen Virol 73(6):1457–1464
Raymundo S, Buddenhagen I (1976) A rice virus disease in West Africa. Int Rice Commiss Newsl 25(1/2)
Raymundo S, Konteh I (1980) Distribution, importance, screening methods and varietal reaction to rice pale yellow mottle disease. Int Rice Commiss Newsl 29(1):51–53
Reckhaus P, Randrianangaly S (1990) Rice yellow mottle virus (RYMV) on rice in Madagascar. International Rice Research Newsletter 15(1):30
Rivera C, Ou S (1965) Leafhopper transmission of "tungro" disease of rice. Plant Dis Report 49(2)
Saito Y, Roechan M (1975) Small bacilliform particles associated with Penyakit Habang (Tungro-like). Phytopathology 65:793–796
Sakamoto S, Putalun W, Vimolmangkang S, Phoolcharoen W, Shoyama Y, Tanaka H, Morimoto S (2018) Enzyme-linked immunosorbent assay for the quantitative/qualitative analysis of plant secondary metabolites. J Nat Med 72(1):32–42
Sarra S, Oevering P, Guindo S, Peters D (2004) Wind-mediated spread of rice yellow mottle virus (RYMV) in irrigated rice crops. Plant Pathol 53(2):148–153
Sasaya T, Nakazono-Nagaoka E, Saika H, Aoki H, Hiraguri A, Netsu O, Ichiki T, Onuki M, Toki S, Saito K, Yatou O (2014) Transgenic strategies to confer resistance against viruses in rice plants. Front Microbiol 4:409
Satoh K, Yoneyama K, Kondoh H, Shimizu T, Sasaya T, Choi IR, Yoneyama K, Omura T, Kikuchi S (2013) Relationship between gene responses and symptoms induced by rice grassy stunt virus. Front Microbiol 4:313
Schwarz DS, Hutvágner G, Haley B, Zamore PD (2002) Evidence that siRNAs function as guides, not primers, in the *Drosophila* and human RNAi pathways. Mol Cell 10(3):537–548
Sharma S, Dasgupta I (2012) Development of SYBR Green I based real-time PCR assays for quantitative detection of rice tungro bacilliform virus and rice tungro spherical virus. J Virol Methods 181(1):86–92
Shen P, Kaniewska M, Smith C, Beachy RN (1993) Nucleotide sequence and genomic organization of rice tungro spherical virus. Virology 193(2):621–630
Shikata E (1972) Rice transitory yellowing virus. CMI/AAB description of plant viruses. No 100
Shikata E (1974) Rice black-streaked dwarf virus. CMI/AAB description of plant viruses. No 135
Shikata E, Senboku T, Ishimizu T (1980) The causal agent of rice grassy stunt disease. Proc Jpn Acad Ser B 56(2):89–94
Shimatani Z, Kashojiya S, Takayama M, Terada R, Arazoe T, Ishii H, Teramura H, Yamamoto T, Komatsu H, Miura K, Ezura H, Nishida K, Ariizumi T, Kondo A (2017) Targeted base editing in rice and tomato using a CRISPR-Cas9 cytidine deaminase fusion. Nat Biotechnol 35(5): 441–443
Shimizu T, Yoshii M, Wei T, Hirochika H, Omura T (2009) Silencing by RNAi of the gene for Pns12, a viroplasm matrix protein of rice dwarf virus, results in strong resistance of transgenic rice plants to the virus. Plant Biotechnol J 7(1):24–32
Shinkai A (1977) Rice waika, a new virus disease, and problems related to its occurrence and control. Jpn Agric Res Q 11:151–155
Shinkai A, Nakano M, Iwasai M (1980) Occurrence of rice ragged stunt disease in Kyushu, Japan. Ann Phytopathol Soc Jpn 46:411
Spetz C, Valkonen JPT (2004) Potyviral 6K2 Protein long-distance movement and symptom-induction functions are independent and host-specific. Mol Plant-Microbe Interact 17(5): 502–510
Su H (1969) Transitory yellowing of rice in Taiwan. In: Symposium on virus diseases rice plant proceeding

Su H-J, Huang J (1965) Intracellular inclusion bodies in the rice plants affected with transitory yellowing. Botan Bull Acad Sinica 6:170–181

Sun F, Yuan X, Zhou T, Fan Y, Zhou Y (2011) *Arabidopsis* is susceptible to rice stripe virus infections. J Phytopathol 159(11–12):767–772

Sun Z, Zhang S, Xie L, Zhu Q, Tan Z, Bian J, Sun L, Chen J (2013) The secretory pathway and the actomyosin motility system are required for plasmodesmatal localization of the P7-1 of rice black-streaked dwarf virus. Arch Virol 158(5):1055–1064

Szittya G, Silhavy D, Molnar A, Havelda Z, Lovas A, Lakatos L, Banfalvi Z, Burgyan J (2003) Low temperature inhibits RNA silencing-mediated defence by the control of siRNA generation. EMBO J 22(3):633–640

Takahashi Y, Omura T, Shohara K, Tsuchizaki T (1991) Comparison of four serological methods for practical detection of ten viruses of rice in plants and insects. Plant Dis 75(5):458–461

Takahashi M, Toriyama S, Hamamatsu C, Ishihama A (1993) Nucleotide sequence and possible ambisense coding strategy of rice stripe virus RNA segment 2. J Gen Virol 74(4):769–773

Tantera DM, Satomi H (1973) Grassy stunt disease of rice in Indonesia. Grassy stunt disease of rice in Indonesia, No 2, pp 8

Teng PS, Hibino H, Leung H (1988) Yield loss due to rice virus diseases in Asian tropics. In: Crop losses due to disease outbreaks in the tropics and countermeasures. Japan Int Research Center for Agricultural Sciences, pp 93–100

Thole V, Hull R (1996) Rice tungro spherical virus: nucleotide sequence of the 3′ genomic half and studies on the two small 3′ open reading frames. Virus Gen 13(3):239–246

Toriyama S, Kimishima T, Takahashi M, Shimizu T, Minaka N, Akutsu K (1998) The complete nucleotide sequence of the rice grassy stunt virus genome and genomic comparisons with viruses of the genus *Tenuivirus*. J Gen Virol 79(8):2051–2058

Traoré O, Pinel A, Fargette D, Konaté G (2001) First report and characterization of rice yellow mottle virus in Central Africa. Plant Dis 85(8):920–920

Tyagi H, Rajasubramaniam S, Rajam M, Dasgupta I (2008) RNA-interference in rice against rice tungro bacilliform virus results in its decreased accumulation in inoculated rice plants. Transgenic Res 17(5):897–904

Upadhyaya NM, Yang M, Kositratana W, Ghosh A, Waterhouse PM (1995) Molecular analysis of rice ragged stunt oryzavirus segment 9 and sequence conservation among isolates from Thailand and India. Arch Virol 140(11):1945–1956

Uyeda I, Kimura I, Shikata E (1995) Characterization of genome structure and establishment of vector cell lines for plant reoviruses. In: Maramorosch K, Murphy FA, Shatkin AJ (eds) Advances in virus research, vol 45. Academic Press, pp 249–279

Van Du P, Cabunagan R, Cabauatan P, Choi H, Choi I, Ho VC, Nguyen HH (2007) Yellowing syndrome of rice: etiology, current status and future challenges. Omonrice 15:94–101

Varga A, James D (2006) Use of reverse transcription loop-mediated isothermal amplification for the detection of Plum pox virus. J Virol Methods 138(1–2):184–190

Wagh SG, Kobayashi K, Yaeno T, Yamaoka N, Masuta C, Nishiguchi M (2016) Rice necrosis mosaic virus, a fungal transmitted Bymovirus: complete nucleotide sequence of the genomic RNAs and subgrouping of bymoviruses. J Gen Plant Pathol 82(1):38–42

Wang G, Zhou Y, Chen Z, Zhou X (2004) Production of monoclonal antibodies to rice stripe virus and application in virus detection. Acta Phytopathol Sinica 34(4):302–306

Wang H-D, Chen J-P, Zhang H-M, Sun X-L, Zhu J-L, Wang A-G, Sheng W-X, Adams MJ (2008) Recent rice stripe virus Epidemics in Zhejiang Province, China, and experiments on sowing date, disease–yield loss relationships, and seedling susceptibility. Plant Dis 92(8):1190–1196

Wang Q, Yang J, Zhou G-H, Zhang H-M, Chen J-P, Adams MJ (2010) The complete genome sequence of two isolates of Southern rice black-streaked dwarf virus, a new member of the genus *Fijivirus*. J Phytopathol 158(11–12):733–737

Wang Z, Yu D, Li X, Zeng M, Chen Z, Bi L, Liu J, Jin L, Hu D, Yang S, Song B (2012) The development and application of a Dot-ELISA assay for diagnosis of southern rice black-streaked dwarf disease in the field. Viruses 4(1):167–183

Waterhouse PM, Wang M-B, Lough T (2001) Gene silencing as an adaptive defence against viruses. Nature 411:834–842

Wei T-Y, Yang J-G, Liao F-L, Gao F-L, Lu L-M, Zhang X-T, Li F, Wu Z-J, Lin Q-Y, Xie L-H, Lin H-X (2009) Genetic diversity and population structure of rice stripe virus in China. J Gen Virol 90(4):1025–1034

Wu S-J, Zhong H, Zhou Y, Zuo H, Zhou L-H, Zhu J-Y, Ji C-Q, Gu S-L, Gu M-H, Liang G-H (2009) Identification of QTLs for the resistance to rice stripe virus in the indica rice variety Dular. Euphytica 165(3):557–565

Xie L, Lin Q (1980) Studies on the epidemic forecasting of rice transitory yellowing and rice dwarf diseases. J Fujian Agric Univ 9:32–43

Xie L, Lv M-F, Zhang H-M, Yang J, Li J-M, Chen J-P (2014) Tumours induced by a plant virus are derived from vascular tissue and have multiple intercellular gateways that facilitate virus movement. J Exp Bot 65(17):4873–4886

Xie K, Minkenberg B, Yang Y (2015) Boosting CRISPR/Cas9 multiplex editing capability with the endogenous tRNA-processing system. Proc Natl Acad Sci U S A 112(11):3570–3575

Xu H, Li Y, Mao Z, Li Y, Wu Z, Qu L, An C, Ming X, Schiemann J, Casper R, Chen Z (1998) Rice dwarf phytoreovirus segment S11 encodes a nucleic acid binding protein. Virology 240(2): 267–272

Xu Q, Ni H, Zhang J, Lan Y, Ren C, Zhou Y (2015) Whole-genome expression analysis of rice black-streaked dwarf virus in different plant hosts and small brown planthopper. Gene 572(2): 169–174

Xu Q, Liu H, Yuan P, Zhang X, Chen Q, Jiang X, Zhou Y (2017) Development of a simplified RT-PCR without RNA isolation for rapid detection of RNA viruses in a single small brown planthopper (*Laodelphax striatellus* Fallén). Virol J 14(1):90

Xue J, Li J, Ta H-A, Zhang H-M, Yang J, Lv M-F, Meng Y, Li P-P, Chen J-P (2014) Complete genomic sequence of southern rice blacked-dwarf virus, a novel fijivirus, from Vietnam. Genome Announc 1(3):e00212–e00213

Yan J, Kudo H, Uyeda I, Lee S-Y, Shikata E (1992) Conserved terminal sequences of rice ragged stunt virus genomic RNA. J Gen Virol 73(4):785–789

Yang X, Huang J, Liu C, Chen B, Zhang T, Zhou G (2017) Rice stripe mosaic virus, a novel cytorhabdovirus infecting rice via leafhopper transmission. Front Microbiol 7:2140–2140

Yassi MNA, Ritzenthaler C, Brugidou C, Fauquet C, Beachy RN (1994) Nucleotide sequence and genome characterization of rice yellow mottle virus RNA. J Gen Virol 75(2):249–257

Ye M, Luo SM, Xie JF, Li YF, Xu T, Liu Y, Song YY, Zhu-Salzman K, Zeng RS (2012) Silencing COI1 in rice increases susceptibility to chewing insects and impairs inducible defense. PLoS One 7(4):e36214

Yu R, Xu X, Liang Y, Tian H, Pan Z, Jin S, Wang N, Zhang W (2014) The insect ecdysone receptor is a good potential target for RNAi-based pest control. Int J Biol Sci 10(10):1171–1180

Zeigler RS, Morales FJ (1990) Genetic determination of replication of rice hoja blanca virus within its planthopper vector, *Sogatodes oryzicola*. Phytopathology 80(6):559–566

Zeigler RS, Pantoja A, Duque MC, Weber G (1994) Characteristics of resistance in rice to rice hoja blanca virus (RHBV) and its vector, Tagosodes orizicolus (Muir). Ann Appl Biol 124(3): 429–440

Zha W, Peng X, Chen R, Du B, Zhu L, He G (2011) Knockdown of midgut genes by dsRNA-transgenic plant-mediated RNA interference in the hemipteran insect *Nilaparvata lugens*. PLoS One 6(5):e20504

Zhang HM, Chen JP, Adams MJ (2001) Molecular characterisation of segments 1 to 6 of rice black-streaked dwarf virus from China provides the complete genome. Arch Virol 146(12):2331–2339

Zhang X, Wang X, Zhou G (2008a) A one-step real time RT-PCR assay for quantifying rice stripe virus in rice and in the small brown planthopper (*Laodelphax striatellus* Fallen). J Virol Methods 151(2):181–187

Zhang H-M, Yang J, Chen J-P, Adams MJ (2008b) A black-streaked dwarf disease on rice in China is caused by a novel fijivirus. Arch Virol 153(10):1893–1898

Zhang C, Pei X, Wang Z, Jia S, Guo S, Zhang Y, Li W (2012) The Rice stripe virus pc4 functions in movement and foliar necrosis expression in Nicotiana benthamiana. Virology 425:113–121

Zhang X, Dong J, Liu H, Wang J, Qi Y, Liang Z (2016) Transcriptome sequencing in response to salicylic acid in salvia miltiorrhiza. PLoS One 11:e0147849

Zhang T, Zhao Y, Ye J, Cao X, Xu C, Chen B, An H, Jiao Y, Zhang F, Yang X, Zhou G (2019) Establishing CRISPR/Cas13a immune system conferring RNA virus resistance in both dicot and monocot plants. Plant Biotechnol J 17(7):1185–1187

Zhao P, Sun X, Li P, Sun J, Yue Y, Wei J, Wei T, Jia D (2019) Infection characteristics of rice stripe mosaic virus in the body of the vector leafhoppers. Front Microbiol 9:3258

Zheng H, Yu L, Wei C, Hu D, Shen Y, Chen Z, Li Y (2000) Assembly of double-shelled, virus-like particles in transgenic rice plants expressing two major structural proteins of rice dwarf virus. J Virol 74(20):9808–9810

Zheng L, Zhang C, Shi C, Yang Z, Wang Y, Zhou T et al (2017) Rice stripe virus NS3 protein regulates primary miRNA processing through association with the miRNA biogenesis factor OsDRB1 and facilitates virus infection in rice. PLoS Pathog 3:e1006662

Zhong B, Kikuchi A, Moriyasu Y, Higashi T, Hagiwara K, Omura T (2003) A minor outer capsid protein, P9, of rice dwarf virus. Arch Virol 148(11):2275–2280

Zhou G, Wen J, Cai D, Li P, Xu D, Zhang S (2008) Southern rice black-streaked dwarf virus: a new proposed Fijivirus species in the family Reoviridae. Chin Sci Bull 53(23):3677–3685

Zhou G, Zhang S, Zou S, Xu Z, Zhou Z (2010) Occurrence and damage analysis of a new rice dwarf disease caused by Southern rice black-streaked dwarf virus. Plant Prot 36(2):144–146

Zhou T, Du L, Fan Y, Zhou Y (2012) Reverse transcription loop-mediated isothermal amplification of RNA for sensitive and rapid detection of southern rice black-streaked dwarf virus. J Virol Methods 180(1):91–95

Zhou G, Xu D, Xu D, Zhang M (2013) Southern rice black-streaked dwarf virus: a white-backed planthopper-transmitted fijivirus threatening rice production in Asia. Front Microbiol 4:270

Zhu Y, Hayakawa T, Toriyama S (1992) Complete nucleotide sequence of RNA 4 of rice stripe virus isolate T, and comparison with another isolate and with maize stripe virus. J Gen Virol 73(5):1309–1312

Zong Y, Wang Y, Li C, Zhang R, Chen K, Ran Y, Qiu J-L, Wang D, Gao C (2017) Precise base editing in rice, wheat and maize with a Cas9-cytidine deaminase fusion. Nat Biotechnol 35(5): 438–440

Botanical Extracts for Rice Fungal Diseases 24

Salman Ahmad, Fazal ur Rehman, Muhammad Adnan, Irfan Ahmad, Shakeel Ahmad, Muhammad Usman Ghazanfar, Ejaz Ashraf, Muhammad Asim, and Maria Kalsoom

Abstract

Frequent and prophylactic applications of fungicides on rice are highly toxic to a broad range of organisms, posing a great risk to aquatic biota. Therefore, there is need to adopt ecofriendly management practices against rice diseases for healthy crop and to obtain high yield. Secondary metabolites of a high diversity are produced by plants as a natural product. These secondary metabolites that are present in botanical extracts have the ability to kill microbial pathogens using their toxicity. It has been investigated that about 2000 higher plant species have pesticidal properties

S. Ahmad (✉) · Fazal ur Rehman · M. Asim
Department of Plant Pathology, College of Agriculture, University of Sargodha, Sargodha, Pakistan
e-mail: salman.ahmad@uos.edu.pk

M. Adnan
Department of Agronomy, College of Agriculture, University of Sargodha, Sargodha, Pakistan

I. Ahmad
Department of Forestry, Range Management and Wildlife, University of Agriculture, Faisalabad, Pakistan

S. Ahmad
Department of Agronomy, Faculty of Agricultural Sciences and Technology, Bahauddin Zakariya University, Multan, Pakistan
e-mail: shakeelahmad@bzu.edu.pk

M. U. Ghazanfar
Department of Agronomy, Bahauddin Zakariya University, Multan, Pakistan

E. Ashraf
Department of Agriculture Extension, College of Agriculture, University of Sargodha, Sargodha, Pakistan

M. Kalsoom
Institute of Food Science and Nutrition, University of Sargodha, Sargodha, Pakistan

N. Sarwar et al. (eds.), *Modern Techniques of Rice Crop Production*,
https://doi.org/10.1007/978-981-16-4955-4_24

against various plant pathogens. Researches on the application of botanical extracts to control fungal pathogens of rice are in progress. Plant extracts have shown antifungal activity against many fungal pathogens that attack rice. In recent years, botanical extracts, mainly neem derivatives, due to their antifungal as well as antibacterial properties, are gaining great importance in the control of rice diseases. A wide range of plants, having potential antifungal substances, are receiving considerable attention throughout the world. This chapter explores the applications of various plant products for the management of devastating fungal diseases of rice. Due to the antifungal activity, biodegradable nature and quick access to availability, the significance of botanical extracts is increasing in modern agriculture for the management of fungal diseases. Botanical extracts are also nonhazardous to environment and induce resistance in plants against the diseases.

Keywords

Botanical extracts · Rice blast · Sheet spot disease · Sheath blight of rice · Stem rot of rice · Leaf scald of rice

Abbreviations

ASS	Aggregate sheath spot disease
BDR	Bakanae disease of Rice
BSR	Brown spot of rice
LSR	Leaf scald of rice
PLS	Phoma leaf spot of rice
RBD	Rice blast disease
SBR	Sheath blight of Rice
SDR	Stackburn disease of rice
SRD	Stem rot disease of rice
SRR	Sheath rot of rice

24.1 Introduction

Rice (*Oryza sativa* L.) is considered as one of the major food crop of the world. More than half of the world's population uses rice as a staple food (Ahmad et al. 2008, 2009, 2012, 2013, 2015, 2019; Ahmad and Hasanuzzaman 2012; Ahmed and Ahmad 2019, 2020; Shahzad and Ahmad 2019). Being a nutritious cereal crop, rice provides 15% of protein and 20% of the calories to world's population (Jayaprakashvel and Mathivanan 2009; Ahmed et al. 2017, 2020, b; Akram et al. 2018, 2019; Hussain et al. 2018; Fahad et al. 2019; Fatima et al. 2020). Plant diseases result in about 20–40% losses in crops annually. It has been reported by Emechebe and Shoyinka (1985) that the pathogenic fungi have the ability to cause up to 100% yield losses in rice crops amounting 60–525 billion US dollars annually

(Sygenta 2012; Hafeez-ur-Rehman et al. 2019; Khan et al. 2019a, b; Naz et al. 2019; Peerzada et al. 2019; Razaq et al. 2019; Sarwar et al. 2013a, b; Tariq et al. 2020; Wasaya et al. 2019; Zahoor et al. 2019).

Fungal diseases of rice are devastating and have been reported to occur in all parts of the world. Rice diseases are causing about 15.6% losses annually (Mondal et al. 2017). Among diseases of rice, the most threatening ones are fungal diseases and their role in yield losses is significant. Narrow brown leaf spot (*Cercospora oryzae*) is one of the major foliar diseases of rice causing yield losses up to 40% (Zhou and Uppala 2015). On the susceptible varieties, it causes severe leaf necrosis and premature ripening that finally results in yield reduction and low grain milling quality (Groth and Hollier 2010). Sheath blight (*Rhizoctonia solani*) is also an important and major soil-borne fungal disease, reducing the grain quality and yield up to 50%. Sheath rot of rice (*Sarocladium oryzae*) is seed-borne fungal disease that is prevalent in all rice growing countries of the world with 85% yield losses (Bigirimana et al. 2015). It results in chaffy and discolored grains that affect the viability and nutritional value of grains. Rice false smut (*Ustilaginoidea virens*, perfect stage *Villosiclava virens*) is one of the most devastating grain diseases which reduces both grain yield and quality. Several epidemics of this disease have been reported in cold weather. The yield losses caused by rice false smut range from 1 to 75% in many rice growing areas. Rice blast disease (*Pyricularia grisea*) caused by a seed-borne fungus is also a major serious disease of rice which is reported to cause major losses in quality and quantity (Ghazanfar et al. 2009).

Synthetic fungicides are usually used to deal with these fungal pathogens. It is an admitted fact that fungicides contribute a lot in order to increase the yield. The application of Mancozeb, Carbendazim, Pyroquilon, Thiophanate methyl, Chlobenthiazone, Dithane M-45, and Tricyclazole are significantly effective in reducing the yield losses in rice caused by fungal diseases. The economic application of fungicides in agriculture helps to improve returns on farm investment. Growers invest millions of dollars on various fungicides to control fungal pathogens of rice and reap billions of dollars in return. Therefore, a very active role is being played by fungicides in the production of rice crops. In different parts of the world, highly bred rice varieties are grown to maintain uniform crop height, crop canopy, grains quality, and quantity as well as overall appearance and quality of produce. Even in these varieties, fungicides are applied extensively to control diseases (Osman and Abdulrahman 2003). In highly advanced countries like the USA, many crops could not be produced on commercial level without the application of fungicides (Croplife 2012).

In recent years, the inappropriate applications of fungicides in agriculture have raised serious concerns which include negative effects on nontarget species; pathogen resistance; pathogen resurgence; and severe impact on environment and human health. Among all these, the pathogen resistance to fungicides is a very serious problem. It has been noted that various fungal pathogens have evolved resistance against fungicides. The evolution of races and biotypes of fungal pathogens is the result of intensive use of these fungicides (Pallant 2011). The applications of synthetic fungicides have residual effects on the treated plant materials as well as on the environment. These undesirable residues then enter into food chain, consumed by nontarget species, and disturb their metabolic processes. These factors have set drive to search for alternatives (Amadioha 2012). Some of the best alternatives are:

- Application of natural enemies of fungal pathogens to control them.
- Re-designing of farming systems by adopting crop rotation practices, good crop density, proper field sanitation, furrow irrigation instead of overhead sprinkler irrigation, and improved field aeration. Obviously, these strategies will minimize predisposal of crops to fungal attacks.
- Use of intercropping to manage pathogens present in smallholding farms.
- Use of various botanical extracts of different plants to combat challenges of fungal diseases.

This chapter mainly focuses on the applications of various botanical pesticides to manage different fungal diseases of rice. Botanical extracts have been known for their antimicrobial and medicinal properties since the ancient times. Many of higher plants of the rain forest have been screened for fungicidal properties of their extracts. Various plant-based extracts have fungitoxic properties that can be utilized to manage phyto-fungal diseases (Opara and Obana 2010). These extracts are readily available in various faming localities of the tropics. They are eco-compatible, cheap, and less harmful to nontarget organisms. These botanical extracts may be valuable in integrated disease management (IDM) programs for the poor farmers having smallholding. A sustainable disease management solution is provided by these extracts especially in organic farming where the use of synthetic fungicides is not allowed. Pre-harvest interval (PHI) is not required for the applications of plant-based pesticides as they produce zero toxic load on treated plants as compared to synthetic chemicals. Plant-based bio-fungicides are systemic and contain multiple bioactive metabolites; hence, very difficult for the pathogens to break them (Enyiukwu et al. 2013). Therefore, it is less likely for the pathogens to become resistant to plant extracts (Adjaye-Gbewonyo et al. 2010; Silva-Aguayo 2013). Thus, plants are possible alternatives of fungicides for the eco-friendly management of fungal rice diseases (Benner 1993).

24.2 Solvent Selection for Extraction of Botanical Extract

The bioactive compounds present in plant extracts are mostly medium to highly polar in nature and can easily be extracted with polar solvents. Methanol is very efficient solvent for the extraction of antimicrobial compounds from plants compared to ethanol, water, and hexane (Lin et al. 1999). The antimicrobial compounds of the plants are not soluble in nonpolar solvents; therefore, hexane extracts have lower inhibitory activities. Various studies have shown that the type of solvents used for extraction considerably affects the successful extraction of active compounds.

The effectiveness of botanical extracts depends on solvents, extraction process, and the composition and structure of antimicrobial compounds present in them (Pinelo et al. 2004). The specific nature of bioactive compounds that are present in plant decides the selection of solvents. Plants of Zingiberaceae family have great antifungal and antibacterial potential in association with methanol extracts (Yusuf et al. 2002). *Inula viscosa* extracts obtained by using various organic solvents including methanol, ethyl acetate, ethanol, acetone, n-hexane, and chloroform have shown to have antifungal activity against *Colletotrichum cucumerinum*, *Phytopthora infestans*, *Plasmophora viticola*, and *Botrytis cinerea* (Cohen et al. 2003). The solvents used for the preparation of botanical extracts do not have any inhibitory activity against the fungal pathogens; these are only the plant extracts which kill them.

24.3 Rice Blast Disease (RBD)

Rice blast is a worldwide serious disease of rice and spread by infected seeds (Ghazanfar et al. 2009) (Fig. 24.2a). The disease causes huge yield losses and also reduces quality of rice. Resistant rice varieties and chemotherapy are the principal practices to control this disease. However, the repeated use of chemical fungicides has made their efficacy limited due the emergence of new races of this fungus. Similarly, these chemicals are too much expensive and also out from the purchase power of small-scale farmers. These chemical fungicides are also detrimental to consumer and farmer health, environment and beneficial predators, and parasitoids (Mbodi et al. 1986). These chemical fungicides also affect the biocontrols used in biological control of plant pathogens (Khan and Nasreen 2010). Amadioha (2000) reported that application of ethanol and water extracts of *Azadirachta indica* leaves and seeds are very effective against *P. grisea*. It reduced the spread and development of blast disease on rice plants in field. *A. indica* is commonly known as Nim, Neem, or Indian lilac and belongs to family Meliaceae. The key ingredient in non-pesticidal management (NPM) is also neem. It provides a natural alternative to synthetic fungicides. It has been demonstrated from various experiments that ethanol and water extracts of seeds and leaves of *A. indica* (neem) significantly inhibited colony growth of *P. oryzae* in vitro, and also considerably reduced spread and development of RBD in greenhouse. Oil extracted from neem seeds exhibits the best controlling activity against fungal pathogens. The oil and water and ethanol extracts of neem control RBD at par to carbendazim in vivo. Neem has a great potential that can be used for the management of RBD in the field. Its seeds are usually ground into powder and soaked overnight in water and then sprayed onto the rice crop. In order to get best results and make its application more effective, it must be applied repeatedly with the interval of 10 days. The spray of *Prosopis juliflora* leaf extracts can effectively reduce the rice blast in field experiments as well as it can also increase the yield (Kamalakannan et al. 2001). The biochemical analysis of rice plants treated with *P. juliflora* showed the increased activities of polyphenol oxidase, peroxidase, and phenylalanine ammonia-lyase. Indeed, the bioactive compounds present in plant

Table 24.1 Botanical extracts effective against rice blast disease (RBD)

Sr. No.	Common name	Scientific name	Family	Plant parts used	References
1	Aloe	*Aloe vera*	Asphodelaceae	Leaves	Uda et al. (2018)
2	Garlic	*Allium sativum*	Amaryllidaceae	Bulbs	Olufolaji et al. (2015)
3	Black jack	*Bidens pilosa*	Asteraceae	Leaves	Hubert et al. (2015)
4	Soursop	*Annona muricata*	Annonaceae	Leaves	Hubert et al. (2015)
5	Coffee	*Coffee arabica*	Coffea arabica	Coffee beans	Hubert et al. (2015)
6	Thorn apple	*Datura stramonium*	Solanaceae	Fresh plants, blossoming	Ganguly (1994)

products act as elicitors and induce resistance in host plants against the disease. Several other botanical extracts have also been reported for the management of RBD (Table 24.1). The commercial products of botanical extracts are cost-effective and freely available in those parts of the world where fungicides are highly expensive.

24.4 Sheath Blight of Rice (SBR)

SBR is a very destructive disease which causes substantial yield losses under favorable conditions of weather in rice growing areas (Fig. 24.1). Due to sclerotia formation, the management of SBR disease is a bit difficult because they remain viable for long time in the soil. The applications of synthetic fungicides for the management of SBR are not recommended owing to developing resistance in fungus against fungicides, pollution in the environment, residual toxicity, and high cost, and also they are carcinogenic.

Dill seed (*Anethum graveolens*) oil has good inhibitory activity against *R. solani.* Plant oils have great potential for controlling several fungal pathogens including *R. solani, L. theobromae*, *Colletotrichum musae*, and *Fusarium proliferatum* while they are also effective against bacterial pathogens (Dorman and Deans 2000). Various plant extracts proved to be very effective in vitro against *R. solani*. Neem oil is effective against sclerotia of *Rhizoctonia* spp., and SBR disease can effectively be controlled by applying 0.5% neem oil (Dohroo and Gupta 1995).

The inhibition activity of plant extracts against SBR disease is also due to their systemic effect which does not allow the further growth of pathogen in tissues beyond certain limits. It has been observed that there are no phytotoxic effects in rice plants on the application of plant extracts up to the concentration of 0.2% under the field conditions. The phytoconstituents of *Clerodendrum infortunatum*, extracted in chloroform, have been found to be most effective in controlling SBR disease (Nayeem and Mehta 2015). Many other botanical extracts are also being used to control SBR disease of rice (Table 24.2).

Fig. 24.1 Sheath blight of rice. (**a**) Sheath blight affecting base of rice culms. (**b**) Sheath blight water-soaked, gray-brown, and oblong lesions on rice. (**c**) Sheath blight water-soaked, gray-brown, oblong lesions on rice leaf (Photographs courtesy of B. Kumar and X. G. Zhou)

Table 24.2 Botanical extracts effective against sheath blight of rice (SBR)

Sr. No.	Common name	Scientific name	Family	Plant parts used	References
1	Arjun	*Terminalia arjuna*	Combretaceae	Leaves	Choudhury et al. (2020)
2	Clerodendrum	*Clerodendrum infortunatum*	Verbenaceae	Leaves	Lin et al. (1995)
3	Dill	*Anethum graveolens*	Umbelliferae	Seeds	Choudhury et al. (2020)
4	Ginger	*Zingiber officinale*	Zingeberaceae	Rhizome	Choudhury et al. (2020)
5	Lantena	*Lantana camara*	Verbenaceae	Tender leaves	Sehajpal et al. (2009)
6	Nicotiana	*Nicotiana tabacum*	Solanaceae	Leaves	Lin et al. (1995)
7	Polygonum	*Polygonum hydropiper*	Polygonaceae	Leaves	Islam and Monjil (2016), Tamuli et al. (2014)
8	Pongamia	*Pongamia pinnata*	Leguminosae	Seeds	Sehajpal et al. (2009)
9	Polyalthia	*Polyalthia longifolia*	Annonaceae	Leaves	Kandhari and Devakumar (2006)
10	Teak	*Tectona grandis*	Verbenaceae	Fallen dried leaves	Choudhury et al. (2020)

24.5 Aggregate Sheath Spot (ASS) Disease

Rhizoctonia oryzae-sativae is the causal agent of ASS disease of rice. The symptoms of this disease are characterized by oval shaped lesions which are gray or green from centers and surrounded by distinct brown margins. Lesions coalesce with each other resulting in an aggregation of lesions. The fungus also attacks panicles and makes the grains discolored (Lanoiselet et al. 2007).

Plant essential oils have been found to be most effective against *R. oryzae-sativae* (Chaijuckam and Davis 2010). Plant essential oils with garlic extracts are mostly used to control *R. oryzae-sativae.* They all have antifungal activity with different degree of efficacy. The cinnamon and lemongrass oils are the most effective plant products and completely inhibit the sclerotia, its production and mycelial growth of *R. oryzae-sativae*. However, cinnamon oil is most effective against *R. oryzae-sativae* compared to lemongrass oil. The similar reports on the relative activity of lemongrass oils and cinnamon oil have been given against *F. proliferatum* (Velluti et al. 2003) and *B. cinerea* (Wilson et al. 1997).

Limonene and citral are the major chemical constituents present in lemongrass oil while eugenal has been reported in cinnamon oil (Wilson et al. 1997; Ranasinghe et al. 2002). At the concentration of 1%, limonene, eugenal, and citral have been found antagonistic against *R. oryzae-sativae*. These chemical constituents of lemongrass and cinnamon oils also have fungicidal effect against fungi *Cercospora arachidicola, Puccinia arachidis*, and *Phaeoisariopsis personata* (Kishore et al. 2007).

The most effective against *R. oryzae-sativae* are garlic extracts and garlic oil. These both have an ability to inhibit the growth and production of sclerotia of *R. oryzae sativae* completely. The garlic oil can significantly reduce the germination of sclerotia of *R. oryzae-sativae*. It has also been reported that garlic oil can also inhibit the sclerotium formation and growth of *R. solani* (Singh and Singh 1980). The extracts of neem, ginger, pepper, and basil also have significant effects on the growth of *R. oryzae-sativae*. It has been reported that garlic and pepper extracts also inhibit spore germination and vegetative growth of *B. cinerea* (Wilson et al. 1997). The effectiveness of neem oil can vary according to different isolates of *R. oryzae-sativae*, that is, partially suppresses sclerotia germination of California isolates in the USA but completely inhibits isolates from India (Banerjee et al. 1989). The same is the case with the cinnamon oil.

The vegetable oil can be used to reduce the concentration of cinnamon oil to increase its spreading and floating, and is also used in fungicides that are active against *R. solani* in rice (Oh and Kim 1988). Generally, the vegetable oil has no effect on rice growth or disease severity of ASS at low concentrations. The medium concentration of cinnamon oil, that is 32.5%, is somehow effective on those rice cultivars that are less vulnerable to *R. oryzae-sativae*. At higher concentration, that is 87.5%, cinnamon oil considerably reduces disease development, but is phytotoxic resulting chlorotic stripes on the leaf sheaths of some tillers. The optimal concentration of cinnamon oil that could be applied and that could reduce the expansion of

Table 24.3 Botanical extracts effective against aggregate sheath spot (ASS) disease

Sr. No.	Common name	Scientific name	Family	Plant parts used	References
1	Pepper	*Capsicum annuum*	Nightshade	Dried peppers	Chaijuckam and Davis (2010)
2	Basil	*Ocimum basilicum*	Lamiaceae	Leaves and flowering tops	Shamim et al. (2017)
3	Garlic	*Allium sativum*	Amaryllidaceae	Fresh garlic bulbs	Jahan et al. (2013)
4	Neem oil	*Azadirachta indica*	Meliaceae	Leaves	Chaijuckam and Davis (2010)
5	Ginger	*Zingiber officinale*	Zingiberaceae	Rhizomes	Chaijuckam and Davis (2010)
6	Lemongrass oil	*Cymbopogon citratus*	Poaceae	Leaves	Chaijuckam and Davis (2010)
7	Cinnamon oil	*Cinnamomum zeylanicum*	Lauraceae	Leaves	Chaijuckam and Davis (2010)

lesions caused by *R. oryzae-sativae* is 60%; this concentration is effective in almost all cultivars and does not affect the rice dry weight.

The biochemical constituent eugenal present in cinnamon oil directly affects *R. oryzae-sativae* by inhibiting vegetative growth and sclerotia germination (Ranasinghe et al. 2002). However, cinnamon leaf oil can also induce systemic resistance in rice plant and inactivate the toxins that are produced by *R. oryzae-sativae*. Because sclerotia are thick-walled structures able to resist unfavorable environments, therefore, to inactivate them relatively high rate of cinnamon oil is required.

Cinnamon oil can also suppress *Phytophthora* blight and reduce the population densities of *P. nicotianae* (Bowers and Locke 2004). Cinnamon oil can also inhibit the spore germination in *Phaeoisariopsis personata*. It can also reduce the mycelial growth of *Aspergillus niger* and minimize the late leaf spot and crown rot in peanut (Kishore et al. 2007). Furthermore, there are many other reports that demonstrate the antagonistic properties of cinnamon oil to plant pathogenic fungi, bacteria, and nematodes (Ranasinghe et al. 2002; Wilson et al. 1997). Thus, the applications of cinnamon oil can prove beneficial to the rice growers managing several pests. Effectiveness of various botanical extracts against ASS disease has been documented in several researches (Table 24.3).

24.6 Brown Spot of Rice (BSR)

Bipolaris oryzae is the cause of BSR (Fig. 24.2b). It is an implacable pathogen, prevalent in all rice growing countries, and is favored by semi-dry climate (Ou 1985). During 1942–1943, this disease caused famine in Bengal which is in history known as “Great Bengal Famine” (Padmanabhan 1973).

Fig. 24.2 Rice blast disease (**a**) and brown spot of rice (**b**) (Photographs courtesy of X. G. Zhou)

The use of various plant products, that is, botanical extracts and other biocontrol agents, has achieved a significant attention in recent years to control BSR. It is due to their antimicrobial activity, easy biodegradability and availability, induction of resistance in host, and non-phytotoxicity. Various botanical extracts have been tested on *B. oryzae* including *Nerium oleander* and *Pithecellobium dulce* having high inhibitory activities against the mycelial growth and spore germination of *B. oryzae*. Similarly, *Solanum nigrum* possess the inhibitory properties against the mycelial growth of *B. oryzae* (Chelven and Sumathi 1994). The inhibition properties of *N. oleander* against *B. oryzae* are due to toxic substances present in their plant extracts. Further, neem extracts are effective reducing the germination and growth of *B. oryzae* in vitro. Ganguly (1994) has also reported that *Velleia rosea* and *Acacia nilotica* have a great efficacy against the mycelial growth of *B. oryzae*. The antifungal activity of *A. indica* considerably reduces the growth of *B. oryzae*. Biocontrol strains are also effective against BSR. For example, *T. viride* and *T. harzianum* can significantly reduce the growth and spore germination of *B. oryzae* through their mycoparasitic activity and cell wall lysis of pathogenic fungi (Geremia et al. 1993; John et al. 2010). These biocontrols, *T. viride* and *T. harzianum,* kill the *B. oryzae* using the same mechanisms (mycoparasitism and lysis).

The sprays of neem extracts on 60 days old and diseased rice plants result in significant reduction in disease that is followed by *N. oleander* leaf extracts and *T. viride.* Same method of the management of diseased rice field by the application of plant products has also been given by Amadioha (2000). The sprays of neem oils are also effective against *B. oryzae,* and *Helminthosporium nodulosum*, the cause of ragi blight (Unnikrishna 1988). Vijayakumar (1998) unraveled that the combined

Table 24.4 Botanical extracts effective against brown spot of rice (BSR)

Sr. No.	Common name	Scientific name	Family	Plant part used	References
1	Kanghi/Atibala	*Abutilon indicum*	Malvaceae	Whole plant	Harish et al. (2008)
2	Slender Amaranth	*Amaranthus viridis*	Amaranthaceae	Leaves	Akbar et al. (2020)
3	Pomegranate	*Punica granatum*	Punicaceae	Fruit bark	Harish et al. (2008)
4	Banyan/ banyan fig	*Ficus bengalensis*	Moraceae	Stem bark	Harish et al. (2008)
5	Butterfly pea	*Clitoria ternatia*	Fabaceae	Flowers	Harish et al. (2008)
6	Great basil	*Ocimum basilicum*	Lamiaceae/ Labiatae	Dried leaves	Sunder et al. (2014)
7	Touch-me-not	*Mimosa pudica*	Fabaceae	Leaves	Harish et al. (2008)
8	Chinese chaste tree	*Vitex negundo*	Verbenaceae	Leaves	Harish et al. (2008)
9	Black night shade	*Solanum indicum*	Solanaceae	Root	Harish et al. (2008)
10	Tamarind	*Tamarindus indica*	Fabaceae	Leaves/ seeds	Harish et al. (2008)

application of neem cake+neem oil gives more effective result against BSR in the field. *N. oleander* contains oleandrin toxin which is responsible for the antimicrobial activity and is also biodegradable (Downer et al. 2003). Thus the residual effects of *N. oleander* plant extracts are very less compared to synthetic fungicides that persist in plant and soil for long time after their application. Recently, the mechanisms through which botanical extracts control BSR has been investigated and unraveled that these extracts induce the treated plants to produce defense enzymes which reduce BSR in field; however, the role of plant products against *B. oryzae* is still obscure. The easy availability of botanical extracts together with less phytotoxicity makes them potential alternatives of chemicals. The applications of plant extracts with biocontrol agents are still needed to be evaluated for the improved protection of rice crop. Thus, the plant extracts in combination with biocontrol agents could be investigated in future for the environment friendly management of BSR. The active ingredients of the botanical extracts could also be isolated and formulated for their commercial products. Many other revelations have shown the effectiveness of various botanical extracts against BSR (Table 24.4).

24.7 Sheath Rot of Rice (SRR)

SRR caused by *Sarocladium oryzae* is the major disease of rice and is seed borne (Shamsi et al. 2010). SRR fungus may infect at all growth stages of rice crop but is most deadly at the panicle stage (Shamsi 1999).

The higher plants are rich with bioactive compounds, that is, phenols, flavanoids, quinones, saponins, sterols, terpenoids, alkaloids, and tannins, and have been found to be effective against SRR. These compounds trigger the defense mechanism of rice plant and make them resistant against SRR. Allelopathic, antimicrobial, bio-regulatory, and antioxidant properties of these plant compounds make them effective against SRR. These natural products are certainly the substitute of harmful fungicides (Patel and Jasrai 2012, 2013; Ahmed et al. 2013). Neem and pungam oils are effective against SSR under field conditions (Narasimhan et al. 1998). Ten percent ethanol extracts of *M. indica* and *Datura metel* can completely inhibit the radial growth of *S. oryzae*. Similarly, 10% ethanol extracts of *A. sativum, A. indica, N. indicum, A. heterophyllus, S. alata, C. medica, Tagetes erecta,* and *A. racemosus* showed significant results inhibiting the mycelial growth of *S. oryzae*; however, the extracts of *Citrus medica* and *A. heterophyllus* were less effective against SRR.

The methanol extracts of *Acacia nilotica, Decalepis hamiltonii, Caesalpinia coriaria, Lawsonia inermis,* and *Emblica officinalis* showed significant antifungal activity against various seed pathogens of rice including *S. oryzae* (Mohana et al. 2011). The seed borne fungal pathogens of rice, that is, *A. flavus, Curvularia oryzae, A. niger, F. oxysporum, B. oryzae, Nigrospora oryzae, F. moniliforme,* and *Penicillium* sp., when treated with Provax and garlic extracts, are controlled up to 100% (Mohana and Raveesha 2007; Yeasmin et al. 2012). The Provax is also found to be best extract against *S. oryzae*. In short, ethanol extracts of *M. indica* and *D. metel* can inhibit completely the radial growth of SRR fungus like the synthetic chemicals. Many other botanical extracts are also being used to control SRR (Table 24.5).

Table 24.5 Botanical extracts effective against SRR

Sr. No.	Common name	Scientific name	Family	Plant part used	References
1	Garlic	*Allium sativum*	Amaryllidaceae	Fresh garlic bulbs	Jose (1997)
2	Jackfruit	*Artocarpus heterophyllus*	Moraceae	Leaves	Shamsi and Chowdhury (2016)
3	Neem tree	*Azadirachta indica*	Meliaceae	Leaves	Shamsi and Chowdhury (2016)
4	Jimsonweed	*Datura metel*	Solanaceae	Crushed seeds	Abbas et al. (1998)
5	Mango	*Mangifera indica*	Anacardiaceae	Fruit stone	Shamsi and Chowdhury (2016)
6	Nerium	*Nerium indicum*	Apocynaceae	Leaves	Shamsi and Chowdhury (2016)
7	Candlesticks	*Senna alata*	Fabaceae	Leaves	Shamsi and Chowdhury (2016)
8	Marigold	*Tagetes erecta*	Asteraceae	Flowers	Shamsi and Chowdhury (2016)

24.8 Stackburn Disease of Rice (SDR)

Alternaria padwickii, the cause of SDR, is asexually reproducing fungus that attacks mainly the seeds of rice and causes seed rot, seed discoloration, and seedling blight resulting huge losses in rice. Fungal pathogens are the main infectious agents in rice. They cause alterations in rice plants pre- and post-harvest. Generally, *A. padwickii* is controlled by synthetic fungicides; however, their applications are causing harmful effects on human health and the environment (Harris and Mantle 2001). Botanical extracts from medicinal plants and their effective concentrations have been tested against *Alternaria* sp. (Wilson et al. 1997). The buffer extracts of *Cynara scolymus* and acid extracts of *Salvia sclarea, S. officinalis,* and *Lippia alba* can significantly inhibit *Alternaria* spp. The antifungal activity of plant extracts can be evaluated depending on the extraction solvent and plant material used. *Schinus sclarea* has also shown significant results against *Alternaria* spp. compared to *S. molle.*

Buffer, aqueous, and acid extracts of various plant species have shown antifungal activity against *Alternaria* sp. The water extracts of *Zea mays* and *Cynara scolymus* are very active against *Alternaria* sp. Similarly, *Cynara scolymus* can inhibit fungal growth of *Alternaria* spp. up to 98%. Whereas the extracts of *Lonicera japonica* also have inhibition activity against *Alternaria* spp. The acid extracts of *Rosmarinus officinalis* have shown highest fungicidal activity against *Alternaria* spp. Solubility of active ingredients of plant extracts varies with solvents which also affect their effectiveness. Sixty percent inhibition of *Alternaria* spp. can be obtained by using *R. officinalis* aqueous extracts compared to their acid extracts which are less effective (Itako et al. 2008).

24.9 Phoma Leaf Spot of Rice (PLS)

This disease is caused by *Phoma oryzae*. Various plant aqueous extracts have been reported as the botanical fungicides for the treatment of PLS of rice and significantly affect the radial growth of *P. oryzae*. Neem extracts have the ability to produce higher inhibitory effect on radial growth of *P. oryzae* in culture. It has also been reported that extracts of *Piper guineensis* also have high efficiency against *P. oryzae*. *Cymbopogon citratus*, *A. indica,* and *Ocimum gratissimum* are very effective against fungal pathogens of rice. The single application of *O. gratissimum* is not very effective against PLS as compared to other plant extracts; however, the repeated applications make it more effective. The leaf extracts of *Venonia amydalina*, *O. basilicun*, *Cymbopogon citratus*, *Carica papaya,* and *A. indica* have the ability to reduce the incidence of seed borne fungi of rice (Singh et al. 1980; Nwachukwu and Umechurba 2001).

Lemon grass extracts are equally effective as neem extracts against *P. oryzae*. The comparative efficacy of plant extracts and synthetic chemicals showed that plant extracts were more effective than chemicals. Ethanol extracts of *O. gratissimum* are also very effective. The extracts of *Piper guineesis* showed best inhibitory effect on *P. oryzae* after the consecutive third day of application. The aqueous extracts of

Table 24.6 Botanical extracts effective against phoma leaf spot (PLS) of rice

Sr. No.	Common name	Scientific name	Plant part used	Discription	References
1	Neem	*Azardiractha indica*	Meliaceae	Leaves	Iwuagwu et al. (2018)
2	Ginger	*Zingiber oficinale*	Piperaceae	Rhizomes	Iwuagwu et al. (2018)
3	Clove basil	*Ocimum gratissimum*	Lamiaceae	Leaves	Iwuagwu et al. (2018)
4	Ashanti pepper	*Piper guinensis*	Piperaceae	Leaves	Iwuagwu et al. (2018)
5	Bitter kola	*Garcinia cola*	Guttiferae	See/leaves	Iwuagwu et al. (2018)

Zingiber oficinale, *A. indica,* and *O. gratissimum* are also very effective against *P. oryzae*, but comparatively less compared to *P. guineesis* and ginger. Many other botanical extracts are also being used to control the PLS of rice (Table 24.6).

24.10 *Aspergillus* spp.

Aspergillus species affect a wide range of host plants including rice and its single species could cause huge yield losses in rice. The most common species attacking on rice are *A. niger*, *A. flavus,* and *A. clavatus*. Bioactive components produced by plants are toxic to *Aspergillus* spp. and thus can be used to control them, and have low mammalian toxicity (Gouka et al. 1997; Machida 2002). Different plant extracts have been screened for their antifungal activity against *Aspergillus* spp. Extracts from the bark of Costa Rican and Australian plants are very effective (Yasui et al. 2020). Bark extracts of *Cupania glabra, Quercus insignis,* and *Neolitsea dealbata* have been proved to be very effective against *A. flavus*.

A. niger is most significantly inhibited by extracts of *Drymonia conchocalyx* vine (Yasui et al. 2020). Botanical extracts including *Cedrela tonduzii* (chloroform bark extracts), *Camellia sinensis* (aqueous leaf extracts), *Ardisia revoluta* (acetone bark extracts), and *Psychotria parviflora* (acetone bark extracts) also show strong inhibition of *A. niger* (Yasui et al. 2020). Similarly, botanical extracts obtained from *Diospyros digyna* (ethanol bark extracts), *Polysoma alangiacea* (ethanol/chloroform bark extracts), *Bocconia frutescens* (acetone leaf extracts), and *Grevillia hilliana* (ethanol/chloroform bark extracts) also give significant inhibition of radial growth of *A. niger*.

Mostly extracts are more effective against *A. niger* compared to *A. flavus* and *A. clavatus*, and also similar results have been seen in case of synthetic fungicides against these two fungal species of *Aspergillus* (Diaz-Godinez et al. 2001; Elinbaum et al. 2002). Therefore, *A. niger* is usually taken as the model fungus to study the inhibition of its cultures by plant extracts and fungicides (Gouka et al. 1997).

24.11 Leaf Scald of Rice (LSR)

LSR is a rice disease caused by the fungal pathogen *Microdochium oryzae* and is potential threat of rice globally. LSR causes scalded appearance of leaves. Different plant species have antifungal activity against *M. oryzae* and they are commonly available. The leaf extracts of *Justicia adhatoda* are found to be highly effective in inhibiting the growth of *M. oryzae*. The leaf extracts of *Lantana camara* and bulb extracts of *Allium sativum* also have a great efficacy against *M. oryzae* (Ahmed et al. 2013). Various plant extracts have been reported to possess different levels of antifungal activities against *M. oryzae*. Seed treatment with *Adhatoda vasica* is most effective to reduce the incidence of LSR. Applications of botanical extracts also significantly increase the yield of rice (Jantasorn et al. 2016).

It has been observed that resistance can be introduced in rice by the applications *A. vasica*. Biological active compounds are present in botanical extracts that act as elicitors and induce resistance in rice plant against pathogenic fungi and thus reduce the disease development (Vidhyasekaran et al. 1992). The applications of leaf extracts of *A. vasica* induce systemic resistance in rice plants against LSR, have significant in vitro inhibition of *M. oryzae*, and also reduce the incidence of bacterial diseases under greenhouse conditions. Therefore, it is highly recommended to use *A. vasica* extracts in order to control *M. oryzae* as well as bacterial pathogens of rice.

24.12 Stem Rot Disease of Rice (SRD)

SRD is caused by *Magnaporthe salvinii* and drastically affects the rice yield (Krause and Webster 1972). Symptoms of this disease first appear during tillering stage with black colored lesions on leaf sheaths. The infected sheaths die and slough off as the disease progresses. The infection may penetrate the culm also. The aqueous and methanol plant extracts of *A. indica*, *Anacardium occidentale*, *Carica papaya,* and *Calotropis procer* have been assessed to control *M. salvinii*. The agar well diffusion method can be used to assess their toxicity. Presence of one or more phytochemicals in each of these plant extracts inhibits the radial growth of *M. salvinii* (Jantasorn et al. 2016).

Phytochemicals tannins, flavonoids, alkaloids, anthocyanin, saponin, and phenol have been identified and inhibit the mycelial growth of *M. salvinii*. The efficacy of plant extracts against *M. salvinii* increases with the increase of their concentration. The efficacy of methanol extracts is higher compared to aqueous extracts against *M. salvinii* (Chowdhury et al. 2015). The higher concentration of *A. occidentale* and *C. procera* showed the highest inhibitions of *M. salvinii* and were more effective compared to *C. papaya* and *A. indica*. However, the methanol and aqueous extracts of *A. occidentale* gave the highest growth inhibition of *Magnaporthe salvinii* at all concentrations levels. The field trials of these plant extracts against *M. salvinii* to control SRD of rice have also proved effective (Ahmed et al. 2013).

24.13 Bakanae Disease of Rice (BDR)

BDR caused by *F. moniliforme* is one of the most important and devastating disease and could cause 25% losses in susceptible varieties of rice (Hossain et al. 2015). BDR is seed-borne disease, and mainly controlled by seed dressing fungicides. Different plant extracts have been documented to be effective against many diseases of rice including this disease (Miah et al. 1990; Hasan et al. 2005, Yang and Clausen 2007). Yasmin et al. (2008) reported that 17 angiosperm plant extracts have inhibitory effects on the growth of *F. moniliforme*. Leaf extracts of *L. inermis* have been found significantly more effective against BDR followed by root extracts of *A. racemosus* and leaf extracts of *Solanum indicum.* The leaf extracts of *Andrographis paniculata*, *L. inermis,* and *Lagerstroemia speciosa* also proved to be very effective against BDR. The extracts of *Vangueria spinosa*, *Avicennia alba,* *Cassia alata*, *Boerhaavia repens*, *Eclipta prostate,* *Mangifera indica*, *Leucas lavendulifolia*, *Eucalyptus citriodora,* *Eupatorium odoratum*, *Smilax macrophylla,* *Cinnamomum camphora,* and *Cuscuta reflexa* plants have shown 20% growth inhibition of *F. moniliforme* (Yasmin et al. 2008).

The *L. inermis* leaf extracts also completely inhibit the growth of different rice fungal pathogens at 20% (w/v) concentration and also have been reported effective against *F. moniliforme.* The antifungal compounds lawsone, 2-hydroxyl-1,4 and napthoquinine, are present in leaf extracts of *L. inermis* and are responsible for inhibition of fungi (Tripathi et al. 1978). The bark extracts of *E. citriodora* and leaf extracts of *Lagerstroemia speciosa* and *Andrographis paniculata* have the ability to reduce 33.53% and 26.82% colony growth of *F. moniliforme*, respectively (Yasmin el al. 2008).

24.14 Conclusion

The use of natural products, specifically plant-derived compounds, is an eco-friendly approach to controlling fungal diseases of rice, and significant success has been obtained to date. Botanical compounds, like essential oils, have antifungal activity against rice fungal pathogens. How these botanical extracts control rice pathogenic fungi is obscure and more research is required to understand the mechanism of action of plant extracts on fungal pathogens of rice. Many studies are in progress to determine the efficacy of various plant extracts to develop their commercial products.

References

Abbas HK, Cartwright RD, Shier WT, Abouzied MM, Bird CB, Rice LG, Ross PF, Sciumbato GL, Meredith FI (1998) Natural occurrence of fumonisins in rice with *Fusarium* sheath rot disease. Plant Dis 82:22–25

Adjaye-Gbewonyo D, Quaye EC, Wubah DA (2010) The effects of extracts of *Piper guineense* seed on insect pest damage to cowpea. J Young Investig 20:264–256

Ahmad S, Hasanuzzaman M (2012) Integrated effect of plant density, N rates and irrigation regimes on the biomass production, N content, PAR use efficiencies and water productivity of rice under irrigated semiarid environment. Not Bot Horti Agrobot Cluj Napoca 40(1):201–211

Ahmad S, Zia-ul-Haq M, Ali H, Shad SA, Ammad A, Maqsood M, Khan MB, Mehmood S, Hussain A (2008) Water and radiation use efficiencies of transplanted rice (*Oryza sativa* L.) at different plant densities and irrigation regimes under semi-arid environment. Pak J Bot 40(1): 199–209

Ahmad S, Ahmad A, Zia-ul-Haq M, Ali H, Khaliq T, Anjum MA, Khan MA, Hussain A, Hoogenboom G (2009) Resources use efficiency of field grown transplanted rice (*Oryza sativa* L.) under irrigated semiarid environment. J Food Agric Environ 7(2):487–492

Ahmad S, Ahmad A, Soler CMT, Ali H, Zia-Ul-Haq M, Anothai J, Hussain A, Hoogenboom G, Hasanuzzaman M (2012) Application of the CSM-CERES-Rice model for evaluation of plant density and nitrogen management of fine transplanted rice for an irrigated semiarid environment. Precis Agric 13(2):200–218

Ahmad S, Ahmad A, Ali H, Hussain A, Garcia y Garcia A, Khan MA, Zia-Ul-Haq M, Hasanuzzaman M, Hoogenboom G (2013) Application of the CSM-CERES-Rice model for evaluation of plant density and irrigation management of transplanted rice for an irrigated semiarid environment. Irrig Sci 31(3):491–506

Ahmad A, Ashfaq M, Rasul G, Wajid SA, Khaliq T, Rasul F, Saeed U, Rahman MH, Hussain J, Baig IA, Naqvi AA, SAA B, Ahmad S, Naseem W, Hoogenboom G, Valdivia RO (2015) Impact of climate change on the rice–wheat cropping system of Pakistan. In: Hillel D, Rosenzweig C (eds) Handbook of climate change and agro-ecosystems: The Agricultural Modeling Intercomparison and Improvement Project (AgMIP) integrated crop and economic assessments. Imperial College Press, London, pp 219–258

Ahmad S, Abbas G, Ahmed M, Fatima Z, Anjum MA, Rasul G, Khan MA, Hoogenboom G (2019) Climate warming and management impact on the change of rice-wheat phenology in Punjab, Pakistan. Field Crop Res 230:46–61

Ahmed M, Ahmad S (2019) Carbon dioxide enrichment and crop productivity. In: Hasanuzzaman M (ed) Agronomic crops, vol 2. Springer Nature Singapore, Singapore, pp 31–46

Ahmed M, Ahmad S (2020) Systems modeling. In: Ahmed M (ed) Systems modeling. Springer Nature, Cham, pp 1–44

Ahmed M, Hossain M, Hassan K, Dash CK (2013) Efficacy of different plant extract on reducing seed borne infection and increasing germination of collected rice seed sample. Univ J Plant Sci 1:66–73

Ahmed M, Fayyaz-ul-Hassan, Ahmad S (2017) Climate variability impact on rice production: adaptation and mitigation strategies. In: Ahmed M, Stockle C (eds) Quantification of climate variability, adaptation and mitigation for agricultural sustainability. Springer, Cham, pp 91–111

Ahmed M, Ahmad S, Raza MA, Kumar U, Ansar M, Shah GA, Parsons D, Hoogenboom G, Palosuo T, Seidel S (2020) Models calibration and evaluation. In: Ahmed M (ed) Systems modeling. Springer Nature Singapore, Singapore, pp 151–178

Ahmed M, Ahmad S, Waldrip HM, Ramin M, Raza MA (2020b) Whole farm modeling: A systems approach to understanding and managing livestock for greenhouse gas mitigation, economic viability and environmental quality. In: Waldrip HM, Pagliari PH, He Z (eds) Animal manure, ASA Special Publication 67. American Society of Agronomy/Soil Science Society of America, Madison, pp 345–371

Akbar M, Sherazi IN, Iqbal MS, Khalil T, Waqas HM (2020) Antibacterial and antioxidant activities of slender amaranth weed. Planta Daninha 38:e020192974

Akram R, Turan V, Hammad HM, Ahmad S, Hussain S, Hasnain A, Maqbool MM, Rehmani MIA, Rasool A, Masood N, Mahmood F, Mubeen M, Sultana SR, Fahad S, Amanet K, Saleem M, Abbas Y, Akhtar HM, Hussain S, Waseem F, Murtaza R, Amin A, Zahoor SA, Sami ul Din M, Nasim W (2018) Fate of organic and inorganic pollutants in paddy soils. In: Hashmi MZ, Varma

A (eds) Environmental pollution of paddy soils. Springer Nature Switzerland, Cham, pp 197–214

Akram R, Fahad S, Masood N, Rasool A, Ijaz M, Ihsan MZ, Maqbool MM, Ahmad S, Hussain S, Ahmed M, Kaleem S, Sultana SR, Mubeen M, Saud S, Kamran M, Nasim W (2019) Plant growth and morphological changes in rice under abiotic stress. In: Hasanuzzaman M, Fujita M, Nahar K, Biswas JK (eds) Advances in rice research for abiotic stress tolerance. Woodhead, Cambridge, pp 69–85

Amadioha AC (2000) Controlling rice blast in vitro and in vivo with extracts of *Azadirachta indica*. Crop Prot 19:287–290

Amadioha AC (2012) Reducing food losses through sustainable methods of plant disease management: an imperative for actualization of food security. A paper presented at the 13th Inaugural Lecture MOUAU, June, 2012

Banerjee S, Bhattacharya I, Mukherjee N (1989) Sensitivity of three sclerotial rice pathogens to plant oils. Int Rice Res Newslett 14:23

Benner JP (1993) Pesticidal compounds from higher plants. Pestic Sci 39:95–102

Bigirimana VDP, Gia HKH, Nyamangyoku OI, Hofte M (2015) Rice sheath rot: an emerging ubiquitous destructive disease complex. Front Plant Sci 6:1066

Bowers JH, Locke JC (2004) Effect of formulated plant extracts and oils on population density of *Phytophthora nicotianae* in soil and control of *Phytophthora* blight in the greenhouse. Plant Dis 88:11–16

Chaijuckam P, Davis RM (2010) Efficacy of natural plant products on the control of aggregate sheath spot of rice. Plant Dis 94:986–992

Chelven PTK, Sumathi L (1994) A fungitoxic substance from *Solanum nigrum*. Ind Phytopathol 47:424–426

Choudhury D, Kundu S, Kole RK, Saha J, Nath R (2020) Efficacy of dill, teak and arjun extract against sheath blight of rice caused by *Rhizoctonia solani*. J Agric Sci Technol 8:27–34

Chowdhury P, Bashar MA, Shamsi S (2015) In vitro evaluation of fungicides and plant extracts against pathogenic fungi of two rice varieties. Bangladesh J Bot 44:251–259

Cohen Y, Baider A, Ben-Daniel BH, Ben-Daniel Y (2003) Fungicidal preparations from Inula viscosa. Plant Prot Sci 38:629–630

Croplife (2012) Value of fungicides in US crop production. Available online at: http//www.croplifefoundation.org/upload/ Retrieved October

Diaz-Godinez G, Soriano-Santos J, Augur C, Viniegra-Gonzalez G (2001) Exopectinases produced by Aspergillus niger in solid-state and sub-merged fermentation: a comparative study. J Ind Microbiol Biotechnol 26:271–275

Dohroo NP, Gupta SK (1995) Neem in plant disease control. Agric Rev 16:133–140

Dorman HJD, Deans SG (2000) Antimicrobial agents from plants: antibacterial activity of volatile oils. J Appl Microbiol 88:308–316

Downer J, Craigmill A, Holstege D (2003) Toxic potential of oleander derived compost and vegetables grown with oleander soil amendments. Vet Hum Toxicol 45:219–221

Elinbaum S, Ferreyra H, Ellenrieder G, Cuevas C (2002) Production of *Aspergillus terreus* alpha-l-rhamnosidase by solid state fermentation. Lett Appl Microbiol 34:67–71

Emechebe AM, Shoyinka SA (1985) Fungal and bacterial diseases of cowpea in Africa. In: Singh RS, Rachie KO (eds) Cowpea research, production and utilization. Wiley, Chichester, pp 173–192

Enyiukwu DN, Awurum AN, Nwaneri JA (2013) Efficacy of plant-derived pesticides in the control of myco-induced postharvest and storage rots of ubers and agricultural products: a review. Neth J Agric Sci 2:30–46

Fahad S, Noor M, Adnan M, Khan MA, Rahman I, Alam M, Khan IA, Ullah H, Mian IA, Hassan S, Saud S, Bakhat HF, Hammad HM, Ahmad S, Nasim W (2019) Abiotic stress and rice grain quality. In: Hasanuzzaman M, Fujita M, Nahar K, Biswas JK (eds) Advances in rice research for abiotic stress tolerance. Woodhead, Cambridge, pp 571–583

Fatima Z, Ahmad M, Hussain M, Abbas G, Ul-Allah S, Ahmad S, Ahmed N, Ali MA, Sarwar G, Ehsan ul Haque, Iqbal P, Hussain S (2020) The fingerprints of climate warming on cereal crops phenology and adaptation options. Sci Rep 10:18013

Ganguly LK (1994) Fungitoxic effect of certain plant extracts against rice blast and brown spot pathogen. Ecol Environ 12:731–733

Geremia RA, Goldman GH, Jacobs D, Ardrtes W, Vila SB, Van Montagu M, Herrera-Estrella A (1993) Molecular characterization of the proteinase-encoding gene, prb1, related to mycoparasitism by *Trichoderma harzianum*. Mol Microbiol 8:603–613

Ghazanfar MU, Habib A, Sahi ST (2009) Screening of rice germplasm against *Pyricularia oryzae*, the cause of rice blast disease. Pak J Phytopathol 21:41–44

Gouka RJ, Punt PJ, van den Hondel CA (1997) Efficient production of secreted proteins by *Aspergillus*: progress, limitations and prospects. Appl Microbiol Biotechnol 47:1–11

Groth D, Hollier C (2010) Narrow brown leaf spot of rice, Identification and management series. Louisiana Plant Pathology, LSU AgCenter, Baton Rouge

Hafeez-ur-Rehman NA, Awan MI, Ijaz M, Hussain M, Ahmad S, Farooq M (2019) Direct seeding in rice: problems and prospects. In: Hasanuzzaman M (ed) Agronomic crops, Production technologies, vol 1. Springer Nature Singapore, Singapore, pp 199–222

Harish S, Saravanakumar D, Radjacommare R, Ebenezar EG, Seetharaman K (2008) Use of plant extracts and biocontrol agents for the management of brown spot disease in rice. BioControl 53: 555

Harris JP, Mantle PG (2001) Biosynthesis of ochratoxins by *Aspergillus ochraceus*. Phytochemistry 58:709–716

Hasan MM, Chowdhury SP, Alam S, Hossain B, Alam MS (2005) Antifungal effects of plant extracts on seed-borne fungi of wheat seed regarding seed germination, seedling health and vigour index. Pakistan J Biol Sci 8(9):1284–1289

Hossain MS, Ali MA, Mollah MI, Khan MA, Islam AS (2015) Evaluation of fungicides for the control of bakanae disease of rice caused by Fusarium moniliforme (Sheldon). Bangladesh Rice 19(1):49–55

Hubert J, Mabagala RB, Mamiro DP (2015) Efficacy of selected plant extracts against *Pyricularia grisea*, causal agent of rice blast disease. Am J Plant Sci 6:602–611

Hussain M, Ahmad S, Hussain S, Lal R, Ul-Allah S, Nawaz A (2018) Rice in saline soils: physiology, biochemistry, genetics and management. Adv Agron 148:231–287

Islam MM, Monjil MS (2016) Effect of aqueous extracts of some indigenous medicinal plants on sheath blight of rice. J Bangladesh Agric Univ 14:7–12

Itako AT, Schwan-Estrada KR, Tolentino Júnior JB, Stangarlin JR, Cruz ME (2008) Atividade antifúngica e proteção do tomateiro por extratos de plantas medicinais. Trop Plant Pathol 33: 241–244

Iwuagwu CC, Onejeme FC, Ononuju CC, Umechuruba CI, Nwogbaga AC (2018) Effects of plant extracts and synthetic fungicides on the radial growth of *Phoma oryzae* on rice (*Oryza sativa* L.) in some rice growing areas of south eastern Nigeria. J Plant Pathol Microbiol 9:468

Jahan B, Ali MA, Alam S, Moni ZR, Alam MA (2013) Invitro evaluation of antifungal activity of plant extracts against *Rhizoctonia oryzae-sativae* causing aggregated sheath spot of rice. Bangladesh J Plant Pathol 29:15–19

Jantasorn A, Moungsrimuangdee B, Dethoup T (2016) In vitro antifungal activity evaluation of five plant extracts against five plant pathogenic fungi causing rice and economic crop diseases. J Biopest 9:1–7

Jayaprakashvel M, Mathivanan N (2009) Biological control and its implications on rice diseases management. Research India Publications, Delhi, pp 440–455

John RP, Tyagi RD, Prévost D, Brar SK, Pouleur S, Surampalli RY (2010) Mycoparasitic Trichoderma viride as a biocontrol agent against *Fusarium oxysporum* f. sp. *adzuki* and *Pythium arrhenomanes* and as a growth promoter of soybean. J Crop Prot 29:1452–1459

Jose B (1997) Integrated management of sheath rot of rice (*Oryza sativa* L.), Doctoral dissertation. Department of Plant Pathology, College of Agriculture, Vellayani. 80p

Kamalakannan A, Shanmugam V, Surendran M (2001) Effect of plant extracts on susceptibility of rice seedlings to blast disease and consequent biochemical changes in rice plants. J Plant Dis Protect 108:536–543

Kandhari J, Devakumar C (2006) Plant extracts for the management of sheath blight (*Rhizoctonia solani* Kuhn) of rice. Oryza 43:293

Khan ZS, Nasreen S (2010) Phytochemical analysis, antifungal activity and mode of action of methanol extracts from plants against pathogens. J Agric Sci Technol 6:793–805

Khan MA, Ahmad S, Raza A (2019a) Integrated weed management for agronomic crops. In: Hasanuzzaman M (ed) Agronomic crops, vol 2. Springer Nature Singapore, Singapore, pp 257–281

Khan SM, Ali S, Nawaz A, Bukhari SAH, Ejaz S, Ahmad S (2019b) Integrated pest and disease management for better agronomic crop production. In: Hasanuzzaman M (ed) Agronomic crops, vol 2. Springer Nature Singapore, Singapore, pp 385–428

Kishore GK, Pande S, Harish S (2007) Evaluation of essential oils and their components for broad-spectrum antifungal activity and control of late leaf spot and crown rot diseases in peanut. Plant Dis 91:375–379

Krause RA, Webster RK (1972) The morphology, taxonomy, and sexuality of the rice stem rot fungus, *Magnaporthe salvinii* (*Leptosphaeria salvinii*). Mycologia 64:103–114

Lanoiselet VM, Cother EJ, Ash GJ (2007) Aggregate sheath spot and sheath spot of rice. J Crop Prot 26:799–808

Lin W, Anuratha CS, Datta K, Potrykus I, Muthukrishnan S, Datta SK (1995) Genetic engineering of rice for resistance to sheath blight. Nat Biotechnol 13:686–691

Lin J, Opoku AR, Geheeb-Keller M, Hutchings AD, Terblanche SE, Jager AK (1999) Preliminary screening of some traditional Zulu medicinal plants for antiinflammatory and anti-microbial activities. J Ethnopharmacol 68:267–274

Machida M (2002) Progress of *Aspergillus oryzae* genomics. Adv Appl Microbiol 51:81–106

Mbodi Y, Gaye S, Diaw S (1986) The role of Tricyclazole in rice protection against blast and cultivar improvement. J Parasitica 43:187–198

Miah MN, Karim MA, Rahman MS, Islam MS (1990) Performance of Nigersail mutants under different row spacings. Bangladesh J Train Develop 3(2):31–34

Mohana DC, Raveesha KA (2007) Anti-fungal evaluation of some plant extracts against some plant pathogenic field and storage fungi. J Agr Technol 4:119–137

Mohana DC, Prasad P, Vijaykumar V, Raveesh KA (2011) Plant extract effect on seed borne pathogenic fungi from seeds of paddy grown in southern India. J Plant Prot Res 51:101–106

Mondal D, Ghosh A, Roy D (2017) Yield loss assessment of rice (*Oryza sativa* L.) due to different biotic stresses under system of rice intensification. J Entomol Zool 5:1974–1980

Narasimhan S, Masilamani S, Muralidharan B (1998) Laboatory evaluation of botanical formulation against blast, brown spot and sheath blight diseases of rice. Pestology 22:19–21

Nayeem N, Mehta KS (2015) A review on family lamiaceae with emphasis on some medicinally important plants of the genus Clerodendrum. Int J Univ Pharm Biol Sci 4:286–302

Naz S, Fatima Z, Iqbal P, Khan A, Zakir I, Noreen S, Younis H, Abbas G, Ahmad S (2019) Agronomic crops: types and uses. In: Hasanuzzaman M (ed) Agronomic crops, Production technologies, vol 1. Springer Nature Singapore, Singapore, pp 1–18

Nwachukwu EO, Umechurba CI (2001) Antifungal activities of some leaf extract on seed-borne fungi of African yambean seeds, seed germination and seed emergence. J Appl Sci Environ Manag 5:29–32

Oh BY, Kim JH (1988) Development feasibility of water-floating fungicide formulation for the control of sheath blight caused by *Rhizoctonia solani* in rice. Agric Chem Biotechnol 31:226–233

Olufolaji DB, Adeosun BO, Onasanya RO (2015) In vitro investigation on antifungal activity of some plant extracts against *Pyricularia oryzae*. Niger J Biotechnol 29:38–43

Opara EU, Obana FT (2010) Performance of some plant extracts and pesticides in the control of bacterial spot disease of *Solanium*. Medwell Agric J 5(2):45–49

Osman KA, Abdulrahman HT (2003) Risk assessment of pesticides to human and the environment. Saudi J Biol Sci 10:81–106

Ou SH (1985) Rice diseases, 2nd edn. Common Wealth Mycological Institute, Kew, pp 61–96

Padmanabhan SY (1973) The great Bengal famine. Annu Rev Phytopathol 11:11–24

Pallant AC (2011) Bioactivity of alkaloidal fraction of *Tabernaemontana elegans* Stapf. Thesis (MagisterScientiae). Department of Pharmacology, Faculty of Health Sciences, University of Pretoria, Pretoria

Patel RM, Jasrai YT (2012) Evaluation of fungitoxic potency of medicinal plant volatile oils (VOs) against plant pathogenic fungi. Pestic Res J 23:168–171

Patel RM, Jasrai YT (2013) Evaluation of fungi toxic potency of *Piper betel* L. (Mysore variety) leaf extracts against eleven phyto-pathogenic fungal strains. Cibtec J Bio Protocols 2:21–28

Peerzada AM, Bukhari SAH, Dawood M, Nawaz A, Ahmad S, Adkins S (2019) Weed management for healthy crop production. In: Hasanuzzaman M (ed) Agronomic crops, vol 2. Springer Nature Singapore, Singapore, pp 225–256

Pinelo M, Rubilar M, Sineiro J, Nunez MJ (2004) Extraction of antioxidant phenolics from almond hulls (*Prunus amygdalus*) and pine sawdust (*Pinus pinaster*). Food Chem 85:267–273

Ranasinghe L, Jayawardena B, Abeywickrama K (2002) Fungicidal activity of essential oils of *Cinnamomum zeylanicum* (L.) and *Syzygium aromaticum* (L.) Merr et LM Perry against crown rot and anthracnose pathogens isolated from banana. Lett Appl Microbiol 35:208–211

Razaq M, Shah FM, Ahmad S, Afzal M (2019) Pest management for agronomic crops. In: Hasanuzzaman M (ed) Agronomic crops, vol 2. Springer Nature Singapore, Singapore, pp 365–384

Sarwar N, Ali H, Ahmad S, EhsanUllah AS, Mubeen K, Hill JE (2013a) Water wise rice cultivation on calcareous soil with the addition of essential micronutrients. J Anim Plant Sci 23(1):244–250

Sarwar N, Ali H, Maqsood M, EhsanUllah SAN, Shahzad M, Mubeen K, Shahid MA, Ahmad S (2013b) Phenological response of rice plants to different micronutrients application under water saving paddy fields on calcareous soil. Turk J Field Crops 18(1):52–57

Sehajpal A, Arora S, Kaur P (2009) Evaluation of plant extracts against *Rhizoctonia solani* causing sheath blight of rice. J Plant Prot Res 1:25–30

Shahzad AN, Ahmad S (2019) Tools and techniques for nitrogen management in cereals. In: Hasanuzzaman M (ed) Agronomic crops, vol 2. Springer Nature Singapore, Singapore, pp 111–126

Shamim MD, Sharma D, Bisht D, Hussain R, Khan NA, Pandey P, Kesari R, Singh KN (2017) Molecular tools for controlling of sheath blight disease of rice and its management. Biotic Stress Manage Rice 6:109

Shamsi S (1999) Colony interaction between *Sarocladium oryzae* and other fungi associated with sheath rot affected rice varieties. J Asiatic Soc Bangladesh Sci 25:369–371

Shamsi S, Chowdhury P (2016) In vitro evaluation of fungicides and some plant extracts against rice sheath rot pathogen *Sarocladium oryzae*. Bangladesh J Sci Res 29:47–54

Shamsi S, Nahar N, Chaowdhury P, Momtaz S (2010) Fungal diseases of three aromatic rice (*Oryza sativa* L.). J Bangladesh Acad Sci 34:63–70

Silva-Aguayo G (2013) Botanical insecticides. In: Radcliff's IPM World Textbook. University of Minnesota, Minneapolis. http://www.ipmworld.umu.edu/chapters/Silva-Aguayo.htm. Retrieved December 10, 2013

Singh HB, Singh UP (1980) Inhibition of growth and sclerotium formation in *Rhizoctonia solani* by garlic oil. Mycologia 72:1022–1025

Singh UP, Singh HB, Singh RB (1980) The fungicidal effect of neem (*A. indica*) extracts on some soil-borne pathogens of grain (*Icer arietinum*). Mycologia 72:1077–1093

Sunder S, Singh R, Agarwal R (2014) Brown spot of rice: an overview. Indian Phytopath 67:201–215

Sygenta (2012) Researches in crop losses (Online). www.sygenta.com/global/corporate/. Retrieved July 15, 2012

Tamuli P, Das J, Boruah P (2014) Antifungal activity of *Polygonum hydropiper* and *Solanum melongena* against plant pathogenic fungi. Plant Arch 14:15–17

Tariq M, Ahmed M, Iqbal P, Fatima Z, Ahmad S (2020) Crop phenotyping. In: Ahmed M (ed) Systems modeling. Springer Nature, Cham, pp 45–60

Tripathi RD, Srivastava HS, Dixit SN (1978) A fungitoxic principle from the leaves of Lawsonia inermis Lam. Experientia 34(1):51–52

Uda MN, Harzana SN, Shamiera SN, Hulwani IN, Akhir MA, Khairul RHM, Salimi MN, Nuradibah MA, Hashim U, Gopinath SC (2018) Antimicrobial activity of plant extracts from Aloe vera, Citrus hystrix, Sabah snake grass and *Zingiber officinale* against *Pyricularia oryzae* that causes rice blast disease in paddy plants. InIOP Conf. Series. Mater Sci Eng 318:012009

Unnikrishna PKP (1988) Studies on the effect of plant extract/products and antagonist bacterium against rice brown spot pathogen *Drechslera oryzae* (Breda de Hann). Subram and Jain MSc (Ag.) Thesis. Tamil Nadu Agricultural University, Coimbatore

Velluti A, Sanchis V, Ramos AJ, Egido J, Marin S (2003) Inhibitory effect of cinnamon, clove, lemongrass, oregano and palmarose essential oils on growth and fumonisin B1 production by *Fusarium proliferatum* in maize grain. Int J Food Microbiol 89:145–154

Vidhyasekaran P, Borromeo ES, Mew TW (1992) Helminthosporium oryzae toxin suppresses phenol metabolism in rice plants and aids pathogen colonization. Physiol Mol Plant Pathol 41:307–315

Vijayakumar S (1998) Use of plant oils and oil cakes in the management of rice diseases. MSc (Ag) Thesis. Tamil Nadu Agricultural University, Coimbatore

Wasaya A, Yasir TA, Ijaz M, Ahmad S (2019) Tillage effects on agronomic crop production. In: Hasanuzzaman M (ed) Agronomic crops, vol 2. Springer Nature Singapore, Singapore, pp 73–99

Wilson CL, Solar JM, Ghaouth AE, Wisniewski ME (1997) Rapid evaluation of plant extracts and essential oils for antifungal activity against *Botrytis cinerea*. Plant Dis 81:204–210

Yang VW, Clausen CV (2007) Antifungal effect of essential oils on southern yellow pine. Int Biodet Biodeg 59:302–306

Yasmin M, Hossain KS, Bashar MA (2008) Effects of some angiospermic plant extracts on in vitro vegetative growth of Fusarium moniliforme. Bangladesh J Bot 37(1):858

Yasui M, Oda K, Masuo S, Hosoda S, Katayama T, Maruyama J, Takaya N, Takeshita N (2020) Invasive growth of Aspergillus oryzae in rice koji and increase of nuclear number. Fungal Biol Biotechnol 7:1–15

Yeasmin F, Ashrafuzzaman M, Hossain I (2012) Effects of garlic extract, Allamanda leaf extract and Provax -200 on seed borne fungi of rice. The Agriculturists 10:46–50

Yusuf NA, Ibrahim H, Khalid N (2002) Antibacterial evaluation and tissue culture studies of selected medicinal *Curcuma* species. University of Malaya, Kuala Lumpur

Zahoor SA, Ahmad S, Ahmad A, Wajid A, Khaliq T, Mubeen M, Hussain S, Sami Ul Din M, Amin A, Awais M, Nasim W (2019) Improving water use efficiency in agronomic crop production. In: Hasanuzzaman M (ed) Agronomic crops, vol 2. Springer Nature Singapore, Singapore, pp 13–29

Zhou XG, Uppala SS (2015) Field efficacy of new fungicides for management of narrow brown leaf spot of rice. Am Phytopathol Soc 105:159–159

Nanotechnology for Rice Fungal Diseases 25

Salman Ahmad, Muhammad Ghayoor Husnain, Zafar Iqbal, Muhammad Usman Ghazanfar, Fazal ur Rehman, Irfan Ahmad, Ejaz Ashraf, Yasir Ali, Mirza Hasanuzzaman, and Shakeel Ahmad

Abstract

Rice is being used as an important source of food by a large number of people. Its production needs to be maintained to feed the increasing number of people on this planet. Various diseases are affecting this important source of food and decreasing both quality and quantity. Fungal diseases contribute a lot in this regard. Many methods have been adopted over the years to combat the diseases of rice. Use of resistant varieties and fertilizers have helped to increase yield. Similarly, chemical fungicides have been used at an immense scale to control the fungal diseases. All these techniques helped to manage the diseases, but on the other hand also raised many concerns. Surplus use of chemicals has always been a hazard for the

S. Ahmad (✉) · M. G. Husnain · Z. Iqbal · M. U. Ghazanfar · Fazal ur Rehman
Department of Plant Pathology, College of Agriculture, University of Sargodha, Sargodha, Pakistan
e-mail: salman.ahmad@uos.edu.pk

I. Ahmad
Department of Forestry, Range Management and Wildlife, University of Agriculture, Faisalabad, Pakistan

E. Ashraf
Department of Agriculture Extension, College of Agriculture, University of Sargodha, Sargodha, Pakistan

Y. Ali
College of Agriculture, Bahauddin Zakariya University, Layyah, Pakistan

M. Hasanuzzaman
Department of Agronomy, Shere-e-Bangla Agriculture University, Dhaka, Bangladesh

S. Ahmad
Department of Agronomy, Faculty of Agricultural Sciences and Technology, Bahauddin Zakariya University, Multan, Pakistan
e-mail: shakeelahmad@bzu.edu.pk

N. Sarwar et al. (eds.), *Modern Techniques of Rice Crop Production*,
https://doi.org/10.1007/978-981-16-4955-4_25

environment. Excessive use of certain chemicals and fertilizers indeed increased the susceptibility of rice to certain fungal diseases. Use of chemical-based fungicides over the years has developed resistance in fungi toward these fungicides. We have now come across many new methodologies to managing the fungal diseases of rice like use of Nanotechnology. In this chapter, needs and benefits of nanotechnology in the management of rice fungal diseases have been discussed. This chapter focuses on major fungal diseases of rice and their management using various nanotechnological methods like use of nanoparticles.

Keywords

Nanotechnology · Nanoparticles · Management · Silver nanoparticles · Copper nanoparticles · Mesoporous silica nanoparticles · Alumino-silicate nanoparticles

25.1 Introduction

Rice (*Oryza sativa* L.) is an important food crop all over the world (Ahmad et al. 2015, 2019, 2012, 2013, 2008, 2009). It is the staple food in most of the Asian countries including Philippines, Bangladesh, and Pakistan. Rice is the most important cereal grain of the world after wheat and maize (Ahmad and Hasanuzzaman 2012; Ahmed and Ahmad 2019, 2020; Shahzad and Ahmad 2019). Out of the total consumption of rice in the world, Asia accounts for the 90% (Ahmed et al. 2017, 2020a, b; Akram et al. 2018, 2019; Hussain et al. 2018; Fahad et al. 2019; Fatima et al. 2020). The total production of rice in the world is approximately 755 million tonnes a year and Asia alone produces annually 677 million tonnes of rice (FAOSTAT 2019). Rice being a very important source of food globally, plays an important role in the food security of the world (Hafeez-ur-Rehman et al. 2019; Khan et al. 2019a, b; Naz et al. 2019; Peerzada et al. 2019; Razaq et al. 2019; Sarwar et al. 2013a, b; Tariq et al. 2020; Wasaya et al. 2019; Zahoor et al. 2019). Many types of fungal diseases affect the yield of rice each year. Fungal diseases have a negative impact on rice both in qualitative and quantitative terms (Ou 1985). Diseases of rice in which fungi are the causal organisms cause about 14% yield loses of rice. Hunger is prevailing in the African countries and in some Asian countries. In sub-Saharan Africa, about 239 million people are suffering from hunger (Sasson 2012). Rice is a major source of food in these countries, especially in the lower-middle class and poor families. Across the globe, daily calories of more than three and a half million people depend upon rice. Thus, fungal diseases of rice are potential threat to the security of food around the globe (Fones et al. 2020).

Human population is increasing at an immense rate (Gilland 2002). The traditional methods and means used for the management of diseases of rice caused by fungi are not very efficient and cannot promise the food security. Further the efficiency of fungicides reduces if pathogen develops resistance (Kongcharoen et al. 2020). Moreover, they are indeed a threat to the environment and sustainability of the world. We use harmful chemical fungicides to manage the fungal diseases, which are either applied in the form of foliar sprays (Groth 1993) or through the irrigation water. This increases the agricultural consumption of water and much of the fungicide is lost in the environment and causes damages to other living organisms including man.

Conventional or traditional ways of controlling fungal diseases include use of chemical fungicides on a vast scale (Morton and Staub 2008). These chemicals are not specific to the diseased tissues rather they affect the whole plant and their residues remain inside them (Liu et al. 2016). Thus, organisms who consume such plants can have health problems if the toxicity levels are high. Due to these factors, the need arises for such an efficient method that helps to prevent the fungal diseases and also has less influence on the environment.

25.2 Nanotechnology

Nanotechnology is the technology of modern era. Nanotechnology utilizes nanoparticles or nanocrystals that are of very minute size. Nanoparticles have the ability to carry the fungicides to the specific tissue of the plant, rather than being applied to the whole plant (Shang et al. 2019). Nanotechnology has the ability to revolutionize the agriculture sector and our food systems (Norman and Hongda 2013). It can help saving precious water being used in foliar sprays of fungicides (Kluge et al. 2017) or in the application of fungicides through irrigation water. The use of nanotechnology to manage the rice fungal diseases can be of immense importance, if used properly and accurately. The uses of nanotechnology in agriculture and the detail of the history of nanotechnology have been given in Figs. 25.1 and 25.2, respectively.

25.3 Need of Nanotechnology

With increasing use of chemical fungicides combating fungal diseases, and their reducing efficiencies due to development of resistance against them by many fungi (Hahn 2014), we need to shift toward a more efficient method. The human population is also increasing day by day, thus each year demand for food is increasing at a faster rate (Tilman et al. 2011). Because of the fungicide resistance, control of plant diseases has become an arduous task (Deising et al. 2008). The conventional practices of disease management are not enough. The conventional practices impose serious threats to biodiversity (Aktar et al. 2009). In order to save the biodiversity, it is required that we review the ways of management of fungal diseases; thus, here comes the need of nanotechnology (Patel et al. 2014).

As nanotechnology can be used to treat and prevent fungal infections by only targeting the specific plant tissues, this helps in reducing the amount of chemical fungicides and eventually preventing the environment from degradation (Shang et al. 2019). Nanotechnology utilizes highly specified nanoparticles. These nanoparticles

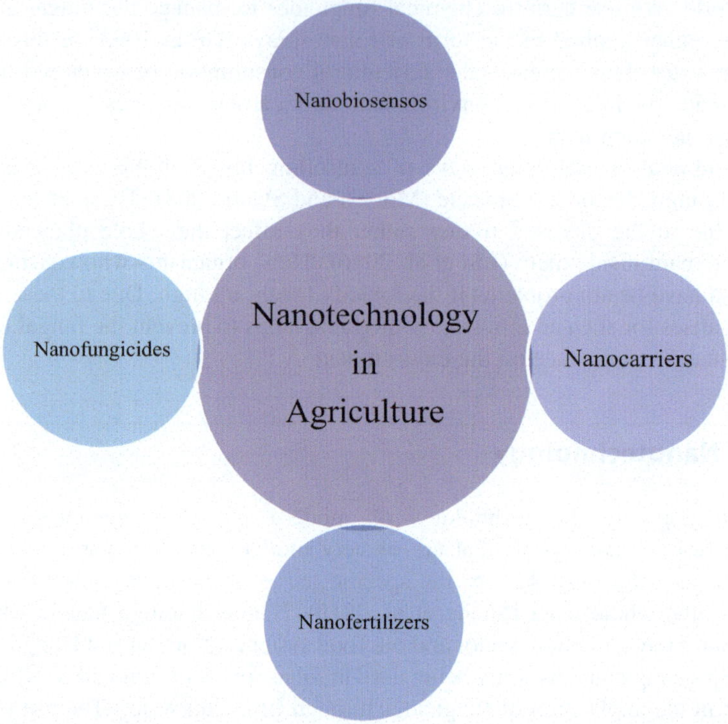

Fig. 25.1 Uses of nanotechnology in agriculture

increase plant vigor and reduce infections by carrying fertilizers and fungicides to the required places. This increases absorption rate of fertilizers and fungicides, and reduces wastage (Shang et al. 2019).

25.4 Nanoparticles

Many types of nanoparticles are in use. Attention of researchers is caught by nanoparticles due to their vast applications in various fields including agriculture (Azharuddin et al. 2019). These nanoparticles include silver nanoparticles, copper nanoparticles, mesoporous silica nanoparticles, and aluminum nanoparticles (Patel et al. 2014). Bio-reduction of nanomaterials achieved by both in vitro and in vivo processes leads to the synthesis of nanoparticles (Singh et al. 2018). Synthesis of nanoparticles involves reducing agents and stabilizing agents.

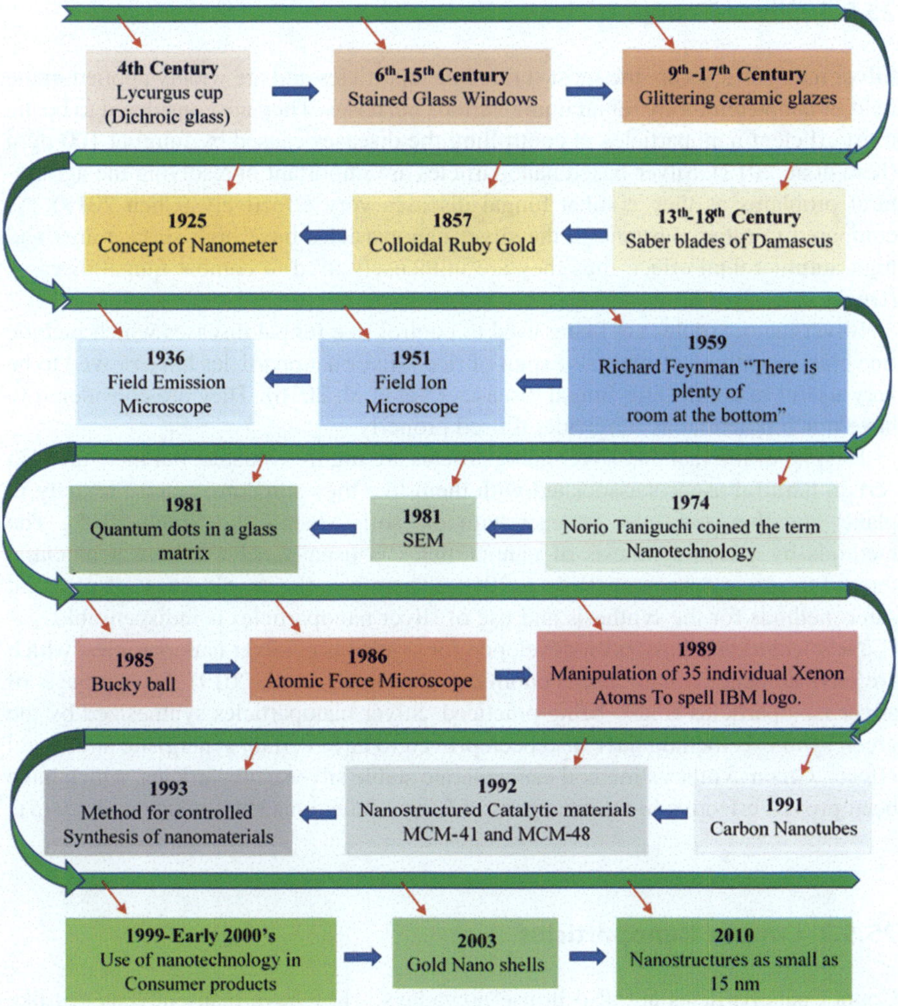

Fig. 25.2 History of nanotechnology

25.5 Nanoparticles and Control of Phytopathogenic Fungi

Few important nanoparticles that are being used in management of plant fungal diseases are described in detail in the following sections.

25.5.1 Silver Nanoparticles

Silver nanoparticles are the most studied nanoparticles and are widely applied in the field of nanotechnology for attaining various purposes. They are considered to be the most efficient nanoparticles in controlling the diseases caused by fungi at 100 ppm (Kim et al. 2012). Silver-based nanoparticles are important in resolving the agricultural problems as they combat fungal diseases very effectively (Chen 2018). As compared to other substances, the silver nanoparticles have high surface area and high antimicrobial effect, thus they are immensely used to combat fungal diseases (Durán et al. 2010).

Silver nanoparticles are being used to control rice fungal diseases which include rice blast, sheath blast, and false smut of rice. Silver nanoparticles have proved to be very useful in controlling fungal diseases (Xia et al. 2016). They are considered to have much potential as fungicide, if used properly.

Despite of the fact that silver nanoparticles are highly valuable, but there are also certain harmful aspects associated with them like they sometimes cause toxicity in plants and they may also interact with the soil (Mishra and Singh 2015). The methods by which synthesis of nanoparticles is usually achieved are a potential threat for our environment (Wilson 2018). Therefore, the development of new and safer methods for the synthesis and use of silver nanoparticles is indispensable.

New techniques have been developed for synthesis of silver nanoparticles, which are environmentally safe and economical (Aritonang et al. 2019). Biosynthesis of silver nanoparticles is also being practiced. Silver nanoparticles synthesized by the green synthesis method have also been proved to have certain synergistic antifungal effects. Green synthesis method can generate stable silver nanoparticles, which have been proved effective in the treatment of fungal infections (Mallmann et al. 2015).

25.5.2 Copper Nanoparticles

Copper nanoparticles are also in use nowadays. They have many advantages like they have high thermal conductivity (Saterlie et al. 2011). They are easily available and are cost efficient. They can act as catalysts (Gawande et al. 2016). Copper nanoparticles can be synthesized by different ways including chemical and electrochemical reduction (Jain et al. 2014). These techniques mostly involve use of expensive equipment and harmful chemicals. A new way of synthesis of copper nanoparticles has been developed called green synthesis method (Rafique et al. 2017). The Green synthesis method involves use of eco-friendly and economical chemicals, which have less harmful effects on environment. The green synthesized nanoparticles have been proved to be efficient in vitro against fungi including *Neofusicoccum* sp., *Fusarium oxysporum*, and *Fusarium solani* (Pariona et al. 2019). Copper nanoparticles have potential of being efficient fungicides in future.

25.5.3 Carbon Nanoparticles

Carbon is the basic brick element for various simple and complex molecular structures due to its unique characteristics. Different properties of materials can be observed at nano-levels. Nanosciences have developed a new hope to solve the existing problems of agriculture sector like extensive and improper use of chemical fertilizers, pesticides, and fungicides. The agriculture research corporations in Brazil are putting a major focus on the use of carbon nanoparticles as pesticide carriers. Carbon nanotubes are getting major attention of the scientists nowadays. Carbon nanotubes are carbon allotropies having cylindrical shaped nanostructure. It is reported that when seeds of tomato were planted in carbon nanotubes containing soil, carbon nanotubes penetrated the hard seed coat of the tomato seeds and enhanced the growth of seedlings (Khodakovsky et al. 2000). The increase in growth was due to enhanced water uptake due to penetration of carbon nanoparticles. Carbon nanoparticles have the potential to prevent seed diseases by delivering the desired molecules to seeds at germinating stage, without any toxicity or adverse effects on the growth of plant (Mukherjee et al. 2016).

25.5.4 Alumino-Silicate Nanoparticles

Nano scale formulation of efficient pesticides is now being done by many well-known chemical companies. Using the loaded aluminum-silicate nanotubes with active ingredients is an important effort in this regard. Alumino-silicate nanotubes are advantageous because they get picked up by insect hairs easily, when sprayed on plant surface. Being environmentally safe, they are also more active biologically (Sharon et al. 2010; El-Naggar et al. 2020).

25.5.5 Mesoporous Silica Nanoparticles

Mesoporous silica nanoparticles have the potential of being a powerful tool for targeted delivery into the cells of the plants, as they have been shown to deliver DNA and chemicals into the cells of the plants (Wang et al. 2002). Spherical shaped and porous silica nanoparticles systems have been developed. These particles have arrays of independent porous channels. Molecules or chemicals can be filled in honeycomb-like structures formed by these channels. Chemical is sealed inside these nanoparticles which have a unique capping strategy (Downing and Jain 2020).

25.5.6 Nano-Emulsions

Emulsions are mixtures of two or more liquids (e.g., oil and water) that are not combined easily. Diameters of dispersed droplets are 500 nm or less in nano-emulsion. Reduction in chemical degradation is facilitated by nano-emulsions, as they can encapsulate functional ingredients within their droplets (McClements and Decker 2000).

25.6 Management of Fungal Diseases of Rice and Nanotechnology

Many types of fungi cause diseases in rice. Fungi affect almost all parts of the rice plant from leaves to roots. Fungal diseases not only reduce the annual yields of rice (Nalley et al. 2016), but also cause huge economic and environmental loses in terms of using various fungicides (Mew et al. 2004).

25.7 Need of Nanotechnology for the Management of Fungal Diseases of Rice

The chemical methods as well as other traditional methods are not enough to control rice diseases. The excessive use of fungicides is also discouraged as this can lead to development of resistance in the fungus against fungicides (Deising et al. 2008). Also, the chemical fungicides pose a serious threat to already degrading environment. Thus, in this scenario we need new and efficient methods to control the fungal diseases of rice. Here comes the need of nanotechnology in controlling these diseases. As in nanotechnology the use of fungicides is reduced to a greater extent as we use nanoparticles (nanocarriers) to carry the fungicides to the desired tissues (Campos et al. 2015), thus having less influence on the environment and the biodiversity.

Important fungal diseases of rice, their conventional management, and uses of nanotechnology in their management are described in detail in this chapter.

25.8 Rice Blast Disease

Rice blast is a major and devastating disease of the rice plants. It affects almost all the parts of the rice plant. It is caused by the fungus *Magnaporthe oryzae* (this fungus was previously known as *Magnaporthe grisea*), a filamentous Ascomycete. It is the first plant pathogenic fungus whose genome was completely sequenced and was released to the public. Even if the infection of this pathogen is moderate in the rice fields, yet it can cause up to 50% yield loses (Choi et al. 2018; Asibi et al. 2019).

25.8.1 Symptomology

The symptoms of this disease include the appearance of lesions and spots on various parts of the rice plant. The morphology of the symptoms depends upon the factors like stage of lesion and resistance of that particular variety, and also the environmental conditions play an important role in this regard (Ou 1985).

25.8.2 Leaf Blast

When this fungus infects the leaves, lesions begin to appear on the leaf blade and this is known as leaf blast. Various factors like age of plant or lesion and resistance of that particular rice variety have effect on shape and size of the lesions. This disease is the most destructive disease for rice plants (Fisher et al. 2012). This disease of rice can cause immense yield losses up to 100% in severe cases. The lesions that develop are oval-shaped or they can be spindle shaped that resembles toothpick (Asibi et al. 2019).

25.8.3 Node Blast

Node is also infected by this fungus. When the infection occurs on the node and the symptoms start to appear on the node it is called as the Node Blast. This causes complete death of the stem above the point of infection. In this type of blast, node of the stem changes its color to black and breaks easily (Devi and Sharma 2010).

25.8.4 Neck Blast

In the neck blast, rice panicle is infected. Infected neck seems to be bordered by the grayish-brown lesions (Devi and Sharma 2010). Neck blast reduces the size of rice grains and has negative effects on yield and seed quality (Khan et al. 2014). It causes falling of panicle when the severity of the infection is intense. If the neck blast occurs before the milking stage, no grain is formed. If the neck blast occurs after the milking stage, then the quality of the grains gets poor. Few varieties of rice have developed resistance against neck blast, but many are still susceptible (Titone et al. 2015).

25.8.5 Collar Blast

When the pathogen infects collar region, then it is known as Collar Blast. In severe cases, it can result in killing of the entire leaf blade. Dark green bordered lesions or spots appear on the collar region (Devi and Sharma 2010). These lesions can be of white to gray-green in color.

25.8.6 Conventional Management

Many traditional or chemical methods are used to control the blast disease of rice. Mostly systemic fungicides including triazoles and strobilurins are being used to control this disease (Kongcharoen et al. 2020). Breeding programs are also in practice to produce resistant varieties, but failing due to high variability of this fungus. Other ways include foliar sprays of Bavistin (0.1%) (Rijal and Devkota 2020) and biological control methods like use of *Trichoderma* spp. to control rice blast disease (Chou et al. 2020).

25.8.7 Use of Nanotechnology for the Management of Rice Blast

25.8.7.1 Silver Nanoparticles against Rice Blast

Silver nanoparticles can be used in low quantities to manage rice blast because they have high surface area and are highly reactive (Sharon et al. 2010). They can penetrate into microbial cells at low concentrations and can act as antimicrobial agents (Lamsal et al. 2011). They cause plasmolysis of fungal hyphae by damaging walls of it (Elamawi and El-Shafey 2013). Silver nanoparticles of usually 100 nm are used to manage rice blast, but nanoparticles of size 20 or 30 nm can also be used (Elamawi and El-Shafey 2013). Nanoparticles of specific size should be used, as the antimicrobial activity of silver nanoparticles is influenced by their size and shape (Mishra et al. 2014). Nanoparticles can be prepared by suspending nano powder in deionized sterile water or by dilution of stock solution using sterile deionized water. Various concentrations can be used like 100, 200, or 400 ppm, but concentration of 100 ppm is proven to be most efficient against the rice blast (Elamawi and El-Shafey 2013). The silver nanoparticles work efficiently in inhibiting the blast disease. But once the infection has occurred in the rice plants, then it becomes very difficult to control this disease. This is the reason that antifungal activity of silver nanoparticles decreases after fungal inoculation. Time of application of nanoparticles has impact on their ability to control disease. It has been reported that when 24 h have been passed after inoculation, antifungal activity of silver nanoparticles reduces (Jo et al. 2009). Moreover, the inhibition of disease is greatly dependent on direct contact of silver nanoparticles with fungal spores or germ tubes (Jo et al. 2009). Thus, while using silver nanoparticles against rice blast we should do their timely application in an accurate manner in order to prevent infection and increase their efficiency.

25.8.7.2 Nano-Particles from Extracts of *Chaetomium* spp. Against Rice Blast

Extracts obtained from *Chaetomium* species have been proved as antifungal agents against number of plant diseases (Soytong et al. 2001). Nanoparticles from *Chaetomium* can be yielded by incorporating crude extracts obtained from *Chaetomium* species into polylactic acid-based nanoparticles through electrospinning (Dar and Soytong 2014). Nanoparticles formed from extracts of *Chaetomium* species can be used to control the rice blast disease. It has been proved by bi-culture tests that nanoparticles from *Chaetomium* species can effectively reduce growth of the fungus that causes blast disease in rice plant (Song and Soytong 2018). They alter the shape of fungal spore and make them abnormal.

25.9 Bakanae Disease of Rice

Bakanae of rice is also very important disease. It is caused by the heterothallic ascomycete fungus *Fusarium fujikuroi*. It causes huge loses to rice crop each year. This disease was first identified in Japan during the year 1828. The indication of elongation of rice seedlings due to the stimuli induced by the hyphae of the *Fusarium hetrosporium* was given by scientists. This disease is called Bakanae disease because of hypertrophic and elongation effect (also called as Bakanae effect), which was demonstrated from culture filtrate of dried rice seedlings and plants by Kurosawa (1926).

This disease was considered as minor disease in the old times, but in the recent times in Asia, it has emerged as a major disease of rice due to its higher severities. The losses in the yield vary with the region in which rice is grown and also with the cultivar. It causes immense losses ranging from 3 to 95.4% (Gupta et al. 2015). Khokhar and Jaffrey (2002) first reported this disease in Pakistan. In Pakistan, the yield losses due to Bakanae disease of rice are 10–50% (Khokhar and Jaffrey 2002).

25.9.1 Symptomology

In this disease, seedlings elongate in an abnormal way and stems become slender. This happens due to excessive secretion of gibberellin by the fungus *Fusarium fujikuroi.* This disease is also called as the Foot rot of rice. The name foot rot was given to this disease in India, because of development of adventitious roots from the lower nodes of stem (Thomas 1931). It also has other names like the "Foolish seedling" disease or "Stupid Rice." It is called so because the diseased plants grow several inches taller than the healthy plants. The plants affected by Bakanae disease are thin and they give a yellowish green appearance.

The stems of the affected plants grow taller and they appear to be chlorotic due to the loss of chlorophyll (Ou 1985). The symptoms on the leaves include their color changing to yellowish green in later stages. Yellowish green flag leaves appear horizontally. Most plants die before reaching the maturity stage (Gupta et al. 2015), but few survive and reach the maturity stage, such plants have only a few tillers that are taller than the normal size. Normal panicles and grains are not produced by the plants infected with this disease at maturity stage.

25.9.2 Conventional Management

Clean seeds should be used for sowing of the rice crop. As this fungus is mainly seed borne, so seed treatment is the most commonly used method for the management. Seed treatment can be done by various ways like using hot water or chemical fungicides like Bavistin or Benlate. Treating the seedling with Benlate (0.1%) for 6–8 h can reduce the occurrence of Bakanae disease up to 92% (Bagga and Sharma 2012). Another practice that can be used to reduce the incidence of this disease is the use of salt water, as it will separate the seeds which are lighter in weight from the normal. Other fungicides like Daconil can also be used (Iqbal et al. 2013), but use of excessive fungicide may have deteriorating impact on the environment.

25.9.3 Use of Silica Nanoparticles for the Management of Bakanae Disease of Rice

Silica nanoparticles can be used as a nanotechnology tool to control the Bakanae disease of rice. They have many benefits for the plants; increase yields and growth (Rastogi et al. 2019). They also increase the rate of germination of seeds (Nair et al. 2011). Silica nanoparticles have the ability to reduce the production and release of spores of Bakanae disease fungus (Elamawi et al. 2020). They increase plant resistance against various biotic factors and thus have proved beneficial in disease control (Shang et al. 2019).

They are economical and eco-friendly for the management of Bakanae disease in rice, as silica nanoparticles can also be prepared by using rice husk (Elamawi et al. 2020). Rice husk is a major waste product of the rice mills. This rice husk contains a large amount of silica, which can be used to prepare silica nanoparticles. Along with large quantity of silica, some impurities are also present in rice husk, which can be removed by various acid treatments (Bakar et al. 2016). Mesoporous silica nanoparticles, due to their porous nature, can be used as potential carriers to carry fungicides and then deliver them to required places (Xu et al. 2017).

Silica nanoparticles can be applied through foliar applications in the fields of rice. After the appearance of Bakanae disease symptoms in rice, foliar applications of silica nanoparticles effectively reduce the disease (Elamawi et al. 2020).

25.10 Brown Leaf Spot of Rice

Brown leaf spot of rice causes huge yield loses globally each year (Barnwal et al. 2013). It is caused by fungus *Bipolaris oryzae* (Quintana et al. 2017). It causes yield loses up to 90%, when it appears in epidemic forms. Great Bengal famine was caused by brown leaf spot of rice in year 1943 (Padmanabhan 1973). *B. oryzae* infects various parts of rice plants including rice leaves and panicles, glumes, and coleoptiles (Iqbal et al. 2015). The infected seeds can be a source of spread of this disease. When infected seedlings are transplanted and they grow into plants, those will be bearing fungal spores. These spores spread to other healthy rice plants and make them infected (Ahmed et al. 2002).

25.10.1 Symptomology

When the infection of fungus starts on rice plant, symptoms like appearance of dark brown spots begin to appear on the leaves (Ou 1985). As the infection continues, brown lesions which are oval-shaped and have grayish centers, appear on leaves and coleoptiles. Glumes and panicles affected by this fungus show dark brown to black spots (Quintana et al. 2017).

25.10.2 Conventional Management

Treatment of seeds with hot water for 10–12 min at 53–54 °C before sowing. Various fungicides are used for seed treatment like Carbendazim and Mancozeb (Faruq et al. 2015). As the incidence of disease is severe at tillering stage, at that stage spray of Amistar (Azoxystrobin) significantly reduces the disease (Hossain et al. 2011.

25.10.3 Use of Nanotechnology in the Management of Brown Leaf Spot

Nanoparticles have also been used against brown leaf spot of rice. Zinc oxide nanoparticles have been proved efficient in controlling the disease at the concentrations of 25–50 ppm by reducing the growth and spore formation of *B. oryzae* (Elamawi et al. 2016). Foliar application of these nanoparticles at certain stages of plant growth can significantly reduce the disease symptoms (Elamawi et al. 2016).

25.11 False Smut of Rice

False smut of rice has become an important disease in all rice growing regions of the world, because it causes significant losses in the yield of rice, thus becoming threat to food security around the globe (Jiehua et al. 2019). This disease caused yield losses of around 158.6 million annually in China alone (Lu et al. 2015). False smut of rice caused by ascomycete fungus *Ustilaginoidea virens* (Cook). Climate change, the immense use of nitrogenous fertilizers for getting higher yields, and growing of high yielding rice cultivars are the major contributing factors in increase of this disease (Jiehua et al. 2019).

25.11.1 Symptomology

The appearance of false smut balls on panicles of rice is typical symptom of this disease (Guo et al. 2012). False smut balls can be yellowish-orange to greenish-black in color (Fan et al. 2016). The fungus mostly attacks the rice floral parts. Studies have shown that rice stamens are necessary for the fungus to form false smut balls (Fan et al. 2020). The interiors of smut balls consist of fungal hyphae and chlamydospores. The chlamydospores when released are orange in color and then become dark green to black with the time.

25.11.2 Conventional Management

Various conventional methods are practiced to control false smut of rice like removing of infected rice panicles and debris after the harvesting, and using certified seeds. Fungicides like trifloxystrobin are also used to manage false smut of rice. Fungicides should be sprayed before the appearance of smut balls, as after the appearance of symptoms the efficiency of fungicides is reduced tremendously (Huang et al. 2019).

25.11.3 Use of Nanotechnology in Managing False Smut Disease of Rice

As the field of nanotechnology is prospering, its use in combating plant fungal diseases is also becoming important (Patel et al. 2014). Various nanoparticles are being used to manage false smut of rice like copper nanoparticles, silicon carbide nanoparticles, silver nanoparticles, and aluminum nanoparticles. But silver nanoparticles have proved to be most efficient nanoparticles at 100 ppm against false smut of rice, causing inhibition of this disease up to 67.11% (Bhargava et al. 2018). Copper sulfide nanoparticles can also be used to manage the false smut of rice as they have significant antifungal effects on *Ustilaginoidea virens* (Barmota et al. 2018).

25.11.4 Sheath Blight of Rice

Sheath Blight is considered one of the most important disease of rice, because this disease has the ability to cause immense yield loses if occurred in epidemic forms (Savary et al. 2001). Sheath blight of rice is caused by fungus *Rhizoctonia solani*. This disease was first reported by Miyake in Japan in the year 1910. Now, this is present in almost all major rice growing areas of the world.

25.11.5 Symptomology

Greenish-gray, elliptical, water-soaked spots having brown margins appear on the leaf sheath. The fungus mainly affects the sheath of leaf and leaf blade, but other parts like emerging panicles can also be affected. Fungal mycelium grows on the surface of the sheath of rice leaf (Ou 1985). The infection occurs through sclerotia which germinate on rice sheaths and form appressoria (Richa et al. 2016).

25.11.6 Conventional Control

Chemical fungicides like Carbenzadim and hexoconazole are used to control this disease. Other fungicides like pencycuron are also considered to be efficient in the control of sheath blight of rice (Lore et al. 2005). Bi-control methods like using certain isolates of antagonistic bacteria can also be used to control this disease (Bashar et al. 2010).

25.11.7 Use of Nanotechnology to Manage Sheath Blight of Rice

Various nanoparticles and nano-chemicals have the potential to control sheath blight of rice. Nano-formulations of azomethines have proved to be efficient against sheath blight of rice and equipotential to fungicide hexaconazole. Polyethylene-based nano forms of azomethines have been proved to have better antifungal effects than chemical fungicides to control the sheath blight of rice under certain conditions (Mondal et al. 2017).

Silver nanoparticles can also be used to control this disease. Silver nanoparticles have been proved to inhibit sclerotia formation of *R. solani* up to 92% at 50 ppm and prevent growth of fungal mycelia up to 85% at the same concentration under certain conditions (Nejad et al. 2017). Silver nanoparticles have the ability to penetrate into microbial cells and can cause disruption of their membranes (Choi et al. 2008). Antifungal effect of silver nanoparticles against causal organism of sheath blight is dose dependent and can significantly reduce the development of lesions on leaves of rice (Nejad et al. 2017). Biologically synthesized silver nanoparticles from *Pseudomonas fluorescens* also have the potential to manage sheath blight of rice (Chiranjeevi et al. 2018).

25.12 Nanotechnology and Diagnosis of Rice Fungal Diseases

Among several other uses of nanoparticles, their use as diagnostic agents is also of great importance (Khiyami et al. 2014). Conventional methods used for disease diagnosis in plants are less efficient and take much time to give results (Martinelli et al. 2015), and also require costly equipment. Nanotechnology-based techniques can help improve the accuracy of disease diagnosis (Hussain 2017). They require less costly equipment and thus are economical (Khiyami et al. 2014). Nanoparticles have the ability to efficiently separate biological molecules like DNA (Khaliq et al. 2012). They can be used in Polymerase Chain Reaction (PCR) to improve the extraction of DNA thus increasing efficiency of PCR technique for the diagnosis of rice diseases, including fungal diseases. Nanoscale technologies can help to give faster results and accurate results in the diagnosis of plant diseases (Li et al. 2020).

25.13 Future Outlook

Nanotechnology has the potential to revolutionize the technologies of pest management which are presently in use and can also be useful in providing solutions for other agricultural issues (Baker et al. 2017). Nano fungicides can offer many advantages; increase the solubility of poorly water-soluble fungicides; increase shelf life; reduce the toxicity produced by fungicides; increase their efficiency; can help to overcome fungicide resistance; can ensure specified delivery of active ingredients to their target; and are safer for environment and humans. Regulatory bodies have not yet clearly defined that what is a nanopesticide and what is not (Kookana et al. 2014). Still very less data is available on whether the nanoparticles can reduce the incidence of fungicide resistance. Only 10 out of 49 fungicides of FRAC groups have been loaded onto nanoparticles and tested up till now. Nanotechnology requires complex analytical tools to generate regulatory parameters for risk assessment (Worrall et al. 2018). Presently the fungicides are applied in a rotatory manner to prevent fungicide resistance; thus, for their use at commercial scale a diverse range of nanofungicides would be required in future. Their safety in field applications and trials and their expensive production and uncertainties in terms of legislation are certain factors that should be considered while developing nanopesticides (Mishra et al. 2017). Approval from regulatory bodies regarding the use of nanofungicides can be attained by using modern ways and methods that can give faster results and provide assessment of the risks associated with the use of nanofungicides. Nanoparticles suitable for a particular purpose can only be selected by having a deep understanding of various structural properties of nanoparticles. These properties include loading capacity, size, and shape. This higher level of understanding can only be attained if experts from various fields of science integrate their skills and efforts. Various studies at different levels like cellular, organism, and ecosystem levels are required to ensure a safe use of nanofungicides. Nanotechnology can help to ensure the reduction in diseases, and could increase production of crops in future. This could happen only if its efficient ways of usage at field level are derived.

25.14 Conclusion

Nanotechnology can play an important part in effective and accurate management of rice diseases. This technology can help in rapid diagnosis of diseases caused by fungi in rice. But, at the present time, nanotechnology is still limited to lab-based tests and requires to be used very accurately at field level. The improper use of nanotechnology can also be a potential hazard for our environment. Integrated efforts of experts from various disciplines are required in this regard. Efforts should be made to increase the use nanotechnology for managing the rice fungal diseases. Governments should acknowledge the need of communicating this important technology to our farmers, so that they can adopt it and use it. This technology has a very big potential to combat fungal diseases of rice and the diseases of other important agricultural crops. Increase in the use of nanotechnology to manage rice fungal diseases, along with its use in a very accurate and professional way, can help to ensure reduction in diseases. It will also reduce usage and wastage of chemicals, and will create lesser impact on our environment.

References

Ahmad S, Hasanuzzaman M (2012) Integrated effect of plant density, N rates and irrigation regimes on the biomass production, N content, PAR use efficiencies and water productivity of rice under irrigated semiarid environment. Not Bot Horti Agrobot Cluj Napoca 40(1):201–211

Ahmad S, Zia-ul-Haq M, Ali H, Shad SA, Ammad A, Maqsood M, Khan MB, Mehmood S, Hussain A (2008) Water and radiation use efficiencies of transplanted rice (*Oryza sativa* L.) at different plant densities and irrigation regimes under semi-arid environment. Pak J Bot 40(1): 199–209

Ahmad S, Ahmad A, Zia-ul-Haq M, Ali H, Khaliq T, Anjum MA, Khan MA, Hussain A, Hoogenboom G (2009) Resources use efficiency of field grown transplanted rice (*Oryza sativa* L.) under irrigated semiarid environment. J Food Agric Environ 7(2):487–492

Ahmad S, Ahmad A, Soler CMT, Ali H, Zia-Ul-Haq M, Anothai J, Hussain A, Hoogenboom G, Hasanuzzaman M (2012) Application of the CSM-CERES-Rice model for evaluation of plant density and nitrogen management of fine transplanted rice for an irrigated semiarid environment. Precis Agric 13(2):200–218

Ahmad S, Ahmad A, Ali H, Hussain A, Garcia y Garcia A, Khan MA, Zia-Ul-Haq M, Hasanuzzaman M, Hoogenboom G (2013) Application of the CSM-CERES-Rice model for evaluation of plant density and irrigation management of transplanted rice for an irrigated semiarid environment. Irrig Sci 31(3):491–506

Ahmad A, Ashfaq M, Rasul G, Wajid SA, Khaliq T, Rasul F, Saeed U, Rahman MH, Hussain J, Baig IA, Naqvi AA, SAA B, Ahmad S, Naseem W, Hoogenboom G, Valdivia RO (2015) Impact of climate change on the rice–wheat cropping system of Pakistan. In: Hillel D, Rosenzweig C (eds) Handbook of climate change and agro-ecosystems: The Agricultural Modeling Intercomparison and Improvement Project (AgMIP) integrated crop and economic assessments. Imperial College Press/American Society of Agronomy, London/Hackensack, pp 219–258

Ahmad S, Abbas G, Ahmed M, Fatima Z, Anjum MA, Rasul G, Khan MA, Hoogenboom G (2019) Climate warming and management impact on the change of rice-wheat phenology in Punjab, Pakistan. Field Crop Res 230:46–61

Ahmed M, Ahmad S (2019) Carbon dioxide enrichment and crop productivity. In: Hasanuzzaman M (ed) Agronomic crops, vol 2. Springer Nature Singapore, Singapore, pp 31–46

Ahmed M, Ahmad S (2020) Systems modeling. In: Ahmed M (ed) Systems modeling. Springer Nature, Cham, pp 1–44

Ahmed MF, Khalequzzaman KM, Islam MN, Anam MK, Islam MT (2002) Effect of fungicides against Bipolaris oryzae of rice under in vitro condition. Plant Pathol J 1:4–7

Ahmed M, Fayyaz-ul-Hassan, Ahmad S (2017) Climate variability impact on rice production: adaptation and mitigation strategies. In: Ahmed M, Stockle C (eds) Quantification of climate variability, adaptation and mitigation for agricultural sustainability. Springer, Cham, pp 91–111

Ahmed M, Ahmad S, Raza MA, Kumar U, Ansar M, Shah GA, Parsons D, Hoogenboom G, Palosuo T, Seidel S (2020a) Models calibration and evaluation. In: Ahmed M (ed) Systems modeling. Springer Nature Singapore, Singapore, pp 151–178

Ahmed M, Ahmad S, Waldrip HM, Ramin M, Raza MA (2020b) Whole farm modeling: A systems approach to understanding and managing livestock for greenhouse gas mitigation, economic viability and environmental quality. In: Waldrip HM, Pagliari PH, He Z (eds) Animal Manure, ASA Special Publication 67. American Society of Agronomy/Soil Science Society of America, Madison, WI, pp 345–371

Akram R, Turan V, Hammad HM, Ahmad S, Hussain S, Hasnain A, Maqbool MM, Rehmani MIA, Rasool A, Masood N, Mahmood F, Mubeen M, Sultana SR, Fahad S, Amanet K, Saleem M, Abbas Y, Akhtar HM, Hussain S, Waseem F, Murtaza R, Amin A, Zahoor SA, Sami ul Din M, Nasim W (2018) Fate of organic and inorganic pollutants in paddy soils. In: Hashmi MZ, Varma A (eds.), Environmental pollution of paddy soils. Springer Nature Switzerland, Cham, pp 197–214

Akram R, Fahad S, Masood N, Rasool A, Ijaz M, Ihsan MZ, Maqbool MM, Ahmad S, Hussain S, Ahmed M, Kaleem S, Sultana SR, Mubeen M, Saud S, Kamran M, Nasim W (2019) Plant growth and morphological changes in rice under abiotic stress. In: Hasanuzzaman M, Fujita M, Nahar K, Biswas JK (eds) Advances in rice research for abiotic stress tolerance. Woodhead, Cambridge

Aktar W, Sengupta D, Chowdhury A (2009) Impact of pesticides use in agriculture: their benefits and hazards. Interdiscip Toxicol 2:1–2

Aritonang HF, Koleangan H, Wuntu AD (2019) Synthesis of silver nanoparticles using aqueous extract of medicinal plants' (*Impatiens balsamina* and *Lantana camara*) fresh leaves and analysis of antimicrobial activity. Int J Microbiol 2019:8642303. https://doi.org/10.1155/2019/8642303

Asibi AE, Chai Q, Coulter JA (2019) Rice blast: a disease with implications for global food security. Agron 9:451–463

Azharuddin M, Zhu GH, Das D, Ozgur E, Uzun L, Turner AP, Patra HK (2019) A repertoire of biomedical applications of noble metal nanoparticles. Chem Commun 55:6964–6996

Bagga PS, Sharma VK (2012) Evaluation of fungicides as seedling treatment for controlling bakanae/foot-rot (*Fusarium moniliforme*) disease in basmati rice. Indian Phytopathol 59:305–308

Bakar RA, Yahya R, Gan SN (2016) Production of high purity amorphous silica from rice husk. Proc Chem 19:189–195

Baker S, Volova T, Prudnikova SV, Satish S, Prasad N (2017) Nanoagroparticles emerging trends and future prospect in modern agriculture system. Environ Toxicol Pharmacol 53:10–17

Barmota H, Sidhu A, Rani R (2018) Microwave oriented sonochemical method for synthesis of copper sulfide nanoparticles as antifungal agents. Int J Chem Stud 6:2090–2093

Barnwal MK, Kotasthane A, Magculia N, Mukherjee PK, Savary S, Sharma AK, Singh HB, Singh US, Sparks AH, Variar M, Zaidi N (2013) A review on crop losses, epidemiology and disease management of rice brown spot to identify research priorities and knowledge gaps. Eur J Plant Pathol 136:443–457

Bashar MA, Hossain MA, Rahman MM, Uddin MN, Begum MN (2010) Biological control of sheath blight disease of rice by using antagonistic bacteria. Bangladesh J Sci Ind Res 45:225–232

Bhargava P, Kumar A, Kumar S, Azad CS (2018) Impact of fungicides and nanoparticles on *Ustilaginoidea virens* causing false smut disease of rice. J Pharmacogn Phytochem 7:1541–1544

Campos EV, De Oliveira JL, Da Silva CM, Pascoli M, Pasquoto T, Lima R, Abhilash PC, Fraceto LF (2015) Polymeric and solid lipid nanoparticles for sustained release of carbendazim and tebuconazole in agricultural applications. Sci Rep 5:1–4. https://doi.org/10.1038/srep13809

Chen H (2018) Metal based nanoparticles in agricultural system: behavior, transport, and interaction with plants. Chem Speciat Bioavailab 30:123–134

Chiranjeevi N, Kumar PA, Jayalakshmi RS, Prasad KH, Prasad TN (2018) Bio efficacy of biogenic silver nanoparticles against rice sheath blight causing pathogen *Rhizoctonia solani* Kuhn. Int J Curr Microbiol App Sci 7:4148–4160

Choi O, Deng KK, Kim NJ, Ross L Jr, Surampalli RY, Hu Z (2008) The inhibitory effects of silver nanoparticles, silver ions, and silver chloride colloids on microbial growth. Water Res 42:3066–3074

Choi J, Lee JJ, Jeon J (2018) Genomic insights into the rice blast fungus through estimation of gene emergence time in phylogenetic context. Mycobiology 46:361–369

Chou C, Castilla N, Hadi B, Tanaka T, Chiba S, Sato I (2020) Rice blast management in Cambodian rice fields using *Trichoderma harzianum* and a resistant variety. Crop Prot 135:104864

Dar J, Soytong K (2014) Construction and characterization of copolymer nanomaterials loaded with bioactive compounds from Chaetomium species. J Agric Technol 10:823–831

Deising HB, Reimann S, Pascholati SF (2008) Mechanisms and significance of fungicide resistance. Braz J Microbiol 39:286–295

Devi S, Sharma GD (2010) Blast disease of rice caused by *Magnaporthe grisea*: a review. Assam Univ J Sci Technol 6:144–154

Downing MA, Jain PK (2020) Mesoporous silica nanoparticles: synthesis, properties, and biomedical applications. In: Chung EJ, Leon L, Rinaldi C (eds) Nanoparticles for biomedical applications. Elsevier, Amsterdam, pp 267–281

Durán N, Marcato PD, Conti RD, Alves OL, Costa F, Brocchi M (2010) Potential use of silver nanoparticles on pathogenic bacteria, their toxicity and possible mechanisms of action. J Braz Chem Soc 21:949–959

Elamawi RMA, El-Shafey RAS (2013) Inhibition effect of silver nanoparticles against rice blast disease caused by *Magnaporthe grisea*. Egypt J Agric Res 91(4)

Elamawi RM, Bassiouni SM, Elkhoby WM, Zayed BA (2016) Effect of zinc oxide nanoparticles on brown spot disease and rice productivity under saline soil. J Plant Prot Pathol 7:171–181

Elamawi RM, Tahoon AM, Elsharnoby DE, El-Shafey RA (2020) Bio-production of silica nanoparticles from rice husk and their impact on rice bakanae disease and grain yield. Arch Phytopathol Plant Protect 53:459–478

El-Naggar ME, Abdelsalam NR, Fouda MM, Mackled MI, Al-Jaddadi MA, Ali HM, Manzer HS, Essam EK, Kandil EE (2020) Soil application of nano silica on maize yield and its insecticidal activity against some stored insects after the post-harvest. Nanomaterials (Basel) 10:739–758

Fahad S, Noor M, Adnan M, Khan MA, Rahman I, Alam M, Khan IA, Ullah H, Mian IA, Hassan S, Saud S, Bakhat HF, Hammad HM, Ahmad S, Nasim W (2019) Abiotic stress and rice grain quality. In: Hasanuzzaman M, Fujita M, Nahar K, Biswas JK (eds) Advances in rice research for abiotic stress tolerance. Woodhead, Cambridge, pp 571–583

Fan J, Yang J, Wang YQ, Li GB, Li Y, Huang F, Wang WM (2016) Current understanding on *Villosiclava virens*, a unique flower-infecting fungus causing rice false smut disease. Mol Plant Pathol 17:1321–1330

Fan J, Liu J, Gong ZY, Xu PZ, Hu XH, Wu JL, Li GB, Yang J, Wang YQ, Zhou YF, Li SC (2020) The false smut pathogen *Ustilaginoidea virens* requires rice stamens for false smut ball formation. Environ Microbiol 22:646–659

FAOSTAT (2019) Food and Agriculture Organization of the United Nations (Source: http://www.fao.org/news/archive/news–by–date/2019/en/)

Faruq AN, Amin MR, Islam MR, Islam MT, Alam MM (2015) Evaluation of some selected seed treatments against leaf blast, brown spot and narrow brown leaf spot diseases of hybrid rice. Adv Agric Biol 4:8–15

Fatima Z, Ahmad M, Hussain M, Abbas G, Ul-Allah S, Ahmad S, Ahmed N, Ali MA, Sarwar G, Ehsan ul Haque, Iqbal P, Hussain S (2020) The fingerprints of climate warming on cereal crops phenology and adaptation options. Sci Rep 10:18013

Fisher MC, Henk DA, Briggs CJ, Brownstein JS, Madoff LC, McCraw SL, Gurr SJ (2012) Emerging fungal threats to animal, plant and ecosystem health. Nature 484:186–194

Fones HN, Bebber DP, Chaloner TM, Kay WT, Steinberg G, Gurr SJ (2020) Threats to global food security from emerging fungal and oomycete crop pathogens. Nat Food 1:332–342

Gawande MB, Goswami A, Felpin FX, Asefa T, Huang X, Silva R, Zou X, Zboril R, Varma RS (2016) Cu and Cu-based nanoparticles: synthesis and applications in catalysis. Chem Rev 116: 3722–3811

Gilland B (2002) World population and food supply: can food production keep pace with population growth in the next half-century? Food Policy 27:47–63

Groth DE (1993) Foliar fungicides for use in the management of rice diseases. LSU Agricultural Experiment Station Reports. 318. Louisiana Agricultural Experiment Station, LSU Agricultural CenterBaton Rouge (Source: http://digitalcommons.lsu.edu/agexp/318)

Guo X, Li Y, Fan J, Li L, Huang F, Wang W (2012) Progress in the study of false smut disease in rice. J Agric Sci Technol 2:1211–1217

Gupta AK, Solanki IS, Bashyal BM, Singh Y, Srivastava K (2015) Bakanae of rice-an emerging disease in Asia. J Anim Plant Sci 25:1499–1514

Hahn M (2014) The rising threat of fungicide resistance in plant pathogenic fungi: Botrytis as a case study. J Chem Biol 7:133–141

Hossain I, Dey P, Hossain MZ (2011) Efficacy of Bion, Amistar and Tilt in controlling brown spot and narrow brown spot of rice cv. BR11 (Mukta). J Bangladesh Agric Univ 9:201–204

Huang S, Liu L, Wang L, Hou Y (2019) Research on advance of rice false smut *Ustilaginoidea virens* (Cooke) takah worldwide: IV. Identification of rice resistance to RFS, management and prospection of RFS. J Agric Sci 11:275–284

Hussain T (2017) Nanotechnology: diagnosis of plant diseases. Agri Res Tech Open Access J 10: 555777. https://doi.org/10.19080/ARTOAJ.2017.10.555777

Hussain M, Ahmad S, Hussain S, Lal R, Ul-Allah S, Nawaz A (2018) Rice in saline soils: physiology, biochemistry, genetics and management. Adv Agron 148:231–287

Iqbal M, Javed N, Yasin SI, Sahi ST, Wakil W (2013) Studies on chemical control of bakanae disease (*Fusarium moniliforme*) of rice in Pakistan. Pak J Phytopathol 25:146–154

Iqbal MF, Hussain M, Waqar MQ (2015) Evaluation of best fungicide for controlling brown leaf spot in transplanted rice. Int J Adv Res Biol Sci 2:44–48

Jain S, Jain A, Devra V (2014) Experimental investigation on the synthesis of copper nanoparticles by chemical reduction method. Int J Sci Eng Res 5:973–978

Jiehua QI, Shuai ME, Yizhen DE, Shiwen HU, Yanjun KO (2019) Ustilaginoidea virens: a fungus infects rice flower and threats world rice production. Rice Sci 26:199–206

Jo YK, Kim BH, Jung G (2009) Antifungal activity of silver ions and nanoparticles on phytopathogenic fungi. Plant Dis 93:1037–1043

Khaliq A, Kafafy R, Salleh HM, Faris WF (2012) Enhancing the efficiency of polymerase chain reaction using graphene nanoflakes. Nanotechnology 23:455106

Khan MA, Bhuiyan MR, Hossain MS, Sen PP, Ara A, Siddique MA, Ali MA (2014) Neck blast disease influences grain yield and quality traits of aromatic rice. C R Biol 337:635–641

Khan MA, Ahmad S, Raza A (2019a) Integrated weed management for agronomic crops. In: Hasanuzzaman M (ed) Agronomic crops, vol 2. Springer Nature Singapore, Singapore, pp 257–281

Khan SM, Ali S, Nawaz A, Bukhari SAH, Ejaz S, Ahmad S (2019b) Integrated pest and disease management for better agronomic crop production. In: Hasanuzzaman M (ed) Agronomic crops, vol 2. Springer Nature Singapore, Singapore, pp 385–428

Khiyami MA, Almoammar H, Awad YM, Alghuthaymi MA, Abd-Elsalam KA (2014) Plant pathogen nanodiagnostic techniques: forthcoming changes? Biotechnol Biotechnol Equip 28: 775–785

Khodakovsky A, Schröder P, Sweldens W (2000) Progressive geometry compression. In: Proceedings of the 27th annual conference on Computer graphics and interactive techniques. Addison-Wesley, New York, pp 271–278

Khokhar LK, Jaffrey AH (2002) Identification of sources of resistance against Bakanae and foot rot disease in rice. Pak J Agric Res 17:176–177

Kim SW, Jung JH, Lamsal K, Kim YS, Min JS, Lee YS (2012) Antifungal effects of silver nanoparticles (AgNPs) against various plant pathogenic fungi. Mycobiology 40:53–58. https://doi.org/10.5941/MYCO.2012.40.1.053

Kluge E, Mendes MC, Faria MV, Santos LA, Santos HO, Szeuczuk K (2017) Effect of foliar fungicide and plant spacing on the expression of lipoxygenase enzyme and grain rot in maize hybrids. Acta Sci Agron 39:407–415

Kongcharoen N, Kaewsalong N, Dethoup T (2020) Efficacy of fungicides in controlling rice blast and dirty panicle diseases in Thailand. Sci Rep 10:1–7

Kookana RS, Boxall AB, Reeves PT, Ashauer R, Beulke S, Chaudhry Q, Cornelis G, Fernandes TF, Gan J, Kah M, Lynch I (2014) Nanopesticides: guiding principles for regulatory evaluation of environmental risks. J Agric Food Chem 62:4227–4240

Kurosawa E (1926) Experimental studies on the nature of the substance secreted by the bakanae fungus. Nat Hist Soc Formosa 16:213–227

Lamsal K, Kim SW, Jung JH, Kim YS, Kim KS, Lee YS (2011) Inhibition effects of silver nanoparticles against powdery mildews on cucumber and pumpkin. Mycobiology 39:26–32

Li Z, Yu T, Paul R, Fan J, Yang Y, Wei Q (2020) Agricultural nanodiagnostics for plant diseases: recent advances and challenges. Nanoscale Adv 2:3083–3094

Liu S, Huang X, He H, Jin Q, Zhu G (2016) Evaluation of selected plant growth regulators and fungicide residues in fruits for dietary risk assessment. Hum Ecol Risk Assess 22:1386–1395

Lore JS, Thind TS, Mohan C (2005) Performance of some new fungicide formulations against sheath blight of rice. Pestic Res J 17:49–51

Lu S, Sun W, Meng J, Wang A, Wang X, Tian J, Fu X, Dai J, Liu Y, Lai D, Zhou L (2015) Bioactive bis-naphtho-γ-pyrones from rice false smut pathogen *Ustilaginoidea virens*. J Agric Food Chem 63:3501–3508

Mallmann EJ, Cunha FA, Castro BN, Maciel AM, Menezes EA, Fechine PB (2015) Antifungal activity of silver nanoparticles obtained by green synthesis. Rev Inst Med Trop São Paulo 57: 165–167

Martinelli F, Scalenghe R, Davino S, Panno S, Scuderi G, Ruisi P, Villa P, Stroppiana D, Boschetti M, Goulart LR, Davis CE (2015) Advanced methods of plant disease detection. A review. Agron Sustain Dev 35:1–25

McClements DJ, Decker EA (2000) Lipid oxidation in oil-in-water emulsions: impact of molecular environment on chemical reactions in heterogeneous food systems. J Food Sci 65:1270–1282

Mew TW, Leung H, Savary S, Vera Cruz CM, Leach JE (2004) Looking ahead in rice disease research and management. Crit Rev Plant Sci 23:103–127

Mishra S, Singh HB (2015) Biosynthesized silver nanoparticles as a nanoweapon against phytopathogens: exploring their scope and potential in agriculture. Appl Microbiol Biotechnol 99:1097–1107

Mishra S, Singh BR, Singh A, Keswani C, Naqvi AH, Singh HB (2014) Biofabricated silver nanoparticles act as a strong fungicide against *Bipolaris sorokiniana* causing spot blotch disease in wheat. PLoS One 9:e97881

Mishra S, Keswani C, Abhilash PC, Fraceto LF, Singh HB (2017) Integrated approach of agri-nanotechnology: challenges and future trends. Front Plant Sci 8:471

Mondal P, Kumar R, Gogoi R (2017) Azomethine based nano-chemicals: development, in vitro and in vivo fungicidal evaluation against *Sclerotium rolfsii*, *Rhizoctonia bataticola* and *Rhizoctonia solani*. Bioorg Chem 70:153–162

Morton V, Staub T (2008) A short history of fungicides. APSnet Features. (Source: https://www.apsnet.org/edcenter/apsnetfeatures/Pages/Fungicides.aspx)

Mukherjee A, Majumdar S, Servin AD, Pagano L, Dhankher OP, White JC (2016) Carbon nanomaterials in agriculture: a critical review. Front Plant Sci 7:172–188

Nair R, Poulose AC, Nagaoka Y, Yoshida Y, Maekawa T, Kumar DS (2011) Uptake of FITC labeled silica nanoparticles and quantum dots by rice seedlings: effects on seed germination and their potential as biolabels for plants. J Fluoresc 21:2057–2068

Nalley L, Tsiboe F, Durand-Morat A, Shew A, Thoma G (2016) Economic and environmental impact of rice blast pathogen (*Magnaporthe oryzae*) alleviation in the United States. PLoS One 11:e0167295

Hafeez-ur-Rehman, Nawaz A, Awan MI, Ijaz M, Hussain M, Ahmad S, Farooq M (2019) Direct seeding in rice: problems and prospects. In: Hasanuzzaman M (ed) Agronomic crops, Production technologies, vol 1. Springer Nature Singapore, Singapore, pp 199–222

Naz S, Fatima Z, Iqbal P, Khan A, Zakir I, Noreen S, Younis H, Abbas G, Ahmad S (2019) Agronomic crops: types and uses. In: Hasanuzzaman M (ed) Agronomic crops, Production technologies, vol 1. Springer Nature Singapore, Singapore, pp 1–18

Nejad MS, Bonjar GH, Khatami M, Amini A, Aghighi S (2017) In vitro and in vivo antifungal properties of silver nanoparticles against Rhizoctonia solani, a common agent of rice sheath blight disease. IET Nanobiotechnol 11:236–240

Norman S, Hongda C (2013) IB in depth special section on nanobiotechnology, Part 2. Ind Biotechnol 9:17–18

Ou SH (1985) Rice diseases. Commonwealth Mycology Institute, Kew. https://www.scirp.org/(S(vtj3fa45qm1ean45vvffcz55))/reference/ReferencesPapers.aspx?ReferenceID=1958442)

Padmanabhan SY (1973) The great Bengal famine. Annu Rev Phytopathol 11:11–24

Pariona N, Mtz-Enriquez AI, Sánchez-Rangel D, Carrión G, Paraguay-Delgado F, Rosas-Saito G (2019) Green-synthesized copper nanoparticles as a potential antifungal against plant pathogens. RSC Adv 9:18835–18843

Patel N, Desai P, Patel N, Jha A, Gautam HK (2014) Agronanotechnology for plant fungal disease management: a review. Int J Curr Microbiol App Sci 3:71–84

Peerzada AM, Bukhari SAH, Dawood M, Nawaz A, Ahmad S, Adkins S (2019) Weed management for healthy crop production. In: Hasanuzzaman M (ed) Agronomic crops, vol 2. Springer Nature Singapore, Singapore, pp 225–256

Quintana L, Gutiérez S, Arriola M, Morinigo K, Ortiz A (2017) Rice brown spot *Bipolaris oryzae* (Breda de Haan) Shoemaker in Paraguay. Trop Plant Res 4:419–420

Rafique M, Shaikh AJ, Rasheed R, Tahir MB, Bakhat HF, Rafique MS, Rabbani F (2017) A review on synthesis, characterization and applications of copper nanoparticles using green method. Nano 12:1750043

Rastogi A, Tripathi DK, Yadav S, Chauhan DK, Živčák M, Ghorbanpour M, El-Sheery NI, Brestic M (2019) Application of silicon nanoparticles in agriculture. 3 Biotech 9:1–11

Razaq M, Shah FM, Ahmad S, Afzal M (2019) Pest management for agronomic crops. In: Hasanuzzaman M (ed) Agronomic crops, vol 2. Springer Nature Singapore, Singapore, pp 365–384

Richa K, Tiwari IM, Kumari M, Devanna BN, Sonah H, Kumari A, Nagar R, Sharma V, Botella JR, Sharma TR (2016) Functional characterization of novel chitinase genes present in the sheath blight resistance QTL: qSBR11-1 in Rice Line Tetep. Front Plant Sci 7:244

Rijal S, Devkota Y (2020) A review on various management method of rice blast disease. Malaysian J Sustain Agric 4:14–18

Sarwar N, Ali H, Ahmad S, EhsanUllah AS, Mubeen K, Hill JE (2013a) Water wise rice cultivation on calcareous soil with the addition of essential micronutrients. J Anim Plant Sci 23(1):244–250

Sarwar N, Ali H, Maqsood M, EhsanUllah SAN, Shahzad M, Mubeen K, Shahid MA, Ahmad S (2013b) Phenological response of rice plants to different micronutrients application under water saving paddy fields on calcareous soil. Turk J Field Crops 18(1):52–57

Sasson A (2012) Food security for Africa: an urgent global challenge. Agric Food Secur 1:1–6

Saterlie M, Sahin H, Kavlicoglu B, Liu Y, Graeve O (2011) Particle size effects in the thermal conductivity enhancement of copper-based nanofluids. Nanoscale Res Lett 6:217

Savary S, Castilla NP, Willocquet L (2001) Analysis of the spatiotemporal structure of rice sheath blight epidemics in a farmer's field. Plant Pathol 50:53–68

Shahzad AN, Ahmad S (2019) Tools and techniques for nitrogen management in cereals. In: Hasanuzzaman M (ed) Agronomic crops, vol 2. Springer Nature Singapore, Singapore, pp 111–126

Shang Y, Hasan M, Ahammed GJ, Li M, Yin H, Zhou J (2019) Applications of nanotechnology in plant growth and crop protection: a review. Molecules 24:2558

Sharon M, Choudhary AK, Kumar R (2010) Nanotechnology in agricultural diseases and food safety. J Phytol 2:83–92

Singh J, Dutta T, Kim KH, Rawat M, Samddar P, Kumar P (2018) 'Green' synthesis of metals and their oxide nanoparticles: applications for environmental remediation. J Nanobiotechnol 16:1–24

Song J, Soytong K (2018) Nano-particles from Chaetomium against rice blast. bioRxiv. https://doi.org/10.1101/339283

Soytong K, Kanokmedhakul S, Kukongviriyapa V, Isobe M (2001) Application of Chaetomium species (Ketomium) as a new broad spectrum biological fungicide for plant disease control. Fungal Divers 7:1–15

Tariq M, Ahmed M, Iqbal P, Fatima Z, Ahmad S (2020) Crop phenotyping. In: Ahmed M (ed) Systems modeling. Springer Nature Singapore, Singapore, pp 45–60

Thomas KM (1931) A new paddy disease in Madras. Madras Agric J 19:34–36

Tilman D, Balzer C, Hill J, Befort BL (2011) Global food demand and the sustainable intensification of agriculture. Proc Natl Acad Sci 108:20260–20264

Titone P, Mongiano G, Tamborini L (2015) Resistance to neck blast caused by *Pyricularia oryzae* in Italian rice cultivars. Eur J Plant Pathol 142(1):49–59

Wang YA, Li JJ, Chen H, Peng X (2002) Stabilization of inorganic nanocrystals by organic dendrons. J Am Chem Soc 124:2293–2298

Wasaya A, Yasir TA, Ijaz M, Ahmad S (2019) Tillage effects on agronomic crop production. In: Hasanuzzaman M (ed) Agronomic crops, vol 2. Springer Nature Singapore, Singapore, pp 73–99

Wilson N (2018) Nanoparticles: environmental problems or problem solvers? Bioscience 68:241–246

Worrall EA, Hamid A, Mody KT, Mitter N, Pappu HR (2018) Nanotechnology for plant disease management. Agronomy 8:285–309

Xia ZK, Ma QH, Li SY, Zhang DQ, Cong L, Tian YL, Yang RY (2016) The antifungal effect of silver nanoparticles on *Trichosporon asahii*. J Microbiol Immunol Infect 49:182–188

Xu Z, Ma X, Gao YE, Hou M, Xue P, Li CM, Kang Y (2017) Multifunctional silica nanoparticles as a promising theranostic platform for biomedical applications. Mater Chem Front 1:1257–1272

Zahoor SA, Ahmad S, Ahmad A, Wajid A, Khaliq T, Mubeen M, Hussain S, Sami Ul Din M, Amin A, Awais M, Nasim W (2019) Improving water use efficiency in agronomic crop production. In: Hasanuzzaman M (ed) Agronomic crops, vol 2. Springer Nature Singapore, Singapore, pp 13–29

Rice Nematodes and Their Integrated Management

26

Salman Ahmad, Fazal ur Rehman, Muhammad Adnan, Irfan Ahmad, Shakeel Ahmad, Zafar Iqbal, Ejaz Ashraf, Maria Kalsoom, and Muhammad Ehetisham ul Haq

Abstract

Nematodes belong to phylum Nematoda, also known as Nemata; due to their moulting nature, they are placed under superphylum Ecdysozoa. They are most ubiquitous multicellular animals on earth. Rice is the vital crop of the world, which attacked by different plant parasitic nematodes (PPNs). Rice parasitic nematodes (RPNs) attack both foliage and roots of rice plants, causing around 10% yield losses and reducing less than 0.2% of the crop value. RPNs such as *Meloidogyne graminicola*, *Ditylenchus angustus*, *Aphelenchoides besseyi*,

S. Ahmad (✉) · Fazal ur Rehman · Z. Iqbal
Department of Plant Pathology, College of Agriculture, University of Sargodha, Sargodha, Pakistan
e-mail: salman.ahmad@uos.edu.pk

M. Adnan
Department of Agronomy, College of Agriculture, University of Sargodha, Sargodha, Pakistan

I. Ahmad
Department of Forestry, Range Management and Wildlife, University of Agriculture, Faisalabad, Pakistan

S. Ahmad
Department of Agronomy, Faculty of Agricultural Sciences and Technology, Bahauddin Zakariya University, Multan, Pakistan
e-mail: shakeelahmad@bzu.edu.pk

E. Ashraf
Department of Agriculture Extension, College of Agriculture, University of Sargodha, Sargodha, Pakistan

M. Kalsoom
Institute of Food Science and Nutrition, University of Sargodha, Sargodha, Pakistan

M. Ehetisham ul Haq
Plant Pathology Research Institute, Ayyub Agricultural Research Institute, Faisalabad, Pakistan

N. Sarwar et al. (eds.), *Modern Techniques of Rice Crop Production*,
https://doi.org/10.1007/978-981-16-4955-4_26

Hemicriconemoides spp., *Pratylenchus* spp., and *Hirschmanniella oryzae* are the major parasites of rice (*Oryza sativa*). The yield losses and damage caused by predominant nematode species of rice crop are increasing with the passage of time in tropical climates. The major reason behind this, is, relatively smaller share of nematological research in the tropics than the most temperate countries. The less awareness among growers, policy-makers, extension agencies, donor funding agencies, and project planners are the other main components which are increasing the yield losses in rice growing regions of the world due to RPNs. The distribution, symptoms of damage, biology and life cycle, effects of environmental factors, and integrated management options of major nematode pests of rice have been reviewed in this chapter.

Keywords

Rice parasitic nematodes · Sheath nematode · Cyst nematode · Root lesion nematodes · Root-knot nematode · Ufra nematode · White tip nematode · Spiral nematode · Lance nematode

Abbreviations

LNs	Lance nematodes
UNs	Ufra nematodes
PPNs	Plant parasitic nematodes
RPNs	Rice parasitic nematodes
RRKNs	Rice root-knot nematodes
RRLNs	Rice root lesion nematodes
RCNs	Rice cyst nematodes
RSNs	Rice sheath nematodes
RRNs	Rice root nematodes
SNs	Spiral nematodes
WTNs	White tip nematodes

26.1 Introduction

About 200 species of PPNs attack rice, but there are only a few that are economically important. The most important PPNs of rice are root knot (*Meloidogyne graminicola*), root infecting (*Hirschmanniella* spp.), white tip (*A. besseyi*), cyst forming (*Heterodera oryzicola*), stem infecting (*Ditylenchus angustus*), and root lesion forming (*Pratylenchus* spp.). Yield losses in rice by major PPNs are well known in different parts of the world. The magnitude of yield losses majorly depends on crop variety, cultivation practices, and type of land. PPNs are present in almost all rice cultivated areas. Cultivation of resistant and tolerant varieties is the most effective way to control them. PPNs are the major constraints in rice-growing countries (Bridge et al. 2005), which demand measurements to control them to

Table 26.1 Yield losses caused by rice nematodes in different countries of the world

Sr. no.	Disease causing nematodes	Mainly affected parts	Yield losses (%)	Country	References
1	Rice root knot nematode	Leaves/ear head	11–73	France	Soriano et al. (2000)
2	Rice root lesion nematode	Reduce plant height/roots	34	Philippines	Prasad and Rao (1978)
3	Cyst nematode of rice	Roots/reduced plant height	17–42	India	Kumari and Kuriyan (1981)
4	Rice root nematode	Roots	2–10	United States	Ichinohe (1972)
5	Lance nematode	Leaves/ear head	2.9	India	Routaray and Das (1982)
6	Spiral nematode	Roots/reduced plant height	30–50	United States	Griffin (1984)
7	White tip nematode	Leaves/reduced plant height	50	China	Lin et al. (2004)
8	Rice stem nematode	Leaves/stem	20–90	Thailand	Latif et al. (2006)

gain high yields (Ramakrishna and Sharma 1998). Resistant cultivars are effective ways to control RPNs (Bos and Parlevliet 1995). Resistant cultivars suppress the disease and do not allow the pathogen to cause disease. One can measure the resistance/susceptibility of the host plants by determining the population densities of nematodes on and in the roots. Precise distribution maps of important nematode species infecting rice, creation of awareness among the farmers, and extension services could also be effective to control RPNs. There is dire need of development of locally feasible, low cost, and sustainable nematode management strategies. The development of sustainable rice-based cropping systems by considering the susceptibility/tolerance/resistance of rice crop against nematodes is very necessary. The exploitation of the antagonistic activities of fungal and bacterial bio-agents against nematode should also be taken into account. The new chemistry nematicides may also be used to control RPNs (He et al. 2020). Silver nanoparticles are also in practice controlling PPNs (Baronia et al. 2020). The current chapter provides an overview of different RPNs and their symptoms and integrated management. The losses caused by RPNs in different countries have been given in Table 26.1.

26.2 Nematode Pests of Rice

RPNs being the microscopic organisms are invisible to the naked eye. Most RPNs feed and develop on roots of rice plants, while some also feed on aerial parts. At the infective stages, these nematodes are worm-like and mobile, while the adults are also mobile. The females of some species of nematodes become sedentary and swollen. These nematodes are insidious, and therefore, the presence of nematodes and their symptoms of damage are often overlooked both by farmers and researchers. The main symptoms shown by infested rice crop include stunting, chlorosis, reduced vigor, and symptoms of water stress. These symptoms could be confused with mineral deficiency, soil physical problems, and low water availability. The RPNs can be divided into two major groups.

26.2.1 Nematodes Infecting Roots of Rice

The RPNs that infect the roots of rice are

- Rice root-knot nematodes (*M. graminicola*)
- Rice root nematodes (*Hirschmanniella* spp.)
- Cyst-forming nematodes (*Heterodera* spp.)
- Sheath nematodes (*Hemicriconemoides* spp.)
- Root lesion nematodes (*Pratylenchus* spp.)
- Lance nematodes (*Hoplolaimus* spp.)
- Spiral nematodes (*Helicotylenchus* spp.)

26.2.2 Nematodes Infecting Aerial Parts of Rice

The PPNs that infect the rice aerial parts are

- Rice stem nematodes/ufra nematodes (*D. angustus*)
- White tip nematodes (*A. besseyi*)

26.3 Nematodes Infecting Roots of Rice

26.3.1 Rice Root-Knot Nematodes (RRKNs)

RRKNs are important and serious pathogens of rice particularly in Asia (Haque 2013; Jain et al. 2012). RRKNs are found on rice grown both in irrigated, upland, and rainfed areas and in deepwater (Bridge et al. 2005). *M. graminicola* is considered as the most damaging on rice grown in lowland, irrigated, upland, and rainfed areas (De Waele and Elsen 2007). However, in flooded conditions, this nematode is also well adapted (Bridge and Page 1982). *M. graminicola* is considered as a main threat to rice production in tropical areas (De Waele and Elsen 2007; Kreye et al. 2009a; Kreye et al. 2009b).

In Myanmar, *M. graminicola* nematode is considered as the most serious nematode attacking all lowland rice varieties (Win et al. 2011). The best way to avoid the occurrence of this nematode is the cultivation of less susceptible or resistant rice varieties.

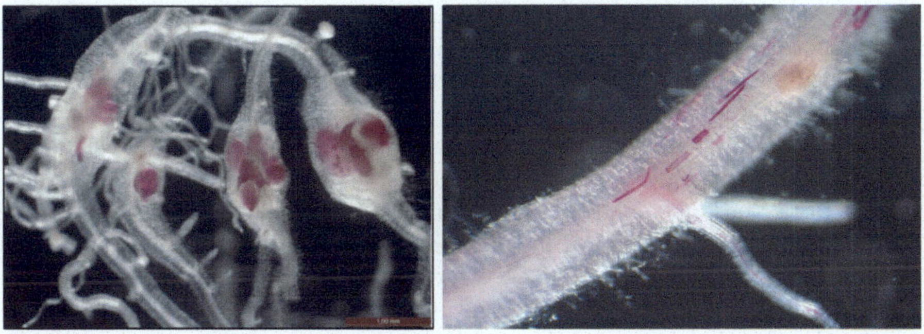

Fig. 26.1 *Meloidogyne graminicola* infected rice roots (Photographs courtesy of Z. Lahari)

26.3.1.1 Taxonomy

- Kingdom: Animalia
- Phylum: Nematoda
- Order: Tylenchida
- Family: Heteroderidae
- Genus: *Meloidogyne*
- Species: *graminicola*
- Binomial name: *M. graminicola*

26.3.1.2 Symptoms of RRKNs

Patches of yellowish and stunted rice plants can easily be seen in nursery beds and main fields. Reduction in leaf size, delayed earhead emergence, and no earheads are produced during high infestation. If earheads are produced, they are poorly filled or have no grains. During moisture stress, infested patches of rice crop are dried early. Terminal hook-like galls are formed on the roots during high infestation of *M. graminicola* (Fig. 26.1). The roots of transplanted crop may become irregularly thickened with no apparent cortical swellings. The plants infested with *M. graminicola* show reduced vigor, yellowing of leaves, and leaf curling along the midribs. Disruption and hypertrophy of cortical cells can be caused by second-stage juveniles. The seedling wilt along with severe reduction in growth parameters can be a result of high initial population of *M. graminicola*, whereas only reduction in growth parameters can be a result of low population (Plowright and Bridge 1990). The other symptoms of high infestation of *M. graminicola* are distortion and crinkling of newly emerged leaves, chlorosis, and stunted growth of infected plants (Dutta et al. 2012; Kavitha et al. 2016).

26.3.1.3 Biology and Life Cycle

The males of *M. graminicola* can reproduce by amphimixis and meiotic parthenogenesis. They deposit about 200–500 partially embryonated or unembryonated eggs in a gelatinous egg matrix (Senthilkumar et al. 2007). About 4–6 days after the embryogenesis, first-stage juveniles (JIs) are produced. JIs molt in 2–3 days and give a second-stage juvenile (J2). The J2 hatches out from egg when there are high moisture and temperature range between 20 and 35 °C. J2 is an infective stage. The hatching rate is slow and can extend over 1 week to several weeks. This difference in egg laying time is due to different ages of females present in a cluster. This may also be due to certain proportion of eggs that do not hatch immediately after laying.

Some eggs only hatch on the stimulation of an extrinsic factor such as the host root diffusate (Fernandez et al. 2014). In relatively dry environmental conditions and absence of rice host, the unhatched J2 can survive for several weeks inside the egg. The hatched J2 finds the root of the plant and enters into it. Generally, they enter from the elongation zone and enhance formation of syncytium and galls. After 3–4 days, J2 swells and molts into J3 and J4 and then differentiated into male and female. The J3 and J4 are enclosed in a cuticular sheath provided by the preceding J2 stage and do not have functional stylet to feed (Dabur et al. 2004). After 3–4 days, the freshly laid eggs can be seen. After 13–15 days of inoculation, the adult males can be seen. The life cycle of *M. graminicola* from egg to egg takes 25–28 days under optimum temperature conditions of 25–30 °C (Rao and Israel 1971a).

26.3.1.4 Effect of Environmental Factors

The hatching of maximum eggs *M. graminicola* in water has been reported at the temperature of 25 and 30 °C (Rao and Israel 1971a), while at 15 and 35 °C hatching is reduced. During December to February when soil temperature is 20.9 °C or less, the larval populations of *M. graminicola* in soil are large. During January to March, the maximum galls on rice roots are observed. The maximum egg masses are usually found during the months of February to March. The most favorable temperature of soil for gall formation is 23.5 °C or less (Rao and Israel 1971b; Ravinndra et al. 2017). At 32% moisture content, the larval invasion is usually highest in soils. The egg mass production and development are highest at 20–30% soil moisture. The highest larval invasion has been found at the pH of 3.5; however, pH has less effect on larval invasion level as well as growth and development of the nematode. Drought conditions at both tillering and flowering stages favor the development and reproduction of *M. graminicola*. The applications of nitrogenous fertilizers increase the reproduction. The nematode infection has been less recorded in clay soils, while sandy or loamy soil favors the development. The population density of *M. graminicola* is usually less in puddled and water-logged soil compared to non-puddled soil.

26.3.1.5 Integrated Management of RRKNs

The soil solarization of nursery beds for 15 days can prove very effective against *M. graminicola*. The application of carbofuran 3G to the nursery at the rate of 15 g/ m^2 is very effective for the management of RRKNs. The soil application and root dipping in carbofuran 3G, carbosulfan 20EC, phorate 10G, and chlorpyriphos 20 EC reduce *M. graminicola* infestation in rice (Khan et al. 2012). Integrated Nematode Management Technology (INMT) can result in reducing the nematode population from 320 J2/200cc soil to nematode population 135 (Somasekhara et al. 2012). The soil application and roots dip in *Trichoderma harzianum* + carbofuran or *Pseudomonas fluorescens* have been found most effective against *M. graminicola* and can suppress the galls formation up to 40–46%. These treatments reduce egg mass up to 45–57% and soil population up to 56–64% and increase plant growth variables by 37–42%.

The application of nematicides against *M. graminicola* in infested rice fields can result in a yield increase of 16–20% or about 1 ton per hectare (Padgham et al. 2004). It has also been demonstrated that in rice fields in Thailand and Indonesia, the applications of nematicide against *M. graminicola* can result in a yield increase of 12–33% and 28–87%, respectively (Arayarungsarit 1987). The root dipping of nursery seedlings of rice in carbofuran + *P. fluorescens* and single spray of carbosulfan or carbofuran on soil are highly effective against *M. graminicola* (Ravinndra et al. 2017). This approach can suppress disease severity to 65–72% and yield can be enhanced to 30–35%.

26.3.2 Rice Root Lesion Nematodes (RRLNs)

After RRKNs and cyst nematodes, *Pratylenchus* spp. are ranked third important PPNs (Jones et al. 2013). *P. zeae* causes heavy yield losses in areas receiving less precipitation. Various PPN invasive species are now becoming a threat to rice crops (Singh et al. 2013). RRLNs have been less studied compared to RRKNs and cyst nematodes; the reason behind this is, species of *Pratylenchus* are migratory and do not develop stable feeding sites; however, cysts and galls can easily be examined due to infections of RRKNs and cyst nematodes. The presence of *Pratylenchus* species, that is, RRLNs, is more challenging to quantify. All stages of RRLNs have the ability to enter and leave the host roots. There are up to 89 putative morphospecies of *Pratylenchus* (Castillo and Vovlas 2007; Subbotin et al. 2008); many of them are widely distributed. The most important species in terms of economic damages are *P. goodeyi*, *P. brachyurus*, *P. coffeae*, *P. loosi*, *P. penetrans*, *P. neglectus*, *P. pratensis*, *P. thornei*, *P. scribneri*, *P. zeae*, and *P. vulnus*. Among them, *P. zeae* attacks important cereal crops, including wheat, rice, maize, vegetables, forages, and fruit crops (Blair and Stirling 2007; Castillo and Vovlas 2007).

26.3.2.1 Taxonomy

- Kingdom: Animalia
- Phylum: Nematoda
- Class: Secernentea
- Subclass: Diplogasteria
- Order: Tylenchida
- Superfamily: Tylenchoidea
- Family: Pratylenchidae
- Subfamily: Pratylenchinae
- Genus: *Pratylenchus*
- Species: *zeae*
- Binomial Name: *P. zeae*

26.3.2.2 Symptoms of Damage

The damage caused by RRLNs can range from minor to severe in terms of infestation severity. The symptoms produced by RRLNs are similar to other diseases and disorders and cause yellowing and stunting of aboveground parts of infected plants. *P. zeae* causes root damage in many crops resulting in moisture stress in these crops. Due to such moisture stress, wilting and stunting occur during hot days. The symptoms are more evident on infested rice plants at early growth stages compared to lateral (Blair and Stirling 2007).

Root damage caused by RRLNs to plants often appears in random patches. The minor to severe necrosis appears on feeding sites and results in the formation of brown and discolored lesions on roots of rice plants. The cracking and rotting of roots also occur by the feeding of RRLNs (Castillo and Vovlas 2007).

26.3.2.3 Biology and Life Cycle

P. zeae is migratory, polyphagous, and intracellular root endoparasite. Its life cycle takes 3–9 weeks for completion and depends on the environmental conditions. When the environmental conditions are favorable together with the susceptible host, the life cycle of RRLNs gets shortened. When *P. zeae* is grown on carrot discs, its life cycle completes in 26 days under controlled conditions, while on red clover, it takes 9 weeks to complete (Turner and Chapman 1972; Chitambar and Raski 1985). J1 of *Pratylenchus* spp. develops within the egg. J1 is then molted into J2 which is then hatched from the egg. All juveniles and adults of *Pratylenchus* spp. are worm-like and mobile in nature. The host plant roots or storage organs are infested by it. They can also enter and leave the roots. They spend most of their life cycle on host plant roots. These nematodes can also be found in adjacent soil. The females lay eggs inside the infested roots when they get matured. They can also deposit eggs singly or in groups in adjacent soil. The reproduction mostly occurs by parthenogenesis. They can survive for longer time under adverse conditions through anhydrobiosis or at the egg stage; in this way, they can survive for several years in soil (Castillo and Vovlas 2007).

26.3.2.4 Effect of Environmental Factors

The abundance of *P. zeae* depends upon mineral components, soil moisture, temperature, organic matter, pH, and aeration (Norton 1979). They are most commonly present in sandy soils (Hallmann et al. 2007). A wide range of environmental conditions can favor the growth of *P. zeae*. The most favorable soils for their breeding and migration are moistened soils, but they could also endure warm and dry environments, however, quiescent under unfavorable moisture levels. The plants can resume their growth after quiescence (Agrios 2005). When soil temperature is less than 15 °C, they remain inactive and again become active when temperature rises above 20 °C.

26.3.2.5 Integrated Management of RRLNs

RRLNs are effectively managed using nematicides, which include aldicarb, fenamiphos, and carbofuran; however, the cost and toxicity of these nematicides do not recommend their applications in rice. Aldicarb at 100 g/100 m, terbufosat 100 g/100 m, and carbofuran at 200 g/100 m are the recommended doses to control of RRLNs. Crop rotation is also not suggested because *P. zeae* have a wide range of hosts. Recognition of *Pratylenchus* at species level could prove helpful in designing crop rotation strategy. Currently, the best method to control RRLNS is the use of resistant cultivars (McDonald et al. 1987).

26.3.3 Rice Cyst Nematodes (RCNs)

RCNs possess habit of sedentary endoparasitism (Fig. 26.2). Cysts are tanned sacs that originate from the female bodies and comprise the eggs. For many years, cysts can endure in soil. Cysts are ambifenestrate and are provided with a small ovoid semifenestra, many bullae, long vulval slit, and a weak underbridge. These structures are not found in some nematodes. *Heterodera oryzycola* has some morphological similarities with *Heterodera oryzae*, *H. elachista*, and *H. sacchari*. These species can be identified using morph-chemical techniques (de Luca et al. 2013). Rice is the primary host of *H. oryzicola*. Other hosts include Bermuda grass (*Cynodon dactylon*), signal grass (*Brachiaria decumbens*), and plantain (*Musa paradisiaca*). RCNs were first identified in Japan in 1974 (Ohshima 1974), but now these are ubiquitous in all Asian rice-growing countries.

26.3.3.1 Taxonomy

- Kingdom: Animalia
- Phylum: Nematoda
- Class: Secernentea
- Subclass: Diplogasteria
- Order: Tylenchida
- Superfamily: Tylenchoidea
- Family: Heteroderidae
- Subfamily: Heteroderinae

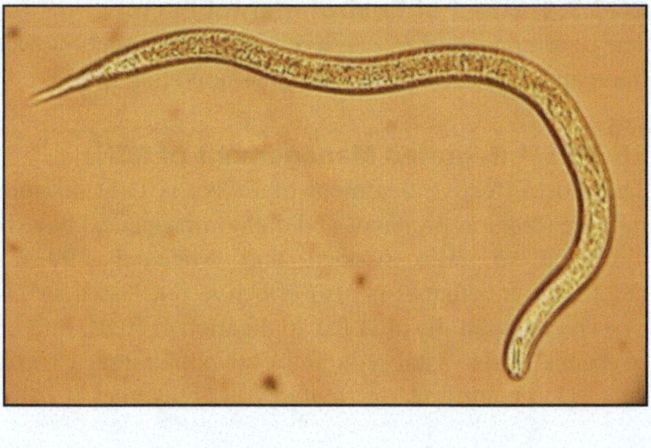

Fig. 26.2 *Heterodera* spp. (Photographs courtesy of R. Smiley)

- Genus: *Heterodera*
- Species: *oryzicola*
- Binomial name: *H. oryzicola*

26.3.3.2 Symptoms of Damage

The attack of RCNs causes severe stunting of rice plants, chlorotic spots on leaves, fewer tillering, reduction in root growth, and blackening and browning of foliage of infected plants. The heavy infestation of RCNs causes death of rice seedlings. Yield reduction of 7–19% has been reported due to the attack of RCNs. On the root surface of affected plants, brown cysts and white lemon-shaped females can be observed. Cysts and females are also present in those soils which are near infected roots (Bridge and Starr 2007).

26.3.3.3 Biology and Life Cycle

RCNs take about 9 days for the completion of embryonic development and protrusion from egg masses. J2 penetrate to the roots of rice within 24 h. After piercing, the endoparasitic juveniles take 14 days to become males. To become an adult female, it takes 20 days. Inclination of females is equal toward root tip (48%) and hypocotyl (42%), while a few of them move in thin secondary roots or small rootlets. The sex ratio among males and females is about 1:4. The virgin females' secretions attract the male nematodes for mating (Job and Thinniyam 1995). Within 22 days, gelatinous matrix is secreted by females in which they lay eggs. Later on, the females are converted into brown cysts within 24 days. A single *H. oryzicola* female can lay about 198 eggs in an egg mass. It can also keep 120 eggs inside the body of the brown cyst. All females pass through matting, lay a single egg mass within 30 days, and complete their life cycle in 30 days, while 12 generations are developed per year (Jayaprakash and Rao 1982). Cysts can persist in soil for many years. J2 originate from the cysts, pierce the host roots, and establish a specialized feeding site called syncytium in the central cylinder of roots. The swollen females keep the eggs that are later converted into large egg masses. After cracking in the root cortex, the females come out from the root surface.

26.3.3.4 Effect of Environmental Factors

The development of *H. oryzicola* in rice fields gets slowed below 20 °C and is high at the temperatures of 30 °C (Athanassiou et al. 2005).

26.3.3.5 Integrated Management of RCNs

The most effective treatment of RCNs is D-D mixture. The applications of D-D mixture (dichloropropanes and dichloropropenes) have been reported to increase the yield up to 30% (Cohen and Michaud 1993). The treatment of EDB (1, 2-Dibromoethane) is also effective but less than D-D mixture; this is because cysts recover rapidly after the application of EDB. Intercropping of sweet potatoes or soybeans is also vital reducing cyst population. Crop rotation is also an effective strategy to control *H. oryzicola* (Van Nguyen et al. 2020).

26.3.4 Rice Sheath Nematodes (RSNs)

RSNs are dimorphic PPNs in terms of sex and mostly present in warm climates. Their cuticle is covered by a loosened outer cuticular sheath which is then further attached to the main body at head and vulva. In males and juveniles, this cuticular sheath is not present. Genus *Hemicriconemoides* have almost 52 species (Geraert 2010) which are not easy to identify due to overlapping of many morphological characters. The molecular data of RSN populations from various localities has been collected, but only few species have been identified including *H. alexis*, *H. chitwoodi*, *H. macrodorus*, and *H. minutus* (Van Den Berg et al. 2014).

26.3.4.1 Taxonomy

- Kingdom: Animalia
- Phylum: Nematoda
- Class: Secernentea
- Subclass: Diplogasteria
- Order: Tylenchida
- Superfamily: Criconematoidea
- Family: Criconematidae
- Subfamily: Criconematinae
- Genus: *Hemicriconemoides*
- Species:
  - *wessoni*
  - *strictathecatus*
- Binomial name:
  - *H. wessoni*
  - *H. strictathecatus*

26.3.4.2 Symptoms of Damage

The infested rice plants with RSNs show stunting, leaf yellowing, necrosis of cortical root tissues, necrosis of leaves and leaf blades, premature wilting, and root malformation (Prasad et al. 1987).

26.3.4.3 Biology and Life Cycle

Major portion of their life cycle related to their developmental stages comprises egg, four juvenile stages (J1–J4) with an absent sheath, and adult stage, that is, female and male. Fassuliotis (1962) made first biological observations on sheathoid nematodes, and subsequent amendments were made by Dasgupta et al. (1969). They recorded the population of *H. chitwoodi* on camellia plants that were grown in pots. They demonstrated that the first two juvenile stages, J1 and J2, of this species were produced inside the egg. Further, they observed that J2 hatched from the embryonated egg and converted into J3 and J4, while the last juvenile stage differentiated into adult males (without sheath) and females (with sheath). Males and juveniles of RSNs mostly live in soil along with females. The fully developed esophagus is not present in males therefore they do not feed. Whereas the juvenile stages (J2–J4) get food from the roots and have cuticle with rows of scales. Loos (1949) and Whitlock and Steele (1960) demonstrated that RSNs are obligate migratory ectoparasites that get nutrition from roots of rice plants. However, there is less information about the parasitic habits of these nematodes.

26.3.4.4 Effect of Environmental Factors

Soil temperature ranges from 20 to 28 °C with appropriate rainfall conducive for the build-up of RSNs while their population is maximum at tillering stage (Curtis 2008).

26.3.4.5 Integrated Management of RSNs

Proper fertilization, balanced irrigation, and the use of nematicides are the best management strategies for RSNs. Drainage of standing water (Bridge and Page 1982), balanced irrigation regimes (Lavini et al. 2008), and appropriate agronomic practices (Porazinska et al. 1999) improve the vigor of infested rice plants and thus make them protected against RSNs. The population of RSNs can effectively be controlled by carbofuran (10% granules), aminofuracarb (10% granules), terbufos (10% granules), carbosulfan (25% emulsifiable concentrate), and aldicarb (15% granules) (Bridge et al. 2005).

26.3.5 Rice Root Nematode (RRNs)

RRNs are the very important and serious pests of rice and are prevalent in irrigated rice (Bridge et al. 2005). Production of rice has significantly increased due to the control of RRNs (Fortuner and Merny 1979; Maung et al. 2010). RRNs belong to family Pratylenchidae, and genus *Hirschmanniella* comprises 35 species. Most of the RRNs are migratory endoparasitic to plant roots (Sher 1968). RRNs are found in rice and non-rice-growing areas of the world. They have also been reported in the USA and tropical and subtropical areas of Asia (Bridge et al. 2005).

26.3.5.1 Taxonomy

- Kingdom: Animalia
- Phylum: Nematoda
- Class: Secernentea
- Subclass: Diplogasteria
- Order: Tylenchida
- Superfamily: Pratylenchidae
- Family: Pratylenchidae
- Subfamily: Pratylenchinae
- Genus: *Hirschmanniella*
- Species: *oryzae*
- Binomial name: *H. oryzae*

26.3.5.2 Symptoms of Damage

The RRN-infested fields do not show clear symptoms except decrease in grain yields. RRNs penetrate through roots of rice plants; however, they are unable to penetrate through root tips or thin lateral roots (Taylor 1969). RRNs may enter

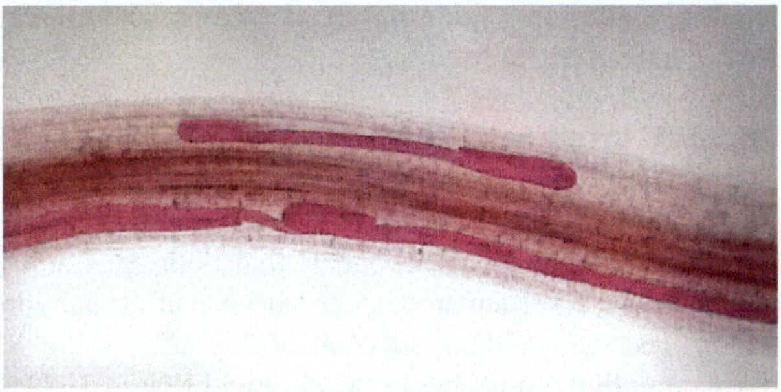

Fig. 26.3 *Hirschmanniella oryzae* inside rice root (Photographs courtesy of S. Nagachandrabose)

completely into the roots or simply insert their heads into the cortex (Fig. 26.3). During migration through the cortex, RRNs nourishment depends on the feeding of cortical cells or vascular bundles inside the root (Mathur and Prasad 1972). However, RRNs also have a tendency toward cellular feeding at the base of root hairs (Mathur and Prasad 1972; Ichinohe 1972). Hirschmanniella species are endoparasites and adapted to aquatic environments. Most species of Hirschmanniella are found in tropical areas, while monocotyledons including oceanic plants are their primary hosts. Some species of *Hirschmanniella* including *H. behningi*, *H. gracilis*, and *H. halophila* are important pests of rice.

The roots that are infested with *H. oryzae* may first show yellowish to brown color and later on they become darkened. Heavily infested roots after turning from brown to black died (Mathur and Prasad 1972; Ichinohe 1972). Below ground symptoms start by the formation of little brown lesions at the points where nematodes have destroyed the surface. After these early symptoms, the epidermal cells become necrotic inside the roots. Cavities are also produced that cause destruction of cortical cells (Mathur and Prasad 1972).

26.3.5.3 Biology and Life Cycle

RRNs are sexually dimorphic, that is, having separate genders. Dimorphic is a type of reproduction in which both sexes are required (Mai et al. 1996). After fertilization, adult females lay eggs that are oval shaped, generally inside the cortex with estimated length of 66–72 μm and width of 26–40 μm. The hatching takes 4–5 days inside the root. Throughout the RRNs life cycle, there are 4 molts, among them the first occurs inside the egg (Mathur and Prasad 1972). The J2 migrate and feed on cells present in the cortex, then complete remaining successive molts. At the last stage, the final molt takes place and immature adult females or males emerge out with underdeveloped gonads (Ichinohe 1972). One cycle from egg to adult is completed in 1 month (Thorne 1937).

26.3.5.4 Effect of Environmental Factors

H. loofi and *H. zostericola* are found in Europe where climatic conditions are highly conducive, while tropical climate favors *H. oryzae*. The *Hirschmanniella* spp. can establish themselves in greenhouses on several crops including tomato. They may also establish in aquaponic systems because their unique lifestyle is adapted to aquatic conditions (Villarroel et al. 2016). Therefore, RRNs are severe threat to hydroponic crops in future.

26.3.5.5 Integrated Management of RRNs

The management of RRNs has two major issues due to nematode's biology. As RRNs are migratory in nature, they leave the main host upon root necrosis and infect neighboring plants. The second is that both eggs and juveniles can overwinter in dead roots. These nematodes can survive in an anhydrobiotic state in response to paddy desiccation. This survival ability allows the RRN populations to remain dormant until the rains begin (Mathur and Prasad 1972; Muthukrishnan et al. 1977).

Such survival of RRNs makes the control measures less successful, and thus applications of nematicides become economically impossible for the small holding farmers (Ichinohe 1972). Therefore, dry fallowing and crop rotations, being the organic control measures, have proven effective; however, this has also an issue that small holding farmers cannot afford to keep their lands out of production for long time to manage RRNs (Muthukrishnan et al. 1977). The use of resistant rice cultivars and weed control are effective and possible control measures (Ichinohe 1972). The applications of nematicides fensulfothion, diazinon, and phorate after transplanting proved effective and also have increased the yield. Further, the nematicides including carbosulfan, oxamyl, benfuracarb, methomyl, and alanycarb at the concentration of 10 μg/mL, 100 μg/mL, 100 μg/mL, 100 μg/mL, and 100 μg/mL have the mortality rate of 100%, 83%, 100%, 77%, and 100%, respectively, against RRNs (Takagi et al. 2020).

26.3.6 Lance Nematode (LNs)

LNs have been reported from Bangladesh, China, India, and Iran, and rice is their main host (CAB International 2001). The three parasitic species of *Hoplolaimus* are pests of agricultural crops. The length of these nematodes is about 1–1.5 mm, while some may reach 2 mm (Bae et al. 2008) (Fig. 26.4). The large stylets equipped with tulips or anchors are present in them. The tail of the males has wing-like folds, and females have a short and rounded tail. Some species of *Hoplolaimus* are amphimictic. LNs reproduce sexually, while few of them are parthenogenetic; however, their females may also produce offsprings without fertilization. LNs are endo- and ectoparasites and semi-endoparasites and feed externally on roots or by inserting their heads into roots (Bae et al. 2008). Some of them have also the ability to enter the roots and feed.

26.3.6.1 Taxonomy

- Kingdom: Animalia
- Phylum: Nematoda
- Class: Secernentea
- Subclass: Diplogasteria
- Order: Tylenchida

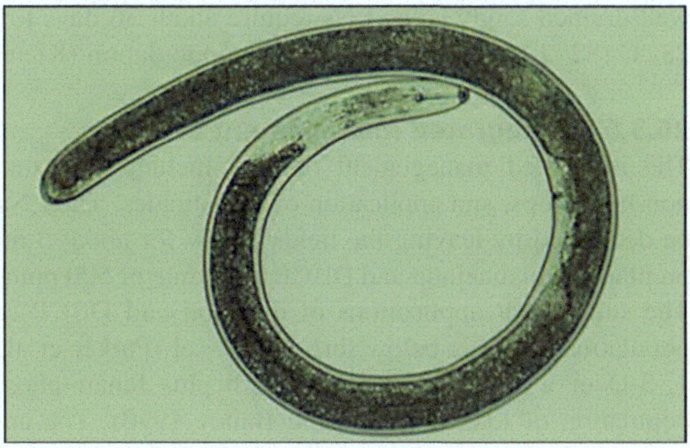

Fig. 26.4 Lance nematode (Photographs courtesy of Jonathan D. Eisenback)

- Superfamily: Tylenchoidea
- Family: Hoplolaimidae
- Subfamily: Hoplolaiminae
- Genus: *Hoplolaimus*
- Species: *indicus*
- Binomial name: *H. indicus*

26.3.6.2 Symptoms of Damage

In rice, the severe attack of LNs results in stunting growth, yellowing and curling of leaves, and suppression of rice ears. The tips of leaves become brittle and turn into an ash color. The severe attack can cause complete crop loss (Prasad et al. 1987).

26.3.6.3 Biology and Life Cycle

For the development of successive post-embryonic stages and oviposition, the prior feeding is necessary. The life cycle of *H. indicus* is completed within 27–36 days from egg to egg and takes 25–27 days from egg to male at 28–32 °C (Nand et al. 1970). Within the egg, the first molt occurs. The three larval stages and the three molts occur outside the eggs. The presence of four specialized ventral cord nuclei around the vagina in females is the only indication of sex differentiation in the second molt. In the details of life cycle, minor variations on different hosts have been reported. The soil moisture content directly determines the population density of LNs; however, soil temperature is also important (Ma et al. 2020). The drought conditions and high soil temperature prevalent during the months of April, May and June, and low temperature in winter have adverse effects on the population of LNs (Vanstone et al. 2008).

26.3.6.4 Effect of Environmental Factors

The best environmental conditions for the maximum growth of LNs population are temperature 30 °C, sandy loam soil with 10–20% clay, soil pH 7, and 16% moisture content. It has been reported that LNs are more prevalent in nurseries and well-drained sandy soils. LNs require about 36 days to complete their life cycle at 28 °C (82 °F) while males are high in population (Khan and Chawla 1975).

26.3.6.5 Integrated Management of LNs

The integrated management of LNs includes resistant host, crop rotation with non-host crops, and application of nematicides. The LNs population can effectively be decreased by leaving the fields fallow for about 3 months. Root dipping before preplanting in diazinon and DBCP at the rate of 500 ppm for 10 min is also effective. The subsequent applications of diazinon and DBCP could also help to keep the populations of LNs below threshold level (Parker et al. 1975). The application of 1, 3-D or a combination of aldicarb plus fenamiphos can profitably reduce the population of LNs (Schmitt and Bailey 1990). The application of vydate® 24SL (oxamyl-DuPont) and mocap® 15GR (ethoprophos-AMVAC) is also effective (Jaramillo et al. 2019).

26.3.7 Spiral Nematodes (SNs)

The nematodes in the genus *Helicotylenchus* are commonly called SNs. The name of SNs is given because of their body structure (Fig. 26.5). When the nematodes are relaxed or dead, their bodies tend to curl into a spiral shape (Fig. 26.6). Sometimes, the name SNs is also applied to other genera in the family Hoplolaimidae, that is, *Scutellonema*, *Rotylenchus*, *Aorolaimus*, and *Peltamigratus*. The most ubiquitous PPNs are among the SNs of *Helicotylenchus* and are associated with turf grasses and various cultivated agronomic and horticultural crops. SNs have been reported in the rice fields of temperate and tropical regions and many islands on all continents except Antarctica. SNs are present in all types of soils, that is, heavy, sandy, and organic soil (Mai et al. 1996).

26.3.7.1 Taxonomy

- Kingdom: Animalia
- Phylum: Nematoda

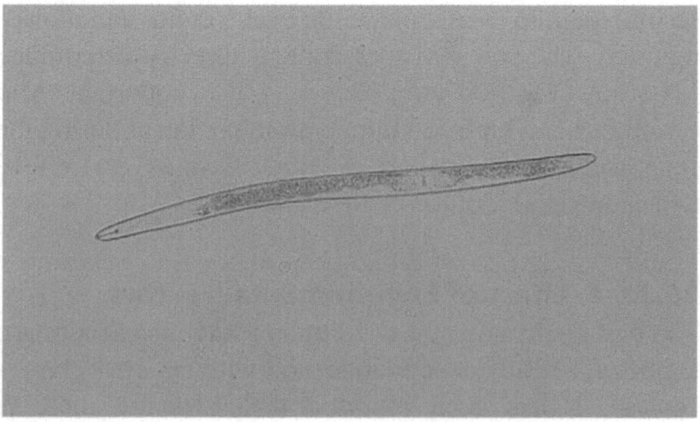

Fig. 26.5 The body of *Helicotylenchus* spp. (Photographs courtesy of William T. Crow)

Fig. 26.6 The spiral body of *Helicotylenchus* spp. (Photographs courtesy of William T. Crow)

- Class: Secernentea
- Subclass: Diplogasteria
- Order: Tylenchida
- Superfamily: Tylenchoidea
- Family: Hoplolaimidae
- Subfamily: Hoplolaiminae
- Genus: *Helicotylenchus*
- Species:
  - *dihystera*
  - *crenacauda*
- Binomial name
  - *H. dihystera*
  - *H. crenacauda*

26.3.7.2 Symptoms of Damage

The small discolored lesions are produced in the root cortex and other underground parts of rice during the infection of SNs, and death of cells occurs at local lesions in the cortex where the nematodes feed. When high populations of SNs feed on roots, they cause severe damage to roots and make them slightly swollen, spongy, and discolored; resultantly the root cortex is sloughed off (Maggenti 1981; Mai et al. 1996). Yellowing of foliage, stunting growth, wilting, and defoliation are the main above ground symptoms and may vary according to the population level of SNs in field.

26.3.7.3 Life Cycle and Biology

Reproduction is most fluctuating in SNs. Some species reproduce sexually in which males and females mate, while self-fertilization also occurs in some species in the same way as in hermaphrodites in which mating is not necessary. Other species also reproduce asexually through parthenogenesis; in this process only females are involved. The life cycle of SNs starts with the egg laying of females in soil. J1 take place in the egg, then before hatching they molt into J2. After hatching, J2 begin feeding on host roots for further development. *H. crenacauda* usually gets its food through cortical cells of host roots (Mondal et al. 2020). The SNs fit their stylet into the epidermis and cortical cells to devour the cellular contents. In some cases, *H. dihystera* causes development of specialized "food cells" that are further used for feeding (Anderson 1968). Before turning into adults, all *Helicotylenchus* spp. pass through two further stages, third and fourth stage juveniles (J3, J4); they look same as the adults but differ only in body size and lack a developed reproductive system. SNs species deposit their eggs into the soil except *H. multicinctus* which deposits within banana roots (Mahfouz and Mohamed 2019).

26.3.7.4 Integrated Management of SNs

Wide range of SNs hosts sometimes make their control failed; however, since the body of *Helicotylenchus* spp. remains uncovered in the soil, nematicides and bionematicides are powerful tools to control them (Mahfouz and Mohamed 2019). Proper sanitation and planting of nematode-free seeds of rice are recommended for the control of SNs. *H. crenacauda* can be controlled in the best way in nurseries of rice and other landscapes by proper sanitation and planting of nematode-free seeds (Bridge et al. 2005). The applications of nematicides including counter® 15GR (terbufos-AMVAC), verango® 50SC (fluopyram-Bayer), and rugby® 10GR (cadusaphos-FMC) can reduce the SNs population ranging from 20 to 49% (Jaramillo et al. 2019).

26.4 Nematodes Infecting Aerial Parts of Rice

26.4.1 White Tip Nematodes (WTNs)

WTNs are one of the most important rice pathogens and are also known by some other names such as spring dwarf nematode, strawberry crimp disease nematode, or flying strawberry nematode (Lin et al. 2004). In rice, *A. besseyi* is a facultative seed-borne parasite, lives as ecto- or endoparasite on both leaves and young tissues, and can survive for several years on stored grains in a state of anhydrobiosis. However, their survival is much less under field conditions. They have large metacorpus and esophageal glands that are dorsally overlapped with cylindrical body having 0.44–0.84 mm length and 14–22 μm width. Near the anterior edge of the nerve ring, excretory pores of females are present (Fig. 26.7). Bursa and spicules are absent in male reproductive structures (Fig. 26.8). *A. besseyi* has been reported in rice fields of Central, North, and South America, Asia, Africa, Eastern Europe, and Pacific islands (McGawley et al. 1984; Hoshino and Togashi 2000; Lin et al. 2004).

26.4.1.1 Taxonomy

- Kingdom: Animalia
- Phylum: Nematoda
- Class: Secernentea

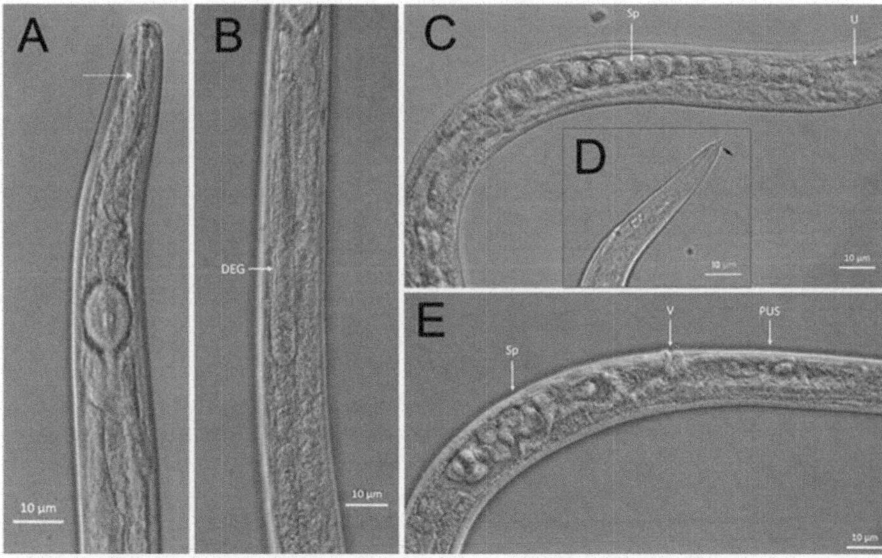

Fig. 26.7 *Aphelenchoides besseyi* female: body with anterior portions (**a**, **b**), arrowed stylet in **a** and dorsal esophageal gland overlapped in intestine in **b**, genital tract filled with round shaped sperm (**c**), tail ending with three pointed processes (**d**), and enlarged oval spermatheca in genital tract (**e**) (Photographs courtesy of J. Desaeger)

- Subclass: Diplogasteria
- Order: Aphelenchida
- Superfamily: Aphelenchoidea
- Family: Aphelenchoididae
- Subfamily: Aphelenchoidinae
- Genus: *Aphelenchoides*
- Species: *besseyi*
- Binomial name: *A. besseyi*

26.4.1.2 Symptoms of Damage

A. besseyi feeds on the plant tissues mostly externally. In *A. besseyi* infected rice plants, the leaves become white from the meristematic regions followed by their necrosis (Adamo et al. 1976; Lin et al. 2004). The infestation of WTNs can also result in stunted growth and sterility, and yield losses up to 50% (Lin et al. 2004; Hoshino and Togashi 1999).

26.4.1.3 Biology and Life Cycle

A. besseyi remains anhydrobiotic in seed till planting. As rice plants show growth, they become agile, attack on meristematic tissues for their food, and propagate amphimicticly in almost 8 days at 23 °C. Parthenogenesis can also occur in them. At reproductive stage of the rice plants, they increase rapidly in their populations, start feeding on reproductive structures, and later on settle in the seeds. With the

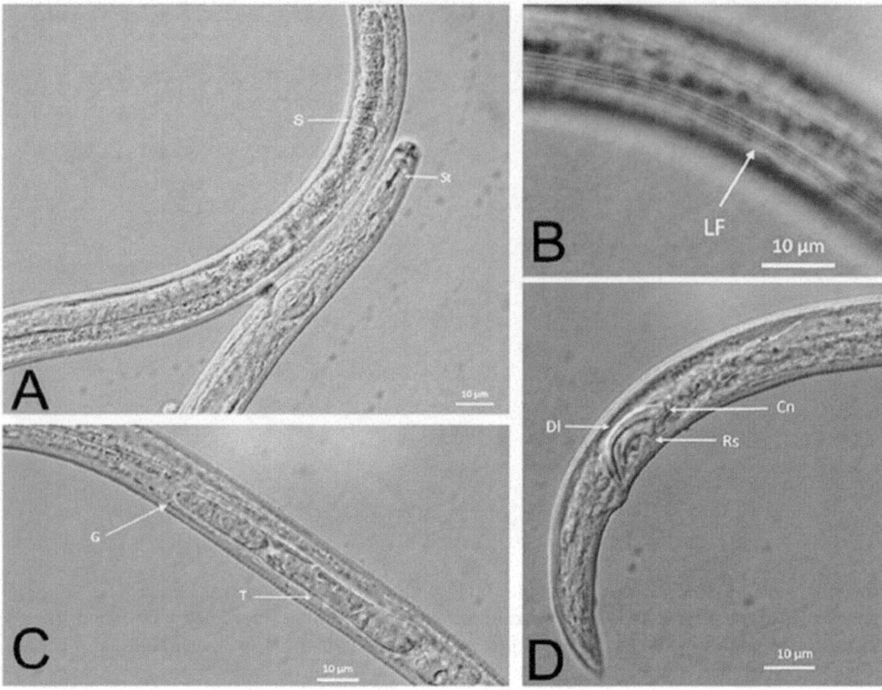

Fig. 26.8 *A. besseyi* male: stylet and sperm in middle and anterior portions of the nematode (**a**), lateral field (**b**), germinal zone in testis anterior portion (**c**), and posterior portion of the nematode (**d**) (Photographs courtesy of J. Desaeger)

dryness of the kernels, the WTNs eventually become quiescent and in dormant form can survive in the kernels for 3 years. The life cycle of *A. besseyi* is not so long and completes within 8–12 days (Lin et al. 2004).

26.4.1.4 Effect of Environmental Factors

A. besseyi is thermophilic, while lower temperature for its development and reproduction is 13 °C. The optimum temperature can vary between 23 °C and 30 °C. The effective temperature for the development of one generation of *A. besseyi* is 80 °C (Anonymous 2008).

26.4.1.5 Integrated Management of WTNs

Maintenance of clean seed is the most effective practice to control *A. besseyi.* The control of this nematode is fairly easy than other rice nematodes because it survives in the seed; hence, their seed treatment is effective. There is another method, in which the seeds are initially soaked into cold water in order to activate the nematodes, after that, these seeds are soaked in hot water to kill them. These hot water treated seeds can then be planted directly or they can also be stored after drying. Presoaking overnight and sun drying prior to sowing can also be effective.

Hot water treatment at 52–53 °C for 15 minutes is also effective to destroy nematodes infecting seeds (Tiwari and Khare 2003). The application of carbofuran 3G at nursery stage before 7 days of transplantation has been found very effective to control the WTNs populations. Chemical treatment of seeds before planting with thiabendazole or benomyl is also effective. The most effective way to control WTNs is the use of resistant varieties. Destruction of infected plant residues like seeds, weeds, and debris is also vital for the IDM of WTNs (Kumar and Sivakumar 1998).

26.4.2 Ufra Nematodes (UNs)

UNs are one of the major pests of deepwater rice; sporadic in irrigated and rainfed lowland rice (Latif et al. 2006). For the first time, *D. angustus* was recorded from East Bengal (Butler 1913). At that time, it was named as *Tylenchus angustus*. It was transferred to the genus *Anguillulina* in 1932 (Goodey 1932) and in 1936 to the genus *Ditylenchus* (Filipjev 1936). *D. angustus* is an obligate ectoparasite that feeds on young tissue of leaves, inflorescence, seeds, and rolled stems of growing plants. It is persistent in crop residues. Reproduction of *D. angustus* is amphimictic, that is, sexual and they can produce three generations in one season. The transmission of *D. angustus* through harvested seeds has also been reported (Prasad and Varaprasad 2002). Although the risk of seed transmission is less and can further be reduced to minimum by a thorough sun drying. Bridge and Starr (2007) reported UNs are not seed-borne and present around the growing points of rice plant seedlings in deepwater, while they are found on all parts of the plants in lowland rice. UNs move upwards for the search of their food and attack on newly forming tissues enclosed in the rolled leaf sheaths.

26.4.2.1 Taxonomy

- Kingdom: Animalia
- Phylum: Nematoda
- Class: Secernentea
- Subclass: Diplogasteria
- Order: Tylenchida
- Superfamily: Tylenchoidea
- Family: Anguinidae
- Subfamily: Anguininae
- Genus: *Ditylenchus*
- Species: *angustus*
- Binomial name: *D. angustus*

26.4.2.2 Symptoms of Damage

The main symptom of the attack of UNs is leaf chlorosis which appears during vegetative growth stage and remains till flag leaf stage. Formation of prominent white patches on different parts of infested plants causes malformation. Speckles also appear on the bases of young leaves during vegetative growth. Necrotic brown stains are also formed on leaves and leaf sheathes of infected rice plants which later become darker. The color of upper internodes of infected rice stem turns to dark brown. The lower nodes of infected plants become swollen with irregular branching. The bases of young leaves become twisted and distortion of leaf sheaths occurs. In case of severe infection, rice plants may die showing light brown appearance. Usually, after the panicle formation, dark brown patches of infected plants can easily be seen within fields (Cox and Rahman 1980; Bridge et al. 1990). In case of highly severe infection, the whole rice field shows dark brown appearance. Crinkling of panicles occurs after heading resulting in empty and shriveled glumes. The head, flag leaf, and panicles become twisted and distorted. The highly infected plants become dead; even if they survive, their grains' quality and quantity are reduced (Ali et al. 1995). At harvesting time, the healthy rice plants bearing tilted panicles while infected remain erect because they are empty.

26.4.2.3 Biology and Life Cycle

D. angustus can multiply on different fungi. The growth and reproduction of *D. angustus* on cultures of *Botrytis cinerea* has been examined and studied (Ali et al. 1995). One day after reaching adulthood, females of *D. angustus* start laying eggs. The two-celled eggs are laid and then they develop into J2 within 3 days which are then hatched out without any stimuli. The total generation time from egg to egg is 10 days. The generation time includes 3 days for the development of embryo, 6 days for the development of larva, and 1 day is required after adulthood for the commencement of egg deposition. When nematodes are cultured on *B. cinerea*, the oviposition period will be 14.8, and single female will produce 4.44 eggs per day. If the nematodes are cultured on *Epicoccum purpurascens* followed by *B. cinerea*, the oviposition period will be 12.8 days and single female will produce 4.26 eggs per day. Totally 65.71 and 54.53 eggs are produced by single female on *B. cinerea* and *E. purpurascens*, respectively (Ali et al. 1995).

26.4.2.4 Effect of Environmental Factors

High level of humidity helps *D. angustus* to migrate from infested plants or residues to healthy plants through stem and leaf contact or through water. UNs spread from plant to plant through irrigation water; minimum 75% humidity is required for *D. angustus* to migrate on the foliage. This nematode is more damaging in wet areas (Cuc and Kinh 1981; Rahman and Evans 1987). *D. angustus* can invade young rice plants within 1 h in water; however, invasion varies with plant age (Bridge et al. 1990). The highest infestation of *D. angustus* occurs at 27–30 °C (81–86 °F). This nematode has a short life cycle of 10–20 days (Bridge et al. 1990; Bridge and Starr 2007).

26.4.2.5 Integrated Management of UNs

The strategies being used against UNs include crop rotation, best cropping systems, solarization, use of resistant cultivars, fallowing of land for some period of time, burning of infected residues, applications of nematicides, and soil amendments/ biological control. Decision-support systems are also facilitating to control UNs.

Recently transgenic plants have been developed which have resistance against *D. angustus* (Khanam et al. 2018). Biological control agents and new safer nematicides are also being evaluated for the environment friendly management of UNs (Bridge 1996). Carbofuran, aminofuracarb, terbufosat, aldicarb, and carbosulfan are common in use for the chemical management of UNs (McDonald et al. 1987).

26.5 Conclusion

Many species of nematodes have been reported in association with rice. Such RPNs have adapted to various cultivation systems and are both foliar and root parasites of rice. The potential of these RPNs as pests of rice has been investigated. Such nematodes are considered as one of the important and limiting factors in rice production in all rice ecosystems. In upland rice, there is a huge reduction in grain yield due to the RPNs attacking young seedlings. The nematicidal compounds including volatile (fumigants) and nonvolatile are applied as soil drenches to cope with the issue of RPNs.

References

Adamo JA, Madamba CP, Chen TA (1976) Extracting the rice white-tip nematode, *Aphelenchoides besseyi*, with match sticks. J Nematol 8:178–179

Agrios GN (2005) Plant pathology, 5th edn. Elsevier Academic, Amsterdam, p 852

Ali R, Ishibashi N, Kondo E (1995) Growth and reproductive parameters of the rice stem nematode *Ditylenchus angustus* on *Botrytis cinerea*. Jpn J Nematol 25:16–23

Anderson RC (1968) The comparative morphology of cephalic structures in the superfamily Filarioidea (Nematoda). Can J Zool 46:181–199

Anonymous 2008. Pests Aphelenchoides besseyi Christie—white tip of rice, rice white tip nematode. Source: http://www.agroatlas.ru/en/content/pests/Aphelenchoides_besseyi/

Arayarungsarit L (1987) Yield ability of rice varieties in fields infested with root-knot nematode. Int Rice Res Newslett 12:14

Athanassiou CG, Vayias BJ, Dimizas CB, Kavallieratos NG, Papagregoriou AS, Buchelos CT (2005) Insecticidal efficacy of diatomaceous earth against *Sitophilus oryzae* (L.) (Coleoptera: Curculionidae) and *Tribolium confusum* du Val (Coleoptera: Tenebrionidae) on stored wheat: influence of dose rate, temperature and exposure interval. J Stored Prod Res 41:47–55

Bae CH, Szalanski AL, Robbins RT (2008) Molecular analysis of the lance nematode, *Hoplolaimus* spp., using the first internal transcribed spacer and the D1-D3 expansion segments of 28S ribosomal DNA. J Nematol 40:20–209

Baronia R, Kumar P, Singh SP, Walia RK (2020) Silver nanoparticles as a potential nematicide against *Meloidogyne graminicola*. J Nematol 52:e2020–e2002

Blair BL, Stirling GR (2007) The role of plant-parasitic nematodes in reducing yield of sugarcane in fine-textured soils in Queensland, Australia. Aust J Exp Agric 47:620–634

Bos L, Parlevliet JE (1995) Concepts and terminology on plant/pest relationships: toward consensus in plant pathology and crop protection. Annu Rev Phytopathol 33:69–102

Bridge J (1996) Nematode management in sustainable and subsistence agriculture. Annu Rev Phytopathol 34:201–225

Bridge J, Page SL (1982) The rice root-knot nematode, *Meloidogyne graminicola*, on deep water rice (*Oryza sativa* subsp. *indica*). Rev Nematol 5:225–232

Bridge J, Starr JL (2007) Plant nematodes of agricultural importance: a color handbook. Elsevier

Bridge J, Luc M, Plowright RA (1990) Nematode parasites of rice. In: Luc M, Sikora RA, Bridge J (eds) Plant parasitic nematodes in subtropical and tropical agriculture. CABI, Wallingford

Bridge J, Plowright RA, Peng D (2005) Nematode parasites of rice. In: Luc M, Sikora RA, Bridge J (eds) Plant parasitic nematodes in subtropical and tropical agriculture, vol 2. CABI Bioscience, Wallingford, pp 87–130

Butler EJ (1913) Ufra disease of rice. Agric J India 8:1–6

CAB International (2001) Hoplolaimus indicusin: crop protection compendium, global module, 3rd edn. CAB International, Wallingford

Castillo P, Vovlas N (2007) *Pratylenchus* (Nematoda: *Pratylenchidae*): diagnosis, biology, pathogenicity and management, vol 6. Brill, Boston. Source. https://brill.com/view/title/13449

Chitambar JJ, Raski DJ (1985) Life history of *Pratylenchus vulnus* on carrot discs. J Nematol 17: 235–236

Cohen SA, Michaud DP (1993) Synthesis of a fluorescent derivatizing reagent, 6-aminoquinolyl-N-hydroxysuccinimidyl carbamate, and its application for the analysis of hydrolysate amino acids via high-performance liquid chromatography. Anal Biochem 211:279–287

Cox PG, Rahman L (1980) Effects of ufra disease on yield loss of deep water rice in Bangladesh. Trop Pest Manage 26:410–415

Cuc NT, Kinh DN (1981) Rice stem nematode in Vietnam. Int. Rice Res Newslett 6:14–15

Curtis RH (2008) Plant-nematode interactions: environmental signals detected by the nematode's chemosensory organs control changes in the surface cuticle and behavior. Parasite 15:310–316

Dabur KR, Taya AS, Bajaj HK (2004) Life cycle of *Meloidogyne graminicola* on paddy and its host range studies. Indian J Nematol 34:80–84

Dasgupta DR, Raski DJ, Van Gundy SD (1969) Revision of the Genus *Hemicriconemoides* Chitwood & Birchfield 1957 (Nematoda: *Criconematidae*). J Nematol 1:126–145

De Waele D, Elsen A (2007) Challenges in tropical plant nematology. Annu Rev Phytopathol 45: 457–485

Dutta TK, Ganguly AK, Gaur HS (2012) Global status of rice root-knot nematode, *Meloidogyne graminicola*. Afr J Microbiol Res 6:6016–6021

Fassuliotis G (1962) Life history of *Hemicriconemoides* chitwoodi Esser. Nematologica 8:110–116

Fernandez L, Cabasan MT, De Waele D (2014) Life cycle of the rice root-knot nematode Meloidogyne graminicola at different temperatures under non-flooded and flooded conditions. Arch Phytopathol Plant Protect 47:1042–1049

Filipjev IN (1936) On the classification of the Tylenchinae. Proc Helminthol Soc Wash 3:80–82

Fortuner R, Merny G (1979) Root-parasitic nematodes of rice. Rev Nematol 2:79–102

Geraert E (2010) The Criconematidae of the World. Identification of the Family *Criconematidae* (Nematoda). Academia Press, Ghent. 615p

Goodey T (1932) The genus Anguillulina Gerv. & v. Ben., 1859, vel Tylenchus Bastian, 1865. J Helminthol 10(2–3):75–180

Griffin GD (1984) Nematodes parasites of alfalfa, cereals, and grasses. In: Nickle WR (ed) Plant and insect nematodes. Marcel Dekker, New York, pp 243–323

Hallmann J, Frankenberg A, Paffrath A, Schmidt H (2007) Occurrence and importance of plant-parasitic nematodes in organic farming in Germany. Nematology 9:869–879

Haque Z (2013) Development of integrated nematode management module for rice rootknot disease caused by *Meloidogyne graminicola*: a success story for endemic area. In: Conference Proceeding: National symposium on nematode: a friend and foe to agri-horticultural crops, at Solan, (HP), India

He Q, Dongya W, Bingxue L, Ambreen M, Haiyan W (2020) Nematicidal evaluation and active compounds isolation of aspergillus japonicus ZW1 against root-knot nematodes Meloidogyne incognita. Agronomy 10:1222

Hoshino S, Togashi K (1999) A simple method for determining *Aphelenchoides besseyi* infestation level of Oryza sativa seeds. J Nematol 31:641–643

Hoshino S, Togashi K (2000) Effect of water-soaking and air-drying on survival of *Aphelenchoides besseyi* in Oryza sativa seeds. J Nematol 32:303–308

Ichinohe M (1972) Nematode diseases of rice. In: Webster JM (ed) Economic Nematology. Academic Press, New York, pp 127–143

Jain RK, Khan MR, Kumar V (2012) Rice root-knot nematode (*Meloidogyne graminicola*) infestation in rice. Arch Phytopathol Plant Protect 45:635–645

Jaramillo J, Vintimilla M, Rubio D, Soto G, Tobar M, Salas E, Araya M (2019) Effect of nematicide rotation on banana (Musa AAA cv. Williams) root nematode control and crop yield. Agron Colomb 37:153–165

Jayaprakash A, Rao YS (1982) Life history and behaviour of the cyst nematode, Heterodera oryzicola Rao and Jayaprakash, 1978 in rice (*Oryza sativa* L.). Proc Anim Sci 91:283–295

Job SK, Thinniyam S (1995) Biology of banana population of *Heterodera oryzicola* (Nemata: *Tylenchina*). Fundam Appl Nematol 18:493–496

Jones JT, Haegeman A, Danchin EG, Gaur HS, Helder J, Jones MG, Kikuchi T, Manzanilla-López R, Palomares-Rius JE, Wesemael WM, Perry RN (2013) Top 10 plant-parasitic nematodes in molecular plant pathology. Mol Plant Pathol 14:946–961

Kavitha PG, Umadevi M, Suresh S, Ravi V (2016) The rice root-knot nematode (*Meloidogyne graminicola*)—life cycle and histopathology. Int J Sci Nat 7:483–486

Khan E, Chawla ML (1975) CIH description of plant parasitic nematodes Set 5, No. 66. Commonwealth Institute of Helminthology, St. Albans. Source: https://nematode.unl.edu/pest43.htm

Khan MR, Zaidi B, Haque Z (2012) Nematicides control rice root-knot, caused by *Meloidogyne graminicola*. Phytopathol Mediterr 51:298–306

Khanam S, Bauters L, Singh RR, Verbeek R, Haeck A, Sultan SM, Demeestere K, Kyndt T, Gheysen G (2018) Mechanisms of resistance in the rice cultivar Manikpukha to the rice stem nematode Ditylenchus angustus. Mol Plant Pathol 19:1391–1402

Kreye C, Bouman BA, Castaneda AR, Lampayan RM, Faronilo JE, Lactaoen AT, Fernandez L (2009a) Possible causes of yield failure in tropical aerobic rice. Field Crop Res 111:197–206

Kreye C, Bouman BA, Reversat G, Fernandez L, Cruz CV, Elazegui F, Faronilo JE, Llorca L (2009b) Biotic and abiotic causes of yield failure in tropical aerobic rice. Field Crop Res 112: 97–106

Kumar S, Sivakumar CV (1998) Management of rice White-tip nematode, *Aphelenchoides besseyi*. Indian J Nematol 28(3):85–87

Kumari U, Kuriyan KJ (1981) Cyst nematode, Heterodera oryzicola, on rice in Kerala I. Estimation of loss in rice due to *H. oryzicola* infestation, in field conditions. Indian. J Nematol 11:106

Latif MA, Akter S, Kabir MS, Ali MA, Hossain MA, Rahman ML (2006) Efficacy of some organic amendments for the control of ufra disease of rice. Bangladesh J Microbiol 23:118–120

Lavini A, Riccardi M, Pulvento C, De Luca S, Scamosci M, d'Andria R (2008) Yield, quality and water consumption of Stevia rebaudiana Bertoni grown under different irrigation regimes in southern Italy. Ital J Agron 2:135–143

Lin M, Ding XF, Wang ZM, Zhou FM, Lin N (2004) Description of Aphelenchoides besseyi from abnormal rice with 'small grains and erect panicles' symptom in China. Rice Sci 12:289–294

Loos CA (1949) Notes on free-living and plant-parasitic nematodes of Ceylon-No. 4. J Zool Soc India 1:17–22

de Luca F, Vovlas N, Lucarelli G, Troccoli A, Radicci V, Fanelli E, Cantalapiedra-Navarrete C, Palomares-Rius JE, Castillo P (2013) Heterodera elachista the Japanese cyst nematode parasitizing corn in northern Italy: integrative diagnosis and bionomics. Eur J Plant Pathol 136:857–872

Ma X, Agudelo P, Richards VP, Baeza JA (2020) The complete mitochondrial genome of the Columbia lance nematode, Hoplolaimus columbus, a major agricultural pathogen in North America. Parasit Vectors 13:1

Maggenti A (1981) General nematology. Springer, New York, 372 p

Mahfouz MM, Mohamed MM (2019) Towards optimization of entomopathogenic nematodes for more service in the biological control. J Nematol 51:1–48
Mai WF, Mullin PG, Lyon HH, Loeffler K (1996) Plant parasitic nematodes–a pictorial key to genera, 5th edn. Comstock, Ithaca, p 277
Mathur VK, Prasad SK (1972) Embryonic development and morphology of larval stages of the rice root nematode, *Hirschmanniella oryzae*. Indian J Nematol 2:146–157
Maung ZT, Kyi PP, Myint YY, Lwin T, Waele DD (2010) Occurrence of the rice root nematode *Hirschmanniella oryzae* on monsoon rice in Myanmar. Trop Plant Pathol 35:3–10
McDonald D, Luis AH, de Waele JH (1987) Chemical control of root-lesion nematodes (Pratylenchus spp.) on maize in South Africa. Phytophylactica 19:479–484
McGawley EC, Rush MC, Hollis JP (1984) Occurrence of Aphelenchoides besseyi in Louisiana rice seed and its interaction with *Sclerotium oryzae* in selected cultivars. J Nematol 16:65–68
Mondal S, Ghosh S, Mukherjee A, Khan MR (2020) Ecological analysis of nematodes associated with vegetable-maize cropping system at Dhapa, Kolkata (India). J Asia Pac Entomol 23:942–949
Muthukrishnan TS, Rajendran G, Ramamurthy VV, Chandrasekaran J (1977) Pathogenicity and control of *Hirschmanniella oryzae*. Indian J Nematol 7:8–16
Nand S, Seshadri AR, Dasgupta DR (1970) Culturing, embryology and life history studies on the lance nematode, *Hoplolaimus indicus*. Nematologica 16:235–248
Norton DC (1979) Ecology of plant-parasitic nematodes. 1978. Soil Sci 127:63
Ohshima Y (1974) Heterodera elachista n. sp., an upland rice cyst nematode from Japan. Jpn J Nematol 4:51–56
Padgham JL, Duxbury JM, Mazid AM, Abawi GS, Hossain M (2004) Yield loss caused by *Meloidogyne graminicola* on lowland rainfed rice in Bangladesh. J Nematol 36:42–48
Parker MB, Minton NA, Brooks OL, Perry CE (1975) Soybean yields and lance nematode populations as affected by subsoiling, fertility, and Nematicide treatments. Agron J 67:663–666
Plowright R, Bridge J (1990) Effect of *Meloidogyne Graminicola* (Nematoda) on the establishment, growth and yield of rice cv. IR36. Nematologica 36:81–89
Porazinska DL, Duncan LW, McSorley R, Graham JH (1999) Nematode communities as indicators of status and processes of a soil ecosystem influenced by agricultural management practices. Appl Soil Ecol 13:69–86
Prasad JS, Rao YS (1978) Potentiality of *Pratylenchus indicus* Das 1960, the root-lesion nematode as a new pest of upland rice. Ann Zool Ecol Animale 10:635–640
Prasad JS, Panwar MS, Rao YS (1987) Nematode problems of rice in India. Int J Pest Manag 33(2): 127–136
Prasad JS, Varaprasad KS (2002) Ufra nematode, *Ditylenchus angustus* is seed borne! Crop Prot 21: 75–76
Rahman ML, Evans AAF (1987) Studies on host-parasite relationships of rice stem nematode, *Ditylenchus angustus* (Nematoda: *Tylenchida*) on rice, *Oryza sativa*, L. Nematologica 4:451–459
Ramakrishna A, Sharma SB (1998) Cultural practices in rice-wheat-legume cropping systems: effect on nematode community. In: Nematode pests in rice-wheat-legume cropping systems: Proceedings of a Regional training course. 1-5 September 1997, Rice-Wheat Consortium Paper Series, vol 4. CCS Haryana Agricultural University, Hisar, Haryana, pp 73–79
Rao YS, Israel P (1971a) Studies on nematodes of rice and rice soils. VI. Influence of soil temperature on the prevalence and incidence of *Meloidogyne graminicola*, the rice root-knot nematode. Oryza 8(2):47–51
Rao YS, Israel P (1971b) Studies on nematodes of rice and rice soils. V. Influence of soil chemical properties on the activity of *Meloidogyne graminicola*, the rice root-knot nematode. Oryza 8: 33–38
Ravinndra H, Sehgal M, Narasimhamurthy HB, Jayalakshmi K, Khan HI (2017) Rice root-knot nematode (*Meloidogyne graminicola*) an emerging problem. Int J Curr Microbiol App Sci 6: 3143–3171

Routaray BN, Das SN (1982) Some growth characteristics of rice as affected by *Hoplolaimus indicus*. Indian J Nematol 12:250–253

Schmitt DP, Bailey JE (1990) Chemical control of Hoplolaimus columbus on cotton and soybean. J Nematol 22(4S):689

Senthilkumar P, Ramakrishnan S, Jonathan EI (2007) Life cycle, varietal reaction, biochemical alteration and histopathology of rice root-knot nematode, *Meloidogyne graminicola*. Indian J Nematol 37:165–171

Sher SA (1968) Revision of the genus *Hirschmanniella* Luc & Goodey, 1963 (Nematoda: *Tylenchoidea*). Nematologica 14:243–275

Singh SK, Hodda M, Ash GJ, Banks NC (2013) Plant-parasitic nematodes as invasive species: characteristics, uncertainty and biosecurity implications. Ann Appl Biol 163:323–350

Somasekhara Y, Mukesh S, Ravichandra NG, Siddegowda DK, Jain RK, Mahadevu P, Ravindra KR, Sardana HR (2012) Validation and economic analysis of adaptable integrated management technology against root-knot nematode (*Meloidogyne graminicola*) in rice (*Oryza sativa*) with farmers' participatory approach. Indian J Agric Sci 82:442–444

Soriano IRS, Prot JC, Matias DM (2000) Expression of tolerance for *Meloidogyne graminicola* in rice cultivars as affected by soil type and flooding. J Nematol 32:309–317

Subbotin SA, Ragsdale EJ, Mullens T, Roberts PA, Mundo-Ocampo M, Baldwin JG (2008) A phylogenetic framework for root lesion nematodes of the genus *Pratylenchus* (Nematoda): evidence from 18S and D2–D3 expansion segments of 28S ribosomal RNA genes and morphological characters. Mol Phylogenet Evol 48:491–505

Takagi M, Goto M, Wari D, Saito M, Perry RN, Toyota K (2020) Screening of nematicides against the lotus root nematode, Hirschmanniella diversa Sher (Tylenchida: Pratylenchidae) and the efficacy of a selected nematicide under lotus micro-field conditions. Agronomy 10:373

Taylor AL (1969) Nematode parasites of rice. In: Peachey JE (ed) Nematodes of tropical crops, vol 40. Commonwealth Bureau of Helminthology, St. Albans, pp 264–268

Thorne G (1937) Superfamily Aphelenchoidea Fuchs (Chapter 15). In: Principles of nematology. McGraw-Hill, New York, pp 391–441

Tiwari SP, Khare MN (2003) White tip caused by *Aphelenchoides besseyi*, an important seed borne disease of rice. Advan Nematol, Trivedi, P.C., Scientific Publishers, pp 103–114

Turner DR, Chapman RA (1972) Infection of seedlings of alfalfa and red clover by concomitant populations of *Meloidogyne incognita* and *Pratylenchus penetrans*. J Nematol 4:280–286

Van den Berg E, Tiedt LR, Inserra RN, Stanley JD, Vovlas N, Rius JE, Castillo P, Subbotin SA (2014) Morphological and molecular characterisation of some *Hemicriconemoides* species (Nematoda: *Criconematidae*) together with a phylogeny of the genus. Nematology 16:519–553

Van Nguyen S, Nguyen PT, Araki M, Perry RN, Tran LB, Chau KM, Min YY, Toyota K (2020) Effects of cropping systems and soil amendments on nematode community and its relationship with soil physicochemical properties in a paddy rice field in the Vietnamese Mekong Delta. Appl Soil Ecol 156:103683

Vanstone VA, Hollaway GJ, Stirling GR (2008) Managing nematode pests in the southern and western regions of the Australian cereal industry: continuing progress in a challenging environment. Australas Plant Pathol 37:220–234

Villarroel M, Junge R, Komives T, König B, Plaza I, Bittsánszky A, Joly A (2016) Survey of aquaponics in Europe. Water 8(10):468

Whitlock LS, Steele AE (1960) Notes on *Hemicriconemoides gaddi* from camellias in Louisiana and Georgia. Plant Dis Rep 44:446–447

Win PP, Kyi PP, De Waele D (2011) Effect of agro-ecosystem on the occurrence of the rice root-knot nematode *Meloidogyne graminicola* on rice in Myanmar. Australas Plant Pathol 40:187–196

Part IV

Advanced Technologies

Managing Greenhouse Gas Emission 27

Sajjad Hussain, Muhammad Mubeen, Syeda Refat Sultana, Ashfaq Ahmad, Shah Fahad, Wajid Nasim, Shakeel Ahmad, Amjed Ali, Hafiz Umar Farid, Hafiz Muhammad Rashad Javeed, Ayman E. L. Sabagh, and Mazhar Ali

Abstract

Rice (*Oryza sativa*) production systems have faced the two opposing challenges all over the world: the need to increase the production to nourish the world's increasing population and reducing the emissions of greenhouse gases (GHG). Nitrous oxide (N_2O), carbon dioxide (CO_2), methane (CH_4), and

S. Hussain · M. Mubeen (✉) · S. R. Sultana · H. M. R. Javeed · M. Ali
Department of Environmental Sciences, COMSATS University, Islamabad, Pakistan
e-mail: muhammadmubeen@cuivehari.edu.pk

A. Ahmad
Climate Change, US—Pakistan Centre for Advanced Studies in Agriculture and Food Security, University of Agriculture, Faisalabad, Pakistan

S. Fahad
Department of Agronomy, The University of Haripur, Haripur, Pakistan

W. Nasim
Department of Agronomy, University College of Agriculture and Environmental Sciences, The Islamia University of Bahawalpur (IUB), Bahawalpur, Pakistan

S. Ahmad
Department of Agronomy, Faculty of Agricultural Sciences and Technology, Bahauddin Zakariya University, Multan, Pakistan
e-mail: shakeelahmad@bzu.edu.pk

A. Ali
University College of Agriculture, University of Sargodha, Sargodha, Pakistan

H. U. Farid
Department of Agricultural Engineering, Bahauddin Zakariya University, Multan, Pakistan

A. E. L. Sabagh
Department of Field Crops, Faculty of Agriculture, Siirt University, Siirt, Turkey

Department of Agronomy, Faculty of Agriculture, Kafrelsheikh University, Kafrelsheikh, Egypt

N. Sarwar et al. (eds.), *Modern Techniques of Rice Crop Production*,
https://doi.org/10.1007/978-981-16-4955-4_27

chlorofluorocarbons (CFCs) are the most significant GHGs because of their global warming mitigation (GWM) and radiative effects on rice. Rice intensive farming system has been producing extreme pressure on fields of rice for producing more rice for the increasing global population, thus declining rice ecosystem balance and soil fertility situation by fluxes of more N_2O, CH_4, and CO_2 to the environment. Many farmers used fertilizer combination and commercial hormone to rice growing. Nowadays, the integrated management system like modifying tillage practices, improving nitrogen fertilization and irrigation patterns, increasing yield potential, and managing organic and fertilizer inputs are set up based on plant physiological needs. These strategies can also increase the yield of rice as well as have benefits on GWM. Satellite-based estimates provide unique opportunities to improve bottom-up and top-down estimate of GHG emissions, and also provide important observations to support the understanding as well as monitoring of environment and earth's surface changes due to human activities. The integrated management system, an eco-farming method, gives the best solution than transgenic plants (in which several problems including field tests and stability of the transgenic lines are inevitable). Adapting drainage systems could be a good option for reducing CH_4 in rice production system.

Keywords

Greenhouse gas emission · Rice · Mitigation prospect · Climate change

27.1 Introduction

Rice (*Oryza sativa*) is a significant economic crop all over the world. Climate change (CC) is an important environmental problem in the twenty-first century, which may meaningfully impact rice productivity side by side accelerating emissions of greenhouse gases (GHGs) from rice ecosystem, which is of great ecological concern (Ahmad et al. 2009a, b). Global population is estimated to increase up to 35% to reach almost nine billion by 2050, demanding 70–100% increase in food demand all over the world. Rice production is projected to increase from 676 million tons in 2010 to 852 M tons in 2035 with a total rise of 26% and 176 M tons in the next 25 years worldwide (Linquist et al. 2012; Tarlera et al. 2016; Ahmed et al. 2017). Fields of rice have been a concern for scientists globally because they produce the three greatest long-lived and potent GHGs, namely, methane (CH_4), nitrous oxide (N_2O), and carbon dioxide (CO_2), due to their positive rises in radiative forcing as well as their influence in global warming (Ali et al. 2009; Lampayan et al. 2015; Ahmed et al. 2020a, b).

Satellite-based observations provide useful information about GHG emissions (Matsunaga and Maksyutov 2018). Different approaches have got advancements to use environmental CO_2 and CH_4 data for approximating CO_2 andCH_4 emissions; these include targeting strong emitters for example the megacities and measurement of other gases that help to differentiate fossil fuel-derived CO_2 (Linquist et al. 2015;

Thu et al. 2016; Ali et al. 2019a). However, soils have the ability to mitigate rising CO_2 concentrations through carbon sequestration, with the maximum potential of global sequestration varying from 0.45 to 0.9 PgC/year. Hence, understanding the impacts of management practices on SOC (soil organic carbon) and GHG emissions is essential to better the management practices to decrease emissions of GHG in rice fields (Wajid et al. 2014). Rice absorbs carbon from the environment, but if the plant cannot use it proficiently, the carbon is disseminated into the soil, where it converts to CH_4. Under these conditions, global warming has become a significant problem nowadays (Wang et al. 2015; Anser et al. 2020).

27.2 Climate Change and GHG Emissions

Climate change (CC) is well-defined as rise in mean temperature of the Earth's near-surface air as well as oceans, with its predicted continuation (Ali et al. 2019b). Increasing the temperature is governed by various components, for example, CFCs, CH_4, N_2O, CO_2, and water vapor that absorb the heat, thus rising temperatures all over the world (Mubeen et al. 2021). The rising temperature is generally due to increase in GHGs and atmospheric and human activities: human needs timber, so they cut down the forests or clear land for agriculture, human residences, and industries. This widespread forest cutting decreases the trees that absorb CO_2. The extensive CH_4 gas production in the environment is due to rice paddies, natural wetlands, as well as livestock (Fahad et al. 2018). Coal mining, termite landfills, natural gas production, and biomass burning also release CH_4. Due to overexploitation of natural resources and fossil fuel, it has become difficult to sustain the worldwide energy balances to and from the earth surface and atmosphere, therefore global temperature is also increasing (Sultana et al. 2014). Agricultural practices like rice-based cropping system are the main GHG emitters, which are added as a main portion of worldwide emissions (Arunrat and Pumijumnong 2017; Wang et al. 2020a).

The GHGs, mostly CH_4, CO_2, and N_2O, have been adding approximately 80% to the present worldwide radiative forcing. Agricultural activities cause about 20% of the current concentrations of atmospheric GHGs, especially N_2O and CH_4emissions from the paddy fields. N_2O and CH_4 are the two most significant GHGs from crop production, with GWP (global warming potentials) of 25 and 298 CO_2 equivalents, respectively, on a hundred years' time horizon. The CC concerns have concentrated to decrease the emission of GHG from agriculture (Wang et al. 2020b). CO_2 is also GHG; yet, on a worldwide scale, fluxes of soil CO_2 are mainly offset by net primary productivity as well as atmospheric CO_2 fixation through crop plants, and therefore contribute approximately 1% to the GWP of agriculture. The N_2O is also a potential GHG with a radiative forcing prospective about 12 times greater than CH_4 (Aslam et al. 2013; Wang et al. 2017).

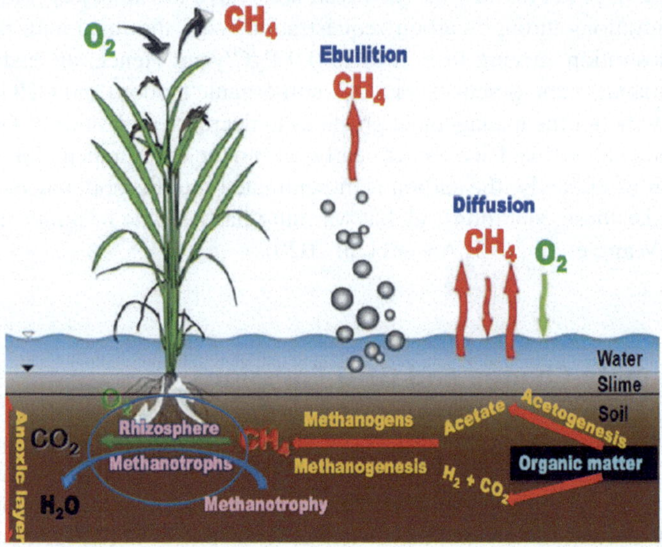

Fig. 27.1 Emissions of GHG from rice

27.3 Characterization and Threats of the Rice System

Rice (*Oryza sativa*) is a vital agricultural staple food for greater than half of the population globally; it is grown in 114 nations to cover an area of around 153 M ha, which is 11% in worldwide arable land. Production of rice must increase by 40% by 2030 to fulfill the rising need of the growing population in the world (Wassmann et al. 2004). Rice production particularly in tropical Asia as well as Southeast Asia is extremely vulnerable to CC. Production of rice is a source of CC globally due to emission of CH_4, CO_2, and N_2O gases emissions and at the same time is impacted due to change in CC (Bakhat et al. 2017; Sabagh et al. 2020a).

The rice occupies about 13.5 M ha in South Asia, including 10 M ha in India, 2.2 M ha in Pakistan, 0.8 M ha in Bangladesh, and 0.5 M in Nepal as shown in Fig. 27.1, spreading across the Indo-Gangetic floodplain into the Himalayan foothills. Rice production systems cover approximately 32% to the whole rice cover area in these four countries (Batool and Chuadhry 2009; Wu et al. 2018).

27.4 Emissions of GHGs from Rice

Rice cropping systems are thought as one of the main anthropogenic causes of N_2O and CH_4. CH_4 emissions from rice paddy soils is estimated about 31–112 Tg y^{-1}, accounting the 19% of overall emissions, whereas 11% of agricultural N_2O

Table 27.1 Highest rice producing countries in Asia

Country	Harvest area (M ha)	Overall production (M tones)	Yield (t ha^{-1})	CH_4 emissions (t CH_4 ha^{-1})
China	29	195	6.5	0.18
India	43	146	3.4	0.11
Indonesia	12	63	4.9	0.21
Bangladesh	11	46	4.3	0.1
Vietnam	7	40	5.3	0.18
Brazil	2	12	4.3	0.06
Colombia	0.5	2	4.6	0.21
Argentina	0.2	1	6.6	0.28

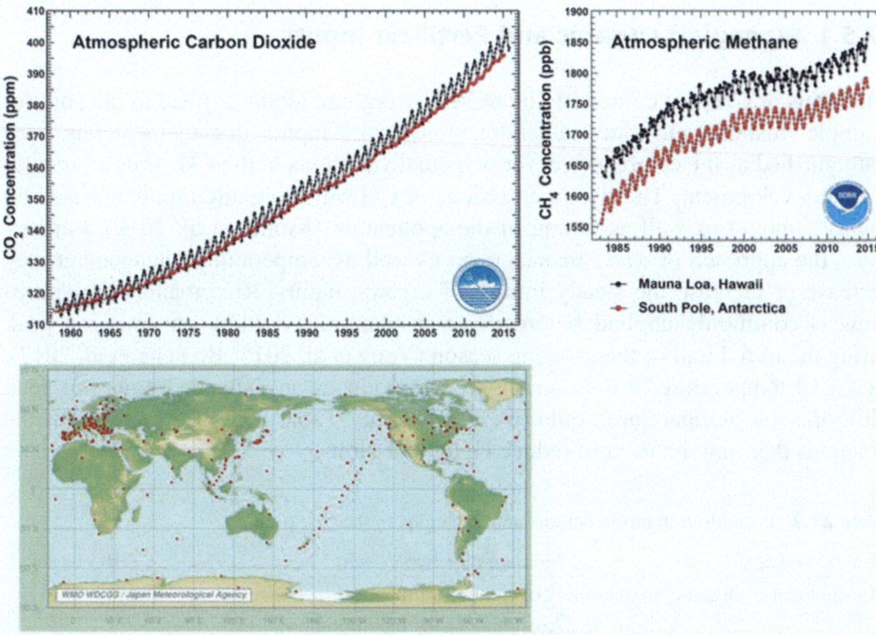

Fig. 27.2 *Top*: Meteorological observations of CH_4 and CO_2. *Bottom*: Map of ground-based quantity sites donating to the World Atmospheric Organization's information Centre for GHG

emissions are derived from rice fields worldwide (Table 27.1). The emissions of GHG from rice have consequently motivated great concerns (Bayer et al. 2014; LaHue et al. 2016).

From a worldwide point of view, the problem of CC is a developing issue, and decreasing non-CO_2 and GHG emissions is a comparatively fast way to decrease the effects of GHG on CC (Kumar et al. 2016). The agricultural sector is a significant cause of GHG emissions, but proper management can decrease agricultural GHG emissions (Fig. 27.2). CH_4 is also an abundant non-CO_2 GHG, and cultivation of rice field is a significant source in agricultural production emissions of CH_4 (Fahad et al. 2015). The warming capability of N_2O is 298 times that of CH_4 and is the greatest significant ozone-removing gas. While the important GHG released from rice fields is CH_4, yet N_2O cannot be discounted. Irregular irrigation to decrease CH_4 emissions will lead to enhanced rice N_2O emissions (Xu et al. 2015). The haphazard application of nitrogen (N) fertilizer in agricultural production is the primary driver of GHG emissions, consequently it is essential to recognize the management measures to decrease the utilization of chemical fertilizers (Bayer et al. 2015).

27.5 Mitigating Strategies to Reduce GHG Emissions in Rice

27.5.1 Managing Organic and Fertilizer Inputs

Emissions of CH_4 are commonly increased by organic inputs applied to the soil, for example, manure and straw amendment. Organic inputs management has been distinguished as the central driver for perpetual variations in the CH_4 source strength of rice development. The rise in emissions of CH_4 after organic inputs is based on quality, amount as well as timing of the application (Kumar et al. 2019). Furthermore, the approach of water management as well as temperature management may decrease or increase the steady impact of organic inputs. Rice manure as well as straw is commonly applied before rice transplanting, yielding an emission peak during the initial half of the growing season (Yang et al. 2015; Boateng et al. 2017). Increased temperature in the coming weeks produces an enhanced emission peak while the low temperature could lower this peak. Table 27.2 shows some of the strategies that may be used to reduce GHG emissions.

Table 27.2 Possible mitigation actions and strategies in the rice rotation

General strategy	Measures for rice system	GHG targeted
Management of organic inputs	Fermentation of manure	CH_4
	Amending straw incorporation	
Modifying nitrogen fertilization	Corresponding N supply with demand	N_2O
	Choosing type of fertilizer and amendment	CH_4
Improving irrigation patterns	Mid-season drainage	CH_4
	Alternative flooding and drying	
	Direct seeding	
Improvement of crop cultivars	Breeding for particular traits	CH_4
	"Aerobic" rice varieties	N_2O
	Improved yield potential	
Improving soil organic C	Managing wetland	CO_2
	Reduced tillage	
	Recycling of residues	

27.5.2 Modifying Tillage Operations

Tillage is a significant factor that determines the flux of oxygen to the topsoil and, consequently, the aerobic decomposition of SOM. Ahmad et al. (2009a, b) modeled the past dynamics of SOM in America's Corn Belt; they calculated depletion of up to 53% of actual value after conversion of the native vegetation toward agriculture area (Kritee et al. 2018). Although this depletion happened in the initial half of twentieth century, present zero-tillage practice has raised the SOC in America's Corn Belt to 61% of its actual. As zero-tillage becomes more and more popular in the Asian rice belt, this development denotes the possibility for reductions, particularly on soils with SOM deficiencies (Fahad et al. 2016).

Bayer et al. (2014) found similar N_2O emissions from the soil under no-till and that under conventional tillage. These results are in line with Zhang et al. (2014), who observed similar N_2O emissions under no-till and conventional tillage for a field experiment in the Jiangsu province of China. However, in a study of Hubei Province (China), N_2O emissions were 33% greater in fertilized no-till fields than fertilized conventional tillage fields.

There are limited statistics on the methods of various types of tillage adopted by rice growers in different countries. Anyhow, about 50% of the world's agricultural area is under no-till, although in various countries about 70% of the overall cultivated land (along with rice fields) is under no-till (Khush 2000; Yang et al. 2017). This means that the region has together the technical capacity and the knowledge to carry out no-tillage cultivation. Since no-tillage systems can decrease the emissions of CH_4 from rice paddies, quantifying the land following this practice as well as other tillage systems will allow improved estimate of local emissions due to tillage from paddy fields (Akram et al. 2018; Mehmood et al. 2020).

27.5.3 Improving Nitrogen (N) Fertilization

The N fertilizer application to soils raises production, but may affect the emissions of GHG from rice crop. In a meta-analysis, N fertilizer-induced emissions of N_2O were stated to be 0.21% under the continuous flooding and 0.40% under AWD (Alternate Wetting and Drying) rice cropping systems (Khan et al. 2013). In this meta-analysis, the influence of type of N fertilizer was also stated; N_2O release rose up to 24% and CH_4 emission decreased up to 40% when ammonium sulfate was applied instead of urea. On the other hand, in soils of irrigated rice, the relationship of N fertilizer with CH_4 is complex. The N fertilizer improves plant growth as well as the amount of root exudates and crop remains which are source of carbon substrate for methanogenesis. Furthermore, likenesses in structure and size of CH_4 and NH_4^+ may affect NH_4^+ constraining CH_4 consumption (Yousaf et al. 2017). In China, during a field experiment, by increasing the amounts of urea and ammonium sulfate fertilizer from 100 to 300 kg N ha^{-1}, CH_4 emissions were reduced by 7% and 30%, respectively. On the other hand, a reduction in the emissions of CH_4 due to greater N application may exhibit trade-offs with a rise in the emissions of N_2O, representing the complication of interactions of N-fertilization and CH_4 emissions (Gan et al. 2014; Khan et al. 2018).

Mitigating N fertilizer-induced N_2O emissions could have the following objectives:

(a) Decreasing the quantity of N_2O production through nitrification and denitrification.
(b) Limiting the concentration of microbial N turnover in the soil. The corresponding emission of N_2O during nitrification and denitrification mainly depends on the form of nitrogen and O_2 presence in the soil (Hardwick and Graven 2016).

27.5.4 Increasing Yield Potential

Production of crop per unit area at farmer level is generally dependent on the "yield potential" of crop genotypes as the rise in the yield potential will pull the recognized yield by farmer level regardless of the magnitude of present gap between them. The assumption of "high yield capacity" is significant as it shows how far human management to raise the crop yields (Deng et al. 2015). Rice production is mainly effected by the yield capacity of cultivated genotypes so increase in the yield capacity of rice is the significant technique to improve rice production worldwide. Studies on yield capacity of crop plants can be helpful in assessing the production downfalls in various cropping system (Cui et al. 2018). Assessments of yield capacity have been used as thumb rule to evaluate progress in plant breeding programs and to find contribution of plant breeding and agronomic advances to crop yields in past (Mingxing and Jing 2002). On contrary, the terms "yield capacity" and "high yield capacity" are interchangeably utilized in the documents to define various yield maximum amount of crop plants without making a clear difference between them (Chen et al. 2015).

27.5.5 Improved Irrigation Patterns

From rice systems, management of water patterns impacts rice production emissions of N_2O and CH_4. Single or alternate drainage for rice are helpful to decrease the emissions of CH_4 from 48 to 93% under to flooding systems. Many researchers have stated a reduction in emissions of CH_4 for single or alternate drainage associated to continuous flooding areas (Mubeen et al. 2019). The decrease in emissions of CH_4 has been credited to variations in soil redox situations (Carrijo et al. 2017; Zahoor et al. 2019).

Cropping system of irrigated rice with pastures and aerobic crops has numerous natural challenges associated to management for variant soil properties (e.g., drainage and compaction) of different countries while emissions of GHG from these systems are yet to be sufficiently quantified (Mubeen et al. 2013c). The possible mitigation of water management practices followed in the area is however to be standardized (Mubeen et al. 2013a, b). For example, farmers are commonly mentioned to suspend irrigation 15–20 days after flower production, which allows farmers to save water as well as money and may also reduction in emissions of CH_4 (Islam et al. 2020).

27.5.6 Selecting Suitable Cultivar

Several earlier studies have observed differences in CH_4 emissions among different varieties from rice paddies. Varietal variation for CH_4 extraction are measured by the amount of degradation and exudates of roots which directly affect the availability of substrate (i.e., carbon); different factors likewise of crop, number of cultivators, area, and number of leaves; grain starch content; length of stay in the field and the structure of aerenchyma affects the transfer of methane from the soil to the atmosphere (Mir et al. 2017). In contrast, the studies carried out did not reveal any racial differences in N_2O emissions into the soil. It is well known that about 80% of N_2O is transported through the rice aerenchyma under flooding situations (low redox possible is related with whole denitrification under these conditions, probably on a varietal basis). However, some of these research found inconsistencies between locations as well as seasons, suggesting greater environmental impacts (Hasan 2013).

27.5.7 Improving Crop Cultivars

Rice and wheat have different effects on the carbon and nitrogen balance in this ecosystem. Plants not only act as a source of carbon balance (complete photosynthesis) but also as important source of methanogenic material by breaking down of different plant parts especially roots. In contrast, some plants act as nitrogen scavengers in the ecosystem like rice and wheat, after removal of crop residues (Houweling et al. 2015). On the other hand, plant material is also one of the main engines of soil N-cycles, providing a substrate for ammonification (Montzka et al. 2011).

The rice plant releases both CH_4 and N_2O gases in to the environment. This property of the rice plant is due to aerenchyma, characteristic of plants modified in response to waterlogging (Hussain et al. 2022).

27.5.8 Increasing Soil Organic Carbon

The rise of CO_2 in environment is mainly due to the burning of fossil fuels which also act as sinks in global carbon budget and are not balanced (Zhang et al. 2016). Recent results have shown that the 1.8 Gt Cyr^{-1} "missing sink" may be due in part to carbon sequestration in terrestrial ecosystems.

Because of the complex issues of emission control, recent research has focused on whether good management can increase the resilience of terrestrial ecosystems to sink (Hsu et al. 2009).

27.6 Reduction of GHG by Improved Management System

One of the main options for reducing GHG emissions from agriculture is by adopting different management through agronomic practices, water management such as (drainage as well as irrigation), nutrient management, sediment/agroforestry, and soil/crop residue management. Management system based on physiological requirements at various stages of plant growth to improve the fertilizer management to attain optimal land use and land cover (LULC) efficiency and fertilizer use efficiency (Hussain et al. 2014; Ahmad et al. 2015; Ahmad et al. 2019; Ahmad et al. 2012, 2013; Ahmad et al. 2008, 2009a, b; Ahmad and Hasanuzzaman 2012; Ahmed and Ahmad 2019, 2020; Shahzad and Ahmad 2019). This management system raises rice yields with three products: by adding pollen extract, root formation, and Act zyme Plus at various growth phases. Pollen extract is a different plant extract that can activate cells to form whole internal organelles for vigorous biochemical reactions leading to overall plant development (Munawar et al. 2020). The root contains organic acids, B vitamins, and special ingredients. This unique formula is formulated to increase root hair density and significantly improves the surface area for nutrient absorption when used correctly (Nasim et al. 2017). Act zyme plus promotes the breakdown of organic matter in the soil. The biological enzymes in this reagent improve plant-friendly microbes as well as inhibit soil pathogens (Hussain et al. 2015; Sathish et al. 2017).

Act zyme Plus leaf spray improves the accumulation of photosynthetic products for the growth of rice grains. The application of Act zyme optimizes the CN ratio, resulting in increased grain yield with quality (Pandey et al. 2014). It differs from transgenic plants or hybridization; they provide rice with additional nutrients depending on the physiological needs of the plant. In the germination phase, they bring in the essential nutrients, nitrogen, phosphate, and potassium, as well as trace elements and rare trace elements, to activate the metabolic reaction of rice (Ravishankara et al. 2009; Sabagh et al. 2020b). With the use of the AgNeeds management system, the yield of rice per unit of agricultural area can increase by 30%. The more biomass is produced; the more CO_2 is converted into rice grains as well as the less carbon is released into the soil (Qin et al. 2010; Revell et al. 2015). This organic farming system from AgNeeds not only increases rice yields but also decreases GHG emissions (Hussain et al. 2020b).

27.7 Reduction of GHG Using Satellite Data

To measure CH_4 and CO_2 satellite spectrometers used a variety of wavelengths that based on their absorption spectra. Satellites observed the smaller wavelengths NIR (near infrared) which are sensitive to near-surface CO_2 concentrations as well as therefore have proven most useful for studying surface emissions (Zhang et al. 2014). Various satellites have been used to identify CO_2 levels in troposphere between 6 and 11 km above the earth's surface, for example, using longer thermal infrared waves (IASI and AIRS satellites). Satellites produce high-resolution Earth's surface which give the data about emissions of GHG.

- Satellites are also used to measure environmental CH_4 and CO_2. This data is used in different models (atmospheric and statistical) for estimation of regional and global sources as well as sinks of these gases (Matsunaga and Maksyutov 2018).
- These satellites provide information about LULC, human population, fires, and infrastructure, as well as the biological and vegetation cover. This information is used to assess the emissions gradient of GHG, biomass burning, spatial circulation of fossil fuel burning, and flows of GHG from marine and terrestrial ecosystems (Hussain et al. 2020c).
- Collaboration on International and National basis is being corresponding to support, promote, and exchange satellite observations.

The LULC change is the second largest source of anthropogenic GHG emissions, after fossil fuel combustion. Earth Satellite images were used to make different LULC maps with forest area. NASA's Landsat program gives a continuous recorded of surface imaging from 1972 to date by using a sequence of satellite missions (Hussain 2018; Hussain et al. 2020a). Satellite measurements determine the concentration of gases in the atmosphere using the properties of gases to absorb electromagnetic radiation of certain wavelengths. These instruments mainly use solar radiation that is reflected off the surface of the earth. However, some may use radiation reflected from the Earth or from lasers on board a satellite (Hussain et al. 2020d).

27.8 The GHG Observing Satellite (GOSAT)

The GOSAT is operated by the Japan Aerospace Exploration Agency (JAXA); the Japan's National Institute for Environmental Studies (NIES); and Ministry of the Environment (MOE). The GOSAT was started in January 2009 and became the first satellite to monitor GHG in the atmosphere. The GOSAT typically registers three to five tracks 10 km in diameter in its 700 km lane and performs worldwide coverage within 3 days (Hardwick and Graven 2016).

27.9 Conclusion

Globally, the cultivated rice production must increase many times over current levels to meet food needs. Nonetheless, agriculture is responsible for 14% and 52% of anthropogenic emissions of CH_4 and CO_2, respectively. The GHG emission in rice can be reduced by increased irrigation, and use of fertilizers as well as the development of new high-yielding rice varieties. On the other hand, this is too expensive for the farmers as they do not want to drain the field or take relevant action by hiring more workers. Consequently, there is an urgent need to develop other strategies to reduce global warming through biotechnology to manipulate genes or through an integrated management system. Satellite-based observations provide valuable information about GHG emissions. Numerous strategies are being developed to estimate CO_2 emissions from fossil fuels by using atmospheric CO_2 data. This includes targeting strong emission sources like megacities as well as measuring other gases that help differentiate CO_2 from fossil fuels. The Integrated Management System can be used in rice paddies to directly mitigate global warming. In this case, an integrated management system appears to be a more effective way to decrease GHG emissions in the short term. If we can combine both the technologies in the upcoming years, there is great potential to stabilize or even decrease GHG emissions from rice paddies, and at the same time increasing rice production without radically changing cultural practices. There is an urgent need to decrease GHG emissions to reduce the adverse influences of CC. Consequently, in the future, rice growing systems will have to combine higher rice yields with reduced GHG emissions.

References

Ahmad S, Hasanuzzaman M (2012) Integrated effect of plant density, N rates and irrigation regimes on the biomass production, N content, PAR use efficiencies and water productivity of rice under irrigated semiarid environment. Not Bot Horti Agrobot Cluj Napoca 40(1):201–211

Ahmad S, Zia-ul-Haq M, Ali H, Shad SA, Ammad A, Maqsood M, Khan MB, Mehmood S, Hussain A (2008) Water and radiation use efficiencies of transplanted rice (*Oryza sativa* L.) at different plant densities and irrigation regimes under semi-arid environment. Pak J Bot 40(1): 199–209

Ahmad S, Ahmad A, Zia-ul-Haq M, Ali H, Khaliq T, Anjum MA, Khan MA, Hussain A, Hoogenboom G (2009a) Resources use efficiency of field grown transplanted rice (*Oryza sativa* L.) under irrigated semiarid environment. J Food Agric Environ 7(2):487–492

Ahmad S, Li C, Dai G, Zhan M, Wang J, Pan S, Cao C (2009b) Greenhouse gas emission from direct seeding paddy field under different rice tillage systems in central China. Soil Tillage Res 106(1):54–61

Ahmad S, Ahmad A, Soler CMT, Ali H, Zia-Ul-Haq M, Anothai J, Hussain A, Hoogenboom G, Hasanuzzaman M (2012) Application of the CSM-CERES-Rice model for evaluation of plant density and nitrogen management of fine transplanted rice for an irrigated semiarid environment. Precis Agric 13(2):200–218

Ahmad S, Ahmad A, Ali H, Hussain A, Garcia y Garcia A, Khan MA, Zia-Ul-Haq M, Hasanuzzaman M, Hoogenboom G (2013) Application of the CSM-CERES-Rice model for evaluation of plant density and irrigation management of transplanted rice for an irrigated semiarid environment. Irrig Sci 31(3):491–506

Ahmad A, Ashfaq M, Rasul G, Wajid SA, Khaliq T, Rasul F, Saeed U, Rahman MH, Hussain J, Baig IA, Naqvi AA, SAA B, Ahmad S, Naseem W, Hoogenboom G, Valdivia RO (2015) Impact of climate change on the rice–wheat cropping system of Pakistan. In: Hillel D, Rosenzweig C (eds) Handbook of climate change and agro-ecosystems: The Agricultural Modeling Intercomparison and Improvement Project (AgMIP) integrated crop and economic assessments. Imperial College Press, London, pp 219–258

Ahmad S, Abbas G, Ahmed M, Fatima Z, Anjum MA, Rasul G, Khan MA, Hoogenboom G (2019) Climate warming and management impact on the change of rice-wheat phenology in Punjab, Pakistan. Field Crop Res 230:46–61

Ahmed M, Ahmad S (2019) Carbon dioxide enrichment and crop productivity. In: Hasanuzzaman M (ed) Agronomic crops, vol 2. Springer Nature Singapore, Singapore, pp 31–46

Ahmed M, Ahmad S (2020) Systems modeling. In: Ahmed M (ed) Systems modeling. Springer Nature, Cham, pp 1–44

Ahmed M, Fayyaz-ul-Hassan, Ahmad S (2017) Climate variability impact on rice production: adaptation and mitigation strategies. In: Ahmed M, Stockle C (eds) Quantification of climate variability, adaptation and mitigation for agricultural sustainability. Springer, Cham, pp 91–111

Ahmed M, Ahmad S, Raza MA, Kumar U, Ansar M, Shah GA, Parsons D, Hoogenboom G, Palosuo T, Seidel S (2020a) Models calibration and evaluation. In: Ahmed M (ed) Systems modeling. Springer Nature Singapore, Singapore, pp 151–178

Ahmed M, Ahmad S, Waldrip HM, Ramin M, Raza MA (2020b) Whole farm modeling: A systems approach to understanding and managing livestock for greenhouse gas mitigation, economic viability and environmental quality. In: Waldrip HM, Pagliari PH, He Z (eds) Animal manure, ASA Special Publication 67. American Society of Agronomy/Soil Science Society of America, Madison, WI, pp 345–371

Akram R, Turan V, Hammad HM, Ahmad S, Hussain S, Hasnain A, Maqbool MM, Rehmani MIA, Rasool A, Masood N, Mahmood F, Mubeen M, Sultana SR, Fahad S, Amanet K, Saleem M, Abbas Y, Akhtar HM, Hussain S, Waseem F, Murtaza R, Amin A, Zahoor SA, Sami ul Din M, Nasim W (2018) Fate of organic and inorganic pollutants in paddy soils. In: Hashmi MZ, Varma A (eds.), Environmental pollution of paddy soils. Springer Nature Switzerland, Cham197–214

Ali MA, Lee CH, Lee YB, Kim PJ (2009) Silicate fertilization in no-tillage rice farming for mitigation of methane emission and increasing rice productivity. Agric Ecosyst Environ 132(1–2):16–22

Ali MA, Inubushi K, Kim PJ, Amin S (2019a) Management of paddy soil towards low greenhouse gas emissions and sustainable rice production in the changing climatic conditions. In: Vázquez-Luna D, del Carmen Cuevas-Díaz M (eds) Soil contamination and alternatives for sustainable development. IntechOpen, London. https://doi.org/10.5772/intechopen.83548

Ali M, Mubeen M, Hussain N, Wajid A, Farid HU, Awais M, Hussain S, Akram W, Amin A, Akram R, Imran M, Ali A, Nasim W (2019b) Role of ICT in crop management. In: Agronomic crops. Springer Nature Singapore, Singapore, pp 637–652. https://doi.org/10.1007/978-981-32-9783-8_28

Anser MK, Hina T, Hameed S, Nasir MH, Ahmad I, Naseer MA (2020) Modeling adaptation strategies against climate change impacts in integrated rice-wheat agricultural production system of Pakistan. Int J Environ Res Public Health 17(7):2522

Arunrat N, Pumijumnong N (2017) Practices for reducing greenhouse gas emissions from rice production in Northeast Thailand. Agriculture 7(1):4

Aslam M, Zamir MSI, Afzal I, Yaseen M, Mubeen M, Shoaib A (2013) Drought stress, its effect on maize production and development of drought tolerance through potassium application. Cercetări Agronomiceîn Moldova 46(2):99–114

Bakhat HF, Zia Z, Fahad S, Abbas S, Hammad HM, Shahzad AN, Abbas F, Alharby H, Shahid M (2017) Arsenic uptake, accumulation and toxicity in rice plants: possible remedies for its detoxification: a review. Environ Sci Pollut Res 24(10):9142–9158

Batool SA, Chuadhry MN (2009) The impact of municipal solid waste treatment methods on greenhouse gas emissions in Lahore, Pakistan. Waste Manag 29(1):63–69

Bayer C, de Souza CF, Pedroso GM, Zschornack T, Camargo ES, de Lima MA, Frigheto RTS, Gomes J, Marcolin E, Macedo VRM (2014) Yield-scaled greenhouse gas emissions from flood irrigated rice under long-term conventional tillage and no-till systems in a humid subtropical climate. Field Crop Res 162:60–69

Bayer C, Zschornack T, Pedroso GM, da Rosa CM, Camargo ES, Boeni M, Marcolin E, dos Reis CES, dos Santos DC (2015) A seven-year study on the effects of fall soil tillage on yield-scaled greenhouse gas emission from flood irrigated rice in a humid subtropical climate. Soil Tillage Res 145:118–125

Boateng KK, Obeng GY, Mensah E (2017) Rice cultivation and greenhouse gas emissions: a review and conceptual framework with reference to Ghana. Agriculture 7(1):7

Carrijo DR, Lundy ME, Linquist BA (2017) Rice yields and water use under alternate wetting and drying irrigation: A meta-analysis. Field Crop Res 203:173–180

Chen Y, Peng J, Wang J, Fu P, Hou Y, Zhang C, Fahad S, Peng S, Cui K, Nie L, Huang J (2015) Crop management based on multi-split topdressing enhances grain yield and nitrogen use efficiency in irrigated rice in China. Field Crop Res 184:50–57

Cui Z, Zhang H, Chen X, Zhang C, Ma W, Huang C, Zhang W, Mi G, Miao Y, Li X, Gao Q, Yang J, Wang Z, Ye Y, Guo S, Lu J, Huang J, Lv S, Sun Y, Liu Y, Peng X, Ren J, Li S, Deng X, Shi X, Zhang Q, Yang Z, Tang L, Wei C, Jia L, Zhang J, He M, Tong Y, Tang Q, Zhong X, Liu Z, Cao N, Kou C, Ying H, Yin Y, Jiao X, Zhang Q, Fan M, Jiang R, Zhang F, Dou Z (2018) Pursuing sustainable productivity with millions of smallholder farmers. Nature 555(7696): 363–366

Deng N, Ling X, Sun Y, Zhang C, Fahad S, Peng S, Cui K, Nie L, Huang J (2015) Influence of temperature and solar radiation on grain yield and quality in irrigated rice system. Eur J Agron 64:37–46

Fahad S, Hussain S, Saud S, Tanveer M, Bajwa AA, Hassan S, Shah AN, Ullah A, Wu C, Khan FA, Shah F, Ullah S, Chen Y, Huang J (2015) A biochar application protects rice pollen from high-temperature stress. Plant Physiol Biochem 96:281–287

Fahad S, Hussain S, Saud S, Khan F, Hassan S, Amanullah NW, Arif M, Wang F, Huang J (2016) Exogenously applied plant growth regulators affect heat-stressed rice pollens. J Agron Crop Sci 202(2):139–150

Fahad S, Ihsan MZ, Khaliq A, Daur I, Saud S, Alzamanan S, Nasim W, Abdullal M, Khan IA, Wu C, Wang D, Huang J (2018) Consequences of high temperature under changing climate optima for rice pollen characteristics-concepts and perspectives. Arch Agron Soil Sci 64(11): 1473–1488

Gan Y, Liang C, Chai Q, Lemke RL, Campbell CA, Zentner RP (2014) Improving farming practices reduces the carbon footprint of spring wheat production. Nat Commun 5(1):5012

Hardwick S, Graven H (2016) Satellite observations to support monitoring of greenhouse gas emissions, Grantham Institute Briefing paper 16. Imperial College, London

Hasan E (2013) Proposing mitigation strategies for reducing the impact of rice cultivation on climate change in Egypt. Water Sci 27(54):69–77

Houweling S, Baker D, Basu S, Boesch H, Butz A, Chevallier F, Deng F, Dlugokencky EJ, Feng L, Ganshin A, Hasekamp O, Jones D, Maksyutov S, Marshall J, Oda T, O'Dell CW, Oshchepkov S, Palmer PI, Peylin P, Poussi Z, Reum F, Takagi H, Yoshida Y, Zhuravlev R (2015) An intercomparison of inverse models for estimating sources and sinks of CO_2 using GOSAT measurements. J Geophys Res Atmos 120(10):5253–5266

Hsu YW, Singh SK, Chiang MY, Wu YY, Chang IF (2009) Strategies to lower greenhouse gas level by rice agriculture. Afr J Biotechnol 8(2):126–132

Hussain S (2018) Land use/land cover classification by using satellite NDVI tool for sustainable water and climate change in Southern Punjab. COMSATS University, Islamabad. https://doi.org/10.13140/RG.2.2.32363.69923

Hussain M, Zaidi SMH, Malik RN, Sharma BD (2014) Greenhouse gas emissions from production chain of a cigarette manufacturing industry in Pakistan. Environ Res 134:81–90

Hussain S, Peng S, Fahad S, Khaliq A, Huang J, Cui K, Nie L (2015) Rice management interventions to mitigate greenhouse gas emissions: a review. Environ Sci Pollut Res 22(5): 3342–3360

Hussain S, Ahmad A, Wajid A, Khaliq T, Hussain N, Mubeen M, Farid HU, Imran M, Hammad HM, Awais M, Ali A, Aslam M, Amin A, Akram R, Amanet K, Nasim W (2020a) Irrigation scheduling for cotton cultivation. In: Ahmad S, Hasanuzzaman M (eds) Cotton production and uses. Springer, Springer, pp 59–80. https://doi.org/10.1007/978-981-15-1472-2_5

Hussain S, Huang J, Huang J, Ahmad S, Nanda S, Anwar S, Shakoor A, Zhu C, Zhu L, Cao X, Jin Q, Zhang J (2020b) Rice production under climate change: Adaptations and mitigating strategies. In: Fahad S, Hasanuzzman M, Alam M, Ullah H, Saeed M, Khan IA, Adnan M (eds) Environment, climate, plant and vegetation growth. Springer, Cham, pp 659–686. https://doi.org/10.1007/978-3-030-49732-3_26

Hussain S, Mubeen M, Ahmad A, Akram W, Hammad HM, Ali M, Masood N, Amin A, Farid HU, Sultana SR, Fahad S, Wang D, Nasim W (2020c) Using GIS tools to detect the land use/land cover changes during forty years in Lodhran District of Pakistan. Environ Sci Pollut Res 27: 39676–39692

Hussain S, Mubeen M, Akram W, Ahmad A, Rahman MH, Ghaffar A, Amin A, Awais M, Farid HU, Farooq A, Nasim W (2020d) Study of land cover/land use changes using RS and GIS: a case study of Multan district, Pakistan. Environ Monit Assess 192(1):2

Hussain S, Amin A, Mubeen M, Khaliq T, Shahid M, Hammad HM, Nasim W (2022) Climate Smart Agriculture (CSA) Technologies. In Building Climate Resilience in Agriculture (pp 319-338). Springer, Cham. https://doi.org/10.1007/978-3-030-79408-8_20

Islam SMM, Gaihre YK, Islam RR, Akter M, Al Mahmud A, Singh U, Sander BO (2020) Effects of water management on greenhouse gas emissions from farmers' rice fields in Bangladesh. Sci Total Environ 734:139382

Khan S, Chao C, Waqas M, Arp HPH, Zhu YG (2013) Sewage sludge biochar influence upon rice (*Oryza sativa* L) yield, metal bioaccumulation and greenhouse gas emissions from acidic paddy soil. Environ Sci Technol 47(15):8624–8632

Khan MTI, Ali Q, Ashfaq M (2018) The nexus between greenhouse gas emission, electricity production, renewable energy and agriculture in Pakistan. Renew Energy 118:437–451

Khush GS (2000) Strategies for increasing the yield potential of rice. Stud Plant Sci 7:207–212

Kritee K, Nair D, Zavala-Araiza D, Proville J, Rudek J, Adhya TK, Loecke T, Esteves T, Balireddygari S, Dava O, Ram K, Abhilash SR, Madasamy M, Dokka RV, Anandaraj D, Athiyaman D, Reddy M, Ahuja R, Hamburg SP (2018) High nitrous oxide fluxes from rice indicate the need to manage water for both long-and short-term climate impacts. Proc Natl Acad Sci 115(39):9720–9725

Kumar A, Nayak AK, Mohanty S, Das BS (2016) Greenhouse gas emission from direct seeded paddy fields under different soil water potentials in Eastern India. Agric Ecosyst Environ 228: 111–123

Kumar A, Nayak AK, Das BS, Panigrahi N, Dasgupta P, Mohanty S, Kumar U, Panneerselvam P, Pathak H (2019) Effects of water deficit stress on agronomic and physiological responses of rice and greenhouse gas emission from rice soil under elevated atmospheric CO_2. Sci Total Environ 650:2032–2050

LaHue GT, Chaney RL, Adviento-Borbe MA, Linquist BA (2016) Alternate wetting and drying in high yielding direct-seeded rice systems accomplishes multiple environmental and agronomic objectives. Agric Ecosyst Environ 229:30–39

Lampayan RM, Rejesus RM, Singleton GR, Bouman BA (2015) Adoption and economics of alternate wetting and drying water management for irrigated lowland rice. Field Crop Res 170:95–108

Linquist BA, Adviento-Borbe MA, Pittelkow CM, van Kessel C, van Groenigen KJ (2012) Fertilizer management practices and greenhouse gas emissions from rice systems: a quantitative review and analysis. Field Crop Res 135:10–21

Linquist BA, Anders MM, Adviento-Borbe MAA, Chaney RL, Nalley LL, Da Rosa EF, Van Kessel C (2015) Reducing greenhouse gas emissions, water use, and grain arsenic levels in rice systems. Glob Chang Biol 21(1):407–417

Mubeen M, Bano A, Ali B, Islam ZU, Ahmad A, Hussain S, Fahad S, Nasim W (2021) Effect of plant growth promoting bacteria and drought on spring maize (Zea mays L.). Pak J Bot 53(2): 731–739. https://doi.org/10.30848/PJB2021-2(38)

Matsunaga T, Maksyutov S (2018) A guidebook on the use of satellite greenhouse gases observation data to evaluate and improve greenhouse gas emission inventories. Satellite Observation

Center, National Institute for Environmental Studies, Onogawa. 129 pp. http://www.nies.go.jp/soc/en/documents/guidebook/GHG_Satellite_Guidebook_1st_12b.~Pdf

Mehmood I, Bari A, Irshad S, Khalid F, Liaqat S, Anjum H, Fahad S (2020) Carbon cycle in response to global warming. In: Fahad S, Hasanuzzman M, Alam M, Ullah H, Saeed M, Khan IA, Adnan M (eds) Environment, climate, plant and vegetation growth. Springer, Cham. https://doi.org/10.1007/978-3-030-49732-3_1

Mingxing W, Jing L (2002) CH_4 emission and oxidation in Chinese rice paddies. Nutr Cycl Agroecosyst 64(1-2):43–55

Mir KA, Purohit P, Mehmood S (2017) Sectoral assessment of greenhouse gas emissions in Pakistan. Environ Sci Pollut Res 24(35):27345–27355

Montzka SA, Dlugokencky EJ, Butler JH (2011) Non-CO_2 greenhouse gases and climate change. Nature 476(7358):43–50

Mubeen M, Ahmad A, Khaliq T, Sultana SR, Hussain S, Ali A, Ali H, Nasim W (2013a) Effect of growth stage-based irrigation schedules on biomass accumulation and resource use efficiency of wheat cultivars. Am J Plant Sci 4(7):1435–1442

Mubeen M, Ahmad A, Wajid A, Bakhsh A (2013b) Evaluating different irrigation scheduling criteria for autumn-sown maize under semi-arid environment. Pak J Bot 45(4):1293–1298

Mubeen M, Ahmad A, Wajid A, Khaliq T, Bakhsh A (2013c) Evaluating CSM-CERES-Maize model for irrigation scheduling in semi-arid conditions of Punjab, Pakistan. Int J Agric Biol 15(1):1–10

Mubeen M, Ahmad A, Hammad HM, Awais M, Farid HU, Saleem M, Sami ul Din M, Amin A, Ali A, Fahad S, Nasim W (2019) Evaluating the climate change impact on water use efficiency of cotton-wheat in semi-arid conditions using DSSAT model. J Water Clim Change 11(4): 1661–1675

Munawar S, Qamar MT, Mustafa G, Khan MS, Joyia FA (2020) Role of biotechnology in climate resilient agriculture. In: Fahad S, Hasanuzzman M, Alam M, Ullah H, Saeed M, Khan IA, Adnan M (eds) Environment, climate, plant and vegetation growth. Springer, Cham. https://doi.org/10.1007/978-3-030-49732-3_14

Nasim W, Ahmad A, Ahmad S, Nadeem M, Masood N, Shahid M, Mubeen M, Hoogenboom G, Fahad S (2017) Response of sunflower hybrids to nitrogen application grown under different agro-environments. J Plant Nutr 40(1):82–92

Pandey A, Mai VT, Vu DQ, Bui TPL, Mai TLA, Bui TPL, Mai TLA, Jensen LS, de Neergaard A (2014) Organic matter and water management strategies to reduce methane and nitrous oxide emissions from rice paddies in Vietnam. Agric Ecosyst Environ 196:137–146

Qin Y, Liu S, Guo Y, Liu Q, Zou J (2010) Methane and nitrous oxide emissions from organic and conventional rice cropping systems in Southeast China. Biol Fertil Soils 46(8):825–834

Ravishankara AR, Daniel JS, Portmann RW (2009) Nitrous oxide (N_2O): the dominant ozone-depleting substance emitted in the 21st century. Science 326(5949):123–125

Revell LE, Tummon F, Salawitch RJ, Stenke A, Peter T (2015) The changing ozone depletion potential of N_2O in a future climate. Geophys Res Lett 42(22):10047–10055

Sabagh AEL, AHossain A, Barutçular C, Iqbal MA, Islam MS, Fahad S, Sytar O, Çiğ F, Meena RS, Erman M (2020a) Consequences of salinity stress on the quality of crops and its mitigation strategies for sustainable crop production: An outlook of arid and semi-arid regions. In: Fahad S, Hasanuzzman M, Alam M, Ullah H, Saeed M, Khan IA, Adnan M (eds) Environment, climate, plant and vegetation growth. Springer, Cham. https://doi.org/10.1007/978-3-030-49732-3_20

Sabagh AEL, Hossain A, Islam MS, Iqbal MA, Fahad S, Ratnasekera D, Azeem F, Wasaya A, Sytar O, Kumar N, Llanes A, Erman M, Ceritoğlu M, Arslan H, Arslan D, Hussain S, Mubeen M, Ikram M, Meena RS, Gharib H, Waraich E, Nasim W, Liu L, Saneoka H (2020b) Consequences and mitigation strategies of heat stress for sustainability of soybean (*Glycine max* L. Merr.) production under the changing climate. In: Hossain A (ed) Plant stress physiology. IntechOpen, London. https://doi.org/10.5772/intechopen.92098

Sathish A, Kumar KA, Reddy PRR, Devi MU (2017) Effect of different crop establishment methods and irrigation regimes on rice (*Oryza sativa* L.) yield and water use efficiency. Int J Curr Microbiol App Sci 6(9):90–95. https://doi.org/10.20546/ijcmas.2017.609.010

Shahzad AN, Ahmad S (2019) Tools and techniques for nitrogen management in cereals. In: Hasanuzzaman M (ed) Agronomic crops, vol 2. Springer Nature Singapore, Singapore, pp 111–126

Sultana SR, Ali A, Ahmad A, Mubeen M, Zia-Ul-Haq M, Ahmad S, Ercisli S, Jaafar HZE (2014) Normalized difference vegetation index as a tool for wheat yield estimation: a case study from Faisalabad, Pakistan. Sci World J 2014:725326

Tarlera S, Capurro MC, Irisarri P, Scavino AF, Cantou G, Roel A (2016) Yield-scaled global warming potential of two irrigation management systems in a highly productive rice system. Sci Agric 73(1):43–50

Thu TN, Phuong LBT, Van TM, Hong SN (2016) Effect of water regimes and organic matter strategies on mitigating Greenhouse Gas Emission from rice cultivation and co-benefits in agriculture in Vietnam. Int J Environ Sci Dev 7:85–90

Wajid A, Ahmad A, Hussain M, ur Rahman MH, Khaliq T, Mubeen M, Rasul F, Bashir U, Awais M, Javed I, Sultana SR, Hoogenboom G (2014) Modeling growth, development and seed-cotton yield for varying nitrogen increments and planting dates using DSSAT. Pak J Agric Sci 51:641–650

Wang W, Guo L, Li Y, Su M, Lin Y, De Perthuis C, Ju X, Lin E, Moran D (2015) Greenhouse gas intensity of three main crops and implications for low-carbon agriculture in China. Clim Chang 128:57–70

Wang ZB, Chen J, Mao SC, Han YC, Chen F, Zhang LF, Li YB, Li CD (2017) Comparison of greenhouse gas emissions of chemical fertilizer types in China's crop production. J Clean Prod 141:1267–1274

Wang H, Zhang Y, Zhang Y, McDaniel MD, Sun L, Su W, Fan X, Liu S, Xiao X (2020a) Water-saving irrigation is a 'win-win' management strategy in rice paddies–With both reduced greenhouse gas emissions and enhanced water use efficiency. Agric Water Manag 228:105889

Wang W, Wang C, Sardans J, Fang Y, Singh BP, Wang H, Huang X, Zeng C, Tong C, Peñuelas J (2020b) Multiple trade-offs between maximizing yield and minimizing greenhouse gas production in Chinese rice croplands. Land Degrad Dev 31(10):1287–1299

Wassmann R, Neue HU, Ladha JK, Aulakh MS (2004) Mitigating greenhouse gas emissions from rice-wheat cropping systems in Asia. In: Eassmann R, PLG V (eds) Tropical agriculture in transition—opportunities for mitigating greenhouse gas emissions? Springer, Dordrecht, pp 65–90

Wu X, Wang W, Xie X, Yin C, Hou H, Yan W, Wang G (2018) Net global warming potential and greenhouse gas intensity as affected by different water management strategies in Chinese double rice-cropping systems. Sci Rep 8:779

Xu Y, Ge J, Tian S, Li S, Nguy-Robertson AL, Zhan M, Cao C (2015) Effects of water-saving irrigation practices and drought resistant rice variety on greenhouse gas emissions from a no-till paddy in the central lowlands of China. Sci Total Environ 505:1043–1052

Yang B, Xiong Z, Wang J, Xu X, Huang Q, Shen Q (2015) Mitigating net global warming potential and greenhouse gas intensities by substituting chemical nitrogen fertilizers with organic fertilization strategies in rice–wheat annual rotation systems in China: A 3-year field experiment. Ecol Eng 81:289–297

Yang Z, Zhang Z, Zhang T, Fahad S, Cui K, Nie L, Peng S, Huang J (2017) The effect of season-long temperature increases on rice cultivars grown in the central and southern regions of China. Front Plant Sci 8:1908

Yousaf M, Li J, Lu J, Ren T, Cong R, Fahad S, Li X (2017) Effects of fertilization on crop production and nutrient-supplying capacity under rice-oilseed rape rotation system. Sci Rep 7(1):1–9

Zahoor SA, Ahmad S, Ahmad A, Wajid A, Khaliq T, Mubeen M, Hussain S, Din MSU, Amin A, Awais M, Nasim W (2019) Improving water use efficiency in agronomic crop production. In: Hasanuzzaman M (ed) Agronomic crops. Springer Nature Singapore, Singapore, pp 13–29. https://doi.org/10.1007/978-981-32-9783-8_2

Zhang X, Yin S, Li Y, Zhuang H, Li C, Liu C (2014) Comparison of greenhouse gas emissions from rice paddy fields under different nitrogen fertilization loads in Chongming Island, Eastern China. Sci Total Environ 472:381–388

Zhang X, Xu X, Liu Y, Wang J, Xiong Z (2016) Global warming potential and greenhouse gas intensity in rice agriculture driven by high yields and nitrogen use efficiency. Biogeosciences 13(9):2701–2714

Applications of Crop Modeling in Rice Production

28

Ghulam Abbas, Mukhtar Ahmed, Ashfaq Ahmad, Aftab Wajid, Fahad Rasool, Shakeel Ahmad, and Gerrit Hoogenboom

Abstract

Rice growth simulation models are being largely applied by researchers along with policy makers all over the world as a vital and effective research tool in rice-growing regions. Different types of rice growth models have been used during previous and present century for the determination of rice crop comebacks to water shortage, climate warming, and macro and micro nutrients stress. Rice growth models are also used for testing the alternate optimal sowing dates for climate changing conditions in rice-growing belts. Rice growth simulated models are valuable researching tool at local, regional, and global levels. Frequently rice crop models are used for the assessment of climatic variations, management practices, as well as irrigation plans on coarse and fine rice production across the world. Wholly rice simulations models are fruitfully applied at local, regional, and national level in world, nevertheless among entirely rice crop growth models. The CSM-CERES-rice model is frequently applied by rice scientists as well as

G. Abbas
Department of Agronomy, Bahauddin Zakariya University, Multan, Pakistan

M. Ahmed
Pir Mehr Ali Shah, Arid Agriculture University, Rawalpindi, Pakistan

A. Ahmad · A. Wajid · F. Rasool
Department of Agronomy, University of Agriculture, Faisalabad, Pakistan

S. Ahmad (✉)
Department of Agronomy, Faculty of Agricultural Sciences and Technology, Bahauddin Zakariya University, Multan, Pakistan
e-mail: shakeelahmad@bzu.edu.pk

G. Hoogenboom
Institute for Sustainable Food Systems, University of Florida, Gainesville, FL, USA

N. Sarwar et al. (eds.), *Modern Techniques of Rice Crop Production*,
https://doi.org/10.1007/978-981-16-4955-4_28

policy planners. For the irrigation management policies, generally ORYZA2000 and AquaCrop models are also used by scientists.

Keywords

Model · Simulation · Climate change · Irrigation management · CERES · Phenology

Abbreviations

CERES	Crop Environment REsource Synthesis
CSM	Cropping system model
DSSAT	Decision support system for agro-technology transfer
GCM	General circulation model
LAI	Leaf area index
NRMSE	Normalized root mean square error
WUE	Water use efficiency

28.1 Introduction

Rice is a main food crop among cereal crops for the people worldwide (Ahmad et al. 2008, 2009a, b, 2012, 2013, 2015, 2019). The role of rice is unavoidable in the existing and forthcoming worldwide food security (Hyun and Kim 2019; Nasir et al. 2020). Rice crop provides almost 20% of worldwide human per capita energy as well as 15% of per capita protein contents. Though rice protein content has position higher in the nourishing quality among cereal crops, protein content is important. Rice crop also gives various types of minerals, vitamins, and fiber contents, although wholly ingredients excluding carbohydrates are declined by rice factories (Swain and Yadav 2009; Ahmad et al. 2013). Rice crop is one of the utmost significant staple foods for more than 3.4 billion people in the world (Ahmad and Hasanuzzaman 2012; Ahmed and Ahmad 2019, 2020; Shahzad and Ahmad 2019). About 17 nations in Asia and the Pacific, 9 nations in North and South America, and 8 nations in Africa continent generally depend on rice crop for the staple diet (Oteng-Darko et al. 2012; Maniruzzaman and Nemes 2014). The Asian states lead the worldwide rice production among all continents (Ahmed et al. 2017, 2020a, b; Akram et al. 2018, 2019; Hussain et al. 2018; Fahad et al. 2019; Fatima et al. 2020). India has the largest rice cultivated area. China is the largest producer of rice in world. While highest productivity of rice is in Japan (Nyang'Au et al. 2014; Phakamas 2015; Hafeez-ur-Rehman et al. 2019; Khan et al. 2019a, b; Naz et al. 2019; Peerzada et al. 2019; Razaq et al. 2019; Sarwar et al. 2013a, b; Tariq et al. 2020; Wasaya et al. 2019; Zahoor et al. 2019).

Crop growth modeling along with cropping system analysis has become imperative researching tool in the contemporary agriculture research studies. Crops cultivated area is a limited natural resource, which cannot be extended to carry the agriculture activities (Dias et al. 2016; Lee et al. 2017). Field crop trials conducted for the evaluation of suitable agronomical crop practices are arduous, time overwhelming, and expensive, particularly when a number of parameters are to be calibrated and entail several years-based data for confirmation (Soundharajan and Sudheer 2009; Liu et al. 2019). The solitary possibility leftward is to upsurge the superiority and quantity of food productions. With the help of technology and decision-making interference scientific tools like crop growth models and yield forecast models are being employed worldwide (Arora 2006; Ahmad et al. 2009a, b). Therefore, crop simulation models are correspondingly beneficial tools for the study of the influence of climatic changes on development, growth, and ultimately economical production under the varied agro-environmental conditions (Amiri 2008; Guo et al. 2019). The crop growth models can be defined as a quantifiable system for simulating the developmental stages, growth, as well as yield of a crop providing a set of genetic characteristics and environment variables like soil and weather data, etc. (Bannayan et al. 2005; Jha et al. 2020). The crop growth simulation model makes our understanding about the physiology and ecology based various processes which administrate growth as well as development of crop in logical algorithm and mathematic set of equations for simulation purpose (Yun 2003; Kadiyala et al. 2015). Rice crop prediction model can animatedly designate the bio-physical and physiology-based progressions of growth, development, and yield and provides a quantifiable tool for the simulation of the productivity level of a rice crop in relation to interaction between genotype, environment, and management practices (Zhang and Tao 2013; Vilayvong et al. 2015). Rice crop modeling techniques have the potential to significantly contribute to food and nutrition security at local, regional, and global level. Novel technology and theoretical advances have contributed to an improved understanding of rice crop performance and grain yield gap and genetic gains, better simulation of pest and insect occurrences, more effective irrigation plans, optimal planting dates, and fertilizer management (Ahmad et al. 2019). Eco-physiological model ORYZA is a rice crop model that represents more than 28 years of worldwide rice crop research at local and regional levels. The growth, development, and grain production formation in the rice crop depends on combined influences of interaction among genotype, environmental, and management practices (Timsina and Humphreys 2006). Growth simulation models for rice crop can bring systematic and quantitative research tools for the prediction of growth, development, and productivity of rice crop under fluctuating environmental circumstances (Larijani et al. 2011). The CSM-CERES-rice model embedded in DSSAT has been developed for supporting the decision-making in agriculture research, crop management practices, proper land-use pattern, and policy making for enhancement of rice crop productivity (Jones et al. 2003; Casanova et al. 2000; Kumar et al. 2009; Hoogenboom et al. 2019a, b) (Figs. 28.1 and 28.2). Rice crop models are physiologically based and predict day-to-day canopy photosynthetic process, respiration process, development, biomass partition, and crop growth as a

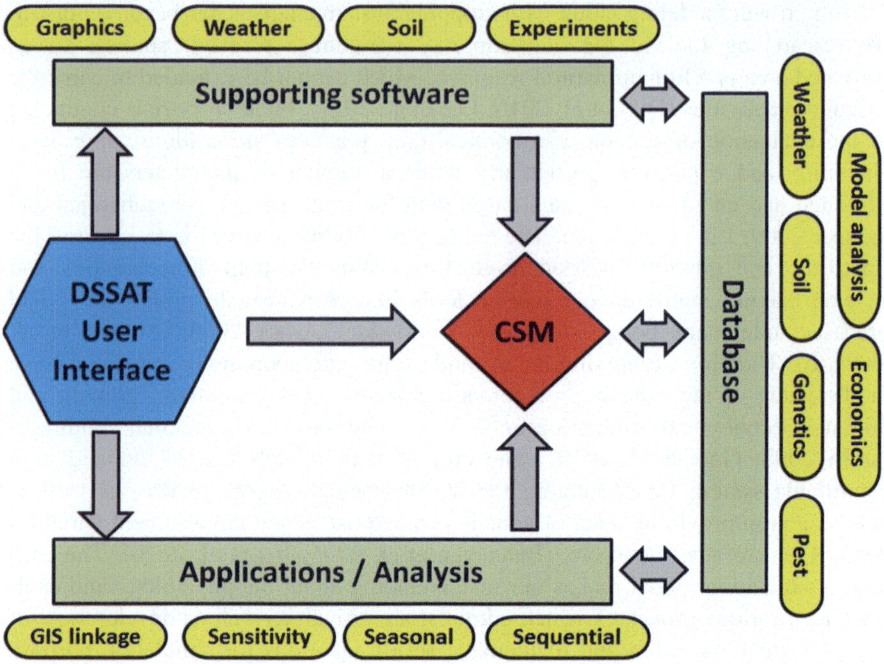

Fig. 28.1 The DSSAT crop modeling ecosystem (**Source**: Modified and adapted from Balwinder-Singh et al. 2009)

function of input information, comprising everyday-based weather data, soil profile characteristics, management practices, as well as genotype features (Radanielson et al. 2018; Zhou et al. 2019). Rice growth simulation models, for example, CERES, ORYZA2000, RiceGrow, CropSyst, RICEMOD, AquaCrop, InfoCrop, SIMRIW, can concomitantly integrate nonlinear relations among soil, water, rice plant, weather data, and crop management practices for the determination of yield, environmental stresses, and water prerequisite in addition to nutrients requirements for the betterment of rice industry across the global level (Bouman, and Van Laar 2006; Maki et al. 2017).

28.1.1 Crop Management Practices

Table 28.1 shows that erraticism of rice grain yield under stressed as well as non-stressed circumstances was predicted with a root mean square error (RMSE) of 191 and 222 kg ha^{-1}, correspondingly, for models ORYZA and APSIM-Oryza, conforming to an RMSEn of 14.8% and 17.3%, respectively. The model index of agreement ranged from 0.86 to 0.99, which indicated well performance of model (Radanielson et al. 2018). Kumar et al. (2009) applied CSM-CERES-rice model that was found to be able to predict the yield of the rice crop well enough with average

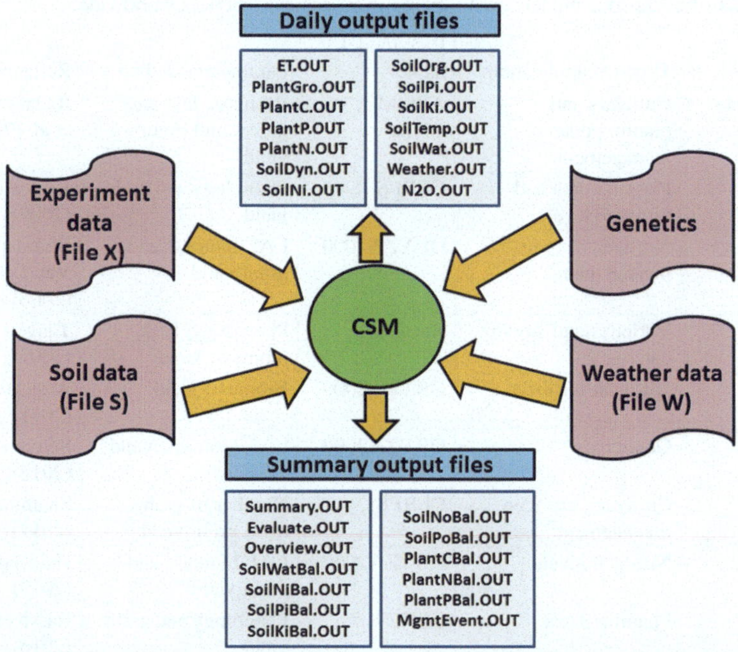

Fig. 28.2 The DSSAT input and output file system (**Source**: Modified and adapted from Hoogenboom et al. 2019a, b)

"RMSE of 148 to 222 kg ha-1" to assist farming community for making broad-scale decisions on the crop management practices like sowing date. Simulated and observed yield was very close to each other. Bouman and Van Laar (2006) used ORYZA2000 model for supporting nitrogen treatments-based field trials and investigation for appropriate nitrogen application regimes along with computed errors of prediction of model. On an average, RMSE value was 690–1280 kg ha^{-1} for biological yield, 350–380 kg ha^{-1} for leaf dry matter, 460–790 kg ha^{-1} for stem dry matter, and 380–580 kg ha^{-1} for panicle dry matter. Economic grain yield was predicted with an RMSE of 840–850 kg ha^{-1} and an RMSEn of 11–13% for rice crop. Tang et al. (2009) used two rice models for comparison to determine optimum sowing date. Comparison of performance of the RiceGrow model was done with the ORYZA2000 model, which resulted in both providing suitable estimations for phenological stages and phases, total dry matter, and grain production. Generally, RiceGrow model can be applied for prediction of growth as well as development with the interaction of different cultivars, environment circumstances, and rice crop husbandry practices (Fig. 28.3). Overall, both rice models can be used for research tools for several uses comprising systematic understandings, policy making, as well as to determine optimal crop management practices for gaining maximum rice productivity. Artacho et al. (2011) used ORYZA2000 model to determine potential rice production and seed yield response to nitrogen fertilizer under prospective

Table 28.1 Various rice models used for crop management practices worldwide

Country	Types of applications	Rice model name	Parameters studied	References
Philippines	Cultivars and planting date management	APSIM-Oryza	Biomass, leaf area index, and grain yield	Radanielson et al. (2018)
India	Planting date and nitrogen levels	CERES	Phenology and grain yield	Kumar et al. (2009)
Netherlands	Nitrogen management	ORYZA2000	LAI, biomass and grain yield	Bouman and Van Laar (2006)
China	Varieties and sowing dates	RiceGrow	Phenology, LAI, biomass, yield	Tang et al. (2009)
Chile	Nitrogen fertilizer	ORYZA2000	Biomass, yield	Artacho et al. (2011)
Vietnam	Genotypes	ORYZA2000	LAI, biomass, yield	Setiyono et al. (2018)
India	Cultivars, nitrogen management	CERES	Number of grains m^{-2}, grain yield	Shamim et al. (2012)
Pakistan	Nitrogen levels	ORYZA2000	LAI, biomass and grain yield	Hameed et al. (2019)
USA	Planting dates	CERES	Phenology and grain yield	Basso et al. (2016)
Pakistan	Seedlings $hill^{-1}$ and nitrogen management	CERES	LAI, biomass and grain yield	Ahmad et al. (2012)
Thailand	Transplanting dates and planting densities	CERES	Phenology, biomass and grain yield	Vilayvong et al. (2015)

weather circumstances in the foremost rice-producing area in the Chile regions. Results indicated that rice simulation model was satisfactorily precise to predict grain production and nitrogen uptake at the culmination of the growing season. Concluding rice nitrogen uptake was predicted with an RMSE of 20 kg ha^{-1} and grain production with an RMSE of 1666 kg ha^{-1} under nitrogen application conditions. Setiyono et al. (2018) applied ORYZA crop growth model in the Red River Delta of Vietnam. Rice growth simulation model facilitated the estimation of grain yield under varied environmental conditions. Sufficiently designated spatial yield variation in the study areas dependably replicated official yield data with RMSE of 0.30 and 0.46 t ha^{-1} (NRMSE of 5% and 8%) in spring and summer season, correspondingly, indicating good performance of the model. ORYZA2000 is a well-established ecophysiological model for the prediction of rice growth and yield in irrigated lowland ecosystem. However, implementation of this model on variable environments of upland, rainfed, and aerobic ecosystems modification is needed (Li et al. 2017). Model ORYZA version 3 (v3) was the newest form with the addition of modules and routines to quantify daily variations in soil temperature, carbon,

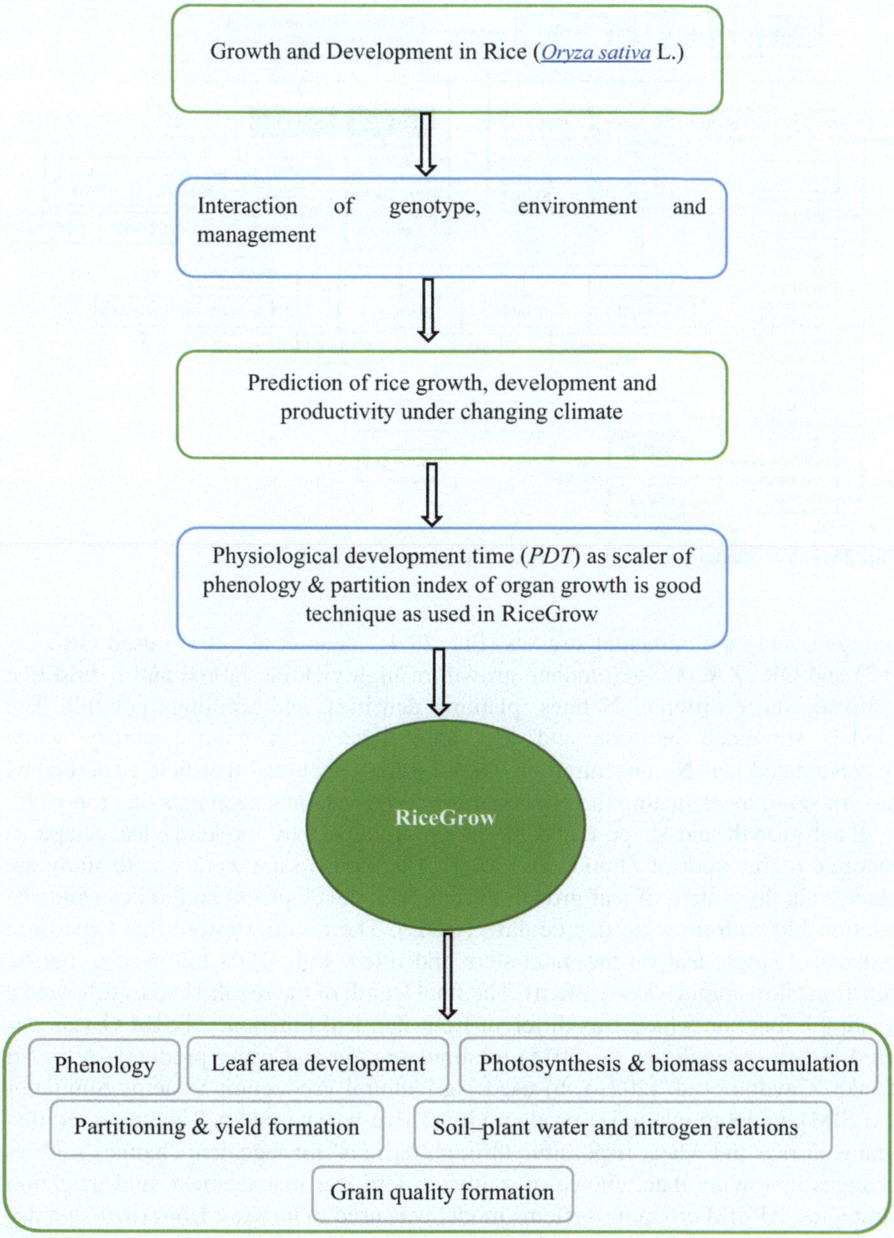

Fig. 28.3 Rice growth and productivity model (**Source**: Tang et al. 2009)

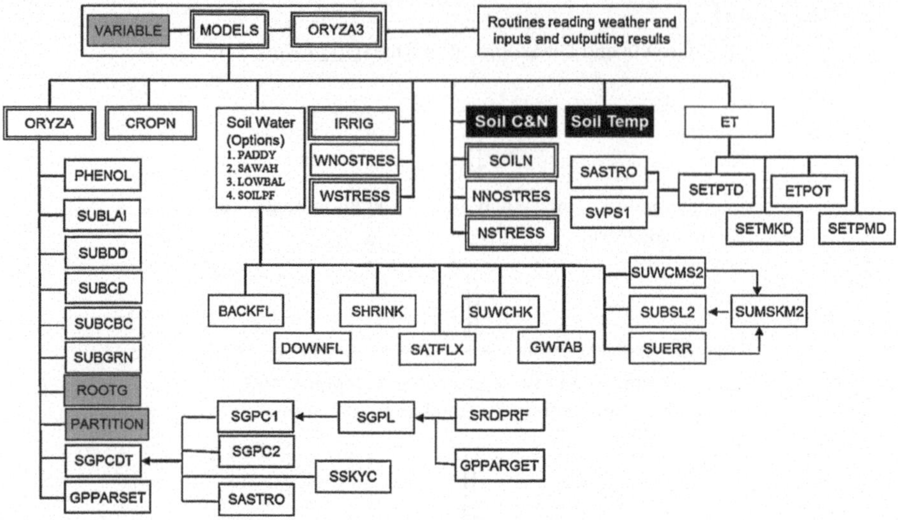

Fig. 28.4 The model structure of ORYZA (v3) (**Source**: Li et al. 2017)

nitrogen, and environmental stresses (Fig. 28.4). Yuan et al. (2017) used ORYZA (v3) and ORYZA2000 to simulate growth of high-yielding inbred and hybrid rice cultivars under different N rates, planting densities, and seedlings per hill. The models simulated biomass and leaf area index with good accuracy while overestimated leaf N concentration. They further concluded that field experiments are crucial in investigating the effects of genotypes and managements on crop yield.

Plant growth and shape could be easily visualized by modeling leaf shape as reported in the work of Zhou et al. (2019). The focus of the work was to study the changes in the pattern of leaf growth during plant development and to determine its relationship with growing degree days (GDD). The results showed that expansion process of single leaf on the main stem and tillers with GDD followed a logistic function (slow–rapid–slow pattern). The final length of the regular leaves followed a quadratic function which was different from flag leaf function. APSIM-Oryza was used by Amarasingha et al. (2015) to simulate crop and water productivity in Sri Lanka. Gaydon et al. (2021a, b) used Agricultural Production Systems Simulator (APSIM) model to maximize productivity of rice-wheat system. They reported that grains in rice and wheat is possible through farmers' management changes such as changes in sowing date, choice of cultivars, fertilizer management, and irrigation strategies. APSIM cropping systems model was used to increase Boro (irrigated dry season) rice production in the saline coastal zone. The work reported that farmers must do early sowing with the use of early-maturing transplanted Aman (T. Aman) rice cultivars (Gaydon et al. 2021a, b).

Agricultural Production Systems Simulator (APSIM)-Oryza was calibrated and validated with field data from research farm of Agricultural and Food Engineering Department, Indian Institute of Technology (IIT) Kharagpur, India, by Biswas et al.

(2021). They reported the model was able to simulate crop yield, evapotranspiration, and consumptive water footprints (CWFs) under different transplanting dates and recommended early transplanting to have good yield with minimum CWF. Dewi et al. (2021) used APSIM model to improve rice productivity by evaluating planting time scenarios. They concluded that modeling is a useful tool in suggesting suitable planting time for rice higher productivity. APSIM was used by Balwinder-Singh et al. (2009) to evaluate the risk-reducing management practices for direct seeded and transplanted rice. Results reported low and variable rice yield due to use of long-duration cultivars under rainfed conditions. Simulation outcomes suggested that to double the yield, it is better to use medium-duration hybrid rice, supplemental post-establishment irrigation, or transplanting appropriately aged seedlings. They also suggested optimum sowing window of early to mid June for irrigated direct seeded rise systems. Khaliq et al. (2019) parameterized APSIM model for local soil and climate in Pakistan. The model was afterward calibrated with rice and wheat growth and yield field data. Calibration was then validated with five years data from the highest and lowest yield producing districts in Punjab. They concluded that N was the primary driver for the larger yield gaps followed by current sowing date. APSIM-Oryza rice growth model was used to study global sensitivity analysis under eight climatic conditions and two CO_2 levels. The work reported that climate has significant influence on sensitivity index of several parameters (Liu et al. 2019).

Shamim et al. (2012) used CSM-CERES-rice model. Extremely positive significant relationship was obtained among models with predicted and field observed phenology ($r = 0.98$). The rice model overestimated total dry matter using predicted test weight and leaf area index. The performance of the model simulated number of grains m^{-2} was poor compared with field observed data. The grain production was predicted in close agreement with the field obtained grain production. Hameed et al. (2019) used ORYZA2000 rice model for the determination of coefficient R^2, whose values were more than 0.85 for wholly treatments, and RMSEn values were acceptable for each treatment variable between field gained and model predicted outcomes. The maximum RMSEn values were ranging from 18.60% to 24.94% and were gained for different nitrogen fertilizer treatments. Basso et al. (2016) used CSM-CERES-rice model to evaluate different field experimental treatments in 43 various countries. Across entirely testing situations, the CSM-CERES model simulated grain yield with a root mean square error (RMSE) of less than 800 kg ha^{-1} with 10% normalized RMSE. Phenological stages were predicted with less than 8 days difference from the field observed data from utmost research studies done worldwide. Ahmad et al. (2012) applied CSM-CERES-rice model in Punjab, Pakistan. The model was capable for the simulation of growth and grain production of irrigated rice crop in semiarid circumstances, with an averaged error percentage of 11% between model predicted and field gained grain yield in Faisalabad. The consequences of the simulation analysis results indicated that two seedlings $hill^{-1}$ along with 200 kg N ha^{-1} produced the maximum grain yield in comparison to other treatment combinations under irrigated conditions. Vilayvong et al. (2015) concluded that CSM-CERES-rice model provided satisfactory accurateness for grain yield with NRMSE error values ranging from 1% to 16% for all experimental treatments. The CERES model is an important alternative research tool for determining suitable crop management practices for rice productivity under both arid and semiarid conditions. Kadiyala et al. (2015) used CERES-rice model in nitrogen experiment. The model accurately simulated phenological stages and phases, grain yield, total biomass, and N uptake, with normalized root mean square of 14.5%, *d*-value of 0.92, that is indicative of acceptable model performance.

28.1.2 Irrigation Management

Table 28.2 shows that different rice simulation models were applied to determine optimum irrigation management. Jha et al. (2020) used CSM-CERES-rice model. The normalized root-mean-square error and degree of agreement d index values were gained to be 2.8% and 0.65%, correspondingly, for the simulation of grain yield with the performance efficiency of model with value of 75% under semiarid conditions. Results indicated that water scarcity during vegetative and maturity phase reduced rice production by 24% and 33%, correspondingly. Nevertheless, the water shortage stress throughout anthesis stage indicated the highest lessening in the grain yield by 43%. Amiri (2008) concluded that the performance of rice model ORYZA2000 was well. Range values of RMSE for model were from 532 to 871 kg ha^{-1} for the total dry matter, from 82 to 246 kg ha^{-1} for grain yield, from 280 to 456 kg ha^{-1} for stem dry matter, from 234 to 473 kg ha^{-1} for panicle dry matter, and from 0.23 to 0.52 for leaf area index. Arora (2006) applied ORYZA2000 rice model for the assessment of influence of water-related decisions on rice grain production and water use strategies for the increasing water use efficiency under irrigated environmental conditions in Punjab, India. Model performance was good because the root mean square of deviations (RMSD) value was gained with 0.52 t ha^{-1} and normalized RMSD value was 7% for grain yield changing from 6.4 to 8.7 t ha^{-1}. Soundharajan and Sudheer (2009) employed ORYZA2000 as the

Table 28.2 Various rice models used for irrigation management worldwide

Country	Rice model name	Parameters studied	References
India	CERES	Phenology and grain yield	Jha et al. (2020)
Iran	ORYZA2000	LAI, biomass, and grain yield	Amiri (2008)
India	ORYZA2000	WUE and grain yield	Arora (2006)
Sri Lanka	ORYZA2000	Soil water contents and grain yield	Soundharajan and Sudheer (2009)
South Korea	AquaCrop	Canopy cover, WUE, grain yield	Lee et al. (2017)
Thailand	CERES	WUE, HI, LAI, biomass, and yield	Phakamas (2015)
Bangladesh	AquaCrop	WUE, canopy cover, and yield	Maniruzzaman, and Nemes (2014)
Pakistan	CERES	LAI, biomass, and yield	Ahmad et al. (2013)

crop growth simulation model. Calibrated crop growth model combined with an optimization algorithm can be used to attain highest water productivity and ultimately more grain yield. RMSE value 165.32 kg ha^{-1} between field observed and model simulated was gained, which indicated good performance of the model. Lee et al. (2017) indicated that the efficiency of AquaCrop model for canopy cover simulation was 0.34–0.82 for rice crop. The model predicted grain production deviated from the field obtained data was ranging from 0.1% to 7.8%. The model efficiency (ME) for grain yield was 0.98. The RMSE for grain yield was t ha^{-1}. Phakamas (2015) used CSM-CERES-rice model to determine water productivity. Model predicted values were in good agreement with d-index 0.91 and 0.93 field observed leaf area index and total biomass values, respectively, while the association between model predicted value and observed value was poor for grain yield and harvest index. The poor association between field obtained and model predicted value for grain yield as well as harvest index was a result of mainly higher infestation of insect pests and diseases attack. Maniruzzaman and Nemes (2014) applied AquaCROP model whose performance was scrutinized by five generally used recital measures that showed realistic fit of the growth simulation model of the recorded irrigation water requirement, canopy cover, total dry matter, and grain yield under various irrigation regimes. When further evaluated for the diverse local field circumstances, this is extensively applicable, user friendly, and easy to parameterizing simulations. Rice growth model is a research tool that assists to upsurge the capacity of rice-growing community to acclimatize climate and potentially other deviations in rice production. Ahmad et al. (2013) evaluated of CSM-CERES-rice model which indicated that the model was capable to precisely predict development, growth, and grain yield under irrigated environmental situations, with a mean error value of 11% between model predicted and field recorded grain production. The consequences of the model indicated that optimum quantity of 1300 mm irrigated water produced the maximum grain yield in comparison to altogether other treatments.

28.1.3 Phenology

Hyun and Kim (2019) applied ORZA2000 rice simulation model. The genotype characteristics have been determined with the help of high-quality data for the crop development, which entail considerable cost and effort. The RMSE value of the panicle emergence was not more than 3 days, when the parameters were associated with phenological stages and phases estimated. In disparity, the coefficient of determination for grain production tended to be less than 0.2. Maki et al. (2017) used a SIMRIW-RS model, which combined remote sensing data with a crop growth simulation model SIMRIW. The model was applied for the estimation of grain yield at different regions. Results indicated that the model has high potential for the estimation of grain yield precisely, when the leaf area index of rice is guesstimated with higher accurateness from satellite data. Zhou et al. (2019) used WOFOST (world food study model) for the improvement in the precision of rice growth simulation at the regional scale under arid environment. The particle swarm optimization algorithm was applied for the optimization of the preliminary phenological stages and phases and rice sowing date in the world food study model by reducing the difference between model predicted and field recorded phenological stages and phases. Performance of model was well. Improvement in simulation of phenology enhances the accurateness of the rice growth simulation, with correlation coefficients equal to 0.80, 083, and 0.82 at three field work dates. Casanova et al. (2000) tested the performance of rice model ORYZA2000 for rice anthesis and maturity date. Results indicated that model predicted rice growth very precisely until maturity date. Subsequently in the anthesis stage, nevertheless, deviations occurred and enhanced particularly at the ripening stage. Larijani et al. (2011) applied ORYZA2000 model, with ranged values of root mean square error from 3 to 4 days, and had higher precision for the simulation of various phenology stages of rice varieties. Considering suitable capability of rice simulation model in simulation of phenological stages of rice cultivars. So, model can be applied as a suitable research tool for planning rice crop management as well as suitable decision support system for adaptation trials of improved as well as introduced genotypes under diverse agroclimatic regions. Timsina and Humphreys (2006) indicated that performance of CSM-CERES-rice model was fair in low nitrogen, water shortage, and lower temperature conditions throughout the phenological stages. Over the three data sets examined, the model simulated the flowering and physiological maturity date good, in which NRMSE value ranged from 3% to 5% and *d*-index from 0.94 to 0.99. Ahmad et al. (2019) used the CSM-CERES-rice model for the standard, field-tested genotypes of rice crop for different regions in Punjab, Pakistan, during 35 years period (1980–2014). Results indicated that the model predicted phenological stages earlier under climate warming conditions in comparison to the field recorded phenological stages. Phenological phases were shortened due to climate warming trend during 1980 to 2014 (Fig. 28.5). A significant share of the adverse influence of climate warming on rice crop (35%) was counterbalance with the help of growing latest varieties which required higher thermal time requirement. Zhang and Tao (2013) applied five rice growth simulation models like CERES-rice, ORYZA2000, RCM, Beta Model, SIMRIW, etc. for prediction of phenological stages and phases (Gao et al. 2020, 2021). Primarily, calibration and evaluation of all models was performed, which was based on a higher number of observations of phenological stages and phases in all over China from 1981 to 2009. The model simulated deviations in phenological stages and phases were usually consistent when air temperatures were less than the appropriate temperature and, nevertheless, varied mostly when temperatures were more than the appropriate temperature in all rice-growing regions.

28.1.4 Climate Change

Table 28.3 showed that the impact of climate change on rice crop was simulated and making adaptation strategies were determined with the help of various growth simulation models. Yun (2003) evaluated CSM-CERES-rice model for climate

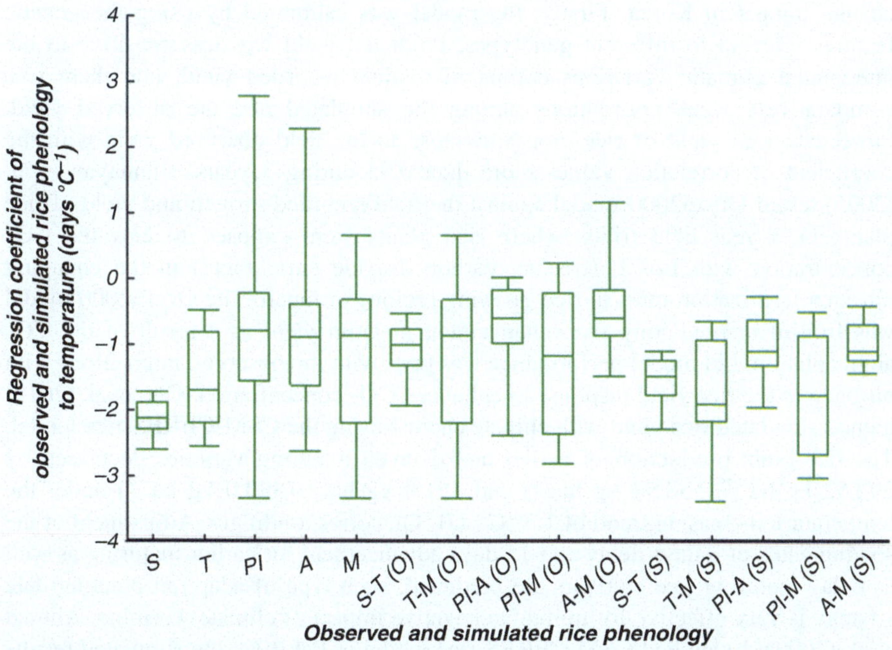

Fig. 28.5 Observed phenology (stages and phases) and simulated phenology (phases) versus temperature trends using a standard rice cultivar at each location in Punjab, Pakistan from 1980 to 2014 (*S* sowing, *T* transplanting, *PI* panicle initiation, *A* anthesis, *M* maturity, *S-T* sowing-transplanting, *T-M* transplanting-maturity, *PI-A* panicle initiation-anthesis, *PI-M* panicle initiation-maturity, *A-M* anthesis-maturity, *O* observed, *S* simulated) (**Source**: Ahmad et al. 2019)

Table 28.3 Various rice models used for assessment of climate change worldwide

Country	Rice model name	Adaptation strategies	References
South Korea	CERES	Heat- and drought-tolerant varieties	Yun (2003)
Iran	ORYZA2000	Efficient irrigation strategies	Bannayan et al. (2005)
China	CERES	Planting date, cultivars	Guo et al. (2019)
Pakistan	CERES	Planting date	Ahmad et al. (2009)
China	APSIM-Oryza	Resources conservation strategies	Liu et al. (2019)
Sri Lanka	CERES	Enhance RUE of cultivars	Dias et al. (2016)
Kenya	CERES	Planting date, cultivars longer duration	Nyang'Au et al. (2014)
Ghana	CERES	Improve soil fertility	Oteng-Darko et al. (2012)
India	CERES	Planting date	Swain and Yadav (2009)
Pakistan	CERES	Heat- and drought-tolerant varieties	Nasir et al. (2020)

change impact in Korea. Firstly, the model was calibrated by using the genetic features relevant to different genotypes. Predicted yield was less sensitive to the interannual climatic variations compared to field recorded yield, and there was comparatively weak correlations among the simulated and the observed yield. Projected grain yield of rice crop was close to the field observed yield with the coefficient of correlation values more than 0.95 during 3 years. Bannayan et al. (2005) tested Oryza2000 model against the field recorded growth and yield of rice plants in 3 year field trials, where rice plants were exposed to elevated CO_2 concentration with FACE (free air carbon dioxide enrichment) in the changing nitrogen fertilization rates in rice-growing regions in Japan. The Oryza2000 model was fruitful in predicting the enhancement in grain yield as a result of the CO_2 augmentation, but model performance was poor with the observed interactions with nitrogen in the rice yield response to enhanced CO_2 concentration. Guo et al. (2019) gained simulated rice yield with climate change using the CSM-CERES-rice model. The rice grain production of earlier and delayed maturing varieties decreased by 292.52 kg ha^{-1} (558.94 kg ha^{-1}) and 151.8 kg ha^{-1} (380.0 kg ha^{-1}) under the temperature increasing trend of 1.5 °C (2.0 °C), correspondingly. Adjustment of the sowing dates of 8 days delay and 15 days advancement for earlier maturing as well as delay maturing rice cultivars are predicted, such type of adaption planting date strategy is very effective for minimize negative impact of climate warming. Ahmad et al. (2009a, b) applied CSM-CERES-rice model in Pakistan, the simulated results of which showed that on an average with the enhancement concentration of CO_2 up to 550 ppm, rice production enhanced by 1.5–4.5% at various sites. Model prediction results also indicated that enhancing air temperature reduced total crop period by 4–5 days at Kala Shah Kaku and Gujranwala locations, correspondingly. Subsequently, rice production reduced at all sites, with the enhancement of temperature. In the future, the end of July will be effectual for transplanting fine cultivars. Liu et al. (2019) used APSIM-Oryza rice model. This model performance was accurate for the simulation in which rice crop does not reach to physiological maturity, which was treated as normal simulations under climate change scenario. Nevertheless, model predictions were a reason for some variations of simulated consequences in rice yields under different climate change scenarios. Dias et al. (2016) applied CERES model, the outcomes of which showed that enhancing air temperature and solar radiations and reducing rainfall in the mid twenty-first century influence the yield and growth of rice crop. Rice production in the middle of the century indicates reduced tendency in all cultivars by 25–35% and sowing to maturity phase shortened compared to baseline period. Nyang'Au et al. (2014) indicated by using CERES-rice model that enhancement in CO_2 level directed an upsurge in rice production of both fine and coarse varieties with changing CO_2 conditions as well as rise in the solar radiations and had an enhancing influence on grain production of all varieties. The consequences of research consequently showed that climate situations significantly impact on rice yield and must be considered for proper planning for the improvement of food security in Kenya. Oteng-Darko et al. (2012) concluded that CSM-CERES-rice model could effectively predict rice production under aerobic condition. Variations in periodic rainfall as well as enhancements in air temperatures particularly throughout dry seasons harmfully influenced rice yield. The decrease in rainfall throughout the wet season favored aerobic rice productivity. Rice yield reduction is double as large as gain. Fluctuations in climate conditions can be a reason for the decline of yield improvement from 83% to 53% and yield lessening to upsurge from 150% to 177% at the end of the present century. Swain and Yadav (2009) applied CERES-rice model for climate change assessment in India. Rice yield will decrease by 28% in future scenario compared to baseline weather data. Proper adaptation strategy like 15 days delay in planting date can be effective for enhancing grain yield under climate warming. Nasir et al. (2020) used CERES-rice model for assessment of climate change in rice-producing areas of Punjab, Pakistan. Grain yield will reduce in future scenario due to climate warming. Rice production of fine cultivars was enhanced by 15% in cool dry as well as 5% in hot dry GCM with adaptation strategies. Rice yield of coarse rice was enhanced by 15–9% under cool dry and hot dry climatic circumstances, correspondingly, with proper adaptation strategies.

28.1.5 Economics and Policy Making

Rice simulation models are very effective tools for the rice crop economic analysis and making proper policy in rice growing areas. Policy makers also apply rice growth models for the finding alternate choices for designing suitable strategies for achieving higher rice productivity under existing agro-ecological conditions (Oteng-Darko et al. 2012; Maniruzzaman and Nemes 2014). Despite some limitations, the rice growth simulation modeling tactic remains very supportive for the economical assessment of the influence of future global climate change; consequently, models provide assistance in preparation of local, regional, and national policies for the proper adaptation strategies determination. Proper planning in the rice industry sector, management operations, and concern of organization decision upon environment concerns are also well supported by using modeling approach across the globe (Soundharajan and Sudheer 2009; Liu et al. 2019).

28.1.6 Conclusion

Rice growth simulation models are very helpful as research tools all over the world. Frequently, rice growth simulation models were applied for determination of best crop husbandry practices, optimum irrigation scheduling, phenological stages and phases, and climate change assessment along with adaptation strategies across the world. All rice simulation models are being effectively used at local, regional, and national levels across the world, nevertheless between all rice growth simulation models. CSM-CERES-rice model was frequently used by rice scientists. For the proper irrigation management planning, generally ORYZA2000 rice model was used by researchers across the world.

References

Ahmad S, Hasanuzzaman M (2012) Integrated effect of plant density, N rates and irrigation regimes on the biomass production, N content, PAR use efficiencies and water productivity of rice under irrigated semiarid environment. Not Bot Horti Agrobot Cluj-Napoca 40(1):201–211

Ahmad S, Zia-ul-Haq M, Ali H, Shad SA, Ammad A, Maqsood M, Khan MB, Mehmood S, Hussain A (2008) Water and radiation use efficiencies of transplanted rice (*Oryza sativa* L.) at different plant densities and irrigation regimes under semi-arid environment. Pak J Bot 40(1): 199–209

Ahmad A, Iqbal S, Hussain A, Wajid S, Khaliq T, Usman M, Nasim W (2009a) Modeling responses of fine rice growth and yield to increased carbon dioxide concentration, temperature and precipitation under current & future climate scenarios. Environ Sci Technol 2:50–59

Ahmad S, Ahmad A, Zia-ul-Haq M, Ali H, Khaliq T, Anjum MA, Khan MA, Hussain A, Hoogenboom G (2009b) Resources use efficiency of field grown transplanted rice (*Oryza sativa* L.) under irrigated semiarid environment. J Food Agric Environ 7(2):487–492

Ahmad S, Ahmad A, Soler CMT, Ali H, Zia-Ul-Haq M, Anothai J, Hasanuzzaman M (2012) Application of the CSM-CERES-Rice model for evaluation of plant density and nitrogen management of fine transplanted rice for an irrigated semiarid environment. Precis Agric 13(2):200–218

Ahmad S, Ahmad A, Ali H, Hussain A, Garcia y Garcia A, Khan MA, Zia-Ul-Haq M, Hasanuzzaman M, Hoogenboom G (2013) Application of the CSM-CERES-Rice model for evaluation of plant density and irrigation management of transplanted rice for an irrigated semiarid environment. Irrig Sci 31(3):491–506

Ahmad A, Ashfaq M, Rasul G, Wajid SA, Khaliq T, Rasul F, Saeed U, Rahman MH, Hussain J, Baig IA, Naqvi AA, SAA B, Ahmad S, Naseem W, Hoogenboom G, Valdivia RO (2015) Impact of climate change on the rice–wheat cropping system of Pakistan. In: Hillel D, Rosenzweig C (eds) Handbook of climate change and agro-ecosystems: The Agricultural Modeling Intercomparison and Improvement Project (AgMIP) integrated crop and economic assessments. Imperial College Press/American Society of Agronomy, London/Hackensack, pp 219–258

Ahmad S, Abbas G, Ahmed M, Fatima Z, Anjum MA, Rasul G, Hoogenboom G (2019) Climate warming and management impact on the change of phenology of the rice-wheat cropping system in Punjab, Pakistan. Field Crop Res 230:46–61

Ahmed M, Ahmad S (2019) Carbon dioxide enrichment and crop productivity. In: Hasanuzzaman M (ed) Agronomic crops, vol 2. Springer Nature Singapore, Singapore, pp 31–46

Ahmed M, Ahmad S (2020) Systems modeling. In: Ahmed M (ed) Systems modeling. Springer Nature, Cham, pp 1–44

Ahmed M, Fayyaz-ul-Hassan, Ahmad S (2017) Climate variability impact on rice production: adaptation and mitigation strategies. In: Ahmed M, Stockle C (eds) Quantification of climate variability, adaptation and mitigation for agricultural sustainability. Springer, Cham, pp 91–111

Ahmed M, Ahmad S, Raza MA, Kumar U, Ansar M, Shah GA, Parsons D, Hoogenboom G, Palosuo T, Seidel S (2020a) Models calibration and evaluation. In: Ahmed M (ed) Systems modeling. Springer Nature Singapore, Singapore, pp 151–178

Ahmed M, Ahmad S, Waldrip HM, Ramin M, Raza MA (2020b) Whole farm modeling: A systems approach to understanding and managing livestock for greenhouse gas mitigation, economic viability and environmental quality. In: Waldrip HM, Pagliari PH, He Z (eds) Animal Manure, ASA Special Publication 67. American Society of Agronomy/Soil Science Society of America, Madison, WI, pp 345–371

Akram R, Turan V, Hammad HM, Ahmad S, Hussain S, Hasnain A, Maqbool MM, Rehmani MIA, Rasool A, Masood N, Mahmood F, Mubeen M, Sultana SR, Fahad S, Amanet K, Saleem M, Abbas Y, Akhtar HM, Hussain S, Waseem F, Murtaza R, Amin A, Zahoor SA, Sami ul Din M, Nasim W (2018) Fate of organic and inorganic pollutants in paddy soils. In: Hashmi MZ, Varma

A (eds) Environmental pollution of paddy soils. Springer Nature Switzerland, Cham, pp 197–214

Akram R, Fahad S, Masood N, Rasool A, Ijaz M, Ihsan MZ, Maqbool MM, Ahmad S, Hussain S, Ahmed M, Kaleem S, Sultana SR, Mubeen M, Saud S, Kamran M, Nasim W (2019) Plant growth and morphological changes in rice under abiotic stress. In: Hasanuzzaman M, Fujita M, Nahar K, Biswas JK (eds) Advances in rice research for abiotic stress tolerance. Woodhead, Cambridge, pp 69–85

Amarasingha RPRK, Suriyagoda LDB, Marambe B, Gaydon DS, Galagedara LW, Punyawardena R, Silva GLLP, Nidumolu U, Howden M (2015) Simulation of crop and water productivity for rice (*Oryza sativa* L.) using APSIM under diverse agro-climatic conditions and water management techniques in Sri Lanka. Agric Water Manag 160:132–143

Amiri E (2008) Evaluation of the rice growth model ORYZA2000 under water management. Asian J Plant Sci 7(3):291–297

Arora VK (2006) Application of a rice growth and water balance model in an irrigated semi-arid subtropical environment. Agric Water Manag 83(1–2):51–57

Artacho P, Meza F, Alcalde JA (2011) Evaluation of the Oryza2000 rice growth model under nitrogen-limited conditions in an irrigated Mediterranean environment. Chil J Agric Res 71(1): 23–33

Balwinder-Singh E, McDonald AJ, Kumar V, Poonia SP, Srivastava AK, Malik RK (2009) Taking the climate risk out of transplanted and direct seeded rice: insight from dynamic simulations in eastern India. Field Crop Res 239:92–103

Bannayan M, Kobayashi K, Kim HY, Lieffering M, Okada M, Miura S (2005) Modeling the interactive effects of atmospheric CO_2 and N on rice growth and yield. Field Crop Res 93(2–3): 237–251

Basso B, Liu L, Ritchie JT (2016) A comprehensive review of the CERES-Wheat, -Maize and -Rice models' performances. Adv Agron. 136:27–132

Biswas A, Mailapalli DR, Raghuwanshi NS (2021) Modelling the effect of changing transplanting date on consumptive water footprints for paddy under the system of rice intensification. J Sci Food Agric 101(13):5378–5390. https://doi.org/10.1002/jsfa.11186

Bouman BAM, Van Laar HH (2006) Description and evaluation of the rice growth model ORYZA2000 under nitrogen-limited conditions. Agr Syst 87(3):249–273

Casanova D, Goudriaan J, Bosch AD (2000) Testing the performance of ORYZA1, an explanatory model for rice growth simulation, for Mediterranean conditions. Eur J Agron 12(3-4):175–189

Dewi ER, Susanti E, Apriyana Y (2021) Planting time options to improve rice productivity based on the Integrated KATAM recommendations. IOP Conf Ser Earth Environ Sci 648:012105

Dias MPNM, Navaratne CM, Weerasinghe KDN, Hettiarachchi RHAN (2016) Application of DSSAT crop simulation model to identify the changes of rice growth and yield in Nilwala river basin for mid-centuries under changing climatic conditions. Proc Food Sci 6:159–163

Fahad S, Noor M, Adnan M, Khan MA, Rahman I, Alam M, Khan IA, Ullah H, Mian IA, Hassan S, Saud S, Bakhat HF, Hammad HM, Ahmad S, Nasim W (2019) Abiotic stress and rice grain quality. In: Hasanuzzaman M, Fujita M, Nahar K, Biswas JK (eds) Advances in rice research for abiotic stress tolerance. Woodhead, Cambridge, pp 571–583

Fatima Z, Ahmad M, Hussain M, Abbas G, Ul-Allah S, Ahmad S, Ahmed N, Ali MA, Sarwar G, Haque EU, Iqbal P, Hussain S (2020) The fingerprints of climate warming on cereal crops phenology and adaptation options. Sci Rep 10:18013

Gao Y, Wallach D, Liu B, Dingkuhn M, Boote KJ, Singh U, Asseng S, Kahveci T, He J, Zhang R, Confalonieri R, Hoogenboom G (2020) Comparison of three calibration methods for modeling rice phenology. Agric For Meteorol 280:107785

Gao Y, Wallach D, Hasegawa T, Tang L, Zhang R, Asseng S, Kahveci T, Liu L, He J, Hoogenboom G (2021) Evaluation of crop model prediction and uncertainty using Bayesian parameter estimation and Bayesian model averaging. Agric For Meteorol 311:108686

Gaydon DS, Khaliq T, Ahmad M-u-D, Cheemad MJM, Gull U (2021a) Tweaking Pakistani Punjab rice-wheat management to maximize productivity within nitrate leaching limits. Field Crop Res 260:107964

Gaydon DS, Radanielson AM, Chaki AK, Sarker MMR, Rahman MA, Rashid MH, Kabir MJ, Khan ASMMR, Gaydon ER, Roth CH (2021b) Options for increasing Boro rice production in the saline coastal zone of Bangladesh. Field Crop Res 264:108089

Guo Y, Wu W, Du M, Liu X, Wang J, Bryant CR (2019) Modeling climate change impacts on rice growth and yield under global warming of 1.5 and 2.0 C in the Pearl River Delta, China. Atmos. 10(10):567–578

Hafeez-ur-Rehman NA, Awan MI, Ijaz M, Hussain M, Ahmad S, Farooq M (2019) Direct seeding in rice: problems and prospects. In: Hasanuzzaman M (ed) Agronomic crops, Production technologies, vol 1. Springer Nature Singapore, Singapore, pp 199–222

Hameed F, Xu J, Rahim SF, Wei Q, Liao Q, Ahmed S (2019) Rice growth and nitrogen uptake simulation by using ORYZA (v3) model considering variability in parameters. Pak J Agric Sci 56(1):245–259

Hoogenboom G, Porter CH, Boote KJ, Shelia V, Wilkens PW, Singh U, White JW, Asseng S, Lizaso JI, Moreno LP, Pavan W, Ogoshi R, Hunt LA, Tsuji GY, Jones JW (2019a) The DSSAT crop modeling ecosystem. In: Boote KJ (ed) Advances in crop modeling for a sustainable agriculture. Burleigh Dodds Science, Cambridge, pp 173–216. https://doi.org/10.19103/AS.2019.0061.10

Hoogenboom G, Porter CH, Shelia V, Boote KJ, Singh U, White JW, Hunt LA, Ogoshi R, Lizaso JI, Koo J, Asseng S, Singels A, Moreno LP, Jones JW (2019b) Decision support system for agrotechnology transfer (DSSAT). Version 4.7.5 (https://DSSAT.net). DSSAT Foundation, Gainesville

Hussain M, Ahmad S, Hussain S, Lal R, Ul-Allah S, Nawaz A (2018) Rice in saline soils: physiology, biochemistry, genetics and management. Adv Agron. 148:231–287

Hyun S, Kim KS (2019) Calibration of cultivar parameters for cv. Shindongjin for a rice growth model using the observation data in a low quality. Kor J Agric For Meteorol 21(1):42–54

Jha RK, Kalita PK, Jat R (2020) Development of production management strategies for a long-duration rice variety: Rajendra Mahsuri—Using crop growth model, DSSAT, for the state of Bihar, India. Paddy Water Environ 18:531–545

Jones JW, Hoogenboom G, Porter CH, Boote KJ, Batchelor WD, Hunt LA, Wilkens PW, Singh U, Gijsman AJ, Ritchie JT (2003) DSSAT cropping system model. Eur J Agron 18:235–265

Kadiyala MDM, Jones JW, Mylavarapu RS, Li YC, Reddy MD (2015) Identifying irrigation and nitrogen best management practices for aerobic rice–maize cropping system for semi-arid tropics using CERES-rice and maize models. Agric Water Manag 149:23–32

Khaliq T, Gaydon DS, Ahmad MUD, Cheema MJM, Gull U (2019) Analyzing crop yield gaps and their causes using cropping systems modelling–A case study of the Punjab rice-wheat system, Pakistan. Field Crop Res 232:119–130

Khan MA, Ahmad S, Raza A (2019a) Integrated weed management for agronomic crops. In: Hasanuzzaman M (ed) Agronomic crops, vol 2. Springer Nature Singapore, Singapore, pp 257–281

Khan SM, Ali S, Nawaz A, Bukhari SAH, Ejaz S, Ahmad S (2019b) Integrated pest and disease management for better agronomic crop production. In: Hasanuzzaman M (ed) Agronomic crops, vol 2. Springer Nature Singapore, Singapore, pp 385–428

Kumar N, Tripathi P, Pal RK (2009) Validation of simulation modeling for growth parameters of rice genotypes using CERES 3.5v for Eastern Uttar Pradesh. Environ Ecol 27(4):1490–1494

Larijani BA, Sarvestani ZT, Nematzadeh GA, Amiri E, Esfahani M (2011) Simulation of phonological development and growth duration of three rice cultivars at different seedling ages using ORYZA2000 model. Iran J Crop Sci 13(3):466–480

Lee KD, Park CW, So KH, Na SI (2017) Selection of optimal vegetation indices and regression model for estimation of rice growth using UAV aerial images. Kor J Soil Sci Fert 50(5):409–421

Li T, Angeles O, Marcaida M, Manalo E, Manalili MP, Radanielson A, Mohanty S (2017) From ORYZA2000 to ORYZA (v3): An improved simulation model for rice in drought and nitrogen-deficient environments. Agric Forest Meteorol 237–238:246–256

Liu J, Liu Z, Zhu AX, Shen F, Lei Q, Duan Z (2019) Global sensitivity analysis of the APSIM-Oryza rice growth model under different environmental conditions. Sci Total Environ. 651:953–968

Maki M, Sekiguchi K, Homma K, Hirooka Y, Oki K (2017) Estimation of rice yield by SIMRIW-RS, a model that integrates remote sensing data into a crop growth model. J Agric Meteorol 73(1):2–8

Maniruzzaman M, Nemes A (2014) Evaluation of the Aquacrop model to simulate rice growth under different water regimes in Bangladesh. Evaluation 159:331–340

Nasir IR, Rasul F, Ahmad A, Asghar HN, Hoogenboom G (2020) Climate change impacts and adaptations for fine, coarse, and hybrid rice using CERES-Rice. Environ Sci Pollut Res. 27(9): 9454–9464

Naz S, Fatima Z, Iqbal P, Khan A, Zakir I, Noreen S, Younis H, Abbas G, Ahmad S (2019) Agronomic crops: types and uses. In: Hasanuzzaman M (ed) Agronomic crops, Production technologies, vol 1. Springer Nature Singapore, Singapore, pp 1–18

Nyang'Au WO, Mati BM, Kalamwa K, Wanjogu RK, Kiplagat LK (2014) Estimating rice yield under changing weather conditions in Kenya using CERES rice model. Int J Agron 2014:1–12

Oteng-Darko P, Kyei-Baffour N, Ofori E (2012) Simulating rice yields under climate change scenarios using the CERES-Rice model. Afr Crop Sci J 20:401–408

Peerzada AM, Bukhari SAH, Dawood M, Nawaz A, Ahmad S, Adkins S (2019) Weed management for healthy crop production. In: Hasanuzzaman M (ed) Agronomic crops, vol 2. Springer Nature Singapore, Singapore, pp 225–256

Phakamas N (2015) Performance of the CSM-CERES-Rice model in evaluating growth and yield of rice in the farm level. J Agric Technol 11(5):1285–1295

Radanielson AM, Gaydon DS, Li T, Angeles O, Roth CH (2018) Modeling salinity effect on rice growth and grain yield with ORYZA v3 and APSIM-Oryza. Eur J Agron. 100:44–55

Razaq M, Shah FM, Ahmad S, Afzal M (2019) Pest management for agronomic crops. In: Hasanuzzaman M (ed) Agronomic crops, vol 2. Springer Nature Singapore, Singapore, pp 365–384

Sarwar N, Ali H, Ahmad S, Ullah E, Ahmad S, Mubeen K, Hill JE (2013a) Water wise rice cultivation on calcareous soil with the addition of essential micronutrients. J Anim Plant Sci 23(1):244–250

Sarwar N, Ali H, Maqsood M, EhsanUllah SAN, Shahzad M, Mubeen K, Shahid MA, Ahmad S (2013b) Phenological response of rice plants to different micronutrients application under water saving paddy fields on calcareous soil. Turk J Field Crops 18(1):52–57

Setiyono TD, Quicho ED, Gatti L, Campos-Taberner M, Busetto L, Collivignarelli F, Holecz F (2018) Spatial rice yield estimation based on MODIS and Sentinel-1 SAR data and ORYZA crop growth model. Remote Sens (Basel) 10(2):293–304

Shahzad AN, Ahmad S (2019) Tools and techniques for nitrogen management in cereals. In: Hasanuzzaman M (ed) Agronomic crops, vol 2. Springer Nature Singapore, Singapore, pp 111–126

Shamim M, Shekh AM, Pandey V, Patel HR, Lunagaria MM (2012) Simulating the phenology, growth and yield of aromatic rice cultivars using CERES-Rice model under different environments. J Agrometeorol 14(1):31–34

Soundharajan B, Sudheer KP (2009) Deficit irrigation management for rice using crop growth simulation model in an optimization framework. Paddy Water Environ 7(2):135–149

Swain DK, Yadav A (2009) Simulating the impact of climate change on rice yield using CERES-rice model. J Environ Inf 13(2):104–110

Tang L, Zhu Y, Hannaway D, Meng Y, Liu L, Chen L, Cao W (2009) RiceGrow: A rice growth and productivity model. NJAS-Wagen J Life Sci 57(1):83–92

Tariq M, Ahmed M, Iqbal P, Fatima Z, Ahmad S (2020) Crop phenotyping. In: Ahmed M (ed) Systems modeling. Springer Nature Singapore, Singapore, pp 45–60

Timsina J, Humphreys EJAS (2006) Performance of CERES-Rice and CERES-wheat models in rice–wheat systems: a review. Agr Syst 90(1–3):5–31

Vilayvong S, Banterng P, Patanothai A, Pannangpetch K (2015) CSM-CERES-Rice model to determine management strategies for lowland rice production. Sci Agric 72(3):229–236

Wasaya A, Yasir TA, Ijaz M, Ahmad S (2019) Tillage effects on agronomic crop production. In: Hasanuzzaman M (ed) Agronomic crops, vol 2. Springer Nature Singapore, Singapore, pp 73–99

Yuan S, Peng S, Li T (2017) Evaluation and application of the ORYZA rice model under different crop managements with high-yielding rice cultivars in central China. Field Crop Res 212:115–125

Yun JI (2003) Predicting regional rice production in South Korea using spatial data and crop-growth modeling. Agr Syst 77(1):23–38

Zahoor SA, Ahmad S, Ahmad A, Wajid A, Khaliq T, Mubeen M, Hussain S, Sami Ul Din M, Amin A, Awais M, Nasim W (2019) Improving water use efficiency in agronomic crop production. In: Hasanuzzaman M (ed) Agronomic crops, vol 2. Springer Nature Singapore, Singapore, pp 13–29

Zhang S, Tao F (2013) Modeling the response of rice phenology to climate change and variability in different climatic zones: comparisons of five models. Eur J Agron. 45:165–176

Zhou G, Liu X, Liu M (2019) Assimilating remote sensing phenological information into the WOFOST model for rice growth simulation. Remote Sens (Basel) 11(3):268–281

Climate Change and Rice Production: Impacts and Adaptations

29

Jamshad Hussain, Sajjad Hussain, Nazia Tahir, Irfan Rasool, Asmat Ullah, and Shakeel Ahmad

Abstract

Changes in climatic conditions is a significantly risk for agronomic crops and food security at local, regional, and global level. Risky weather circumstances and shifting patterns of rainfall resulted in reduction of rice crop productivity. Heat stress and indefinite rainfall decline the grain production of rice crop through shortening in the phenological phases in rice-growing areas. Forthcoming predictions indicated that temperature would be enhanced by 5 °C up to end of the current century. The predicted intensification in climate warming would result in the higher frequent and extend heat waves which could be serious decline in the rice crop production in rice track regions. The increase in climate warming resulted in earlier occurrence of rice crop phenological stages. Sustaining the rice grain yield under climate change conditions is a significant challenge at local,

J. Hussain
Adaptive Research Farm, Karor Lal Esan, Pakistan

S. Hussain
Institute of Environment and Sustainable Environment in Agriculture, Chinese Academy of Agricultural Sciences, Beijing, China

N. Tahir
Adaptive Research Farm, DG Khan, Pakistan

I. Rasool
Agronomic Research Station, Khanewal, Pakistan

A. Ullah
Department of Agriculture, Abdul Wali Khan University, Mardan, Pakistan

S. Ahmad (✉)
Department of Agronomy, Faculty of Agricultural Sciences and Technology, Bahauddin Zakariya University, Multan, Pakistan
e-mail: shakeelahmad@bzu.edu.pk

N. Sarwar et al. (eds.), *Modern Techniques of Rice Crop Production*,
https://doi.org/10.1007/978-981-16-4955-4_29

regional, and global level. So, adaptation strategies are mandatory for reduction of the climate susceptibilities. The adversative impact of climate warming can be lessened by evolving of heat tolerance, demanding higher growing degree days cultivars and some modifications in current rice crop management practices and technologies. Adaptation strategies deliver the valuable information for the researchers, academia, as well as farming community for the mitigation of the adverse impact of climate warming.

Keywords

Paddy · Climate change · Adaptation · Phenology · Methane · Crop model

Abbreviations

CERES	Crop Environment Resource Synthesis
CSM	Cropping system model
DSSAT	Decision Support System for Agrotechnology Transfer
FATE	Free-air Temperature Enhancement
GCM	General circulation model
IPCC	Intergovernmental Panel on Climate Change
WUE	Water use efficiency

29.1 Introduction

Currently, climate change has threatened food security and, in future, would be big challenge for sustainable development of agriculture (Nasir et al. 2020). Weather and climate are still decisive factors in crop production in the era of advanced technology. Climate change impacts on agriculture and food production are aggravating with passage of time in recent past (IPCC 2007).

Hiking demand for rice with increasing world population can only be met by improving current management and discovering new options for it. Rice is being cultivated and grown around 100 countries around the globe. Among different major sources of food, half of world population is dependent on rice for food and nutrition (Ahmad et al. 2008a, b, 2009a, b, c, d, 2012, 2013, 2015, 2019a, b; Ahmad and Hasanuzzaman 2012; Ahmed and Ahmad 2019, 2020; Shahzad and Ahmad 2019). Pace of rice production worldwide is not parallel to increasing pace of world population. Rice around the globe is facing many problems of harsh environment, low productivity, and high input cost (Ahmed et al. 2017, 2020, b; Akram et al. 2018, 2019; Hussain et al. 2018; Fahad et al. 2019; Fatima et al. 2020). Among different rice species, fine rice is the best and favorite for taste. But aromatic rice is facing different problems elevating temperature, low water use efficiency, water shortage, changing rainfall patterns, and low productivity (Endo et al. 2009; Hafeez-ur-Rehman et al. 2019; Khan et al. 2019a, b; Naz et al. 2019; Peerzada et al. 2019; Razaq et al. 2019; Sarwar et al. 2013a, b; Tariq et al. 2020; Wasaya et al. 2019; Zahoor et al. 2019). Climate change-related risks such as high temperature, drought spells, changing rainfall patterns, and water availability are major natural factors, which are extremely affecting rice production (Bhuvaneswari et al. 2014; Ahmed et al. 2015).

Average temperatures have risen by 0.6 °C globally since the Industrial Revolution as described by the Intergovernmental panel on climate change (IPCC). There is expectation to rise in global average temperature by 1.4–5.8 °C in future with current emission scenarios (Houghton et al. 2001). Carbon dioxide has elevated from 360 ppm during 1995 to 416 ppm up to date, which is predicted to be 571 ppm during 2055 under RCPs 8.5 scenario (Rosenzweig et al. 2013). Brohan et al. (2006) also reported that the twentieth century has the highest average temperature as compared to previous centuries, while 1990s was the warmest decade during the nineteenth century. It has been predicted that there would be rise in temperature of Indus Delta by 5 °C by the end of twenty-first century. It is expected that rise in temperature would multiply the water demand of agriculture and domestic and livestock sectors by 1.5 times as compared to current requirement in Pakistan (Rasul et al. 2012).

The relationship between rice production system and climate change is negatively correlated around the globe by different studies. Sustainable production of rice is mainly challenged by climate change (Hina et al. 2019). There is a dire need of climate change mitigation and adaptations strategies in rice sector to sustain the rice production system (Zhou et al. 2003). The greenhouse gas methane (CH_4) is produced during anaerobic degradation of organic matter in wetlands and rice fields by methanogenic archaea. Methane is the second most contributor of the global warming and 18% of total radiative forcing (IPCC 2007). Rice field contributes 5–19% of total CH_4 emission (IPCC 2007). Top ten emitters of methane are given in Fig. 29.1. There should be adoption of different climate change mitigation technologies for reduction of methane emission from paddy field.

Physiology, growth, and development are significantly affected by climate change and variability. Different studies have reported that higher CO_2 concentration would affect the photosynthesis positively to certain extent, but rice plant development is very temperature sensitive. Rosenzweig and Hillel (1995) concluded that fine paddy productivity is reduced because of reduced photosynthesis due to increased respiration at high temperatures. Rice is a C_3 crop and great responsive to elevated CO_2. Hunsaker et al. (2000) concluded that likewise, water use efficiency (WUE) also enhanced plants' salinity resistance besides increased nutrient uptake at higher rates of CO_2 concentration (Kaya et al. 2001). These positives effects of elevated CO_2 have no significance with elevated temperature.

Impacts of changing climate have been well studied in different studies (Ahmed and Ahmad 2020). However, studies on impacts of adaptations under changing climate are the key ways to minimize the negative impacts on paddy field like many other crops (Wassmann et al. 2009). There is need of an integrative approach for adaptations such as policy, management, awareness, and breeding adaptations for

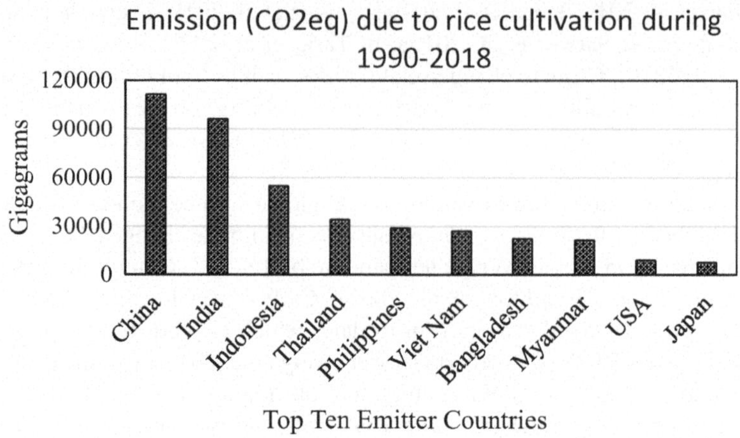

Fig. 29.1 Top ten emitters of methane (CO_2eq) due to rice cultivation during 1990–2018

sustainable production of rice. Results of different studies on adaptations have significant impacts on rice production (Ahmad et al. 2015). Thus, there are evidences that proper adaptation strategies according to requirements can enhance rice yield under changing climate (Howden et al. 2007). Alternate wetting drying, introduction of drought-resistant cultivars, enhanced water use efficiency, and minimum tillage practices should be advocated among farmers for maximum adoptability. Furthermore, strategies are required to cope with effects of climate change on rice production by application of anaerobic methanotrophs to oxidize the CH_4, and the development of high-yielding and abiotic stresses-tolerant (temperature, drought) and resistant rice cultivars by using different new breeding, genetic engineering, and genomic tools. Besides that, other management options such as development of weather-proofed farm equipment, shifting of planting and adjustments in cropping dates and use of climate forecasting by using remote sensing and modeling can also be used to sought out the climatic issues.

29.2 Future Projections of Climate Changes

During the twenty-first century, different unprecedented events have proved serious issue of climate change. Temperature around the globe is elevating since the Industrial Revolution due to accumulation of greenhouse gases (GHGs) and predicted to rise at higher pace in future (IPCC 2007). Concentration of GHGs would be increased with passage of time as predicted by Rosenzweig et al. (2013). Elevating temperature and GHG concentration have affected badly all other natural cycles of natural resources and factors affecting these cycles. Agriculture is open environment industry mainly affected by climatic factors such as temperature, carbon dioxide concentration, precipitation, and water availability. Globally,

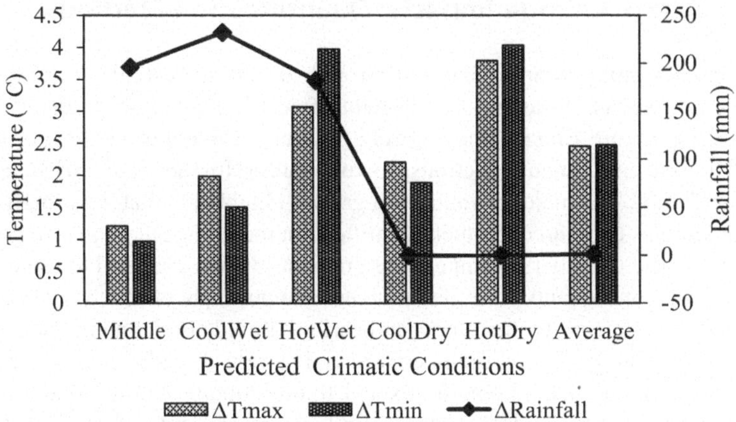

Fig. 29.2 Predicted temperature and rainfall changes of Middle, CoolWet, HotWet, CoolDry, HotDry, and average during rice season (July–October) under RCPs 8.5 for mid-century (2040–2069) as compared to baseline (1980–2010) at Central Punjab, Pakistan

impacts of climate change on agriculture are visible and evident, and world food security is an emerging challenge.

Pakistan is an agricultural country and severely affected by climate change. Among different weather components, temperature is the most important factor affecting the crop yield significantly. Different studies have predicted the rise in temperature and change in rainfall pattern in Pakistan separately and in combination (Hussain et al. 2020). Rise of temperature is basically dependent upon the amount of GHGs prevailing in the environment. Hussain et al. (2020) predicted the weather condition in five combinations (CoolWet, CoolDry, HotWet, HotDry, and Middle) using 29 GCMs of two arid and semiarid environments under RCPs 8.5 for mid-century (2040–2069). There would be increase in average temperature (2–3 °C) at both environments while rainfall of arid environment would be higher as compared to semiarid environment during mid-century. This shows that arid environment would be semiarid and semiarid environment would be arid environment during mid-century under RCPs 8.5 (Hussain et al. 2020). Shabbir et al. (2020) predicted the temperature and rainfall changes of Middle, CoolWet, HotWet, CoolDry, HotDry, and average during rice season (July–October) under RCPs 8.5 for mid-century (2040–2069) as compared to baseline (1980–2010) at Central Punjab Pakistan (Fig. 29.2). Ahmad et al. (2015) predicted the rise of temperature by in rice region during study the rice-wheat cropping system under climate change. Nasir et al. (2020) also predicted the rise in average temperature and rainfall of rice zone by 3 °C and 107.66 mm, respectively, during mid-century under RCPs 8.5.

29.3 Climate Change Impacts: Quantification Methods

Climate change impacts are quantified through different methods such as growth chambers rain shelter to assess the drought affects, temperature gradient tunnels, Free-air Temperature Enhancement (FATE) systems (White et al. 2011), Free Air Carbon Dioxide enrichment; locations, sowing dates (Hussain et al. 2018), and crop growth modeling. The major advantage of growth chamber is that it has high control on temperature as compared to other quantification methods, but plant root growth is restricted in these chambers. Temperature gradient tunnels are good to represent the field conditions and quantify the real climate change impacts but they obstruct the solar radiations. Free air temperatures and carbon dioxide enrichment are important techniques to quantify the high temperature and carbon dioxide level which are the most representatives of field conditions and future climate, but these are not cost-effective and have not an ideal control over temperature and carbon dioxide especially during windy environment. Crop growth models are important cost- and time-effective tools to quantify the climate change impacts like rising temperature and CO_2 and changing rainfall pattern. For example, Fang et al. (2006) used statistical regression, while process-based model was used by Ahmad et al. (2015).

Crop growth models are important tool to understand the change in environment factors around the globe. Crop modeling has enabled the researchers to simulate the effects of different adaptations and mitigation strategies for future. This quantification method has the ability to quantify the individual and cumulative effects of different factors such as temperature, CO_2, rainfall, management options, soil, and varietal characters (Hussain et al. 2018), but there is uncertainty in the results of models (Asseng et al. 2013).

Usage of crop models is an emerging technique for making decisions for climate smart agronomy. These models provide useful facts to scientists, growers, and policy makers across the world (Hoogenboom et al. 2015). DSSAT-CERES-Rice model using sensitivity analysis approach was employed to quantify the impact of climate change on aromatic rice. It was observed that rice productivity was most vulnerable to temperature, and adverse impacts of temperature counterweigh the positive impacts of elevated CO_2 rates which increase crop productivity (Ahmed and Hassan 2011). It is a dire need to apply these models under diverse agro-environmental conditions and devise adaptation approaches like temperature-resilient varieties, optimum management options for growers for coping issues of climate change.

29.4 Interaction of Changing Climate and Rice

Climatic conditions significantly and directly affect crop development and productivity and indirectly via variations in the availability of irrigation water. Being input of photosynthesis, an increase in CO_2 is fruitful for increasing rate of photosynthesis of C_3 and C_4 crop plants. But regardless of the positive effects, integrated effect of increasing CO_2 and increasing temperature is very detrimental for crop production (Aggarwal 2008). Rice, wheat, cotton, and maize crops are hard hit by the climate

Table 29.1 Calculation of mean (SE) minimum temperature (T_{min}), optimum temperature (T_{opt}), and maximum temperature (T_{max}) for different stages of rice

Stages	T_{min} ± SE (*n*)	T_{opt} ± SE (*n*)	T_{max} ± SE (*n*)
Germination/emergence	11.35 ± 1.15 (9)	28 ± 2.85 (7)	40 ± 1.29 (6)
Leaf initiation	10.6 ± 0.6 (6)	29.7 ± 0.75 (10)	42.5 ± 2.4 (2)
Shoot growth	13.73 ± 2.1 (5)	28.5 ± 1 (6)	35.5 ± 0.5 (2)
Root growth	15.82 ± 0.8 (8)	27.6 ± 0.3 (13)	35.8 ± 0.56 (9)
Tillering	16.4 ± 0.75 (10)	28.35 ± 1.2 (9)	35.25 ± 1 (6)
Panicle initiation	15.75 ± 0.3 (6)	26.7 ± 4.3 (2)	33.1 ± 1.7 (3)
Anthesis	16.3 ± 1.5 (9)	26.25 ± 1.31 (9)	37 ± 1.1 (10)
Grain filling	20.6 ± 0.65 (15)	24.5 ± 1.6 (10)	31.1 ± 0.65 (11)
Whole plant	<13.45 ± 2 (8)	27.63 ± 2.1 (5)	>35.5 ± 2 (8)

n number of literature sources
Source: Modified and adapted from Sánchez et al. (2013)

change in the country due to erratic changes in environmental variables. Rice crop is negatively affected by the erratic rainfall during anthesis and maturity and yield could be reduced up to 15% (Mahamood et al. 2003); fungal diseases and insect pest boosted up by the delaying of monsoon spell in the rice wheat cropping zone of the country (Saseendran et al. 2000). Peng et al. (2004) observed that there will be 10% reduction in paddy production with 1% increase of minimum temperature during dry season. Higher temperature promoted grain sterility in paddy. Up to 30 °C sterility increased, however at 36 °C spikelet sterile completely. Rice plant requires specific optimum temperatures at different phenology stages and phases from transplanting to maturity (Table 29.1). However, temperature after threshold level reduces phases that resulted in lesser productivity and quality (Kim et al. 2011; Ahmad et al. 2015).

29.4.1 Phenology

Among the important factors affecting rice physiological maturity are changes in temperatures and solar radiations (Basak et al. 2009). Transplanting dates also affect rice productivity as delayed transplanting was found to reduce physiological maturity with consequent impact on significant reduction in yield. Ahmad et al. (2009a, b) applied CSM-CERES-Rice model in Pakistan which simulated results showed that on an averaged bases of the enhancement concentration of CO_2 up to 550 ppm, rice production was enhanced by 1.5–4.5% at various sites. Model prediction results also indicated that enhancing air temperature reduced total crop period by 4–5 days at Kala Shah Kaku and Gujranwala locations, correspondingly. Subsequently, rice production was reduced at wholly sites, with enhancement of temperature. Transplanting of fine cultivars in the end of July will be the furthermost effectual transplanting date in future scenario at entirely locations. The study conducted by Otengdarko and coworkers (2012) indicated 26% greater productivity at lower temperature, while at the same time, yield was reduced by 60% due to 4 °C increase in temperature. Increase in temperature reduced the crop growth duration and vice versa with resulting effects on yield. Ahmad et al. (2019a, b) used the CSM-CERES-Rice model for the standard, field-tested genotypes of rice crop for different regions during 35 years period. Results indicated that the model-predicted phenological stages were earlier under climatic warming condition in comparison to the field-recorded phenological stages (Fig. 29.3). A significant share of the adverse influence of climate warming on rice crop (35%) was counterbalanced with the help of growing latest varieties which required higher thermal time requirement.

29.4.2 Grain Formation

According to some previous studies, solar radiations play an important part on paddy production; thus, paddy productivity might be improved by harvesting greater solar radiations on the canopy. These findings are in line with the works of Otengdarko et al. (2012) and Mahamood et al. (2003), who depicted that by enhancing solar radiation MJ m^{-2} day^{-1}, paddy productivity would be increased by 33%. Low temperature results in longer vegetative and grain filling durations besides finally producing more productivity; however, shorter duration and lower yield were observed under high temperature. Similarly, late transplanting of rice has the benefit of producing more assimilates by capturing higher solar radiation MJ m^{-2} on canopy surface. A significant reduction in the rice yield was observed by Mahamood et al. (2003) when planting was delayed afar June 1. Consequently, climatic change will not merely reduce the paddy productivity significantly, but it could make productivity more susceptible to transplanting times.

29.4.3 Grain Yield (kg ha^{-1})

Climate change is an important long-term challenge for sustained crop productivity (Vaghefi et al. 2011). Temperature after critical threshold level reduces development period of paddy along with increases sterility of spikelet's, shortens grain filling phase, besides intensify the loss of respiratory producing lesser productivity and seed quality (Kim et al. 2011; Ahmad et al. 2015). Different impacts of heat stress on paddy development have been observed. Nasir et al. (2020) used CERES-Rice model for assessment of climate change in rice-producing areas of Punjab, Pakistan. Grain yield will be reduced during future scenario due to climate warming. Rice production of fine cultivars was enhanced by 15% in cool dry as well as 5% in hot dry GCM with adaptation strategies. Rice yield of coarse rice was enhanced by 15% and 9% under cool dry and hot dry climatic circumstances, correspondingly, with proper adaptation strategies. According to Welch et al. (2010), 1 °C rise in temperature beyond critical threshold of rice (which is 24 °C) will bring about 10% reduction in grain yield and biomass. Higher temperature during nighttime leads to increase in respiration consequently, reducing the grain yield. Sudden rise in the air temperature during early spring (when wheat and other winter crops were at their reproductive stage) caused notable reductions in the grain yield despite affecting the

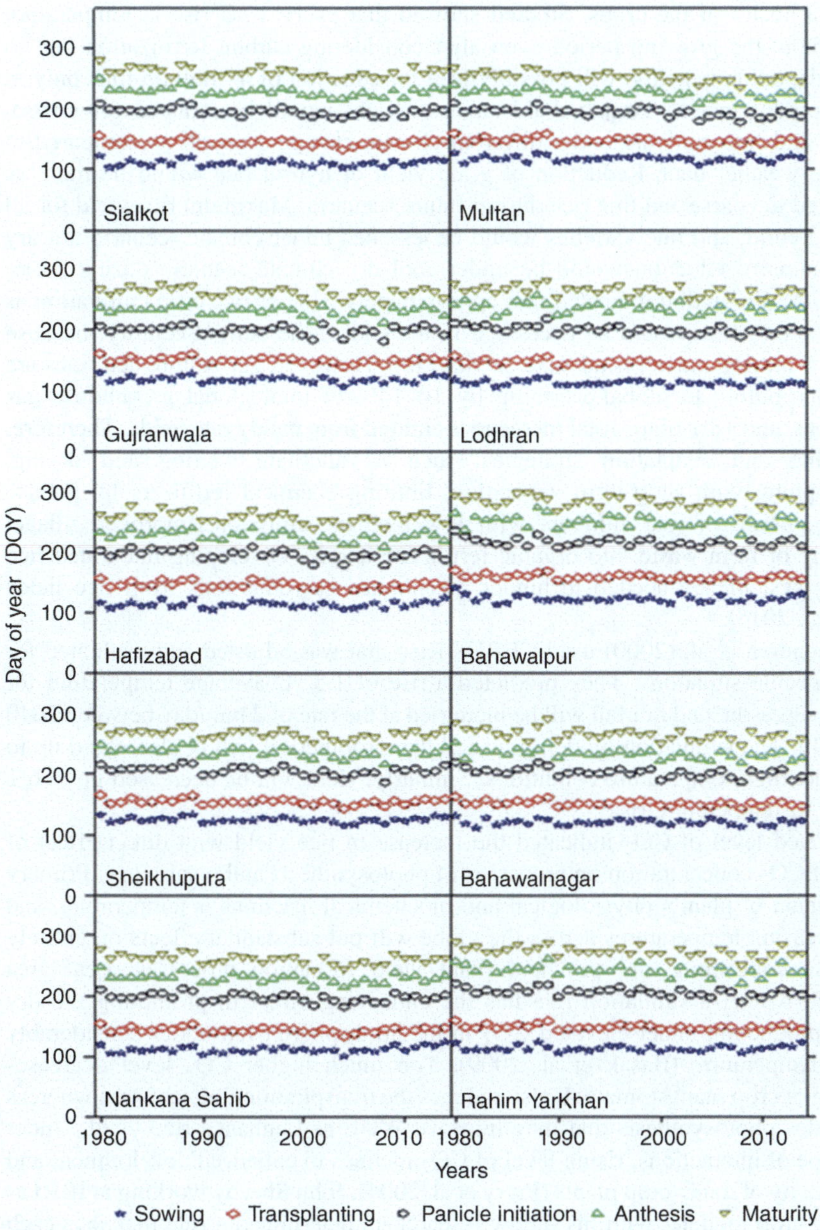

Fig. 29.3 Time series plots of farmer field observed dates of sowing, transplanting, panicle initiation, anthesis, and maturity of rice in Punjab, Pakistan (**Source**: Ahmad et al. 2019a, b)

apparent health of the crops. Studied showed that every 1 °C rise in temperature (throughout the growing period even after considering carbon fertilization but no other adaptation benefits) will possibly lead to reduction of 4– million tons only in wheat production. In Punjab, Pakistan, rice yield would be reduced for coarse, hybrid, and fine cultivars under different climate change scenario as compared to baseline weather data. Reduction of grain yield of hybrid rice would be more as compared to coarse and fine rice during future scenario. Maximum rice yield for all coarse, hybrid, and fine varieties would be lessened under climate scenario hot dry and minimum reduction would be under cool dry climate scenario during future (Fig. 29.4). Prediction indicated that approximately 51% of rice crop cultivation as well as production would be decreased in the end of the current century, because climate warming impact is adverse on rice crop. However, agriculture activities are also contributing to global warming by 10–14% of total global greenhouse gas emissions, and 18% of the total methane is emitted from paddy rice fields. Therefore, mitigating and adaptation strategies, such as alternate wetting and drying, intercropping with short-term vegetation, limiting chemical fertilizers by precise farming, usage of rice cultivars with low methane emission, improved tillage, recycling of farm waste into organic fertilizers, and by developing integrated rice farming system, are needed to hinder greenhouse gas emissions from rice fields (Xu et al. 2015).

Saseendran et al. (2000) used CERES-Rice that was adjusted and evaluated for those specific situations. They predicted a rise of 1.5 °C average temperature for monsoon season, and rainfall will be increased at the rate of 2 mm/day between 2040 and 2049. The simulation model indicated that crop period will be decreased up to 9% under high temperature conditions. Similarly, yield will be decreased up to 6% upon 1 °C temperature rise.

Elevated level of CO_2 indicated the increase in rice yield with direct effect of ambient CO_2 concentration on processes of photosynthesis and respiratory. Primary determinant of plant's physiological and biochemical reactions is temperature, and the increasing temperatures across the globe will put substantial effects on entirely all these controlling and principal plant yield formation processes. Leaf area followed by net assimilation rate has the enhancing effect in promoting the dry matter production under elevated CO_2 in the atmosphere even under considerably higher temperature (Basak et al. 2009). Too much higher CO_2 level decreases opening of crop plant stomata. It also reduces the transpiration/unit leaf area whereas aggravates photosynthesis that may improve WUE and enhance rice yield. Under such type of interactions, rising level of CO_2 inclines to enhanced development and productivity of many crop plants (Parry et al. 2004). John Sheehy working at IRRI as scientist crop modeler with his fellow researchers determined a rule that says each 75 ppm rise in CO_2 conc. Will increase the paddy yield (0.5 t ha^{-1}) while an increase in temperature by 1 °C will reduce the yield by 0.6 t ha^{-1} (Parry et al. 2004). Modeling results from this study are in agreement with the above discussion. Guo et al. (2019) indicated that rice grain production for earlier maturing varieties and delay maturing varieties are decreased by 292.52 kg ha^{-1} (558.94 kg ha^{-1}) and 151.8 kg ha^{-1} (380.0 kg ha^{-1}) under the temperature increasing trend 1.5 °C

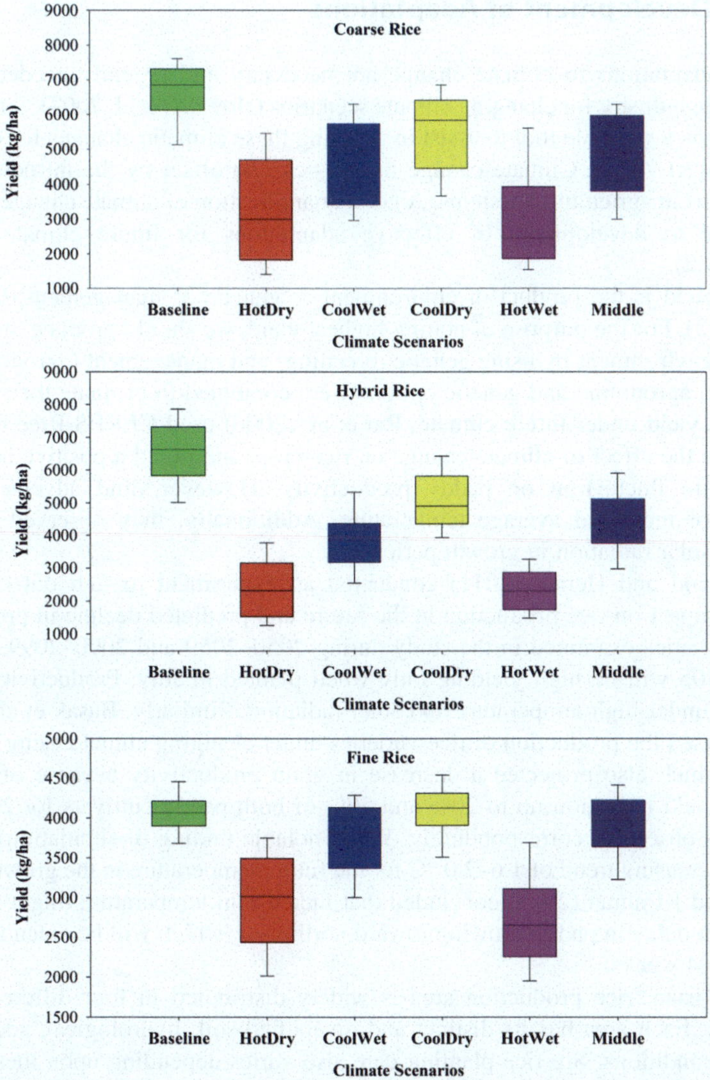

Fig. 29.4 Simulated climate change impacts on fine, coarse, and hybrid rice in Pakistan under different climatic conditions: Baseline, HotDry, CoolWet, CoolDry, HotWet, and Middle simulated under RCPs 8.5 for mid-century scenarios (2040–2069)

(2.0 °C) under climate change, correspondingly. Adjustment of the sowing dates of 8 days delay and 15 days advancement for earlier maturing as well as delay maturing rice cultivars is predicted; such type of adaption planting date strategy is very effective to minimize negative impact of climate warming.

29.5 Development of Adaptations

Timely adaptations to climate change are necessary and urgently needed by the agriculture industry for changing climate scenarios (Howden et al. 2007). Simulation modeling is a valuable tool to assist in adapting these climatic changes for different crops (Ewert 2012). Climate change impacts can be offset by the introduction of adaptations in agricultural systems. A good quantification of climate change impacts may lead to development of effective adaptations for future climate change (Table 29.2).

Rice yield is the product of environment × genetic × management (Craufurd et al. 2013). For the purpose of getting highest yield, we should produce the best fit with the environment by using genetic (breeding) and management (agronomic). In our study, agronomic and genetic options were combined to evaluate their impacts on wheat yield under future climate. Rai et al. (2006) used CERES-Rice model to determine the effect of climate change on rice crop, and noted a positive impact of temperature fluctuation on paddy productivity. However, find adverse impact because of increased average temperature. Additionally, they observed positive effect of solar radiation in growth periods.

Kawasaki and Herath (2011) conducted an experiment to find out changing climate impact on rice production in the future and predicted decline in production for all varieties examined in the study during 2050–2059 and 2090–2099. Variety KDML 105 will be high yielding only when planted in July. Productivity would decrease under high temperature and solar radiation. Similarly, Basak et al. (2010) also assessed the production of rice varieties under changing climate using DSSAT model, which also projected a decrease in grain productivity because of climate change; yield reduction up to 20% and 50% of both paddy cultivars for 2050 and 2070 was observed, correspondingly. While notable finding of simulation result is that an increasing trend of 1.6–2.0 °C for the future temperature in the growth cycle. Shunji and Kimuraa (2007) concluded that increase in temperature might result in significant delay in paddy growing toward northward, which will be extended up to around 3–4 weeks.

In Pakistan, rice production area is widely distributed in four different agro-ecologies. Each area has its distinct and diversified soil, hydrological, social, and weather conditions. So, rice planting date also varies depending upon these zones (Ahmad et al. 2019a, b). The selection of the planting date by the farmers is pertinent to have desirable life cycle of rice well adapted to prevailing temperature and other climatic factors (Ahmad et al. 2019a, b). Optimum transplanting time for rice crop is very important due to the following two reasons: (1) appropriate growth and development through vegetative phase regarding temperature besides solar radiation, and (2) Ensuring that grain filling will happen or coincides when temperature is appropriate, thus good quality seeds are attained (Farrell et al. 2003).

Ahmad et al. (2019a, b) revealed that any deviation from recommended sowing or transplanting times might affect rice productivity by upsetting rice phenology. In the upcoming years, heat stress may also cause remarkable loss to rice yield (Semenov 2009). Therefore, optimal transplanting time for approved cultivars for certain

Table 29.2 Proposed adaptations for rice production for different regions

Area of study	Factors identified as adaptation strategies	References
Ebonyi state, Nigeria	Minimum tillage, bond and drainage, fertilizer, crop diversification, livelihood diversification, improved rice varieties, pesticide, nursery, and adjusting planting and harvesting dates	Onyeneke (2021)
India	Short duration varieties, intercropping, changing cropping pattern, investment in irrigation, and agroforestry	Tripathi and Mishra (2017)
Central Vietnam	Late transplanting of rice (20–27%); supplementary irrigation (42%); increasing the fertilizer application rate (0.3–30%); number of doses of fertilizer (1.8–5.1%); heat-tolerant varieties	Shrestha et al. (2016)
India	Modified system of rice intensification (MSRI), supplemental irrigation, alternate wetting and drying, improved management practices	Palanisami et al. (2017)
Pakistan	Increase in 15% nitrogen application; 15% increase in sowing density; reducing no. of irrigation by 15%; early transplanting of nursery by 15 days; developed heat-tolerant cultivars up to 3 °C	Ahmad et al. (2015); Nasir et al. (2020); Shabbir et al. (2020)
Bangladesh	System of rice intensification (SRI); alternate wetting and draying	Latif et al. (2005)
Sri Lanka	Direct planting	Weerakoon et al. (2011)
Sri Lanka	Varietal improvement, and coping with water scarcity mechanization in transplanting	Zubair et al. (2015)
China	Adequate irrigation and suitable sowing dates; development of new rice varieties resistant for high temperature ranges; strengthen farmers' awareness	Ding et al. (2020); Yanan Hu et al. (2019)
Thailand	Use of advanced irrigation methods	Polthanee and Promkhambut (2014)
World	Abiotic stresses tolerance (temperature, drought), shifting of planting, and adjustments in cropping dates and use of climate forecasting	Hussain et al. (2020)
Indonesia	Planned adaptation policy, and prioritizes farmers with insecure land tenure	Rondhi et al. (2019)

locality or region is a key element liable for paddy productivity. Farmers should take into account this imperative factor; otherwise, the potential of a high-yielding variety will be lost (Laborte et al. 2012).

29.6 Conclusions

Climate warming and significant variations in climate pattern have adverse impacts on rice crop phenological stages, phases, and ultimately yield. Changes in rainfall frequency and intensity and intensification in extreme events like heat waves, flood, and drought have harmful influence on rice crop phenology and productivity. Future climate change prediction data indicated that temperature would be augmented by 5 °C in the end of the current century. Number of hot days and warm night will be enhanced in rice-growing areas, and less rainfall will occur in rice crop-growing period. Thus, climate warming condition would be resulted in an enormous lessening in grain yield of rice crop during future scenario. The harmful impacts of climate warming can be minimized by adopting adaptation strategies like growing heat-tolerant varieties and modification in current production technologies and crop management practices.

References

Aggarwal PK (2008) Global climate change and Indian agriculture: impacts adaptation and mitigation. Indian J Agric Sci 78:911–919

Ahmad S, Hasanuzzaman M (2012) Integrated effect of plant density, N rates and irrigation regimes on the biomass production, N content, PAR use efficiencies and water productivity of rice under irrigated semiarid environment. Not Bot Horti Agrobot Cluj-Napoca 40(1):201–211

Ahmad S, Zia-ul-Haq M, Ali H, Shad SA, Ahmad A, Maqsood M, Khan MB, Mehmood S, Hussain A (2008a) Water and radiation use efficiencies of transplanted rice (*Oryza sativa* L.) at different plant densities and irrigation regimes under semi-arid environment. Pak J Bot 40:199–209

Ahmad S, Zia-ul-Haq M, Ali H, Shad SA, Ammad A, Maqsood M, Khan MB, Mehmood S, Hussain A (2008b) Water and radiation use efficiencies of transplanted rice (*Oryza sativa* L.) at different plant densities and irrigation regimes under semi-arid environment. Pak J Bot 40(1): 199–209

Ahmad A, Iqbal S, Ahmad S, Khaliq T, Nasim W, Husnain Z, Hussain A, Zia-Ul-Haq M, Hoogenboom G (2009a) Seasonal growth, radiation interception, its conversion efficiency and biomass production of *Oryza sativa* L. under diverse agro-environments in Pakistan. Pak J Bot 41:1241–1257

Ahmad A, Iqbal S, Hussain A, Wajid S, Khaliq T, Usman M, Nasim W (2009b) Modeling responses of fine rice growth and yield to increased carbon dioxide concentration, temperature and precipitation under current & future climate scenarios. Environ Sci Technol 2:50–59

Ahmad S, Ahmad A, Zia-ul-Haq M, Ali H, Khaliq T, Anjum MA, Khan MA, Hussain A, Hoogenboom G (2009c) Resources use efficiency of field grown transplanted rice (*Oryza sativa* L.) under irrigated semiarid environment. J Food Agric Environ 7(2):487–492

Ahmad S, Zia-ul-Haq M, Ali H, Ahmad A, Khan MA, Khaliq T, Husnain Z, Hussain A, Hoogenboom G (2009d) Morphological and quality parameters of *Oryza sativa* L. as affected by population dynamics, nitrogen fertilization and irrigation regimes. Pak J Bot 41:1259–1269

Ahmad S, Ahmad A, Soler CMT, Ali H, Zia-Ul-Haq M, Anothai J, Hussain A, Hoogenboom G, Hasanuzzaman M (2012) Application of the CSM-CERES-Rice model for evaluation of plant density and nitrogen management of fine transplanted rice for an irrigated semiarid environment. Precis Agric 13(2):200–218

Ahmad S, Ahmad A, Ali H, Hussain A, Garcia y Garcia A, Khan MA, Zia-Ul-Haq M, Hasanuzzaman M, Hoogenboom G (2013) Application of the CSM-CERES-Rice model for evaluation of plant density and irrigation management of transplanted rice for an irrigated semiarid environment. Irrig Sci 31(3):491–506

Ahmad A, Ashfaq M, Rasul G, Wajid SA, Khaliq T, Rasul F, Saeed U, Rahman MH, Hussain J, Baig IA, Naqvi AA, SAA B, Ahmad S, Naseem W, Hoogenboom G, Valdivia RO (2015) Impact of climate change on the rice–wheat cropping system of Pakistan. In: Hillel D, Rosenzweig C (eds) Handbook of climate change and agro-ecosystems: the Agricultural Modeling Intercomparison and Improvement Project (AgMIP) integrated crop and economic assessments. Imperial College Press, London, pp 219–258

Ahmad S, Abbas G, Ahmed M, Fatima Z, Anjum MA, Rasul G, Hoogenboom G (2019a) Climate warming and management impact on the change of phenology of the rice-wheat cropping system in Punjab, Pakistan. Field Crops Res 230:46–61

Ahmad S, Abbas G, Ahmed M, Fatima Z, Anjum MA, Rasul G, Khan MA, Hoogenboom G (2019b) Climate warming and management impact on the change of rice-wheat phenology in Punjab, Pakistan. Field Crops Res 230:46–61

Ahmed M, Ahmad S (2019) Carbon dioxide enrichment and crop productivity. In: Hasanuzzaman M (ed) Agronomic crops, vol 2. Springer Nature Singapore, Singapore, pp 31–46

Ahmed M, Ahmad S (2020) Systems modeling. In: Ahmed M (ed) Systems modeling. Springer Nature, Cham, pp 1–44

Ahmed M, Hassan FU (2011) APSIM and DSSAT models as decision support tools. In 19th International Congress on Modelling and Simulation, Perth, Australia, 12–16 December. http://mssanz.org.au/modsim

Ahmed M, Fayyaz-ul-Hassan, Ahmad S (2017) Climate variability impact on rice production: adaptation and mitigation strategies. In: Ahmed M, Stockle C (eds) Quantification of climate variability, adaptation and mitigation for agricultural sustainability. Springer, Cham, pp 91–111

Ahmed M, Ahmad S, Raza MA, Kumar U, Ansar M, Shah GA, Parsons D, Hoogenboom G, Palosuo T, Seidel S (2020) Models calibration and evaluation. In: Ahmed M (ed) Systems modeling. Springer Nature Singapore, Singapore, pp 151–178

Ahmed M, Ahmad S, Waldrip HM, Ramin M, Raza MA (2020b) Whole farm modeling: a systems approach to understanding and managing livestock for greenhouse gas mitigation, economic viability and environmental quality. In: Waldrip HM, Pagliari PH, He Z (eds) Animal manure, ASA Special Publication 67. American Society of Agronomy/Soil Science Society of America, Madison, pp 345–371

Akram R, Turan V, Hammad HM, Ahmad S, Hussain S, Hasnain A, Maqbool MM, Rehmani MIA, Rasool A, Masood N, Mahmood F, Mubeen M, Sultana SR, Fahad S, Amanet K, Saleem M, Abbas Y, Akhtar HM, Hussain S, Waseem F, Murtaza R, Amin A, Zahoor SA, Sami ul Din M, Nasim W (2018) Fate of organic and inorganic pollutants in paddy soils. In: Hashmi MZ, Varma A (eds) Environmental pollution of paddy soils. Springer Nature Switzerland, Cham, pp 197–214

Akram R, Fahad S, Masood N, Rasool A, Ijaz M, Ihsan MZ, Maqbool MM, Ahmad S, Hussain S, Ahmed M, Kaleem S, Sultana SR, Mubeen M, Saud S, Kamran M, Nasim W (2019) Plant growth and morphological changes in rice under abiotic stress. In: Hasanuzzaman M, Fujita M, Nahar K, Biswas JK (eds) Advances in rice research for abiotic stress tolerance. Woodhead, Cambridge, pp 69–85

Asseng S, Ewert F, Rosenzweig C et al (2013) Uncertainty in simulating wheat yields under climate change. Nat Clim Chang 3:827–832

Basak JK, Ali MA. Islam MN, Alam MJB (2009) Assessment of the effect of climate change on boro rice production in Bangladesh using CERES-Rice model. In: Proceedings of the international conference on climate change impacts and adaptation strategies for Bangladesh, February, pp 18–20

Basak JK, Ali MA, Islam MN, Rashid MA (2010) Assessment of the effect of climate change on Boro-rice production in Bangladesh using DSSAT model. J Civil Eng 2:95–108

Bhuvaneswari K, Geethalakshmi V, Lakshmanan A, Anbhazhagan R, Sekhar DNU (2014) Climate change impact assessment and developing adaptation strategies for rice crop in western zone of Tamil Nadu. J Agromet 16:38–43

Brohan P, Kennedy JJ, Harris I, Tett SFB, Jones PD (2006) Uncertainty estimates in regional and global observed temperature changes: a new data set from 1850. J Geophys Res 111:D12106

Craufurd PQ, Vadez V, Jagadish SVK, Prasad PVV, Zaman-Allah M (2013) Crop science experiments designed to inform crop modelling. Agric For Meteorol 170:8–18

Ding Y, Wang W, Zhuang Q, Luo Y (2020) Adaptation of paddy rice in China to climate change: the effects of shifting sowing date on yield and irrigation water requirement. Agric Water Manag 228:105890

Endo M, Tsuchiya T, Hamada K, Kawamura S, Yano K, Ohshima M, Higashitani A, Watanabe M, Kawagishi-Kobayashi M (2009) High temperatures cause male sterility in rice plants with transcriptional alterations during pollen development. Plant Cell Physiol 50(11):1911–1922

Ewert F (2012) Opportunities in climate change? Nat Clim Chang 2:153–154

Fahad S, Noor M, Adnan M, Khan MA, Rahman I, Alam M, Khan IA, Ullah H, Mian IA, Hassan S, Saud S, Bakhat HF, Hammad HM, Ahmad S, Nasim W (2019) Abiotic stress and rice grain quality. In: Hasanuzzaman M, Fujita M, Nahar K, Biswas JK (eds) . Woodhead, Advances in rice research for abiotic stress tolerance, pp 571–583

Fang K-T, Li R, Sudjianto A (2006) Design and modeling for computer experiments. Chapman & Hall/CRC

Farrell TC, Fox K, Williams RL, Fukai S, Lewin LG (2003) Avoiding low temperature damage in Australia's rice industry with photoperiod sensitive cultivars. Proceedings of the Australian Agronomy Conference, Australian Society of Agronomy

Fatima Z, Ahmad M, Hussain M, Abbas G, Ul-Allah S, Ahmad S, Ahmed N, Ali MA, Sarwar G, Ehsan ul Haque, Iqbal P, Hussain S (2020) The fingerprints of climate warming on cereal crops phenology and adaptation options. Sci Rep 10:18013

Hina T, Adil SA, Ashfaq M, Ahmad A (2019) Economic impact assessment of climatic change sensitivity in rice-wheat cropping system of Pakistan. Indian J Sci Technol 12:37

Guo Y, Wu W, Du M, Liu X, Wang J, Bryant CR (2019) Modeling climate change impacts on rice growth and yield under global warming of 1.5 and 2.0 C in the Pearl River Delta, China. Atmos 10(10):567–578

Hafeez-ur-Rehman, Nawaz A, Awan MI, Ijaz M, Hussain M, Ahmad S, Farooq M (2019) Direct seeding in rice: problems and prospects. In: Hasanuzzaman M (ed) Agronomic crops, Production technologies, vol 1. Springer Nature Singapore, Singapore, pp 199–222

Hoogenboom G, Jones JW, Wilkens PW, Porter CH, Boote KJ, Hunt LA, Singh U, Lizaso JI, White JW, Uryasev O, Ogoshi R, Koo J, Shelia V, Tsuji GY (2015) Decision Support System for Agrotechnology Transfer (DSSAT). DSSAT Foundation, Prosser 4(7):5. Washington Version (http://dssat.net)

Houghton JT, Ding Y, Griggs DJ, Nouger M, van der Linden PJ, Dai X, Maskell K, Johnson CA (eds.) (2001), IPCC, 2001: Climate Change 2001: The scientific Basis. Contribution of Working Group I to the Third Assessment Report of the Intergovernmental Panel on Climate Change, Cambridge

Howden SM, Soussana JF, Tubiello FN, Chhetri N, Dunlop M, Meinke H (2007) Adapting agriculture to climate change. Proc Natl Acad Sci 104:19691–19696

Hu Y, Fan L, Liu Z, Yu Q, Liang S, Chen S, You L, Wu W, Yang P (2019) Rice production and climate change in Northeast China: evidence of adaptation through land use shifts. Environ Res Lett 14:024014

Hunsaker DJ, Kimball BA, Pinter PJ, Wall GW, LaMorte RL, Adamsen FJ, Leavitt SW, Thompson TL, Matthias AD, Brooks TJ (2000) CO enrichment and soil nitrogen effects on wheat evapotranspiration and water use efficiency. Agric For Meteorol 104:85–105

Hussain M, Ahmad S, Hussain S, Lal R, Ul-Allah S, Nawaz A (2018) Rice in saline soils: physiology, biochemistry, genetics and management. Adv Agron 148:231–287

Hussain S, Huang J, Huang J, Ahmad S, Nanda S, Anwar S, Shakoor A, Zhu C, Zhu L, Cao X, Jin Q, Zhang J (2020) Rice production under climate change: adaptations and mitigating strategies. In: Fahad S, Hasanuzzaman M, Alam M, Ullah H, Saeed M, Ali Khan I, Adnan M (eds) Environment, climate, plant and vegetation growth. Springer, Cham

IPCC (2007) Contribution of Working Group II to the Fourth Assessment Report of the Intergovernmental Panel on Climate Change. In: M. L. Parry, O. F. Canziani and P. J. Palutikof, eds

EhsanUllah, Jabran K, Asghar G, Hussain M, Rafiq M (2012) Effect of nitrogen fertilization and seedling density on fine rice yield in Faisalabad, Pakistan. Soil Environ 31:152–156

Kawasaki J, Herath S (2011) Impact assessment of climate change on rice production in Khon Kaen province, Thailand. Journal of ISSAAS 17:14–28

Kaya C, Higgs D, Kirnak H (2001) The effects of high salinity (NaCl) and supplementary phosphorus and potassium on physiology and nutrition development of spinach. J Plant Physiol 27:3–4

Khan MA, Ahmad S, Raza A (2019a) Integrated weed management for agronomic crops. In: Hasanuzzaman M (ed) Agronomic crops, vol 2. Springer Nature Singapore, Singapore, pp 257–281

Khan SM, Ali S, Nawaz A, Bukhari SAH, Ejaz S, Ahmad S (2019b) Integrated pest and disease management for better agronomic crop production. In: Hasanuzzaman M (ed) Agronomic crops, vol 2. Springer Nature Singapore, Singapore, pp 385–428

Kim J, Shon J, Lee CK, Yang W, Yoon Y, Yang WH, Kim YG, Lee BW (2011) Relationship between grain filling duration and leaf senescence of temperate rice under high temperature. Field Crops Res 122:207–213

Laborte A, Nelson A, Jagadish K, Aunario J, Sparks A, Ye C, Redona ED (2012) Rice feels the heat. Rice Today, July–September 30–31

Latif MA, Islam MR, Ali MY, Saleque MA (2005) Validation of the system of rice intensification (SRI) in Bangladesh. Field Crops Res 93(2–3):281–292

Mahamood R, Meo M, Legates DR, Morrissey ML (2003) The CERES-Rice model based estimates of potential monsoon season rainfed rice productivity in Bangladesh. Prof Geogr 55:259–273

Nasir IR, Rasul F, Ahmad A, Asghar HN, Hoogenboom G (2020) Climate change impacts and adaptations for fine, coarse, and hybrid rice using CERES-Rice. Environ Sci Pollut Res 27(9): 9454–9464

Naz S, Fatima Z, Iqbal P, Khan A, Zakir I, Noreen S, Younis H, Abbas G, Ahmad S (2019) Agronomic crops: types and uses. In: Hasanuzzaman M (ed) Agronomic crops, Production technologies, vol 1. Springer Nature Singapore, Singapore, pp 1–18

Onyeneke R (2021) Does climate change adaptation lead to increased productivity of rice production? Lessons from Ebonyi state, Nigeria. Renew Agric Food Syst 36(1):54–68

Otengdarko P, Kyeibaffour N, Ofori E (2012) Simulating rice yields under climate change scenarios using the CERES-rice model. Afr Crop Sci J 20:401–408

Palanisami K, Kakumanu KR, Nagothu US, Ranganathan CR, Senthilnathan S (2017) Climate change and India's future rice production: evidence from 13 major rice growing states of India. SF J Glob Warm 1(2):1–12

Parry ML, Rosenzweig C, Iglesias A, Livermore M, Fischer G (2004) Effects of climate change on global food production under SRES emissions and socio-economic scenarios. Glob Environ Chang 14:53–67

Peerzada AM, Bukhari SAH, Dawood M, Nawaz A, Ahmad S, Adkins S (2019) Weed management for healthy crop production. In: Hasanuzzaman M (ed) Agronomic crops, vol 2. Springer Nature Singapore, Singapore, pp 225–256

Polthanee A, Promkhambut A (2014) Impact of climate change on rice-based cropping systems and farmers' adaptation strategies in Northeast Thailand. Asian J Crop Sci 6(3):262–272

Rai YK, Ale BB, Alam J (2006) Impact assessment of climate change on paddy yield: a case study of Nepal Agriculture Research Council (NARC), Tarahara, Nepal. J Inst Eng 8(3):147–167

Rasul G, Chaudhry QZ, Mahmood A (2012) Diagnostic analysis of heavy rainfall and temperature events in South Asia. J Advances Atmospheric Sci 22:375–391

Razaq M, Shah FM, Ahmad S, Afzal M (2019) Pest management for agronomic crops. In: Hasanuzzaman M (ed) Agronomic crops, vol 2. Springer Nature Singapore, Singapore, pp 365–384

Rondhi M, Khasan AF, Mori Y, Kondo T (2019) Assessing the role of the perceived impact of climate change on national adaptation policy: the case of rice farming in Indonesia. Land 8:81–92

Rosenzweig C, Hillel D (1995) Potential impacts of climate change on agriculture and world food supply. Consequences 1(2):23–32

Rosenzweig C, Jones JW, Hatfield JL, Ruane AC, Boote KJ, Thorburn P, Antle J, Nelson G, Porter C, Janssen S, Asseng S, Basso B, Ewert F, Wallach D, Baigorria G, Winter JM (2013) Agricultural Model Inter-comparison and Improvement Project (AgMIP): Protocols and pilot studies. Agric For Meteorol 170:166–182

Safdar ME, Ali A, Muhammad S, Sarwar G, Awan TH (2008) Effect of transplanting dates on paddy yield of fine grain rice genotypes. Pak J Bot 40:2403–2411

Sánchez B, Rasmussen A, Porter J (2013) Temperatures and the growth and development of maize and rice: a review. Glob Chang Biol 20(2):408–417

Sarwar N, Ali H, Ahmad S, EhsanUllah, Ahmad S, Mubeen K, Hill JE (2013a) Water wise rice cultivation on calcareous soil with the addition of essential micronutrients. J Anim Plant Sci 23(1):244–250

Sarwar N, Ali H, Maqsood M, EhsanUllah, Shahzad AN, Shahzad M, Mubeen K, Shahid MA, Ahmad S (2013b) Phenological response of rice plants to different micronutrients application under water saving paddy fields on calcareous soil. Turk J Field Crops 18(1):52–57

Saseendran AA, Singh KK, Rathore LS, Singh SV, Sinha SK (2000) Effects of climate change on rice production in the tropical humid climate of Kerala, India. Clim Chang 44:495–514

Semenov MA (2009) Impacts of climate change on wheat in England and Wales. J R Soc Interface 6:343–350

Shabbir G, Khaliq T, Ahmad A et al (2020) Assessing the climate change impacts and adaptation strategies for rice production in Punjab, Pakistan. Environ Sci Pollut Res 27:22568–22578

Shahzad AN, Ahmad S (2019) Tools and techniques for nitrogen management in cereals. In: Hasanuzzaman M (ed) Agronomic crops, vol 2. Springer Nature Singapore, Singapore, pp 111–126

Shrestha S, Deb P, Bui TTT (2016) Adaptation strategies for rice cultivation under climate change in Central Vietnam. Mitig Adapt Strateg Glob Change 21:15–37

Shunji O, Kimuraa A (2007) Impacts of climate changes on the temperature of paddy waters and suitable land for rice cultivation in Japan. Agric For Meteorol 147:196–198

Tariq M, Ahmed M, Iqbal P, Fatima Z, Ahmad S (2020) Crop phenotyping. In: Ahmed M (ed) Systems modeling. Springer Nature Singapore, Singapore, pp 45–60

Tripathi A, Mishra AK (2017) Knowledge and passive adaptation to climate change: an example from Indian farmers. Clim Risk Manag 16:195–207

Vaghefi N, Samsudin MN, Makmom A, Bagheri M (2011) The economic impact of climate change on the rice production in Malaysia. Int J Agric Res 6:67–74

Wasaya A, Yasir TA, Ijaz M, Ahmad S (2019) Tillage effects on agronomic crop production. In: Hasanuzzaman M (ed) Agronomic crops, vol 2. Springer Nature Singapore, Singapore, pp 73–99

Wassmann R, Jagadish K, Heuer S, Ismail A, Redona E, Serraj R, Singh R, Howell G, Pathak S, Sumfleth K (2009) Climate change affecting rice production. Advan Agron 101:59–122

Weerakoon WMW, Mutunayake MMP, Bandara C, Rao AN, Bhandari D, Ladha J (2011) Direct-seeded rice culture in Sri Lanka: lessons from farmers. Field Crops Res 121:53–63

Welch JR, Vincent JR, Auffthammer M, Moya PF, Dobermann A (2010) Rice yields in tropical/subtropical Asia exhibit large but opposing sensitivities to minimum and maximum temperatures. Proc. Natl Acad. Sci. USA

White J, Kimball B, Wall G, Ottman M, Hunt L (2011) Responses of time of anthesis and maturity to sowing dates and infrared warming in spring wheat. Energy Fuels 124:213–222

Xu Y, Ge J, Tian S, Li S, Nguy-Robertson AL, Zhan M, Cao C (2015) Effects of water-saving irrigation practices and drought resistant rice variety on greenhouse gas emissions from a no-till paddy in the central lowlands of China. Sci Total Environ 505:1043–1052

Zahoor SA, Ahmad S, Ahmad A, Wajid A, Khaliq T, Mubeen M, Hussain S, Sami Ul Din M, Amin A, Awais M, Nasim W (2019) Improving water use efficiency in agronomic crop production. In: Hasanuzzaman M (ed) Agronomic crops, vol 2. Springer Nature Singapore, Singapore, pp 13–29

Zhou YL, Chen H, Wang ZG, Shen JF (2003) Effects of planting density, sowing date and N fertilizer application on yield and quality of rice. Acta Agric Shanghai 19:28–32

Zubair L, Nissanka S, Weerakoon WMW, Herath DI, Karunaratne AS, ASM P, Agalawatte MB, Herath RM, Yahiya SZ, BVR P, Vishwanathan J (2015) Climate change impacts on rice farming systems in northwestern Sri Lanka. In: Rosenzweig C, Hillel D (eds) Handbook of climate change and agroecosystems: the Agricultural Model Intercomparison and Improvement Project (AgMIP) Integrated Crop and Economic Assessments, Part 2, ICP series on climate change impacts, adaptation, and mitigation, vol 3. Imperial College Press, London, pp 315–352

Rice Production and Crop Improvement Through Breeding and Biotechnology

30

Ali Hassan, Ahmad Naeem Shahzad, and Muhammad Kamran Qureshi

Abstract

Rice is world leading food crop having basic chromosome number 12. It is a self-pollinated crop. The genus *Oryza* contains 20 rice species. Cultivated rice species such as *O. sativa* and *O. glabirrema* have evolved in geographical areas such as Asia and Africa. Rice is a major source of food for half of the world population. It is produced all over the world with major share from Asia. Similarly, about 85% of the total produced rice is consumed by humans. On commercial scale, rice is divided into three classes based upon grain length, maturity, and aroma. Common breeding method of rice are pure line selection, mass selection, pedigree, and backcross breeding. Moreover, recently hybrid rice technology substantially increased rice production in major rice-producing regions. Additionally, blast resistance, salinity tolerance, and many traits related to grain quality have been improved in rice by the use of biotechnological applications such as in vitro culture and marker-assisted selection (MAS). Similarly, functional genomic studies also helped in the determination of variability for many metabolites and drought tolerance in rice cultivars. Genetic engineering played a significant role in rice improvement. By using gene delivery systems, rice crop is transformed with various agronomically important genes. This chapter aims to cover origin of rice, world picture of rice production, consumption, morphology, and recent advancements in rice crop improvement by using conventional breeding and modern technologies.

A. Hassan · M. K. Qureshi (✉)
Department of Plant Breeding and Genetics, Faculty of Agricultural Sciences & Technology, Bahauddin Zakariya University, Multan, Pakistan

A. N. Shahzad
Department of Agronomy, Faculty of Agricultural Sciences & Technology, Bahauddin Zakariya University, Multan, Pakistan

N. Sarwar et al. (eds.), *Modern Techniques of Rice Crop Production*,
https://doi.org/10.1007/978-981-16-4955-4_30

Keywords

Rice · Productivity · Production · Breeding · Improvement

30.1 Introduction

Rice (*Oryza sativa*) is a member of family Poaceae, and is a leading food crop worldwide. About half of the global population consumes rice as staple food, with major share from Asia. After wheat, rice crop is cultivated on large areas as a cereal crop. Rice meets 20% of per capita energy demands and 13% of protein intake by individuals in the world (Zhang and Wu 1988; Juliano 1994). It is perceived as a semiaquatic grass and grown in different soils and water cultures such as upland, lowland, and flood inclined and rainfed. Mostly, cultivated rice is grown in irrigated and flooded fields impounded by bunds (Li et al. 2003; Wei et al. 2018; Ma et al. 2020; Beighley 2010).

30.2 World Rice Picture

30.2.1 Production

Rice is a leading crop, along with wheat and corn, and hence constitutes a major food source for half of the world's population. In some regions of the world, it is a strategic commodity as it contributes in economic growth, political stability, as well as food security of many countries (FAO 2014). Rice crop is produced all over the world. Major share in rice production was from Asia during 1994 to 2019 as it holds 90.6% rice production globally. Second largest rice-producing region is America with contribution of 5.2%, followed by Africa 3.5%, Oceania 0.1%, and Europe 0.6% (Fig. 30.1). About 80% of rice global production comes from eight different countries which are concentrated in Asia such as China, India, Indonesia, Bangladesh, Vietnam, Thailand, Myanmar, and the Philippines. Among these countries, China leads in rice production with a share of 209.6 metric tons (MT), whereas world's second largest rice producer is India with production of 177.65 MT (Fig. 30.2).

Globally, rice consumption is 480 million MT (MMT) per year, and nearly 408 MMT (85%) of globally produced rice is consumed by humans. Countries such as China and India consume 50% of the world's total produced rice. Asia is the top most continent with highest rice consumption on daily per capita basis. Countries like Bangladesh, Myanmar, Lao People's Democratic Republic, Cambodia, Thailand, Vietnam, Indonesia, and the Philippines reported intake of more than 300 g rice-based food on a daily basis (FAO 2013).

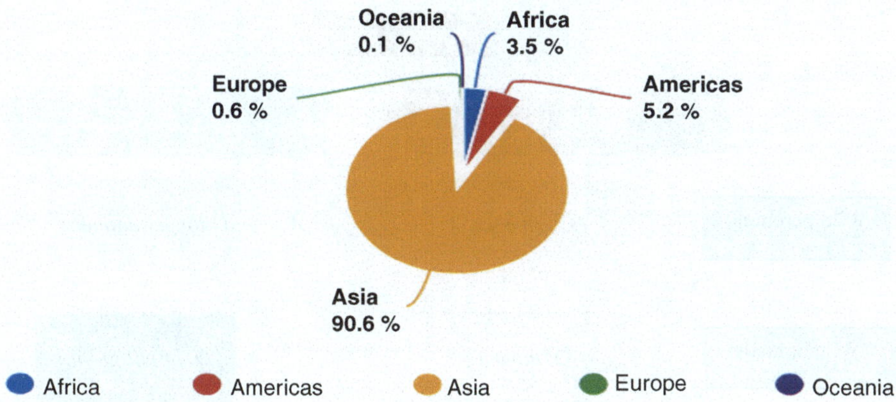

Fig. 30.1 Continental share of rice-producing regions in percentage (**Source**: FAO 2019)

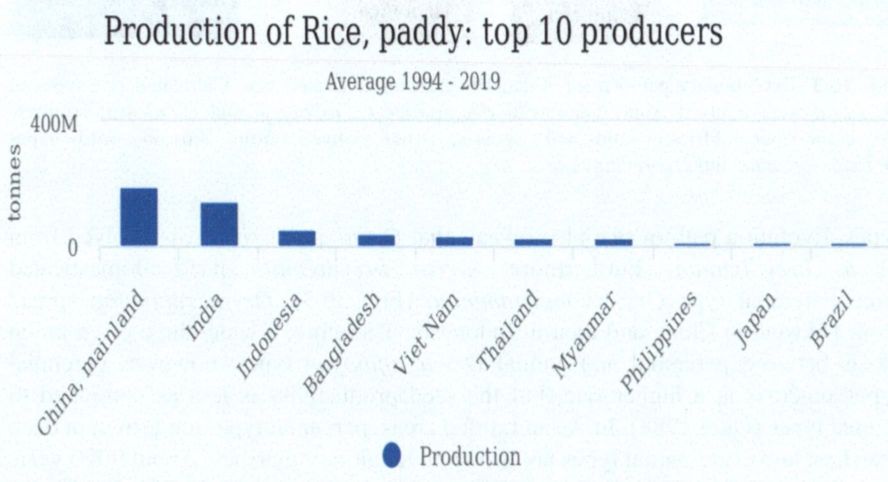

Fig. 30.2 Countries with major share in rice production (**Source**: FAO 2019)

30.3 Origin and Domestication of Cultivated Rice

The genus *Oryza* contains 20 species with a basic chromosome number of 12. Genus include both diploid and tetraploid species with six genomes such as A, B, C, D, E, and F. Cultivated species of rice *O. sativa* ($2n = 2x = 24$) has AA genome, while *O. glaberrima* ($2n = 2x = 24$) has AgAg genome. Six species of *Oryza* are annuals, while remaining species are perennial. The common rice, *Oryza sativa*, and the African rice, *Oryza glaberrima*, are considered to be an example of parallel evolution in crop plants. *Oryza rufipogon* is an Asian wild rice species having a wide range of variations from annual to perennial types and is considered as ancestor of *Oryza sativa*. Similarly, *Oryza sativa* was domesticated from *Oryza nivara* annual

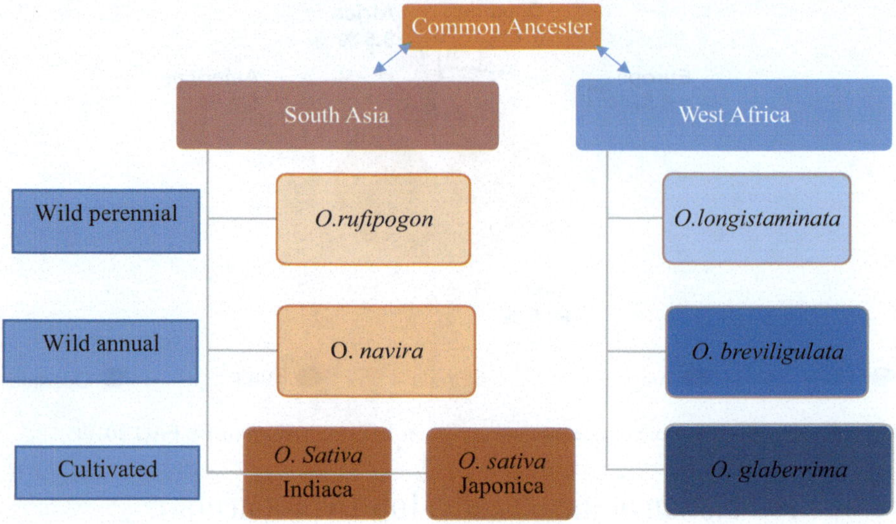

Fig. 30.3 Evolutionary pathway of Asian and African cultivated rice. Cultivated rice types of *O. sativa* were evolved from Asian wild rice species *O. rufipogon* and *O. nivara;* similarly *O. glaberrima*, African cultivated species, was evolved from African wild types *O. longistaminata* and *O. breviligulata*

types. Evolution path of rice also reveals that *Oryza glaberrima* was evolved from *Oryza breviligulata*. Furthermore, *Oryza breviligulata* itself domesticated from perennial type *Oryza longistaminata* (Fig. 30.3). *Oryza rufipogon* spread from Pakistan to China and then in Indonesia. Therefore, a wide range of variation exists between perennial and annual *Oryza rufipogon* types; however, perennial types outcross at a higher rate, but the seed productivity is less as compared to annual types (Okra 1988). In Asian rainfed areas, perennial types are grown in deep marshes; however, annual types are grown in temporary marshes. About 9000 years ago, domestication of wild rice probably started, where evolution in rice is also associated with the environmental and abiotic stress such as alternating periods of prolong drought and variations in temperatures during 10,000–15,000 years ago enhancing the development of annuals types under different mutative evolutions and natural selection in east India, north southeast Asia, and south west China (Whyte 1972).

30.3.1 Domestication in Africa

Oryza glaberrima is being cultivated from 3000 years in West Africa (Linares 2002). African cultivated rice, as compared with Asian rice, has some abominable characteristics, such as a seed shattering, hard grain, and less yields. Currently, Asian species are replacing *Oryza glaberrima* in West Africa which were introduced into the Africa by the Portuguese as early in the mid-sixteenth century (Sweeney and McCouch 2007; Van et al. 2016). African rice was subjected to population genomic analyses by utilizing 20 *Oryza glaberrima* and 94 *Oryza barthii* accessions which reveal the developmental history of artificial selected and domesticated rice in Africa providing evidence backed by the hypothesis that *Oryza glaberrima* was domesticated from *Oryza barthii* subgroup and individually domesticated from *Oryza sativa* (Wang et al. 2014).

30.3.2 Domestication in Asia

In Asia, rice domestication started individually and concurrently at several sites that extend from the plains below the eastern foothills of the Himalayas in India through Upper Myanmar, northern Thailand, Laos, and Vietnam to southwest or south China (Chang 1976; Ramiah 1937; Roschevitz 1931). Initially, rice was cultivated by direct seed sowing and also without standing water. Later on soil puddling and transplanting of rice seedlings was introduced in China for efficient rice production. With the advent of this system, rice became truly domesticated. In Southeast Asia, rice crop was grown in dry-land conditions in uplands and in vast river deltas. *Oryza sativa* cultivation begun in Indian subcontinent which is indeed an oldest rice cultivation region as grain samples were collected from Mohanjo Daro region which date back to about 2500 BC (Andrus and Mohammed 1958).

The species of *Oryza sativa* has evolved in different types or eco-graphic races such as Indica, Japonica, Tropical Japonica, and Javanica.

(a) *Indica*
This is a tropical type of *O. sativa* and mainly characterized as tall plants with weak stems, long leaves, vulnerable to low temperature and photoperiod, slender grains, seed remain dormant for long period.

(b) *Japonica*
It is a temperate type with short leaves, short stems, moderate tillering, resistance to low temperature, short and round grains, low amylose contents, usually sticky nature after cooking.

(c) *Tropical Japonica*
It is a tropical type having unique germplasm pool, mostly used in upland condition, long grain, photoperiod insensitive, dry and fluffy, and good milling yield.

(d) *Javanica*
This type is characterized with broad leaves, low tillers, thick and tall stems, long panicles, resistance to shattering, and large and bold grains.

30.4 Rice Cultures

Rice crop is basically grown in different types of culture such as irrigated or flood rice, rainfed or lowland rice, and deepwater or floating rice.

30.4.1 Irrigated or Flood Type Rice Culture

In irrigated type culture, rice crop is grown in standing water impounded by bunds or levees. In this rice culture, water is required in much higher quantities, where more than 15 cm water depth is needed.

30.4.2 Rainfed or Lowland Rice Culture

Rice is grown in standing water impounded by bunds and levees from natural rainfalls. In this culture, supplementary irrigation is not mandatory. Efficiency of this rice culture depends upon monsoon season. In this culture, water depth is usually from 50 to 100 cm (Mackill et al. 2010).

30.4.3 Deepwater or Floating Rice Culture

In this rice culture, crop is cultivated in low-lying areas that may come to drench at the depth of 1–5 m during rainy season. In floating rice, stems of the plant draw out in water as the water level raises and leaves of the plant float on the water surface (Dobermann and Fairhurst 2000).

30.4.4 Upland or Dry-Land Rice Culture

Rice crop is grown with natural rainfall, mostly in hilly or newly cleared areas where topography does not permit impounding water for irrigation.

Among all rice culture, irrigated rice culture is the principal method for growing rice. This culture had occupied over half of the world's total rice-growing areas. Globally, about one-fourth of rice is cultivated under rainfed rice culture. This culture is applicable in regions with rainy climate such as in Asia and Africa. Moreover, world's 11% rice-growing areas are under deepwater rice culture, whereas dry-land rice culture is strictly rainfed and crop is grown without supplementary irrigations (Arraudeau 1995).

30.5 Commercial Rice Classes

On the basis of commercial or marketing scale, rice is divided into following three classes.

30.5.1 Grain Length

Rice cultivars are principally characterized by grain length. Grain length in rice is categorized as short grain (5.50 mm), medium grain (5.51–6.60 mm), long grain (6.61–7.50 mm) and very long grain (>7.50 mm). Japonica type varieties usually have short grains, short stature, lodge resistance ability, which make them more receptive for fertilizer applications. The varieties with long grains are Indica types, having taller and more fragile stems which results in lodging in case of fertilizer applications (Cruz and Khush 2000; Acquaah 2009).

30.5.2 Maturity

Rice is classified on the basis of earliness as early maturity, mid-season maturity, and late maturity. Early maturing rice varieties take about 120 to 129 days to mature; whereas, mid-season rice varieties need 130–139 days, and late maturing rice varieties somehow take 140 days or more to mature (Acquaah 2009).

30.5.3 Aromatic Rice

Rice may be aromatic or nonaromatic. Basmati and jasmine are the two most aromatic types of rice. Basmati rice is characterized with extraordinary aroma; grains extend double in size after cooking and remain nonsticky. It is usually grown mainly in central Punjab of Pakistan and Northern India (Cruz and Khush 2000; Acquaah 2009). Whereas Jasmine rice type is cultivated in Thailand. Jasmine rice is characterized as soft, moist, and sticky after cooking which is due to starch in grains (Singh et al. 2000).

30.6 Development and Growth Stages in Rice

Rice crop is divided into three growth phases such as vegetative phase, reproductive phase, and ripening phase (Fig. 30.4).

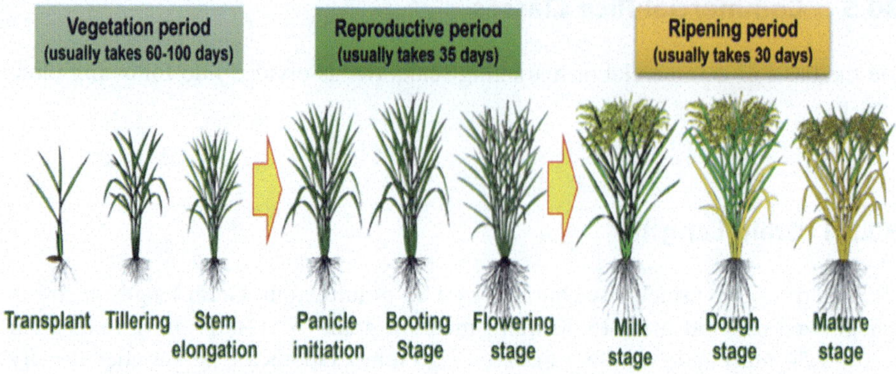

Fig. 30.4 Growth stages of rice after transplantation of rice seedlings. Reproductive stage starts with the panicle initiation and ends at flowering stage. Finally, ripening stage starts after the termination of flowering and ends at maturity of crop (**Source**: Mosleh et al. 2015)

30.6.1 Vegetative Phase

30.6.1.1 Seedling

Germination and seedling growth begin when seed dormancy is broken and seed absorbs adequate water and become soft as well as exposed to a temperature between 10 and 40 °C. From embryo, coleoptile is the first to emerge, along with the root development. Seminal roots develop from coleorhiza followed by coleoptile emergence (Moldenhauer and Slaton 2001).

30.6.1.2 Stem and Tillering

Stem development in rice comprised of a progression of nodes and internodes. Much variation exists in rice varieties for intermodal length; however, this trait is influenced by the environment. Usually, rice plant stem bears 13–16 nodes separated by internodes. In some deepwater rice varieties, internodal length can increase up to 30 cm under increase in water level. At node, leaf blade is clinched by leaf sheath that encloses stem. There is a combination of claw-like projections at the point where leaf blade and the leaf sheath meet, and this junction is known as auricle, which surrounds the stem. Auricles are covered by boorish hairs. Along with auricles, upright delicate membrane which is called ligule is also present. After the emergence of 3/4th leaf, tillering initiates, which ends at panicle initiation. The first tiller in rice appears when the seedling is at 5 leaf stage. The first tiller is established at the base of plant among main stem and second leaf. Main stem tillers are known as primary tillers from which secondary tillers develop followed by tertiary tillers. In rice, tillers are produced within synchronous manner. Tillers remain attached to plant at initial stages of development, whereas at later stages, tillers become independent by producing their roots (Pawar et al. 2016).

30.6.1.3 Panicle and Spikelets

Primary structure of rice panicle contains base, axis, underlying branches, pedicel, rudimentary glumes, and spikelets. The panicle axis development starts from panicle base to the apex; it has 8–10 nodes after the interval of 2–4 cm. Primary branches develop followed by secondary branches. From the nodes of these branches, pedicels develop; and on pedicels, spikelets are positioned and each spikelet in rice hold single flower. Lemma and palea enclosed the flower, which may be either awned or awnless depending on varietal character. Moreover, rice flower is a complete flower having male and female parts in a single spikelet (Mohapatra et al. 2011).

30.6.2 Reproductive Growth Phase

Reproductive phase of rice is recognized by panicle initiation, booting, and anthesis.

30.6.2.1 Panicle Initiation

This stage is initiated before heading with the development of panicle. Spikelet anthesis starts with panicle heading. Heading in rice crop takes 2 weeks to complete because of variation in tillers development (Mosleh et al. 2015).

30.6.2.2 Booting

Subsequently, after the initiation of panicle, spikelets differentiate and panicles extend upward inside the flag leaf sheath. The increase in size of panicle extends upward inside the flag leaf sheath. The increase in size of panicle and its upward growth in the flag leaf sheaths are recognizable due to swelling in the sheath (Moldenhauer and Slaton 2001).

30.6.2.3 Anthesis

In tropical environments, anthesis in rice occurs between 10 and 13 h followed by fertilization process which compete in 1–2.5 h. It takes 4–7 days for all spikelets to complete anthesis within the same panicle (Moldenhauer and Slaton 2001).

30.6.3 Ripening Phase

Ripening stage in rice crop begun after fertilization, which is divided into different stages, such as milky stage in which contents of developing caryopsis turn milky, dough stage in which milky portion turns into hard dough from jelly, yellow-ripe stage, and maturity. Maturity is primarily based upon the grain texture and grain color. Usually, it takes 15–40 days to complete the ripening stage; however, this stage may vary among different genotypes (Vergara 1991).

30.7 Flowering and Pollination

Rice inflorescence is of determinate type, which is known as a panicle having primary and secondary floral branches. Spikelet is a floral unit of rice present on the panicle. Spikelet has a pair of glumes containing single floret. Inflorescence of rice is perfect type, as it contains both pistil and stamens (Yoshida and Nagato 2011). In rice flowers, reproductive parts are confined by glume, lemma, and palea. Pistil comprises of stigma, style, and ovary. Stigma is feathery in nature. There are six stamens, and each stamen contains two-celled filament and anther, respectively (Ikeda et al. 2004). The emergence of rice panicle from flag leaf is called heading. Normally, flowering stage occurs right after the heading stage, which is the result of different events such as opening and closing of spikelets, and generally it continues up to 1–2.5 h. Opening of lemma and palea, elongation of filaments, anther exertion from the glumes, and dehisce are the events which occur during flowering and end with the closure of glumes. Usually, anther dehiscence takes place after the opening of lemma and palea (Ikeda et al. 2004). Pollen grains of rice are viable for only 5 min, whereas stigma of rice can remain receptive for 3–7 days for pollen grains. Environmental factors such as soil fertility, light, temperature, and humidity are much important in flower development and seed setting (Fehr 1987). Rice is a self-pollinated crop. However, there are 5% chances of outcrossing which generally depends upon rice variety and environment. Usually, pollen shedding in rice starts with the opening of flower along with spikelets blooming. In rice, spikelets blooming starts from the top of panicle and proceeds downward (Singh and Kumar 2009). In order to make crossing in rice, it is important to know when to make emasculation along with maximum availability of viable pollens. Therefore, plant breeders should be aware about the blooming time of the spikelets.

30.8 Breeding Methods in Rice

30.8.1 Pure Line Selection

In self-pollinating species like rice, certain landraces can be considered as a combination of pure lines, including some heterozygous individuals which were developed from a low frequency of cross-fertilization. In such populations, choosing single plants and developing inbreed progenies perpetually bring about certain lines that can perform better as compared to their original landraces under accustomed conditions. Nonetheless, such dominance includes some significant downfalls, because pure lines are ordinarily less steady than diverse populations. Therefore, pure lines cannot stand with most of the biotic and abiotic stresses; moreover, there is less scope for long-term adaptations in pure lines due to monogenic inheritance of different traits. Pure lines are used as parents in hybrid development (Priyadarshan 2019).

30.8.2 Mass Selection

Mass selection is the oldest methods of crop improvement by which local and existing varieties of crop are improved for particular traits. In rice, plants of same height, maturity, and grain types are bulked. In the next growing season, selection and evaluation of few hundred to few thousand plants is made from bulk populations on phenotypic basis for yield and morphological traits, and superior population is selected and released for general cultivation. In mass selection, varieties having considerable genetic variations can be improved. However, this method of crop improvement can only be used in specific environments with the expression of specific traits. Therefore, this constrain limits its application in offseason crop improvement (Acquaah 2009).

30.8.3 Pedigree Selection

This technique was described by H.H. Lowe in 1927. Pedigree selection is a broadly utilized strategy for breeding self-pollinated crops. A basic characteristic of pedigree method of selection is the hybridization which put this method at superiority as compared to pure line and mass selection. This technique is confined with the crossing of parents and bringing out segregating populations with further repetition of self-pollination and selection cycles. This cycle continues until the desired combination of both parents is obtained in new line. It is an efficient method for breeding of monogenically controlled traits, like disease resistance, plant architecture, color, and shape. This method of selection is used more by researches because it allows the creation of superior varieties by putting together good characteristics, in the single plant. In pedigree method, record is maintained of segregating populations; therefore, it makes possible to have the genealogy of each variety or cultivar developed by this technique (Breseghello and Alexandre 2013).

30.8.4 Backcross

Backcross is normally considered as a procedure suitable for the transfer of one or more genes in a single cultivar mainly for biotic stress resistance. By using backcross method of crop improvement, plant breeders could make improvement in already existing elite or adapted variety by keeping it as a recurrent parent. Donor parent having genes for specific traits which were missing in the elite variety is for the introduction of one of more genes. Backcrossing technique could be utilized to make crosses followed by the generation of segregating populations. These segregating populations at the end will contain special characteristic in high frequency as compared to adapted or preexisting elite recurrent parent (Singh et al. 2012).

30.8.5 Hybrid Rice

In 1930, successful development of maize hybrid laid the foundation of crop improvement by using hybrid technology. Therefore, later on hybrids of many self- and cross-pollinated crops were developed and commercially exploited. In developing countries, maintaining of rice self-sufficiency, its stability, and price are the most important objectives (Hossain 1995). With the increase in population, most of the Asian countries have made remarkable achievement in terms of sustainable food production to meet the food needs.

During 1960 to 1975 in China, rice yield per hectare was increased up to 2–3.5 tons per hectare by the utilization of semi dwarf rice cultivars. However, in 1970, hybrid rice was introduced in China which was fully commercialized in 1983. This rice hybrid technology brings a revolutionary change in rice production. By using hybrid rice cultivars per hectare, yield of rice was increased up to 6 tons as compared to nonhybrid cultivars. However, hybrid rice cultivars were developed mostly in the regions having more resources for rice research (Lin 1992).

30.8.5.1 Methods for Hybrid Rice Development

Rice is a self-pollinated crop. Hybrid development in any crop needs emasculation and pollination procedures to produce F_1 progeny. In case of self-pollinated crops such as rice, it is much difficult to attempt the emasculation and pollination on a large scale; moreover, it is time-consuming and labor-intensive. Therefore, male sterility system was introduced in rice to bypass the process of hand emasculation. Rice plant with male sterile panicle bears nonviable pollens. Therefore, self-pollination in such panicles does not occur and such panicles are utilized as female parent in hybrid development.

(a) *Cytoplasmic Male Sterile (CMS) System*

Among all the sterility methods in rice, CMS system is proven effective for hybrid rice production. This system of hybrid rice development involves three lines as A line (CMS), B line (Normal), and R line (Restorer). This system is also known as ABR system. The hybrid which is produced by this ABR system is also called three-line hybrid. This system has widely been used in hybrid rice production in China and other Asian countries (Li et al. 2007). A CMS line and maintainer line are crossed with each other (A × B) for multiplication of CMS seed. This process is usually carried out by hand or by simple outcrossing depending upon the seed quantity to be produced. These two lines, A and B lines, have same morphology but with differences in fertility, as A line is sterile and B line is male fertile line. R line is fertility restore line. In R line, there are dominant fertility restorer genes. When (A × B) CMS is crossed with R line, it restores the fertility in F_1 as (A × B) R. Schematic diagram of CMS is given in Fig. 30.5.

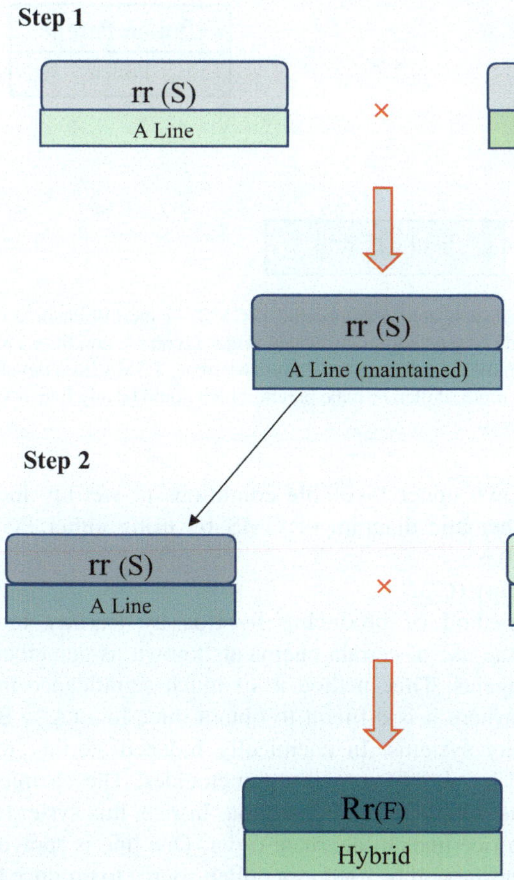

Fig. 30.5 ABR line system. In the first step, A line (sterile) with rr (non-restorer genes) is crossed to B line with normal cytoplasm instead of sterile to maintain A line. B line also has rr (non-restorer genes). In the second step maintained, A line is crossed with R line (restorer) with RR (fertility restorer genes) to restore the fertility in hybrid seed. Thus, hybrid between the A line and R line is fertile which upon selfing produces fertile seed

(b) *Genic Male Sterility*

This sterility system for hybrid production in rice is also known as two-line hybrid system. Basically, in genic male sterility system, expression of sterility is determined by the environmental factors. These factors involve photoperiod and temperature. Therefore, genic male sterility is also known as photoperiod male sterility (PGMS) and thermosensitive male sterility (TGMS). PGMS lines remain male sterile in long days and TGMS lines remain male sterile at temperature above 30 °C. Hybrid development by using such lines is much simple as it does not require any maintainer line as in CMS system. Simple selfing procedures are followed in order to produce hybrid seeds when both PGMS

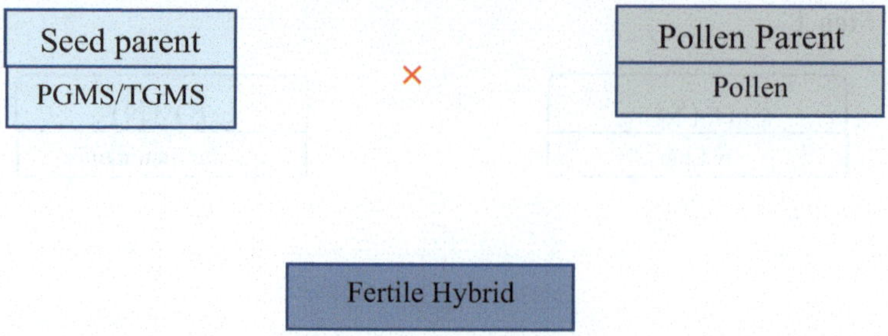

Fig. 30.6 GMS or two- line sterility system. Photoperiod (PGMS) or thermosensitive (TGMS) male sterile lines are used as female or seed parent cultivated under favorable conditions to induce male sterility. PGMS lines develop male sterility in long days; whereas, TGMS lines develop male sterility at temperature of 30 °C or more. Pollen or male parent which could be any line or variety is used to cross PGMS or TGMS line

and TGMS lines are grown under favorable conditions of sterility induction (Zhang et al. 2013). Schematic diagram of GMS by using either PGMS or TGMS is given in Fig. 30.6.

(c) *Chemically Induced Sterility (CIM)*

This is a nongenetic method of producing hybrids by sterility induction mechanisms along with the use of certain chemicals known as gametocides or chemical hybridization agents. This method is of much significance in crops having bisexual flowers where it is difficult to obtain three-line (A × B) R or two-line (A × B) sterility systems. In chemically induced sterility method, hybrids can be produced by simply spraying gametocides. The chemical will kill pollen grains of a line without effecting stigma. In rice, this system can be operated by planting two rice lines in alternate order. One line is sprayed with gametocides and the other line will be used as a pollen source to produce hybrid.

30.8.5.2 Prospects of Hybrid Rice

This technology of hybrid rice development is now introduced in more than 40 countries including China. In most of hybrid rice-producing countries, initially parents were supplied by the IRRI. Later, different countries developed their own hybrid rice breeding programs. In India, hybrid rice technology was adapted in 1989 after China. Similarly, USA adopted this hybrid technology in 2000. Significant progress has been made in USA in hybrid rice breeding. Many hybrid rice cultivars have been developed which include weedicide resistance, late maturity, and new plant types. Overall yield of hybrid rice is 16–39% higher as compared to simple inbreed lines (Rout et al. 2020).

30.9 Rice Biotechnology

Expression of a phenotype depends upon the genetic constitution of genotype and its environment, and this phenomenon is also known as genotype by environmental interaction G × E. In agriculture crop system, agronomy of crops is different under different environments. Therefore, genotype of a crop plant is considered as the deciding factor for crop development and productivity. Crop improvement and development of desired crop varieties cannot be achieved by using simple or traditional techniques. However, use of biotechnological tools made it easier to engineer desired genes into plant species. Thus, various biotechnological resources and planning have been undertaken with the aim of enhanced rice production. Broadly, this included in vitro culture techniques, marker-assisted selection, functional genomics, and genetic engineering of plants (Biswal et al. 2017).

30.9.1 In Vitro Culture

30.9.1.1 Somaclonal Variations

Morphological and genetic variations which are present in clonally inseminated plants are called somaclonal variations. A wide range of genetic variation is present in undifferentiated cells, protoplasts, callus, and tissues of regenerated plants. Therefore, the term somaclonal variation is basically used for in vitro culture in which genetic variability is present among all types of cells and organs of plants (Lesteri 2006). Thus, such somaclonal variation could be utilized to genetically manipulate the polygenic traits in crop species (Jain 2001). In rice, attempts were made for the development of somaclones by irradiation process for submergence tolerance. Similarly, variations in some regenerated plants from somaclones of rice were studied for grain and quality characters (Le et al. 1989). Moreover, variations for blast disease resistance salt tolerance in rice genotypes were also accessed by using somaclonal culture, and remarkable variation for salt tolerance was detected in regenerated plants (Mandal et al. 1999).

30.9.1.2 Anther Culture

Anther culture is an in vitro culture technique in which microspores are utilized to regenerate the whole plant. Callus produced from microspores generally contain a haploid genome set after meiosis in the parental plant, which can aid the characterization of genetic factors controlling the cell viability of male gametes (Niizeki and Oono 1968). Generally, variation which is developed by culturing of haploid somatic organs is greater than that of radiations. Fully homozygous line could be developed rapidly by using anther culture technique. Therefore, this technique provides an efficient alternative to the conventional inbreed line development. This technique has been used efficiently as supplementary breeding tool in Japonica rice varieties; however, due to inherent recalcitrant in genetic background, this technique was not established in Indica rice varieties. Success of anther culture is dependent on factors such as genotype and physiological status of donor parents, development of pollen, anther wall, culture media composition, and anther pretreatments (Zhu et al. 1998). In rice, anther culture has been used in the selection of superior DHLs for the improvement of grain quality characters (Xa and Lang 2011). Similarly, blast resistance gene was also identified in rice by using anther culture along with some RFLP markers (Araujo et al. 2010). Moreover, by using anther culture, many salinity-tolerant lines of rice were developed in Bangladesh, Myanmar, Egypt, Dominican Republic, the Philippines, Thailand, and Mexico (Senadhira et al. 2002).

30.9.1.3 Protoplast Culture

Cell fusion or protoplast fusion is an in vitro technique which is used by plant breeders to make hybridization where normal hybridization is not possible. It could be used to overcome the obstacles of fertilization associated with interspecific crossing. First successful application of protoplast fusion was carried out in 1975. In rice, this technique has been used in Japonica cultivars. Similarly, seed progeny derived from protoplasts was also evaluated for morphological traits in rice, and it was reported that such progenies reflected delayed flowering, shorter flag leaves, and small panicles, and it was recommended that proto-clones derived from protoplast culture with acceptable variation could be selected and utilized in crop improvement (Ogura et al. 1987). A reproducible plant regeneration system has been developed for protoplasts from embryogenic cell suspension cultures of the commercial Asian long-grain Javanica rice. Similarly, hybridization between *Oryza sativa* and *P. coractata* was made by protoplast culture for salt tolerance (Tang et al. 2000). Many useful characters have been introgressed into rice from wild relatives.

30.9.2 Molecular Markers

In rice breeding, molecular markers had provided much potential in case of biotic and abiotic stress breeding. No doubt there are more advantages of molecular markers as compared to morphological markers. Molecular markers can be used for screening of various traits without the limitation of plant growth stages. This technique is routinely used in the screening of difficult and time-consuming traits. Moreover, a grater edge of molecule markers is that progeny testing is not required for the determination of heterozygous condition. By utilizing different techniques like Southern blotting and PCR-based molecular markers such as RAPD, RFLPs, AFLPs, SSRs, and SNPs have been developed. Among PCR-based markers, SSRs are preferred for tagging various genes of interest, because such markers have higher polymorphism, versatility, higher reproducibility, and amenability (Gao et al. 2016). In rice, about 20,000 SSR markers have been developed, and their locations on chromosome and polymorphism have also been determined so far. Due to such reasons, marker-assisted selection is mostly based on SSRs for rice crop improvement. Besides SSRs, recently another class of markers like SNPs is gaining much importance in rice breeding. It is a variation in DNA sequence which occur by a single nucleotide such as A, T, G, or C change in genome between the member of a species and paired chromosome in an individual (Lema 2018).

30.9.2.1 Marker-Assisted Breeding

In rice breeding over the past decade, scientists have developed and applied MAS analysis techniques. Such approaches promote identification of new germplasm and replacing old traditional method of varietal development. Therefore, MAS is an efficient, reliable, and effective method as compared to morphological based selection. Furthermore, it is a cost-effective way of selection as well as shortens the time of variety development. MAS method has a much higher crop improvement potential as compared to genetic engineering in plants. In rice, pyramiding of genes by marker-assisted breeding has been done; this technique is now extensively being used in Asian countries like India, China, the Philippines, Thailand, and Indonesia. Genes for blast resistance were introduced in rice through marker assisted backcrossing (Singh et al. 2012). Similarly, by using marker-assisted backcrossing, resistance for blast and bacterial blight is also introduced (Fu et al. 2012). Moreover, submergence tolerance gene and QTL for salinity tolerance have also been introgressed into rice by the use of marker-assisted backcrossing. Some examples of genes by MAS for blast tolerance in rice are mentioned in Table 30.1.

30.9.3 Functional Genomics

Globally, in rice functional genomics, most of the rapid advances have been made in recent years which could be summarized in categories as (1) development of technological platforms for gene identification; (2) isolation of functional genes; and (3) functional genomics and biological analysis of agronomic traits (Li et al. 2018). Main aim of functional genomics study in crop species is to know the genome function by analyzing the information preserved in the nucleotide sequences, genes, and regulatory elements. Challenges which need to be addressed for sustainable rice production are: (a) insect pests and diseases; (b) uncontrolled pesticide applications; (c) overuse of fertilizers; (d) drought; and e) extensive cultivation in marginal lands. These challenges could be addressed by the utilization of combination of approaches and techniques which are based on recent advancements in genomic research. Moreover, efforts have been made on rice germplasm for the identification of genes for resistance to insect pest and diseases,

Table 30.1 Important genes tagged for blast resistance in rice by using marker-assisted selection (Tanweer et al. 2015)

No.	Gene	Trait	Marker
1	Pi1	Blast resistance	RFLP
2	Pi2	Blast resistance	SSR
3	Pi54	Blast resistance	SSR
4	Pi9 (t)	Blast resistance	PB8
5	Pi-ta	Blast resistance	Gene specific
6	Pi1, Piz-5, Pi2, Pita	Blast resistance	SSR, ISSR, and RFLP
7	Pid1, Pib, and Pita	Blast resistance	SSR

drought resistance, nitrogen and phosphorus use efficiency, quality, and yield (Jiang et al. 2013; Fujino et al. 2008; Huang et al. 2016).

30.9.4 Metabolomics

Plants can generate up to one million metabolites (Dixon and Strack 2003). Recently, with the improvement of metabolomics, especially development in metabolic profiling, mass spectra, and magnetic imaging, the metabolomics field is contributing significantly in crop improvement (Saito 2013). Identification of genes, metabolic pathway analysis, and genetic investigation of natural diversity through coordination with other omics advances like genome-wide association study (GWAS) has been utilized to distinguish a few many loci that regulate natural diversity in metabolite substances. More than 160 novel metabolites, together with flavonoids, nutrients, and terpenes, were reported on the basis of genome-wide association study (Chen et al. 2014).

30.9.5 Proteomics

Proteomics research concentrates around the identification of proteins, their quantification, activity and stability determination, confinement, and functions which play fundamental role in cell signaling and other pathways (Wilkins et al. 1996; Lodha et al. 2013). Recently, incredible improvements have been accomplished in rice proteomics that provide comprehensive preview on the comprehension of yield improvement, stress resistance, organelle, secretome, and protein posttranslational modification (Tappiban et al. 2021). Proteomics studies in rice have been carried out generally by utilizing gel-based (1DE, 2DE, and 2DIGE) and gel-free (LC-MS/MS or MudPIT) approaches. Moreover, recently isobaric tags for relative and absolute quantitation (iTRAQ) technique for protein was also performed to study proteomics under high temperature in various rice cultivars, and results suggested that heat resistance rice cultivars under high temperature produce heat-shock proteins along with expansins and lipids transfer proteins (Mu et al. 2017). Additionally, PEG-induced drought stress responsiveness in resistance cultivars was also studied, and it was observed that during drought stress, roots of resistant plants exhibited differentially expressed genes that appear to be involved in stress regulations (Agarwal et al. 2016).

30.10 Genetic Engineering in Rice Breeding/Transgenic Rice

Genetically modified (GM) technology is one of the most important tools which promises to recast rice production scheme. Prominent advantage of this technology is to assemble useful genes from non-rice gene pool into the rice. In early 1980s with the advent of tissue culture technique and plant regeneration strategies, initial studies to develop GM were conducted. In japonica rice, gene transformation is performed routinely in many laboratories; however, this system in indica rice is more complex (Yookongkaew et al. 2007). To date, many copies of a gene(s) are inserted, but the expression pattern of such introduced genes is different among individuals. Thus, in order to perform selection for desirable transformants to define the expression pattern of transformed genes, there is a need to develop large number of transgenic plants.

Two most important and efficient methods of gene delivery are particle bombardment and *Agrobacterium tumefaciens*-mediated transformation. Most of the *Agrobacterium* transformations were carried out for dicot species. Monocot species including important cereals like wheat, rice, and maize may not be transformed for considerable time. The reasons for less ability of transformation of *Agrobacterium* into cereal crops such as rice, wheat, and maize were bacterial contamination; inconsistency of results was overcome by the use of particle bombardment technique. By using particle bombardment technique, transgenic plants were generated with genes for blast and brown plant hopper (Tang et al. 1999).

For irrigated tropical lowlands, IRRI had released IR-8 first transgenic rice high yielding cultivar. Since then, tremendous efforts are being continually made to improve rice plant for higher yield and desirable features; however, more innovations are required to breakdown yield barrier (Basu et al. 2014).

30.10.1 Controversies on Transgenic Rice

Transgenic rice no doubt holds much significance in crop improvement against insect pest and diseases. However, a few concerns with regards to GM rice have been raised. A portion of these issues include natural environmental, economic, and ecological issues. On a natural and biological point of view, the development of modified insect- and weed-resistant rice cultivar raises an issue on the outcomes of GM rice as gene flow from GM to non-GM crop or weedy relatives which can cause the development of supper weeds or insects (Bawa and Anilakumar 2013).

30.11 Conclusion

Rice is a most important cereal crop grown all over the world since its origin. Many improvements such as biotic and abiotic stress tolerance have been developed in rice crop by using conventional breeding techniques. Rice production is uplifted by the use to hybrid technology in rice crop. Hybrid rice played a significant role in food security, and up to now, this technology has been adopted in more than 40 countries. However, to fulfill the future rice demand, there is a need to upgrade hybrid technology. This objective could be achieved by enhancing seed production capability of parents by the use of both conventional and molecular tools. There is a wide scope for biotechnology in crop improvement. Apart from the understanding of economically important traits, biotechnology plays a significant role in trait development and incorporation. MAS and transgenic technologies are considered two basic tools for crop improvement. By using both of these technologies, numerous achievements have been made in rice crop. But in future, in order to meet global climate change and food security, there is a need to involve other techniques such as recombinant DNA technology along with some genome editing tools like TALENs and CRISPER/Cas, as such techniques are used to either modify the nonfunctional alleles or to knock out unwanted genes. Moreover, biotechnological tools aim with the creation of useful variability which could be efficiently utilized by plant breeders for rice improvement.

References

Acquaah G (2009) Principles of plant genetics and breeding. John Wiley & Sons

Agarwal P, Parida SK, Raghuvanshi S, Kapoor S, Khurana P, Khurana JP, Tyagi AK (2016) Rice improvement through genome-based functional analysis and molecular breeding in India. Rice 9:1–17

Andrus J, Mohammed AF (1958) The economy of Pakistan. Oxford University Press, Oxford

Araujo LG, Prabhu AS, Pereiral PA (2010) Marker-assisted selection for the rice blast resistance gene Pi-ar in a backcross population. Crop Breed Appl Biotechnol 10:23–31

Arraudeau MA (1995) Upland rice: challenges and opportunities in a less favorable ecosystem. Geo J 35(3):325–328

Basu A, Ray S, Sarkar S, Chadhuri RT, Kundu S (2014) Agrobacterium mediated genetic tansformation of popular indica rice Ratna (IET1411). Afr J Biotechnol 13:4187–4197

Bawa AS, Anilakumar KR (2013) Genetically modified foods: safety, risks and public concerns a review. J Food Sci Technol 50:1035–1046

Beighley DH (2010) Growth and production of rice. In: Verheye WH (ed) Soils, plant growth and crop production, vol 2. Encyclopedia of Life Support Systems, Paris, p 349

Biswal AK, Shamim MD, Cruzado K, Soriano G, Ghatak A, Toleco M, Vikram P (2017) Role of biotechnology in rice production. In: Chauhan B, Jabran K, Mahajan G (eds) Rice production worldwide. Springer, Cham, pp 487–547

Breseghello F, Alexandre SG (2013) Traditional and modern plant breeding methods with examples in rice (*Oryza sativa* L.). J Agric Food Chem 35:8277–8286

Chang TT (1976) The origin, evolution, cultivation, dissemination and diversification of Asian and African rice. Euphytica 25:435–444

Chen W, Gao Y, Xie W, Gong L, Lu K, Wang W, Li Y, Liu X, Zhang H, Dong H (2014) Genome-wide association analyses provide genetic and biochemical insights into natural variation in rice metabolism. Nat Genet 46:714–721

Cruz ND, Khush GS (2000) Rice grain quality evaluation procedures. Aromat Rices 3:15–28

Dixon RA, Strack D (2003) Phytochemistery meets genome analysis, and beyond. Phytochemistery 62:815–816

Dobermann A, Fairhurst TH (2000) Nutrient disorders and nutrient management. Potash and Phosphate Institute/PPI of Canada and International Rice Research Institute, Singapore, p 192

Fehr WR (1987) Principles of cultivar development, Theory and technique, vol 1. Macmillan Publishing Company, New York

Food and Agricultural Organisation (2013) FAOSTAT Database. FAO rice market monitoring, Rome. http://www.fao.org/economic/est/ricepublications/rice%20market/monitorrmm/en/

Food and Agriculture Organization of the United Nations (2014) A regional rice strategy for sustainablefood security in Asia and the Pacific. http://www.fao.org/3/a-i3643e

Food and Agriculture Organization of United Nations (2019) Production of rice paddy by region, production of rice peddy. http://www.fao.org/faostat/en/#data/QC/visualize

Fu C, Wu T, Liu W, Wang F, Li J, Zhu X, Huang Y (2012) Genetic improvement of resistance to blast and bacterial blight of the elite maintainer line Rongfeng B in hybrid rice (*Oryza sativa* L.) by using marker-assisted selection. Afr J Biotechnol 11:13104–13114

Fujino K, Sekiguchi H, Matsuda Y, Sugimoto K, Ono K, Yano M (2008) Molecular identification of a major quantitative trait locus, qLTG3-1, controlling low-temperature germinability in rice. Proc Natl Acad Sci U S A 105:12623–12628

Gao L, Jia J, Kong X (2016) A SNP-based molecular barcode for characterization of common wheat. PLoS One 11:e0150947

Hossain M (1995) Sustaining food security for fragile environments in Asia: achievements, challenges and implications for rice research. In: Fragile lives in fragile ecosystems. Proc. International Rice Research Conference, Manila, pp 3–23

Huang XY, Deng F, Yamaji N, Pinson SR, Fujii M, Danku J, Douglas A, Guerinot ML, Ma JF (2016) A heavy metal P-type ATPase OsHMA4 prevents copper accumulation in rice grain. Nat Commun 7:12138

Ikeda K, Sunohara H, Nagato Y (2004) Developmental course of in florescence and spikelet in rice. J Breed Sci 54:147–156

Jain SM (2001) Tissue culture-derived variation in crop improvement. Euphytica 118:153–166

Jiang L, Liu X, Xiong G, Liu H, Chen F, Wang L, Meng X, Liu G, Yu H, Yuan Y (2013) DWARF 53 acts as a repressor of strigolactone signaling in rice. Nature 504:401–405

Juliano BO (1994) Rice chemistry and technology. The American Association of Cereal Chemists, St. Paul. 2nd Printing

Le K, Zong ZG, Yu L, Zhen MX (1989) Somatic cell culture of rice cultivars with different grain types: somaclonal variation in some grain and quality characters. Plant Cell Tissue Organ Cult 18:201–208

Lema M (2018) Application of biotechnology on rice (Oryza sativa) improvement review article. Mod Concepts dev Agron 2(2):121–127

Lesteri EG (2006) Invitro selection and somaclonal variation for biotic and abiotic stress tolerance. Biodiverse J 7:3

Li X, Qian Q, Fu Z, Wang Y, Xiong G, Zeng D, Li J (2003) Control of tillering in rice. Nature 422(6932):618–621

Li SQ, Yang DC, Zhu YG (2007) Characterization and use of male sterility in hybrid rice breeding. J Integr Plant Biol 49:791–804

Li Y, Xiao J, Chen L, Huang X, Cheng Z, Han B, Wu C (2018) Rice functional genomics research: past decade and future. Mol Plant 11(3):359–380

Lin JY (1992) Hybrid rice innovation in China: a study of market-demand induced technological innovation in acentrally planned economy. Rev Econ Stat 74:14–20

Linares OF (2002) African rice (*Oryza glaberrima*) history and future potential. PNAS 25:16360–16365

Lodha TD, Hembram P, NitileTep JB (2013) Proteomics: a successful approach to understand the molecular mechanism of plantpathogen interaction. Am J Plant Sci 4(6):1212–1226

Ma Z, Wang L, Zhao M, Gu S, Wang C, Zhao J, Xu Z (2020) iTRAQ proteomics reveals the regulatory response to Magnaportheoryzae in durable resistant vs. susceptible rice genotypes. PLoS One 15(1):e0227470

Mackill DJ, Ismail AM, Kumar A, Gregorio GB (2010) The role of stress-tolerant varieties foradapting to climate change. Based on a paper from the CURE Workshop on Climate Change, Siem Reap, Cambodia, 4 May 2010

Mandal AB, Pramanik SC, Chowdhury B, Bandyopadhyay AK (1999) Salt-tolerant Pokkali somaclones: performance under normal and saline soils in Bay Islands. Field Crops Res 61: 13–21

Mohapatra PK, Panigrahi R, Turner NC (2011) Physiology of spikelet development on the rice panicle: is manipulation of apical dominance crucial for grain yield improvement? Adv Agron 110:333–359

Moldenhauer KEWC, Slaton N (2001) Rice growth and development. Rice Prod Handb 192:7–14

Mosleh MK, Hassan QK, Chowdhury EH (2015) Application of remote sensors in mapping rice area and forecasting its production: a review. Sensors 15:769–791
Mu Q, Zhang W, Zhang Y, Yan H, Liu K, Matsui T, Yang P (2017) iTRAQ-based quantitative proteomics analysis on rice anther responding to high temperature. Int J Mol Sci 18:1811
Niizeki H, Oono K (1968) Induction of haploid rice plant from anther culture. Proc Jpn Acad 44: 554–557
Ogura H, Kyozuka J, Hayashi Y, Koba T, Shimamoto K (1987) Field performance and cytology of protoplast-derived rice (*Oryza sativa*): high yield and low degree of variation of four japonica cultivars. Theor Appl Genet 74:670–676
Okra HI (1988) Origin of cultivated rice. Japan Scientific Societies Press, Tokyo
Pawar SY, Radhakrishnan VV, Mohanan KV (2016) The importance of Optimum Tillering in rice-an overview. SIJBS 2:125–127
Priyadarshan PM (2019) Breeding self-pollinated crops. In: Plant breeding: classical to modern. Springer, Singapore, pp 223–241
Ramiah K (1937) Rice in Madras—popular handbook. Government Press, Madras
Roschevitz RJ (1931) A contribution to the knowledge of rice. Appl Bot Genet Plant Breed 27: 1–133
Rout D, Jena D, Singh V, Ahlavat MK, Arsode P, Singh P, Verma R (2020) Hybrid rice research: current status and prospects. In: Recent advances in rice research. IntechOpen, London
Saito K (2013) Phytochemical genomics a new trend. Curr Opin Plant Biol 16:373–380
Senadhira O, Zapata AF, Gregorio GB, Alejar MS, Cruz HC, Padolina TF, Galvez AM (2002) Development of the first salt-tolerant rice cultivar through indica/indica anther culture. Field Crop Res 76:103–110
Singh MP, Kumar S (2009) Genetics and plant breeding. APH Publishing Corporation
Singh RK, Gautam PL, Saxena S, Singh S (2000) Scented rice germplasm: conservation, evaluation and utilization. In: Aromatic rices. Kalyani, New Delhi, pp 107–133
Singh VK, Singh A, Singh S, Ellur RK, Choudhary V, Sarkel S, Vinod K (2012) Incorporation of blast resistance into "PRR78", an elite Basmatiice restorer line, through marker-assisted back-cross breeding. Field Crops Res 128:8–16
Sweeney M, McCouch S (2007) The complex history of the domestication of rice. Ann Bot 100: 951–957
Tang K, Tinjuangjun P, Xu Y, Sun X, Gatehouse JA, Ronald PC, Kohli A (1999) Particle-bombardment-mediated co-transformation of elite Chinese rice cultivars with genes conferring resistance to bacterial blight and sap-sucking insect pests. Planta 208:552–563
Tang K, Sun X, Power JB, Cocking EC, Davey MR (2000) Plant regeneration from protoplasts of a commercial Asian long-grain javanica rice. Plant Cell Tissue Organ Cult 60:79–82
Tanweer FA, Rafii MY, Sijam K, Rahim HA, Ahmed F, Latif MA (2015) Current advance methods for the identification of blast resistance genes in rice. C R Biol 338:321–334
Tappiban P, Ying Y, Xu F, Bao J (2021) Proteomics and post-translational modifications of starch biosynthesis-related proteins in developing seeds of rice. Int J Mol Sci 22(11):5901
Van TR, Meyer RS, Aflitos SA, Carney JA, Veltman MA (2016) Tracing ancestor rice of Suriname Maroons back to its African origin. Nature Plants 10:1–5
Vergara BS (1991) Rice plant growth and development. In: Rice. Springer, Boston, MA, pp 13–22
Wang M, Yu Y, Haberer G, Marri PR, Fan C, Goicoechea JL, Wing RA (2014) The genome sequence of African rice (*Oryza glaberrima*) and evidence for independent domestication. Nat Genet 46:982–988
Wei J, Wang A, Li R, Qu H, Jia Z (2018) Metabolome-wide association studies for agronomic traits of rice. Heredity 120(4):342–355
Whyte RO (1972) The Gramineae, wild and cultivated plants of monsoonal and equatorial. Asia. I. Southeast Asia. Asian Perspect 15:127–151
Wilkins MR, Pasquali C, Appel RD, Golaz O, Sanchez JC, Hochstrasser DF (1996) From proteins to proteomes: large scale protein identification by two-dimensional electrophoresis and amino acid analysis. Biotechnology 14:61–65

Xa TT, Lang NT (2011) Rice breeding for high grain quality through anther culture. Omonrice 18: 68–72

Yookongkaew N, Srivatanakul M, Narangajavana J (2007) Development of genotype-independent regeneration system for transformation of rice (*Oryza sativa* ssp. indica). J Plant Res 120: 237–245

Yoshida H, Nagato Y (2011) Flower development in rice. J Exp Bot 62(14):4719–4730

Zhang W, Wu R (1988) Efficient regeneration of transgenic plants from rice protoplasts and correctly regulated expression of the foreign gene in the plants. Theor Appl Genet 76:835–840

Zhang H, Xu C, Zong CH, Yang X (2013) Mutation in CSA creates a new photoperiod- sensitive genic male sterile line applicable for hybrid rice seed production. Proc Natl Acad Sci 110:76–81

Zhu DY, Sun ZX, Pan XG, Ding XH, Shen XH, Won Y, Datta SK (1998) Use of anther culture in hybrid rice breeding. In: Advances in hybrid rice technology. IRRI, Manila, pp 265–281

Hybrid Rice Production

31

Zhicai Huang and Wenqiang Liu

Abstract

Rice is one of the most important food crops in the world. Since the successful application of three-line hybrid rice in 1970s, it has been proved that hybrid rice has more than 20% yield advantage over conventional rice in China. Subsequently, two-line hybrid rice was developed successfully and further improve the yield of hybrid rice in 1980s. But it seems to be more difficult for improving rice yield due to yield ceiling. With rapid progress of molecular biology and bioinformation, more and more QTLs/genes were cloned and their functions were further investigated. These findings facilitate to integrate QTLs/genes to restore lines for improving grain yield, grain quality, disease and pest resistance, abiotic stress, and so on. Accordingly, some super-hybrid rice varieties with elite agronomic traits and breakthrough records in yield were produced. Although huge achievement has been made in hybrid rice breeding in the past decades in China, some restrains on utilization of hybrid rice remain to be overcome, such as limited restorer-maintainer pattern for three-line and fluctuated fertility/sterility by environmental factor for two-line. In order to solve the above problems, Professor Yuan Longping claimed that the third-generation hybrid rice technology was studied preliminarily successfully. The achievement of hybrid rice production in China has greatly encouraged many countries to develop their own hybrid rice, which would be beneficial for ensuring the world food security.

Keywords

Hybrid rice · Quantitative trait locus (QTL) · Gene · Resistance

Z. Huang (✉) · W. Liu
Hunan Rice Research Institute, Hunan Academy of Agricultural Sciences, Changsha, China

N. Sarwar et al. (eds.), *Modern Techniques of Rice Crop Production*,
https://doi.org/10.1007/978-981-16-4955-4_31

Abbreviations

CMS	Cytoplasmic male sterility
GA3	Gibberellic acid
GMS	Genic male sterility
Ha	Hectare
I-KI	iodine and potassium iodide
MAS	Marker-assisted selection
NGS	Next-generation sequencing
PGMS	Photoperiod-sensitive genetic male sterile
QTL	Quantitative trait locus
RFLP	Restriction fragment length polymorphism
SNP	Single-nucleotide polymorphism
SPT	Seed production technology
TGMS	Thermosensitive genetic male sterility
WA	Abortive type

31.1 Introduction

Rice is one of the staple food crops for more than half of the world's population and is grown over an area of 161 M ha with an annual production of about 764 M ton. About 90% of the world's rice is produced in Asia (Ahmad et al. 2008, 2009, 2012, 2013, 2015, 2019; Ahmad and Hasanuzzaman 2012; Ahmed and Ahmad 2019, 2020; Shahzad and Ahmad 2019). China is the leading rice producer, accounting for more than 28% of overall worldwide rice production (Cheng et al. 2004; Ahmed et al. 2017, 2020a, b; Akram et al. 2018, 2019; Hussain et al. 2018; Fahad et al. 2019; Fatima et al. 2020). With the growth of world population, humans will face huge challenges in providing rice energy to feed the world due to shortage of arable land. Fortunately, some attentions have been gained in meeting these challenges and ensuring food security (Hafeez-ur-Rehman et al. 2019; Khan et al. 2019a, b; Naz et al. 2019; Peerzada et al. 2019; Razaq et al. 2019; Sarwar et al. 2013a, b; Tariq et al. 2020; Wasaya et al. 2019; Zahoor et al. 2019).

In recent years, huge progresses in hybrid rice breeding have been attained by China. Productivity has a remarkable improvement compared to that in several decades ago, but the yield ceiling has also been encountered and further increase seems to be more difficult. Some attentions begin to seek the exploitation of *indica*/*japonica* heterosis. With rapid development of molecular biology in rice, hybrids of *indica*/*japonica* with proper proportional genotypes can be produced. Furthermore, molecular genetics has made it possible to pinpoint besides studying effects of an individual quantitative trait locus (QTL). In combination with molecular marker-assisted selection (MAS), some elite lines carrying beneficial genes/QTLs for yield and disease/pest resistance are developing for super-hybrid rice breeding program.

31.2 History and Need for Hybrid Rice

Heterosis of rice was initially researched by Prof. Yuan in China in 1964. In 1970, a pollen abortive plant from natural population of wild rice was found, which was used later as donor to develop cytoplasmic male sterility (CMS) lines. In 1973, the first CMS line was developed in China, and subsequently was released commercially in 1976 (Yuan et al. 2003). To date, CMS has shown that it is an effective system for the development of rice hybrids in the world. CMS system is effective for taking advantage of heterosis in rice, but it involves three lines (CMS line, maintainer line, and restorer line) and was complex. Its utilization is still restricted to some degree in practice.

In 1981, Shi Mingsong, a Chinese rice scientist, first reported that a male sterile mutant plant from the variety of Nongken58 was a photoperiod-sensitive genetic male sterile (PGMS) line. Development of PGMS line laid down the foundation for developing a two-line hybrid rice. In 2002, growing areas of two-line hybrids reached up 2.745 M ha in China.

Recently, a super-rice variety, Y-U-2, realized the yield breakthrough with joint efforts of rice breeders. In 2011, mean yield reached 13.9 t/ha in demonstration trial. In addition, super-rice, Y-U-900, reached 14.8 and 15.4 t/ha in Longhui county, Hunan province in 2013 and in Xupu county, Hunan in 2014, respectively, in demonstration trials. A star variety of super-hybrid rice, super-1000, reached 16.0 t/ha in similar above-mentioned trials in Gejiu county, Yunnan province in 2015 (Yuan 2017) (Fig. 31.1).

31.2.1 Development of Hybrid Rice

31.2.1.1 Development of Parental Lines

Exploiting heterosis in rice includes cytoplasmic genetic male sterility (CMS) for three-line system and photo- and/or thermosensitive genetic male sterility (PGMS or TGMS) for two–line system.

31.2.1.2 Definition of Three Lines

The three lines refer to the cytoplasmic genetic male sterile line (CMS line), the maintainer line (B line), and the restorer line (R line).

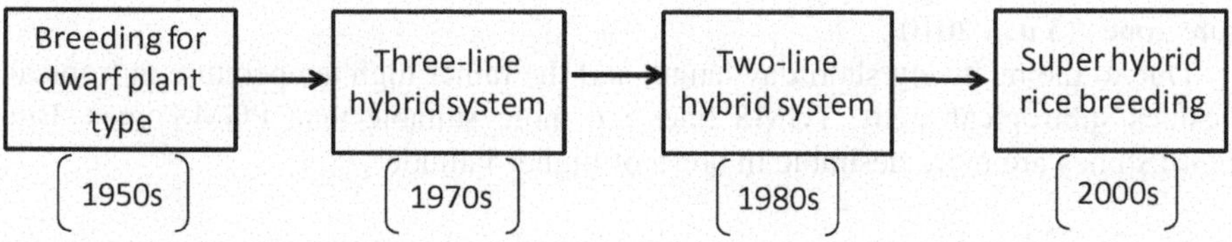

Fig. 31.1 Development process of hybrid rice in China

A CMS line is abnormal in anthers; there was only abortive pollen or no pollen; no seed can be borne by selfing. A maintainer line is a specific pollinator variety used to pollinate CMS line, and the progenies produced still show male sterility. The role of B line is to multiply a CMS line. Major characteristics of a CMS line are determined by its corresponding maintainer line. Actually, genotype of a CMS line and its corresponding maintainer line was identical in nucleus rather than in cytoplasm. Thus, they are similar to each other in appearance, differing only in few traits such as heading date, pollen, anther shape, and flowering habit. A restorer line is a pollinator variety used for pollinating a CMS line to produce F_1 hybrids that become fertile and thus can produce normal seeds by selfing. As an elite restorer line, it should have strong restoring ability, good agronomic characteristics and combining ability, well-developed anthers with heavy pollen load, good flowering habits, and normal dehiscence (Yuan 2010).

31.2.1.3 Definition of Two Lines

Fertility/sterility alteration of thermosensitive genetic male sterile (TGMS) lines and photoperiod-sensitive genetic male sterile (PGMS) lines is induced by day length and temperature. PGMS and TGMS lines are foundation for developing two-line system hybrid rice.

A rice plant that became male sterile in case of longer days length and recovered fertility under shorter days length was first discovered in China. The plant was a spontaneous mutant of a japonica variety, Nongkeng58, and was called PGMS rice. PGMS rice have the following characteristics: the photoperiod-sensitive stage for fertility/sterility alteration is from the development of secondary branch primordial to the formation of pollen mother cell; the critical day length for inducing sterility generally is 13.75 h and the critical light intensity is above 50 lux; temperature has a slight influence on fertility/sterility alteration. Beyond a certain temperature, range of day length will have no effect on fertility/sterility alteration (Fig. 31.2).

Fertility/sterility alteration of TGMS lines is induced mainly by temperature. The existing TGMS lines become completely male sterile under higher temperature. The day length has very little influence on fertility. PGMS rice have the following characteristics: the thermosensitive stage for fertility/sterility alteration is the period from pollen mother cell formation to beginning of meiotic division; the critical male sterility-inducing temperature is 23–29 °C, varying from line to line; under the critical temperature for about two consecutive days, the TGMS line resumes male fertility. The most important point in the practical utilization of TGMS lines is that critical male-inducing temperature must be relatively lower, especially for temperature zones (Yuan 2010).

Due to the relatively short day length and the rather high temperature in tropical besides subtropical areas, TGMS lines are more suitable than PGMS lines. But PGMS lines are more desirable in areas of higher latitudes.

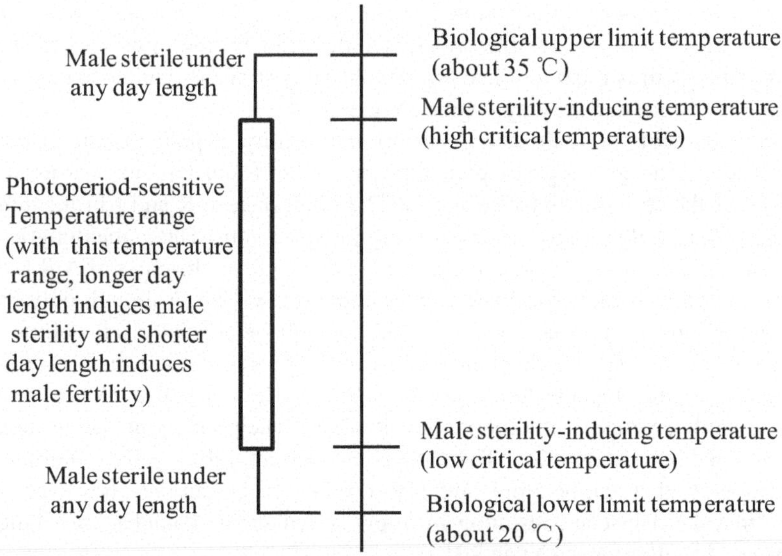

Fig. 31.2 The pattern of fertility/sterility alteration of PGMS lines in relation to day length and temperature

31.2.2 Developmental Biology of Male Sterile Rice

With the rapid development of hybrid rice, more and more CMS lines have been bred. These CMS lines can be grouped into various types based on the following criteria.

31.2.2.1 Based on the Genetic Behavior of Sterile Genes, CMS Lines Are Divided into Two Types: Sporophytic and Gametophytic

In the sporophytic male sterile system, the pollen sterility or fertility is determined by the genotypes of sporophyte (plant body), and the genotype of pollen grains (gametophyte) has no effect on it. When the sporophytic genotype is S(rr), all the pollen grains will be abortive. If the genotype is N(RR)o S(RR), all the pollen grains will be fertile. As for the sporophytic genotype S(Rr), although it produces two kinds of male gametes, namely S(R) and S(r), still all pollen grains are fertile because the fertility of pollen grains is determined by the dominant fertile gene, R, present in the sporophyte (Yuan 2010).

The CMS lines of the wild abortive type (WA) and Gambiaca (Gam) type belong to sporophytic sterility, which has the following properties: the F_1 hybrid from cross of A/R has normal pollen grains with normal fertility. Segregation in fertility will occur in F_2 and a certain proportion of male sterile plants will appear in the population. Abortion of pollen grains occurs at the earlier stage of microspore development. Most pollen grains look wrinkled and irregular (less pollen grains are round in shape) and are unstained with iodine and potassium iodide (I-KI) solution. The anthers are milky white in color, water-soaked in feature, and indehiscent. The male sterility is stable and less influenced by the environment, but its restoring spectrum is relatively narrow. The basal part of the panicle is enclosed in the flag leaf sheath to a varying degree (Yuan 2010).

In the gametophytic male sterile system, the fertility of pollen grains is directly determined by the genotype of gametophyte (pollen) and has no bearing on the genotype of the sporophyte (plant body). The nuclear gene R and r in the gametophyte (pollen) results in fertility and sterility, respectively. The gametophytic male sterile has the following characteristics: the F1 hybrid from the cross of A/R has two pollen genotypes, namely S(R) and S(r) in equal proportion. Although only half of the pollen grains are normal, the F_1 hybrid can still be normally self-pollinated to produce seeds. No male sterile plants appear in F_2. This is because the S(r) pollen of F_1 is abortive and only the fertile pollen S(R) is able to fertilize. Pollen abortion occurs at the later stage of micropore development. The pollen grains are round and lightly stained by I-KI solution. The anthers are slender, milky yellow, and indehiscent. Under higher temperature and lower moisture conditions, however, some anthers may be dehiscent, resulting in a few selfed seeds. Panicles are slightly or not enclosed in the sheath (Yuan 2010).

31.2.2.2 Based on the Genetic Behavior of Sterile Genes, CMS Lines Are Divided into Two Types: Sporophytic and Gametophytic

According to the difference in restorer lines and maintainer lines to a given male sterility or simply called maintaining-restoring patterns, the existing CMS lines can be classified into three basic groups, namely CMS lines of WA type, CMS lines of Honglian type, and CMS lines of BT type.

31.2.2.3 Classification Based on Morphology of Sterile Pollen Grains

Male sterility in rice could be due to abnormality at any stage of microsporogenesis. Failure of the mechanism to form microspores from sporogenous cells results in male sterility of pollen-free type, while failure of the mechanism during microspore development and pollen maturation results in various types of pollen abortion. The classification of male sterility based on the morphology of pollen grains stained by I-KI solution contains typical abortion types, spherical abortion types, and stained abortion types.

31.2.3 Molecular Breeding

Conventional breeding methods were used by selecting phenotypes that are generally effective for qualitative traits. It is not workabe for quantitative traits because of their small effects. Progress of molecular biology and genomics begins to allow applications of molecular technology in rice breeding by marker-assisted selection.

The process of using MAS involves selection of interest of genes and construction of target populations, besides molecular marker screening. During recent years, pronounced progresses have been made in gene clone and functional genomics.

Table 31.1 Examples of marker-assisted selection in rice

Varieties/lines	Types	Genes involved	Donors	Tolerance to/exploited traits
Zhonghui8006	Indica	Xa21, GM6	Duoxil, Minghui63	BB, GM
Zhonghui218	Indica	Xa21	IRBB21	BB
Guodao1	Indica	Xa21	Zhonghui8006	BB
Guodao3	Indica	Xa21	Zhonghui8007	BB
IIYou8006	Indica	Xa21	Zhonghui8008	BB
IIYou-218	Indica	Xa21	Zhonghui218	BB
Yuanhui611	Indica	Yld1.1, yld2.1	*O. rufipogon*	High yield
Y-You7	Indica	Yld1.1, yld2.1	Yuanhui611	High yield
Shuihui527	Restorer	Xa4, Xa21	1318/88-R3360	BB
Zhunlinagyou527	Indica	Xa4, Xa21	Shuihui527	BB
D-You 527	Indica	Xa4, Xa21	Shuihui527	BB
XieYou527	Indica	Xa4, Xa21	Shuihui527	BB
RGB-7S/RGD-8S	CMS	Pi1, Pi2	BL122	RB
Yueza 746/763	Indica	Pi1, Pi2	RGD-7S	RB
W3660	Japonica	Lgc-1	LGC-1	Low glutelin content
W017	Japonica	Lox-3	DawDam	Prolonged storage seed
W025	Japonica	ge	Haiminori	Huge embryo
Zhonghui161	Restorer	Pita, xa13, wx	IRBB51, Teqing	RB, BB, good quality
Bph68S/ Luohong4A	CMS	Bph14, Bph15	B5	BPH
Ning9108	Indica	Stv-bi, Wx-mq	Guandong194	Strip blight, good quality

Abbreviations: *BB* bacterial blight, *RB* rice blast, *GM* gall midge, *BPH* blight planthopper

These cloned QTL/genes provide sources for breeding application in rice. So far, molecular breeding was mainly applied on high productivity, resistance to biotic stress plus abiotic stress, grain quality, and improved physiological traits (Rao et al. 2014) (Table 31.1).

31.2.4 Application of Molecular Markers

In essence, molecular markers are nucleotide sequences that were developed on the basis of polymorphic insertion, deletion, point mutation, duplication, translocation, and so on. Ideal molecular markers should be codominant, reproducible, and detectable. Molecular markers are grouped into several groups based on their characteristic and detection method. Multiple types of DNA molecular markers have been developed and applied in genetics and breeding in rice. During the past 30 years, the molecular marker technology was developed from restriction fragment length polymorphism (RFLP) to single-nucleotide polymorphisms (SNPs) and a diversity of array-technology-based markers. Developments in sequencing skills have led to improvement of next-generation sequencing (NGS) platforms which are cost-effective with higher throughput.

Application of molecular markers in rice is very wide, including genetic diversity assessment, identification of haploid, investigation of heterosis, target gene selection in backcrossing, QTL mapping and clone, genetic linkage map construction, detection of F_1 purity, and so on.

31.3 Production Technology of Hybrid Rice

Heterosis of hybrid rice is the base of its high-yielding potential. Some cultivation techniques in accordance with these characteristics are required.

31.3.1 Seed Treatment and Germination

Seeds should be disinfected by mercury compounds or formalin or trichloroisocyanurate before sowing. After disinfecting, these seeds should be rinsed carefully and soaked alternatively at day and night until seeds get germinated.

31.3.2 Raising Seedling

Sowing time and seedling age for transplanting are determined according to the cropping system, variety character, temperature, and so on. The methods of raising seedling are irrigated nursery, upland nursery, wet nursery, and tray nursery. The healthy seedlings were of dark green leaves, with tillers, white and vigorous roots, uniform in growth, and free from disease and insect pests.

31.3.3 Transplanting

The paddy field should be well prepared before transplanting. The transplant space and number of seedlings transplanted varies with hybrid variety, growth season, growth duration, and seedling age. In general, transplanting space for hybrid rice is wider than that for inbred rice. Two seedlings with tillers per hill are transplanted.

31.3.4 Fertilizer and Water Management

Amount of fertilizer for nitrogen, phosphorus, and potassium is about 10–12 kg of N, 5–6 kg of P_2O_5, and 12–15 kg of K_2O, respectively, to produce 500 kg of rice grains. The commonly used fertilizers are organic manure and chemical fertilizer, such as urea, super phosphate, KCL, and compound fertilizer. All of the organic manure and phosphate, and 50%–60% of nitrogen, are usually applied as basal dressing, and the others as top-dressing.

In order to recover from transplant shock, a thin layer of standing water is kept for 3–5 days after transplanting. If herbicide is used, the standing water should be kept for 5–7 days. From the beginning of tillering to the initiation of panicle differentiation, it is important to irrigate and drain the water alternatively. When the number of tillers reaches 80–120% of the expected panicle numbers, the field should be drained and kept dry to the point of surface cracking for control of nonproductive tillers and promotion of root growth. After panicle initiation, the field should be kept with a shallow layer of water and wet alternatively until heading.

31.3.5 Pest and Weed Control

Major disease and pests in hybrid rice are similar to those in inbred rice, such as rice blast, sheath blight, stem borers, and so on. Different methods are taken to control these diseases and pests.

Weed control includes preventive, cultural, manual, mechanical, biological, and chemical methods. Among them, the chemical control is the most effective and economical way.

31.4 Adaptation of Hybrid Rice

The huge challenge that humans face in the twenty-first century includes population increase, environment resource degeneration, and food supplies scarceness, which made the problem of searching methods for enhancing rice productivity. Hybrid rice has shown its powerful productivity advantage all over the world. Some studies on adaption of hybrid rice show that hybrid rice needs more fertilizer and possesses strong photosynthetic capacity.

31.4.1 Abiotic and Biotic Stress

Compared with inbred rice, hybrid rice has some obvious superiority in agronomic traits such as tillering capacity, root activity, and so on. Correspondingly, hybrid rice has high rate of metabolism in physiological performance. Thus, under abiotic stress treatment, hybrid rice tends to show high tolerance capacity and better recovery capacity. Many studies indicated that hybrid rice showed better performance than inbred rice in dry, flooding, low/high temperature condition. Of course, there are also some exceptions. Recent studies showed that cadmium level in hybrid rice did not necessarily show high cadmium concentration from the parents in polluted high cadmium field.

At present, number of genes for pest/disease resistance has been cloned, most of them were dominant. Therefore, it is not significant difference between hybrid rice and inbred rice for biotic stress of pest and disease.

In a word, more attentions were paid on rice high yield rather than adaption of hybrid rice previously. The effort should be intensively conducted presently, which will be beneficial for introducing some varieties among countries.

31.5 Biotechnology in Hybrid Rice

Two breakthrough records in rice production were utilization of semidwarf gene besides heterosis. At present, application of heterosis in rice contains CMS for three-line and PTGMS for two-line system in China. With rapid improvement of molecular biology, molecular MAS contributes to improve disease and pest resistance of hybrid rice. The MAS technology in combination with conventional means is becoming a popular approach in rice breeding. A number of genes involving CMS/Rf system and PTGMS were cloned and the mechanism was further investigated; some PTGMS lines, CMS lines, and R lines have been bred using molecular-assisted breeding.

However, compared to PGTMS lines, CMS line is safe in hybrid seed productivity due to its stability besides complete sterility. But application of CMS/Rf system requires restore genes from restore line, which could compensate impairment caused by the CMS in F_1 hybrids. The PTGMS line could be restored by any other rice cultivars; the system could contribute to a wider exploration of heterosis. But fertility of PTGMS line is easily affected by environment factor, causing hybrid seed productivity quite vulnerable. Therefore, a new sterile line both capable of safe hybrid seed productivity and amenable to free combination for developing hybrid rice is imperative.

In 1993, Plant Genetic Systems Company proposed that introducing into fertility restoration gene, pollen abortion gene and selection marker gene could produce a male sterile plant as a maintainer line. In 2002, a sterile line was achieved by introducing fertility restoration gene and selection marker gene into mutant lines. During 2006, US DuPont–Pioneer Company finished seed production technology (SPT) of genic male sterility (GMS) in maize for the first time. In 2016, professor Yuan Longping claimed that the third-generation hybrid rice technology was studied initially fruitful. The third-generation hybrid rice refers to hybrid rice with a genetically engineered male sterile line as a genetic tool. It contains advantages both stabel sterility of three system and free combination of two-line system.

31.6 Hybrid Rice Seed Production

Hybrid rice seed productivity includes foundation seed productivity of parental lines, multiplication of parental line, and F_1 hybrid rice seed productivity. Actually, hybrid rice seed productivity is called outcrossing cultivation of rice. It includes three-line hybrid rice seed productivity and two-line hybrid rice seed productivity. The three-line hybrid rice seed productivity is taken as the female parental lines for the development of three-line system rice hybrids. Whereas two-line hybrid rice seed productivity is taken as PGMS or TGMS lines as female parental lines for developing two-line system rice hybrids.

31.6.1 Outcross Character of Rice

Rice is a typical self-pollinated and hybrid rice seed production is the process of getting the seeds by outcross pollination. The outcross characters of CMS/PTGMS lines and restorer lines have to be studied and utilized. It includes character of heading and flowering, flowering time, exsertion and vitality of stigma, and so on.

31.6.2 Optimum Season for Heading and Flowering

Favorable climatic condition is very important. The daily mean temperature for heading and flowering of the parental lines is around 26–28 °C, the daily highest temperature is not more than 35 °C, and the daily lowest temperature is not more less 21 °C and sufficient sunlight. The relative humidity in the field is around 70–90% without dry and hot wind. There is no continuous rain for more than 3 days in the flowering stage.

31.6.3 Synchronization in Flowering

Rice is a crop with 10–15 days period of flowering time; during the period, both parental lines are required to flower at the same time for outcrossing. The synchronization in flowering is judged by the late or early initial heading of parental lines. The degree of synchronization in flowering of parental lines determines the seed production yield. In order to ensure the synchronization in flowering, sowing dates and sowing intervals between parental lines must be done. If necessary, prediction and adjustment of flowering date would be required.

31.6.4 The Establishment of the Parental Line Population

Hybrid rice seed productivity is composed of two parental line populations. The male parent provides pollens, and the female parent receives pollen for outcrossing and developing seeds, seeds harvested from female lines. The grains from the male parent are usually not used for production. Therefore, the female population should be more than that of the male population; the area ration between male and female lines is 1:3–4. The planting methods of male parental lines include single row, narrow double row, small double row, and big double row. The row ratio between

Fig. 31.3 Small double-row male parental line (*left*) and *big double-row* male parental line

male and female parental lines varies with the planting methods of male parental lines (Fig. 31.3).

31.6.5 Field Management of Hybrid Rice Seed Production

Field management is very important to achieve higher productivity. Such management includes raising healthy and tillered seedling, planting one tillered seedling per hill for restorer line and two tillered seedlings per hill for CMS lines, applying fertilizer during the early growth stage after transplanting, and fertilizing and irrigating less during the middle and late stages.

31.6.6 Techniques for Improving Outcrossing Characters

In order to improve outcrossing posture, raising healthy plants of female lines is a basic condition, spraying gibberellic acid (GA3) in proper dosage n proper time is the key point, and cutting leaves is taken as supplementary.

The methods for spraying GA3 are different with different sterile lines, different synchronization in flowering, and different growing condition of the plants. For GA3 usage per unit area, the fields with sensitive sterile lines is 120–150 g/ha, with not sensitive sterile lines 500 g/ha, and medium sensitive sterile lines 300 g/ha; The first spraying time is 5%, 40%, and 15% heading rate for not sensitive sterile lines, sensitive sterile lines and medium sensitive sterile lines, respectively. If synchronization in flowering is different, spraying GA 2–3 times in the cotinuous 2–3 days is required (Fig. 31.4).

Fig. 31.4 Spraying of GA3 in hybrid rice seed production

31.6.7 Supplementary Pollination

Supplementary pollination is necessary for getting higher outcrossing seed-setting rate. It contains rope-pulling method, single long-rod driving method, single long-rod pushing method, and double short-rod pushing method.

Rope-pulling method is adopted with the transplanting ways as single row, narrow double row, and small double row of male parental lines. With the rope-pulling method, panicles of restorer line are shaken by a long rope (about 0.5 cm in diameter besides 20–30 m in length) which is pulled by two persons on both ends of the rope and runs against the wind; the rope should run parallel to R-line row direction. The rope flies across the panicle level quickly and shakes the panicles to cause the pollens flying out from the anthers of R line to the female spikelets.

Single long-rod driving method is adopted with the transplanting ways as single row, narrow double row, and small double row of male parental lines. While walking in the rows among the parental lines, one person holds one end of a long rod (length about 6 m and diameter in 2–3 cm) in both hands, and puts the rod at the base of the panicle canopy of R-line in level direction and sweep the panicle canopy left and right in fan shape to shake the panicles of male parental line, so as to cause the pollens to fly out from the anthers of R line to the female spikelets (Fig. 31.5).

Single long-rod pushing method is adopted with all transplanting ways of male parental lines. Before driving the pollens, the plots with 5 m length as the length of the rod are divided in the direction vertical to the R-line row in the fields, the distance (30 cm) between each plot is used as the walking path. When walking in the walking path of the field, one person holds the middle part of the rod and puts it at the up-middle part of the R-line plants, then pushes and shakes the R-line plants with power row by row. The advantages are with good effect, quick speed, and without driving the female plants.

Double short-rod pushing method is adopted with the transplanting ways as big double row and small double row of male parental lines. While walking between the rows of the R lines, one person holds a short rod in each hand (length about 2 m and

Fig. 31.5 Seed production of three-line hybrid rice

diameter in 3 cm) in both hands and puts the two rods at the middle-up part of R-line plants and pushes and shakes the R-line plants with power for 2–3 times at the two sides so as to cause the pollens to fully fly out from the anthers of R line to the female spikelets.

Under good weather condition and when most of the male plants flowering, supplementary pollination can be carried out from about 10:00–12:00 AM. If meeting the raining and cloudy weather with low temperature, the flowering time will be changed and then should wait for good weather until the R line blooming. From the date on which male and female flower together to the end of flowering of male, it needs 8–12 days for supplementary pollination.

31.7 Restraints in Hybrid Rice Production

As accelerating urbanization in China, many youths leave their hometowns and seek to earn money in city. Currently there is a big labor shortage for rice production in village, which cause that many fertile rice paddy fields are barren. Suitable machinery should be designed and distributed among the farmers for transplanting and other agricultural operations.

In addition, the hybrid price is very high while the grain rice is much cheap in market, which leads to loss of motivation of rice cultivation for farmers. The government should take some stringent measures to control markets and provide subsidy to encourage farmers for rice production.

In recent year, some extreme weather occurs frequently. Rice has long growth duration and was prone to suffer the weather. As a result, the rice yield was reduced or even nothing. Some insurance company should be encouraged to join in the situation. Once the extreme climate happens, farmer will be reduce some loss to some degree.

Three-line hybrid is limited by restorer-maintainer pattern, while two-line hybrid is risked in seed production due to external temperature fluctuation. The third-generation hybrid that combines the advantages of three-line and avoids the disadvantage of two line is a promising line and it is still an early begin (Song et al. 2020). We believe that it will be a great breakthrough in the future.

31.8 Conclusions

Great achievement has been made in rice breeding program. How then to increase further grain yield and improve quality of rice next step? Firstly, more attentions should be paid to develop and exploit wild rice. At present, many parental materials have similar genetic background, while wild rice can provide abundant genetic resources such as disease/pest resistance and stress tolerance. Secondly, wide compatible resources are very important in the utilization of *indica–japonica* heterosis and should be further discovered and applied widely, as they could overcome hybrid impediments. In addition, intermediate materials with a proportional compatibility between cultivated and wild rice should be developed. Recently, molecular design breeding was highly recommended in rice breeding. Molecular design breeding is a novel breeding method in combination platform of bioinformatics with databases of genomics and proteomics. According to breeding objectives and growth environment in rice, molecular design breeding optimizes proper scheme besides then carries out trials (Cheng et al. 2007a, b). The transmission from conventional breeding to molecular design cultivars will be a general inclination, ensuring new varieties with elite traits of high yield, high grain quality, disease/pest resistance, besides stress tolerance. This will contribute toward national food security.

References

Ahmad S, Hasanuzzaman M (2012) Integrated effect of plant density, N rates and irrigation regimes on the biomass production, N content, PAR use efficiencies and water productivity of rice under irrigated semiarid environment. Not Bot Horti Agrobot Cluj-Napoca 40(1):201–211

Ahmad S, Zia-ul-Haq M, Ali H, Shad SA, Ammad A, Maqsood M, Khan MB, Mehmood S, Hussain A (2008) Water and radiation use efficiencies of transplanted rice (*Oryza sativa* L.) at different plant densities and irrigation regimes under semi-arid environment. Pak J Bot 40 (1):199–209

Ahmad S, Ahmad A, Zia-ul-Haq M, Ali H, Khaliq T, Anjum MA, Khan MA, Hussain A, Hoogenboom G (2009) Resources use efficiency of field grown transplanted rice (*Oryza sativa* L.) under irrigated semiarid environment. J Food Agric Environ 7(2):487–492

Ahmad S, Ahmad A, Soler CMT, Ali H, Zia-Ul-Haq M, Anothai J, Hussain A, Hoogenboom G, Hasanuzzaman M (2012) Application of the CSM-CERES-Rice model for evaluation of plant density and nitrogen management of fine transplanted rice for an irrigated semiarid environment. Precis Agric 13(2):200–218

Ahmad S, Ahmad A, Ali H, Hussain A, Garcia y Garcia A, Khan MA, Zia-Ul-Haq M, Hasanuzzaman M, Hoogenboom G (2013) Application of the CSM-CERES-Rice model for evaluation of plant density and irrigation management of transplanted rice for an irrigated semiarid environment. Irrig Sci 31(3):491–506

Ahmad A, Ashfaq M, Rasul G, Wajid SA, Khaliq T, Rasul F, Saeed U, Rahman MH, Hussain J, Baig IA, Naqvi AA, SAA B, Ahmad S, Naseem W, Hoogenboom G, Valdivia RO (2015) Impact of climate change on the rice–wheat cropping system of Pakistan. In: Hillel D,

Rosenzweig C (eds) Handbook of climate change and agro-ecosystems: the agricultural Modeling Intercomparison and improvement project (AgMIP) integrated crop and economic assessments. Imperial College Press, London, pp 219–258

Ahmad S, Abbas G, Ahmed M, Fatima Z, Anjum MA, Rasul G, Khan MA, Hoogenboom G (2019) Climate warming and management impact on the change of rice-wheat phenology in Punjab, Pakistan. Field Crops Res 230:46–61

Ahmed M, Ahmad S (2019) Carbon dioxide enrichment and crop productivity. In: Hasanuzzaman M (ed) Agronomic crops, vol 2. Springer Nature Singapore, Singapore, pp 31–46

Ahmed M, Ahmad S (2020) Systems modeling. In: Ahmed M (ed) Systems modeling. Springer Nature, Cham, pp 1–44

Ahmed M, Fayyaz-ul-Hassan, Ahmad S (2017) Climate variability impact on rice production: adaptation and mitigation strategies. In: Ahmed M, Stockle C (eds) Quantification of climate variability, adaptation and mitigation for agricultural sustainability. Springer, Cham, pp 91–111

Ahmed M, Ahmad S, Raza MA, Kumar U, Ansar M, Shah GA, Parsons D, Hoogenboom G, Palosuo T, Seidel S (2020a) Models calibration and evaluation. In: Ahmed M (ed) Systems modeling. Springer Nature Singapore, Singapore, pp 151–178

Ahmed M, Ahmad S, Waldrip HM, Ramin M, Raza MA (2020b) Whole farm modeling: A systems approach to understanding and managing livestock for greenhouse gas mitigation, economic viability and environmental quality. In: Waldrip HM, Pagliari PH, He Z (eds) Animal Manure, ASA Special Publication 67. American Society of Agronomy/Soil Science Society of America, Madison, WI, pp 345–371

Akram R, Turan V, Hammad HM, Ahmad S, Hussain S, Hasnain A, Maqbool MM, Rehmani MIA, Rasool A, Masood N, Mahmood F, Mubeen M, Sultana SR, Fahad S, Amanet K, Saleem M, Abbas Y, Akhtar HM, Hussain S, Waseem F, Murtaza R, Amin A, Zahoor SA, Sami ul Din M, Nasim W (2018) Fate of organic and inorganic pollutants in paddy soils. In: Hashmi MZ, Varma A (eds) Environmental pollution of paddy soils. Springer Nature Switzerland, Cham, pp 197–214

Akram R, Fahad S, Masood N, Rasool A, Ijaz M, Ihsan MZ, Maqbool MM, Ahmad S, Hussain S, Ahmed M, Kaleem S, Sultana SR, Mubeen M, Saud S, Kamran M, Nasim W (2019) Plant growth and morphological changes in rice under abiotic stress. In: Hasanuzzaman M, Fujita M, Nahar K, Biswas JK (eds) Advances in rice research for abiotic stress tolerance. Woodhead, Cambridge, pp 69–85

Cheng SH, Cao LY, Yang SH, Zhai HQ (2004) Forty years' development of hybrid rice: China's experience. Rice Sci 11(5–6):225–225

Cheng SH, Cao LY, Zhuang JY, Chen SG, Zhan XD, Fan YY, Zhu DF, Min SK (2007a) Super hybrid rice breeding in China: achievements and prospects. J Integr Plant Biol 49(6):805–810

Cheng SH, Zhuang JY, Fan YY, Du JH, Cao LY (2007b) Progress in research and development on hybrid rice: a super-domesticate in China. Ann Bot 100:959–966

Fahad S, Noor M, Adnan M, Khan MA, Rahman I, Alam M, Khan IA, Ullah H, Mian IA, Hassan S, Saud S, Bakhat HF, Hammad HM, Ahmad S, Nasim W (2019) Abiotic stress and rice grain quality. In: Hasanuzzaman M, Fujita M, Nahar K, Biswas JK (eds) Advances in rice research for abiotic stress tolerance. Woodhead, Cambridge, pp 571–583

Fatima Z, Ahmad M, Hussain M, Abbas G, Ul-Allah S, Ahmad S, Ahmed N, Ali MA, Sarwar G, Haque EU, Iqbal P, Hussain S (2020) The fingerprints of climate warming on cereal crops phenology and adaptation options. Sci Rep 10:18013

Hafeez-ur-Rehman NA, Awan MI, Ijaz M, Hussain M, Ahmad S, Farooq M (2019) Direct seeding in rice: problems and prospects. In: Hasanuzzaman M (ed) Agronomic crops, Production technologies, vol 1. Springer Nature Singapore, Singapore, pp 199–222

Hussain M, Ahmad S, Hussain S, Lal R, Ul-Allah S, Nawaz A (2018) Rice in saline soils: physiology, biochemistry, genetics and management. Adv Agron 148:231–287

Khan MA, Ahmad S, Raza A (2019a) Integrated weed management for agronomic crops. In: Hasanuzzaman M (ed) Agronomic crops, vol 2. Springer Nature Singapore, Singapore, pp 257–281

Khan SM, Ali S, Nawaz A, Bukhari SAH, Ejaz S, Ahmad S (2019b) Integrated pest and disease management for better agronomic crop production. In: Hasanuzzaman M (ed) Agronomic crops, vol 2. Springer Nature Singapore, Singapore, pp 385–428

Naz S, Fatima Z, Iqbal P, Khan A, Zakir I, Noreen S, Younis H, Abbas G, Ahmad S (2019) Agronomic crops: types and uses. In: Hasanuzzaman M (ed) Agronomic crops, Production technologies, vol 1. Springer Nature Singapore, Singapore, pp 1–18

Peerzada AM, Bukhari SAH, Dawood M, Nawaz A, Ahmad S, Adkins S (2019) Weed management for healthy crop production. In: Hasanuzzaman M (ed) Agronomic crops, vol 2. Springer Nature Singapore, Singapore, pp 225–256

Rao Y, Li Y, Qian Q (2014) Recent progress on molecular breeding of rice in China. Plant Cell Rep 33(4):551–564

Razaq M, Shah FM, Ahmad S, Afzal M (2019) Pest management for agronomic crops. In: Hasanuzzaman M (ed) Agronomic crops, vol 2. Springer Nature Singapore, Singapore, pp 365–384

Sarwar N, Ali H, Ahmad S, EhsanUllah AS, Mubeen K, Hill JE (2013a) Water wise rice cultivation on calcareous soil with the addition of essential micronutrients. J Anim Plant Sci 23(1):244–250

Sarwar N, Ali H, Maqsood M, EhsanUllah SAN, Shahzad M, Mubeen K, Shahid MA, Ahmad S (2013b) Phenological response of rice plants to different micronutrients application under water saving paddy fields on calcareous soil. Turk J Field Crops 18(1):52–57

Shahzad AN, Ahmad S (2019) Tools and techniques for nitrogen management in cereals. In: Hasanuzzaman M (ed) Agronomic crops, vol 2. Springer Nature Singapore, Singapore, pp 111–126

Song S, Wang T, Li Y, Hu J, Kan R, Qiu M, Deng Y, Liu P, Zhang L, Dong H, Li C, Yu D, Li X, Yuan D, Yuan L, Li L (2020) A novel strategy for creating a new system of third-generation hybrid rice technology using a cytoplasmic sterility gene and a genic male-sterile gene. Plant Biotechnol J 19:251–260

Tariq M, Ahmed M, Iqbal P, Fatima Z, Ahmad S (2020) Crop phenotyping. In: Ahmed M (ed) Systems modeling. Springer Nature Singapore, Singapore, pp 45–60

Wasaya A, Yasir TA, Ijaz M, Ahmad S (2019) Tillage effects on agronomic crop production. In: Hasanuzzaman M (ed) Agronomic crops, vol 2. Springer Nature Singapore, Singapore, pp 73–99

Yuan LP (2010) Yuan Longping's collection works. Science Press, Beijing

Yuan LP (2017) Progress in super-hybrid rice breeding. Crop J 5:100–102

Yuan LP, Wu XJ, Liao FM, Ma GH, Xu QS (2003) Hybrid rice technology. China Agriculture Press, Beijing

Zahoor SA, Ahmad S, Ahmad A, Wajid A, Khaliq T, Mubeen M, Hussain S, Sami Ul Din M, Amin A, Awais M, Nasim W (2019) Improving water use efficiency in agronomic crop production. In: Hasanuzzaman M (ed) Agronomic crops, vol 2. Springer Nature Singapore, Singapore, pp 13–29

Rice Biotechnology

32

Batool Fatima, Dilshad Hussain, Maryam Jamil,
and Mohibullah Shah

Abstract

Rice crop has been cultivated since very long to provide food for the humans as well as animals. Gene discovery and functional genomics of rice are being used for the study of the complete gene profiling of rice that can be further used for obtaining more sustainable transcripts for the rice. Main purpose of the whole biotechnological study of the rice is to enhance the yield of this crop. For this, transgenic plants are being introduced that give abundant yield. Similarly, C4 plants marker-assisted breeding to get more number of grains per plant is also in use. Another approach to deal with the high demand of rice is to make them more resistant against the viral attacks, environmental hazards, and herbaceous as well as bacterial attack. Besides, the amount of nutritional quality of the rice crop is also a target of the study. This is also achieved by biotechnological approaches. Targeted genome editing technologies like TALENs and CRISPR/Cas system could be exploited for modifying defunct alleles of beneficial genes or to knock out unwanted genes in plants without any trace of transgene. Biotechnological instrumentation and practices provide a chance for creating novel variations for the rice breeders.

Keywords

Rice · Yield · Production · Improvement · Technology

B. Fatima · M. Jamil · M. Shah (✉)
Department of Biochemistry, Bahauddin Zakariya University, Multan, Pakistan
e-mail: mohib@bzu.edu.pk

D. Hussain
International Center for Chemical and Biological Sciences, HEJ Research Institute of Chemistry, University of Karachi, Karachi, Pakistan

N. Sarwar et al. (eds.), *Modern Techniques of Rice Crop Production*,
https://doi.org/10.1007/978-981-16-4955-4_32

32.1 Introduction

Rice, being the world's most important staple food crop, feeds more than two billion people where 40–70% of the entire food calories being consumed are provided by rice (Ahmad et al. 2008, 2009, 2012, 2013, 2015, 2019; Ahmad and Hasanuzzaman 2012; Ahmed and Ahmad 2019, 2020; Shahzad and Ahmad 2019). Rice has also been used for animal graze and hence has been the prime source of earnings for agrarian people; high-quality rice produce additional revenues (Durand-Morat et al. 2018; Ahmed et al. 2017, 2020a, b; Akram et al. 2018, 2019; Hussain et al. 2018; Fahad et al. 2019; Fatima et al. 2020). On account of the Green Revolution, a quantum surge in the yield of rice cropped up throughout the last three decades; however, poverty, as well as hunger, was not abolished by elevated production of food (Hafeez-ur-Rehman et al. 2019; Khan et al. 2019a, b; Naz et al. 2019; Peerzada et al. 2019; Razaq et al. 2019; Sarwar et al. 2013a, b; Tariq et al. 2020; Wasaya et al. 2019; Zahoor et al. 2019). The yield growth precluded famine and halted a substantial obstruction of the food provision in Asia—contrary to some African countries, where a scarcity of infrastructure and the legislative decision came about the malfunction to manipulate Green Revolution techniques (Datta 2004).

Hence, scientists see it as a firmly established model plant. Our proficiency has been altered efficaciously for discovery of gene besides functional genomics of rice by contemporary developments in molecular biology and omics technologies (Rana et al. 2020). Transcript and metabolite profiling can provide us a comprehensive picture of one cell of rice. Current breakthroughs in phenomics research have enabled us to simply unmask germplasm at varied specialty conditions in an additional combinatory method revealing the whole-plant modification besides trait variegation. Eventually, we can scrutinize discrete component attributes having genetic control in controlled environments using a reductionist approach. On account of radical betterment in systems biology and molecular techniques, we can forecast roles of gene cohorts and foretell, authenticate, and enforce their regulation for refinement of rice cultivars (Biswal et al. 2017).

A variety of crops in the world are being ameliorated by manipulating biotechnology; rice is one of them. Now rice production has been shot up quantitatively and qualitatively by shifting of cost-effective crucial attributes from genus/species fence into the rice gene pool, utilization of target attribute without wrecking of the non-target loci of the rice genome, and truncating the breeding cycle (Pathak et al. 2018). The execution of genetic engineering on rice has focused on tolerance to cold, salinity, and drought; resistance to insects, pests, and diseases; nutritional traits (iron improvement, to reduce vitamin A deficiency); and herbicide tolerance but gave more prominence to the two promising areas of nutrition improvement and plant protection for instant use (Peng et al. 2020; Rasool et al. 2020) (Fig. 32.1).

The increased accuracy and rate of rice breeding, being the potential benefits, has been proposed using molecular markers over phenotypic markers. The research of rice in the International Rice Research Institute (IRRI) and National Agriculture Systems (NARSs) has resulted in genetically ameliorated seed, refined rice culture exercises, and better methods for soil management, biotic factors, and water.

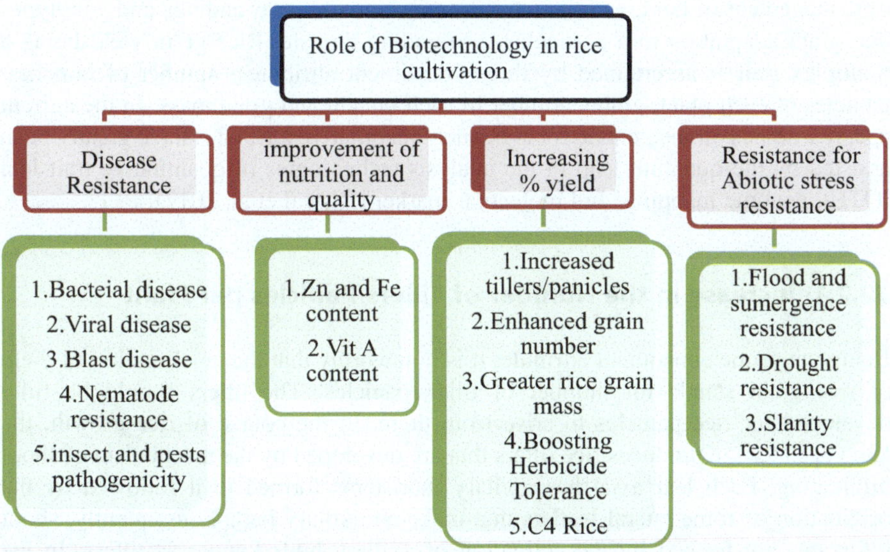

Fig. 32.1 Amendments made by rice biotechnology

Consequently, rice research has fundamental influences on certifying food security, environment protection, handling climate change, and poverty reduction (Reardon et al. 2019).

Biotechnology comprises of the use of living organisms (e.g., plants, animals, microorganisms), either wholly or in part, to prepare, remold, or ameliorate a product, animal, or plant, and developing the microorganisms for peculiar ambitions. Examples include soy sauce, bread, and beer (foods synthesized using biotechnology). Antibiotics including penicillin, insulin for treating diabetes, and vaccines for rabies, hepatitis B, and measles (Panda et al. 2020). Being a valuable tool, it can amend the standard of life in innumerable approaches, now and also in the future, by the synthesis of life-protracting medicines, enhanced food nutrition, and improved plant varieties (Frewer et al. 1994).

32.2 Ameliorating the Rice Yield

Yield is the most crucial and multiplex feature in genetic amendment of rice. Across previous years, immense efforts are made in genomic research of rice functions. Cloning besides characterization of function-related genes which might be correlated or directly linked to productivity attributes has guided to substantial advancements to understand the molecular as well as biological pathways rudimentary to traits in rice yield (Joshi et al. 2020).

Varieties of rice vary substantially for the productivity. This discrepancy is impelled through their colossal genetic variety, atmospheric conditions, practices of management of field, and the interlinkages between surroundings and genotypes that grant adaptation to a particular surrounding climate. Rice grain yield being a multiplex trait is ascertained by three constituent attributes: number of bunches/panicles of each plant, grains number in each bunch, and grain mass. In the current age, researches on the genetic roots of such quantitative attributes have greatly been assisted by the quantum leap in the analysis technologies of quantitative trait loci (QTL), genome mapping, and molecular marker (Biswal et al. 2017).

32.2.1 Increase in the Number of Tillers/Panicles per Plant

Being the prime constituent attribute, it is noteworthy that the amelioration of yield of rice grain stands for number of tillers/panicles. The tillers should be fully developed for rice panicles to arise from them. In the course of rice growth, the late vegetal stage produces rice tillers that are developed by the mechanism of shoot bifurcating. Each leaf axial has axillary meristems formed in it followed by the origination of some lateral leaflets that make an axillary bud. Consequently, shoot bifurcates are formed by the activation of axillary buds known as tillers. In the favorable conditions, secondary as well as higher-ordered tillers may arise by further development (Baral et al. 2020).

Once fully developed, as tillers develop completely, they cause the growth of panicles and further grains development after pollination. Axillary bud activity is controlled via a systematic network of signals in consequence of multiplex interconnections of phytohormones. The environmental, growth, and genetic signals regulate the signaling through phytochromes (Shimizu-Sato and Mori 2001). These phytohormones including auxins (Aux), cytokinins (CKs), and strigolactones along with their derivatives facilitate flow of information among various organs of the plant required for the development (Domagalska and Leyser 2011).

Role of Aux and CK in modulation of shoot apical-meristem of rice shoot shad has been shown by a famed notion of apical dominance (Azizi et al. 2015). It has been witnessed that CKs promote branching, which leads to a rise in the rice spikelet number, whereas Aux halt the growth of axillary bud, being synthesized by the apex of primary shoot (Azizi et al. 2015). CK lessens suppression and facilitates lateral branches to grow, consequently the low CK level of turn out in the boost of apical dominance and arrest of growth of axillary bud (Azizi et al. 2015).

Agronomically, it is recommended that attempts to modulate the activity of axillary bud via genetic engineering and biotechnology would be profitable in the trials for rice yield elevations (Xing and Zhang 2010).

32.2.2 Marker-Assisted Breeding

Currently, numerous significant QTLs are identified for panicle number modulation. In a cross of ingressed indica IR71033-121-15 and Junambyeo, chromosome numbers 4 and 6 were found to possess two QTLs that were accountable to influence panicle number in both residentiaries, denominated as pn4 and pn6, respectively. It was found in the same research that revealed the certainty that IR71033 conferred a provoking effect on chromosome 4; contrarily, chromosome 6 had a declining impact.

32.2.3 Transgenic Research

Certain plant hormones have also been ascertained that affect the branching as well as tillering, along with the multiple genes being influenced by these phytohormones subsequentially. Rice OsTB1 (*Oryza sativa* Teosinte Branched 1) is an example of such genes. OsTB1 encrypts for a recognized transcription protein that carries a DNA-binding motif of the type helix-loop-helix, and it suppresses the sidewise ramification in rice. Currently studied, transformation mediated by Agrobacterium had upregulated this gen. Furthermore, it was distinguished that in RNAi-mediated knockdown OsTB1 and upregulated rice plants, the number of panicles was ameliorated and reduced, respectively, in contrast to the wild-type rice plants (Choi et al. 2012).

32.3 Increase in Grain Number in each Bunch/Panicle

Being the most multiplex quantitative feature of the rice plant, grain number has assuredly correspond to grain yield per plant and hence is important in the betterment of rice yield. The grain number in each panicle is classified into three constituent traits: development of panicle, rate of formation of spikelet, and panicle differentiation duration (Tripathi et al. 2012). The development of panicle points to the shift from late vegetal stage to the reproductive stage and is affected through the cooperation of phytohormones including several genetic agents. Two significant modulating genes for axillary meristem production in rice are identified in mutant analyses including the SMALL PANICLE and the LAX PANICLE1 (Lax1) genes. Lax1 and SPA genes were demonstrated to play interconnected function, which is conspicuous by the shifting from the vegetative stage to the reproductive stage (Huang et al. 2019).

32.3.1 Marker-Assisted Breeding

Ghd7 encrypts for a protein family called the CCT motif family and acts as one of the crucial modulators in control of various traits pleiotropically, comprising grain number per panicle, height of a plant, and also the heading date (hence named Ghd). Currently, it has been revealed that Ghd7 upregulates both panicles besides tiller branches through a density-dependent mechanism, pointing out that Ghd7 impacts on branch development control responding to atmospheric context (Weng et al., 2014). Moreover, upregulated Ghd7 downregulates the Hd3a which elongates the panicle differentiation.

Ghd7-3, Ghd7-1, besides related Ghd7 operative alleles with intense impacts permit the rice to use temperature and light by deferring flowering in case of long-day conditions in the areas having prolonged growing seasons, hence produce big panicles causing increased grain yield (Xue et al. 2008). Likewise, Ghd8/DTH8 is capable to downregulate early heading date 1 (Ehd1) as well as Hd3a in a state of long-day conditions resulting in a postponed heading date along with elevated grains per plant (Yan et al. 2011).

The central molecular and genetic mechanisms that modulate GN remained greatly unknown. Another recent study detected the quantitative trait loci (QTLs) for traits related to grain yield utilizing a series of chromosomal segment substitution lines (CSSLs) that were produced through a cross between the Indica cultivar 9311 being the donor of the trait, and the japonica cultivar "Nipponbare" being the recipient. A total of 25 QTLs were identified on eight chromosomes that encrypt the traits related to panicle, including secondary branch number, primary branch number, length of panicle, and grain number (GN) (Huang et al. 2018). Among the QTLs, it was revealed that qGN1c for grain number (GN) is situated near Gn1a that had been previously recognized as the main QTL for several grains on chromosome 1. Ultrafine mapping was carried out to place qGN1c inside the region of ~ 379 kb in an interval of chromosome 1 that was flanking Gn1a, revealing qGN1c as an unclosed gene. Besides, assessment of agronomic features in a near-isogenic line (NIL-qGN1c9311) devised the qGN1c not to have auxiliary effects on agronomic features excluding GN and grain weight in thousands. Crucially, the NIL-qGN1c9311 grain yields in each plant were greatly escalated by 14.46% and 13.34% in two planting loci. Consequently, the determination of qGN1c gives a beneficial genetic tool to upgrade the rice grain yield during its breeding (Xu et al. 2019).

32.3.2 Transgenic Approach

Various studies have manipulated the transgenic approach method to characterize some genes functionally to reveal either they affect the grain number per panicle. For this, a study reports the LRK1 (leucine-rich repeat receptor-like kinase 1) to come about with 27.09% increase in overall productivity per plant as the consequence of its overexpression causing a boost in panicle number, the weight of each grain, spikelet number per panicle, and cellular escalation that collectively contribute to the total grain yield (Zha et al. 2009).

32.4 Greater Rice Grain Mass

Grain weight is an important trait about the quality and yield potency of rice grain. To identify the net weight of grains, there are three frequently followed criteria including grain width, grain length, and grain filling. Present studies on cloning and QTL delineation have made great breakthroughs on the determination of genes and crucial QTLs modulating the net grain weight. The significant QTLs/genes that have been stated to modulate these criteria are GW2, GS3, and grain incomplete filling 1 (GIF1) (Tripathi et al. 2012). These all put light on role of biotechnology for elevating the potential of rice yield, that is, having knowledge and by regulating these genes, being agronomically important, modulate significant yield features, and put the master plans into an effort to magnify the rice productivity.

32.5 Boosting Herbicide Tolerance

Herbicide tolerance (HT), being an agronomically significant trait, has been utilized to supervise weeds efficaciously for numerous decenniums. A huge number of approaches have been utilized to generate HT crops that involve in vitro selection, cell culture, and genetic engineering strategies. Genes conferring the resistance against herbicides are being manipulated to synthesize herbicide-impervious rice. Commercially, nowadays, three central HT systems are rooted in generating resistance to the herbicides by inhibiting the metabolic production of amino acids. Systems included are glyphosate, imidazolinone, and glufosinate resistance (Duke 2005). The three HT rice systems have been established in the rice plant (Wang et al. 2014a). IMI-tolerant rice offers resistance against the potential herbicides having the imidazolinone group (i.e., imazapic, imazamox, imazapyr, imazethapyr), controlling a vast spectrum of weeds and grass, attaining the conducive environmental consequences (Tan et al. 2005).

Herbicides with the imidazolinone group govern the weeds by negative regulation of the enzyme acetolactate synthase. Because of this, transgenic strategies including traditional breeding for gene mutation have been treating ALS as their crucial target. Bispyribac sodium (BS) is the ALS suppressor. Tolerance against BS is integrated by two ALS gene·point mutations (Endo et al. 2007). Besides, a greatly efficacious system of gene-targeting (GT) mediated by T-DNA was discovered to cause S627I and W548 L mutations in the ALS gene that came up with the development of highly tolerant plants of rice against BS. The transgenic strategy has also been manipulated for the insertion of genes resistant to herbicides into rice plants obtained from another organism. It involves the gene that encodes the herbicides having glyphosate as well as the bar gene that is obtained from *Streptomyces hygroscopicus*, which is capable of abrupt metabolic withdrawal of glufosinate (Kumar et al. 2008).

32.6 C4 Rice

C4 photosynthesis being the most significant adaptation of the flowering plants comprises about 20–30% of terrene carbon fixation is comprised by C4 photosynthesis despite being utilized by hardly about 3% of the entire angiosperm families (Kellogg 2013). Contrary to the common C3 plants including rice, most of the efficacy of energy conversion of C4 photosynthesis greatly relies on its carbon dioxide concentration procedure.

The enzyme named "ribulose-1,5-bisphosphate carboxylase/ oxygenase or Rubisco" is the central catalyst to carry out the carbon fixation in all plants. Formation of an unstable intermediate having six carbons by carboxylation (addition of CO_2) of ribulose-1,5-bisphosphate (RuBP) is catalyzed by Rubisco (Döring 2017). Subsequently, two molecules of 3-phosphoglycerate (3-PGA), a three-carbon compound, being the first stable product (hence the name C3), are resulted by the cleavage followed by hydration of the above intermediate. The leaf surface has localized mesophyll (M) cells where the process takes place in C3 plants. The aberrant interface of the enzyme with oxygen evokes Rubisco's oxygenase role, which involves the covalent attachment of O_2 at Carbon 2 of the precursor RuBP which results in one 3-PGA and one 2-phosphoglycolate (2-PG) molecules. A greater concentration of 2-PG is noxious for the plant, hence, must be detoxified in mitochondria and peroxisome through the process known as photorespiration involving the regeneration of 2-PG to 3-PGA. Although it comprises removal of previously imbibed CO_2 besides NH_3, the equivalents of energy consumption as well as reduction (Peterhansel et al. 2010), ultimately it suppresses the efficacy of photosynthesis up to 30% (Mallmann et al. 2014). Due to this, photorespiration is usually thought to be a fruitless phenomenon.

Initially, occurring in C4 plants, CO_2 is fixed resulting in the generation of four-carbon compound oxaloacetate (OAA) (hence C4) by phosphoenolpyruvate carboxylase (PEPC) which is a cytosolic enzyme. The PEPC significantly interacts with CO_2 only, being inconsiderate to O_2. A large number of trials are being made either to introduce this C4 pathway into the rice or to get a whole understanding of the complete biomolecular reaction of C4 photosynthesis.

32.6.1 Marker-Assisted Breeding

As a matter of fact, due to the existence of C3–C4 intermediates and the gradual conversion from C3 to C4, the C4 approach has been reported as a polygenic quantitative trait (Westhoff and Gowik 2010). Since the 1970s, trials to map the QTLs among the species related closely have been carried out. Malcolm Nobs and Olle Björkman made the earlier crosses between Atriplexrosea (C4) and Atriplexprostrata (C3), the generated F2 population exhibited individual C4 segregation, but it did not progress due to anomaly. Likewise, crosses have been carried out between C3, C4, besides C3–C4 intermediates of *Flaveria* species by biotechnological approaches (Hernández-Prieto et al. 2019).

32.6.2 Transgenic Research

In the first instance, it was assumed that a single-cell C4 system was thought to prosper faster in C3 plants, and there have been earlier attempts to genetic engineering of single-cell C4 system in rice has been endeavored formerly (Miyao et al. 2011). Though, the only inefficacious cycle has been produced in previous trials (Zhu et al. 2010). Another attempt for the more efficacious consolidation of the C4 system of Kranz-type into rice is the monopolypioneered genetic engineering of familiar genes into the rice. C4 genes that have been engineered by cloning from maize and incorporated into rice comprise PPDK, PEPC, NADP-ME, NADP-MDH, and CA (Kajala et al. 2011). At present, phase III in the project of C4 has been commenced legalizing for a more rectified genetic instrumentation which has been amassed during previous phases and an enhanced perception of modulatory processes that inaugurate mechanism in C4 crops. Such attempts are being oriented for the complete recognition for engineering the C4 pathway genetically into the rice.

32.7 Improvement of Nutrition and Quality

A worldwide serious problem of health, peculiarly prevailing the successful countries, is hidden hunger or malnutrition of micronutrients. Minute amounts of minerals and vitamins, often designated to be micronutrients, are crucial for development and growth. The absence of micronutrients poses severe consequences in overall growth besides development, and growth of the human body is severely affected by micronutrient deficiency. Food chains are responsible for the provision of micronutrients to the human body. No access to costly foods depleted with micronutrients, that is, fortified items of food, animal beneficial products, vegetables, and fruits cause deficiencies among people. Eight African countries, nine north and south American countries, and 17 Asian countries take rice as a staple or the diet with a primary energy source (30%). Depleted levels of vitamin E, provitamin A, zinc, folate, and iron occur in milled rice. Trials for micronutients·fortification and supplementation of cereals including related crops have been attempted previously. Biofortification is directed to generate vitamin and nutrients rich rice cultivars via transgenic or traditional breeding approaches for plants. Complete functional profiles for the contestant are being tried to explore via genomic rooted efforts to evolve the nutraceutical rice cultivars. Multiplex genes for endosperm storage with micronutrients can be induced to a single cultivar to produce increased micronutrient concentration diversity. Additionally, biofortification can also be manipulated for the generation of superior varieties of rice with integrated increased micronutrient concentration and large yield features. Current advances in genetic engineering facilitate such innovative and cost-effective outcomes (Hasegawa et al. 2019).

32.7.1 Zinc and Iron Content

Vitamin A, zinc, and iron scarcity are perceived as most recurrent structures of micronutrient scarcity. Iron depletion only influences approximately two billion individuals, whereas vitamin A scarcity influences at the minimum hundred million children annually, where about 1.3–2.5 million get perished, while above 0.25 million get suffered from persistent health errors. Iodine scarcity may be encountered by almost 30% of the world's whole population. Deficient food on account of impoverishment is thought to be the most leading cause of zinc scarcity that declines hunger and deleteriously influences the body's defense system. Averagely zinc and iron contents in polished rice are 12 and 2 parts per million, respectively. Multiplex assorting strategies have been undergone to recognize the high zinc and iron level producing rice cultivars (Sardar et al. 2015).

Early germ plasmas sortation and field estimation of procreation lines by Philippines, India, Indonesia, Bangladesh, and Korea has exhibited the strategy to escalate the zinc and iron content of polished rice by the factor of 2–4. Nearly double iron and approximately 40% increased zinc contents have been observed in the floating rice type in India called, Jalmagna. Many diverse rice types that are proficient for zinc can be manipulated for zinc·iron levels.

XuaBueNuo, KuatikPutih, BilleKagga, Zuchem, and Getu are some of the classes of rice that are a very efficient source of zinc or iron having an abundant amount of zinc in it. It was revealed that the high-iron trait is controlled by three types of genes located over 7, 8, and 9 chromosomes. Alteration in iron levels in different varieties of rice is also described following these genes. Permanent mapping populations of F8 c were developed to map high-iron besides high-zinc traits (Anuradha et al. 2012).

Levels of zinc and iron are suggested to be associated with each other (Anuradha et al. 2012). However, the amount of the minerals is partly determined by the climate and characteristics of the soil as well. For experimentation three genes, for example, *Phaseolus vulgaris* (common bean) ferritin (pfe) gene, cysteine-efficient metallothionein-like (rgMT) protein, and *Aspergillus fumigatus* phytase (phyA) were injected in rice. The digestive system produces phytase and cysteine peptides that hydraulically break phytic acid and increase absorption of iron respectively. Iron and zinc are multivalent metal ions that are chelated by phytic acid found in plant seeds. This chelating produces insoluble salts thus shoeing lesser bioavailability. Rice that is non-transgenic kind have seeds that possess elevated phytase level and cysteine residues 130 and seven times respectively. Genes from a ruminal bacterium such as *Escherichia coli* (appA) and *Selenomonas ruminantium* (SrPf6) were introduced by Hong et al. (2004) in Taiwan in transgenic rice to describe the elevation of phytase activity. Phytase functionality was found to be increased at pH ranges, for example, 2.0–6.0 at the optimized temperature of 55 and 60 °C by expression of these genes. MxIRT1 enhanced expression causes the transportation of iron by utilizing ATP (Tan et al. 2015). Processed and unprocessed rice have enhanced iron due to the strong expression of PvFERRITIN, AtIRT1, and AtNAS1 (Boonyaves et al. 2016) (Fig. 32.2).

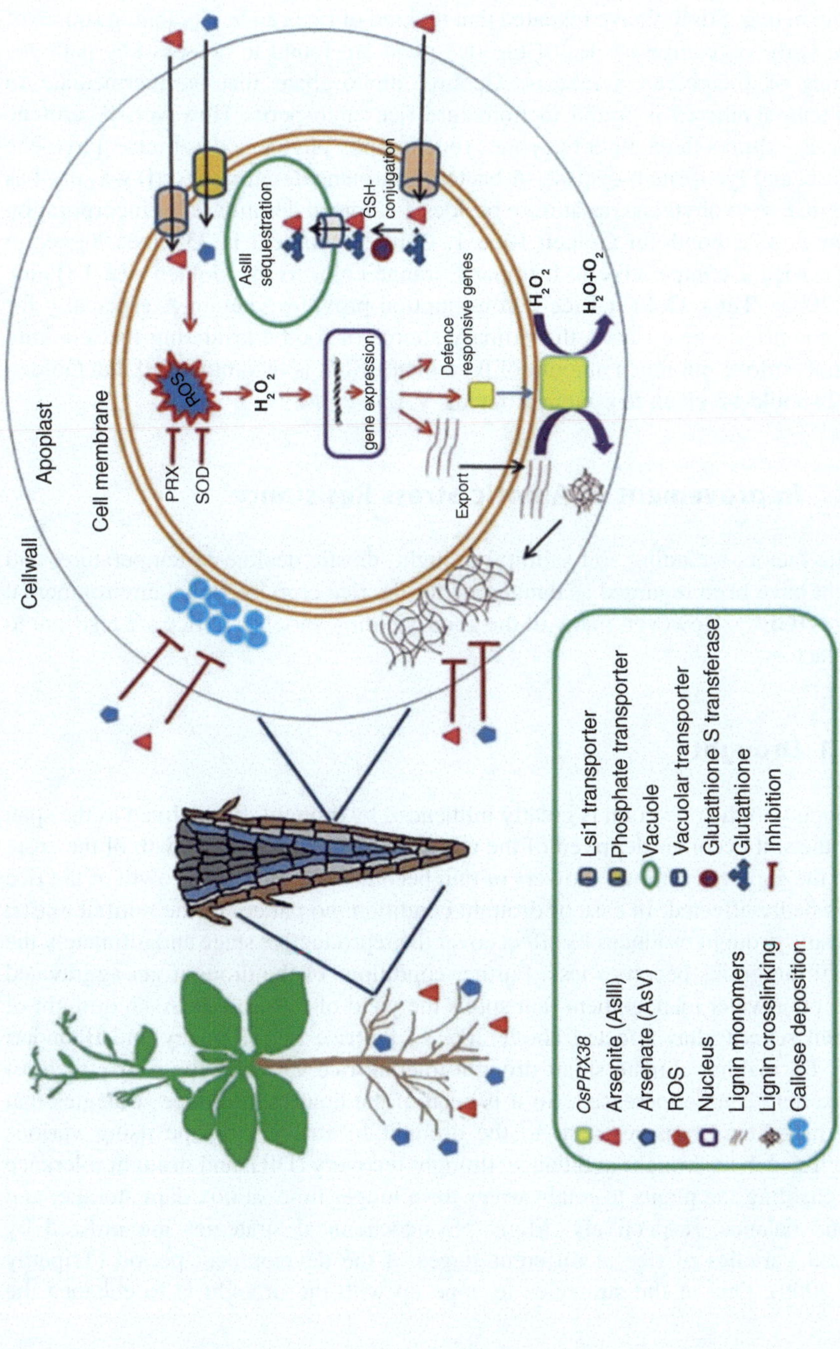

Fig. 32.2 Transgenic rice development

32.7.2 Vitamin A

Rice crop is considered as a chief source of energy particularly in those countries facing vitamin A deficiency (VAD). β-carotene, which is the precursor of Vit A, is missing in rice. Studies have revealed that no kind of rice can be a potential source of Vit A. Only vegetative tissues of the rice plant are found to possess key pathway enzymes of β-carotene synthesis. Geranyl diphosphate that is intermediate to β-carotene syntheses is found in immature rice endosperm. However, β-carotene synthesis requires three other enzymes, for example, phytoene desaturase, phytoene synthase, and lycopene β-cyclase. A bacterial carotene desaturase (crtI) was used as an alternative to phytoene desaturase besides ξ-carotene desaturase for incorporation of four double bonds in Golden Rice 1. Carotene content is 23 times higher in Golden Rice 2 comparatively. It imparts orange color to the Golden rice 1 (Paine et al. 2005). Thus, Golden Rice 2 consumption provides vitamin A especially for those countries where rice is the primary source of food considering that cooking does not remove out much amount of β-carotene. So it is recommended that Golden Rice 2 should be given to countries facing VAD.

32.8 Improvement of Abiotic Stress Resistance

Abiotic factors including soil salinity, drought, drastic decline of temperature, and drought have been regarded as dangerous for the rice crop by global environmental changes (GEC). However, many of the good-yielding varieties of rice are resistant to these factors.

32.8.1 Drought

The quality of the rice grain is greatly influenced by drought. It is defined as the span when the soil becomes deprived of the moisture required for the growth of the crop. When the gap between two showers of rain becomes too long, the growth of the rice crop is badly affected. In case of drought condition, no pattern of the rainfall exists; as a result, drought produces its effects over the reproductive stage and ultimately the yield of the grains becomes less. Further conditions of the drought get aggravated due to poor water management that spoils the yield of the crop. In Asia, drought of different severity has affected above 20% of the rice field (Pandey and Bhandari 2009). By making climate-smart drought-tolerant rice varieties, the yearly yield of the rice field can be increased. So it is need of the hour to introduce strategies that could make the crops resistant to the drought by drought escape using various strategies such as drought avoidance, drought recovery (DR), and drought tolerance (DT) enabling the plants to retain water for a longer time, antioxidant storage, and osmotic balance, respectively. Many physicochemical strategies are utilized by different varieties of rice at different stages of the development period (Tripathy et al. 2000). One of the strategies to cope up with the drought is to enhance the uptake of water by deep and thick roots to store extra water. Studies have revealed that this deep rooting has produced good results more than the less deep root system. High-density SNP genotyping array and molecular markers have been used to identify root-specific gene expression, and parental polymorphism has been used to explain drought-tolerant rice plant (Chen et al. 2014).

There is a huge variety of drought depending upon the severity, span, time, and hydrology, which is a leading cause to the development of drought-resistant plants. The drought condition gets aggravated due to nutrients insufficiency and elevated temperature. Birsagora, Dular, and Laloo 14 are some of the known drought-resistant varieties. Vandana and Annada in India and some of the varieties in the Philippines such as PSBRc82 and Apo have shown the best drought resistance. So a bit of achievement has been achieved using the conventional mode of breeding. For the sake of the development of the drought-resistant varieties of rice, various national as well as international works are in progress. Direct application of higher grain yield varieties of the rice is becoming common for the last decade. The advanced studies have revealed the importance of the varieties that could withstand drought conditions as well as normal availability of the water and produce the same yield in both conditions. But much of the improvements are required in introducing proper drought conditions for the experiments designed to study phenotypic behavior of drought-resistant and drought-sensitive cell lines (Fig. 32.3).

Though it is difficult, to attain a phenotypic selection of drought has been an ideal characteristic for the betterment of MAS. Many aspects regarding the drought plants have been unfolded. At the, first secondary level, influencing traits, was focused on the selection. Structural characteristics of the root such as depth of root and anchorage were found to be influenced by genetic variability. The morphology of rice roots is found to be influenced by root dry biomass tiller^{-1}, root length, and maximum root length that were linked to QTLs (Steele et al. 2006). An inverse relation between the morphology of root helpful in drought resistance and osmotic adjustment and dehydration tolerance was also revealed (Shen et al. 2001).

Allele from Azucena deep root was transferred to the elite high-yielding cultivar IR64 to study marker-assisted backcross. The influence of the other characteristics was also studied to found the role in drought resistance. It was suggested that 42 QTLs are related to drought in rice. Chief QTLs are described as the most efficient factors for the resistance against drought as the polygenic nature of these resistant plants is much complex (Obara et al. 2010).

At the time of the reproductive stage, the grain yield was estimated for the evaluation of the first large-effect QTL, qDTY12.1, in the Way Rarem population. It depicted the genetic variance (Dixit et al. 2015). Reorganization of marker-assisted backcross breeding and large-effect QTLs (qDTY2.2, qDTY4.1, qDTY9.1, qDTY10.1, and qDTY12.1) has helped in the progressive development of the rice varieties Vandana and IR64 even in the state of drought at the reproductive stage (Mishra et al. 2013). In drought stress, directly choosing the reproductive stage has played a beneficial role in the development of drought-resistant and good-yielding varieties in Africa, South Asia, and Southeast Asia (Kumar et al. 2014).

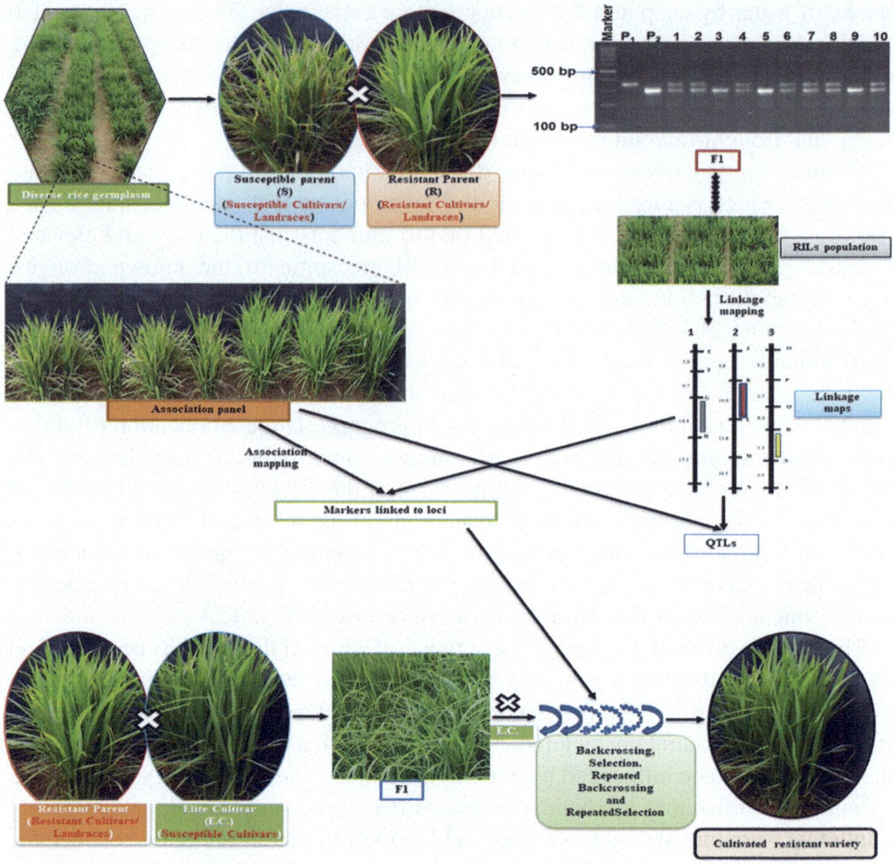

Fig. 32.3 Mechanism of marker-assisted breeding (**Source**: Mehta et al. 2019)

Studies have confirmed that transgenic rice lines can better withstand drought conditions. The mechanism of transgenicity has been the center of discussion for molecular biologists because of many ambiguities. Transgenic cell lines have been attained by various gene modifications such as genes targeting secondary traits (water use efficiency (WUE)), root system architecture, or by showing transcription factors. These factors are involved in regulation of gene expression that helps to cope up with stressed conditions. For example, elevated expression of NAC (NAM, ATAF, and CUC) transcription factor helps in the development of the resistance against drought in these transgenic varieties of rice. These transgenic rice are obtained with the same yield as well as produce resistance at reproductive and vegetative stages.

Extra grain yield due to higher tiller numbers and larger panicle sizes have also been reported with transgenic rice in drought conditions (Todaka et al. 2015). An increased expression of OsNAC6 in rice plants also produces salinity resistance as well as drought resistance. Similarly, transcriptional factors such as OsbZIP16 and

OsbZIP23 are also found to be involved in the ABA-dependent regulatory pathway and ultimately tolerance against these stresses (Chen et al. 2012). Cellular protection to fight with such stress is attained by important stress-inducible proteins of late embryogenesis abundant. Enhanced stress resistance against drought and salinity is obtained by different LEA proteins of transgenic rice (Duan and Cai 2012). An enzyme known as pea manganese superoxide dismutase (MnSOD) is an antioxidant that is found to be involved in producing drought resistance (Gu et al. 2013).

Another transgenic cell line of rice was developed with maize C4-specific phosphoenolpyruvate carboxylase (PCK) and individually as well. The combined cell lines gave more yield of grain as compared to wild type even in drought (Kader and Lindberg 2010). Overexpression of OsCPK4, OsSIK1, and OsSIK2 was also involved in transgenic rice plants that can withstand drought and salinity by good water storage capacity and lowered lipid peroxidation in membranes (Campo et al. 2014; Guo et al. 2013). Much research work is the need of the hour for improved transgenic rice varieties that can explain the mechanisms working behind the development of drought-resistant rice plants.

32.8.2 Salinity

Salinity is an important factor to determine the growth of rice plants. Rice plants exhibit an altered level of capacity to cope with salinity. At the time of germination, active tillering, and maturity, rice plant is more resistant to salinity as compared to the seedling and reproductive stage (Singh et al. 2010). Soil salinity has affected one-fifth of the total irrigated arable lands worldwide (Negrão et al. 2011). Brackish water gets intuited in the coastal region rice crops especially at the start of the wet and dry seasons. Bad quality irrigation water and improper drainage make the salinity condition more adverse. Different varieties of rice show genetic variation for resistance in salinity. Some old varieties such as Damodar, Dasal, and Pokkali have become resistant to salinity while some new varieties are still suitable to salinity. Independent genes exhibit different behavior toward salinity. QTLs are found to be linked with salinity resistance in rice plants. For example, RIL from this population is found to be helpful in the development of salinity resistance in BT7 variety utilizing marker-assisted backcross (Linh et al. 2012). The reproductive stage of robust QTLs is not found to be resistant to salinity (Jena and Mackill 2008). Highly robust QTL is Saltol/SKC1 present on short arm of chromosome 1, which is one of the best varieties. Genetic makeup has been widely studied to explore genomic reasoning behind salinity resistance. All of the efforts are made to produce improved varieties of rice that are more resistant to salinity even at their reproductive stage during dry seasons (Hossain et al. 2015). Some of the factors such as Na+/K+ ratio, pollen fertility, and Na+ exclusion are also found to be influential against salinity at reproductive stage. Utilizing embryo rescue technique, a new cross of exotic wild-rice species *Oryza coarctata* besides cultivated rice cultivar IR56 has been made. It has given two folds more salt resistance in an obtained generation.

The transgenic study has been conducted to evaluate the genes of late embryogenesis proteins and aquaporins (water channel proteins) to correlate it with drought and salinity stress. However, much work is still needed to help breeders in this case. It was revealed that *Arthrobacter globiformis* processing gene choline oxidase (codA) were salt-resistant (Mohanty et al. 2002). This resistance is attained by catalyzing the conversion of choline to glycine betaine. The obtained form is proved to play a protective role, particularly in photosynthetic organelles. Those transgenic varieties having codA gene in the chloroplast were found to be more resistant than those having this gene in cytosol (Kathuria et al. 2009). Overexpression of superoxide dismutase (SOD), calcium-dependent protein kinase OsCDPK7 gene, and HVA1 are also helpful in salinity resistance. Increased growth was observed in two rice varieties by the expression of Barley LEA protein genes under drought and salt stress (Babu et al. 2004).

32.8.3 Flood and Submergence

Lowlands of South and Southeast Asian cultivators of rice have been facing the problem of flooding and submergence of the crop. So, the low-yielding varieties that can hold with flooding are preferred over high-yielding varieties that cannot withstand floods (Fig. 32.4). It is the third most affecting abiotic factor to the rice crop in Eastern India and only overcome by weeds and drought. The genes that can impart the tolerance against submergence are found to be complex genetically and environment-dependent. "Quiescence" strategy has been adopted in recent times to make the rice field resistant to flooding (Bailey-Serres and Voesenek 2008). This survival is attained by an ethylene-responsive factor (ERF), Sub1A, that gives

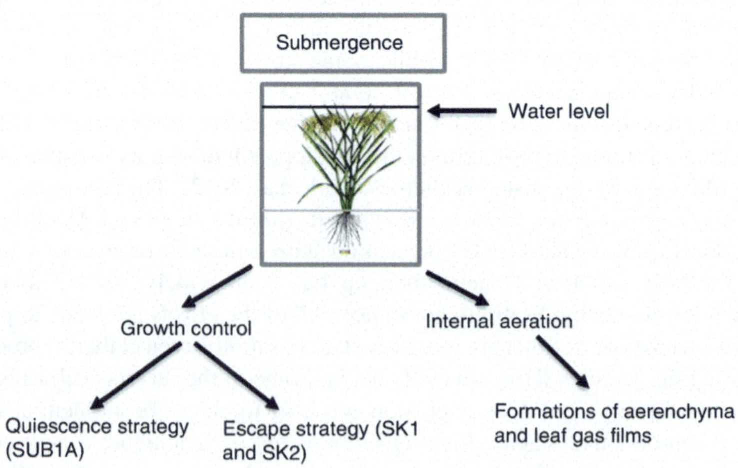

Fig. 32.4 The stress of submergence and flooding in rice plant (**Source**: Mahmood et al. 2019)

hormonal trigger to submerged plants in the form of gaseous ethylene (Xu et al. 2006).

Simple and quantitative inheritance can be useful for obtaining submergence tolerance. FR13A, a tolerant landrace from Odisha, is being used to cope up with submergence in India. A cross between susceptible japonica line ("P1543851") and submergence-tolerant indica line (IR40931-26) was designed that has given Submergence 1 (Sub1). Submergence tolerance was varied 70% phenotypically (Hattori et al. 2009). The stress of 2 weeks submergence was given to tolerant variety FR13A having gene Sub1. Variation in the resistance to submergence was studied by a backcross of IR64-Sub1 cultivar carrying QTL/Sub1 (Lang et al. 2013). These marker-assisted backcrossings were used to combine QTL with Sub1 to increase the resistivity against submergence. Waterlogging response has been studying by identifying two factors, for example, overexpression of transcription and lowered expression of candidate genes such as ethanol synthesis using sense and antisense constructs (Dennis et al. 2000). Long-term tolerance to less oxygen availability was found to be attained by these two factors (Quimio et al. 2000).

32.9 Improvement of Biotic Stress Resistance

The 37% of rice productivity gets disturbed by pests and diseases. This damage can be up to 24% besides 41% depending upon condition of production (Sparks et al. 2012).

32.9.1 Disease Resistance

A pathological condition that causes maximum loss of rice is rice blast. It is carried by the fungus known as *Magnaporthe oryzae*, which is an ascomycete of filamentous nature having a genome distributed into seven chromosomes (Dean et al. 2005). It damages Sesma and Osbourn at their adult stages, panicles, nodes, and seedling stage as well (Sesma and Osbourn 2004) (Fig. 32.5).

A large number of blast pathogen resistance genes in *M. oryzae* have been isolated using advanced molecular analysis techniques (Sharma et al. 2012), for example, Pib, PiaPik, Pi1-Pi62, Pii, Pi-kh (same as Pi54), Pish, Pit, Pita, Pita 2, Pitp, etc. (Wang et al. 2014b). Marker-assisted backcross breeding (MABB) approach has played its role in introducing the resistance genes in the genetic background of PRR78. This incorporation has resulted in the formation of resistant Pusa1603 (PRR78 + Pi54) and Pusa1602 (PRR78 + Piz5 (Singh et al. 2012).

Resistance against neck blast was found to be developed in the transgenic rice having blast resistance gene Pi-d2 (CHEN et al. 2010). This resistance was found to be developed due to transcription factor-related genes, large expression of the enzyme triggering the fight system against fungus (Gupta et al. 2012). Basmati rice was also made resistant by this gene transfer (Asghar et al. 2007). Inhibition of rice radical growth and wilting of seedlings of rice was observed due to the Rice

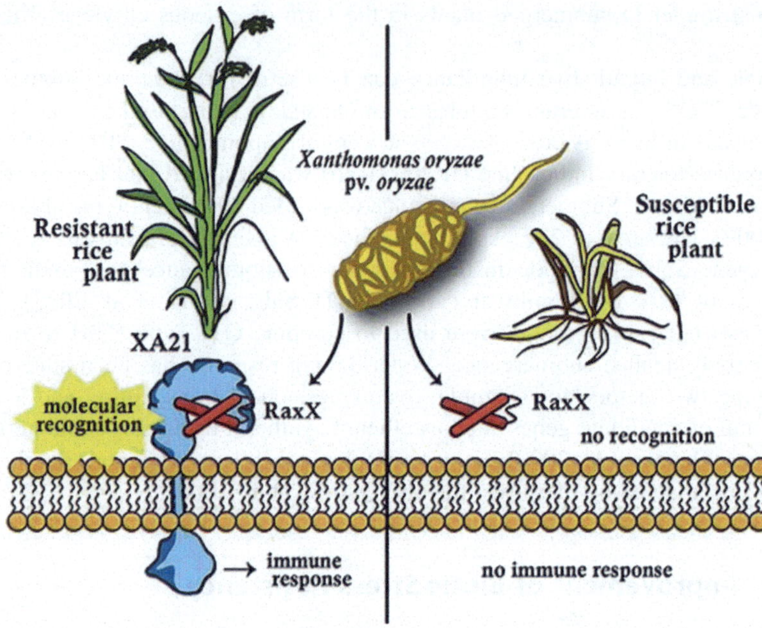

Fig. 32.5 Disease resistance in rice plants (**Source**: Pruitt et al. 2015)

ShB pathogen. Studies are continued on resistance to ShB. Rice of different varieties has shown resistant genotypes of varying levels of resistance (Fig. 32.6).

ShB resistance-related studies were made in different species of *Oryza*. High-level resistance was found in more number of species, while some also have shown moderate resistance to ShB. It was revealed that mixed genetic groups were found to be involved in high-level resistance (Ram et al. 2008). When pathogens attack a plant, specific kind of pathogenesis-related proteins are produced. For example, β-1,3-glucanases (PR-2), plant- or microbe-derived antifungal proteins, and chitinase (PR-3) overexpression help in the development of transgenic plants against fungal infection. ShB and many other disease resistance have been observed in rice plants having overexpression of Dm-AMP1 chattiness and Ace-AMP1 (Jha et al. 2009). The collaboration of resistance genes in a single rice plant has been under study. Different genes such as Chinsurah Boro II, Swarna, IR64, and Basmati 122 has been transferred in rice to get resistance against ShB (Kalpana et al. 2006). Kalpana et al. (2006) have developed rice cultivars having overexpression of IR50, ADT38, and Pusa Basmati1. These genes were found to be involved in the encoding of thaumatin-like protein. These transformed progenies exhibited a better level of resistance against *Rhizoctonia solani* and sheath blight pathogen. *R. solani* resistance was in rice chi11 having a combination of chitinase and tlp.

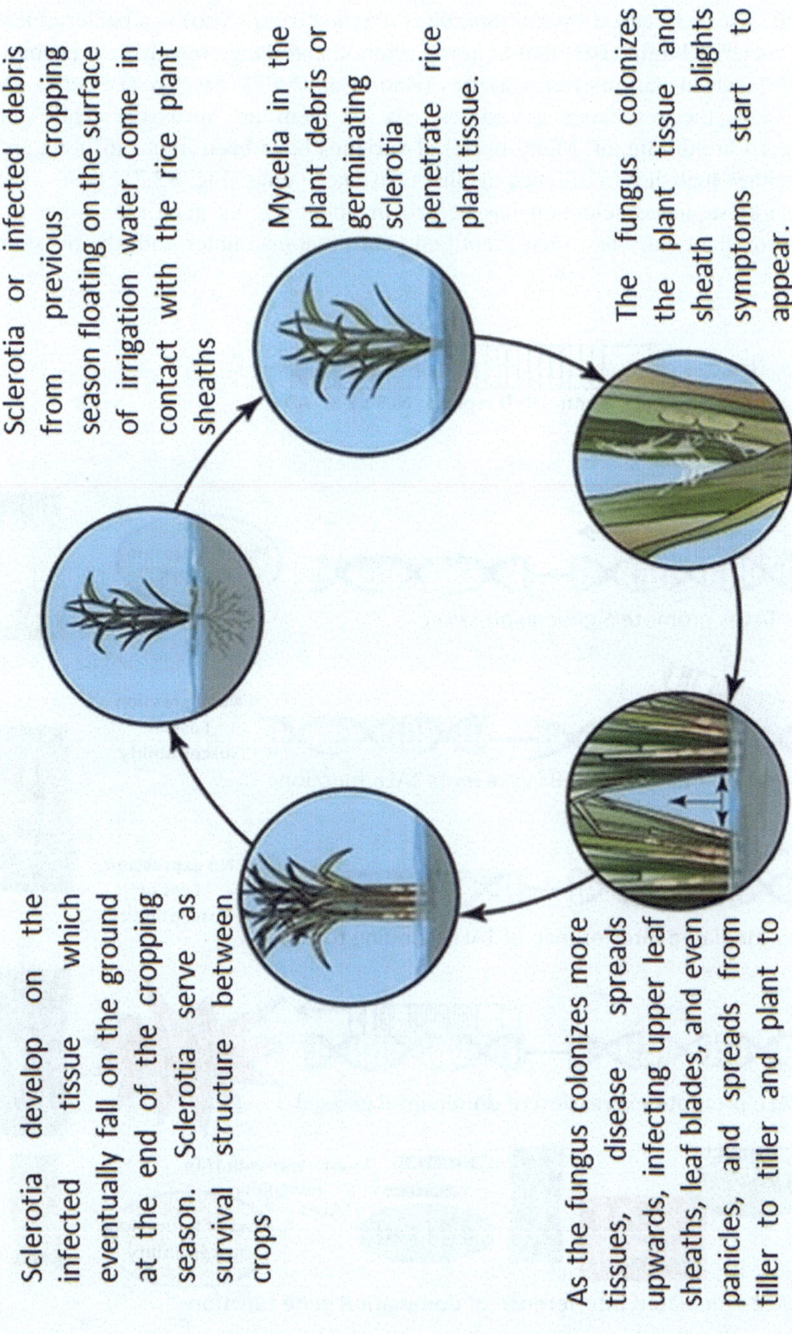

Fig. 32.6 Rice sheath blight

32.9.2 Bacterial Disease

Pathogenic diseases caused by *Xanthomonas oryzae Oryza* (Xoo) is a bacterial leaf blight or bacterial blight (BB) that is most common and dangerous disease encountering resistance in various rice varieties (Rao et al. 2002). Twenty-six genes are involved with the resistance encounter. Six of them are recessive while the remaining 14 are dominant. Many bacterial varieties have been found in India and the Philippines that show virulence against resistance genes (Fig. 32.7).

The marker-assisted selection has helped in obtaining Xa gene clones for the release of resistant cultivars. First identified genes that encounter with the resistant

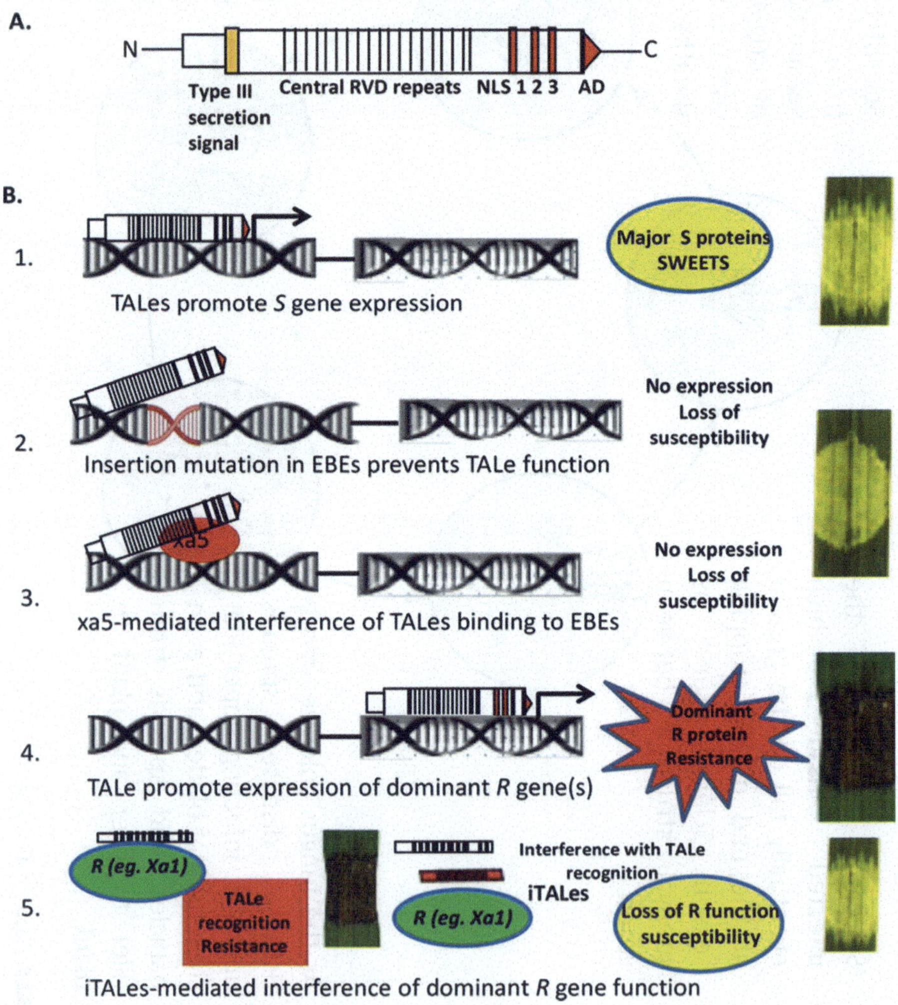

Fig. 32.7 Xoo TALe-dependent resistance in rice plant

genes Japanese Xoo race I was gene Xa-1. It was reported in Japan. RFLP markers were studied in correlation to the Xa-1 gene at chromosome 4. In Japan, a gene known as positional cloning of the gene was revealed. From *O. longistaminata* into *O. sativa* cross, a broad-spectrum bacterial blight resistance gene Xa-21 was incorporated. Three bacterial blight resistance genes, xa5, xa13, and xa21, were successfully transferred to the three NPT lines by a marker assisted backcrossing process to increase their resistance to xoo (Sanchez et al. 2000). Bacterial blight resistance genes were transferred by this cloning and resulted in novel plant production. This set of genes was also found in Samba Mahsuri (BPT5204), a most cultivated and well yielding variety in Indian Punjab. Xoo resistance was observed in rice cultivar Minghui 63 and Xa3/Xa26 in MRKa (Cao et al. 2007). Chitinase gene and the Bt fusion gene have lacked for insect resistance besides sheath blight in reciprocal crossing rice (Datta et al. 2002).

Bacterial blight besides sheath blight resistance were found to be increased in transgenic Pusa Basmati1 line pyramided with Xa21, chi11, and tlp. *Agrobacterium*-mediated transformation system was used for getting Xa21 transformed with five commonly used Chinese rice varieties. Enhance blight disease resistance was observed in these transgenic plants (Zhai et al. 2004).

32.9.3 Virus Resistance

Rice yield has been widely affected by the viral attacks. More than 828,000 tons of rice gets deteriorated by the attack of grassy stunt virus (RGSV), rice ragged stunt virus (RRSV), and coinfection by RGSV (Sasaya et al. 2014). Resistance allele Stvb-i was found in indicia paddy varieties. This allele is partially dominant and at chromosome 11 it is located in the multi-allelic locus *Stvb*. To obtain rice stripe virus (RSV), some japonica varieties such as Chugoku 31, Aichi 97, and Aichi 6 were incorporated with Stvb-i from Modan (Maeda et al. 2006). Stable resistance to RSV was observed in varieties cultivated in China and Japan that possess Stvb-i (Wang 2006). A QTL for RSV resistance in Milyang 23 was revealed at chromosome 11 on the interval between markers C1172 and XNpb202 and found to be allelic with Stvb-i (Ding et al. 2004). In Japan, two QTLs of upland rice variety were developed using markers such as SSR and RFLP. Decreased RSV infection by QTL on chromosome 11 was dedicated to Stv-b rather than chromosome 2 (Maeda et al. 2006). Indica variety and DV85 also exhibited RSV resistance due to the presence of two major QTLs located on chromosome 7 and the other was the same as Stvb-i (Ding et al. 2004).

Indian landrace Dular has three QTLs that were mapped, two of them on chromosome 11 at the gaps of RM209–RM21 and RM287–RM209 and one on chromosome 3 (Wu et al. 2009). Two Hoja Blanca Virus (RHBV) resistant rice varieties namely Fedearroz 2000 & 50 were crossed with non resistant variety WC366 for producing segregating lines (Romero et al. 2014). A major QTL on chromosome 4 short arm was found to be involved in resistance against (RHBV) in this population (Romero et al. 2014). Similarly, resistance against *T. orizicolus* in two major QTLs at chromosomes 5 and 7 were also found. This study permitted an improved understanding of the genetic control of RHBV resistance mechanism. A cross between resistant variety Kasalath and vulnerable Nipponbare using 98 back-cross inbred lines was conducted in QTL for visualization of RSV resistance (Zhang et al. 2011). RSV resistance was observed in two QTLs i.e. qSTV11KAS and qSTV7 which were located on chromosomes 11 and 7, respectively. Both QTLs were detected under artificial inoculation while one of these was detected under natural inoculation in the same location on chromosome 11.

A prominent reduction in disease development was observed in the transformed rice with the RHBV nucleocapsid protein (N) gene (Lentini et al. 2003). Development of complete resistance was observed in plants. It was due to the formation of a recovery phenotype known as local lesions coupled with appearing of symptomless new leaves. In these resistant rice lines, the N gene RNA level was below the detection limit of Northern blots and resolved by RT-PCR only. ELISA tests and Western blot were used for nucleocapsid protein detection in transgenic plants. It was concluded that RNA helps in the development of the resistance with the help of the N gene. RHBV-resistant transgenic lines exhibited elevated grain production, the number of tillers increased, and improved yield as compared to nonresistant lines. These resistant lines also showed some similar characteristics to the uninucleated, nontransgenic Cica 8 control. Thus, the development of the transgenic rice by utilizing DNA constructs can prove to help get RHBV-resistant cultivars (Verma et al. 2012). Plants having resistance against RTSV and RTBV have also been developed using a transgenic approach (Dai and Beachy 2009). Transgenic plants of indicia rice and Basmati-1 with coat protein (CP) gene of an Indian isolate of RTBV are also developed (Ganesan et al. 2009). Under controlled conditions, transgenic rice line through RNA interference (RNAi) much of resistance against many pathological conditions was observed particularly in RTBV infection (Tyagi et al. 2008).

32.9.4 Nematodes

Thirty-five genera from 300 nematode species are found to be involved in the infectious diseases in rice. Some nematode species are also involved in the economic production of rice. Depending upon the land type, different types of rice are affected by different nematodes. For example, *Aphelenchoides besseyi* and *Hirschmanniella* spp. attack irrigated rice while root-knot (*Meloidogyne* spp.) and Ufra (*Ditylenchus angustus*) attack deepwater rice. White tip nematodes and ufra are classified as regulatory pests (Biswal et al. 2017). In India, the yield loss caused by ufra nematodes is enormous. This problem was ignored due to subsistence farming of the crop, poor economic condition, and unawareness. But now the conditions are changing such as change of ecosystem of many rice-producing areas, and altered cropping systems has enhanced the value of nematode pests. Plant-parasitic nematode (PPN) species are more than 200 linked with rice up till now. For example, stem nematode (*Ditylenchus angustus*), rice root nematode (*Hirschmanniella oryzae*), and

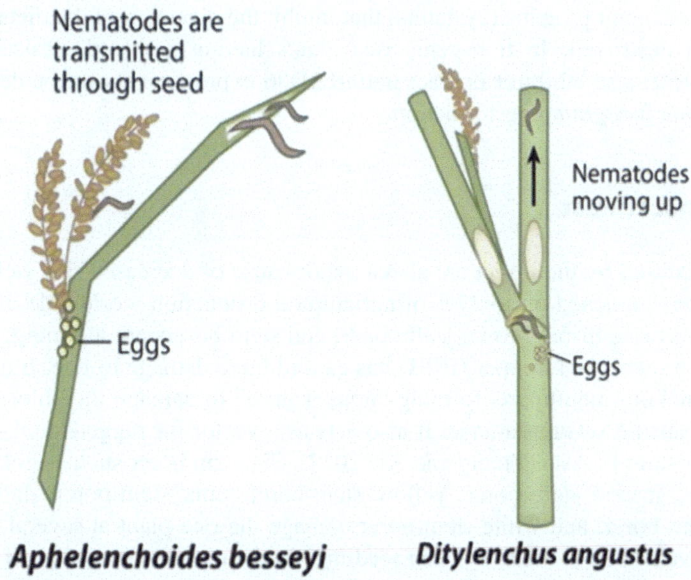

Fig. 32.8 Rice and the parasitic nematode *Ditylenchus angustus* (**Source**: Khanam 2016)

white tip nematode (*Aphelenchoides besseyi*), are widely known PPNs of rice. Rainfed areas are mostly encountered by rice root-knot nematodes (*Meloidogyne* sp.). Resistance against *M. graminicola* is trying to be obtained. Naggardhan and Achhoo are recognized to have good resistance against *M. graminicola* (Srivastava et al. 2011). Ranbir Basmati and Hasan Sarai are most vulnerable to nematodes. MAS-breeding approaches can be made by using these contrasting phenotypes (Ravindra et al. 2015). It was observed that development of the resistant lines has been hurdled because of lack of uniform resistance evaluation procedure (Pokharel et al. 2012) (Fig. 32.8).

Indica rice variety yield is not much affected by *M. graminicola* as compared to the yield of japonica rice variety Azucena that is decreased significantly. It refers to the semi-tolerance against nematodes. A cross between Bala (tolerant) × Azucena (susceptible) has helped in the identification of semi-resistance against *M. graminicola* (Shrestha et al. 2007). Six QTLs were found to be resistant against nematodes. Better tolerance than Bala was observed in two QTLs. This tolerance was given by the tolerance alleles of Azucena making a new hope for the development of more tolerant varieties. Different drought-tolerant rice lines were tested to find lines harboring drought tolerance and root-knot nematode resistance (Prasad et al. 2006). The perennial teqing and the donors cvs Type 3, Zihui 100, and Shwe Thwe Yin Hyv were found resistant while the donors cvs Binam, Khaza, Haonnong, and parent IR 64 were susceptible to the said nematode (Prasad et al. 2006).

Transgenic plants having nematode resistance have been developed adopting several strategies such as feeding-cell attenuation, migration strategies, anti-nematode feeding besides development, and anti-invasion. In obtaining nematode-resistant rice, plant proteins (cystatins) that inhibit the digestion of the dietary plants have been destroyed. In transgenic rice plants having nematode resistance, the cysteine proteinase inhibitor oryzacystatin-IΔD86 expression showed a decrease in *Meloidogyne incognita* egg formation.

32.9.5 Insect Pest

Damages caused by the insect are also a major cause of decreased rice yield. These damages have reached up to 37% of agricultural production worldwide. Egregious pests such as rice hispa, brown leafhopper, and stem borers are alarming. In Asian countries, brown plant hopper (BPH) has caused huge damage to rice. It utilizes its piercing-sucking mouthparts forming "hopper burn" to damage the phloem part of the plant causing severe damage. It also acts as a vector for ragged stunt virus and rice grassy stunt in Asia (Zhang and Xie 2014). Six stem borer such as gold-fringed stem-borer, striped stem-borer, yellow stem-borer, pink stem-borer, dark-headed striped stem-borer, and white stem-borer damage the rice plant at several stages of their life cycle such as maturity even seedling. At the reproductive stage, it damages the panicle stalk by residing inside it making a hollow grain known as "whiteheads." In the vegetative stage, larvae of the insect live in the stem making young leaf whorl causing "dead heart" and ultimately death (Biswal et al. 2017). Damages caused by insects are ten million tons per year throughout the world.

Rice crop grown in a temperate environment is mostly attacked by insects brown plant hopper (BPH). A cross between vulnerable variety C418 (japonica restorer line) and resistant parent ASD7 possessing a resistant gene bph2 was mapped on long arm of chromosome 12 between SSR markers RM7102 and RM463 (Li-Hong et al. 2006). Another cross between IR65482-17 (developed from wild rice species *Oryza australiensis*) and Zhenshan 97 (ZS97) showed three QTLs for feeding rate to BPH and seedling resistance (Hu et al. 2015). QTLs and the genes Bph10 and Bph18 (t) were recognized by using australiensis (Jena et al. 2006). Another cross of Kasalath (Indica) and Nipponbare (Japonica) were also found to have resistant genes on chromosomes 2, 10, and 12 imparting partial tolerant parents (Su et al. 2002). A normal level of resistance against stem borer was also obtained in rice varieties by conventional strategies. But the elevated level of resistance is still to be achieved in donor varieties. Stem borer resistance genes were not recognized even after screening of 30,000 rice accessions.

Tolerance against yellow stem borer, lepidopteran insects, and leaf folder was found to be developed in transgenic elite rice lines having Bt fusion gene-derived actinI promoter from cryIA(b) and cryIA without decreasing average yield. Many other transgenic lines also have given good results (Tang et al. 2006). For the production of two cry1Ab/Ac (Bt) lines, Chinese Ministry of Agriculture has given biosafety certificates (Chen et al. 2011). These cell lines are suggested to be safe for the environment by Rigorous testing even as food. Successful control over lepidopteran complex on rice can prove to be beneficial economically using these two Bt rice lines with lesser harms to the ecosystem (Chen et al. 2011).

Many QTLs have functions irrespective of their genetic background with small impact on the phenotype are important challenges for the cultivators in choosing the better lines (Gowda et al. 2011). Intraspecific crosses for the selection of QTL mapping are also a challenge. Many efforts are still necessary for interspecific crosses for exploring novel alleles besides their effective incorporation into breeding programs of drought tolerance in rice.

32.10 Future Research and Conclusion

Biotechnology is an emerging area of scientific research and is playing a vital role in different domains of life sciences. Plant biotechnology has a broad scope in improving the agriculture sector. Improving the existing quality of crops and plants, disease treatment, and development of new products with desired characteristics are some key areas where biotechnology has been played its part since the last decade. Tremendous research has been done to use biotechnology for the improvement of rice quality, as rice is one of the primary food all over the world. To meet the ever-increasing demand for rice, different techniques including transgenic deployment and marker-assisted breeding are two potential areas in which focus is being laid. Rapid progress is seen over the marker-assisted rice breeding. By developing the resistance mechanism and studies on gene function, rice varieties with improved characteristics are being prepared. By employing modern biotechnological tools including recombinant DNA techniques and tissue culture technology, different modification in rice has been achieved. Latest gene-editing tools such as CRISPR/Cas system and TALENs can lead to the removal of unwanted genes from the plant species and introduce the desired characteristics. Rice species with high production yield, improved disease and stress resistance, better nutritional quality, and appropriate mineral contents can be achieved by modern biotechnological tools with a unique prospect to create a novel beneficial variation for rice breeders. Further work on plant biotechnology not only can revolutionize horticulture, forestry, and agriculture, but also can meet food requirements of world.

References

Ahmad S, Hasanuzzaman M (2012) Integrated effect of plant density, N rates and irrigation regimes on the biomass production, N content, PAR use efficiencies and water productivity of rice under irrigated semiarid environment. Not Bot Horti Agrobot Cluj Napoca 40(1):201–211

Ahmad S, Zia-ul-Haq M, Ali H, Shad SA, Ammad A, Maqsood M, Khan MB, Mehmood S, Hussain A (2008) Water and radiation use efficiencies of transplanted rice (*Oryza sativa* L.) at different plant densities and irrigation regimes under semi-arid environment. Pak J Bot 40(1):199–209

Ahmad S, Ahmad A, Zia-ul-Haq M, Ali H, Khaliq T, Anjum MA, Khan MA, Hussain A, Hoogenboom G (2009) Resources use efficiency of field grown transplanted rice (*Oryza sativa* L.) under irrigated semiarid environment. J Food Agric Environ 7(2):487–492

Ahmad S, Ahmad A, Soler CMT, Ali H, Zia-Ul-Haq M, Anothai J, Hussain A, Hoogenboom G, Hasanuzzaman M (2012) Application of the CSM-CERES-Rice model for evaluation of plant density and nitrogen management of fine transplanted rice for an irrigated semiarid environment. Precis Agric 13(2):200–218

Ahmad S, Ahmad A, Ali H, Hussain A, Garcia y Garcia A, Khan MA, Zia-Ul-Haq M, Hasanuzzaman M, Hoogenboom G (2013) Application of the CSM-CERES-Rice model for evaluation of plant density and irrigation management of transplanted rice for an irrigated semiarid environment. Irrig Sci 31(3):491–506

Ahmad A, Ashfaq M, Rasul G, Wajid SA, Khaliq T, Rasul F, Saeed U, Rahman MH, Hussain J, Baig IA, Naqvi AA, SAA B, Ahmad S, Naseem W, Hoogenboom G, Valdivia RO (2015) Impact of climate change on the rice–wheat cropping system of Pakistan. In: Hillel D, Rosenzweig C (eds) Handbook of climate change and agro-ecosystems: The Agricultural Modeling Intercomparison and Improvement Project (AgMIP) integrated crop and economic assessments. Imperial College Press/American Society of Agronomy, London/Hackensack, pp 219–258

Ahmad S, Abbas G, Ahmed M, Fatima Z, Anjum MA, Rasul G, Khan MA, Hoogenboom G (2019) Climate warming and management impact on the change of rice-wheat phenology in Punjab, Pakistan. Field Crop Res 230:46–61

Ahmed M, Ahmad S (2019) Carbon dioxide enrichment and crop productivity. In: Hasanuzzaman M (ed) Agronomic crops, vol 2. Springer Nature Singapore, Singapore, pp 31–46

Ahmed M, Ahmad S (2020) Systems modeling. In: Ahmed M (ed) Systems modeling. Springer Nature, Cham, pp 1–44

Ahmed M, Fayyaz-ul-Hassan, Ahmad S (2017) Climate variability impact on rice production: adaptation and mitigation strategies. In: Ahmed M, Stockle C (eds) Quantification of climate variability, adaptation and mitigation for agricultural sustainability. Springer, Cham, pp 91–111

Ahmed M, Ahmad S, Raza MA, Kumar U, Ansar M, Shah GA, Parsons D, Hoogenboom G, Palosuo T, Seidel S (2020a) Models calibration and evaluation. In: Ahmed M (ed) Systems modeling. Springer Nature Singapore, Singapore, pp 151–178

Ahmed M, Ahmad S, Waldrip HM, Ramin M, Raza MA (2020b) Whole farm modeling: a systems approach to understanding and managing livestock for greenhouse gas mitigation, economic viability and environmental quality. In: Waldrip HM, Pagliari PH, He Z (eds) Animal Manure, ASA Special Publication 67. American Society of Agronomy/Soil Science Society of America, Madison, WI, pp 345–371

Akram R, Turan V, Hammad HM, Ahmad S, Hussain S, Hasnain A, Maqbool MM, Rehmani MIA, Rasool A, Masood N, Mahmood F, Mubeen M, Sultana SR, Fahad S, Amanet K, Saleem M, Abbas Y, Akhtar HM, Hussain S, Waseem F, Murtaza R, Amin A, Zahoor SA, Sami ul Din M, Nasim W (2018) Fate of organic and inorganic pollutants in paddy soils. In: Hashmi MZ, Varma A (eds) Environmental pollution of paddy soils. Springer Nature Switzerland, Cham, pp 197–214

Akram R, Fahad S, Masood N, Rasool A, Ijaz M, Ihsan MZ, Maqbool MM, Ahmad S, Hussain S, Ahmed M, Kaleem S, Sultana SR, Mubeen M, Saud S, Kamran M, Nasim W (2019) Plant growth and morphological changes in rice under abiotic stress. In: Hasanuzzaman M, Fujita M, Nahar K, Biswas JK (eds) Advances in rice research for abiotic stress tolerance. Woodhead, Cambridge, pp 69–85

Anuradha K, Agarwal S, Rao YV, Rao K, Viraktamath B, Sarla N (2012) Mapping QTLs and candidate genes for iron and zinc concentrations in unpolished rice of Madhukar× Swarna RILs. Gene 508:233–240

Asghar A, Rashid H, Ashraf M, Khan MH, Chaudhry Z (2007) Improvement of basmati rice against fungal infection through gene transfer technology. Pak J Bot 39:1277

Azizi P, Rafii M, Maziah M, Abdullah S, Hanafi M, Latif M, Rashid A, Sahebi M (2015) Understanding the shoot apical meristem regulation: a study of the phytohormones, auxin and cytokinin, in rice. Mech Dev 135:1–15

Babu RC, Zhang J, Blum A, Ho THD, Wu R, Nguyen H (2004) HVA1, a LEA gene from barley confers dehydration tolerance in transgenic rice (*Oryza sativa* L.) via cell membrane protection. Plant Sci 166:855–862

Bailey-Serres J, Voesenek L (2008) Flooding stress: acclimations and genetic diversity. Annu Rev Plant Biol 59:313–339

Baral BR, Pande KR, Gaihre YK, Baral KR, Sah SK, Thapa YB, Singh U (2020) Increasing nitrogen use efficiency in rice through fertilizer application method under rainfed drought conditions in Nepal. Nutr Cycl Agroecosyst 118:103–114

Biswal AK, Shamim M, Cruzado K, Soriano G, Ghatak A, Toleco M, Vikram P (2017) Role of biotechnology in rice production. In: Chauhan BS, Jabran K, Mahajan G (eds) Rice production worldwide. Springer, Cham

Boonyaves K, Gruissem W, Bhullar NK (2016) NOD promoter-controlled AtIRT1 expression functions synergistically with NAS and FERRITIN genes to increase iron in rice grains. Plant Mol Biol 90:207–215

Campo S, Baldrich P, Messeguer J, Lalanne E, Coca M, San Segundo B (2014) Overexpression of a calcium-dependent protein kinase confers salt and drought tolerance in rice by preventing membrane lipid peroxidation. Plant Physiol 165:688–704

Cao Y, Duan L, Li H, Sun X, Zhao Y, Xu C, Li X, Wang S (2007) Functional analysis of Xa3/Xa26 family members in rice resistance to Xanthomonasoryzae pv. oryzae. Theor Appl Genet 115: 887–895

Chen DX, Chen XW, Lei CL, Wang YP, Li SG (2010) Rice blast resistance of transgenic rice plants with Pi-d2 gene. Rice Sci 17:179–184

Chen M, Shelton A, Ye GY (2011) Insect-resistant genetically modified rice in China: from research to commercialization. Annu Rev Entomol 56:81–101

Chen H, Chen W, Zhou J, He H, Chen L, Chen H, Deng XW (2012) Basic leucine zipper transcription factor OsbZIP16 positively regulates drought resistance in rice. Plant Sci 193:8–17

Chen H, Xie W, He H, Yu H, Chen W, Li J, Yu R, Yao Y, Zhang W, He Y (2014) A high-density SNP genotyping array for rice biology and molecular breeding. Mol Plant 7:541–553

Choi MS, Woo MO, Koh EB, Lee J, Ham TH, Seo HS, Koh HJ (2012) Teosinte branched 1 modulates tillering in rice plants. Plant Cell Rep 31:57–65

Dai S, Beachy RN (2009) Genetic engineering of rice to resist rice tungro disease. In Vitro Cell Dev Biol Plant 45:517–524

Datta SK (2004) Rice biotechnology: a need for developing countries. AgBioForum 7(1–2):31–35

Datta K, Baisakh N, Thet KM, Tu J, Datta S (2002) Pyramiding transgenes for multiple resistance in rice against bacterial blight, yellow stem borer and sheath blight. Theor Appl Genet 106:1–8

Dean RA, Talbot NJ, Ebbole DJ, Farman ML, Mitchell TK, Orbach MJ, Thon M, Kulkarni R, Xu JR, Pan H (2005) The genome sequence of the rice blast fungus Magnaporthe grisea. Nature 434:980–986

Dennis ES, Dolferus R, Ellis M, Ranman M, Wu Y, Hoeren F, Grover A, Ismond K, Good A, Peacock W (2000) Molecular strategies for improving waterlogging tolerance in plants. J Exp Bot 51:89–97

Ding XL, Jiang L, Liu SJ, Wang CM, Chen LM, Chang ZB, Fan YJ, Zhou YJ, Wan JM (2004) QTL analysis for rice stripe disease resistance gene using recombinant inbred lines (RILs) derived from crossing of Kinmaze and DV85. Yi Chuan Xue Bao 31(3):287–292

Dixit S, Biswal AK, Min A, Henry A, Oane RH, Raorane ML, Longkumer T, Pabuayon IM, Mutte SK, Vardarajan AR (2015) Action of multiple intra-QTL genes concerted around a co-localized transcription factor underpins a large effect QTL. Sci Rep 5:15183

Domagalska MA, Leyser O (2011) Signal integration in the control of shoot branching. Nat Rev Mol Cell Biol 12:211–221

DÖring F (2017) Understanding the bundle sheath in C_4 evolution—forward genetic and transcriptomic approaches. Universitäts-und Landesbibliothek der Heinrich-Heine-Universität Düsseldorf

Duan J, Cai W (2012) OsLEA3-2, an abiotic stress induced gene of rice plays a key role in salt and drought tolerance. PLoS One 7:e45117

Duke SO (2005) Taking stock of herbicide-resistant crops ten years after introduction. Pest Manag Sci 61:211–218

Durand-Morat A, Chavez E, Wailes E (2018) World and US rice baseline outlook, 2018–2028, B.R. Wells Arkansas Rice Research Studies, vol 2018. Agricultural Experiment Station, Fayetteville, p 412

Endo M, Osakabe K, Ono K, Handa H, Shimizu T, Toki S (2007) Molecular breeding of a novel herbicide-tolerant rice by gene targeting. Plant J 52:157–166

Fahad S, Noor M, Adnan M, Khan MA, Rahman I, Alam M, Khan IA, Ullah H, Mian IA, Hassan S, Saud S, Bakhat HF, Hammad HM, Ahmad S, Nasim W (2019) Abiotic stress and rice grain quality. In: Hasanuzzaman M, Fujita M, Nahar K, Biswas JK (eds) Advances in rice research for abiotic stress tolerance. Woodhead, Cambridge, pp 571–583

Fatima Z, Ahmad M, Hussain M, Abbas G, Ul-Allah S, Ahmad S, Ahmed N, Ali MA, Sarwar G, Ehsan ul Haque, Iqbal P, Hussain S (2020) The fingerprints of climate warming on cereal crops phenology and adaptation options. Sci Rep 10:18013

Frewer LJ, Shepherd R, Sparks P (1994) Biotechnology and food production: knowledge and perceived risk. Br Food J 96:26–32

Ganesan U, Suri SS, Rajasubramaniam S, Rajam MV, Dasgupta I (2009) Transgenic expression of coat protein gene of rice tungro bacilliform virus in rice reduces the accumulation of viral DNA in inoculated plants. Virus Genes 39:113–119

Gowda VR, Henry A, Yamauchi A, Shashidhar H, Serraj R (2011) Root biology and genetic improvement for drought avoidance in rice. Field Crop Res 122:1–13

Gu JF, Qiu M, Yang JC (2013) Enhanced tolerance to drought in transgenic rice plants overexpressing C_4 photosynthesis enzymes. Crop J 1:105–114

Guo C, Ge X, Ma H (2013) The rice OsDIL gene plays a role in drought tolerance at vegetative and reproductive stages. Plant Mol Biol 82:239–253

Gupta SK, Rai AK, Kanwar SS, Chand D, Singh NK, Sharma TR (2012) The single functional blast resistance gene Pi54 activates a complex defence mechanism in rice. J Exp Bot 63:757–772

Hafeez-ur-Rehman NA, Awan MI, Ijaz M, Hussain M, Ahmad S, Farooq M (2019) Direct seeding in rice: problems and prospects. In: Hasanuzzaman M (ed) Agronomic crops, Production technologies, vol 1. Springer Nature Singapore, Singapore, pp 199–222

Hasegawa T, Shibata A, Wainaina CM, Lucob-Agustin N, Makihara D, Samejima H, Kikuta M, Menge DM, Kimani JM, Yamauchi A, Inukai Y (2019) Evaluation of a newly identified mutation gene that promotes root elongation for improvement of drought avoidance in rice. J Int Cooper Agric Dev 17:34–40

Hattori Y, Nagai K, Furukawa S, Song XJ, Kawano R, Sakakibara H, Wu J, Matsumoto T, Yoshimura A, Kitano H (2009) The ethylene response factors SNORKEL1 and SNORKEL2 allow rice to adapt to deep water. Nature 460:1026–1030

HernÁndez-Prieto MA, Foster C, Watson-Lazowski A, Ghannoum O, Chen M (2019) Comparative analysis of thylakoid protein complexes in the mesophyll and bundle sheath cells from C_3, C_4 and C_3 –C_4 Paniceae grasses. Physiol Plant 166:134–147

Hong CY, Cheng KJ, Tseng TH, Wang CS, Liu LF, Yu SM (2004) Production of two highly active bacterial phytases with broad pH optima in germinated transgenic rice seeds. Transgenic Res 13(1):29–39. https://doi.org/10.1023/b:trag.0000017158.96765.67

Hossain H, Rahman M, Alam M, Singh R (2015) Mapping of quantitative trait loci associated with reproductive-stage salt tolerance in rice. J Agron Crop Sci 201:17–31

Hu J, Xiao C, Cheng MX, Gao GJ, Zhang QL, He YQ (2015) A new finely mapped Oryza australiensis-derived QTL in rice confers resistance to brown planthopper. Gene 561:132–137

Huang Y, Zhao S, Fu Y, Sun H, Ma X, Tan L, Liu F, Sun X, Sun H, Gu P (2018) Variation in the regulatory region of FZP causes increases in secondary inflorescence branching and grain yield in rice domestication. Plant J 96:716–733

Huang Y, Bai X, Luo M, Xing Y (2019) Short panicle 3 controls panicle architecture by upregulating APO2/RFL and increasing cytokinin content in rice. J Integr Plant Biol 61:987–999

Hussain M, Ahmad S, Hussain S, Lal R, Ul-Allah S, Nawaz A (2018) Rice in saline soils: physiology, biochemistry, genetics and management. Adv Agron 148:231–287

Jena K, Mackill D (2008) Molecular markers and their use in marker-assisted selection in rice. Crop Sci 48:1266–1276

Jena K, Jeung J, Lee J, Choi H, Brar D (2006) High-resolution mapping of a new brown planthopper (BPH) resistance gene, Bph18 (t), and marker-assisted selection for BPH resistance in rice (*Oryza sativa* L.). Theor Appl Genet 112:288–297

Jha S, Tank HG, Prasad BD, Chattoo BB (2009) Expression of Dm-AMP1 in rice confers resistance to Magnaporthe oryzae and Rhizoctonia solani. Transgenic Res 18:59–69

Joshi B, Chaudhary A, Singh H, Kumar PA (2020) Prospective evaluation of individual and consortia plant growth promoting rhizobacteria for drought stress amelioration in rice (*Oryza sativa* L.). Plant Soil 457:225–240

Kader MA, Lindberg S (2010) Cytosolic calcium and pH signaling in plants under salinity stress. Plant Signal Behav 5:233–238

Kajala K, Covshoff S, Karki S, Woodfield H, Tolley BJ, Dionora MJA, Mogul RT, Mabilangan AE, Danila FR, Hibberd JM (2011) Strategies for engineering a two-celled C_4 photosynthetic pathway into rice. J Exp Bot 62:3001–3010

Kalpana K, Maruthasalam S, Rajesh T, Poovannan K, Kumar KK, Kokiladevi E, Raja JA, Sudhakar D, Velazhahan R, Samiyappan R (2006) Engineering sheath blight resistance in elite indica rice cultivars using genes encoding defense proteins. Plant Sci 170:203–215

Kathuria H, Giri J, Nataraja KN, Murata N, Udayakumar M, Tyagi AK (2009) Glycinebetaine-induced water-stress tolerance in codA-expressing transgenic indica rice is associated with up-regulation of several stress responsive genes. Plant Biotechnol J 7:512–526

Kellogg EA (2013) C_4 photosynthesis. Curr Biol 23:R594–R599

Khan MA, Ahmad S, Raza A (2019a) Integrated weed management for agronomic crops. In: Hasanuzzaman M (ed) Agronomic crops, vol 2. Springer Nature Singapore, Singapore, pp 257–281

Khan SM, Ali S, Nawaz A, Bukhari SAH, Ejaz S, Ahmad S (2019b) Integrated pest and disease management for better agronomic crop production. In: Hasanuzzaman M (ed) Agronomic crops, vol 2. Springer Nature Singapore, Singapore, pp 385–428

Khanam S (2016) Characterisation of the interaction between rice and the parasitic nematode Ditylenchus angustus. Ghent University, Ghent. http://hdl.handle.net/1854/LU-8098941

Kumar V, Bellinder R, Gupta R, Malik R, Brainard D (2008) Role of herbicide-resistant rice in promoting resource conservation technologies in rice–wheat cropping systems of India: a review. Crop Prot 27:290–301

Kumar A, Dixit S, Ram T, Yadaw R, Misra K, Mandal N (2014) Breeding high-yielding drought-tolerant rice: genetic variations and conventional and molecular approaches. J Exp Bot 65: 6265–6278

Lang NT, Thi P, Ha T, Nha C, Van NH, Van HD (2013) Introgression of Sub1 gene into local popular varieties and newly developed elite breeding lines in the Mekong Delta adapt to the climate change. Omonrice 19:27–39

Lentini Z, Lozano I, Tabares E, Fory L, DomÍnguez J, Cuervo M, Calvert L (2003) Expression and inheritance of hypersensitive resistance to rice hoja blanca virus mediated by the viral nucleocapsid protein gene in transgenic rice. Theor Appl Genet 106:1018–1026

Li-Hong S, Chun-Ming W, Chang-Chao S, Yu-Qiang L, Hu-Qu Z, Jian-Min W (2006) Mapping and marker-assisted selection of a brown planthopper resistance gene bph2 in rice (*Oryza sativa* L.). Acta Genet Sin 33:717–723

Linh LH, Linh TH, Xuan TD, Ham LH, Islamail AM, Khanh TD (2012) Molecular breeding to improve salt tolerance of rice (*Oryza sativa* L.) in the red River Delta of Vietnam. Int J Plant Genomics 2012:949038. https://doi.org/10.1155/2012/949038

Maeda H, Matsushita K, Iida S, Sunohara Y (2006) Characterization of two QTLs controlling resistance to rice stripe virus detected in a Japanese upland rice line, Kanto 72. Breed Sci 56: 359–364

Mahmood U, Hussain HA, Hussain S, Ashraf U, Khaliq A, Hussain S (2019) Submergence stress in rice: physiological disorders, tolerance mechanisms, and management. In: Hasanuzzaman M,

Hakeem KR, Nahar K, Alharby HF (eds) Plant abiotic stress tolerance: agronomic, molecular and biotechnological approaches. Springer International Publishing, Cham

Mallmann J, Heckmann D, BrÄutigam A, Lercher MJ, Weber AP, Westhoff P, Gowik U (2014) The role of photorespiration during the evolution of C_4 photosynthesis in the genus Flaveria. elife 3:e02478

Mehta S, Singh B, Dhakate P, Rahman M, Islam MA (2019) Rice, marker-assisted breeding, and disease resistance. In: Wani SH (ed) Disease resistance in crop plants: molecular, genetic and genomic perspectives. Springer International Publishing, Cham

Mishra KK, Vikram P, Yadaw RB, Swamy BM, Dixit S, Cruz MTS, Maturan P, Marker S, Kumar A (2013) qDTY 12. 1: a locus with a consistent effect on grain yield under drought in rice. BMC Genet 14:12

Miyao M, Masumoto C, Miyazawa SI, Fukayama H (2011) Lessons from engineering a single-cell C4 photosynthetic pathway into rice. J Exp Bot 62:3021–3029

Mohanty A, Kathuria H, Ferjani A, Sakamoto A, Mohanty P, Murata N, Tyagi A (2002) Transgenics of an elite indica rice variety Pusa Basmati 1 harbouring the codA gene are highly tolerant to salt stress. Theor Appl Genet 106:51–57

Naz S, Fatima Z, Iqbal P, Khan A, Zakir I, Noreen S, Younis H, Abbas G, Ahmad S (2019) Agronomic crops: types and uses. In: Hasanuzzaman M (ed) Agronomic crops, Production technologies, vol 1. Springer Nature Singapore, Singapore, pp 1–18

NegrÃo S, Courtois B, Ahmadi N, Abreu I, Saibo N, Oliveira M (2011) Recent updates on salinity stress in rice: from physiological to molecular responses. Crit Rev Plant Sci 30:329–377

Obara M, Tamura W, Ebitani T, Yano M, Sato T, Yamaya T (2010) Fine-mapping of qRL6. 1, a major QTL for root length of rice seedlings grown under a wide range of NH 4+ concentrations in hydroponic conditions. Theor Appl Genet 121:535–547

Paine JA, Shipton CA, Chaggar S, Howells RM, Kennedy MJ, Vernon G, Wright SY, Hinchliffe E, Adamd JL, Silverstone AL (2005) Improving the nutritional value of Golden Rice through increased pro-vitamin A content. Nat Biotechnol 23:482

Panda AK, Mishra R, Mahanty A, Lenka S, Panda KK, Bisht SS (2020) Genomics and genetic engineering of rice for tolerance to ozone and anoxia. In: Roychoudhury A (ed) Rice research for quality improvement: genomics and genetic engineering, Breeding techniques and abiotic stress tolerance, vol 1. Singapore, Springer Singapore

Pandey S, Bhandari H (2009) Drought, coping mechanisms and poverty. IFAD, Rome. Occasional Papers

Pathak RK, Baunthiyal M, Pandey D, Kumar A (2018) Augmentation of crop productivity through interventions of omics technologies in India: challenges and opportunities. 3 Biotech 8:454

Peerzada AM, Bukhari SAH, Dawood M, Nawaz A, Ahmad S, Adkins S (2019) Weed management for healthy crop production. In: Hasanuzzaman M (ed) Agronomic crops, vol 2. Springer Nature Singapore, Singapore, pp 225–256

Peng H, Wang K, Chen Z, Cao Y, Gao Q, Li Y, Li X, Lu H, Du H, Lu M (2020) MBKbase for rice: an integrated omics knowledgebase for molecular breeding in rice. Nucleic Acids Res 48: D1085–D1092

Peterhansel C, Horst I, Niessen M, Blume C, Kebeish R, Kürkcüoglu S, Kreuzaler F (2010) Photorespiration. Arabidopsis Book 8:e0130. https://doi.org/10.1199/tab.0130

Pokharel RR, Duxbury JM, Abawai G (2012) Evaluation of protocol for assessing the reaction of rice and wheat germplasm to infection by Meloidogyne graminicola. J Nematol 44:274

Prasad JS, Vijayakumar CHM, Sankar M, Varaprasad KS, Prasad MS, Rao YK (2006) Root-knot nematode resistance in advanced back cross populations of rice developed for water stress conditions. Nematol Medit 34:3–8

Pruitt RN, Schwessinger B, Joe A, Thoas N, Liu F, Albert M, Robinson MR, Chan LJG, Luu DD, Chen H, Bahar O, Daudi A, De Vleesschauwer D, Caddell D, Zhang W, Zhao X, Li X, Heazlewood JL, Ruan D, Majumder D, Chern M, Kalbacher H, Midha S, Patil PB, Sonti RV, Petzold CJ, Liu CC, Brodbelt JS, Felix G, Ronald PC (2015) The rice immune receptor XA21 recognizes a tyrosine-sulfated protein from a Gram-negative bacterium. Sci Adv 1:e1500245

Quimio CA, Torrizo LB, Setter TL, Ellis M, Grover A, Abrigo EM, Oliva NP, Ella ES, Carpena AL, Ito O (2000) Enhancement of submergence tolerance in transgenic rice overproducing pyruvate decarboxylase. J Plant Physiol 156:516–521

Ram T, Majumder N, Laha G, Ansari M, Kar C, Mishra B (2008) Identification of donors for sheath blight resistance in wild species of rice. Ind J Genet Plant Breed 68:317

Rana N, Rahim MS, Kaur G, Bansal R, Kumawat S, Roy J, Deshukh R, Sonah H, Sharma TR (2020) Applications and challenges for efficient exploration of omics interventions for the enhancement of nutritional quality in rice (*Oryza sativa* L.). Crit Rev Food Sci Nutr 60:3304–3320

Rao KK, Lakshminarasu M, Jena K (2002) DNA markers and marker-assisted breeding for durable resistance to bacterial blight disease in rice. Biotechnol Adv 20:33–47

Rasool A, Akbar F, Rehman A, Jabeen H (2020) Genetic engineering of rice for resistance to insect pests. In: Roychoudhury A (ed) Rice research for quality improvement: genomics and genetic engineering. Springer Singapore, Singapore, pp 129–148

Ravindra H, Sehgal M, Narasimhamurthy H, Khan HI, Shruthi S (2015) Evaluation of rice landraces against rice root-knot nematode, Meloidogyne graminicola. Afr J Microbiol Res 9: 1128–1131

Razaq M, Shah FM, Ahmad S, Afzal M (2019) Pest management for agronomic crops. In: Hasanuzzaman M (ed) Agronomic crops, vol 2. Springer Nature Singapore, Singapore, pp 365–384

Reardon T, Echeverria R, BerdeguÉ J, Minten B, Liverpool-Tasie S, Tschirley D, Zilberman D (2019) Rapid transformation of food systems in developing regions: highlighting the role of agricultural research & innovations. Agric Syst 172:47–59

Romero LE, Lozano I, Garavito A, Carabali SJ, Triana M, Villareal N, Reyes L, Duque MC, Martiniz CP, Calvert L, Lorieus M (2014) Major QTLs control resistance to rice hoja blanca virus and its vector Tagosodes orizicolus. G3 (Bethesda) 4:133–142

Sanchez A, Brar D, Huang N, Li Z, Khush G (2000) Sequence tagged site marker-assisted selection for three bacterial blight resistance genes in rice. Crop Sci 40:792–797. https://doi.org/10.2135/cropsci2000.403792x

Sardar G, Mallick G, Jana K, Ghosh S (2015) Screening of high iron and zinc rice genotypes in red and lateritic areas of West Bengal. Int J Appl Biores 23:1–3

Sarwar N, Ali H, Ahmad S, EhsanUllah AS, Mubeen K, Hill JE (2013a) Water wise rice cultivation on calcareous soil with the addition of essential micronutrients. J Anim Plant Sci 23(1):244–250

Sarwar N, Ali H, Maqsood M, EhsanUllah SAN, Shahzad M, Mubeen K, Shahid MA, Ahmad S (2013b) Phenological response of rice plants to different micronutrients application under water saving paddy fields on calcareous soil. Turk J Field Crops 18(1):52–57

Sasaya T, Nakazono-Nagaoka E, Saika H, Aoki H, Hiraguri A, Netsu O, Ichiki TU, Onuki M, Toki S, Saito K (2014) Transgenic strategies to confer resistance against viruses in rice plants. Front Microbiol 4:409

Sesma A, Osbourn AE (2004) The rice leaf blast pathogen undergoes developmental processes typical of root-infecting fungi. Nature 431:582–586

Shahzad AN, Ahmad S (2019) Tools and techniques for nitrogen management in cereals. In: Hasanuzzaman M (ed) Agronomic crops, vol 2. Springer Nature Singapore, Singapore, pp 111–126

Sharma T, Rai A, Gupta S, Vijayan J, Devanna B, Ray S (2012) Rice blast management through host-plant resistance: retrospect and prospects. Agric Res 1:37–52

Shen L, Courtois B, Mcnally KL, Robin S, Li Z (2001) Evaluation of near-isogenic lines of rice introgressed with QTLs for root depth through marker-aided selection. Theor Appl Genet 103: 75–83

Shimizu-Sato S, Mori H (2001) Control of outgrowth and dormancy in axillary buds. Plant Physiol 127:1405–1413

Shrestha R, Uzzo F, Wilson MJ, Price AH (2007) Physiological and genetic mapping study of tolerance to root-knot nematode in rice. New Phytol 176:665–672

Singh R, RedoÑa E, Refuerzo L (2010) Varietal improvement for abiotic stress tolerance in crop plants: special reference to salinity in rice. In: Pareek A, Sopory SK, Bohnert HJ (eds) Abiotic stress adaptation in plants: physiological, molecular and genomic foundation. Springer Netherlands, Dordrecht, pp 387–415

Singh VK, Singh A, Singh S, Ellur RK, Choudhary V, Sarkel S, Singh D, Krishnan SG, Nagarajan M, Vinod K (2012) Incorporation of blast resistance into "PRR78", an elite basmati rice restorer line, through marker assisted backcross breeding. Field Crop Res 128:8–16

Sparks A, Nelson A, Castilla N (2012) Where rice pests and diseases do the most damage. Rice Today 11:26–27

Srivastava A, Rana V, Rana S, Singh D, Singh V (2011) Screening of rice and wheat cultivars for resistance against root knot nematode, (Golden and Birchfield) in rice wheat cropping system. J Rice Res 4:2

Steele K, Price AH, Shashidhar H, Witcombe J (2006) Marker-assisted selection to introgress rice QTLs controlling root traits into an Indian upland rice variety. Theor Appl Genet 112:208–221

Su CC, Cheng XN, Zhai HQ, Wan J (2002) Detection and analysis of QTL for resistance to the brown planthopper, Nilaparvata lugens (Stål), in rice (*Oryza sativa* L.), using backcross inbred lines. Yi Chuan Xue Bao 29:332–338

Tan S, Evans RR, Dahmer ML, Singh BK, Shaner DL (2005) Imidazolinone-tolerant crops: history, current status and future. Pest Manag Sci 61:246–257

Tan S, Han R, Li P, Yang G, Li S, Zhang P, Wang WB, Zhao WZ, Yin LP (2015) Over-expression of the MxIRT1 gene increases iron and zinc content in rice seeds. Transgenic Res 24:109–122

Tang W, Chen H, Xu C, Li X, Lin Y, Zhang Q (2006) Development of insect-resistant transgenic indica rice with a synthetic cry1C* gene. Mol Breeding 18:1. https://doi.org/10.1007/s11032-006-9002-9

Tariq M, Ahmed M, Iqbal P, Fatima Z, Ahmad S (2020) Crop phenotyping. In: Ahmed M (ed) Systems modeling. Springer Nature Singapore, Singapore, pp 45–60

Todaka D, Shinozaki K, Yamaguchi-Shinozaki K (2015) Recent advances in the dissection of drought-stress regulatory networks and strategies for development of drought-tolerant transgenic rice plants. Front Plant Sci 6:84

Tripathi AK, Pareek A, Sopory SK, Singla-Pareek SL (2012) Narrowing down the targets for yield improvement in rice under normal and abiotic stress conditions via expression profiling of yield-related genes. Rice (N Y) 5:37

Tripathy J, Zhang J, Robin S, Nguyen TT, Nguyen H (2000) QTLs for cell-membrane stability mapped in rice (*Oryza sativa* L.) under drought stress. Theor Appl Genet 100:1197–1202

Tyagi H, Rajasubramaniam S, Rajam MV, Dasgupta I (2008) RNA-interference in rice against Rice tungro bacilliform virus results in its decreased accumulation in inoculated rice plants. Transgenic Res 17:897–904

Verma V, Sharma S, Devi SV, Rajasubramaniam S, Dasgupta I (2012) Delay in virus accumulation and low virus transmission from transgenic rice plants expressing Rice tungro spherical virus RNA. Virus Genes 45:350–359

Wang C (2006) The advance of the resistant breeding of rice stripe disease in Jiangsu province. Jiangsu Agric Sci 3:1–5

Wang L, Czedik-Eysenberg A, Mertz RA, Si Y, Tohge T, Nunes-Nesi A, Arrivault S, Dedow LK, Bryant DW, Zhou W (2014a) Comparative analyses of C 4 and C 3 photosynthesis in developing leaves of maize and rice. Nat Biotechnol 32:1158–1165

Wang X, Lee S, Wang J, Ma J, Bianco T, Jia Y (2014b) Current advances on genetic resistance to rice blast disease. In: Yan WG (ed) Rice-germplasm, genetics and improvement. IntechOpen, London, pp 195–217

Wasaya A, Yasir TA, Ijaz M, Ahmad S (2019) Tillage effects on agronomic crop production. In: Hasanuzzaman M (ed) Agronomic crops, vol 2. Springer Nature Singapore, Singapore, pp 73–99

Weng X, Wang L, Wang J, Hu Y, Du H, Xu C, Xing Y, Li X, Xiao J, Zhang Q (2014) Grain number, plant height, and heading date7 is a central regulator of growth, development, and stress response. Plant Physiol 164:735–747

Westhoff P, Gowik U (2010) Evolution of C4 photosynthesis—looking for the master switch. Plant Physiol 154:598–601

Wu SJ, Zhong H, Zhou Y, Zuo H, Zhou LH, Zhu JY, Ji CQ, Gu SL, Gu MH, Liang GH (2009) Identification of QTLs for the resistance to rice stripe virus in the indica rice variety Dular. Euphytica 165:557–565

Xing Y, Zhang Q (2010) Genetic and molecular bases of rice yield. Annu Rev Plant Biol 61:421–442

Xu K, Xu X, Fukao T, Canlas P, Maghirang-Rodriguez R, Heuer S, Ismail AM, Bailey-Serres J, Ronald PC, Mackill DJ (2006) Sub1A is an ethylene-response-factor-like gene that confers submergence tolerance to rice. Nature 442:705–708

Xu Z, Miao Y, Chen Z, Gao H, Wang R, Zhao D, Zhang B, Zhou Y, Tang S, Zhang H (2019) Identification and fine mapping of qGN1c, a QTL for grain number per panicle, in rice (*Oryza sativa*). Mol Breed 39:129

Xue W, Xing Y, Weng X, Zhao Y, Tang W, Wang L, Zhou H, Yu S, Xu C, Li X (2008) Natural variation in Ghd7 is an important regulator of heading date and yield potential in rice. Nat Genet 40:761–767

Yan WH, Wang P, Chen HX, Zhou HJ, Li QP, Wang CR, Ding ZH, Zhang YS, Yu SB, Xing YZ (2011) A major QTL, Ghd8, plays pleiotropic roles in regulating grain productivity, plant height, and heading date in rice. Mol Plant 4:319–330

Zahoor SA, Ahmad S, Ahmad A, Wajid A, Khaliq T, Mubeen M, Hussain S, Sami Ul Din M, Amin A, Awais M, Nasim W (2019) Improving water use efficiency in agronomic crop production. In: Hasanuzzaman M (ed) Agronomic crops, vol 2. Springer Nature Singapore, Singapore, pp 13–29

Zha X, Luo X, Qian X, He G, Yang M, Li Y, Yang J (2009) Over-expression of the rice LRK1 gene improves quantitative yield components. Plant Biotechnol J 7:611–620

Zhai W, Chen C, Zhu X, Chen X, Zhang D, Li X, Zhu L (2004) Analysis of T-DNA-Xa21 loci and bacterial blight resistance effects of the transgene Xa21 in transgenic rice. Theor Appl Genet 109:534–542

Zhang F, Xie J (2014) Genes and QTLs resistant to biotic and abiotic stresses from wild rice and their applications in cultivar improvements. Rice-Germplasm, Genetics and Improvement, pp 59–78

Zhang YX, Wang Q, Jiang L, Liu LL, Wang BX, Shen YY, Cheng XN, Wan JM (2011) Fine mapping of qSTV11 KAS, a major QTL for rice stripe disease resistance. Theor Appl Genet 122:1591–1604

Zhu XG, Shan L, Wang Y, Quick WP (2010) C4 rice—an ideal arena for systems biology research. J Integr Plant Biol 52:762–770

New Breeding Techniques (NBTs) and Biotechnology for Boosting Rice Grain Yield to Feed 5 Billion in 2050

33

Babar Hussain, Qasim Raza, Rana Muhammad Atif, and Muhammad Qadir Ahmad

Abstract

Rice is the staple food of over 50% world population that accounts for around three billion people. Human population is estimated to swell to ten billion by 2050, and feeding an extra four billion population then will be a challenging task. Classical breeding requires 10–12 years to release a variety making it hard to improve the grain yield at required pace. Therefore, innovative solutions in crop improvement programs are required to ensure the food security. New breeding techniques (NBTs) such as molecular breeding, genomic selection, speed breeding, DNA recombinant technology, and Clustered Regularly Interspaced Short Palindromic Repeats (CRISPR)/CRISPR-associated protein 9 (CRISPR/Cas9) system are promising tools to speed up the breeding process. Therefore, we discuss the practical applications of these NBTs for improving the grain yield

B. Hussain (✉)
Department of Biological Sciences, Middle East Technical University, Ankara, Turkey

Faculty of Life Sciences, University of Central Punjab, Lahore, Pakistan
e-mail: babar.hussain@ucp.edu.pk

Q. Raza
Molecular Breeding Laboratory, Rice Research Institute, Sheikhupura, Pakistan

R. M. Atif
Department of Plant Breeding and Genetics, University of Agriculture Faisalabad, Faisalabad, Pakistan

Center for Advanced Studies in Agriculture and Food Security (CAS-AFS), University of Agriculture Faisalabad, Faisalabad, Pakistan

M. Q. Ahmad
Department of Plant Breeding and Genetics, Bahauddin Zakariya University, Multan, Punjab, Pakistan

N. Sarwar et al. (eds.), *Modern Techniques of Rice Crop Production*,
https://doi.org/10.1007/978-981-16-4955-4_33

of rice to ensure food security. We hope this chapter will be a useful resource for future rice breeding programs.

Keywords

Rice · Grain yield · Molecular breeding · Genomic selection · CRISPR/Cas9 · DNA recombinant technology · Speed breeding

33.1 Introduction

Rice (*Oryza sativa* L.) is consumed by around three billion or 50% of the world's population (Zhang et al. 2014). The World population is expected to reach 9.7–11 billion in 2050 (https://population.un.org/wpp/) that will significantly increase the demand for food including rice. It is estimated that there will be around five billion rice consumers in 2050; thus, significant increase in rice yield will be required to cope with surging demand. Additionally, climate change results in increased drought spells and soil salinization worldwide (Budak et al. 2015; Hussain et al. 2015) in addition to increased occurrence of plant diseases and insect pests (Hussain 2015). Abiotic stresses such as drought can cause 60–78% reduction in rice grain yield (Swamy et al. 2017), while cold stress poses serious threat to rice grain production in upland areas of Japan, China, Korea, etc. (Zhang et al. 2014). Improving crop plants through classical breeding is a long and hectic process as it involves selection of desired plants for 6–7 years to achieve desired homozygosity and could take 9–10 years to release cultivars for farmers (Budak et al. 2015). Therefore, new breeding techniques (NBTs) such as molecular breeding, genomic selection, speed breeding, haploid breeding, DNA recombinant technology, CRISPR/Cas9 technology, genetic tuning, etc. that could fast forward the selection process and shorten the breeding cycles could help in increasing the rice production to feed ten billion population in 2050. With the advances in sequencing technologies and bioinformatics tools (Raza et al. 2020a), molecular breeding or marker-assisted selection (MAS) that employs DNA markers particularly single-nucleotide polymorphisms (SNPs) for selection of desired parents and plants during breeding cycles to improve the selection efficiency is a routinely used technique for rice improvement (Yu et al. 2017; Raza et al. 2020b). Similarly, genomic selection is an advanced form of molecular breeding or MAS that utilizes large numbers of molecular marker and phenotyping data for a collection of lines/varieties to estimate the genomic estimated breeding values (GEBVs). This helps to establish parameters for fast-forwarding the selection of the progenies in the absence of additional phenotyping (Hussain et al. 2021), thus making the selection process more reliable and powerful. Speed breeding is a powerful NBT that utilizes the procedures used at space station, that is, growing the plants under extended photoperiods through provision of supplementary lighting to produce grains in 2 months, thus making it possible to develop six breeding progenies in single year that could require 6 years in classical breeding (Ghosh et al. 2018; Hussain et al. 2021). Although DNA recombinant technology or transgenic method dates back to the 1990s, however, its ability to transfer genes from unrelated plant species, a major bottleneck of classical breeding, makes it an exciting procedure (Babar et al. 2020; Hussain and Mahmood 2020). Recently, development of a simple, robust, efficient, and powerful genome editing system, that is, CRISPR/Cas9 (Jinek et al. 2012), has opened new horizons in crop improvement (Hussain et al. 2018). This revolutionary and powerful tool utilizes single-guide RNA (sgRNA) and Cas9 nuclease to make double-strand breaks (DSBs) in DNA that are repaired later with ability to edit any DNA sequence in any species with high accuracy and specificity (Jinek et al. 2012; Budak et al. 2015; Hussain et al. 2018, 2021). CRISPR/Cas9 system has been routinely used to improve yield, herbicide resistance, biotic and abiotic stress tolerance, and quality in several crops such as barley, corn, citrus, cucumber, rice, soybean, tobacco, tomato, and wheat in addition to model plants including *Arabidopsis thaliana*, *Camelina sativa*, *Nicotiana benthamiana*, *Physcomitrella patens,* etc. (Hussain et al. 2018). Indeed, CRISPR/Cas9 is one of the most exciting crop improvement tools developed in human history.

Keeping in view the potential of these NBTs to achieve the target of feeding the ten billion population in 2050, we review the latest developments and applications of these NBTs for improving the grain yield of rice that feeds around three billion people across the globe (Zhang et al. 2014). This chapter will be a useful guide for the future application of NBTs in crop improvement programs.

33.2 NBTs for Improving Grain Yield of Rice

Several NBTs have been applied for increasing rice grain yield and are described below.

33.2.1 Molecular Breeding for Improving Rice Grain Yield

Molecular breeding or MAS refers to the scheme of identification, characterization, and genetic exploitation of trait-linked genomic regions to fulfill a particular breeding objective. It involves the characterization of germplasm, selection of parents and traits, and gene mapping based on DNA marker data. Among various types of DNA markers, SNPs are predominantly used for the purpose as these are the most prevalent types of markers in genome, reliable, cost-effective, and suitable for high-throughput genotyping and high-density linkage map construction (Hussain et al. 2017, 2021; Majeed et al. 2019). Additionally, molecular markers are not influenced by environmental factors in contrast to morphological traits. Thus, MAS is a robust, efficient, and reliable tool that could reduce the selection cycles and time to release cultivars by 3–4 years (Budak et al. 2015). Identification and validation of genomic regions are the two most important components of MAS. With the advent of next-generation sequencing (NGS) technologies, mapping of genomic regions controlling genetic diversity in qualitative and quantitative characters has been

Table 33.1 Examples of application of molecular breeding for improvement of grain yield and related traits in rice

Associated traits	Study type	Mapping material/ cross	Lines used (n)	QTL/ SNPs mapped (n)	Candidate genes	References
Agronomic traits	QTL	DHs	111, 146	15	*OsD10, OsHd3a, OsRFP, OSVIL1*	Descalsota-Empleo et al. (2019)
		IR64/IR69428, BR29/IR75862				
Grain yield and yield-related	QTL	*Indica* and *japonica* accessions	60	86	*Os02g0120800 Os04g0619500 Os05g0474600 Os06g0162800 Os06g0560300*	Donde et al. (2020)
Grain quality and yield-related	QTL	RILs	182	15	*GW5, chalk5*	Gao et al. (2016)
		Gang46B/K1075				
Yield and yield-related	QTL	RILs	105	22	*Os03t0263600-01*	Kulkarni et al. (2020)
					Os03t0308200-01	
					Os03t0333200-01	
		IR58025A/KMR3R			*Os03t0764900-01*	
					Os03t0809900-01	
					Os03t0815700-01	
Grain yield and yield-related	QTL	RILs	190	39	*Os02g0826050*	Lei et al. (2018)
					Os04g0482300	
					Os05g0588600	
		Dongnong422/ Kongyu131			*Os08g0105100*	
					Os10g0524700	
Total spikelets number per panicle	QTL	NILs	3	2	*Os12g0618800*	Sasaki et al. (2017)
					Os12g0619000	
					Os12g0619700	
		YTH63/IR 64, YTH83/IR 64			*Os12g0620000*	
					Os12g0620400	
					Os12g0620600	
					Os12g0621000	

Yield and yield-related	QTL	Cultivars, breeding lines, landraces	75	20	*OsMADS29, OsSPL4, OsFBX76, OsSub37, OsSCP48*	Swamy et al. (2017)
Grain yield-related	QTL	NILs 9311/Nipponbare	125	25	*Gn1a*	Xu et al. (2019)
Grain number and yield	QTL	NILs Zhonghui8006/ Wuyunjing8	200	1	*GN4-1*	Zhou et al. (2018)
Grain weight	QTL-seq	*Indica*, *japonica*, *aus*, and wild accessions	11	6	*LOC_Os05g15880* *LOC_Os05g18604*	Daware et al. (2016)
Grain length and weight	QTL-seq	NILs Hui 12-29/Fuhui 212	176	1	—	Yaobin et al. (2018)
TGW, grain length, grain width, and grain thickness	SLAF-Seq/BSA	RILs M201/JY293	60	3	*GS3*	Xu et al. (2015)
Grain length	GWAS	Cultivars	504	99	*OsLG3, GS3, SSG6, GW5, GW6a, TGW6*	Yu et al. (2017)
Yield and related traits	GWAS	NILs PA64s/9311, Nipponbare/9311, PA64s/Nipponbare	81, 55, 61	71	*sd1, LSCHL4, LAX1, GW5*	Zhang et al. (2019)
Grain yield-related	GPWAS, GWAS	Diverse accessions	471	106	*Os01g0140100* *Os02g0809800* *Os06g0133000* *Os07g0669700* *Os08g0159900*	Zhong et al. (2021)
Grain yield and yield-related	GWAS	Breeding lines	363	52	*OsMADS50, Hd1, RFT1/FT-3, FT-2*	Begum et al. (2015)
Heading date, panicle traits, and panicle number	GWAS	Accessions	1568	138	*SD1, EP3/LP, OsMADS47, OsKS1, CYP90D3, GID1, OsGA2 oxidase-5, OsBZR1, FZP, WRKY2*	Crowell et al. (2016)

(continued)

Table 33.1 (continued)

Associated traits	Study type	Mapping material/ cross	Lines used (n)	QTL/ SNPs mapped (n)	Candidate genes	References
Grain yield-related	GWAS	Cultivars	266	294	*OsCCD7, CYP97A4*	Li et al. (2018)
Plant height, grain yield, and drought resistance	GWAS	Cultivars and landraces	270	29	*OsGA2ox3, OsGH3-2, sd-1, OsGNA1, OsSAP11/OsDOG, OsCYP51G3, OsRRMh, OsPYL2, OsGA2ox9, OsRLK5*	Ma et al. (2016)
Yield, yield components, and other related traits	GWAS	MAGIC populations	128	26	*GS3, Nal1, Psd1, sd1, DTH3, Hd3a*	Meng et al. (2016)
Grain yield under drought	GWAS	Upland rice accessions	175	13	*LOC_Os02g09650 LOC_Os03g50570*	Pantalião et al. (2016)
					LOC_Os06g41700	
					LOC_Os06g41820	
					LOC_Os08g13840	
Yield and yield-related	Meta-QTL analysis	Diverse biparental populations	122	114	*OHK4, TMK, ZHD11, RLCK115, MYB52, SCL7, WDR5a*	Khahani et al. (2020)
Yield under reproductive stage heat stress	Meta-QTL analysis	BCs, DHs, RILs, CSSLs	19	35	*OsBiP2, OsMed37_1, OsNAS3, OsTEF1, OsWRKY10, OsWRKY21*	Raza et al. (2020b)
High grain number per panicle	MABB	Cultivars	12	1	*LOC_Os04g52479*	Singh et al. (2018)
					LOC_Os04g52590	

QTL: quantitative trait loci mapping; GWAS: genome-wide association study; *MABB*: marker-assisted backcross breeding; (*n*): number, *MAGIC*: Multiparent advanced generation intercross population accelerated. In recent years, a plethora of quantitative trait loci (QTL) and marker-trait associations (MTAs) for grain yield improvement have been reported in rice (Table 33.1). Exploitation of these genomic regions using modern plant breeding strategies including MAS, marker-assisted backcrossing, and genome-wide association (GWAS) or genomic selection could bolster the grain yield potential of rice to feed the growing population.

Grain yield is a quantitatively controlled and most complex character. It is the product of number of tillers/panicles per plant, number of grains per panicle, and grain weight traits (Li et al. 2018). During recent past, significant progress has been made in mapping and functional characterization of QTL/genes which are directly or indirectly related to these yield contributing traits. Construction of a genetic linkage map using the genotyping data is a prerequisite for the QTL mapping, and phenotypic data are used to tag the genomic regions controlling a particular trait (Hussain et al. 2017; Xu et al. 2019). Meng and colleagues (Meng et al. 2016) identified a QTL for panicle number and several other QTL for yield-related traits through GWAS involving a multiparent advanced generation intercross (MAGIC) population derived from elite *indica* lines. Similarly, another group (Liu et al. 2017) fine mapped a major tiller inhibition gene *tiller suppression 1* (*ts1*) from an F_2 population of a cross between an EMS-induced rice tiller suppression mutant and an *indica* cultivar. Candidate gene analysis revealed that the causal gene harbored a point mutation in *ts1* mutant plants. They also developed a co-dominant SNP marker which could facilitate MAS. In another study (Lei et al. 2018), a total of ten QTL affecting tiller number in rice were identified from a recombinant inbred line (RIL) population derived from a cross between a high-yielding variety and a *japonica* variety.

Chromosomal segment substitution lines were used to identify 25 QTL tightly linked with the panicle-related traits (Xu et al. 2019). They also fine mapped a new QTL (*qGN1c*) controlling grain number per panicle and 1000-grain weight in rice. The identification and fine mapping of *qGN1c* would accelerate molecular breeding efforts to improve rice yield. Several other groups have reported major QTL for chalkiness, grain weight, and length (Xu et al. 2015a, b; Gao et al. 2016; Yaobin et al. 2018), spikelet number per panicle (Sasaki et al. 2017), grain number and grain yield (Zhou et al. 2018; Descalsota-Empleo et al. 2019; Donde et al. 2020; Kulkarni et al. 2020), etc., and further details about mapping populations, cross combination, number of lines, and QTL; and candidate genes are detailed in Table 33.1.

Unlike QTL mapping, construction of linkage map is not required for GWAS, and combination of phenotypic and phenotypic data is sufficient for identification of MTAs controlling a particular trait. Additionally, development of populations is not required, large collections of cultivars and accessions can be characterized, and MTAs reported for a trait can be used for MAS or genomic selection (Hussain et al. 2021). For example, Li and colleagues (Li et al. 2018) conducted a GWAS to understand the genetic basis of correlations among yield-related traits. They studied a diverse and large collection of cultivated rice accessions under short- and long-day environments. Their GWAS identified 15 pleiotropic QTL, 45 pleiotropic genes, and 35 pleiotropic SNPs that may contribute to the underlying genetic correlations among panicle number per plant, grains per panicle, and kilo-grain weight traits. These interactive QTL, genes, SNPs, and associated molecular markers could be utilized through molecular breeding for genetic improvement in rice yield. Similarly, another study (Zhong et al. 2021) explored genetic mechanisms underlying panicle architecture in rice using genome–phenome and genome-wide association studies. They investigated 471 homozygous and diverse rice accessions to discover the relationships between genes and multi-traits. A total of 106 tightly linked QTL with six panicle architecture traits were identified, including 21 QTL for grain number per panicle and 17 QTL for panicle number per plant. Additionally, 23 candidate genes possibly regulating these traits were also mapped on rice chromosomes.

Khahani and colleagues (Khahani et al. 2020) reported a comprehensive genome-wide meta-analysis of yield and yield-related traits in rice. Their analysis included 1052 QTL from 122 diverse rice populations and identified 114 stable meta-QTL (mQTL) controlling yield and yield-related traits under diverse genetic backgrounds and environments. Genetic markers associated with these mQTL and the underlying candidate genes could pave the way for molecular breeding-based yield improvement in rice. Similarly, another meta-QTL analysis reported 35 stable mQTL for heat tolerance including the ones for grain yield (Raza et al. 2020b). With the advances in NGS technologies and bioinformatics tools, high-throughput genotyping of rice through genotyping arrays and genotyping by sequencing (GBS) results in deep coverage of genome. Therefore, most of the QTL, GWAS, and mQTL studies have identified a plethora of candidate genes such as *OsHd3a, GW5, OsMADS29, OsMADS47, OsMADS50, OsSPL4, OsFBX76, Gn1a, GN4-1, GS3, OsLG3, SSG6, GW6a, TGW6, SD1, CYP90D3, OsGA2 oxidase-5, OsGA2ox3, OsGA2ox9, WRKY2, OsWRKY10, OsWRKY21, OHK4, MYB52, OsNAS3,* etc. among various others and are listed in Table 33.1. Characterization and utilization of these candidate genes in future breeding programs could boost the efforts to increase rice production.

High-throughput genotyping and next-generation sequencing technologies have identified a metadata of trait-linked SNPs in rice. However, a large gap exists between SNP markers development and their application in molecular breeding. Recently, Yang and colleagues (Yang et al. 2019) developed a core set of 467 SNP arrays (including 244 and 87 markers for agronomic traits and yield, respectively) based on the Kompetitive Allele-Specific PCR (KASP) method. These KASP markers when combined with molecular breeding are a valuable resource for the improvement of yield and other agronomic traits in rice.

During recent years, genotypic and phenotypic data in crop genetics and genomics fields are accumulating at an exponential rate. However, the bottleneck to utilizing these huge amounts of data in molecular breeding is lack of data integration and tools development for unrevealing the genotype to phenotype relationships. To address these subordinate issues, researchers from Chinese Academy of Sciences have developed an integrated omics knowledgebase (MBKbase) for molecular breeding in rice (Peng et al. 2020). This online knowledgebase integrates information from rice germplasms, reference genomes, and omics data, and could facilitate researchers in molecular breeding of targeted traits.

33.2.2 Genomic Selection for Improving Rice Grain Yield

With the dramatic increase in human population, the food security has become a serious issue in the world. In the next 30 years to keep up the pace with the global population growth, there is a need to double the production of staple cereals grain including rice, wheat, and maize. However, our agriculture system is damaged by human-induced climate change, and both farmers and plant breeders have to fight against these environmental stresses to overcome this damage. In Asian tropics, the breeding of rice varieties is challenging and aspire-intensive resources. The Asian public sectors of rice breeding program mostly use conventional breeding schemes like pedigree selection which involves selection and screening of desired traits for several generations which is laborious and time-consuming (Budak et al. 2015). With the advancement in genomics and molecular genetics, the faster breeding methods like MAS are being adopted to improve the specific traits, but its effect on increasing the efficiency of breeding has been limited (Hussain et al. 2021). The effect of genetic background and epistasis effect in rice make molecular breeding more complicated. In 2001, Meuwissen and colleagues introduced genomic selection which is another marker-based strategy that is alternative to MAS and use genome-wide DNA marker data to predict the breeding value of breeding progenies and increase the effectiveness of breeding quantitative traits (Meuwissen et al. 2001). In genomic selection (GS), the identification and selection of individuals having superior breeding values is based on prediction models constructed (Hussain et al. 2021) by correlating genotype and phenotype in breeding population of interest. The prediction accuracy depends upon various factors including size of training population, trait heritability, genetic relationship between training population and selection candidates, and marker density (Wang et al. 2018). The advantage of genomic selection over traditional pedigree method is efficiency. The benefit of selection during genomic selection is proportional to GEBVs accuracy. If the genome-estimated breeding value accuracy is higher, GS can reduce the breeding time by increasing the proportion of high-performing offspring in breeding population and increase the gain from selection (Hussain et al. 2021). Genomic selection is being successfully implemented in dairy cattle, but in plant breeding it has the potential to improve the breeding efficiency.

In hybrid rice breeding, GS offers the opportunity to predict the hybrid performance before the measurement of their phenotype. Studies show that the average yield of top 100 hybrids from predicted 21,945 hybrids give a 16% higher yield and indicates the potential of genomic selection for yield improvement (Xu et al. 2014). Similarly, GS for hybrid breeding predicts relatively higher value of 0.54 for yield and 0.92 for grain length for potential hybrid of rice varieties (Cui et al. 2020b). In rice, GWAS and QTL mapping studies show the large effect of QTL for agronomic traits including grain yield, plant height, grain yield under drought stress, and flowering time (Table 33.1). The variation in genetic architecture of different rice traits may affect the effectiveness of GS statistical methods. The GWAS concurrence with fivefold GS cross-validation on 363 elite breeding lines from the International Rice Research Institute (IRRI) was used to build genome prediction model. The prediction accuracy ranged from 0.31 for grain yield, 0.34 for plant height to 0.63 for flowering time (Spindel et al. 2015). The analysis by using subset of the full marker set recommends that the use of 1 SNP every 0.2 cm is sufficient for genomic selection in collection of rice breeding material (Thomson 2014). Ridge-regression best linear unbiased prediction (RR-BLUP) is the most computationally efficient of best statistical method for grain yield where there was no large effect QTL detected by GWAS. For plant height, four mid-sized QTL were identified by GWAS and random forest produced the most consistently accurate GS models. Whereas, for time to flowering, a single large effect QTL was detected, the non-GS multiple linear regression method performs better GS model (Spindel et al. 2015). This shows that genomic selection informed by GWAS interpretations of population structure and genetic architecture can be proved an effective tool for increasing the efficiency of rice breeding as the costs of genotyping continue to decline.

33.2.3 Speed Breeding for Improving Rice Grain Yield

Rice is one of the major food crops on earth, and it is also a staple food of about half of the world's population. Both the increasing global population and changing climate are putting heavy pressure on the crop plants for increased grain harvest. So, there is utmost need to enhance the yield for rapidly increasing population through the development of climate-resilient rice varieties. The conventional breeding techniques are time-consuming as they can take at least 10–12 years from the crossing of desirable parents till the release of approved variety including 6–7 years of selection cycles to achieve desired homozygosity (Budak et al. 2015). On contrary, the marker-assisted breeding methods offer shortcuts in the genotypic selection of desirable plants with the best allelic combinations, but it can still take 4–6 years for the release of desirable variety. This pace of variety development must be accelerated to achieve the food security challenge posed by the ever-increasing population (Hickey et al. 2019). To cope with this scenario, speed breeding is an evolving strategy for the rapid development of desired cultivars by accelerating their life cycle. So, plants are grown in greenhouses with specific light wavelength and intensity, standard day and night temperature, and photoperiod (Ghosh et al. 2018). All these parameters speed up the plant's physiological processes such as photosynthesis and flowering by reducing generation time span (Hickey et al. 2019). This approach is helps to obtain 4–6 generations within a year instead of 1–2 generations under normal environment (Ghosh et al. 2018; Hickey et al. 2019; Hussain et al. 2021). Speed breeding can be used as a complementary tool for integrating genome editing, genomic/MAS, and high-throughput genotyping and phenotyping approaches for the development of suitable traits in crops in less time period. The efficiency of speed breeding has been successfully demonstrated in peanut (*Arachis hypogaea*) (O'Connor et al. 2013), barley (*Hordeum vulgare*), wheat (*Triticum aestivum*), oats (*Avena sativa*), field mustard (*Brassica rapa*), rapeseed (*Brassica napus*), etc. (Ghosh et al. 2018). Now, this is being applied in other perennial crops such as apple for obtaining more generations per year. Speed breeding protocols for long- as well as short-day plants are now available (Wanga et al. 2021), and this application is highly efficacious instead of other breeding tools.

Quite recently, speed breeding has also been successfully applied for generation advancements in combination with MAS for generating salt-tolerant rice (Rana et al. 2019). Before applying speed breeding, a *hitomebore salt-tolerant 1* (*hst1*) gene conferring the salinity tolerance in "Kaijin" rice variety (used s donor parent) was introgressed into "Yukinko-mai," a high-yielding rice variety (used as recipient parent). Subsequently, backcross breeding populations were developed aided with SNP-based MAS. The vegetative growth of the populations was accelerated in a Biotron with longer photoperiod (14 h a day) for initial 30 days. The reproductive phase was then induced by a shorter photoperiod (10 h a day). In this way, six generations till BC3F3 were achieved in a record period of 17 months. The whole-genome sequencing of this BC population revealed the recovery of 93.5% of the genome of the recurrent recipient parent "Yukinko-mai" (Rana et al. 2019). This study revealed the potential of speed breeding coupled with MAS for the development of rice varieties against abiotic stresses. Thus, speed breeding can be used as a valuable approach not only for rapid generations but also for desirable alleles to improve plant's genetics against biotic and abiotic stresses as well as quality- and yield-related parameters.

33.2.4 Transgenic Approaches for Improving Rice Grain Yield

DNA recombinant technology or transgenic approach is a powerful tool for gene transfer between unrelated plant species and wild/distant relatives of crop plants. Therefore, it breaks the genetic barriers by allowing gene transfer otherwise not possible through classical breeding techniques (Babar et al. 2020; Hussain and Mahmood 2020). Since the beginning of twenty-first century, much effort has been dedicated to develop C4 rice plants with high grain yield. For this purpose, genes derived from C4 plants involved in C4 cycle were overexpressed, for example, four transgenic rice lines overexpressing maize pyruvate *orthophosphate dikinase* (*PK*), *phosphoenolpyruvate carboxylase* (*PC*), *PC* + *PK* (CK), and rice *NADP-malic* enzymes were developed (Jiao et al. 2002). Among these, PC and CK transgenic lines showed 22–24% increase in grain yield as compared to wild-type plants. Similarly, transfer and overexpression of C4 pathway enzyme, *phosphoenolpyruvate carboxylase* (*PEP carboxylase* or *PEPC*) from foxtail millets (Ding et al. 2013) and sugarcane (Lian et al. 2014) increased the photosynthesis rate, grain yield, and related traits in transgenic rice plants as compared to wild-type (Table 33.2).

Gene transfer from wild relatives of crops and unrelated species is one of the most prevalent strategies for crop improvement, for example, transfer of *SPIKELET NUMBER* (*SPIKE*) from a japonica rice landrace to cultivated Indica rice increased the grain yield by improving the plant architecture (Fujita et al. 2013). Map-based cloning revealed that SPIKE gene was identical to *NARROW LEAF 1* (*NAL1*) gene which controls vein pattern in leaf. Indica lines overexpressing *SPIKE* revealed increased spikelet number and enhancement in source and sink size as well as

Table 33.2 Examples of Agrobacterium-mediated gene transformation/transgenic rice for improving grain yield of rice

Transgene/s	Origin of transgene/s	Recipient line/s	Improved trait/s	References
Pyruvate *orthophosphate dikinase, phosphoenolpyruvate carboxylase*	Maize	*Japonica* cv. Kitaake	Grain yield (22–24%)	Jiao et al. (2002)
Putative protein phosphatase with Kelch-like repeat domain (*OsPPKL1*)	Rice (*Oryza sativa*)	*Japonica* cv. Zhongua11	Grain yield (13–16%), grain weight (37%), grain length (20%), panicle length (12%)	Zhang et al. (2012)
Positive regulator of grain length 1 (*PGL1*)	*Japonica* cv. Nipponbare	*Japonica* cv. Nipponbare	Grain length, grain weight	Heang and Sassa (2012)
NAL1	Landrace	*Indica* cv. IR64	Grain yield (13–36%), plant architecture	Fujita et al. (2013)
Dehydroascorbate reductase (*DHAR*)	*Japonica* rice	*Japonica* cv. Ilmi	Grain yield, biomass	Kim et al. (2013)
SiPEPC	Foxtail millet (*Setaria italica* cv. Gufeng1)	*Japonica* cv. Zhonghua 8	Biological yield, harvest index, photosynthesis rate panicle length, number of spikes, grains per spike	Ding et al. (2013)
SoPEPC	Sugarcane (*Saccharum officinarum*)	*Indica* cv. Hang2	Grain yield, photosynthesis rate	Lian et al. (2014)
Ammonium transporter (OsAMT1;1)	Rice (*Oryza sativa*)	*Indica* cv. Kaybonnet	Grain yield, nitrogen assimilates, starch, sugars, chlorophyll	Ranathunge et al. (2014)
PcGDH	Fungus *Pleurotus cystidiosus*	*Japonica* cv. Kitaake	Nitrogen assimilation, grain weight, number of panicles	Zhou et al. (2014)
GS2	Rice (*Oryza sativa*)	*Japonica* cv. Zhonghua11	Grain yield, panicle length, number of grains, grain length, width, and thickness, 1000-grain weight	Hu et al. (2015)

(continued)

Table 33.2 (continued)

Transgene/s	Origin of transgene/s	Recipient line/s	Improved trait/s	References
CYP78A13 or *GL3.2*	Rice (Oryza sativa)	*Japonica* cv. Hejiang19	Grain yield, grain weight	Xu et al. (2015)
Plant architecture and yield 1 (PAY1)	Rice (*Indica*)	*Indica* cv. Teqing and 9311	Grain yield (38%), number of grains (200%), plant architecture	Zhao et al. (2015)
GNP1	Rice (*Japonica*)	*Japonica* cv. Zhonghua 11 and Lemont	Grain yield, grain number	Wu et al. (2016)
OsMPH1	Rice (*Oryza sativa*)	*Japonica* cv. Kitaake	Grain yield (45–50%)	Zhang et al. (2017)
OsSUS1-6; *OsSUS3*	Rice (*Japonica*)	*Japonica* cv. Zhonghua11	Grain weight (15–19%), hull size, starch accumulation	Fan et al. (2019)
OsRBCS2 or rubisco	Rice (*Oryza sativa*)	*Japonica* cv. Notohikari	Grain yield, biomass, nitrogen use efficiency	Yoon et al. (2020)

translocation capacity. Furthermore, IRRI146 cultivar of Indica rice showed 13–36% increase in grain yield. In another study (Zhou et al. 2014), overexpression of *NADP(H)-dependent glutamate dehydrogenase (GDH)* gene (*PcGDH*) from fungus *Pleurotus cystidiosus* resulted in improved nitrogen assimilation, grain weight, and number of panicles in transgenic rice plants.

Grain size and weight are the two important parameters associated with grain yield in rice. *Sucrose synthase* genes (*SUS*) gene has been found associated with grain yield enhancement in rice. Fan and colleagues (Fan et al. 2019) overexpressed six *SUS* genes in rice. Transgenic lines showed enhanced hull size and grain weight. It was also observed that transgenic lines contained much larger endosperm volume than those of grains of non-transgenic lines. These transgenic lines depicted a 15–19% increase in weight of brown grains. Similarly, overexpression of *OsMPH1* gene belonging to MYB family of transcription factors leads to the increased grain yield by 45–50% and plant height, whereas knockdown of this gene showed opposite results (Zhang et al. 2017). Among the various transformation methods, Agrobacterium-mediated gene transfer is one of the most prevalent methods (Hussain and Mahmood 2020) and has been routinely used for overexpression of several genes to improve grain yield and related traits in rice (Table 33.2). However, transgenic or genetically modified organisms undergo several biosafety checks before being approved for cultivation, and this is probably one of the most challenging aspects of transgenic rice technology.

33.2.5 CRISPR/Cas9 for Improving Grain Yield of Rice

Increase in rice yield is an essential part of ensuring global food security as it is consumed by billions. Classical breeding takes 7–8 years to get the desired level of homozygosity and 10–12 years to release a variety. On the other hand, CRISPR/Cas9 could deliver transgene-free complete homozygous lines within a cropping season and help to release a variety within 3–4 years (Hussain et al. 2018, 2021). Therefore, CRISPR/Cas9 system has become one of the most predominant systems for genome editing in several crop plants, and this is due to the fact that it is a simple, robust, and powerful tool that utilizes only sgRNA and Cas9 nuclease for multiplexed gene editing in any species without any off-target activity (Hussain et al. 2018).

Several genes and QTL have been edited/knocked out to improve the grain yield and yield-related traits in rice (Tabassum et al. 2021) (Table 33.3). In addition to being a simple and robust genome editing system, one of the most important advantages of CRISPR technology is the multiplexed genome editing. For example, CRISPR-based simultaneous knockout of four genes/QTL, that is, *grain number 1a* (*Gn1a*), *Grain Size 3* (*GS3*), *Ideal Plant Architecture1* (*IPA1*), and *Dense and Erect*

Table 33.3 Examples of CRISPR/Cas9-mediated improvement in grain yield and yield-related traits of rice

Targeted gene/s	Improved trait/s in mutants	References
IPA1, GS3, DEP1, Gn1a	Grain yield, number of tillers, grain size, plant architecture, dense erect panicles, grain number	Li et al. (2016)
GS3, OsGW2, Gn1a	Grain yield, grain size, grain weight, grain number	Zhou et al. (2019)
OsPIN5b, GS3	Panicle length, grain size	Zeng et al. (2020)
GW2, 5 and 6	Grain yield, grain weight	Xu et al. (2016)
OsSPL16/ qGW8	Grain yield, grain size	Usman et al. (2021)
Gn1a, GS3, DEP1	Grain yield, grain number, grain size, panicle architecture	Shen et al. (2018); Huang et al. (2018)
OsFWL	Grain yield, grain length, number of tillers, flag leaf area, number of cells in flag leaf	Gao et al. (2020)
Cytochrome P450, OsBADH2	Grain yield, grain size, aroma (2-acetyl-1-pyrroline (2AP) content)	Usman et al. (2020a)
PYL1, PYL4, PYL6	Number of grains, grain yield	Miao et al. (2018)
OsPYL9	Grain yield under normal and limited water availability	Usman et al. (2020b)
SD1	Grain yield, dwarf plants, resistance to lodging	Hu et al. (2019)
OsGA20ox2	Grain yield, semi-dwarf plants, reduced gibberellins, and flag leaf length	Han et al. (2019)
SCM1, SCM2, SD2	Number of tillers, stem cross-sectional area	Cui et al. (2020a)

Panicle (*DEP1*), improved the number of grains, grain size, number of tillers, and panicle architecture that lead to improved grain yield (Li et al. 2016). Similarly, simultaneous knockout of *GS3*, *OsGW2*, and *Gn1a* improved the rice grain yield by increasing the grain size, grain weight, and number of grains (Zhou et al. 2019). CRISPR-based multiplexed editing of three genes, *Gn1a, GS3,* and *DEP1*, increased the number of grains, grain size, panicle architecture, and grain yield (Shen et al. 2018; Huang et al. 2018). Additionally, rice panicle length and aroma or 2-acetyl-1-pyrroline (2AP) content has been enhanced through CRISPR-based mutagenesis (Zeng et al. 2020; Usman et al. 2020a).

Grain size is an important contributor of grain yield, and editing of *grain width/weight 2, 5, 6,* and *8* (*GW2*, *GW5, GW6,* and *GW8)* (Xu et al. 2016; Zhou et al. 2019; Usman et al. 2021) through CRISPR/Cas9 technology increased the rice grain yield. In another study (Xu et al. 2016), CRISPR/Cas9 technology was used to edit *GW2*, *GW5*, *GW6,* and *GS3* that increased the grain yield through enhancement of grain weight, width, and size. Similarly, CRISPR technology was used to create mutations in abscisic acid receptor genes, *Pyrabactin Resistance 1-Like 1* (*PYL1*), *PYL4*, and *PYL6* (Miao et al. 2018) that resulted in 31% increase in number of grains in rice, while CRISPR mutation in *PYL9* (Usman et al. 2020b) increased the grain yield as well. Furthermore, grain yield in rice was also improved through creating mutations in genes coding for plant height such as *SD1* (Hu et al. 2019), *OsGA20ox2* (Han et al. 2019), and *SD2* (Cui et al. 2020a). The mutant plants showed short or semi-dwarf plant height, resistance to lodging, and increased grain yield. In conclusion, CRISPR/Cas9 gene-editing system emerged as a powerful plant breeding tool for rice crop improvement to ensure global food security.

33.3 Conclusions and Prospects

Human population will swell to ten billion by 2050, and feeding them will be a daunting challenge. Therefore, out-of-box solutions in crop improvement programs are required to ensure sustainable agriculture and food security. As we discussed above, NBTs are being successfully employed utilized by several research groups to identify and utilize important genes for significantly improving the rice grain yield. Among the NBTs, CRISPR/Cas9 is the most powerful and promising tool till date and has been extensively used to improve grain yield of rice. The success of CRISPR system is attributed to its simplicity, robustness, higher efficiency, specificity, precision, and multiplexed gene editing. Although CRISPR/Cas9 has the potentail to be a predominant tool to improve grain yield of rice in near future, combination of different NBTs will more useful for the purpose. There are already some examples where a combination of different NBTs has been used to increase the grain yield in rice. For example, QTL identified through molecular breeding have been edited through CRISPR/Cas9 system to improve the grain yield of rice (Li et al. 2016; Shen et al. 2018; Huang et al. 2018; Zhou et al. 2019; Zeng et al. 2020; Usman et al. 2021). Similarly, speed breeding has been used in combination with molecular breeding to improve yield under salt stress (Rana et al. 2019). With the advances in sequencing technologies and reduction in cost of genotyping, MAS is being replaced with genomic selection, and we hope that GS will be combined with speed breeding to improve the rice production in the years to come. We also expect that the identification of novel QTL and genes through MAS and GS will lead to their application in overexpression through DNA recombinant technology and CRISPR/Cas9 genome editing technology to improve the grain yield of rice.

References

Babar U, Nawaz MA, Arshad U et al (2020) Transgenic crops for the agricultural improvement in Pakistan: a perspective of environmental stresses and the current status of genetically modified crops. GM Crops Food 11:1–29. https://doi.org/10.1080/21645698.2019.1680078

Begum H, Spindel JE, Lalusin A et al (2015) Genome-wide association mapping for yield and other agronomic traits in an elite breeding population of tropical rice (Oryza sativa). PLoS One 10: e0119873. https://doi.org/10.1371/journal.pone.0119873

Budak H, Hussain B, Khan Z et al (2015) From genetics to functional genomics: improvement in drought signaling and tolerance in wheat. Front Plant Sci 6:1–13. https://doi.org/10.3389/fpls.2015.01012

Crowell S, Korniliev P, Falcão A et al (2016) Genome-wide association and high-resolution phenotyping link Oryza sativa panicle traits to numerous trait-specific QTL clusters. Nat Commun 7:1–14. https://doi.org/10.1038/ncomms10527

Cui Y, Hu X, Liang G et al (2020a) Production of novel beneficial alleles of a rice yield-related QTL by CRISPR/Cas9. Plant Biotechnol J 18:1987–1989. https://doi.org/10.1111/pbi.13370

Cui Y, Li R, Li G et al (2020b) Hybrid breeding of rice via genomic selection. Plant Biotechnol J 18: 57–67. https://doi.org/10.1111/pbi.13170

Daware A, Das S, Srivastava R et al (2016) An efficient strategy combining SSR markers- and advanced QTL-seq-driven QTL mapping unravels candidate genes regulating grain weight in rice. Front Plant Sci 7:1535. https://doi.org/10.3389/fpls.2016.01535

Descalsota-Empleo GI, Amparado A, Inabangan-Asilo MA et al (2019) Genetic mapping of QTL for agronomic traits and grain mineral elements in rice. Crop J 7:560–572. https://doi.org/10.1016/j.cj.2019.03.002

Ding Z-S, Huang S-H, Zhou B-Y et al (2013) Over-expression of phosphoenolpyruvate carboxylase cDNA from C4 millet (Seteria italica) increase rice photosynthesis and yield under upland condition but not in wetland fields. Plant Biotechnol Rep 7:155–163. https://doi.org/10.1007/s11816-012-0244-1

Donde R, Mohapatra S, Baksh SKY et al (2020) Identification of QTLs for high grain yield and component traits in new plant types of rice. PLoS One 15:e0227785. https://doi.org/10.1371/journal.pone.0227785

Fan C, Wang G, Wang Y et al (2019) Sucrose synthase enhances hull size and grain weight by regulating cell division and starch accumulation in transgenic rice. Int J Mol Sci 20:4971. https://doi.org/10.3390/ijms20204971

Fujita D, Trijatmiko KR, Tagle AG et al (2013) NAL1 allele from a rice landrace greatly increases yield in modern indica cultivars. Proc Natl Acad Sci 110:20431–20436. https://doi.org/10.1073/pnas.1310790110

Gao Y, Liu C, Li Y et al (2016) QTL analysis for chalkiness of rice and fine mapping of a candidate gene for qACE9. Rice 9:41. https://doi.org/10.1186/s12284-016-0114-5

Gao Q, Li G, Sun H et al (2020) Targeted mutagenesis of the rice FW 2.2-like gene family using the CRISPR/Cas9 system reveals OsFWL4 as a regulator of tiller number and plant yield in rice. Int J Mol Sci 21:809. https://doi.org/10.3390/ijms21030809

Ghosh S, Watson A, Gonzalez-Navarro OE et al (2018) Speed breeding in growth chambers and glasshouses for crop breeding and model plant research. Nat Protoc 13:2944–2963. https://doi.org/10.1038/s41596-018-0072-z

Han Y, Teng K, Nawaz G et al (2019) Generation of semi-dwarf rice (Oryza sativa L.) lines by CRISPR/Cas9-directed mutagenesis of OsGA20ox2 and proteomic analysis of unveiled changes caused by mutations. 3 Biotech 9:387. https://doi.org/10.1007/s13205-019-1919-x

Heang D, Sassa H (2012) Antagonistic actions of HLH/bHLH proteins are involved in grain length and weight in rice. PLoS One 7:e31325. https://doi.org/10.1371/journal.pone.0031325

Hickey LT, Hafeez AN, Robinson H et al (2019) Breeding crops to feed 10 billion. Nat Biotechnol 37:744–754. https://doi.org/10.1038/s41587-019-0152-9

Hu J, Wang Y, Fang Y et al (2015) A rare allele of GS2 enhances grain size and grain yield in Rice. Mol Plant 8:1455–1465. https://doi.org/10.1016/j.molp.2015.07.002

Hu X, Cui Y, Dong G et al (2019) Using CRISPR-Cas9 to generate semi-dwarf rice lines in elite landraces. Sci Rep 9:19096. https://doi.org/10.1038/s41598-019-55757-9

Huang L, Zhang R, Huang G et al (2018) Developing superior alleles of yield genes in rice by artificial mutagenesis using the CRISPR/Cas9 system. Crop J 6:475–481. https://doi.org/10.1016/j.cj.2018.05.005

Hussain B (2015) Modernization in plant breeding approaches for improving biotic stress resistance in crop plants. Turk J Agric For 39:515–530. https://doi.org/10.3906/tar-1406-176

Hussain B, Mahmood S (2020) Development of transgenic cotton for combating biotic and abiotic stresses. In: Ahmad S, Hasanuzzaman M (eds) Cotton production and uses: agronomy, crop protection, and postharvest technologies. Springer, Singapore, pp 527–545

Hussain B, Khan AS, Ali Z (2015) Genetic variation in wheat germplasm for salinity tolerance at seedling stage: improved statistical inference. Turk J Agric For 39:182–192. https://doi.org/10.3906/tar-1404-114

Hussain B, Lucas SJ, Ozturk L, Budak H (2017) Mapping QTLs conferring salt tolerance and micronutrient concentrations at seedling stagein wheat. Sci Rep 7:15662. https://doi.org/10.1038/s41598-017-15726-6

Hussain B, Lucas SJ, Budak H (2018) CRISPR/Cas9 in plants: at play in the genome and at work for crop improvement. Brief Funct Genomics 17:319–328. https://doi.org/10.1093/bfgp/ely016

Hussain B, Akpınar BA, Alaux M et al (2021) Wheat genomics and breeding: bridging the gap. AgriRxiv:1–57. https://doi.org/10.31220/agriRxiv.2021.00039

Jiao D, Huang X, Li X et al (2002) Photosynthetic characteristics and tolerance to photo-oxidation of transgenic rice expressing C4 photosynthesis enzymes. Photosynth Res 72:85–93. https://doi.org/10.1023/A:1016062117373

Jinek M, Chylinski K, Fonfara I et al (2012) A programmable dual-RNA-guided DNA endonuclease in adaptive bacterial immunity. Science 337:816–822. https://doi.org/10.1126/science.1225829

Khahani B, Tavakol E, Shariati V, Fornara F (2020) Genome wide screening and comparative genome analysis for meta-QTLs, ortho-MQTLs and candidate genes controlling yield and yield-related traits in rice. BMC Genomics 21:294. https://doi.org/10.1186/s12864-020-6702-1

Kim Y-S, Kim I-S, Bae M-J et al (2013) Homologous expression of cytosolic dehydroascorbate reductase increases grain yield and biomass under paddy field conditions in transgenic rice (Oryza sativa L. japonica). Planta 237:1613–1625. https://doi.org/10.1007/s00425-013-1862-8

Kulkarni SR, Balachandran SM, Ulaganathan K et al (2020) Molecular mapping of QTLs for yield related traits in recombinant inbred line (RIL) population derived from the popular rice hybrid KRH-2 and their validation through SNP genotyping. Sci Rep 10:13695. https://doi.org/10.1038/s41598-020-70637-3

Lei L, Zheng HL, Wang JG et al (2018) Genetic dissection of rice (Oryza sativa L.) tiller, plant height, and grain yield based on QTL mapping and metaanalysis. Euphytica 214:109. https://doi.org/10.1007/s10681-018-2187-2

Li M, Li X, Zhou Z et al (2016) Reassessment of the four yield-related genes Gn1a, DEP1, GS3, and IPA1 in rice using a CRISPR/Cas9 system. Front Plant Sci 7:377. https://doi.org/10.3389/fpls.2016.00377

Li F, Xie J, Zhu X et al (2018) Genetic basis underlying correlations among growth duration and yield traits revealed by GWAS in rice (Oryza sativa L.). Front Plant Sci 9:650. https://doi.org/10.3389/fpls.2018.00650

Lian L, Wang X, Zhu Y et al (2014) Physiological and photosynthetic characteristics of indica Hang2 expressing the sugarcane PEPC gene. Mol Biol Rep 41:2189–2197. https://doi.org/10.1007/s11033-014-3070-4

Liu L, Meng F, He Y et al (2017) Fine mapping and candidate gene analysis of the tiller suppression gene ts1 in rice. PLoS One 12:e0170574. https://doi.org/10.1371/journal.pone.0170574

Ma X, Feng F, Wei H et al (2016) Genome-wide association study for plant height and grain yield in rice under contrasting moisture regimes. Front Plant Sci 7:1801. https://doi.org/10.3389/fpls.2016.01801

Majeed S, Rana IA, Atif RM et al (2019) Role of SNPs in determining QTLs for major traits in cotton. J Cotton Res 2:5. https://doi.org/10.1186/s42397-019-0022-5

Meng L, Zhao X, Ponce K et al (2016) QTL mapping for agronomic traits using multi-parent advanced generation inter-cross (MAGIC) populations derived from diverse elite indica rice lines. F Crop Res 189:19–42. https://doi.org/10.1016/j.fcr.2016.02.004

Meuwissen TH, Hayes BJ, Goddard ME (2001) Prediction of total genetic value using genome-wide dense marker maps. Genetics 157:1819–1829

Miao C, Xiao L, Hua K et al (2018) Mutations in a subfamily of abscisic acid receptor genes promote rice growth and productivity. Proc Natl Acad Sci 115:6058–6063. https://doi.org/10.1073/pnas.1804774115

O'Connor DJ, Wright GC, Dieters MJ et al (2013) Development and application of speed breeding technologies in a commercial peanut breeding program. Peanut Sci 40:107–114. https://doi.org/10.3146/PS12-12.1

Pantalião GF, Narciso M, Guimarães C et al (2016) Genome wide association study (GWAS) for grain yield in rice cultivated under water deficit. Genetica 144:651–664. https://doi.org/10.1007/s10709-016-9932-z

Peng H, Wang K, Chen Z et al (2020) MBKbase for rice: an integrated omics knowledgebase for molecular breeding in rice. Nucleic Acids Res 48(D1):D1085–D1092. https://doi.org/10.1093/nar/gkz921

Rana MM, Takamatsu T, Baslam M et al (2019) Salt tolerance improvement in rice through efficient SNP marker-assisted selection coupled with speed-breeding. Int J Mol Sci 20:2585. https://doi.org/10.3390/ijms20102585

Ranathunge K, El-kereamy A, Gidda S et al (2014) AMT1; 1 transgenic rice plants with enhanced NH4+ permeability show superior growth and higher yield under optimal and suboptimal NH4+ conditions. J Exp Bot 65:965–979. https://doi.org/10.1093/jxb/ert458

Raza Q, Riaz A, Atif RM et al (2020a) MADS-box genes galore in wheat genome: phylogenomics, evolution and stress associated functions. BioRxiv:1–33. https://doi.org/10.1101/2020.10.23.351635

Raza Q, Riaz A, Bashir K, Sabar M (2020b) Reproductive tissues-specific meta-QTLs and candidate genes for development of heat-tolerant rice cultivars. Plant Mol Biol 104:97–112. https://doi.org/10.1007/s11103-020-01027-6

Sasaki K, Fujita D, Koide Y et al (2017) Fine mapping of a quantitative trait locus for spikelet number per panicle in a new plant type rice and evaluation of a near-isogenic line for grain productivity. J Exp Bot 68:2693–2702. https://doi.org/10.1093/jxb/erx128

Shen L, Wang C, Fu Y et al (2018) QTL editing confers opposing yield performance in different rice varieties. J Integr Plant Biol 60:89–93. https://doi.org/10.1111/jipb.12501

Singh VK, Ellur RK, Singh AK et al (2018) Effect of qGN4.1 QTL for grain number per panicle in genetic backgrounds of twelve different mega varieties of rice. Rice 11:8. https://doi.org/10.1186/s12284-017-0195-9

Spindel J, Begum H, Akdemir D et al (2015) Genomic selection and association mapping in rice (Oryza sativa): effect of trait genetic architecture, training population composition, marker number and statistical model on accuracy of rice genomic selection in elite, tropical rice breeding lines. PLoS Genet 11:e1004982. https://doi.org/10.1371/journal.pgen.1004982

Swamy BPM, Shamsudin NAA, Rahman SNA et al (2017) Association mapping of yield and yield-related traits under reproductive stage drought stress in rice (Oryza sativa L.). Rice 10:21. https://doi.org/10.1186/s12284-017-0161-6

Tabassum J, Ahmad S, Hussain B, Mawia AM, Zeb A, Ju L (2021) Applications and potential of genome-editing systems in rice improvement: current and future perspectives. Agronomy 11(7): 1359. https://doi.org/10.3390/agronomy11071359

Thomson MJ (2014) High-throughput SNP genotyping to accelerate crop improvement. Plant Breed Biotechnol 2:195–212. https://doi.org/10.9787/PBB.2014.2.3.195

Usman B, Nawaz G, Zhao N et al (2020a) Generation of high yielding and fragrant rice (Oryza sativa L.) lines by CRISPR/Cas9 targeted mutagenesis of three homoeologs of cytochrome P450 gene family and OsBADH2 and transcriptome and proteome profiling of revealed changes triggered by mutations. Plants 9:788. https://doi.org/10.3390/plants9060788

Usman B, Nawaz G, Zhao N et al (2020b) Precise editing of the OsPYL9 gene by RNA-guided Cas9 nuclease confers enhanced drought tolerance and grain yield in rice (Oryza sativa L.) by regulating circadian rhythm and abiotic stress responsive proteins. Int J Mol Sci 21:7854. https://doi.org/10.3390/ijms21217854

Usman B, Nawaz G, Zhao N et al (2021) Programmed editing of rice (Oryza sativa l.) osspl16 gene using crispr/cas9 improves grain yield by modulating the expression of pyruvate enzymes and cell cycle proteins. Int J Mol Sci 22:1–19. https://doi.org/10.3390/ijms22010249

Wang X, Xu Y, Hu Z, Xu C (2018) Genomic selection methods for crop improvement: current status and prospects. Crop J 6:330–340. https://doi.org/10.1016/j.cj.2018.03.001

Wanga MA, Shimelis H, Mashilo J, Laing MD (2021) Opportunities and challenges of speed breeding: a review. Plant Breed 140:185–194. https://doi.org/10.1111/pbr.12909

Wu Y, Wang Y, Mi X-F et al (2016) The QTL GNP1 encodes GA20ox1, which increases grain number and yield by increasing cytokinin activity in rice panicle meristems. PLoS Genet 12: e1006386. https://doi.org/10.1371/journal.pgen.1006386

Xu S, Zhu D, Zhang Q (2014) Predicting hybrid performance in rice using genomic best linear unbiased prediction. Proc Natl Acad Sci 111:12456–12461. https://doi.org/10.1073/pnas.1413750111

Xu F, Fang J, Ou S et al (2015a) Variations in CYP 78 A 13 coding region influence grain size and yield in rice. Plant Cell Environ 38:800–811. https://doi.org/10.1111/pce.12452

Xu F, Sun X, Chen Y et al (2015b) Rapid identification of major QTLs associated with rice grain weight and their utilization. PLoS One 10:e0122206. https://doi.org/10.1371/journal.pone.0122206

Xu R, Yang Y, Qin R et al (2016) Rapid improvement of grain weight via highly efficient CRISPR/Cas9-mediated multiplex genome editing in rice. J Genet Genomics 43:529–532. https://doi.org/10.1016/j.jgg.2016.07.003

Xu Z, Miao Y, Chen Z et al (2019) Identification and fine mapping of qGN1c, a QTL for grain number per panicle, in rice (Oryza sativa). Mol Breed 39:129. https://doi.org/10.1007/s11032-019-1039-7

Yang G, Chen S, Chen L et al (2019) Development of a core SNP arrays based on the KASP method for molecular breeding of rice. Rice 12:21. https://doi.org/10.1186/s12284-019-0272-3

Yaobin Q, Peng C, Yichen C et al (2018) QTL-Seq identified a major QTL for grain length and weight in rice using near isogenic F2 population. Rice Sci 25:121–131. https://doi.org/10.1016/j.rsci.2018.04.001

Yoon D-K, Ishiyama K, Suganami M et al (2020) Transgenic rice overproducing rubisco exhibits increased yields with improved nitrogen-use efficiency in an experimental paddy field. Nat Food 1:134–139. https://doi.org/10.1038/s43016-020-0033-x

Yu J, Xiong H, Zhu X et al (2017) OsLG3 contributing to rice grain length and yield was mined by Ho-LAMap. BMC Biol 15:28. https://doi.org/10.1186/s12915-017-0365-7

Zeng Y, Wen J, Zhao W et al (2020) Rational improvement of rice yield and cold tolerance by editing the three genes OsPIN5b, GS3, and OsMYB30 with the CRISPR–Cas9 system. Front Plant Sci 10:1663. https://doi.org/10.3389/fpls.2019.01663

Zhang X, Wang J, Huang J et al (2012) Rare allele of OsPPKL1 associated with grain length causes extra-large grain and a significant yield increase in rice. Proc Natl Acad Sci 109:21534–21539. https://doi.org/10.1073/pnas.1219776110

Zhang Q, Chen Q, Wang S et al (2014) Rice and cold stress: methods for its evaluation and summary of cold tolerance-related quantitative trait loci. Rice 7:24. https://doi.org/10.1186/s12284-014-0024-3

Zhang Y, Yu C, Lin J et al (2017) OsMPH1 regulates plant height and improves grain yield in rice. PLoS One 12:e0180825. https://doi.org/10.1371/journal.pone.0180825

Zhang B, Shang L, Ruan B et al (2019) Development of three sets of high-throughput genotyped rice chromosome segment substitution lines and QTL mapping for eleven traits. Rice 12:33. https://doi.org/10.1186/s12284-019-0293-y

Zhao L, Tan L, Zhu Z et al (2015) PAY 1 improves plant architecture and enhances grain yield in rice. Plant J 83:528–536. https://doi.org/10.1111/tpj.12905

Zhong H, Liu S, Meng X et al (2021) Uncovering the genetic mechanisms regulating panicle architecture in rice with GPWAS and GWAS. BMC Genomics 22:86. https://doi.org/10.1186/s12864-021-07391-x

Zhou Y, Liu H, Zhou X et al (2014) Over-expression of a fungal NADP(H)-dependent glutamate dehydrogenase PcGDH improves nitrogen assimilation and growth quality in rice. Mol Breed 34:335–349. https://doi.org/10.1007/s11032-014-0037-z

Zhou Y, Tao Y, Yuan Y et al (2018) Characterisation of a novel quantitative trait locus, GN4-1, for grain number and yield in rice (*Oryza sativa* L.). Theor Appl Genet 131:637–648. https://doi.org/10.1007/s00122-017-3025-y

Zhou J, Xin X, He Y et al (2019) Multiplex QTL editing of grain-related genes improves yield in elite rice varieties. Plant Cell Rep 38:475–485. https://doi.org/10.1007/s00299-018-2340-3

CRISPR/Cas9 for Rice Crop Improvement: Recent Progress, Limitations, and Prospects

34

Babar Hussain and Shakeel Ahmad

Abstract

The Clustered Regularly Interspaced Short Palindromic Repeats (CRISPR)/CRISPR-associated protein 9 (CRISPR/Cas9) system utilizes a single-guide RNA (sgRNA) for genome editing that makes it a simple, efficient, precise, and powerful crop improvement. In addition to knockout mutagenesis, it can also be used for knock-in/replacement, base editing, and transcriptional regulation. There is strong evidence that CRISPR/Cas9 has emerged as a powerful plant breeding tool in several crop species, in particular, for rice improvement. In this book chapter, we describe the progress made in CRISPR-based approaches for single and multiplexed gene editing in rice for improving grain yield, herbicide resistance, disease resistance, and resilience against abiotic stresses such as drought, osmotic stress, salinity, cold, and metal toxicity stresses. This chapter will provide useful resource for future rice breeding through CRISPR/Cas9-mediated genome editing.

B. Hussain (✉)
Department of Biological Sciences, Middle East Technical University, Ankara, Turkey

Faculty of Life Sciences, University of Central Punjab, Lahore, Pakistan
e-mail: babar.hussain@ucp.edu.pk

S. Ahmad
Maize Research Station, Ayub Agricultural Research Institute, Faisalabad, Pakistan

National Centre for Genome Editing for Crop Improvement and Human Health (NCGE), Center for Advanced Studies in Agriculture and Food Security (CAS-AFS), University of Agriculture Faisalabad, Faisalabad, Pakistan

N. Sarwar et al. (eds.), *Modern Techniques of Rice Crop Production*,
https://doi.org/10.1007/978-981-16-4955-4_34

Keywords

Rice · CRISPR/Cas9 · Crop improvement · Grain yield · Climate resilience · Abiotic stresses · Drought · Disease resistance

34.1 Introduction

Rice (*Oryza sativa* L.) is produced and consumed in all parts of the world (Ahmad et al. 2008, 2009, 2012, 2013; Ahmad and Hasanuzzaman 2012). It is staple food of around 3 billion people worldwide (Zhang et al. 2014). In 2019, it was grown on 162 million hectares with total global production of 755 million (http://www.fao.org/faostat/en). The world population will reach 9.7–11 billion in 2050 as compared to 7.7 billion in 2020 (https://population.un.org/wpp/). Although global rice production increased 31.92% during the last 30 years from 514 to 755 million tonnes (Fig. 34.1), we shall need at least 30% more rice (annual 1% increase) to feed almost 10 billion people in 2050. However, only 9% increase in rice yield was reported during the last 10 years (Fig. 34.1). Therefore, new genetic tools are needed to increase the grain yield. The situation becomes more complicated when effects of climate change on rice production are taken into account (Fatima et al. 2020; Ahmad et al. 2015, 2019; Ahmed and Ahmad 2019, 2020). Indeed, rice grain yield is severely affected by abiotic stresses (drought, heat stress, and salinity) in addition to diseases such rice blast and bacterial leaf blight. These stresses cause as high as 50–70% yield losses in rice (Tabassum et al. 2021). Keeping in view the growing population and the growing impacts of climate change, several crop improvment methods have been employed in rice to produce high yielding and stress resilient plants.

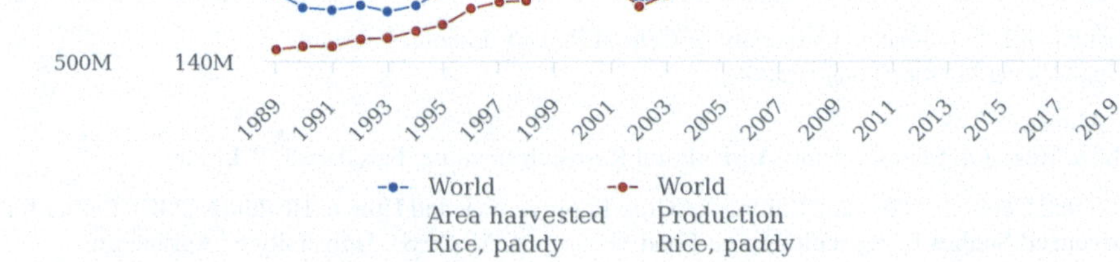

Fig. 34.1 World area under cultivation (ha) and production (Tonnes) of rice during the last 30 years (1989–2019) (**Source**: FAOSTAT 2021)

The narrow genetic diversity in crop plants makes them more vulnerable to biotic and abiotic stresses (Budak et al. 2015; Hussain 2015). Therefore, various mutagenesis strategies such as antisense RNA, TILLING, RNA interference (RNAi), and T-DNA insertion have been used to create novel variation. However, large mutant screens, partial decrease in gene expression, and undesirable mutagenesis events have limited their efficacy (Alonso and Ecker 2006). Therefore, development of a simple, cost-effective, robust, and powerful genome editing system, Clustered Regularly Interspaced Short Palindromic Repeats (CRISPR)/CRISPR-associated protein 9 (CRISPR/Cas9), in 2012 (Jinek et al. 2012) has provided unique oppertunities for crop improvement through plant genome editing (Hussain et al. 2018). The success of CRISPR/Cas9 system as an efficient genome editing tool is due to the fact that it uses single-guide RNA (sgRNA), protospacer adjacent motif (PAM), and Cas9 protein to target and edits specific DNA sequences with high accuracy and specificity (Jinek et al. 2012). The PAM is essential for Cas9 nuclease to cut the DNA and is situated 3–4 nucleotide bases downstream from the cut site (Fig. 34.2). These characteristics make it a simple, robust, efficient, and powerful genome editing tool as compared with other mutagenesis approaches (Hussain et al. 2018). Further reading about the mechanism, components, and procedure of the CRISPR/Cas9 system can be found in our review papers on the applications of the system for genome editing and crop improvement (Hussain et al. 2018; Tabassum et al. 2021).

CRISPR/Cas9 system has been extensively utilized for improvement of yield, quality, herbicide resistance, biotic, and abiotic stress tolerance in several plant and crop species including but not limited to *Arabidopsis thaliana*, Barley, *Camelina sativa*, corn, citrus, cucumber, *Nicotiana benthamiana*, *Physcomitrella patens*, soybean, tobacco, tomato, and wheat as reviewed by Hussain and colleagues (Hussain et al. 2018) and rice (Li et al. 2016b, a, 2017; Xu et al. 2016, 2017; Wang et al. 2016; Ma et al. 2017; Minkenberg et al. 2017; Sun et al. 2017; Hussain et al. 2018;

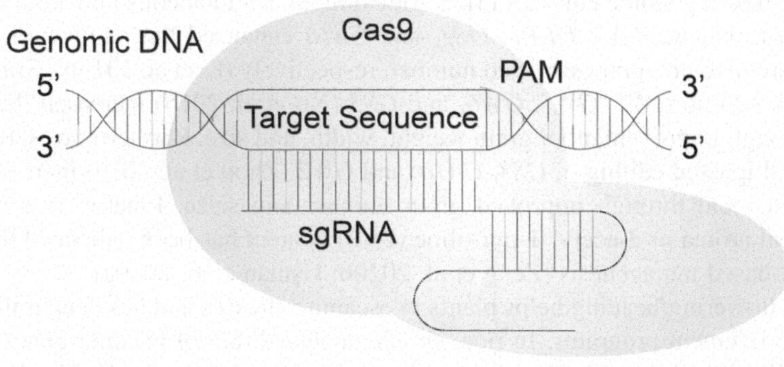

Fig. 34.2 A schematic presentation of CRISPR/Cas9 system

Tabassum et al. 2021). Keeping in view of the widespread applications of the CRISPR/Cas9 system in crop improvement, we review the latest developments in the field of CRISPR/Cas9 for plant genome editing and improving grain yield, disease resistance, herbicide resistance, and tolerance against abiotic stresses such as drought, salinity, cold, osmic, and metal toxicity in rice. This chapter will be a useful resource for future genome editing-based crop improvement programs.

34.2 CRISPR/Cas9 for Improving Grain Yield of Rice

Rice is the major staple crop that is consumed by billions of people worldwide, and increasing its yield to feed growing population is a major challenge. Classical breeding involves selection of parents, hybridization, and multiple selection cycles to improve quantitative traits like yield, and it takes 7–8 years to release a variety while this can be achieved in less than a year through CRISPR/Cas9 technology (Hussain et al. 2018, 2021). Therefore, CRISPR/Cas9 system has been used to knock out plethora of negative regulator genes to improve yield and related traits (Hussain et al. 2018). Rice, being the model crop among the cereals, has been the most important target of CRISPR-based genome editing sytem for its improvement. Several genes and quantitative trait loci (QTL) have been knocked out to improve grain yield and related traits (Tabassum et al. 2021) (Table 34.1).

Grain yield in rice has been improved by CRISPR/Cas9-based editing of genes/QTL coding for grain yield components such as *Ideal Plant Architecture1* (*IPA1*) (Li et al. 2016b), *Grain Size 3* (*GS3*) (Li et al. 2016b; Zhao et al. 2016; Zhou et al. 2019; Zeng et al. 2020b), *grain width/weight 2, 5, 6,* and *8* (*GW2, GW5, GW6, and GW8)* (Xu et al. 2016; Zhou et al. 2019; Usman et al. 2021), *Dense and Erect Panicle* (*DEP1*) (Li et al. 2016b; Huang et al. 2018), *Grain Number 1a* (*Gn1a*) (Shen et al. 2018; Huang et al. 2018; Zhou et al. 2019), *Fruit Weight Like, OsFWL* (Gao et al. 2020), etc. Being a powerful genome editing tool, CRISPR/Cas9 has been used for multiplexed genome editing in rice, for example, simultaneous knockout of four rice genes, that is, *IPA1*, *DEP1*, *GS3,* and *Gn1a* enhanced the number of tillers, panicle architecture, grain size, and number, respectively (Li et al. 2016b). Similarly, multiplex editing *GW2*, *GW5*, *GW6,* and *GS3* (Xu et al. 2016) increased the grain yield through improvement of grain weight, width, and size. Furthermore, CRISPR-based multiplexed editing of *GS3, Gn1a,* and *GW2* (Zhou et al. 2019) increased the rice grain yield through improved grain number and seize. Finally, rice panicle length and aroma or 2-acetyl-1-pyrroline (2AP) content has been enhanced through CRISPR-based mutagenesis (Zeng et al. 2020b; Usman et al. 2020a).

Early flowering/heading helps plants in escaping stresses and has been utilized to crop improvement programs. In rice, simultaneous editing of heading genes (*Hd2*, *Hd4*, *Hd5*) altered the flowering time and produced plants with early flowering and maturity (Li et al. 2017). Similarly, CRISPR–Cas9-based mutagenesis in genes coding for the abscisic acid (ABA) receptor genes, *Pyrabactin Resistance 1-Like 1* (*PYL1*), *PYL4*, and *PYL6* (Miao et al. 2018) resulted in 31% increase in number of grains in rice, while CRISPR mutation in *PYL9* (Usman et al. 2020b) increased the

Table 34.1 Examples of CRISPR/Cas9-mediated improvement in yield and yield-related traits of rice

Targeted gene/s	Improved trait/s in mutants	References
IPA1	Number of tillers, plant architecture	Li et al. (2016b)
GS3	Grain yield, grain size	Li et al. (2016b)
GS3	Grain yield, grain size	Zhao et al. (2016)
DEP1	Dense erect panicles	Li et al. (2016b)
Gn1a	Grain number, grain yield	Li et al. (2016b)
GW2, 5 and *6*	Grain yield, grain weight	Xu et al. (2016)
Hd2, 4 and *5*	Flowering time; early flowering, heading, and maturity	Li et al. (2017)
OsSWEET11	Sugar transport, grain filling	Ma et al. (2017)
Gn1a, DEP1	Grain yield	Huang et al. (2018)
Gn1a, GS3	Grain yield; contrasting yield contribution	Shen et al. (2018)
PYL1, PYL4, PYL6	Number of grains, grain yield	Miao et al. (2018)
SD1	Grain yield, dwarf plants, resistance to lodging	Hu et al. (2019)
OsGA20ox2	Grain yield, semi-dwarf plants, reduced gibberellins, and flag leaf length	Han et al. (2019)
GS3	Grain yield, grain size	Zhou et al. (2019)
OsGW2	Grain yield, grain weight	Zhou et al. (2019)
Gn1a	Grain number, grain yield	Zhou et al. (2019)
OsPYL9	Grain yield under normal and limited water availability	Usman et al. (2020b)
Cytochrome P450, OsBADH2	Grain yield, grain size, aroma (2-acetyl-1-pyrroline (2AP) content)	Usman et al. (2020a)
OsPIN5b, GS3	Panicle length, grain size	Zeng et al. (2020b)
SCM1, SCM2, SD2	Number of tillers, stem cross-sectional area	Cui et al. (2020)
OsFWL	Grain yield, grain length, number of tillers, flag leaf area, number of cells in flag leaf	Gao et al. (2020)
OsSPL16/qGW8	Grain yield, grain size	Usman et al. (2021)

grain yield as well. Furthermore, grain yield in rice was also improved through creating mutations in genes coding for plant height such as *SD1* (Hu et al. 2019), *OsGA20ox2* (Han et al. 2019), and *SD2* (Cui et al. 2020). The mutant plants showed short or semi-dwarf plant height, resistance to lodging, and increased grain yield. Interestingly, CRISPR-based editing of the same QTL in five different genetic backgrounds or cultivars showed diverse even contrasting phenotypes (plant height) in rice (Shen et al. 2018) highlighting genetic variation. In conclusion, CRISPR/Cas9 gene editing system is emerged as a powerful plant breeding tool for rice crop improvement to ensure global food security.

34.3 CRISPR/Cas9 for Improving Abiotic Stress Tolerance of Rice

Climate change is predicted to increase the global temperature, drought spells, and soil salinity (Budak et al. 2015; Hussain and Mahmood 2020) and thus poses a potent threat to crop production. The abiotic stress tolerance is often a complex mechanism, for example, drought tolerance is conferred by a combination of genes, microRNAs (miRNAs), proteins, transcription factors (TFs), hormones, metabolites, and ions (Budak et al. 2015). The complexity of abiotic stress tolerance mechanisms has required more precise and powerful genomic tools for crop improvement. During the past decade, CRISPR/Cas9 has become a powerful tool for improving abiotic stress resilience in several plant and crop species such as *Arabidopsis thaliana,* maize/corn, wheat, rice, tomato, and *Physcomitrella patens* (Hussain et al. 2018). Several genes, miRNAs, TFs, and genes coding for transporter proteins have been edited/mutated for improving tolerance against drought, salinity, osmotic, cold, and metal toxicity stresses in rice (Table 34.2).

Table 34.2 Examples of application of CRISPR/Cas9 system for improving abiotic stress tolerance of rice

Stress	Edited gene/s	Improved traits	References
Drought	*OsSAPK2*	Drought tolerance, stomatal, and ABA signaling	Lou et al. (2017)
	OsSRL1, OsSRL2	Drought tolerance; reduced stomatal conductance and transpiration rate; increased panicle number, abscisic acid (ABA), and antioxidants	Liao et al. (2019)
	OsPYL9	Drought tolerance; higher grain yield, antioxidants, chlorophyll content, ABA content, and leaf cuticle; reduced stomatal conductance and transpiration rate	Usman et al. (2020b)
	OsERA1	Drought tolerance, stomatal conductance, increased sensitivity to ABA	Ogata et al. (2020)
	DST	Drought tolerance, leaf architecture, reduced stomatal density	Kumar et al. (2020)
	OsmiR535	Drought tolerance, ABA insensitivity, number of lateral roots	Yue et al. (2020)
Salinity	*OsSAPK2*	Salinity tolerance, osmotic tolerance	Lou et al. (2017)
	OsRR22	Salinity tolerance, shoot length, shoot weight	Zhang et al. (2019)
	DST	Salinity tolerance, osmotic tolerance	Kumar et al. (2020)
	OsmiR535	Salinity tolerance, osmotic tolerance, shoot length, number of lateral roots, root length	Yue et al. (2020)
Metal toxicity	*OsHAK1*	Production of low-Cs^{+} (radioactive) rice	Nieves-Cordones et al. (2017)
	OsNramp5	Reduced cd accumulation	Tang et al. (2017)
Cold stress	*OsAnn3*	Response to cold tolerance	Shen et al. (2017)
	OsMYB30	Cold tolerance	Zeng et al. (2020b)

34.3.1 CRISPR/Cas9 for Improving Drought Tolerance of Rice

Drought is considered one of the most eminent threats to global food production (Ali et al. 2011; Budak et al. 2015; Hussain and Mahmood 2020). For the production of 1 kg of rice grains, 3000 L of water is required and rice is one of the most drought susceptible plants owing to thin cuticle wax and small root system. Hence, drought can result in up to 100% yield losses depending on the growth stage (Oladosu et al. 2019). Resultantly, CRISPR technology has been used frequently to alter drought tolerance response in rice. For example, CRISPR/Cas9-based knockout of *OsSAPK2* revealed its involvement in germination, drought, salinity, and osmotic stress tolerance. Drought tolerance was conferred through upregulation of genes coding for enzymes responsible for reactive oxygen species-scavenging, stomatal closure, and ABA accumulation (Lou et al. 2017). This highlights power of CRISPR system as an efficient functional genomics tool. CRISPR-based editing of genes, TFs, and miRNAs such as *OsSRL1, OsSRL2* (multiplexed editing) (Liao et al. 2019), *OsPYL9* (Usman et al. 2020b), *OsERA1* (Ogata et al. 2020), *DST* (Kumar et al. 2020), and *OsmiR535* (Yue et al. 2020), significantly enhanced the drought tolerance in mutant rice plants through reduced stomatal conductance and stomatal density transpiration rate (Liao et al. 2019; Kumar et al. 2020; Yue et al. 2020; Usman et al. 2020b; Ogata et al. 2020). On the other hand, most of these mutants had higher grain yield, leaf cuticle, number of panicles and lateral roots in addition to increased chlorophyll, antioxidants (CAT, SOD), and ABA content.

34.3.2 CRISPR/Cas9 for Improving Salt Tolerance of Rice

Soil salinity is a complex trait (Hussain et al. 2015), and salt tolerance in plants is controlled by multiple genes, TFs, transporter proteins, ions, and ion transporters responding to osmotic and sodium toxicity stresses (Hussain et al. 2017). It is well known that rice is more susceptible to soil salinity as compared to other cereals and threatens to reduce global rice production by 50% (Ma et al. 2018). Therefore, CRISPR/Cas9 editing system has been utilized to edit the genes such as *OsRR22* (Zhang et al. 2019), *DST* (Kumar et al. 2020), *OsmiR535* (Yue et al. 2020), and resultant mutants had significantly higher salt and osmotic stress tolerance. The improved salinity and osmotic stress of the mutants were measured in terms of higher root length, shoot length, and root and shoot weights under salinity as compared to wild-type plants.

34.3.3 CRISPR/Cas9 for Improving Cold Tolerance of Rice

Low-temperature or cold stress occurring at reproductive stages in highlands of Japan, China, Korea, and other regions of the world adversely affects the grain yield and/or grain quality of rice (Zhang et al. 2014). CRISPR/Cas9-based knockout of rice annexin gene, *OsAnn3,* revealed an important role of this gene under cold stress

4–6 °C for 3 days. The mutant plants had significantly less survival rate after cold stress (Shen et al. 2017), thus indicating a critical role of *OsAnn3* in conferring cold tolerance in rice. In another study, CRISPR/Cas9 technology-based multiplexed editing of three rice genes, that is, *GS3, OsPIN5b* (encoding panicle length), and *OsMYB30* generated mutants with 42–66% mutagenesis efficiency (Zeng et al. 2020b) and showed improved grain size, panicle length, grain yield, and cold tolerance in mutants as compared to wild-type plants.

34.3.4 CRISPR/Cas9 for Improving Metal Toxicity Response of Rice

CRISPR/Cas9 has been successfully utilized to reduce the accumulation of toxic metals in plants. This has been achieved through knockout of metal transporter genes, for example, CRISPR-based knockout of potassium transporter 1 (*OsHAK1*) produced mutant plants with significantly low-Cs^+ rice plants as compared with wild-type plants (Nieves-Cordones et al. 2017), thus reducing radioactive element in plant system. Similarly, CRISPR/Cas9 was utilized to knockout metal transporter *OsNramp5,* and the resultant mutant rice grain has reduced Cd accumulation (Tang et al. 2017). These studies show the utility of CRISPR technology for reducing the accumulation of toxic metals in plant parts that are very hazardous for human health if consumed and could be utilized as an efficient breeding strategy in the future.

34.4 CRISPR/Cas9 for Improving Herbicide Resistance of Rice

Herbicide-resistant are able to tolerate the herbicides (chemicals that kill herbs/ weeds) so that the main crop is not injured by application of herbicides. Herbicides such as glyphosate (N-phosphonomethylglycine) and Basta are routinely used to control weeds in several cropping systems (Hussain and Mahmood 2020). Several herbicides are broad-spectrum in nature and could be absorbed by crop plant leaves in addition to weed leaves. For example, glyphosate is used in over 130 countries, and its use has doubled since the 1970s (Sang et al. 2021). Traditionally, DNA recombinant technology has been used to develop genetically engineered crops for improving herbicide resistance in crop plants such as cotton, soybean, and maize (Hussain and Mahmood 2020; Sang et al. 2021). However, biosafety concerns about the transgenic plants have shifted focus on development of herbicide resistance through CRISPR/Cas9 system (Table 34.3). For example, a novel approach based on knock-in/replacement of *OsEPSPS* (*5-enolpyruvylshikimate-3-phosphate synthase*) gene enhanced the glyphosate resistance in mutant rice plants as compared to wild types (Li et al. 2016a). Similarly, knockout of rice *Acetolactate Synthase* (*OsALS*) gene resulted in increased herbicide resistance against 100 μM bispyribac sodium herbicide at the five-leaf stage (Sun et al. 2016). Another novel approach involving multiplexed base editing of *OsALS* and *FTIP1e* genes resulted in enhanced resistance against imazamox pesticide (Shimatani et al. 2017).

Table 34.3 Examples of application of CRISPR/Cas9 system for improving herbicide resistance in rice

Edited gene/s	Improved traits	References
OsEPSPS	Herbicide (glyphosate) resistance	Li et al. (2016a)
OsALS	Herbicide (Bispyribac-sodium) resistance	Sun et al. (2016)
OsALS, OsFTIP1e	Herbicide (Imazamox) resistance	Shimatani et al. (2017)
OsALS	Herbicide (Imazethapyr-IMT and Imazapic-IMP) resistance	Wang et al. (2021)
OsALS	Herbicide (Bispyribac-sodium) resistance	Kuang et al. (2020)
OsALS	Herbicide (Nicosulfuron, Imazapic-IMP, Pyroxsulam, Flucarbazonesodium, and bispyribac-sodium) resistance	Zhang et al. (2020)

CRISPR/Cas9-based knockout (Wang et al. 2021) and base editing (Kuang et al. 2020; Zhang et al. 2020) of *OsALS* gene have emerged as a novel breeding strategy for improving herbicide resistance of the rice crop. Importantly, one group reported broad-spectrum resistance against five herbicides (Zhang et al. 2020). Several countries consider CRISPR-edited plants as non-transgenic ones, but European Union considers otherwise (Hussain et al. 2018). However, it is a developing field, and we expect that genome editing-based approach will completely replace transgenic approach in near future.

34.5 CRISPR/Cas9 for Improving Disease Resistance of Rice

Diseases caused by bacteria, viruses, and fungi cause severe losses to agricultural crop production. Historical plant epidemic events such as maize leaf bight in the USA (1969–1970), potato blight in Ireland (1845–1852, also known as the Irish Potato Famine, the Great Famine, the Great Hunger, or the Great Starvation), and coffee rust in Brazil are examples of complete or partial crop failure due to plant disease. Similarly, the Great Bengal Famine (1943) was also prompted by wars and plant diseases. The Irish Potato Famine and the Great Bengal Famine resulted in millions of deaths and large population migrations (Hussain 2015), thus highlighting the threat of plant diseases for food production. Plant diseases alone cause an average of 10% loss to global food production and put 800 million people underfed (Hussain 2015) and thus are an eminent threat to global food security. In this regard, to develop healthy plants and overcome this issue, plant breeding, genetics, genomics, and application of high-throughput technologies have played a critical role. For example, identification of resistance (R) gene/s from plants and their incorporation in other disease susceptible plants has been a widely used system for developing specific and/or broad-spectrum disease-resistant plants (Pandolfi et al. 2017; Hussain et al. 2021). However, the disease development via R genes is considered a time-consuming and labor-intensive process. On the other hand, the identification of different susceptible (S) gene/s and factors in plants has opened new avenues of resistance development in plants (Xu et al. 2019; Oliva et al. 2019; Zafar et al. 2020; Zeng et al. 2020a). Actually, the expression of a particular S gene/s facilitates the proliferation of any particular pathogen/s in the plants. Thus, the S gene/s interruption via any gene editing tools such as Clustered Regularly Interspaced Palindromic Repeats (CRISPR) systems has developed resistance against different biotic stresses in plants (Hussain et al. 2018; Ahmad et al. 2020).

CRISPR/Cas9 technology has been utilized to edit *OsMPK1*, *OsMPK2*, *OsMPK5*, and *OsMPK6* genes in rice to improve disease resistance and disease resistance signaling pathways (Xie and Yang 2013; Minkenberg et al. 2017). Similarly, CRISPR/Cas9-based genome editing has led to improvement in resistance against several bacteria, viruses, and fungal pathogens in rice (Table 34.4).

Rice blast, caused by a fungus *Magnaporthe oryzae*, is the most devastating disease of cultivated rice that reduces the grain yield up to 30% annually but could cause 100% yield loss in favorable conditions for pathogen (Hussain 2015; Asibi et al. 2019) and thus pose a serious threat to global food security.

Table 34.4 Examples of application of CRISPR/Cas9 system for improving disease resistance in rice

Pathogen	Improved disease/pathogen resistance	Targeted gene	References
General disease resistance	Disease resistance signaling pathway	*OsMPK5*	Xie and Yang (2013)
		OsMPK1, *OsMPK2*, *OsMPK5*, *OsMPK6*	Minkenberg et al. (2017)
Fungi	Rice blast (*Magnaporthe oryzae*)	*OsERF922*	Wang et al. (2016)
		OsALB1, *OsRSY1*,	Foster et al. (2018)
		OsPi21	Li et al. (2019)
		OsPi21	Nawaz et al. (2020)
	False smut (*Ustilaginoidea virens*)	*OsUSTA*, *OsUvSLT2*	Liang et al. (2018)
Bacteria	Bacterial leaf blight (*Xanthomonas oryzae pv. Oryzae*)	*OsSWEET14*, *OsSWEET11*	Jiang et al. (2013)
		OsSWEET11 or *Os8N3*	Kim et al. (2019)
		OsXa13	Li et al. (2019)
		OsSWEET11, *OsSWEET13*, *OsSWEET14*	Oliva et al. (2019)
		OsSWEET11, *OsSWEET14*	Xu et al. (2019)
		OsSWEET14	Zeng et al. (2020a)
		OsSWEET14	Zafar et al. (2020)
		OsXa13	Li et al. (2020)
Virus	*Rice tungro spherical virus (RTSV)*	*eIF4G*	Macovei et al. (2018)

CRISPR/Cas9-based deletion of *Ethylene Response Factor, OsERF922,* enhanced the resistance against rice blast as measured by less blast lesions and improved agronomic performance when compared to wild-type plants (Wang et al. 2016). Unlike classical breeding that takes at least 6–7 years for variety development (Budak et al. 2015; Hussain et al. 2021), this technology generated the disease resistance in less than 1 year, and thus could be applied to locally adapted, high-yielding rice varieties for increasing the farmers' income (Hussain et al. 2018). Similarly, editing of *OsALB1* and *OsRSY1,* genes via CRISPR/Cas9 generated homozygous mutant plants, which showed resistance against rice blast (Foster et al. 2018). Recently, CRISPR-based editing in *Pi21* gene generated such mutants which exhibited resistance against rice blast, *Magnaporthe oryzae* (Li et al. 2019; Nawaz et al. 2020), as indicated by less lesions and improved the agronomic traits. Moreover, CRISPR-based editing of two genes, *OsUSTA* and *OsUvSLT2,* resulted in increased resistance against false smut (*Ustilaginoidea virens*) (Liang et al. 2018).

Bacterial leaf blight (BLB) is one of the most serious diseases of rice that can cause as high as 70% grain yield loss under favorable conditions for the causal bacteria, *Xanthomonas oryzae pv. Oryzae* (http://www.knowledgebank.irri.org). Rice plants resistant to BLB have been routinely generated by editing three well-known host S genes *Sugar Will Eventually be Exported Transporters* (SWEET), that is, *OsSWEET11* (also called as *Os8N3*), *OsSWEET13,* and *OsSWEET14* via CRISPR/CRISPR-associated protein 9 (Cas9) by multiple groups (Jiang et al. 2013; Xu et al. 2019; Oliva et al. 2019; Kim et al. 2019; Zafar et al. 2020; Zeng et al. 2020a). Resultant homozygous mutant plants showed less or no symptoms of BLB and improved agronomic traits. Moreover, promotor sequence of *Xa13* gene was disrupted by using CRISPR/Cas9 system to generate BLB-resistant mutants (Li et al. 2019, 2020). Mutant plants showed truncated proteins of *Xa13* and better agronomic traits than wild type.

Likewise, rice resistance has also been improved against viruses, for example, eIF4G, a host S gene, and has been knocked out, and rice mutants resistant against rice tungro spherical virus (RTSV) were generated (Macovei et al. 2018). These applications of CRISPR/Cas9 system show that this technology has immense potential to improve rice crop against different biotic stresses. Applications of CRISPR/Cas9 system in rice are summarized in Table 34.4. However, for further study, we refer our readers to reviews published on the applications of CRISPR/Cas9 systems for improving biotic stress resistance in plants (Hussain et al. 2018; Ahmad et al. 2020).

34.6 Advantages and Limitations of CRISPR/Cas9 System

34.6.1 Advantages of CRISPR/Cas9 System

CRISPR/Cas9-based genome editing has several advantages over other approaches for crop improvement as described by Hussain and colleagues (Hussain et al. 2018):

- CRISPR/Cas9 is a simple and robust genome editing tool as compared to previously used methods.
- CRISPR/Cas9 has higher genome editing efficiency as compared to previously used methods.
- It has higher accuracy as compared to other mutagenesis tools, and no widespread off-target editing has been reported.
- It has ability to target and edit any gene in any organism, thus making it a powerful tool.
- Unlike other mutagenesis tools, it does not require mutagenesis screen involving thousands of individuals to identify mutant plants, thanks to its higher efficiency.
- In addition to knockout mutations, this system can be used for knock-in and replacement of genes that make it a desirable tool for gene recombination.
- Its multiplexed gene editing ability makes it a preferable breeding tool to improve multiple traits at the same time.
- Due to precise gene editing, it does not create non-desirable mutagens in non-targeted genes and thus becomes a preferable breeding tool.
- CRISPR-based crop improvement takes less than a year as compared to 6–7 years required by classical breeding, thus making it a powerful plant breeding tool.

34.6.2 Limitations of CRISPR/Cas9 System

Despite being a powerful genome editing and crop improvement tool, CRISPR/Cas9 system has some limitations too as described by our recent reviews (Hussain et al. 2018; Ahmad et al. 2020).

- One of the major limitations of CRISPR/Cas9 system is the limited availability of PAM sequences.
- Regulatory oversight over CRISPR-edited crops despite the absence of any transgene by regions like EU can limit the application of the system.
- A gene of interest can only be edited if we completely know its sequence and possible role in controlling the trait of the interest.
- Low efficiency of Agrobacterium-based transformation, required for transferring gene editing constructs into plant cells, is a major limiting factor of CRISPR/Cas9 system.

34.7 Conclusions and Prospects

CRISPR/Cas9 has been widely utilized in many rice breeding programs to improve grain yield, grain quality, disease resistance, herbicide resistance, and to improve resistance against abiotic stresses such as drought, salinity, cold, osmic, and metal toxicity stresses. Despite being used to edit plant genes in 2013 for the first time, the success of CRISPR system has been attributed to several factors such as its simplicity, higher efficiency, and precise editing of the targeted genes. The amount of success achieved by the application of this system in crop breeding programs is enormous and predicts more applications of this system in the future. We hope that the advances in next-generation sequencing technologies (Hussain et al. 2021) and subsequent characterization of genes (Raza et al. 2020) and identification of novel PAM sequences will pave the way for the incorporation of CRISPR system in many more rice crop improvement programs. Similarly, improvement in transformation methods will also be vital for enhancing the efficiency of the system. Additionally, the applications of the CRISPR system for improving insect resistance in plants have not been achieved yet, and it will be interesting to see the selection of the targeted genes for the purpose. Another aspect that lacks the application of the CRISPR system is the improvement of the heat stress tolerance.

References

Ahmad S, Hasanuzzaman M (2012) Integrated effect of plant density, N rates and irrigation regimes on the biomass production, N content, PAR use efficiencies and water productivity of rice under irrigated semiarid environment. Not Bot Horti Agrobot Cluj Napoca 40(1):201–211

Ahmad S, Zia-ul-Haq M, Ali H, Shad SA, Ammad A, Maqsood M, Khan MB, Mehmood S, Hussain A (2008) Water and radiation use efficiencies of transplanted rice (*Oryza sativa* L.) at different plant densities and irrigation regimes under semi-arid environment. Pak J Bot 40(1): 199–209

Ahmad S, Ahmad A, Zia-ul-Haq M, Ali H, Khaliq T, Anjum MA, Khan MA, Hussain A, Hoogenboom G (2009) Resources use efficiency of field grown transplanted rice (*Oryza sativa* L.) under irrigated semiarid environment. J Food Agric Environ 7(2):487–492

Ahmad S, Ahmad A, Soler CMT, Ali H, Zia-Ul-Haq M, Anothai J, Hussain A, Hoogenboom G, Hasanuzzaman M (2012) Application of the CSM-CERES-Rice model for evaluation of plant density and nitrogen management of fine transplanted rice for an irrigated semiarid environment. Precis Agric 13(2):200–218

Ahmad S, Ahmad A, Ali H, Hussain A, Garcia y Garcia A, Khan MA, Zia-Ul-Haq M, Hasanuzzaman M, Hoogenboom G (2013) Application of the CSM-CERES-Rice model for evaluation of plant density and irrigation management of transplanted rice for an irrigated semiarid environment. Irrig Sci 31(3):491–506

Ahmad A, Ashfaq M, Rasul G, Wajid SA, Khaliq T, Rasul F, Saeed U, Rahman MH, Hussain J, Baig IA, Naqvi AA, SAA B, Ahmad S, Naseem W, Hoogenboom G, Valdivia RO (2015) Impact of climate change on the rice–wheat cropping system of Pakistan. In: Hillel D, Rosenzweig C (eds) Handbook of climate change and agro-ecosystems: The Agricultural Modeling Intercomparison and Improvement Project (AgMIP) integrated crop and economic assessments. Imperial College Press, London, pp 219–258

Ahmad S, Abbas G, Ahmed M, Fatima Z, Anjum MA, Rasul G, Khan MA, Hoogenboom G (2019) Climate warming and management impact on the change of rice-wheat phenology in Punjab, Pakistan. Field Crop Res 230:46–61

Ahmad S, Wei X, Sheng Z, Hu P, Tang S (2020) CRISPR/Cas9 for development of disease resistance in plants: recent progress, limitations and future prospects. Brief Funct Genomics 19:26–39. https://doi.org/10.1093/bfgp/elz041

Ahmed M, Ahmad S (2019) Carbon dioxide enrichment and crop productivity. In: Hasanuzzaman M (ed) Agronomic crops, vol 2. Springer Nature Singapore, Singapore, pp 31–46

Ahmed M, Ahmad S (2020) Systems modeling. In: Ahmed M (ed) Systems modeling. Springer Nature, Cham, pp 1–44

Ali Q, Elahi M, Hussain B, Khan NH, Ali F, Elahi F (2011) Genetic improvement of maize (*Zea mays* L.) against drought stress: an overview. Agric Sci Res J 1:228–237

Alonso JM, Ecker JR (2006) Moving forward in reverse: genetic technologies to enable genome-wide phenomic screens in Arabidopsis. Nat Rev Genet 7:524–536. https://doi.org/10.1038/nrg1893

Asibi AE, Chai Q, Coulter JA (2019) Rice blast: a disease with implications for global food security. Agronomy 9:451. https://doi.org/10.3390/agronomy9080451

Budak H, Hussain B, Khan Z, Ozturk NZ, Ullah N (2015) From genetics to functional genomics: improvement in drought signaling and tolerance in wheat. Front Plant Sci 6:1–13. https://doi.org/10.3389/fpls.2015.01012

Cui Y, Hu X, Liang G, Feng A, Wang F, Ruan S, Dong G, Shen L, Zhang B, Chen D, Zhu L, Hu J, Lin Y, Guo L, Matsuoka M, Qian Q (2020) Production of novel beneficial alleles of a rice yield-related QTL by CRISPR/Cas9. Plant Biotechnol J 18:1987–1989. https://doi.org/10.1111/pbi.13370

Fatima Z, Ahmad M, Hussain M, Abbas G, Ul-Allah S, Ahmad S, Ahmed N, Ali MA, Sarwar G, Ehsan ul Haque, Iqbal P, Hussain S (2020) The fingerprints of climate warming on cereal crops phenology and adaptation options. Sci Rep 10:18013

Foster AJ, Martin-Urdiroz M, Yan X, Wright HS, Soanes DM, Talbot NJ (2018) CRISPR-Cas9 ribonucleoprotein-mediated co-editing and counterselection in the rice blast fungus. Sci Rep 8:14355. https://doi.org/10.1038/s41598-018-32702-w

Gao Q, Li G, Sun H, Xu M, Wang H, Ji J, Wang D, Yuan C, Zhao X (2020) Targeted mutagenesis of the rice FW 2.2-like gene family using the CRISPR/Cas9 system reveals OsFWL4 as a regulator of tiller number and plant yield in rice. Int J Mol Sci 21:809. https://doi.org/10.3390/ijms21030809

Han Y, Teng K, Nawaz G et al (2019) Generation of semi-dwarf rice (*Oryza sativa* L.) lines by CRISPR/Cas9-directed mutagenesis of OsGA20ox2 and proteomic analysis of unveiled changes caused by mutations. 3 Biotech 9:387. https://doi.org/10.1007/s13205-019-1919-x

Hu X, Cui Y, Dong G, Feng A, Wang D, Zhao C, Zhang Y, Hu J, Zeng D, Guo L, Qian Q (2019) Using CRISPR-Cas9 to generate semi-dwarf rice lines in elite landraces. Sci Rep 9:19096. https://doi.org/10.1038/s41598-019-55757-9

Huang L, Zhang R, Huang G, Li Y, Melaku G, Zhang S, Chen H, Zhao Y, Zhang Y, Hu F (2018) Developing superior alleles of yield genes in rice by artificial mutagenesis using the CRISPR/Cas9 system. Crop J 6:475–481. https://doi.org/10.1016/j.cj.2018.05.005

Hussain B (2015) Modernization in plant breeding approaches for improving biotic stress resistance in crop plants. Turkish J Agric For 39:515–530. https://doi.org/10.3906/tar-1406-176

Hussain B, Mahmood S (2020) Development of transgenic cotton for combating biotic and abiotic stresses. In: Ahmad S, Hasanuzzaman H (eds) Cotton production and uses. Springer Singapore, Singapore, pp 527–545

Hussain B, Khan AS, Ali Z (2015) Genetic variation in wheat germplasm for salinity tolerance at seedling stage: improved statistical inference. Turkish J Agric For 39:182–192. https://doi.org/10.3906/tar-1404-114

Hussain B, Lucas SJ, Ozturk L, Budak H (2017) Mapping QTLs conferring salt tolerance and micronutrient concentrations at seedling stagein wheat. Sci Rep 7:15662. https://doi.org/10.1038/s41598-017-15726-6

Hussain B, Lucas SJ, Budak H (2018) CRISPR/Cas9 in plants: at play in the genome and at work for crop improvement. Brief Funct Genomics 17:319–328

Hussain B, Akpınar BA, Alaux M et al (2021) Wheat genomics and breeding: bridging the gap. AgriRxiv:1–57. https://doi.org/10.31220/agriRxiv.2021.00039

Jiang W, Zhou H, Bi H, Fromm M, Yang B, Weeks DP (2013) Demonstration of CRISPR/Cas9/sgRNA-mediated targeted gene modification in Arabidopsis, tobacco, sorghum and rice. Nucleic Acids Res 41:e188–e188. https://doi.org/10.1093/nar/gkt780

Jinek M, Chylinski K, Fonfara I, Hauer M, Doudna JA, Charpenier E (2012) A programmable dual-RNA-guided DNA endonuclease in adaptive bacterial immunity. Science 337:816–821. https://doi.org/10.1126/science.1225829

Kim Y-A, Moon H, Park C-J (2019) CRISPR/Cas9-targeted mutagenesis of Os8N3 in rice to confer resistance to Xanthomonas oryzae pv. oryzae. Rice 12:67. https://doi.org/10.1186/s12284-019-0325-7

Kuang Y, Li S, Ren B, Yan F, Spetz C, Li X, Zhou X, Zhou H (2020) Base-editing-mediated artificial evolution of OsALS1 in planta to develop novel herbicide-tolerant rice germplasms. Mol Plant 13:565–572. https://doi.org/10.1016/j.molp.2020.01.010

Kumar VVS, Verma RK, Yadav SK, Yadav P, Watts A, Rao MV, Chinnusamy V (2020) CRISPR-Cas9 mediated genome editing of drought and salt tolerance (OsDST) gene in indica mega rice cultivar MTU1010. Physiol Mol Biol Plants 26:1099–1110. https://doi.org/10.1007/s12298-020-00819-w

Li J, Meng X, Zong Y, Chen K, Zhang H, Liu J, Li J, Gao C (2016a) Gene replacements and insertions in rice by intron targeting using CRISPR–Cas9. Nat Plants 2:16139. https://doi.org/10.1038/nplants.2016.139

Li M, Li X, Zhou Z, Wu P, Fang M, Pan X, Lin Q, Luo W, Wu G, Li H (2016b) Reassessment of the four yield-related genes Gn1a, DEP1, GS3, and IPA1 in rice using a CRISPR/Cas9 system. Front Plant Sci 7:377. https://doi.org/10.3389/fpls.2016.00377

Li X, Zhou W, Ren Y, Tian X, Lv T, Wang Z, Fang J, Chu C, Yang J, Bu Q (2017) High-efficiency breeding of early-maturing rice cultivars via CRISPR/Cas9-mediated genome editing. J Genet Genomics 44:175–178. https://doi.org/10.1016/j.jgg.2017.02.001

Li S, Shen L, Hu P, Liu Q, Zhu X, Qian Q, Wang K, Wang Y (2019) Developing disease-resistant thermosensitive male sterile rice by multiplex gene editing. J Integr Plant Biol 61:1201–1205. https://doi.org/10.1111/jipb.12774

Li C, Li W, Zhou Z, Chen H, Xie C, Lin Y (2020) A new rice breeding method: CRISPR/Cas9 system editing of the Xa13 promoter to cultivate transgene-free bacterial blight-resistant rice. Plant Biotechnol J 18:313–315. https://doi.org/10.1111/pbi.13217

Liang Y, Han Y, Wang C, Jiang C, Xu JR (2018) Targeted deletion of the USTA and UvSLT2 genes efficiently in Ustilaginoidea virens with the CRISPR-Cas9 system. Front Plant Sci 9:699. https://doi.org/10.3389/fpls.2018.00699

Liao S, Qin X, Luo L, Han Y, Wang X, Usman B, Nawaz G, Zhao N, Liu Y, Li R (2019) CRISPR/Cas9-induced mutagenesis of semi-rolled Leaf1,2 confers curled leaf phenotype and drought tolerance by influencing protein expression patterns and ROS scavenging in rice (*Oryza sativa* L.). Agronomy 9:728. https://doi.org/10.3390/agronomy9110728

Lou D, Wang H, Liang G, Yu D (2017) OsSAPK2 confers abscisic acid sensitivity and tolerance to drought stress in rice. Front Plant Sci 8:1–15. https://doi.org/10.3389/fpls.2017.00993

Ma L, Zhang D, Miao Q, Yang J, Xuan Y, Hu Y (2017) Essential role of sugar transporter OsSWEET11 during the early stage of rice grain filling. Plant Cell Physiol 58:863–873. https://doi.org/10.1093/pcp/pcx040

Ma NL, Lah WAC, Kadir NA, Mustaqim M, Rahmat Z, Ahmad A, Lam SD, Ismail MR (2018) Susceptibility and tolerance of rice crop to salt threat: physiological and metabolic inspections. PLoS One 13:e0192732. https://doi.org/10.1371/journal.pone.0192732

Macovei A, Sevilla NR, Cantos C, Jonson GB, Slamet-Loedin I, Čermák T, Voytas DF, Choi IR, Chadha-Mohanty P (2018) Novel alleles of rice eIF4G generated by CRISPR/Cas9-targeted mutagenesis confer resistance to Rice tungro spherical virus. Plant Biotechnol J 16:1918–1927. https://doi.org/10.1111/pbi.12927

Miao C, Xiao L, Hua K, Zou C, Zhao Y, Bressan RA, Zhu JK (2018) Mutations in a subfamily of abscisic acid receptor genes promote rice growth and productivity. Proc Natl Acad Sci 115:6058–6063. https://doi.org/10.1073/pnas.1804774115

Minkenberg B, Xie K, Yang Y (2017) Discovery of rice essential genes by characterizing a CRISPR-edited mutation of closely related rice MAP kinase genes. Plant J 89:636–648. https://doi.org/10.1111/tpj.13399

Nawaz G, Usman B, Peng H, Zhao N, Yuan R, Liu Y, Li R (2020) Knockout of Pi21 by CRISPR/Cas9 and iTRAQ-based proteomic analysis of mutants revealed new insights into M. oryzae resistance in elite rice line. Genes (Basel) 11:735. https://doi.org/10.3390/genes11070735

Nieves-Cordones M, Mohamed S, Tanoi K, Kobayashi NI, Takagi K, Vernet A, Guiderdoni E, Périn C, Sentenac H, Véryn AA (2017) Production of low-Cs + rice plants by inactivation of the K + transporter Os HAK 1 with the CRISPR –Cas system. Plant J 92:43–56. https://doi.org/10.1111/tpj.13632

Ogata T, Ishizaki T, Fujita M, Fujita Y (2020) CRISPR/Cas9-targeted mutagenesis of OsERA1 confers enhanced responses to abscisic acid and drought stress and increased primary root growth under nonstressed conditions in rice. PLoS One 15:e0243376. https://doi.org/10.1371/journal.pone.0243376

Oladosu Y, Rafii MY, Samuel C, Fatai A, Magaji U, Kareem I, Kamarudin ZS, Muhammad I, Kolapo K (2019) Drought resistance in rice from conventional to molecular breeding: a review. Int J Mol Sci 20:3519. https://doi.org/10.3390/ijms20143519

Oliva R, Ji C, Atienza-Grande G, Huguet-Tapia JC, Perez-Quintero A, Li T, Eom JS, Li C, Nguyen H, Liu B, Auguy F, Sciallano C, Luu VT, Dossa GS, Cunnac S, Schmidt SM, Slamet-Loedin IH, Cruz CV, Szurek B, Frommer WB, White FF, Yang B (2019) Broad-spectrum resistance to bacterial blight in rice using genome editing. Nat Biotechnol 37:1344–1350. https://doi.org/10.1038/s41587-019-0267-z

Pandolfi V, Neto JRCF, da Silva MD, Amorim LLB, Wanderley-Nogueira AC, de Oliveira Silva RL, Kido EA, Crovella S, Iseppon AMB (2017) Resistance (R) genes: applications and prospects for plant biotechnology and breeding. Curr Protein Pept Sci 18:323–334. https://doi.org/10.2174/1389203717666160724195248

Raza Q, Riaz A, Atif RM, Hussain B, Ali Z, Budak H (2020) MADS-box genes galore in wheat genome: phylogenomics, evolution and stress associated functions. BioRxiv:1–33. https://doi.org/10.1101/2020.10.23.351635

Sang Y, Mejuto J-C, Xiao J, Simal-Gandara J (2021) Assessment of glyphosate impact on the agrofood ecosystem. Plan Theory 10:405. https://doi.org/10.3390/plants10020405

Shen C, Que Z, Xia Y, Tang N, Li D, He R, Cao M (2017) Knock out of the annexin gene OsAnn3 via CRISPR/Cas9-mediated genome editing decreased cold tolerance in rice. J Plant Biol 60:539–547. https://doi.org/10.1007/s12374-016-0400-1

Shen L, Wang C, Fu Y, Wang J, Liu Q, Zhang X, Yan C, Qian Q, Wang K (2018) QTL editing confers opposing yield performance in different rice varieties. J Integr Plant Biol 60:89–93. https://doi.org/10.1111/jipb.12501

Shimatani Z, Kashojiya S, Takayama M, Terada R, Arazoe T, Ishii H, Teramura H, Yamamoto T, Komatsu H, Miura K, Ezura H, Nishida K, Ariizumi T, Kondo A (2017) Targeted base editing in rice and tomato using a CRISPR-Cas9 cytidine deaminase fusion. Nat Biotechnol 35:441–443. https://doi.org/10.1038/nbt.3833

Sun Y, Zhang X, Wu C, He Y, Ma Y, Hou H, Guo X, Du W, Zhao Y, Xia L (2016) Engineering herbicide-resistant rice plants through CRISPR/Cas9-mediated homologous recombination of acetolactate synthase. Mol Plant 9:628–631. https://doi.org/10.1016/j.molp.2016.01.001

Sun Y, Jiao G, Liu Z, Zhang X, Li J, Guo X, Du W, Du J, Francis F, Zhao Y, Xia L (2017) Generation of high-amylose rice through CRISPR/Cas9-mediated targeted mutagenesis of starch branching enzymes. Front Plant Sci 8:1–15. https://doi.org/10.3389/fpls.2017.00298

Tabassum J, Ahmad S, Hussain B, Mawia AM, Zeb A, Ju L (2021) Applications and potential of genome-editing systems in rice improvement: current and future perspectives. Agronomy 11(7):1359. https://doi.org/10.3390/agronomy11071359

Tang L, Mao B, Li Y, Lv Q, Zhang L, Chen C, He H, Wang W, Zeng X, Shao Y, Pan Y, Hu Y, Peng Y, Fu X, Li H, Xia S, Zhao B (2017) Knockout of OsNramp5 using the CRISPR/Cas9 system produces low Cd-accumulating indica rice without compromising yield. Sci Rep 7:14438. https://doi.org/10.1038/s41598-017-14832-9

Usman B, Nawaz G, Zhao N, Liu Y, Li R (2020a) Generation of high yielding and fragrant rice (*Oryza sativa* L.) lines by CRISPR/Cas9 targeted mutagenesis of three homoeologs of Cytochrome P450 gene family and OsBADH2 and transcriptome and proteome profiling of revealed changes triggered by mutations. Plants 9:788. https://doi.org/10.3390/plants9060788

Usman B, Nawaz G, Zhao N, Liao S, Liu Y, Li R (2020b) Precise editing of the OsPYL9 gene by RNA-guided Cas9 nuclease confers enhanced drought tolerance and grain yield in rice (*Oryza sativa* L.) by regulating circadian rhythm and abiotic stress responsive proteins. Int J Mol Sci 21: 7854. https://doi.org/10.3390/ijms21217854

Usman B, Nawaz G, Zhao N, Liao S, Qin B, Liu F, Liu Y, Li R (2021) Programmed editing of rice (Oryza sativa l.) osspl16 gene using CRISPR/Cas9 improves grain yield by modulating the expression of pyruvate enzymes and cell cycle proteins. Int J Mol Sci 22:1–19. https://doi.org/10.3390/ijms22010249

Wang F, Wang C, Liu P, Lei C, Hao W, Gao Y, Liu YG, Zhao K (2016) Enhanced rice blast resistance by CRISPR/Cas9-targeted mutagenesis of the ERF transcription factor gene OsERF922. PLoS One 11:1–18. https://doi.org/10.1371/journal.pone.0154027

Wang F, Xu Y, Li W, Chen Z, Wang J, Fan F, Ao Y, Jiang Y, Zhu QH, Yang J (2021) Creating a novel herbicide-tolerance OsALS allele using CRISPR/Cas9-mediated gene editing. Crop J 9: 305–312. https://doi.org/10.1016/j.cj.2020.06.001

Xie K, Yang Y (2013) RNA-guided genome editing in plants using a CRISPR–Cas system. Mol Plant 6:1975–1983. https://doi.org/10.1093/mp/sst119

Xu R, Yang Y, Qin R, Li H, Qiu C, Li L, Wei P, Yang J (2016) Rapid improvement of grain weight via highly efficient CRISPR/Cas9-mediated multiplex genome editing in rice. J Genet Genomics 43:529–532. https://doi.org/10.1016/j.jgg.2016.07.003

Xu R, Qin R, Li H, Li D, Li L, Wei P, Yang J (2017) Generation of targeted mutant rice using a CRISPR-Cpf1 system. Plant Biotechnol J 15:713–717. https://doi.org/10.1111/pbi.12669

Xu Z, Xu X, Gong Q, Li Z, Li Y, Wang S, Yang Y, Ma W, Liu L, Zhu B, Zou L, Chen G (2019) Engineering broad-spectrum bacterial blight resistance by simultaneously disrupting variable TALE-binding elements of multiple susceptibility genes in rice. Mol Plant 12:1434–1446. https://doi.org/10.1016/j.molp.2019.08.006

Yue E, Cao H, Liu B (2020) OsmiR535, a potential genetic editing target for drought and salinity stress tolerance in *Oryza sativa*. Plan Theory 9:1337. https://doi.org/10.3390/plants9101337

Zafar K, Khan MZ, Amin I, Mukhtar Z, Yasmin S, Arif M, Ejaz K, Mansoor S (2020) Precise CRISPR-Cas9 mediated genome editing in super basmati rice for resistance against bacterial blight by targeting the major susceptibility gene. Front Plant Sci 11:575. https://doi.org/10.3389/fpls.2020.00575

Zeng X, Luo Y, Vu NTQ, Shen S, Xia K, Zhang M (2020a) CRISPR/Cas9-mediated mutation of OsSWEET14 in rice cv. Zhonghua11 confers resistance to Xanthomonas oryzae pv. oryzae without yield penalty. BMC Plant Biol 20:313. https://doi.org/10.1186/s12870-020-02524-y

Zeng Y, Wen J, Zhao W, Wang Q, Huang W (2020b) Rational improvement of rice yield and cold tolerance by editing the three genes OsPIN5b, GS3, and OsMYB30 with the CRISPR–Cas9 system. Front Plant Sci 10:1663. https://doi.org/10.3389/fpls.2019.01663

Zhang Q, Chen Q, Wang S, Hong Y, Wang Z (2014) Rice and cold stress: methods for its evaluation and summary of cold tolerance-related quantitative trait loci. Rice 7:24. https://doi.org/10.1186/s12284-014-0024-3

Zhang A, Liu Y, Wang F, Li T, Chen Z, Kong D, Bi J, Zhang F, Luo X, Wang J, Tang J, Yu X, Liu G, Luo L (2019) Enhanced rice salinity tolerance via CRISPR/Cas9-targeted mutagenesis of the OsRR22 gene. Mol Breed 39:47. https://doi.org/10.1007/s11032-019-0954-y

Zhang R, Chen S, Meng X, Chai Z, Wang D, Yuan Y, Chen K, Jiang L, Li J, Gao C (2020) Generating broad-spectrum tolerance to ALS-inhibiting herbicides in rice by base editing. Sci China Life Sci. https://doi.org/10.1007/s11427-020-1800-5

Zhao Y, Zhang C, Liu W, Gao W, Liu C, Song G, Li WX, Mao L, Chen B, Xu Y, Li X, Xie C (2016) An alternative strategy for targeted gene replacement in plants using a dual-sgRNA/Cas9 design. Sci Rep 6:23890. https://doi.org/10.1038/srep23890

Zhou J, Xin X, He Y, Chen H, Li Q, Tang X, Zhong Z, Deng K, Zheng X, Akher SA, Cai G, Qi Y, Zhang Y (2019) Multiplex QTL editing of grain-related genes improves yield in elite rice varieties. Plant Cell Rep 38:475–485. https://doi.org/10.1007/s00299-018-2340-3

Part V

Nutritional Aspects

Rice Nutritional Aspects

35

Shahneel Shafaq and Abrar Hussain

Abstract

Rice being a first- or second-order staple food has major role in ensuring food security in the world. It is mainly produced and consumed in Asia. However, its nutritional quality aspect is very important in this regard, as the health of rice-consuming population is linked with its quality. Keeping in view the importance of its quality, this chapter focuses on the nutritional aspect of rice grain.

Keywords

Rice · Yield · Quality · Nutrition

35.1 Origin and History of Rice

In the early time, the west scholars recognized the importance of rice; at that time, Asians did not even know the rice. Various stories were published in China and India. The main objective was about cultivation and use of rice. It should be known that the artificial hybridization was documented only after the rediscovery of Mendel's law in the twentieth century. Russian botanist Roschevicz (1931), after a systematic study of 19 species, created a clear statement that species in the section Sativa Roschev had a common origin in Africa. *Australiensis*, *brachyantha*, *glaberrima*, *grandiglumis*, *latifolia*, *officinalis*, and *sativa* has covered all genomes of A, B, C, D, E, and F. The geographic origin and later disintegration of the genes continued to be a mystery.

S. Shafaq · A. Hussain (✉)
COMSAT Institute of Information and Technology, Sahiwal, Pakistan
e-mail: abrar.hussain@cuisahiwal.edu.pk

N. Sarwar et al. (eds.), *Modern Techniques of Rice Crop Production*,
https://doi.org/10.1007/978-981-16-4955-4_35

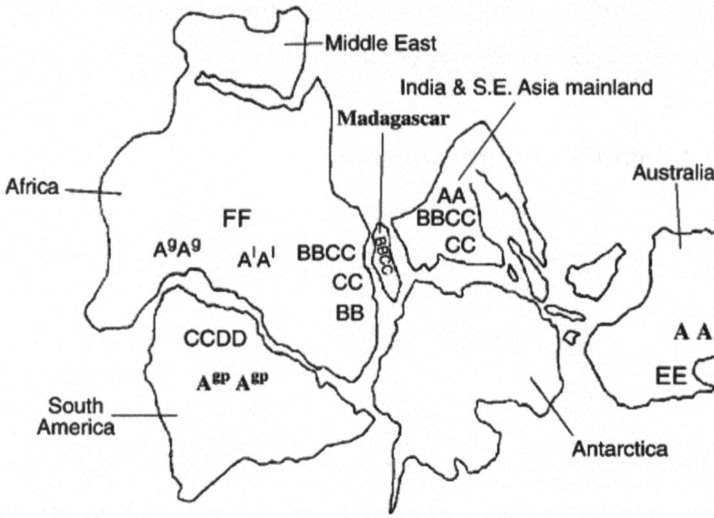

Fig. 35.1 The Gondwana origin of the *Oryza* species

During 1970s Chang saw a new and craw map of the Gondwana supercontinent, it targeted the Chang to the solution of mystery. When large genome was placed on map, he got a perfect fit before it broke or swept away. The Gondwana origin of the *Oryza* species was then proposed as shown in Fig. 35.1.

Chang concise the whole genomic account in 1985. This plate signified the best assimilation of results from tectonics, biosystematics, and crop geography among economically important plants. About 130 million years ago in the new system of plate tectonics, the breakage of the Gondwana supercontinent was seen. The initiation was from South America, and after 20 years, it was the Australia and Antarctica. The last to break were South and Southeast Asia from the large plate, and it stuck with the Central Asia about 45 million years BP. This impact and the continuation of South Asia plate to the north led to the rise of the Himalayas and associated mountains in Burma and Malaysia. Madagascar, the main islands of Indonesia, Papua, and New Guinea were fragments of Gondwana. The above summary is an amazing account of these huge landmarks, the expansion of early forms of *Oryza* plant to the corners of southern hemisphere.

Two main species of rice are grown by humans: *Oryza sativa* in Asia and *Oryza glaberrima* in Africa. *Oryza sativa* is now also grown in other parts of the world, but the other is not spread. Both of them have a lot of similarities, which an ordinary person cannot differentiate. Asian rice came to Africa in the sixteenth century and became more popular.

35.2 Major Rice-Consuming Countries

The current global usage of rice is 480 million metric tons (MMT) per annum, and of this, about 85% (408 MMT) is consumed by humans. About 50% of globally produced rice is consumed by India and China. In Asia, the daily utilization of rice is high. In Bangladesh, Cambodia, Myanmar, Indonesia, Laos, Vietnam, Thailand, and the Philippines, it is declared that there is higher annual consumption of rice which is higher than 110 kg per capita annually. It is also reported that in some countries like Caribbean and Latin America, including Guyana, Suriname, Panama, Cuba, Guyana, Peru, Costa Rica, Nicaragua, and Ecuador, there is a high consumption of rice. Rice consumption is increasing in countries like Solomon Island, Fiji, and Vanuatu. A major portion of rice is eaten as rice kernel and broken kernel (Fig. 35.2).

35.3 Nutritional Content of Rice and their Importance

35.3.1 Starch

A polymer of dextrorotatory glucose is termed starch. It is a polymer which attaches a-(1–4). It typically contains two types of molecules, a linear molecule called amylose and a branched molecule called amylopectin. The branched ones are attached via a-(1–6) links. Inventive methods have revealed that amylose present in rice occupies two to four chains. They have a DP n, which is termed degree of polymerization of about 900–1100 glucose units (Tester et al. 2004). The rice

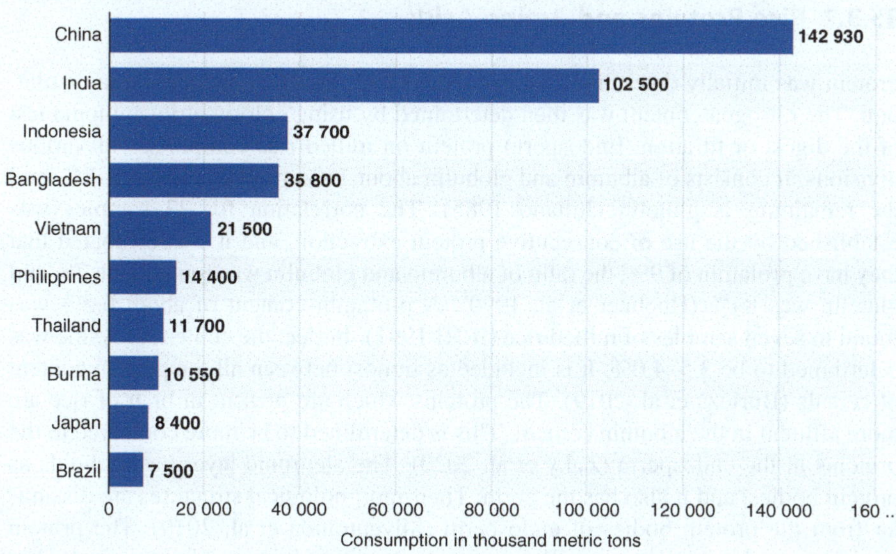

Fig. 35.2 Major rice-consuming countries

amylopectin has an iodine affiliation of 0.4–0.9%; this value is present in rice having low and intermediate amylose. In rice having higher amylose, it is about 2–3%. In samples having high content of amylose, the disbranched isoamylase amylopectin has shown DP n of 100. They have longer chain and high affinity for iodine about 9–14% (Hizukuri et al. 1989).

A classification of milled rice is presented by a study of colorimetric color absorption for starch and iodine done at 590–620 nm. This classification is given below:

- Waxy rice having amylose (1–2%)
- Rice having amylose in very low amount (2–12%)
- Rice having amylose in low amount (12–20%)
- Rice having amylose in moderate amount (20–25%)
- Rice having higher amylose content (25–33%) (Juliano 1985)

A recent study has revealed the committed amount of amylose to be 20%. That supplementary capacity for the binding of iodine is because of longer and linearized chains present in the amylopectin (Takeda et al. 1987). The "apparent amylose content" is recognized by colorimetric amylose values. The endosperm is waxy and opaque. It illustrates gaps between the starch granules. They appear to have lower density than the non-waxy granules. The morphology of the starch granule is not complete and well understood. But it is revealed that crystallinity and staling are associated with the fractions of amylopectin.

35.3.2 Rice Proteins and Amino Acids

Protein was initially determined by micro-Kjeldahl digestion and ammonia distillation. The nitrogen content was then determined by using colorimetric ammonia test of the digest or titration. Endosperm protein on milled rice comporises of various divisions. It consists of albumin and globulin about 15%, prolamin about 5–8%, and the remaining is glutelin (Juliano 1985). The correlation for 33 samples was established by the use of consecutive protein extraction, and it was estimated that they have prolamin of 9%; the ratio of albumin and globulin was found to be 7% and glutelin was 84% (Huebner et al. 1990). A prolamin content of about 6.5% was found in seven samples of milled rice (IRRI 1991). In rice, the content of lysine was determined to be 3.5–4.0%. It is included as utmost between all the protein content of cereals (Burieza et al. 2019). The proteins which are present in bran of rice are more affluent in the albumin content. This is determined to be more compared to the proteins in the endosperm (Zaky et al. 2020). The aleuronic layer has globoids as protein bodies, and it also has the germ. Their morphological structures are dissimilar from the protein bodies of endosperm (Silventoinen et al. 2019). The protein present in endosperm is generally confined to protein bodies. The protein bodies which are termed PB-1 are richer in prolamin content, and they are large and

Table 35.1 Amino gram (g/16 g N) of the acidic and basic subunits of rice glutelin and the major and minor subunits of prolamin

	Glutelin subunit		Prolamin subunit		
Amino acid	3039 kd (acidic)	19–25 kd (basic)	13 kd	10 kd	6 kd
Histidine	2.2–2.5	2.6–2.7	2–2.4	1.7	4.2
Isoleucine	3.2–3.3	4.1–4.9	3.8–5.4	1.6	3.6
Lucien	6.4–7.5	7–8.5	17.9–26.4	4.7	8.1
Lysine	2.2–3	3–4.1	4–5.5	1.0	3.3
Methionine + cysteine	0.2–1.9	0.1–2.4	0.7–1.2	22.5	5.3
Phenylalanine + tyrosine	10–10.5	10.1–10.8	12.7–21.6	4.3	7.6
Threonine	2.8–3.7	2.5–3.7	1.8–2.8	6.8	2.7
Valine	5.1–5.7	5.7–7	2.7–3.9	4.4	3.9
Amino acid scores %	38–52	52–71	7–8	18	57

Sources: Juliano and Boulter (1976), Villareal and Juliano (1978) (glutelin subunits), Hibino et al. (2015) (prolamin subunits)

spherical in shape. On the other hand, the protein bodies referred to as PB-2 are higher in glutelin, and they are crystalline in nature (Saito et al. 2012).

The starch granules of rice, amylose, fetters up to 0.7% protein. Amylose the rice starch granule fetters up to 0.7% (Villareal and Juliano 1978). Glutelins of rice have three acidic and two basic subunits (Kagawa et al. 1988). The subunits of two kinds were obtained when there occurs a cleavage. This cleavage was of a polypeptide precursor of 57 kd (Sugimoto et al. 1986). The amino acid content of some acidic and basic subunits of glutelin and prolamin is given in Table 35.1.

In the areas where rice is the main dietary consumption, 25% of children are affected by protein-energy malnutrition. If rice is consumed as a staple food, it provides insufficient essential amino acids (Gearing 2015). So the efforts are in progress to enhance the nutritional aspects of rice. For this purpose, the main focus is on PC which is called protein content. Moreover, the other standards for nutrition are under consideration. In rice, the amount of protein is comparatively less than the grains of other kinds (8.5%). The combination of rice seed protein consists of 60–80% glutelin and 20–30% prolamin which are controlled by 15 and 34 genes, respectively (Kawakatsu et al. 2008; Xu and Messing 2009) (Table 35.2).

35.3.3 Lipid

They occur precisely in the form of lipid bodies or spherosomes in the aleuronic layer and bran. The rice lipid or fat content is present principally in the bran fraction (20%, dry basis). However, around 1.5–1.7% occurs in milled rice, in which they chiefly occur as non-starch lipids (Chuan and Jinsong 2019). They are extracted by ether, chloroform–methanol, and cold water-saturated butanol (Cruz et al. 2016). Protein bodies, specifically the core, are rich in lipids (Saito et al. 2012). About 34.65% linoleic acid and 0.19% linolenic acid are the main essential fatty acids in rice oil (Tahira and Ata-ur-Rehman 2007).

Table 35.2 Amino acid content of rough rice and its milling fractions at 14% moisture (g per 16 g N)

Rice	Rough rice	Brown rice	Milled rice	Rice bran	Rice hull
Histidine	1.5–2.8	2.3–2.5	2.2–2.6	2.7–3.3	1.6–2
Isoleucine	3–4.8	3.4–4.4	3.5–4.6	2.7–4.1	3.2–4
Lucien	6.9–8.8	7.9–8.5	8–8.2	6.9–7.6	8–8.2
Lysine	3.2–4.7	3.7–4.1	3.2–4	4.8–5.4	3.8–5.4
Methionine + cysteine	4.5–6.2	4.4–4.6	4.3–5	4.2–4.6	3.5–3.7
Phenylalanine + tyrosine	9.3–10.8	8.6–9.3	9.3–10.4	7.7–8	6.6–7.3
Threonine	3–4.5	3.7–3.6	3.5–3.7	3.6–4.2	4.2–5
Tryptophan	1.2–2	1.2–1.4	1.2–1.7	0.6–1.2	0.6
Valine	4.6–7	4.8–6.3	4.7–6.5	4.9–6	5.5–7.5
Amino acid score (%)	55–81	64–71	55–69	83–93	66–93

Sources: Juliano (1985), Eggum et al. (1982), Pedersen and Eggum (1983)

During the development of grain, the content of essential fatty acids might be enhanced with temperature. It leads to decrease in whole oil content (Lakkakula et al. 2004). Starch lipids are mostly monoacyl lipids (fatty acids and lysophosphatides) compiled with amylose (Pardo et al., 2017). For waxy starch granule, the starch lipid content is lowermost about 0.2%. For intermediate amylose rice, it is maximum of about 1.0%. It can be a little lower in high-amylose rice (Colussi et al. 2014). Waxy milled rice has more non-starch lipids than non-waxy rice. Starch lipids are endangered from oxidative rancidity, and the amylose–lipid complex is digested by growing rats (Park et al. 2007). Though, starch lipids pay slightly to the energy content of the rice grain. The major fatty acids of starch lipids are palmitic and linoleic acids, with lesser amounts of oleic acid (Annor et al. 2015).

35.3.4 Vitamins and Minerals

To attain a proper growth and development, about 42 nutrients are necessary. To get these nutrients, rice is the main source and it fulfills the nutrient requirement (Mahender et al. 2016).

Potassium: The main need of potassium is for regulation of water balance. It also plays a vital role in maintaining blood pressure and for the proper functioning of neuromuscular system. Potassium aids in the proper functioning of heart and kidney. It has a significant role in transmitting the heart's electrical potential (Pohl et al. 2013).

Magnesium: It is an important nutrient to carry out a number of functions in our body. It plays an important role in bone health. It is a key factor for the synthesis of proteins and fatty acids. It also has a role in energy production by Adenosine Tri Phosphate (ATP). It is important for relaxing muscles and also for lowering blood pressure. Functions like formation of new cells and blood clotting are also

Table 35.3 Vitamin and mineral contents of rough rice and its milling fractions at 14% moisture

Rice type	Rough rice	Brown rice	Milled rice	Rice bran	Rice hull
Thiamine (mg)	0.26–0.33	0.29–0.61	0.02–0.1 1	1.20–2.40	0.09–0 21
Riboflavin (mg)	0 0.06–0.1 1	0.04–0.14	0.02–0.06	0.18–0.43	0.05–0.07
Niacin (mg)	2.9–5.6	3.5–5.3	1.3–2.4	26.7–49.9	1.6–4.2
α-Tocopherol (mg)	0.90–2.00	0.90–2.50	75–0.30	2.60–13.3	0
Calcium (mg)	10–80	10–50	10–30	30–120	60–130
Phosphorous (g)	0.17–0.39	0 17–0.43	0.08–0.15	1.1–2.5	0.03–0.07
Phytin P (g)	0.18–0.21	0.13–0.27	0.02–0.07	0.9–2.2	0
Iron (mg)	1.4–6.0	0.2–5.2	0.2–2.8	8.6–43.0	3.9–9.5
Zinc (mg)	1.7–3.1	0.6–2.8	0.6–2.3	4.3–25.8	0.9–4.0

Sources: Juliano (1985), Pedersen and Eggum (1983)

performed by magnesium. It is also important for insulin functioning (Swaminathan 2003).

Thiamine or Vitamin B1: The processing of macronutrients like carbohydrates, proteins, and fat is done by vitamin B1. For the formation of ATP, this vitamin is necessary for every cell (Fitzpatrick and Chapman 2020).

Niacin: It is also termed vitamin B3, for the synthesis of energy from carbohydrates body utilized vitamin B3. It also helps in regulation of cholesterol (Păucean et al. 2010).

Vitamin B6: It is the main and key vitamin in amino acid processing. For making hormones like serotonin, dopamine, and melatonin, this vitamin is required. For the regulation of mental process, it is a key nutrient because it is involved in the formation of many neurotransmitters (Parra et al. 2018).

Pantothenic Acid: It is also termed vitamin B5. In the process of energy synthesis, it is a key nutrient. It is also essential for making acetylcholine; this is a neurotransmitter. The production, transport, and release of energy from fat also require pantothenic acid. Cholesterol synthesis also depends on pantothenic acid (Tahiliani and Beinlich 2008).

Iron: It is part of hemoglobin. It is the component of blood which is responsible for carrying oxygen. If the iron is not insufficient in body, the body gets starved for oxygen, and because of this person tires easily. Iron is a part of myoglobin, which helps muscle cells to store oxygen (Toyokuni 2009).

Phosphorous: For bone and teeth health, phosphorous is important. It plays a key role in process of metabolism of fats, proteins, and carbohydrates (Takeda et al. 2004) (Table 35.3).

35.4 Improving Rice Yield and Quality

Breeders conventionally emphasized enhancing the paddy yield and the ability to tolerate the stress. It is basically to make sure the rice ability to sustain and security. Yet, it is not enough to only focus on paddy yield because during processing the poor

Fig. 35.3 Rice grain quality steps

milling and cooking deteriorate the quality. It also leads to quality reduction and economic loss. Such things can happen on any step. It can also occur at post-harvest stage (Hodges et al. 2011). Rice has to pass through the mechanical stress throughout the processing. It causes the breakage of rice, and it is demonstrated that 30% loss can occur due to this breakage (Buggenhout et al. 2013). The amount of rough rice left after the process of milling is termed HRY. It is standard use to identify the loss in kernel (Fig. 35.3).

In general, it is estimated that HRY in rice is 24–74%, and rice produced after milling is between 70–80% (Bell et al. 1998). There are various methods that researchers have revealed to get better HRY properties (Buggenhout et al. 2013). All these factors are discussed here.

35.4.1 Recognizing Best Post-Harvest Conditions

In rice, the methods which are used to manage their quality after production are understood very well (Bell et al. 1998). On the basis of a broad and summarized study, the post-harvest and harvest practices are provided to farmers. During harvesting, the moisture should be carefully monitored within 22–24%. Throughout the process of drying, temperature should not be more than 43 °C. The temperature should not be more than 50 °C if the rice milling process is to be carried out in a dryer that is of mixing type. If doing the sun drying, the grains must have depth of 2 and 4 cm, and there must be mixing every 30 min. The basic goal is to dry deliberately to get 14% MC. It makes sure the minimal breakage. There should be proper storage and protection from pests and water after drying.

35.4.2 Reducing Rice Fissuring

A major problem which is seen in rice industry is rice fissuring. It is a very common problem, and it leads to a decrease in market value of rice; moreover, it also causes a reduction in HRY. Along the length of grain, huge internal cracks occur which are termed rice fissures. These fissures cause weakness of grain structure, and when it is subjected to milling and hulling the grain undergoes breakage. The increased amount of grain breakage leads to a decrease in HRY. Environmental factors greatly influence the grain fissuring. It may occur before, after, and during processing of grains. In the last phase of grain filling, the water supply from plant tissues in the rice is cut off to begin the ripening of grain. During this phase, the MC is no longer under the control of moisture transfer. Its variation occurs because of differences in environmental conditions.

Three main environmental conditions which affect the occurrence of fissure are air, temperature, and rain. At this stage on the basis of temperature, humidity, and MC, the rice grain properties change relatively. Typically, the fissure in grains is developed in the fields, and it happens as the MC falls under 15% (Otto and David 2004) (Fig. 35.4).

Fissures in grains lead to break the grain, but it is not compulsory that all fissured grains will be broken away (Zhang et al. 2005). Various types of rice grains are typically obtained by farmers or grain processors while doing harvesting and storage. If the rice kernels are subjected to intense dry process, it will lead to a higher number of fissures. It is seen that generally grains are not fissured instantly after drying (Sharma and Kunze 1982). As the rice gets cooled after drying, the fissures develop (Cnossen et al. 2003).

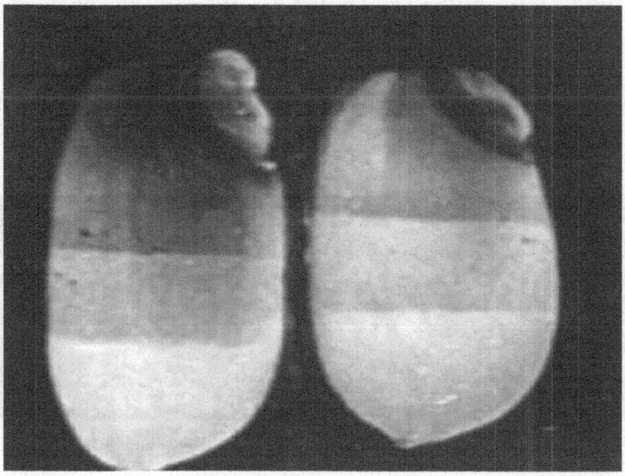

Fig. 35.4 Rice fissuring issue

35.4.3 Methods for Tracing Fissures

An innovative practice for recognizing the fissures in grains is magnetic resonance imaging analyses (Hwang et al. 2009). The range of temperature on which polymers are transferred from a rubber texture to glass-like material and from glass material to rubber-like is termed Tg. For measuring, various methods can be used which include dynamic mechanical analysis, differential scanning calorimetric, and thermomechanical analysis (Sun et al. 2002). While doing the tempering and drying of rice, it is very important to take Tg range under consideration. It is critical because milling quality is affected by it (Cnossen and Siebenmorgen 2000).

It is favorable to dry out the rice in glassy area because it does not reduce HRY. An increase in MC is seen if the process of drying occurs in rubber form and is cooled down instantly without tempering. This increase in MC leads to a substantial decrease in HRY (Yang et al. 2003). By reducing the moisture content inside the rice, the viscous stress component can be reduced (Zhang et al. 2003). If it is desirable to enhance milling production, infrared drying with tempering is more appreciable than outdated drying (Ding et al. 2015).

35.4.4 Reducing Grain Chalkiness

Chalk is determined to be a cloudy whitish band in a translucent grain of rice (Buggenhout et al. 2013). Disturbed translocation of assimilates is associated with the existence of chalk (Fitzgerald and Resurreccion 2009). It also leads to starch breakdown biosynthesis of proteins in developing grain (Sreenivasulu et al. 2015). It leads to the production of partially filled granules of starch. Along the transparent grain, they appear as opaque spots. A process called X-ray micro-tomography can be used to visualize these gaps (Zhu et al. 2012). Chalked grains are fragile; they tend to break easily than translucent grains (Lyman et al. 2013). The reason for this breakage is the softness and increase in fissures (Ashida et al. 2009) (Bhattacharya et al. 1982). While doing hulling and polishing of grains, the chalked grains are more subjected to mechanical stress. If the chalked grains resist breakage and pass through the milling process, still their presence decreases their market value and consumer preference (Fitzgerald et al. 2009). The reason for this is because it affects their cooking quality (Chun et al. 2009) (Fig. 35.5).

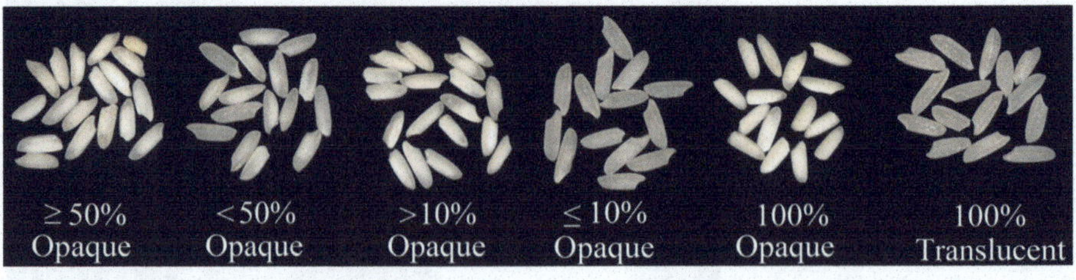

Fig. 35.5 Rice quality parameter

Chalkiness has negative effect on rice so it is necessary to study them deeply. It is required to develop a highly correct method, because the traditional method by use of Cervitec is limited. This is because grains which undergo breakage are removed before analyzing grain. Cervitec just determines the chalkiness which only affects the quality of appearance of grain. It does not determine the chalkiness which has an effect on HRY. On the basis of seed count, a new practice is determined. This method determines chalkiness, color, whiteness, as well as broken and immature grain analysis.

35.4.5 Determining Grain Maturity and Kernel Dimension

The maturity of asynchronous grains causes an increase in immature rice content. It not only destroys the appearance; it also reduces milling quality. This is because while doing the processing of grain, the immature grain is highly subjected to breakage. For photosensitive rice variety, the incidence of braking is considerably less. The reason for this is in such varieties the chances of grain breakage are less. In non-photosensitive varieties, there are more immature grains in early grains of 90–100 days as compared to medium mature rice grain of 130–140 days. To reduce HRY and chalkiness, DTF abbreviated for tailoring days to flowering and synchronizing anthesis are main and vital methods. In rice, they show a connection among the development of various rice grain parts like flowering, panicle, and grain yield (Endo-Higashi and Izawa 2011). It is now considered a very important area for research. It is very challenging as it includes very complicated system genetics and epigenetics (Sun et al. 2014a, b). In rice, recently developed methods of genomics can identify this complexity in a better way (Butardo et al. 2017). For this reason, it is currently testing non-destructive control of rice quality. It is also considering the grading and milling quality to be tested under laboratory conditions (Cheng et al. 2006). Evaluation of grain quality through computer visualization and other hyperspectral and multispectral methods are being used for other grains (Shahin et al. 2012). To achieve this goal, the use of hyperspectral imaging technique was to determine the quality of rice (Wang et al. 2015). In this research, Principle Component Analysis (PCA) and back-propagation neural networks are used to perform dimensionality reducing methods on spectrum and image information (Kumar and Bal 2007), but conventional rice quality laboratories are usually unable to because they are still not verified and commercialized by international cooperative testing systems. In the area of milling and physical quality of kernel maturation, dimensional analysis has another innovative technique. It involves the separation of grain on the basis of their quality. It involves the separation of defective and superior grains. Now a new machine is available which can sort 50 kernels per second. This machine is called Qsorter (Qualisense). To attain fast and accurate results of milling quality and HRY, the sorting is done automatically. Such revolutionary technologies are aimed to surpass traditional inspection by humans. The inspection by humans is not accurate and is highly subjected to errors. Rice exporting requires a specific grading. These developing technologies can help in satisfying the exporting grading and hence will improve the rice export.

35.5 Future Prospect of Rice

For determining the properties influencing the quality of grains, rice cereal chemists are developing advanced methods. These methods help in determining the constantly emerging and spreading of consumers' preferences. In this part, future procedures are being involved for assessment of quality. It is discussed by keeping the past methods and newly developed methods in consideration.

35.5.1 Correct Assessment of Consumer Preference

In rice breeding, sensory evaluation is crucial. It is crucial for the identification of parental line; this is before the release of newly formed rice cultivators. For sensory evaluation, it is important to take instrumental measures. This is important to do in the early phases of breeding. To achieve this goal, the mixing of various methods is interesting (Concepcion et al. 2015). Determining the sensory characteristics is getting crucial due to market demand. Understanding the sensory and quality attributes of huge rice varieties by various testing is important. To recognize the consumer preference and demand, it is important to search market. This is to determine their preference for a number of sensory and quality attributes.

35.5.2 Using Nutritional Quality in Rice Improvement

The amylose in rice apparently ranges from 0 to 30%. It makes rice a unique and divergent crop (Butardo et al. 2017). Rice plays a very important role in lessening the diseases which are non-communicable (Butardo et al. 2011). The processes like cooking of rice can alter the digesting property, so it should be studied. Along with this, industrial processing should also be performed to change digesting property of rice. It includes parboiling of rice and the development of amylose-containing products. With this, the attributes according to religion and demand should also be kept in mind. It is also important to know that cooking rice can alter the nutritional aspects of rice. So it should be a concern to recognize the factors which influence the cooking quality of rice grain. And hence it should be ensured to give health benefits.

35.5.3 Associating Additional Molecular Markers to Rice Grain Quality

An in-depth knowledge of genetic basis of rice quality parameters may lead to aid in helping the breeding programs to determine the quality of newly developed breeding at a fast rate. This can be done by using the genomic selection (GS) and marker-assisted breeding. The quantitative trait loci (QTL) control the quality attributes of rice. They can be major as well as minor (Stokes et al. 2013). Attributes like storage, processing, and biochemical composition affect them. In this decade for determining the sequenced genome of rice, next-generation sequencing is used (2014). It helps to access the genomic study to indicate the major regions which regulate the quality and nutritional value of rice grain. In this case, the progression of robust and precise grain nutritional quality phenotyping techniques explained must have a pace to the quick genomic developments and the genomic tools to describe the rice variety. The use of molecular markers for studying grain quality attributes also assists to predict the rice sorting at genotype level. It also helps rice cultivators to get upper ranked and increased yield by using molecular genetics. MAS can assist in rice breeding, marker-assisted selection. But still for rice quality the use of robust GS is at the initial level because there is a lack of knowledge about association of biochemical and physical characters to traits like cooking and processing. So it is important for rice cultivators to develop the additional markers for studying the rice quality. They should have novel combined attributes to make sure the consumers' acceptance. Recently developed phenotyping techniques to understand starch morphology directly affect the cooking quality. There is a need for a collaborative effort by breeders, food scientists, molecular geneticists, and cereal chemists for achieving the goal. They should have a line of attack to imitate the attributes of different kinds to differentiate the segments of low, medium, and high quality. It is expected that there will be a considerable enhancement in identity and use of SNPs linked to rice nutrition and quality by using various methods of genomics.

35.5.4 Integrating the Quality of Hybrid Rice

Accomplishment of hybrid rice is usually measured in terms of production and the ability to withstand and resist certain stress conditions. Rice breeders should work to enhance rice quality in a more effective way because now rice consumers concern about rice quality and its impact on health. In the first step, there should be an indication of cooking quality associated with texture. Gradually by using genome-wide association studies, they should reveal the genetic variations of eating and cooking quality attributes. Variety of rice collection and pre-breeding materials should be studied for this purpose. The main focus is to develop effective knowledge of key molecular and physiological methods for grain quality attributes by discovering gene through QTL cloning and genomic approaches. Then this knowledge should be used in breeding sectors to produce rice of superior quality. It will lead to the development of diverse hybrid rice. If we summarize it, there is a need for various techniques to enhance the rice quality. To achieve this goal, a recently done study revealed that high-yielding, high-quality rice lines can have a variety of complex traits to ensure superior grain quality (Zeng et al. 2017). In the future, the high yield and superior quality will not only be based on quick innovations in genetics. It will also depend on present and future phenotyping techniques for rice grain quality.

References

Annor GA, Marcone M, Corredig M, Bertoft E, Seetharaman K (2015) Effects of the amount and type of fatty acids present in millets on their in vitro starch digestibility and expected glycemic index (eGI). J Cereal Sci 64:76–81

Ashida K, Iida S, Yasui T (2009) Morphological, physical, and chemical properties of grain and flour from chalky rice mutants. Cereal Chem 86(2):225–231. https://doi.org/10.1094/cchem-86-2-0225

Bell M, Bakker R, De Padua D, Rickman J (1998) Rice quality management-principles and some lessons. In: ACIAR proceedings, 2000. ACIAR. pp 255–263

Bhattacharya KR, Sowbhagya CM, Swamy YI (1982) Quality profiles of rice: a tentative scheme for classification. J Food Sci 47(2):564–569

Buggenhout J, Brijs K, Celus I, Delcour JA (2013) The breakage susceptibility of raw and parboiled rice: a review. J Food Eng 117(3):304–315

Burieza HP, Rizzo AJ, Vale EM, Silveira V, Maldonado S (2019) Shotgun proteomic analysis of quinoa seeds reveals novel lysine-rich seed storage globulins. Food Chem 293:299–306

Butardo VM Jr, Fitzgerald MA, Bird AR, Gidley MJ, Flanagan BM, Larroque O, Resurreccion AP, Laidlaw HK, Jobling SA, Morell MK, Rahman S (2011) Impact of down regulation of starch branching enzyme IIb in rice by artificial microRNA- and hairpin RNA-mediated RNA silencing. J Exp Bot 62(14):4927–4941

Butardo VM Jr, Anacleto R, Parween S, Samson I, de Guzman K, Alhambra CM, Misra G, Sreenivasulu N (2017) Systems genetics identifies a novel regulatory domain of amylose synthesis. Plant Physiol 173(1):887–906

Cheng F, Ying YB, Li YB (2006) Detection of defects in rice seeds using machine vision. Trans ASABE 49(6):1929–1934

Chuan T, Jinsong B (2019) Rice lipids and rice bran oil. Rice fourth edition chemistry and technology. In: Rice, chemistry and technology, 4th edn. AACC International Press, Washington, DC, pp 131–168

Chun A, Song J, Kim KJ, Lee HJ (2009) Quality of head and chalky rice and deterioration of eating quality by chalky rice. J Crop Sci Biotechnol 12(4):239–244. https://doi.org/10.1007/s12892-009-0142-4

Cnossen AG, Siebenmorgen TJ (2000) The glass transition temperature concept in rice drying and tempering: effect on milling quality. Trans ASAE 43(6):1661–1667

Cnossen AG, Jimenez MJ, Siebenmorgen TJ (2003) Rice fissuring response to high drying and tempering temperatures. J Food Eng 59(1):61–69

Colussi R, Pinto VZ, Mello El Halal SL, Vanier NL, Villanova FA, Marques e Silva R, Zavareze E, Dias ARG (2014) Structural, morphological, and physicochemical properties of acetylated high-, medium-, and low-amylose rice starches. Carbohydr Polym 103:405–413

Concepcion JCT, Ouk M, Zhao D, Fitzgerald MA (2015) The need for new tools and investment to improve the accuracy of selecting for grain quality in rice. Field Crop Res 182(0):60–67

Cruz M, Wang M, Frisch-Daiello J, Han X (2016) Improved butanol–methanol (BUME) method by replacing acetic acid for lipid extraction of biological samples. Lipids 51:887–896. https://doi.org/10.1007/s11745-016-4164-7

Ding C, Khir R, Zhongli P, Zhang J, El-Mashad H, Tu K, El-Mashad H (2015) Effect of infrared and conventional drying methods on physicochemical characteristics of stored white rice. Cereal Chem 92(5):441–448

Eggum BO, Juliano BO, Maningat CC (1982) Protein and energy utilization of rice milling fractions by rats. Plant Foods Hum Nutr 31(4):371–376

Endo-Higashi N, Izawa T (2011) Flowering time genes heading date 1 and early heading date 1 together control panicle development in rice. Plant Cell Physiol 52(6):1083–1094

Fitzgerald MA, Resurreccion AP (2009) Maintaining the yield of edible rice in a warming world. Funct Plant Biol 36(12):1037–1045. https://doi.org/10.1071/fp09055

Fitzgerald MA, McCouch SR, Hall RD (2009) Not just a grain of rice: the quest for quality. Trends Plant Sci 14(3):133–139. https://doi.org/10.1016/j.tplants.2008.12.004

Fitzpatrick TB, Chapman LM (2020) The importance of thiamine (vitamin B1) in plant health: from crop yield to biofortification. J Biol Chem 295(34):12002–12013, ISSN 0021-9258. https://doi.org/10.1074/jbc.REV120.010918

Gearing RE, MacKenzie MJ, Ibrahim RW, Brewer KB, Batayneh JS, Schwalbe CS (2015) Stigma and mental health treatment of adolescents with depression in Jordan. Community Ment Health J 51(1):111–117

Hibino K, Kidzu T, Masumura T, Ohtsuki K, Tanaka K, Kawabata K, Fujii S, Gearing ME (2015) Good as gold: can golden rice and other biofortified crops prevent malnutrition? In: Science in the news. Harvard University, Boston. http://sitn.hms.harvard.edu/

Hizukuri S, Takeda Y, Maruta N, Juliano BO (1989) Molecular structures of rice starch. Carbohydr Res 189:227–235

Hodges RJ, Buzby JC, Bennett B (2011) Postharvest losses and waste in developed and less developed countries: opportunities to improve resource use. J Agric Sci 149:37–45. https://www.statista.com/statistics/255971/top-countries-based-on-rice-consumption-2012-2013/

Huebner FR, Bietz JA, Webb BD, Juliano BO (1990) Rice cultivar identification by high-performance liquid chromatography of endosperm proteins. Cereal Chem 67:129–135

Hwang SS, Cheng YC, Chang C, Lur HS, Lin TT (2009) Magnetic resonance imaging and analyses of tempering processes in rice kernels. J Cereal Sci 50(1):36–42. https://doi.org/10.1016/j.jcs.2008.10.012

IRRI (1991) Program report for 1990. IRRI, Manila

Juliano BO (1985) Factors affecting nutritional properties of rice protein. Trans Natl Acad Sci Technol 7:205–216

Juliano BO, Boulter D (1976) Extraction and composition of rice endosperm glutelin. Phytochemistry 15(11):1601–1606

Kagawa H, Hirano H, Kikuchi E (1988) Variation in glutelin seed storage protein in rice (*Oryza sativa* L.). Jpn J Breed 38:327–332

Kawakatsu T, Yamamoto MP, Hirose S, Yano M, Takaiwa F (2008) Characterization of a new rice glutelin gene *GluD-1* expressed in the starchy endosperm. J Exp Bot 59:4233–4245. https://doi.org/10.1093/jxb/ern265

Kumar PA, Bal S (2007) Automatic unhulled rice grain crack detection by X-ray imaging. Trans ASABE 50(5):1907–1911

Lakkakula NR, Lima M, Walker T (2004) Rice bran stabilization and rice bran oil extraction using ohmic heating. Bioresour Technol 92:157–161

Lyman NB, Jagadish KSV, Nalley LL, Dixon BL, Siebenmorgen T (2013) Neglecting rice milling yield and quality underestimates economic losses from high-temperature stress. PLoS One 8(8): e72157. https://doi.org/10.1371/journal.pone.0072157

Mahender A, Anandan A, Pradhan SK, Pandit E (2016) Rice grain nutritional traits and their enhancement using relevant genes and QTLs through advanced approaches. Springerplus 5: 2086. https://doi.org/10.1186/s40064-016-3744-6

Otto RK, David LC (2004) Rough-rice drying-moisture adsorption and desorption. In: Champagne ET (ed) Rice: chemistry and technology, Grain science references. American Association of Cereal Chemists, St. Paul, pp 223–268. https://doi.org/10.1094/1891127349.009

Pardo IM, Shanks RA, Adhikari B, Adhikari R (2017) Thermoplastic starch-nanohybrid films with polyhedral oligomeric silsesquioxane. Carbohydr Polym 173:170–177

Park IM, Ibáñez AM, Zhong F, Shoemaker CF (2007) Gelatinization and pasting properties of waxy and non-waxy rice starches. Starch 59:388–396. https://doi.org/10.1002/star.200600570

Parra M, Stahl S, Hellmann H (2018) Vitamin B6 and its role in cell metabolism and physiology. Cell 7(7):84. https://doi.org/10.3390/cells7070084

Păucean A, Pârlog MR, Vodnar CD, Socaciu C, Mudura E (2010) HPLC analysis of vitamin B3 and vitamin C from a dairy product containing brewer's yeast. J Agroaliment Process Technol 16: 136–140

Pedersen B, Eggum BO (1983) The influence of milling on the nutritive value of flour from cereal grains. 4. Rice. Plant Foods Hum Nutr 33(4):267–278

Pohl HR, Wheeler JS, Murray HE (2013) Sodium and potassium in health and disease. In: Sigel A, Sigel H, Sigel R (eds) Interrelations between essential metal ions and human diseases, Metal ions in life sciences, vol 13. Springer, Dordrecht. https://doi.org/10.1007/978-94-007-7500-8_2

Roschevicz RJ (1931) A contribution to the knowledge of rice. Bull Appl Bot Genet Plant Breed 27:1–33

Saito Y, Shigemitsu T, Yamasaki R, Sasou A, Goto F, Kishida K, Kuroda M, Tanaka K, Morita S, Satoh S, Masumura T (2012) Formation mechanism of the internal structure of type-I protein bodies in rice endosperm: relationship between the localization of prolamin species and the expression of individual genes. Plant J 70:1043–1055. https://doi.org/10.1111/j.1365-313X.2012.04947.x

Shahin MA, Hatcher DW, Symons SJ (2012) 17-Development of multispectral imaging systems for quality evaluation of cereal grains and grain products. In: Sun D-W (ed) Computer vision technology in the food and beverage industries. Woodhead, Cambridge, pp 451–482. https://doi.org/10.1533/9780857095770.3.451

Sharma AD, Kunze OR (1982) Post-drying fissure developments in rough rice. Trans ASAE 25(2): 465–468

Silventoinen P, Rommi K, Holopainen-Mantila U, Poutanen K, Nordlund E (2019) Biochemical and techno-functional properties of protein- and fibre-rich hybrid ingredients produced by dry fractionation from rice bran. Food Bioprocess Technol 12:1487–1499. https://doi.org/10.1007/s11947-019-02307-w

Sreenivasulu N, Butardo VM, Misra G, Cuevas RP, Anacleto R, Kavi Kishor PB (2015) Designing climate-resilient rice with ideal grain quality suited for high-temperature stress. J Exp Bot 66(7): 1737–1748. https://doi.org/10.1093/jxb/eru544

Stokes JR, Boehm MW, Baier SK (2013) Oral processing, texture and mouthfeel: from rheology to tribology and beyond. Curr Opin Colloid Interface Sci 18(4):349–359

Sugimoto T, Tanaka K, Kasai Z (1986) Molecular species in the protein body II (PB-II) of developing rice endosperm. Agric Biol Chem 50:3031–3035

Sun ZH, Yang WD, Siebenmorgen T, Stelwagen A, Cnossen A (2002) Thermomechanical transitions of rice kernels. Cereal Chem 79(3):349–353. https://doi.org/10.1094/Cchem.2002.79.3.349

Sun CH, Chen D, Fang J, Wang PR, Deng XJ, Chu CC (2014a) Understanding the genetic and epigenetic architecture in complex network of rice flowering pathways. Protein Cell 5(12): 889–898

Sun CM, Liu T, Ji CX, Jiang M, Tian T, Guo DD, Wang LJ, Chen YY, Liang XM (2014b) Evaluation and analysis the chalkiness of connected rice kernels based on image processing technology and support vector machine. J Cereal Sci 60(2):426–432

Swaminathan R (2003) Magnesium metabolism and its disorders. Clin Biochem Rev 24(2):47–66

Tahiliani AG, Beinlich CJ (2008) Pantothenic acid in health and disease. Vitam Horm 46:165–228

Tahira R, Ata-ur-Rehman BMA (2007) Characterization of rice bran oil. J Agric Res 45:225–230

Takeda Y, Hizukuri S, Juliano BO (1987) Structures of rice amylopectin with low and high affinities for iodine. Carbohydr Res 168:79–88

Takeda E, Taketani Y, Sawada N, Sato T, Yamamoto H (2004) The regulation and function of phosphate in the human body. Biofactors 21:345–355. https://doi.org/10.1002/biof.552210167

Tester RF, Karkalas J, Qi X (2004) Starch—composition, fine structure and architecture. J Cereal Sci 39(2):151–165. https://doi.org/10.1016/j.jcs.2003.12.001

The 3000 rice genomes project (2014) The 3,000 rice genomes project. GigaScie 3:7

Toyokuni S (2009) Role of iron in carcinogenesis: cancer as a ferrotoxic disease. Cancer Sci 100:9–16. https://doi.org/10.1111/j.1349-7006.2008.01001.x

Villareal RM, Juliano BO (1978) Properties of glutelin from mature and developing rice grain. Phytochemistry 17:177–182

Wang L, Liu D, Pu HB, Sun DW, Gao WH, Xiong ZJ (2015) Use of hyperspectral imaging to discriminate the variety and quality of rice. Food Anal Methods 8(2):515–523

Xu JH, Messing J (2009) Amplification of prolamin storage protein genes in different subfamilies of the Poaceae. Theor Appl Genet 119(8):1397–1412

Yang W, Jia CC, Siebenmorgen TJ, Pan Z, Cnossen AG (2003) Relationship of kernel moisture content gradients and glass transition temperatures to head rice yield. Biosyst Eng 85(4):467–476

Zaky AA, Abd El-Aty AM, Ma A, Jia Y (2020) An overview on antioxidant peptides from rice bran proteins: extraction, identification, and applications. Crit Rev Food Sci Nutr. https://doi.org/10.1080/10408398.2020.1842324

Zeng D, Tian Z, Rao Y, Dong G, Yang Y, Huang L, Leng Y, Xu J, Sun C, Zhang G, Hu J, Zhu L, Gao Z, Hu X, Guo L, Xiong G, Wang Y, Li J, Qian Q (2017) Rational design of high yield and superior-quality rice. Nat Plants 3:17031

Zhang Q, Yang W, Jia C (2003) Preservation of head rice yield under high-temperature tempering as explained by the glass transition of rice kernels. Cereal Chem 80(6):684–688. https://doi.org/10.1094/CCHEM.2003.80.6.684

Zhang Q, Yang W, Sun Z (2005) Mechanical properties of sound and fissured rice kernels and their implications for rice breakage. J Food Eng 68(1):65–72

Zhu LJ, Dogan H, Gajula H, Gu MH, Liu QQ, Shi YC (2012) Study of kernel structure of high-amylose and wild-type rice by X-ray microtomography and SEM. J Cereal Sci 55(1):1–5

Rice Grain Quality

36

Muhammad Mazhar Iqbal, Tayyaba Naz, Shazia Iqbal, Mazhar Iqbal Zafar, Omer Farooq, Atique-ur-Rehman, and Muhammad Akram Qazi

Abstract

Grain quality of rice (*Oryza sativa* L.) is more complex than other cereals, since it is mostly consumed as whole grain in countries where it serves as a staple food. Grain quality of rice is more complex than other cereals, since it is mostly consumed as whole grain in countries where it serves as a staple food. The price on the market is determined by qualitative factors which include metabolism, physical appearance, cooking, sensitivity, and the amount of healthy food. It would be beneficial to have a deeper understanding of the variables that influence these quality factors. We will review the progress made in the production of the critical components of grain quality and its genetic makeup in this chapter. In the current context of declining natural resources and biodiversity, this chapter will also provide fresh insights into knowledge obtained from modern instruments, including grain quality and high yields. Rice is a distinctive and very significant crop that is believed to help feed about half of the world every day. For the supply of high-quality goods, knowing their properties and their meaning is important. This is particularly true today as, in recent years, international trade in rice has

M. M. Iqbal (✉)
Soil and Water Testing Laboratory Chiniot, Department of Agriculture, Government of Punjab, Chiniot, Pakistan

T. Naz · S. Iqbal
Institute of Soil and Environmental Sciences, University of Agriculture, Faisalabad, Pakistan

M. I. Zafar
Department of Environmental Science, Quaid-i-Azam University, Islamabad, Pakistan

O. Farooq · Atique-ur-Rehman
Department of Agronomy, Bahauddin Zakariya University, Multan, Pakistan

M. A. Qazi
Rapid Soil Fertility Survey and Soil Testing Institute, Lahore, Pakistan

N. Sarwar et al. (eds.), *Modern Techniques of Rice Crop Production*,
https://doi.org/10.1007/978-981-16-4955-4_36

risen rapidly. This substantial chapter of the book reviews the difference in the characteristics of rice and their impact on rice quality. Rice is a unique and very important crop, which is thought to help feed about half of the planet every day. Understanding its properties and its importance is essential to the provision of high-quality products. This is especially true today as the international trade in rice has grown rapidly in recent years. This important chapter of the book reviews the variation in rice characteristics and their impact on rice quality.

Keywords

Rice · Quality parameters · Milling recovery · Cooking and eating quality

36.1 Introduction

The other name for rice (*Oryza sativa* L.) is contrariness. A high-value one and a subsistence crop are both scattered and concentrated. A relatively undeveloped area is connected with it. It has accurate characteristics but is also unpredictable. It is a cultural and survival food (Ahmad et al. 2008, 2009, 2012, 2013, 2015, 2019; Ahmad and Hasanuzzaman 2012; Ahmed and Ahmad 2019, 2020; Shahzad and Ahmad 2019). In Asia, some 90 percent of total of the world's rice is cultivated (Ahmed et al. 2017, 2020a, b; Akram et al. 2018, 2019; Hussain et al. 2018; Fahad et al. 2019; Fatima et al. 2020). Yet rice feeds the world's largest number of individuals. Rice is extremely versatile and grown virtually in the whole world (Hafeez-ur-Rehman et al. 2019; Khan et al. 2019a, b; Naz et al. 2019; Peerzada et al. 2019; Razaq et al. 2019; Sarwar et al. 2013a, b; Tariq et al. 2020; Wasaya et al. 2019; Zahoor et al. 2019). The rice country accounts for 25% of arable land but about 14% of the land and accounts for 54% of the world's population. Another characteristic of rice is its great variety, first of all between the three zones in the "rice country" and second between varieties.

Widely cultivated in limited amounts, aromatic rice is highly prized and used for special occasions. These rice, historically grown in South Asia, belong to category V of the genetic classification of Glaszmann and type IV of the consistency classification of Bhattacharya. Of these, India's and Pakistan's basmati rice and Thailand's jasmine rice are now well known due to international trade. Basmati is probably the world's longest and slenderest rice grain, displays extreme grain elongation and formation of "ring" during cooking, and has a distinctive distal-end grain shape (Bhattacharya 2011a).

For at least half the people of the world, rice (Oryza sativa) is an important food (Wei and Huang 2019). Because of need for more food for the population's growth and simultaneous decline in arable land, new high yield potential varieties, biotic and abiotic stress resistance varieties, are required. Rice buyers and breeding schemes are now mainly concentrated on improving rice quality.

36.2 Concepts of Rice Quality

The International Standard Organization 8402 1986 defines quality as a total of the features and characteristics of a product or service capable of meeting the requirements set out or specified. Quality characteristics are established characteristics for better rice grains. Rice quality parameters usually affect the market rate as well as approval by consumer. In the recent era, grain quality and yield are the most important factors to grow new rice varieties. If the taste, texture, taste, smell, appearance, cooking, or processing of freshly made rice is not appreciated by consumers, any other excellent quality of this type may be in vain.

The word rice grain quality is defined in many ways and depends on the end uses, the area of interest, experience, ethnic context, etc. The quality of appearance is primary in marketing; producers and mills emphasize the quality of milling; food manufacturing firms focus on the quality of processing; food suppliers are in need of nutritional quality and customers demand a large range of qualities in cooking and eating. There is also a high degree of good quality or bad quality, and if choice varies, the same rice, considered to be good by one, maybe considered poor by another.

Quality can be divided into four wide range of fields: (1) quality of milling; (2) quality of cooking, food, and processing; (3) quality of nutrition; and (4) basic quality, cleanliness, soundness, and purity standards. To judge the adequacy of rice for a specific use, all categories are significant. The quality characteristics of rice, however, differ greatly and are ultimately related to the final acceptance of consumers of any rice-containing food or rice product (Fig. 36.1).

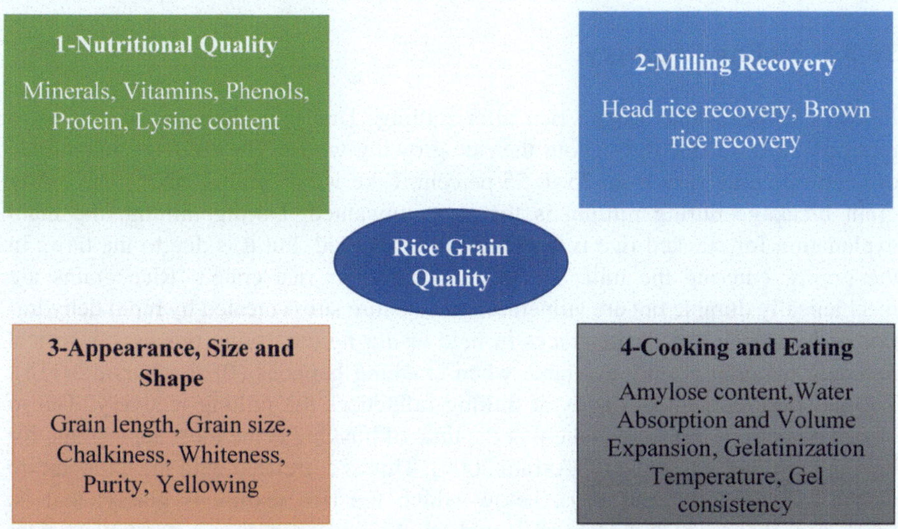

Fig. 36.1 Rice quality parameters

All rice is used in one or more different ways, as a food for human consumption. The boiled rice prepared in institutions is the best pattern of consumption. Parboiled rice, fast food rice, cereals for dry breakfast, value-added rice, soups, rice flour, foods, snack food, clothes, confections, and brewing are also essential and constantly growing applications of rice. The quality of rice must therefore be assessed based on proposed use and defined wholeness requests.

Factors influencing rice quality include those under genetic regulation, such as sanitation and hygiene, and those that are independent of genetic control. These factors are mainly the role of management and retention and are therefore well represented in the US Standards for Rice for raw rice, brown rice processed, and milled rice and the Rice Inspection Handbook and the Rice Inspection Handbook. A significant factor in rice quality is the genetic composition of grain. Modern breeding systems constantly optimize and enhance the consistency of genetic traits to produce the most desirable product. New rice cultivars are produced for all important quality attributes in the United States through intense genetic selection. Critical elements in a responsible varietal improvement program include breeding and selection for suitable milling procedure, cooking, consumption, and processing traits in the hybrid selection of new rice varieties. These programs are conducted by the US Department of Agriculture (USDA) in cooperation with agricultural experiment stations in Mississippi, Arkansas, Louisiana, Texas, and California. Before being released into commercial production, the new varieties produced in these systems meet the critical requirements of the refining, food preparation, catering, and quality processing industries. A comprehensive list of quality assessment methods for different parameters is given in Table 36.1.

36.3 Components of Rice Grain Quality

36.3.1 Milling Recovery

It refers to head and broken rice after milling. Due consideration is offered to recovery of head rice throughout the rice-growing world. The recovery magnitude of head rice may vary from 25 to 55 percent. Like whole grains, rice is used. Any grain breakage during milling is therefore unwanted. During milling, the main explanation for cracked rice is not the milling method, but it is due to the flaws in the grains, entering the mill. The primary cause is rice cracks. Rice grains are mechanically durable but are vulnerable to moisture stress created by rapid dehydration or hydration and create cracks in field or during the drying process. A critical moisture material is only available when cracking happens (Bhattacharya 2011b). The moisture content of paddy at milling influences the milling recovery. Ten to eleven percent of moisture content at the time of husking is the most appropriate for having high milling recovery (Akram 2009). This is correlated with the temperature of the transition to glass (Tg), below which the rice texture is glassy, that is, breakable, above which the texture is rubbery. Moisture serves as a starch plasticizer and decreases the Tg. Thus, moisturized grains do not crack. Cracking by warm

Table 36.1 Different valuation methods used for determining rice quality

Rice grain quality parameters		Recent quality evaluation method (s)	References
Nutritional quality	Lipid content	Metabolomic approach using LC-MS or GC-MS	Liu et al. (2017)
	Protein content	NIRS	Lapchareonsuk and Sirisomboon (2015)
	Resistant starch content	None	
	Non-starch, polysaccharide content, and dietary fiber content	CE, HPLC coupled with mass spec detector	Mantovani et al. (2018)
	Micronutrients	AAS, ICP-OES, ICP-MS, XRF	Hansen et al. (2009), Wheal et al. (2011)
	Digestibility	Time-resolved NMR	Dona et al. (2009)
Appearance, size, and shape	Texture profiling	Texturometer, hardness tester, extrusion and back extrusion, consistometer, viscoelastograph, surface tensiometer, tensipresser, Kramer shear or texture press, puncture test	Sharma and Khanna (2019)
	Aroma profiling	Detection and quantification of 2-acetyl-1-pyrroline by GC-MS Detection of total volatile metabolome by GC-MS	Fitzgerald et al. (2008)
	Rancidity test	Detection of free fatty acids by titration or colorimetry	Wang et al. (2006)
Cooking and eating quality	Pasting properties	Brabender visco-amylograph, micro-visco-analyzer	Champagne et al. (1999), McKenzie and Rutger (1983)
	Gelatinization Temperature	Measurement of starch gelatinization by DSC, Photometric method, alkali photometry, or RVA pasting curve	Dang and Bason (2014)
	Kernel elongation and grain volume expansion	–	Sharma and Khanna (2019)
	Cooking time	Measured indirectly by estimating gelatinization temperature using DSC	Biliaderis et al. (1986)
	Apparent amylose Content	HPLC-SEC, DSC, NIRS	Jane and Chen (1992), Biliaderis et al. (1986), Villareal et al. (1994), Delwiche et al. (1995)

Source: Modified and adopted from Sharma and Khanna (2019)

tempering to overcome moisture gradient before cooling can be avoided during drying. Milling machinery only affects the defeat/damaged grain in rice breakage. Small friction defends some faulty grains from failure, whereas harsh friction causes more and more of them to fail (Bhattacharya 2011b).

The degree of milling (DM) of rice is called the degree to which bran is separated during milling. The greater the extent of bran removal, the lower is the yield of polished rice and vice versa (Akram 2009). Paddy, or rough rice, gives brown rice when it is dehusked, which is called bran in an outer layer. The DM has a lot to do with rice consistency. As the layer of the rice has various intricacies, the chemical composition of rice is influenced by the DM. During partial friction, fat is flushed over the grain surface; hence, DM affects rice fluid, packaging properties, and storage stability. Many micronutrients are present in the bran layer, and rice DM, therefore, has an effect on their nutritional value. DM also affects the cooking quality as the bran layer provides some resistance to cooking (Bhattacharya 2011c).

36.3.2 Varietal Purity

Varietal mixtures cause difficulties at milling and usually result in low milling recovery of rice, excessive grain breakage, reduced capacity, and slashed head rice. Difference in shapes and sizes of rice grains makes it problematic to accommodate polishers, hullers, and whiteners to make whole grains. This causes a higher percentage of re-circulated paddy, low initial husking efficiencies, lower grade of milled rice, and non-uniform whitening (Mackill et al. 1996).

36.3.2.1 Degree of Purity

The dockage presence in the grain is linked to purity. Dockage is created by the materials like stones, chaff, seeds of weed, dirt, stalks, and rice straw. Such materials are typically mixed from drying floor or field. Un-clean paddy increased the amount of time needed to cleanse the grain. External grain stuff decreases the milling recovery and rice quality (Mackill et al. 1996).

36.3.3 Grain Size, Appearance, and Shape

Grain appearance, size, and shape influence consumer's satisfactoriness to a larger degree. The long, translucent, and willowy rice is preferred usually in Pakistan by the consumers. A quality index (length/width $\times$ thickness) is another parameter which classifies rice into fine and coarse groups. If a computed value of quality index for a particular variety is more than 2.0, it is fine grain; otherwise, it is coarse (Akram 2009).

36.3.3.1 Whiteness

Whiteness is a mixture of natural characteristics of the variety and the degree of milling. The whiteness of the grain is influenced greatly by milling, whitening, and polishing. The bran layer and silver skin from the brown rice are separated during whitening. Polishing after whitening is carried out to increase the whiteness of rice. In the process of polishing, some bran particles stick to the rice surface and give a brighter appearance (Mackill et al. 1996).

36.3.3.2 Chalkiness

It is sometimes described as "chalky" if the color of milled rice kernel is opaque, not translucent. Too much chalkiness was produced due to the disturbance during the final grain filling stages. Chalkiness downgrades milled rice, but when cooked, it disappears and has no effect on taste or flavor. Even while cooking, chalkiness disappears and does not affect cooking. Excessive chalkiness decreases consistency and lessens milling recovery (Mackill et al. 1996).

36.3.3.3 Grain Dimensions

Grain shape and size are a varietal property (length–width ratio). Long grains usually undergo breakage more compared to small grains, and thus a lower recovery of milled rice occurs. Grain dimension often determines the form of milling equipment to some degree (Mackill et al. 1996).

Length

Asian rice "Indica varieties" are long rice variety and typically cultivated in warm climates, while Asian rice "Japonica" varieties are short grains and cultivated in both tropical and temperate climates. Broad- and short-grain varieties are also available for African rice as well as glutinous rice (Asian rice).

Short Grains

Grains of short-grain rice varieties prefer to stay together when cooked, like Japonica varieties of Asian rice. Short-grain rice is up to 5.2 mm in length. These are not like sticky or glutinous rice. Uruchimai or "sushi rice" and Arborio are a variety of short-grained Japanese rice.

Long Grain

When cooked, long-grain rice prefers to remain separate and fluffy rather than to remain together. Two common varieties of long-grain rice are Jasmine rice and Basmati rice; both are aromatic and fragrant. Indica (long-grain) rice is produced in Southern Asia. Rice with a grain length of more than 6.0 mm refers to long-grain rice.

Medium-Sized Grain

Rice with grain length from 5.2 to 6.0 mm falls under the category of medium-sized grains.

Unripe Grains

The quantity of immature paddy grains has a significant influence on head rice yield and consistency. Quite slender and chalky, the immature rice kernels result in broken grains, unsustainable bran production, and rice production from the brewer. The optimum moisture stage for harvesting is 20–24 percent humidity of grain. Drying or shattering causes loss of many grains if the harvest is too late. During threshing, grains are cracked.

Cracked Grains

Over-exposure to changing temperature and humidity conditions lead to production of individual kernel cracks. Most significant element in breakage during rice milling is cracks in the kernel. This results in the recovery of milled rice and head rice yields being decreased.

Damaged Grains

Due to biochemical changes, production of odors, and appearance changes, paddy deteriorates. Water, insects, and exposure to heat cause harm to rice.

Yellowing/Staining

Yellowing is caused prior to drying, due to moist conditions. A mixture of chemical and microbiological activity that overheats the grain results in yellowing. Sometimes, these fermented grains have partially gelatinized cells of starch and usually reduce the applied compressions during the milling of grain. Although fermented grains' presence does not influence yield, due to the unattractive appearance of yellowing, the consistency of the milled rice degrades (Mackill et al. 1996).

Aroma of Rice

The aroma is a value-adding and favorite feature of rice customers. It is a valuable sensory element of cooked rice that enhances its value in market (Chen et al. 2006). However, it is uncommon for consumers to identify the aroma of rice. Scientists have therefore invented ways to describe the scent of rice. Nice aromas appear to be connected with good aromatics present in the varieties of Jasmine as well as Basmati rice. Out of the 100 known volatile compounds, the main chemical is 2-acetyl-1-pyrroline (2-AP) that contributes to the aroma of jasmine rice and basmati rice, with popcorn and cracker-like, toasted odor (Czerny et al. 2008; Chen et al. 2006). This 2-AP aroma is also associated with milky aroma, sweet nutty odor, and pandan aroma (Paraskevopoulou et al. 2014; Kobayashi and Nishimura, 2014). Chemical 2-AP has a low point of scent tolerance in water and, as a result, can be found usually not in a tasteful variety (Buttery et al. 1982). The 2-AP aroma is primarily determined by the genetic history and by the agronomic and post-harvest conditions for the plants and subsequently, the cereals undergo (Wakte et al. 2017). The taste has historically been identified by smelling after a 0.1 M KOH (Potassium Hydroxide) reaction. But with a constant and long exposure, this technique is subjective and damaged the analyst's nasal cavity (Grimm et al. 2001). Apart from 2-AP, other volatile compounds also have specified rice flavors (Anacleto et al. 2015). The existence of these other compounds and the varying concentrations differentiate between aromas of various rice varieties, that is, Basmati and Jasmine rice. The contribution of these volatiles to the fragrance of rice, however, is not clear, and aromatic rice description is still unfinished.

36.3.4 Cooking and Eating Characteristics

These characteristics include a wide range of tests. But some common and reproducible tests which have a direct effect on cooking and eating are as follows.

36.3.4.1 Amylose Contents

Thousands of varieties of rice around the world show significant differences in the quality of their food. Preliminary studies had repeatedly shown that its texture was the only major determinant of amylose starch. Recent research has shown that what was considered to be amylose or its waterproof component was a branch, especially the longer branches of starch amylopectin. Rheological studies show that these long branches affect the strength of starch granules by their internal and internal intramolecular interactions. The texture was revealed by the hardness/fragility of starch granules caused by the large/small amount of these amylopectin chains. A second role may have been played by protein substances (Bao 2019; Bhattacharya 2011e).

Milled rice makes up to 90% starch. Starch contains amylose and amylopectin. Ratings for amylose to amylopectin are the main choice for cooking and eating qualities of milled rice. Amylose content is also used as a condition of classification rice such as waxy (1–2% amylose), non-waxy (>2% amylose), high (25–35% amylose), medium (20–25% amylose), low (9–20% amylose), and very low (2–9% amylose) (Akram 2009). The amylose starch content generally varies from 15 to 35 percent. Rice with high amylose content indicates a high increase in volume (not really stretching) and a high degree of bitterness. High amylose grains are cooked dry, slightly porous, and difficult to cool. Low-amylose rice, on the other hand, makes cookies moist and sticky. In many rice-growing regions, medium-sized rice is preferred, with the exception of low-grade amylose Japonicas (Mackill et al. 1996).

36.3.4.2 Gelatinization Temperature

It has been reported in the literature that the temperature of gelatinization (GT) affects the cooking time of rice positively (Akram 2009). The time required to cook milled rice is ascertained by the temperature of gelatinization or by GT. Environmental factors affect GT, such as temperature during maturation. By testing the Alkali distribution value, the GT of milled rice is tested. There is a different option for rice with a moderate temperature of gelatinization in many rice-growing countries (Mackill et al. 1996). Usually, the first-served rice is cooked; it takes 5–5 min to cook in boiling water vigorously. GT is a substance starch and can also be distinguished as the range of temperatures within it. Ninety percent of starch granules in hot water begin to swell irreversibly. It varies between 55 and 79 °C. This feature also classifies rice varieties as low (55–69 °C), medium (70–74 °C), and high (74.5–79 °C) try GT rice. This is in the middle GT rice that are the most popular. GT rice is estimated to absorb six whole grains of milled rice at 1.7 percent of 23% potassium hydroxide (KOH) hours at 30 °C temperature. Rice is numbered numerically (1–7) based on dispersion. Prices with low GT are completely different (6–7), medium GT rice show partial separation (4–5); and high GT rice remain intangible (1–3) (Akram 2009).

36.3.4.3 Gel Consistency

There are many varieties of rice that have the same amylose content as well as body sizes such as grain size, shape, and appearance. But it gets worse. It is difficult to distinguish well. A reliable measure that distinguishes such rice is the consistency of the gel at the end of the substance. The movement characteristics and the flat length of rice gel (mm) can be described as being specified on the 13 × 100 ml test tube. This experiment divides the rigid rice (gel length 36 mm or less) in the middle (gel length 36–59 mm) and soft gel steadiness (gel length above 60 mm) in the middle. The books reveal that medium-sized amylose rice has a low gel consistency and are popular because of their generosity (Akram 2009).

36.3.4.4 Parboiled Rice

Parboiled rice is nothing but pre-cooked rice with a husk on it. The method leads to various variations in the grain components and different grain properties. During the cooking of rice, kernel cracks and calcinating are cured, which significantly reduces the rice breakage during the frying of parboiled rice. During the milling of parboiled rice, vitamins and micronutrients cannot be extracted easily making it much superior in nutrition. Cooking in reduced humidity induces the development of novel starch polymorphs, namely, starch annealed, amylose–lipid complexes I and II, and amylopectin and amylose retrograde complexion. For this reason, parboiled rice takes a lot more cooking time, and the cooked grains are hard and distinctive. The disturbance in the aleuronic layer of the fat global components pushes out the fat, resulting in poor flow, packaging, and frying properties for undermilled parboiled rice (Bhattacharya 2011d).

36.3.4.5 Elongation Ratio

After frying, some rice types are longer in size than others. The "stretching ratio" is defined as the average length of completely cooked grains and raw lengths. This is highly attractive aspect of rice grain quality. High-quality rice is stated to reflect a 100 percent rating (Akram 2009).

36.3.4.6 Water Absorption and Volume Expansion

Water absorption and volume expansion ratios are also determinants of good cooking qualities. For this purpose usually, 2.0 grams of whole rice are doused in slightly excess water for about 30 min in a tube. The tube along with its contents is heated for 45 min at 77–82 °C. Water absorption ratio is calculated as the amount of water absorbed/gram of rice. While volume expansion is measured as the difference between raw rice volume and fully cooked rice volume, the ratio is computed as the cooked rice volume to raw rice volume (Akram 2009). During direct boiling of rice in water, traditional intake of water (g of water absorbed at a time by g of rice) is attached to the grain surface. Whole rice, when cooked in a grain bowl, absorbs water twice as much as 2.5 times its weight, regardless of its composition and characteristics. Rice hydration at moderate temperatures, on the other hand, is influenced in contrast to its amylose content, GT, and is facilitated by grain coating. Hydration is different in relation to GT at 70–80 °C. Immersion before cooking decreases the cooking time of grain expansion helps and affects grain cracking and splitting. Immersion before cooking decreases the cooking time of grain expansion and helps and affects grain cracking and splitting. The cooking method contributes to the formation and other parameters of cooked rice. Standardization is also ideal for laboratory cooking (Bhattacharya 2011f).

The staple food in northern Thailand and Laos is waxy rice. It is also commonly cultivated and used for snack foods and special dishes. Colored rice is considered to have physical and medicinal properties (Bhattacharya 2011a).

36.3.5 Nutritional Quality

Not only does rice feed the greatest number of people on Earth, but it also makes up the bulk of the diet of a great number of poor people. Therefore, its nutritional value is of primary importance (Bhattacharya 2011g). Protein is the second-highest component of milled rice. It acts as a tributary aspect that gives a vital contribution to nutritional quality (Akram 2009). Compared to many cereal grains, protein contents in rice are quite low; yet its quality is good (Kawakatsu and Takaiwa 2019). Phenolics have many health-promoting effects in rice and help minimize the risk of many chronic diseases, for instance, inflammation, cardiovascular diseases, obesity, type II diabetes, and certain forms of tumors (Shao and Bao 2019). Many processes lead to the loss of the nutritional value of rice. Washing and cooking lead to Vitamin B loss. Milling also causes loss of Vitamin B. Iron and calcium deficiency are also main defects of poor rice eater's diet. From a broad perspective, excessive refining of grains especially rice is now considered a basic contributor to diseases of affluence. For proper and healthy diet balance of rice, the content of resistant starch and glycemic index are considered (Bhattacharya 2011g).

36.3.6 Preferred Quality Parameters for International Trade

Rice is used and traded internationally and as consumers are from different regions, traditions, and different eating habits, a standard rice quality definition is needed. As an interdisciplinary analysis of farmer, consumer, and industry views, literature and genetic information show that the quality of rice is perspective based. Consumer differentiations are heterogeneous among regions, countries, cities, towns, and urbanizations. There remains unanimous discussion among food technicians about how rice quality, especially in terms of grain size and form, chalkiness, whiteness, aroma, nutrients, quality of cooking, and GT, should be measured and ranked. The rice division currently does not have unanimity on how "rice quality" should be described and calculated. The misalignment of food scientists and quality class in the value chain actors will run to a potential problem in trading and will avoid rice farmers from putting appropriate targets on propagating traits.

In the concept of low-, good-, or high-quality rice, consumers of different economic and economic classes seem to agree surprisingly in a specific context, while these quality categories can be seen as distinct but rather as softening. It is unfinished as the aroma is established by the concentration of different 2-AP and many other flexible rice grain compounds.

The focus is on the appearance of uncooked grains in South Asia (shape and size, thinness, whiteness), and premium rice in South Asia has the benefits of nutritious food, softness and scent, and satiety of cooked products, and characteristics that Southeast Asian consumers can easily take into account in any rice product other than "low-quality rice."

The patterns are in line with the views of farmers and industry, indicating that the value chain of rice effectively conveys ideas for the use of Southeast Asian and South Asian line players. These regional and national features have a profound effect on the allocation of labor between national and regional breeding programs. The rice chain players, particularly in countries, where customers usually purchase rice in packages, can improve the quality of rice by labeling, branding, and packaging (Custodio et al. 2019).

36.3.7 Effect of Climate Change in Rice Quality

The steady production of high-quality goods poses a major threat to change of climate. Concentrations of atmospheric CO_2 are growing at a speed not unprecedented, with other greenhouse gasses, and we see rising temperatures, different precipitation levels, and more common extreme weather conditions. Climate change is posing grave threats to the production and livelihood of rice farmers because of extreme temperatures, extreme rain, osmotic stress, drought, salinity, frost damage, and floods (Atique-ur-Rehman 2018). The effect of the change of grain filling time is comparatively greater than the effect of the increase in temperature. Quality of rice is highly sensitive to the date of transplantation than yields; therefore, the later shifts in date may be a suitable measure for climate change adaptation (Okada et al. 2011).

Increasing daily temperature may reduce rice spikelet fertility, which will help reduce production while increasing CO_2 concentration patterns will increase rice output (Dharmarathna et al. 2012). Threshold beyond limit temperatures increased respiratory rate (Mohammed and Tarpley 2009), decreased crop length, increased spikelet sterility, and reduced grain-fill time (Kim et al. 2011), resulting in lower yields and smaller grains of rice (Fitzgerald and Resurreccion 2009).

36.4 Conclusion

Quality of rice grain is a complex trait as a whole. This trait includes appearance qualities of grain like whiteness, chalkiness, yellowing, size and shapes, aroma, milling quality, milling recovery, and nutritional quality. Aroma is the most dominant and important quality parameter for trade purpose. Appearance, size, and shapes perform a critical part in quality of rice. Shorter sized grains of rice cooked early but usually stuck together, however, absorb an equal water volume as large-sized rice grain absorbed. Large-sized, white, and transparent or translucent rice are considered of good quality. Milling recovery and extent of milling are important in rice quality parameters. Milling causes cracks in rice if moisture content is low. Starch, amylose, gel consistency, and GT affect the texture of rice, and rice is graded on the basis of these contents. Different nutrients such as phytochemicals, vitamins, etc. are usually present in the upper layer of bran. These vitamins can be present in very low quantity in milled grains. Protein, calcium, zinc, and iron are low in rice, and these results in enhancing the benefits of rice by improving these nutrients will benefit human health. Genetic control research for quality indicators has greatly improved, especially in terms of appearance, cooking, and food quality. With the quality of digestion and the quality of healthy eating, genetic studies required a new theme for community growth, and improving living conditions that need researchers' attention will be the basis for food and intake worth rice. Priorities among the studies are food recipes, genetics, and cell reproduction.

References

Ahmad S, Hasanuzzaman M (2012) Integrated effect of plant density, N rates and irrigation regimes on the biomass production, N content, PAR use efficiencies and water productivity of rice under irrigated semiarid environment. Not Bot Horti Agrobot Cluj-Napoca 40(1):201–211

Ahmad S, Zia-ul-Haq M, Ali H, Shad SA, Ammad A, Maqsood M, Khan MB, Mehmood S, Hussain A (2008) Water and radiation use efficiencies of transplanted rice (*Oryza sativa* L.) at different plant densities and irrigation regimes under semi-arid environment. Pak J Bot 40(1): 199–209

Ahmad S, Ahmad A, Zia-ul-Haq M, Ali H, Khaliq T, Anjum MA, Khan MA, Hussain A, Hoogenboom G (2009) Resources use efficiency of field grown transplanted rice (*Oryza sativa* L.) under irrigated semiarid environment. J Food Agric Environ 7(2):487–492

Ahmad S, Ahmad A, Soler CMT, Ali H, Zia-Ul-Haq M, Anothai J, Hussain A, Hoogenboom G, Hasanuzzaman M (2012) Application of the CSM-CERES-Rice model for evaluation of plant density and nitrogen management of fine transplanted rice for an irrigated semiarid environment. Precis Agric 13(2):200–218

Ahmad S, Ahmad A, Ali H, Hussain A, Garcia y Garcia A, Khan MA, Zia-Ul-Haq M, Hasanuzzaman M, Hoogenboom G (2013) Application of the CSM-CERES-Rice model for evaluation of plant density and irrigation management of transplanted rice for an irrigated semiarid environment. Irrig Sci 31(3):491–506

Ahmad A, Ashfaq M, Rasul G, Wajid SA, Khaliq T, Rasul F, Saeed U, Rahman MH, Hussain J, Baig IA, Naqvi AA, Bokhari SAA, Ahmad S, Naseem W, Hoogenboom G, Valdivia RO (2015) Impact of climate change on the rice–wheat cropping system of Pakistan. In: Hillel D, Rosenzweig C (eds) Handbook of climate change and agro-ecosystems: the agricultural modeling Intercomparison and improvement project (AgMIP) integrated crop and economic assessments. Imperial College Press and the American Society of Agronomy, pp 219–258

Ahmad S, Abbas G, Ahmed M, Fatima Z, Anjum MA, Rasul G, Khan MA, Hoogenboom G (2019) Climate warming and management impact on the change of rice-wheat phenology in Punjab, Pakistan. Field Crops Res 230:46–61

Ahmed M, Ahmad S (2019) Carbon dioxide enrichment and crop productivity. In: Hasanuzzaman M (ed) Agronomic crops, vol 2. Springer, Singapore, pp 31–46

Ahmed M, Ahmad S (2020) Systems modeling. In: Ahmed M (ed) Systems Modeling. Springer, Singapore, pp 1–44

Ahmed M, Fayyaz-ul-Hassan, Ahmad S (2017) Climate variability impact on rice production: adaptation and mitigation strategies. In: Ahmed M, Stockle C (eds) Quantification of climate variability, adaptation and mitigation for agricultural sustainability. Springer, Cham, pp 91–111

Ahmed M, Ahmad S, Raza MA, Kumar U, Ansar M, Shah GA, Parsons D, Hoogenboom G, Palosuo T, Seidel S (2020a) Models calibration and evaluation. In: Ahmed M (ed) Systems Modeling. Springer, Singapore, pp 151–178

Ahmed M, Ahmad S, Waldrip HM, Ramin M, Raza MA (2020b) Whole farm modeling: a systems approach to understanding and managing livestock for greenhouse gas mitigation, economic viability and environmental quality. In: Waldrip HM, Pagliari PH, He Z (eds) Animal manure. ASA special publication 67. American Society of Agronomy and Soil Science Society of America, Madison, WI, pp 345–371

Akram M (2009) Aromatic rice's of Pakistan. Pak J Agric Sci 22(3–4):154–160

Akram R, Turan V, Hammad HM, Ahmad S, Hussain S, Hasnain A, Maqbool MM, Rehmani MIA, Rasool A, Masood N, Mahmood F, Mubeen M, Sultana SR, Fahad S, Amanet K, Saleem M, Abbas Y, Akhtar HM, Hussain S, Waseem F, Murtaza R, Amin A, Zahoor SA, Sami ul Din M, Nasim W (2018) Fate of organic and inorganic pollutants in paddy soils. In: Hashmi MZ, Varma A (eds) Environmental pollution of Paddy soils. Springer, Cham, pp 197–214

Akram R, Fahad S, Masood N, Rasool A, Ijaz M, Ihsan MZ, Maqbool MM, Ahmad S, Hussain S, Ahmed M, Kaleem S, Sultana SR, Mubeen M, Saud S, Kamran M, Nasim W (2019) Plant growth and morphological changes in rice under abiotic stress. In: Hasanuzzaman M, Fujita M, Nahar K, Biswas JK (eds) Advances in Rice Research for Abiotic Stress Tolerance. Woodhead Publishing, pp 69–85

Anacleto R, Cuevas RP, Jimenez R, Llorente C, Nissila E, Henry R, Sreenivasulu N (2015) Prospects of breeding high-quality rice using post-genomic tools. Theor Appl Genet 128: 1449–1466. https://doi.org/10.1007/s00122-015-2537-6

Atique-ur-Rehman M, Farooq A, Rashid F, Nadeem S, Stürz FA, Bel RW (2018) Boron nutrition of rice in different production systems. A review. Agron Sustain Dev 38:25

Bao J (2019) Rice starch. In: Bao J (ed) Rice (fourth edition) chemistry and technology. AACC International Press, Pages, pp 55–108

Bhattacharya KR (2011a) Introduction: rice in historical and social perspectives. In: Bhattacharya KR (ed) Rice quality, Woodhead publishing series in food science, technology and nutrition. Elsevier Science, pp 1–25

Bhattacharya KR (2011b) Milling quality of rice. In: Bhattacharya KR (ed) Rice quality, Woodhead publishing series in food science, technology and nutrition. Elsevier Science, pp 61–99

Bhattacharya KR (2011c) Degree of milling (DM) of rice and its effect. In: Bhattacharya KR (ed) Rice quality, Woodhead publishing series in food science, technology and nutrition. Elsevier Science, pp 100–115

Bhattacharya KR (2011d) Cooking quality of rice. In: Bhattacharya KR (ed) Rice quality, Woodhead publishing series in food science, technology and nutrition. Elsevier Science, pp 164–192

Bhattacharya KR (2011e) Eating quality of rice. In: Bhattacharya KR (ed) Rice quality, Woodhead publishing series in food science, technology and nutrition. Elsevier Science, pp 193–246

Bhattacharya KR (2011f) Effect of parboiling on rice quality. In: Bhattacharya KR (ed) Rice quality, Woodhead publishing series in food science, technology and nutrition. Elsevier Science, pp 247–297

Bhattacharya KR (2011g) Nutritional quality of rice. In: Bhattacharya KR (ed) Rice quality, Woodhead publishing series in food science, technology and nutrition. Elsevier Science, pp 377–409

Biliaderis CG, Page CM, Maurice TJ, Juliano BO (1986) Thermal characterization of rice starches: polymeric approach to phase transition of granular starch. J Agric Food Chem 34:6–24

Buttery RG, Ling LC, Juliano BO (1982) 2-acetyl-1-pyrroline: an important aroma component of cooked rice. Chem Ind 23:958–959

Champagne ET, Bett KL, Vinyard BT, McClung AM, Barton FE II, Moldenhauer K, Linscombe S, Mckenzie K (1999) Correlation between cooked rice texture and rapid visco analyses measurements. Cereal Chem 76:764–771

Chen S, Wu J, Yang Y, Shi W, Xu M (2006) The fgr gene responsible for rice fragrance was restricted within 69 kb. Plant Sci 171:505–514

Custodio MC, Cuevas RP, Ynion J, Laborte AG, Velasco ML, Demont M (2019) Rice quality: how is it defined by consumers, industry, food scientists, and geneticists? Trends Food Sci Tech 92: 122–137

Czerny M, Christlbauer M, Christlbauer M, Fischer A, Granvogl M, Hammer M, Hartl C, Hernandez NM, Schieberle P (2008) Re-investigation on odor thresholds of key food aroma compounds and development of an aroma language based on odor qualities of defined aqueous odorant solutions. Eur Food Res Technol 228:265–273. https://doi.org/10.1007/s00217-008-0931-x

Dang JMC, Bason ML (2014) AACCI approved methods technical committee report: collaborative study on a method for determining the gelatinization temperature of milled rice flour using the rapid visco analyser. Cereal Foods World 59:31–34

Delwiche SR, Bean MM, Miller RE, Webb BD, Williams PC (1995) Apparent amylose content of milled rice by near infrared reflectance spectroscopy. Cereal Chem 72:182–187

Dharmarathna WR, Weerakoon SB, Rathnayaka UR, Herath S (2012) Variation of irrigated rice yield under the climate change scenarios. In: SAITM research symposium on engineering advancements, pp 31–34

Dona AC, Pages G, Gilbert RG, Gaborieau M, Kuchel PW (2009) Kinetics of in vitro digestion of starches monitored by time-resolved 1H nuclear magnetic resonance. Biomacromolecules 10: 638–644

Fahad S, Noor M, Adnan M, Khan MA, Rahman I, Alam M, Khan IA, Ullah H, Mian IA, Hassan S, Saud S, Bakhat HF, Hammad HM, Ahmad S, Nasim W (2019) Abiotic stress and rice grain quality. In: Hasanuzzaman M, Fujita M, Nahar K, Biswas JK (eds) Advances in Rice Research for Abiotic Stress Tolerance. Woodhead Publishing, pp 571–583

Fatima Z, Ahmad M, Hussain M, Abbas G, Ul-Allah S, Ahmad S, Ahmed N, Ali MA, Sarwar G, Ehsan ul Haque, Iqbal P, Hussain S (2020) The fingerprints of climate warming on cereal crops phenology and adaptation options. Sci Rep 10:18013

Fitzgerald MA, Resurreccion AP (2009) Maintaining the yield of edible rice in a warming world. FPB 36:1037–1045

Fitzgerald MA, Hamilton NRS, Calingacion MN, Verhoeven HA, Butardo VM (2008) Is there a second fragrance gene in rice? Plant Biotechnol J 6:416–423

Grimm CC, Bergman CJ, Delgado JT, Bryant RJ (2001) Screening for 2- acetyl-1-pyrroline in the headspace of rice using SPME/GC-MS. J Agric Food Chem 49:245–249. https://doi.org/10.1021/jf0008902

Hafeez-ur-Rehman, Nawaz A, Awan MI, Ijaz M, Hussain M, Ahmad S, Farooq M (2019) Direct seeding in rice: problems and prospects. In: Hasanuzzaman M (ed) Agronomic crops, volume 1: production technologies. Springer, Singapore, pp 199–222

Hansen TH, Laursen KH, Persson DP, Pedas P, Husted S, Schjoerring JK (2009) Microscaled high-throughput digestion of plant tissue samples for multi-elemental analysis. Plant Methods 5:12

Hussain M, Ahmad S, Hussain S, Lal R, Ul-Allah S, Nawaz A (2018) Rice in saline soils: physiology, biochemistry, genetics and management. Adv Agron 148:231–287
Jane JL, Chen JF (1992) Effect of amylose molecular size and amylopectin branch chain length on paste properties of starch. Cereal Chem 69:60–65
Kawakatsu T, Takaiwa F (2019) Rice proteins and essential amino acids. In: Bao J (ed) Rice (fourth edition) chemistry and technology. AACC International Press, Pages, pp 109–130
Khan MA, Ahmad S, Raza A (2019a) Integrated weed management for agronomic crops. In: Hasanuzzaman M (ed) Agronomic crops, vol 2. Springer, Singapore, pp 257–281
Khan SM, Ali S, Nawaz A, Bukhari SAH, Ejaz S, Ahmad S (2019b) Integrated pest and disease management for better agronomic crop production. In: Hasanuzzaman M (ed) Agronomic crops, vol 2. Springer, Singapore, pp 385–428
Kim J, Shon J, Lee CK, Yang W, Yoon Y, Yang WH, Kim YG, Lee BW (2011) Relationship between grain filling duration and leaf senescence of temperate rice under high temperature. Field Crop Res 122:207–213
Kobayashi N, Nishimura O (2014) Availability of detection frequency method using three-port gas chromatography-olfactometry for rapid comparison of whole milk powders. Food Sci Technol Res 20:809–814. https://doi.org/10.3136/fstr.20.809
Lapchareonsuk R, Sirisomboon P (2015) Sensory quality evaluation of rice using visible and shortwave near-infrared spectroscopy. Int J Food Prop 18:1128–1138
Liu J, Chen J, Zheng X, Wu F, Lin Q, Heng Y, Tian P, Cheng Z, Yu X, Zhou K, Zhang X, Guo X, Wang J, Wang H, Wan J (2017) GW5 acts in the brassinosteroid signalling pathway to regulate grain width and weight in rice. Nat Plants 3:17043
Mackill DJ, Coffman WR, Garrity DP (1996) Rainfed lowland rice improvement. International Rice Research Institute, Manila, p 242
Mantovani V, Galeotti F, Maccari F, Volpi N (2018) Recent advances in capillary electrophoresis separation of monosaccharides, oligosaccharides, and polysaccharides. Electrophoresis 39:179–189
McKenzie KS, Rutger JN (1983) Genetic analysis of amylose content, alkali spreading score, and grain dimensions in rice. Crop Sci 23:306–311
Mohammed AR, Tarpley L (2009) Impact of high nighttime temperature on respiration, membrane stability, antioxidant capacity, and yield of rice plants. Crop Sci 49:313–322
Naz S, Fatima Z, Iqbal P, Khan A, Zakir I, Noreen S, Younis H, Abbas G, Ahmad S (2019) Agronomic crops: types and uses. In: Hasanuzzaman M (ed) Agronomic crops. Volume 1: production technologies. Springer, Singapore, pp 1–18
Okada M, Iizumi T, Hayashi Y, Yokozawa M (2011) Projecting climate change impacts both on rice quality and yield in Japan. J Agric Meteorol 67(4):285–295. https://doi.org/10.2480/agrmet.67.4.10
Paraskevopoulou A, Amvrosiadou S, Biliaderis CG, Kiosseoglou V (2014) Mixed whey protein isolate-egg yolk or yolk plasma heat-set gels: rheological and volatile compounds characterisation. Food Res Int 62:492–499. https://doi.org/10.1016/j.foodres.2014.03.056
Peerzada AM, Bukhari SAH, Dawood M, Nawaz A, Ahmad S, Adkins S (2019) Weed management for healthy crop production. In: Hasanuzzaman M (ed) Agronomic crops, vol 2. Springer, Singapore, pp 225–256
Razaq M, Shah FM, Ahmad S, Afzal M (2019) Pest management for agronomic crops. In: Hasanuzzaman M (ed) Agronomic crops, vol 2. Springer, Singapore, pp 365–384
Sarwar N, Ali H, Ahmad S, Ehsan Ullah, Ahmad S, Mubeen K, Hill JE (2013a) Water wise rice cultivation on calcareous soil with the addition of essential micronutrients. J Anim Plant Sci 23(1):244–250
Sarwar N, Ali H, Maqsood M, Ehsan Ullah, Shahzad AN, Shahzad M, Mubeen K, Shahid MA, Ahmad S (2013b) Phenological response of rice plants to different micronutrients application under water saving paddy fields on calcareous soil. Turk J Field Crops 18(1):52–57
Shahzad AN, Ahmad S (2019) Tools and techniques for nitrogen management in cereals. In: Hasanuzzaman M (ed) Agronomic crops, vol 2. Springer, Singapore, pp 111–126

Shao Y, Bao J (2019) Rice phenolics and other natural products. In: Bao J (ed) Rice (fourth edition) chemistry and technology. AACC International Press, Pages, pp 221–271

Sharma N, Khanna R (2019) Rice grain quality: current developments and future prospects. In: Shah F, Khan Z, Iqbal A, Turan M, Olgun M (eds) Recent advances in grain crops research rice. Intech Open. https://doi.org/10.5772/intechopen.89367

Tariq M, Ahmed M, Iqbal P, Fatima Z, Ahmad S (2020) Crop phenotyping. In: Ahmed M (ed) Systems Modeling. Springer, Singapore, pp 45–60

Villareal CP, De La Cruz NM, Juliano BO (1994) Rice amylose analysis by near-infrared transmittance spectroscopy. Cereal Chem 71:292–296

Wakte K, Zanan R, Hinge V, Khandagale K, Nadaf A, Henry RJ (2017) Thirtythree years of 2-acetyl-1-pyrroline, a principal basmati aroma compound in scented rice (*Oryza sativa* L.): a status review. J Sci Food Agric 97:384–395. https://doi.org/10.1002/jsfa.7875

Wang HL, Wan XY, Bi JC, Wang JK, Jiang L, Chen LM, Zhai HQ, Wan JM (2006) Quantitative analysis of fat content in rice by near-infrared spectroscopy technique. Cereal Chem 83:402–406

Wasaya A, Yasir TA, Ijaz M, Ahmad S (2019) Tillage effects on agronomic crop production. In: Hasanuzzaman M (ed) Agronomic crops, vol 2. Springer, Singapore, pp 73–99

Wei X, Huang X (2019) Origin, taxonomy, and phylogenetics of rice. In: Bao J (ed) Rice (fourth edition) chemistry and technology. AACC International Press, pp 1–29

Wheal MS, Fowles TO, Palmer LT (2011) A cost-effective acid digestion method using closed polypropylene tubes for inductively coupled plasma optical emission spectrometry (ICP-OES) analysis of plant essential elements. Anal Methods 3:2854–2863

Zahoor SA, Ahmad S, Ahmad A, Wajid A, Khaliq T, Mubeen M, Hussain S, Sami Ul Din M, Amin A, Awais M, Nasim W (2019) Improving water use efficiency in agronomic crop production. In: Hasanuzzaman M (ed) Agronomic crops, vol 2. Springer, Singapore, pp 13–29

Phytochemical Composition of Rice

37

Haq Nawaz, Huzaifa Rehman, Momna Aslam, Hina Gul, Iqra Zakir, Zartash Fatima, Pakeeza Iqbal, Amna Khan, and Kamrun Nahar

Abstract

Rice is a widely grown cereal crop all over the world. All parts of rice including grain, husk, and straw are used as nutrition for humans, fodder for animals, and fertilizer for soil, respectively. Rice grain due to its good nutritional value is a frequently used staple food all over the world. This chapter provides a comprehensive review of the bioactive phytochemical components present in different parts of different rice varieties, their antioxidant potential, and their medicinal importance. All types of rice are a good source of bioactive phytochemical compounds with strong antioxidant potential. Principal phytochemicals occuring in various parts of different types of rice include tocopherols, tocotrienols, gamma oryzanol, phenolic acids, flavonoids, anthocyanins, proanthocyanidins, phytosterols, and gamma-aminobutyric acid. The presence of these

H. Nawaz (✉) · M. Aslam
Department of Biochemistry, Bahauddin Zakariya University, Multan, Pakistan
e-mail: haqnawaz@bzu.edu.pk

H. Rehman
Department of Biochemistry and Biotechnology, The Women University Multan, Multan, Pakistan

H. Gul
Pir Mehr Ali Shah, Arid Agriculture University, Rawalpindi, Pakistan

I. Zakir · Z. Fatima
Department of Agronomy, Bahauddin Zakariya University, Multan, Pakistan

P. Iqbal
Department of Botany, University of Agriculture, Faisalabad, Pakistan

A. Khan
Department of Agronomy, University of Sargodha, Sargodha, Pakistan

K. Nahar
Department of Agricultural Botany, Sher-e-Bangla Agricultural University, Dhaka, Bangladesh

N. Sarwar et al. (eds.), *Modern Techniques of Rice Crop Production*,
https://doi.org/10.1007/978-981-16-4955-4_37

phytochemicals is associated with strong antioxidant potential and diverse biological activities including antitumor, anti-hypercholesterolemic, anti-hyperglycemic, anti-hyperuricemic, anti-mutagenic, anticancer, anti-metastasis, cytotoxic, anti-inflammatory, immunologic, mitogenic, antimicrobial, anti-proliferation, antioxidant, antiradical, antidiabetic, and enzyme inhibitory activities. Based on these biological activities, rice possesses great medicinal importance and can be a good candidate for various pharmaceutical formulations.

Keywords

Rice · Grain · Quality · Phytochemical

37.1 Introduction

Rice (*Oryza sativa*) is the second-largest staple food in the world particularly in Asia and Africa (Palombini et al. 2013; Ahmad et al. 2008, 2009, 2012, 2013, 2015, 2019; Ahmad and Hasanuzzaman 2012; Ahmed and Ahmad 2019, 2020; Shahzad and Ahmad 2019). The whole grain, bran, white rice, husk, and straw are the main parts of rice utilized as nutrition for humans, fodder for animals, and fertilizer for the soil (Ahmed et al. 2017, 2020a, b; Akram et al. 2018, 2019; Hussain et al. 2018; Fahad et al. 2019; Fatima et al. 2020; Hafeez-ur-Rehman et al. 2019; Khan et al. 2019a, b; Naz et al. 2019; Peerzada et al. 2019; Razaq et al. 2019; Sarwar et al. 2013a, b; Tariq et al. 2020; Wasaya et al. 2019; Zahoor et al. 2019). Rice grain is a rich source of macro- and micronutrients particularly carbohydrates, protein, vitamins, and minerals. The major carbohydrates found in rice bran include cellulose, hemicellulose, β-glucan, and starch. Palmitic, oleic, and linoleic acids are the major fatty acids with small amounts of waxes and unsaponifiable lipids in rice bran crude oil (Saunders 1985). The rice bran oil also contains vitamin E (tocopherol) as a major vitamin. Moreover, rice bran is also rich in vitamin B-complex and minerals including sodium, potassium, calcium, magnesium, iron, aluminum, manganese, zinc, phosphorus, silicon, and chlorine (Lu and Luh 1991). The presence of vitamins, minerals, and bioactive phytochemical antioxidants confers the medicinal significance of rice. In addition to its nutritional value, the rice bran is also a good source of non-nutritional bioactive phytochemical components, which possess antioxidant potential, show diverse biological activities, and confer the medicinal significance of rice (Xu et al. 2001; Ghasemzadeh et al. 2015; Muttagi and Ravindra 2020).

37.2 Phytochemical Composition

Phytochemicals are non-nutritional components present in plants that perform different biological functions for the maintenance and survival of the plant in unfavorable besides harsh environmental circumstances. The compounds help the plant to fight against pathogens and stress conditions such as nutritional and environmental stresses. Polyphenols, phenolic acids flavonoids, anthocyanins, proanthocyanidins, tocotrienols, tocopherols, γ-oryzanol, and phytic acid are the major phytochemicals present in various parts of rice that possess potent biological activities and help the plant to resist microbial attack and stress conditions (Scavariello and Arellano 1998; Pietta 2000; Hernández et al. 2009; Niu et al. 2013; Wang et al. 2018a). Although not being part of nutrition, these compounds are equally beneficial for humans due to their pharmaceutical and medicinal importance. These phytochemical compounds are known to possess antimicrobial, anticancer, anti-obesity, antiaging, antitumor, anti-hypercholesterolemic, anti-hyperglycemic, anti-hyperuricemic, anti-mutagenic, anti-metastasis, cytotoxic, anti-inflammatory, anti-proliferative, antioxidant, antiradical, antidiabetic, immunomodulatory, and enzyme inhibitory activities (Nam et al. 2005; Manosroi et al. 2011; Khanthapoka et al. 2015; Park et al. 2016; Gao et al. 2018; Minh et al. 2019; Meselhy et al. 2020; Verma and Srivastav 2020). These activities are effective in fighting against environmental stress, microbial infections, and other diseases in humans. The phytochemical quality of various parts of different varieties of rice is presented in Table 37.1.

The phytochemical quantity in various parts of different rice varieties is presented in Table 37.2. The whole rice grain of various parts of different varieties consisted of total phenolic content (Ferulic acid equivalent FAE: 364.8–492.8 mg/100 g, gallic acid equivalent GAE: 7.40–398 ± 0.23 mg/100 g, and tannic acid equivalent TAE: 0.676 ± 0.078–0.90 ± 0.057 mg/g fresh weight), total flavonoid content (Epicatechin equivalent ECE: 9.48 ± 0.19 mg/100 g, quercetin equivalent QE: 1308 ± 0.04 mg/100 g, and TAE: 5.36 ± 0.75–6.38 ± 0.82 mg/g fresh weight), total anthocyanin content (FAE: 109.5–256.6 mg/100 g and cyanidin-3-glucoside equivalent CGE: 11.6–16.5 μg/g), condensed tannins content (TAE: 0.078 ± 0.015–0.104 ± 0:017 mg/g fresh weight), γ-oryzanol (37–55 mg/100 g), and phytic acid content (9.94–236.94–558.67 mg/100 g). The rice bran of various parts of different varieties consisted of total phenolic content (GAE: 0.51–3.97 mg/g extract 0.25–3.88 mg/g dry weight), total flavonoid content (Catechin equivalent CE: 3596–12,448 mg/100 g dw), total anthocyanin content (CGE: 1231–5101 mg/100 g dw), and γ-oryzanol (1.6–1.8 mg/kg and 3861.93–5911.12 μg/g bran). The rice bran oil contained total phenolic content (GAE: 0.8931–1.2239 mg/mg of bran) and γ-oryzanol (17.54 ± 0.75–18.49 ± 0.52 mg/g), while rice husk contained total phenolic content (GAE: 1.39 ± 0.04–3.75 ± 0.13 mg/g), total flavonoids (CE: 153.63 ± 15.92–656.23 ± 27.19 μg/g), and γ-oryzanol (46.06 ± 6.41–124.55 ± 3.36 μg/g fresh weight). The rice leaf essential oil showed 3.70–7.35% of phytoalexin momilactones A and B.

37.3 Antioxidant Potential

The majority of the phytochemicals possess antioxidant potential due to the presence of donate-able hydrogens in their chemical structure. These antioxidant compounds reduce oxidative stress, prevent oxidative damage to biomolecules, and protect the cellular structures by different mechanisms. Antioxidants perform their function by

Table 37.1 Phytochemical quality of different types and parts of rice

Source	Class of phytochemicals	Phytochemicals components	References
Rice grain (black and purple-black rice)	Anthocyanins	Cyanidin 3- O-b-D-glucoside, peonidin 3-O-glucoside, malvidin 3-O-glucoside, pelargonidin 3-O-glucoside, and delphinidin 3-O-glucoside	Ichikawa et al. (2001), Park et al. (2008), Sompong et al. (2011)
Rice hulls	Phenolic compounds	p-coumaric acid, 3-vinyl-1-oxybenzene, p-hydroxybenzaldehyde, vanillin, p-hydroxybenzoic acid, and 4,7-dihydroxy vanillic acid	Lee et al. (2003)
Rice grain (red rice)	Phenolic acids	Syringic and chlorogenic acids	Talcott et al. (2003)
Rice grains (different types)	Phenolic acids and flavones	Ferulic acid, ρ-coumaric acids, sinapic acid, protocatechuic acid, chlorogenic acid, hydroxybenzoic acid, vanillic acid, syringic acid, caffeic acid, gallic acid, tricin (flavone), the esters 6'-O-(E)-feruloyl sucrose, and 6'-O(E)-sinapoyl sucrose	Tian et al. 2004, Zhou et al. (2004), Walter and Marchesan (2011), Ndolo and Beta (2014)
	Anthocyanins	Cyanidin-3-O-β-D-glucoside and peonidin-3-O-β-D-glucoside, anthocyanidins cyanidin and malvidin, the anthocyanins pelargonidin3,5-diglucoside and cyanidin-3,5-diglucoside	Oki et al. (2002), Awoyinka et al. (2007), Yawadio and Morita (2007), Walter and Marchesan (2011)
Rice bran (defatted)	Oryzanols	Oryzanols, tools, and ferulic acid	Devi et al. (2007)
Rice flour	Carotenoids	β-Carotene and lutein	Lamberts and Delcour (2008)
Rice bran (black rice)		Anthocyanin	Park et al. (2008)
Rice bran (brown rice)	Carotenoids (pigments)	β-Carotene, lutein, and zeaxanthin	Lamberts and Delcour (2008)
Rice bran	Tocopherols and oryzanols	α-Tocopherol, α-tocotrienol, γ-tocopherol, γ-tocotrienol, tocopherol, tocotrienol, and oryzanols	Xu et al. (2001), Lai et al. (2009)
	Phenolic acids	Cycloartenyl ferulate, 24-methylene cycloartenyl ferulate, and campesteryl ferulate	
Rice bran (unpolished red rice)		c-glycosyl apigenin	Hirawan et al. (2011)

Rice beans	Phenolic compounds	p-coumaric acid, ferulic acid, vitexin, isovitexin, sinapic acid, and quercetin, catechin, epicatechin	Yao et al. (2012)
Rice grain (red and brown rice)	Phenolic acids	Ferulic, isoferulic, trans-ferulic acid, trans-trans-8-O-4′ diferulic acid, syringic, chlorogenic, chlorogenic, sinapic, p-coumaric, 4-hydroxybenzoic, syringic, and vanillic acids	Niu et al. (2013), Gong et al. (2017)
Rice grain (white, black, red, and brown rice)		Gamma-oryzanols, including 24-methylene cycloartenol, campesterol, cycloartenol, and ß-sitosterol ferulates, carotenoids in rare	Pereira-Caro et al. (2013)
Rice seed (Japanese black-purple rice)	Anthocyanins	Cyanidin-3-O-glucoside and peonidin-3-O-glucoside predominating	Pereira-Caro et al. (2013)
	Flavonol glycosides and flavones	Quercetin-3-O-glucoside and quercetin-3-O-rutinoside	
	Carotenoids	Lutein, zeaxanthin, lycopene, and b-carotene	
	c-Oryzanol	24-methylene cycloartenol ferulate, campesterol ferulate, cycloartenol ferulate, and b-sitosterol ferulate	
Rice bran (red, brown, and black rice)	Phenolic acids	Protocatechuic acid, syringic acid, ferulic acid, cinnamic acid, and p-coumaric acid	Ghasemzadeh et al. (2015)
	Flavonoids	Quercetin, apigenin, catechin, luteolin, and myricetin	
Brown rice bran and its processed products		*Trans*-ferulic acid was the most abundant monomeric phenolic acid with *trans-trans*-8-O-4′ diferulic acid being the most abundant diferulic acid	Gong et al. (2017)
Rice bran/seed		Phytic acid, gallic acid, tannic acid, ascorbic acid, p-hydroxybenzoic acid, caffeic acid, protocatechuic acid, ferulic acid, p-coumaric acid sinapic acid, syringic acid, vanillic acid, 4-hydroxybenzoic, and isoferulic acids	Manosroi et al. (2010), Vichapong et al. (2010), Niu et al. (2013), Wang et al. (2018b), Muttagi and Ravindra (2020)
Rice leaf oil		Methyl ricinoleic acid, palmitic acid	Minh et al. (2019)
Rice bran oil		γ-Oryzanol	Munarko et al. (2020)
Rice bran oil		Cholesterol, β-sitosterol, and other 4-desmethylsterols	Vissers et al. (2000)

Table 37.2 Phytochemical quantity in different types and parts of rice

Rice type	Extracting solvent	Component	Quantity	References
Rice bran	Methanol/acetone	TPC	GAE: 250–397 mg/100 g extract	Chatha et al. (2006)
Flour of different rice varieties		Carotenoids	β-Carotene and lutein (100 ng/g), zeaxanthin (30 ng/g)	Lamberts and Delcour (2008)
Japonica rice bran	Methanol/ethyl acetate/hexane	TPC	GAE: 2.5 mg/g dw	Lai et al. (2009)
		GOC and Tocols	1.6–1.8 and 126–130 mg/kg bran, respectively	
Rice bran	Ethanol/carbon dioxide fluid	TPC	GAE: 0.51 ± 0.07–0.65 ± 0.05 mg/g extract	Manosroi et al. (2010)
Rice bran (white, red, and black rice)		TPC	GAE: 0.8931–0.9884, 1.0103–1.0494, 1.0810–1.2239 mg/g	Manosroi et al. (2010)
Rice bran (Thai white, red, and black rice)	Methanol	TPC	GAE: 0.89–1.223 mg/g extract	Muntana and Prasong (2010)
Rice bran (Vasumathi, Yamini, Jyothi, and Njavara rice)	Methanol	TPC	GAE: 3.2–12.4 mg/g bran QE: 1.68–8.5 mg/g bran	Rao et al. (2010)
Rice bran (black rice)		TPC	– Free phenolics: 2086–7043 mg GAE/100 g dw – Bound phenolics: 221.2–382.7 mg GAE/100 g dw – Total phenolics: 2365–7367 mg GAE/100 g dw	Zhang et al. (2010)
		TFC	Free: Bound and total flavonoids ranged from 3462 to 12,061, 126.7 to 386.9, and 3596 to 12,448 mg CE/100 g dw, respectively	
		TAC	Free, bound, and total anthocyanins ranged from 1227 to 5096, 4.89 to 8.23, and 1231 to 5101 mg of CGE/100 g dw, respectively	
Rice husk	Aqueous ethanol	TPC	GAE: 1.39 ± 0.04–3.75 ± 0.13 mg/g sample	Kim et al. (2011a)
		TFC	CE: 153.63 ± 15.92–656.23 ± 27.19 μg/g sample	
		GOC	46.06 ± 6.41–124.55 ± 3.36 μg/g sample (wet weight)	
Fermented rice sap	Water	TPC	GAE: 0.13–8.38 mg/g sap	Manosroi et al. (2011)
		TAC	CGE: 0.04–0.10 mmol/g dw	

Rice bran (white, light brown, brown, purple, and red rice)	Polar to non-polar fractions	Tocopherols and GOC	Tocopherols: 319.67–443.73 μg/g bran γ-Oryzanols: 3861.93–5911.12 μg/g bran	Min et al. (2011)
Rice grain (red and black rice)	Acidified methanol	TAC	FAE: 0.3–1.4 mg/100 g in red rice and 109.5–256.6 mg/100 g in black rice	Sompong et al. (2011)
	Aqueous methanol	TPC	FAE: 364.8 mg/100 g in red rice and 492.8 mg/100 g	
Rice bran (red and black rice)		TPC	GAE: 0.25–50.32 mg/g black rice bran of different pericarp grains	Walter and Marchesan (2011)
Rice bran (red and black rice)	Aqueous methanol	TAC	CGE: 0.3–1.4 for red and 109.5–256.6 mg/100 g for black rice	Sompong et al. (2011)
Rice seed (colored and non-colored rice)	Methanol	TPC	GAE: 7.40 mg/100 g	Chakuton et al. (2012)
		Tannic acid	ME: 1045.12 mg/100 g	
		PAC	9.94 mg/100 g	
Rice bran layer, rice bran, and rice germ (black rice)		TPC	GAE: 7.14 ± 0.60 mg/g (bran layer), 6.65 ± 0.93 mg/g (Rice bran), and 1.35 ± 0.01 mg/g (Rice germ)	Moongngarm et al. (2012)
Rice bran layer, rice bran, and rice germ (red rice)		TPC	GAE: 4.01 ± 0.04 mg/g (bran layer), 4.39 ± 0.09 mg/g (Rice bran), and 1.18 ± 0.01 mg/g (Rice germ)	
Rice bean (13 varieties)	Aqueous ethanol	TPC	GAE: 3.27 ± 0.04–43 ± 0.25 mg/g	Yao et al. (2012)
		TFC	CE:55.95 ± 11.16–320.39 ± 31.77 mg/g	
Rice grain (22 red rice varieties)		TPC	FAE: 433–2213 μg/g	Niu et al. (2013)
		TAC	CGE: 11.6–16.5 μg/g	
Rice grain (7 cultivars of Brazilian rice)		TPC	GAE: 24.45–37.93 mg/100 g	Palombini et al. (2013)
Rice bran oil (dehusked black and white rice)	Hexane	TPC	GAE: 0.8931–1.2239 mg/mg of bran	Bopitiya and Madhujith (2014)
Rice bran oil	Hexane	GOC	17.54 ± 0.75–18.49 ± 0.52 mg/g	Pengkumsri et al. (2015)

(continued)

Table 37.2 (continued)

Rice type	Extracting solvent	Component	Quantity	References
Rice bran (Hashemi rice)	Ethanol–water	TTEC	56.23 mg/100 g dw	Ghasemzadeh et al. (2015)
		TPC	288.40 mg/100 g dw	
		TFC	156.20 mg/100 g dw	
Rice grain (giant embryo brown rice)	Ethanol	TPC	GAE: 44.88 ± 0.83 mg/100 g	Im Chung et al. (2016)
		TFC	ECE: 9.48 ± 0.19 mg/100 g	
Rice leaf essential oil	Hexane	Phytoalexin momilactones A and B	A: 9.80 ng/g fresh weight and 7.35% of essential oil B: 4.93 ng/g fresh weight and 3.70% of essential oil	Minh et al. (2019)
Rice grain (Indonesian brown rice)	Ethanol	TPC	GAE: 47–70 mg/100 g	Munarko et al. (2020)
		GOC	37–55 mg/100 g	
Rice grain (Indian rice)		TPC	GAE: 47.82–160.71 mg/100 g	Muttagi and Ravindra (2020)
		TAC	CGE: 25.74–56.88 mg/100 g	
		PAC	236.94–558.67 mg/100 g	
Rice grain (pigmented Indian rice)		TPC	GAE: 398 ± 0.23 mg/100 g	Nayeem et al. (2021)
		TFC	QE: 1308 ± 0.04 mg/100 g	
		TAC	CGE:13.36 ± 0.00 mg/100 g	
Rice grain	Water	TPC	TAE: 0.676 ± 0.078–0.90 ± 0.057 mg/g fw	Priyanthi and Sivakanesan (2021)
		TFC	TAE: 5.36 ± 0.75–6.38 ± 0.82 mg/g fw	
		MAC	CGE: 0.02 ± 0.005–0.0292 ± 0.009 mg/g fw	
		CTC	TAE: 0.078 ± 0.015–0.104 ± 0:017 mg/g fw	

CGE Cyanidin-3-glucoside equivalent, *CE* Catechin equivalent, *CTC* Condensed tannin content, *dw* dry weight, *ECE* (−)-epicatechin equivalent, *FAE* Ferulic acid equivalents, *fw* fresh weight, *GAE* Gallic acid equivalent, *GOC* γ-oryzanol content, *MAC* Monomeric anthocyanin content, *PAC* Phytic acid content, *QE* Quercetin equivalent, *TPC* Total phenolic content, *TFC* Total flavonoid content, *TAC* Total anthocyanins content, *TTEC* Total tocotrienol content, *TAC* Tannic acid reducing the reactive species and terminating the free radical chain reaction or inhibiting oxidation reactions by giving the electrons (Pietta 2000). In pharmacy and medicine, food quality measurements are based on compounds that perform the scavenging activity for oxidative stress. Antioxidants are the compounds that capture the reactive oxygen and reactive nitrogen species and terminate the free radical chain reactions initiated by these species. The antioxidants perform their action by either donating hydrogen or transferring the electron to the reactive species. These substances can reduce oxidative stress and protect the body from many diseases (Xu et al. 2001; Shad et al. 2012; Lee et al. 2013; Palombini et al. 2013). All rice portions are a good source of phytochemical compounds possessing strong antioxidant potential. Antioxidant properties of various parts of different varieties of rice are presented in Table 37.3. The whole rice grain and rice bran of different rice varieties possess significant antioxidant potential in terms of Trolox equivalent antioxidant capacity, free radical scavenging capacity (DPPH, 2, 2-diphenyl picrylhydrazyl, hydroxyl, and ABTS: 2,2-azino-bis(3-ethylbenzothiazoline-6-sulfonic acid) radical scavenging), ferrous ion chelating activity, inhibition of lipid peroxidation, ferric reducing antioxidant power, and reducing power.

37.4 Biological Activities of Rice

Based on phytochemical composition and antioxidant potential, numerous components of different varieties of rice have been found to possess diverse biological activities that have been found effective in the treatment of microbial infections, cancer, diabetes, cardiovascular abnormalities, and other diseases associated with oxidative stress. The phytochemical antioxidant compounds present in rice show antitumor, cardioprotective, anti-hypercholesterolemic, anti-hyperglycemic, anti-hyperuricemic, anti-mutagenic, anticancer, anti-metastasis, cytotoxic, anti-inflammatory, immunologic, mitogenic, antimicrobial, anti-proliferation, antioxidant, antiradical, antidiabetic, and enzyme inhibitory activities. Biological activities of different parts of diverse varieties of rice are mentioned in Table 37.4. The presence of antioxidant potential, bioactive phytochemical compounds, and other biological activities highlights the medicinal and pharmaceutical significance of different parts of rice. Based on its biological activities, various parts of pigmented rice varieties may be a good choice for the treatment of different diseases such as cardiovascular and hepatic abnormalities, cancer, diabetes, and infectious diseases.

37.5 Factors Affecting Phytochemical Profile and Antioxidant Potential of Rice

Table 37.5 presents different factors that have been described to affect phytochemical besides antioxidant quality of rice. The storage such as extrusion temperature, post-extrusion holding time, storage temperature, and storage time result in a decrease in vitamin E and oryzanol retention. The processing conditions such as

Table 37.3 Antioxidant potential of different types and parts of rice

Source	Extracting solvent	Activity	Response	References
Rice grain (processed rice)	Methanol	Reducing power, radical scavenging capacity, ferrous ion chelating activity	Monascus-fermented polished rice showed comparatively highest antioxidant activity	Yang et al. (2006)
Rice bran	Methanol/ acetone	Inhibition of lipid peroxidation, loss of β-carotene	Aqueous methanolic extract (80%) showed comparatively higher antioxidant activity than acetone extracts	Chatha et al. (2006)
Defatted rice bran	Methanol	DPPH and superoxide radical scavenging activity	The acetone-extract polar fraction of crude methanolic extract of defatted rice bran showed strong activity against DPPH and superoxide radicals	Devi and Arumughan (2007)
Defatted rice bran	Methanol	Inhibition of oxidation in a linoleic acid system	The antioxidant efficacy of acetone-polar fraction of crude methanolic extract of defatted rice bran was closer to that of TBHQ and far greater than that of BHT	Devi et al. (2007)
Rice grain (black rice)	Ethanol with 0.1% trifluoroacetic acid	Inhibition of lipid peroxidation and free radical scavenging capacity	100 μg/ml concentration of anthocyanins extracted from black rice inhibited lipid peroxidation (88.3%), DPPH radical (55.2%), superoxide anion radical 54.96%, and hydrogen peroxide 72.77%	Park et al. (2008)
Japonica rice bran	Methanol/ethyl acetate/hexane	Inhibition of lipid peroxidation and free radical scavenging capacity, ferrous ion chelating activity, and reducing power	Inhibition peroxidation in linoleic acid (57%), scavenging of DPPH radicals (93%), ferrous ion chelating activity (EDTA Eqv: ~1300 μg/g), and reducing power (78%)	Lai et al. (2009)
Defatted rice endosperm protein hydrolysate	Water	Inhibition of lipid peroxidation and free radical scavenging capacity, ferrous ion chelating activity, and reducing power	The antioxidant peptides extracted from protein hydrolysate of rice endosperm exhibited hydroxyl, and superoxide radical scavenging activity (EC_{50}: 2.0 and >6 mg/ml, respectively), inhibition of lipid peroxidation (82.09%), and ferrous ion chelating activity 89.15%	Zhang et al. (2009)

Rice bran (Thai white, red, and black rice)	Methanol	Inhibition of lipid peroxidation and free radical scavenging capacity	The extracts showed inhibition of peroxidation (10.15–38.80%) and DPPH radical scavenging capacity (IC_{50}: 0.0057–0.2582 g/ml)	Muntana and Prasong (2010)
Rice bran (Vasumathi, Yamini, Jyothi, and Njavara rice)	Methanol	Free radical scavenging capacity	The extracts showed DPPH and nitric oxide scavenging capacities (IC_{50}: 30.85–87.72 and 52.25–107.18 μg/ml, respectively)	Rao et al. (2010)
Rice husk	Aqueous ethanol	Free radical scavenging capacity and reducing power	The samples showed DPPH and ABTS radical scavenging capacities (26.81–58.46% and 2.67 mg AAE/g sample)	Kim et al. (2011a)
Fermented (rice sap)	Water	Free radical scavenging	The sap at different fermentation days showed DPPH radical scavenging activity (SC_{50}: 91.60–169.78 mg/ml)	Manosroi et al. (2011)
Rice grain (red and black rice)	Aqueous methanol	Ferric reducing antioxidant power (FRAP), free radical scavenging capacities	FRAP: 0.9–8.1 mmol Fe(II)/100 g in red rice and 3.7–7.6 mmol Fe(II)/100 g in black rice DPPH radical inhibition from 23.6 to 87% ABTS radical inhibition: TEAC = 2.1–12.6 mmol/100 g	Sompong et al. (2011)
Rice bran (brown, red, and black pericarp)		Total antioxidant activity	TE: 10–345.3 μM/g bran for different pericarp grains	Walter and Marchesan (2011)
Rice bean (13 varieties)	Aqueous ethanol	DPPH radical scavenging capacities	TE: 39.87 ± 1.37–46.40 ± 2.18 μM/g	Yao et al. (2012)
Rice grain (pigmented rice)	Aqueous ethanol	Antioxidant activity	Pigmented rice showed greater antioxidant activity in terms of free radical scavenging, reducing power, iron chelation, inhibition of lipid peroxidation, and superoxide dismutase activity	Kang et al. (2013)
Rice grain (7 cultivars of Brazilian rice)		DPPH radical scavenging capacity	TE: 794.51–1461.66 μmol/g	Palombini et al. (2013)
Rice grain (different rice types)		Trolox equivalent antioxidant capacity	TEAC: 794.51–1461.66 μmol/g	Palombini et al. (2013)

(continued)

Table 37.3 (continued)

Source	Extracting solvent	Activity	Response	References
Rice bran oil (white and red rice)		Free radical scavenging activity	DPPH radical scavenging: IC_{50} = 0.027 ± 0.007 mg/ml for white and 0.025a ± 0.001 mg/ml for red rice	Bopitiya and Madhujith (2014)
Rice bran oil (red and white rice)	Hexane	Free radical scavenging capacity	DPPH radical scavenging (IC_{50}: 0.025 ± 0.001–0.027 ± 0.007 mg/ml) ABTS radical scavenging (IC_{50}: 0.017 ± 0.001–0.018 ± 0.006 mg/ml)	Bopitiya and Madhujith (2014)
Rice bran (red, brown, and black rice)	Acidified methanol	Free radical scavenging capacity	DPPH radical scavenging: 10.7–87.9% Nitric oxide radical scavenging: 4.0–89.2%	Ghasemzadeh et al. (2015)
Rice bran oil		Trolox equivalent antioxidant capacity	TEAC: 12.42 ± 0.58–18.80 ± 0.57 mg/g bran oil extract	Pengkumsri et al. (2015)
Rice grain (gaint embryo brown rice)	Ethanol	Free radical scavenging capacity	DPPH radical: 36.10 ± 0.85% ABTS radical: AAE= 67.54 ± 1.28 μg/g Hydroxyl radical: 40.68 ± 1.76a %	Im Chung et al. (2016)
		Reducing power	OD at 700 nm: 0.17 ± 0.00	
		Ferrous ion chelating ability	64.98 ± 1.00%	
		Enzyme inhibitory activity	XOD inhibition: 50.50 ± 0.42% SOD-like activity: 16.70 ± 0.75%	
Rice bran		DPPH radical scavenging capacity	IC_{50}: 6.94 mg/ml	Lakshmi et al. (2017)
Rice leaf essential oil	Steam distillation	Free radical scavenging capacity	DPPH radical scavenging (IC_{50}: 73.1 μg/ml) ABTS radical scavenging (IC_{50}: 198.3 μg//ml)	Minh et al. (2019)
		Ferric reducing antioxidant power (FRAP) and β-carotene oxidation (BCO)	FRAP (IC_{50}: 700.8 μg/ml) BCO (79%)	

Rice grain (Indian rice varieties)		Total antioxidant capacity	AAE: 257.93–800.25 μM/100 g	Muttagi and Ravindra (2020)
Rice grain (pigmented Indian rice)		Total antioxidant activity	AAE: 316 ± 0.03 mg/100 g	Nayeem et al. (2021)
Rice grain (Sri Lankan red rice)	Water	Free radical scavenging capacity and ferric reducing antioxidant power	FRAP: 0.561 ± 0.113–0.695 ± 0.077 mmol/100 g fw DPPH radical scavenging: 26.07 ± 3.08–53.66 ± 7.61 mg/ml	Priyanthi and Sivakanesan (2021)

AAE Ascorbic acid equivalent, *ABTS* 2,2-azino-bis(3-ethylbenzothiazoline-6-sulfonic acid), *BHT* Butylated hydroxyl toluene, *DPPH* 2, 2-diphenyl picrylhydrazyl, EC_{50} Effective concentration for 50% activity, *FRAP* Ferric reducing antioxidant power, *dw* dry weight, *fw* fresh weight, IC_{50} Inhibitory concentration for 50% radical inhibition, *LPI* Lipid peroxidation inhibition, *OD* Optical density, SC_{50} Scavenging concentration for 50% radical inhibition, *SOD* Superoxide dismutase, *TBHQ* Tertiary butylated hydroxyl quinolone, *TE* Trolox equivalent, *TEAC* Trolox equivalent antioxidant capacity, *XOD* Xanthine oxidase

Table 37.4 Biological activities of different types and parts of rice

Source	Extracts/ Compounds	Activity	References
Rice bran	Saccharides	Antitumor activity (rice bran saccharides reduced the immunocompetence in the course of carcinogenesis, suppressed carcinogenesis, and prolonged the survival of rats with gastrointestinal cancer)	Takeshita et al. (1992)
Rice bran oil	Sterols and terpenes alcohols	Anti-hypercholesterolemic activity (reduced serum total cholesterol by 0.19 mmol/l)	Vissers et al. (2000)
Rice bran	Tocopherols and γ-oryzanol	γ-Oryzanol showed a strong antioxidant potential than tocopherols in a cholesterol oxidation system.	Xu et al. (2001)
Rice bran oil		Improved catalase activity and reduction of lipid peroxidation in rat	Rana et al. (2004)
Rice bran (blackish-purple and brown rice)	Ethanol–water	– Antioxidant activity in terms of xanthine oxidase inhibition, ferrous ions chelation, potassium ferricyanide reduction, intracellular peroxides, superoxide anion, and hydroxyl radical scavenging activities – Anti-mutagenic activity in terms of inhibition of 4-nitroquinoline N-oxide-induced mutagenesis – Antitumor activity in terms of inhibition of phorbol ester-induced tumor promotion	Nam et al. (2005)
Rice bran	Ethanol–water	Antimicrobial, anti-inflammatory anti-mutagenic	Nam et al. (2005)
Rice bran	Arabinoxylan	Anti-oncogenic activity, modulation of lipid peroxidation, augmentation of the antioxidant defense system	Noaman et al. (2008)
Rice bran	Ethanol/carbon dioxide fluid	– Antioxidant activity in terms of DPPH radical scavenging capacity, ferrous ions chelation, and inhibition of lipid peroxidation – Tyrosinase inhibitory activity	Manosroi et al. (2010)
Rice bran (different rice varieties)	Methanol	Antioxidant potential in terms of total antioxidant activity, DPPH and nitric oxide radical scavenging capacity, and reducing power Cell cytotoxic properties	Rao et al. (2010)
Rice husk	Aqueous ethanol	Antioxidant, mitogenic, and anti-proliferative activities	Kim et al. (2011a)
Germinated brown rice grain	Aqueous ethanol	Anti-proliferative activity against breast cancer cells (MCF-7) and immunological activity (macrophage and mitogenic activities)	Kim et al. (2011b)

(continued)

Table 37.4 (continued)

Source	Extracts/ Compounds	Activity	References
Fermented rice sap	Water	Antioxidant, cytotoxic, and tyrosinase, and MMP-2 inhibitory activities	Manosroi et al. (2011)
Brown red and black bran		Cardioprotective, antidiabetic, inhibitory effect on lens opacity of rat, reduction in total cholesterol concentration	Walter and Marchesan (2011)
Rice beans	Ethanol	Antiradical and antidiabetic activity (alpha-glucosidase inhibition and glycation end product inhibition)	Yao et al. (2012)
Rice grain (22 red rice varieties)	Aqueous alcohol	Antioxidant (DPPH and ABTS radical scavenging capacity and ferric reducing ability power) and anti-inflammatory activities (suppression of LPS stimulated IL-1β, IL-6, and COX-2 mRNA expressions in mouse macrophage cells)	Niu et al. (2013)
Rice bran	Protein hydrolysate	Enzyme inhibition activity	Khanthapoka et al. (2015)
Rice bran	Gramisterol	Antitumor and immune-enhancing activity	Somintara et al. (2016)
Rice hull	Enzyme/water	Antioxidant, anti-inflammatory, and antiadipogenic activities	Park et al. (2016)
Rice grain (Nigerian local rice)	Methanol	Antioxidant and anti-cholinesterase activity	Salawu et al. (2016)
Rice bran (Gramisterol)	Hexane and dichloromethane	– Anticancer activity against white blood cell tumors by the production of anticancer immune-related cytokines – Immune-enhancing activity by increasing the amount of immune function-related cells, interferons, and cytokines	Somintara et al. (2016)
Glutinous and non-glutinous brown rice	Bound fraction	Antioxidant, cellular antioxidant, and anti-proliferative activities	Gao et al. (2018)
Rice leaf oil	Steam distillation	Growth inhibition and anti-hyperuricemic activity	Minh et al. (2019)
Rice grain		Antioxidant, anti-inflammatory, anticancer, and antidiabetic activities	Verma and Srivastav (2020)
Rice bran	Alcoholic	Cytotoxic, anticancer, antioxidant, anti-metastasis activity	Meselhy et al. (2020)
Rice grain (red rice)		Anti-hyperglycemic: Inhibitory effect against α-amylase and α-glucosidase	Krishnan et al. (2021)

Table 37.5 Factors affecting the phytochemical composition and biological activities of rice

Source	Factor	Effect on biological activity	References
Rice bran	Storage conditions (extrusion temperature, post-extrusion holding time, microwave heating, and gamma irradiation)	Reduced retention of vitamin E and oryzanol concentration during storage	Shin (1995)
Rice bran	Gamma irradiation	An increase in free fatty acids and a decreased E vitamers and oryzanol in rice bran	Shin and Godber (1996)
Rice leaf	Salinity stress	The stress decreased superoxide dismutase activity and increased the peroxidase activity, lipid peroxidation, electrolyte leakage higher Na+ accumulation in the leaves	Dionisio-Sese and Tobita (1998)
Rice bran	Commercial milling	Heavy milling increased the concentration of tocopherol, tocotrienol in outer bran layers, and that of oryzanol in long-grain rice bran	Lloyd et al. (2000)
Rice hulls	Far-infrared irradiation	Increase in free radical scavenging activity from 47.74 to 79.63%	Lee et al. (2003)
Rice grain (dehulled white and red rice)	Milling and cooking	Milling resulted in the loss of antioxidant capacity while cooking as risotto retained the activity	Finocchiaro et al. (2007)
Rice grain (germinated rough rice)	Germination	Germination increased the total phenolic content and antioxidant activity of rice	Lee et al. (2007)
Rice grain (brown rice)	Parboiling and milling	Parboiling and milling reduced the carotenoid pigments in rice grain	Lamberts and Delcour (2008)
Rice grain (10 varieties of Pakistani rice)	Extracting solvent	A decrease in polarity of extracting solvent increased the extract yield, phytochemical content, and antioxidant activity of rice	Zubair et al. (2012)
Rice grain (Brewer's rice)	Stabilization techniques (microwave heating, gamma irradiation, and acid treatment)	Significant reduction in γ-oryzanol and α-tocopherol contents of rice	Nordin et al. (2014)
Rice bran and rice husk	Far-infrared radiation	Significant increase in total phenolic content, α- and γ-tocopherols, DPPH radical scavenging activities, and ferric reducing antioxidant power of rice bran	Wanyo et al. (2014)
	Enzymatic treatments (cellulose)	Significant increase in vanillic acid and γ-oryzanol but a decrease in ferulic acid of rice husk	

(continued)

Table 37.5 (continued)

Source	Factor	Effect on biological activity	References
Rice flour (Nigerian rice varieties)	Germination	An increase in physicochemical and antioxidant properties with a decrease in phytic acid content	Chinma et al. (2015)
Rice bran (Hashemi rice)	Extraction techniques (Ultrasonication with ethanol–water solvent)	Increase in total phenolics, total flavonoids, total tocotrienols, and antioxidant activity	Ghasemzadeh et al. (2015)
Rice bran oil	Extraction techniques: Solvent extraction (hexane), hot-pressed, cold-pressed, and supercritical fluid extraction	Solvent extraction increased the phytochemical contents and antioxidant properties	Pengkumsri et al. (2015)
Rice grain (black rice)	Thermal cooking	Decrease in total anthocyanin, cyanidin-3-glucoside contents, and antioxidant activity but no effect on anti-inflammatory activities of rice	Bhawamai et al. (2016)
Red yeast rice	Mixed fermentation with *Monascus purpureus*	Increase in the antioxidant activity of red yeast rice	Huang et al. (2017)
Rice flour (purple rice)	Blending/mixing of purple rice flour with wheat flour	Improved nutritional and antioxidant properties	Klunklin and Savage (2018)
Rice flour (black rice)	Gamma radiation, storage temperature, and storage time	– Gamma irradiation increase in antioxidant activity but decrease in anthocyanin content – Storage temperature and time decreased the antioxidant activity and anthocyanin content	Ito et al. (2019)
Rice bran oil	Refining process (neutralization)	Losses of phytochemicals and antioxidant activities	Liu et al. (2019)
Rice grain (black rice)	Thermal treatments (roasting, frying, and cooking)	Thermal decomposition of anthocyanin and cyanidin-3-glucoside, increased levels of protocatechuic acid, inhibition of α-glucosidase, and α-amylase, and reduction in glycation and glycemic index	Aalim et al. (2021)

commercial milling increase tocopherol, tocotrienol, and oryzanol but decrease the antioxidant activity and anthocyanin content of rice. The parboiling and milling decrease the carotenoid pigments while roasting, frying, and cooking result in the thermal decomposition of anthocyanin and cyanidin-3-glucoside and increase the levels of protocatechuic acid, inhibition of α-glucosidase, and α-amylase, and reduction in glycation and glycemic index of rice. Microwave heating and gamma irradiation cause a reduction in γ-oryzanol, α-tocopherol, and oryzanol in rice.

However, gamma irradiation increases the free fatty acids and antioxidant activity but decreases the anthocyanin content in rice. The infrared irradiation increases the phenolic content and antioxidant activity of rice. The extraction techniques and extraction solvents also increase the phytochemical contents and antioxidant properties of rice as a function of solvent polarity. The salinity stress decreases the antioxidant potential, while germination and fermentation increase the total phenolic content and antioxidant activity of rice.

37.6 Conclusion

All rice plant parts are good sources of bioactive phytochemical compounds that have the antioxidant potential and several other biological activities. The presence of bioactive phytochemical compounds, antioxidant potential, and other biological activities highlights the medicinal and pharmaceutical significance of different parts of rice. Based on the diverse biological activities, the grain and bran of pigmented rice varieties could be a good choice for the treatment of various diseases such as cardiovascular and hepatic abnormalities, cancer, diabetes, and infectious diseases.

References

Aalim H, Wang D, Luo Z (2021) Black rice (Oryza sativa L.) processing: evaluation of physicochemical properties, in vitro starch digestibility, and phenolic functions linked to type 2 diabetes. Food Res Int 141:109898

Ahmad S, Hasanuzzaman M (2012) Integrated effect of plant density, N rates and irrigation regimes on the biomass production, N content, PAR use efficiencies and water productivity of rice under irrigated semiarid environment. Not Bot Horti Agrobot Cluj-Napoca 40(1):201–211

Ahmad S, Zia-ul-Haq M, Ali H, Shad SA, Ammad A, Maqsood M, Khan MB, Mehmood S, Hussain A (2008) Water and radiation use efficiencies of transplanted rice (*Oryza sativa* L.) at different plant densities and irrigation regimes under semi-arid environment. Pak J Bot 40(1): 199–209

Ahmad S, Ahmad A, Zia-ul-Haq M, Ali H, Khaliq T, Anjum MA, Khan MA, Hussain A, Hoogenboom G (2009) Resources use efficiency of field grown transplanted rice (*Oryza sativa* L.) under irrigated semiarid environment. J Food Agric Environ 7(2):487–492

Ahmad S, Ahmad A, Soler CMT, Ali H, Zia-Ul-Haq M, Anothai J, Hussain A, Hoogenboom G, Hasanuzzaman M (2012) Application of the CSM-CERES-Rice model for evaluation of plant density and nitrogen management of fine transplanted rice for an irrigated semiarid environment. Precis Agric 13(2):200–218

Ahmad S, Ahmad A, Ali H, Hussain A, Garcia y Garcia A, Khan MA, Zia-Ul-Haq M, Hasanuzzaman M, Hoogenboom G (2013) Application of the CSM-CERES-Rice model for evaluation of plant density and irrigation management of transplanted rice for an irrigated semiarid environment. Irrig Sci 31(3):491–506

Ahmad A, Ashfaq M, Rasul G, Wajid SA, Khaliq T, Rasul F, Saeed U, Rahman MH, Hussain J, Baig IA, Naqvi AA, Bokhari SAA, Ahmad S, Naseem W, Hoogenboom G, Valdivia RO (2015) Impact of climate change on the rice–wheat cropping system of Pakistan. In: Hillel D, Rosenzweig C (eds) Handbook of climate change and agro-ecosystems: the agricultural modeling Intercomparison and improvement project (AgMIP) integrated crop and economic assessments. Imperial College Press and the American Society of Agronomy, pp 219–258

Ahmad S, Abbas G, Ahmed M, Fatima Z, Anjum MA, Rasul G, Khan MA, Hoogenboom G (2019) Climate warming and management impact on the change of rice-wheat phenology in Punjab, Pakistan. Field Crops Res 230:46–61

Ahmed M, Ahmad S (2019) Carbon dioxide enrichment and crop productivity. In: Hasanuzzaman M (ed) Agronomic crops, vol 2. Springer, Singapore, pp 31–46

Ahmed M, Ahmad S (2020) Systems modeling. In: Ahmed M (ed) Systems Modeling. Springer, Singapore, pp 1–44

Ahmed M, Fayyaz-ul-Hassan, Ahmad S (2017) Climate variability impact on rice production: adaptation and mitigation strategies. In: Ahmed M, Stockle C (eds) Quantification of climate variability, adaptation and mitigation for agricultural sustainability. Springer, Cham, pp 91–111

Ahmed M, Ahmad S, Raza MA, Kumar U, Ansar M, Shah GA, Parsons D, Hoogenboom G, Palosuo T, Seidel S (2020a) Models calibration and evaluation. In: Ahmed M (ed) Systems Modeling. Springer, Singapore, pp 151–178

Ahmed M, Ahmad S, Waldrip HM, Ramin M, Raza MA (2020b) Whole farm modeling: a systems approach to understanding and managing livestock for greenhouse gas mitigation, economic viability and environmental quality. In: Waldrip HM, Pagliari PH, He Z (eds) Animal manure. ASA special publication 67. American Society of Agronomy and Soil Science Society of America, Madison, WI, pp 345–371

Akram R, Turan V, Hammad HM, Ahmad S, Hussain S, Hasnain A, Maqbool MM, Rehmani MIA, Rasool A, Masood N, Mahmood F, Mubeen M, Sultana SR, Fahad S, Amanet K, Saleem M, Abbas Y, Akhtar HM, Hussain S, Waseem F, Murtaza R, Amin A, Zahoor SA, Sami ul Din M, Nasim W (2018) Fate of organic and inorganic pollutants in paddy soils. In: Hashmi MZ, Varma A (eds) Environmental pollution of Paddy soils. Springer, Cham, pp 197–214

Akram R, Fahad S, Masood N, Rasool A, Ijaz M, Ihsan MZ, Maqbool MM, Ahmad S, Hussain S, Ahmed M, Kaleem S, Sultana SR, Mubeen M, Saud S, Kamran M, Nasim W (2019) Plant growth and morphological changes in rice under abiotic stress. In: Hasanuzzaman M, Fujita M, Nahar K, Biswas JK (eds) Advances in Rice Research for Abiotic Stress Tolerance. Woodhead Publishing, pp 69–85

Awoyinka OA, Balogun IO, Ogunnowo AA (2007) Phytochemical screening and in vitro bioactivity of Cnidoscolus aconitifolius (Euphorbiaceae). Journal of Medicinal Plants Research 1:63–65

Bhawamai S, Lin S-H, Hou Y-Y, Chen Y-H (2016) Thermal cooking changes the profile of phenolic compounds, but does not attenuate the anti-inflammatory activities of black rice. Food Nutr Res 60:32941

Bopitiya D, Madhujith T (2014) Antioxidant potential of rice bran oil prepared from red and white rice. Tropical Agricultural Research 26:1–11

Chakuton K, Puangpronpitag D, Nakornriab M (2012) Phytochemical content and antioxidant activity of colored and non-colored Thai rice cultivars. Asian J Plant Sci 11:285–293

Chatha SAS, Anwar F, Manzoor M, ur-Rehman Bajwa J (2006) Evaluation of the antioxidant activity of rice bran extracts using different antioxidant assays. Grasas Aceites 57:328–335

Chinma CE, Anuonye JC, Simon OC, Ohiare RO, Danbaba N (2015) Effect of germination on the physicochemical and antioxidant characteristics of rice flour from three rice varieties from Nigeria. Food Chem 185:454–458

Devi RR, Arumughan C (2007) Antiradical efficacy of phytochemical extracts from defatted rice bran. Food Chem Toxicol 45:2014–2021

Devi RR, Jayalekshmy A, Arumughan C (2007) Antioxidant efficacy of phytochemical extracts from defatted rice bran in the bulk oil system. Food Chem 104:658–664

Dionisio-Sese ML, Tobita S (1998) Antioxidant responses of rice seedlings to salinity stress. Plant Sci 135:1–9

Fahad S, Noor M, Adnan M, Khan MA, Rahman I, Alam M, Khan IA, Ullah H, Mian IA, Hassan S, Saud S, Bakhat HF, Hammad HM, Ahmad S, Nasim W (2019) Abiotic stress and rice grain quality. In: Hasanuzzaman M, Fujita M, Nahar K, Biswas JK (eds) Advances in Rice Research for Abiotic Stress Tolerance. Woodhead Publishing, pp 571–583

Fatima Z, Ahmad M, Hussain M, Abbas G, Ul-Allah S, Ahmad S, Ahmed N, Ali MA, Sarwar G, ul Haque E, Iqbal P, Hussain S (2020) The fingerprints of climate warming on cereal crops phenology and adaptation options. Sci Rep 10:18013

Finocchiaro F, Ferrari B, Gianinetti A, Dall'asta C, Galaverna G, Scazzina F, Pellegrini N (2007) Characterization of antioxidant compounds of red and white rice and changes in total antioxidant capacity during processing. Mol Nutr Food Res 51:1006–1019

Gao Y, Guo X, Liu Y, Zhang M, Zhang R, Abbasi AM, You L, Li T, Liu RH (2018) Comparative assessment of phytochemical profile, antioxidant capacity and anti-proliferative activity in different varieties of brown rice (Oryza sativa L.). LWT 96:19–25

Ghasemzadeh A, Jaafar HZ, Juraimi AS, Tayebi-Meigooni A (2015) Comparative evaluation of different extraction techniques and solvents for the assay of phytochemicals and antioxidant activity of hashemi rice bran. Molecules 20:10822–10838

Gong ES, Luo S, Li T, Liu C, Zhang C, Chen J, Zeng Z, Liu RH (2017) Phytochemical profiles and antioxidant activity of processed brown rice products. Food Chem 232:67–78

Hafeez-ur-Rehman, Nawaz A, Awan MI, Ijaz M, Hussain M, Ahmad S, Farooq M (2019) Direct seeding in rice: problems and prospects. In: Hasanuzzaman M (ed) Agronomic crops, volume 1: production technologies. Springer, Singapore, pp 199–222

Hernández I, Alegre L, Van Breusegem F, Munné-Bosch S (2009) How relevant are flavonoids as antioxidants in plants? Trends Plant Sci 14:125–132

Hirawan R, Diehl-Jones W, Beta T (2011) Comparative evaluation of the antioxidant potential of infant cereals produced from purple wheat and red rice grains and LC-MS analysis of their anthocyanins. J Agric Food Chem 59:12330–12341

Huang YJ, Kuo MC, Dunn P et al (2017) 75 - Spliceosome mutations in chronic Myelomonocytic leukemia: prognostic impact and their co-existed mutation profile. Leuk Res 55(Supplement 1): S49. https://doi.org/10.1016/S0145-2126(17)30188-1

Hussain M, Ahmad S, Hussain S, Lal R, Ul-Allah S, Nawaz A (2018) Rice in saline soils: physiology, biochemistry, genetics and management. Adv Agron 148:231–287

Ichikawa H, Ichiyanagi T, Xu B, Yoshii Y, Nakajima M, Konoshi T (2001) Antioxidant activity of anthocyanin extract from purple black rice. J Med Food 4:211–218

Im Chung S, Lo LMP, Nam SJ, Jin X, Kang MY (2016) Antioxidant capacity of giant embryo rice Seonong 17 and Keunnunjami. J Adv Agric Technol 3:94–98

Ito VC, Zielinski AAF, Demiate IM, Spoto M, Nogueira LG (2019) Effects of gamma radiation on the stability and degradation kinetics of phenolic compounds and antioxidant activity during storage of (Oryza sativa L.) black rice flour. Braz Arch Biol Technol 62:e19180470

Kang MY, Rico CW, Bae HJ, Lee SC (2013) Antioxidant capacity of newly developed pigmented rice cultivars in Korea. Cereal Chem 90:497–501

Khan MA, Ahmad S, Raza A (2019a) Integrated weed management for agronomic crops. In: Hasanuzzaman M (ed) Agronomic crops, vol 2. Springer, Singapore, pp 257–281

Khan SM, Ali S, Nawaz A, Bukhari SAH, Ejaz S, Ahmad S (2019b) Integrated pest and disease management for better agronomic crop production. In: Hasanuzzaman M (ed) Agronomic crops, vol 2. Springer, Singapore, pp 385–428

Khanthapoka P, Muangpromb A, Sukronga S (2015) Antioxidant activity and DNA protective properties of rice grass juices. Sci Asia 41:119–129

Kim D-J, Oh S-K, Chun A, Yoon M-R, Hong H-C, Choi I-S, Lee J-S, Yu K-W, Kin Y-K (2011a) Evaluation of biological activities of rice husk extracts. J Food Sci Nutr 16:179–183

Kim D-J, Oh S-K, Yoon M-R, Chun A, Choi I-S, Lee D-H, Lee J-S, Yu K-W, Kim Y-K (2011b) The change in biological activities of brown rice and germinated brown rice. J Korean Soc Food Sci Nutr 40:781–789

Klunklin W, Savage G (2018) Physicochemical, antioxidant properties and in vitro digestibility of wheat–purple rice flour mixtures. Int J Food Sci Technol 53:1962–1971

Krishnan V, Rani R, Awana M, Pitale D, Kulshreshta A, Sharma S, Bollinedie H, Archana SA, Singh B, Singh AK, Praveen S (2021) Role of nutraceutical starch and proanthocyanidins of pigmented rice in regulating hyperglycemia: enzyme inhibition, enhanced glucose uptake and hepatic glucose homeostasis using in vitro model. Food Chem 335:127505

Lai P, Li KY, Lu S, Chen HH (2009) Phytochemicals and antioxidant properties of solvent extracts from japonica rice bran. Food Chem 117:538–544

Lakshmi NB, Jessie SW, Maheswari KU, Kumari BA, Prabhakar BN (2017) Antioxidant potential of rice bran and vegetable waste powders incorporated extrudates. Pharm Innov J 6:12–16

Lamberts L, Delcour JA (2008) Carotenoids in raw and parboiled brown and milled rice. J Agric Food Chem 56:11914–11919

Lee N, Hui D, Wu A, Chan P, Cameron P, Joynt GM, Ahuja A, Yung MY, Leung CB, To KF, Lui SF, Szeto CC, Chung S, Sung JJY (2003) A major outbreak of severe acute respiratory syndrome in Hong Kong. N Engl J Med 348:1986–1994

Lee Y-R, Woo K-S, Kim K-J, Son J-R (2007) Antioxidant activities of ethanol extracts from germinated specialty rough rice. Food Sci Biotechnol 16:765–770

Lee C, Lee SM, Namgung R (2013) Bilirubin metabolism and bilirubin encephalopathy. Neonatal Med 20:268–275

Liu R, Liu R, Shi L, Zhang Z, Zhang T, Lu M, Chang M, Jin Q, Wang X (2019) Effect of refining process on physicochemical parameters, chemical compositions and in vitro antioxidant activities of rice bran oil. LWT 109:26–32

Lloyd BJ, Siebenmorgen TJ, Beers KW (2000) Effects of commercial processing on antioxidants in rice bran. Cereal Chem 77:551–555

Lu S, Luh BS (1991) Properties of the rice caryopsis. In: Rice. Springer, pp 389–419

Manosroi A, Ruksiriwanich W, Abe M, Sakai H, Manosroi W, Manosroi J (2010) Biological activities of the rice bran extract and physical characteristics of its entrapment in niosomes by supercritical carbon dioxide fluid. J Supercrit Fluids 54:137–144

Manosroi A, Ruksiriwanich W, Kietthanakorn B, Manosroi W, Manosroi J (2011) Relationship between biological activities and bioactive compounds in the fermented rice sap. Food Res Int 44:2757–2765

Meselhy KM, Shams MM, Sherif NH, El-Sonbaty SM (2020) Phytochemical study, potential cytotoxic and antioxidant activities of selected food byproducts (pomegranate peel, Rice bran, Rice straw & mulberry bark). Nat Prod Res 34:530–533

Min B, McClung AM, Chen M-H (2011) Phytochemicals and antioxidant capacities in rice brans of different color. J Food Sci 76:C117–C126

Minh TN, Xuan TD, Van TM, Andriana Y, Viet TD, Khanh TD, Tran H-D (2019) Phytochemical analysis and potential biological activities of essential oil from rice leaf. Molecules 24:546

Moongngarm A, Daomukda N, Khumpika S (2012) Chemical compositions, phytochemicals, and antioxidant capacity of rice bran, rice bran layer, and rice germ. Apcbee Procedia 2:73–79

Munarko H, Sitanggang AB, Kusnandar F, Budijanto S (2020) Phytochemical, fatty acid and proximal composition of six selected Indonesian brown rice varieties. CyTA-J Food 18: 336–343

Muntana N, Prasong S (2010) Study on total phenolic contents and their antioxidant activities of Thai white, red and black rice bran extracts. Pak J Biol Sci 13:170–174

Muttagi GC, Ravindra U (2020) Phytochemical and antioxidant capacity of traditional rice varieties of Karnataka, India. Int J Curr Microbiol App Sci 9:67–75

Nam SH, Choi SP, Kang MY, Kozukue N, Friedman M (2005) Antioxidative, antimutagenic, and anticarcinogenic activities of rice bran extracts in chemical and cell assays. J Agric Food Chemi 53:816–822

Nayeem S, Sundararajan S, Ashok A, Abusaliya A, Ramalingam S (2021) Effects of cooking on phytochemical and antioxidant properties of pigmented and non-pigmented rare Indian rice landraces. Biocatal Agric Biotechnol:101928

Naz S, Fatima Z, Iqbal P, Khan A, Zakir I, Noreen S, Younis H, Abbas G, Ahmad S (2019) Agronomic crops: types and uses. In: Hasanuzzaman M (ed) Agronomic crops. Volume 1: production technologies. Springer, Singapore, pp 1–18

Ndolo VU, Beta T (2014) Comparative studies on composition and distribution of phenolic acids in cereal grain botanical fractions. Cereal Chem 91:522–530

Niu Y, Gao B, Slavin M, Zhang X, Yang F, Bao J, Shi H, Xie Z, Yu L (2013) Phytochemical compositions, and antioxidant and anti-inflammatory properties of twenty-two red rice samples grown in Zhejiang. LWT-Food Sci Technol 54:521–527

Noaman E, El-Din NKB, Bibars MA, Mossallam AAA, Ghoneum M (2008) Antioxidant potential by arabinoxylan rice bran, MGN-3/biobran, represents a mechanism for its oncostatic effect against murine solid Ehrlich carcinoma. Cancer Lett 268:348–359

Nordin N, Karim R, Ghazali HM, Adzahan NM, Sultan MT (2014) Effects of various stabilization techniques on the nutritional quality and antioxidant potential of brewer's rice. J Eng Sci Technol 9:347–363

Oki T, Masuda M, Kobayashi M, Nishiba Y, Furuta S, Suda I, Sato T (2002) Polymeric procyanidins as radical-scavenging components in red-hulled rice. J Agric Food Chem 50: 7524–7529

Palombini SV, Maruyama SA, Claus T, Carbonera F, de Souza NE, Visentainer JV, Gomes STM, Matsushita M (2013) Evaluation of antioxidant potential of Brazilian rice cultivars. Food Sci Technol 33:699–704

Park Y-S, Kim S-J, Chang H-I (2008) Isolation of anthocyanin from black rice (Heugjinjubyeo) and screening of its antioxidant activities. Microbiol Biotechnol Lett 36:55–60

Park C-Y, Kim S, Lee D, Park DJ, Imm J-Y (2016) Enzyme and high pressure assisted extraction of tricin from rice hull and biological activities of rice hull extract. Food Sci Biotechnol 25:159–164

Peerzada AM, Bukhari SAH, Dawood M, Nawaz A, Ahmad S, Adkins S (2019) Weed management for healthy crop production. In: Hasanuzzaman M (ed) Agronomic crops, vol 2. Springer, Singapore, pp 225–256

Pengkumsri N, Chaiyasut C, Sivamaruthi BS, Saenjum C, Peerajan S, Suwannalert P, Sirisattha S, Chaiyasut K, Kesika P (2015) The influence of extraction methods on composition and antioxidant properties of rice bran oil. Food Sci Technol 35:493–501

Pereira-Caro G, Watanabe S, Crozier A, Fujimura T, Yokota T, Ashihara H (2013) Phytochemical profile of a Japanese black–purple rice. Food Chem 141:2821–2827

Pietta P-G (2000) Flavonoids as antioxidants. J Nat Prod 63:1035–1042

Priyanthi C, Sivakanesan R (2021) The total antioxidant capacity and the total phenolic content of rice using water as a solvent. Int J Food Sci 2021:5268584

Rana P, Vadhera S, Soni G (2004) In vivo antioxidant potential of rice bran oil (RBO) in albino rats. Indian J Physiol Pharmacol 48:428–436

Rao AS, Reddy SG, Babu PP, Reddy AR (2010) The antioxidant and antiproliferative activities of methanolic extracts from Njavara rice bran. BMC Complement Altern Med 10:4

Razaq M, Shah FM, Ahmad S, Afzal M (2019) Pest management for agronomic crops. In: Hasanuzzaman M (ed) Agronomic crops, vol 2. Springer, Singapore, pp 365–384

Salawu SO, Alao OF, Faloye OF, Akindahunsi AA, Boligon AA, Athayde ML (2016) Antioxidant potential of phenolic-rich two varieties of Nigerian local rice and their anti-cholinesterase activities after in vitro digestion. Nutr Food Sci 46:171–189

Sarwar N, Ali H, Ahmad S, EhsanUllah AS, Mubeen K, Hill JE (2013a) Water wise rice cultivation on calcareous soil with the addition of essential micronutrients. J Anim Plant Sci 23(1):244–250

Sarwar N, Ali H, Maqsood M, EhsanUllah SAN, Shahzad M, Mubeen K, Shahid MA, Ahmad S (2013b) Phenological response of rice plants to different micronutrients application under water saving paddy fields on calcareous soil. Turk J Field Crops 18(1):52–57

Saunders RM (1985) Rice bran: composition and potential food uses. Food Rev Int 1:465–495

Scavariello EM, Arellano DB (1998) Gamma-oryzanol: an important component in rice brain oil. Arch Latinoam Nutr 48:7–12

Shad MA, Zafar ZI, Nawaz H, Anwar F (2012) Effect of sorbic acid and some other food preservatives on human serum cholinesterase activity. Afr J Biotechnol 11:11280–11286

Shahzad AN, Ahmad S (2019) Tools and techniques for nitrogen management in cereals. In: Hasanuzzaman M (ed) Agronomic crops, vol 2. Springer, Singapore, pp 111–126

Shin T (1995) *Changes of antioxidants and fatty acid composition in stabilized rice bran during storage*. PhD Dissertation University of King Sejong
Shin T-S, Godber JS (1996) Changes of endogenous antioxidants and fatty acid composition in irradiated rice bran during storage. J Agric Food Chem 44:567–573
Somintara S, Leardkamolkarn V, Suttiarporn P, Mahatheeranont S (2016) Anti-tumor and immune enhancing activities of rice bran gramisterol on acute myelogenous leukemia. PLoS One 11: e0146869
Sompong R, Siebenhandl-Ehn S, Linsberger-Martin G, Berghofer E (2011) Physicochemical and antioxidative properties of red and black rice varieties from Thailand, China and Sri Lanka. Food Chem 124:132–140
Takeshita M, Nakamura S, Makita F, Ohwada S, Miyamoto Y, Morishita Y (1992) Antitumor effect of RBS (rice bran saccharide) on ENNG-induced carcinogenesis. Biotherapy 4:139–145
Talcott ST, Percival SS, Pittet-Moore J, Celoria C (2003) Phytochemical composition and antioxidant stability of fortified yellow passion fruit (Passiflora edulis). J Agric Food Chem 51:935–941
Tariq M, Ahmed M, Iqbal P, Fatima Z, Ahmad S (2020) Crop phenotyping. In: Ahmed M (ed) Systems Modeling. Springer, Singapore, pp 45–60
Tian S, Nakamura K, Kayahara H (2004) Analysis of phenolic compounds in white rice, brown rice, and germinated brown rice. J Agric Food Chem 52:4808–4813
Verma DK, Srivastav PP (2020) Bioactive compounds of rice (Oryza sativa L.): review on paradigm and its potential benefit in human health. Trends Food Sci Technol 97:355–365
Vichapong J, Sookserm M, Srijesdaruk V, Swatsitang P, Srijaranai S (2010) High performance liquid chromatographic analysis of phenolic compounds and their antioxidant activities in rice varieties. LWT-Food Sci Technol 43:1325–1330
Vissers MN, Zock PL, Meijer GW, Katan MB (2000) Effect of plant sterols from rice bran oil and triterpene alcohols from sheanut oil on serum lipoprotein concentrations in humans. Am J Clin Nutr 72:1510–1515
Walter M, Marchesan E (2011) Phenolic compounds and antioxidant activity of rice. Braz Arch Biol Technol 54:371–377
Wang L, Shuai T, Wang Y-Y, Cao H (2018a) The effect of traditional Chinese medicine washing combined with massage for neonatal jaundice: a meta-analysis. TMR Integrative Nursing 1:36–44
Wang W, Li Y, Dang P, Zhao S, Lai D, Zhou L (2018b) Rice secondary metabolites: structures, roles, biosynthesis, and metabolic regulation. Molecules 23:3098
Wanyo P, Meeso N, Siriamornpun S (2014) Effects of different treatments on the antioxidant properties and phenolic compounds of rice bran and rice husk. Food Chem 157:457–463
Wasaya A, Yasir TA, Ijaz M, Ahmad S (2019) Tillage effects on agronomic crop production. In: Hasanuzzaman M (ed) Agronomic crops, vol 2. Springer, Singapore, pp 73–99
Xu Z, Hua N, Godber JS (2001) Antioxidant activity of tocopherols, tocotrienols, and γ-oryzanol components from rice bran against cholesterol oxidation accelerated by 2, 2 '-azobis (2-methylpropionamidine) dihydrochloride. J Agric Food Chem 49:2077–2081
Yang J-H, Tseng Y-H, Lee Y-L, Mau J-L (2006) Antioxidant properties of methanolic extracts from monascal rice. LWT-Food Sci Technol 39:740–747
Yao Y, Cheng X-Z, Wang L-X, Wang S-H, Ren G (2012) Major phenolic compounds, antioxidant capacity and antidiabetic potential of rice bean (Vigna umbellata L.) in China. Int J Mol Sci 13: 2707–2716
Yawadio R, Morita N (2007) Color enhancing effect of carboxylic acids on anthocyanins. Food Chem 105:421–427
Zahoor SA, Ahmad S, Ahmad A, Wajid A, Khaliq T, Mubeen M, Hussain S, Sami Ul Din M, Amin A, Awais M, Nasim W (2019) Improving water use efficiency in agronomic crop production. In: Hasanuzzaman M (ed) Agronomic crops, vol 2. Springer, Singapore, pp 13–29

Zhang J, Zhang H, Wang L, Guo X, Wang X, Yao H (2009) Antioxidant activities of the rice endosperm protein hydrolysate: identification of the active peptide. Eur Food Res Technol 229: 709–719

Zhang MW, Zhang RF, Zhang FX, Liu RH (2010) Phenolic profiles and antioxidant activity of black rice bran of different commercially available varieties. J Agric Food Chem 58:7580–7587

Zhou Z, Robards K, Helliwell S, Blanchard C (2004) The distribution of phenolic acids in rice. Food Chemi 87:401–406

Zubair M, Anwar F, Shahid SA (2012) Effect of extraction solvents on phenolics and antioxidant activity of selected varieties of Pakistani rice (Oryza sativa). Int J Agric Biol 14:935–940

Rice-Based Products

38

Ehsan Ul Haque, Sohaib Afzaal, Akbar Hayat, Mian Anjum Murtaza, Ahmad Din, Shinawar Waseem Ali, and Shakeel Ahmad

Abstract

Rice is a cereal grain, staple food, feeding half of the world population. Rice is low fibre and rich caloric food providing one fifth of the calories consumed by the world's human population. Several value-added rice products and by-products can be made in which stickiness of rice usually works to hold its shape when prepared as sweet dishes, steamed products, cooked preparations, savoury dishes, rice pudding and various products in various regions of the world through instant rice and parboiled rice.

Keywords

Rice · Grain · Human · Products

Ehsan Ul Haque · S. Afzaal · A. Hayat
Citrus Research Institute, Sargodha, Pakistan

M. A. Murtaza
Institute of Food Science & Nutrition, University of Sargodha, Sargodha, Pakistan

A. Din
National Institute of Food Science and Technology, University of Agriculture, Faisalabad, Pakistan

S. W. Ali
Department of Food Sciences, University of Punjab, Lahore, Pakistan

S. Ahmad (✉)
Department of Agronomy, Faculty of Agricultural Sciences and Technology, Bahauddin Zakariya University, Multan, Pakistan
e-mail: shakeelahmad@bzu.edu.pk

N. Sarwar et al. (eds.), *Modern Techniques of Rice Crop Production*,
https://doi.org/10.1007/978-981-16-4955-4_38

38.1 Introduction

Rice is consumed as a staple diet by greater than 50% of world's population as variety of processed and semi-processed products (Ahmad et al. 2008, 2009, 2012, 2013, 2015, 2019; Ahmad and Hasanuzzaman 2012; Ahmed and Ahmad 2019, 2020; Shahzad and Ahmad 2019; Fatima et al. 2020). Maximum of the milled rice in various countries including Bangladesh, Philippines, Nigeria etc., is consumed as cooked whole grain. As a food, it is basically energetic in nature due to its rich caloric value and low fibre content, as its most important components are carbohydrates (Fairhurst and Dobermann 2002). A lot of rice-based value-added products are being consumed and liked worldwide. Value addition in rice products is basically the objective to improve the profitability of rice industry as well as to diversify its taste by developing wide range of products through various food engineering technologies. A lot of processing techniques to develop unique rice-based products including canning, flaking, flavouring, puffing, fortification and supplementation are being employed to rice industry. Broken rice is also being subjected to developing various sweet products and extraction and utilization of rice starch. Rice products enriched with various vitamins and minerals are available in market and are getting popular nowadays. Value-added products of organic rice along with medicinal rice cultivars having therapeutic value have good role in national and foreign markets (Juliano and Hicks 1996).

By-products developed from the rice milling have been reported to have very high nutrients compared to white rice itself. Rice by-products may include rice straw, rice hulls, broken rice (traditionally known as Tota), rice bran and rice bran oil and various waxes utilized in food industry (Perretti et al. 2003). These by-products are normally applied as their original form, but nowadays researches are continued to utilize them for value addition and extraction of their bioactive compounds to impart various functional properties. Rice products development and rice processing mainly depend upon the cultivar and variety of rice, quality and texture of rice grains, types of the end-product and method of processing (Mestres et al. 2011; Roy et al. 2011).

Rice was normally consumed as main coarse meal after simple boiling or cooking with some salt and spices, but since last three decades various rice snacks including flaked products, puffs, popped or expanded products prepared from fermented wet-ground rice paste or dry-ground rice paste followed by deep frying, have got huge popularity. Canned rice-based products are also available in markets including soups with rice, meat with rice, boiled rice, flavoured rice, rice puddings and fried rice.

Various factors affect the cooking or processing time for rice products especially protein content of rice. Protein content in rice products affects the processing or cooking time of rice products, as high proteinaceous rice variety take longer time for cooking due to presence of physical blockade against H_2O absorption formed by protein matrix and rice starch. Lesser proteinaceous rice products impart sweet taste and are more tender and cohesive.

Rice parboiling is a common practice to improve the rice grain texture and parboiled rice is subjected to canning due to stability and firmness of kernel and maintenance of its shape without disintegration. Additional aspects which might affect canned rice quality comprise fat contents, pH and salt concentration along with blanching time (Sharp et al. 1985). Alkaline solution affects rice resulting in development of yellow colour.

38.2 Processing Techniques

Type of processing method is a crucial factor for developing rice products. A variety of processing techniques and methods has been employed in rice industry. The following processing/cooking methods for rice products are commercially beneficial and quick-cooking processes (Luh 1991):

38.2.1 Soak Boil Steam Dry Method

This method is based upon softening of rice grains by soaking in water up to 30% moisture absorption prior processing. Rice grains are boiled in water till 50–60% moisture with or without steam is achieved. After slight gap, product is boiled more to enhance till a moisture content of 60–70% is achieved, followed by careful drying to 8–14% moisture content, which is needed for maintaining porous structure.

38.2.2 Expanded and Pre-Gelatinization Method

This method is based on increasing the rice expansion ration and slight gelatinization prior processing. Briefly, rice grains are soaked for 40–50 min and processed by boiling/steaming or pressure cooking for gelatinization of grains. Finally, grains are dried at lower temperatures to produce dense glassy structured seeds followed by expansion or puffing at high temperature to produce final product having desirable porous texture.

38.2.3 Rolling or Bumping Method

This method is quite similar to expanded and pre-gelatinization method, as gelatinization is done accordingly. After gelatinization, rolling or bumping is done to flatten the rice grains followed by drying to produce a relatively hard, glassy textured product.

38.2.4 Dry Heat Treatment Method

Dry heat treatment is applied in a hot air oven at temperature of 65–82 °C for 10–30 min or at 272 °C for 16–18 s, depending upon final product. This caused the dextrinization and fissuring or expansion of rice grains. Steaming or boiling is not applied in this method and product is cooked in comparatively lesser time.

38.2.5 The Freeze-Thaw-Drying Process

This is an expensive method for rice processing because high efficiency equipment and process are required to impart specific thawed taste to the final product. Pre-cooked rice is instantly frozen using blast freezing method, followed by thawing and drying.

38.2.6 Gun Puffing

Gun puffing is basically grouping of three methods, that is, pre-conditioning of rice till 20–22% moisture is achieved, followed by steaming (165 °C) in a pressure cooker for 5–10 min followed by puffing. Final product normally contains up to 25% of moisture content.

38.3 Rice Products

Rice products can be divided into various groups including:

1. Regular rice meals in Indo-Pak subcontinent
2. Rice flakes
3. Rice flour products
4. Liquefied rice-based products
5. Puffed rice/paddy
6. Fortified products
7. Infant and baby products
8. Fermented products
9. Rice-based products at Pakistan
10. Rice by-products

38.3.1 Regular Rice Meals in Indo-Pak Subcontinent

38.3.1.1 Plain Cooked (Boiled)

Cooked rice refers to rice that has been cooked either by steaming or boiling (Fig. 38.1). Plain cooked rice grains are referred to boiled rice grains in boiling water for 7–10 min; after soaking in water for about 45–60 min; without addition of any salt, spices, or condiments. This is common for any variants of Asian rice varieties including Indica and Japonica. This method is specific for white milled

Fig. 38.1 Plain cooked (boiled) rice

Fig. 38.2 Saffron rice

rice grains of any length. Normally, this product is consumed along with some lentils or fish in India and Pakistan.

38.3.1.2 Saffron

Saffron rice dish is commonly used in the cuisines of South Asia, China, Burma, Vietnam and Thailand, etc. White milled rice grains are cooked along with saffron and simmered vegetables and meats including bones (Fig. 38.2). Processing method is almost similar to plain cooked rice except addition of ingredients.

38.3.1.3 Tehri

Tehri may be called as potato rice, widely liked and consumed in North India and Pakistan. These are plain cooked rice with potatoes and some minor spices and salt (Fig. 38.3). Various other vegetables including peas, carrots, garlic etc., may also be added and it is dum-baked in sealed clay-pot after cooking topped with garnish layer.

Fig. 38.3 Tehri rice

Fig. 38.4 Zarda (cooked with edible *yellow* colour)

It is consumed with Raita (minsed herbs in semi-liquid yogurt), mango and lemon pickles and *Papad.* While in Bangladesh some meat is also added to Tehri to develop taste and aroma.

38.3.1.4 Zarda (Cooked with Edible Yellow Colour)

Zarda refers to traditionally boiled sweet rice dish in South Asia, native to India and Pakistan. It is prepared by addition of various ingredients including food colour (yellow), milk, sugar, cardamoms, raisins, saffron, pistachios, almonds etc. Zarda word is derived from Urdu word Zard which mean yellow colour, and due its yellow colour, the name Zarda is given to this (Fig. 38.4).

38.3.1.5 Biryani

Biryani is native to Persia, as word Biryani was derived from Persian word Biran meaning 'fried before cooking' is a meat-based rice dish, which was initially prepared by adding rice in to meat chops and then cooking on flame. But nowadays, lot of styles of Biryani are available in market. This is the most liked dish in Asia and Middle East. Normally, basmati rice variety is considered fit for preparation of

Fig. 38.5 Biryani rice

Fig. 38.6 Pulao rice

Biryani. Some vegetables may also be added in meat and rice with spices and salt (Fig. 38.5).

Biryani is prepared by half frying mutton in ghee and half frying of rice in ghee separately. Then, both half-cooked dishes are combined and layered in a clay-based cookware commonly known as *Handi*. Various types of Biryani include Lucknow Biryani, Hyderabad Biryani, Sindhi Biryani, Malabar Biryani, Bombay Biryani, Iran Biryani and Myanmar Biryani and are being liked and consumed by whole world, even in European countries.

38.3.1.6 Pulao

Pulao, *Pilaf or Pilav* is native to middle-east Turkish populated areas (today's Turkey, Greek, Kosovo etc.,) and Persia. Almost similar words are available in different languages like *Pulaka* in Sanskrit, *Palaw or Pulav* in Persian having same meaning of the rice-based dish with some vegetables, meat and nuts browned in cooking oil and with herbal spices (Fig. 38.6). Basic difference in *Biryani* and *Pulao* is layering, for example, layering is not done in Pulao while Biryani is layered with meat and rice. Subject to local cuisine, it might contain meat and vegetables diversity. Pulao along with dishes having mixture of meat and rice are common in cuisines of Central, South Asian, Middle Eastern, Latin American and Caribbean. While in Indian subcontinent, *Pulao* generally refers to vegetable rice.

38.3.1.7 Zeera Rice

Zeera is Urdu word for cumin seeds and in Hindi Zeera is pronounced as Jeera. Zeera rice (native to India and Pakistan) is easy to prepare rice dish cooked along-with cumin seeds and it is popular.

38.3.1.8 Khichri

Khichuṛī is a common dish in homes along with cuisines in South Asia, comprising rice and lentils, however, further variants comprise millet grains and or mung dal in rice (Fig. 38.7). In India and Pakistan, it is considered as first solid foods for babies and for patients during disease or medical treatment and after recovery.

38.3.2 Rice Flakes

Rice flakes are prepared by flaking of rice grains after parboiling (Fig. 38.8). Primary rice flakes are used as brewery adjuncts in beer industries. Chiwra or Poha is a traditionally flaked rice product in India and Pakistan. The unit operations involved in preparation of Chiwra include cleaning, soaking, roasting, shelling, polishing, flaking, sieving and drying.

38.3.3 Rice Flour Products

38.3.3.1 Rice Noodles

Rice noodles, second main rice product being consumed in the world, normally prepared from Indica rice varieties, are common in Asia and consumed in the form of soups (or) snack foods (Fig. 38.9). They are also called bifun (bihon) or vermicelli in

Fig. 38.7 Khichri rice

Fig. 38.8 Rice flakes

Fig. 38.9 Bifun or vermicelli rice noodles

Taiwan, China and east southern Asia and harusame in Japan. They are usually prepared from high amylose rice flour. Sometimes rice noodles are fermented with *Lactobacilli* and *Streptococci* strains, which imparts specific acidic taste by lowering their pH. Rice dough is kneaded and passed through extruders and noodles are steam cooked followed by oven drying (Li et al. 2020).

Taking advantage of the established technology for rice noodle production from Asia, low-grade broken rice fractions from local mills have been converted into noodles and sensory evaluation data indicated satisfactory consumer preference. Traditionally, rice noodles are made from long grain rice flour with high amylose content >25 g/100 g (Fu 2008; Juliano and Sakurai 1985), because amylose content plays a critical role in the formation of gel networks and setting of the noodle structure (Mestres et al. 1988).

38.3.3.2 Instant Rice Noodles

Instant rice noodles are named because these are ready to serve within 3–8 min. It's a commercial product widely consumed in Asia (Fig. 38.10). Instant rice noodles are manufactured in two step extruders, rest of process is same as traditional noodles. Critical factor in instant rice noodles is their diameter that should not be more than 0.68 mm. Because if diameter exceeds from 0.70 mm, then it will not be easy to prepare instantly and bursting ratio will be increased (Hatcher et al. 2008).

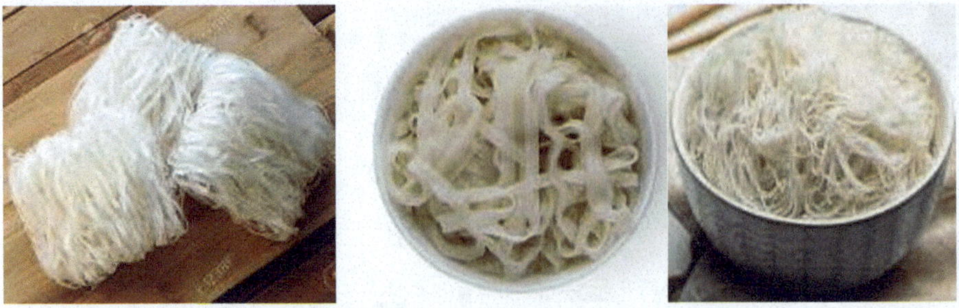

Fig. 38.10 Instant rice noodles

Fig. 38.11 Mitaimu or bilabial snack food noodles popular in Taiwan

38.3.3.3 Snack Food Noodles (Mitaimu)

Snack foods noodles or commonly known as Mitaimu are basically short, wet-type noodles having large diameter. Mitaimu are prepared from Indica varieties only, having high amylose contents. Some other starches are also added to adjust and maintain its texture including sweet potato to increase its elasticity, corn and tapioca to decrease its hardness. Mitaimu are also known as Bilabial in Taiwan (Fig. 38.11). Mitaimu is normally consumed in summer with syrup and ice, sometimes with meat and green onions in Taiwan. Monosodium glutamate, pepper and salt are fundamental ingredients for Mitaimu. Manufacturing process is quite similar to conventional rice noodles except a few unit operations, that is, during mixing of flour, hot water is used for starch gelatinization. These are generally shorter than 10 cm. Main issues are rehydration and drying of Mitaimu, which is still much difficult, so no dried Mitaimu product is commercially available.

38.3.3.4 Sheeted or Flat Noodles

Sheeted or flat noodles are prepared preferably by Indica rice with high amylose contents using wet, milled rice flour and widely liked in Asia especially in Thailand, Japan and Taiwan. Briefly, a milk layer is formed using drum and steamed followed by gelatinizing and drying of the sheets and cutting into noodle strips (Fig. 38.12). Some advanced methods also use mechanical extruders for cooking and formation of sheets. Some Japanese non-waxy rice (amylose content 20%) has been successfully

Fig. 38.12 Sheeted or flat noodles popular in Asia

Fig. 38.13 Egg roll rice rappers popular in Vietnam

used to prepare sheeted or flat noodles. Corn or tapioca starch can be used to adjust the texture of the final products. Only thin flat rice noodles (thickness <15 mm) are dried and packed as commercial products. The wet-type product is used in most restaurants.

38.3.3.5 Egg Roll Rice Rappers

Egg roll rice wrappers commonly used in Vietnam, contain eggs and rice flour. Wrappers are prepared from a mixture of eggs, rice flour, water and salt (Fig. 38.13).

38.3.3.6 Tteok Mandu Guk

Tteok Mandu Guk or rice cake with dumplings is most likely dish by Koreans. Sometime they use Kimchi along with rice cakes. Small pieces of rice cakes are consumed with soups (Fig. 38.14).

Fig. 38.14 Tteok Mandu Guk—rice cake with soup popular in Korea

Table 38.1 A comparison of various oils with rice bran oil

Oil type	Vitamin E tocopherol (ppm)	Vitamin E tocotrienol (ppm)	Oryzanol (ppm)	Total natural antioxidants (ppm)
Rice bran	81	336	2000	2417
Olive	51	0	0	51
Canola	650	0	0	650
Peanut	487	0	0	487
Soybean	1000	0	0	1000
Grape seed	256	149	0	405

ppm parts per million
Source: Available from; www.californiariceoil.com/nutrition.htm

38.3.4 Liquified Rice-Based Products

38.3.4.1 Rice Bran Oil

Rice bran oil (RBO) has an attractive nut-like flavour. Its higher smoke point (254 °C) is considered as appropriate for cooking techniques at higher temperature like stir or deep frying. Many exceptional properties mark it interesting for markets (Juliano 1985; Zigoneanu et al. 2008). Rogers et al. (1993) mentioned its significant features of higher nutraceutical value like gamma-oryzanol along with tocotrienols and notable reduction for cholesterol absorption. Ausman et al. (2005) and Wilson et al. (2000) found that its capacity to lower cholesterol is much better than canola, corn, peanut and coconut oils. Chou et al. (2009) mentioned that its medicinal behaviour is due to oryzanol that decreases plasma cholesterol and absorption of cholesterol, thus reducing earlier atherosclerosis. Moreover, its superior qualities than other oils are good shelf life, oxidative and thermal stability, higher antioxidants besides nutraceuticals, best composition of fatty acids and lesser absorption from processed foods (Table 38.1).

38.3.4.2 Rice Malt Beverages

Rice beverages are produced either wholly or in part from malt, either fermented or unfermented, that contains <1% of alcohol by volume. Various rice-based beverages include Andong soju, Awamori, Apo, Genmaicha, Horchata, Huangjiu, Jūrokucha, Kokkoh, Mijiu, Rice baijiu, Sikhye, Soju etc. Many types of alcohol, shakes and cocktail are also based on rice (Dung 2013).

38.3.4.3 Rice Vinegar

Rice vinegar is prepared from fermented rice in East Asian countries like Japan, China and Korea along with Vietnam in Southeast Asia. Rice vinegar is used as sweetener for fries. Rice vinegar prepared in different countries differ from each other in terms of strength, colour range, pH and sweetness (Shibayama et al. 2020).

38.3.4.4 Rice Milk

A liquid derived from wet milling of rice, referred to as 'rice milk' has been used as a substitute for cow's milk in various part of the world (Fig. 38.15). It is often served as breakfast cereals and weaning foods in Southeast Asia, especially in China and Taiwan (Russell and Delahunty 2004). The allergic response that many consumers have towards soy beverages, the bean-like flavour, the after taste of soy products and lactose intolerance (condition resulting from inability to fully utilize lactose) have created demand for rice milk as an alternative to cow's milk (Mitchell et al. 1988; www.milk.org). Because of its milk-like texture and functionality, rice milk has been used as substitute for milk and milk forms in many preparations of various food products including beverages and non-dairy puddings.

Rice milk can be prepared by soaking white milled rice in distilled water for 2 h in a ratio of 1:3 at room temperature. Grind with distilled water using and attrition mill until a smooth slurry is produced. The slurry is then filtered and sweetener added

Fig. 38.15 Rice milk

before homogenized for 5 min. Each batch is then heated at 70–80 °C separately on a stove before packaging.

38.3.4.5 Rice Syrups

Rice syrup is a sweetener rich in sugars that is derived through steeping cooked rice starch by saccharifying enzymes of starch breakdown and followed by straining off liquid till desired consistency (Fig. 38.16). Enzymes for saccharification are attained from sprouted barley seeds or through addition of fungal-derived purified enzymes isolates (Ofoedu et al. 2020; Zeng et al. 2008).

38.3.5 Puffed Rice

Puffed rice is very popular in Indo-Pak Subcontinent as a low cost ready-to-eat breakfast cereal as well as snack because of its crispness and lightness (Fig. 38.17). It is also a favourite food product made in different forms like puffed rice balls, bars and fatty pastes, chocolate (or) boiled sugar confectioneries in many countries. Puffing of rice grains results from sudden expansion of water vapour (steam) in interstices of starch granules during high-temperature-short-time heating of grains. Hoke et al. (2005) discovered that particle is fixed in its expanded state by dehydration resulting from rapid diffusion of water vapour out of it. Puffed product should

Fig. 38.16 Rice syrups

Fig. 38.17 Puffed rice

Fig. 38.18 Sand puffing

be maintained with around 3% moisture in order to achieve desired crispness. Various puffing methods for rice are given below:

38.3.5.1 Sand Puffing

This method is traditionally followed in India where parboiled rice grains having initial moisture content of 11–12% are mixed with 2% salt and limited water so that moisture content of rice increases up to 16–29%. Moistened rice grains are heated (tempering) for 30 min–8 h for moisture equilibration within the grains. The treated rice grains are dried in the bright sun or by conduction heating in a hemispherical metallic container with continuous stirring until grains get dried (Fig. 38.18). Small quantity of the rice is roasted at an optimum sand temperature for 6–20 s to give puffed rice with expansion ratio of 8–10 (Swarnakar et al. 2019).

38.3.5.2 Roller Puffing

In roller puffing, dough with moisture content of 8–18% is fed into the rolls with temperature of 190–440 °C. Puffed products are obtained at 6–7% moisture content. The rolls are heated by radiant heat or by circulation of high temperature fluid media inside the cylinder.

38.3.5.3 Oil Puffing

Pre-heated rice (parboiled) is puffed in vegetable oil at 200–220 °C to give expansion ratio of 5–7.

38.3.5.4 Air Puffing

Blast of air at 200–300 °C for 7–10 s is given to pre-treated rice at 10–12% moisture content which gives puffed rice with expansion ratio of 8–10.

38.3.5.5 Gun Puffing

In this process, raw milled rice and other grains can be puffed and do not need pre-treatment (parboiled) before puffing which is an essential step for puffing with other technique. Pre-moistened pearled or unpearled grains are fed into a pressure vessel, which is continuously rotated and externally heated. As the optimum pressure varies from grain to grain, the sudden release of chamber pressure causes the superheated water to flash into steam resulting in porous structure of the puffed products. In some cases, preheated grains at 272–337 °C (521–638 °F) are fed to the pressure vessel in which pressure is built by superheated steam at a temperature of 241.6 °C at the gun. After a short cooking time, the gun is suddenly opened to produce puffed rice. Initial moisture content of the grains and attaining optimum pressure within minimum time are critically important for gun puffing.

38.3.6 Fortified Rice Products

Enrichment commonly deals with refurbishment of vitamins along with minerals which disappears in processing. Fortification commonly deals in addition of vitamin and minerals in foods in greater quantities than pre-processing limits. Added nutrients must be stable while processing and storage, however, a significant quantity is being added to compensate the losses while processing and storage.

Rice can be fortified by adding a micronutrient powder that adheres to grains or spraying surface of ordinary rice grains in many layers with vitamins along with mineral mix forming a protective coat. Rice kernels may also be fortified with various micronutrients, that is, Fe, folic acid, B-complex vitamins, vitamin A and zinc. But still biofortification is considered as the best way for micronutrient fortification in rice (Muthayya et al. 2012).

38.3.7 Rice-Based Infant and Baby Foods

A baby has exclusive nutritional requirement, rice is the first food being introduced in the baby's diet plan as it is being digested easily, hypoallergenic and has lesser fat. Rice as flour or granulated rice is utilized from making baby foods. It blends well by addition of ingredients like proteins and vegetables.

Medium along with short seed cultivars having lesser amylose contents are preferably used to avoid retrogradation during storage, which results in formation of a very rigid gel and separation of water. Largest usage of rice in baby food industry is in manufacture of pre-cooked infant rice cereals.

38.3.7.1 Pre-Cooked Rice Cereal

For preparation of pre-cooked foods especially targeted for infants and toddlers, various cereals can be used but rice flour is the first preferred grain for this purpose based on various characteristics. Flour from rice is smooth textured, non-allergic, easy to digest and well balanced. Baby rice cereal often called rice porridge can be prepared at home but in market, it is offered readymade where cereals are processed after precooking and dehydration. The pre-cooked rice cereal is enriched for nutritional losses during processing sometimes, and therefore additionally fortified with some vital minerals and vitamins. These balanced nutrition preparations have stable shelf life, easy to store and transport. Sometimes antioxidant agents are incorporated in moisture proof packing to enhance storage stability of the product. But special permission for such practices from the regulatory bodies like Food and Drug Administration (FDA) may be necessary because baby foods require special attention for safety reasons.

For commercial preparation of instant pre-cooked rice-based cereal, rice powder is pre-cooked in water to produce slurry which is aimed to digest starch from crystalline form to unstructured form. The whole process of starch digestion is conducted in controlled conditions for temperature and time in presence of amylase enzyme which aids the process of starch digestibility. Basic process remains similar but the recipe and manufacturing protocol vary among producing facilities operating in the production of pre-cooked rice cereals. Generally, formulations consist of rice flour, milk powder, sugars, salt, flavours, minerals and vitamins, rice oil, emulsifier and antioxidant agent. Some producers use puree or pulp of the choice fruits like banana, apple, pear or strawberries to enhance nutritional value and flavour according to the market demand. All the given ingredients are dehydrated through drum drier after cooking with digested slurry. The dried product is collected in the shape of flakes which is packed in moisture free packaging to extend shelf life. Pre-cooked rice cereals are instant and quick dissolving creamy preparations when mixed with water or milk. These products are widely popular and used as a first choice solid food for infants and toddlers ranging from 6 months to 3 years.

38.3.7.2 Extrusion Cooked Baby Foods

Extrusion is a cooking process for snack foods adopted to get product in some regular shape and size. Process involves forcing of formulation mix through a barrel and screw against a die fitted with cutters to attain the required shape and size of extruded products. The product mix according to the formulation of the producer usually consisting of rice powder mixed with fruit purees, sugars and fats is added to the pre-conditioner where steam is injected to start cooking. All ingredients should be properly ground to suitable particles size to get good consistency of the mix with proper moisture contents. This mix is added to the extruder consisting of a barrel fitted with screw used to push the product mix against a dye to shape the output according to the desired structure and cut in proper size through blades adjacent to the dye. Output efficiency is increased after the introduction of the twin screw extruders in the market. Key factors defining process efficiency and product quality include water content, temperature and residence time of the mix. Snack production process is revolutionized by the advent of extruder technology especially used for high starch products. Various extruded products are available in the market, Kasetsart University; Bangkok, Thailand has developed an extruded baby food product containing rice flour 72% mixed with 13% soy flour.

38.3.7.3 Formulated Rice-Based Baby Foods

Formulated baby foods often called baby formula used to feed infants solely, as a supplement to nursing or babies require some special diets due to digestive system illnesses. All grocery and pharmacy stores have usually a specified desk for baby foods or infant formula which indicates the market expansion of this special food category. According to data of Nestle investor seminar in 2019, baby formula or infant nutrition is expanding and by 2023, it is expected to touch 92 billion USD. Researchers have designed the formulation of baby formula to provide the full nutritional requirement of the babies under guidelines encoded by the regulatory bodies like Codex Alimentarius, FAO and WHO. Almost 30 important nutritional components are listed for baby formula including carbohydrates, fats, proteins, vitamins and minerals. The basic ingredients of the baby formula remain same but a variety of formulations is marketed under different brands usually in powder form which is easily dissolvable in water. Traditionally, rice cereal products are utilized in formulated baby food preparations which also help to maintain the consistency. Rice cereal is the important component of the balanced food diet for babies, which is less allergic and easily digestible being gluten free. Rice-based foods being rich in carbohydrates can provide required energy to keep babies active enough. Rice cereals could provide babies with enough calcium and magnesium to build healthy bones. Several body functions require essential micronutrients like copper, zinc and selenium which are also provided through rice-based formulations.

38.3.8 Fermented Rice Products

38.3.8.1 Fermented Rice

A paste like sweet and sour tasting fermented food product made from cooked rice often consumed directly or as a part of various East and Southeast Asian recipes is called Tapai or fermented rice. Cook the soaked rice and then cool to the body temperature to assist the fermentation process. Incubate the mixture overnight after mixing the started culture to ensure mixing well in each layer. Sugar water mixture can be added to quickly start the fermentation process. Fermented rice can be used with salt and curd preferred as breakfast snack. The probiotic effect of healthy bacteria in Tapai helps in bowel movement acting as laxative. Fermentation increases bioavailability of several nutrients of rice and helps to reduce fatigue as it is being packed with vitamin B12. The magnesium and selenium in this fermented product help to strengthen bone build up and potassium works to regulate blood pressure.

38.3.8.2 Idli, Dhokla and Dosa

Idli is a popular rice-based fermented snack food made with combination of black lentil (Urad dal) with rice. Mixed dough made from rice and black lentil is steamed and acid leavened resulting in small fermented cake. Product is seasoned with curd, spices and ginger paste give a pleasing taste and flavour. Fermentation breaks down the starch converting it ready for digestion in the body. This healthy snack food of Indian subcontinent has zero fats and quickly absorbable proteins with enhanced vitamin B availability.

Dhokla is also popular culinary food in Indian subcontinent especially in Gujarat state and its areas, usually used as breakfast snack. Formulation, preparation and product shape are very similar to idli as stated above but it has additionally flour of chickpea (gram powder called Besan) in formulation. Dhokla is probiotic, fermented, high protein, rice-based product suitable for weight conscious consumers. Dhokla should not be mixed with khaman which is made from chickpeas flour and appears yellow in colour.

Dosa is also a similar crispy fermented product like idli made from combination of soaked rice, black lentil and spices like fenugreek. This pancake-type savoury food is also popular in Indian subcontinent especially in south Indian region. Mixture is blended with water using a processor and shaped into thin layer, after completion of fermentation process, the batter is cooked using hot pan to a crispy and delicious final product. Dosas can be taken solely, dipped in curries or filled with spicy mix of potato or other vegetables.

38.3.8.3 Ambali or Ragi

Ambali or millet ragi is a fermented probiotic food prepared by combination of rice and millet flour. Finger millet flour termed as ragi is mixed with water and fermented which is mixed with cooked rice, later on mixed with buttermilk or yoghurt to serve. Ragi flour should be mixed well with salt then fermentation process require at least 8–10 h. For formulation, raw rice is usually selected but any kind of rice can opted which is ground to granule size before cooking. This savoury mix is made spicy by adding chilli, ginger and other spices according to choice. If we want a sweet taste then milk and brown sugar is added while stirring into cooked rice. Through acid fermentation some other bread-like products are prepared by utilizing dough made from rice flour. Kichudok and Puto respectively in Korea and Philippine are two products in which cooked rice cake is fermented to produce bread-like products similar to Indian idli. These products are rice-based cereals having no legumes or chickpea powder.

38.3.9 Rice-Based Snacks in Pakistan

38.3.9.1 Kheer (Sweet Rice with Boiled Milk)

Kheer or sweet pudding is a well-known traditional rice dish made by boiling milk, rice and sugar. This rice dessert is locally known with various regional names like kheer, kheeri, phirni, payasam and kannada. Cardamom, coconut, saffron and various dry fruits, nuts like almonds, pistachios, cashews and raisins can be used to enhance flavour during serving. For sweetness sugar, brown sugar or jaggery (Gur) can be used. However, in some regions of the subcontinent same dish is prepared by replacing rice with wheat, millet, sweet corn or some other cereal according to the taste buds.

38.3.9.2 Rice Balls

Rice balls are stuffed quick snacks made through boiled plain or mixed rice popular in Japan. Rice ball is a popular food available in Japanese restaurants often ball, triangular or cylindrical in shape (Fig. 38.19). It is known by various local names like nigirimeshi, omusubi or onigiri which is a traditional sandwich type food often filled with chicken karaage, shrimp tempura, tarako, kombu, pickled plum, salted salmon or other salty or sour filling which has a preservative effect on product, making it palatable and portable. There was no refrigeration system when formulation of onigiri was started so these salty and sour fillings were introduced for introduction of preservative effect to keep fresh rice. This product should not be confused with sushi which a fish is preserving method made of rice with vinegar.

A similar shaped product is popular in various regions of Punjab called rice ball (Penni). This is different from above-stated product as it is a dessert food, very

Fig. 38.19 Rice balls (Onigiri—a salty and sour stuffed snack popular in Japan)

Fig. 38.20 Rice ball (Penni—a sweet dessert food popular in Pakistan; Photo by Muhammad Junaid)

Fig. 38.21 Marunda (traditional sweet cracker popular in Pakistan)

energetic made from rice flour, jiggery or sugar mixed with various nuts and oil or ghee, and rolled into balls (Fig. 38.20) and served as a snack food usually in winter.

38.3.9.3 Marunda (Puffed Rice)

Marunda is a traditional sweet, snack food made by using puffed rice coated with jaggery, popular in Punjab both in India and Pakistan. Sometimes jaggery is replaced with sugar to have a bright colour product which is garnished with almond or coconut slices. During Marunda production, rice grains are puffed and mixed with hot syrup made from sugar or jaggery (Fig. 38.21). Marunda also known as Kheel marunda or Marmura chikki is shaped like a cake or in ball shape.

38.3.9.4 Rice Bread and Biscuits

Rice flour is difficult to bake due to absence of gluten, that's why rice flour is mixed with wheat flour to prepare traditional bread or roti or biscuits (Fig. 38.22). Rice biscuits are also baked in the same way, imparting highly nutritious characteristics.

Fig. 38.22 Rice biscuits

Fig. 38.23 Rice flour and bread (Chapati—popular rice bread in Pakistan; Photo by Aysha Shakeel and Afifa Shakeel)

Traditional bread (chapatti) is made from rice flour in various rice growing regions of Pakistan. Although it is tricky to make bread from rice flour as it is gluten free having no elasticity to roll and spread the loaf but domestic women make it possible by mixing with some cereal flour of their choice, flour mixed with water to make dough which is spread manually into a thin rounded sheet or layer called chapatti, roti or bread (Fig. 38.23). This bread is served to eat after making spicy within dough, used to dip in curry or served with butter and traditional Saag made from different vegetables.

38.3.10 Rice by-Product

Main by-products attained from rice crop may include rice straw, rice hull or husk and rice bran. The by-products of rice are broken rice, rice starch, liquified glucose, high fructose syrup and wild rice stem. Nutritional value, cook time and sensory characteristics are improved through rice milling. By-products of rice had better nutritional value and various functional properties in comparison to the polished rice product. Rice husk or hull had major amount of dietary fibre consisting of lignin, cellulose and hemicelluloses. In bakery products, dietary fibre from different sources, is added to enhance nutritional value and their rheological properties for which rice bran or rice hull is opted as a preferred source. Moreover, rice bran is utilized to extract oil called as bran oil and remaining flakes can be utilized in various traditional food preparations.

38.3.10.1 Rice Starch

Rice starch is the main component of rice prepared from broken rice called endosperm of rice grain consisting of approximately 90% of milling dry weight. Starch is mainly used in laundry works and as thickening agent in food preparations, sauces, desserts, syrups and baby formulas. Starch is the main component of rice grain which is chemically a polymeric carbohydrate consisting of amylose and amylopectin. Starch is naturally a white powder varying in composition (Amagliani et al. 2016).

Rice starch can be used as fat replacer in various servings as it bears fine granularity alike fat globules. Starch converts into a creamy, smooth white-coloured gel when heated with water. Gel produced from rice starch has no flavour and taste which assist various food preparations helping in formulations to retain natural colour and taste of the food mix. These characteristics of the rice starch like low in allergens, gluten-free, easily digestible, creamy texture, white colour and neutral in taste made its wider acceptability for various industrial uses (Bao and Bergman 2018; Eliasson 2004).

Various uses of rice starch could be enlisted below:

- Baby formulas
- Organic rice-based meals
- Ready-to-eat rice-based preparations
- Thickening agent in syrups, soups and sausages
- Rice-based cereals and cereal bars
- Layer as confectionery coatings
- Stiffening agent in Laundry
- Pudding or custard starch
- Cold starching of fabrics
- Dusting powder in cosmetics

38.3.10.2 Liquefied Glucose

Liquid glucose, glucose syrup or rice syrup is a colourless or mild yellow in colour, odourless sweetener syrup obtained through partial hydrolysis of starch carried out by enzymatic process. This thick, nutritive syrup contains water, dextrose maltose used in various food preparations. Liquid glucose can be further purified and concentrated to the required percentage of solids at the manufacturing process.

38.3.10.3 Rice Fructose Syrup

Rice starch is converted by enzymes into a natural sweetener which is called rice fructose syrup. Chen and Chang (1984) stated that rice fructose syrup is sweet syrup used as sweetener obtained from rice which is non-genetically modified organisms (GMO) and free of allergens. Plain rice or organically grown rice can be used for production of fructose rice syrup. Rice fructose syrup has low glycemic index and is a favourable choice for vegetarian foods. Rice fructose works as bulking agent and is a natural alternate to high fructose corn syrup (Zhang et al. 2010, 2011).

Another product derived from brown rice is called rice syrup which is free of fructose. High fructose corn syrup has high amount of fructose as its name indicates but brown rice syrup instead have maltose, glucose and maltotriose sugars in it. Brown rice syrup is free of fructose, so converse to other sweeteners and it is safe for use without effecting liver and metabolic functions of the body.

Rice fructose syrup finds wide application in health foods and beverages, due to its inherent nature of being 'natural', 'non-GMO', 'allergen-free', 'gluten-free' 'organic' and has 'low glycemic index' (fewer calories). Being a plant source, rice fructose is suitable for 'Vegan' and 'vegetarian' foods. Rice fructose makes a great base for pollen-free table top sweeteners, such as pancake syrup, breakfast sweetener, honey substitute, etc. Rice fructose is a good bulking agent for natural and blended sweeteners. Rice fructose is a perfect healthy alternate to high fructose corn syrup (Zhang et al. 2010, 2011).

Rice fructose syrup is a pollen-free natural sweetener widely used in food industry having fructose ranging from 40 to 49%. It has higher sweetness index than commonly used sucrose which makes its wide acceptance in various food manufacturing operations like beverages, cold drinks, dairy industry, canned fruits, confectionary, juice industry preserved fruit and salad dressings. In beverage industry, sucrose can be totally replaced by high fructose rice syrup (Jackson et al. 2012).

38.3.10.4 Wild Rice Stems

Wild rice is in the same family as common bamboo called Poaceae. It is known by names like wild rice, Kuw-sun, water bamboo, Coba or Jiao-Bai. Wild rice is widely eaten as vegetable and has popularity in Japan and China as medicinal usage. Wild

Fig. 38.24 Wild rice stem

rice is considered best source of vitamin C, A, iron calcium and other vital minerals. Wild rice shoot is used as vegetable by stir frying to preserve its natural light sweet flavour and crunch (Fig. 38.24).

References

Ahmad S, Hasanuzzaman M (2012) Integrated effect of plant density, N rates and irrigation regimes on the biomass production, N content, PAR use efficiencies and water productivity of rice under irrigated semiarid environment. Not Bot Horti Agrobot Cluj-Napoca 40(1):201–211

Ahmad S, Zia-ul-Haq M, Ali H, Shad SA, Ammad A, Maqsood M, Khan MB, Mehmood S, Hussain A (2008) Water and radiation use efficiencies of transplanted rice (*Oryza sativa* L.) at different plant densities and irrigation regimes under semi-arid environment. Pak J Bot 40(1): 199–209

Ahmad S, Ahmad A, Zia-ul-Haq M, Ali H, Khaliq T, Anjum MA, Khan MA, Hussain A, Hoogenboom G (2009) Resources use efficiency of field grown transplanted rice (*Oryza sativa* L.) under irrigated semiarid environment. J Food Agric Environ 7(2):487–492

Ahmad S, Ahmad A, Soler CMT, Ali H, Zia-Ul-Haq M, Anothai J, Hussain A, Hoogenboom G, Hasanuzzaman M (2012) Application of the CSM-CERES-Rice model for evaluation of plant density and nitrogen management of fine transplanted rice for an irrigated semiarid environment. Precis Agric 13(2):200–218

Ahmad S, Ahmad A, Ali H, Hussain A, Garcia y Garcia A, Khan MA, Zia-Ul-Haq M, Hasanuzzaman M, Hoogenboom G (2013) Application of the CSM-CERES-Rice model for evaluation of plant density and irrigation management of transplanted rice for an irrigated semiarid environment. Irrig Sci 31(3):491–506

Ahmad A, Ashfaq M, Rasul G, Wajid SA, Khaliq T, Rasul F, Saeed U, Rahman MH, Hussain J, Baig IA, Naqvi AA, SAA B, Ahmad S, Naseem W, Hoogenboom G, Valdivia RO (2015) Impact of climate change on the rice–wheat cropping system of Pakistan. In: Hillel D, Rosenzweig C (eds) Handbook of climate change and agro-ecosystems: the agricultural modeling Intercomparison and improvement project (AgMIP) integrated crop and economic assessments. Imperial College Press and the American Society of Agronomy, pp 219–258

Ahmad S, Abbas G, Ahmed M, Fatima Z, Anjum MA, Rasul G, Khan MA, Hoogenboom G (2019) Climate warming and management impact on the change of rice-wheat phenology in Punjab, Pakistan. Field Crops Res 230:46–61

Ahmed M, Ahmad S (2019) Carbon dioxide enrichment and crop productivity. In: Hasanuzzaman M (ed) Agronomic crops, vol 2. Springer, Singapore, pp 31–46

Ahmed M, Ahmad S (2020) Systems modeling. In: Ahmed M (ed) Systems Modeling. Springer, Singapore, pp 1–44

Amagliani L, O'Regan J, Kelly AL, O'Mahony JA (2016) Chemistry, structure, functionality and applications of rice starch. J Cereal Sci 70:291–300

Ausman LM, Rong NI, Nicolosi RJ (2005) Hypocholesterolemic effect of physically refined rice bran oil: studies of cholesterol metabolism and early atherosclerosis in hypercholesterolemic hamsters. J Nutr Biochem 16(9):521–529

Bao J, Bergman CJ (2018) Rice flour and starch functionality. In: Starch in food. Woodhead Publishing, pp 373–419

Chen WP, Chang YC (1984) Production of high-fructose rice syrup and high-protein rice flour from broken rice. J Sci Food Agric 35(10):1128–1135

Chou TW, Ma CY, Cheng HH, Chen YY, Lai MH (2009) A rice bran oil diet improves lipid abnormalities and suppress hyperinsulinemic responses in rats with streptozotocin/nicotinamide-induced type 2 diabetes. J Clin Biochem Nutr 45(1):29–36

Dung NTP (2013) Vietnamese rice-based alcoholic beverages. Int Food Res J 20(3):1035–1041

Eliasson AC (ed) (2004) Starch in food: structure, function and applications. CRC press

Fairhurst T, Dobermann A (2002) Rice in the global food supply. WORLD 5.7(502):454–349
Fatima Z, Ahmad M, Hussain M, Abbas G, Ul-Allah S, Ahmad S, Ahmed N, Ali MA, Sarwar G, Ehsan ul Haque, Iqbal P, Hussain S (2020) The fingerprints of climate warming on cereal crops phenology and adaptation options. Sci Rep 10:18013
Fu BX (2008) Asian noodles: history, classification, raw materials, and processing. Food Res Int 41(9):888–902
Hatcher DW, Edwards NM, Dexter JE (2008) Effects of particle size and starch damage of flour and alkaline reagent on yellow alkaline noodle characteristics. Cereal Chem 85(3):425–432
Hoke K, Houšová J, Houška M (2005) Optimum conditions of rice puffing. Czech J Food Sci 23(1): 1–11
Jackson BP, Taylor VF, Karagas MR, Punshon T, Cottingham KL (2012) Arsenic, organic foods, and brown rice syrup. Environ Health Perspect 120(5):623–626
Juliano BO (ed) (1985) Rice: chemistry and technology. American Association of Cereal Chemists, St. Paul, MN, pp 569–618
Juliano BO, Hicks PA (1996) Rice functional properties and rice food products. Food Rev Int 12(1): 71–103
Juliano BO, Sakurai J (1985) Miscellaneous rice products. In: Rice: chemistry and technology, pp 569–618
Li N, Guo Y, Zhao S, Kong J, Qiao D, Lin L, Zhang B (2020) Amylose content and molecular-order stability synergistically affect the digestion rate of indica rice starches. Int J Biol Macromol 144: 373–379
Luh BS (1991) Quick-cooking rice. In: Rice. Springer, Boston, MA, pp 540–565
Mestres C, Colonna P, Buléon A (1988) Characteristics of starch networks within rice flour noodles and mungbean starch vermicelli. J Food Sci 53(6):1809–1812
Mestres C, Ribeyre F, Pons B, Fallet V, Matencio F (2011) Sensory texture of cooked rice is rather linked to chemical than to physical characteristics of raw grain. J Cereal Sci 53(1):81–89
Mitchell CR, Mitchell PR, Nissenbaum R (1988) Nutritional rice milk production. US4744992 17
Muthayya S, Hall J, Bagriansky J, Sugimoto J, Gundry D, Matthias D, Maberly G (2012) Rice fortification: an emerging opportunity to contribute to the elimination of vitamin and mineral deficiency worldwide. Food Nutr Bull 33(4):296–307
Ofoedu CE, Osuji CM, Omeire GC, Ojukwu M, Okpala COR, Korzeniowska M (2020) Functional properties of syrup from malted and unmalted rice of different varieties: a comparative study. J Food Sci 85(10):3081–3093
Perretti G, Miniati E, Montanari L, Fantozzi P (2003) Improving the value of rice by-products by SFE. J Supercrit Fluids 26(1):63–71
Rogers EJ, Rice SM, Nicolosi RJ, Carpenter DR, McClelland CA, Romanczyk LJ Jr (1993) Identification and quantitation of γ-oryzanol components and simultaneous assessment of tocols in rice bran oil. J Am Oil Chem Soc 70(3):301–307
Roy P, Orikasa T, Okadome H, Nakamura N, Shiina T (2011) Processing conditions, rice properties, health and environment. Int J Env Res Pub He 8(6):1957–1976
Russell K, Delahunty C (2004) The effect of viscosity and volume on pleasantness and satiating power of rice milk. Food Qual Prefer 15(7–8):743–750
Shahzad AN, Ahmad S (2019) Tools and techniques for nitrogen management in cereals. In: Hasanuzzaman M (ed) Agronomic crops, vol 2. Springer, Singapore, pp 111–126
Sharp RN, Kattan MW, Sharp C (1985) Effects of toasting on the quality of canned rice. J Food Sci 50(2):340–341
Shibayama Y, Kanouchi H, Fujii A, Nagano M (2020) A review of Kurozu, amber rice vinegar made in pottery jars. Functional Foods in Health and Disease 10(6):254–264
Swarnakar AK, Srivastav PP, Das SK (2019) Optimization of preconditioning process of pressure parboiled brown rice (unpolished) for microwave puffing and its comparison with hot sand bed puffing. J Food Process Eng 42(3):e13007
Wilson TA, Ausman LM, Lawton CW, Hegsted DM, Nicolosi RJ (2000) Comparative cholesterol lowering properties of vegetable oils: beyond fatty acids. J Am Coll Nutr 19(5):601–607

Zeng JX, Huang WF, Zhou XS (2008) The development of a brown rice syrup. Modern Food Sci Technol 7

Zhang QZ, Wang ZN, Chen C, Wu HY, Li Y, Liu P (2010) Preparation optimization of high fructose syrup from broken rice. Food Sci 18:195–198

Zhang QZ, Wang ZN, Li JJ, Li Y, Wu HY, Liu P (2011) Technology of high fructose syrup using broken rice glucose by borate separation. Food Sci Technol 06:207–211

Zigoneanu IG, Williams L, Xu Z, Sabliov CM (2008) Determination of antioxidant components in rice bran oil extracted by microwave-assisted method. Bioresour Technol 99(11):4910–4918

Part VI
Miscellaneous

Advances Approached to Mitigate Abiotic Stresses in Rice (*Oryza sative* L.) Crop

39

Sibgha Noreen, Seema Mahmood, Kausar Hussain Shah, Shahzadi Saima, Muhammad Salim Akhter, Nawishta Saleem, Muhammad Rashid, Fahd Rasul, Hassan Munir, Kamrun Nahar, Mirza Hasanuzzaman, Muhammad Azam Khan, and Shakeel Ahmad

Abstract

The food insecurity has been haunting the mankind since time immemorial and even continuing to the present age. In the beginning of twentieth century, the greatest efforts were made in the areas of plant breeding, agronomy, and development of synthetic fertilizers, pesticides, and weedicides, resultantly the productivity of wheat yield increased from 0.5 to ≥6.0 tons by 1950s. This research endeavors witnessed the dawn of green revolution across the developing countries. This travel has a long run to meet the expected increase in population of 9.5 billion by the year 2050. Thereby, it necessitates to enhance the agricultural productivity by more than 70% over the current ratio of production. At the advent of climate change with concurrent deterioration in natural land and water resources, it is imminent to gear up research endeavors toward food and insecurity in the long run. The sustainability

S. Noreen (✉) · S. Mahmood · K. H. Shah · S. Saima · M. S. Akhter · N. Saleem · M. Rashid
Institute of Pure and Applied Biology, Bahauddin Zakariya University, Multan, Pakistan
e-mail: sibgha.noreen@bzu.edu.pk

F. Rasul · H. Munir
Department of Agronomy, University of Agriculture Faisalabad, Faisalabad, Pakistan

K. Nahar
Department of Agricultural Botany, Sher-e-Bangla Agricultural University, Dhaka, Bangladesh

M. Hasanuzzaman
Department of Agronomy, Sher-e-Bangla Agricultural University, Dhaka, Bangladesh

M. A. Khan
In-Service Agriculture Training Institute, Sargodha, Pakistan

S. Ahmad
Department of Agronomy, Faculty of Agricultural Sciences and Technology, Bahauddin Zakariya University, Multan, Pakistan
e-mail: shakeelahmad@bzu.edu.pk

N. Sarwar et al. (eds.), *Modern Techniques of Rice Crop Production*,
https://doi.org/10.1007/978-981-16-4955-4_39

in food crops production is challenged by a number of biotic (insect pests and pathogens) and abiotic (salinity, drought, extreme temperatures, heavy metals, flooding, and nutrient stress) stresses. The crop plants being resilient in nature make some metabolic adjustments to adapt to the environment. Among cereal crops, rice holds key position in providing food security for billions of peoples around the globe. Rice crop is being cultivated widely in different ecological environments and its yields are reduced to a greater proportion in response to abiotic stresses. The shortfall in productivity of rice crop may revert to severe shortage of food. Among the abiotic stresses, high temperature, drought, salinity, and heavy metals significantly impact yield and productivity of rice crop. The occurrence of these stresses results in change in protein profile, mRNAs, transcriptome, transcriptional factors, and epigenetic changes. During the changes at cellular level, certain genes are regulated. To cope with abiotic stress, the genes are ought to be upregulated. Therefore, there is a need to breed or develop different rice varieties/cultivars/lines that could be efficient for upregulation of genes at the highest proportion. Other than conventional plant breeding technology, the development of transgenic rice plant by employing genetic engineering could be more prospective and efficacious in protecting the rice crop against abiotic stresses.

Keywords

Oryza sativa L. · Cereal · Abiotic · Biotic · Stress · Crop

39.1 Introduction

The agriculture sector is essential for food and nutrient security for billions of inhabitants across the continents, particularly among developing countries. To feed the expected 9.5 billion mouths by the year 2050, we ought to enhance the productivity of agricultural crops by more than 70% over the ratio of production level. However, there are many challenges that are to be addressed adequately to meet the nutrient need of 9.5 billion people by 2050. The most important is the livelihood of billions of farmers, who will face a high amount of threat related to agricultural profitability and stability. The threat of advancing footprints of climate change with concurrent deterioration in land and water resources and loss of soil and plant diversity represents one key challenge and may adversely affect the functioning, stability, resilience, and adaptability of increased productivity of food crops and associated ecosystems services, both within and beyond farm level.

Rice is the member of family Gramineae (Poaceae) under subfamily Pooideae, tribe Oryzeae, and further placed in series sativa in section Sativae as *Oryza sativa*. Rice is a diploid (2n) plant species with 24 chromosomes having genomic formula AA (Bardenas 1965). It is an annual grass with round, hollow, and jointed clums with sessile leaf blades having a terminal panicle (Bardenas 1965).

Unlike other plant species, the rice plants have the ability to tolerate submerged and waterlogged conditions through formation of longitudinal interconnected gas spaces (aerenchyma) that enable internal aeration across roots and shoots (Colmer and Pedersen 2008a). Furthermore, internal aeration continues through leaf gas films, these are microlayers of trapped air present between submerged leaves (Colmer and Pedersen 2008b; Pedersen et al. 2009). Rice is being cultivated in 129.6 mha, with rice production of 497.7 million tons. Among rice growing countries, China ranks first with 146.7 million tons, followed by Bangladesh with 118.9 million tons (FAO 2021). While Pakistan ranks tenth in production by producing 7.4 million tons paddy rice gathered from 3 million hectare, mainly composed of basmati and IRRI varieties (Ahmed and Schmitz 2011; Rahman et al. 2016).

Among the agricultural crops, cereals constituting wheat, rice, and maize crops provide more than 50% energy and essential nutrients to the humans around the globe (Gnanamanickam 2009). The nutrient contents of rice vary widely due to agroecological environment. The nutritional value (protein and fat) of brown rice is much higher as compared with white rice (Seki et al. 2005). It contains higher content of glutamic and aspartic amino acids while limited in lysine amino acid. It is a good source of vitamins like thiamine (B1), riboflavin (B2), and niacin (B3) (Kim et al. 2018a). The risk of diabetes mellitus increases due to its high glycemic index, which is associated with diabetes mellitus (Rohman et al. 2014).

39.2 Constraints for Rice Production

A number of environmental constraints severely affect the growth and productivity of rice crop. These limitations could be mitigated by employing biotechnological approaches and cultural measures as an effort to sustain its yield at higher level.

39.2.1 Abiotic Constraints

The rapid changes in climatic conditions are impacting negatively on the productivity of rice crop across different agroecological environments. The farming community have little capability to cope with potential threats of water shortage, extreme temperatures, brackish ground water, floods, sea-level rise near coastal areas, and intense tropical cyclones. The yield of rice crop is expected to decrease by 15% under irrigated and 12% under climatic change scenario by 2050 (IFRI). A greater proportion of population will face risk of food insecurity due to major food shortage.

39.2.1.1 Edaphic and Climatic Conditions

The changes in climatic conditions due to anthropogenic intervention pose a great challenge toward realization of yield potential of crops (Bellard et al. 2012; Alley et al. 2003) Among cereal crops, rice crop is highly prone to climatic changes (Rosenzweig et al. 2014). The cumulative effect of abiotic stresses could lower crop yield because, in most of the cases, plants are subjected to multiple stresses at a time, which are eventually hazardous to the crop. Therefore, it is impossible to check

Table 39.1 Major edaphic constrains

Components	Stress factors
Natural stress	
Internal soil process	
Chemical condition	Salinity; nutrient deficiency; heavy metal toxicity; and low pH
Physical condition	Higher soil erosion; low soil structure stability; low water-holding capacity; and root restricting layers
Biological condition	Imbalanced organic matter
Ecosystem conditions	Soil degradation and reductions in soil resilience
External condition and process	
Climatic condition	Reduction in soil moisture; water logging; atmospheric temperature fluxions; and disturbance in length of growing seasons
Biological condition	Diseases and pests
Catastrophic condition	Drought and floods; volcanic and seismic activity; and landslides
Ecosystem condition	Loss of soil quality and health
Anthropic stresses	
Chemical condition	Acidification of soils by acid rains, extensive fertilizers, and chemicals
Physical condition	Increase in soil erosion and compaction of soil
Biological condition	Higher incidences of diseases and pest attacks; loss of predators; and allelopathy

independent effects of drought or temperature stress during the growth period (Suzuki et al. 2014; Paul and Roychoudhury 2019). Invariably, the ecoedaphic factors produce far-reaching effects on the sustainability or deterioration of natural resources and greatly prompt anthropogenic activities (Table 39.1).

39.2.2 Adaptation of Rice to Abiotic Stresses

The plants can adopt themselves against changing climatic conditions (Yoshida et al. 2014), still multiple stresses can reduce or completely destroy crop production (Des Marais et al. 2013). However, under stressful conditions, the plants can initiate series of internal and external determinants and make certain morphological, biochemical, and molecular adaptations, as to coincide with the external changes in most of the cases in order to protect plant's metabolic machinery.

39.2.2.1 Soil Salinity

The rice crop has greater tolerance to salinity at 4 dSm^{-1} (40 $mmoll^{-1}$ NaCl) in germination stage, however, sensitive in flowering stage (Dobermann and Fairhurst 2000). The tolerance of rice seedlings at germination stage is independent of sensitivity to salinity at flowering/reproductive stage (Singh and Flowers 2010). The growth and development of rice crop are affected to a greater proportion due to uptake of Na^+ ion in a larger amount, while depressing the absorption of other mineral nutrients (Rajendran et al. 2009; Horie et al. 2012). The reduction in the uptake of essential nutrients, namely phosphorus, zinc, iron, and boron, leads to slowing down of the photosynthetic and metabolic functions in the plant (Dobermann and Fairhurst 2000). The cumulative effects of salinity stress result in lowering yield potential of rice, in terms of decrease in panicle length, number of spikelets per panicle, and increased spikelets sterility (Zeng and Shannon 2000).

Ion exclusion is an important mechanism for salt tolerance in rice at organ or tissue level (Munns and Tester 2008; Roy and Chakraborty 2014). It is attributed to reduction in loading and transport of Na^+ and Cl^- from roots and decreasing the build-up of these ions in leaves (Roy and Chakraborty 2014). However, the tissue tolerance is achieved by transport of Na^+ into vacuoles. Similarly, the salt tolerance is an expression of complex qualitative traits, which are controlled by multiple genes (Reddy et al. 2017).

The transport and uptake of selective salts by root cells in rice plants take place via symplastic as well as apoplastic pathways (Das et al. 2017). In the apoplastic pathway, the transport of water takes through cell walls of adjacent cells of roots but, at endodermis level, the presence of casparian strip forces the water from apoplastic to symplastic pathway. So the salts have to cross the plasma membrane into the cytoplasm of endodermis as highly suberized endodermal layer in rice provides a major resistance to radial flow of water and dissolved substances (Miyamoto et al. 2012). Resultantly, the concentration of Na^+ in the shoot system is reduced in rice plants subjected to salinity (Enstone et al. 2002; Cai et al. 2011). When the transport of water and salts takes place through symplastic pathway, the channels and protein carriers of plasma membrane transport Na^+ outside the endodermis. Therefore, majority of ions moving through apoplastic pathway have to cross plasma membrane at casparian bands, until discharged to the xylem (Das et al. 2015). In rice plant, apoplastic pathway is the major route for transportation of Na^+ ions in the shoots (Krishnamurthy et al. 2009).

39.2.2.2 Drought Stress

The water-deficit condition affects plant, germination, vegetative growth, and developmental stages of crop species via disturbance in a number of physiological processes, that is, decrease in leaf relative water contents and water potential (Halder and Burrage 2003; Farooq et al. 2009). It is determined that effect is more pronounced in reducing the yield-contributing determinants, resultantly causing greater reduction in rice yield (Hasanuzzaman et al. 2013). The biochemical changes are also brought about by increasing the ratio of abscisic acid (ABA) biosynthesis,

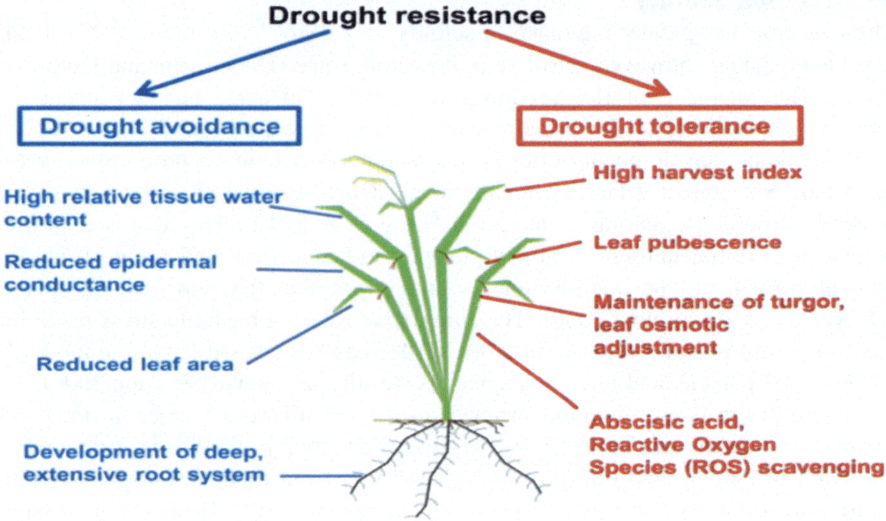

Fig. 39.1 Drought resistance in rice

diminishing stomatal efficiency, conductance, and transpirational efficiency (Yamaguchi-Shinozaki and Shinozaki 2005).

The rice plants make in-built and/or imminent strategy to avoid the salt and drought stress, either drought avoidance or drought tolerance, to complete the life span. The plants make certain adjustments in retaining greater amount of water by maintaining relatively higher leaf tissue water potential during the drought stress. The mechanism of avoidance of drought by plants is associated with reduction of epidermal transpiration rate to lower loss of water. Moreover, drought-stressed plants develop deep root system to further enhance the water uptake to meet the needs of normal metabolism (Farooq et al. 2009; Gowda et al. 2011) (Fig. 39.1).

The drought stress is controlled by genetic factors and interacted by simulation of certain mechanisms in cellular adjustments, physiological acclimation, and morphological adaptation at different stages of growth. The physiological acclimation is comprised of higher density and conductance of stomata, decline in transpiration rates, and delay in reproductive maturity of male and female parts of floret. While morphological adaptation includes increase in root thickness, greater length and proliferation, presence of waxy and/ or thick leaf coating, small leaf size, increased green leaf area, and delayed leaf senescence (Gnanamanickam 2009), a wide set of processes (biochemical and molecular) are activated to protect the plants subjected to moisture-deficit stress (Fujita et al. 2006). The important adaptation in rice plants under drought stress includes osmotic adjustment (Wei et al. 2014), regulation in stomatal conductance (gs) (Price et al. 2002), enhancement in biomass accumulation (Xangsayasane et al. 2014), and activation of drought-responsive genes (Degenkolbe et al. 2009).

39.2.2.3 Heavy Metal Toxicity

Agricultural food safety is on serious risk due to polluted air and soils with heavy metals (HM), including arsenic (As), cadmium (Cd), lead (Pb), copper (Cu), and zinc (Zn), triggered by natural sources (volcano eruption) and human activities (mining and industrialization) (Duan et al. 2016; Jacob et al. 2018). Plant growth and yield significantly reduced due to presence of heavy metals (HMs) in soil media (Seneviratne et al. 2017). The common symptoms of HM stress include reduction in plant height and leaf area (Pant and Tripathi 2014), damaging photosynthetic apparatus including degradation of photosynthetic pigments, and slowing down of xanthophyll cycle and carbon reduction cycle (Rai et al. 2016).

Plants have developed certain processes to tolerate HM stress through metal exclusion, accumulation of metal in older parts of plant, and binding of metal ions by strong ligands such as metallothioneins (MTs), cysteine-rich proteins, phytochelatins (PCs), and thiol-rich peptides (Datta 2004; Zibaee et al. 2009). For example, As (III) is compartmentalized in vacuoles via shuttling of As from cytoplasm as As-PC complexes (Shri et al. 2009). Other alterations include anatomical adjustments adopted by plants in their physiological and morphological processes, and decrease in parenchymatous and mesophyll cells with concurrent reduction in the number of xylem vessels and root and leaf diameter (Batool et al. 2015).

39.2.2.4 High Temperature

Heat stress arises due to increase in temperature beyond the tolerance limit of crops which causes irreversible damage to germination, growth, reproduction, yield, and grain quality. The impact of heat stress is obvious even if the stress lasts for a few hours (Jagadish et al. 2007). If the temperature rises to 39 °C on the onset of flowering stage, it results in low anther dehiscence (Matsui and Omasa 2002). However, the high temperature at ripening stage of fruit reduces the quality and yield process, resulting in low yield (Zinn et al. 2010).

Normally, the rice is cultivated in areas where the temperature is nearly optimum ranging from 27 to 32 °C (Yin et al. 1996). However, any further increases in temperature even for a short period during sensitive stage could reduce grain yield (Ceccarelli et al. 2010). The pollination stage is most sensitive to high temperature as the temperatures higher than 35 °C lead to reduction in pollination, resulting in spikelet sterility and, finally, low yield (Matsui et al. 1997). Similarly, spikelet sterility is also high if the temperature at night is higher the normal (Zinn et al. 2010) (Fig. 39.2).

Generally, the high temperature tolerance is the ability of the plants to express better growth and optimum yield produce (Wahid et al. 2007). Being sessile in nature, the plants have a single option to defend temperature stress through structural and metabolic adjustments (Yamanouchi et al. 2002). When plants are continuously exposed to lethal heat stress, they induce internally the heat stress tolerance cascade as a cell autonomous phenomenon (Larkindale and Vierling 2008). The tolerance mechanism mainly depends on initiation of particular pathways during the period of acclimation and subsequent achievement of temperature tolerance (Klueva et al. 2001). Under heat stress, the rice plant adjusts the of panicle emergence timing and opening of spikelet/floret in response to stressful conditions (Weerakoon et al. 2008).

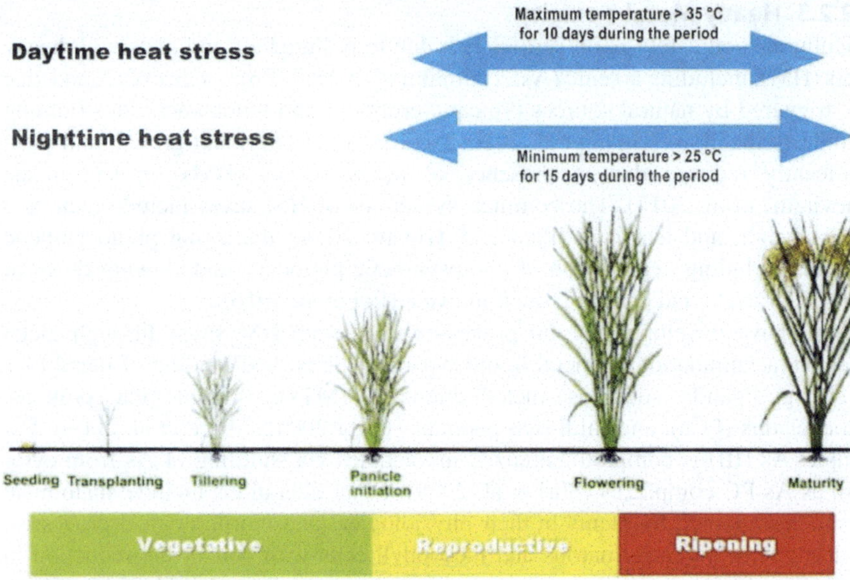

Fig. 39.2 Heat stress thresholds at critical growth stages of rice. (Source: Laborte et al. 2012)

39.2.2.5 Ultra Violet (UV) Radiations

Ultraviolet (UV) radiations from sun significantly decrease crop growth and yield attributes (Mmbando et al. 2020). The exposure of rice plants to UV radiations, even for a shorter duration, has serious biological effects, resulting into damage to biological membranes, DNA, and proteins leading to low growth and production (Chen et al. 2020). Moreover, UV radiations stimulate the reactive oxygen species (ROS) synthesis that disturbs normal metabolism and disturbance in photosynthetic electron transport chain (Yu et al. 2013; Tripathi et al. 2017). UV radiation encourages the plants to generate ROS, which results in decrease of photosynthetic pigments, light harvesting, and chlorophyll-binding proteins and protective proteins upregulation, for example, pigment-dispersing factor 1.2 and pathogen-related protein-1 (A.-H.-Mackerness et al. 2001; Kalbina and Strid 2006; Du et al. 2011). Plants have established various protective mechanisms against UV radiations through the augmentation of defense system (antioxidant) (Brosché and Strid 2003) and compounds accumulation which can absorb UV radiations (Frohnmeyer and Staiger 2003).

39.2.3 Biotic Constrains

The rice crop is prone to a number of biotic stresses. Among these constrains, pests and diseases are major cause of concern leading to loss in rice production. Attack of insects on rice include rice stink bug, rice water weevil, chinch bug, Mexican rice borer, fall armyworm, grasshoppers, sugarcane borer leafhoppers, and blister beetles. Various pathogens cause serious damages such as *Magnaporthe oryzae* and rice yellow mottle virus (Reissig 1985; Zibaee 2013).

Several biotechnological approaches are developed to escalate improvement in rice crop and also its resistance to pests, diseases, and various abiotic stresses (Datta 2004; Zibaee 2013). These approaches have remained a success story as a result of (1) transfer of economically important traits in rice genome, (2) manipulation of target trait, and (3) shortening of the breeding cycle (Wing et al. 2018).

Moreover, research is undergoing to prevent any eventual yield loss (Wanger et al. 2014; Rijal and Devkota 2020). Farmers use insecticides and pesticides to protect their crops from attack of pests and insects (Dent and Binks 2020). Due to increase in environmental pollution and development of resistance in insects/pests against certain insecticides, the researchers and farmers are looking for safer and more efficient tactics such as using insect growth regulators, biocontrol agents, sanitation, and many more.

39.3 Induced Stress Tolerance

Besides internal defense and signaling system, the plants have several adaptive mechanisms by increasing or decreasing some endogenous synthesized compounds (antioxidant compounds and osmolytes). The plant growth regulators, that is, abscisic acid, gibberellic acid, indole acetic acid, salicylic acid, jasmonates, and cytokinins, are being used to maximize the plant growth and yield especially under stressful conditions (Hamdia and Shaddad 2010). Similarly, application of micro- and macronutrients, such as Ca, K, N, Si, etc., has considerable attention as a shotgun approach to ameliorate the adverse effects of stresses. The brief summary for stress tolerance in plants under different stresses is as follows (Table 39.2).

39.4 Omics Approaches to Produce Tolerant Varieties of Rice

Abiotic stresses are the key factors negatively affecting the crop growth and productivity worldwide. To cope and mitigate the abiotic stresses, plants adopt various strategies. In these strategies, the production of antioxidants, osmoprotectants, proteins, and expression of genes are included. To understand the responses of plants to environmental stresses, the advances in plants physiology, genetics, and molecular biology have greatly improved (Munns 2002; Singh et al. 2010; Rashid et al. 2020). Rice is one of the important cereal crops for the human being food. It is also facing different challenges for productivity. Rice plants are sensitive to several abiotic stresses. The understanding of physiology and molecular biology helps us to improve the stress-tolerant varieties of rice. The breeding, gene regulation, transcriptions factors, and transcriptome approaches are included to improve the rice under abiotic stresses (Zou et al. 2011; Zhu et al. 2015).

Table 39.2 The response rice to different application under different stresses

Stress	Application	Effect	References
Salinity	Apigenin	Regulated selective ion uptake; and maintaining higher K^+/Na^+ ratios Triggering the induction of the antioxidant defense system	Mekawy et al. (2018)
Salinity	Na_2SeO_4	Enhanced antioxidant enzymes (SOD, APX, CAT, and GSH-Px); and reduced H_2O_2 and MDA concentrations in plants under NaCl stress	Subramanyam et al. (2019)
Salinity	Salicylic acid	Biomass accumulation and higher chlorophyll contents	Kim et al. (2018b)
Salinity	Melatonin	Activated salt stress-responsive genes; and increased the expression of K^+ uptake transporters in the root tip	Liu et al. (2020)
Salinity	Zinc	Enhancements in indole-3-acetic acid and reductions in abscisic acid	Nadeem et al. (2020)
Salinity	Proline	Increased photosynthetic pigment; protects macromolecules; and stabilized protein structures	Wutipraditkul et al. (2015)
Salinity	Salicylic acid	Improved photosynthetic pigments andreduced antioxidant enzymes activity	Khan et al. (2019)
Salinity	Nanosilica	Upregulation of phytohormones and antioxidant defense system; and induced osmolyte production	Abdel-Haliem et al. (2017)
Salinity	Melatonin	Increased rice dry weight; inhibited the uptake of Na^+; and enhanced K^+ content	Yan et al. (2021)
Salinity	$Na_2SiO_3{\cdot}5H_2O$	Low electrolytic leakage and lipid peroxidation; regulating phytohormonal; and enzymatic antioxidants' responses	Kim et al. (2014)
High temperature	Biochar	Increased in pollen fertility and retention, anther dehiscence, and rate of pollen germination	Fahad et al. (2015)
High temperature	Abscisic acid	Enhanced sucrose metabolism to prevent ATP depletion and maintain energy homeostasis	Rezaul et al. (2019)
High temperature	Salicylic acid	Enhanced pollen viability; lowered H_2O_2 levels as well as MDA levels	Feng et al. (2018)
Heavy metal stress	Hemin	Upregulation of antioxidant activities of SOD, APX, and GR in roots; and enhanced biomass accumulation	Chen et al. (2017)
Arsanilic acid	Sodium silicate	Decreased lipid peroxidation and proteins decomposition; and improved protein synthesis and protein metabolism	Geng et al. (2018)
Arsenic stress	Glutathione	Enhanced antioxidant enzyme's activities and nonenzymatic antioxidants	Jung et al. (2019)
Cadmium stress	$CaCl_2$	Enhanced antioxidant defense system and reversed the overproduction of ROS	Rahman et al. (2016)

(continued)

Table 39.2 (continued)

Stress	Application	Effect	References
Cadmium stress	Aspartic acid	Increased shoot length and dry weight; altering antioxidant enzymes	Rizwan et al. (2017)
Drought	Polyethylene glycol	Decreased leaf water potential and water contents	Ding et al. (2014)
Drought	Salicylic acid	Improved photosynthetic and protein content; reduced antioxidant enzymes activity; and lowered ROS accumulation	Ahmed et al. (2019)

39.4.1 Proteins Expression

Proteins profiling is an important and powerful tool to analyze the plant response toward the stress condition. Under submerged conditions, the chloroplastic (protein ID: A2XZ01) and photosystem II 10 kDa polypeptide proteins were downregulated in rice (Xiong et al. 2019b). Under heat stress, the protection proteins, protein biosynthesis, proteins related to proteins degradation, carbohydrate metabolism, and redox homeostasis were isolated and detected in rice (Zou et al. 2011). Dirigent (DIR) gene protein gives relief to rice plant against extreme temperatures, heavy metals, drought, and salinity (Singh et al. 2020). The high level of expression of cluster-binding, iron-sulfur protein (OsFd), which is an integral part of electron transport chain, was isolated from root tissue of rice under heat stress (Sailaja et al. 2014).

The molecular chaperons (heat shock proteins—Hsps) play important role in proteins intracellular localization, secretion, protein folding, and truncated or misfolded proteins degradation. Under cold, salinity, and drought stress, these Hsps were highly expressed in rice. However, the expression level is more in drought in comparison to other stresses in rice (Hu et al. 2009). In endosperm of rice plants, five of the genes (*OsHsp72.90, OsHsp72.57, OsHsp71.18, OsHsp24.15*, and *OsHsp18.03*) exhibited high expression, representing that *OsHsp* genes play significant role in seed development of rice. By exposing to heat, PEG, and abscisic acid, the nine *OsHsps* genes were upregulated. However, six genes were upregulated due to salt stress, that is, *OsHsp71.10, OsHsp93.04, OsHsp72.57, OsHsp71.18, OsHsp24.15*, and *OsHsp18.03*. In contrast to salinity stress, two genes, namely *OsHsp93.04* and *OsHsp72.57*, were downregulated under cold stress. These various expressions of *Hsp* gene family imply potential functional diversity in rice (Ye et al. 2012). OsHSP genes proteins were induced strongly by exposing the plants to heat.

The *OsHSP80.2, OsHSP23.7*, and *OsHSP71.1* transcripts were enhanced under salinity stress. The high level of expression, that is, OsHSP80.2 and OsHSP24.1 gene, was detected under 10% PEG. One OsHSP71.1 was encouraged by the application of ABA. However, the OsHSP24.1 was downregulated by ABA. These annotations indicated that the nine *OsHSP* genes play diverse roles in plant development under abiotic stress (Zou et al. 2009). The rice proteins, namely gamma subunits of the G-protein (RGG1 and RGG2), were highly expressed under abiotic stresses. The upregulation of RGG1 and RGG2 occurred in cold, heat, and salinity stress. However, only RGG1 was upregulated under moisture-deficit conditions in rice plants (Yadav et al. 2012). Calcineurin B-like protein-interacting protein kinases (CIPKs) are important proteins. The *OsCIPK31* was upregulated by mannitol and salt stress. However, long-term exposure downregulated these proteins.

39.4.2 Epigenetic Approaches

Epigenetic alterations are (characteristically) chemically heritable reversible modifications in structure of chromatin and DNA sequences in plants and animals. The DNA methylation and histone acetylation might facilitate certain changes in the physical structure of DNA, which causes the regulation of gene expression (Joel 2013). The methylation of DNA cytosine in plants looks in sequences, that is, CpG, CpNpG, and CpNpN, where N means A, C, or T. After DNA replication, the CpNpG and CpG methylation may occur because of symmetrical nature, while CpNpN nonsymmetrical methylation recognition is de novo (Sahu et al. 2013). Four rice HAT (*OsHAC703*, *OsHAG703*, *OsHAF701*, and *OsHAM701*) genes were expressed under drought due to epigenetic modifications. Salinity stress induces the rice HDAC genes that were regulated by other abiotic stresses. Under salinity and osmotic stress, the overexpression of *OsHDT701* was noted (Deng et al. 2016).

The cultivars internal methylation (5'-CmCGG-3') was observed dominant in rice leaves under drought stress signifying a high frequency of mCpG dinucleotide in comparison to mCpC dinucleotide in the 5'-CCGG-3' sequence. The drought susceptible plants exhibited relatively higher level of MspI and HpaII expression as compared with non-stressed plants, indicating that demethylation of these genes might have occurred under stress condition and the expression was enhanced. While, the drought-tolerant cultivars showed less digestion of HpaII under stress conditions than the control plants (Suji and Joel 2010).

Joel (2013) reported hypermethylation in substantive rice plant, while hypomethylation in tolerant rice plant under drought stress. Under salinity stress, the DNA methylation was also observed in rice that may give tolerance to rice plant against salinity stress. The methylated changes were more in shoot than root (Karan et al. 2012). DNA methylation also favors the tolerance of rice plant against nitrogen stress deficiency (Kou et al. 2011). The rice seedling showed 48 and 51.1% of annotated genes (with H3K4me3 modification) under water-deficit and control conditions, respectively.

Moreover, 4837 genes (910 genes with reduced H3M and 3927 genes with enhanced H3M) and 5866 genes were differentially expressed (3721 downregulated and 2145 upregulated) in water-deficit condition. The differential pattern of the H3K4me3 methylation does not affect stress-responsive genes but only up to a minor proportion, and the H3K4me3 modification level was correlated with transcripts under water-deficit condition. Additionally, for the H3K4me3-regulated stress-responsive genes, the level of H3K4me3 modification was mostly better in genes with truncated expression and decreased in genes expression under water-deficit condition. The data of H3K4me3 and gene expression profiles under water-deficit condition provide a good resource for studying the plants' epigenetic regulation (Zong et al. 2013).

The expression profiles (genome wide) of 10 cytosine DNA-MTase genes (belonging to four subfamilies: DNMT2, MET1, DRM, and CMT) were analyzed in rice plants. The diversity in the expression and tissue specification is found in all family members that might play a role in some elementary metabolic pathways. The expression of rice cytosine DNA-MTase genes especially *OsDRM1a* and *OsDRM1b* was decreased by jasmonic acid application. Under salinity and osmotic stress, the low level of transcription for 10 genes was observed in rice roots and shoots. They play important role in stress tolerance mechanism (Ahmad et al. 2014).

DNA demethylation and histone modification changes occurred in *OsMYB91* gene locus that leads toward the salt tolerance in rice under salinity stress. The overexpression of *OsMYB91* gene enhanced the proline level and ROS scavenging mechanism. This gene enhances tolerance in rice under abiotic stress by the regulation of SLR1 gene expression (Zhu et al. 2015). Under stress conditions, DNA methylation modifies the gene expression. Rice plants were analyzed by using MeDIP-chip hybridization under salinity and drought stress. Then, 14 genes related to CYP family were known, which showed variable marks of methylation in rice genome. The responses of CYPs genes were checked under different abiotic stresses, like salinity, drought, and cold. Under phytohormones, the expression level of different CYPs has been checked, which indicated the role of CYPs in hormone regulation (Waseem et al. 2021). Modification of Snf2 methylation increased resistance against abiotic stresses (Hu et al. 2013).

39.4.3 Transcription Factors

The MYBS plays a significant role in short-term and long-term chilling stress adaptation by suppressing the CBF-dependent signaling in rice plants (Su et al. 2010).

Under abiotic stress, the rice MYB TFs, that is, *OsMYB2P-1, OsMYBS3, OsMYB3R-2, OsMYB4*, and *OsMYB2*, were regulated (Huang et al. 2018). The upregulation of TF *OVP1* boosted the transgenic rice tolerance against the cold stress by enhancing proline contents and cell membrane integrity while lowering the monoaldehyde contents (Li et al. 2010). The expression level of *OsLEA3*, *OsRab16*, and *OsP5CS1* genes was significantly increased in rice overexpression lines with *OsNCED1* and *OsNCED3 mOsNAC2* (Jiang et al. 2019). Under sugar starvation, salinity, and cold stress, the expression of *OsMAPK4* results in activation of MAPK (Huang et al. 2002). ERF was upregulated and showed tolerance against the abiotic stresses (Thamilarasan et al. 2014).

OsSRO1c is a rice SRO gene that functions as downregulation of the transcription factors under stress conditions. SNAC1 is the key stress-responding gene in SRO family of rice. Under abiotic stresses, it responded differentially by binding with different transcription factors (TF) (You et al. 2014). The OsNAC6 functions as a transcriptional activator and play a role in tolerance against abiotic stresses

(Nakashima et al. 2007). The expression of certain stress-inducible genes such as *OsLEA3* leads to the enhanced expression of *OsNAC5*, *OsNAC6*, and *SNAC1* genes in rice plants exhibiting better stress management (Takasaki et al. 2010). The mutant *Osabf2* exhibited significant reduction to sensitivity, which enhanced the levels of abscisic acid at germination and postgermination. The rice *OsABF2* gene regulates transcriptional control of stress genes with ABA-dependent pathway (Hossain et al. 2010a). The sDREB2B exhibited nuclear-specific localization and the maximum transactivation activity. It has some nonfunctional and functional orthologues and analogous in Poaceae. The expression of its functional form was better under stress. The *OsDREB2B* gene is a key gene encoding a stress-inducible transcription factor (DREB2 type) in rice plants (Matsukura et al. 2010). The *ONAC045* was upregulated in drought and salinity stress in rice plant (Zheng et al. 2009). The *OsABF1* gene, that is, *Oryza sativa* ABA-responsive element-binding factor 1, encodes a transcription factor (bZIP). It was isolated from rice plants under abiotic stresses. Its expression in seedlings (roots and shoots) was induced by different abiotic stresses, for example, salinity, oxidative stress, drought, anoxia, abscisic acid (ABA), and cold (Hossain et al. 2010b).

39.4.4 Transcriptomic Profiling

The transcriptomic data indicated that the chilling stress-related genes were upregulated in cold-tolerant genotypes of rice connected with the upregulation of the transcription factors (TFs) of those genes and their signal transduction. However, in prolonged chilling stress, the regulation of these genes was diverse. TF genes and enriched cis elements explained that various regulatory pathways, including CBF and MYBS3 regulons, have important role in chilling stress relief (Zhang et al. 2012).

The transcriptomic data of the rice plants under cold stress indicated that the 241 and 244 M readings were aligned with 25.703 and 26.963 genes in Oro and Tio Taka genotypes, respectively. Furthermore, variations in gene expression were observed in 259 genes in Oro genotype and 5579 genes in Tio Taka genotype. Some of these genes are related to phytohormones metabolism, signal transduction in biotic stress, and antioxidant system (da Maia et al. 2017). The transcriptome profile indicated that a diverse group of genes encoding enzymatic antioxidents (*CAT*, *APX*, *GR*, and *SOD*), osmolytes regulatory enzymes (*PDH*, *BADH1*, and *P5CS*), polyamines regulatory enzymes (*SPDS*, *SAMDC*, *DAO*, and *SPMS*), and transcription factors (*WRKY*-71 and TRAB-1) were differentially regulated under cadmium stress in sensitive and tolerant varieties of rice. The regulation of above-mentioned genes plays an important role under cadmium stress (Paul and Roychoudhury 2018). The overexpression of calmodulin gene, *OsCam1–1*, gave relief against the salinity stress (Yuenyong et al. 2018).

Transcriptome profiling in rice under mercy stress indicated the functional genes responsible for chemical detoxification, cell wall formation, secondary metabolism, and signal transduction. In addition to these, several genes involved in aromatic amino acids level were also increased by upregulation of certain genes (Chen et al. 2014). Transcriptomic profiling showed various types of genes that were upregulated in rice under drought stress. In these genes, zinc finger domain encompassing proteins, a bZIP transcription factor, a MYB factor, and an OsMADS7 are included. Moreover, genes related to ABA and brassinosteroids were also upregulated by drought stress (Muthurajan et al. 2018).

The *OsCDPK21* and *OsCDPK23*, cytochrome P450 monooxygenase *CYP76M8*, were upregulated under low temperature in rice (Buti et al. 2018). Under zinc deficiency, the membrane transport genes, organic acid biosynthesis, and phytosiderophore activity exhibited high variation in their expression (Bandyopadhyay et al. 2017). Transcriptomic profile of rice under arsenic stress indicated that certain genes such as photosynthesis putative genes, lipid biosynthesis, and electron transport genes were downregulated in wild-type plants, while the genes of cysteine biosynthesis were enhanced significantly (Hwang et al. 2017). The *OsWRKYs* showed different ways of expression. The *OsWRK24, OsWRKY69*, and *OsWRKY21* were with higher expression. However, the genes *OsWRK47, OsWRKY10, OsWRKY72, OsWRKY77*, and *OsWRKY62* were found to be chilling stress tolerance-negative regulators. The 12 different *OsMAPKs* genes were expressed at different levels in rice (Viana et al.).

In two rice cultivars, 801 and 507 transcripts showed diverse expressions in stress environment. Gene ontology investigation recommended the transcripts enrichment intricate under abiotic stress and gene expression regulation in tolerant cultivars. The upregulation of the thioredoxin transcripts responsible for phenylpropanoid metabolism was noted in one cultivar, while the terpenoid and wax metabolism transcripts were upregulated in some cultivars (Shankar et al. 2016).

The genome-wide rice roots transcriptome profiling indicated that genes were related to defense system, accumulation of different type of stress-related molecules, and tolerance against chromium. The orthologous promoters database has 13 and 14 promoter motifs that were found to be highly significant, however, their functions are unknown (Dubey et al. 2010).

39.4.5 DNA Marker and QTLs Mapping

Quantitative trait loci (QTL) mapping is an important and evolving technique in plant breeding. Through using the QTL mapping, the plant tolerance mechanism can be understood. Under abiotic stress, the crop productivity can be achieved through these QTL. The agronomic and physiological attributes in plants can be improved under abiotic stresses. For translational genomics and the breeding of quantitative trait, QTL mapping works as an excellent tool particularly in comparative genomic approaches. By using the QTL mapping, the yield of crop can be improved under abiotic stress (Ganie et al. 2019).

Under salinity stress, seven QTL have been discovered that were linked with salinity stress in rice plant. By using the F2 population, from salt-tolerant to wild-type rice plant, a key gene has been isolated which was present in chromosome 7. On the chromosome 1, a major *Saltol* QTL has been mapped that were flanked by RM23 and RM140 markers. These QTL were involved in more than 70% of salinity tolerance. It was discovered that a specific QTL in chromosome 1 enhanced the potassium uptake in shoot of rice when subjected to salinity stress. In salt-tolerant cultivar, the potassium ion homeostasis was determined by the mapping of SKC1 present in chromosome 1 (Das et al. 2015).

39.4.6 MicroRNA (miRNAs) in Abiotic Stress Tolerance

In RNAs, there are small noncoding endogenous RNAs present known as small RNAs (sRNAs), which control the expression of genes at the levels of transcription and posttranscription. These sRNAs are classified on the basis of their function and biogenesis in plants. One class of these sRNAs is MicroRNAs (miRNAs) that have hairpin-like structure, derived from single-stranded precursors, and are primarily 21-nucleotide (nt)-long molecules. The posttranscriptional gene silencing (PTGS) mechanism involves miRNA as depicted in the variable expression of stress-related genes (Guan et al. 2010). Jones-Rhoades and Bartel showed a high degree of conservation among miRNAs and their mRNA targets while working on rice (*Oryza sativa*) and arabidopsis (*Arabidopsis thaliana*). Many miRNAs and their targets are species specific. Genome-wide analysis and development of high-throughput sequencing technologies have improved the identification of new variants of miRNA from different plant species. Under abiotic and biotic stresses, plant miRNAs are frequently explored as key regulators of specific gene expression. The overexpression of Osa-miR319a increased the tolerance against the salinity and drought stresses by increasing the wax contents in leaves to reduce the water and sodium ion uptake. In addition to this, the leaf size and number of tillers were also increased by overexpression of Osa-miR319a. The upregulation of miR319 also increased the cold tolerance in rice. The miR393 (OsAFB2 and OSTIR1) increased tolerance against the salinity and drought stresses (Zhou 2012). The overexpression of *mOsNAC2* conferred enhanced tolerance to drought stress (Jiang et al. 2019). The *miR820*—a rice-specific miRNA—was downregulated in water-deficit environment (Jeong and Green 2013).

Under abiotic stress, 227 miRNAs of 127 families were identified. About 70 miRNAs were not listed in the miRBase. The 62 miRNAs (10 novel) were validated with the previously published data of small RNA expression in RNAi lines of *DCL1*, *RDR2,* and *DCL3*. About 210 targets were set from 86 miRNAs with the help of degradome published data. The miRNAs 18, 15, and 10 were regulated in response to cold, salt, and drought stresses (Barrera-Figueroa et al. 2012). Shen et al. (2010) quantified 41 microRNAs in rice in drought, salinity, cold, and abscisic acid stress. Additional analysis suggests that *cis*-elements (stress responsive) were enhanced in the promoters region of stress-responsive microRNA genes. Under low nitrogen, 32 miRNAs were differentially expressed in root and leaves. Six miRNAs (miR1318, miR156, miR528, miR164, miR820, and miR821 were tested in leaves. While the miR167, miR164, miR528, and miR168 were reported in roots (Nischal et al. 2012).

Under salinity stress, the rice miRNA Osa-miR12477 was isolated and identified that encodes L-ascorbate oxidase (Bankole et al. 2017), and oxidative stress by DAB staining was indicated. Thus, salt tolerance mechanism might involve (miRNA-mediated) regulation of the hormone signaling components, for example, ARF; synthesis of the abiotic stress-related transcription factors and antioxidative components to mitigate the oxidative damages (Parmar et al. 2020). The extreme temperatures, heavy metals, drought, and salinity have triggered differential regulation of DIR in rice (Singh et al. 2020) (Table 39.3).

39.4.7 Metabolomics Approaches

The metabolites are the final products of cellular actions and reactions. Metabolomic analysis describes the cellular biochemical status at specific stage. The metabolomics tools provide further strength to the proteomics and transcriptomics findings to exhibit more clear figures of the cellular study. The rice metabolomics studies have been done to investigate the differences between the wild-type, transgenic lines and varieties under normal growth conditions as well as under stress conditions. Generally, in response to abiotic stresses, the level of sugar, sugar alcohol, and amino acids gets enhanced. The proline and raffinose sugar was increased by exposing to salinity, cold, and drought stresses in rice plants (Kaplan et al. 2004; Das et al. 2015).

Under dry cultivation system, the rice plants accumulate higher level of choline, lactose, and nicotinoylcholine in grains, which improves the protein contents. The higher rate of serine, threonine, glycine, sucrose, starch, and galactose metabolism was detected in rice through metabolomic approach. The biosynthesis of phenylpropanoids, anthocyanins, flavonoids, and subsidiary proteins and starch accumulation was increased in the endosperms (Chen et al. 2020). The nanoparticles of TiO_2 caused toxicity in rice plants. A study revealed that profile of ~105 different metabolites was significantly different in plants grown under normal conditions as compared with those grown under stress conditions. For instance, the levels of glucose-1-phosphate, glucose-6-phosphate, isocitric acid, and succinic were enhanced while the levels of glyoxylic acid and sucrose were reduced. The synthesis of carbohydrates and metabolism of sucrose, starch, dicarboxylate, and glyoxylate was inhibited while the secondary metabolites, amino acids, and fatty acids biosynthetic were enhanced (Wu et al. 2017).

The rice plants grown under drought stress exhibited variation in pathways of the carbon fixation and carbohydrates reactive oxygen species (ROS) and energy metabolism (Xiong et al. 2019b). Under salinity stress, the rice varieties differentially expressed some metabolites such as polyols, sugars, amino acids, certain

Table 39.3 Rice microRNAs involved in abiotic stress

MicroRNA	Stress conditions	Response	Validated/putative targets	References
miR169	Drought	Upregulated	CBF/DREBs transcription factors (TFs)	Zhao et al. (2007)
miR393	Salt and drought	Upregulated	Auxin receptors TIR1, AFB2, and AFB3	Xia et al. (2012)
miR398	Oxidative stress	Upregulated	Copper SOD enzyme	Li et al. (2011)
miR169	Oxidative stress	Upregulated	HAP2-like transcription factor	Li et al. (2011)
miR397	Oxidative stress	Upregulated	Laccase	Li et al. (2011)
miR827	Oxidative stress	Upregulated	SPX domain protein	Li et al. (2011)
miR1425	Oxidative stress	Upregulated	Pentatricopeptide repeat (PPR) protein	Li et al. (2011)
miR528	Oxidative stress cadmium	Down regulated/ upregulated	F-box containing protein, dicer-like 1	Li et al. (2011), Ding et al. (2011)
miR167	Auxin signaling	Upregulated	Auxin response factor	Meng et al. (2010)
miR160	Auxin signaling	Upregulated	Auxin response factor	Meng et al. (2010)
miR162	Cadmium	Downregulated	Dicer-like 1	Ding et al. (2011)
miR168	Cadmium	Downregulated	Argonaute	Ding et al. (2011)
miR166	Cadmium	Downregulated	HD-zip TFs	Ding et al. (2011)
miR171	Cadmium	Downregulated	Scarecrow-like TFs	Ding et al. (2011)
miR396	Cadmium	Downregulated	Rhodenase-like protein and kinesin-like protein B	Ding et al. (2011)
miR390	Cadmium	Downregulated	Leucine-rich repeat receptor-like protein kinase	Ding et al. (2011)
miR156	Cadmium	Downregulated	Squamosa promoter binding protein TFs	Ding et al. (2011)
miR1432	Cadmium	Downregulated	EF-hand proteins	Ding et al. (2011)
miR444	Cadmium	Downregulated	MADS-box TFs	Ding et al. (2011)

purine derivatives, and organic acids. The salt induced the higher level of serotonin in the salt-tolerant varieties. The signaling molecules such as ferulic acid, vanillic acid, and gentisic acid production were significantly increased under stress conditions. The salt-sensitive varieties showed relatively higher level of 4-hydroxybenzoic acid, cinnamic acid derivatives, and 4-hydroxycinnamic acid (Gupta and De 2017).

The metabolome analysis of the rice plants grown under salt stress with ABA application revealed certain variations in different processes in the plant cells such as the metabolic processes of cellular lipids and fatty acids, cytoplasmic transport, cell-wall remodeling, detoxification, and vacuolar sequestration in shoots and redox reactions in roots (Wang et al. 2015). Under drought-flooding alternate conditions, for rice plants, metabolites and proteins expressions indicated changes in metabolic pathways of photosynthesis, reactive oxygen species, and energy resulting in reduced grains yield (Xiong et al. 2019a). Under low temperature, the rice plants showed differences in phenylpropanoids, amino acids biosynthesis, and inositol phosphate and glutathione metabolism (Yang et al. 2019). Higher levels of amino acids, sorbitol, melezitose, and pipecolic acid play important role in salt tolerance in rice under salinity stress (Xie et al. 2020).

Under drought stress, the higher concentration of proline, lysine, arginine, ABA, aromatic and branched chain amino acids, 2-aminoadipate, GABA, allantoin, and saccharopine was noted (You et al. 2019). Under salinity stress, many secondary metabolites, valine, proline, glyceric acid, ascorbic acid, and phosphoenol pyruvic acid, were increased conferring salinity tolerance in rice plants (Wang et al. 2021).

39.5 Conclusions

The plants have to face a variety of abiotic stresses in their environment, which make their development, growth, and productivity difficult and challenging. These stresses bring loss to the productivity of crops such as the rice crop, one of the major food crops having significant importance in meeting the human food requirement all around the world, and eventually, lowering the economy of a state. To find out a better solution to this problem it is necessary to understand molecular mechanism of the cellular damage in plants due to abiotic stresses and the molecular mechanism of defense in naturally resistant plants. The molecular genetics serves as excellent tool to investigate stress-responsive- and stress tolerance-related genes and proteins/ peptides along with the regulatory machinery of their expression. Various molecular approaches have been followed to understand pathways and to confer abiotic stress tolerance in crop plants. The development of stress-related molecular markers, QTL mapping, transcription factors, microRNAs, siRNAs, genetic cloning and expression, molecular breeding, and genome editing, especially by state-of-the-art CRISPR/Cas technique, are the in-practice strategies to increase abiotic stress tolerance in rice crop. In spite of the fact that identification, isolation, and validation of a variety of stress-responsive genes have been done in different varieties of the rice crop, there is a need for more efforts and continuous research to further explore the effectiveness of stress-related genes, proteins/peptides, and metabolites in future molecular breeding programs.

References

A.-H.-Mackerness S, John CF, Jordan B, Thomas B (2001) Early signaling components in ultraviolet-B responses: distinct roles for different reactive oxygen species and nitric oxide. FEBS Lett 489:237–242

Abdel-Haliem ME, Hegazy HS, Hassan NS, Naguib DM (2017) Effect of silica ions and nano silica on rice plants under salinity stress. Ecol Eng 99:282–289

Ahmad F, Huang X, Lan H, Huma T, Bao Y, Huang J, Zhang H (2014) Comprehensive gene expression analysis of the DNA (cytosine-5) methyltransferase family in rice (Oryza sativa L.). Genet Mol Res 13:5159–5172

Ahmed M, Schmitz M (2011) Economic assessment of the impact of climate change on the agriculture of Pakistan. Bus Econ Horiz (BEH) 4:1–12

Ahmed MS, Soundhararajan R, Akther T, Kashif M, Khan J, Waseem M, Srinivasan H (2019) Biogenic AgNps synthesized via endophytic bacteria and its biological applications. Environ Sci Pollut Res 26:26939–26946

Alley RB, Marotzke J, Nordhaus WD, Overpeck JT, Peteet DM, Pielke RA, Pierrehumbert RT, Rhines PB, Stocker TF, Talley LD (2003) Abrupt climate change. Science 299:2005–2010

Bandyopadhyay T, Mehra P, Hairat S, Giri J (2017) Morpho-physiological and transcriptome profiling reveal novel zinc deficiency-responsive genes in rice. Funct Integr Genomics 17: 565–581

Bankole F, Menkir A, Olaoye G, Crossa J, Hearne S, Unachukwu N, Gedil M (2017) Genetic gains in yield and yield related traits under drought stress and favorable environments in a maize population improved using marker assisted recurrent selection. Front Plant Sci 8:808

Bardenas EA (1965) Morphology and varietal characteristics of the rice plant. The Int Rice Res Inst

Barrera-Figueroa BE, Gao L, Wu Z, Zhou X, Zhu J, Jin H, Liu R, Zhu JK (2012) High throughput sequencing reveals novel and abiotic stress-regulated microRNAs in the inflorescences of rice. BMC Plant Biol 12:132

Batool R, Hameed M, Ashraf M, Ahmad MSA, Fatima S (2015) Physio-anatomical responses of plants to heavy metals phytoremediation for green energy. Springer, pp 79–96

Bellard C, Bertelsmeier C, Leadley P, Thuiller W, Courchamp F (2012) Impacts of climate change on the future of biodiversity. Ecol Lett 15:365–377

Brosché M, Strid Å (2003) Molecular events following perception of ultraviolet-B radiation by plants. Physiol Plant 117:1–10

Buti M, Pasquariello M, Ronga D, Milc JA, Pecchioni N, Ho VT, Pucciariello C, Perata P, Francia E (2018) Transcriptome profiling of short-term response to chilling stress in tolerant and sensitive Oryza sativa ssp. Japonica seedlings. Funct Integr Genomics 18:627–644

Cai X, Chen T, Zhou Q, Xu L, Qu L, Hua X, Lin J (2011) Development of Casparian strip in rice cultivars. Plant Signal Behav 6:59–65

Ceccarelli S, Grando S, Maatougui M, Michael M, Slash M, Haghparast R, Rahmanian M, Taheri A, Al-Yassin A, Benbelkacem A (2010) Plant breeding and climate changes. J Agric Sci 148:627–637

Chen YA, Chi WC, Trinh NN, Huang LY, Chen YC, Cheng KT, Huang TL, Lin CY, Huang HJ (2014) Transcriptome profiling and physiological studies reveal a major role for aromatic amino acids in mercury stress tolerance in Rice seedlings. PLoS One 9:e95163

Chen Q, Zhang X, Liu Y, Wei J, Shen W, Shen Z, Cui J (2017) Hemin-mediated alleviation of zinc, lead and chromium toxicity is associated with elevated photosynthesis, antioxidative capacity; suppressed metal uptake and oxidative stress in rice seedlings. Plant Growth Regul 81:253–264

Chen Z, Gao W, Reddy K, Chen M, Taduri S, Meyers S, Shankle M (2020) Ultraviolet (UV) B effects on growth and yield of three contrasting sweet potato cultivars. Photosynthetica 58:37–44

Colmer TD, Pedersen O (2008a) Oxygen dynamics in submerged rice (Oryza sativa). New Phytol 178:326–334

Colmer TD, Pedersen O (2008b) Underwater photosynthesis and respiration in leaves of submerged wetland plants: gas films improve CO_2 and O_2 exchange. New Phytol 177:918–926

da Maia LC, Cadore PR, Benitez LC, Danielowski R, Braga EJ, Fagundes PR, Magalhães AM, de Oliveira AC (2017) Transcriptome profiling of rice seedlings under cold stress. Funct Plant Biol 44:419–429

Das P, Nutan KK, Singla-Pareek SL, Pareek A (2015) Understanding salinity responses and adopting 'omics-based' approaches to generate salinity tolerant cultivars of rice. Frontiers in Plant Sci 6:712

Das G, Patra JK, Baek KH (2017) Insight into MAS: a molecular tool for development of stress resistant and quality of rice through gene stacking. Front Plant Sci 8:985

Datta SK (2004) Rice biotechnology: a need for developing countries

Degenkolbe T, Do PT, Zuther E, Repsilber D, Walther D, Hincha DK, Köhl KI (2009) Expression profiling of rice cultivars differing in their tolerance to long-term drought stress. Plant Mol Biol 69:133–153

Deng X, Song X, Wei L, Liu C, Cao X (2016) Epigenetic regulation and epigenomic landscape in rice. Natl Sci Rev 3:309–327

Dent D, Binks RH (2020) Insect pest management Cabi

Des Marais DL, Hernandez KM, Juenger TE (2013) Genotype-by-environment interaction and plasticity: exploring genomic responses of plants to the abiotic environment. Annu Rev Ecol Evol Syst 44:5–29

Ding Y, Chen Z, Zhu C (2011) Microarray-based analysis of cadmium-responsive microRNAs in rice (Oryza sativa). J Exp Bot 62:3563–3573

Ding L, Li Y, Li Y, Shen Q, Guo S (2014) Effects of drought stress on photosynthesis and water status of rice leaves. Chin J Rice Sci 28:65–70

Dobermann A, Fairhurst T (2000) Nutrient disorders and nutrient management. Potash and Phosphate Institute, Potash and Phosphate Institute of Canada and International Rice Research Institute, Singapore

Du H, Liang Y, Pei K, Ma K (2011) UV radiation-responsive proteins in rice leaves: a proteomic analysis. Plant Cell Physiol 52:306–316

Duan Q, Lee J, Liu Y, Chen H, Hu H (2016) Distribution of heavy metal pollution in surface soil samples in China: a graphical review. Bull Environ Contam Toxicol 97:303–309

Dubey S, Misra P, Dwivedi S, Chatterjee S, Bag SK, Mantri S, Asif MH, Rai A, Kumar S, Shri M, Tripathi P, Tripathi RD, Trivedi PK, Chakrabarty D, Tuli R (2010) Transcriptomic and metabolomic shifts in rice roots in response to Cr (VI) stress. BMC Genomics 11:648

Enstone DE, Peterson CA, Ma F (2002) Root endodermis and exodermis: structure, function, and responses to the environment. J Plant Growth Regul 21:335–351

Fahad S, Hussain S, Saud S, Tanveer M, Bajwa AA, Hassan S, Shah AN, Ullah A, Wu C, Khan FA (2015) A biochar application protects rice pollen from high-temperature stress. Plant Physiol Biochem 96:281–287

FAO (2021) Food and agriculture organization of the United Nations. http://www.fao.org/home/en/

Farooq M, Wahid A, Kobayashi N, Fujita D, Basra S (2009) Plant drought stress: effects, mechanisms and management. Sustain Agric:153–188

Feng B, Zhang C, Chen T, Zhang X, Tao L, Fu G (2018) Salicylic acid reverses pollen abortion of rice caused by heat stress. BMC Plant Biol 18:1–16

Frohnmeyer H, Staiger D (2003) Ultraviolet-B radiation-mediated responses in plants. Balancing damage and protection. Plant Physiol 133:1420–1428

Fujita M, Fujita Y, Noutoshi Y, Takahashi F, Narusaka Y, Yamaguchi-Shinozaki K, Shinozaki K (2006) Crosstalk between abiotic and biotic stress responses: a current view from the points of convergence in the stress signaling networks. Curr Opin Plant Biol 9:436–442

Ganie SA, Molla KA, Henry RJ, Bhat K, Mondal TK (2019) Advances in understanding salt tolerance in rice. Theor Appl Genet 132:851–870

Geng A, Wang X, Wu L, Wang F, Wu Z, Yang H, Chen Y, Wen D, Liu X (2018) Silicon improves growth and alleviates oxidative stress in rice seedlings (Oryza sativa L.) by strengthening antioxidant defense and enhancing protein metabolism under arsanilic acid exposure. Ecotoxicol Environ Saf 158:266–273

Gnanamanickam SS (2009) Rice and its importance to human life biological control of rice diseases. Springer, pp 1–11

Gowda VR, Henry A, Yamauchi A, Shashidhar H, Serraj R (2011) Root biology and genetic improvement for drought avoidance in rice. Field Crops Res 122:1–13

Guan Y, Serraj R, Liu S, Xu J, Ali J, Wang W, Venus E, Zhu L, Li Z (2010) Simultaneously improving yield under drought stress and non-stress conditions: a case study of rice (Oryza sativa L.). J Exp Bot 61:4145–4156

Gupta P, De B (2017) Metabolomics analysis of rice responses to salinity stress revealed elevation of serotonin, and gentisic acid levels in leaves of tolerant varieties. Plant Signal Behav 12: e1335845

Halder K, Burrage S (2003) Drought stress effects on water relations of rice grown in nutrient film technique. Pak J Biol Sci 6:441–444

Hamdia M, Shaddad M (2010) Salt tolerance of crop plants. J Stress Physiol Biochem 6:64–90

Hasanuzzaman M, Nahar K, Gill SS, Fujita M (2013) Drought stress responses in plants, oxidative stress, and antioxidant defense. Clim Change Plant Abiot Stress Toler:209–250

Horie T, Karahara I, Katsuhara M (2012) Salinity tolerance mechanisms in glycophytes: An overview with the central focus on rice plants. Rice 5:1–18

Hossain MA, Cho JI, Han M, Ahn CH, Jeon JS, An G, Park PB (2010a) The ABRE-binding bZIP transcription factor OsABF2 is a positive regulator of abiotic stress and ABA signaling in rice. J Plant Physiol 167:1512–1520

Hossain MA, Lee Y, Cho JI, Ahn CH, Lee SK, Jeon JS, Kang H, Lee CH, An G, Park BP (2010b) The bZIP transcription factor OsABF1 is an ABA responsive element binding factor that enhances abiotic stress signaling in rice. Plant Mol Biol 72:557–566

Hu W, Hu G, Han B (2009) Genome-wide survey and expression profiling of heat shock proteins and heat shock factors revealed overlapped and stress specific response under abiotic stresses in rice. Plant Sci 176:583–590

Hu Y, Zhu N, Wang X, Yi Q, Zhu D, Lai Y, Zhao Y (2013) Analysis of rice Snf2 family proteins and their potential roles in epigenetic regulation. Plant Physiol Biochem 70:33–42

Huang DD, Huang HJ, Fu SF, Chou WC (2002) Transcriptional regulation of a Rice mitogen-activated protein kinase gene, OsMAPK4, in response to environmental stresses. Plant Cell Physiol 43:958–963

Huang Y, Zhao H, Gao F, Yao P, Deng R, Li C, Chen H, Wu Q (2018) A R2R3-MYB transcription factor gene, FtMYB13, from Tartary buckwheat improves salt/drought tolerance in Arabidopsis. Plant Physiol Biochem 132:238–248

Hwang S-G, Chapagain S, Lee JW, Han AR, Jang CS (2017) Genome-wide transcriptome profiling of genes associated with arsenate toxicity in an arsenic-tolerant rice mutant. Plant Physiol Biochem 120:40–51

Jacob JM, Karthik C, Saratale RG, Kumar SS, Prabakar D, Kadirvelu K, Pugazhendhi A (2018) Biological approaches to tackle heavy metal pollution: a survey of literature. J Environ Manag 217:56–70

Jagadish SK, Craufurd PQ, Wheeler TR (2007) High temperature stress and spikelet fertility in rice (Oryza sativa L.). J Exp Bot 58:1627–1635

Jeong DH, Green PJ (2013) The role of rice microRNAs in abiotic stress responses. J Plant Biol 56: 187–197

Jiang D, Zhou L, Chen W, Ye N, Xia J, Zhuang C (2019) Overexpression of a microRNA-targeted NAC transcription factor improves drought and salt tolerance in Rice via ABA-mediated pathways. Rice 12:76

Gayacharan, Joel AJ (2013) Epigenetic responses to drought stress in rice (Oryza sativa L.). Physiol Mol Biol Plants 19:379–387

Jung HI, Kong MS, Lee BR, Kim TH, Chae MJ, Lee EJ, Jung GB, Lee CH, Sung JK, Kim YH (2019) Exogenous glutathione increases arsenic translocation into shoots and alleviates arsenic-induced oxidative stress by sustaining ascorbate–glutathione homeostasis in rice seedlings. Front Plant Sci 10:1089

Kalbina I, Strid Å (2006) Supplementary ultraviolet-B irradiation reveals differences in stress responses between Arabidopsis thaliana ecotypes. Plant Cell Environ 29:754–763

Kaplan F, Kopka J, Haskell DW, Zhao W, Schiller KC, Gatzke N, Sung DY, Guy CL (2004) Exploring the temperature-stress metabolome of Arabidopsis. Plant Physiol 136:4159–4168

Karan R, DeLeon T, Biradar H, Subudhi PK (2012) Salt stress induced variation in DNA methylation pattern and its influence on gene expression in contrasting Rice genotypes. PLoS One 7: e40203

Khan MS, Akther T, Ali DM, Hemalatha S (2019) An investigation on the role of salicylic acid alleviate the saline stress in rice crop (Oryza sativa (L)). Biocatal Agric Biotechnol 18:101027

Kim YH, Khan AL, Waqas M, Shim JK, Kim DH, Lee KY, Lee IJ (2014) Silicon application to rice root zone influenced the phytohormonal and antioxidant responses under salinity stress. J Plant Growth Regul 33:137–149

Kim DS, Kim HS, Hong SJ, Cho JJ, Choi M, Heo SU, Lee J, Chung H, Shin EC (2018a) Investigation of water-soluble vitamin (B 1, B 2, and B 3) content in various rice, soups, and stews produced in Korea. Korean J Food Sci Technol 50:362–370

Kim Y, Mun BG, Khan AL, Waqas M, Kim HH, Shahzad R, Imran M, Yun BW, Lee IJ (2018b) Regulation of reactive oxygen and nitrogen species by salicylic acid in rice plants under salinity stress conditions. PLoS One 13:e0192650

Klueva N, Maestri E, Marmiroli N, Nguyen H (2001) Mechanisms of thermotolerance in crops. Crop responses and adaptations to temperature stress. Food Products Press, Binghamton, pp 177–217

Kou HP, Li Y, Song XX, Ou XF, Xing SC, Ma J, Von Wettstein D, Liu B (2011) Heritable alteration in DNA methylation induced by nitrogen-deficiency stress accompanies enhanced tolerance by progenies to the stress in rice (Oryza sativa L.). J Plant Physiol 168:1685–1693

Krishnamurthy P, Ranathunge K, Franke R, Prakash H, Schreiber L, Mathew M (2009) The role of root apoplastic transport barriers in salt tolerance of rice (Oryza sativa L.). Planta 230:119–134

Larkindale J, Vierling E (2008) Core genome responses involved in acclimation to high temperature. Plant Physiol 146:748–761

Li Z, Baldwin CM, Hu Q, Liu H, Luo H (2010) Heterologous expression of Arabidopsis H+-pyrophosphatase enhances salt tolerance in transgenic creeping bentgrass (Agrostis stolonifera L.). Plant Cell Environ 33:272–289

Li T, Li H, Zhang YX, Liu JY (2011) Identification and analysis of seven H2O2-responsive miRNAs and 32 new miRNAs in the seedlings of rice (Oryza sativa L. ssp. indica). Nucleic Acids Res 39:2821–2833

Liu J, Shabala S, Zhang J, Ma G, Chen D, Shabala L, Zeng F, Chen ZH, Zhou M, Venkataraman G (2020) Melatonin improves rice salinity stress tolerance by NADPH oxidase-dependent control of the plasma membrane K+ transporters and K+ homeostasis. Plant Cell Environ 43:2591–2605

Matsui T, Omasa K (2002) Rice (Oryza sativa L.) cultivars tolerant to high temperature at flowering: anther characteristics. Ann Bot 89:683–687

Matsui T, Namuco OS, Ziska LH, Horie T (1997) Effects of high temperature and CO_2 concentration on spikelet sterility in indica rice. Field Crops Res 51:213–219

Matsukura S, Mizoi J, Yoshida T, Todaka D, Ito Y, Maruyama K, Shinozaki K, Yamaguchi-Shinozaki K (2010) Comprehensive analysis of rice DREB2-type genes that encode transcription factors involved in the expression of abiotic stress-responsive genes. Mol Gen Genomics 283:185–196

Mekawy AMM, Abdelaziz MN, Ueda A (2018) Apigenin pretreatment enhances growth and salinity tolerance of rice seedlings. Plant Physiol Biochem 130:94–104

Meng Y, Chen D, Ma X, Mao C, Cao J, Wu P, Chen M (2010) Mechanisms of microRNA-mediated auxin signaling inferred from the rice mutant osaxr. Plant Signal Behav 5:252–254

Miyamoto T, Ochiai K, Takeshita S, Matoh T (2012) Identification of quantitative trait loci associated with shoot sodium accumulation under low potassium conditions in rice plants. Soil Sci Plant Nutr 58:728–736

Mmbando GS, Teranishi M, Hidema J (2020) Very high sensitivity of African rice to artificial ultraviolet-B radiation caused by genotype and quantity of cyclobutane pyrimidine dimer photolyase. Sci Rep 10:1–14

Munns R (2002) Comparative physiology of salt and water stress. Plant Cell Environ 25:239–250

Munns R, Tester M (2008) Mechanisms of salinity tolerance. Annu Rev Plant Biol 59:651–681

Muthurajan R, Rahman H, Manoharan M, Ramanathan V, Nallathambi J (2018) Drought responsive transcriptome profiling in roots of contrasting rice genotypes. Indian J Plant Physiol 23: 393–407

Nadeem F, Azhar M, Anwar-ul-Haq M, Sabir M, Samreen T, Tufail A, Awan HUM, Juan W (2020) Comparative response of two rice (Oryza sativa L.) cultivars to applied zinc and manganese for mitigation of salt stress. J Soil Sci Plant Nutr 20:2059–2072

Nakashima K, Tran LSP, Van Nguyen D, Fujita M, Maruyama K, Todaka D, Ito Y, Hayashi N, Shinozaki K, Yamaguchi-Shinozaki K (2007) Functional analysis of a NAC-type transcription factor OsNAC6 involved in abiotic and biotic stress-responsive gene expression in rice. Plant J 51:617–630

Nischal L, Mohsin M, Khan I, Kardam H, Wadhwa A, Abrol YP, Iqbal M, Ahmad A (2012) Identification and comparative analysis of microRNAs associated with low-N tolerance in rice genotypes. PLoS One 7:e50261

Pant PP, Tripathi A (2014) Impact of heavy metals on morphological and biochemical parameters of Shorea robusta plant. Ekológia 33:116

Parmar S, Gharat SA, Tagirasa R, Chandra T, Behera L, Dash SK, Shaw BP (2020) Identification and expression analysis of miRNAs and elucidation of their role in salt tolerance in rice varieties susceptible and tolerant to salinity. PLoS One 15:e0230958

Paul S, Roychoudhury A (2018) Transcriptome profiling of abiotic stress-responsive genes during cadmium chloride-mediated stress in two indica rice varieties. J Plant Growth Regul 37:657–667

Paul S, Roychoudhury A (2019) Transcript analysis of abscisic acid-inducible genes in response to different abiotic disturbances in two indica rice varieties. Theor Exp Physiol 31:249–272

Pedersen O, Rich SM, Colmer TD (2009) Surviving floods: leaf gas films improve O2 and CO2 exchange, root aeration, and growth of completely submerged rice. Plant J 58:147–156

Price AH, Cairns JE, Horton P, Jones HG, Griffiths H (2002) Linking drought-resistance mechanisms to drought avoidance in upland rice using a QTL approach: progress and new opportunities to integrate stomatal and mesophyll responses. J Exp Bot 53:989–1004

Rahman A, Mostofa MG, Nahar K, Hasanuzzaman M, Fujita M (2016) Exogenous calcium alleviates cadmium-induced oxidative stress in rice (Oryza sativa L.) seedlings by regulating the antioxidant defense and glyoxalase systems. Rev Bras Bot 39:393–407

Rai R, Agrawal M, Agrawal S (2016) Impact of heavy metals on physiological processes of plants: with special reference to photosynthetic system plant responses to xenobiotics. Springer, pp 127–140

Rajendran K, Tester M, Roy SJ (2009) Quantifying the three main components of salinity tolerance in cereals. Plant Cell Environ 32:237–249

Rashid M, Ejaz S, Shah KH (2020) Regulatory role of transcription factors in abiotic stress responses in plants plant ecophysiology and adaptation under climate change: mechanisms and perspectives II. Springer, pp 543–565

Reddy INBL, Kim BK, Yoon IS, Kim KH, Kwon TR (2017) Salt tolerance in rice: focus on mechanisms and approaches. Rice Sci 24:123–144

Reissig W (1985) Illustrated guide to integrated pest management in rice in tropical Asia. Int Rice Res Inst

Rezaul IM, Baohua F, Tingting C, Weimeng F, Caixia Z, Longxing T, Guanfu F (2019) Abscisic acid prevents pollen abortion under high-temperature stress by mediating sugar metabolism in rice spikelets. Physiol Plant 165:644–663

Rijal S, Devkota Y (2020) A review on various management method of rice blast disease. Malay J Sustain Agric 4:14–18

Rizwan M, Ali S, Akbar MZ, Shakoor MB, Mahmood A, Ishaque W, Hussain A (2017) Foliar application of aspartic acid lowers cadmium uptake and cd-induced oxidative stress in rice under cd stress. Environ Sci Pollut Res 24:21938–21947

Rohman A, Helmiyati S, Hapsari M, Setyaningrum DL (2014) Rice in health and nutrition. Int Food Res J 21

Rosenzweig C, Elliott J, Deryng D, Ruane AC, Müller C, Arneth A, Boote KJ, Folberth C, Glotter M, Khabarov N (2014) Assessing agricultural risks of climate change in the 21st century in a global gridded crop model intercomparison. Proc Natl Acad Sci 111:3268–3273

Roy S, Chakraborty U (2014) Salt tolerance mechanisms in salt tolerant grasses (STGs) and their prospects in cereal crop improvement. Bot Stud 55:1–9

Sahu PP, Pandey G, Sharma N, Puranik S, Muthamilarasan M, Prasad M (2013) Epigenetic mechanisms of plant stress responses and adaptation. Plant Cell Rep 32:1151–1159

Sailaja B, Anjum N, Prasanth VV, Sarla N, Subrahmanyam D, Voleti S, Viraktamath B, Mangrauthia SK (2014) Comparative study of susceptible and tolerant genotype reveals efficient recovery and root system contributes to heat stress tolerance in rice. Plant Mol Biol Report 32:1228–1240

Seki T, Nagase R, Torimitsu M, Yanagi M, Ito Y, Kise M, Mizukuchi A, Fujimura N, Hayamizu K, Ariga T (2005) Insoluble fiber is a major constituent responsible for lowering the post-prandial blood glucose concentration in the pre-germinated brown rice. Biol Pharm Bull 28:1539–1541

Seneviratne M, Weerasundara L, Ok YS, Rinklebe J, Vithanage M (2017) Phytotoxicity attenuation in Vigna radiata under heavy metal stress at the presence of biochar and N fixing bacteria. J Environ Manag 186:293–300

Shankar R, Bhattacharjee A, Jain M (2016) Transcriptome analysis in different rice cultivars provides novel insights into desiccation and salinity stress responses. Sci Rep 6:23719

Shen J, Xie K, Xiong L (2010) Global expression profiling of rice microRNAs by one-tube stem-loop reverse transcription quantitative PCR revealed important roles of microRNAs in abiotic stress responses. Mol Gen Genomics 284:477–488

Shri M, Kumar S, Chakrabarty D, Trivedi PK, Mallick S, Misra P, Shukla D, Mishra S, Srivastava S, Tripathi RD, Tuli R (2009) Effect of arsenic on growth, oxidative stress, and antioxidant system in rice seedlings. Ecotoxicol Environ Saf 72:1102–1110

Singh, R. and T. Flowers. 2010. 36 the physiology and molecular biology of the effects of salinity on Rice

Singh A, Giri J, Kapoor S, Tyagi AK, Pandey GK (2010) Protein phosphatase complement in rice: genome-wide identification and transcriptional analysis under abiotic stress conditions and reproductive development. BMC Genomics 11:435

Singh DK, Mehra S, Chatterjee S, Purty RS (2020) In silico identification and validation of miRNA and their DIR specific targets in Oryza sativa Indica under abiotic stress. Non-coding RNA Res 5:167–177

Su CF, Wang YC, Hsieh TH, Lu CA, Tseng TH, Yu SM (2010) A novel MYBS3-dependent pathway confers cold tolerance in rice. Plant Physiol 153:145–158

Subramanyam K, Du Laing G, van Damme EJ (2019) Sodium selenate treatment using a combination of seed priming and foliar spray alleviates salinity stress in rice. Front Plant Sci 10:116

Suji K, Joel AJ (2010) An epigenetic change in rice cultivars under water stress conditions. Electron J Plant Breed 1:1142–1143

Suzuki N, Rivero RM, Shulaev V, Blumwald E, Mittler R (2014) Abiotic and biotic stress combinations. New Phytol 203:32–43
Takasaki H, Maruyama K, Kidokoro S, Ito Y, Fujita Y, Shinozaki K, Yamaguchi-Shinozaki K, Nakashima K (2010) The abiotic stress-responsive NAC-type transcription factor OsNAC5 regulates stress-inducible genes and stress tolerance in rice. Mol Gen Genomics 284:173–183
Thamilarasan SK, Park JI, Jung HJ, Nou IS (2014) Genome-wide analysis of the distribution of AP2/ERF transcription factors reveals duplication and CBFs genes elucidate their potential function in Brassica oleracea. BMC Genomics 15:422–422
Tripathi DK, Singh S, Singh VP, Prasad SM, Dubey NK, Chauhan DK (2017) Silicon nanoparticles more effectively alleviated UV-B stress than silicon in wheat (Triticum aestivum) seedlings. Plant Physiol Biochem 110:70–81
Viana VE, da Maia LC, Busanello C C. Pegoraro and A. Costa de Oliveira. When rice gets the chills: comparative transcriptome profiling at germination shows WRKY transcription factors responses. Plant Biol
Wahid A, Gelani S, Ashraf M, Foolad MR (2007) Heat tolerance in plants: an overview. Environ Exp Bot 61:199–223
Wang WS, Zhao XQ, Li M, Huang LY, Xu JL, Zhang F, Cui YR, Fu BY, Li ZK (2015) Complex molecular mechanisms underlying seedling salt tolerance in rice revealed by comparative transcriptome and metabolomic profiling. J Exp Bot 67:405–419
Wang Y, Huang L, Du F, Wang J, Zhao X, Li Z, Wang W, Xu J, Fu B (2021) Comparative transcriptome and metabolome profiling reveal molecular mechanisms underlying OsDRAP1-mediated salt tolerance in rice. Sci Rep 11:5166
Wanger TC, Darras K, Bumrungsri S, Tscharntke T, Klein AM (2014) Bat pest control contributes to food security in Thailand. Biol Conserv 171:220–223
Waseem M, Huang F, Wang Q, Aslam MM, Abbas F, Ahmad F, Ashraf U, Hassan W, Fiaz S, Ye X, Yu L, Ke Y (2021) Identification, methylation profiling, and expression analysis of stress-responsive cytochrome P450 genes in rice under abiotic and phytohormones stresses. GM Crops & Food:1–13
Weerakoon W, Maruyama A, Ohba K (2008) Impact of humidity on temperature-induced grain sterility in rice (Oryza sativa L). J Agron Crop Sci 194:135–140
Wei S, Hu W, Deng X, Zhang Y, Liu X, Zhao X, Luo Q, Jin Z, Li Y, Zhou S (2014) A rice calcium-dependent protein kinase OsCPK9 positively regulates drought stress tolerance and spikelet fertility. BMC Plant Biol 14:1–13
Wing RA, Purugganan MD, Zhang Q (2018) The rice genome revolution: from an ancient grain to Green super Rice. Nat Rev Genet 19:505–517
Wu B, Zhu L, Le XC (2017) Metabolomics analysis of TiO2 nanoparticles induced toxicological effects on rice (Oryza sativa L.). Environ Pollut 230:302–310
Wutipraditkul N, Wongwean P, Buaboocha T (2015) Alleviation of salt-induced oxidative stress in rice seedlings by proline and/or glycinebetaine. Biol Plant 59:547–553
Xangsayasane P, Jongdee B, Pantuwan G, Fukai S, Mitchell J, Inthapanya P, Jothiyangkoon D (2014) Genotypic performance under intermittent and terminal drought screening in rainfed lowland rice. Field Crops Res 156:281–292
Xia K, Wang R, Ou X, Fang Z, Tian C, Duan J, Wang Y, Zhang M (2012) OsTIR1 and OsAFB2 downregulation via OsmiR393 overexpression leads to more tillers, early flowering and less tolerance to salt and drought in rice. PLoS One 7:e30039
Xie Z, Wang C, Zhu S, Wang W, Xu J, Zhao X (2020) Characterizing the metabolites related to rice salt tolerance with introgression lines exhibiting contrasting performances in response to saline conditions. Plant Growth Regul 92:157–167
Xiong Q-Q, Shen T-H, Zhong L, Zhu C-L, Peng X-S, He X-P, Fu J-R, Ouyang L-J, Bian J-M, Hu L-F, Sun X-T, Xu J, Zhou H-Y, He H-H, Chen X-R (2019a) Comprehensive metabolomic, proteomic and physiological analyses of grain yield reduction in rice under abrupt drought–flood alternation stress. Physiol Plant 167:564–584

Xiong Q, Cao C, Shen T, Zhong L, He H, Chen X (2019b) Comprehensive metabolomic and proteomic analysis in biochemical metabolic pathways of rice spikes under drought and submergence stress. Biochim Biophys Acta (BBA) Prote Proteom 1867:237–247

Yadav DK, Islam SMS, Tuteja N (2012) Rice heterotrimeric G-protein gamma subunits (RGG1 and RGG2) are differentially regulated under abiotic stress. Plant Signal Behav 7:733–740

Yamaguchi-Shinozaki K, Shinozaki K (2005) Organization of cis-acting regulatory elements in osmotic-and cold-stress-responsive promoters. Trends Plant Sci 10:88–94

Yamanouchi U, Yano M, Lin H, Ashikari M, Yamada K (2002) A rice spotted leaf gene, Spl7, encodes a heat stress transcription factor protein. Proc Natl Acad Sci 99:7530–7535

Yan F, Wei H, Ding Y, Li W, Chen L, Ding C, Tang S, Jiang Y, Liu Z, Li G (2021) Melatonin enhances Na+/K+ homeostasis in rice seedlings under salt stress through increasing the root H+-pump activity and Na+/K+ transporters sensitivity to ROS/RNS. Environ Exp Bot 182:104328

Yang M, Yang J, Su L, Sun K, Li D, Liu Y, Wang H, Chen Z, Guo T (2019) Metabolic profile analysis and identification of key metabolites during rice seed germination under low-temperature stress. Plant Sci 289:110282

Ye S, Yu S, Shu L, Wu J, Wu A, Luo L (2012) Expression profile analysis of 9 heat shock protein genes throughout the life cycle and under abiotic stress in rice. Chin Sci Bull 57:336–343

Yin X, Kroff MJ, Goudriaan J (1996) Differential effects of day and night temperature on development to flowering in rice. Ann Bot 77:203–213

Yoshida T, Mogami J, Yamaguchi-Shinozaki K (2014) ABA-dependent and ABA-independent signaling in response to osmotic stress in plants. Curr Opin Plant Biol 21:133–139

You J, Zong W, Du H, Hu H, Xiong L (2014) A special member of the rice SRO family, OsSRO1c, mediates responses to multiple abiotic stresses through interaction with various transcription factors. Plant Mol Biol 84:693–705

You J, Zhang Y, Liu A, Li D, Wang X, Dossa K, Zhou R, Yu J, Zhang Y, Wang L, Zhang X (2019) Transcriptomic and metabolomic profiling of drought-tolerant and susceptible sesame genotypes in response to drought stress. BMC Plant Biol 19:267

Yu G, Li W, Yuan Z, Cui H, Lv C, Gao Z, Han B, Gong Y, Chen G (2013) The effects of enhanced UV-B radiation on photosynthetic and biochemical activities in super-high-yield hybrid rice Liangyoupeijiu at the reproductive stage. Photosynthetica 51:33–44

Yuenyong W, Chinpongpanich A, Comai L, Chadchawan S, Buaboocha T (2018) Downstream components of the calmodulin signaling pathway in the rice salt stress response revealed by transcriptome profiling and target identification. BMC Plant Biol 18:335

Zeng L, Shannon MC (2000) Salinity effects on seedling growth and yield components of rice. Crop Sci 40:996–1003

Zhang T, Zhao X, Wang W, Pan Y, Huang L, Liu X, Zong Y, Zhu L, Yang D, Fu B (2012) Comparative transcriptome profiling of chilling stress responsiveness in two contrasting rice genotypes. PLoS One 7:e43274

Zhao B, Liang R, Ge L, Li W, Xiao H, Lin H, Ruan K, Jin Y (2007) Identification of drought-induced microRNAs in rice. Biochem Biophys Res Commun 354:585–590

Zheng X, Chen B, Lu G, Han B (2009) Overexpression of a NAC transcription factor enhances rice drought and salt tolerance. Biochem Biophys Res Commun 379:985–989

Zhou M (2012) Genetic engineering of turfgrass for enhanced performance under environmental stress

Zhu N, Cheng S, Liu X, Du H, Dai M, Zhou D-X, Yang W, Zhao Y (2015) The R2R3-type MYB gene OsMYB91 has a function in coordinating plant growth and salt stress tolerance in rice. Plant Sci 236:146–156

Zibaee A (2013) Rice: importance and future. J Rice Res 1:e102

Zibaee A, Jalali Sendi J, Ghadamyari M, Alinia F, Etebari K (2009) Diazinon resistance in different selected strains of Chilo suppressalis (Lepidoptera: Crambidae) in northern Iran. J Econ Entomol 102:1189–1196

Zinn KE, Tunc-Ozdemir M, Harper JF (2010) Temperature stress and plant sexual reproduction: uncovering the weakest links. J Exp Bot 61:1959–1968

Zong W, Zhong X, You J, Xiong L (2013) Genome-wide profiling of histone H3K4-tri-methylation and gene expression in rice under drought stress. Plant Mol Biol 81:175–188

Zou J, Liu A, Chen X, Zhou X, Gao G, Wang W, Zhang X (2009) Expression analysis of nine rice heat shock protein genes under abiotic stresses and ABA treatment. J Plant Physiol 166:851–861

Zou J, Liu C, Chen X (2011) Proteomics of rice in response to heat stress and advances in genetic engineering for heat tolerance in rice. Plant Cell Rep 30:2155–2165

Is Geographical Indication System an Opportunity for Developing-8 (D-8) Countries? An Evaluation of Registered Rice Production

40

Mustafa Kan, Arzu Kan, and Muhammad Ashfaq

Abstract

GI, which is an intellectual property right, is perceived as quality product, differentiated product, and local product, identified with its production origin, produced with the contribution of human elements and at the same time a part of culture. GI, which is becoming more and more accepted in the international arena, adds a multi-dimensional value to it both as a local development tool and is being used for the protection and promotion of cultural values. At the same time, it can be used to have an important market segment in the trade of products in the international arena. GI, which provide competitive advantage for certain products with a perception of quality in trade, has started to find a place in the legal legislation of many countries. D-8 OEC, which dates back to 1996, is one of the important economic cooperation organizations formed by eight Muslim countries. Countries within this organization have the potential to make the most of the benefits of the GI system with their biodiversity and cultural diversity. One of these products is rice. D-8 countries, which have approximately 20% of the world production, also have 15 GI registered rice that they have registered in their national legislation. Bangladesh, Indonesia, Nigeria, and Pakistan are among the leading countries in the D-8 OEC, which are an important production center especially for aromatic rice. Among the D-8 countries, the GI system has been used for a long time in some countries (Turkey and Malaysia), while in others there is either no specific national legal basis or the system is in a new

M. Kan (✉) · A. Kan
Agricultural Faculty, Department of Agricultural Economics, Kirsehir Ahi Evran University, Kırşehir, Turkey
e-mail: mustafa.kan@ahievran.edu.tr

M. Ashfaq
Muhammad Nawaz Shareef University of Agriculture, Multan, Pakistan

N. Sarwar et al. (eds.), *Modern Techniques of Rice Crop Production*,
https://doi.org/10.1007/978-981-16-4955-4_40

development state. With this study, the general situation of the GI system in D-8 countries has been presented and a general evaluation has been made on rice. As a result, it is revealed that the efforts of the D-8 countries to carry out joint studies in many fields should also be demonstrated on the issue of intellectual property rights and that the studies should turn into more action.

Keywords

Geographical indication · Developing-8 · Economic cooperation · Rice

Abbreviations

ASEAN	Association of Southeast Asian Nations
D-8 OEC	Developing-8 Organization for Economic Cooperation
EU	European Union
GDP	Gross domestic product
GI	Geographical indication
GIRPA	Geographical Indications Registration and Protection Act
SDG-2030	Sustainable Development Goals-2030
TRIPS	Trade-Related Aspects of Intellectual Property
UN	United Nations
WIPO	World Intellectual Property Organization
WTO	World Trade Organization

40.1 Introduction

The agricultural sector is more prominent due to the role it plays in the development process, especially in underdeveloped and developing countries, the population and workforce it hosts, its importance in terms of food security and security, and its interaction with other sectors (Ahmad et al. 2008, 2009, 2012, 2013, 2015, 2019). Especially with the prevalence of Covid-19 pandemic, we once again clearly understood how important and vital agriculture sector is. The road to development is primarily through food security and food safety (Ahmad and Hasanuzzaman 2012; Ahmed and Ahmad 2019, 2020; Shahzad and Ahmad 2019; Fatima et al. 2020). For this reason, it cannot be denied that regional integrations should enter a new era that will help each other both commercially, economically, and humanly. In evaluating the development parameters, it should be an important criterion to have the ability to be resilient against the food security risk from now on.

According to the scenarios made for the world population, it is predicted that the people will reach 9.5 billion with an overall increase of 32% by the year 2050 from its current number of 7.2 billion, and will reach to 11 billion by increasing 53% in 2100. Today, the fact is that arable agricultural lands are on the border, the limitations in water and energy require a significant increase in efficiency to meet the increasing food demand (Reynolds et al. 2015). This increase will be possible with the joint efforts and coordination among countries.

Reducing poverty, ending hunger, and raising awareness of responsible consumption and production constitute 3 out of 17 Sustainable Development Goals-2030 (SDG-2030; UN 2015). One of the keys to achieving these goals is shown as partnerships and collaboration in the international arena. In the literature, many studies focus on the importance of global partnerships on essential issues such as poverty reduction, ending hunger, combating climate change, food security, food safety, and the environment (Röder 1985; UN 2015, 2018; Peker et al. 2019). For this reason, within the scope of efforts to increase cooperation, countries can enter into economic integrations to increase their productivity by expanding their production capacity economically and consequently, growing the level of social welfare and developing a culture of working together to resolve regional and global problems.

Developing-8 Organization for Economic Cooperation (D-8 OEC) consists of eight Muslim countries (Bangladesh, Indonesia, Iran, Malaysia, Egypt, Nigeria, Pakistan, and Turkey). The foundations were laid in October 1996 by the former Prime Minister of the Republic of Turkey, Prof. Dr. Necmettin Erbakan, with the representatives of these countries at the "Development Cooperation Conference" held in Istanbul. After a series of preparatory meetings following the "Conference on Cooperation in Development" on October 22, 1996, the establishment of D-8 was officially declared at the Heads of State and Government Summit held in Istanbul on June 15, 1997 (Istanbul Declaration). The purpose of D-8 is to increase the trade and cooperation among member countries. The main goal of launching the D-8 initiative is to create and diversify new opportunities in trade relations among member countries that represent a tremendous economic potential, diverse resources, a broad population, and geographical area, to increase participation in the decision-making process at the international level, to provide better living conditions, to develop economic cooperation around projects, and to strengthen the status of developing countries in the world economy. Agriculture and food security and renewable energy sources, industry, transportation, and tourism are some of the main issues expected in cooperation (D-8 OEC 2021a).

D-8, targeting global partnership rather than regional integration, has created a common platform for defending the rights of developing countries against developed countries in World Trade Organization (WTO) decisions. In addition, they have the fields of joint work under the titles of "Agriculture and Food Security," "Trade," "Transportation," "Industrial Partnership," "Energy and Minerals," "Tourism," and "Other Partnerships" (D-8 OEC 2021b). With a population of more than 650 million, the agricultural sector has a significant share in the total gross domestic product (GDP) in D-8 countries, which contain approximately 11% of the world's population. As of 2019, Pakistan ranked first with 22.04% of the share of agriculture in total GDP, followed by Nigeria (21.91%) and Indonesia (12.72%). It can be seen here that the country with the lowest proportional shares of agriculture is Turkey (6.42%) (The Global Economy 2021).

One of the most important assets of D-8 countries is their genetic resources. The importance of the subject was identified at the first D-8 Agricultural Ministerial Meeting on Food Security which was held on 25–27 February 2009 in Kuala Lumpur, Malaysia, as establishing a Seed Bank was one of the areas under agricultural cooperation. In the fourth Working Group Meeting on D-8 Seed Bank, which was held on 3 October 2012 in Mataram, West Nusa Tenggara-Indonesia, member countries agreed to hold workshops, construct standard rules in material exchange, harmonization of the evaluation method for breeding, etc. (D-8 2021c). Although the decisions taken and the importance of the issue have been emphasized so far, it is seen that there is no significant progress so far.

D-8 countries stand out in terms of cultural richness around these resources as well as the diversity of genetic resources in both plant and animal products. Especially coffee, rice, different types of fruits and vegetables, and the gastronomic diversity make these countries destination centers in terms of cuisine and halal tourism. The D-8 organization, formed by the combination of countries with such a significant diversity, needs to make more efforts to become a global player and to promote its wealth and use it for their benefits. Therefore, one of these efforts is geographical indication (GI), known as an intellectual property right, which is one of the labeling systems used to promote and protect countries' values.

In the studies on GI, both in Turkey and in the international arena, the concept of geographical indication has been examined with different dimensions. It is one of the common ideas of these studies that GIs create an important dynamic in local development (Wilson et al. 1999; Treager 2003; Treager et al. 2007; Lopez and Martin 2005; Kan and Gülçubuk 2008; Kan et al. 2010; Orhan 2010; Çalışkan and Koç 2012; Kan and Gülçubuk 2012, 2013; Kan et al. 2013; UNCTAD 2015; Çukur and Çukur 2017; Belletti et al. 2017; FAO 2018; ARISE+IPR 2019). The feature that makes geographical signs important is that local products and flavors can be protected within this system, which can be used as a tool in regional economic development. For this reason, GIs appear as a developing trend in many countries, especially in European countries, and they do not fall off the agenda in terms of both protecting and maintaining cultural and local heritage (Kan et al. 2012). GIs, which are evaluated in terms of advantages and disadvantages in many ways today, have started to show significant developments in Europe and Asia, and African countries in recent years (Kan 2011; Marescotti 2003; Drahos 2017).

In areas where there is a lot of genetic and cultural diversity, many products that may be subject to GIs are likely to emerge. In this context, paddy is one of those products, among the D-8 countries, which constitute 20.09% of the world production area and 18.71% of the production, especially Bangladesh, Indonesia, Nigeria, and Pakistan are the major contributors. In addition, because some of these countries have a geographical indication system and some do not, the D-8 framework can provide an important opportunity to work together. In today's world, where competition is increasing day by day, introducing and evaluating existing values under a common platform means gaining significant global power. One of the most critical characteristics of D-8 countries is their rich biodiversity. This wealth makes these countries a vital production center in agriculture. One of these production areas is the rice production, which is the primary food of people in many countries.

In this chapter, the status of the geographical indication system in D-8 countries and paddy production in this system are explained. This chapter has tried to reveal the existing potentials of the geographical indication system or similar systems in the possible future working areas of the D-8 union in terms of joint action in many products, including paddy.

40.2 Rice Production and Trade in D-8 Countries

Paddy is grown significantly in Asia, on the border between Russia and China, up to the shores of the Amur River (53°N) in the north and south of central Argentina (40°S). It is grown in cool climates in the mountains of Nepal and India and the hot climatic conditions of Pakistan, Iran, and Egypt by applying the irrigation. It is a crop being grown in highlands in Asia, Africa, and some parts of Latin America. In addition to this, in areas such as the Mekong in Vietnam, Chao Phraya in Thailand, Irrawady in Myanmar, and Ganges-Brahmaputra in eastern India, the areas that remain in seasonally formed deep water beds (Patra et al. 2016). Paddy spread from South India to China in 3000 BC and to Java in 1000 BC, and it was introduced to Europe in 300 BC at the end of Alexander, the Great's Asian expeditions (Kün 1997). The debates are still going on where the paddy was first cultivated (UCL 2021).

Paddy is one of the oldest cultivated plants. It is estimated that the gene center is India and China in Southeast Asia. It is a grain that can germinate in water and its roots can utilize the oxygen in the water. As a food source for humans, it is the most important product after wheat among grains. At the same time, rice is the main food source for more than half of the people in the world (UHK 2011).

India, Bangladesh, Burma, Thailand, Vietnam, and Southern China are the regions with the greatest genetic variation. It is accepted that it spread from here to other areas for the first time. Recent genetic evidence show that all forms of Asian rice, both indica and japonica, come from a single domestication event that occurred 8200–13,500 years ago in the Pearl River valley region of China (Ricepedia 2021). Among the eight gene centers determined by Vavilov in his study in 1935, Hindustani gene center (including tropical India, Indo-china, southern China, and the islands of southeastern Asia) is shown as the gene center of paddy (Hummer and Hancock 2015).

There are many types of rice which are being produced around the world. Rice can be divided into the following categories, depending on where it is grown (PACRA 2020):

- **Indica**: Grown in tropical and sub-tropical regions and accounts for ~75% of the global rice trade.
- **Aromatic (Jasmine & Basmati)**: Thailand, Vietnam, India, and Pakistan are major producer countries. It accounts for ~16 to 18% of the world rice trade.

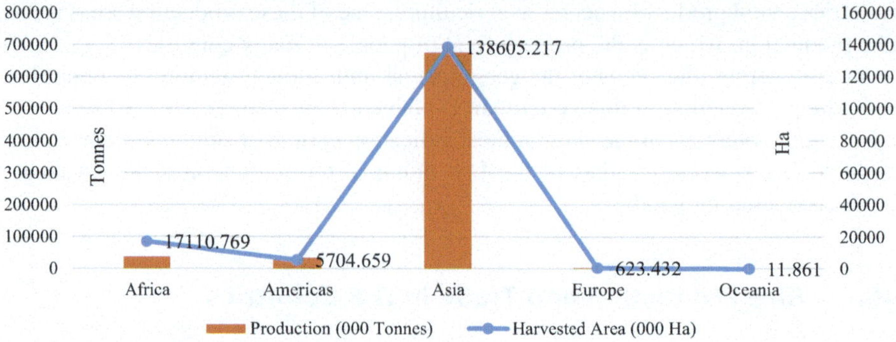

Fig. 40.1 Paddy production and harvested area by the Continentals (FAOSTAT 2020)

- **Japonica**: It grows in colder climates and accounts for ~5 to 6% of global trade.
- **Glutinous**: Majorly grown in Southeast Asia, and contributes to the remaining share of world trade.

As of 2019, there were approximately 1.44 billion hectares of agricultural land in the world. Grains are cultivated on approximately 724 million hectares of this area. Paddy is approximately 22% of the world's total grain cultivation and has 25% share in its production. Approximately, 90% of the world rice production of 755 million tons in 2019 was grown in Asia. China ranks first in terms of world paddy production (approximately 210 million tonnes), followed by India (178 million tonnes), Indonesia (55 million tonnes), and Bangladesh (55 million tons) (Fig. 40.1).

In Fig. 40.2, the leading countries in rice production and trade in the world are shown. When the figures are examined, the countries with the highest output are China, India, Bangladesh, and Indonesia. While the most exporting countries were India, Thailand, and Vietnam, the most importing countries were the Philippines, China, and Iran.

When the place of D-8 countries in world production and trade is examined, it is seen that they have approximately 20% of the world cultivation area, 18% of the total output, 11% of the world export, and 10% of its imports. Especially Bangladesh and Indonesia are the leaders in production, while Pakistan is at the forefront with its exports and Iran and Malaysia with its imports (Table 40.1).

40.3 Geographical Indication System in D-8 Countries

In today's world, diversity in the markets has led to a tremendous increase in competition. In an increasingly competitive environment, different methods and strategies have been applied to ensure sustainability. In this context, product differentiation is an essential strategy. The process of directing and managing the consumer's perception is also vital to ensure selectivity in the structure that becomes homogeneous with the developing technologies. In recent years, healthy and reliable

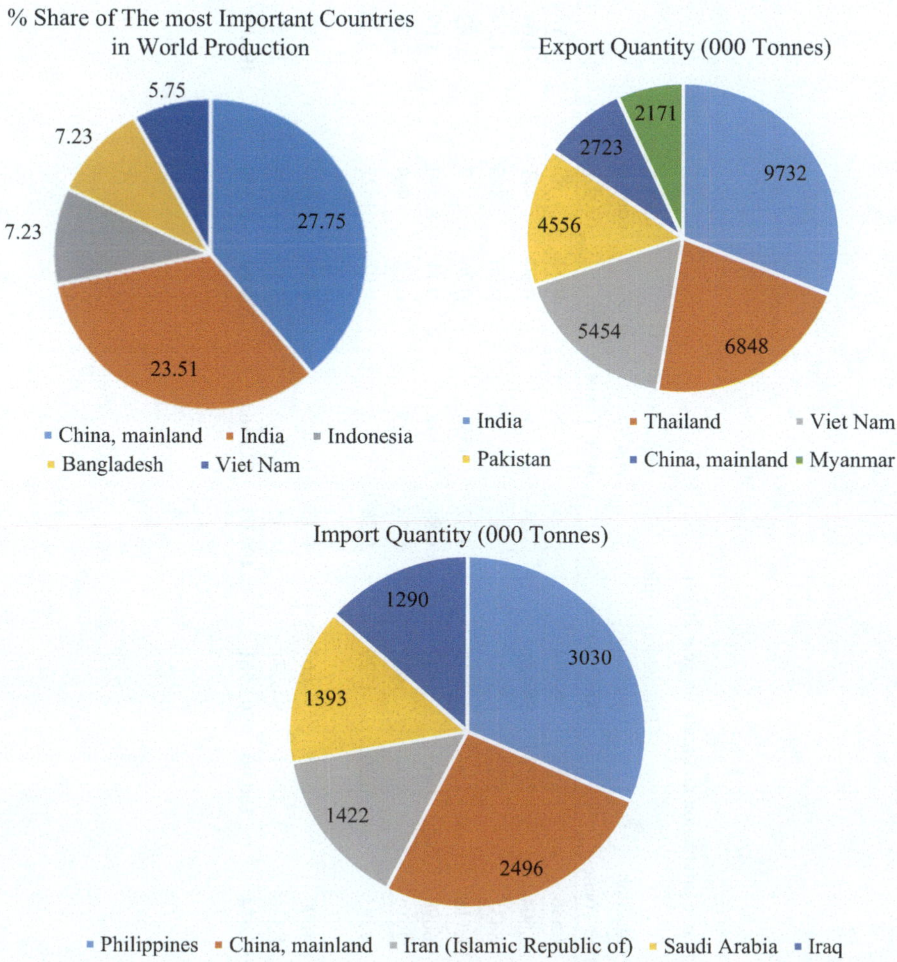

Fig. 40.2 Top countries on production, export, and import on rice, paddy (rice milled equivalent) (year 2019) (FAOSTAT 2020)

food has an essential place in the choice list of consumer. Especially with globalization, foods that become more homogeneous and industrialized create an important risk perception for consumers. This has led to increased consumer interest in differentiated products (Bramley et al. 2009; Stolzenbach et al. 2013; Kılıç and Kurnaz 2010; Asif et al. 2018), the preference of quality, health friendly (Akbay and Eugene 2005; Verbeke 2013; Tomescu 2015), and environmental friendly products (Schlegelmilch et al. 1996; Hopkins and Roche 2009; Chaudhary and Bisai 2018) rather than quantity, as well as an increased interest in authentic flavors (Bessière 1998; Trichopoulou et al. 2007; Löker et al. 2013; Lee et al. 2017). Within this structure, traditional, local, geographical indication, etc., product descriptions, and labeling try to meet the consumer's needs in this market segment.

Table 40.1 The Share of D-8 countries in the world rice production and trade (in year 2019)

	Area		Production		Production (milled equivalent)		Export (milled equivalent)		Import (milled equivalent)	
Countries	(000 Ha)	%	(000 Tonnes)	%	(000 Tonnes)	%	(000 Tonnes)	%	(000 Tonnes)	%
Bangladesh	11,517	7.11	54,586	7.23	36,409	7.23	8	0.02	83	0.18
Indonesia	10,678	6.59	54,604	7.23	36,421	7.23	1	0.00	471	1.04
Nigeria	5281	3.26	8435	1.12	5626	1.12	0	0.00	845	1.87
Pakistan	3034	1.87	11,115	1.47	7414	1.47	4556	10.76	7	0.02
Egypt	799	0.49	6690	0.89	4462	0.89	48	0.11	493	1.09
Malaysia	684	0.42	2912	0.39	1942	0.39	25	0.06	969	2.15
Iran (Islamic Republic of)	437	0.27	1993	0.26	1329	0.26	2	0.01	1422	3.15
Turkey	126	0.08	1000	0.13	667	0.13	202	0.48	484	1.07
D-8	32,557	20.09	141,336	18.71	94,271	18.71	4843	11.43	4774	10.58
World	**162,056**	**100.00**	**755,474**	**100.00**	**503,901**	**100.00**	**42,356**	**100.00**	**45,130**	**100.00**

Source: FAOSTAT (2020)

GIs are a vital tool in meeting the needs and expectations of consumers in this global market. GIs, which are evaluated within the scope of the protection of intellectual property rights, are also crucial in the international arena to show the countries' values and protect their values. In particular, the WTO rules to be followed in the international arena and efforts to abolish protectionism policies in domestic and foreign trade, along with practices such as product differentiation to gain competitive advantage, bring along the efforts of countries to create a "Special for Me" perception. For this reason, it means that the expectations from this concept increase in countries implementing this system.

The fact that the concept of a geographical sign is multi-dimensional, brings the conclusion that it should be multi-dimensional and diverse in its expectations from this concept. Geographical signs are not just tools that provide a commercial advantage or provide legal protection. Their multi-functional structure makes them different from other intellectual property rights, enabling them to exist in a broader context. In this context, GIs, on the one hand, activate local values such as the intended use of the environment, tradition, and culture; on the other hand, as an integrated form of rural development, they vigorously develop and advance commercial and economic interests. GIs are an excellent example of the new term "globalization." These types of products both take place in the global market and support local cultures and economies.

From the point of view of supporting local economies, it has been observed that GIs are increasingly creating a higher quality rural workforce. These indications can confirm and protect unique intellectual or socio-cultural property embodied in local knowledge or traditional and artisanal skills, which are valuable forms of expression for a particular community (Giovannucci et al. 2009).

Products with GIs are market-oriented products. Because they tend to have standards for quality, traceability, and food safety, they are often in line with the emerging trade demands. Geographical signs have many of the characteristics of a top-class brand. They can also influence a supply chain and even other products and services in a region, promoting business clustering and rural integration.

Traceability and control are of great importance in GIs. This situation should be transparent and accessible. For this, the necessary database should be established, audits should be made regularly, and the results should be open to everyone. A superior and independent authority to ensure this is important. Therefore, good governance is essential in GIs.

GIs reveal traditional production and processing methods that arise in a region and are difficult to imitate in other areas or countries. From this point of view, they play an essential role in preserving culture and passing it on to future generations. Thus, it can be said that the region, which has different cultures and traditions, has more advantages in cultural tourism to be known.

One of the most important outputs of GIs is forming a union between the producers and even the input providers, intermediaries, and sellers that provide the primary input for the product subject to the geographical indication. This unity can be established through a cooperative and an union or by creating a brand (especially a collective brand). Most of the applications for GI in the world are made through producer organizations. Thus, GIs provide a culture of acting together, which helps create a society with high organizational commitment.

GIs are not a static process, and on the contrary, they have a dynamic structure. To see this process as a closed box in itself, turning it off to innovation will not go beyond misinterpreting the system. Studies on this subject indicate that the geographical indication system is a dynamic process, that it is open to innovation and must have the ability to develop itself according to changing market conditions, and it is stated that the adaptation capabilities and sustainability of geographical signs with this structure are high (Gugerell et al. 2017; Quiñones-Ruiz et al. 2018).

In an European Union (EU)-wide study on the theoretical benefits of the GIs as well as their performances in practice, it was revealed that the development of the European Commission GIs strengthened the following issues (Giovannucci et al. 2009):

- Regional cooperation between local authorities, commercial, and social partners
- Positive recognition of the regions (especially their culture, market, etc.)
- The improvement of general infrastructure and rural services
- Increasing the attractiveness of the region as a business center
- Improvements in environmental quality and related resource use

The advantages of GIs locally and therefore in rural development can be grouped under the following five main headings (Wilson et al. 1999; Armesto-Lopez and Martin 2005; Treager et al. 2007; Kan and Gülçubuk 2008).

- **Protection tool**: It can be used as a tool to prevent the usurpation of the rights of the producers due to counterfeiting and the deceit of the consumers.
- **Marketing tool**: It positively affects the image, reputation, and profile of the product in the market.
- **Rural development tool**: It is a different approach to production and can be used to ensure local businesses' sustainability and protect cultural heritage and biodiversity.
- **A tool to create an economic balance**: It can eliminate the financial difference between less developed and developed areas. In addition, protecting by the state means less expense for producers.
- **Information tool**: Geographical signs are an essential information tool that enables sharing knowledge and culture between the producer and the consumer. In particular, it allows consumers to have information about the product they consume and the region's culture.

As can be seen, the studies show that GIs will provide essential contributions in many areas. The transformation of such potential positive contributions into cooperation in the region and the international arena will be advantageous for the cooperating countries.

Table 40.2 The legal base and definition of geographical indication in D-8 countries

D-8 countries	Legal status	GI definition
Bangladesh	• Geographical Indication Act on November 10, 2013. Act No. 54 of 2013	Geographical indication of goods means any agricultural, natural, and manufactured goods having geographical indication that indicates that such goods have originated or manufactured in a country or territory or a locality or region of such country or territory, where a given quality, reputation, or other characteristics of such goods are essentially attributable to its geographical origin and in case such goods are manufactured goods, one of the activities of either the production, processing, or preparation of such goods takes place in such territory, region, or locality (Zahur 2017)
Egypt	In Egypt, there is no specific system for the protection of GIs; however, in practice, the Egyptian Trademark Office has been known to reject any trademark that includes a geographical indication, city name, or national/international identifier in accordance with Law No. 82 of 2002. Law No. 82 of 2002 on the Protection of Intellectual Property Rights, Act 67(8) provides that, "The following shall not be registered as trademarks or components thereof ... Marks and geographical indications which are likely to mislead or confuse the public or which contain false descriptions as to the origin of products, whether goods or services, or their other qualities, as well as the signs that contain an indication of a fictitious, imitated, or forged trade name"	Geographical indications are defined by the WIPO as "A sign used on products that have a specific geographical origin and possess qualities or a reputation that are due to that origin. A sign must identify a product as originating in a given place. In addition, the qualities, characteristics, or reputation of the product should be essentially due to the place of origin" (Eldib 2021)
Indonesia	The Republic of Indonesia has developed the concept of "geographical indication" since regulated in Law No. 14 of 1997 concerning Marks, and in the years 2001, The Law of the Republic of Indonesia regulated Law n°15/2001 on TM stipulated some rules on GIs. Then, regulated further in the government regulation n°52/2007 that developed the GI system as	Geographical indication is defined as "a sign, which indicates the geographical origin of a good and/or a product with this origin giving a certain reputation, quality, and characteristic to the good and/or product because of the geographical environment, the human factor, or a combination of the two

(continued)

Table 40.2 (continued)

D-8 countries	Legal status	GI definition
	enforceable and applicable in the country. Now two laws that are interested with GI in Indonesia (WIPO 2021) are as follows: • Law of the Republic of Indonesia Number 20 of 2016 on Trade Marks and Geographical Indications • Elucidation of Law of the Republic of Indonesia Number 20 of 2106 on Marks and Geographical Indications	
Iran	Geographical indications in Iran are protected under the provisions of the Act on Protecting Geographical Indications (Islamic Consultative Assembly of Iran, 2005) that are in consistent with provisions of the Paris Convention for the Protection of Industrial Property	Article 1 of the latter Act defines a geographical indication as "a sign which correlates with the origin of a product to a territory or a region of the country, provided that the quality, reputation, or other features of the product is attributable to that geographical region" (Hekmat 2017)
Malaysia	• Geographical Indications Act 2000	Geographical indication is an indication which identifies any goods as originating in a country or territory or a region or locality in that country or territory, where a given quality, reputation or other characteristics of the goods are essentially attributable to their geographical origin (MyIPO 2000)
Nigeria	There is no specific legal regulation on this subject. The only legislation that bears a resemblance to GI legislation is the Trademarks Act (Cap T 13 Laws of the Federation of Nigeria, 2004. Section 43 of the Act provides for registration of certification marks) (Gwom 2017)	
Pakistan	• 31 March 2020, Pakistan the Geographical Indications (Registration and Protection) Act, 2020 (GIRPA'20) (National Assembly of Pakistan 2020)	The term "Geographical Indication" or "GI" has been defined in GIRPA'20, in relation to goods includes an indication which identifies such goods as agricultural goods, natural goods, or manufactured goods originating or manufactured or produced in a territory or a region or locality as determined by the country, where a given quality, reputation or other characteristics of the goods or the ingredients or components, is essentially attributable to its geographical origin, and in the case of manufactured goods, one of the

(continued)

Table 40.2 (continued)

D-8 countries	Legal status	GI definition
		activities of either the production or processing or preparation of the goods concerned takes place in such territory, region, or locality as the case may be. There are three subcategories of registered GI. These are • Pakistan GI • Foreign GI • Homonymous GI
Turkey	• Regulation on the protection of geographical indications numbered 555 and the Decree Law on the Protection of geographical indications numbered 555 in 1995 (The protection started with the regulation in 1995) • The Industrial Property Law No. 6769 on 22.12.2016 (It is the current law and has replaced the Statutory Decree No. 555 of 1995 year) (TURKPATENT 2021)	Geographical indications are signs which show a product identified with a district, area, region, or country of origin in terms of a specific reputation, distinction, or other characteristics. Geographical indications are registered as name of origin or geographical indication according to the characteristics. There are two terms in the law under the concept of geographical indication: • Appellation of origin (Menşei) • Designation of origin (Mahreç)

The existence of a system related to GI protection in D-8 countries is different. Table 40.2 shows the legal infrastructures and GI definitions of the existing geographical indication system in D-8 countries.

The concept of a GI is increasingly accepted in the international arena, and countries integrate this concept with the concepts of both development and competitive advantage. The popularity of the idea in D-8 countries is also increasing. Studies on this subject show how important geographical indication can be for D-8 countries. Islam (2013) tried to reveal the difference between the geographical indication system and trademarks in Bangladesh. He stated that GIs provide tremendous benefits, whether in developed, developing, or least developed countries. Hyder (2016) examined the status of the geographical indication system in Bangladesh and stated that the system is late for Bangladesh. Still, if steps are taken in line with the international arena, this system can significantly contribute to Bangladesh.

In his study, Zahur (2017) evaluated the "GI Act of 2013," which constitutes the legal infrastructure of the geographical indication system in Bangladesh. It states that this law is vital for protecting cultural heritage in Bangladesh, but it is not sufficient by itself. However, he said that the first step was taken to preserve and promote cultural heritage in the international arena with this regulation. Gwom (2017) states that Nigeria's lack of a geographical indication system is a significant deficiency. This system is essential for Nigeria and other countries in the African continent in economic development. However, the lack of legal legislation and the acceptance of this system and the establishment of technical and financial infrastructure are stated. In the webinar organized by World Intellectual Property (WIPO) and Africa International Trade & Commerce Research (WIPO and AITCR 2020), it was noted that GIs could play an essential role in the increase of Nigeria's non-oil product exports and economic development. Antons (2017) focused on developing the geographical indication system in Indonesia and the relationship of this development with the localization of the administration in Indonesia. It is stated that GIs are a marketing tool. Their positive effects on growth are emphasized since they are not tools that exclude competitors at the national or international level. It is stated that the impact on cultural heritage and the environment will only become visible in the long term. In the study conducted by Durand and Fournier (2017), the contributions of GIs to modernization in agriculture in Indonesia and Vietnam were evaluated through the case studies. Contribution of GIs to localization efforts has been evaluated and it has been stated that government intervention is required to activate and financially support the GI dynamics as long as the awareness of the producers on GIs remains low. Earterasarun and Nursiah (2015) evaluated the Geographical Indication System in Indonesia and stated that GI registration not only creates a niche market for existing products, but also increases the value of the product. They also noted that GIs create job opportunities for local people, while also helping to learn about traditions and support other industries such as tourism. This is why they strongly recommended the provision of GI protection. Tianprasit (2016) states in his study that the Association of Southeast Asian Nations (ASEAN) country community consisting of 10 countries closely follows the advantages of geographical indication protection. He stated that each of the countries within this community has geographic areas that can make a difference on their own products, so they care about systems such as geographical signs that can turn this geographical advantage into an opportunity and add value to their products. He stated that one of the best answers to increase product value and improve the livelihoods of its people is the geographical indication system, yet there are many problems that the ASEAN Community has to solve with regard to GIs.

Another country among the D-8 countries is Egypt. Although Egypt does not have a special legal infrastructure for GIs, it is obliged to fulfill the minimum obligations regarding protection of GIs within the scope of Trade-Related Aspects of Intellectual Property (TRIPS) due to its membership in the WTO. The studies conducted are based on two items that Egypt needs to do more, studies especially with geographical indication protection. The first of these is that Egypt is an important agricultural country and agriculture has an important share in its export. The Geographical Indication System has the potential to add value to export products in this regard. Another factor is that Egypt is an important cotton growing country and the textile industry has an important place in Egypt. Until the "Egypt Cotton" is protected as a geographical indication, Egypt is vulnerable to the loss of the benefit of obtaining a geographical indication for this valuable product (USAID 2009). Among the D-8 countries, Turkey is one of the countries with a high potential for geographically indication registered products. As of April 2021, 728 products have been registered, while applications have been made for the registration of 744 products in Turkey, which has the oldest geographical indication system among D-8 countries (1995) (TURKPATENT 2021). It is stated that GIs are an important opportunity for Turkey in terms of both contributions to local economic development, importance for rural development, rural tourism, and gastronomy tourism (Kan and Gülçubuk 2008; Kan et al. 2010, 2011; Kan and Kan 2020).

As can be seen from the studies carried out, the subject of GI is still an evolving process among the D-8 countries. The positive effects of the Geographical Indication System, especially for agricultural products, are a means of gaining an important competitive advantage, protecting cultural heritage, and providing economic benefit for D-8 countries with good agricultural potential and cultural heritage. For this reason, the D-8 Association is an important platform, where joint work can be done, for the development of this system and the promotion and trade of the products included in the system. It is important for a stronger society that the development of GIs and/or similar systems that can be established within the D-8 countries where the agricultural product potential is higher than ever. Islamic culture is a common value in terms of cultural values, and the advantages of unity in trade can be achieved.

40.4 The Situation of Geographical Indication Registered Rice in D-8 Countries

In the whole world, there are mostly coarse rice in the market. Approximately, 18% of the world rice market consists of aromatic rice. These rice come from a limited number of countries in some specific areas where pedoclimatic conditions and human knowledge are combined. Aromatic rice is usually protected by a GI. In this market, Basmati rice is supplied from Pakistan and India, while both countries are registered in GI systems. Jasmine rice, which carries the Khao Hom Mali GI in Thailand, is also one of the important aromatic rice in this market. The export prices of these rice are also changing and it is stated that the average price of rice with GI is two times higher than coarse varieties and one and a half times more than higher quality but without GI rice (Giraud et al. 2018).

Especially, aromatic type rice with high value is grown in D-8 countries to a large extent. It is seen that important paddy producers, such as Indonesia, Malaysia, and Pakistan have registered aromatic rice in their Geographical Indication Systems. In addition, Turkey is one of the countries with significant potential in this field, although it is not a high percentage of producers and exporters with 8 rice varieties (registered geographical indication and candidate for registration) (Table 40.3). While this situation reveals that GIs are important even for a single product (rice) in D-8 countries, it also provides an opportunity for further utilization of this potential within D-8 countries. When Table 40.3 is examined, among the D-8 countries, the most GI registered rice is found in Turkey. Turkey is followed by Indonesia and Malaysia. D-8 countries, which produce about 20% of the world rice harvest, have many paddy landraces, geographical differences, and reflection of the long-standing relationship between rice and humans on the culture, although it has

Table 40.3 Some geographical indication registered and candidate rice names in D-8

No	GI names	Country	Status	Registration date
1	Kataribhog Rice of Dinajpur	Bangladesh	Registered	02/01/2018
2	Kalijira Rice of Bangladesh	Bangladesh	Registered	02/01/2018
3	Bareh Solok (Solok Rice)	Indonesia	Registered	08/10/2018
4	Beras Adan Krayan	Indonesia	Registered	06/01/2012
5	Beras Pandanwangi Cianjur (Pandawangi Cianjur rice)	Indonesia	Registered	16/10/2015
6	Beras Raja Uncak Kapuas Hulu (Kapuas Hulu Raja Uncak Rice)	Indonesia	Registered	28/02/2017
7	Bario Rice	Malaysia	Registered	10/03/2008
8	Sarawak Beras Bajong	Malaysia	Registered	17/02/2009
9	Sarawak Beras Biris	Malaysia	Registered	17/02/2009
10	Basmati Rice	Pakistan	Registered	27/01/2020
11	Bolu Kıbrıscık Pirinci[a]	Turkey	Registered	28/07/2020
12	Karacadağ Pirinci	Turkey	Registered	28/03/2018
13	Konuralp Pirinci	Turkey	Registered	25/06/2019
14	Tosya Pirinci	Turkey	Registered	08/11/2017
15	İpsala Pirinci	Turkey	Registered	02/05/2016
16	Gönen Pirinci	Turkey	Candidate (At the evaluation stage)	
17	Terme Pirinci	Turkey	Candidate (At the evaluation stage)	
18	Yusufeli Pirinci	Turkey	Candidate (At the evaluation stage)	

Sources: ASEAN GI Database (2021), TURKPATENT (2021)
[a]"Pirinç" is Turkish Word. The English meaning is "Rice"

been noticed in gastronomy tourism in this case, it is an important deficiency that the desired advantages in rice could not be obtained.

Among the D-8 countries, there are superior quality rice varieties and landraces, some of which have received a GI. Studies on this subject show that rice is one of the important products with GI potential among D-8 countries. Gwom (2017) mentioned the potential of important products in studies conducted in Nigeria. It is stated that rice production is exclusive, especially in the North-Central region of Nigeria. In the webinar workshop organized by WIPO and AITCR (2020), Dr. Yauri stated that there are approximately 10,000 potential geographically indication registered products in Nigeria. Among these products, three varieties of rice are mentioned, especially in paddy. These are the Kebbi Rice (Lake Rice), Abakaliki Rice, and Ofada Rice. In the same webinar, Dr. Abani states that Ofada rice, a rice unique to Ogun from Southwest Nigeria, should also be protected as GI.

In his study, Antons (2017) tried to establish a relationship between "Adan Krayan Rice," one of the 32 GI products registered in Indonesia, and the people living in the area where this product is produced. It is stated that Adan Krayan Rice is known as "Bario Rice" in Malaysia and this product was registered in Indonesia in

Solok Regency Solok Rice	Nunukan Regency Beras Adan Krayan	Cianjur Regency Pandawangi Cianjur rice	Kapuas Hulu Regency Kapuas Hulu Raja Uncak Rice

Fig. 40.3 Geographical distribution of primary rice cultivation area (USDA 2016) and geographical indication registered rice places

2012 and there is a competition for this product between Indonesia and Malaysia. Indonesia has five and Malaysia has three GI registered rice. The grown places of GI registered rice in Indonesia and Malaysia are shown in Figs. 40.3 and 40.4.

Siddique et al. (2018) tried to characterize GI registered rice varieties grown in Bangladesh at the Bangladesh Rice Research Institute in Gazipur according to their morphological characteristics. In this study, 20 potentially different geographically indication (GI) registered rice are mentioned, and this shows that there are different paddy varieties unique to Bangladesh. Now there are two GI registered rice in Bangladesh and both of them are being produced in Dinajpur district of Bangladesh (Fig. 40.5).

Turkey is a region with a very high geographical indication potential. Turkey has five rice varieties registered as a geographical indication, and awaits registration of geographical indications for three rice varieties. The geographical distribution of rice with geographical indication in Turkey is presented in Fig. 40.6. It is widely believed that rice, which has an important place in Turkish cuisine, and its plant, paddy, were

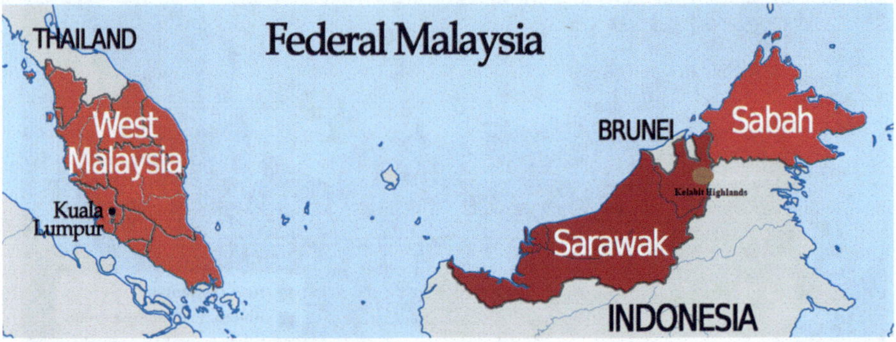

Fig. 40.4 Geographical distribution of geographical indication registered rice places in Malaysia

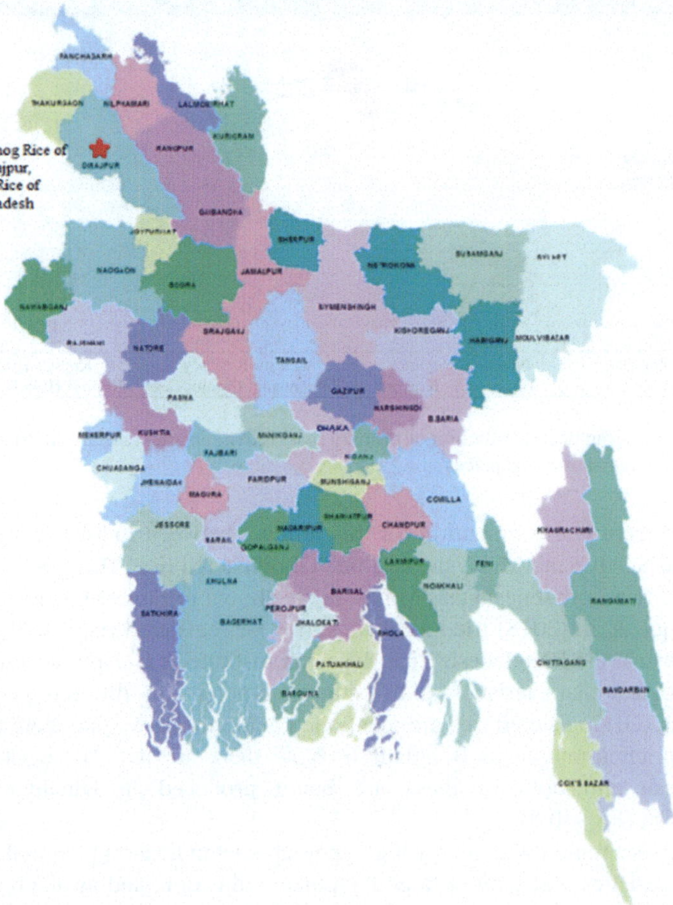

Fig. 40.5 Geographical distribution of geographical indication registered rice places in Bangladesh

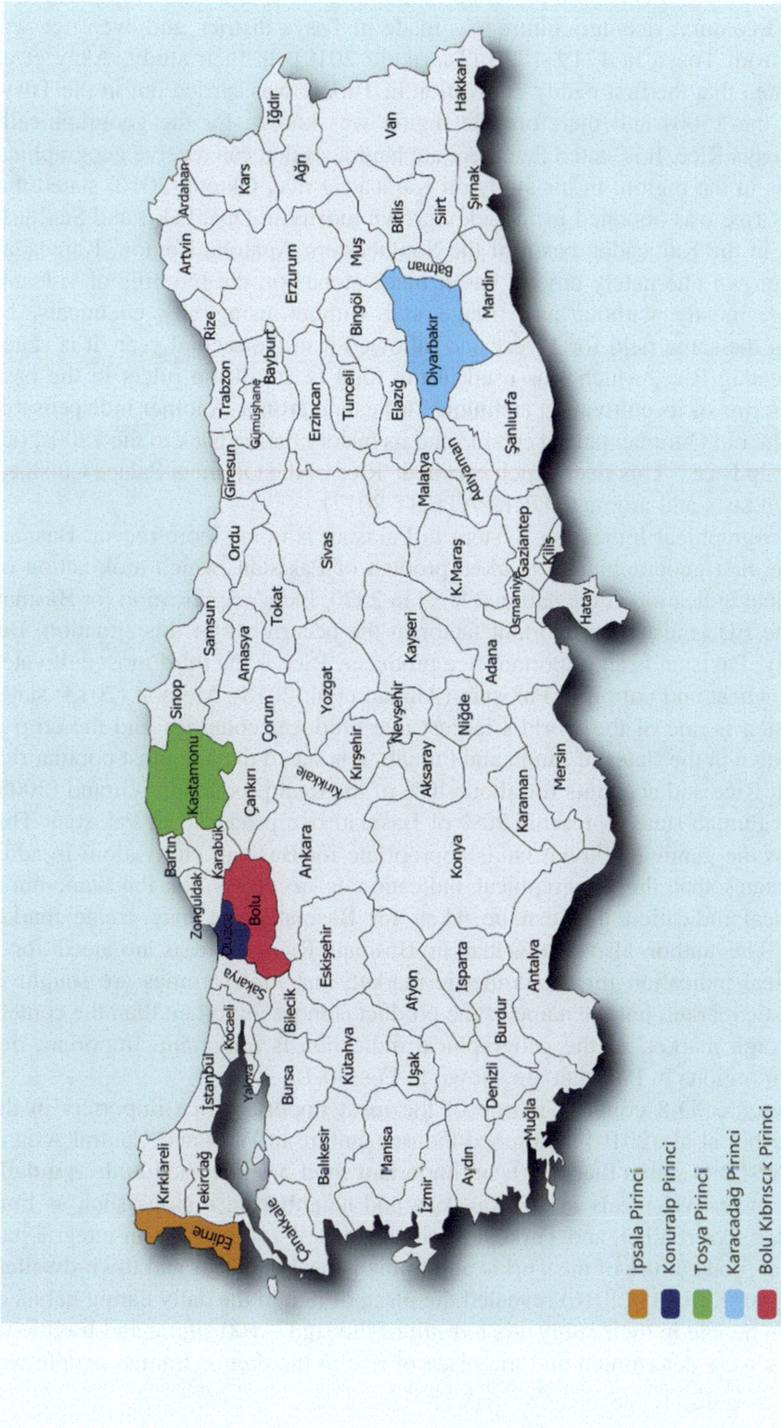

Fig. 40.6 Geographical distribution of geographical indication registered rice in Turkey (Edited on the map at www.paintsmap.com over TURKPATENT records by the author)

brought to Anatolia from Egypt in the XV century and the first plantings were made in Tosya district of Kastamonu province. According to the Ottoman Empire records; in the XVI century, rice agriculture was made in Tosya district, and even rice was exported from Tosya in 1719–1720 (Tanrıkulu 2019). In their study, Akay et al. (2018) stated that the first paddy cultivation in Turkey was carried out in the Tosya region in the 1500s and therefore the region was known for the geographically marked Tosya Rice. It is stated that there are landraces that can receive geographical indications in the region. In his study on Karacadağ rice, Öktem (2016) stated that Karacadağ rice was obtained from paddy grown mostly in Diyarbakır and Şanlıurfa provinces in the Karacadağ basin of the Southeastern Anatolia Region. Karacadağ paddy is grown alternately among basalt black stones on the foothills of volcanic Karacadağ, mostly without any tillage and without using any chemicals, by cultivating the same field for 1 year and fallowed it in the second year. It is stated that Karacadağ Rice, which was used in the food feasts of the rulers in the past, differs in terms of its cultivation technique, taste, and aroma. Another indispensable flavor of the old Ottoman palace cuisine and its history dating back to the 1800s, rice is “Konuralp Rice.” This rice, which is called “Rice of the Ottoman Palace Cuisine,” has its own taste and aroma (TURKPATENT 2017).

The Geographical Indication System in Pakistan is much more recent. Basmati Rice is the first geographically marked product of Pakistan, which took action on geographical indication at the national level in 2020. India’s application for Basmati Rice at the EU is also an important factor in the occurrence of this situation. But historically, Pakistan is an important rice producer. Rice is the third most cultivated area after wheat and cotton in Pakistan (Ahmad et al. 2018). Memon (2013) states that Pakistan is one of the world’s largest rice producer countries and the crop is mostly grown in the states of Sindh and Punjab. It is stated that the most popular rice is Basmati Rice and accounts for about 40% of the total production. Giraud (2008) states that Punjab state represents 90% of Basmati rice produced in Pakistan. This area forms the genuine alluvial lands appropriate for Basmati cultivation. In addition, he states that the geographical indication is necessary for Pakistan, but a geographical indication that can be taken for Basmati Rice may create market shortage. The author also states that in Basmati Rice, there is no need for a geographical indication for the domestic market, that more aromas are sought in the domestic market, but the name of the product is more important than the content in the foreign market, so the geographical indication is important. Important rice production regions in Pakistan are shown in Fig. 40.7.

Iran, another D-8 country, is one of the most important rice importers in the world. Nesbitt et al. (2010) mentioned the rice culture in West and Central Asia in their studies. It is stated that rice is an important food product used both as a daily meal and in festival meals in Central Asia and neighboring regions such as Iran, Afghanistan, Azerbaijan, and Turkey. Various kinds of “pilovs” are an integral part of the local food culture of the Turkic and Iranian-speaking oasis and town-dwellers of the region. Karizaki (2016) revealed the place of rice in the daily eating habits of the Iranian people in their study. As a result of the study, 100 ethnic and traditional rice dishes were determined and the place of rice in the diet of Iranian people was

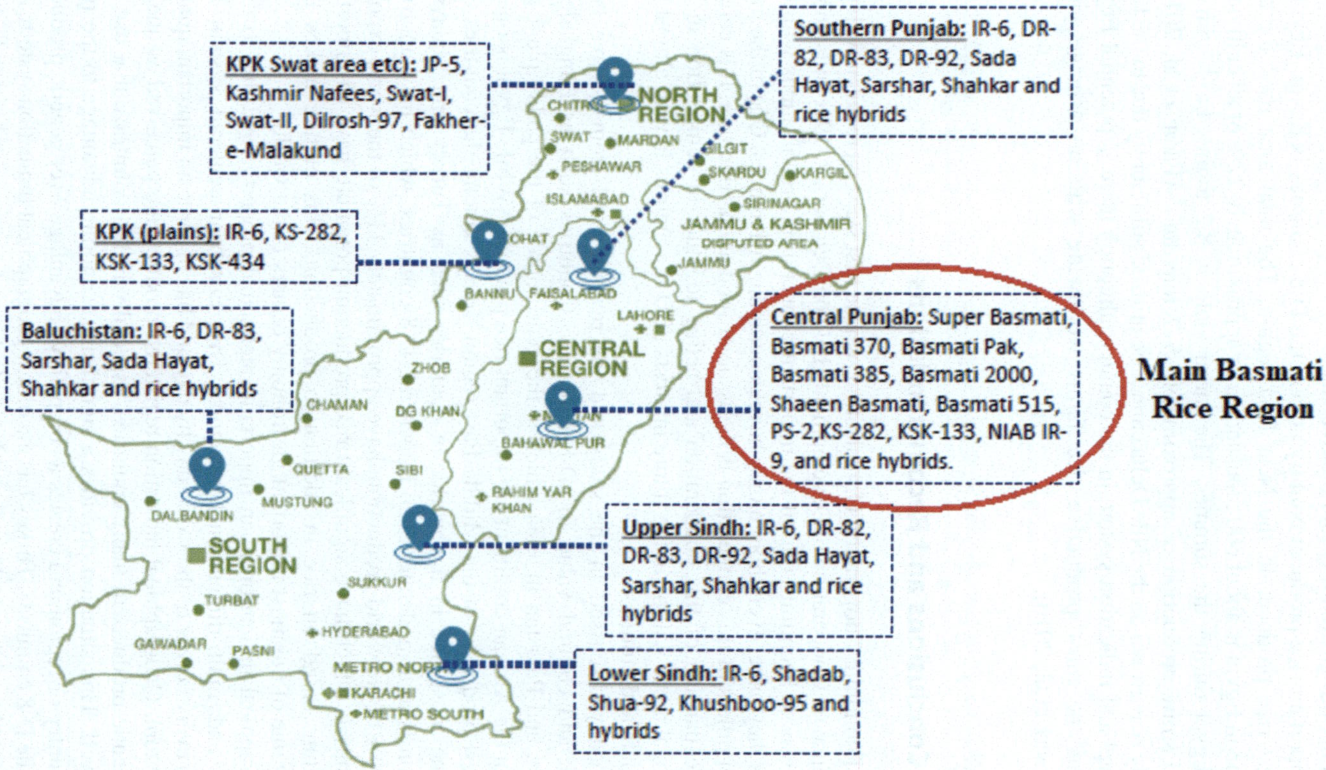

Fig. 40.7 Geographical distribution of main rice varieties in Pakistan (PACRA 2020)

revealed. Although there are no GI registered products in Iran, it is one of the rich countries in both gastronomy and biodiversity.

An important rice producer country among the D-8 countries is Egypt. Egypt is the largest rice producer in the Near East region. The total area used for rice cultivation in Egypt is about 600 thousand ha or approximately 22% of all cultivated area in Egypt during the summer. The mean yield is 8.2 tonnes ha^{-1} with an estimated straw production of approximately 5–7 tons ha^{-1} (Farag et al. 2013). Especially rice grown on the Nile Delta cannot be utilized sufficiently due to the lack of geographical indication system in the country, although there is potential Egypt which has the most productive rice farms worldwide with an average yield (El-Shahway et al. 2016).

40.5 Conclusions and Recommendations

The D-8 Economic Cooperation Organization, whose first steps were taken with the "Conference on Cooperation in Development" held on October 22, 1996 in Istanbul, is on its way to becoming one of the important cooperation organizations with the resources they have, even though they have geographic differences. One of the most important pillars of this cooperation is the issue of food and agriculture. It is certain that the expansion of this cooperation and acting together in competition in the international arena will be in the future interests of D-8 countries having almost the same beliefs and cultures.

The paddy discussed in the study and the rice obtained from it have both economic and cultural value for D-8 countries. Countries such as Bangladesh, Indonesia, and Pakistan in Asia which is the gene source of the paddy plant and Nigeria in the African continent are important rice producer. While Egypt is one of the biggest producers of the Middle East region, Iran is among the countries that import the most rice in the world. In countries such as Turkey and Malaysia, production is at a moderate level. In general, D-8 countries, which have approximately 20% of the world production, have approximately 11% of the world exports. D-8 countries, which have significant rice production potential, have an important food culture based on rice in addition to production. This culture makes D-8 countries one of the most important gastronomy centers in the world.

Geographical indication is an important intellectual property right due to their meaning and multi-dimensional structure. This European-centered system has spread to every corner of the world day by day and has become an important quality indicator tool. Geographical indication registered products are perceived as quality and authentic products in many countries, especially in EU countries, that use the system well. This situation provides a significant economic advantage to the producer country or region and creates a competitive advantage over generic products. Among the D-8 countries, Turkey has the oldest geographical indication system, and countries such as Nigeria, Egypt, and Iran still do not have geographical indication protection in a special legal structure. This situation is an important loss for D-8 countries, which have a rich structure in terms of both biodiversity and cultural diversity.

With regard to paddy, Turkey has the most GI registered products/goods in D-8 countries, followed by Indonesia, Malaysia, and Bangladesh. The geographical indication system in Pakistan is much more recent. Pakistan, which is an especially important rice producer, is known as the largest Basmati Rice producer and exporter in the world along with India. The latest incident regarding geographical indications, which is an important tool for gaining competitive advantage in the international arena, is happening between India and Pakistan. The activities of both countries in terms of Basmati Rice are known to have superiority to each other in both production and commercial activities. In particular, India's application for Basmati Rice at the European Union level is an example of India's efforts to create a competitive advantage in exports to the European Union under the name of Basmati Rice. In the registration and use of GIs, first of all, it is important to create fair trade conditions. Conditions such as geographical origin, popularity, historical background, and connection between product and humans exist for both India and Pakistan for Basmati Rice. For this reason, issues that will cause unfair competition should not be allowed in the registration of the GIs in the international arena, and it should be taken into account that more than one country may have a geographical origin in the registration of a product in the international area. The same situation is valid in Turkey and GI registration should not give unfair advantage to any country in terms of competing products in international competition.

As a result, the evaluation of GIs and similar potentials not only for rice but for many other products is important for D-8 countries to take a more important role in the international arena and become an important player. For this reason, the D-8 Economic Cooperation organization is an important platform for the evaluation, development, and expansion of such activities. In today's world where a culture of cooperation and collaborative work are needed more than ever, more action should be done and action should be taken together in this structure where we have common beliefs, cultures, and values rather than political concerns. This may be the case with regard to GIs as well as in all other matters.

References

Ahmad S, Hasanuzzaman M (2012) Integrated effect of plant density, N rates and irrigation regimes on the biomass production, N content, PAR use efficiencies and water productivity of rice under irrigated semiarid environment. Not Bot Horti Agrobot Cluj-Napoca 40(1):201–211

Ahmad S, Zia-ul-Haq M, Ali H, Shad SA, Ammad A, Maqsood M, Khan MB, Mehmood S, Hussain A (2008) Water and radiation use efficiencies of transplanted rice (*Oryza sativa* L.) at different plant densities and irrigation regimes under semi-arid environment. Pak J Bot 40(1): 199–209

Ahmad S, Ahmad A, Zia-ul-Haq M, Ali H, Khaliq T, Anjum MA, Khan MA, Hussain A, Hoogenboom G (2009) Resources use efficiency of field grown transplanted rice (*Oryza sativa* L.) under irrigated semiarid environment. J Food Agric Environ 7(2):487–492

Ahmad S, Ahmad A, Soler CMT, Ali H, Zia-Ul-Haq M, Anothai J, Hussain A, Hoogenboom G, Hasanuzzaman M (2012) Application of the CSM-CERES-Rice model for evaluation of plant density and nitrogen management of fine transplanted rice for an irrigated semiarid environment. Precis Agric 13(2):200–218

Ahmad S, Ahmad A, Ali H, Hussain A, Garcia y Garcia A, Khan MA, Zia-Ul-Haq M, Hasanuzzaman M, Hoogenboom G (2013) Application of the CSM-CERES-Rice model for evaluation of plant density and irrigation management of transplanted rice for an irrigated semiarid environment. Irrig Sci 31(3):491–506

Ahmad A, Ashfaq M, Rasul G, Wajid SA, Khaliq T, Rasul F, Saeed U, Rahman MH, Hussain J, Baig IA, Naqvi AA, Bokhari SAA, Ahmad S, Naseem W, Hoogenboom G, Valdivia RO (2015) Impact of climate change on the rice–wheat cropping system of Pakistan. In: Hillel D, Rosenzweig C (eds) Handbook of Climate Change and Agro-ecosystems: The Agricultural Modeling Intercomparison and Improvement Project (AgMIP) Integrated Crop and Economic Assessments. Imperial College Press and the American Society of Agronomy, pp 219–258

Ahmad S, Rehman A, Ejaz M et al (2018) Agricultural land-use change of major field crops in Pakistan (1961-2014). Sci Technol Dev 37(3):113–121

Ahmad S, Abbas G, Ahmed M, Fatima Z, Anjum MA, Rasul G, Khan MA, Hoogenboom G (2019) Climate warming and management impact on the change of rice-wheat phenology in Punjab, Pakistan. Field Crops Res 230:46–61

Ahmed M, Ahmad S (2019) Carbon dioxide enrichment and crop productivity. In: Hasanuzzaman M (ed) Agronomic Crops, vol 2. Springer, Singapore, pp 31–46

Ahmed M, Ahmad S (2020) Systems modeling. In: Ahmed M (ed) Systems Modeling. Springer, Singapore, pp 1–44

Akay H, Sezer İ, Mut Z (2018) Coğrafi İşaretler ve Tosya Yerel Çeltik Genotipleri Örneği. Adnan Menderes Üniversitesi Ziraat Fakültesi Dergisi 15(2):81–85

Antons C (2017) Geographical indications, heritage, and decentralization policies: the case of Indonesia. In: Calboli I, Loon W, Loy NG (eds) Geographical Indications at the crossroads of trade, development, and culture; focus on Asia-Pacific. Cambridge University Press. Chapter 18, p 485–507

Armesto-Lopez XA, Martin BG (2005) Tourism and quality agro-food products: An opportunity for the Spanish countryside. Tijdschr Econ Soc Geogr 97(2):166–177

ARISE+IPR (2019) Geographical indications in the AESEA region, a booklet on ASEAN geographical indications procedure and products. https://euipoeuf.eu/sites/default/files/arisedocs/2019/ASEAN_GI-Booklet.pdf

ASEAN GI Database (2021) All eASEAN GIs. http://www.asean-gidatabase.org/gidatabase/gi-search-2

Asif M, Xuhui W, Nasiri A et al (2018) Determinant factors influencing organic food purchase intention and the moderating role of awareness: a comparative analysis. Food Qual Prefer 63: 144–150

Beletti G, Marescotti A, Brazzini A (2017) Old world case study: the role of protected geographical indications to foster rural development dynamics: the case of Sorana Bean PGI, pp.253-276. In: W. Van Caenegem and J. Cleary (eds.), The Importance of Place: Geographical Indications as a Tool for Local and Regional Development, Springer International Publishing AG

Bessière J (1998) Local development and heritage: traditional food and cuisine as tourist attractions in rural areas. Sociol Rural 38(1):21–34

Bramley C, Biénabe E, Kirsten J (2009) The economics of geographical indications: towards a conceptual framework for geographical indication research in developing countries. The Economics of İntellectual Property 109, WIPO

Çalışkan V, Koç H (2012) Türkiye'de coğrafi işaretlerin dağılış özelliklerinin ve coğrafi işaret potansiyelinin değerlendirilmesi. Eastern Geog Rev 28:193–214

Chaudhary R, Bisai S (2018) Factors influencing green purchase behavior of millennials in India. Manag Environ Qual 29:798–812

Çukur F, Çukur T (2017) Coğrafi İşaretli ürünlerin kırsal kalkınma açısından değerlendirilmesi: Muğla ili örneği. Turkish J Agr Econ 23:187–194

D-8 OEC (D-8 Organisation for Economic Cooperation) (2021a) Brief history of D-8. http://developing8.org/about-d-8/brief-history-of-d-8/
D-8 OEC (D-8 Organisation for Economic Cooperation) (2021b) Areas of cooperation. http://developing8.org/d-8-baseindbasic-social-and-economic-indicators/
D-8 OEC (D-8 Organisation for Economic Cooperation) (2021c) Agriculture and food security: seed security and trade. https://developing8.org/areas-of-cooperation/seed-security-and-trade/
Durand C, Fournier S (2017) Can geographical indications modernize Indonesian and Vietnamese agriculture? Analyzing the role of national and local governments and producers' strategies. World Dev 98:93–104
Drahos P (2017) Sunshine in a bottle? Geographical indications, the Australian wine industry, and the promise of rural development. The promise and problems of geographical indications for local and rural development. In: Calboli I, Ng-Loy W (eds) Geographical indications at the crossroads of trade, development, and culture: focus on Asia-Pacific. Cambridge University Press, Cambridge, pp 257–436
Earterasarun S, Nursiah I (2015) Registering geographical indications in Indonesia. p 9. www.tilleke.com
Eldib (2021) Geographical Indications in Egypt and its effects on trademark registration. https://eldib.com/news/geographical-indications-in-egypt-and-its-effects-on-trademark-registration/
El-Shahway AS, Mahmoud MMA, Udeigwe TK (2016) Alterations in soil chemical propertiesinduced by continuous rice cultivation: a study on the arid Nile delta soils of Egypt. Land Degrad Dev 27:231–238
FAO (2018) Strengthening sustainable food systems through geographical indications; An analysis of economic impacts. http://www.fao.org/3/I8737EN/i8737en.pdf
FAOSTAT (2020) Live animals data-2018. http://www.fao.org/faostat/en/#data/QA
Farag AA, Radwan HA, Abdrabbo MAA, Heggi MAM, McCarl BA (2013) Carbon footprint for paddy rice production in Egypt. Nat Sci 11(12):36–45
Fatima Z, Ahmad M, Hussain M, Abbas G, Ul-Allah S, Ahmad S, Ahmed N, Ali MA, Sarwar G, Ehsan ul Haque, Iqbal P, Hussain S (2020) The fingerprints of climate warming on cereal crops phenology and adaptation options. Sci Rep 10:18013
Giovannucci D, Josling T, Kerr W, O'Connor B, Yeung MT (2009) Guide to geographical indications linking products and their origins, ITC/P228.E/DMD/ SC/09-II, 2009, Geneva. http://legacy.intracen.org/publications/Free-publications/Geographical_Indications.pdf
Giraud G (2008) Range and Limit of Geographical Indication Scheme: The Case of Basmati Rice from Punjab, Pakistan. International Food and Agribusiness Management Review 11(1):51–76
Giraud G, Gallo JL, Boucher H (2018) Is geographical indication acting on rice export price? Proceedings in system dynamics and innovation in food networks 2018. pp 131–132. http://centmapress.ilb.uni-bonn.de/ojs/index.php/proceedings/article/view/1809/762
Gugerell K, Uchiyama Y, Kieninger PR, Penker M, Kajima S, Kohsaka R (2017) Do historical production practices and culinary heritages really matter? Food with protected geographical indications in Japan and Austria. Journal of Ethnic Foods 4(2):118–125
Gwom S (2017) Protection of Geographical Indications in Nigeria: A Legal and Policy Deficit. WIPO-WTO Colloquium Papers 2017:1–13
Hekmat V (2017) Protecting the textile carpet industry by intellectual property rights. Master's thesis of law, Graduate School of Law, Seoul National University, Intellectual Property Law Major. https://s-space.snu.ac.kr/bitstream/10371/128756/1/000000140890.pdf
Hopkins MS, Roche C (2009) What the 'green' consumer wants. MIT Sloan Management Review, https://sloanreview.mit.edu/article/what-thegreen-consumer-wants/
Hummer KE, Hancock JF (2015) Vavilovian centers of plant diversity: Implications and impacts. HortScience 50(6):780–783
Hyder Md NA (2016) Geographical indications protection in Bangladesh: An overview. Int J Bus Soc Sci Res 4(2):95–102. http://www.ijbssr.com/currentissueview/14013133
Islam MT (2013) Introducing Geographical Indications in Bangladesh. Dhaka University Law Journal 24(1):51–82

Kan M (2011) Yerel Düzeyde Ekonomik Kalkınmada Coğrafi İşaretlerin Kullanımı ve Etkisi: Akşehir Kirazı Araştırması. Ph.D. diss. Ankara Üniversity, Grad. Sch. Nat. and Appl. Sci. Dept. of Agr. Econ., Ankara-Turkey

Kan M, Gülçubuk B (2008) Geographical indications for recovery and local owned of rural economy. Journal of Agricultural Faculty of Uludag University 22(2):57–66

Kan M, Kan A (2020) Qualitative and quantitative analysis of the geographical indication system in Turkey. Journal of Global Innovations in Agricultural and Social Sciences 8(2):114–123

Kan M, Gülçubuk B (2012) Are geographical indications an opportunity for local economic development; a case of Aksehir Kirazı (Cherry of Aksehir). International Interdisciplinary Social Inquiry Conference, 17-21 June 2012, Bursa-Turkey

Kan M, Gülçubuk B (2013) Coğrafi işaretlerin yerel kalkınma potansiyelinin değerlendirilmesi; Akşehir Kirazı araştırması bulguları. Ulusal KOP Bölgesel Kalkınma Konferansı, 14-16 November

Kan M, Gülçubuk B, Kan A, Küçükçongar M (2010) Karaman Divle Skinbag Cheese As a Geographical Indication. Karamanoğlu Mehmetbey Üniversitesi Sosyal ve Ekonomik Araştırmalar Dergisi 12(19):15–23

Kan M, Gülçubuk B, Küçükçongar M (2011). Coğrafi İşaretlerin Kırsal Turizmde Kullanılma İmkanları. 1.Ulusal Kırsal Turizm Sempozyumu 13-15 Eylül 2011, Karaman

Kan M, Gülçubuk B, Küçükçongar M (2012) The Usage Opportunities of Geographical Indications on Rural Tourism. Karamanoğlu Mehmetbey Üniversitesi Sosyal Ve Ekonomik Araştırmalar Dergisi 2012(1):93–101

Kan M, Gülçubuk B, Peker K et al (2013) Türkiye'de Yerel Kalkınma Dinamiklerini Harekete Geçirmede Bir Araç; Yerel Ürünler. 2.Uluslararası Bölgesel Kalkınma Konferansı, 16-17 Mayıs 2013, Elazığ

Karizaki VM (2016) Ethnic and traditional Iranian rice-based foods. Journal of Ethnic Foods 2016: 1–11. https://doi.org/10.1016/j.jef.2016.05.002

Kılıç B, Kurnaz A (2010) Alternatif turizm ve ürün çeşitliliği oluşturmada ekolojik çiftlikler: pastoral vadi örneği. İşletme Araştırmaları Dergisi 2(4):39–56

Kün E (1997) Tahıllar-II (Sıcak İklim Tahılları). Ankara Üniversitesi Zir.Fak. Yayınları:1452 Ders Kitabı: 432 A.Ü. Basımevi. Ankara, 97.S

Lee S, Ko E, Chae H et al (2017) A study of the authenticity of traditional cultural products: focus on Korean, Chinese, and Japanese consumers. J Glob Scholars Market Sci 27(2):93–110

Löker BG, Amoutzopoulos B, Özge Özkoç S et al (2013) A pilot study on food composition of five Turkish traditional foods. Br Food J 115(3):394–408

Lopez XAA, Martin BG (2005) Tourism and quality agro-food products: an opportunity for the Spanish countryside. Tijdschr Econ Soc Geogr 97:166–177

Marescotti A (2003) Typical products and rural development; who benefits from PDO/PGI recognition? pp:1-17. 83rd EAAE Seminar. Food Quality Products in the Advent of the 21st Century: Production, Demand and Public Policy 4-7 September, 2003 - Chania, Greece

Memon NA (2013) Rice: important cash crop of Pakistan. http://lcwu.edu.pk/ocd/cfiles/Pakistan%20Studies/Maj/Pak-St-206/Exclusive-article-RiceImportanceofCashCropinPakistan.pdf

MyIPO (2000) Laws of Malaysia act 602 geographical indications act 2000. http://www.myipo.gov.my/wp-content/uploads/2016/09/GEOGRAPHICAL-INDICATIONS-ACT-2000-ACT-602.pdf

National Assembly of Pakistan (2020) The geographical indications. http://www.na.gov.pk/uploads/documents/1588593635_938.pdf

Nesbitt M, Simpson SJ, Svanberg I (2010) History of rice in Western and Central Asia. In: Sharma SD (ed) Rice: origin, antiquity and history. Science Publishers. Book details: Science Publishers (USA), Enfield, NH. p 308–340, 535–541

Öktem A (2016) Şanlıurfa Koşullarında Karacadağ Çeltiği (Oryza sativa L.) Yetiştiriciliği Üzerine Anket Çalışması. Türkiye Tarımsal Araştırmalar Dergisi 3(2):102–108

Orhan A (2010) Yerel değerlerin turizm ürününe dönüştürülmesinde "coğrafi işaretlerin" kullanımı: İzmit Pişmaniyesi örneği, Anatolia. J Tourism Res 21:243–254

PACRA (2020) Rice sector; an overview. The Pakistan Credit RatingAgency Limited. http://pacra.pk/sector_research/Rice%20Sector%20PACRA_1604759631.pdf
Patra BC, Ray S, Ngangkham U, Mohapatra T (2016) Rice. In: Singh M, Upadhyaya HD (eds) Genetic and Genomic Resources for Grain Cereals Improvement. Academic Press, San Diego, pp 1–80
Peker K, Kan M, Nadeem M (2019) Corporate governance of climate change adaptation. J Glob Innov Agric Soc Sci 7(1):1–5
Quiñones-Ruiz XF, Forster H, Penker M, Belletti G, Marescotti A, Scaramuzzi S, Broscha K, Braito M, Altenbuchner C (2018) How are food Geographical Indications evolving?—an analysis of EU GI Amendments. Br Food J 120(8):1876–1887
Reynolds LP, Wulster-Radcliffe MC, Aaron DK, Davis TA (2015) Importance of Animals in Agricultural Sustainability and Food Security. J Nutr 145(7):1377–1379. https://doi.org/10.3945/jn.115.212217
Ricepedia (2021) History of rice cultivation. https://ricepedia.org/culture/history-of-rice-cultivation
Röder KH (1985) Global problems: a challenge to cooperation between states of different social systems. Int Polit Sci Rev 6(1):35–43. https://www.jstor.org/stable/pdf/1600966.pdf
Schlegelmilch BB, Bohlen GM, Diamantopoulos A (1996) The link between green purchasing decisions and measures of environmental consciousness. Eur J Mark 30:35–55
Shahzad AN, Ahmad S (2019) Tools and techniques for nitrogen management in cereals. In: Hasanuzzaman M (ed) Agronomic Crops, vol 2. Springer, Singapore, pp 111–126
Siddique A, Khalequzzaman M, Islam MZ, Bhuiya A, Baktiar MHK, Ara A (2018) Agro-Morphological Characterization of Geographical Indication Rice Cultivars Grown in Bangladesh. The Experiment 45(4):2598–2610
Stolzenbach S, Bredie WL, Byrne DV (2013) Consumer concepts in new product development of local foods: traditional versus novel honeys. Food Res Int 52(1):144–152
Tanrıkulu M (2019) Tosya'da Pirinç Üretiminin Dünü, Bugünü ve Geleceği. Akademik Bakış Dergisi 71:229–250
The Global Economy (2021) GDP share of agriculture (2019)—country rankings. https://www.theglobaleconomy.com/rankings/Share_of_agriculture/
Tianprasit T (2016) The Protection of Geographical Indications in ASEAN Community. NIDA Development Journal 56(4):1–17
Trichopoulou A, Soukara S, Vasilopoulou E (2007) Traditional foods: a science and society perspective. Trends Food Sci Technol 18(8):420–427
Treager A (2003) From Stilton to Vimto: using food history to re-think typical products in rural development. Sociol Rural 43:91–100
Treager A, Arfini F, Belletti G, Marescotti A (2007) Regional foods and rural development: The role of product qualification. J Rural Stud 23:12–22
TURKPATENT (2017) Konuralp Pirinci. https://ci.turkpatent.gov.tr/Files/GeographicalSigns/e07ca1ed-092a-4fb1-98bf-341b3fbc51ce.pdf
TURKPATENT (2021) Geographical indications of Turkey. Turkish Patent and Trademark Office. https://ci.turkpatent.gov.tr/
UCL (2021) Debating the origins of rice. University College London (UCL). https://www.ucl.ac.uk/rice/history-rice/debating-origins-rice
UHK (2011) Ulusal Hububat Konseyi (UHK) Çeltik Raporu / Aralık 2011. http://uhk.org.tr/dosyalar/UHK_celtikraporu.pdf
UN (2015) Sustainable development goals. https://sdgs.un.org/goals
UN (2018) Economic commission for Latin America and the Caribbean (ECLAC)/Organisation for Economic Cooperation and Development (OECD), Emerging challenges and shifting paradigms: new perspectives on international cooperation for development (LC/PUB.2018/16), Santiago
UNCTAD (2015) Why Geographical Indications for Least Developed Countries (LDCs)? https://unctad.org/en/PublicationsLibrary/aldc2015d4_en.pdf

USAID (2009) International approaches to the protection of geographical indications. Prepared by Judy Winegar Goans. https://pdf.usaid.gov/pdf_docs/PA00WFJ3.pdf

USDA (2016) INDONESIA: rice production prospects reduced by El Nino. United States Department of Agriculture, Commodity Intelligence Report, 9 March. https://ipad.fas.usda.gov/highlights/2016/03/Indonesia/Index.htm

Wilson N, Van Ittersum K, Fearne A (1999) Co-operation and co-ordination in the supply chain: a comparison between the Jersey royal and the opperdoezer Ronde Potato. Contribution appeared in Sylvander B, Barjolle D, Arfini (eds) The socio-economics of origin labelled products: spatial, institutional and co-ordination aspects. Proceedings of the 67th EAAE Seminar, p 96–102

WIPO (2021) Geografis indication in Indonesia. https://www.wipo.int/edocs/mdocs/geoind/en/wipo_geo_yty_17/wipo_geo_yty_17_21.pdf

WIPO and AITCR (2020) The importance of geographical indications to the sustainable development of Nigeria. Webinar organized by the World Intellectual Property Organization (WIPO) Nigeria Office, in collaboration with the Africa International Trade & Commerce Research (AITCR). https://www.wipo.int/meetings/en/doc_details.jsp?doc_id=522734

Zahur M (2017) The geographical indication act 2013: protection of traditional knowledge in Bangladesh with special reference to Jamdani, In: Calboli I, Loon W, Loy NG (eds) Geographical indications at the crossroads of trade, development, and culture; focus on Asia-Pacific. Cambridge University Press. Chapter 18, p 439–460

The manufacturer's authorised representative in the EU is Springer Nature Customer Service Centre GmbH, Europaplatz 3, 69115 Heidelberg, Germany. If you have any concerns regarding our products, please contact ProductSafety@springernature.com

Printed and bound by CPI Group (UK) Ltd, Croydon, CR0 4YY
06/10/2026
02225252-0020